ON SOLAR HYDROGEN & NANOTECHNOLOGY

Editor
Lionel Vayssieres
National Institute for Materials Science, Japan

John Wiley & Sons (Asia) Pte Ltd

Visit our Home Page on www.wiley.com

Other Wiley Editorial Offices

John Wiley & Sons, Ltd, The Atrium, Southern Gate, Chichester, West Sussex, PO19 8SQ, UK

John Wiley & Sons Inc., 111 River Street, Hoboken, NJ 07030, USA

Jossey-Bass, 989 Market Street, San Francisco, CA 94103-1741, USA

Wiley-VCH Verlag GmbH, Boschstrasse 12, D-69469 Weinheim, Germany

John Wiley & Sons Australia Ltd, 42 McDougall Street, Milton, Queensland 4064, Australia

John Wiley & Sons Canada Ltd, 5353 Dundas Street West, Suite 400, Toronto, ONT, M9B 6H8, Canada

Wiley also publishes its books in a variety of electronic formats. Some content that appears in print may not be available in electronic books.

Library of Congress Cataloging-in-Publication Data

On solar hydrogen & nanotechnology / editor, Lionel Vayssieres.
p. cm.
Includes bibliographical references and index.
ISBN 978-0-470-82397-2 (cloth)
1. Solar energy. 2. Nanotechnology. 3. Hydrogen as fuel. 4. Water oxidation. 5. Photocatalysis.
I. Vayssieres, Lionel, 1968-
TJ810.O52 2010
621.47–dc22

2009040144

ISBN 978-0-470-82397-2 (HB)

Typeset in 10/12pt Times by Thomson Digital, Noida, India.
Printed and bound in Singapore by Markono Print Media Pte Ltd, Singapore.
This book is printed on acid-free paper responsibly manufactured from sustainable forestry in which at least two trees are planted for each one used for paper production.

Contents

PART FIVE—NEW DEVICES FOR SOLAR THERMAL HYDROGEN GENERATION

List of Contributors

Bruce D. Alexander, University of Greenwich, UK.

Shamindri M. Arachchige, Virginia Polytechnic Institute and State University, USA.

Jan Augustynski, Warsaw University, Poland. Email: Jan.Augustynski@unige.ch

Marcus Bär, Helmholtz-Zentrum Berlin für Materialien und Energie, Berlin, Germany. Email: marcus.baer@helmholtz-berlin.de

Flemming Besenbacher, Aarhus University, Denmark. Email: fbe@inano.dk

Karen J. Brewer, Virginia Polytechnic Institute and State University, USA. Email: kbrewer@vt.edu

Lin X. Chen, Argonne National Laboratory, USA; Northwestern University, USA. Email: lchen@anl.gov

Chinglin Chang, Tamkang University, Taiwan.

Sahab Dass, Dayalbagh Educational Institute, India.

Per-Anders Glans, Lawrence Berkeley National Laboratory, USA.

Jinghua Guo, Lawrence Berkeley National Laboratory, USA. Email: jguo@lbl.gov

Clemens Heske, University of Nevada, Las Vegas, USA. Email: heske@unlv.nevada.edu

Yongsheng Hu, University of California, Santa Barbara, USA.

Hicham Idriss, University of Aberdeen, UK; Robert Gordon University, UK. Email: h.idriss@abdn.ac.uk

Yasunobu Inoue, Nagaoka University of Technology, Japan. Email: inoue@analysis.nagaokaut.ac.jp

Hideyuki Kamisaka, The University of Tokyo, Japan.

Alan Kleiman-Shwarsctein, University of California, Santa Barbara, USA.

Athanasios G. Konstandopoulos, CERTH/CPERI, Greece; Aristotle University, Greece. Email: agk@cperi.certh.gr

Stuart Licht, The George Washington University, USA. Email: slicht@gwu.edu

Yi-Sheng Liu, Tamkang University, Taiwan.

Souzana Lorentzou, CERTH/CPERI, Greece; Aristotle University, Greece.

Eric W. McFarland, University of California, Santa Barbara, USA. Email: mcfar@engineering.ucsb.edu

Eric L. Miller, Hawaii Natural Energy Institute, USA. Email: ericm@hawaii.edu

Manoranjan Misra, University of Nevada, Reno, USA. Email: misra@unr.edu

Torsten Oekermann, Leibniz Universität Hannover, Germany. Email: oekermann@pci.uni-hannover.de

Frank Osterloh, University of California, Davis, USA. Email: osterloh@chem.ucdavis.edu

Lars Österlund, Totalförsvarets forskningsinstitut (FOI), Sweden; Uppsala University, Sweden. Email: lars.osterlund@foi.se

Krishnan S. Raja, University of Nevada, Reno, USA.

Vibha Rani Satsangi, Dayalbagh Educational Institute, India. Email: vibhasatsangi@gmail.com

Rohit Shrivastav, Dayalbagh Educational Institute, India.

Yasuhiro Tachibana, Osaka University, Japan; RMIT Australia. Email: y.tachibana@chem.eng.osaka-u.ac.jp

Ronnie T. Vang, Aarhus University, Denmark.

Lionel Vayssieres, National Institute for Materials Science, Japan. Email: vayssieres.lionel@nims.go.jp

Geoff Waterhouse, University of Auckland, New Zealand.

Lothar Weinhardt, Universität Würzburg, Germany.

Stefan Wendt, Aarhus University, Denmark.

Abraham Wolcott, University of California, Santa Cruz, USA.

Koichi Yamashita, The University of Tokyo, Japan. Email: yamasita@chemsys.t.u-tokyo.ac.jp

Jin Z. Zhang, University of California, Santa Cruz, USA. Email: zhang@chemistry.ucsc.edu

Peng Zhang, University of California, Santa Barbara, USA.

Preface

Finding new ways to power the future by making cleaner and safer energy without creating additional CO_2 in the atmosphere is one, if not simply the most important, problem facing humanity today. Indeed, meeting the current and future demand for energy, which according to various sources should double by 2050, while managing the environmental consequences of energy production and consumption, is of crucial importance nowadays, with the industrial and economic rise of highly populated countries such as India and China. Reducing urban air pollution and the build-up of greenhouse gases that threaten severe climate change due to global warming, and lowering dependence on foreign oil is of utmost importance. Indeed, the transition from fossil fuels to hydrogen is of revolutionary importance, not only for its societal impact, but also for the new discovery of materials for renewable energy. Energy is not just a part of the world economy, it is the economy.

If one considers the largest and geographically balanced free resource available on earth, that is seawater, and that more sunlight energy is striking our blue planet in one hour than all of our annual energy consumption, the direct solar-to-hydrogen conversion by photo-oxidation of water is a very straightforward and attractive solution for the production of hydrogen, as it is clean, sustainable and renewable and so without the use of a sacrificial agent. It offers an alternative solution to fossil-fuel-based energy sources and explains the tremendous interest in renewable, sustainable energy sources and materials for energy conversion.

Indeed, *materials* – natural or manmade, that is, in raw or engineered forms – are the key to technological development and industrial revolution and thus, to stable, stronger and more secure societies. The ages of ancient civilization are remembered by their materials. Prehistoric eras are named after the materials that defined their technological status, for example, the *Stone Age*, the *Bronze Age*, the *Iron Age*. Truly, the present age will be remembered as the *Nanomaterials Age*, as they should provide our society with the necessary tools to supply renewable and sustainable energy sources and energy-source carriers, while protecting the environment and providing better health, if scientists manage to unravel their full potential while understanding and managing their eventual toxicity.

New materials bring new knowledge and new fields of science, which in turn evolve into new technologies. In many respects, *materials* can be considered as the parents of almost all technologies. Most technological breakthroughs have been achieved by the development of new materials (and technologies based on them). New technologies create new industries, which in turn provide jobs, as well as better and more secure living conditions and societies. It is well established by economists (and not just scientists) that long-term economic growth depends on innovation and technological progress (which account for about 90% of the economic growth).

However, to address these new challenges, current materials and conventional technologies are simply not good enough. The necessity of materials development which is not limited to materials that can achieve their intrinsic theoretical limits, but makes it possible to raise those limits by changing the fundamental underlying physics and chemistry is crucial. The demand of novel multifunctional materials is a major challenge for scientists to address to solve crucial contemporary issues related to energy, environment and health. Truly, the transition of energy resources from their fossil-fuel-based beginnings to clean and renewable technologies relies on the widespread implementation of solar-related energy systems, however the high cost of energy production and the relatively low efficiency of currently used material combinations pose an intrinsic limitation. Indeed, (r)evolutionary development is required to achieve the necessary increases in efficiency and decrease in cost of materials for energy conversion. The need for low-cost functional materials purposely built from optimized building blocks with controlled size, morphology, orientation and aspect ratio, fabricated by cost-effective large-scale manufacturing methods, will play a decisive role in the successful large-scale implementation of solar-related energy sources. However, fabricating and manufacturing large areas of such functional materials still represents a tremendous challenge. Novel smarter and cheaper fabrication techniques and, just as important, better fundamental knowledge and comprehensive understanding of the structure–property relationships using materials chemistry and nanoscale phenomena such as quantum confinement, for instance, to create multifunctional structures and devices, is the key to success.

The materials requirements for water splitting and thus the direct solar-to-hydrogen generation are drastic. First, the materials must be stable in water. Second, they must be stable (upon illumination) against photocorrosion and their bandgap must be small enough to absorb visible light, but large enough not to "dissolve" once illuminated. Finally, their band edges must be positioned below and above the redox potential of hydrogen and oxygen, respectively. Bandgap energy and band-edge positions, as well as the overall band structure of semiconductors are of crucial importance in photoelectrochemical and photocatalytic applications. The energy position of the band edges can be controlled by the electronegativity of the dopants and solution pH, as well as by new concepts such as quantum confinement effects and the fabrication of novel nanostructures. Fulfilling the abovementioned requirements while keeping the cost of the materials low is a tremendously difficult challenge, which explains why direct solar-to-hydrogen generation is still in its infancy, compared to photovoltaics.

This book written by leading experts in major fields of physical sciences from USA, Europe and Asia covers the fundamentals of photocatalysis at oxide interfaces for direct solar-to-hydrogen conversion, the latest developments in materials discovery and their in-depth characterization by high-resolution electron scanning probe microscopy and synchrotron-radiation-based X-ray spectroscopy, as well as the latest development of devices for solar thermal generation of hydrogen. It consists of five distinctive parts and 21 chapters and addresses in detail: the fundamentals, modeling and experimental investigation of photocatalytic reactions for direct solar hydrogen generation (Part I); the electronic structure, energetics and transport dynamics of photocatalyst nanostructures (Part II); the development of advanced nanostructures for efficient solar hydrogen production from classical large bandgap semiconductors (Part III); the new design and approaches to bandgap profiling and visible-light-active nanostructures (Part IV) and the development of new devices for solar thermal hydrogen generation (Part V). While covering the entire spectrum of studies involved in solar hydrogen, the main focus was intentionally given to novel materials development leading to stable and cost-effective visible-light-active semiconductors and efficient (sea)water splitting, the holy grail of photocatalysis.

Hopefully, political leaders, policy makers and program managers of governmental funding agencies, as well as industrial and private investors will soon realize that it is actually not enough to provide funding for existing devices by just multiplying the current technology, but rather investment in R&D of new technologies/devices. Moreover, the world electricity infrastructure needs to be modernized to involve cost-effective and more efficient distribution, for instance by investing massively in smart-grid technology and significantly reducing the cost of renewable technologies. In addition, given that the world is in urgent need of reinventing its energy sources, changing and improving technology policies (new laws could also actually help address global warming), and using tax on carbon-related energy to fund innovative spin-offs and small businesses to induce a shift to noncarbon energy sources, as well as contributing to the necessary market momentum for helping them to survive the world recession by increasing productivity and creating new markets. Finally, a better perception and understanding by the public would certainly help to support and embrace the very urgent shift to renewable energy sources and I sincerely hope this book will contribute to it.

Lionel Vayssieres,
Berkeley, USA, July 12, 2009

Editor Biography

Born in 1968, he obtained his BSc and MSc in physical chemistry in 1989 and 1990, respectively, and a PhD in inorganic chemistry in November 1995 from the Université Pierre et Marie Curie, Paris, France for his research work on the interfacial and thermodynamic growth control of metal-oxide nanoparticles in aqueous solutions. Thereafter, he joined Uppsala University, Sweden as a postdoctoral researcher for the Swedish Materials Consortium on Clusters and Ultrafine Particles to extend his concepts and develop purpose-built metal-oxide nanomaterials for photoelectrochemical applications, as well as to characterize their electronic structure by X-ray spectroscopies at synchrotron radiation facilities. He has been invited as a visiting scientist to: the University of Texas at Austin; the UNESCO Center for Macromolecules & Materials, Stellenbosch University, and iThemba Laboratory for Accelerator-Based Sciences, Cape Town, South Africa; the Glenn T. Seaborg Center, Chemical Sciences Division, at Lawrence Berkeley National Laboratory; Texas Materials Institute; the Ecole Polytechnique Federale de Lausanne, Switzerland; the University of Queensland, Australia and Nanyang Technological University, Singapore, developing novel metal-oxide nanorod-based structures and devices.

He has (co-)authored 60 refereed publications in major international journals which have already generated over 3000 citations since the year 2000; Essential Science Indicators (as of November 1, 2009) shows 114 citations per paper for Materials Science and 71 for All Fields; five ISI highly cited papers (four as first author) for the last 10 years, a single-author 2003 paper ranked no. 1 in the Top 10 hot papers in chemistry (Jul-Aug 05), no. 2 (Sep-Dec 05) and no. 3 (May-June 05) in the Top 3 hot papers in materials science and the most cited paper in materials science for the country of Sweden for the last 10 years as identified by Essential Science Indicators. He has been interviewed by In-Cites and by ScienceWatch in 2006 for this single-authored 2003 paper which has now been cited over 800 times. Two other first-and-corresponding author 2001 papers have already been cited 400 times, five other papers of his (three as the first author) have been cited over 100 times during the last 10 years. He has presented over 125 seminars at universities, governmental and industrial research institutes and 90 talks at international conferences including 70 invited lectures (20 plenaries) in 26 countries and acted as an organizer, chairman, executive program committee member and advisory member for major international conferences (ACerS, ICTP, IEEE, IUPAC, MRS, SPIE and stand-alones) in the fields of Nanoscience and Nanotechnology and projects worldwide. He is, for instance, the founder (2006) of the SPIE Optics and Photonics symposium entitled *Solar Hydrogen and Nanotechnology*, which has been held every year since its creation.

He is currently an independent scientist at the World Premier International Center for Materials NanoArchitectonics, National Institute for Materials Science, in Tsukuba, Japan; a guest scientist at Lawrence Berkeley National Laboratory, USA and an R&D consultant. He is also the founding editor-in-chief of the *International Journal of Nanotechnology* (Impact Factor 2008: 1.184) and a referee for 63 SCI scientific journals, as well as for major funding agencies in USA, Europe, Asia and Africa.

Part One

Fundamentals, Modeling, and Experimental Investigation of Photocatalytic Reactions for Direct Solar Hydrogen Generation

1

Solar Hydrogen Production by Photoelectrochemical Water Splitting: The Promise and Challenge

Eric L. Miller

Hawaii Natural Energy Institute, University of Hawaii at Manoa, Honolulu, HI, USA, Email: ericm@hawaii.edu

1.1 Introduction

Photoelectrochemical (PEC) water splitting, using sunlight to break apart water molecules into constituent hydrogen and oxygen gases, remains one of the "holy grail" technologies for clean and renewable hydrogen production. Hydrogen is an extremely valuable chemical commodity, not only in today's industrial marketplace, but even more so in the emerging *green economies* so vital to the future of our planet and its people. "Green" futures will need to rely less and less on fossil fuels, and more and more on solar, wind, geothermal and other renewable energy resources. Hydrogen is envisioned as a primary media for the storage and distribution of energy derived from this renewable portfolio. To achieve the vision, much work is needed in the development of new, more practical technologies and infrastructures for hydrogen production, storage, delivery and utilization. Non-polluting technologies for large-scale hydrogen production utilizing renewable energy are of particular importance.

Among the viable renewable hydrogen-production approaches, PEC water splitting remains one of the most intriguing, yet one of the most elusive. A PEC system combines the harnessing of solar energy and the electrolysis of water into a single semiconductor-based device. Sunlight plus water gives us clean hydrogen plus oxygen. It sounds good, but it's not all that easy. When a PEC semiconductor device is immersed in a water-based solution, solar energy can be

On Solar Hydrogen & Nanotechnology Edited by Lionel Vayssieres

converted directly to electrochemical energy for splitting the water. This will only happen, however, if key criteria are all met. The semiconductor material must efficiently absorb sunlight and generate sufficient photovoltage to split water, while the semiconductor interface must be favorable to sustaining the hydrogen and oxygen gas evolution reactions. In addition, the PEC system needs to remain stable in solution, and must be cheap for any large-scale deployment.

This sounds like a very tall, complicated order, which in fact it is! No known semiconductor system achieves *all of the above criteria*, though some have come close. Multijunction PEC devices based on III–V semiconductor technology have been demonstrated at National Renewable Energy Laboratory (NREL) with impressive solar-to-hydrogen conversion efficiencies exceeding 16% [1]. Unfortunately, these devices lack long-term stability, and moreover III–V semiconductors are *very expensive* [2]! Lower-cost PEC devices based on thin-film semiconductors have been demonstrated, with stable hydrogen conversion efficiencies in the 3–5% range [3,4], but progress has plateaued at this efficiency level for quite some time. Important breakthroughs are needed in the development of new PEC materials and devices before practical PEC hydrogen production becomes a reality. Fortunately, the scientific community seems unable to resist the daunting challenge. Across the United States and around the world, the newest scientific techniques in materials theory, synthesis and characterization are being brought to the table, and powerful synergies among researchers in the PEC, photovoltaics and nanotechnology fields are emerging in collaborative pursuit of the needed breakthroughs.

Is this really a "holy grail?" PEC offers the potential for efficiently harnessing solar energy to produce high-purity hydrogen from water, at low operating temperatures, with no carbon emissions and using low-cost materials – definitely worth the quest! This chapter presents a brief overview of the PEC hydrogen-production research goals, progress and ongoing hurdles. A broad palette of topics, including hydrogen, solar-energy conversion, semiconductor materials and electrochemistry, are brought together to illustrate the promise and the challenge of PEC.

1.2 Hydrogen or Hype?

Hydrogen has become a hot topic in recent years, both in political and scientific circles. A national spotlight was cast on hydrogen in Former President George W. Bush's State of the Union Speech in 2004, which featured such memorable quotes as: "America can lead the world in developing clean, hydrogen powered automobiles" and "With a new national commitment, our scientists and engineers will overcome obstacles to taking these cars from laboratory to showroom" [5]. From a completely different perspective, energy expert Joseph J. Romm in his book *The Hype About Hydrogen* asserts: "Neither government policy nor business investment should be based on the belief that hydrogen cars will have meaningful commercial success in the near- or medium-term" [6].

Whatever spin you take on the future importance of hydrogen, the world's most abundant element is in fact an extremely valuable chemical commodity today. In the contemporary industrial marketplace, hydrogen is a high-volume chemical with US production exceeding 5 000 000 kg annually. Important industry uses include the production of chemicals, processing of materials, semiconductor manufacturing, generator cooling and fertilizer production, among others. Hydrogen's low density, high thermal conductivity and strong chemical reducing properties make it ideal for such applications. To satisfy the industrial demand, current hydrogen production relies primarily on fossil-fuel technologies. Worldwide, over 95% of hydrogen is produced from natural gas, oil or coal. US production relies mainly on steam–methane reforming.

Of course, fossil fuels are vulnerable to dwindling availability and rising cost, and result in carbon emissions and other forms of environmental contamination. All current industries, including those utilizing hydrogen, suffer the risks and disadvantages of our current *fossil-fuel economy*. In any event, new, cleaner and long-term approaches to hydrogen production will need to be considered. The motivation increases exponentially if *hydrogen economy* proponents have their way [7–10]!

In a future hydrogen economy, hydrogen is envisioned as the ideal energy-carrier for the storage and distribution of renewable energy resources such solar, wind, geothermal, hydroelectric and others. Using fuel-cell or combustion-engine technologies, hydrogen can be converted simply and cleanly to power or heat with no carbon emissions, and with water as the primary by-product. As an added bonus, hydrogen is nature's most abundant element. Unfortunately, it exists primarily in strongly bonded chemical compounds, and extracting it is a difficult and energy-intensive process. With current technologies, it would be difficult to economically produce, store or utilize this ideal energy carrier. An enormous amount of technology and infrastructure development would be needed to attain a pure hydrogen economy.

Realistically, a green economy will emerge comprising a broad portfolio of alternative energy sources and carriers, including hydrogen. As our present reliance on petroleum-based fossil fuels become increasingly difficult to sustain, both economically and environmentally, this will become inevitable. As a result, new, more distributed approaches to national and world energy management will take hold. Different locations rich in their own renewable resources can manufacture energy currencies such as electricity or hydrogen for large-scale distribution to the broader energy marketplace. Electricity is a key energy carrier today, and will remain so long into the future. Hydrogen, however, will also emerge in an important complementary role, providing important benefits in large-scale energy storage and long-distance distribution.

Bottom line, without hype: hydrogen is valuable today, and will become increasingly valuable as an energy carrier with the future development of new renewable-energy production and distribution infrastructures. In the process, new and improved technologies will emerge for the economical and environmentally friendly production, storage, delivery and utilization of hydrogen. Production technologies using solar energy to split water are enormously attractive, motivating, for example, accelerated PEC research and development.

1.3 Solar Pathways to Hydrogen

1.3.1 The Solar Resource

In discussing the world energy situation of the early twentieth century, Thomas Edison once said: "I'd put my *money* on the sun and solar energy. What a source of power! I hope we don't have to wait 'til oil and coal run out before we tackle that" [11]. It's almost 100 years later, and we are still hoping, perhaps now with a little more urgency. The sun is, in fact, the ultimate renewable energy resource, continuously bombarding earth with about 180 000 000 000 000 000 W (or 180 000 TW) of radiant power, enough to power 3 quadrillion 60 W light bulbs [12,13]! About 50 000 TW of this is directly reflected back to space, and 82 000 TW is absorbed by earth and re-emitted as heat. Of this, 36 000 TW is absorbed at the earth's land masses, where terrestrial-based solar-energy conversion plants could be installed practically.

To put this in perspective, our society on average consumes 13–15 TW, with some predictions doubling this consumption rate by the year 2060 [14]. Although these numbers are staggering, they still represent a small fraction of the sun's influx of radiant power. It might seem that solar energy alone could satisfy our insatiable hunger for energy. Of course, it is not that simple. The planet relies on the sun for many things, including sustaining plant-life and driving its weather patterns, and our voracious energy demands are relatively low on nature's priority list. Still, despite the abundance of spare solar energy at our disposal, large-scale conversion is currently quite costly and somewhat problematic. At peak times of daylight, the solar intensity available for terrestrial conversion scales to approximately 1000 W m^{-2}. Large collection areas and significant landmass would therefore be needed for commercial-scale power production. Such expansive commercial deployment requires an enormous capital investment. For example, commercial photovoltaic technologies today can convert sunlight to electricity at efficiencies between 10 and 20%, at $2–5 per installed watt [15]; a single Gigawatt plant would cost billions of dollars and span over 2500 acres! Even worse, this gargantuan installation would be a sleeping giant at night and under severe cloud cover.

There are certainly practical difficulties and challenges, yet our sun is still the most generous renewable resource, and the most underutilized in modern society. Currently, less than 0.05% of the world energy production is from solar-energy plants, though this number is on the rise of late [16]. Encouragingly, improved technologies for solar-energy conversion, storage and utilization are emerging to make their impact on the world energy scene. New and improved solar-to-electric and solar-to-hydrogen conversion technologies are all poised to be part of the new energy mix.

1.3.2 Converting Sunlight

In converting sunlight, whether to electricity or to hydrogen, fundamental thermodynamic principles govern the energy-conversion process. As illustrated in Figure 1.1, the sun can be viewed as a black body radiating at a temperature of 5780 K, while the earth, as a *black body*, radiates at 300 K. The Carnot limit between these source and sink temperatures is readily calculated to 95%, representing the amount of radiant energy that can be converted into other more useable energy forms. This is very encouraging! A lot of solar energy available, and in theory most of it can be converted for practical end-uses. Unfortunately, however, actually converting sunlight is always further limited by unavoidable losses associated with available energy-conversion routes. Thermodynamically, efficiency is lost with every added conversion step in the process.

The sun transmits energy radiatively via photons, quantum particles of light with discrete energy content. Figure 1.2 shows the standard $AM1.5_{global}$ atmosphere-filtered solar spectrum [17] indicating the range of photon energies comprising sunlight, and the distribution of energy transmitted by these photons. The solar photons (γ) reaching earth readily interact with electrons, energizing them to excited states (e^-), as illustrated in Figure 1.1. Two basic routes for energy conversion of the photoexcited electrons are also depicted. In the *solar-thermal route*, the energized electrons thermalize to their surroundings, converting the energy to heat (ν). This thermal energy can be converted further, for example, using heat-engines to produce work, though now restricted by a lower Carnot limit based on an intermediate source temperature.

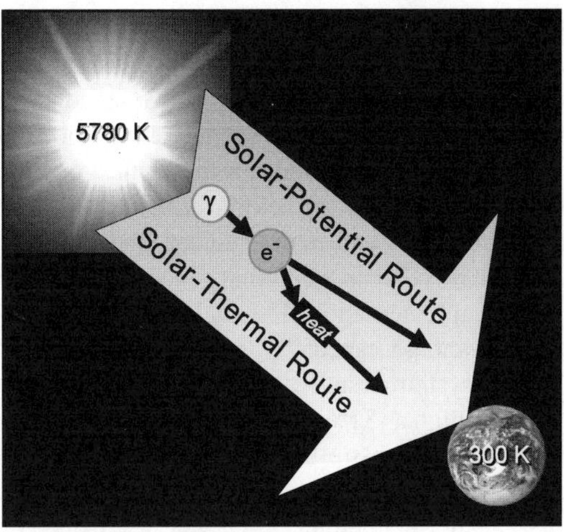

Figure 1.1 Black-body representation of the sun and earth showing solar-energy conversion routes.

Alternatively, in the *solar-potential route*, the elevated electrochemical potential of the energized electrons can directly drive further conversion processes, for example, producing electricity or chemical products. Thermal energy is not being converted, so no additional Carnot limits are imposed.

1.3.3 Solar-Thermal Conversion

In solar-thermal conversion systems, concentrated sunlight produces high temperatures to drive heat-engines for generating mechanical work, electrical energy, or chemical products.

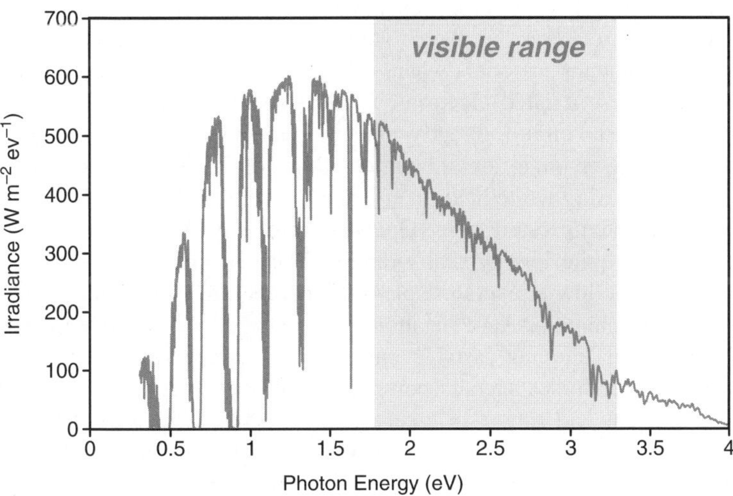

Figure 1.2 The AM1.5_{global} spectrum shown as a function of photon energy. The range of photons in visible light is highlighted.

A good example of this route is the solar-thermal production of electricity. Concentrating solar thermal (CST) systems employ mirrored troughs, dishes or heliostats for focusing sunlight to heat working fluids of a gas- or liquid-phase turbine cycle. Solar concentration up to 1000× and working-fluid temperatures in the 250–1100 °C range are common. Conversion efficiencies, governed by the Carnot limit, can be quite high for the highest operational temperatures, but significant materials issues arise. Exotic refractory materials are needed, adding significant cost to plant production, operations and maintenance. One important advantage of solar-thermal production of electricity is that conventional generators and infrastructure can be used, facilitating plant design and implementation. Thermal storage of the energy via storage of the heated fluid can also be an advantage, especially in the higher-temperature systems. Recently, an experimental CST 25 kW system installed at Sandia National Laboratories has reported solar-to-electric efficiencies as high as 31% [18]. Larger-scale installations, such as the 50 MW AndaSol plant in Spain operate at gross efficiencies closer to 3% [19].

Another example of solar-thermal energy conversion is the production of hydrogen as a chemical by-product of solar-thermochemical cycles (STC). Concentrated sunlight provides the net heat for driving a multistep thermochemical process involving the splitting of water into hydrogen and oxygen gases. Many STC chemical cycles are known, including the sulfur–iodine [20,21] and copper–chlorine [22] cycles with reaction temperatures ranging up to 1200 K. High solar-to-hydrogen conversion efficiencies are possible, reported between 42–57% in the sulfur–iodine cycles, with a high-temperature step at 1123 K. The high-temperature, corrosive operating environment of all STC reactors, however, can be problematic, requiring specialized, and usually expensive materials, components and maintenance.

1.3.4 Solar-Potential Conversion

In the solar-potential route, photons in the incident solar light energize electrons, which can be converted directly to electrical or electrochemical energy. The primary example of the solar-potential conversion process is the photovoltaic (PV) production of electricity. Photons are absorbed in semiconductor materials where they excite electrons from the valence into the conduction band. These excited electrons, at elevated electrochemical potentials, can be extracted into an external circuit, directly converting the photon energy into electric energy. Though direct, the conversion is not without loss. Some of the excited electrons thermalize to their surroundings, causing waste heat. In efficient PV cells, however, this waste is minimal, resulting in moderate temperature rises. Operating temperatures in PV installations without concentration can be quite low, typically ranging from 30 to 80 °C. This is a particularly attractive feature, since low-temperature plants do not require specialized materials, and are easy to operate and maintain. Another attractive feature of PV-generated electricity is the absence of mechanical "moving parts" common to the turbine systems in CST generation. Large-scale power electronics such as power inverters are needed, but these systems have become more efficient and robust in recent years. On the down side, PV semiconductor materials and systems are still relatively expensive. Although cumulative global installations of PV generation has reached over 1500 MW [15], the per-installed-watt price still exceeding $3 is somewhat prohibitive in may economic sectors.

Other examples of solar-potential conversion include photoelectrochemical processes such as waste-water remediation, and the industrial synthesis of chemicals and synthetic fuels.

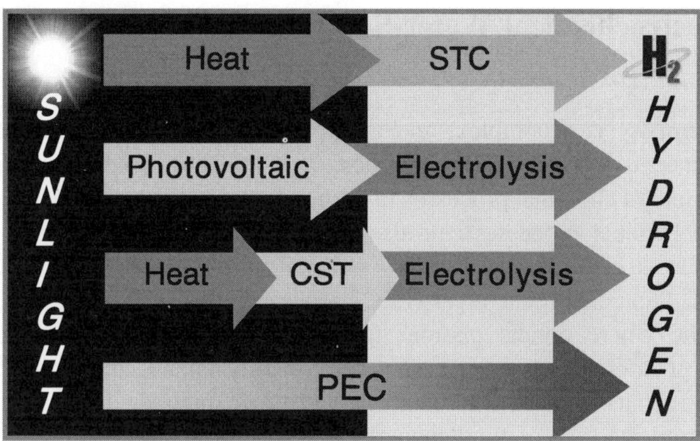

Figure 1.3 Solar-to-hydrogen conversion pathways. (STC stands for solar-thermochemical, CST for concentrating solar-thermal, and PEC for photoelectrochemcial).

Hydrogen production by PEC water splitting, an attractive low-temperature alternative to solar-to-hydrogen water splitting, falls into this category.

1.3.5 Pathways to Hydrogen

Using sunlight to split water for hydrogen production can follow several different conversion routes, as shown in Figure 1.3. The solar-thermal route is essentially a two-step process, with a photon-to-thermal energy-conversion step followed by a thermal-to-chemical conversion step. The other two-step process shown in the figure represents PV-electrolysis, where a photon-to-electric conversion step is followed by an electric-to-chemical conversion process. The three-step process represents a CST-electrolysis route, involving photon-to-heat, heat-to-electricity and electricity-to-chemical conversion steps. The final pathway depicted, representing a single-step direct conversion from photon-to-chemical energy, is the PEC water-splitting process. Other solar-to-hydrogen pathways are possible, including photobiological routes [23,24] and the ultra-high-temperature thermolysis route [25]. All pathways can contribute to renewable hydrogen production for future "green economies," but economics will determine which will predominate.

From an economic viewpoint, it is important to remember that both hydrogen and electricity will be valuable as renewable-energy carriers in the future. Processes capable of producing both, such as PV-electrolysis and CST-electrolysis, could be advantageous. In fact, PV- and CST-electrolysis systems can be assembled today using off-the-shelf components. The electricity and the hydrogen produced would not be inexpensive, but this will change with further maturing of the technologies. It will remain vital to keep an eye on the alternative, even less-mature approaches. The solar-electrolysis routes comprise multiple conversion steps, with efficiency loss at each step. In terms of hydrogen production, the most direct conversion processes, such as PEC water splitting, could have some inherent performance advantages. PEC hydrogen production as a low-temperature single-stage process remains one of the front-running alternatives.

1.4 Photoelectrochemical Water-Splitting

1.4.1 Photoelectrochemistry

Photoelectrochemistry is a complex, and extremely rich scientific field drawing together fundamental concepts from chemistry, physics, optics, electronics and thermodynamics. In contrast with standard chemical processes involving interactions between chemical and ionic species, electrochemical processes can also involve interfacial interactions between ionic conductors, such as electrolyte solutions, and solid-state electronic conductors, such as semiconductors. Photoelectrochemical (PEC) processes comprise electrochemical systems exposed to light, where optical photons interact with the electrochemical reactions. In semiconductor photoelectrochemistry, photons typically create electron–hole pairs within the semiconductor that can react with redox chemistry at semiconductor/electrolyte interfaces. Although a complicated set of fundamental electrochemical and solid-state optoelectronic principles govern the behavior of such systems, some useful simplifications can be helpful in providing a broad overview of the PEC water-splitting process.

1.4.2 PEC Water-Splitting Reactions

Most texts on PEC water splitting will start with the simple two-electrode setup shown in Figure 1.4. In this canonical model, a light-sensitive semiconductor photoelectrode is immersed in an aqueous solution, with electrical wiring connected to a metallic counter-electrode. With exposure to sunlight, photogenerated electron–hole pairs in the semiconductor interact electrochemically with ionic species in solution at the solid/liquid interfaces. Photoexcited holes drive the oxygen-evolution reaction (OER) at the anode surface, while photoexcited electrons drive the hydrogen-evolution reaction (HER) at the cathode surface. Figure 1.4 depicts a *photoanode* system where holes are injected into solution at the semiconductor surface for evolving oxygen, while photoexcited electrons are shuttled to the counter-electrode where hydrogen is evolved. Conversely, in *photocathode* systems, electrons are injected into solution and hydrogen is evolved at the semiconductor surface, while oxygen is evolved at the counter-electrode. Similar to solid-state pn-junction solar cells, PEC photoelectrodes typically act as *minority carrier* devices [26,27]. The semiconductor/liquid junction, like the pn junction, allows the flow of minority carriers, while blocking majority-carrier flow. For this reason,

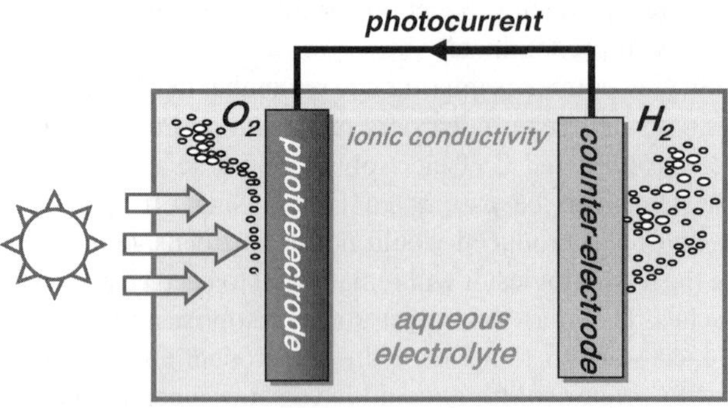

Figure 1.4 Standard two-electrode setup for PEC water splitting, shown in the photoanode configuration with a separated counter-electrode.

n-type semiconductors allowing minority-carrier hole injection are better suited as photoanodes, while p-type semiconductors are used as photocathodes.

In the PEC water-splitting process, oxygen evolution at the anode and hydrogen evolution at the cathode can be modeled as two electrochemical "half-reactions." Both must be sustained simultaneously, coupled by their exchange of electrons in the solid state, and ions in solution. A simplified equation set describing the half-reactions in addition to the net conversion process can be written:

$$2\gamma \rightarrow 2e^- + 2h^+ \quad \text{Photon-induced eletron-hole pair generation} \tag{1.1a}$$

$$H_2O + 2h^+ \rightarrow 2H^+ + {}^1\!/_2 O_2(\text{gas}) \quad \text{OER : anodic water-oxidation half-reaction} \tag{1.1b}$$

$$2H^+ + 2e^- \rightarrow H_2(\text{gas}) \quad \text{HER : cathodic } H^+ \text{ reduction half-reaction} \tag{1.1c}$$

$$H_2O + 2\gamma \rightarrow H_2(\text{gas}) + {}^1\!/_2 O_2(\text{gas}) \quad \text{Net PEC water splitting reaction} \tag{1.1d}$$

$$\Delta G^\circ = +237.18\ \text{kJ mol}^{-1} \quad \text{Standard Gibbs free energy} \tag{1.1e}$$

$$V^\circ_{rev} = \Delta G^\circ / nF = 1.23\ V \quad \text{Standard reversible potential} \tag{1.1f}$$

$$V_{op} = V^\circ_{rev} + \eta_a + \eta_c + \eta_\Omega + \eta_{sys} \quad \text{Operating voltage with overpotential losses} \tag{1.1g}$$

where γ is photon energy, e^- is an electron, h^+ is a hole, ΔG^o is the standard Gibbs free energy, V°_{rev} is the standard reversible potential, n (=2) is the number of electrons exchanged, F is the Faraday constant, V_{op} is the operational voltage, η_a, η_c, η_Ω, and η_{sys}, are overpotentials associated with anode, cathode, ionic-conductivity and system losses, respectively.

Implicit in Equation Set 1.1, solid-state electrons/holes are exchanged between the anode and cathode through a conductive pathway (such as a wire), while H^+ ion migration from anode to cathode is through the aqueous media. It is clear from the equation set that PEC water splitting is a delicate balancing act, where photon-energized electron–hole pairs under the right conditions can simultaneously drive the electrochemical half-reactions. In steady-state, the reactions in Equations 1.1b and 1.1c, must be sustained at the same reaction rate. Since H^+ ions are generated at the anode surface and consumed at the cathode surface, unless these events are proceeding concurrently at identical rates, charge build-up will impede or even stop the entire process. A similar situation exists with the charge carriers in the solid state. The anodic half-reaction consumes two holes (i.e., supplies two electrons) while the cathode half-reaction consumes two electrons. These electrons must be shuttled from anode to the cathode via electrical current (e.g., through the interconnecting wire shown in Figure 1.4), and steady state cannot be maintained if anode and cathode reaction rates are not the same.

There are several additional key points that should be emphasized regarding the thermodynamic parameters included in Equation Set 1.1:

- $\Delta G^\circ = +237.18\,\text{kJ}\,\text{mol}^{-1}$ is the standard Gibbs free energy change, representing a thermodynamic minimum for splitting water into the constituent gases at standard conditions of 25 °C and 1 bar. Since ΔG° is positive, energy needs to be supplied to the drive electrolysis process.

- $V^{\circ}_{rev} = 1.23\,eV$ is the corresponding reversible potential, indicating the minimum electrical potential needed to sustain reversible photoelectrolysis.
- Water splitting will *not* occur at a bulk potential of 1.23 V. This value does not take into account unavoidable process losses, including overpotential loss at the anode (η_a), overpotential loss at the cathode (η_c) or ionic conductivity losses in solution (η_Ω), in addition to other balance-of-system losses (η_{sys}).
- The half-reactions described in Equations 1.1b and 1.1c are simplifications of more complex multistep electrochemical reaction pathways [28,29]. The electrode overpotential losses of these multistage reactions, including the effects of activation energy, kinetics and mass-transport can be substantial, commonly several tenths of volts. The water-oxidation reaction at the anode is the more complex and less facile reaction, so the anodic overpotential losses are more severe.
- Overpotential losses due to ionic conductivity in the solution can also be severe. These losses depend on solution properties, as well as electrode geometry and spacing. The splitting of *pure water* is particularly difficult, since the ionic conductivity, typically less than $0.05\,S\,m^{-1}$, is prohibitively low. Weak acid or alkaline solutions with conductivities exceeding $10\,S\,m^{-1}$ are typically used to compensate, although this creates a more corrosive environment for the electrodes.
- V_{op}, the operating voltage for water splitting, must exceed V°_{rev} to compensate for all the losses, as indicated in Equation 1.1g. In practice, water electrolysis systems typically require operating voltages of 1.6–1.9 V, depending on gas-production rates [30,31].

The focus of Equation Set 1.1 is on electrochemical behavior and the losses in solution. Losses in the solid-state electrodes, including electron–hole-pair recombination losses and electronic conductivity losses, among others, also degrade system performance. To drive the water-splitting process, including all solution and electrode losses, the absorbed photons must induce sufficient electrochemical potential to the electron–hole pairs. The photoelectrolysis *balancing act* can be set into motion *only* if the photopotential requirement is met. Once in operation, the hydrogen evolution will be proportional to electron consumption, as per Equation 1.1c.

During steady-state operations, the solid-state shuttling of charges between anode and cathode represents a photon-induced current, or photocurrent, that is integrally tied to the hydrogen-producing performance of the PEC system. Explicitly from Equation 1.1c, two electrons are consumed in the evolution of one H_2 molecule. The rate of hydrogen production is therefore half the rate of electron flow, in other words, half the photocurrent. This is technically written:

$$R_{H_2} = \frac{I_{ph}}{2e} = \frac{(J_{ph} \times \text{Area})}{2e} \Rightarrow J_{ph} = \left(\frac{R_{H_2}}{\text{Area}}\right) \times 2e \qquad (1.2)$$

where R_{H_2} is the hydrogen production rate (s^{-1}), I_{ph} is the photocurrent (A), e is the electronic charge (C), Area is the illuminated photoelectrode area, J_{ph} is photocurrent density ($A\,m^{-2}$).

In Equation 1.2, the photocurrent density J_{ph} is normalized to the illuminated area of the photoelectrode, and is therefore inversely proportional to the incident photon flux. Upon closer look, J_{ph} is proportional to the ratio between the hydrogen production rate and the solar energy input. As a result, this parameter becomes particularly important when evaluating the solar-to-hydrogen conversion performance of a PEC system.

1.4.3 Solar-to-Hydrogen Conversion Efficiency

The chemical solar-to-hydrogen (STH) conversion efficiency of any solar-based hydrogen production system is defined as the ratio of the useable chemical energy in the generated hydrogen gas to the total solar energy delivered to the system. For steady-state operations, this is equivalent to the ratio of the *power output* to the *power input*. In words, this can be expressed:

$$\frac{P_{\text{out}}}{P_{\text{in}}} = \frac{(\text{hydrogen production rate}) \times (\text{hydrogen energy density})}{\text{solar flux integrated over illuminated area}} \tag{1.3}$$

Using the hydrogen-production rate from Equation 1.2, the Gibbs energy as the useful energy density of the hydrogen and an integrated solar flux of $1000\,\text{W m}^{-2}$ for AM1.5_{global} solar irradiation, the STH efficiency for a PEC system can be expressed:

$$\text{STH}(\%) = \frac{\Delta G R_{\text{H}_2}}{P_{\text{solar}} \times \text{Area}} = \frac{\Delta G\left(\frac{J_{\text{ph}} \times \text{Area}}{2\text{e}}\right)}{P_{\text{solar}} \times \text{Area}} \approx \underbrace{0.123 \times J_{\text{ph}}(\text{Am}^{-2}) = 1.23 \times J_{\text{ph}}(\text{mA cm}^{-2})}_{\text{for AM1.5}_g \text{ solar irradiation}} \tag{1.4}$$

The first ratio in Equation 1.4 is generic for any solar-to-hydrogen production system, while the second is derived specifically for PEC hydrogen processes. The third term, explicitly relating conversion efficiency to the photocurrent density, is calculated for a PEC system under AM1.5_{global} solar illumination. In Equation 1.4, the use of the Gibbs free energy reflects chemical energy in the hydrogen that can be retrieved using an ideal fuel cell. This, in effect, calculates the lower-heating value (LHV), which is standard in practical comparisons between different fuels.

On the subject of standard efficiency terminology, numerous types of efficiency have been defined and employed throughout PEC literature [32–35], but extreme care needs to be taken in the appropriate application and interpretation of each. For valid side-by-side comparisons with other solar-to-hydrogen conversion technologies, the STH definitions in Equation 1.4 must be used. To be strictly correct, the hydrogen gas evolved should be collected and certified in any efficiency determination, since parasitic effects cannot be quantified in volume or photocurrent measurements alone. Alternative efficiency definitions have included three-terminal efficiencies, efficiencies specific to a limited range of photon wavelengths, and energy-saving efficiencies for externally biased systems. These can be extremely useful in characterization of PEC materials and interfaces, but cannot be equated with a system STH efficiency. STH calculations using the higher heating value (HHV) for hydrogen have been reported, founded on novel utilization schemes recovering the water's heat of condensation. These, however, do not conform to industry-standard reporting practices based on the LHV.

Independent of the heating value used, it is clear from Equation 1.4 that the PEC STH efficiency is *all about the photocurrent.* In contrast to solid-state solar cells, operated at the maximum power point (i.e., maximum product of photocurrent and photovoltage) for the best solar-to-electric conversion efficiency [36], the PEC cell should be operated at maximum photocurrent for best hydrogen-production performance. This becomes extremely important in the design and optimization of PEC semiconductor materials and devices. It is the *saturated photocurrent density* limit of a semiconductor that ultimately constrains the hydrogen production rate. For peak efficiency, sufficient photopotential must be generated in the device to drive the photocurrent into saturation. With all the built-in losses, this can be a difficult challenge.

1.4.4 Fundamental Process Steps

Losses affecting STH conversion efficiency are inherent in all processes occurring in the solid state, in solution and, often most importantly, at the interface. To help keep track of the loss mechanisms, it is useful to break down the PEC water-splitting dynamics into fundamental process steps, tracking events all along the way from the photon collection in the photoelectrode to the hydrogen release in solution. These basic steps can be summarized as follows:

1. **Photon Absorption/Charge Generation (solid-state):** Solar photons are absorbed in the semiconductor, creating excited charge carriers in the form of electron–hole pairs. Absorption losses are related to semiconductor bulk properties.
2. **Charge Separation and Transport (solid-state/interface):** Photoexcited electron–hole pairs must be separated spatially, and before recombining to lower energy states, must be transported to opposite contact surfaces for extraction. The separation mechanisms are tied to charge distributions in the solid state and at the interface. Transport losses are related to semiconductor defects and other mobility-limiting effects.
3. **Charge Extraction/Electrochemical Product Formation (interface):** Charge carriers transported to anode/cathode surfaces can be extracted into the water-oxidation/hydrogen-reduction half-reactions, respectively. Oxygen/hydrogen gas is produced, while hydrogen ions are consumed/formed. Interface losses are many, including poorly aligned energetics, reaction overpotentials and slow reaction kinetics.
4. **Electrochemical Product Management (solution):** Hydrogen and oxygen gas need to be removed from solution, while the hydrogen ion concentrations need to redistribute. Ionic conductivity losses as well as "bubble" losses (e.g., related to disruption of mass transport in solution and to possible light blockage) are present.

In a semiconductor photoelectrode system, photons are absorbed in the semiconductor bulk, and the photogenerated charge carriers (in the form of electron–hole pairs) are separated, transported and extracted due to the rectifying nature of the semiconductor/electrolyte junction. Not surprisingly, hydrogen production performance is strongly influenced both by semiconductor material properties and by junction characteristics.

1.5 The Semiconductor/Electrolyte Interface

1.5.1 Rectifying Junctions

In semiconductor-based PEC water-splitting systems, the semiconductor/electrolyte interface can form a *rectifying* junction, similar to the solid-state pn junctions or Schottky diode junctions used in solar cells. Such rectifying junctions exhibit built-in electric fields capable of separating excited charge carriers (i.e., electron–hole pairs) created by absorbed solar photons. In solid-state solar cells, this charge-separation mechanism drives photocurrents to produce electricity, while in the PEC case, the charge separation can drive the HER and OER half-reactions for water-splitting. In both cases, illumination creates extra photoexcited electron–hole pairs, which need to be separated and extracted before they recombine. Extraction of the photogenerated charge carriers with elevated electrochemical potentials in effect converts the solar energy to electricity or hydrogen. Many good sources of information are available, detailing semiconductor material properties, solid-state junctions and solar cells [13,26,27,37–39], and these provide an excellent background for understanding fundamentals of rectifying junction formation and behavior.

1.5.2 A Solid-State Analogy: The np$^+$ Junction

Before tackling the semiconductor/electrolyte junction, it is interesting to consider the analogous solid-state np$^+$ junction represented in Figure 1.5. In this device, the *n-region* is doped with *donor* atoms to provide an excess of free electrons, and the *p$^+$-region* is more heavily doped with *acceptor* atoms for a high concentration of holes. Carrier densities at equilibrium typically would be, for example, 10^{15} cm^{-3} for the n-region and 10^{16} cm^{-3} for the p$^+$-region. Figure 1.5 includes the classic band-diagram representation of the np$^+$ device, with Figure 1.5a showing the separated semiconductors in thermal equilibrium. Consistent with the relative level of doping, Fermi levels (F) are close to the conduction band (CB) in the n-region, and very close to valence band (VB) on the p+ side.

Figure 1.5b depicts the np$^+$ junction formation as the two doped materials are brought together. At equilibrium, the Fermi levels align across the device, resulting in the band-bending

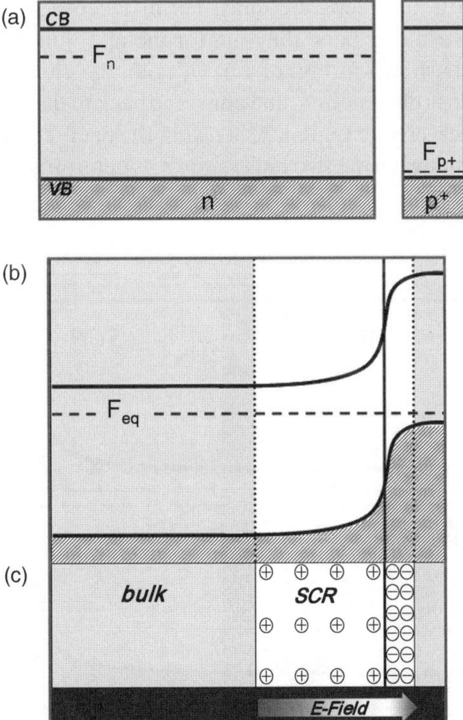

Figure 1.5 Formation of the np$^+$ solid-state rectifying junction showing: (a) materials before contact; (b) band diagram of junction formation and equilibration and (c) charge distributions and built-in electric field. (F_n, F_{p+} and F_{eq} represent Fermi levels in the isolated n and p$^+$ regions, and in equilibrium after contact, respectively).

observed in the figure. Near the junction, free electrons from the n-side diffuse into the p^+-side, leaving behind the fixed positive charges of ionized donor atoms. Conversely, holes from the p^+-side diffuse into the n-region, exposing fixed negatively charged acceptor atoms. A *depletion region*, also known as a *space-charge region* (SCR), forms where free carrier diffusion is counter-balanced by the built-in electric field generated by the fixed charges flanking the junction. As seen in the fixed-charge distribution shown in Figure 1.5c, the space-charge region extends further into the less-conductive n-region, to the point where the net space charges in the n- and p^+-sides balance. Typical space-charge widths for such a device would extend 1 micron into the n-region, but only 0.1 micron into the p^+-region. The built-in electric field established by the fixed charge distribution is also indicated in Figure 1.5c. This built-in field is the critical mechanism for separating electron–hole pairs generated under illumination.

When sunlight shines in the vicinity of the np^+ junction, some of the solar photons are absorbed, creating elevated concentrations of electron–hole pairs. Electron–hole pairs generated within or near the space-charge region are separated by the built-in field, with the electrons and holes pushed toward opposite sides of the device. The photocharges successfully extracted are converted to electricity, while the remainder will recombine, losing energy to radiation or heat. Under solar illumination, thermal equilibrium at the junction is disturbed, and a single Fermi level cannot be defined. Instead, quasi-Fermi analysis can be applied, where the original Fermi level splits into separate quasi-Fermi levels for electrons (F_e) and holes (F_h) [26,27,39,40]. The resulting band diagram for the np^+ junction under illumination at open-circuit conditions is shown in Figure 1.6. Of note, the electron and hole quasi-Fermi levels are continuous across the junction, and converge back to the bulk n-region and p-region levels away from the space-charge region. As a result, the open-circuit potential, represented by the Fermi level offset between the two bulk regions, is determined by the quasi-Fermi level separation in the device.

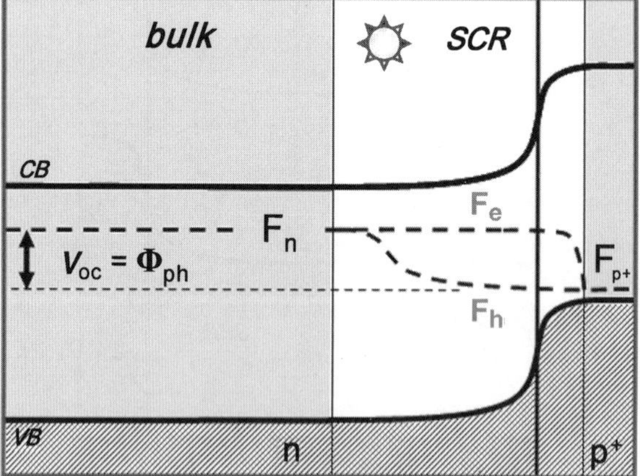

Figure 1.6 Band diagram of the np^+ junction under solar illumination, illustrating the development of usable photopotential through quasi-Fermi level separation. (SCR stands for space-charge region, Φ_{ph} for photopotential, V_{oc} for open-circuit voltage, and F_e and F_h for electron and hole quasi-Fermi levels, respectively).

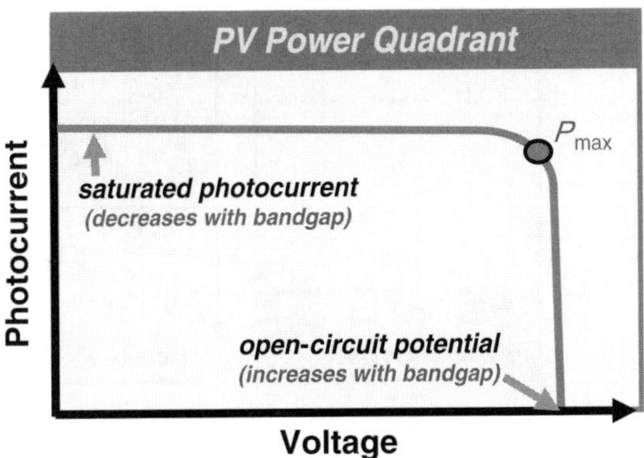

Figure 1.7 Current–voltage performance curve of solid-state solar cell showing open-circuit potential, the saturated photocurrent level and the maximum power point.

The effect of this quasi-Fermi level split is extremely important to how much *useable* photopotential the device can generate, specifically in relation to the bandgap energy. In crystalline silicon, for example, the bandgap energy is 1.1 eV, while single-junction silicon solar cells have open-circuit voltages typically between 0.6 and 0.7 V. Thermodynamically, semiconductor band-diagram representations reflect the *internal energy* of electrons and holes, not the *useable energy*. Electricity can be extracted from a solar cell at potentials below the open-circuit voltage, which is typically 50–75% of the semiconductor bandgap energy. The output voltage can be increased using higher-bandgap cells, but as a tradeoff, fewer solar-spectrum photons (refer to Figure 1.2) would be absorbed, limiting the saturated photocurrent. The effects of bandgap on photopotential and photocurrent are indicated in the generic solar-cell performance curve of Figure 1.7. These effects remain extremely relevant to the performance of PEC rectifying junctions under sunlight.

1.5.3 PEC Junction Formation

To describe rectifying PEC junctions formed at semiconductor/electrolyte interfaces, key principles from solid-state physics and electrochemistry need to be combined. For reference, there is a wealth of literature on fundamental electrochemical principles [41–46], in addition to the previous citations covering semiconductor physics. Of particular interest to PEC studies are the models developed by Gerischer, which make the important connections between the in-solution electrochemical potentials of electrons and solid-state Fermi levels [40,47,48]. Using the Gerischer models, descriptions of semiconductor/electrolyte junctions follow closely the solid-state junction analogies. Photoanodes using n-type semiconductors form PEC junctions similar to an np^+ junction, as depicted in Figure 1.8, while p-type photocathode junction formation, illustrated in Figure 1.9, would be more analogous to a solid-state pn^+ device. Since there are clear symmetries in the development of photoanode versus photocathode junctions, as a starting point it is instructional to focus initially on one, for example the photoanode from Figure 1.8.

Figure 1.8a depicts the photoanode and electrolyte solution before contact. The Fermi level in the n-type semiconductor (F_n) is close to the CB, and the Fermi level in solution (F_s) falls between the redox (reduction/oxidation) levels for hydrogen reduction (H^+/H) and water

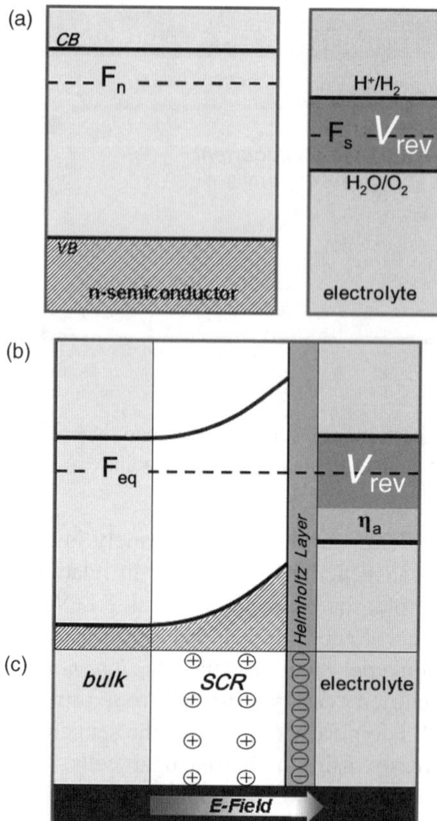

Figure 1.8 Formation of a semiconductor/electrolyte PEC junction based on an n-type photoanode, showing: (a) semiconductor and electrolyte before contact; (b) band diagram of junction formation and equilibration; and (c) charge distributions in the space-charge region and Helmholtz double layer, and the built-in electric field. (F_s represents solution Fermi level, $V_{rev} = 1.23\,\text{eV}$ is the reversible potential for water splitting, and η_a is the anodic overpotential).

oxidation (H_2O/O_2). After immersion, the electrode and electrolyte Fermi levels must align to reach thermal equilibrium, as shown in Figure 1.8b. Since the initial electrode Fermi level is higher than the electrolyte Fermi level, free electrons in the n-type semiconductor will migrate to the solid–liquid interface exposing positively charged fixed donor sites, similar to the np^+ example. In the PEC case, however, the electrons form a surface charge layer at the interface, which induces a thin Helmholtz double layer in the electrolyte. The charge distributions including the fixed space charges in the solid-state and the Helmholtz layer charges in solution are shown in Figure 1.8c. Typically, Helmholtz layers are on the order of a few nm, compared with several microns for the semiconductor space-charge region.

As in the np^+ case, electron–hole pairs generated by photon absorption in the space-charge region can be separated by the built-in electric field. Photoexcited electrons are driven toward the electrode's back contact, where they can be extracted, for example to a counter-electrode in solution. On the other hand, photogenerated holes will be driven toward the interface, where

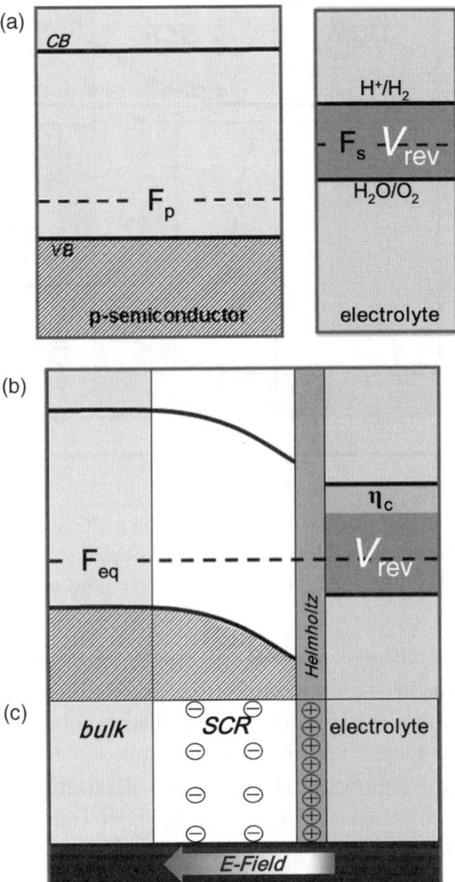

Figure 1.9 Formation of a semiconductor/electrolyte PEC junction based on a p-type photocathode, showing: (a) semiconductor and electrolyte before contact; (b) band diagram of junction formation and equilibration; and (c) charge distributions in the space-charge region and Helmholtz double layer, and the built-in electric field. (Here, η_c is the cathodic overpotential).

(with the appropriate energetics and kinetics) they can drive oxidation reactions in the electrolyte. Before focusing on the dynamics of the illuminated photoanode, it is worth noting that the equilibrium charge-distribution process is the same for p-type photocathodes, but as shown in Figure 1.9 the charges and band-bending are reversed.

1.5.4 Illuminated Characteristics

Coming back to the photoanode, the PEC junction response to solar illumination is detailed in Figure 1.10. In the figure, the back contact of the photoelectrode is connected by external wiring to a counter-electrode also in solution; and added in the band diagram are η_a and η_c, the overpotentials for water oxidation and hydrogen reduction associated with photoanode and counter-electrode interfaces, respectively. With the addition of sunlight, there is again

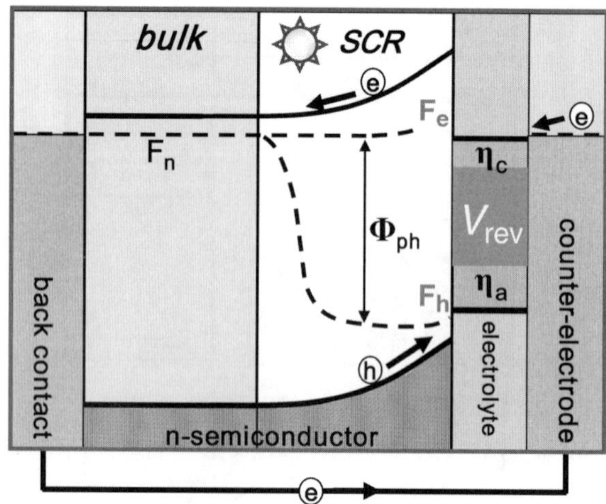

Figure 1.10 Band diagram of the photoanode PEC junction under solar illumination, illustrating the development of sufficient photopotential (i.e., quasi-Fermi-level split) to overcome the reversible potential plus the anodic and cathodic overpotentials for water splitting.

a quasi-Fermi-level split resulting from the excess concentration of photogenerated electron–-hole pairs. With respect to the thermal equilibrium populations of carriers, the excess hole population significantly alters the minority-carrier distribution, while the excess electrons barely affect the majority-carrier numbers. As a result, the hole quasi-Fermi level shifts substantially in contrast to insignificant change in the electron quasi-Fermi level.

As in the solid-state case, the quasi-Fermi separation determines the *useable* energy in the device. In Figure 1.10, the electron potential at the counter-electrode, tied to the photoanode's back contact potential, is sufficiently high to drive the hydrogen-reduction half-reaction (including η_c). Simultaneously, the quasi-Fermi hole energy at the solution interface is sufficiently *low* to drive water-oxidation (including η_a). In consequence, this configuration is capable of sustaining the net PEC water-splitting process, driven by *useable* energy in the quasi-Fermi split (*not* by the *internal* energy in the bandgap.). It is often misreported that semiconductors with bandgaps "straddling" the redox potentials can photosplit water. Counter examples are illustrated in Figures 1.11 and 1.12. In both cases the conduction and valence-band edges clearly straddle the redox levels, including the overpotentials. In Figure 1.11, however, the hole quasi-Fermi level at the PEC interface is too high, while in Figure 1.12, the electron potential in the counter-electrode is too low. In either case PEC water splitting is not sustainable. An interesting variation to Figure 1.12 is seen in Figure 1.13: an external voltage source has been added between the photoanode and counter-electrode, boosting electron energies enough to support the hydrogen reduction. This can split water, but no longer qualifies as a simple solar-to-hydrogen conversion process.

1.5.5 Fundamental Process Steps

From the previous examples, it becomes evident that high-bandgap semiconductors are needed just to meet the energetic requirements for single-junction PEC water splitting.

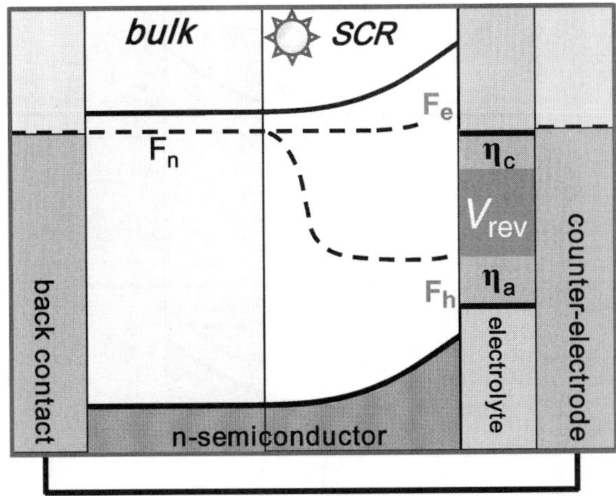

Figure 1.11 Non-operational photoanode junction with insufficient F_h for driving the water oxidation half-reaction.

As in the solid-state case, a large bandgap increases the photopotential, but also reduces the photocurrent. This is particularly unfortunate, since hydrogen production rates and STH conversion efficiencies, as shown in Equations 1.2 and 1.3, are directly proportional to the photocurrent. Quantitative effects of this current–voltage tug-of-war will be evident in some real-world examples of single-junction PEC water splitting. Beforehand, to focus on the important performance-limiting loss mechanisms, it's worth reviewing the

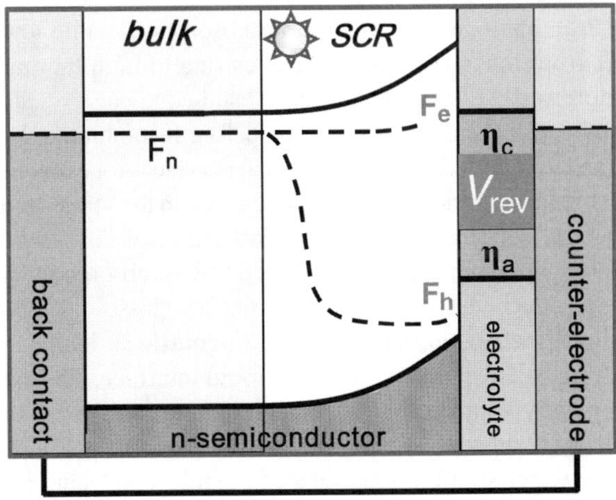

Figure 1.12 Non-operational photoanode junction with insufficient F_e for driving the hydrogen reduction half-reaction.

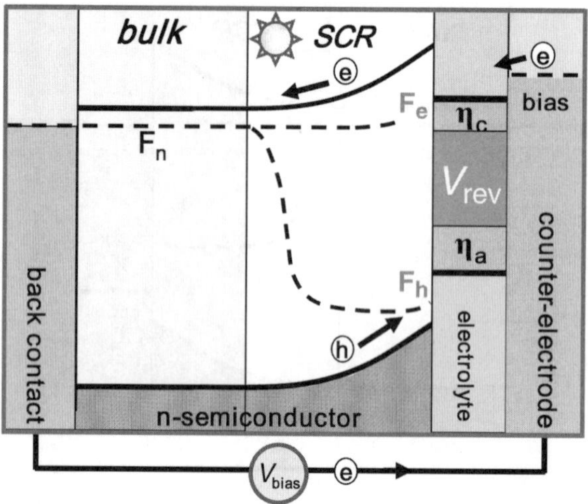

Figure 1.13 Addition of an external bias to the photoanode configuration shown in Figure 1.12, in effect raising the counter-electrode electron energy to enable PEC water-splitting.

fundamental solar-to-hydrogen process steps specifically applied to single-junction photoelectrode systems:

1. **Photon-Absorption and Charge Generation:** In single-junction absorbers, photons with energies below the semiconductor bandgap cannot be absorbed or converted. Photons with energies exceeding the bandgap are absorbed at rates dependent on the allowed transitions in the semiconductor. Direct bandgap materials absorb more efficiently than indirect bandgap materials. Photogenerated electron–hole pairs rapidly thermalize (usually within picoseconds) to band-edge energy levels, losing energy to heat. High-bandgap semiconductors generate little photocurrent due to poor absorption, while low bandgap semiconductors can suffer from low conversion efficiency due to high thermalization losses.
2. **Charge Separation and Transport:** While at band-edge energy states, the electron–hole pairs can often survive for several microseconds before recombining. During this time, they must be separated and transported to electrochemical interfaces for extraction. This separation is assisted by the electric fields set up by charge distributions in the semiconductor and at the solid/liquid interface. Defects in the bulk and at the interface can adversely affect the separation fields, and also result in poor mobility for charge transport. If wide absorption widths are needed (for example, in indirect semiconductors) the charge transport losses can be severe.
3. **Charge Extraction/Electrochemical Product Formation:** Ideally, charge is extracted via the water splitting half-reaction at the solid/liquid interface. The extraction process can be slowed or completely inhibited by poor energetic alignment or poor surface kinetics at the photoelectrode or counter-electrode surfaces. Moreover, parasitic or corrosion reactions competing with the water-splitting reactions can result in substantial loss. Surface treatments can be employed to kinetically and/or energetically favor water splitting over the parasitic processes, but such treatments could also block sunlight. Surface incorporation of nanoparticle catalysts is one approach. Since PEC water splitting is a low-current-density

process (typically operating below 20 mA cm^{-2}), non-precious-metal catalysts can be used. Additionally, nanostructuring of electrode surfaces can increase effective surface area for enhanced charge extraction, although this can also lead to higher surface recombination loss. On the solution side, the electrolyte is an important factor determining stability, efficiency of the charge-extracting reactions, and the electrochemical byproducts. Splitting seawater, for example, is a challenge, since it is difficult to electrochemically suppress the production of chlorine gas from the Cl^- ions [49].

4. **Electrochemical Product Management:** During PEC water splitting, the evolved hydrogen or oxygen gas must be efficiently removed from the photoelectrode surface to avoid mass-transport losses in the surface reactions, and to minimize adverse optical effects. Surfactants added to the electrolyte have been successful in promoting rapid bubble formation and dissipation. In solution, ionic conductivity losses tend to be a bigger problem. High electrolyte concentrations can be used to minimize this loss, but the tradeoff is in higher corrosivity. Photoelectrode geometry and counter-electrode proximity are critical parameters to the redistribution of ions. In some geometries, gas-separating membranes are needed, introducing further ionic-transport loss.

1.6 Photoelectrode Implementations

1.6.1 Single-Junction Performance Limits

With inherent electrochemical and solid-state losses, it is difficult to achieve high STH conversion efficiencies in single-junction PEC photoelectrode systems. The bandgap tradeoff between photopotential and photocurrent is particularly detrimental for single junctions. For example, if the minimum water-splitting potential (based on redox separation with overpotentials) amounts to 1.6 eV, and the quasi-Fermi-level separation in the semiconductor can achieve 50% of the bandgap level, then the minimum bandgap for the onset of photoelectrolysis would be 3.2 eV. For any appreciable level of hydrogen production, overpotential and other system losses increase, requiring even higher bandgaps.

How efficient would such high-bandgap single junctions be? It is easy to establish an upper bound based on optical absorption limits. Figure 1.14 plots the maximum attainable $AM1.5_{global}$ photocurrent densities in a semiconductor as a function of bandgap. The derivation assumes that every photon in the solar spectrum with an energy exceeding the bandgap will create an electron–hole pair, and that all of these electron–hole pairs are converted to photocurrent. For bandgaps greater than 3.2 eV, photocurrent density is limited to approximately 1 mA cm^{-2}. This, according to Equation 1.3, places an upper STH efficiency limit of 1.23%. STH values based on Equation 1.3 are listed in parentheses on the right vertical axis of the Figure 1.14, but these only apply to standalone configurations capable of water splitting. For any bandgap, if the photoelectrode system cannot sustain photoelectrolysis, there is no photocurrent density, and the conversion efficiency is 0% STH.

To date, the only demonstrations of single-junction water splitting have utilized very-high-bandgap materials, such as $SrTiO_3$ and $KTaO_4$ [50,51]. Based on poor photon absorption, the demonstrated STH values have been very small, consistent with the predictions in Figure 1.14. Also indicated in this figure are the bandgap positions for polycrystalline Fe_2O_3, WO_3 and TiO_2, three commonly studied in PEC materials. The potential photocurrent densities look encouraging, especially for iron oxide, but none of these materials develop enough

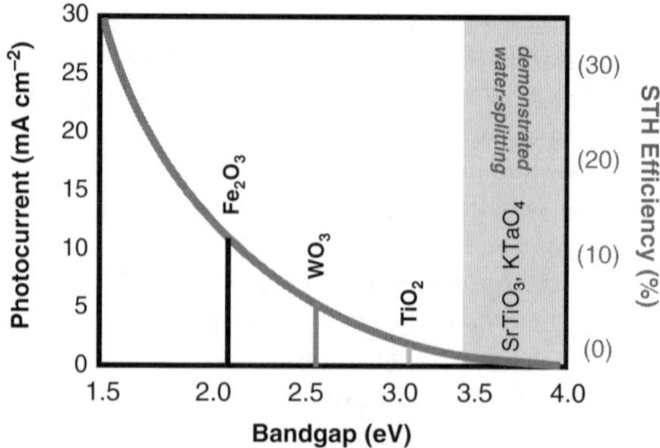

Figure 1.14 Maximum attainable AM1.5_{global} photocurrent densities (in mA cm^{-2}) for single-junction PEC devices shown as a function of semiconductor bandgap. Also shown are the corresponding STH efficiencies applicable only to operational water-splitting junctions.

photopotential under sunlight to split water, not even TiO_2, with a bandgap over 3.0 eV. To enhance the photopotential, these and other promising semiconductors can be incorporated in multijunction PEC schemes.

1.6.2 Multijunction Performance Limits

Multijunction devices are well-known in the PV community. In fact, the use of multiple junctions to improve the solar-to-electricity conversion efficiency is one of the cornerstones of *third generation research* [52]. For PEC devices, a similar approach can be taken to enhance photopotential and increase absorption efficiency in a PEC. The concept is illustrated in Figure 1.15 for a two-junction (tandem) system. Sunlight is partly absorbed in the higher-bandgap top junction, while the remaining filtered light is absorbed in the lower-bandgap bottom junction. Both junctions generate photovoltage and photocurrent. Since they are stacked in a series-connected configuration, the photovoltages V_1 and V_2 will add, but the photocurrents will not. In fact, the net photocurrent will be the minimum of J_1 and J_2, which is the bottle-neck for current flow across the device. For optimal performance, it is therefore critically important to current-match J_1 and J_2 by optical tuning of the two junctions.

In PEC water-splitting applications, multiple junctions can be stacked to take advantage of the photopotential enhancement. The tradeoff is reduced photocurrent, which directly limits hydrogen production rates. The device design must strike the right balance to maximize conversion efficiency. Two different design approaches using tandem junctions to photosplit water are shown in Figure 1.16. Figure 1.16a depicts a PEC/PEC tandem [53], where a photoanode and photocathode, both deposited onto transparent substrates, are stacked one in front of the other, and electrically connected by a wire. The configuration in Figure 1.16b represents a PEC/PV hybrid electrode [1,54–56], with a PEC top junction monolithically stacked with a solid-state PV bottom junction, and connected to a counter-electrode. Both designs have their own merits and disadvantages. The PEC/PV tandem entails extra fabrication complexity, but takes advantage of synergies with PV technology. Many good device-quality thin-film PV materials are available for the bottom cell, so the challenge is to develop a

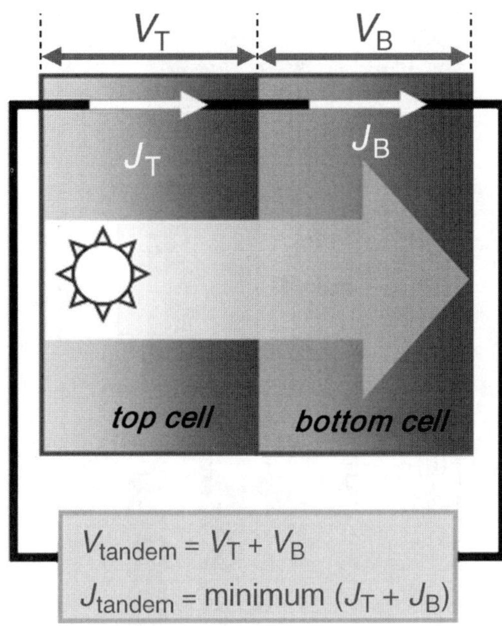

Figure 1.15 Block diagram of a stacked, series-connected tandem device under solar illumination. Photocurrents, J, and photovoltages, V, are shown for the top cell, the bottom cell and the combined tandem device.

compatible PEC material for the top junction. The PEC/PEC tandem has better potential for low-cost electrode synthesis, but involves external interconnections and requires development of two PEC materials (as if one wasn't enough!). The real bottom line will rest with actual relative device performance in the different tandem configurations.

As with the single-junction cases, efficiency bounds can be placed on such tandem configurations based on optical absorption limits. Figure 1.17 is essentially a two-dimensional extension of Figure 1.15 for tandem devices, where maximum photocurrent and the corresponding STH levels are calculated as a function of both top- and bottom-junction bandgaps. The assumptions are as follows: (1) the top cell absorbs all photons with energies exceeding the top-cell bandgap; (2) the bottom cell is illuminated with the top-cell-filtered light, and absorbs all of the remaining photons with energies exceeding its bandgap; (3) photocurrent density (in $mA\,cm^{-2}$) is calculated based on the minimum electron–hole pair count from the two junctions, assuming all of these pairs are converted to current; and (4) STH efficiencies, shown in parentheses, are calculated using Equation 1.3, and only apply to systems with sufficient photopotentials to split water.

The numbers from Figure 1.17 are definitely encouraging for the viability of tandem PEC hydrogen production. Despite the split spectrum and reduced photocurrents, a wide range of bandgap combinations yield high enough photocurrents for $>10\%$ STH conversion. An important key is to identify the possible combinations capable of the necessary photopotential levels. A quick-and-dirty rule of thumb can be derived based on typical quasi-Fermi level splits, for example about 50% of the bandgap for many semiconductors under solar illumination: the average of the top- and bottom-cell bandgaps needs to exceed the water-splitting potential, including the overpotentials, for example around 1.6 eV in typical

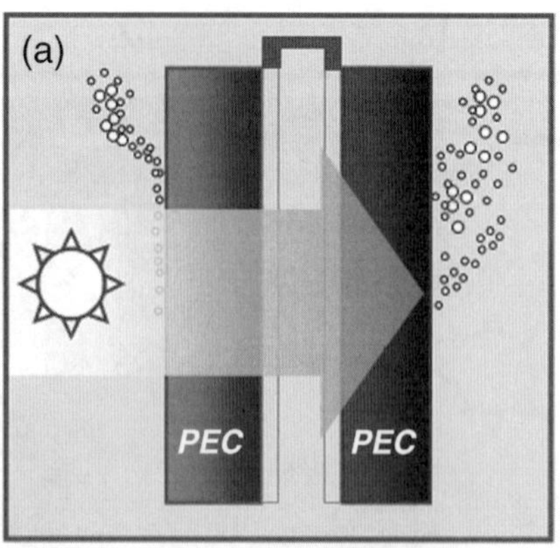

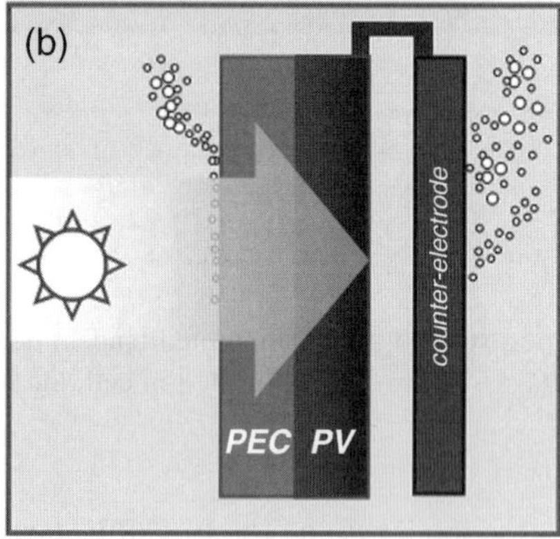

Figure 1.16 Design configurations for PEC water-splitting tandem devices showing: (a) a PEC/PEC photoanode/photocathode tandem with the two electrically connected photoelectrodes fabricated on transparent substrates for light transmission; and (b) a hybrid PEC/PV monolithically stacked device connected to a counter-electrodes.

photoelectrode configurations. There are many pitfalls and extra losses in real-world multijunction implementations, so this *rule* is by no means a guarantee of success. It is, however, quite useful in brainstorming sessions about device design.

As an illustration, TiO_2 with bandgaps of 3.1 eV could be combined with a low-bandgap semiconductor such as silicon (1.1 eV) in a hybrid tandem to meet the photopotential requirement for PEC water-splitting. As seen in Figure 1.17, however, independent of the bottom junction, the device performance will never exceed 5% STH. In this, and any series-connected tandem configuration, the performance is bound by the photocurrent limits of the PEC semiconductor. To achieve STH efficiencies above 10%, we can start back at Figure 1.14 and select semiconductors with bandgaps less than 2.2 eV for a top cell. For an average exceeding 1.6 eV, the bottom cell must be at least 1.0 eV. Using Figure 1.17, STH just straddles the 10% mark for bottom-cell bandgaps ranging from the 1.0 eV minimum up to 1.7 eV. The bandgaps should be somewhat lower than 2.2 eV for the top cell, and correspondingly higher than 1.0 eV in the bottom cell for more robust hydrogen production rates. In fact, such a device has already been designed, fabricated and successfully demonstrated in the laboratory.

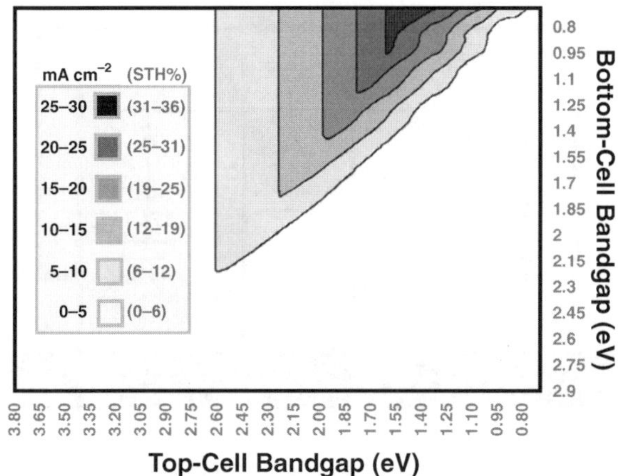

Figure 1.17 Maximum attainable AM1.5$_{global}$ photocurrent densities (in mA cm^{-2}) for tandem PEC devices shown as a function of top-cell and bottom-cell bandgaps. Also shown are the corresponding STH efficiencies applicable only to operational water-splitting devices.

1.6.3 A Shining Example

The NREL, using a multijunction hybrid photoelectrode, holds the STH efficiency world record for PEC water splitting [1]. The hybrid tandem device incorporates high-quality III–V crystalline semiconductor materials, with a 1.44 eV GaAs PV junction as the bottom cell, and a 1.83 eV Ga-InP PEC junction as the PEC top cell. As seen in Figure 1.18, this is a tandem photocathode configuration, where hydrogen is evolved at the PEC interface and oxygen is evolved at the separated counter-electrode. As seen in the device band diagram representation in Figure 1.19, there is a characteristic downward band-bending at the solid/liquid interface, consistent with the photocathode formulation in Figure 1.9. Tracking the combined quasi-Fermi level splits across the tandem, it can be seen that sufficient photopotential is developed to

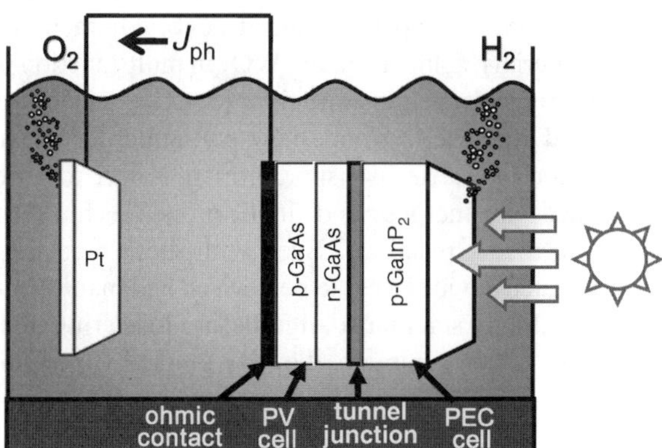

Figure 1.18 Two-electrode setup for the NREL hybrid PEC/PV photocathode demonstrating a 16% world-record STH efficiency.

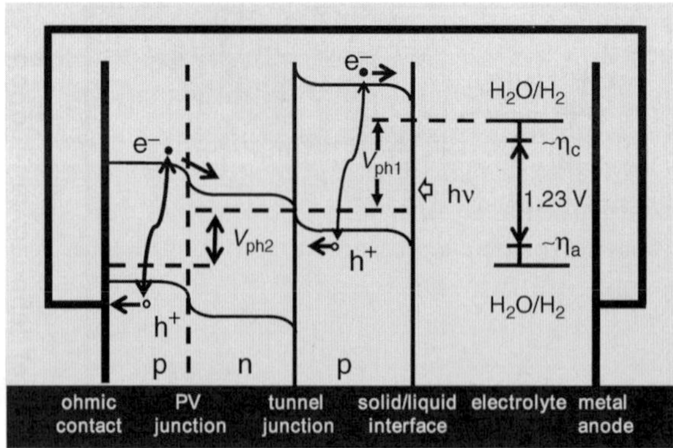

Figure 1.19 Band diagram of the PEC water-splitting process in the NREL tandem photocathode implemented in a two-electrode configuration.

drive both the hydrogen-reduction and water-oxidation reactions. Cross-referencing the top- and bottom-cell bandgaps in Figure 1.17, we would expect a solar-to-hydrogen conversion efficiency in the 10–20% range. In fact, this tandem device has demonstrated an astounding 16% STH efficiency! This laboratory-scale device is fabricated using extremely expensive III–V materials (the kind only NASA can afford!) and suffers from limited durability, it's not quite the grail. Still, the NREL hybrid tandem remains the best shining example of *what PEC can achieve.*

1.7 The PEC Challenge

1.7.1 What's Needed, Really?

In a nutshell, can the PEC performance levels of the NREL tandem device be reproduced using low-cost, high-durability devices and systems? The usual suspects under investigation for decades, including particulate and thin-film forms of metal oxides such as TiO_2, WO_3 and Fe_2O_3, haven't been able to cut it in the performance category. In fact, the best-demonstrated STH efficiencies in these materials, achieved using WO_3 in multijunction photoanode devices, has plateaued in the 3–5% range for quite some time [3,4].

Breakthroughs are needed, on several fronts: new particulate or thin-film semiconductor materials with good absorption and carrier-transport properties need to be developed; and novel interfaces need to be designed with energetic and kinetic properties favoring the water-splitting reactions, and inhibiting corrosion reactions. Even with these breakthrough materials and interfaces, very clever integrated devices need to be devised and manufactured using low-cost, commercial-scale processes. Many stars must align before mastering that delicate balancing act of PEC hydrogen production. In summary, the holy grail PEC system will be held to the following decrees:

- **Thou shall generate sufficient photopotential to split water.** In other words, the net quasi-Fermi level split must be large enough to overcome the reversible potential plus the overpotentials for water-splitting, typically over 1.6 eV in practical systems.

- **Thou shall generate sufficient photocurrent for efficient hydrogen production.** In other words, semiconductor and solution properties, in addition to the interface energetics and kinetics, must support the efficient generation, separation, transport and collection of photogenerated charges to drive the water-splitting reactions. Photocurrents exceeding $10\,mA\,cm^{-2}$, for example, are needed for STH conversion efficiencies over 12.3%.
- **Thou shall not corrode.** High hydrogen production rates must be sustained over long durations with minimal corrosion and degradation. Interface energetics/kinetics must favor water-splitting reactions over corrosion reactions.
- **In the long term, Thou shall cost next to nothing!**

1.7.2 Tradeoffs and Compromises

Again, this is not an easy task. In efforts to achieve practical PEC hydrogen production, tradeoffs abound and compromises are many. Cost versus performance will remain the central tradeoff in PEC research pending major breakthroughs in materials and interfaces. In the meantime, researchers remain preoccupied searching for pathways forward, based on the right compromises along numerous avenues. The fundamental photocurrent/photovoltage tradeoff will not go away, but pathways around this are available, including multijunctions and further down the road, hot-carrier collection [52,57,58]. The basic implementation of PEC semiconductors in commercial hydrogen production plants has not been settled. What are the tradeoffs between large-scale photoelectrode reactors (e.g., panel- or tubular- types) using low-cost thin-film materials and large-scale slurry-bed reactors utilizing functionalized semiconductor particles [59]. Assuming the right thin-film or particulate materials can be found, more techno-economics analyses will be needed to resolve this debate. The path forward needs to navigate all the economic, as well as the scientific obstacles.

1.7.3 The Race with PV-Electrolysis

If the PEC holy grail is found, how would it stack up against, for example, PV-electrolysis systems for hydrogen production? At first glance, PEC might look pretty good next to PV-electrolysis. Take for example what a *low-cost* PV-electrolysis system might look like today. A *relatively cheap* 10% amorphous silicon PV system coupled with a 60% PEM electrolyzer gives us only 6% STH. Higher-cost more-efficient components are available, but in the *low-cost/high-performance* race, much R&D is needed here as well. Interestingly, the semiconductor materials research efforts in PV and PEC are highly synergistic, and breakthroughs in either can greatly benefit both. The real race is toward the practical solar production of hydrogen from water, no matter how we get there, everyone wins!

1.8 Facing the Challenge: Current PEC Materials Research

Across the United States and around the world, the latest and greatest scientific advances are being brought to the table to face the PEC challenge. Key patrons in the quest include the US Department of Energy's (DOE) Working Group on PEC Hydrogen Production [60], and the International Energy Agency's Hydrogen Implementation Agreement (IEA-HIA) [61]. In recent years, powerful synergies developed in the PEC community have become infectious, drawing together researchers from diverse fields, including photovoltaics, nanotechnology and the *other* solar-hydrogen routes. Globally, *renewable hydrogen* research is getting some new life, as evidenced in numerous recent publications, including several informative books [62,63]. There is an important unifying theme linking all renewable-energy research fields, including PEC hydrogen, and that theme is *materials, materials, materials!*

Fundamental materials and materials-interface properties hold the keys to successful PEC research and development. In response, the worldwide PEC network is constructing a specialized "tool-chest", including state-of-the art techniques in materials theory, synthesis and characterization, to facilitate research progress, and to help inspire fundamental breakthroughs. This tool-chest is already proving invaluable in the investigation of promising PEC semiconductors and interfaces. As collaborative activities expand, it is expected to become an even more powerful asset.

To foster collaboration, the DOE PEC Working Group and the IEA-HIA PEC Annex-26, are encouraging the formation of informal international "task forces" to coordinate important PEC research activities. While some of the collaborative task forces center on the R&D of specific PEC materials classes, others focus on critical activities to advance the supporting science and technologies in the PEC tool-chest. These include:

- Development of standardized testing and reporting protocols for evaluating candidate PEC materials systems on a level playing field. In the past, the lack of standardized conditions and procedures for reporting PEC results has greatly hampered research progress across the board [64].
- Development of advanced characterization techniques to enhance understanding of PEC materials and interfaces, and promote breakthrough discoveries [65]. The most advanced methods in evaluating optoelectronic properties of semiconductor materials and interfaces, *in situ* as well as *ex situ*, are being deployed.
- Development of new theoretical models of PEC materials and interfaces critical to the design and engineering of brand new semiconductor systems [66,67]. Sophisticated models of band states and bandgaps are needed, including effects of surface, interfaces and grain boundaries.
- Development and implementation of innovative synthesis techniques, including combinatorial synthesis methods, to facilitate the PEC materials discovery process [68,69]. Innovative synthesis routes can make or break the viability of a semiconductor system, a fact well-appreciated by the PV community.
- Development and refinement of *techno-economics analyses* of PEC hydrogen-production systems incorporating performance and processing-cost feedback from the broader materials R&D efforts. This will provide a basis for evaluating the long-term feasibility of large-scale PEC production technologies in comparison with other renewable approaches.

On an impressive scale, the PEC research community is applying its tool-chest to investigate a broad spectrum of promising materials classes. Some of the materials task forces are concentrating on modification of traditional materials, others are focusing on more practical synthesis approaches for expensive, high-performance materials, and yet others are attempting to discover completely new material systems. For all of the different materials, task-force researchers must carefully consider the benefits, barriers and approaches for addressing the barriers; all in close coordination with the ongoing advances in theory, synthesis and characterization. Some of the important PEC materials classes under current investigation worldwide include:

- **Tungsten-oxide and related modified compounds:** Tungsten oxide, particularly in thin-film and nanoparticle forms, has been a workhorse in photoelectrochemical [70–73] and electrochromic [74] applications for years. It is inexpensive and stable, but its high bandgap ($\sim$2.6 eV) is limiting to PEC performance. Photocurrent densities around 3 mA cm^{-2} have been achieved [4,75,76], with STH efficiencies over 3% in tandem configurations. To break the performance barrier, current research is focused on reducing bandgap through ion incorporation into the WO_3 structure [66,67,77], and further integration in multijunction devices.
- **Iron-oxide and related modified compounds:** Iron-oxide is abundant, stable, inexpensive and has a near-ideal bandgap ($\sim$2.1 eV) for PEC applications. Unfortunately, its poor absorption, photocarrier lifetime and transport properties have been prohibitive to practical water splitting. Current research to overcome these barriers has been encouraging, with recent progress in thin films [78–81] and nanostructured materials [82]. Iron oxide in tandem configurations may also be of interest.
- **Amorphous silicon compounds, including silicon carbides and nitrides:** Amorphous silicon compounds have recently demonstrated interesting performances in PEC applications [56,83–87]. The progress of this material class in photoelectrochemical applications has benefited from decades of research in the PV community. Technical barriers remain in PEC stability and interface properties, but electrolyte and surface modification studies could help overcome these barriers. With material and interface improvements, monolithically fabricated multijunction devices using amorphous silicon compound films can have practical appeal for PEC water-splitting.
- **Copper chalcopyrite compounds:** Copper chalcopyrite thin films are among the best absorbers of solar energy. As a result, chalcopyrite alloys formed with copper and gallium, indium, sulfur and selenium have been widely characterized in the PV world [88,89]. A great advantage of this material class for PEC applications is the bandgap tailoring based on composition, with bandgaps ranging from 1.0 eV in $CuInSe_2$ to 1.6 eV in $CuGaSe_2$, and up to 2.43 eV in $CuGaS_2$ [90]. The $CuGaSe_2$ bandgap is attractive for PEC applications, and photocurrent density exceeding 13 mA cm^{-2} have been demonstrated with this material [91] in biased PEC cells. Stability, surface kinetics and surface energetics remain as current barriers, but if current research can successfully address these, high STH efficiency could be achievable in low-cost thin-film copper chalcopyrite systems.
- **Tungsten and molybdenum sulfide nanostructures:** As bulk materials, tungsten and molybdenum sulfides are excellent hydrogen catalysts, but their bandgaps (below 1.2 eV) are too low for PEC water splitting. Quantum confinement using nanostructuring, however, can increase the bandgap up to 2.5 eV. Current studies on nanostructured MoS_2 are focused on stabilized synthesis routes and integration of the nanostructures into practical bulk PEC devices [92].
- **III–V semiconductor classes:** High-quality crystalline semiconductor compounds of gallium, indium, phosphorous and arsenic have been studied for decades [93]. In PEC experiments to date, STH efficiencies between 12–16% have been demonstrated in $GaInP_2$/GaAs hybrid tandem photocathodes [1,94]. High cost and limited durability are the barriers to practical PEC hydrogen production, and breakthroughs in synthesis and in surface stabilization are being pursued.

Although there is still much work ahead, research in these promising candidate materials, among others, has seen significant recent progress. Fortunately, in today's broad-based network of PEC collaborators, progress in any one area can greatly benefit all. The needed scientific breakthroughs are still on the way, but once they get here, watch out! The implications of new low-cost, high-efficiency semiconductors will be enormous, not only to PEC solar water splitting for hydrogen production, but also to PV and other solar-energy conversion pathways. The challenge is great, but the promise even greater.

Acknowledgments

First, I would like to applaud the worldwide network of PEC researchers who carry out the quest with such dedication and enthusiasm, and would like to thank especially all participants in the US DOE PEC Working Group and of the IEA-HIA PEC Annex-26 for their excellent work and fruitful collaborative efforts. This list is long, and it includes (but is by no means limited to) the following: my good friend Dr. Clemens Heske and his excellent cohorts at UNLV; Dr. Arun Madan and his incredibly dedicated research team at MVSystems Incorporated; Dr. Eric McFarland and his amazing pool of talented students at UCSB; the brilliant Dr. Tom Jaramillo and his brand new laboratory at Stanford; and of course Drs. John Turner, Mowafak Al-Jassim, Todd Deutsch, Huyen Dinh and all the other superb NRELians! My sincere appreciation and gratitude goes out to the Hawaii Natural Energy Institute (HNEI) at the University of Hawaii at Manoa; Its director Dr. Richard Rocheleau and excellent staff, including Dr. Michael Antal, Dr. Bor Yann Liaw and Mitch Ewan have been instrumental in the advancement of hydrogen and renewable Energy. My very special thanks go out to all researchers and assistants, past and present, of the HNEI Thin Films Laboratory (TFL) who have carried the PEC torch with dignity and grace for many years From the recent past, Drs. Bjorn Marsen, Brian Cole and Daniela Paluselli have made significant contributions; The current TFL staff of Dr. Nicolas Gaillard, Jess Kaneshiro, Alex DeAngelis, Xi Song and Stewart Mallory will have my eternal gratitude, not only for their good work, but also for their tireless efforts in the preparation of this chapter. I would like to send a special Mahalo to Mary-Rose Valladares and Andreas Luzzi of the IEA, and Richard Farmer and David Peterson of the US DOE for their invaluable support of PEC research. Finally, I humbly give my thanks and Aloha to Dr. Robert Perret of Nevada Technical Services, LLC, and Roxanne Garland of the US DOE for their vision, inspiration, and unwavering moral support!

References

[1] Khaselev, O., Bansal, A. and Turner, J.A. (2001) High-efficiency integrated multijunction photovoltaic/electrolysis systems for hydrogen production. *International Journal of Hydrogen Energy*, **26**, 127–132.

[2] Andreev, V.M. (2003) *Practical Handbook of Photovoltaics: Fundamentals and Applications* (eds T. Markvart and L. Castañer), Elsevier.

[3] Grätzel, M. (2001) Photoelectrochemical Cells. *Nature*, **414**, 338.

[4] Marsen, B., Miller, E.L., Paluselli, D. and Rocheleau, R.E. (2007) Progress in sputtered tungsten trioxide for photoelectrode applications. *International Journal of Hydrogen Energy*, **32**, 3110–3115.

[5] Bush, G.W.(January 2003) State of the Union. Presented in Washington, D.C.

[6] Romm, J.J. (2004) *The Hype About Hydrogen*, Island Press.

[7] Ball, M. and Wietschel, M. (2009) *The Hydrogen Economy: Opportunities and Challenges*, Cambridge Press.

[8] Rifkin, J. (2003) *The Hydrogen Economy*, Tarcher.

[9] Yürüm, Y. (1995) *Hydrogen Energy System: Production and Utilization of Hydrogen and Future Aspects*, Kluwer Academic Publishers.

[10] Turner, J.A. (1999) A realizable renewable energy future. *Science*, **285**, 687–689.

[11] Newton, J. (1989) *Uncommon Friends: Life with Thomas Edison, Henry Ford, Harvey Firestone, Alexis Carrel, & Charles Lindbergh*, Mariner Books.

[12] US Department of Energy, Office of Science (2005) Basic Research Needs for Solar Energy Utilization.

[13] Green, M.A. (1982) *Solar Cells: Operating Principles, Technology, and System Applications*, Prentice-Hall, Inc.

[14] US Department of Energy, Energy Information Administration (2008) International Energy Outlook 2008 (DOE/EIA-0484).

[15] Greentech Media, the Prometheus Institute (2008) PV Technology, Production and Cost, 2009 Forecast: The Anatomy of a Shakeout.

[16] Solarbuzz (2009) Marketbuzz 2009: Annual World Solar PV Market Report. San Francisco, CA.

[17] Emery, K. (2003) *Handbook of Photovoltaic Science and Engineering* (eds A. Luque and S. Hegedus), John Wiley & Sons, Ltd.

[18] Sandia National Laboratories (2008) Sandia Stirling Energy Systems set new world record for solar-to-grid conversion efficiency. Press Release. 2 Feb. 2008.

[19] CSP Project Developments in Spain (2002) *IEA SolarPACES Implementing Agreement*. http://www.solarpaces.org/News/Projects/Spain.htm.

[20] Kasahara, S., Hwang, G.J., Nakajima, H. *et al.* (2003) Effects of process parameters of the IS process on total thermal efficiency to produce hydrogen from water. *Journal of Chemical Engineering of Japan*, **36**, 887–889.

[21] Onuki, K., Inagaki, Y., Hino, R. and Tachibana, Y. (2005) Research and development on nuclear hydrogen production using HTGR at JAERI. *Progress in Nuclear Energy*, **47**, 496–503.

[22] Lewis, M.A., Serban, M. and Basco, J.K. (2004) A progress report on the chemistry of the low temperature Cu-Cl thermochemical cycle. *Transactions of the American Nuclear Society*, **91**, 113–114.

[23] Akkerman, I., Janssen, M., Rocha, J. and Wijffels, R.H. (2002) Photobiological hydrogen production: photochemical efficiency and bioreactor design. *International Journal of Hydrogen Energy*, **27**, 1195–1208.

[24] Zaborsky, O.R. (1998) *Biohydrogen*, Plenum Press.

[25] Funk, J.E. and Reinstrom, R.M. (1966) Energy requirements in production of hydrogen from water. *Industrial & Engineering Chemistry Process Design and Development*, **5**, 336–342.

[26] Sze, S.M. (2006) *Physics of Semiconductor Devices*, John Wiley and Sons, New York.

[27] Neamen, D.A. (2002) *Semiconductor Physics and Devices: Basic Principles*, McGraw-Hill Science/Engineering/Math.

[28] Marcus, R.J. (1965) Chemical conversion of solar energy. *Science*, **123**, 399–405.

[29] Bockris, J.O.M. (1956) Kinetics of activation controlled consecutive electrochemical reactions: anodic evolution of oxygen. *Journal of Chemical Physics*, **24**, 817–827.

[30] Dutta, S. (1990) Technology assessment of advanced electrolytic hydrogen production. *International Journal of Hydrogen Energy*, **15**, 379–386.

[31] LeRoy, R.L. (1983) Industrial water electrolysis: present and future. *International Journal of Hydrogen Energy*, **8**, 401–417.

[32] Parkinson, B. (1984) On the efficiency and stability of photoelectrochemical devices. *Accounts of Chemical Research*, **17**, 431–437.

[33] Dohrmann, J.K. and Schaaf, N.S. (1992) Energy conversion by photoelectrolysis of water: determination of efficiency by *in situ* photocalorimetry. *The Journal of Physical Chemistry*, **96**, 4558–4563.

[34] Heller, A. (1982) Electrochemical solar cells. *Solar Energy*, **29**, 153–162.

[35] Khan, S.U.M., Al-shahry, M. and Ingler, W.B. Jr. (2002) Efficient photochemical water splitting by a chemically modified n-TiO_2. *Science*, **297**, 2243–2245.

[36] Luther, J. (2003) *Handbook of Photovoltaic Science and Engineering* (eds A. Luque and S. Hegedus), John, Wiley & Sons Ltd.

[37] Balandin, A.A. and Wang, K.L. (2006) *Handbook of Semiconductor Nanostructures and Nanodevices*, (5-Volume Set), American Scientific Publishers.

[38] Muller, R.S. and Kamins, T.I. (2002) *Device Electronics for Integrated Circuits*, John Wiley & Sons, Ltd.

[39] Yu, P.Y. and Cardona, M. (2004) *Fundamentals of Semiconductors: Physics and Materials Properties*, Springer; Fonash, S. (1982) *Solar Cell Device Physics*, Academic Press.

[40] Gerischer, H. (1979) *Topics in Applied Physics Volume 31: Solar Energy Conversion, Solid-State Physics Aspects* (ed. B.O. Seraphin), Springer-Verlag.

[41] Bard, A.J. and Faulknerk, L.R. (2000) *Electrochemical Methods: Fundamentals and Applications*, John Wiley & Sons, Ltd.

[42] Bockris, J.O.M., Reddy, A.K.N. and Gamboa-Aldeco, M.E. (2001) *Modern Electrochemistry: Fundamentals of Electrodics v. 2a*, Springer.

[43] Memming, R. (2001) *Semiconductor Electrochemistry*, Wiley-VCH.

[44] Lipkowski, J. and Ross, P.N. (1994) *Electrochemistry of Novel Materials*, VCH Publishers.

[45] Gellings, P.J. and Bouwmeester, H.J.M. (1997) *The CRC Handbook of Solid State Electrochemistry*, CRC Press.

[46] Nozik, A.J. and Memming, R. (1996) Physical chemistry of semiconductor-liquid interface. *The Journal of Physical Chemistry*, **100**, 13061–13078.

[47] Gerischer, H. (1970) *Physical Chemistry: An Advanced Treatise*, **9a**, Academic Press.

[48] Gerischer, H. (1990) The impact of semiconductors on the concept of electrochemistry. *Electrochim Acta*, **35**, 1677–1690.

[49] Mussini, T. and Longhi, P. (1985) *Standard Potentials in Aqueous Solution* (eds A.J. Bard, R. Parsons and J. Jordan), IUPAC.

[50] Mavroides, J.G., Kafalas, J.A. and Kolesar, D.F. (1976) Photoelectrolysis of water in cells with $SrTiO_3$ anodes. *Applied Physics Letters*, **28**, 241–243.

[51] Ellis, A.B., Kaiser, S.W. and Wrighton, M.S. (1976) Semiconducting potassium tantalate electrodes. *The Journal of Physical Chemistry*, **80**, 1325–1328.

[52] Green, M.A. (2003) *Third Generation Photovoltaics: Advanced Solar Energy Conversion*, Springer-Verlag.

[53] Ingler, W.B. Jr. and Khan, S.U.M. (2006) A self-driven p/n-Fe_2O_3 tandem photoelectrochemical cell for water splitting. *Electrochemical and Solid – State Letters*, **9**, G144–G146.

[54] Miller, E.L., Rocheleau, R.E. and Deng, X.M. (2003) Design considerations for a hybrid amorphous silicon/ photoelectrochemical multijunction cell for hydrogen production. *International Journal of Hydrogen Energy*, **28**, 615–623.

[55] Miller, E.L., Marsen, B., Paluselli, D. and Rocheleau, R.E. (2005) Optimization of hybrid photoelectrodes for solar water-splitting. *Electrochemical and Solid – State Letters*, **8**, A247–A249.

[56] Zhu, F., Hu, J., Kunrath, A. *et al.* (2007) a-SiC:H films used as photoelectrodes in a hybrid, thin-film silicon photoelectrochemical (PEC) cell for progress toward 10% solar-to hydrogen efficiency. Solar Hydrogen and Nanotechnology – Proceedings of SPIE, 6650, p. 66500.

[57] Archer, M.D. (1981) *Photochemical Conversion and Storage of Solar Energy* (ed. J.S. Connolly), Academic Press.

[58] Hanna, M.C., Lu, Z. and Nozik, A.J. (1997) Hot carrier solar cells. Proceedings of 1st NREL Conference on Future Photovoltaic Generation. Technology – AIP Conference Proceedings, 404, pp, 309–316.

[59] Linkous, C.A., Muradov, N.Z. and Ramser, S.N. (1995) Consideration of reactor design for solar hydrogen production from hydrogen sulfide using semiconductor particulates. *International Journal of Hydrogen Energy*, **20**, 701–709.

[60] Miller, E.L., Garland, R. and Perret, R.(June (2008)) Photoelectrochemical hydrogen production: DOE PEC working group overview & UNLV-SHGR program subtask. Presented at the DOE Hydrogen Program AMR, Arlington, VA.

[61] International Energy Agency. (1977) Hydrogen Implementing Agreement. http://www.ieahia.org/.

[62] Grimes, C.A., Varghese, O.K. and Ranjan, S. (2008) *Light, Water, Hydrogen*, Springer.

[63] Krishnan, R., Licht, S. and McConnell, R. (eds) (2008) *Solar Hydrogen Generation, Toward a Renewable Energy Future*, Springer.

[64] Murphy, A.B., Barnes, P.R.F., Randeniya, L.K. *et al.* (2006) Efficiency of solar water splitting using semiconductor electrodes. *International Journal of Hydrogen Energy*, **31**, 1999–2017.

[65] Heske, C. (2007) Soft X-ray and electron spectroscopy studies of oxide semiconductors for photoelectrochemical hydrogen production. SPIE proceedings Solar Hydrogen and Nanotechnology II, San Diego, USA, 26–30 August 2007.

[66] Huda, M.N., Yanfa, Y., Moon, C.Y. *et al.* (2008) Density-functional theory study of the effects of atomic impurity on the band edges of monoclinic WO3. *Physical Review B*, **77**, 195102.

[67] Yan, Y. and Wei, S.-H. (2008) Doping asymmetry in wide-bandgap semiconductors: Origins and solutions. *Physica Status Solidi B-Basic Research*, **245**, 641.

[68] Jaramillo, T.F., Baeck, S.-H., Kleiman-Shwarsctein, A. *et al.* (2005) Automated electrochemical synthesis and photoelectrochemical characterization of $Zn_{1-x}Co_xO$ thin films for solar hydrogen production. *Journal of Combinatorial Chemistry*, **7**, 264–271.

[69] Woodhous, M., Herman, G. and Parkinson, B.A. (2005) A combinatorial approach to identification of catalysts for the photoelectrolysis of water. *Chemistry of Materials*, **17**, 4318.

[70] Yoon, K.H., Seo, D.K., Cho, Y.S. and Kang, D.H. (1998) Effect of Pt layers on the photoelectrochemical properties of a WO_3/p-Si electrode. *Journal of Applied Physics*, **84**, 3954–3959.

[71] Santato, C., Ulmann, M. and Augustynski, J. (2001) Photoelectrochemical properties of nanostructured tungsten trioxide films. *The Journal of Physical Chemistry B*, **105**, 936–940.

[72] Solarska, R., Alexander, B.D. and Augustynski, J. (2006) Electrochromic and photoelectrochemical characteristics of nanostructured WO_3 films prepared by a sol–gel method. *Comptes Rendus Chimie*, **9**, 301–306.

[73] Weinhardt, L., Blum, M., Bär, M. *et al.* (2008) Electronic surface level positions of WO_3 thin films for photoelectrochemical hydrogen production. *The Journal of Physical Chemistry C*, **112**, 3078–3082.

[74] Washizu, E., Yamamoto, A., Abe, Y. *et al.* (2003) Optical and electrochromic properties of RF reactively sputtered WO_3 films. *Solid State Ionics*, **165**, 175–180.

[75] Marsen, B., Cole, B. and Miller, E.L. (2007) Influence of sputter oxygen partial pressure on photoelectrochemical performance of tungsten oxide films. *Solar Energy Materials and Solar Cells*, **91**, 1954–1958.

[76] Alexander, B.D., Kulesza, P.J., Rutkowska, I. *et al.* (2008) Metal oxide photoanodes for solar hydrogen production. *Journal of Materials Chemistry*, **18**, 2298–2303.

[77] Cole, B., Marsen, B., Miller, E.L. *et al.* (2008) Evaluation of nitrogen doping of tungsten oxide for photoelectrochemical water splitting. *The Journal of Physical Chemistry C*, **112**, 5213–5220.

[78] Miller, E.L., Paluselli, D., Marsen, B. and Rocheleau, R.E. (2004) Low-temperature reactively sputtered iron oxide for thin film devices. *Thin Solid Films*, **466**, 307–313.

[79] Duret, A. and Graetzel, M. (2005) Visible light-induced water oxidation on mesoscopic a-Fe_2O_3 films made by ultrasonic spray pyrolysis. *The Journal of Physical Chemistry. B*, **109**, 17184–17191.

[80] Hu, Y.-S., Kleiman-Shwarsctein, A., Forman, A.J. *et al.* (2008) Pt-doped α-Fe_2O_3 thin films active for photoelectrochemical water splitting. *Chemistry of Materials*, **20**, 3803–3805.

[81] Kleiman-Shwarsctein, A., Hu, Y.-S., Forman, A.J. *et al.* (2008) Electrodeposition of α-Fe_2O_3 doped with Mo or Cr as photoanodes for photocatalytic water splitting. *The Journal of Physical Chemistry C*, **112**, 15900–15907.

[82] Kay, A., Cesar, I. and Graetzel, M. (2006) New benchmark for water photooxidation by nanostructured alpha-Fe_2O_3 films. *Journal of the American Chemical Society*, **128**, 15714–15721.

[83] Zhu, F., Hu, J., Matulionis, I. *et al.* (2009) Amorphous silicon carbide photoelectrode for hydrogen production directly from water using sunlight. *Philosophical Magazine*, **89**, 1478–6443.

[84] Matulionis, I., Zhu, F., Hu, J. *et al.* (2008) Development of a corrosion-resistant amorphous silicon carbide photoelectrode for solar-to-hydrogen photovoltaic/photoelectrochemical devices. SPIE Solar Energy and Hydrogen 2008, San Diego, USA, 10–14 August 2008.

[85] Stavrides, A., Kunrath, A., Hu, J. et al. (2006) Use of amorphous silicon tandem junction solar cells for hydrogen production in a photoelectrochemical cell. SPIE proceedings Solar Hydrogen and Nanotechnology, San Diego, USA, 13–17 August 2007.

[86] Yae, S., Kobayashi, T., Abe, M. et al. (2007) Solar to chemical conversion using metal nanoparticle modified microcrystalline silicon thin film photoelectrode. *Solar Energy Materials and Solar Cells*, **91**, 224–229.

[87] Sebastian, P.J., Mathews, N.R., Mathew, X. et al. (2001) Photoelectrochemical characterization of SiC. *International Journal of Hydrogen Energy*, **26**, 123–125.

[88] Bär, M., Bohne, W., Röhrich, J. *et al.* (2004) Determination of the band gap depth profile of the penternary $Cu(In_{(1(x)}Ga_x)(S_ySe_{(1(y)})_2$ chalcopyrite from its composition gradient. *Journal of Applied Physics*, **96**, 3857–3860.

[89] Bär, M., Weinhardt, L., Pookpanratana, S. *et al.* (2008) Depth-resolved band gap in $Cu(In,Ga)(S,Se)_2$ thin films. *Applied Physics Letters*, **93**, 244103.

[90] Bär, M., Weinhardt, L., Heske, C. *et al.* (2008) Chemical structures of the $Cu(In,Ga)Se_2$/Mo and Cu(In,Ga) $(S,Se)_2$/Mo interfaces. *Physical Review B-Condensed Matter*, **78**, 075404.

[91] Marsen, B., Cole, B. and Miller, E.L. (2008) Photoelectrolysis of water using thin copper gallium diselenide electrodes. *Solar Energy Materials & Solar Cells*, **92**, 1054–1058.

[92] Jaramillo, T.F., Jørgensen, K.P., Bonde, J. *et al.* (2007) Identification of active edge sites for electrochemical H_2 evolution from MoS_2 nanocatalysts. *Science*, **317**, 100–102.

[93] Deutsch, T.G., Koval, C.A. and Turner, J.A. (2006) III–V nitride epilayers for photoelectrochemical water splitting: GaPN and GaAsPN. *The Journal of Physical Chemistry B*, **110**, 25297–25307.

[94] Khaselev, O. and Turner, J.A. (1998) A monolithic photovoltaic photoelectrochemical device for hydrogen production via water splitting. *Science*, **280**, 425–427.

2

Modeling and Simulation of Photocatalytic Reactions at TiO_2 Surfaces

Hideyuki Kamisaka and Koichi Yamashita
Department of Chemical System Engineering, School of Engineering, The University of Tokyo, Tokyo 113-8656, Japan, Email: yamasita@chemsys.t.u-tokyo.ac.jp

2.1 Importance of Theoretical Studies on TiO_2 Systems

Since the discovery of the Fujishima–Honda effect [1], the surface chemistry of TiO_2 has been expected to play a central role in solar hydrogen generation processes. In addition, strong photocatalytic activity [2] and photoinduced hydrophilic conversion phenomena [3] have been observed on TiO_2 surfaces. A dye-sensitized TiO_2 surface can be used as a building block for solar-cell devices [4]. Introduction of some impurities in TiO_2 thin films induces significant changes in their magnetic [5] and conducting properties [6]. All these characteristic features show the large number of possible industrial applications that are yet to materialize; for example, the fabrication of new catalysis and optoelectronic devices and solar cells. Some of these features have already been utilized in industrial applications [7]. Because TiO_2 is nontoxic and has a low cost, it will improve such processes. Most of the applications of TiO_2 are related to clean chemical processes and reusable-energy resources. From these favorable features, the importance of TiO_2 in the next-generation chemical industry cannot be overemphasized.

However, despite intensive research into TiO_2, from both experimental and theoretical aspects, the understanding of the fundamentals of TiO_2 chemistry is still limited. Most of the microscopic mechanisms of the properties mentioned above have not been clarified. Even some basic quantities, such as the effective mass of the electrons and the stoichiometry of bulk TiO_2,

On Solar Hydrogen & Nanotechnology Edited by Lionel Vayssieres

have not been resolved satisfactorily. Several chemically active species on the surface of TiO_2 are speculated to exist in photocatalytic reactions, but these have not been clearly identified. Without knowledge of such mechanisms, no guiding principle can be followed to direct improvements.

In this chapter, we review our recent theoretical studies on TiO_2. The following topics are discussed: (i) the structure and optical properties of carbon-doped (C-doped) TiO_2 [8], (ii) the structure and stoichiometry of niobium-doped TiO_2 (TNO) [9], (iii) water molecules and hydroxyl groups on the surface of TiO_2 and their effect on the surface stress [10] and (iv) the timescale of the decoherence between the electronic states of PbSe quantum dots (QDs) [11]. C-doped TiO_2 and TNO are newly developed materials that have a high photocatalytic activity under visible light and a high electrical conductivity, respectively. These two materials are important from an engineering perspective. The third topic is related to the photoinduced hydrophilic conversion process on a TiO_2 surface. An understanding of this phenomenon is necessary for industrial applications and also for fundamental surface science. PbSe QDs exhibit a multiexciton generation (MEG) phenomenon when irradiated with light. The MEG process is expected to improve the efficiency of solar cells [12]. QDs can be used as sensitizers for TiO_2 surfaces. These four topics are representative research fields in TiO_2 chemistry.

Among the several theoretical methodologies that can be used, we emphasize the use of the *first-principle* method. This terminology refers to a method that solves a wavefunction, Green's function, or the electron density of a system directly from the Schrödinger or Kohn–Sham equations. A proper description of the chemical reaction process requires a rigorous treatment of the chemical bonds. During the course of a reaction, bonds break and are newly formed, and the reliable prediction of the changes to chemical bonds and their dynamics requires that the total energy of the system be calculated to an accuracy of $1\,\mathrm{kcal\,mol^{-1}}(=4.2\,\mathrm{kJ\,mol^{-1}} = 0.043\,\mathrm{eV})$, which is sometimes referred to as the *chemical accuracy*. Usually, the chemical accuracy can only be obtained using the first-principle method. Other methodologies, such as the molecular dynamics (MD) method or an approximate calculation of electronic states, cf. the tight-binding (TB) method, rely on empirical parameters and preliminary prescriptions that only work for foreseen situations.

However, the importance of approximate approaches should not be underestimated. In the last section of this chapter, recent trends and applications in such directions are reviewed, and their fundamental ideas are outlined in brief. Despite their limited ability to describe chemical reaction processes, these methodologies have a great advantage in reducing computational costs in actual applications. The size of a system that can be solved using the first-principle method is limited. To facilitate a realistic simulation of a system involving a large number of atoms, both methods should be incorporated in a complementary way. From this viewpoint, the future direction of theoretical studies on TiO_2 is addressed. The essential difficulties originating from a semi-infinite surface system are also discussed, along with potential remedies.

This review consists of six sections. Section 2.1 is the Introduction. In Section 2.2, theoretical calculations on bulk TiO_2 are reviewed, and our recent theoretical studies on C-doped TiO_2 and TNO are discussed. In Section 2.3 our recent studies on the surface stress of TiO_2 and the effect of water molecules are introduced. These calculations are related to the photoinduced hydrophilic conversion on the surface. In Section 2.4, our theoretical research into PbSe QDs is reviewed. The potential importance of QDs as sensitizers for TiO_2 surfaces is emphasized, and the controversy over the mechanism of multiexciton generation (MEG) is explained. In Section 2.5, recent progress in theoretical modeling and simulation schemes for TiO_2 surface chemistry is reviewed. A brief discussion is given regarding the validity of combined approaches that consist of the first-principle method and molecular dynamics. The future direction of research in this field is discussed in this section.

2.2 Doped TiO_2 Systems: Carbon and Niobium Doping

Several properties of semiconductors can be modified by doping them with impurities. Such processes are especially important for fabricating semiconductor materials that have some functionality with a desired value. As a number of impurities are utilized in the silicon industry, several of these impurities have been used in TiO_2 bulk systems. In this section, we focus on our calculations on C-doped TiO_2 [8] and TNO [9].

Pure bulk TiO_2 is transparent in the visible-light region. On the other hand, photocatalytic reactions on TiO_2 surfaces are initiated by the charge separation that occurs on the substrate owing to the adsorption of photons. This means that a large fraction of sunlight is not utilized in the photocatalytic reaction processes on TiO_2. The use of dopants may cause a narrowing of the bandgap and may improve the efficiency of such photocatalytic reactions. Experimentally, several dopants have been investigated, and some of these have been reported to facilitate visible-light absorption and photocatalytic reactions. Nitrogen [13,14], carbon [15–17], sulfur [18,19] and iodine [20] atoms are examples of such dopants. Improvement of the photocatalytic efficiency is not the only purpose of doping, as modification of the magnetic [5] and electronic [6] properties of TiO_2 is also achieved by doping. In particular, the recent discovery of a high electronic conductivity in Nb-doped anatase thin films of TiO_2 means that they are a potential alternative transparent conducting oxide (TCO) and are currently receiving much attention [6]. Similarly, Ta-doped [21] anatase TiO_2 has been theoretically surveyed as an alternative to TCO. Besides, Co- [5], Cr- [22] and V-doped [23,24] TiO_2 have been studied because of their ferromagnetic character.

2.2.1 First-Principle Calculations on TiO_2

First-principle calculations can be categorized into two groups: molecular orbital (MO)-based theory and density functional theory (DFT)-based theory. The former method solves the wavefunction of electrons, and the latter method calculates the electron density of a system. Theoretical calculations and the prediction of the structure, electronic structure and optical properties of a bulk material are usually provided by the DFT-based first-principle method. This method is based on the well-known Hohenberg–Kohn [25] theorem and Kohn–Sham (KS) [26] ansatz. In this framework, the effect of electron correlation is treated as an effective potential for the noninteracting electrons. The prescription for the effective potential from the electron density constitutes a functional, which is called the *DFT functional*. Several types of functional have been advocated; for example, PW91(Perdew-Wang 91) [27] and PBE (Perdew-Burke-Enzerhof) [28] functionals.

As far as the electronic ground state is concerned, the DFT method predicts the structure of the bulk, the relative energetic stability among the polymorphic phases, the chemical bond length and the elastic properties, quantitatively. However, theoretical prediction of the bandgap energy in semiconductors is difficult. The estimate taken from the energies of the KS orbitals usually underestimates the width of the bandgap. For example, the estimate using the PW91 functional gives bandgaps of 1.8 and 2.1 eV for the rutile and anatase TiO_2, respectively. The corresponding experimental values are 3.0 and 3.2 eV. Existing remedies to this will be introduced in a later section.

The primitive unit cell of rutile has tetragonal symmetry, with a space group of $D_{4h}^{14}-P4_2/mnm$ [29]. The experimental lattice constants of rutile are: $a=b=4.584$ Å and $c=2.953$ Å [30]. In the case of anatase, the primitive unit cell is tetragonal body-centered, with the symmetry of space group of $D_{4h}^{19}-I4_1/amd$ [29]. Usually a conventional unit cell is constructed for the anatase structure by extending the third translational vector as $\mathbf{t}'_3 = -\mathbf{t}_1-\mathbf{t}_2+2\mathbf{t}_3$. This cell is tetragonal, and the experimental lattice constants are $a=b=$ 3.782 Å and $c=9.502$ Å [31]. In either form, all the titanium and oxygen atoms are equivalent under the symmetry operation of the system.

Early first-principle theoretical calculations on TiO_2 were performed by Glassford and Chelikowsky [32]. In their calculations, the DFT functional used was the local density approximation (LDA) functional, and the norm-conserving pseudopotential of Trouiller and Matrins (TM) was employed. Mo *et al.* used the LDA functional and a linear combination of atomic orbital (LCAO) basis set to calculate the electronic and optical properties of three TiO_2 phases: rutile, anatase and brookite [33]. The electronic and optical properties of anatase TiO_2 were calculated by Asahi *et al.* using the full-potential linearized augmented plane-wave (FPLAPW) method [34].

Theoretical calculations on doped systems usually use the DFT-based band-structure method. In this approach, the doped system is modeled with large unit cells with a periodic boundary condition (PBC). In reality, impurities are supposed to be spread over the material in a random manner. In such circumstances, one cannot assume any specific space symmetry. However, if the change in chemical structure is localized in the vicinity of the dopants, and the size of the calculated unit cell is large, then the theoretical calculations should be able to describe the effect of the dopants. When some impurities are placed into the cell, then their mutual interactions in the bulk can be clarified. The energy of formation is then calculated from the total energy calculations of the system and the subsystems [35]. Using the calculated energy of formation, and counting the number of possible doping sites, statistical analysis can be performed to estimate the doping rate [35,36]. Several physical properties in the equilibrium state can be analyzed as a function of the external chemical environment, most typically the chemical potential of a component and the temperature.

Naturally occurring rutile is an n-type semiconductor. The color of anatase can be changed by annealing it in an oxygen or hydrogen atmosphere [37]. The cause of this nonstoichiometry may be the formation of defect states, such as interstitial oxygen (O_i) and titanium (Ti_i) atoms, or their corresponding vacancies (V_O and V_{Ti}), or antisite defects (Ti_O and O_{Ti}). Shallow donor states have been suggested to form in n-type anatase [38], and the defect center of $[Ti^{3+}-V_O]$ has been suggested to occur in rutile [39]. Calculations on natural defects in rutile have been carried out by several researchers independently [36,40,41]. Calculations of the energy of formation of the defect sites mentioned above suggest the dominance of the V_O structures [41]. Natural defects in the anatase form of TiO_2 have been calculated by Na-Phattalung *et al.* [42]. It was found that Ti_i, O_i, V_{Ti} and V_O have low energies of formation, and that Ti_i is a quadruple donor in p-type samples, whereas V_{Ti} is a quadruple acceptor defect in n-type samples. The O_i defect in anatase spontaneously bonds to the lattice oxygen atoms, resulting in an O_2 dimer.

2.2.2 C-Doped TiO_2

The visible-light photocatalytic activity of a doped TiO_2 system was first pointed out in nitrogen-doped samples [13,43]. Sulfur-doped TiO_2 in either anion-doping or cation-doping form was found to be active under visible light [18,19]. However, S-doped TiO_2 suffers from catalytic poisoning by the sulfate ions generated [44]. Carbon-doped TiO_2 was then fabricated using several methods [45], and visible-light photocatalytic activity was reported for these C-doped samples [15,16]. However, the structure of the carbon dopant has not been clarified experimentally. The performance of C-doped TiO_2 as a photocatalytic material strongly depends on the fabrication process. This implies the possibility of several doped structures, and only some of the doped structures may contribute to the photocatalytic reactions. In the case of S-doping, the S_{Ti} structure shows a high visible-light response [18]. The IR spectrum of a photocatalytic active C-doped TiO_2 sample has been measured [16]. A peak corresponding to the carbon oxide anion (CO_3^{2-}) was detected, and the carbon atoms were speculated to be substituting for oxygen atoms. Recently, in addition to N-doped and C-doped TiO_2 samples, iodine-doped (I-doped) TiO_2 has also been found to be a catalyst under visible light [20].

We carried out first-principle calculations to survey the structure and optical properties of C-doped TiO_2 [8]. The calculations used the periodic boundary condition, and the PW91 DFT functional [27] was employed. The KS orbital for the electron density in the valence and conduction bands was expanded using the plane-wave basis set. The effect of the core electrons was treated implicitly as a pseudopotential for the valence electrons in the framework of Vanderbilt's ultrasoft pseudopotential (USPP) technique [46], as supplied in Ref. [47]. The Brillouin zone integral was carried out using the Monkhorst–Pack **k**-point sampling scheme [48]. All the computations were carried out using the Vienna *Ab Initio* Simulation Package (VASP) software package [49].

We extended the primitive unit cell to describe the dilute dopant. The new lattice vectors were: $\mathbf{t}'_1 = 2\mathbf{t}_1, \mathbf{t}'_2 = 2\mathbf{t}_2$ and $\mathbf{t}'_3 = -\mathbf{t}_1 - \mathbf{t}_2 + 2\mathbf{t}_3$ for anatase, and $\mathbf{t}'_1 = 2\mathbf{t}_1, \mathbf{t}'_2 = 2\mathbf{t}_2$ and $\mathbf{t}'_3 = 2\mathbf{t}_3$ for rutile. The unit cell was eight times larger than the primitive unit cell. The chemical composition of the unit cell was $Ti_{16}O_{32}$. Using the computational scheme described above, the maximum number of atoms feasible with our current computational facilities was limited to around 100. Several cells with different impurities – carbon atoms at the oxygen (C_O) or titanium (C_{Ti}) sites, oxygen vacancies (V_O), and both types of impurity (C_O–V_O and C_{Ti}–V_O) – were calculated for both the rutile and anatase cells.

Figure 2.1 shows the optimized structure of the C_{Ti} cell. In both forms of TiO_2, the doped carbon atoms were displaced from the center of the six original coordinating oxygen atoms and were orientated toward the center of a triangular plane composed of three oxygen atoms. The internuclear distance between the carbon atoms and the coordinating oxygen atoms were: r(C–O) = 1.30 Å in rutile and r(C–O) = 1.26–1.33 Å in anatase. The usual internuclear distance of carbonate ions is r(C–O) = 1.28 Å, and the usual internuclear distance of carbon dioxide molecules is r(C–O) = 1.16 Å. The formation of carbonate ions is reasonable from this internuclear distance. This finding is consistent with the infrared structure obtained from measurements on anatase [16]. In contrast, the C_O cells showed no significant structural change from that of pure TiO_2 in either form.

In contrast with the optimized C_{Ti} structures, where no significant change was observed in the electronic density of states (DOS) plots in the bandgap region, a change in the DOS was found in the C_O cells. Figure 2.2a and b shows the DOS of the C_O cells and the pure cells around

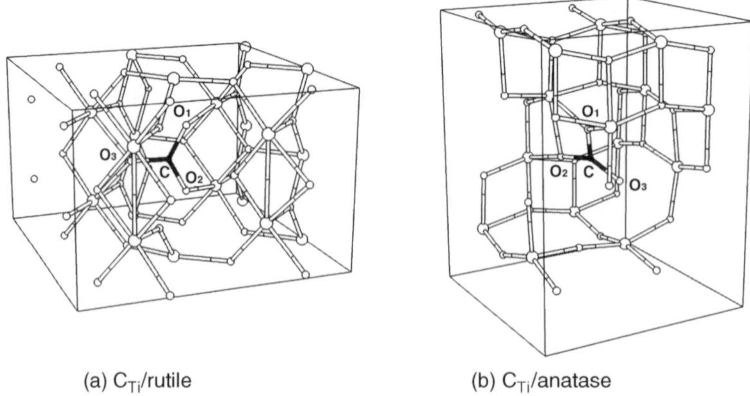

Figure 2.1 Optimized structure of C_{Ti}-doped cells. The apparent number of atoms is larger because the atoms on the boundary of the unit cell are drawn on both sides.

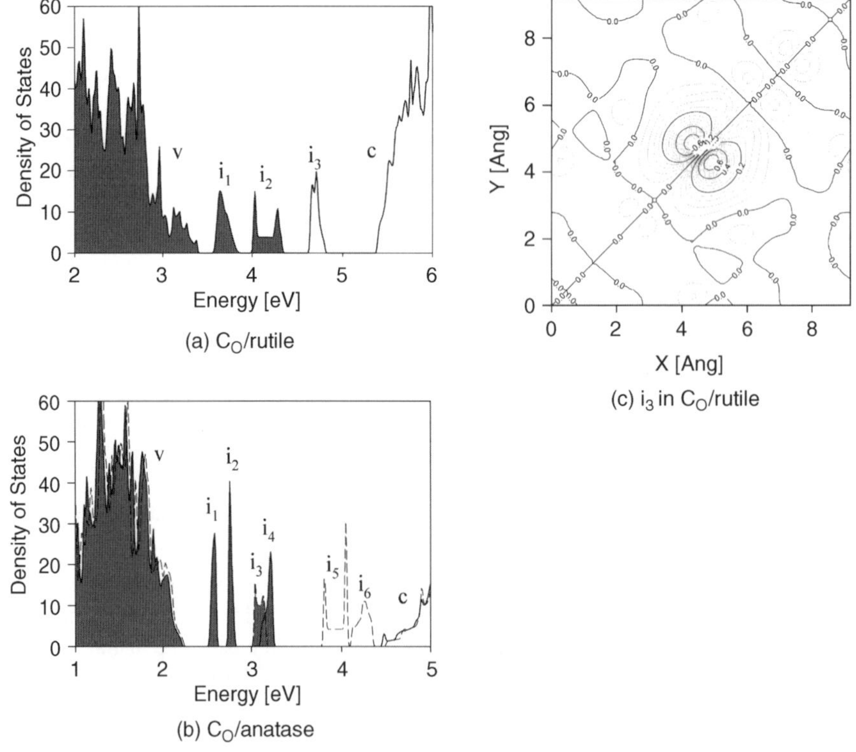

Figure 2.2 (a), (b) Density of states of C_O-doped cells around the bandgap region. Electronically occupied states are depicted in gray. In (b), the plots of the up- and down-spin electrons are represented by the solid and dashed lines. (c) The wavefunction of i_3 state in (a). The x and y axes denote the a, b lattice vectors. The doped C_O is located at the center. (d), (e) The optical absorbance of C_O-doped cells. The thick line denotes the total absorbance and the thin lines denote the decomposed absorbance according to their transitions. The thick dashed lines represent the corresponding absorbance of pure TiO_2.

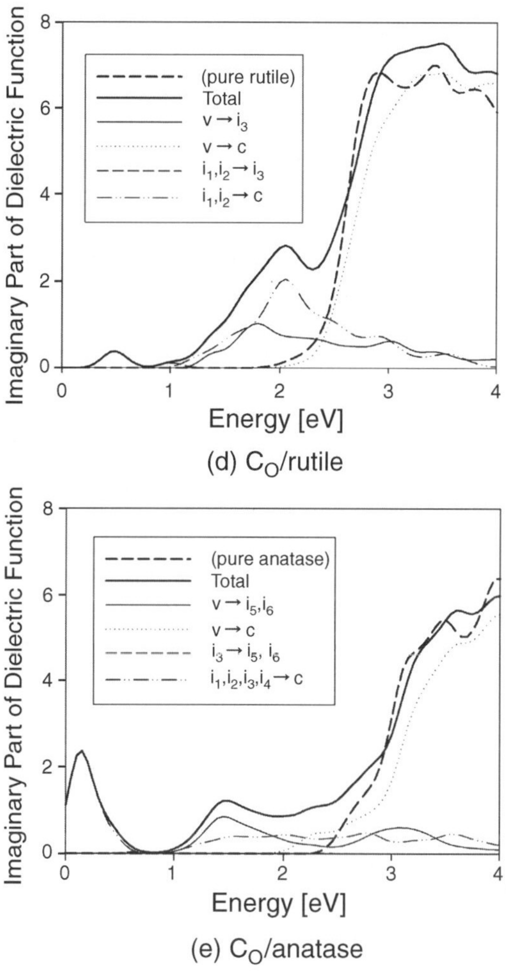

Figure 2.2 *(Continued)*

the bandgap region. As discussed in the previous subsection, the width of the gap of rutile (anatase) was underestimated to be 1.8 (2.1) eV in our present calculations. In either form, three peaks appear in the original bandgap region. Hereafter, we will denote these three states as: i_1, i_2 and i_3 in energetically ascending order. The Fermi energy is located between these states, and the occupied and unoccupied states are distinguished by the gray area in Figure 2.2. All these impurity states posses a 2p orbital shape around the carbon atoms. Figure 2.2c shows the wavefunction of the i_3 state at the Γ point in the rutile form. The spatial size of the wavefunction is the same size as the unit cell. The unit cell is considered to have the minimum size needed to describe the doped carbon atoms using the band-structure calculation method with a periodic boundary condition.

The optical absorbance was calculated for the C_O cells. In addition to the total absorbance, the decomposed absorbance with respect to the transition under consideration is also depicted in Figure 2.2d and e. The visible light absorbance was resolved from two distinguishable components: (i) transitions from the valence band to unoccupied impurity states, and (ii) transitions from occupied impurity states to the conduction band. Transitions between impurity states only contributed to the small energy region located around 0.2–0.3 eV. As we will discuss in the next section, the most important active species in the photocatalytic reaction is supposed to be surface-trapped electron holes. Thus, this visible-light response is considered to be able to contribute directly to the photocatalytic mechanism.

We will now discuss the effect of oxygen vacancies in C-doped TiO_2. In this study, a small peak at the edge of the conduction band was observed in the DOS plots. Because the Fermi level of the V_O cells is higher than the Fermi level of the corresponding C_O cells, two electrons are expected to transfer from a V_O cell to a C_O cell, when they are connected electronically. These transferred electrons will fill the unoccupied impurity states and suppress the formation of electron holes in the conduction band under visible light. As a result, the V_O cell is expected to inhibit the photocatalytic reactions on the surface.

However, this situation could be different when the two cells are located close to each other, and when changes are made in the chemical-bonding framework around them. To assess the effect of oxygen vacancies lying close to a carbon dopant, we carefully checked the results of the C_{Ti}–V_O and C_O–V_O cells. The left side of Figure 2.3 shows the total energy of the system plotted as a function of the distance between the cells. In either form, a significant energy stabilization was observed at a specific distance, and so the formation of a paired C_{Ti}–V_O structure may be possible.

The existence of paired C_O–V_O structures can be assessed using a statistical analysis approach. Construction of the partition function consists of isolated C_O and V_O cells and the several types of paired C_O–V_O cells with different bond distances, and the fraction of C_O cells forming the paired C_O–V_O structure can be estimated. The total energy of the system can be evaluated from the sum of several factors: (i) the energy of the independent C_O and V_O cells, the energy of the C_O–V_O pairs, and the energy of the conducting electrons. Electron transfer from the V_O to the C_O cells was also taken into consideration. The Fermi energy of the system was then determined according to the conservation of the number of electrons. The entropy factor was calculated by counting the number of possible doping sites, and the contribution from the conductive electrons was then added. The formation of the C_O–V_O pairs was assessed as a minimizing parameter for the total free energy at a given temperature.

Figure 2.3 shows the result of carbon doping and an oxygen vacancy rate of 0.01% on the right. The carbon atoms are assumed to replace the oxygen atoms (C_O). The experimental doping rate reported varies in the range 0.03–7.5% [15,16,44]. The oxygen vacancy rate in TiO_2 is controversial, but an experimental estimate is 0.02–0.2% [50]. Both these experimental values are greater than 0.01%, the value used in our analysis. The data in Figure 2.3 show that the C_O–V_O pairs are thermodynamically stable under these assumptions. In fact, no significant change was found in this structure by incorporating the C_O cells, nor were there any changes in the structure of the DOS plot. Nevertheless, the present results indicate the possibility of other complex structures that were not included in our analysis. Thus, for complete clarification of C-doped TiO_2, a systematic sampling of the structure is necessary. In V-doped [23] and Co-doped [51] systems, the essential role of oxygen vacancies has also been discussed.

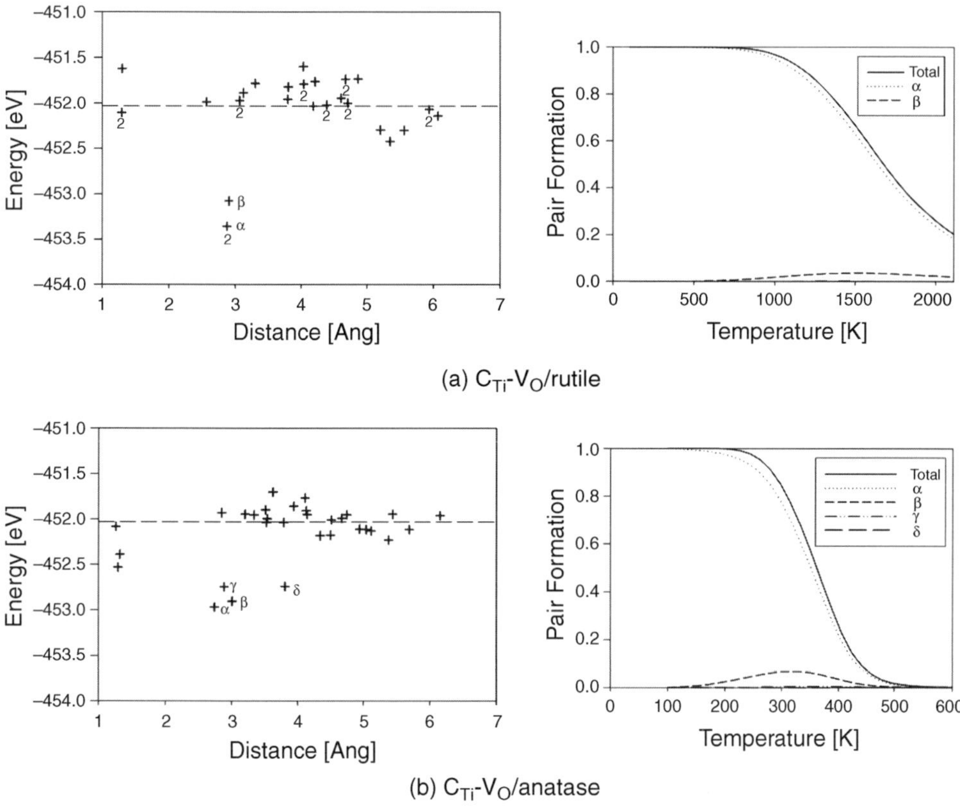

Figure 2.3 The energy levels of the optimized C_{Ti}–V_O pair structures on the left. The horizontal axis represents the distance between C_{Ti} and V_O. Pair formation is plotted on the right when both the C_{Ti} rate and the V_O rate were 0.01% as a function of temperature.

2.2.3 Nb-Doped TiO_2

Modern optoelectronic devices, such as light-emitting diodes (LEDs), thin-film transistor (TFT) displays and solar cells, rely on the availability of a transparent electrode. The class of oxide compounds that are transparent and exhibit electronic conductivity is called *transparent conducting oxides* (TCOs). Currently, the most widely used TCO is indium tin oxide (ITO), which is an alloy of tin oxide (SnO_2) and indium oxide (InO_2). However, the estimated amount of indium in the earth's upper continental crust is very small (50 ppb) [52], and the exhaustion of available indium ores is a serious concern. Recently, Furubayashi *et al.* discovered that an alternative TCO could be made from TiO_2 [6]. When niobium atoms are doped into anatase thin films of TiO_2, these films exhibit a high electrical conductivity and are transparent, which means that they can compete with ITO. This system was first fabricated on $SrTiO_3$ substrates using pulsed laser deposition, and later it was synthesized on glass substrates [53]. The estimated amount of Nb in the Earth's upper continental crust is 25 ppm, which is much larger than the amount of In. Thus, Nb-doped anatase TiO_2 is considered to be a promising alternative

TCO, and its improvement and the clarification of the microscopic operating mechanism are highly desirable.

In this section, our recent first-principle calculation results on TNO are presented for the doped structure, and we also discuss the effect of nonstoichiometry in TNO. Experimentally, the performance of TNO is strongly influenced by the ambient atmosphere during fabrication [53]. A reducing atmosphere is necessary, and the electrical conductivity deteriorates when the samples are exposed to O_2 gas [54]. The charge state of the niobium atoms was confirmed using Nb 3d core-level X-ray photoelectron spectroscopy (XPS) [55]. The ratio of Nb^{5+} to Nb^{4+} changed on annealing in O_2, and a strong correlation to the conductivity was observed. The results show the importance of nonstoichiometry, which occurs close to the niobium atoms.

The site of the Nb dopant was experimentally suggested as being at the Ti sites, from X-ray absorption spectroscopy (XAS) [56] and Rutherford backscattering data [57]. The properties of a cell with a single Nb atom in place of a Ti atom, designated as an Nb_{Ti} cell (see below), were calculated, as were the properties of cells with an oxygen vacancy or an interstitial oxygen atom. These cells are designated as V_O and O_i cells, respectively. In addition to these simple doped cells, the properties of cells with an Nb_{Ti} and a V_O (or O_i) site were also calculated. These cells were denoted as Nb_{Ti}–V_O and Nb_{Ti}–O_i cells, respectively. In the cells with two impurities, all symmetrically independent combinations of the doping sites in the enlarged cell were considered.

The computational scheme used was almost identical to the earlier study involving C-doped TiO_2. The same size of unit cell was constructed from the primitive cell. The PW91 functional [27], the plane-wave basis set, the USPP [46], and the Monkhorst–Pack **k**-point sampling scheme [48] were used again.

First, we will discuss the results of the Nb_{Ti} cells. The optimized structure of the Nb_{Ti} cell was similar to that of the pure cell. A small enlargement of the chemical bonds around the Nb atom relative to the Ti atom was observed. The internuclear distance was r(Nb–O) = 1.979 Å in the *a,b*-plane and r(Nb–O) = 2.037 Å along the *c*-axis. These values were increased by +1.8% and +2.1% measured from the pure value, r(Ti–O) = 1.944 Å and r(Ti–O) = 1.995 Å. In addition, the density of states did not show any significant changes from the pure cell, and the only apparent change was seen in the Fermi level. An electron was emitted from the Nb dopant, and at a doping rate of 6.25%, the Fermi level was raised to $E_F \sim 0.16\,eV$, measured from the conduction band minimum (CBM). The CBM at the Γ point is composed of the Ti $3d_{xy}$ orbitals [34]. A DOS plot of the Nb_{Ti} cell is shown in Figure 2.4, where the projection of the DOS to each atomic component is also shown. The contributions from the Nb 4d and Ti 3d orbitals are similar. A spatial plot of the CBM state also reveals that the difference between these two orbital contributions is minor. The effective mass of an electron in the CBM was calculated for the Nb_{Ti} and pure cells. The estimated value was the same for both structures within the computational accuracy, $m^*_{xx} = 0.42$ and $m^*_{zz} = 4.05\, m_0$. These results suggest that the orbital mixing of the two cells is almost complete, and they confirm the validity of the rigid-band model for electronic structure calculations of this simple Nb_{Ti} cell.

Second, the results of the V_O and O_i cells will be explained. In the optimized V_O cell, no significant structural changes were seen, except for the removed oxygen atoms. The Fermi energy was raised, $E_F \sim 0.30\,eV$, which corresponds to two electrons per unit cell in the conduction band. The electrons are weakly localized in place of the removed oxygen, and the

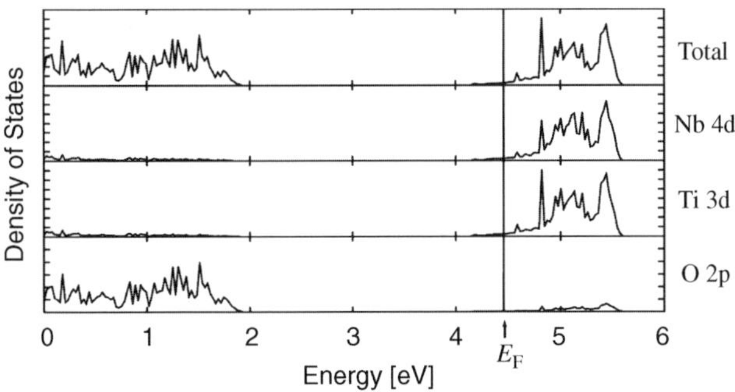

Figure 2.4 The DOS and PDOS plots of the Nb_{Ti}-doped cell. For comparison, the heights of the projected DOS (PDOS) are normalized according to the number of atoms.

density is composed of the Ti 3d orbitals of the three surrounding Ti atoms. In the band-structure plot, the density has the lowest energy at the A point, which is close to CBM at the Γ point. The optimized O_i cell has an interstitial oxygen atom located next to another lattice oxygen atom [42]. The two oxygen atoms are structurally identical, with an internuclear distance of r(O–O) = 1.471 Å. No novel impurity state was found in the bandgap in these structures. The energy of formation of these cells was calculated from the total energy of the cell, and the energy of an O_2 molecule $(^3\Sigma_g^-)$ was obtained using the same method. The energy of formation, $E(O_i) = 1.38$ eV and $E(V_O) = 4.85$ eV, is consistent with the other reported values [35].

So far, we have focused on the results of independent Nb_{Ti}, V_O and O_i cells. None of these structures shows any significant electronic structural changes from the pure cells, besides changes to the Fermi energy. This finding indicates the importance of their mutual interactions. Hereafter, we will discuss the results of Nb_{Ti}–V_O and Nb_{Ti}–O_i cells. These structures are also important for the clarification of the role of nonstoichiometry. In the data shown in Figure 2.5, the total energy of these cells is plotted as a function of their internuclear distances, r(Nb_{Ti}–V_O) and r(Nb_{Ti}–O_i). The data in Figure 2.5 show that the interaction between an Nb_{Ti} and a V_O cell is repulsive. The total energy increases about 0.1 eV when the two cells are neighbors. In contrast, neighboring Nb_{Ti} and O_i cells show an energy stabilization of about 0.8 eV. Hereafter, we will designate the structures in which the Nb_{Ti} dopant is placed most closely to a V_O/O_i site as “adjacent” Nb_{Ti}–V_O (Nb_{Ti}–O_i) cells. These structures were employed in the following analysis to rationalize the experimental results. Other structures with an Nb_{Ti} cell separated from a V_O/O_i cell did not show any novel results, based on expectations from the independent Nb_{Ti}, V_O and O_i cells.

The adjacent Nb_{Ti}–V_O cells did not show any changes in structure from the V_O cells. In contrast, in the adjacent Nb_{Ti}–O_i structure, an increase of the internuclear distance between the two oxygen atoms was observed, and the distance became r(O–O) = 1.946 Å, which is an increase of 32% on the original O_i cell. Impurity states were observed in the gap in these adjacent structures. Figure 2.6 shows a DOS plot and a spatial plot of the electron density of this state. In the case of an adjacent Nb_{Ti}–V_O structure, the electron density observed in the place of the V_O cell is inclined toward the Nb_{Ti} dopant. The larger cationic character of the $Nb_{Ti}{}^{5+}$ dopant relative to Ti^{4+} explains this and is probably the

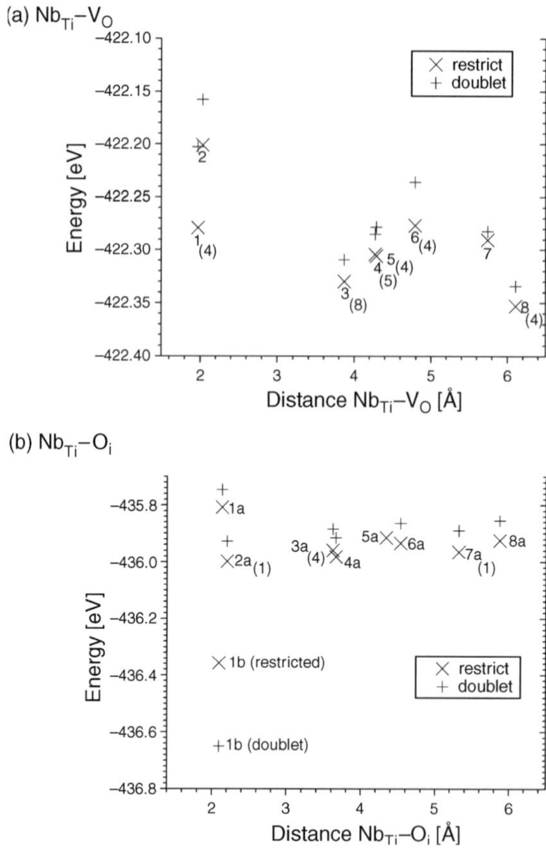

Figure 2.5 Energy levels of the Nb_{Ti}–V_O and Nb_{Ti}–O_i structures with different spatial separations. The numbers in parentheses show the number of structural degeneracies. Energies obtained from restricted and unrestricted (doublet) calculations were plotted.

reason why the state appears in the bandgap, while the corresponding states are located close to the CBM in the V_O cell. In the case of the Nb_{Ti}–O_i structure, the electron density of the state is localized on the two oxygen atoms and has the shape of a σ^* antibonding orbital. This orbital was obtained using the restricted method and is occupied by an electron. When the unrestricted method was employed, the up/down states were split, and only one of them was occupied.

The characteristic electronic structure in the adjacent Nb_{Ti}–V_O/O_i cells can be used to rationalize the experimental results. The charge state of the Nb dopant was determined from the Nb 3d core-level XPS experimental data [55]. According to this, both the Nb^{4+} and Nb^{5+} charge states exist in the initial TNO sample, with a ratio of $Nb^{4+} : Nb^{5+} = 4 : 1$. The sample had been fabricated in a reducing atmosphere. After annealing the sample at 400–600 °C in an oxygen atmosphere at a partial pressure of 3.5×10^2 torr, the Nb^{4+} peak disappeared and the carrier concentration decreased from $n_e = 9 \times 10^{20}\,\mathrm{cm}^{-3}$ to $n_e = 4 \times 10^{19}\,\mathrm{cm}^{-3}$.

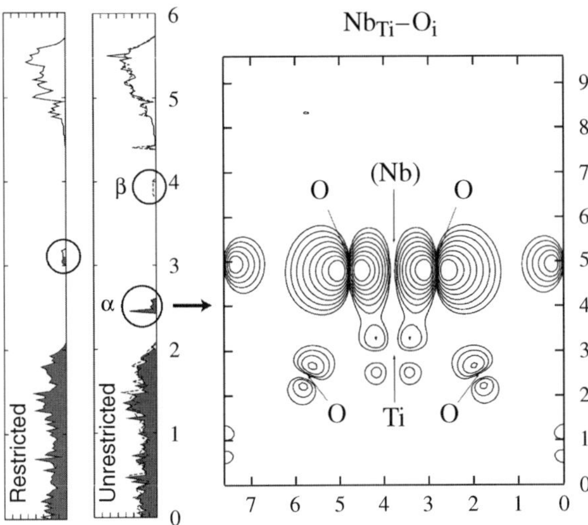

Figure 2.6 Contour plot of the electron density of the impurity state found in the adjacent Nb_{Ti}–O_i structure. Two oxygen atoms are on the plane, while the Nb_{Ti} dopant protrudes. The contours start from 0.005 electrons $Å^{-3}$ and change successively by a factor of two.

The effect of the oxygen annealing can be quantitatively analyzed using the calculated results and a statistical technique. The Gibbs free energy of the O_2 molecules in the ambient gas phase can be easily expressed in terms of the mass and rotational momentum of the molecules and the temperature. The free energy of the bulk can be expressed as the sum of the energy of the nuclear and valence electrons and the conductive electrons, and their entropy factor. The entropy for the structure can be assessed by counting the number of possible doping sites and using Starling's formula. When the DOS in the conduction band is expressed in particular empirical form, the statistics of the conducting electrons [6,58] can be described by a Fermi–Dirac distribution. By comparing the chemical potential of the oxygen atoms in the two phases, the thermal equilibrium O_i doping rate can be determined, and its effect on the spectroscopic measurements can be clarified by referring to this result.

Figure 2.7 shows a comparison of the two free energies for a 6% niobium-doping rate and at room temperature. The dash and solid lines represent the results with and without counting the change in free energy in conducting electrons, respectively. Note that this statistical analysis does not include the adjacent Nb_{Ti}–V_O structure. The adjacent Nb_{Ti}–V_O structure will supposedly suppress any change in electronic free energy in the conduction band because the structure will trap excess electrons in its impurity state. In the data shown in Figure 2.7, the solid line crosses over the experimental oxygen pressure range in which the change in conducting properties was observed. This coincidence indicates the existence of the adjacent Nb_{Ti}–V_O structure in the initial TNO samples.

The existence of the adjacent Nb_{Ti}–V_O and Nb_{Ti}–O_i structures can be corroborated from the lattice constant of the system. The experimental results show an expansion of the lattice in all three directions [6,58]. Table 2.1 shows the optimized lattice cell. Despite the increase in bond length between the Nb dopant and all six coordinating oxygen atoms, a minor shrinkage

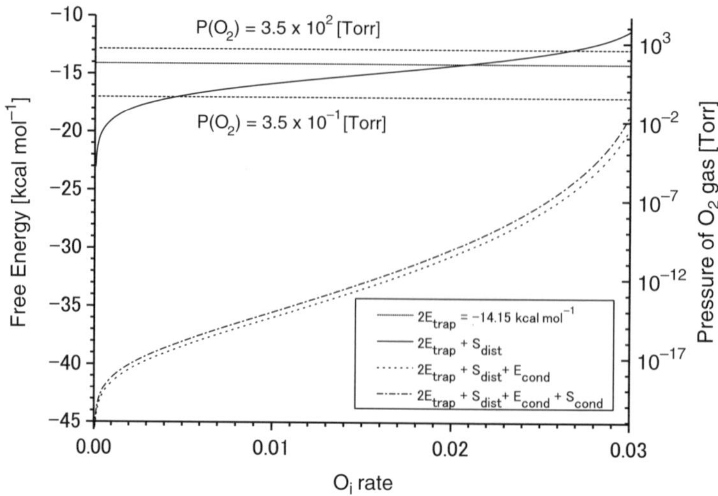

Figure 2.7 Free energy of O_2 molecules in the gas phase and $2O_i$ absorbed in TNO. The doping rate of TNO is 6%. On the right of the figure, the pressure of O_2 gas is plotted in axis. The horizontal dashed lines represent the experimental pressure range where changes in XPS signals were observed. Three curves show the contribution from the distribution of O_i and that from electrons in the conduction band.

(0.20%) of the lattice is observed along the *c*-axis. In contrast, an expansion is observed for all three axes in the adjacent Nb_{Ti}–V_O and Nb_{Ti}–O_i structures. If a small number of Nb dopant atoms form this structure, then the experimental results are reasonable.

In our calculations, the Nb_{Ti}–V_O cells exhibit a repulsive interaction. Nevertheless, the discussion above explains the experimental results consistently if the existence of the adjacent Nb_{Ti}–V_O structure is assumed. In reality, a more complex Nb_{Ti}–V_O structure may exist, and an aggregation of Nb_{Ti} and V_O structures will possibly exhibit energy stabilizations because of the electrostatic attractions between the trapped electrons in the V_O cell and the Nb^{5+} cations. The electrons in the V_O cell are only lightly trapped, and a conduction path for the electrons may be provided by such an aggregated structure.

Table 2.1 Optimized lattice constants of several 8-times cells [Å].

	Pure*1	Pure*2	TNO(6%)*2	Pure	Nb_{Ti}	V_O	O_i	Nb_{Ti}–V_O	Nb_{Ti}–O_i
a	7.564	7.564	7.592	7.6331	7.6725	7.6508	7.6486	7.6409	7.6374
			+0.37%		+0.517%	+0.231%	+0.202%	+0.101%	+0.055%
b				7.6331	7.6741	7.6422	7.6368	7.6529	7.6027
					+0.537%	+0.120%	+0.048%	+0.258%	−0.390%
c	9.502	9.544	9.560	9.7151	9.6957	9.6884	9.7571	9.7335	9.8342
			+0.17%		−0.200%	−0.275%	+0.432%	+0.189%	+1.225%

*1 and *2 designate data taken from Refs. [6,34], respectively.

2.3 Surface Hydroxyl Groups and the Photoinduced Hydrophilicity of TiO_2

2.3.1 Speculated Active Species on TiO_2 – Superoxide Anion (O_2^-) and the Hydroxyl Radical ($OH^•$)

Complete understanding of the photocatalytic reaction process on a TiO_2 surface has not been established. Initially, the reaction was conjectured to involve several reactive intermediates [2,59–61]: a superoxide anion (O_2^-), hydroxyl radicals ($OH^•$), and surface-trapped electron holes (h^+). The $OH^•$ radical was considered to be a product of the reaction between a surface hydroxyl group and a photogenerated hole:

$$[\mathrm{Ti{-}OH}]_s + h^+ \rightarrow [\mathrm{Ti^g OH}]_s^+ .$$

However, several researchers denied the direct production of hydroxyl radicals. In reports, there was confusion with regards to the electron paramagnetic resonance (EPR) signals that originate from $OH^•$ radicals and those that originate from photogenerated holes trapped at the lattice oxygen sites [62].

In 2004, Nakamura and Nakato carefully observed the Fourier-transform infrared spectroscopy (FTIR) adsorption *in situ* and confirmed a signal that corresponded to a bridge-type surface peroxo species $[\mathrm{Ti{-}O{-}O{-}Ti}]_s$ [60]. From this, they advocated a reaction mechanism that begins with a nucleophilic attack of a H_2O molecule onto a surface-trapped hole. In the same year, Kubo and Tatsuma detected H_2O_2 molecules released from TiO_2 in air [63]. They suggested that the oxygen molecules adsorbed on the TiO_2 will be either reduced directly:

$$O_2 + 2H^+ + 2e^- \rightarrow H_2O_2,$$

or indirectly:

$$O_2 + e^- \rightarrow O_2^-,$$

$$2O_2^- + 2H^+ \rightarrow H_2O_2 + O_2.$$

Then, the H_2O_2 molecules will be decomposed into two $OH^•$ radicals by UV irradiation.

2.3.2 Theoretical Calculations of TiO_2 Surfaces and Adsorbents

Among the possible rutile surfaces, the surface designated with a Miller index of (110) is the most stable [29]. This surface can assume either stoichiometric or reduced forms. The stoichiometric TiO_2 (110) surface has oxygen atoms placed on the titanium atoms protruding from the outermost layer. The oxygen atoms bridge two titanium atoms and thus are called *bridging oxygen atoms*. The bridging oxygen atoms are loosely bonded to the surface and tend to be missing, depending on the external conditions. A surface lacking any bridging oxygen atoms is considered to be the main structure of a reduced (110) surface. The sites with missing bridging oxygen atoms are known to act as adsorption sites for several chemical species [29,59,64]. The first step of a theoretical elucidation of the photocatalytic reactions on TiO_2 needs to address such adsorbed molecules on these surfaces.

In theoretical calculations, the surface of the bulk TiO_2 is typically modeled using a periodic slab. Because the system has periodic boundary conditions in all three directions, the methodology of the standard DFT-based band-structure calculations is applicable. Theoretical calculations on TiO_2 using the first-principle method began in the early 1990s. Reinhardt *et al.* calculated the structure of (001), (100) and (110) rutile surfaces using a periodic Hartree–Fock method and discussed their relative stability and electronic structures [65]. Ramamoorthy *et al.* calculated several stoichiometric rutile surfaces systematically and reproduced the Wulff diagram [66]. As mentioned above, the bridging oxygen atoms are loosely bound to the surface. Early calculations on the rutile (110) surface showed discrepancies between experimental values with regard to this position. Later, the discrepancy was ascribed to the soft vibrational mode of these surfaces [67]. Compared with the mature rutile (110) calculations, the number of publications dealing with anatase surfaces is currently small.

Calculations of a reduced rutile (110) surface are more difficult than calculations of a stoichiometric surface. On this type of surface, excess electrons exist, and the energy level of these excess electrons is identified by photoelectron spectroscopy and electron energy loss spectroscopy (EELS) [68]. This energy level has been observed to be 2 eV above the valence band edge. This state is called the *gap state*. The early theoretical calculations using the extended Hückel method or the DV–Xα method did not successfully reproduce the energy level, nor did the first-principle calculations by Ramamoorthy *et al.* [69]. Lindan *et al.* calculated the reduced (110) surface using either the spin restricted or unrestricted DFT method [70]. According to their report, the effect of spin polarization is essential to reproduce the state. Their calculation utilized the periodic slab model. Naturally, the oxygen defects were introduced periodically. To elucidate an isolated oxygen defect, Bredow *et al.* calculated a cluster model using the DFT method [71]. In their report, the effect of the spin polarization was also pointed out.

Needless to say, water molecules are the most fundamental adsorbent on a TiO_2 surface. However, theoretical calculations of this structure have not reached a definite conclusion as yet. From experimental data, some authors have reported that a molecularly adsorbed form is dominant at low temperatures and that the dissociated form only occurs when the surface defects are mediated [72,73]. On the other hand, theoretical calculations have predicted both molecular and dissociated adsorption [74]. After a careful convergence check and controversy, it turns out that the preferred form of adsorbed water on TiO_2 relies on the choice of the DFT functional [75–77].

The hydroxyl radical has been proposed as the important active species on a TiO_2 surface. This molecule was calculated by Shapavalov *et al.* using a standard quantum chemistry method [78]. They discussed the formation of the $OH^{\bullet}$ radicals from adsorbed water molecules. A cluster model of the (110) surface, Ti_4O_{14}, was constructed and a water molecule was placed on it. The water molecule was adsorbed in either a molecular or dissociated form. The electronic ground state and excited states were variationally calculated using the configuration interaction (CI) method. From the results, the electron density between the electronic ground and excited states was compared, and the depletion of electron density on the water was found for the excited states, in both the molecular and dissociated forms. Shapavalov *et al.* concluded that this feature would lead to the production of $OH^{\bullet}$ radicals, especially for the dissociated case. The transferred electron density appeared on the titanium atom in the third layer. Shapavalov *et al.* related these findings to the proposed mechanism where the electron holes migrate to the surface and trigger the production of radical species.

The O_2^- species on a TiO_2 surface was calculated by de Lara-Castells *et al.* [79]. A periodic slab model was constructed for oxygen molecules adsorbed on a reduced (110) surface. The surface was reconstructed to form a $(3 \times 2)-(110)$ surface, and the periodic Hartree–Fock method was applied. According to their calculation, oxygen molecules are adsorbed on an oxygen vacancy site, and a triplet state of O_2^- is formed. The density of the excess electron in the O_2^- extended to the adjacent titanium atoms. De Lara-Castells *et al.* discussed that any further adsorption of two oxygen molecules would be induced by these electrons. According to an EELS experimental report, the O_2^- molecule consists of two distinguishable oxygen atoms [80]. Measurement of the photoinduced desorption of this molecule suggests the existence of three types of adsorbent [81]. The calculation by de Lara-Castells *et al.* rationalized these two experimental findings. However, these calculations were essentially for the ground state, and neither the excited states nor the photodesorption process was treated explicitly.

2.3.3 Surface Hydroxyl Groups and Photoinduced Hydrophilic Conversion

The hydrophilicity of a TiO_2 surface increases significantly on irradiation with UV light, and this change lasts for a period of several hours in the dark [3]. This phenomenon is called *photoinduced hydrophilic conversion* and is used in many industrial applications. Initially, the cause of this hydrophilic conversion was the dissociative adsorption of water molecules on the oxygen vacancies that are generated by UV light. Using XPS, Sakai *et al.* showed that a strong correlation existed between the water contact angle and the intensity of the O(1s) spectra of the surface hydroxyl groups [82].

However, several groups have proposed a counter argument. Photocatalytic decomposition of hydrocarbon molecules on a TiO_2 surface was suggested by White *et al.* [83] and Zubkov *et al.* [84]. In their experiments, trimethyl acetate (TMA) and hexane were the hydrocarbons used, respectively. Zubkov *et al.* also surveyed the effect of oxygen in the air, and the adsorbed Ti–OH was measured using infrared spectroscopy. In addition to these arguments, a mixed mechanism involving the surface hydroxyl groups and hydrocarbons has also been advocated [85].

To aid this controversy, Shibata *et al.* measured the dynamic hardness of the (100), (110) and (001) surfaces of rutile both before and after UV irradiation [86]. In all three cases, the irradiation enhanced the dynamic hardness of the surface. Independently, a tensile stress was introduced on the surface of a thin TiO_2 layer from the difference in the thermal expansion coefficients of the film and the substrate [87]. Suppression of the reverse process, that is, the conversion from a hydrophilic to a hydrophobic surface in the dark, was induced by the tensile stress.

From the data from the XPS and surface stress experiments, it is reasonable to expect that the hydrophilic conversion process is associated with the generation of hydroxyl groups and a compressive stress on the surface. The generation of a hydrophilic surface is a reversible process that is triggered by UV irradiation, and the hydrophilicity lasts for a period of hours in the dark. A structural change in the surface must be arise because of the participation of electronic excited states, and this change is trapped as a metastable structure, even after the de-excitation of the electronic state.

To assess the validity of the corroboration using the data measurements of the dynamic hardness, we studied the correlation between the adsorbed water molecules, the number of surface hydroxyl groups and the surface stress from first-principle calculations. The metastable

Table 2.2 The surface stress of a clean slab [Nm^{-1}].

		8,6 layers		4,3 layers		extended 4,3 layers	
		xx	*yy*	*xx*	*yy*	*xx*	*yy*
PW91	(110)	−23.77	−14.24	−23.02	−15.12	−23.10	−16.90
		(−24.64)	(−15.28)	(−22.48)	(−14.87)		
	(100)	20.99	−13.46	20.63	−13.92	16.24	−17.35
		(19.80)	(−14.47)	(19.83)	(−14.83)		
PBE	(110)	−25.55	−11.35	−22.50	−12.31	−21.92	−12.07
		(−26.44)	(−12.30)	(−22.40)	(−11.79)		
	(100)	20.57	−6.89	19.15	−8.73	18.84	−9.08
		(18.96)	(−9.47)	(18.61)	(−9.17)		

The numbers of layers are 8 and 4 on the (110) surface, and 6 and 3 on the (100). The *xx*, *yy*-components correspond to the $[1\bar{1}0]$ and [0 01] directions on the (110) surface, and the [010] and [001] on the (10 0). The numbers in parenthesis are results obtained with increased parameters. The extended 4,3 layers have non-primitive surface translational vectors.

structure discussed above should be found using DFT-based calculations on the electronic ground state. DFT-based first-principle calculations could be used to obtain the stress of the cell. Using the optimized lattice constant of the bulk system, the surface stress can be extracted from the results using a periodic slab model. In general, the hardness of the surface is related to the surface stress.

We calculated the surface stress of rutile (110) and (100) surfaces. Four types of adsorbents were considered: (i) molecularly adsorbed water (H_2O(mol)), (ii) dissociatively adsorbed water (H_2O(dis)), (iii) dissociatively adsorbed water at an oxygen vacancy, and (iv) adsorbed hydrogen atoms. The size of the surface translational vector was extended to describe an isolated water molecule. In this study, two DFT functionals, the PW91 and PBE functionals, were employed.

As a preliminary measure, we checked the convergence of the surface stress with respect to the thickness of the slab. The surface stress calculated using a different number of slab layers is summarized in Table 2.2, which shows that a good convergence of the stress with respect to the slab thickness was achieved. Unlike the surface energy and adsorption energy of the adsorbents, which is known to oscillate as a function of the number of layers [75,88], the surface stress is localized on the top few layers. For example, the surface stress of a diamond (111) surface is determined by the top layer, and the contribution of the second layer is less than 1% [89].

The optimized structure and the calculated surface stress are shown in Figure 2.8. The surface stress tensor was diagonalized, and the eigenvalues are depicted along the principle axis. The horizontal arrows facing each other show the tensile stress. The other arrows represent the compressive stress. The magnitude of the stress is denoted in units of Nm^{-1}.

The molecularly adsorbed structure is considered in many experimental and theoretical results. One indicator of the reliability of the optimized structure is the internuclear distance between the terminal titanium atom and the oxygen atom in the adsorbed water. Recent experimental measurements using scanned energy mode photoelectron diffraction have reported r(Ti−O) = 2.21 ± 0.02 Å [90]. Preceding theoretical values were in the range r(Ti−O) 2.25–2.41 Å [91,92]. Our result of r(Ti−O) = 2.23 Å is in good agreement with these values.

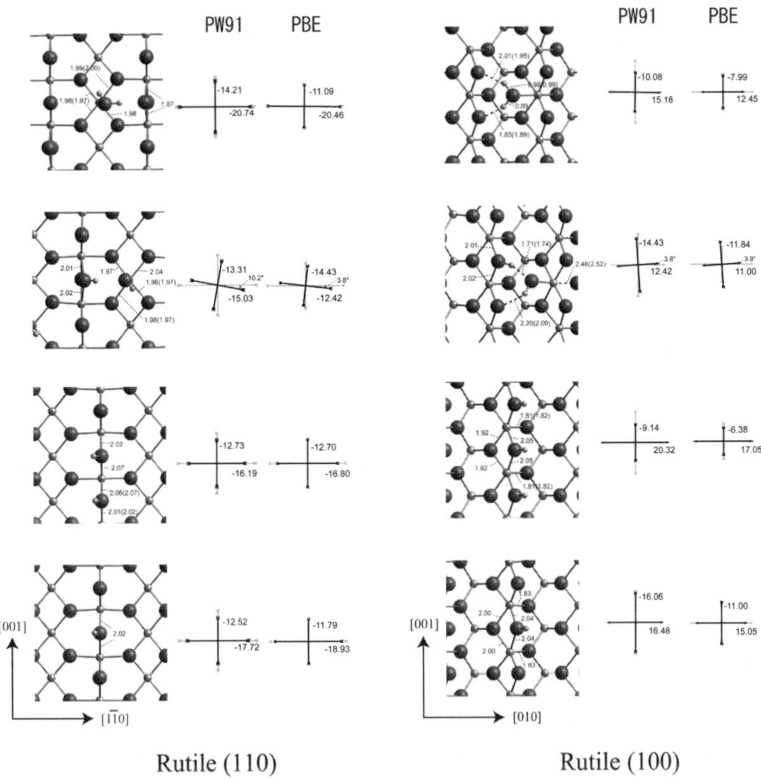

Figure 2.8 The optimized structures of the adsorbents on (110) and (10 0) surfaces. The small, medium and large spheres represent the hydrogen, titanium and oxygen atoms, respectively. The calculated stress tensors are drawn on the right-hand side in Nm^{-1}. The arrows facing each other show the tensile stress. The other arrows represent the compressive stress. The light-gray arrows on the right-hand side represent the stress on the clean surface.

The H_2O(dis) structure has a similar structure to the H_2O(mol) adsorbent. The position of one of the hydrogen atoms is transferred from the molecular water to the vicinity of the bridging oxygen atom. As a result, a terminal hydroxyl group and a bridging hydroxyl group are formed on the surface. A small shift in the titanium atom is also observed. On both surfaces, the bonded titanium atoms protrude from the surface toward the hydroxyl groups. The $Ti-O_{br}$ bond distance of the bridging hydroxyl group has been measured in the case of a formic acid reaction [93]. A significantly enlarged bond length of $r(Ti-O_{br}) = 2.02 \pm 0.05$ Å was reported, whereas values of 2.01–2.02 Å arise in our calculations.

The adsorption energy of both structures can be calculated from the total energy of the water molecules and the clean surface. The adsorption energy is listed in Table 2.3. In our calculations, the H_2O(mol) adsorbent showed a higher adsorption energy than H_2O(dis) did by ~0.25 eV. Note that preference of the molecular form in these calculations was obtained at 1/3 ML. This result may change under high coverage conditions and the choice of DFT functional. The water molecules adsorbed on the oxygen vacancies show a higher adsorption

Table 2.3 Adsorption energy of H_2O on (110) and (100) surfaces [eV].

	PW91		PBE	
	(110)	(100)	(110)	(100)
H_2O(mol)	0.80	1.01	0.70	0.91
H_2O(dis)	0.58	0.73	0.46	0.66
$H_2O@V_O$	1.39	1.53	1.30	1.39

energy on both stoichiometric surfaces. This result agrees with earlier theoretical reports [91] and the conclusions from temperature programmed desorption (TPD) spectra [73]. However, it should be stressed that the possibility of another structure has been pointed out experimentally [94], and the interpretation of the TPD data requires a kinetic analysis of the process and the effect of reconstructing the water layers.

Hereafter, the discussion will concentrate on the calculated results. First, we will discuss the surface stress on clean surfaces. On a clean (110) surface, a tensile stress is observed in both directions. An emergence of a tensile stress is usually found on metal and semiconductor surfaces [95]. When a bulk material undergoes cleavage, electrons transfer from the missing bonds between the layers to the remaining bonds on the surface. This transfer decreases the equilibrium bond length on the surface, and so a tensile stress is observed. On a (110) surface, the decrease in bond length between the Ti and O atoms is observed, and the internuclear distance is r(Ti−O) = 1.83 Å, while the corresponding value in the bulk is 2.00 Å in the z-direction. This displacement will attract the surrounding oxygen atoms in the x,y-plane. The short r(Ti−O) distance for the fivefold Ti may also come about from an imbalance in the electrostatic force due to the missing apex oxygen atom as well. The increase in bond strength due to the missing apex oxygen atom will be induced, not only the vertical Ti−O bond, but also in the lateral Ti−O bonds.

A compressive stress is observed in the [010] direction on a clean (100) surface. The cause of a compressive stress on a surface consists of several competing factors [96], and, in general, no simple explanation can be proposed. However, one explanation is as follows. Unlike the (110) surface discussed above, the ends of the Ti−O bond chains are pendant on a (100) surface, as seen in Figure 2.8. Therefore, the influence of the decreased bond length can be compensated for without inducing a tensile stress. To clarify the origin of the compressive stress further, a one-layer slab model of a (100) surface was optimized for the structure and the lattice constants. The one-layer slab expanded to 112.3% in the [010] direction and shrank to 98.1% in the [001] direction, as expected from the surface stress. In the optimized one-layer slab, the bond length was found to be very close to the corresponding bond length of a thicker slab with a bulk lattice constant, and the maximum difference was around 0.01 Å. In contrast, significant changes were observed in the bond angles. Therefore, unfolding of the zigzag structure in the [010] direction in the top layer induces a compressive stress.

The effect of the adsorbent will now be discussed. On metal surfaces, adsorbents generally induce a compressive stress relative to the clean surface [95]. The electron density on the surface is depleted because of the adsorbent, which is in contrast to the cleavage process discussed above. In our calculations, the same mechanism was observed for the −H and −H × 2 cases. Figure 2.9 shows the electron depletion in the case of −H × 2 on a (100) surface. The attachment of hydrogen atoms on the bridging oxygen increases the electron density

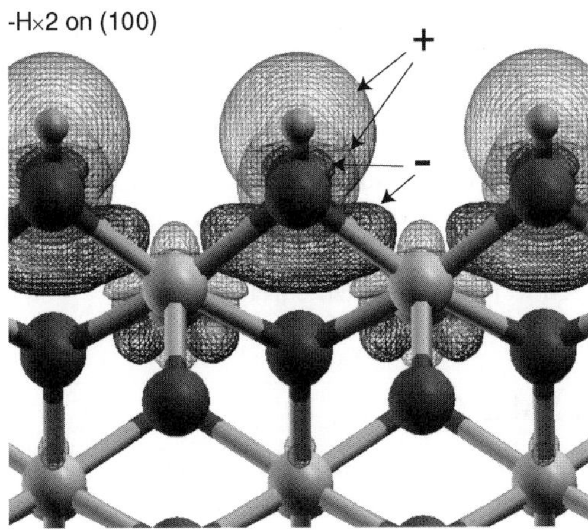

Figure 2.9 The changes in electron density induced by attaching a hydrogen atom to the bridging oxygen on a (10 0) surface. The increase (+) and depletion (−) of the density is depicted by the light and dark gray balloons, respectively.

around the hydrogen atoms, while the electron density in the Ti−O−Ti chains decreases. A similar depletion is observed on a (110) surface as well.

The effect of H_2O(mol) and H_2O(dis) is more complicated. On a (100) surface, hydrogen bonds are formed for both adsorbents. The internuclear distances between the H and O atoms are r(H−O) = 1.71–1.85 Å. The bonds reduce the magnitude of compressive stress along the [010] direction relative to a clean surface. On a (110) surface, the corresponding distances are 1.93–2.04 Å, which are longer than the internuclear distances on a (100) surface. Therefore, the strength of the hydrogen bonds should be weaker. A reduction in the tensile stress is observed for H_2O(dis) and H_2O(mol) adsorbates on a (110) surface. In these structures, the fivefold coordinated titanium is displaced from its position on a clean surface toward the original position in the bulk. The recovery of this position is related to the sixfold coordinated character of the atom and can be regarded as the reason for the reduction in the tensile surface stress.

These results show the complexity of the surface stress induced by hydroxyl groups. For example, the H_2O(dis) structure can cause opposite surface stress, depending on the surface. The morphology of the surface affects the internuclear distances of the hydrogen and oxygen atoms, and the formation of hydrogen bonds between the surface and the adsorbents is influenced strongly. The termination of chains of bonds also depends on the surface. These terminations show the different nature of the surface stress at the level of a clean surface, and the magnitude of this character is significantly changed by the sixfold character of titanium atoms, which is influenced by the attachment of water or hydroxyl groups, as discussed above.

In reality, at high coverage condition, or for a water–TiO_2 interface, there must be a hydrogen-bond network present involving adsorbed water molecules, bulk water molecules, hydroxyl groups on the surface, and the bridging oxygen atoms on the TiO_2. The formation of a hydrogen network has already been discussed theoretically [77]. In such circumstances, the results of this study may not be applicable.

In experiments, only the magnitude of the surface hydroxyl groups can be inferred from the XPS signals, and the type is not resolved [82]. The effect of artificially introduced surface stress was investigated on a composite surface with a complex morphology [87]. Considering the complex results obtained from these calculations, a corroboration of the role of surface hydroxyl groups in hydrophilic conversion, and any changes in dynamic hardness, cannot be determined. The type of surface hydroxyl group needs to be identified to further our understanding of these phenomena. The effect of coverage rate, surface morphology and hydrogen bonds also need to be investigated carefully.

2.4 Dye-Sensitized Solar Cells

The TiO_2 system can be used as a component of solar-cell devices. When a given dye molecule on a TiO_2 surface is electronically excited by irradiation of light, electrons are injected from the dye into the substrate. This electron injection process is very fast, on the femtosecond timescale. A solar-cell device can contain such a surface, an electrolysis component (I^-/I_3^-) and a transparent electrode (TCO). This type of cell was first fabricated by O'Regan and Grätzel and is known as a dye-sensitized solar cell (DSSC) or a *Grätzel cell* [4]. A number of combinations using a wide variety of dye and electrolyte have been investigated. The use of ruthenium compounds (Blackdye [97], N3 dye [98]) or organic coumarin dyes [99] (C343) and the I^-/I_3^- electrolyte is commonly seen in actual applications. This combination gives the maximum efficiency of about 8–10%. However, the performance is still lower than that of solar cells made of polysilicon, and further improvement in the efficiency is required. In the following text, we will overview some preceding theoretical studies of this injection process and review our recent analysis of PbSe quantum dots. As stated in the Introduction, QDs can be used as sensitizers of TiO_2 surfaces, and a significant improvement in efficiency is expected.

2.4.1 Conventional Sensitizers: Ruthenium Compounds and Organic Dyes

Electron injection from the dye into the TiO_2 substrate has been the subject of several theoretical studies. Some pioneering work was carried out by Ramakrishna *et al.* [100] and Thoss *et al.* [101]. In their studies, the system was represented using a model Hamiltonian and several phenomenological parameters. The analysis was intended to clarify the qualitative behavior of the process rather than to provide an elucidation of charge transfer in a molecular system.

An atomistic simulation was carried out by Stier *et al.* [102]. In their study, a model system that consisted of an anatase (101) surface and isonicotic acid was considered. An approximate simulation scheme based on DFT calculations was adapted, and the effect of a nonadiabatic transition was included, referring to the overlap of the KS orbitals along a trajectory calculated using a finite time step. Their findings were that both the adiabatic and nonadiabatic processes contribute to the electron injection. However, attention should be paid to the qualitative reliability of their results. In their scheme, restricted KS orbitals were employed, and the electronic excited states were represented by simple fractional occupation numbers for the KS orbital. In reality, the dye injects an electron into the substrate and so will form an open-shell structure.

Other atomistic simulations have been carried out by Rego *et al.* [103] and Kondov *et al.* [104]. The former employed an extended Hückel approach for mixed quantum-classical simulations of the catechol/anatase system and revealed a superexchange hole tunneling between the dye and the titanium atoms. The latter applied a partitioning scheme to define donor and acceptor states and the coupling between them. A theoretical approach was applied to the coumarin/TiO_2 nanoparticle system, and quantum dynamical simulations were performed employing the multilayer multiconfiguration time-dependent Hartree (MCTDH) method. These simulations solved the electronic structure of the system in the LCAO representation, but both of them relied on a single-particle picture of the electrons.

2.4.2 Multiexciton Generation in Quantum Dots: A Novel Sensitizer for a DSSC

Recently, QDs have received much attention as an important component for solar-cell devices. In QD systems, electrons are confined to a small structure of the order of a nanometer, which is comparable to the Bohr length in a semiconductor. Because of this strong confinement, several novel quantum phenomena emerge, such as phonon bottlenecks and Coulomb or spin blockades. In 2001, Nozik *et al.* predicted a novel phenomenon in QDs, MEG [105]. When irradiated with light, several excitons can form per photon in a semiconductor QD.

This phenomenon suggests the possibility of solar-cell devices that would have significantly higher efficiency than at present. In a conventional solar-cell device, there is an upper limit to the efficiency, known as the *Schockley–Queisser* limit [106]. When a photon is adsorbed in a bulk semiconductor, an electron is promoted from the valence band (VB) to the conduction band (CB). The difference in energy between the adsorbed photon and the bandgap quickly dissipates as heat. This process cools the hot electron and is called *Auger cooling*. At first sight, an inefficiency due to Auger cooling is inevitable, because the spectral range of sunlight encompasses this intrinsic range. According to the analysis of Schockley and Queisser, the estimated maximum efficiency is 31%. However, this analysis depends on the following four assumptions: (i) a single p–n junction exists, (ii) a single electron–hole pair is generated per single photon, (iii) the excess energy beyond the bandgap is relaxed to form heat, and (iv) the sample is excited with nonconcentrated solar radiation. If one or more of these conditions is not present, then the system will be free from this limitation, and a significant increase in efficiency may be possible.

In 2004, experimental evidence of MEG in PbSe QDs was provided by Schaller *et al.* [107]. The MEG process apparently negates condition (ii) above. This discovery shed light on the importance of QDs as promising building blocks of solar cells. Since then, several groups have confirmed MEG in QDs [107–112]. At the time of writing this article, MEG of up to seven excitons per single phonon has been reported [112]. Not only confirmation of this phenomenon, but also intensive research into the fabrication of DSSCs is now in progress, and DSSCs made from TiO_2 and CdSe (or PbS) QDs have already been reported in the literature [12]. Their efficiency is around 3% [113], which is still smaller than that of conventional DSSCs.

However, the microscopic mechanism of MEG is still an open question. Several mechanisms have been suggested to be applicable to this process, including impact ionization [105], an inverse Auger process, direct carrier multiplication [108,114] and MEG [109]. One essential difference between these hypotheses can be seen in the role of the decoherence timescale between the electronic states. In MEG, the excitons are generated in a decoherent process. The Auger mechanism assumes an incoherent process and requires a faster decoherence time than the creation of multiple excitons. Direct carrier multiplication relies on multiexciton coupling to virtual single-exciton states and the coherent superposition of these during an optical pulse.

2.4.3 Theoretical Estimation of the Decoherence Time between the Electronic States in PbSe QDs

In an attempt to solve this controversy, we estimated the theoretical timescale of the decoherence between the electronic states in PbSe QDs [11]. PbSe is a IV–VI (14–16) semiconductor with a rock-salt structure. Its narrow bandgap is 0.27 eV at ambient temperature [115], which means the absorption spectra of the QDs is in the visible-light region. The conduction and valence bands of PbSe QDs are formed from the p-orbital chains of Pb and Se, respectively, and the optical transition from the VB to the CB is almost symmetrical [116]. PbSe is also known for its strong dielectric screening. The Bohr radius of an exciton in PbSe is as large as 46 nm [116]. Therefore, the effect of quantum confinement is prominent in PbSe QDs.

We performed atomistic calculations on two PbSe clusters: $Pb_{16}Se_{16}$ and $Pb_{68}Se_{68}$. Figure 2.10 shows the structure of the QDs. The initial geometries were generated from the bulk and were fully optimized at 0 K. The surface effects were relatively small in the PbSe QDs, and no surface ligands or artificial terminations on the surface were included. Each cluster was brought up to 300 K by preliminary MD simulations using repeated velocity rescaling. A microcanonical trajectory was generated for each cluster in the electronic ground state. The simulation was continued for a period of 2 and 4 ps for $Pb_{16}Se_{16}$ and $Pb_{68}Se_{68}$ clusters, respectively.

The exciton (biexciton) states were represented in the Kohn–Sham orbital scheme by promoting an (two) electron(s) from the occupied to the unoccupied orbitals. The excitation energy, $\Delta E(t)$, was estimated from the orbital energy and the occupation number. The large

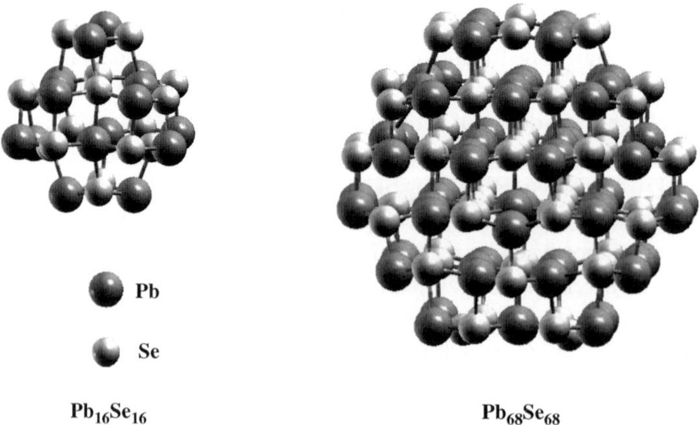

Figure 2.10 Typical geometry of the $Pb_{16}Se_{16}$ and $Pb_{68}Se_{68}$ clusters at room temperature.

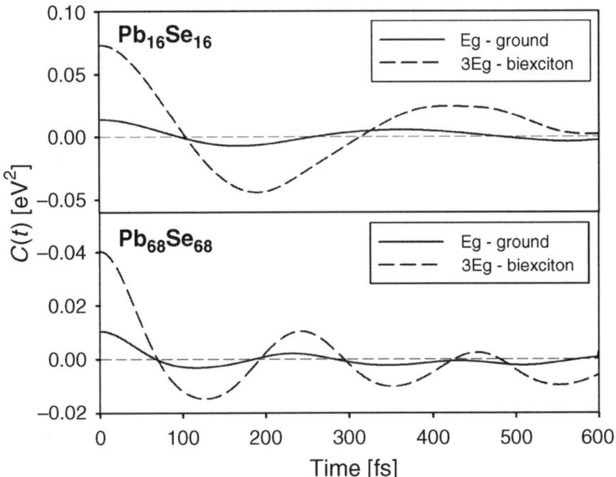

Figure 2.11 Autocorrelation function of the bandgap (solid line) and of the energy difference between the triple-gap exciton (3Eg) and biexciton states (dashed line).

Bohr radius in PbSe rationalizes this treatment [117]. The autocorrelation function, $C(t)$, was calculated from the trajectory:

$$C(t) = \langle \Delta E(t) \cdot \Delta E(0) \rangle_T.$$

Figure 2.11 shows $C(t)$ for the Eg/ground and 3Eg/biexciton pairs of states. The amplitude was notably greater for the 3Eg/biexciton pair. The larger cluster allowed for a wider range of vibrational frequencies, and the corresponding autocorrelation functions oscillated faster. In all cases, the correlation decayed within a period of several hundred femtoseconds.

The decoherence function, $D(t)$, was calculated by exponentiation of the double integral of $C(t)$ [118]:

$$D(t) = \exp\left(\int_0^t d\tau_1 \int_0^{\tau_1} C(\tau_2) d\tau_2\right).$$

The same decoherence time can be estimated from semiclassical theory [119,120]. In this approach, wavefunctions for each nuclear coordinate are represented by a Gaussian function whose width is related to the thermal de Broglie wavelength. The decoherence time is estimated by considering the evolution of the Gaussians correlated with each of the two electronic states to the lowest order:

$$\tau_D = \left[\left\langle \sum_n \frac{1}{2a_n h^2} |F_{1n} - F_{2n}|^2 \right\rangle_T\right]^{-1/2},$$

where F denotes the force experienced by the nth atom in the first and second electronic states. The forces on each atom in different electronic states were calculated using the appropriate orbital occupations and the Hellmann–Feynman theorem. The thermal de Broglie width

Table 2.4 Dephasing time [fs] between pairs of electronic states.

Cluster	State Pair	Response Function	Semiclassical
$Pb_{16}Se_{16}$	Eg/ground	7.1 (8.4)	12.5
	2Eg/Eg	6.2 (7.7)	7.4
	3Eg/Eg	4.0 (5.0)	8.1
	2Eg/biexciton	3.5 (4.2)	5.0
	3Eg/biexciton	3.2 (2.3)	5.4
$Pb_{68}Se_{68}$	Eg/ground	9.3 (8.8)	5.8
	2Eg/Eg	9.8 (9.5)	5.0
	3Eg/Eg	7.0 (7.5)	5.7
	2Eg/biexciton	4.7 (4.8)	2.8
	3Eg/biexciton	3.5 (3.7)	3.0

parameter, a_n, was defined according to atomic mass and temperature [119]:

$$a_n = \frac{6mk_BT}{\mathrm{h}^2}.$$

The estimated decoherence times from $D(t)$ and the semiclassical analysis are listed in Table 2.4. Despite the crudeness of the approximation made by both methods, a good correspondence was observed.

The semiclassical approach has an advantage in that the decoherence time is expressed as the sum of each atomic contribution. Thus, one can identify the atomic motions that are dominant in inducing the decoherence. Figure 2.12 shows these contributions as a function of the distance from the center of the QD. The data were averaged over equivalent atoms. The largest atomic

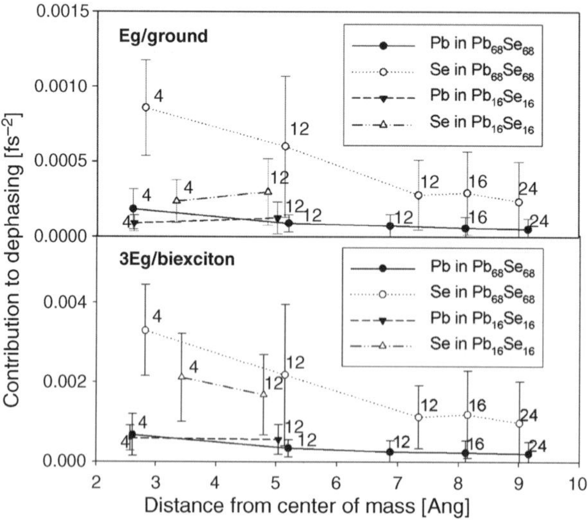

Figure 2.12 Contributions of individual atoms to the decoherence time as a function of distance from the quantum dot center. The number of equivalent atoms accompanies each data point.

contribution to the decoherence came from the core atoms. Even though individual core atoms contribute more strongly to decoherence than individual surface atoms do, the overall contribution from the surface is significant and agrees with the proposed incoherent mechanism for the creation of multiple excitons in the surface region [108,114]. The Se atoms had a stronger effect on the decoherence for two reasons. First, Se atoms are lighter than Pb atoms, and their value of a_n is larger than that of Pb atoms by a factor of 2.6. Second, the difference of the force on atoms between electronic states are larger for Se, by approximately a factor of two. This analysis indicates that the decoherence time should increase with the square root of the atomic mass of the lighter atom. The decoherence time for PbS and PbTe QDs should be different from the decoherence time of PbSe QDs, roughly in proportion to the square root of the mass of the lighter atom.

Because of the small size of the PbSe QDs considered in this study, definitive conclusions cannot be reached for the size dependency of the decoherence time. It is known both experimentally [121,122] and theoretically [123] that the decoherence time decreases in larger clusters. The calculated decoherence time should be regarded as being the lower limit of the PbSe experimental data. The calculated times agree with the extrapolation of the decoherence times measured in larger CdSe clusters ($\geq$2 nm) [122] and with large experimental line widths observed in CdSe clusters [124], which are comparable in size with the present PbSe system.

The ultrafast decoherence time has direct implications for the proposed mechanisms of MEG. On timescales longer than a few femtoseconds, the MEG process must be incoherent and is described properly by rate expressions [108,114]. A short, but finite, coherence time may be sufficient for the direct photogeneration of multiple excitons, provided that the Coulomb interaction that mixes the single- and multiexciton states is of the order of at least few tens of electron volts [110,112].

MEG in PbSe has also been studied by Franceschetti *et al.* [125]. In their analysis, approximately 2000 single-particle electronic states were solved using atomistic pseudopotential calculations. The density of the excitonic and biexcitonic states was estimated from these single-particle states in a similar way to our method. The matrix elements of the screened Coulomb interactions between these states were calculated, and the timescale of the MEG was assessed using the Fermi golden rule. The research of Franceschetti *et al.* successfully explains the experimental results without introducing any peculiar mechanism. In their analysis, the electronic decoherence timescale was assumed to occur on a very fast timescale. This assumption is consistent with our estimation presented above.

2.5 Future Directions: *Ab Initio* Simulations and the Local Excited States on TiO_2

In this chapter so far, we have reviewed our recent studies on the bulk and surface chemistry of TiO_2, and related works by other researchers. The structure and the electronic states of the bulk system, a clean surface and surfaces with adsorbents have been solved using DFT-based methods. The optical properties and electronic conductivity have been discussed for doped TiO_2 systems and interpretation of the experimental results has been provided. In reviewing some papers, the nature of the excited state of adsorbed molecules has been discussed using a cluster model. The power of the first-principle method was fully exploited in these studies.

Nevertheless, it should be stressed that none of these works has treated the dynamics of the electron, or the chemical-reaction dynamics, directly. Essentially, the electronic structure of the system was solved for the ground state, or was solved as an excited state in a finite system with a fixed nuclear coordinate. Statistical analysis only works for the thermal equilibrium state. However, in reality, photochemical reactions occur for an adsorbent on the surface with the participation of electronically excited states. The surface of the bulk is a *semi-infinite* open system with an infinite number of electrons. Therefore, the physics occurring on the surface are essentially different from those of the isolated molecule. Without a proper treatment of these features, a reliable understanding or prediction of surface phenomena is not possible. In the next section, we review the importance of these missing factors and reviewed some recent progress made in research in this field.

2.5.1 Improvement of the DFT Functional

The DFT-based first-principle method provides reliable predictions for the structure, chemical bond length, stability between several polymorphic phases, surface energy and bulk properties. However, the width of the bandgap is difficult to predict using the DFT method, and the use of conventional DFT functionals, either at the LDA or the GGA level, means that the bandgap is usually underestimated when it is inferred from the energy level of the KS orbitals. Needless to say, the bandgap is one of the most important physical quantities of a semiconductor, as the width of the bandgap determines the optical properties and the dynamics of the excited electrons.

The cause of this discrepancy is complex, and several remedies have been advocated. One strategy is the use of screened Coulomb interactions. In the GW approximation, the screening effect is incorporated into the framework of many-body perturbation theory [126]. The interaction between the excited electron and other electrons in the VB are included in the self-energy term for the excited electron. However, the methodology is computationally expensive and has a disadvantage in that the gradient of the excitation energy with respect to the structure of the system cannot be calculated. Alternatively, the effect of screening can be plugged into the KS formalism more directly. Recently, Heyd *et al.* advocated a novel DFT functional, HSE3 [127], which includes the effect of screening, and this functional has been applied to bulk silicon and TiO_2 [128]. A significant improvement in the width of the bandgap was reported for both cases.

The reproduction of the bandgap can be improved using different approaches. In standard DFT formalism, because of the crudeness of the exchange-correlation kernel, a spurious self-interaction effect appears for electrons. This situation is especially serious in metal oxide systems, or in systems with electrons confined to the 4*f* orbital of a transition metal. The self-interaction correction (SIC) formalism eliminates this spurious effect [129]. Another improvement in calculating the bandgap arises from considering electron correlations. When a given orbital is highly localized, then the electrons experience a strong on-site Coulomb interaction. In the LDA + U (local density approximation + Hubbard coefficient) approach, the on-site Coulomb interaction is added to the Kohn–Sham equation, as is done for the Hubbard model [130].

In the DFT method, van der Waals (VdW) interactions between molecules are usually absent. The lack of any VdW interactions is especially problematic in π-stacking systems and the rare gases. Several combined DFT schemes have been used recently and have been applied to molecules on surfaces that are adsorbed by VdW interactions [131].

2.5.2 *Molecular Mechanics and* Ab Initio *Molecular Dynamics*

In the Introduction, we emphasized the importance of nonempirical first-principle calculations. However, the applicability of this method is limited to small systems. For calculations on large systems, one has to resort to an approximate or empirical methodology.

The simplest description of a bulk system can be made using two-body potential functions for atoms. Despite the simplicity of this method, some systems, such as ionic compounds, can be described qualitatively, and several physical quantities can be calculated. For example, elastic properties, surface reconstruction, phonon dispersion and the velocity of sound can be reproduced. Many two-body potential function forms have been advocated. For the TiO_2 bulk system, Matsui *et al.* defined a parameter set in the form of a Buckingham potential [132]. The potential consisted of a formal charge at the nucleus, and parameters for the Lennard–Jones potential and the VdW interactions between the atoms. The parameter set reproduced the relative stability of the rutile and anatase forms, and was applied in several simulations of TiO_2 nanoparticles [133] and the TiO_2–H_2O interface [134]. In the latter calculations, a modified parameter was used for atoms on the surface. Some researchers have tried to improve the two-body function of Matsui *et al.* For example, Swamy *et al.* adapted the charge equilibration (QEq) method for the charges on the nuclei [28]. However, no significant improvement has been made so far, and the Buckingham parameters of Matsui *et al.* are still in use.

More accurate simulations can be carried out if the electronic structure of a system is solved at each time step during a run. An empirical calculation method, known as the tight-binding DFT (TB-DFT), is used for relatively large systems. In this method, the Hamiltonian matrix is composed from a LCAO representation using predefined interactions between atomic orbitals. The predefined interactions are prepared either empirically or according to preliminary calculations. A similar treatment is known as the Hückel method in molecular physics. In the TB-DFT method, the effect of electron correlation is partially included in the Hamiltonian by using a DFT functional [135]. An example of TB-DFT-based *ab initio* dynamics can be seen in simulations of TiO_2 nanoparticles in water [136].

A similar simulation can be seen when using the DFT-based first-principle method. This approach is called *ab initio* dynamics. In this simulation scheme, the electron density of the system is directly solved at each simulation time step, and the forces on the nuclei are evaluated using the gradient of the total energy of the system or using the Hellman–Feynman theorem. Using this method, Alvarez-Ramirez *et al.* simulated the formation of titanase-type nanosheets from anatase [137]. A combination of *ab initio* dynamics and a molecular mechanism is also possible. Schiffmann *et al.* simulated an acetonitrile/anatase (101) system using a combined method [138]. The combination of an *ab initio* method and molecular mechanics was also used in the optimization of the surface structure [134].

A simplification of the *ab initio* dynamics was made by Car and Parinello [139]. In their scheme, the Hamiltonian for the electrons and nuclei was integrated by introducing a fictional mass for electronic motion. This enabled a simultaneous treatment for the electron system and for nuclear motion, and the dynamics of the system could be calculated at a lower computational cost than for any other method. The Car–Parinello method is powerful and elegant, and with the rapid increase in computational power and better implementation techniques, its range of application is expanding. However, one has to remember that this method is only applicable to the electronic ground state, and quantum effects on the nuclei and the nonadiabaticity of the electronic states are not covered.

2.5.3 Description of Local Excited States

As we emphasized in the Introduction, photocatalytic reactions involve electronically excited states. In condensed matter physics, the straightforward way to calculate excited electrons and excitonic states in periodic systems is given by a diagrammatic expansion of Green's function in many-body perturbation theory. The GW approximation is frequently used for the description of excited electrons. Similarly, excitonic states can be elucidated by solving the Bethe–Salpeter equation (BSE) [140]. For example, the excited electrons and excitonic states in carbon nanotubes (CNTs) are well reproduced using these schemes [141]. However, such a rigorous treatment is computationally highly intensive and is not applicable to photocatalytic reaction systems. In this case, the chemical reaction occurs when molecules are adsorbed on a surface, and the system does not have a clear periodicity. Not only the excitation of electrons but also nuclear dynamics occur, which are associated with the breaking and formation of chemical bonds.

In the field of *ab initio* calculations of molecules, the excited states of a system are usually solved using a variational method. The configuration interaction (CI) and multiconfigurational self-consistent field (MCSCF) methods are representative examples. However, direct application of such methods to photocatalytic systems involves theoretical difficulties. The number of electrons involved is infinite, and some of the excited states are metastable, for example the excited state of a dye in a DSSC. Such a metastable state cannot be calculated variationally. The excited state, which is localized in a given region and has a finite lifetime, is referred to as a *local excited state*. The local excited state can be solved using the variational method when the system is decomposed into a finite reaction center and its surroundings. The coupling between two regions determines the lifetime of the state.

Separation of the system into a central region and its surroundings has been adapted for *ab initio* calculations. This method is known as an *embedded* calculation scheme. Many embedding schemes have been advocated, differing in the way of plugging in the effect that the surroundings have on the central region [142,143]. However, the majority of embedded schemes are intended for the electronic ground state.

The use of embedded calculations for local excited states can be seen in the papers of Whitten and coworkers, who have applied an embedded cluster model and a configuration interaction method to study the photoinduced dissociation of methyl nitrite and formaldehyde on an Ag (111) surface [144]. Schulger *et al.* combined a classical shell model and molecular-orbital-based *ab initio* calculations [145], and calculated several types of defects on an MgO surface, where they calculated the excitation energy of the surface exciton to analyze the optical absorption and photoinduced desorption of atomic O and Mg species [146].

In the DFT community, the excited-state molecules were initially discussed using Kohn–Sham orbitals and using nonaufbau occupation numbers for these orbitals. Later, a rigorous treatment was made using Runge–Gross theorem [147], which is an extension of the Hohenberg–Kohn theorem to the frequency domain. The excitation energy of a system is extracted as singular points of the dynamic response to an external perturbation. Casida and Wesołowski extended the TDDFT scheme to an embedded system [148]. In their formalism, the electron density is divided into two parts, and only one of these parts is solved for the excited state. The theory does not rely on an orbital picture for the surrounding part, but needs an expression for the kinetic energy density in the form of a contribution from two independent electron densities. This theory is known as *frozen-density embedding* or Kohn–Sham orbitals with a constrained electron density (KSCED). Successful application of this theory can be seen in the calculations on solvatochromic shifts [149].

Govind *et al.* proposed a combined approach [143]. In their theory, the local potential, calculated using a DFT functional and the electron density of the surroundings, is directly applied to molecular-orbital calculations of the central region to form an effective potential. The application of this method can be seen in the prediction of the excited state of CO on a Pd(111) surface [150].

2.5.4 Nonadiabatic Behavior of a System and Interfacial Electron Transfer

When a molecule is excited to one of its electronically excited states, the nuclear motion is accelerated by the steep gradient of the surface. In such cases, the breakdown of the Born–Oppenheimer approximation occurs, and a nonadiabatic transition between the electronic states is facilitated. The coupling term between the electronic states with respect to the displacement of the nuclei is called the nonadiabatic coupling (NAC) term, and NACs are present in both isolated molecules in a gas and in condensed materials. In the theory of gas dynamics, because this strong transition usually occurs in a given small region in the nuclear configuration space where the electronic states are quasi-degenerate, this is known as a *curve-crossing* point.

However, in the case of adsorbents on surfaces, the situation is not so simple. Because the number of electrons involved is infinite in the substrate, simple partition into the electronic degree of freedom and the nuclear degree of freedom is questionable. In particular, on a metal surface, the adsorbed molecules experience strong friction against nuclear motion. The vibrationally excited states of the adsorbents are quickly quenched because of coupling with the conducting electrons [151]. A similar friction is observed in the photoinduced desorption process from a metal surface as well [152]. The conducting electrons have a continuous energy spectrum, and thus, they dissipate the vibrational energy efficiently by creating electron–hole pairs [153]. In general, the number of electrons in an adsorbent or embedded cluster system is not necessarily an integer. Head-Gordon *et al.* theoretically analyzed the vibrational relaxation of CO/Cu(100) [154] using *molecular dynamics with electronic friction* (MDEF) theory, which has been applied to several other simulations [155]. The interfacial electron transfer triggered by the photoabsorption also treats the change in the number of electrons within the system of interest. In case of a photoinduced desorption process, hot electrons are generated in the bulk and are transported to the surface region [156]. The photocatalytic reactions on a TiO_2 surface undergo similar transport, either as excited electrons or as holes.

Quantum dynamics in contact with an open system, or with a dissipation term, are usually solved using a density matrix representation [152,157]. However, such a treatment is computationally difficult for large systems, and the use of quantum classical theory is necessary [158]. In such theory, the nuclear degree of freedom is treated stochastically, and the electronic states collapse to the eigenfunctions for a given structure. At the same time, the simple band picture of conducting electrons needs to be reconsidered. Electrons and holes strongly couple with lattice deformations and the surface structure, and the formation of a polaron or trapping at the surface occurs [159]. Such electronic states, and the competition between the localized and delocalized states, need to be incorporated in simulations.

All these factors may be necessary for calculations on photocatalytic reactions and interfacial electron transfer, and detailed assessments of these factors need to be addressed carefully. Any essential factors identified need to be incorporated into the simulation scheme.

Acknowledgments

This research was supported by a grant from the Elements Science and Technology Project and the Global COE Program, Chemical Innovation, from the Ministry of Education, Culture, Sports, Science and Technology of Japan. The authors are grateful to Dr. Svetlana Kilina and Prof. Oleg Prezhdo for investigating the decoherence mechanism in quantum dots.

References

[1] Fujishima, A. and Honda, K. (1972) Electrochemical photolysis of water at a semiconductor electrode. *Nature*, **238**(5358), 37.

[2] Fujishima, A., Rao, T.N., and Tryk, D.A. (2000) Titanium dioxide phytocatalysis. *Journal of Photochemistry and Photobiology C: Photochemistry Reviews*, **1**, 1.

[3] Wang, R., Hashimoto, K., Fujishima, A. *et al.* (1997) Light-induced amphiphilic surfaces. *Nature*, **388**, 431.

[4] O'Regan, B. and Grätzel, M. (1991) A low-cost, high-efficiency solar cell based on dye-sensitized colloidal TiO_2 films. *Nature*, **353**, 737.

[5] Matsumoto, Y., Murakami, M., Shono, T. *et al.* (2001) Room-temperature ferromagnetism in transparent transition metal-doped titanium dioxide. *Science*, **291**, 854.

[6] Furubayashi, Y., Hitosugi, T., Yamamoto, Y. *et al.* (2005) A transparent metal: Nb-doped anatase TiO_2. *Applied Physics Letters*, **86**, 252101.

[7] Irie, H., Sunada, K., and Hashimoto, K. (2004) Recent developments in TiO_2 Photocatalysis: Novel applications to interior ecology materials and energy saving systems. *Electrochemistry*, **72**, 807.

[8] Kamisaka, H., Adachi, T., and Yamashita, K. (2005) Theoretical study of the structure and optical properties of carbon-doped rutile and anatase titanium oxides. *Journal of Chemical Physics*, **123**, 084704.

[9] Kamisaka, H., Hitosugi, T., Suenaga, T. *et al.* (2009) Density functional theory based first-principle calculation of Nb-doped anatase TiO_2 and its interactions with oxygen vacancies and intersitital oxygen. *Journal of Chemical Physics*, **131**, 034702.

[10] Kamisaka, H. and Yamashita, K. (2007) The surface stress of the (110) and (100) surfaces of rutile and the effect of water adsorbents. *Surface Science*, **601**, 4824.

[11] Kamisaka, H., Kilina, S.V., Yamashita, K., and Prezhdo, O.V. (2006) Ultrafast vibrationally-induced dephasing of electronic excitations in PbSe quantum dots. *Nano Letters*, **6**, 2295.

[12] Kamat, P.V. (2008) Quantum dot solar cells. semiconductor nanocrystals as light harvesters. *The Journal of Physical Chemistry. C*, **112**, 18737.

[13] Asahi, R., Morikawa, T., Ohwaki, T. *et al.* (2001) Visible-light photocatalysis in nitrogen-doped titanium oxides. *Science*, **293**, 269; Ihara, T., Ando, M., and Sugihara, S. (2001) Preparation of visible light active TiO_2 photocatalysis using wet method. *Photocatalysis*, **5**, 19; Sakatani, Y., Okusako, K., Koike, H., and Ando, H. (2001) Development of TiO_2 photocatalysis with visible light responce. *Photocatalysis*, **4**, 51.

[14] Tachikawa, T., Takai, Y., Tojo, S. *et al.* (2006) Visible light-induced degradation of ethylene glycol on nitrogen-doped TiO_2 powders. *Journal of Physical Chemistry B*, **110**, 13158.

[15] Khan, S.U.M., Al-Shahry, M., and Ingler, W.B. Jr. (2002) Efficient photochemical water splitting by a chemically modified n-TiO_2. *Science*, **297**, 2243. Ohno, T., Tsubota, T., Nishijima, K., and Miyamoto, Z. (2004) Degradation of metylene blue on carbonate species-doped TiO_2 photocatalysis under visible light. *Chemistry Letters*, **33**, 750.

[16] Sakthivel, S. and Kisch, H. (2003) Daylight photocatalysis by carbon-modified titanium dioxide. *Angewandte Chemie-International Edition*, **42**, 4908.

[17] Shaban, Y.A. and Khan, S.U.M. (2008) Visible light active carbon modified n-TiO_2 for efficient hydrogen production by photoelectrochemical splitting of water. *International Journal of Hydrogen Energy*, **33**, 1118.

[18] Ohno, T., Mitsui, T., and Matsumura, M. (2003) TiO_2-photocatalyzed oxidation of adamantane in solutions containing oxygen or hydrogen peroxide. *Chemistry Letters*, **32**, 364.
[19] Umebayashi, T., Yamaki, T., Itoh, H., and Asai, K. (2002) Band gap narrowing of titanium dioxide by sulfur doping. *Applied Physics Letters*, **81**, 454.
[20] Tojo, S., Tachikawa, T., Fujitsuka, M., and Majima, T. (2008) Iodine-Doped TiO_2 photocatalysts: correlation between band structure and mechanism. *Journal of Physical Chemistry C*, **112**, 14948.
[21] Osorio-Guillén, J., Lany, S., and Zunger, A. (2008) Atomic control of conductivity versus ferromagnetism in wide-gap oxides via selective doping: V, Nb, Ta in anatase TiO_2. *Physical Review Letters*, **100**, 036601.
[22] Kaspar, T.C., Droubay, T., Shutthanandan, V. *et al.* (2006) Ferromagnetism and structure of epitaxial Cr-doped anatase TiO_2 thin films. *Physical Review B-Condensed Matter*, **73**, 155327.
[23] Du, X., Li, Q., Su, H., and Yang, J. (2006) Electronic and magnetic properties of V-doped anatase TiO_2 from first principles. *Physical Review B-Condensed Matter*, **74**, 233201.
[24] Wang, Y. and Doren, D.J. (2005) Electronic structures of V-doped anatase TiO_2. *Solid State Communications*, **136**, 142.
[25] Hohenberg, P. and Kohn, W. (1964) Inhomogeneous electron gas. *Physical Review*, **136**, B864.
[26] Kohn, W. and Sham, L.J. (1965) Self-consistent equations including exchange and correlation effects. *Physical Review*, **140**(4A), A1133.
[27] Perdew, J.P., Chevary, J.A., Vosko, S.H. *et al.* (1992) Atoms, molecules, solids, and surfaces – applications of the generalized gradient approximation for exchange and correlation. *Physical Review B-Condensed Matter*, **46** (11), 6671.
[28] Perdew, J.P., Burke, K., and Ernzerhof, M. (1996) Generalized gradient approximation made simple. *Physical Review Letters*, **77**(18), 3865.
[29] Diebold, U. (2003) The surface science of titanium dioxide. *Surface Science Reports*, **48**, 53.
[30] Nakahara, M. (ed.) (1997) *Dictionary of Inorganic Compounds and Complexes*, Kodansha, Tokyo.
[31] Samsonov, G.V. (1982) *The Oxide Handbook*, IFI/Plenum Press, New York.
[32] Glassford, K.M. and Chelikowsky, J.R. (1992) Structural and electronic properties of titanium dioxide. *Physical Review B-Condensed Matter*, **46**, 1284.
[33] Mo, S.-D. and Ching, W.Y. (1995) Electronic and optical properties of three phases of tiatnium dioxide: Rutile, anatase, and brookite. *Physical Review B-Condensed Matter*, **51**, 13023.
[34] Asahi, R., Taga, Y., Mannstadt, W., and Freeman, A.J. (2000) Electronic and optical properties of anatase TiO_2. *Physical Review B-Condensed Matter*, **61**, 7459.
[35] Zhang, S.B. and Northrup, J.E. (1991) Chemical potential dependence of defect formation energies in GaAs: application to ga self-diffusion. *Physical Review Letters*, **67**, 2339.
[36] He, J., Behera, R.K., Finnis, M.W. *et al.* (2007) Prediction of high-temperature point defect formation in TiO_2 from combined ab initio and thermodynamics calculations. *Acta Materialia*, **55**, 4325.
[37] Sekiya, T., Yagisawa, T., Kamiya, N. *et al.* (2004) Defects in anatase TiO_2 single crystal controlled by heat treatments. *Journal of the Physical Society of Japan*, **73**, 703.
[38] Forro, L., Chauvet, O., Emin, D., and Zuppiroli, L. (1994) High mobility n-type charge carriers in large single crystals of anatase (TiO_2). *Journal of Applied Physics*, **75**, 633.
[39] Lu, T.-C., Wu, S.-Y., Lin, L.-B., and Zheng, W.-C. (2001) Defects in the reduced rutile single crystal. *Physica B*, **304**, 147.
[40] Cho, E., Han, S., Ahn, H.-S. *et al.* (2006) First-principles study of point defects in rutile TiO_{2-x}. *Physical Review B-Condensed Matter*, **73**, 193202.
[41] Iddir, H., Öğüt, S., Zapol, P., and Browning, D. (2007) Diffusion mechanisms of native point defects in rutile TiO_2: Ab initio total-energy calculations. *Physical Review B-Condensed Matter*, **75**, 073203.
[42] Na-Phattalung, S., Smith, M.F., Kim, K. *et al.* (2006) First-principles study of native defects in anatase TiO_2. *Physical Review B-Condensed Matter*, **73**, 125205.
[43] Sato, S. (1986) Photocatalytic activity of NO_x-doped TiO_2 in the visible light region. *Chemical Physics Letters*, **123**, 126.
[44] Irie, H., Watanabe, T., and Hashimoto, K. (2003) Carbon-doped anatase TiO_2 powders as a visible-light sensitive photocatalyst. *Chemistry Letters*, **32**, 772.
[45] Lettman, C., Hindenbrad, K., Kisch, H. *et al.* (2001) Visible light photodegradation of 4-chlorophenol with a coke-containing titanium dioxide photocatalyst. *Applied Catalysis B-Environmental*, **32**, 215.
[46] Vanderbilt, D. (1990) Soft self-consistent pseudopotentials in a generalized eigenvalue formalism. *Physical Review B-Condensed Matter*, **41**(11), 7892.

[47] Kresse, G. and Hafner, J. (1994) Norm-conserving and ultrasoft pseudopotentials for first-row and transition-elements. *Journal of Physics: Condensed Matter*, **6**(40), 8245.

[48] Monkhorst, H.J. and Pack, J.D. (1976) Special points for Brillouin-zone integrations. *Physical Review B-Condensed Matter*, **13**(12), 5188.

[49] Kresse, G. and Furthmüller, J. (1996) Efficient iterative schemes for ab initio total-energy calculations using a plane-wave basis set. *Physical Review B-Condensed Matter*, **54**(16), 11169; Kresse, G. and Hafner, J. (1993) Ab initio molecular-dynamics for liquid-metals. *Physical Review B-Condensed Matter*, **47**(1), RC558.

[50] Breckenridge, R.G. and Hosler, W.R. (1953) Electrical properties of titanium dioxide semiconductors. *Physical Review*, **91**, 793.

[51] Anisimov, V.I., Korotin, M.A., Nekrasov, I.A. *et al.* (2006) The role of transition metal impurities and oxygen vacancies in the formation of ferromagnetism in Co-doped TiO_2. *Journal of Physics: Condensed Matter*, **18**, 1695.

[52] Taylor, S.R. and McLennan, S.M. (1995) The geochemical evolution of the continental-crust. *Reviews of Geophysics*, **33**(2), 241.

[53] Hitosugi, T., Ueda, A., Nakao, S. *et al.* (2007) Fabrication of highly conductive $Ti_{1-x}Nb_xO_2$ polycrystalline films on glass substrates via crystallization of amorphous phase grown by pulsed laser deposition. *Applied Physics Letters*, **90**, 212106.

[54] Dabney, M.S., van Hest, M.F.A.M., Teplin, C.W. *et al.* (2008) Pulsed laser deposited Nb doped TiO_2 as a transparent conducting oxide. *Thin Solid Films*, **516**, 4133.

[55] Hitosugi, T., Kamisaka, H., Yamashita, K. *et al.* (2008) Electronic band structure of transparent conductor: Nb-doped anatase TiO_2. *Applied Physics Express*, **1**, 111203.

[56] Sacerdoti, M., Dalconi, M.C., Carotta, M.C. *et al.* (2004) XAS investigation of tantalum and niobium in nanostructured TiO_2 anatase. *Journal of Solid State Chemistry*, **177**, 1781.

[57] Zhang, S.X., Kundaliya, D.C., Yu, W. *et al.* (2007) Niobium doped TiO_2: Intrinsic transparent metallic anatase versus highly resistive rutile phase. *Journal of Applied Physics*, **102**, 013701.

[58] Kurita, D., Ohta, S., Sugiura, K. *et al.* (2006) Carrier generation and transport properties of heavily Nb-doped anatase TiO_2 epitaxial films at high temperatures. *Journal of Applied Physics*, **100**, 096105.

[59] Linsebigler, A.L., Lu, G.Q., and Yates, J.T.J. (1995) Photocatalysis on TiO_2 surfaces: principles, mechanisms, and selected results. *Chemical Reviews*, **95**, 735.

[60] Nakamura, R. and Nakato, Y. (2004) Primary intermediates of oxygen photoevolution reaction on TiO_2 (Rutile) particles, revealed by *in situ* FTIR absorption and photoluminescence measurements. *Journal of the American Chemical Society*, **126**, 1290.

[61] Thompson, T.L. and Yates, J.T.J. (2006) Surface science studies of the photoactivation of TiO_2 – new photochemical processes. *Chemical Reviews*, **106**, 4428.

[62] Nosaka, Y., Komori, S., Yawata, K. *et al.* (2003) Photocatalytic $^{\bullet}OH$ radical formation in TiO_2 aqueous suspension studied by several detection methods. *Physical Chemistry Chemical Physics*, **5**, 4731.

[63] Kubo, W. and Tatsuma, T. (2004) Detection of H_2O_2 released from TiO_2 photocatalyst to air. *Analytical Sciences*, **20**, 591; Kubo, W. and Tatsuma, T. (2005) Photocatalytic remote oxidation with various photocatalysts and enhancement of its activity. *Journal of Materials Chemistry*, **15**, 3104.

[64] Henderson, M.A. (2002) The interaction of water with solid surfaces: fundamental aspects revisited. *Surface Science Reports*, **46**, 1.

[65] Reinhardt, P. and Heß, B.A. (1994) Electronic and geometrical structure of rutile surfaces. *Physical Review B-Condensed Matter*, **50**, 12015.

[66] Ramamoorthy, M., Vanderbilt, D., and King-Smith, R.D. (1994) First-principles calculations of the energetics of stoichiometric TiO_2 surfaces. *Physical Review B-Condensed Matter*, **49**, 16721.

[67] Harrison, N.M., Wang, X.-G., Muscat, J., and Scheffler, M. (1999) The influence of soft vibrational modes on our understanding of oxide surface structure. *Faraday Discussions*, **114**, 305.

[68] Egdell, R.G., Eriksen, S., and Flavell, W.R. (1986) Oxygen deficient SnO_2 (110) and TiO_2 (110): A comparative study by photoemission. *Solid State Communications*, **60**, 835; Eriksen, S. and Egdell, R.G. (1987) Electronic excitations at oxygen deficient TiO_2 (110) surfaces: A study by EELS. *Surface Science*, **180**, 263.

[69] Ramamoorthy, M., King-Smith, R.D., and Vanderbilt, D. (1994) Defects on TiO_2 (110) surfaces. *Physical Review B-Condensed Matter*, **49**, 7709.

[70] Lindan, P.J.D., Harrison, N.M., Gillan, M.J., and White, J.A. (1997) First-principles spin-polarized calculations on the reduced and reconstructed TiO_2 (110) surface. *Physical Review B-Condensed Matter*, **55**, 15919.

[71] Bredow, T. and Pacchion, G. (2002) Electronic structure of an isolated oxygen vacancy at the TiO_2 (110) surface. *Chemical Physics Letters*, **355**, 417.

[72] Henderson, M.A. (1996) An HREELS and TPD study of water on TiO_2 (110): the extent of molecular versus dissociative adsorption. *Surface Science*, **355**, 151; Schaub, R., Thostrup, R., Lopez, N. *et al.* (2001) Oxygen vacancies as active sites for water dissociation on rutile TiO_2(110). *Physical Review Letters*, **87**, 266104.

[73] Hugenschmidt, M.B., Gamble, L., and Campbell, C.T. (1994) The interaction of H_2O with a TiO_2(110) surface. *Surface Science*, **302**(3), 329.

[74] Bandura, A.V., Sykes, D.G., Shapovalov, V. *et al.* (2004) Adsorption of water on the TiO_2 (rutile) (110) surface: A comparison of periodic and embedded cluster calculations. *The Journal of Physical Chemistry B*, **108**(23), 7844; Goniakowski, J. and Gillan, M.J. (1996) The adsorption of H_2O on TiO_2 and SnO_2(110) studied by first-principles calculations. *Surface Science*, **350**(1–3), 145; Lindan, P.J.D., Harrison, N.M., and Gillan, M.J. (1998) Mixed dissociative and molecular adsorption of water on the rutile (110) surface. *Physical Review Letters*, **80**(4), 762.

[75] Harris, L.A. and Quong, A.A. (2004) Molecular chemisorption as the theoretically preferred pathway for water adsorption on ideal rutile TiO_2(110). *Physical Review Letters*, **93**(8), 086105.

[76] Harris, L.A. and Quong, A.A. (2005) Comment on "Molecular chemisorption as the theoretically preferred pathway for water adsorption on ideal rutile TiO_2(110)" – Harris and Quong reply. *Physical Review Letters*, **95** (2), 029602; Lindan, P.J.D. and Zhang, C.J. (2005) Comment on "Molecular chemisorption as the theoretically preferred pathway for water adsorption on ideal rutile TiO_2(110)". *Physical Review Letters*, **95**(2), 029601.

[77] Lindan, P.J.D. and Zhang, C.J. (2005) Exothermic water dissociation on the rutile TiO_2(110) surface. *Physical Review B-Condensed Matter*, **71**(7), 075439.

[78] Shapovalov, N., Wang, Y., and Truong, T.N. (2003) Theoretical analysis of the electronic spectra of water adsorbed on the rutile TiO_2 (110) and MgO (100) surfaces. *Chemical Physics Letters*, **375**, 321.

[79] de Lara-Castells, M.P. and Krause, J.L. (2003) Theoretical study of the UV-induced desorption of molecular oxygen from the reduced TiO_2 (110) surface. *Journal of Chemical Physics*, **118**, 5098.

[80] Henderson, M.A., Epling, W.S., Perkins, C.L. *et al.* (1999) Interaction of molecular oxygen with the vacuum-annealed TiO_2 (110) surface: molecular and dissociative channels. *Journal of Physical Chemistry B*, **103**, 5328; Oviedo, J. and Gillan, M.J. (2001) First-principles study of the interaction of oxygen with the SnO_2 (110) surface. *Surface Science*, **490**, 221.

[81] Lu, G., Linsebigler, A.L., and Yates, J.T. Jr. (1995) Molecular oxygen-mediated vacancy diffusion on TiO_2 (110) – new studies of the proposed mechanism. *Chemical Physics Letters*, **102**, 4657; Rusu, C.N. and Yates, J.T. Jr. (1997) Defect Sites on TiO_2(110). Detection by O_2 Photodesorption. *Langmuir*, **13**, 4311.

[82] Sakai, N., Fujishima, A., Watanabe, T., and Hashimoto, K. (2003) Quantitative evaluation of the photoinduced hydrophilic conversion properties of TiO_2 thin film surfaces by the reciprocal of contact angle. *Journal of Physical Chemistry B*, **107**(4), 1028.

[83] White, J.M., Szanyi, J., and Henderson, M.A. (2003) The photon-driven hydrophilicity of titania: A model study using TiO_2(110) and adsorbed trimethyl acetate. *Journal of Physical Chemistry B*, **107**(34), 9029.

[84] Zubkov, T., Stahl, D., Thompson, T.L. *et al.* (2005) Ultraviolet light-induced hydrophilicity effect on TiO_2(110) (1×1). Dominant role of the photooxidation of adsorbed hydrocarbons causing wetting by water droplets. *Journal of Physical Chemistry B*, **109**(32), 15454.

[85] Gao, Y.F., Masuda, Y., and Koumoto, K. (2004) Light-excited superhydrophilicity of amorphous TiO_2 thin films deposited in an aqueous peroxotitanate solution. *Langmuir*, **20**(8), 3188; Wang, C.Y., Groenzin, H., and Shultz, M.J. (2003) Molecular species on nanoparticulate anatase TiO_2 film detected by sum frequency generation: Trace hydrocarbons and hydroxyl groups. *Langmuir*, **19**(18), 7330.

[86] Shibata, T., Irie, H., and Hashimoto, K. (2004) High sensitization of the photo-induced hydrophilicity on TiO_2 surface by photo-etching. *Photocatalysis*, **14**, 24.

[87] Shibata, T., Irie, H., and Hashimoto, K. (2003) Enhancement of photoinduced highly hydrophilic conversion on TiO_2 thin films by introducing tensile stress. *Journal of Physical Chemistry B*, **107**(39), 10696.

[88] Bredow, T., Giordano, L., Cinquini, F., and Pacchioni, G. (2004) Electronic properties of rutile TiO_2 ultrathin films: Odd-even oscillations with the number of layers. *Physical Review B-Condensed Matter*, **70**(3), 035419.

[89] Halicioglu, T. (1995) Stress Calculations on Diamond Surfaces. *Thin Solid Films*, **260**(2), 200.

[90] Allegretti, F., O'Brien, S., Polcik, M. *et al.* (2005) Adsorption bond length for H_2O on TiO_2(110): A key parameter for theoretical understanding. *Physical Review Letters*, **95**, 226104; Allegretti, F., O'Brien, S., Polcik, M. *et al.* (2006) Quantitative determination of the local structure of H_2O on TiO_2(110) using scanned-energy mode photoelectron diffraction. *Surface Science*, **600**(7), 1487.

[91] Menetrey, M., Markovits, A., and Minot, C. (2003) Reactivity of a reduced metal oxide surface: hydrogen, water and carbon monoxide adsorption on oxygen defective rutile TiO_2(110). *Surface Science*, **524**(1–3), 49.

[92] Stefanovich, E.V. and Truong, T.N. (1999) Ab initio study of water adsorption on TiO_2(110): molecular adsorption versus dissociative chemisorption. *Chemical Physics Letters*, **299**(6), 623; Zhang, C.J. and Lindan, P. J.D. (2004) A density functional theory study of the coadsorption of water and oxygen on TiO_2(110). *Journal of Chemical Physics*, **121**(8), 3811.

[93] Sayago, D.I., Polcik, M., Lindsay, R. *et al.* (2004) Structure determination of formic acid reaction products on TiO_2(110). *Journal of Physical Chemistry B*, **108**(38), 14316.

[94] Wendt, S., Matthiesen, J., Schaub, R. *et al.* (2006) Formation and splitting of paired hydroxyl groups on reduced TiO_2(110). *Physical Review Letters*, **96**(6), 066107; Zhang, Z., Bondarchuk, O., Kay, B.D. *et al.* (2006) Imaging water dissociation on TiO_2(110): Evidence for inequivalent geminate OH groups. *Journal of Physical Chemistry B*, **110**(43), 21840.

[95] Haiss, W. (2001) Surface stress of clean and adsorbate-covered solids. *Reports on Progress in Physics*, **64**(5), 591.

[96] Meade, R.D. and Vanderbilt, D. (1989) Origins of stress on elemental and chemisorbed semiconductor surfaces. *Physical Review Letters*, **63**(13), 1404.

[97] Nazeeruddin, M.K., Pechy, P., and Grätzel, M. (1997) Efficient panchromatic sensitization of nanocrystalline TiO_2 films by a black dye based on a trithiocyanato-ruthenium complexes. *Chemical Communications*, 1705.

[98] Nazeeruddin, M.K., Kay, A., Rodicio, I. *et al.* (1993) Conversion of light to electricity by cis-X2bis (2,2′-bipyridyl-4,4′-dicarboxylate) ruthenium (II) charge-transfer sensitizers ($X = Cl^-$, Br^-, I^-, CN^-, and SCN^-) on nanocrystalline titanium dioxide electrodes. *Journal of the American Chemical Society*, **115**, 6382.

[99] Rehm, J.M., McLendon, G.L., Nagasawa, Y. *et al.* (1996) Femtosecond electron-transfer dynamics at a sensitizing dye-semiconductor (TiO_2) interface. *The Journal of Physical Chemistry*, **100**, 9577.

[100] Ramakrishna, S., Willig, F., May, V., and Knorr, A. (2003) Femtosecond spectroscopy of heterogeneous electron transfer: extraction of excited-state population dynamics from pump-probe signals. *Journal of Physical Chemistry B*, **107**, 607.

[101] Thoss, M., Kondov, I., and Wang, H. (2004) Theoretical study of ultrafast heterogeneous electron transfer reactions at dye-semiconductor interfaces. *Chemical Physics*, **304**, 169.

[102] Stier, W.M. and Prezhdo, O.V. (2002) Nonadiabatic molecular dynamics simulation of light-induced electron transfer from anchored molecular electron donor to a semiconductor acceptor. *Journal of Physical Chemistry B*, **106**, 8047.

[103] Rego, L.G.C. and Batista, V.S. (2003) Quantum dynamics simulations of interfacial electron transfer in sensitized TiO_2 semiconductors. *Journal of the American Chemical Society*, **125**, 7989.

[104] Kondov, I., Čížek, M., Benesch, C. *et al.* (2007) Quantum dynamics of photoinduced electron-transfer reactions in dye-semiconductor systems: first-principles description and application to coumarin 343-TiO_2. *Journal of Physical Chemistry C*, **111**, 11970.

[105] Nozik, A.J. (2001) Spectroscopy and hot electron relaxation dynamics in semiconductor quantum wells and quantum dots. *Annual Review of Physical Chemistry*, **52**, 193.

[106] Shockley, W. and Queisser, H.J. (1961) Detailed balance limit of efficiency of p-n junction solar cells. *Journal of Applied Physics*, **32**, 510.

[107] Schaller, R.D. and Klimov, V.I. (2004) High efficiency carrier multiplication in PbSe nanocrystals: implications for solar energy conversion. *Physical Review Letters*, **92**, 186601.

[108] Califano, M., Zunger, A., and Franceschetti, A. (2004) Direct carrier multiplication due to inverse Auger scattering in CdSe quantum dots. *Applied Physics Letters*, **84**(13), 2409; Califano, M., Zunger, A., and Franceschetti, A. (2004) Efficient inverse Auger recombination at threshold in CdSe nanocrystals. *Nano Letters*, **4**(3), 525.

[109] Ellingson, R.J., Beard, M.C., Johnson, J.C. *et al.* (2005) Highly efficient multiple exciton generation in colloidal PbSe and PbS quantum dots. *Nano Letters*, **5**, 865; Murphy, J.E., Beard, M.C., Norman, A.G. *et al.* (2006) PbTe colloidal nanocrystals: synthesis, characterization, and multiple exciton generation. *Journal of the American Chemical Society*, **128**, 3241.

[110] Schaller, R.D., Agranovich, V.M., and Klimov, V.I. (2005) High-efficiency carrier multiplication through direct photogeneration of multi-excitions via virtual single-exciton states. *Nature Physics*, **1**, 189.

[111] Schaller, R.D., Sykora, M., Jeong, S., and Klimov, V.I. (2006) High-efficiency carrier multiplication and ultrafast charge separation in semiconductor nanocrystals studied via time-resolved photoluminescence. *Journal of Physical Chemistry B*, **110**(50), 25332.

[112] Schaller, R.D., Sykora, M., Pietryga, J.M., and Klimov, V.I. (2006) Seven excitions at a cost of one: redefining the limits for conversion efficiency of photons into charge carriers. *Nano Letters*, **6**, 424.
[113] Diguna, L.J., Shen, Q., Kobayashi, J., and Toyoda, T. (2007) High efficiency of CdSe quantum-dot-sensitized TiO_2 inverse opal solar cells. *Applied Physics Letters*, **91**, 023116; Niitsoo, O., Sarkar, S.K., Pejoux, C., Rühle, S., Cahen, D., and Hodes, G. (2006) Chemical bath deposited CdS/CdSe-sensitized porous TiO_2 solar cells. *Journal of Photochemistry and Photobiology A*, **181**, 306.
[114] Wang, L.-W., Califano, M., Zunger, A., and Franceschetti, A. (2003) Pseudopotential theory of auger processes in CdSe quantum dots. *Physical Review Letters*, **91**(5), 056404.
[115] Baleva, M., Georgiev, T., and Lashkarev, G. (1990) On the temperature dependence of the energy gap in PbSe and PbTe. *Journal of Physics: Condensed Matter*, **1**, 2935.
[116] Wise, F.W. (2000) Lead salt quantum dots: the limit of strong quantum confinement. *Accounts of Chemical Research*, **33**, 773.
[117] Kayanuma, Y. (1988) Quantum-size effects of interacting electrons and holes in semiconductor microcrystals with spherical shape. *Physical Review B-Condensed Matter*, **38**, 9797.
[118] Mukamel, S. (1995) *Principles of Nonlinear Optical Spectroscopy*, Oxford University Press, New York.
[119] Neria, E. and Nitzan, A. (1993) Semiclassical evaluation of nonadiabatic rates in condensed phases. *Journal of Chemical Physics*, **99**(2), 1109; Schwartz, B.J., Bittner, E.R., Prezhdo, O.V., and Rossky, P.J. (1996) Quantum decoherence and the isotope effect in condensed phase nonadiabatic molecular dynamics simulations. *Journal of Chemical Physics*, **104**(15), 5942.
[120] Prezhdo, O.V. and Rossky, P.J. (1997) Evaluation of quantum transition rates from quantum-classical molecular dynamics simulations. *Journal of Chemical Physics*, **107**(15), 5863; Prezhdo, O.V. and Rossky, P.J. (1998) Relationship between quantum decoherence times and solvation dynamics in condensed phase chemical systems. *Physical Review Letters*, **81**(24), 5294.
[121] Colonna, A.E., Yang, X., and Scholes, G.D. (2005) Photon echo studies of biexcitons and coherences in colloidal CdSe quantum dots. *Physica Status Solidi B-Basic Research*, **242**(5), 990; Salvador, M.R., Hines, M.A., and Scholes, G.D. (2003) Exciton-bath coupling and inhomogeneous broadening in the optical spectroscopy of semiconductor quantum dots. *Journal of Chemical Physics*, **118**(20), 9380.
[122] Mittleman, D.M., Schoenlein, R.W., Shiang, J.J. *et al.* (1994) Quantum size dependence of femtosecond electronic dephasing and vibrational dynamics in CdSe nanocrystals. *Physical Review B-Condensed Matter*, **49**, 14435.
[123] Takagahara, T. (1996) Electron-phonon interactions in semiconductor nanocrystals. *Journal of Luminescence*, **70**, 129.
[124] Soloviev, N.V., Eichhöfer, A., Frenske, D., and Banin, U. (2000) Molecular limit of a bulk semiconductor: size dependence of the "Band Gap" in CdSe cluster molecules. *Journal of the American Chemical Society*, **122**, 2673.
[125] Franceschetti, A., An, J.M., and Zunger, A. (2006) Impact ionization can explain carrier multiplication in PbSe quantum dots. *Nano Letters*, **6**, 2191.
[126] Hedin, L. (1965) New method for calculating the one-particle Green's function with application to the electron-gas problem. *Physical Review*, **139**, A796.
[127] Heyd, J., Scuseria, G.E., and Ernzerhof, M. (2003) Hybrid functionals based on a screened Coulomb potential. *Journal of Chemical Physics*, **118**, 8207.
[128] Nakai, H., Heyd, J., and Scuseria, G.E. (2006) Periodic-boundary-condition calculation using Heyd-Scuseria-Ernzerhof screened coulomb hybrid functional: electronic structure of anatase and rutile TiO_2. *Journal of Computer Chemistry, Japan*, **5**, 7.
[129] Perdew, J.P. and Zunger, A. (1981) Self-interaction correction to density-functional approximations for many-electron systems. *Physical Review B-Condensed Matter*, **23**, 5048.
[130] Dudarev, S.L., Botton, G.A., Savrasov, S.Y. *et al.* (1998) Electron-energy-loss spectra and the structural stability of nickel oxide: An LSDA + U study. *Physical Review B-Condensed Matter*, **57**, 1505.
[131] Chakarova-Käck, S.D., Borck, Ø., Schröder, E., and Lundqvist, B.I. (2006) Adsorption of phenol on graphite (0001) and α-Al_2O_3 (0001): Nature of van der Waals bonds from first-principles calculations. *Physical Review B-Condensed Matter*, **74**, 155402; Lazić, P., Crljen, Ž., Brako, R., and Gumhalter, B. (2005) Role of van der Waals interactions in adsorption of Xe on Cu(111) and Pt(111). *Physical Review B-Condensed Matter*, **72**, 245407.
[132] Matsui, M. and Akaogi, M. (1991) Molecular dynamics simulation of the structural and physical properties of the four polymorphs of TiO_2. *Molecular Simulation*, **6**, 239.
[133] Koparde, V.N. and Cummings, P.T. (2005) Molecular dynamics simulation of titanium dioxide nanoparticle sintering. *Journal of Physical Chemistry B*, **109**, 24280; Naicker, P.K., Cummings, P.T., Zhang, H., and Banfield,

J.F. (2005) Characterization of titanium dioxide nanoparticles using molecular dynamics simulations. *Journal of Physical Chemistry B*, **109**, 15243.

[134] Bandura, A.V. and Kubicki, J.D. (2003) Derivation of force field parameters for TiO_2-H_2O Systems from *ab initio* caculations. *Journal of Physical Chemistry B*, **107**, 11072.

[135] Seifert, G. (2007) Tight-binding density functional theory: an approximate Kohn-Sham DFT scheme. *Journal of Physical Chemistry A*, **111**, 5609.

[136] Erdin, S., Lin, Y., Halley, J.W. *et al.* (2007) Self-consistent tight binding molecular dynamics study of TiO_2 nanocluster in water. *Journal of Electroanalytical Chemistry*, **607**, 147.

[137] Alvarez-Ramirez, F. and Ruiz-Morales, Y. (2007) *Ab initio* molecular dynamics calculations of the phase transformation mechanism for the formation of TiO_2 titanate-type nanosheets from anatase. *Chemistry of Materials*, **19**, 2947.

[138] Schiffmann, F., Hutter, J., and VandeVondele, J. (2008) Atomistic simulations of a solid/liquid interface: a combined force field and first principles approach to the structure and dynamics of acetonitrile near an anatase surface. *Journal of Physics: Condensed Matter*, **20**, 064206.

[139] Car, R. and Parrinello, M. (1985) Unified approach for molecular dynamics and density-functional theory. *Physical Review Letters*, **55**, 2471.

[140] Rohlfing, M. and Louie, S.G. (2000) Electron-hole excitations and optical spectra from first principles. *Physical Review B-Condensed Matter*, **62**, 4927.

[141] Deslippe, J., Spataru, C.D., Prendergast, D., and Louie, S.G. (2007) Bound excitons in metallic single-walled carbon nanotubes. *Nano Letters*, **7**, 1626.

[142] Ahlrichs, R., Scharf, P., and Ehrhardt, C. (1985) The coupled pair functional (CPF). A size consistent modification of the CI(SD) based on an energy functional. *Journal of Chemical Physics*, **82**, 890; Barandiarán, Z. and Seijo, L. (1988) The *ab initio* model potential representation of the crystalline environment. Theoretical study of the local distortion on NaCl: Cu^{+}. *Journal of Chemical Physics*, **89**, 5739; DiLabio, G.A., Hurley, M. M., and Christiansen, P.A. (2002) Simple one-electron quantum capping potentials for use in hybrid QM/MM studies of biological molecules. *Journal of Chemical Physics*, **116**, 9578; Poteau, R., Ortega, I., Alary, F., Solis, A.R., Barthelat, J.-C., and Daudey, J.-P. (2001) Effective group potentials. 1. method. *The Journal of Physical Chemistry A*, **105**, 198; Shidlovskaya, E.K. (2002) Improved embedded molecular cluster model. *International Journal of Quantum Chemistry*, **89**, 349; Stefanovich, E.V. and Truong, T.N. (1996) Embedded density functional approach for calculations of adsorption on ionic crystals. *Journal of Chemical Physics*, **104**, 2946; Whitten, J.L. and Yang, H. (1995) Theoretical-studies of surface-reactions on metals. *International Journal of Quantum Chemistry*, **29**, 41; Yasuike, T. and Nobusada, K. (2007) Open-boundary cluster model for calculation of adsorbate-surface electronic states. *Physical Review B-Condensed Matter*, **76**, 235401.

[143] Govind, N., Wang, Y.A., da Silva, A.J.R., and Carter, E.A. (1998) Accurate ab initio energetics of extended systems via explicit correlation embedded in a density functional environment. *Chemical Physics Letters*, **295**, 129.

[144] Sremaniak, L.S. and Whitten, J.L. (2002) Photoinduced dissociation of methylnitrite on Ag(111). *Surface Science*, **516**, 254.

[145] Shluger, A.L. and Gale, J.D. (1996) One-center trapping of the holes in alkali halide crystals. *Physical Review B-Condensed Matter*, **54**, 962.

[146] Sushko, P.V., Shluger, A.L., and Catlow, C.R.A. (2000) Relative energies of surface and defect states: *ab initio* calculations for the MgO (001) surface. *Surface Science*, **450**, 153.

[147] Runge, E. and Gross, E.K.U. (1984) Density-functional theory for time-dependent systems. *Physical Review Letters*, **52**, 997.

[148] Casida, E. and Wesołowski, T.A. (2004) Generalization of the Kohn-Sham equations with constrained electron density formalism and its time-dependent response theory formalism. *International Journal of Quantum Chemistry*, **96**, 577.

[149] Neugebauer, J., Louwerse, M.J., Baerends, E.J., and Wesolowski, T.A. (2005) The merits of the frozen-density embedding scheme to model solvatochromic shifts. *Journal of Chemical Physics*, **122**, 094115.

[150] Klüner, T., Govind, N., Wang, Y.A., and Carter, E.A. (2001) Prediction of electronic excited states of adsorbates on metal surfaces from first principles. *Physical Review Letters*, **86**, 5954.

[151] Persson, M. and Hellsing, B. (1982) Electronic damping of adsorbate vibrations on metal surfaces. *Physical Review Letters*, **49**, 662.

[152] Abe, A. and Yamashita, K. (2003) Effects of vibrational relaxation on the photodesorption of NO from Pt(111): A density matrix study. *Journal of Chemical Physics*, **119**, 9710.

[153] Persson, B.N.J. and Persson, M. (1980) Vibrational lifetime for CO adsorbed on Cu(100). *Solid State Communications*, **36**, 175.
[154] Head-Gordon, M. and Tully, J.C. (1992) Vibrational relaxation on metal surfaces: Molecular-orbital theory and application to CO/Cu(100). *Journal of Chemical Physics*, **96**, 3939.
[155] Tully, J.C. (2000) Chemical dynamics at metal surfaces. *Annual Review of Physical Chemistry*, **51**, 153.
[156] Nakamura, H. and Yamashita, K. (2005) Electron tunneling of photochemical reactions on metal surfaces: Nonequilibrium Green's function-density functional theory approach to photon energy dependence of reaction probability. *Journal of Chemical Physics*, **122**, 194706.
[157] Abe, A., Yamashita, K., and Saalfrank, P. (2003) STM and laser-driven atom switch: An open-system density-matrix study of H/Si (100). *Physical Review B-Condensed Matter*, **67**, 235411.
[158] Prezhdo, O.V. (1999) Mean field approximation for the stochastic Schrödinger equation. *Journal of Chemical Physics*, **111**, 8366.
[159] Deskins, N.A. and Dupuis, M. (2007) Electron transport via polaron hopping in bulk TiO_2: A density functional theory characterization. *Physical Review B-Condensed Matter*, **75**, 195212. Ke, S.C., Wang, T.C., Wong, M.S., and Gopal, N.O. (2006) Low temperature kinetics and energetics of the electron and hole traps in irradiated TiO_2 nanoparticles as revealed by EPR spectroscopy. *Journal of Physical Chemistry B*, **110**, 11628. Kerisit, S., Deskins, N.A., Rosso, K.M., and Dupuis, M. (2008) A shell model for atomistic simulation of charge transfer in titania. *Journal of Physical Chemistry C*, **112**, 7678.

3

Photocatalytic Reactions on Model Single Crystal TiO_2 Surfaces

G.I.N. Waterhouse[1] and H. Idriss[2,3]

[1]*Department of Chemistry, University of Auckland, Auckland, New Zealand*
[2]*Department of Chemistry, University of Aberdeen, Aberdeen, UK*
[3]*School of Engineering, Robert Gordon University, Schoolhill, Aberdeen, UK,*
Email: h.idriss@abdn.ac.uk

Light from the sun is by far the most abundant source of energy on earth. Yet, at present, less than 0.05% of the total power (15 000 GW annual) used by humans is generated from the sun (excluding solar heating, which contributes around 0.6%). The estimated practical and convertible power that the earth surface receives is equivalent to that provided by 600 000 nuclear reactors (one nuclear power plant generates, on average, 1 GW power) or about 40 times the present global need.[1] One mode of solar energy utilization is the use of sunlight to generate energy carriers, such as hydrogen, from renewable sources (e.g., ethanol and water) using semiconductor photocatalysts.

The photoassisted splitting of water into hydrogen and oxygen was first achieved by Fujushima and Honda [1], who showed that hydrogen and oxygen could be generated in an electrochemical cell containing a titania photoelectrode, provided an external bias was applied. Since that time, numerous researchers have explored ways of achieving direct water dissociation without the need for an external bias. Much work has been conducted, a large fraction of which is discussed in a recent review [2]. Among the many issues affecting direct water splitting is the need to separate hydrogen from oxygen and the relatively low hydrogen evolution rates so far achieved. These, in addition to the need for using UV light (>3eV) to excite TiO_2 and other related materials, has been one of the main obstacles for practical applications. Many authors have sought modified photocatalysts which, unlike pure TiO_2, respond to visible (sunlight) excitation, with limited success to date; see some of these materials in ref. [2].

[1] The total amount of sunlight reaching the earth surface is orders of magnitude higher than the quoted figure.

On Solar Hydrogen & Nanotechnology Edited by Lionel Vayssieres

Due to its widespread use as a photocatalyst, and its participation in many other processes and applications, including biomaterials, sensors and as catalyst support for transition metals and metal oxides, TiO_2 powder and single-crystal surfaces have been studied extensively. The reactions over TiO_2 single-crystal surfaces of chemical compounds with different functional groups, including carboxylic acids, alcohols, aldehydes, aromatics and alkynes, in addition to smaller molecules such as CO, NO, NO_2, CO_2, H_2S and H_2O, among others, have been studied in the last two decades [3–5]. Results of these studies are of some interest when dealing with photoreactions, in particular those related to adsorption geometry and energetics. The photoreactions of organic compounds over these single-crystal surfaces have received less attention, but interest in these reactions are increasing [6–10]. The most-studied photoreactions on rutile $TiO_2(110)$ single-crystal surfaces include those of ethanol [6], acetic acid [11], trimethyl acetic acid [8,9] and acetone [10]. Sporadic work has been conducted on anatase TiO_2 single crystals, although, to date, no published work dealing with photoreactions is known. Before discussing some of these results, it is worthwhile giving a brief introduction on the surfaces of rutile TiO_2. This will be followed by another brief introduction about the chemical and physical processes thought to be involved in photocatalytic reactions over a semiconductor.

3.1 TiO_2 Single-Crystal Surfaces

The surfaces of single crystals of rutile TiO_2, such as the (001), (110) and (100), have been studied for decades as models for metal oxide. TiO_2 exists in several different polymorphic forms, such as the rutile, anatase and brookite polymorphs. While brookite is not a common TiO_2 phase, the anatase phase is often found in nature. In addition, because the anatase phase can be easily synthesized with high surface area many TiO_2-based catalytic materials contain substantial amounts of the anatase phase [12]. However, the anatase phase is considerably less stable than the rutile phase; it can transform to the rutile phase at temperatures as low as 500 K. While this may not be very critical to some catalytic reactions, it has rendered their study by surface-science methods more complex, because high temperatures are required to prepare clean surfaces for chemical reactions. Some single-crystal surfaces of anatase TiO_2 have been successfully studied by scanning tunneling microscopy (STM) [13,14] although most studies are conducted by computation modeling [15–17]. Surface energy is one of the main parameters determining the stability of a surface and possibly reactivity. One of the crucial factors determining surface stability is the coordination number of surface atoms. The closer the coordination number of a given surface atom to that of the bulk, the more stable the surface. In both the rutile and anatase structures the coordination number of Ti atoms in the bulk is six, and three for oxygen. Table 3.1 presents the computed surface energies for some low-Miller-index rutile TiO_2 surfaces, while Figure 3.1 shows the most common surface structures.

The (110) rutile surface contains alternating rows of Ti and O atoms five- and twofold coordinated, respectively. Ti atoms underneath the twofold coordinated oxygen atoms are sixfold coordinated. Because of the high coordination numbers of Ti atoms, this surface is the most stable, as evidenced by the value of its surface energy (Table 3.1). Because of its high stability, the (110) surface has received most of the attention in surface-science studies and has been often considered as a bench mark for metal oxides. The (100) surface is second most stable surface, where all surface Ti atoms are fivefold coordinated and all O atoms are twofold

Table 3.1 Computed surface energy by density functional theory (DFT) for rutile TiO_2 surfaces [18].

Surface	Surface energy		DFT method[a]
	meV/(a.u.)^2	J m^{-2}	
(110)	15.6	0.89	LDA
(100)	19.6	1.12	LDA
(011)	24.4	1.39	LDA
(011)[b] 1 × 1	19.3	1.10	GGA
(011)[c] 2 × 1	7.4	0.42	GGA
(001)	28.9	1.65	LDA

[a] LDA: local density approximation; GGA: generalized gradient approximation.
[b] From [19].
[c] From [20].

coordinated. The (001) surface with fourfold coordinated Ti atoms is the least stable of the low-index rutile surfaces. However, this surface is particularly interesting as it readily facets to the {011} or {114} surfaces at high temperature [21]. These facets are highly reactive, and the reactivity of a number of small molecules, including numerous oxygen- and nitrogen-containing compounds have been studied [22–24].

The exact structures of the (011) and (114) surfaces are still debated. Models derived from STM studies demonstrate that the (114) surface is composed of steps, each containing one fourfold-coordinated Ti atom. The shortest distance between two fourfold-coordinated Ti atoms is 0.65 nm (along the [110] direction) while the distance between two successive steps is equal to 0.30 nm along the [$1\bar{1}0$] direction). This model agrees with Ti2p core level X-ray photoelectron spectroscopy (XPS) results, indicating that the surface is composed of stoichiometric TiO_2 units [25]. On the other hand, low-energy electron diffraction (LEED) [21] and STM [28] data suggest that the stable (011) surface is a (2 × 1) reconstructed surface. Initially a model for the (011) surface containing terminal Ti = O groups was proposed [26]. Based on surface X-ray diffraction, a new model has been recently introduced [20,27]. This model, unlike the previous one, does not contain Ti = O structures.

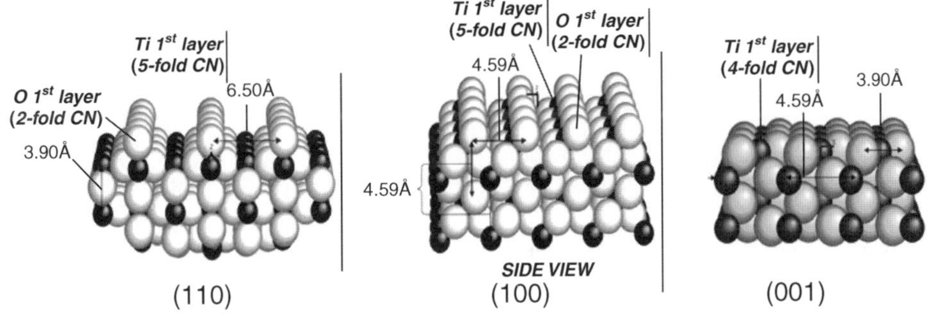

Figure 3.1 Side view of the unreconstructed (bulk terminated) rutile TiO_2 surfaces. The black circles represent Ti atoms, and the gray circles are O atoms. The coordination number (CN) for the Ti atoms is five for the (110) and (1 0 0) and four for the (001) structure.

3.2 Photoreactions Over Semiconductor Surfaces

Figure 3.2 is a schematic description of the complex processes involved in photocatalysis over a material composed of a metal deposited on a wide bandgap semiconductor. Excitation of a semiconductor, such as TiO_2, with photons of appropriate energy produces hole–electron pairs, which may either recombine or react with appropriate adsorbed species. Photocatalysis with TiO_2 is a well-established field, which exploits these phenomena and which has been extensively reviewed [28]. Despite the enormous number of published papers in this field (particularly in the context of environmental clean-up), many aspects of reaction mechanisms remain poorly understood and therefore catalyst optimization has proceeded in a largely empirical manner.

Much higher hydrogen production rates are reported for the photocatalytic reforming of alcohols when compared to those for water. This is mainly because alcohols are very active hole scavengers. The loading of titania photocatalysts with metal particles dramatically enhances the rate of hydrogen evolution, presumed to be due to trapping of conduction-band electrons within the metal clusters. The ambient temperature reforming of methanol over a variety of metal-loaded titania photocatalysts can be described as [29]

$$CH_3OH + H_2O \rightarrow CO_2 + 3\,H_2 \tag{3.1}$$

However, methanol is made from fossil fuel ($CO + H_2$) and therefore contributes to increasing carbon dioxide emissions. This method is, in addition, largely nonefficient, because hydrogen is consumed to make methanol in the first place.

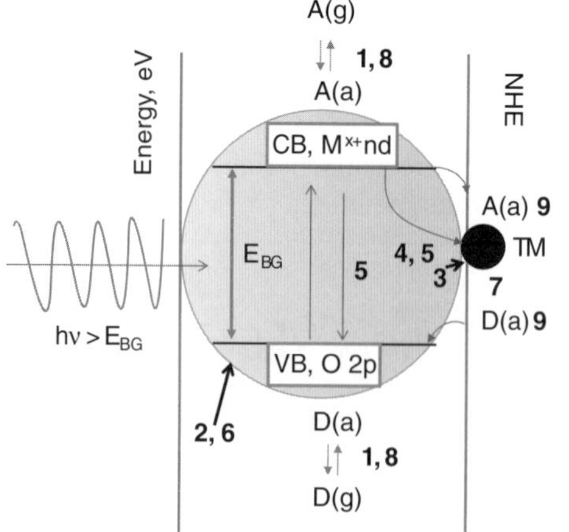

Figure 3.2 A schematic of the different photoreaction/photoexcitation processes relevant to catalytic reactions. The figure presents a schematic of the main chemical and physical reactions involved on the surfaces of a semiconductor under photon irradiation (legend at the right-hand side of the figure). Large sphere: semiconductor, small sphere: transition metal (TM) such as Au, Ru or Rh; VB: valence band (mainly O2p in the case of TiO_2); CB: conduction band (mainly Ti3d levels in TiO_2) A: acceptor; D: donor; NHE: normal hydrogen electrode; E_{BG}: Bandgap = 3–3.2 eV for TiO_2; (a): adsorbed; (g): gas phase.

In contrast, ethanol is a biorenewable hydrogen source because it is obtained from crop fermentation, and the release of CO_2 during the reaction to ethanol does not add to CO_2 emissions because of the biological origin of ethanol [30].

$$CH_3CH_2OH + 3\,H_2O \rightarrow 2\,CO_2 + 6\,H_2 \tag{3.2}$$

3.3 Ethanol Reactions Over TiO_2(110) Surface

There is no reported hydrogen production from ethanol or any other alcohol over single-crystal surfaces via photocatalysis. However, the photo-oxidation of ethanol has been studied over the rutile TiO_2(110) surface. Ethanol is dissociatively adsorbed via its oxygen lone pair on fivefold coordinated Ti atoms to produce adsorbed ethoxide species (Figure 3.3). STM studies of the adsorption of ethanol on TiO_2(110) demonstrated the presence of both alkoxides and surface hydroxyls [31], confirming the adsorption is dissociative. The adsorption on this surface is similar to that of an acid–base type, whereby the O atom of ethanol acts as a Lewis base and the surface Ti acts as a Lewis acid. Because of this strong interaction, the OH bond is broken, even though it is the strongest bond of the molecule (see bond energies of an ethanol molecule at the left-hand side of Figure 3.3). The adsorption energy of ethanol on this surface has not been computed yet, although that of methanol, at $\theta = 0.5$, has been found equal to 100 and 120 kJ mol^{-1} for the molecular and dissociated adsorptions, respectively [32].

Figure 3.4a at 0 minutes displays the XPS C1s spectra after saturation exposure of ethanol at 300 K; the surface coverage θ was found to be 0.5 with respect to surface Ti atoms, and this

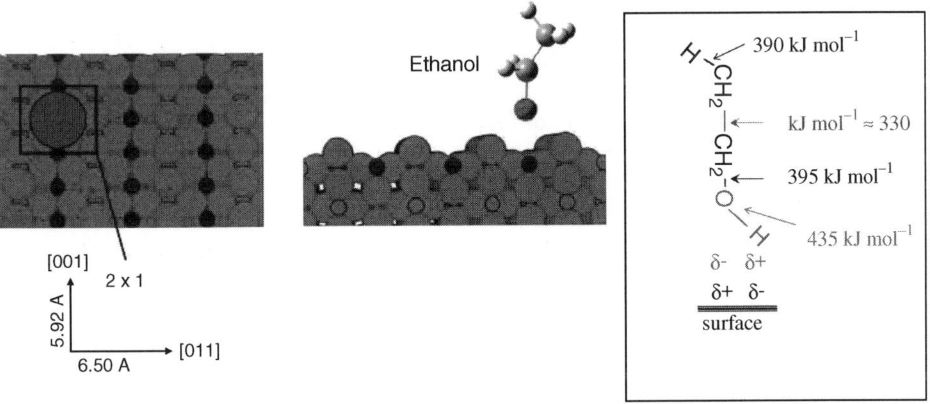

Figure 3.3 The left-hand side of the figure: top and side views of the (110) surface of the rutile TiO_2. Small (blue) circles represent Ti atoms and large (red) circles represent O atoms. The right-hand side of the figure gives the bond energies of an ethanol molecule. A representation of the adsorbed ethoxide species on top of a fivefold coordinated Ti is also seen. The H resulting from the dissociative adsorption is presumably on top of the twofold coordinated surface O atoms (bridging O atoms), not shown. The atomic scale between ethanol and TiO_2 is not unified. The very large circle (pink) represents the van der Waals dimension of an ethanol molecule. Data for this figure are provided from spectroscopic measurement. (Adapted with permission from P.M. Jayaweera *et al.*, Photoreaction of ethanol on TiO_2(110) single-crystal surface, *Journal of Physical Chemistry C*, **111**(4), 2007. © 2007 American Chemical Society.) See also color-plate section.

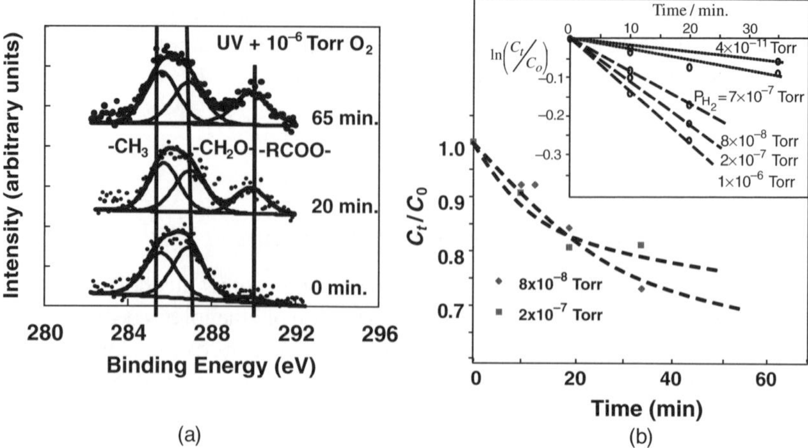

Figure 3.4 (a) XPS C1s of adsorbed ethanol on $TiO_2(110)$ at 300 K at saturation before (0 min) and after UV illumination (Hg lamp with main line at 360 nm) in the presence of 10^{-6} Torr O_2 at different time intervals. (b) The peak area of the XPS C1s peak as a function of illumination time at different O_2 partial pressures. The inset is the initial linear decrease of the computed peak areas at different partial pressures of O_2. Included in the figure is an additional run under H_2 (7×10^{-7} Torr) for comparison with the one in the absence of O_2 ($P_{O_2} = 4 \times 10^{-11}$ Torr is estimated as the upper value for oxygen pressure at a vacuum of 2×10^{-10} Torr). (Adapted with permission from P.M. Jayaweera *et al.*, Photoreaction of ethanol on $TiO_2(110)$ single-crystal surface, *Journal of Physical Chemistry C*, **111**(4), 2007. © 2007 American Chemical Society.)

might be due to the van der Waals radius of the ethanol molecule, as depicted in the left-hand side of Figure 3.3. As seen in Figure 3.4, the C1s structure consists of two species, because the two carbon atoms of ethoxides are in different chemical environments (-CH_3 and -CH_2O-).

Upon exposure to UV light in presence of 10^{-6} Torr O_2, two main changes in the XPS spectra are observed [6]: (i) a slight decrease in the C1s peak area for both functional groups (-CH_2O- and -CH_3) with increasing irradiation time, and (ii) the formation of a peak at ~290 eV, which is attributed to an $RCOO^-$(a) species. The absorption of a photon by a surface-bound ethoxide results in the transfer of electrons from the valence band (occupied states containing contributions from the O2p level) to the conduction band (empty states containing contributions from Ti3d level). The decrease in the ethoxide C1s peaks is due to its reaction with $^{\bullet}O_2$ and/or $^{\bullet}O$, which form according to the following reactions:

$$TiO_2 \xrightarrow{UV} e^- + h^+ \tag{3.3}$$

$$e^- + O_2 \rightarrow O_2^{\bullet -} \tag{3.4}$$

$$\begin{aligned} O_2^{\bullet -} + CH_3CH_2O_{(a)} &\rightarrow HO_2 + CH_3{-}CHO^{\bullet}_{(ad)} \\ &\rightarrow + O_{(a)} \quad \text{or} \quad O_{2,(g)} \rightarrow CH_3COO_{(a)} \\ &\rightarrow \rightarrow CO_2 + H_2O \end{aligned} \tag{3.5}$$

The two successive arrows ($\rightarrow \rightarrow$) indicate multiple reaction steps of unidentified intermediates.

The change in ethoxide concentration (expressed as the ratio, C_t/C_0, where C_t is the area of the two C1s peaks of the ethoxide species at time t and C_0 is the area of the two peaks before irradiation) is plotted as a function of time (Figure 3.4b). These data are similar to those reported in Figure 3.4a, except each line is the result of a photoreaction conducted at a different O_2 pressure. Assuming a simple exponential decay of adsorbed ethoxide with time, the time-dependent concentration is written as:

$$C_t = C_0 \exp(-kt) \quad \text{or} \quad C_t = C_0 \exp(-FQt) \tag{3.6}$$

and rearranging, the photoreaction cross-section, Q (in cm^2) becomes

$$Q = \frac{1}{Ft} \ln\left(\frac{C_0}{C_t}\right) \tag{3.7}$$

where k is a pseudo first-order rate constant, t is the illumination time and F is the UV flux. Q is found to be equal to $\sim 2 \times 10^{-18}\,cm^2$ at a pressure of 10^{-6} Torr O_2.

Some of the ethoxide species are converted to acetate ($RCOO^-$, where $R = CH_3$). In the photo-oxidation of ethanol, the reaction products are CO_2 and water. Detailed studies of reaction intermediates on model metal oxides are scarce, however more work has been conducted on high-surface-area powders. The reaction intermediates on TiO_2 have been studied by temperature programmed desorption (TPD) [33] and IR [34] and the formation of CH_3CHO by the removal of an electron from the radical of Equation 3.8 has been reported

$$CH_3{-}CHO^{\bullet}_{(a)} + h^+ \rightarrow CH_3CHO_{(a)} \tag{3.8}$$

The acetaldehyde radical may further react in a complex set of reactions, which are not well understood, to yield the final product CO_2 [35].

3.4 Photocatalysis and Structure Sensitivity

The steady-state photocatalytic reactions in ultra-high-vacuum conditions, using acetic acid and rutile TiO_2 (001) single crystals, have been studied in considerable detail. The rutile TiO_2 (001) surface reconstructs to two stable facets, depending on the annealing temperature: the low-temperature phase (011) and the high-temperature phase (114). Because photocatalysis directly depends on both bulk and surface properties, studying these single-crystal surfaces has helped in decoupling surface from bulk contributions.

Acetic acid reacts photocatalytically at RT over TiO_2 to give methane, ethane and CO_2, as depicted by the two equations at the bottom of Figure 3.5. The left-hand side of the figure shows the formation of ethane (m/z 30), methane (m/z 16) and CO_2 (m/z 44) upon UV illumination over the (011) reconstructed TiO_2 surface. While CO_2 and methane production is constant and tracks UV light exposure, that of ethane decreases with time. The reason has been found to be depletion of surface oxygen; that is why the surface oxygen atoms in the second equation of the figure are in bold as they are removed from the surface in the form of water. Co-feeding O_2 in the gas phase with acetic acid was found to maintain ethane production, as it restores surface oxygen atoms. The right-hand side of the figure shows the activity towards ethane formation under UV light over the (011) and (114) reconstructed surfaces. It was found that the (011) reconstructed surface of rutile TiO_2(001) is more active than the (114) reconstructed surface of the same crystal [36,37]. While the reasons for the different reactivity between the two surfaces

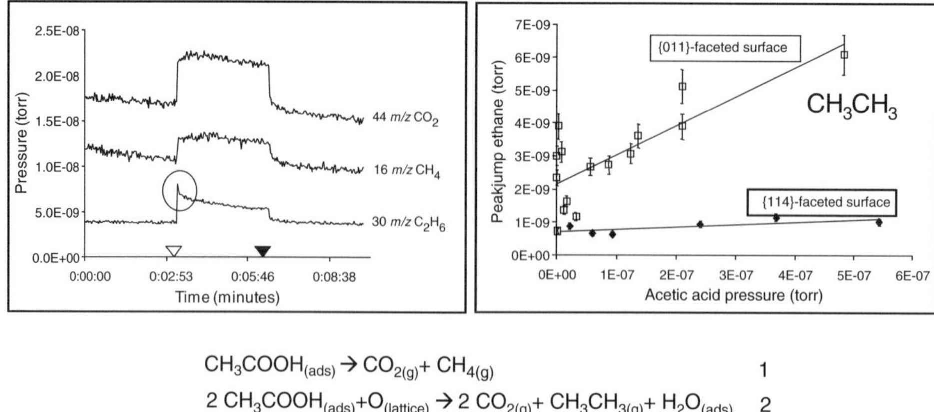

Figure 3.5 Structure sensitivity studied by photocatalytic reaction of rutile TiO_2 surfaces: acetic acid photoreactions over TiO_2(001) reconstructed surfaces ((011) and (114)). Acetic acid reacts under photons with the surface to form CO_2 and methane or CO_2, ethane and water. Reaction 2 consumes surface O atoms, while reaction 1 does not. The (011) surface is easier to reduce than the (114) surface and is therefore more active for reaction 2. Injecting O_2 together with acetic acid regenerates surface O atoms and makes both surfaces behave similarly.

are not understood, the fact that both surfaces share the same bulk structures removes the bulk electron–hole recombination rate as a factor, and therefore different surface different electron–hole recombination rates would be one of the main contributing factors.

One accurate method for monitoring adsorbate evolution on model surfaces is shown in Figure 3.6, where the signal of a dosed compound, in this case acetic acid, on a surface of a single crystal, TiO_2(011) rutile, is monitored as a function of exposure to photon time, as well as gas-phase oxygen pressure. From such measurements one can estimate the photoionization cross-section, as well as the reaction quantum yield, in a similar way to that presented for ethanol.

3.5 Hydrogen Production from Ethanol Over Au/TiO_2 Catalysts

Figure 3.7 presents some recent data for the photocatalyzed dehydrogenation of ethanol using 1.5 Au wt% on TiO_2 Degussa P25. Degussa P25 is composed of about 80% anatase and 20% rutile and is one of the most photocatalytically active TiO_2 materials known to date. The inset at the right-hand side of the figure is a TEM image, where anatase TiO_2 particles are 30 nm in size, while the larger particles of 70–80 nm size are those of the rutile phase. The small dark particles of about 3 nm size are those of Au particles. The left-hand side of the figure presents XPS Au4f lines of Au particles of the catalyst, together with a comparison with an Au foil metal. The shift of the Au4f lines to lower binding energy with respect to that of the Au foil is due to negative charge transfer from the support to the metal. The atomic percentage of Au as determined from Au4f XPS is about 0.7%. Dark catalytic oxidation of CO to CO_2 on Au/TiO_2 has been shown to be most active when the particle size of Au is close to 3 nm [38,39]. Moreover, the upward shift of the Fermi level has been determined (with respect to the normal hydrogen electrode (NHE)) to increase with decreasing Au particle size, with a shift of 290 mV for 3 nm particles [40]. It is

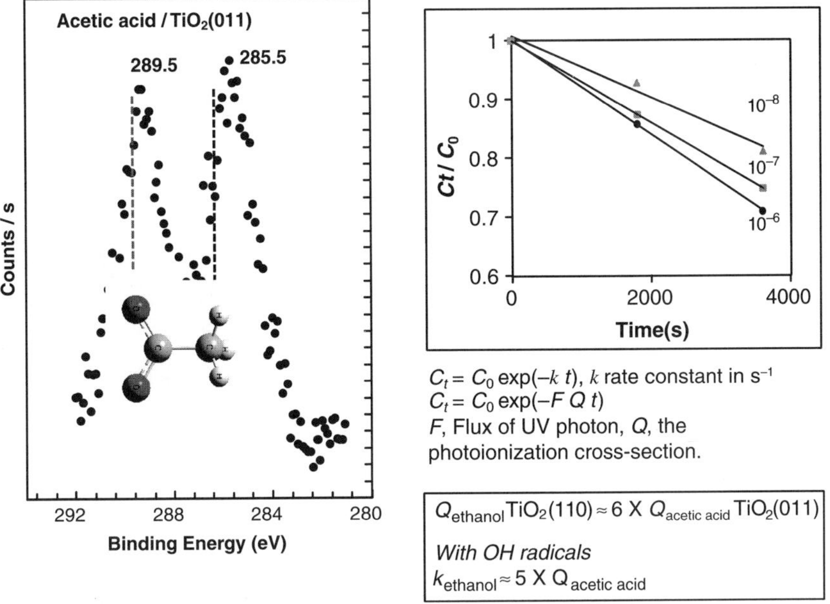

Figure 3.6 XPS C1s following acetic acid adsorption on rutile TiO_2 (011) surface. Acetates are formed from dissociative adsorption of acetic acid on a TiO_2(011) surface. The left-hand side of the figure presents the signal of the two groups as studied by their XPS C1s signal. From monitoring the surface population with increasing photons and exposure time, as well as under different oxygen pressures, the intrinsic parameters of the reaction can be extracted. It was found that acetic acid photodecomposition is about six times slower than that of ethanol on TiO_2(110). The second-order rate constant for ethanol and acetic acid reactions with OH radicals in the gas phase indicates similar numbers.

seen in the figure that after an initial (not well-understood) "induction" period, close to linear production of hydrogen occurs. The calculated quantum yield was found to be 0.06 (or 6%). Details of the reaction mechanism is not well understood, but involves in part a two-step electron transfer to make one mole of hydrogen per mole of ethanol and the associated production of acetaldehyde, as indicated below.

Within a semiconductor (such as TiO_2), upon excitation with photons of energy equal or higher than its bandgap (E_{BG}), electron transfers from the valence band (VB) to the conduction band (CB) occur, consequently creating electron (e^-) and hole (h^+) pairs.

$$\text{Photoexcitation}: TiO_2 + 2\,\text{UV photons} \rightarrow 2\,e^- + 2\,h^+ \quad (3.9)$$

If we consider ethanol as the reactant, Equations 3.10–3.15 describe the formation of one hydrogen molecule per one ethanol molecule, using two photons. The presence of a transition metal (acting as an electron trap) decreases the electron–hole recombination rate (the reverse of Equation 3.9). If an adsorbed species (an acceptor such as H^+) is at the interface M/TiO_2, as depicted in Figure 3.2, it may react with the electron and is therefore reduced ($H^+ + e^- \rightarrow {}^1/_2\, H_2$). Similarly, regeneration of electrons in the VB may be possible by electron injection

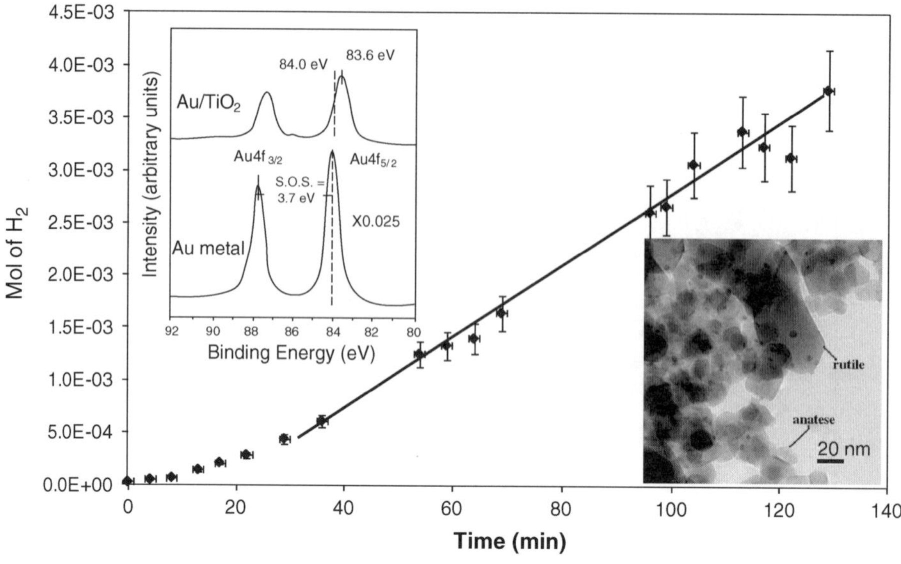

Figure 3.7 Hydrogen production from ethanol over 1.5 wt% Au/TiO_2 P25 (80% anatase – 20% rutile). y-axis presents mol l^{-1} of hydrogen evolved as a function of time (x-axis) from a solid–liquid batch reactor containing 50 mg of catalyst and 15 ml of ethanol (total reactor volume about 100 ml). Prior to reaction, the catalyst was reduced using hydrogen at 573 K overnight. Top inset: XPS Au4f of the catalyst and of gold foil. Inset bottom TEM of the catalyst. Au particles are seen as dark spots with an average diameter of 3 nm, larger particles (about 30 nm size) are anatase, while the largest particles (70–80 nm size) are rutile. Reaction temperature 315 K.

from another surface intermediate ($CH_3CH_2O_{(a)} \rightarrow CH_3CHO^{\bullet}_{(a)} + H^+$). The latter reaction repeated a second time gives CH_3CHO.

Dissociative adsorption of ethanol on Ti–O sites:

$$CH_3CH_2OH + \mathbf{Ti\text{-}O} \rightarrow CH_3CH_2O(\mathbf{Ti}) + OH \tag{3.10}$$

First electron reduction at the CB:

$$OH_{(a)} + e^- \rightarrow O_{(a)} + {}^1\!/\!{}_2H_2 \tag{3.11}$$

First hole trapping at the VB:

$$CH_3CH_2O(\mathbf{Ti}) + h^+ \rightarrow CH_3CHO^{\bullet}(\mathbf{Ti}) + H^+ \tag{3.12}$$

Second hole trapping at the VB:

$$CH_3CHO^{\bullet}(\mathbf{Ti}) + h^+ \rightarrow CH_3CHO(g) + \mathbf{Ti} \tag{3.13}$$

Second electron reduction at the CB:

$$H^+ + e^- \rightarrow {}^1\!/\!{}_2H_2 \tag{3.14}$$

Overall reaction:

$$TiO_2 + 2\,\text{UV photons} + CH_3CH_2OH \rightarrow CH_3CHO(g) + H_2(g) \quad (3.15)$$

For the above equation, the time intervals between two absorbed photons is considered too small compared to electron (hole) transfer reactions. This is true for large particles (bulk materials), but breaks down for nanosize particles, where large time intervals may prevent the formation of acetaldehyde, and other products can be formed [41].

3.6 Conclusions

An overview of the photoreactions of organic compounds on TiO_2 single-crystal surfaces has been presented. The photoreactions of ethanol and acetic acid were used as specific examples. The acetic acid photoreaction over reconstructed TiO_2 (001) single crystal surfaces (the (011) and (114)) displays a structure-sensitive reaction with different yields attributed to changes in the electron–hole recombination rates. The ethanol photo-oxidation reaction yield on the rutile TiO_2(110) surface has been shown to be a function of O_2 pressure. The build up of surface acetates appears to slow the oxidation reaction with time, which is explained by the fact that the latter have a slower photo-oxidation reaction rate. The photoreduction of ethanol over transition metals and TiO_2 can be successfully conducted. The example presented here taking 1.5 wt.% Au/TiO_2 P25 shows that uniform 3 nm size gold particles supported on titania give a photoreaction quantum yield close to 10%, depending on the ethanol to water ratio.

References

[1] Fujishima, A. and Honda, K. (1972) Electrochemical photolysis of water at a semiconductor electrode. *Nature*, **238**, 37.

[2] Kudo, A. and Miseki, Y. (2009) Heterogeneous photocatalyst materials for water splitting. *Chemical Society Reviews*, **38**, 253.

[3] Idriss, H. and Barteau, M.A. (2000) Active sites on oxides: from single crystals to catalysts. *Advances in Catalysis*, **45**, 261.

[4] Diebold, U. (2003) The surface science of titanium dioxide. *Surface Science Reports*, **48**, 53.

[5] Qiu, H., Idriss, H., Wang, Y., and Wöll, C. (2008) Carbon-carbon bond formation on model titanium oxide surfaces: identification of surface reaction intermediates by high-resolution electron energy loss spectroscopy. *Journal of Physical Chemistry C*, **112**, 9828.

[6] Jayaweera, P.M., Quah, E.L., and Idriss, H. (2007) Photoreaction of ethanol on TiO_2(110) single-crystal surface. *Journal of Physical Chemistry C*, **111**, 1764.

[7] Henderson, M.A., White, J.M., Uetsukab, H., and Onishi, H. (2006) Selectivity changes during organic photooxidation on TiO_2: Role of O_2 pressure and organic coverage. *Journal of Catalysis*, **238**, 153.

[8] Lyubinetsky, I., Yu, Z.Q., and Henderson, M.A. (2007) Direct observation of adsorption evolution and bonding configuration of TMAA on TiO_2(110). *Journal of Physical Chemistry C*, **111**, 4342.

[9] White, J.M., Szanyi, J., and Henderson, M.A. (2004) Thermal chemistry of trimethyl acetic acid on TiO_2(110). *Journal of Physical Chemistry B*, **108**, 3592.

[10] Henderson, M.A. (2005) Photooxidation of acetone on TiO_2(110): conversion to acetate via methyl radical ejection. *Journal of Physical Chemistry B*, **109**, 12062.

[11] Wilson, J.N. and Senanayake, S.D. (2004) Carbon coupling on titanium oxide at surface oxygen defects, H. Idriss. *Surface Science*, **L231**, 562.

[12] Ni, M., Leung, M.K.H., Leung, D.Y.C., and Sumathy, K. (2007) A review and recent developments in photocatalytic water-splitting using TiO_2 for hydrogen production. *Renewable and Sustainable Energy Reviews*, **11**, 401.

[13] Ruzycki, N., Herman, G.S., Boatner, L.A., and Diebold, U. (2003) Scanning tunneling microscopy study of the anatase (1 0 0) surface. *Surface Science*, **529**, L239.

[14] Tanner, R.E., Sasahara, A., Liang, Y., Altman, E.I., and Onishi, H. (2002) Formic acid adsorption on anatase TiO_2(001)–(1 × 4) thin films studied by NC-AFM and STM. *Journal of Physical Chemistry B*, **106**, 8211.

[15] Lazzeri, M. and Selloni, A. (2001) Stress-driven reconstruction of an oxide surface: the anatase TiO_2(001)-(1 × 4) surface. *Physical Review Letters*, **87**, 266105.

[16] Bonapasta, A.A. and Filippone, F. (2005) Photocatalytic reduction of oxygen molecules at the (100) TiO_2 anatase surface. *Surface Science*, **577**, 59.

[17] Czekaj, I., Piazzesi, G., Kröcher, O., and Wokaun, A. (2006) DFT modeling of the hydrolysis of isocyanic acid over the TiO_2 anatase (101) surface: Adsorption of HNCO species. *Surface Science*, **600**, 5158.

[18] Ramamoorthy, M. and Vanderbilt, A. (1994) First principles calculations of the energetic of stoichiometric TiO_2 surfaces. *Physical Review B-Condensed Matter*, **49**, 16721.

[19] McGill, P.R. and Idriss, H. (2008) Plane wave density functional theory computation of the molecular adsorption of ammonia on the (011) surface of titanium dioxide. *Langmuir*, **24**, 97.

[20] Gong, X.-Q., Khorshidi, N., Stierle, A., Vonk, V., Ellinger, C., Dosch, H., Cheng, H., Selloni, A., He, Y., Dulub, O., and Diebold, U. (2009) The 2 × 1 reconstruction of the rutile TiO_2(011) surface: A combined density functional theory, X-ray diffraction, and scanning tunneling microscopy study. *Surface Science*, **603**, 138.

[21] Firment, L.E. (1982) Thermal faceting of the rutile TiO_2(001) surface. *Surface Science*, **116**, 205.

[22] Kim, K.S. and Barteau, M.A. (1990) Structure and composition requirements for deoxygenation, dehydration, and ketonization reactions of carboxylic acids on TiO_2(001) single-crystal surfaces. *Journal of Catalysis*, **125**, 353.

[23] Idriss, H., Kim, K.S., and Barteau, M.A. (1991) Structure-activity and structure-selectivity relations for reactions of carboxylic acids on TiO_2(001) surfaces. *Studies in Surface Science and Catalysis*, **64**, 327.

[24] Senanayake, S.D. and Idriss, H. (2006) Photo-catalyses and the origin of life. The synthesis of nucleotides from formamide on TiO_2(001) single surfaces. *Proceedings of the National Academy of Sciences of the United States of America*, **103**, 1194.

[25] Ariga, H., Taniike, T., Morikawa, H., Tero, R., Kondoh, H., and Iwasawa, Y. (2008) Lattice-work structure of a TiO_2(001) surface studied by STM, core-level spectroscopies and DFT calculations. *Chemical Physics Letters*, **454**, 350.

[26] Valentin, C., Tilocca, A., Selloni, A., Beck, T.J., Klust, A., Batzill, M., Losovyj, Y., and Diebold, U. (2005) Adsorption of water on reconstructed rutile TiO_2(011)-(2 × 1): TiO double bonds and surface reactivity. *Journal of the American Chemical Society*, **127**, 9895.

[27] Torrelles, X., Cabailh, G., Lindsay, R., Bikondoa, O., Roy, J., Zegenhagen, J., Teobaldi, G., Hofer, W.A., and Thornton, G. (2009) Geometric structure of TiO_2(011) (2 × 1). *Physical Review Letters*, **101**, 185501.

[28] Fujishima, A., Zhang, X., and Tryk, D.A. (2008) TiO_2 photocatalysis and related surface phenomena. *Surface Science Reports*, **63**, 515 and reference therein.

[29] Al-Mazroai, L.S., Bowker, M., Davies, P., Dickinson, A., Greaves, J., James, D., and Millard, L.M. (2007) The photocatalytic reforming of methanol. *Catalysis Today*, **122**, 46.

[30] Yang, Y.Z., Chang, C.H., and Idriss, H. (2006) Photo-catalytic production of hydrogen from ethanol over M/TiO_2 catalysts (M = Pd, Pt or Rh). *Applied Catalysis B-Environmental*, **67**, 217.

[31] Zhang, Z., Bondarchuk, O., Kay, B.D., White, J.M., and Dohnàlek, Z. (2007) Direct visualization of 2-butanol adsorption and dissociation on TiO_2(110). *The Journal of Physical Chemistry C*, **111**, 3021.

[32] Bates, S.P., Kresse, G.K., and Gillan, M.J. (1998) The adsorption and dissociation of ROH molecules on TiO_2(110). *Surface Science*, **409**, 336.

[33] Reztova, T., Chang, C.-H., Koresh, J., and Idriss, H. (1999) Dark and photoreactions of ethanol and acetaldehyde over TiO_2/carbon molecular sieve fiber. *Journal of Catalysis*, **185**, 223.

[34] Yu, Z. and Chuang, S.S.C. (2007) In situ IR study of adsorbed species and photogenerated electrons during photocatalytic oxidation of ethanol on TiO_2. *Journal of Catalysis*, **246**, 118.

[35] Zehr, R.T. and Henderson, M.A. (2008) Acetaldehyde photochemistry on TiO_2(110). *Surface Science*, **602**, 2238.

[36] Wilson, J.N. and Idriss, H. (2002) Structure-sensitivity and photo-catalytic reactions of semiconductors. Effect of the last layer atomic arrangement. *Journal of the American Chemical Society*, **124**, 11284.

[37] Wilson, J.N. and Idriss, H. (2003) Effect of surface reconstruction of TiO_2(001) single crystal on the photoreaction of acetic acid. *Journal of Catalysis*, **214**, 46.

[38] Goodman, D.W. and Lai, X. (2000) Structure–reactivity correlations for oxide-supported metal catalysts: new perspectives from STM. *Journal of Molecular Catalysis A-Chemical*, **162**, 33.
[39] Haruta, M. (1997) Size- and support-dependency in the catalysis of gold. *Catalysis Today*, **36**, 153.
[40] Kamat, P.V. (2008) Quantum dot solar cells. semiconductor nanocrystals as light harvesters. *Journal of Physical Chemistry C*, **112**, 18737 and references therein.
[41] Mller, B.R., Majoni, S., Memming, R., and Meissner, D. (1997) Particle size and surface chemistry in photoelectrochemical reactions at semiconductor particles. *Journal of Physical Chemistry B*, **101**, 2501.

4

Fundamental Reactions on Rutile $TiO_2(110)$ Model Photocatalysts Studied by High-Resolution Scanning Tunneling Microscopy

Stefan Wendt, Ronnie T. Vang, and Flemming Besenbacher
Interdisciplinary Nanoscience Center (iNANO), Department of Physics and Astronomy, Ny Munkegade, Building 1521, Aarhus University, DK-8000 Aarhus C, Denmark, Email: fbe@inano.dk

4.1 Introduction

As discussed in Chapter 1, the holy grail of photocatalytic water splitting is the development of an efficient, robust and inexpensive catalyst that allows one to produce hydrogen in a sustainable way. However, so far the method of direct water splitting by solar light has not matured into a commercially viable technology, despite intensive research into new photocatalyst materials [1–5]. Probably, one of the main reasons for this missing success is a lack of atomic-scale insight into the fundamental processes of photocatalytic reactions, which has hampered the development of the ideal photocatalyst. Although numerous papers have been published on water-splitting photocatalysis, a detailed atomic-scale understanding of the basic principles and concepts governing the photocatalyzed chemical transformations have not yet been established.

"Real", industrial heterogeneous catalysts, including photocatalysts, are structurally rather complex materials, and photocatalysts typically consist of highly porous nanoparticles or films modified with bulk and/or surface dopants; the photocatalytic processes involved are often highly influenced by a wide range of external parameters (temperature, pH etc.). In photocatalysis, the reactants are temporarily bound to the solid surface of the catalyst that serves as light absorber and provides a reaction route with suitable energy barriers to enable the reaction to

On Solar Hydrogen & Nanotechnology Edited by Lionel Vayssieres

proceed at a high rate. A catalyzed chemical reaction is therefore the outcome of a complex interplay between a number of sequential elementary processes that involve, for example, absorption of light, excitation of electron–hole pairs, adsorption of reactants, diffusion to the active sites, reaction and a final desorption step of the products. Being inherently atomic-scale processes, fundamental understanding of catalytic reactions thus requires the ability to characterize surface structures at the atomic level. In particular the so-called catalytically active sites, for example, step edges, kinks, vacancies or single adatoms, are important for a detailed understanding, and thus studies of the catalyst and single reactants with high spatial and temporal resolution are required to obtain detailed fundamental insight into catalytic reactions.

The structural complexity of most photocatalysts limits the possibilities for detailed structural characterization of the catalysts. This is one of the reasons why in the past catalysts, in many cases, have been developed by trial-and-error methods often based on "research intuition". However, the need for better and more efficient catalysts has recently led to the development of a plethora of specialized experimental nanoscience techniques that provide detailed structural and chemical insight into complex catalysts. To gain such detailed fundamental insight into the elementary steps involved in catalytic reactions, many researches have adopted the so-called *surface-science approach*, which has proven extremely successful [6]. In the surface-science approach the considerations of reactions at surfaces are simplified by studying idealized model systems consisting of either flat single-crystal surfaces or well-defined nanoclusters at surfaces under clean and well-controlled, often ultrahigh vacuum (UHV), conditions [7]. Within this surface-science approach, an impressive variety of techniques have been developed and employed, which has resulted in a remarkable insight into a variety of fundamental processes on surfaces. The studies have, however, typically been based on diffraction or spectroscopy signals averaged over macroscopic areas of the surface, and generally do not provide direct information about individual atoms and particles.

The area of surface-science was, however, revolutionized by the development of scanning probe microscopes in the 1980s [8,9]. These scanning probe microscope (SPM) techniques are a family of direct, real-space, local probe techniques that are capable of exploring the atomic-scale realm of surfaces with atomic resolution. With the best and most stable SPM microscopes, it has become possible to use SPM as an outstanding tool for monitoring atomic and molecular dynamics on surfaces, the adsorption, diffusion and reaction of gas molecules, which are the basic steps involved in reactions on surfaces with direct relevance to heterogeneous catalysis. Furthermore, the ultimate resolution of the SPMs has led to unprecedented new atomic-scale insight into catalytic active sites, and these atomic-scale studies have emphasized the importance of the catalytic active sites such as edges, kinks, atom vacancies and other defect sites, which are often extremely difficult to detect when using more traditional averaging techniques [10]. The knowledge base hereby obtained has provided valuable information about a number of catalytic reactions, and several examples now exist where this basic insight has been used to design new high-surface-area catalysts [11]. One of the challenges today is to apply the surface-science approach also for the study of photocatalyzed reactions, in order to improve our understanding of the processes going on in a photoelectrochemical cell.

A photoelectrochemical cell utilizing TiO_2 for water splitting consists of a TiO_2 photoanode and a metal (often Pt) counterelectrode as the cathode immersed in a suitable electrolyte. At the TiO_2 photoanode, light is absorbed, which leads to the generation of electron–hole pairs. The holes oxidize water molecules resulting in the formation of molecular oxygen and H^+ ions. The electrons are subsequently transported through the electric circuit to the cathode, where they reduce the H^+ ions whereby molecular hydrogen, H_2, is formed. The reactions at both the photoanode and the cathode involve several elementary steps, and, for the reaction at the photoanode in a neutral medium, the following reaction mechanism has been suggested [2,5,12]:

$$
\begin{aligned}
(TiO_2) + h\nu &\rightarrow h^+ + e^- \\
h^+ + H_2O &\rightarrow OH + H^+_{(aq)} \\
OH + OH &\rightarrow H_2O_2 \\
OH + H_2O_2 &\rightarrow HO_2^{\bullet} + H_2O \\
HO_2^{\bullet} &\rightarrow O^{\bullet -}_{2(aq)} + H^+_{(aq)} \\
h^+ + O^{\bullet -}_{2(aq)} &\rightarrow O_2 \\
O^{\bullet -}_{2(aq)} + O^{\bullet -}_{2(aq)} &\rightarrow O_2^{\bullet 2-} + O_2(\text{at pH} < 6) \\
HO_2^{\bullet} + O_2^{\bullet 2-} &\rightarrow HO_2^- + O_2 \\
HO_2^{\bullet} + HO_2^{\bullet} &\rightarrow H_2O_2 + O_2(\text{at}\, pH < 3)
\end{aligned}
$$

From this set of intermediate reaction steps it is obvious that the surface chemistry of TiO_2 is extremely important for the understanding of its photocatalytic properties, and, in particular, the fundamental understanding of the chemistry of adsorbed water, oxygen and their intermediate species needs to be improved. By means of high-resolution, fast-scanning STM, the elementary steps in reactions involving water and oxygen on TiO_2 single crystals can indeed be monitored with atomic resolution in real time, thus unfolding its complex surface chemistry.

For fundamental studies of photocatalysts, including water-splitting catalysts, the most appropriate model system is a single-crystal surface of the active material, and, in particular, rutile $TiO_2(110)$ has been the subject of many surface-science studies [4,13–17]; in our group this surface has also been intensely studied [18–30]. In this chapter, we will not provide a comprehensive review of surface-science studies on TiO_2 single crystals, but rather the aim is to illustrate what potential exists in the characterization of TiO_2 model photocatalysts at the atomic scale. In this respect we will discuss some examples from our own recent work. After a brief summary of the peculiar structural features of the $TiO_2(110)$–(1×1) surface, we will first address fundamental reactions taking place on $TiO_2(110)$ involving either water or oxygen, before we discuss more complex reactions wherein both these molecules play a role and where a series of intermediate species can be identified. In particular, for those reactions where oxygen is involved, it becomes more and more evident that the surface O vacancies alone are not sufficient to explain the observed phenomena and the identified reaction pathways. In fact, for reactions where electronegative species are involved, the reduction of the bulk needs to be taken into account in order to reach satisfying agreement with density functional theory (DFT) calculations. Finally, we will present recent STM results addressing the nucleation of gold nanoclusters, since finely dispersed gold supported on metal oxides has shown very interesting catalytic and photocatalytic properties.

4.2 Geometric Structure and Defects of the Rutile TiO_2 (110) Surface

Among numerous investigations addressing TiO_2-based materials, many studies have been carried out using TiO_2 single crystals under UHV conditions [4,13–17,20,31], applying the

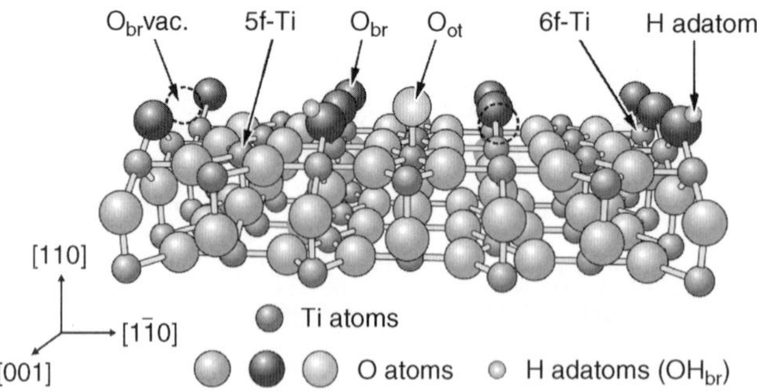

Figure 4.1 Ball-and-stick model of the $TiO_2(110)$–(1 × 1) surface with some of its known point defects. Large gray balls represent O atoms and medium-sized red balls sixfold coordinated Ti (6f-Ti) and fivefold coordinated surface Ti atoms (5f-Ti). Small light gray balls represent H adatoms. The bridge-bonded O species (O_{br}), single oxygen vacancies (O_{br} vac.), and the on-top bonded O species (O_{ot}) are also indicated. See also color-plate section.

surface-science approach described above. For model studies of this kind the rutile $TiO_2(110)$ surface has often been preferred (Figure 4.1). Today this anisotropic surface is still by far the most frequently studied model system, since it is the thermodynamically most stable surface of rutile, and $TiO_2(110)$ crystals of high quality can be purchased from several suppliers. The $TiO_2(110)$–(1 × 1) surface consists of alternating rows of fivefold coordinated Ti (5f-Ti) atoms (the Ti troughs) and protruding, twofold coordinated O_{br} atoms, as depicted in Figure 4.1. Like bulk Ti atoms in stoichiometric rutile TiO_2, the Ti atoms underneath the O_{br} atoms are sixfold coordinated. Ion-sputtering and vacuum-annealing of $TiO_2(110)$ crystals result in reduced n-type semiconductors, with a bulk conductivity that allows for the use of STM and other electron spectroscopy techniques.

Figure 4.2 shows a large-scale STM image obtained on a clean, reduced $TiO_2(110)$ crystal [r-$TiO_2(110)$] that was finally annealed *in vacuo* at a temperature as high as ~1000 K. Generally, STM images reflect the local density of states at the Fermi level, projected to the tip apex position, and thus the STM contrast results from a convolution of the geometrical *and* the electronic structure of the surface and *not* solely from the geometrical structure [32–34]. Today it is generally accepted that the STM images acquired on $TiO_2(110)$ surfaces are dominated by electronic effects, that is, the bright rows correspond to the Ti troughs, and geometrically protruding O_{br} atoms appear as the dark troughs in the STM images [13,14,20,35]. As can be seen from the STM image shown in Figure 4.2, steps are preferentially running parallel to the [$1\bar{1}1$] and the [001] directions. Steps running parallel to the [$1\bar{1}1$] direction are very uniform and smooth, whereas steps running parallel to the [001] direction are often rough, as also observed in previous work [13]. The terrace widths are typically in a range between 100 and 200 Å. The alternating rows of 5f-Ti atoms and O_{br} atoms are clearly resolved, as can be seen from the zoom-in STM image that was acquired from the same crystal (see the inset of Figure 4.2). The predominant point defects on vacuum-annealed rutile $TiO_2(110)$–(1 × 1) surfaces – the O_{br} vacancies, that is, missing surface O atoms – are also immediately revealed in the STM image depicted as inset. The O_{br} vacancies appear as faint protrusions along the dark rows [13,14,20,36]. In Figure 4.2, the density of O_{br} vacancy was estimated to be ~10% ML, where one ML (monolayer) is defined as the density of the (1 × 1) units, $5.2 \times 10^{14}\,cm^{-2}$. In the STM images acquired on ion-sputtered and vacuum-annealed $TiO_2(110)$ crystals, we

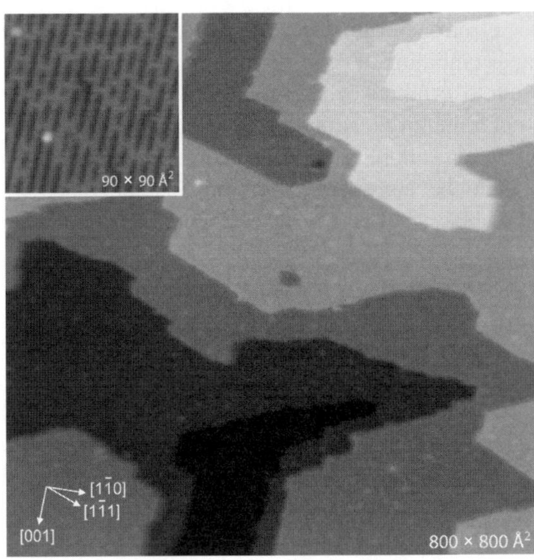

Figure 4.2 STM image of a clean $TiO_2(110)$–(1 × 1) surface obtained after applying cycles of sputtering and final annealing at ~1000 K in UHV. Step edges are predominantly running parallel to the [$1\bar{1}1$] and the [001] directions. The inset shows a zoom-in STM image obtained within the same experiment. The prevailing point defects on the surface – O_{br} vacancies – are seen as faint features connecting the bright rows, that is, the Ti troughs, that are running parallel to the [001] direction. Before starting the STM measurement, a short flash up to 600 K was applied to free the sample from residual water and hydroxyls. The STM images were acquired with a tunneling current $(I_t) \leq 0.1$ nA and a tunneling voltage $(V_t) = 1.2$ V, and with the sample at ~120 K.

exclusively observed single O_{br} vacancies, but no paired O_{br} vacancies or clusters of O_{br} vacancies. This can be explained by repulsion between the negative charges associated with the vacancies [37–41].

In most previous surface-science work, the O_{br} vacancies have been proposed as being the most important point defects that determine the chemistry on the $TiO_2(110)$ surface [4,13,14,36,39,41,42]. Within this picture it has long been anticipated that the defect state located within the ~3.1 eV wide bandgap of reduced TiO_2 is caused by the O_{br} vacancies [4,13,14,39,41–44]. Indeed, at first glance, the O_{br} vacancies appear to be very important, since, as will be discussed in the following, O_{br} vacancies serve as dissociation sites for water, oxygen and also for other molecules. Moreover, evidence has been provided that the O_{br} vacancies serve as nucleation sites for small gold clusters [19,22]. Nevertheless, it can be questioned whether O_{br} vacancies are relevant for "real" catalysis with high pressures of water and other gases. The fact alone that it is rather difficult to produce $TiO_2(110)$ surfaces with clean O_{br} vacancies [r-$TiO_2(110)$] even under UHV, that is, clean and well-controlled conditions [20,45,46] points to a view wherein surface O_{br} vacancies are of only minor relevance for "real" catalysts – simply because they may not persist under the high pressure conditions. From this point of view, the study of r-$TiO_2(110)$ surfaces with O_{br} vacancies may be mainly of academic interest.

It is thus of utmost importance to clarify what other structural features of reduced TiO_2 samples may cause the defect state that is located within the bandgap, since for most of the applications the excess electrons originating from this state play an important role. Recent studies from our group, combining several surface-science techniques in combination with DFT calculations, point to a model wherein surplus Ti – most likely in the form of Ti interstitials – are a structural entity leading, or, at least, contributing to a large extent, to the defect state [23]. This scheme implies that for a detailed understanding of photocatalytic reactions, the bulk has to be considered, not only as the place where the creation of electron–hole pairs takes place, but also as a "reservoir" for defect centers that provide the excess electrons.

4.3 Reactions of Water with Oxygen Vacancies

In the field of surface science, the reaction of water with the $TiO_2(110)$–(1×1) surface has become one of the most-studied systems, both experimentally [18,20,21,36,45,47–57] and theoretically [20,21,58–64]. In these studies the most frequently addressed question is regarding water dissociation, since this issue has high relevance with respect to the splitting of water for the production of hydrogen. Generally, it has long been assumed that defect sites on oxide surfaces play an important role for the dissociation of water. More specifically, it has been assumed that water dissociates in surface oxygen vacancies [47,48]. Addressing the water interaction with the $TiO_2(110)$–(1×1) surface, Kurtz *et al.* reported evidence, based on synchrotron radiation experiments, for a small portion of dissociatively adsorbed water and suggested water dissociation in O_{br} vacancies [49]. Similar conclusions have been drawn by Henderson [50,51], who used a combination of high-resolution electron energy loss measurements (HREELS) and temperature-programmed desorption (TPD), and also in a modulated molecular beam study by Brinkley *et al.* [52]. However, the portion of water dissociation was (and still is) somewhat disputed [53,54,58–60]. Furthermore, based on the utilized averaging, spectroscopic techniques applied in the studies by Kurtz *et al.* [49], Henderson *et al.* [50,51] and others [52–54], direct evidence for the proposed dissociation mechanism could not be provided.

The atomistic mechanism for water dissociation in O_{br} vacancies suggested previously by Kurtz *et al.* [49] implies the formation of H adatom pairs in the rows of O_{br} atoms (or OH_{br} pairs in an alternative notation), meaning that, for each water molecule that heals a vacancy, one of the protons is released to one of the O_{br} atoms next to the vacancy site. measurements with the fast-scanning Aarhus STM have made it possible to directly reveal the water-dissociation mechanism, and, in high-resolution STM images, features have indeed been detected that can be ascribed to such H adatom pairs in the rows of O_{br} atoms [20,21,36]. An example of a high-resolution STM image showing paired H adatoms is depicted in Figure 4.3. From this STM image, it is clear that various point-like defects do exist on the O_{br} rows after exposing the crystal to water. In order to distinguish between all the possible point-like defects in the O_{br} rows and those defects that are localized within the Ti troughs, we will, in the following, use the term "type-A defect" for defects in the O_{br} rows (and formally, defects in the Ti troughs are labeled as "type-B defect"). The predominant new features in the O_{br} rows, that is, type-A defects, appearing in the STM images after a low water exposure at room temperature (RT) – some of which are indicated by large squares in Figure 4.3 – are clearly brighter than the faint spots arising from O_{br} vacancies (indicated by small squares in Figure 4.3). Thus, these two different type-A defects can be easily distinguished from each other in the STM images.

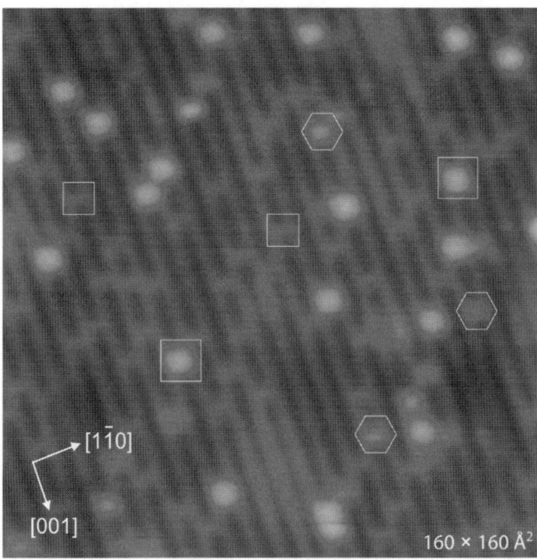

Figure 4.3 High-resolution STM image of the *r*-$TiO_2(110)$ surface with a number of paired H adatoms and a few single H adatoms obtained after water exposure at RT. Symbols indicate O_{br} vacancies (small squares), single H adatoms (hexagons) and pairs of H adatoms (large squares). The STM image has been acquired at a sample temperature of ~120 K.

However, in addition to the predominant new bright features on the O_{br} rows, new features have also appeared (indicated by hexagons in Figure 4.3) that are less bright than the predominant new features in the O_{br} rows but brighter than the faint spots arising from O_{br} vacancies. By superimposing lattice grids on the STM images, we have checked the positions of the various type-A defects in the [001] direction [20]. Aligning a grid at the O_{br} vacancies yields that the predominant new features in the O_{br} rows defects lie exactly between two intersection points of the lattice, whereas the medium bright type-A defects lie exactly on the intersection points. This result is consistent with a description wherein the brightest type-A defects are assigned to H adatom pairs and the medium bright defects to isolated H adatoms [in the literature often also termed as bridging hydroxyls (OH_{br})] [20].

That these assignments of the type-A defects are indeed correct has been proven by means of time-lapsed STM studies performed in our laboratory at temperatures below 200 K, in combination with DFT calculations [21]. In the work by Wendt *et al.* [21], first a freshly prepared *r*-$TiO_2(110)$ sample was quenched to ~110 K and then exposed to water corresponding to ~0.1 ML. Subsequently, the $TiO_2(110)$ sample was slowly warmed up, while recording STM movies. Snapshots from one of these movies "dissociation" (2.7 s per image) acquired at ~187 K are depicted in Figure 4.4 (The full-length movie can be found under *additional information* at www.phys.au.dk/spm). In addition to the O_{br} vacancies and H adatoms in the dark O_{br} rows, a protruding species is seen in Figure 4.4a on the bright Ti troughs. Its identity was revealed when tracking the dynamics of this species. Focusing on the Ti trough marked by a dashed line, it is evident that this species is adsorbed on 5f-Ti sites, and diffuses along the Ti trough (Fig. 4.4(a)–(c)). When this species reaches a 5f-Ti site next to an O_{br} vacancy

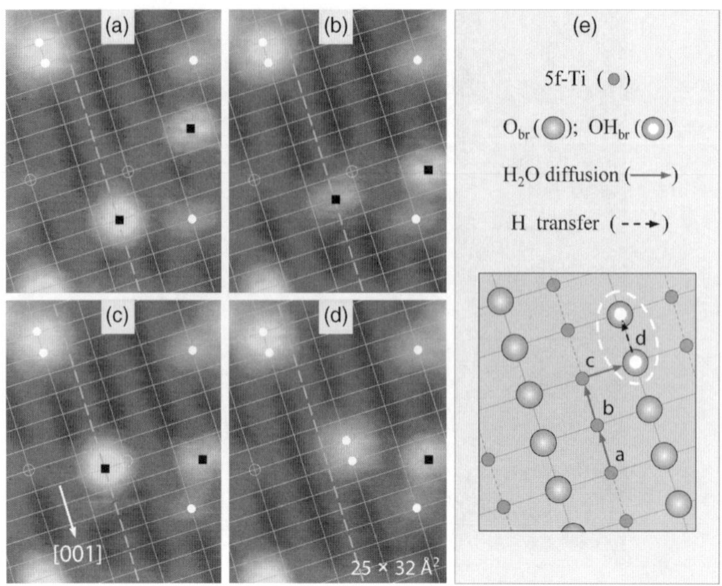

Figure 4.4 (a)–(d): Dissociation of monomeric water in an O_{br} vacancy on $TiO_2(110)$ at ~187 K as obtained in STM movie "dissociation" (see *additional information* at www.phys.au.dk/spm). Protrusions are labeled as follows: O_{br} vacancies (open white circles), H adatoms (filled white circles), monomeric water on 5f-Ti sites (filled black squares). (e) Schematic: Letters **a** to **d** mark oxygen positions of the water monomer. (Reprinted with permission from S. Wendt *et al.*, Formation and splitting of paired hydroxyl groups on reduced $TiO_2(110)$, *Physical Review Letters*, **96**, 066107, 2006. © 2006 American Physical Society.)

(Figure 4.4c), both the species on the Ti trough, as well as the vacancy in the O_{br} row disappeared and were replaced by a pair of adjacent H adatoms (Figure 4.4d). The reaction monitored in STM movie "dissociation" can be explained straightforwardly when ascribing the species in the Ti trough to an isolated water molecule that dissociates in the O_{br} vacancy (Figure 4.4e). This interpretation is fully consistent with the TPD data in the literature [50,51,53], since the species in the Ti troughs are absent after flashing the sample to 350 K. In addition, the water dissociation in O_{br} vacancies at ~180 K, is in line with vibrational spectroscopy data from Henderson *et al.* [51]. Moreover, DFT calculations strongly support this interpretation, because all the calculated barriers for the diffusion of water monomers, as well as for the proton transfers are low, in line with the low temperatures at which we observed the reactions in the STM movies [21]. Also, DFT calculations have predicted that water monomers are adsorbed on-top on 5f-Ti sites [20,21,55,58–62], which is also consistent with our STM observations.

4.4 Splitting of Paired H Adatoms and Other Reactions Observed on Partly Water Covered $TiO_2(110)$

As discussed in the previous section, three different type-A defects have been identified in the high-resolution STM images acquired on *r*-$TiO_2(110)$ surfaces: O_{br} vacancies, single H adatoms and paired H adatoms. Whereas paired H adatoms often can be found directly after water exposure to a clean r-$TiO_2(110)$ surface (sometimes the water background from the residual gas in the UHV chamber is already sufficient), isolated H adatoms are predominant for a longer waiting time after the exposure, in particular if the crystal was kept at temperatures high enough for water monomers to diffuse along the Ti troughs, that is, $\geq$170 K [20]. In fact, after a waiting time of 24 h we found only a very few paired H adatoms on the surface, but many single H adatoms [20]. Apparently, with time, the paired H adatoms somehow split apart from each other, leading to single H adatoms.

To unravel the atomic-scale mechanism of the H adatom pair splitting, again STM movies were recorded (see Figure 4.5 and movie "splitting" and *additional information* at www.phys.au.dk/spm). From STM movies obtained at cryogenic temperatures, we found that water plays an important role, since the pairs of H adatoms split into two single H adatoms when interacting with water monomers [21]. In the example presented in Figure 4.5, several changes can be discerned after a water molecule has diffused to a 5f-Ti site next to an H adatom pair (Figure 4.5a and b). Instead of the H adatom pair in the center of Figure 4.5a and b, two single H adatoms are observed the subsequent image (Figure 4.5c), one residing at the same place as in Figure 4.5b, but the other in the adjacent O_{br} row on the left. In addition, the water molecule also appears at a different adsorption site, namely shifted one Ti trough to the left. The changes seen

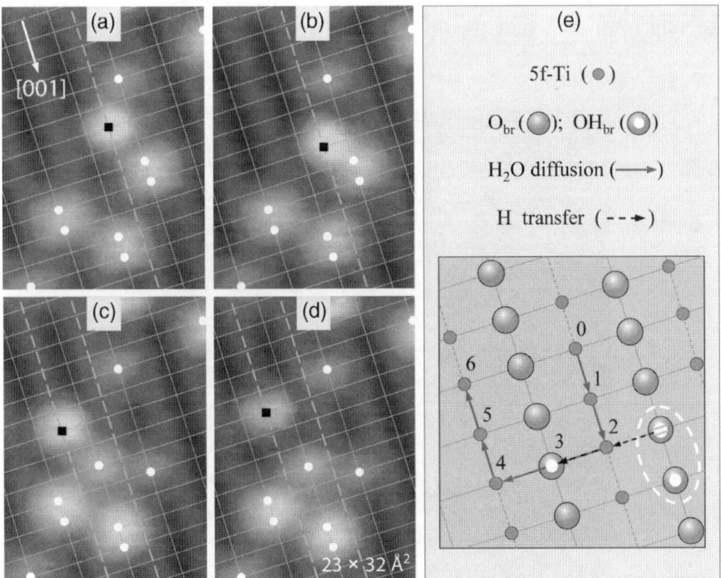

Figure 4.5 (a)–(d): Splitting of an H adatom (OH_{br}) pair mediated by a water molecule at ~187 K, as obtained in STM movie "splitting" (see *additional information* at www.phys.au.dk/spm). Protrusions are labeled as in Figure 4.4 (e) Schematic: Numerals 0 to 6 mark oxygen positions of the water for selected configurations. The hatched white disk indicates the H adatom of the pair that was transferred. (Reprinted with permission from S. Wendt *et al.*, Formation and splitting of paired hydroxyl groups on reduced $TiO_2(110)$, *Physical Review Letters*, **96**, 066107, 2006. © 2006 American Physical Society.)

in the STM movie "splitting" shows that water molecules mediate H transfer from one O_{br} row to another, and further that water molecules can easily diffuse over the H adatoms in the O_{br} rows. In Figure 4.5e, a schematic is depicted denoting the intermediate configurations corresponding to the STM movie "splitting".

In addition to the reactions seen in the STM movie "splitting" (Figure 4.5), it was observed in further STM movies that the diffusion of single H adatoms across the O_{br} rows is mediated by water molecules. However, the reverse reaction as compared to Figure 4.5, that is, the formation of H adatom pairs from two single H adatoms, was observed only very rarely and only for high coverage of H adatoms. Thus, it can be concluded that water dissociation in O_{br} vacancies is the major channel for the formation of H adatom pairs. Furthermore, it was revealed from numerous STM movies that water monomers can diffuse over the ~1.1 Å protruding O_{br} atoms at H adatoms, but not at regular O_{br} atoms. The latter finding was strongly supported by DFT calculations, since at an O_{br} atom capped by an H adatom, the diffusion barrier for the jump of the water molecule from one O_{br} row to another was calculated to be ~0.19 eV, whereas the barrier is as high as ~0.6 eV in the absence of the H adatom [21]. The explanation for this large difference in the diffusion barrier is the formation of a strong H bond between the oxygen of the water molecule and the capping H adatom on the O_{br} atom [21].

In Figure 4.6 we illustrate additional interesting results addressing the water dynamics on the $TiO_2(110)$ surface, which were also revealed by high-resolution STM studies at temperatures around 200 K and below. The STM image in Figure 4.6 was acquired after a water exposure corresponding to 9% ML at ~150 K to a clean r-$TiO_2(110)$ surface, that is, after a preparation similar to the one applied when the movies "dissociation" and "splitting" were recorded.

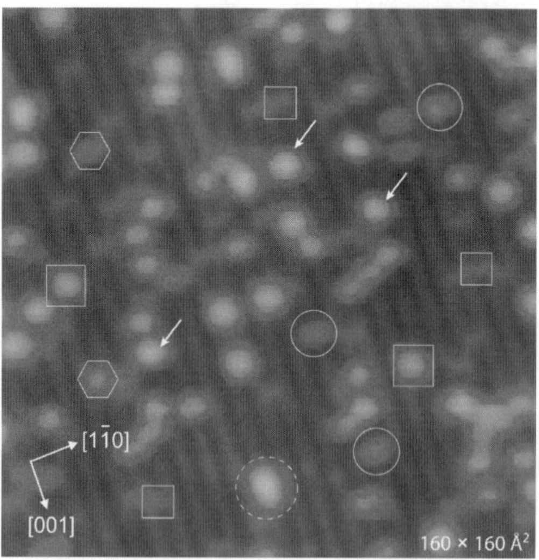

Figure 4.6 High-resolution STM image of the r-$TiO_2(110)$ surface obtained after water exposure at ~150 K. Type-A defects are indicated as follows: O_{br} vacancies (small squares), single H adatoms (hexagons) and pairs of H adatoms (large squares). Additional symbols indicate water monomers (small circles, solid line), water dimers (white arrows), and a water trimer (big circle, dashed line). The STM image was acquired at a sample temperature of ~120 K.

Within the O_{br} rows, again the three type-A defects discussed above can be recognized. Within the Ti troughs, however, there is not only evidence for monomeric water (small circles, solid line), but also for clustering of water species. Specifically, in Figure 4.6, protrusions are observed in the Ti troughs that are ascribed to water dimers (white arrows) and to water trimers (big circle, dashed line). By means of STM movies in combination with DFT calculations, we have recently shown that the formation of water dimers is energetically favorable on the $TiO_2(110)$ surface [25]. For the diffusion of water dimers, a roll-over mechanism was proposed that is characterized by an intermolecular H bond maintained during the diffusion. In addition, the O_{br} atoms of the rutile $TiO_2(110)$ surface were found to play an active role through H bond formation [25]. Because of these H bonds, the diffusion of water dimers along the Ti troughs is more facile than the diffusion of water monomers. In addition to dimers, we also observed the formation of larger water clusters; however, these clusters were found to be rather unstable [25].

4.5 O_2 Dissociation and the Role of Ti Interstitials

From the data presented above it is obvious that the O_{br} vacancies are very active sites for the dissociation of water on the *r*-$TiO_2(110)$ surface. Therefore, it may be logical to assume that the O_{br} vacancies are also active sites for the dissociation of O_2 molecules. As can be seen from the STM images depicted in Figure 4.7, the O_{br} vacancies are indeed active centers for the dissociation of O_2 molecules. The STM image depicted in Figure 4.7b was acquired after 5 L O_2 exposure [1 L (Langmuir) $= 1.33 \times 10^{-6}$ mbar s] at ~120 K to the *r*-$TiO_2(110)$ surface from

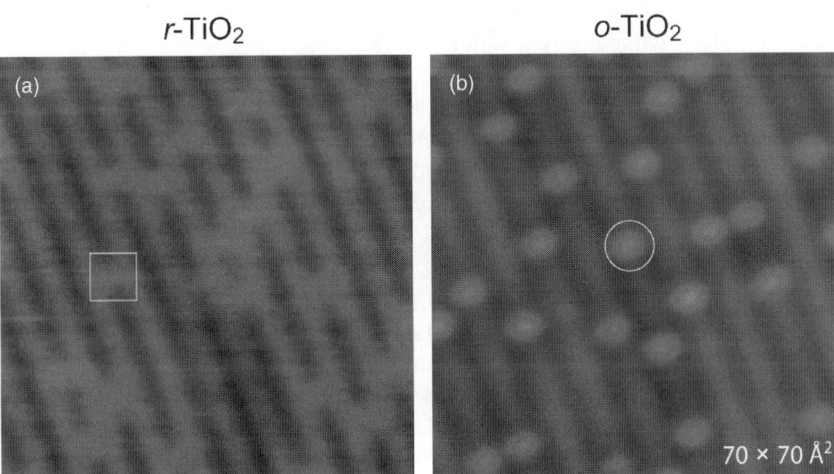

Figure 4.7 High-resolution STM images obtained on $TiO_2(110)$ before (a) and after 5 L O_2 exposure at 120 K (b). The *r*-$TiO_2(110)$ crystal used for this experiments was characterized by an O_{br} vacancy density of $5.5 \pm 0.2\%$ ML. Symbols indicate O_{br} vacancies (square) and O_{ot} adatoms (circle). (Reprinted from S. Wendt, R. Schaub, J. Matthiesen *et al.*, Oxygen vacancies on $TiO_2(110)$ and their interaction with H_2O and O_2: A combined high-resolution STM and DFT study, *Surface Science*, **598**, 226–245, 2005, Elsevier.)

which we acquired the STM image depicted in Figure 4.7a. Since only single O_{br} vacancies are observed on r-$TiO_2(110)$, one O atom of a dissociating O_2 molecule at an O_{br} vacancy heals the vacancy and one O atom is left over. Therefore, the new protrusions in the Ti troughs observed in the STM image depicted in Figure 4.7b arise from the "left over" single oxygen atoms, O_{ot} adatoms, located on-top on 5f-Ti sites [20,45,65]. Upon O_2 exposure (5 L in case of the discussed experiment) at ~120 K, the vast majority of the O_{ot} adatoms are adsorbed on 5f-Ti sites right next to the healed O_{br} vacancy sites, that is, it is very unlikely that the O_{ot} adatoms diffuse along the Ti trough away from the healed O_{br} vacancy sites, as was observed by Du *et al.* for O_2 dissociation in O_{br} vacancies at RT [65]. The assignment of the new protrusions in the Ti troughs to O_{ot} adatoms is strongly supported by the fact that the density of the new protrusions is about as high as the density of O_{br} vacancies before the O_2 exposure. The interaction of O_2 with the r-$TiO_2(110)$ surface is, however, more complex than the water–$TiO_2(110)$ interaction. Whereas all the O_{br} vacancies can be filled through the dissociation of water, regardless of the reduction state of the crystal, the filling of all O_{br} vacancies through O_2 dissociation is only possible for very low reduced r-$TiO_2(110)$ crystals that are characterized by O_{br} vacancy densities of ~5%ML or less. The STM images depicted in Figure 4.7 were acquired on a $TiO_2(110)$ crystal that was characterized by an O_{br} vacancy density of ~5.5%ML. The same type of experiment as shown in Figure 4.7 was also performed for more reduced $TiO_2(110)$ crystals that were characterized by O_{br} vacancy densities of ~9.5%ML [66]. In these experiments usually 2–3%ML unfilled O_{br} vacancies were observed, essentially independent of the amount of O_2 exposed. The explanation for this phenomenon is that the O_{ot} adatoms are electronegative species that can only be stabilized on the surface as long as electronic charge is available from the TiO_2 bulk. DFT calculations have shown that O_2 dissociation is energetically feasible for slabs that either contain an O vacancy [20,38] or a Ti interstitial [23], both of which are electron donors. However, O_2 adsorption and dissociation in O_{br} vacancies is energetically unfavorable for slabs without any additional defects, such as bulk oxygen vacancies or H adatoms [20,38,67].

In the following, we compare STM results from our own laboratory with previously published TPD data by M. A. Henderson *et al.* At first glance, the STM results presented in Figure 4.7 appear to be consistent with previous TPD experiments, from which the presence of O_{ot} adatoms on the $TiO_2(110)$ surface has been inferred [68,69]. For example, we found that the O_{ot} adatoms still reside on the $TiO_2(110)$ surface when water was co-adsorbed at cryogenic temperatures, followed by a flash to 375 K [20], as has been reported in Ref. [68]. However, we found the mechanism where one O of the O_2 molecules fills the O_{br} vacancy and the second O adsorbs on one of the 5f-Ti sites right next to the healed O_{br} vacancy to be operative at low temperatures (100–180 K) rather than at RT, as proposed by Epling *et al.* [68]. Furthermore, the fact that O_2 dissociation at O_{br} vacancy sites is feasible at ~120 K is at variance with the mechanisms proposed by Henderson *et al.* [69], where a 1 : 1 correlation between density of O_{br} vacancies before and O_{ot} adatoms after the O_2 exposure has been suggested only for temperatures above ~150 K. In fact, we find a ratio close to 1 : 1 only in the low-temperature range, say for temperatures below ~180 K, whereas the situation is different for higher adsorption temperatures [66]. It should be noted here that at temperatures lower than ~180 K, O_2 molecules can also be stabilized on reduced $TiO_2(110)$ [15,51,69–71], in addition to the O_{ot} adatoms, as evidenced by TPD measurements. Whereas the O_{ot} adatoms can be easily observed in high-resolution STM images (Figures 4.7–4.9), the O_2 molecules have not been detected so far with STM [66]. In this context it appears that the TPD and STM techniques are

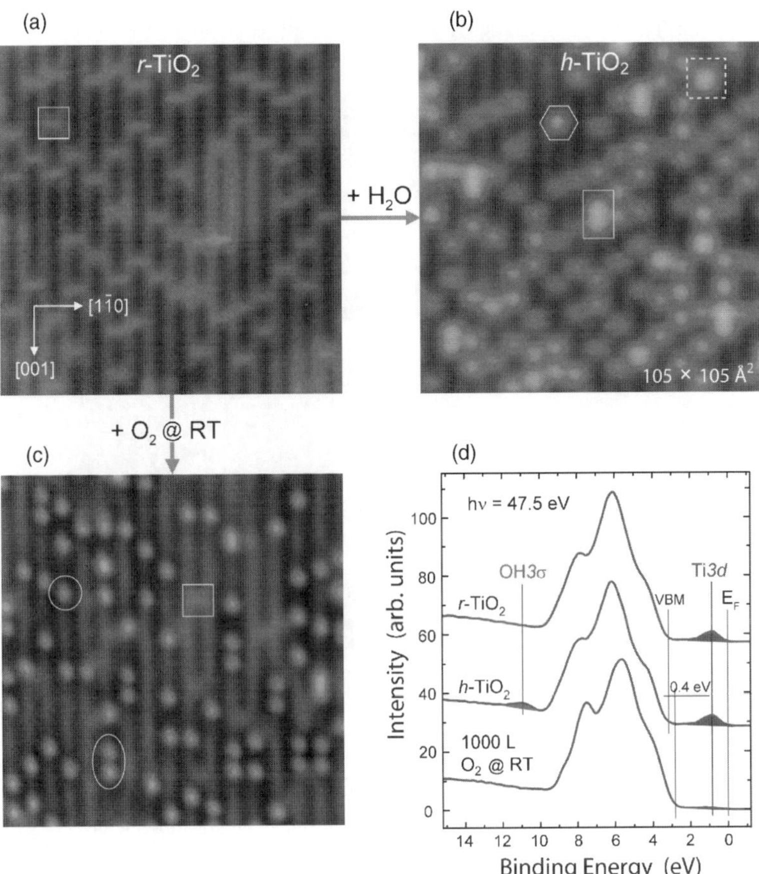

Figure 4.8 STM images of *r*-$TiO_2(110)$ (a), *h*-$TiO_2(110)$ (b) and an at RT O_2-saturated $TiO_2(110)$ surface (c) acquired with a sample in a reduction state corresponding to an O_{br} vacancy density of $11.4 \pm 0.3\%$ ML. Symbols indicate O_{br} vacancies (square), single H adatoms (hexagon), paired H adatoms (square, dotted line), next-nearest H adatoms (rectangle, solid line) in the O_{br} rows, and O_{ot} adatoms (circle), as well as pairs of next-nearest O_{ot} adatoms (ellipse) in the Ti troughs. Corresponding PES valence-band spectra are shown in (d). PES spectra and STM images were recorded at sample temperatures between 100 and 130 K. The position of the valence band maximum is indicated by the label "VBM" and the Ti3d defect state is observed within the ~3.2 eV wide band gap. (Reprinted from S. Wendt, P.T. Sprunger, E. Lira *et al.*, The role of interstitial sites in the Ti3d defect state in the band gap of titania, *Science*, **320**, 1755–1759, 2008, American Association for the Advancement of Science.)

complementary for the study of O_2 exposed *r*-$TiO_2(110)$ surfaces. Whereas high-resolution STM allows one to reliably estimate the density of O_{ot} adatoms, TPD is the technique of choice to determine the amount of adsorbed O_2 molecules on the $TiO_2(110)$ surface. It is evident from this discussion that the interaction of O_2 with *r*-$TiO_2(110)$ surface is much more complex than the water–$TiO_2(110)$ interaction.

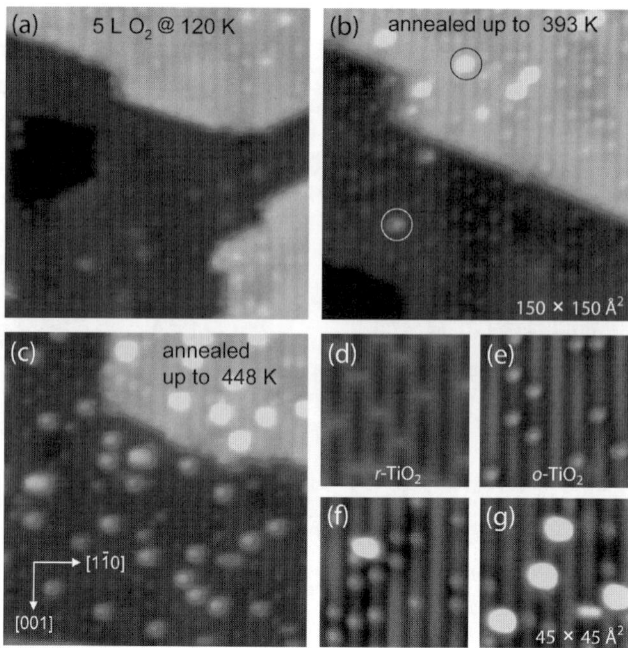

Figure 4.9 (a)–(c) STM images acquired on a $TiO_2(110)$ crystal characterized by an O_{br} vacancy density of ~4.5%ML. (a) Exposed to 5 L O_2 at 120 K, (b) subsequently annealed up to 393 K, and (c) further annealed up to 448 K. A circle in (b) indicates one of the newly formed TiO_x islands on the terraces with an STM height of ~0.22 nm. (d) Zoom-in STM images showing a clean *r*-$TiO_2(110)$ surface with O_{br} vacancies (d), and enlarged regions (e) to (g) of oxidized and flash-annealed $TiO_2(110)$ surfaces corresponding to the image shown in (a)–(c). (Reprinted from S. Wendt, P.T. Sprunger, E. Lira *et al.*, The role of interstitial sites in the Ti3d defect state in the band gap of titania, *Science*, **320**, 1755–1759, 2008, American Association for the Advancement of Science.)

To improve our understanding of the O_2–*r*-$TiO_2(110)$ interaction, we performed experiments combining high-resolution STM studies with photoelectron spectroscopy (PES) measurements to systematically explore the origin of the Ti3d-derived state in the bandgap (see Figure 4.8). Finding the origin of the Ti3d-derived state is not only important for a more correct description of the O_2–*r*-$TiO_2(110)$ interaction, but also for a better understanding of the adsorption of many other molecules, and the redox chemistry on titania in general. Starting from an *r*-$TiO_2(110)$ surface (Figure 4.8a), we first studied the effect of hydration by letting water dissociate in the O_{br} vacancies (Figure 4.8b). As discussed above, the brighter protrusions in the O_{br} rows appearing on the resulting hydrated $TiO_2(110)$ surface [*h*-$TiO_2(110)$], originate from capping H atoms. In the PES valence-band spectra corresponding to *r*- and *h*-$TiO_2(110)$ (Figure 4.8d), the only obvious difference is the OH3σ feature at a binding energy (BE) of ~10.8 eV, that is evident in the valence band after hydration [49]. However, the Ti3d-derived state in the bandgap [13,42,43] at ~0.85 eV below the Fermi level (E_F) is only minimally affected by hydration, in spite of the complete absence of O_{br} vacancies on *h*-$TiO_2(110)$. In an attempt to "heal" the O_{br} vacancies on *r*-$TiO_2(110)$ without producing capping H atoms on the

O_{br} rows, we studied the interaction of O_2 with clean *r*-$TiO_2(110)$ at RT. As can be seen in Figure 4.8c, the number of O_{br} vacancies decreased after the O_2 exposure, and a number of O_{ot} adatoms appeared simultaneously on the Ti troughs, some of which are next-nearest neighbors. Note, however, that O_2 exposure at RT does not lead to healing of all O_{br} vacancies, even in the case of saturation (Figure 4.8c and Ref. [66]). In the valence-band spectrum corresponding to the O_2 exposed $TiO_2(110)$ surface, the most striking feature is the strongly suppressed defect state at ~0.85 eV (Figure 4.8d). Both of these adsorption experiments, hydration of *r*-$TiO_2(110)$, as well as oxidation of *r*-$TiO_2(110)$ at RT, are at variance with the conventional model in which the Ti3d defect state has been fully ascribed to O_{br} vacancies [13,14,39,41–44]. Since additional experiments showed that the Ti3d defect state remains almost unchanged on surfaces without O_{br} vacancies and H adatoms [23], we have recently suggested that the Ti3d defect state essentially stems from Ti interstitials on octahedral sites in the near-surface region, rather than from O_{br} vacancies [23]. This model is consistent with previous experimental studies addressing the Ti3d defect state, including the pioneering work by Henrich *et al.* [43,44], electron paramagnetic resonance spectroscopy [72–74], bulk photonic [75] and transport [76] studies, recent photoelectron diffraction studies [77], and thin-film growth studies [78]. Moreover, recent DFT calculations confirmed that surplus Ti on interstitial sites indeed leads to the appearance of a defect state within the bandgap [23,79–81], whereas for slabs containing O_{br} vacancies, the majority of previous DFT calculations did not indicate a defect state within the bandgap [82]. It is noted that hybrid DFT methods [39] may be biased toward a gap state and the finding of an O_{br} vacancy related gap state in Ref. [39] may therefore be an artifact of the method. Of course, it could be argued that standard DFT is biased against a gap state because, with standard DFT, the bandgap is underestimated and the hybridization of localized states is overestimated [82]. However, since calculations using standard DFT methods [23,79,80] *and* those using hybrid DFT methods [81] do predict a gap state for slabs with Ti interstitials, good arguments exists that indeed the Ti interstitials lead to a gap state, in agreement with the experimental evidence discussed above and in Ref. [23].

That Ti charge donors indeed exist in the near-surface region of vacuum-annealed TiO_2 crystals is evident from the STM data depicted in Figure 4.9. After exposing the surface of a clean $TiO_2(110)$ crystal characterized by an O_{br} vacancy density of ~4.5%ML to 5 L O_2 at 120 K (Figure 4.9a), all the O_{br} vacancies disappeared and O_{ot} adatoms appeared in the Ti troughs, exactly as discussed above (cf. Figure 4.7). After annealing of this O_2-exposed $TiO_2(110)$ crystal to 393 K (Figure 4.9b), the density of O_{ot} adatoms increased by a factor of ~2, indicating that a nonvacancy-related O_2 dissociation channel exists that is energetically more activated than the one associated with the O_{br} vacancies. In addition, a number of small islands appeared on the terraces, one of which is indicated in Figure 4.9b by a circle. These newly formed islands are positioned in between two Ti troughs and show up in the STM images with a height of ~2.2 Å. Subsequent annealing to an even higher temperature (448 K) led to an increase in the density of the islands, and, in this case, islands with a height of ~3.2 Å were observed, in addition to the islands of ~2.2 Å height Figure 4.9c. In this context, it should be noted that the height of ~3.2 Å closely resembles the step height on $TiO_2(110)$ crystals, which is ~3.25 Å. Because of the step height expected for $TiO_2(110)$ and because exclusively Ti and O features are evident in the PES core-level spectra acquired in experiments following identical experimental procedures [23], the islands appearing on previously flat terraces have been ascribed to newly formed TiO_x structures with $x \sim 2$. This finding implies that, upon annealing,

Ti ions diffuse from the near-surface region to the top-most surface layer where reactions with O_{ot} adatoms and possibly also with O_2 molecules occurr.

The results depicted in Figure 4.9 show that the interaction of O_2 with r-$TiO_2(110)$ surfaces is more complex than previously anticipated in the experimental studies by Epling *et al.* [68], Henderson *et al.* [69] and Kimmel *et al.* [71]. In addition to O_{br} vacancies, surplus Ti ions on interstitial sites in the near-surface are crucial to describe reduced TiO_2 surfaces in a more complete and coherent picture [23]. This conclusion is in accordance with several previous experimental findings, such as the temperature-programmed static secondary-ion mass spectrometry data reported by Henderson [83], where the onset of Ti interstitial diffusion was found to occur at ~400 K, in good agreement with the STM data depicted in Figure 4.9. Furthermore, the STM results presented in Figure 4.9 are in good agreement with previous STM studies, where O_2-induced growth of new surface structures has been observed as well [84–87]. That Ti interstitials are indeed an important feature of reduced $TiO_2(110)$ samples is also supported by studies addressing the interaction of O_2 with the clean r-$TiO_2(110)$ surface at RT (Figure 4.8c and Figure 4.9f). In addition to the generally accepted O_2 dissociation channel associated with the O_{br} vacancies, a nonvacancy-related O_2 dissociation channel in the Ti troughs is evident from the high number of *paired* O_{ot} adatoms on next-nearest 5f-Ti sites appearing upon RT exposure [23,88]. Both, the second O_2 dissociation channel, as well as the PES valence-band spectra discussed above are in accord with the presence of Ti interstitials in the near-surface region. Within this model, electronegative adsorbates such as O_2 molecules and O_{ot} adatoms are stabilized on the $TiO_2(110)$ surfaces through charge transfer from Ti interstitials to the adsorbates.

4.6 Intermediate Steps of the Reaction Between O_2 and H Adatoms and the Role of Coadsorbed Water

As outlined in the introduction, photocatalyzed reactions are intimately connected to the chemistry of adsorbed water, oxygen and their intermediate species. Therefore, there is a huge interest within the field of photocatalysis and related fields in exploring the intermediate steps of the reactions leading from water to O_2 and vice versa. However, it is very challenging to unravel the intermediate steps by means of averaging spectroscopic techniques, since the concentrations of intermediates are usually quite low and the corresponding features in the spectra are often difficult to resolve. Aiming at an improved understanding of the chemistry of adsorbed water, oxygen and their intermediate species, high-resolution STM studies were recently undertaken by two research groups to follow the reaction of O_2 molecules with H adatoms on the O_{br} rows of the $TiO_2(110)$ surface [24,57,89]. Matthiesen *et al.* unraveled unprecedented details about this reaction by means of time-lapsed, high-resolution STM movies in conjunction with DFT calculations [24]. A series of intermediate H-transfer reaction steps was identified, that finally leads to the formation of water dimers [24].

In the reaction studies by Matthiesen *et al.*, the initial configuration was an h-$TiO_2(110)$ surface characterized by a density of H adatoms of ~14%ML (Figure 4.10a). In the STM image depicted in Figure 4.10a, protrusions are observed within the dark O_{br} rows arising from single H adatoms (hexagon, solid line), pairs of neighboring H adatoms (square) and H adatoms on next-nearest O_{br} atoms (rectangle, solid line). The paired H adatoms and likewise the H adatoms on next-nearest O_{br} atoms appear in the STM images brighter than single

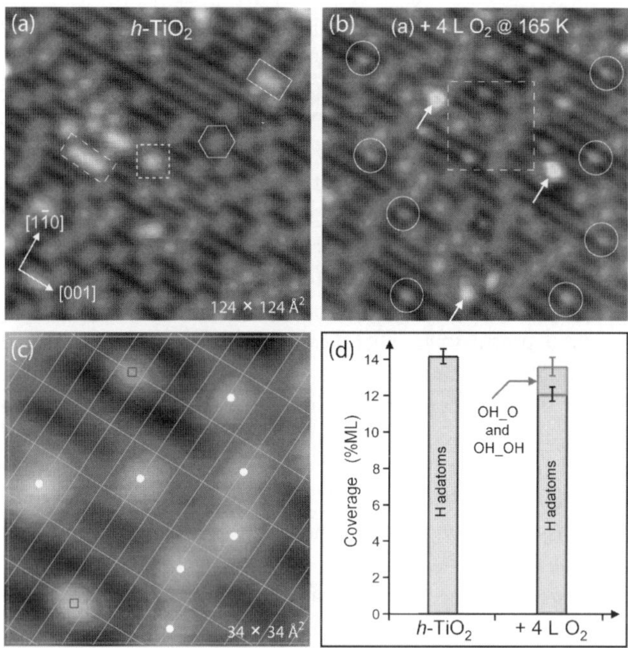

Figure 4.10 STM images of the h-$TiO_2(110)$ surface before (a) and after (b) 4 L O_2 exposure at ~165 K. Symbols in (a) indicate single H adatoms (hexagon), paired H adatoms (square, dotted line), next-nearest H adatoms (rectangle, solid line) and a chain of next-nearest H adatoms (rectangle, dashed line). The circles in (b) indicate the newly formed species in the Ti troughs and arrows indicate larger protrusions in the Ti troughs of STM heights, as typically observed for water species. The indicated square area in (b) (white, dashed line) is shown in (c), enlarged. A lattice grid was centered on top of 5f-Ti sites in (c) using single H adatoms (white small dots) for aligning in the [001] direction. Note that the z contrast in the STM image depicted in (c) is higher than in the STM images shown in (a) and (b). The STM images were acquired at ~110 K. (d) Bar graph of measured coverages in %ML of the obtained surface species before and after O_2 exposure. To extract the densities of surface species areas of ~2000 nm^2 were scanned and analyzed. (Reprinted with permission from J. Matthiesen *et al.*, Observation of all the intermediate steps of a chemical reaction on an oxide surface by scanning tunneling microscopy, *ACS Nano*, **3**, 517–526, 2009. © 2009 American Chemical Society.) See also color-plate section.

H adatoms [20,21]. The STM image depicted in Figure 4.10b was acquired after 4 L O_2 exposure at ~165 K to the h-$TiO_2(110)$ surface from which the image depicted in Figure 4.10a was obtained. Upon O_2 exposure, new species with an elongated shape in the [001] direction have formed in the Ti troughs, some of which are indicated by white circles. In addition, as a result of the reaction, a few very bright features (arrows) have appeared in the Ti troughs that possess STM heights typically observed for water species [21,40]. Figure 4.10c depicts a zoom in on the indicated area in Figure 4.10b, whereon a grid has been superimposed. Because the centers (small blue squares) of the most abundant new protrusions in the Ti troughs lie between two intersection points along the [001] direction, it can be concluded that the new species with an elongated shape occupy two 5f-Ti sites in the Ti troughs. Therefore, the new species with an elongated shape are most likely composed from two oxygen atoms. However, from the results presented in Figure 4.10, the number of H atoms in the new species cannot be determined. This information was gathered by means of STM movies recorded during the inlet of O_2 gas, an example of which will be discussed in the following. From the atomic-scale insights into the reactions, it was revealed that the new species with an elongated shape in the [001] direction arise from intermediate species with HO_2 and H_2O_2 stoichiometry [24]. This conclusion is consistent with an analysis of the full data set corresponding to the STM images shown in Figure 4.10a and Figure 4.10b, since, assigning the new protrusions to species with HO_2 and H_2O_2 stoichiometry best fulfils the H balance, that is, the total amounts of hydrogen before and after the O_2 exposure are equal in this case (Figure 4.10d). Thus, O_2 exposure at $\sim$165 K to an h-TiO_2(110) surface leads to at least two intermediate species in the Ti troughs, and both of them arise as elongated protrusions. Therefore, the different appearances of the two new protrusions in the Ti troughs (indicated in Figure 4.10c by small blue squares) is caused by their different stoichiometry. The bigger protrusion in Figure 4.10c (lower left corner) arises from a species which has H_2O_2 stoichiometry, whereas the smaller protrusion (upper middle part in the image) is associated with a species with HO_2 stoichiometry [24]. In the following we denote these intermediate species as OH_O and OH_OH, respectively, because, according to the DFT calculations discussed below, the O—O bonds are broken in both of these intermediate species.

To identify the new species in the Ti troughs more directly and to unravel their further fate an STM movie was recorded on the h-TiO_2(110) surface at $\sim$190 K during the inlet of O_2 gas at a background pressure at $\sim 10^{-8}$ Torr and in the presence of co-adsorbed water species (see Figure 4.11 and STM movie "O2_h-TiO2_reaction" that can be accessed at www.phys.au.dk/spm/movies/O2_h-TiO2_reaction.gif and as Supporting Material from Ref. [24]). Water corresponding to $\sim$6% ML was dosed directly after hydration of the TiO_2(110) surface to facilitate the diffusion of H adatoms (see above and Ref. [21]). Without the co-adsorbed water, the reaction may stop after the formation of the OH_O and OH_OH species, since thermally activated diffusion of the H adatoms is not feasible at $\sim$190 K [21,90]. In Figure 4.11, water species in the form of water dimers are denoted using the labels A_n, B_n, C_n and D, respectively, where n indicates the estimated number of containing H atoms, that is, A_n with $n = 4$ denotes a stoichiometric water dimer. In Figure 4.11a, the formation of an OH_O species is observed, since this protrusion appeared in the Ti trough concomitant with the disappearance of one H adatom. In the STM movie, three additional H-uptake reactions can be identified: (i) OH_OH formation (Figure 4.11b), (ii) OH_OH_2 formation (Figure 4.11c) and finally (iii) the creation of a water dimer, $[H_2O]_2$ (Figure 4.11d). That water dimers are produced in the reaction was confirmed by comparing the number of reacted O_2 molecules with the number of reacted H adatoms in several STM movies [24].

It is possible from the high-resolution STM images to identify the individual reaction intermediates upon H uptake, partly by following the transfer of H adatoms from the O_{br} rows and from other adsorbates to new species in the Ti troughs, and partly from the apparent STM heights and the diffusivity of the intermediate species. It is easy to differentiate in the STM movies when OH_O changes to OH_OH and when OH_OH changes to OH_OH_2, respectively, solely based on the apparent heights. However, solely from the measured apparent STM heights it is very difficult to distinguish between OH_OH_2 and $[H_2O]_2$ species [24]. For the latter transition, it appeared that the diffusion rate of the intermediates along the Ti trough differs markedly depending on how many H atoms the intermediate contains. More specifically, it was found that the mobility of the OH_OH_2 species is a factor of $\sim$30 lower than that of the water

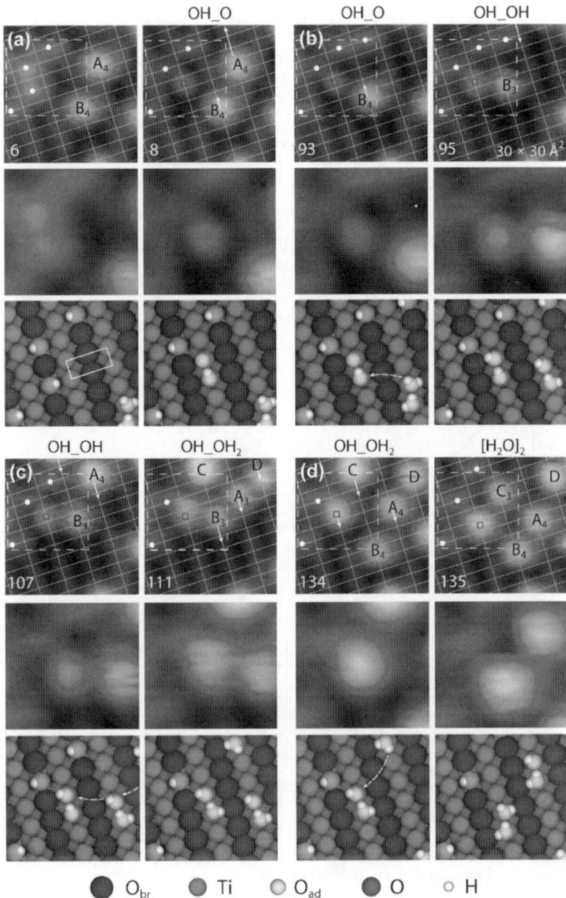

Figure 4.11 Snapshots extracted from an STM movie "O_2_h-TiO_2_reaction" recorded at ~190 K on an *h*-$TiO_2(110)$ surface with a water coverage of ~6 %ML and in an O_2 background of 1×10^{-8} Torr. STM images were acquired with a rate of 2.76 s per frame. OH_O formation is shown in (a), and the three following H uptake reactions are identified in (b), OH_OH formation, (c), OH_OH_2 formation, and (d), $[H_2O]_2$ formation, respectively. Numbers at the lower left edges indicate the appearance of the selected STM images within the movie. White small dots indicate H adatoms, and water species in the form of dimers are denoted using the labels A_n, B_n, C_n and D, respectively, where n indicates the estimated number of containing H atoms, that is, A_n with $n = 4$ denotes a stoichiometric water dimer. In (c) and (d) the index n was omitted for some water dimers. The lattice grid is cantered on 5f-Ti sites. White arrows along the Ti troughs indicate the diffusion directions of the adsorbates. Indicated areas of $18 \times 16.5\,Å^2$ size in the STM images in the first row (white dashed rectangles) are shown enlarged in the second row and explained in the third row using ball models (top views). The bent arrows indicate from which side in the STM images the H adatoms were provided. (Reprinted with permission from J. Matthiesen *et al.*, Observation of all the intermediate steps of a chemical reaction on an oxide surface by scanning tunneling microscopy, *ACS Nano*, **3**, 517–526, 2009. © 2009 American Chemical Society.) See also color-plate section.

dimer species [24]. Thus, in spite of their similar STM heights OH_OH_2 and $[H_2O]_2$ species can be distinguished in STM movies because of their different diffusivity along the Ti troughs.

The reaction pathway revealed from the STM movie has been corroborated by first-principles DFT calculations [24]. In order to take the bulk reduction of the samples into account a crystallographic shear-plane (CSP) [91,92] parallel to the (110) surface was introduced by removing every second O atom between the fourth and fifth layer of Ti atoms and by shifting the upper four tri-layers in the $\frac{1}{2}[0\bar{1}1]$ direction. In Figure 4.12a, the calculated energetics is depicted, whereas the adsorption configurations of all reaction intermediates are shown in Figure 4.12b–h. Starting from the r-TiO_2(110) surface ($E = 0$) with two O_{br} vacancies

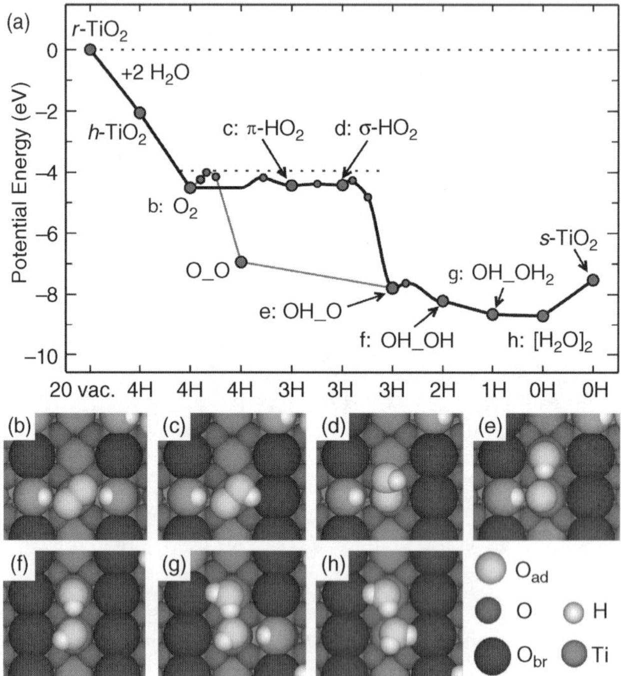

Figure 4.12 (a) DFT-based potential energy diagram. Big red dots correspond to calculated local potential energy minima, whereas small red dots indicate values deduced using the climbing NEB procedure. The initial configuration is a reduced TiO_2(110) surface [r-TiO_2(110)] with two O_{br} vacancies and two H_2O molecules in the gas phase. The first step is dissociation of the H_2O molecules in O_{br} vacancies, thereby forming four H adatoms [h-TiO_2(110)] (the number of O_{br} vacancies or H adatoms on the TiO_2(110) surface are denoted on the bottom axis). The second step is the adsorption of an O_2 molecule onto h-TiO_2(110). The pathway in which O_2 becomes HO_2 before dissociation into OH_O is shown as a thick black line, whereas the thin red line accounts for a pathway in which O_2 dissociates ($E_a = 0.51$ eV) before the reduction. The former pathway is kinetically preferred. Desorption of the $[H_2O]_2$ leads to a clean TiO_2(110) surface, s-TiO_2(110), with no point defects in the uppermost surface layer. (b)–(h) Ball models (top views) of selected reaction intermediates. O_{ad} (yellow balls) indicate the oxygen atoms of the adsorbates. (Reprinted with permission from J. Matthiesen *et al.*, Observation of all the intermediate steps of a chemical reaction on an oxide surface by scanning tunneling microscopy, *ACS Nano*, **3**, 517–526, 2009. © 2009 American Chemical Society.) See also color-plate section.

and two H_2O molecules in the gas phase, energy is released upon water dissociation in O_{br} vacancies, leading to two H adatoms instead of one O_{br} vacancy. The potential energy curve continues to go "downhill" upon subsequent O_2 adsorption, Figure 4.12b, OH_O and OH_OH formation, and further, all the way to the final reaction product – a water dimer, $[H_2O]_2$ (Figure 4.12h), consistent with the STM results and previously published spectroscopic evidence of water formation [51]. However, the energetics presented in Figure 4.12 are at variance with previous theoretical work addressing the O_2 adsorption on the *h*-$TiO_2(110)$ surface [67,93]. This discrepancy can be traced back primarily to the fact that the authors of Ref. [67] and Ref. [93] assumed a stoichiometric TiO_2 bulk without extra electrons available.

Concerning the activation energy barriers, E_a, it can be seen from Figure 4.12a that the transfer of the first H adatom from a position on an O_{br} atom to O_2 is hindered by only $E_a = 0.34$ eV. The resulting π-bonded HO_2 (hydroperoxyl) species (Figure 4.12c) instantly ($E_a = 0.05$ eV) transforms into a σ-bonded HO_2 species (Figure 4.12d) that dissociates very easily ($E_a = 0.15$ eV). Due to the low barriers for the latter two reaction steps, any HO_2 species is very short lived, even at the low temperatures during the STM measurements, and thus these DFT calculations strongly suggest that all reaction intermediates observed in the STM images are species in which the O–O bond has been broken [24]. In light of the calculations discussed in Figure 4.12, the imaging of stable, adsorbed HO_2 species by STM at RT, as reported by Du *et al.* [57], can be debated. Again, the discrepancy is probably related to the questionable assumption of a stoichiometric bulk without extra electrons available, as in Ref. [57].

From the interplay of STM movies and DFT calculations, where the bulk reduction was taken into account, we revealed the following reaction pathway:

$$\begin{array}{ll} O_2 + H \rightarrow \overset{(DFT)}{HO_2} \rightarrow \overset{(STM)}{OH\_O} & (O-O \text{ bond broken}) \\ OH_O + H \rightarrow OH\_OH & (\text{the H can be supplied from } [H_2O]_2) \\ OH\_OH + H \rightarrow OH\_OH_2 & (\text{H transfer mediated by } [H_2O]_2) \\ OH\_OH_2 + H \rightarrow [H_2O]_2 & (\text{H transfer mediated by } [H_2O]_2) \end{array}$$

The discussed results of the O_2 + H reaction on $TiO_2(110)$ are interesting for several reasons. First, the observed reaction mechanism has presumably universal character for TiO_2 surfaces, that is, that the reaction proceeds similarly on other faces of rutile and maybe also on other reducible transition metal oxides. Comparing the available spectroscopic data for $TiO_2(110)$ with those acquired on the (110) face of RuO_2 [94,95] – an oxide surface that is isomorphous to the $TiO_2(110)$ surface – some striking similarities are revealed. On both these surfaces the transfer of H adatoms from O_{br} sites to oxygen species in the troughs is facile, and in both cases the formation of water has been observed. Second, the role of co-adsorbed water is remarkable with a view to the promoting role of moisture for catalytic reactions on supported gold clusters [96,97]. Thus, co-adsorbed water may be crucial for the activation of molecular oxygen, which is considered as the most important reaction step on supported Au nanocatalysts [67,97], because traces amounts of water facilitate the diffusion of hydrogen to adsorbed O_2. Finally, with a view to photocatalysis, it is highly interesting to unravel how UV light illumination influences the O_2 + H reaction and to figure out under what circumstances the direction of the reaction can be reversed.

4.7 Bonding of Gold Nanoparticles on $TiO_2(110)$ in Different Oxidation States

Transition-metal nanoparticles are interesting from the perspective of both fundamental studies and applications, and in recent years specifically gold (Au) nanoparticles have attracted great attention. The main reason for this is the surprising catalytic activity of small Au nanoparticles dispersed on oxide supports for numerous reactions [97–102], but other potential uses, such as in photocatalysis [3,103–110] and solar energy conversion [111,112] have also triggered intense research activity into Au nanoparticles on TiO_2 supports.

For applications in catalysis, it is generally accepted that the size of the Au particles significantly affects the activity, and a high catalytic activity exists only on nm sized Au particles [113–118]. Furthermore, indications exist that there is a "support effect" in addition to the "size effect" [22,97,119–121], that is, the choice of the support material (normally oxides) also influences the catalysis on gold nanoclusters. However, the mechanism(s) of catalyzed reactions on dispersed Au catalysts and the nature of the active Au species still remain(s) open for dispute [22,97,122–124]. One suggestion is that the low-coordinated Au atoms on the perimeter of the Au nanoclusters are especially active in O_2 activation [101,125], whereas in the case of CO oxidation, the CO molecules preferentially are adsorbed on Au atoms at corners and on edges of the nanoclusters [126]. However, other suggestions to explain the distinct catalytic properties, such as the special electronic structure of small oxide-supported Au clusters, metal-support charge transfer and the specific shape [127] of the Au particles have been suggested in the literature in the past.

In photocatalysis the situation is even more complex than in ordinary heterogeneous catalysis without light-induced enhancement of the reactivity, since the oxide material is often a high-surface-area mesoporous network, or is itself composed from oxide nanoparticles, a fact which also influences the efficiency of the photocatalytic system [2,128]. Generally, enhanced photocatalytic activity through the addition of transition metals has been traced back to a reduced rate of electron–hole recombination, since the metal particles will act as electron traps, whereas large metal nanoparticles may act as centers of electron–hole recombination [2]. However, other factors also play a role, such as charge equilibration between photoirradiated TiO_2 and Au nanoparticles [3,105] and intense visible-light absorption by Au [109] through surface plasmon resonance [129,130] that enhance the photocatalytic activity. Moreover, a so-called antenna mechanism [131,132] has recently been proposed wherein the three-dimensional mesoporous TiO_2 network has been suggested to act as an antenna system, transferring the initially generated electrons from the location of light absorption to a suitable interface with the noble metal catalyst and subsequently, the electrons are transferred to the the metal nanoparticle where the actual electron-transfer reaction takes place [131,132]. Nevertheless, the mechanism(s) of charge transport in metal/TiO_2 nanostructures still remain(s) an issue that needs to be further explored.

Also, with respect to the applications of the Au/TiO_2 system in (photo)catalysis, the stable adhesion of the Au nanoclusters on the TiO_2 support is of fundamental importance. Atomic-scale information on this issue can best be obtained by using the surface-science approach, and small gold clusters on the rutile $TiO_2(110)$ surface have indeed become one of the most-studied model systems [97,114,123,133–139]; we recently have studied this system in great detail as well [22]. Whereas in most previous studies step edges have been reported to be nucleation centers [19,114,135,136], Matthey *et al.* [22] recently reported that point defect sites are crucial for strong Au cluster adhesion, which allows the stabilization of Au nanoclusters on the terraces.

In the study by Matthey *et al.* [22], three well-defined $TiO_2(110)$ surfaces, *r*-$TiO_2(110)$ with O_{br} vacancies, *h*-$TiO_2(110)$ with H adatoms and *o*-$TiO_2(110)$ with O_{ot} adatoms, were exposed to Au (3% ML) at RT, and quite different Au cluster morphologies were obtained. In case of the *r*-$TiO_2(110)$ surface, numerous rather small Au clusters (mainly gold monomers) are distributed homogeneously on the terraces (Figure 4.13a and d). In contrast, on the *h*-$TiO_2(110)$ surface, only fairly large Au clusters (containing up to about 20 gold atoms), that preferentially decorate the step edges of the substrate, were observed (Figure 4.13b and e), whereas on the *o*-$TiO_2(110)$ surface, Au clusters were again found to nucleate homogeneously on the terraces (Figure 4.13c and f). From the results depicted in Figure 4.13 it was concluded that the

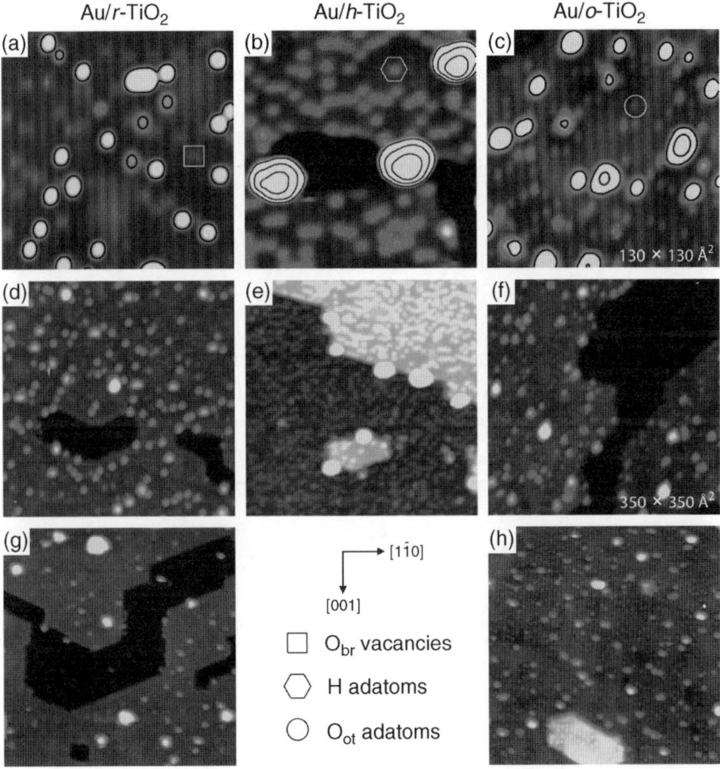

Figure 4.13 (a)–(f) STM images of $TiO_2(110)$ surfaces in different oxidation states, *r*-TiO_2, *h*-TiO_2 and *r*-TiO_2, respectively, after 3% ML Au exposure at RT [1 ML is defined as 1.387×10^{15} atoms per cm^2 corresponding to Au(111)]. In (a)–(c) the heights of the Au nanoclusters are given by contour lines at 1.2, 3.2 and 5.2 Å above the terrace. (g) STM image of the Au/*r*-$TiO_2(110)$ surface (350×350 Å^2) prepared by 3% ML Au deposition at RT followed by heating up to 68 °C. (h) Like (g), but for the Au/*o*-$TiO_2(110)$ surface. The STM images were acquired at ~110 K. (Reprinted from D. Matthey, J.G. Wang, S. Wendt *et al.*, Enhanced bonding of gold nanoparticles on oxidized $TiO_2(110)$, *Science*, **315**, 1692–1696, 2007, American Association for the Advancement of Science.)

interaction between Au clusters and the *h*-TiO_2(110) surface is weaker than that on the *r*- and the *o*-TiO_2(110) surfaces, suggesting that indeed the point defects (O_{br} vacancies and O_{ot} adatoms) are required to stabilize small Au clusters on the terraces.

To compare the strength of the gold attachment on *r*- and *o*-TiO_2(110) surfaces, Matthey *et al.* explored the effect of heating on the Au cluster morphologies on these two TiO_2 surfaces [22]. Starting with *r*- and *o*-TiO_2(110) surfaces, as shown in Figure 4.13d and f, respectively, the samples were heated to 68 °C. Whereas Au sintering was clearly evident for Au/*r*-TiO_2(110), no effect was observed for the Au/*o*-TiO_2(110) system (Figure 4.13g and h). These results clearly reveal that the Au clusters bind more strongly to the *o*-TiO_2(110) surface than on the *r*-TiO_2(110) surface. This trend was found irrespective of the precise size of the Au nanoclusters.

To rationalize the different adhesion strength of the Au clusters on the three TiO_2(110) surfaces considered, the experimental results were compared with DFT calculations [22]. In the calculations, also the stoichiometric TiO_2(110) [*s*-TiO_2(110)] was considered, and, as depicted in Figure 4.14, the adhesion of small Au_n clusters ($1 \leq n \leq 4$) is very low on *s*- and *h*-TiO_2(110) surfaces. For these Au_n clusters, the adhesion strength increases on *r*-TiO_2(110), because covalent bonds between Ti and Au can form more readily when not all of the Ti atoms are in their fully oxidized state. However, an adhesion mechanism involving cationic Au is triggered when the surface is oxidized [*o*-TiO_2(110)] [22]. When the TiO_2 support is oxidized, the Au nanoclusters are not only stabilized via covalent bonds, but also through ionic bonding. As a consequence, the Au_n–support adhesion strength is larger on the *o*-TiO_2(110) surfaces compared to the situation on the *r*-TiO_2(110) surface. However, in the calculations by Matthey *et al.* (Figure 4.14) the bulk reduction of the TiO_2(110) crystal was not considered. Therefore, we anticipate that the difference in Au_n–support adhesion strength between *o*-TiO_2(110) and the other three TiO_2(110) surfaces is somewhat overestimated. More recent calculations by Madsen and Hammer have revealed that the introduction of a Ti interstitial in a TiO_2(110) slab without other defects leads to a stabilization of small Au clusters [80].

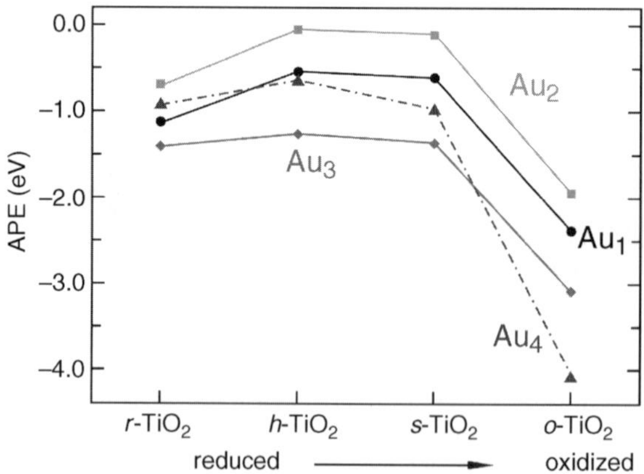

Figure 4.14 Plot of the calculated adhesion potential energies (APEs) for the most stable small Au_n clusters ($n = 1$–4) as a function of the "TiO_2(110) oxidation state". The different oxidation states of the TiO_2(110) surfaces were modeled via introduction of one point defect (O_{br} vacancy, H adatom and O_{ot} adatom, respectively) in a (4 × 2) surface unit cell. For details see Ref. [22].

In light of the presented experimental and theoretical results, it can be argued that O vacancies are not relevant for the stabilization of dispersed Au clusters under real reaction conditions with usually high oxygen (and water) pressures. Rather, the study by Matthey *et al.* suggests that O-rich Au–support interfaces are important to stabilize the Au nanoclusters. Therefore, the oxidation state of the supporting oxide is highly relevant and may also indicate that the perimeter of the Au nanoclusters is of special interest for catalytic reactions.

Within the Ti interstitial model, a more complete picture emerges of how the stability of Au nanoclusters on the various $TiO_2(110)$ surfaces depicted in Figure 4.13 can be generalized. Since the O_{ot} adatoms serve as "glue" between the Au nanoclusters and the oxide support, reducible oxides are proposed to be better catalysts than those supported on nonreducible ones, because reducible oxides are capable of forming O-rich Au–support interfaces, while non-reducible ones are not. The latter comes about since only reducible oxides can supply the electronic charge that is needed to stabilize and dissociate O_2 molecules on the surface. In the case of rutile $TiO_2(110)$ surfaces, the electronic charge probably stems from donor sites in the near-surface region and the bulk, that is from the Ti interstitials [23,80,81]. O-rich surface terminations have been identified also for several other reducible oxide supports, including vanadium and iron oxide [140], which underlines the availability of electronic charge on these oxide materials and the above statement addressing the capability of forming of O-rich Au–support interfaces. It should finally be mentioned that additional factors may also contribute to the higher catalytic activity of Au nanoclusters supported on reducible oxides compared to those supported on nonreducible oxides [120,141–143]. For example, Overbury *et al.* suggested that the support may influence the activity through the stabilization of subnanometer particles, the formation of active oxygen-containing intermediates, or the stabilization of confined, optimal Au structures [141].

4.8 Summary and Outlook

The enormous interest in the surface chemistry of $TiO_2(110)$ surfaces traces back to numerous applications, wherein, among others, TiO_2-based materials are used as photocatalysts and as support materials for catalytic reactions. Furthermore, TiO_2 continues to play an important role in the present search for new synthetic strategies to design nanostructure architectures for solar energy conversion. However, the underlying reaction mechanisms of the present and future applications of TiO_2-based materials are today still not well understood. Using the surface-science approach, the considerations of reactions at surfaces are simplified and fundamental insights can be achieved. In this chapter we have revealed how high-resolution STM studies of the $TiO_2(110)$ surface can provide new insight into TiO_2-based photocatalytic systems. The studies emphasize that atomic-scale insight is essential in order to understand important aspects of catalysis and photocatalysis, such as identification of active sites and the importance of electronic defects of the oxide. Whereas in most previous surface-science studies the O_{br} vacancies have been proposed as being the most important point defects that determine the chemistry on the $TiO_2(110)$ surface, we have presented compelling evidence for a new model, in which the importance of surplus Ti is highlighted. This new model is crucial for an improved understanding of the interaction between O_2 and reduced $TiO_2(110)$ surfaces, which is more complex than previously anticipated. In addition to O_2 dissociation in O_{br} vacancies, we have shown that a second, nonvacancy-assisted O_2 dissociation channel exists that is associated with the Ti interstitials and that leads to characteristic pairs of O_{ot} adatoms in the Ti troughs. For sputtered and vacuum-annealed $TiO_2(110)$ crystals strong evidences exists that the defect state located within the ~3.1 eV wide bandgap of reduced TiO_2 is caused by Ti interstitials in the near-surface region rather than by O_{br} vacancies. The redox chemistry on TiO_2 surfaces is strongly influenced by these Ti interstitials, as we have exemplified for the O_2 + H reaction on $TiO_2(110)$, where the bulk reduction needs to be taken into account to reach good agreement between experimental and theoretical results. For the reaction between O_2 and H adatoms on $TiO_2(110)$, we have identified intermediate species of HO_2 and H_2O_2 stoichiometry, and unraveled how they further react on the surface in the presence of co-adsorbed water. In addition to the studies where O_2 was involved, we have also presented new insight in relation to the interaction of water with the $TiO_2(110)$ surface: (i) water molecules mediate the diffusion of H adatoms on the surface, (ii) they form dimers, and (iii) they were found to dissociate in O_{br} vacancies. The latter reaction – although interesting from a fundamental point of view – is probably only of minor relevance for "real" applications, simply because O_{br} vacancies cannot persist under high-pressure conditions. Finally, we have addressed the nucleation of gold nanoclusters on $TiO_2(110)$ surfaces in different reduction states. We discussed why a stronger gold–oxide support adhesion exists on O-rich gold–support interfaces than on O-poor oxide support surfaces. For catalytic and photocatalytic applications, this means that O_{ot} adatoms and/or OH species are crucial to stabilize well-dispersed Au nanoclusters of the right size. On $TiO_2(110)$ surfaces with O_{br} vacancies, the stabilization of gold monomers and gold trimers is feasible, but larger gold clusters cannot be sufficiently stabilized. Whereas the surface-science approach allows one to gather atomic-scale information that cannot be gained by any other means, the presented examples show once more that the results obtained in surface-science work cannot naively be transferred to the situation in "real" catalysis. Both the so-called structure gap and the pressure gap have to be taken into account when the results obtained from the surface-science studies are transferred to the situation for high-surface-area catalysts and photocatalysts.

For future work, we anticipate that much work will be focused on reactions and photoreactions of organic compounds over TiO_2 single-crystal surfaces. Besides the studies on rutile, we need to improve our understanding of the chemistry on anatase, which is generally recognized to be the most active phase in photocatalytically active TiO_2-based materials. While the first studies exist for photoreactions on rutile TiO_2 single-crystal surfaces, no such studies have been published to date for photoreactions on anatase TiO_2 single crystals. Additionally, for both TiO_2 polymorphs, the effects of metal and nonmetal doping needs to be better understood, in order to gather insights into how the efficiency of TiO_2-based photocatalysts can be improved. Although the fundamental adsorption processes and reactions discussed in this chapter do not involve light exposure, these data represent a solid basis for surface-science studies addressing photocatalysis. It is anticipated that a similar scheme to that discussed here may also be applied to similar model systems under the illumination of UV/visible light to obtain detailed insight into the elementary steps involving photons. Therefore, the potential of the surface-science approach within the field of photocatalysis is enormous. In conjunction with traditional surface-science techniques, high-resolution STM is the technique of choice to shed light on the underlying processes at the atomic scale. High-resolution dynamic STM studies coupled with state-of-the-art DFT calculations makes it today possible to unravel reaction pathways and to test their relevance for "real" catalysts.

References

[1] Grimes, G.A., Varghese, O.K. and Ranjan, S. (2008) *Light, Water, Hydrogen*, Springer, New York.

[2] Carp, O., Huisman, C.L. and Reller, A. (2004) Photoinduced reactivity of titanium dioxide. *Progress in Solid State Chemistry*, **32**, 33–177.

[3] Kamat, P.V. (2002) Photophysical, 'Photochemical and photocatalytic aspects of metal nanoparticles'. *Journal of Physical Chemistry B*, **106**, 7729–7744.

[4] Thompson, T.L. and Yates, J.T. Jr. (2006) Surface science studies of the photoactivation of TiO_2 – new photochemical processes. *Chemical Reviews*, **106**, 4428–4453.

[5] Fujishima, A., Zhang, X.T. and Tryk, D.A. (2008) TiO_2 photocatalysis and related surface phenomena. *Surface Science Reports*, **63**, 515–582.

[6] Ertl, G. (2008) Reactions at surfaces: from atoms to complexity (Nobel Lecture). *Angewandte Chemie-International Edition*, **47**, 3524–3535.

[7] Ertl, G. (2001) Heterogeneous catalysis on the atomic scale. *Chemical Record*, **1**, 33–45.

[8] Binnig, G. and Rohrer, H. (1999) In touch with atoms. *Reviews of Modern Physics*, **71**, S324–S330.

[9] Binnig, G., Rohrer, H., Gerber, C. and Weibel, E. (1982) Surface studies by scanning tunneling microscopy. *Physical Review Letters*, **49**, 57–61.

[10] Besenbacher, F., Lauritsen, J.V. and Vang, R.T. (2008) *Handbook of Heterogeneous Catalysis*, vol. **2** (eds G. Ertl, H. Knözinger, F. Schüth and J. Weitkamp), Wiley, Weinheim, 833–870.

[11] Lauritsen, J.V., Vang, R.T. and Besenbacher, F. (2006) From atom-resolved scanning tunneling microscopy (STM) studies to the design of new catalysts. *Catalysis Today*, **111**, 34–43.

[12] Getoff, N. (1990) Photoelectrochemical and photocatalytic methods of hydrogen production: a short review. *International Journal of Hydrogen Energy*, **15**, 407–417.

[13] Diebold, U. (2003) The surface science of titanium dioxide. *Surface Science Reports*, **48**, 53–229.

[14] Pang, C.L., Lindsay, R. and Thornton, G. (2008) Chemical reactions on rutile $TiO_2(110)$. *Chemical Society Reviews*, **37**, 2328–2353.

[15] Thompson, T.L. and Yates, J.T. Jr. (2005) TiO_2-based photocatalysis: surface defects, oxygen and charge transfer. *Topics in Catalysis*, **35**, 197–210.

[16] Bowker, M. (2006) The surface structure of titania and the effect of reduction. *Current Opinion in Solid State & Materials Science*, **10**, 153–162.

[17] Iwasawa, Y. (1998) Dynamic oxide interfaces by microscopic techniques at high resolutions. *Surface Science*, **402**-404, 8–19.

[18] Schaub, R., Thostrup, P., Lopez, N. *et al.* (2001) Oxygen vacancies as active sites for water dissociation on rutile $TiO_2(110)$. *Physical Review Letters*, **87**, 266104.

[19] Wahlström, E., Schaub, R., Africh, C. *et al.* (2003) Bonding of gold nanoclusters to oxygen vacancies on rutile $TiO_2(110)$. *Physical Review Letters*, **90**, 026101.

[20] Wendt, S., Schaub, R., Matthiesen, J. *et al.* (2005) Oxygen vacancies on $TiO_2(110)$ and their interaction with H_2O and O_2: A combined high-resolution STM and DFT study. *Surface Science*, **598**, 226–245.

[21] Wendt, S., Matthiesen, J., Schaub, R. *et al.* (2006) Formation and splitting of paired hydroxyl groups on reduced $TiO_2(110)$. *Physical Review Letters*, **96**, 066107.

[22] Matthey, D., Wang, J.G., Wendt, S. *et al.* (2007) Enhanced bonding of gold nanoparticles on oxidized $TiO_2(110)$. *Science*, **315**, 1692–1696.

[23] Wendt, S., Sprunger, P.T., Lira, E. *et al.* (2008) The role of interstitial sites in the Ti3d defect state in the band gap of titania. *Science*, **320**, 1755–1759.

[24] Matthiesen, J., Wendt, S., Hansen J.Ø. *et al.* (2009) Observation of all the intermediate steps of a chemical reaction on an oxide surface by scanning tunneling microscopy. *ACS Nano*, **3**, 517–526.

[25] Matthiesen, J., Hansen J.Ø., Wendt, S. *et al.* (2009) Formation and Diffusion of Water Dimers on Rutile $TiO_2(110)$. *Physical Review Letters*, **102**, 226101.

[26] Lauritsen, J.V., Foster, A.S., Olesen, G.H. *et al.* (2006) Chemical identification of point defects and adsorbates on a metal oxide surface by atomic force microscopy. *Nanotechnology*, **17**, 3436–3441.

[27] Enevoldsen, G.H., Foster, A.S., Christensen, M.C. *et al.* (2007) Noncontact atomic force microscopy studies of vacancies and hydroxyls of $TiO_2(110)$: experiments and atomistic simulations. *Physical Review B-Condensed Matter*, **76**, 205415.

[28] Enevoldsen, G.H., Glatzel, T., Christensen, M.C. *et al.* (2008) Atomic scale kelvin probe force microscopy studies of the surface potential variations on the $TiO_2(110)$ surface. *Physical Review Letters*, **100**, 236104.

[29] Enevoldsen, G.H., Pinto, H.P., Foster, A.S. *et al.* (2009) Imaging of the hydrogen subsurface site in rutile TiO_2. *Physical Review Letters*, **102**, 136103.

[30] Kibsgaard, J., Clausen, B.S., Topsøe, H. *et al.* (2009) Scanning tunneling microscopy studies of TiO_2-supported hydrotreating catalysts: anisotropic particle shapes by edge-specific MoS_2-support bonding. *Journal of Catalysis*, **263**, 98–103.

[31] Barteau, M.A. (1996) Organic reactions at well-defined oxide surfaces. *Chemical Reviews*, **96**, 1413–1430.

[32] Wiesendanger, R. (1994) *Scanning Probe Microscopy and Spectroscopy*, 1st edn, Cambridge University Press, Cambridge, p. 637.

[33] Besenbacher, F. (1996) Scanning tunnelling microscopy studies of metal surfaces. *Reports on Progress in Physics*, **59**, 1737–1802.

[34] Bonnell, D.A. and Garra, J. (2008) Scanning probe microscopy of oxide surfaces: atomic structure and properties. *Reports on Progress in Physics*, **71**, 044501.

[35] Onishi, H. and Iwasawa, Y. (1994) STM-imaging of formate intermediates adsorbed on a $TiO_2(110)$ surface. *Chemical Physics Letters*, **226**, 111–114.

[36] Zhang, Z., Bondarchuk, O., Kay, B.D. *et al.* (2006) Imaging water dissociation on $TiO_2(110)$: evidence for inequivalent geminate OH groups. *Journal of Physical Chemistry B*, **110**, 21840–21845.

[37] Paxton, A.T. and Thien-Nga, L. (1998) Electronic structure of reduced titanium dioxide. *Physical Review B-Condensed Matter*, **57**, 1579–1584.

[38] Rasmussen, M.D., Molina, L.M. and Hammer, B. (2004) Adsorption, diffusion, and dissociation of molecular oxygen at defected $TiO_2(110)$: A density functional theory study. *Journal of Chemical Physics*, **120**, 988–997.

[39] Di Valentin, C., Pacchioni, G. and Selloni, A. (2006) Electronic structure of defect states in hydroxylated and reduced rutile $TiO_2(110)$ surfaces. *Physical Review Letters*, **97**, 166803.

[40] Teobaldi, G., Hofer, W.A., Bikondoa, O. *et al.* (2007) Modelling STM images of $TiO_2(110)$ from first-principles: defects, water adsorption and dissociation products. *Chemical Physics Letters*, **437**, 73–78.

[41] Minato, T., Sainoo, Y., Kim, Y. *et al.* (2009) The electronic structure of oxygen atom vacancy and hydroxyl impurity defects on titanium dioxide (110) surface. *Journal of Chemical Physics*, **130**, 124502.

[42] Thomas, A.G., Flavell, W.R., Mallick, A.K. *et al.* (2007) Comparison of the electronic structure of anatase and rutile TiO_2 single-crystal surfaces using resonant photoemission and X-ray absorption spectroscopy. *Physical Review B-Condensed Matter*, **75**, 035105.

[43] Henrich, V.E. and Cox, P. (1996) *The Surface Science of Metal Oxides*, Cambridge Univ. Press, Cambridge.

[44] Henrich, V.E., Dresselhaus, G. and Zeiger, H.J. (1976) Observation of 2-dimensional phases associated with defect states on surface of TiO_2. *Physical Review Letters*, **36**, 1335–1339.

[45] Bikondoa, O., Pang, C.L., Ithnin, R. *et al.* (2006) Direct visualization of defect-mediated dissociation of water on $TiO_2(110)$. *Nature Materials*, **5**, 189–192.

[46] Suzuki, S., Fukui, K., Onishi, H. and Iwasawa, Y. (2000) Hydrogen adatoms on $TiO_2(110)$-(1×1) characterized by scanning tunneling microscopy and electron stimulated desorption. *Physical Review Letters*, **84**, 2156–2159.

[47] Henderson, M.A. (2002) The interaction of water with solid surfaces: fundamental aspects revisited. *Surface Science Reports*, **46**, 5–308.

[48] Thiel, P.A. and Madey, T.E. (1987) The interaction of water with solid surfaces-fundamental aspects. *Surface Science Reports*, **7**, 211–385.

[49] Kurtz, R.L., Stockbauer, R., Madey, T.E. *et al.* (1989) Synchrotron radiation studies of H_2O adsorption on $TiO_2(110)$. *Surface Science*, **218**, 178–200.

[50] Henderson, M.A. (1996) An HREELS and TPD study of water on $TiO_2(110)$: The extent of molecular versus dissociative adsorption. *Surface Science*, **355**, 151–166.

[51] Henderson, M.A., Epling, W.S., Peden, C.H.F. and Perkins, C.L. (2003) Insights into photoexcited electron scavenging processes on TiO_2 obtained from studies of the reaction of O_2 with OH groups adsorbed at electronic defects on $TiO_2(110)$. *Journal of Physical Chemistry B*, **107**, 534–545.

[52] Brinkley, D., Dietrich, M., Engel, T. *et al.* (1998) A Modulated molecular beam study of the extent of H_2O dissociation on $TiO_2(110)$. *Surface Science*, **395**, 292–306.

[53] Hugenschmidt, M.B., Gamble, L. and Campbell, C.T. (1994) The interaction of H_2O with a $TiO_2(110)$ surface. *Surface Science*, **302**, 329–340.

[54] Ketteler, G., Yamamoto, S., Bluhm, H. *et al.* (2007) The nature of water nucleation sites on $TiO_2(110)$ surfaces revealed by ambient pressure X-ray photoelectron spectroscopy. *Journal of Physical Chemistry C*, **111**, 8278–8282.

[55] Allegretti, F., O'Brien, S., Polcik, M. *et al.* (2005) Adsorption bond length for H_2O on $TiO_2(110)$: A key parameter for theoretical understanding. *Physical Review Letters*, **95**, 226104.
[56] Brookes, I.M., Muryn, C.A. and Thornton, G. (2001) Imaging water dissociation on $TiO_2(110)$. *Physical Review Letters*, **87**, 266103.
[57] Du, Y.G., Deskins, A., Zhang, Z. *et al.* (2009) Imaging consecutive steps of O_2 reaction with hydroxylated $TiO_2(110)$: Identification of HO_2 and terminal OH intermediates. *Journal of Physical Chemistry C*, **113**, 666–671.
[58] Zhang, Z., Fenter, P., Cheng, L. *et al.* (2004) Ion adsorption at the rutile-water interface: linking molecular and macroscopic properties. *Langmuir*, **20**, 4954–4969.
[59] Lindan, P.J.D. and Zhang, C.J. (2005) Exothermic water dissociation on the rutile $TiO_2(110)$ Surface. *Physical Review B-Condensed Matter*, **72**, 075439.
[60] Harris, L.A. and Quong, A.A. (2004) Molecular chemisorption as the theoretically preferred pathway for water adsorption on ideal rutile $TiO_2(110)$. *Physical Review Letters*, **93**, 086105.
[61] Zhang, W.H., Yang, J.L., Luo, Y. *et al.* (2008) Quantum molecular dynamics study of water on $TiO_2(110)$ surface. *Journal of Chemical Physics*, **129**, 064703.
[62] Oviedo, J., Sanchez-De-Armas, R., Miguel, M.A.S. and Sanz, J.F. (2008) Methanol and water dissociation on $TiO_2(110)$: The role of surface oxygen. *Journal of Physical Chemistry C*, **112**, 17737–17740.
[63] Du, Y., Deskins, A., Zhang, Z. *et al.* (2009) Two pathways for water interaction with oxygen adatoms on $TiO_2(110)$. *Physical Review Letters*, **102**, 096102.
[64] Kowalski, P.M., Meyer, B. and Marx, D. (2009) Composition, structure, and stability of the rutile $TiO_2(110)$ surface: oxygen depletion, hydroxylation, hydrogen migration, and water adsorption. *Physical Review B-Condensed Matter*, **79**, 115410.
[65] Du, Y.G., Dohnalek, Z. and Lyubinetsky, I. (2008) Transient mobility of oxygen adatoms upon O_2 dissociation on reduced $TiO_2(110)$. *Journal of Physical Chemistry C*, **112**, 2649–2653.
[66] Wendt, S. *et al.* Dissociative and molecular oxygen chemisorption channels on reduced rutile TiO2(110): A high-resolution STM and DFT study (to be submitted in 2009. *Journal of Physical Chemistry C*).
[67] Liu, L.M., McAllister, B., Ye, H.Q. and Hu, P. (2006) Identifying an O_2 Supply Pathway in CO oxidation on Au/$TiO_2(110)$: A density functional theory study on the intrinsic role of water. *Journal of the American Chemical Society*, **128**, 4017–4022.
[68] Epling, W.S., Peden, C.H.F., Henderson, M.A. and Diebold, U. (1998) Evidence for oxygen adatoms on $TiO_2(110)$ resulting from O_2 dissociation at vacancy sites. *Surface Science*, **413**, 333–343.
[69] Henderson, M.A., Epling, W.S., Perkins, C.L. *et al.* (1999) Interaction of molecular oxygen with the Vacuum-annealed $TiO_2(110)$ surface: molecular and dissociative channels. *Journal of Physical Chemistry B*, **103**, 5328–5337.
[70] Perkins, C.L. and Henderson, M.A. (2001) Photodesorption and trapping of molecular oxygen at the $TiO_2(110)$-water ice interface. *Journal of Physical Chemistry B*, **105**, 3856–3863.
[71] Kimmel, G.A. and Petrik, N.G. (2008) Tetraoxygen on reduced $TiO_2(110)$: Oxygen adsorption and reactions with bridging oxygen vacancies. *Physical Review Letters*, **100**, 196102.
[72] Li, M., Hebenstreit, W., Diebold, U. *et al.* (2000) The influence of the bulk reduction state on the surface structure and morphology of rutile $TiO_2(110)$ single crystals. *Journal of Physical Chemistry B*, **104**, 4944–4950.
[73] Attwood, A.L., Murphy, D.M., Edwards, J.L. *et al.* (2003) An EPR study of thermally and photochemically generated oxygen radicals on hydrated and dehydrated titania surfaces. *Research on Chemical Intermediates*, **29**, 449–465.
[74] Aono, M. and Hasiguti, R.R. (1993) Interaction and ordering of lattice-defects in oxygen-deficient rutile $TiO_{2(X}$. *Physical Review B-Condensed Matter*, **48**, 12406–12414.
[75] Ghosh, A.K., Wakim, F.G. and Addiss, R.R. (1969) Photoelectronic processes in rutile. *Physical Review*, **184**, 979–988.
[76] Yagi, E., Hasiguti, R.R. and Aono, M. (1996) Electronic conduction above 4 K of slightly reduced oxygen-deficient rutile $TiO_{2(X}$. *Physical Review B-Condensed Matter*, **54**, 7945–7956.
[77] Krüger, P., Bourgeois, S., Domenichini, B. *et al.* (2008) Defect states at the $TiO_2(110)$ surface probed by resonant photoelectron diffraction. *Physical Review Letters*, **100**, 055501.
[78] Chambers, S.A., Cheung, S.H., Shuttanandan, V. *et al.* (2007) Properties of structurally excellent N-doped TiO_2 rutile. *Chemical Physics*, **339**, 27–35.

[79] Cho, E., Han, S., Ahn, H.-S. *et al.* (2006) First-principles study of point defects in rutile TiO_{2-x}. *Physical Review B-Condensed Matter*, **73**, 193202.

[80] Madsen, G.K.H. and Hammer, B. (2009) Effect of subsurface Ti-interstitials on the bonding of small gold clusters on rutile TiO_2(110). *Journal of Chemical Physics*, **130**, 044704.

[81] Finazzi, E., Di Valentin, C. and Pacchioni, G. (2009) Nature of Ti interstitials in reduced bulk anatase and rutile TiO_2. *Journal of Physical Chemistry C*, **113**, 3382–3385.

[82] Ganduglia-Pirovano, M.V., Hofmann, A. and Sauer, J. (2007) Oxygen vacancies in transition metal and rare earth oxides: current state of understanding and remaining challenges. *Surface Science Reports*, **62**, 219–270.

[83] Henderson, M.A. (1999) Surface perspective on self-diffusion in rutile TiO_2. *Surface Science*, **419**, 174–187.

[84] Onishi, H. and Iwasawa, Y. (1996) Dynamic visualization of a metal-oxide-surface/gas-phase reaction: time-resolved observation by scanning tunneling microscopy at 800 K. *Physical Review Letters*, **76**, 791–794.

[85] Bennett, R.A., Stone, P., Price, N.J. and Bowker, M. (1999) Two (1×2) reconstructions of TiO_2(110): Surface rearrangement and reactivity studied using elevated temperature scanning tunneling microscopy. *Physical Review Letters*, **82**, 3831–3834.

[86] Li, M., Gross, L., Diebold, U. *et al.* (1999) Oxygen-induced restructuring of the TiO_2(110) surface: a comprehensive study. *Surface Science*, **437**, 173–190.

[87] Smith, R.D., Bennett, R.A. and Bowker, M. (2002) Measurement of the surface-growth kinetics of reduced TiO_2(110) during reoxidation using time-resolved scanning tunneling microscopy. *Physical Review B-Condensed Matter*, **66**, 035409.

[88] Henderson, M.A., White, J.M., Uetsuka, H. and Onishi, H. (2003) Photochemical charge transfer and trapping at the interface between an organic adlayer and an oxide semiconductor. *Journal of the American Chemical Society*, **125**, 14974–14975.

[89] Zhang, Z., Du, Y., Petrik, N.G. *et al.* (2009) Water as a catalyst: imaging reactions of O_2 with partially and fully hydroxylated TiO_2(110) surfaces. *Journal of Physical Chemistry C*, **113**, 1908–1916.

[90] Li, S.C., Zhang, Z., Sheppard, D. *et al.* (2008) Intrinsic diffusion of hydrogen on rutile TiO_2(110). *Journal of the American Chemical Society*, **130**, 9080–9088.

[91] Bennett, R.A. (2000) The re-oxidation of the substoichiometric TiO_2(110) surface in the presence of crystallographic shear planes. *PhysChemComm*, **3**, doi: 10.1039/b309812p.

[92] Rohrer, G.S., Henrich, V.E. and Bonell, D.A. (1990) Structure of the reduced TiO_2(110) surface determined by scanning tunneling microscopy. *Science*, **250**, 1239–1241.

[93] Tilocca, A., Di Valentin, C. and Selloni, A. (2005) O_2 Interaction and reactivity on a model hydroxylated rutile (110) surface. *Journal of Physical Chemistry B*, **109**, 20963–20967.

[94] Knapp, M., Crihan, D., Seitsonen, A.P. and Over, H. (2005) Hydrogen transfer reaction on the surface of an oxide catalyst. *Journal of the American Chemical Society*, **127**, 3236–3237.

[95] Knapp, M., Crihan, D., Seitsonen, A.P. *et al.* (2006) Unusual process of water formation on RuO_2(110) by hydrogen exposure at room temperature. *Journal of Physical Chemistry B*, **110**, 14007–14010.

[96] Daté, M., Okumura, M., Tsubota, S. and Haruta, M. (2004) Vital role of moisture in the catalytic activity of supported gold nanoparticles. *Angewandte Chemie-International Edition*, **43**, 2129–2132.

[97] Meyer, R., Lemire, C., Shaikhutdinov, S.K. and Freund, H.-J. (2004) Surface chemistry of catalysis by gold. *Gold Bulletin*, **37**, 72–124.

[98] Haruta, M., Yamada, N., Kobayashi, T. and Iijima, S. (1989) Gold catalysts prepared by coprecipitation for low-temperature oxidation of hydrogen and of carbon-monoxide. *Journal of Catalysis*, **115**, 301–309.

[99] Hayashi, T., Tanaka, K. and Haruta, M. (1998) Selective vapor-phase epoxidation of propylene over Au/TiO_2 catalysts in the presence of oxygen and hydrogen. *Journal of Catalysis*, **178**, 566–575.

[100] Boccuzzi, F., Chiorino, A., Manzoli, M. *et al.* (1999) FTIR study of the low-temperature water-gas shift reaction on Au/Fe_2O_3 and Au/TiO_2 catalysts. *Journal of Catalysis*, **188**, 176–185.

[101] Haruta, M. (2004) Gold as a novel catalyst in the 21st century: preparation, working mechanism and applications. *Gold Bulletin*, **37**, 27–36.

[102] Thompson, D.T. (2006) An overview of gold-catalysed oxidation processes. *Topics in Catalysis*, **38**, 231–240.

[103] Subramanian, V., Wolf, E.E. and Kamat, P.V. (2001) Semiconductor-metal composite nanostructures. To what extent do metal nanoparticles improve the photocatalytic activity of TiO_2 films? *Journal of Physical Chemistry B*, **105**, 11439–11446.

[104] Subramanian, V., Wolf, E.E. and Kamat, P.V. (2003) Influence of metal/metal ion concentration on the photocatalytic activity of TiO_2–Au composite nanoparticles. *Langmuir*, **19**, 469–474.

[105] Subramanian, V., Wolf, E.E. and Kamat, P.V. (2004) Catalysis with TiO_2/Gold nanocomposites. Effect of metal particle size on the fermi level equilibration. *Journal of the American Chemical Society*, **126**, 4943–4950.

[106] Daniel, M.C. and Astruc, D. (2004) Gold nanoparticles: assembly, supramolecular chemistry, quantum-size-related properties applications toward biology, catalysis, and nanotechnology. *Chemical Reviews*, **104**, 293–346.

[107] Tada, H., Mitsui, T., Kiyonaga, T. *et al.* (2006) All-solid-state Z-scheme in CdS-Au–TiO_2 three-component nanojunction system. *Nature Materials*, **5**, 782–786.

[108] Sonawane, R.S. and Dongare, M.K. (2006) Sol-gel synthesis of Au/TiO_2 thin films for photocatalytic degradation of phenol in sunlight. *Journal of Molecular Catalysis A-Chemical*, **243**, 68–76.

[109] Li, H.X., Bian, Z., Zhu, J. *et al.* (2007) Mesoporous Au/TiO_2 nanocomposites with enhanced photocatalytic activity. *Journal of the American Chemical Society*, **129**, 4538–4539.

[110] Chen, X., Zhu, H.Y., Zhao, J.C. *et al.* (2008) Visible-light-driven oxidation of organic contaminants in air with gold nanoparticle catalysts on oxide supports. *Angewandte Chemie-International Edition*, **47**, 5353–5356.

[111] Gur, I., Fromer, N.A., Geier, M.L. and Alivisatos, A.P. (2005) Air-stable all-inorganic nanocrystal solar cells processed from solution. *Science*, **310**, 462–465.

[112] McFarland, E.W. and Tang, J. (2003) A photovoltaic device structure based on internal electron emission. *Nature*, **421**, 616–618.

[113] Bamwenda, G.R., Tsubota, S., Nakamura, T. and Haruta, M. (1997) The influence of the preparation methods on the catalytic activity of platinum and gold supported on TiO_2 for CO oxidation. *Catalysis Letters*, **44**, 83–87.

[114] Valden, M., Lai, X. and Goodman, D.W. (1998) Onset of catalytic activity of gold clusters on titania with the appearance of nonmetallic properties. *Science*, **281**, 1647–1650.

[115] Yoon, B., Häkkinen, H., Landman, U. *et al.* (2005) Charging effects on bonding and catalyzed oxidation of CO on Au_8 clusters on MgO. *Science*, **307**, 403–407.

[116] Lee, S.S., Fan, C.Y., Wu, T.P. and Anderson, S.L. (2004) CO oxidation on Au_n/TiO_2 catalysts produced by size-selected cluster deposition. *Journal of the American Chemical Society*, **126**, 5682–5683.

[117] Kung, M.C., Davis, R.J. and Kung, H.H. (2007) Understanding Au-catalyzed low-temperature CO oxidation. *Journal of Physical Chemistry C*, **111**, 11767–11775.

[118] Herzing, A.A., Kiely, C.J., Carley, A.F. *et al.* (2008) Identification of active gold nanoclusters on iron oxide supports for CO oxidation. *Science*, **321**, 1331–1335.

[119] Bond, G.C. and Thompson, D.T. (2000) Gold-catalysed oxidation of carbon monoxide. *Gold Bulletin*, **33**, 41–51.

[120] Schubert, M.M., Hackenberg, S., van Veen, A.C. *et al.* (2001) CO oxidation over supported gold catalysts-"Inert" and "Active" support materials and their role for the oxygen supply during reaction. *Journal of Catalysis*, **197**, 113–122.

[121] Chou, J., Franklin, N.R., Baeck, S.H. *et al.* (2004) Gas-phase catalysis by micelle derived Au nanoparticles on oxide supports. *Catalysis Letters*, **95**, 107–111.

[122] Hashmi, A.S.K. and Hutchings, G.J. (2006) Gold catalysis. *Angewandte Chemie-International Edition*, **45**, 7896–7936.

[123] Wang, J.G. and Hammer, B. (2006) Role of Au^+ in supporting and activating Au_7 on $TiO_2(110)$. *Physical Review Letters*, **97**, 136107.

[124] Chen, M.S. and Goodman, D.W. (2006) Structure-activity relationships in supported Au catalysts. *Catalysis Today*, **111**, 22–33.

[125] Molina, L.M. and Hammer, B. (2004) Theoretical study of CO oxidation on Au nanoparticles supported by MgO (100). *Physical Review B-Condensed Matter*, **69**, 155424.

[126] Lopez, N., Janssens, T.V.W., Clausen, B.S. *et al.* (2004) On the origin of the catalytic activity of gold nanoparticles for low-temperature CO oxidation. *Journal of Catalysis*, **223**, 232–235.

[127] Xu, Y. and Mavrikakis, M. (2003) Adsorption and dissociation of O_2 on gold surfaces: effect of steps and strain. *Journal of Physical Chemistry B*, **107**, 9298–9307.

[128] Chen, X. and Mao, S.S. (2007) Titanium dioxide nanomaterials: synthesis, properties, modifications, and applications. *Chemical Reviews*, **107**, 2891–2959.

[129] Mulvaney, P. (1996) Surface plasmon spectroscopy of nanosized metal particles. *Langmuir*, **12**, 788–800.

[130] Yonezawa, T., Matsune, H. and Kunitake, T. (1999) Layered nanocomposite of close-packed gold nanoparticles and TiO_2 gel layers. *Chemistry of Materials*, **11**, 33–35.

[131] Friedmann, D., Hansing, H. and Bahnemann, D. (2007) Primary processes during the photodeposition of Ag clusters on TiO_2 nanoparticles. *Zeitschrift Für Physikalische Chemie-International Journal of Research in Physical Chemistry & Chemical Physics*, **221**, 329–348.

[132] Ismael, A.A., Bahnemann, D., Bannat, I. and Wark, M. (2009) Gold nanoparticles on mesoporous interparticle networks of titanium dioxide nanocrystals for enhanced photonic efficiencies. *Journal of Physical Chemistry C*, **113**, 7429–7435.

[133] Cosandey, F. and Madey, T.E. (2001) Growth, morphology, interfacial effects and catalytic properties of Au on TiO_2. *Surface Review and Letters*, **8**, 73–93.

[134] Molina, L.M., Rasmussen, M.D. and Hammer, B. (2004) Adsorption of O_2 and oxidation of CO at Au nanoparticles supported by $TiO_2(110)$. *Journal of Chemical Physics*, **120**, 7673–7680.

[135] Tong, X., Benz, L., Kemper, P. *et al.* (2005) Intact size-selected Au_n clusters on a $TiO_2(110)$-(1×1) surface at room temperature. *Journal of the American Chemical Society*, **127**, 13516–13518.

[136] Chen, M., Cai, Y., Yan, Z. and Goodman, D.W. (2006) On the origin of the unique properties of supported Au nanoparticles. *Journal of the American Chemical Society*, **128**, 6341–6346.

[137] Chretien, S. and Metiu, H. (2007) Density functional study of the interaction between small Au clusters, Au_n ($n =$ 1–7) and the rutile TiO_2 surface. I. adsorption on the stoichiometric surface. *Journal of Chemical Physics*, **127**, 084704.

[138] Chretien, S. and Metiu, H. (2007) Density functional study of the interaction between small Au clusters, Au_n ($n =$ 1–7) and the rutile TiO_2 surface. II. adsorption on a partially reduced surface. *Journal of Chemical Physics*, **127**, 244708.

[139] Wang, J.G. and Hammer, B. (2007) Oxidation state of oxide supported nanometric gold. *Topics in Catalysis*, **44**, 49–56.

[140] Abu Haija, M., Guimond, S., Romanyshyn, Y. *et al.* (2006) Low temperature adsorption of oxygen on reduced $V_2O_3(0001)$ surfaces. *Surface Science*, **600**, 1497–1503.

[141] Overbury, S.H., Ortiz-Soto, L., Zhu, H. *et al.* (2004) Comparison of Au catalysts supported on mesoporous titania and silica: investigation of au particle size effects and metal-support interactions. *Catalysis Letters*, **95**, 99–106.

[142] Comotti, M., Li, W.C., Spliethoff, B. and Schüth, F. (2006) Support effect in high activity gold catalysts for CO oxidation. *Journal of the American Chemical Society*, **128**, 917–924.

[143] Boccuzzi, F., Chiorino, A., Tsubota, S. and Haruta, M. (1996) FTIR study of carbon monoxide oxidation and scrambling at room temperature over gold supported on ZnO and TiO_2. *The Journal of Physical Chemistry*, **100**, 3625–3631.

Part Two

Electronic Structure, Energetics, and Transport Dynamics of Photocatalyst Nanostructures

5

Electronic Structure Study of Nanostructured Transition Metal Oxides Using Soft X-Ray Spectroscopy

Jinghua Guo[1], Per-Anders Glans[1], Yi-Sheng Liu[2], and Chinglin Chang[2]
[1]*Advanced Light Source, Lawrence Berkeley National Laboratory, Berkeley, CA 94720, USA, Email: jguo@lbl.gov*
[2]*Department of Physics, Tamkang University, Tamsui, Taiwan, 250, R.O.C.*

5.1 Introduction

Solar energy can be converted to heat for warming space and water, to electricity and chemical fuels for energy use and storage [1–4]. However, the conversion efficiency has hampered the potential use of solar energy. There are emerging technologies using semiconductors for light-harvesting assemblies and charge-transfer processes for solar cells. Sunlight in the near infrared, visible and near ultraviolet regions has considerable energy (about 0.9–3.2 eV per photon) and intensity. It could provide a significant contribution to our electrical and chemical resources if efficient and inexpensive systems utilizing readily available materials could be devised for the conversion process.

Upon absorption of sunlight, the electron–hole pair formation that occurs at the interface between a semiconductor and a solution leads to oxidation or reduction of solution species. The fabrication of artificial photosynthetic systems for the conversion of H_2O and CO_2 to fuels (for examples, H_2 and CH_3OH) has become a field of much research interest and has encouraged new fundamental investigations between the interactions of sunlight, electron flow and chemical reactions.

On Solar Hydrogen & Nanotechnology Edited by Lionel Vayssieres

Synchrotron radiation-based soft X-ray spectroscopy has become a powerful tool to determine the bandgap properties of semiconductors [5,6]. X-rays originate from an electronic transition between a localized core state and a valence state. Soft X-ray absorption (XAS) probes the local *unoccupied* electronic structure (conduction band), soft X-ray emission (XES) probes the *occupied* electronic structure (valence band) and the addition of resonant inelastic soft X-ray scattering (Raman spectroscopy with soft X-rays) can identify the energy levels that reflect the chemical and physical properties of semiconductors. Recently, quantum size effects on the exciton and bandgap energies were observed in single-walled carbon nanotubes (SWNTs) [7].

In particular, it was possible to apply the detailed knowledge obtained in soft X-ray spectroscopy of bulk rutile and anatase TiO_2, and hematite Fe_2O_3 to the study of the interaction of adsorbates on undoped and doped TiO_2 and Fe_2O_3 surfaces, and the bonding that can occur as a result. This is potentially important, for instance, for photolysis, in relation to both solar cells and to the production of hydrogen.

5.2 Soft X-Ray Spectroscopy

5.2.1 Soft X-Ray Absorption and Emission Spectroscopy

Soft X-ray absorption spectra provide information about the unoccupied states. For example, in oxygen *K*-edge absorption, the oxygen 1*s* electron is excited to empty electronic states in the carbon allotrope conduction band, and the dipole selection rule provides a tool to study locally the O2*p* character of these unoccupied valence bands (Figure 5.1). The atomic nature of the core hole implies elemental and site selectivity. The probability of such a transition is related to the X-ray absorption cross-section. The intensity of these secondary electrons or the photons can be measured as a function of incoming photon energy. This will reflect the absorption cross-section as the intensity of the secondary electrons/emitted photons are proportional to the

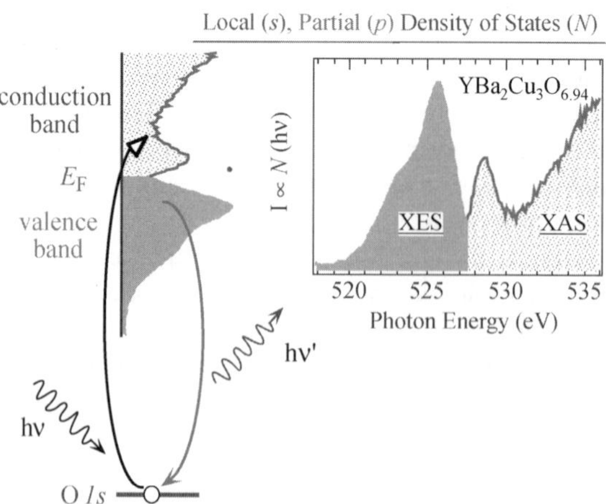

Figure 5.1 Schematic representation of X-ray absorption and emission processes that gives an example of the electronic structure study of a complex system.

absorbed intensity. Because of the short mean free path of electrons, the electron yield detection method is very surface sensitive. If the out-coming photons are detected (fluorescence yield), the X-ray absorption becomes bulk probing (about 100–200 nm) due to the comparatively larger attenuation lengths. Figure 5.1 gives an example of an XAS study of a high-Tc superconductor $YBa_2Cu_3O_7$ and the XAS spectrum reflects the partial density of states (DOS) of the conduction band.

The core vacancy left by the excited 1s electron is filled by an electron from a valence orbital; thereby soft X-ray emission also gives direct information about the chemical bonding. Figure 5.1 also gives an example of an XES study of a high-Tc superconductor YBCO and the XES spectrum reflects the density of states (DOS) of the valence band. In addition to the inherent elemental selectivity of X-ray spectra, energy-selective excitation allows separation of features that pertain to different atoms of a sample. Emission from chemically nonequivalent sites of the same atomic species can be separated. Interest in the technique is presently booming, due to the advent of third-generation synchrotron radiation sources.

5.2.2 Resonantly Excited Soft X-Ray Emission Spectroscopy

The introduction of synchrotron radiation did not immediately lead to great progress in soft X-ray emission spectroscopy, in the way photoemission and X-ray absorption spectroscopies developed when synchrotron radiation became available. The first soft X-ray emission spectroscopic study using monochromatized synchrotron radiation was carried out in 1987 [8]. X-ray absorption and emission have traditionally been treated as two independent processes, with the absorption and emission spectra providing information on the unoccupied and occupied electronic states, respectively. The formulations of resonant inelastic X-ray scattering (RIXS) lead to a Kramers–Heisenberg-type dispersion formula for the cross-section with generally only the resonant part of the scattering process taken into account [9]. Second-order perturbation theory for the RIXS process leads to the Kramers–Heisenberg formula for the resonant X-ray scattering amplitude. Using this starting point, RIXS has been analyzed in periodic solids as a momentum-conserving process, suggesting that it can be used as a novel "band-mapping" technique [10]. The same starting point was adopted to unravel the symmetry-selective properties of RIXS in theoretical work focused on molecules [11–14]. Resonant inelastic X-ray scattering at core resonances has become a new tool for probing the optical transitions in transition-metal oxides [15,16]. Final states probed via such a channel, RIXS or XRS, are related to the eigenvalues of the ground state Hamiltonian. The core-hole lifetime is not a limit on the resolution in this spectroscopy. According to the many-body picture, the energy of a photon, scattered on a certain low-energy excitation, should change by the same amount as the change in the excitation energy of the incident beam, so that inelastic scattering structures have constant energy losses and follow the elastic peak on the emitted-photon energy scale. Figure 5.2 shows an example of such an energy loss originating from the *dd* excitations observed at the Ti *L*-edge.

5.3 Experiment Set-Up

Photon-in/photon-out soft X-ray spectroscopy offers a number of unique features, including element and chemical site-specific probing, based on the energy tunability of synchrotron

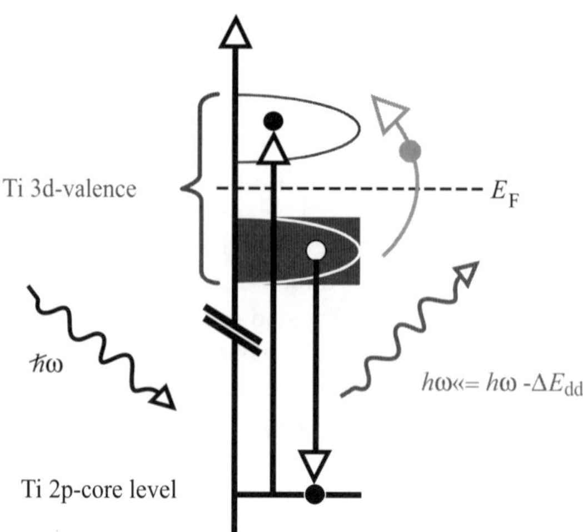

Figure 5.2 Schematic representation of RIXS process at Ti *L*-edge to illustrate the dd excitations.

radiation. This is a valuable general asset in the study of complex materials, and is often important for the study of nanostructured materials. Photons are insensitive to external fields, such as electric and magnetic fields, and can also be made partially insensitive to a gaseous and aqueous environment. This allows the investigation of materials in real photoelectrochemical reactions or under a controlled gas atmosphere and selected temperature. Resonant inelastic scattering implies independence of the core-hole lifetime broadening, and thus offers a unique opportunity to utilize very high resolution in the study of changes in the electronic structure, for example the opening of a gap, associated with phase transitions, doping or nanofabrications. In short, soft X-ray spectroscopy offers a high degree of information selectivity on electronic properties with respect to atomic species, chemical site, symmetry and local order.

The increasingly improved performance of synchrotron radiation during the last two decades has opened new areas of application for SXES. The most dramatic improvement came with the Advanced Light Source (ALS) at Berkeley, where the high brightness of the first soft X-ray third-generation source was combined with high quality optical systems for refocusing the monochromatized beam in the1990s [16–18]. At the same period of time, there were also developments in SXES instrumentation at ESRF in France, Spring-8 in Japan and the newly operational facility: the Swiss Light Source (SLS).

5.3.1 Beamline

At the ALS, two different beamlines were taken into operation for soft X-ray fluorescence spectroscopy at approximately the same time, BL7.0 [20] and BL8.0. The two beamlines are essentially of the same type, based on identical undulators and similar spherical grating monochromators. Beamline 7.0 at the ALS is equipped with a spherical grating monochromator (SGM), providing resolving power of 5000 in the energy range 80–1000 eV, available for studying the fundamental electronic properties of materials in solid, gaseous and aqueous phases [21]. The beamline and endstation are capable for performing XAS measurements, and

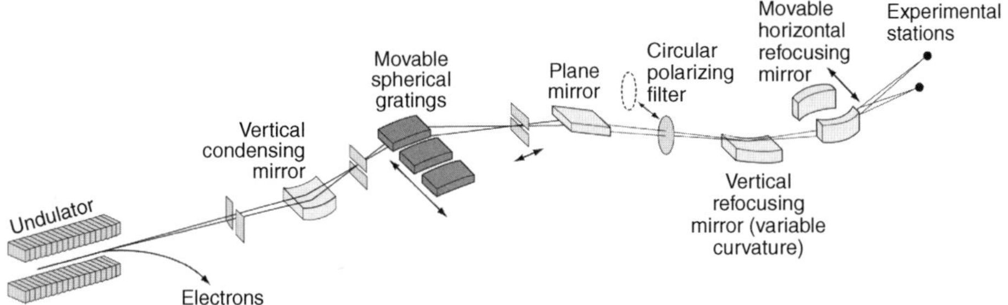

Figure 5.3 The schematic diagram of beamline 7.0.1 at the ALS.

a high-resolution grating spectrometer is used for XES and RIXS studies. Beamline 8.0 is a sister beamline at the ALS with the same performance as BL7.0.

Figure 5.3 shows an outline of ALS BL7.0 with its principle elements. The first element of the beamline is a horizontal beam-defining aperture (HBDA), comprised of two water-cooled copper blocks forming an aperture of horizontal width. The second element is a spherical mirror that images the source onto the entrance slit of the monochromator. A piezo-activated adjustment of these optics is provided in order to optimize the flux into the monochromator.

The monochromator has an entrance slit, a set of gratings on a movable carriage and a movable exit slit. The entrance slit is made of a block of water-cooled Glidcop, which has been cut to produce a structure of flexure elements providing parallel and precise movements of the slit jaws. The slit can be set at widths from a few microns to 100 microns. There is a provision to move the entrance slit, but this is usually not used for soft X-ray fluorescence experiments with this beamline. Three gratings are available in the grating tank, mounted on a carriage that provides water cooling to the gratings and facilitates grating rotation for wavelength scanning. Changing the grating is readily accomplished by sliding the carriage horizontally. In order to account for the fact that the focal length of the spherical grating changes with angle, the exit slit is movable over a 700 mm translation stage. The exit slit width can be set from a few microns up to 250 microns.

The post-monochromator optics are designed to deliver a focused vertical beam of monochromatized photons to the experiment. To accomplish this, the system has one focusing vertical mirror. Fine-tuning of the post-focusing mirrors by means of piezo actuators allows the beam to be electronically controlled with respect to focal length (only vertically) and alignment. This feature has turned out to be of great value in several cases, such as aligning the beam in certain gas- and liquid-phase experiments, where one needs to direct the beam through the small apertures.

5.3.2 Spectrometer and Endstation

The advancement of soft X-ray fluorescence spectroscopy in the 1990s has been accomplished mainly based on the compact spherical grating spectrometers and modern synchrotron radiation sources. A number of the grating spectrometers used (Rubensson *et al.*; Hague *et al.*, private communications) [22–26] are basically similar and thus the inherent performance of the various spectrometers is not very different from the instrument that was first used for these types of studies [22,23]. One difference exists in the sense that a few spectrometers are mounted on rotatable chambers. This arrangement has allowed polarization studies to be carried out in a more extensive and detailed manner.

The basic strategy behind the design of the compact soft X-ray spectrometer is one of adaptability to various experiment chambers and minimization of the source-to-detector distance for maximum angular acceptance. This latter objective stands in opposition to resolution, obviously, since resolution is a matter of simple geometric scaling. Thus, the bigger the instrument, the higher the resolution for a given ruling density. However, by using multiple gratings of different radii and ruling density, one can obtain an optimization of the overall performance of the spectrometer.

The small size of this instrument has allowed it to be mounted on ultra-high vacuum (UHV) rotatable chambers. Figure 5.4 (left) illustrates the optical components of the grating spectrometer. This is a prerequisite for utilizing the polarization properties of synchrotron radiation in an effective manner. One needs to be able to select independently the angle between the polarization of the incident photons and the sample orientation on the one hand, and the sample orientation and the emission direction on the other.

The optical components consist of an entrance slit, two movable shutters for grating selection, three fixed mounted gratings and an area detector that can be moved in a three-axis coordinate system (two translations and one rotation). The optical arrangement follows Rowland geometry, and the concept can be seen as three different Rowland spectrometers merged into each other to have a common entrance slit and a detector that can be aligned to the focal curve of the grating in use. Gratings with different ruling densities as well as radii are mixed in order to accomplish optimum performance. This instrument is currently on the market from the Gammadata/Scienta company. The plane of dispersion of the instrument is vertical or horizontal depending on choice. For synchrotron radiation applications, a vertical dispersion plane has advantages, as it aligns the entrance slit to be parallel with the synchrotron radiation beam.

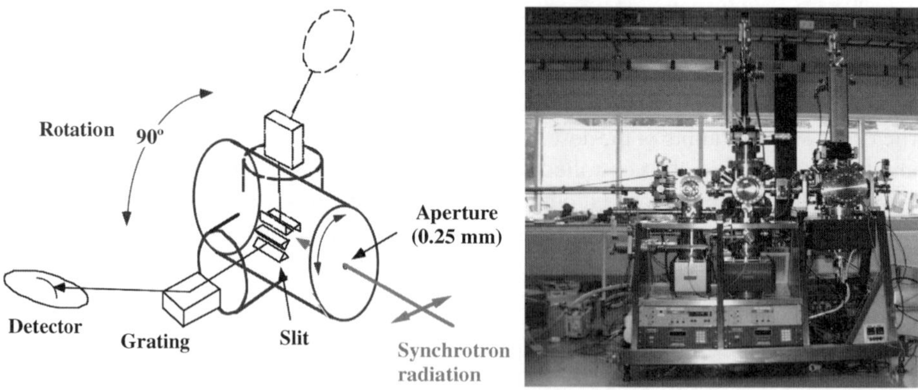

Figure 5.4 Outline of the grating spectrometer, which rotates around the center axis of the experimental chamber with respect to the incoming synchrotron radiation (left) and a photograph of the endstation.

The endstation (see Figure 5.4) is comprised of three main sections: analysis, preparation and load-lock stages, separated by UHV valves. The load-lock chamber has a rubber-sealed quick-open flange and much less volume compared to the other two stages, so it can be pumped down very quickly. Typically it takes less than 30 minutes to get chamber pressure down from atmospheric pressure to the low 10^{-7} torr range at which the load lock can be opened to the preparation and analysis chambers for transferring samples. The transfer arm, with linear and rotation motions controlled mechanically, is able to take the samples from storage in the load lock and transfer them onto the manipulators in the preparation and analysis chambers, and vice versa. The sample storage can store seven 20-mm diameter sample holders. The preparation chamber is designed for *in situ* sample cleaning and preparation. For example, films of C_{60}, K-doped C_{60} and C_{70} [27–29] were evaporated in the preparation chamber shortly before the soft X-ray spectroscopy measurements. The analysis chamber, reaching UHV ($<5 \times 10^{-10}$ torr), is directly attached to the output of the synchrotron radiation beamline. The analysis chamber can be rotated up to 90° under UHV conditions, so allowing the X-ray fluorescence parallel and perpendicular to the polarization plane of the incident photons to be recorded.

5.3.3 *Sample Arrangements*

Soft X-ray fluorescence spectroscopy is essentially a bulk-sensitive technique, since the attenuation length of photons in this energy range is typically 100–200 nm in solid matter. Thus the penetration depth offers a few experimental opportunities not possible in electron-based spectroscopy. Apart from the obvious advantage of relaxed requirements for sample surface treatment, one can address scientific issues involving buried structures down to 10 nm in depth [30]. Furthermore, windows of considerable transmittance can be used to separate the sample environment from the vacuum system required for the soft X-ray spectroscopy experiment. This offers the possibility to study samples in the liquid or gas phase, confined in a closed volume, such as, for example, a solid exposed to an ambient gas [31]. For gaseous samples, one can use high sample pressures, tens of mbar or higher [32–34], so that intensities can become close to the corresponding solid sample intensities. The gas cell is illustrated in Figure 5.5a.

To exploit the possibility of *in situ* characterization, liquid and chemical-reaction cells have been constructed to allow monitoring of the chemical reaction under a range of temperatures and pressures. The liquid cell was built based on our experience of gas-phase and buried-layer SXES studies. It has a window for compatibility with the UHV conditions of the spectrometer and beamline. We have demonstrated the soft X-ray spectroscopic studies of liquids, gases and solid samples in a gaseous environment [18,19,35,36], for example to examine the influence of the intermolecular interaction on the local electronic structure of liquid water and methanol [37,38]. The cell is capable of holding atmospheric pressure. [39,40].

The static liquid cell uses a thin membrane window for compatibility with the UHV conditions of the fluorescence spectrometer and synchrotron radiation beamline. A thin window (100 nm silicon nitride), separating the liquid from the surrounding vacuum, is penetrated both by the incident photon beam and the X-ray emission. The thin silicon nitride window is commercially available. The test showed that a 100-nm-thick Si_3N_4 membrane of 1×1 mm^2 could hold a nitrogen gas pressure of more than 18 bars. The static liquid cell (see in Figure 5.5b) consists of a metal container and 1×1 mm^2 and this 100 nm Si_3N_4 membrane,

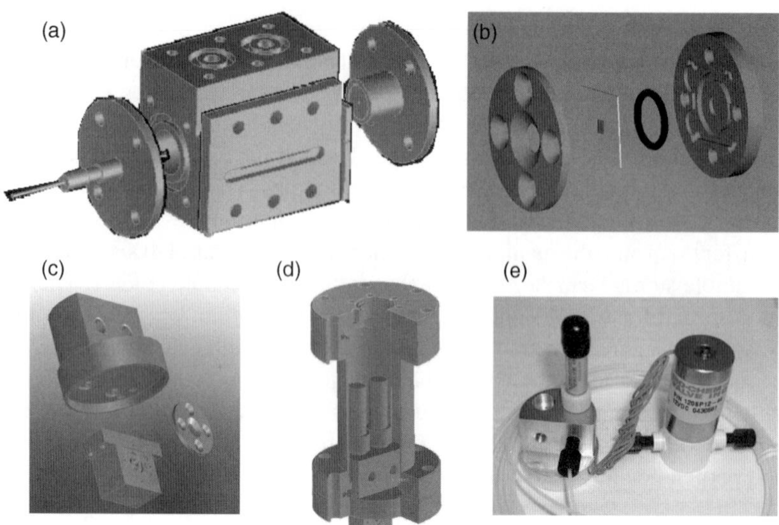

Figure 5.5 Cells for soft X-ray spectroscopic studies: (a) gas cell, (b) static liquid cell, and (c,d,e) liquid flow cell.

which can withstand the differential pressure between the liquid on one side and UHV on the other side. The transmissions of X-rays at the C *K*-, O *K*- and Fe *L*-edges for a 100 nm thick window are 46%, 66% and 82%, respectively. X-ray absorption spectra of liquids were also measured in fluorescence-yield mode, using a photon diode and a fluorescence spectrometer. The experiments were performed in vacuum, with a base pressure of 1×10^{-9} Torr.

When X-ray photon-induced sample damage becomes a problem in a soft X-ray spectroscopic study, a flow liquid cell is used. Figure 5.5c–e shows drawings and a photograph of a flow liquid cell assemblage and a few important components, including valves and a pump for the liquid flow cell. The central part of the liquid cell is similar to the earlier static cell, while liquid flow allows sample refresh to eliminate any sample damage problems. The flow can refresh liquid samples in the rate of 50–200 nl s^{-1}. The valves are located near to the vacuum side to minimize the volume of liquid exposed to vacuum chamber. In the event of the window being broken, the impact to the vacuum chamber will be minimal.

5.4 Results and Discussion

The research interest in nanostructured TiO_2 is based on the possibilities of using the material in various applications, such as Li-ion batteries, [41] displays [42] and dye-sensitized solar cells. [43] ZnO has nearly the same bandgap and electron affinity as TiO_2, making it a possible candidate as an effective dye sensitized solar cell (DSSC) semiconductor. While little work has been done on large-scale, template-free growth of TiO_2 nanowires, ZnO can readily be grown in a variety of morphologies and by several different processing methods. There have already been many soft X-ray spectroscopic experiments performed on ZnO crystals, nanocrystals and films, and also doped on ZnO [44,45].

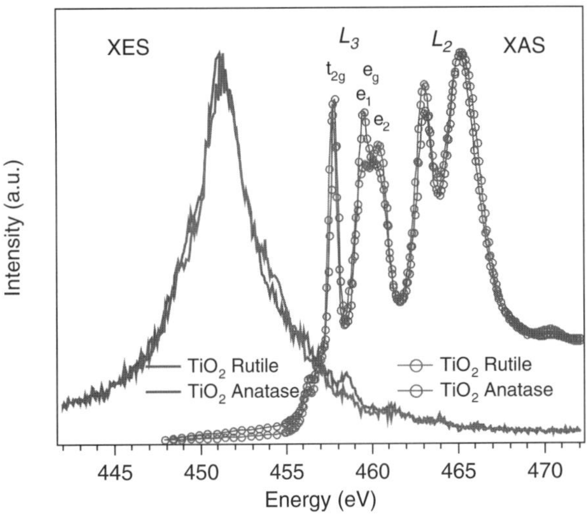

Figure 5.6 Ti2p-XAS and L-XES spectra of TiO_2 in crystal structure of anatase and rutile.

Figure 5.6 shows the Ti2p XAS spectra of nanoporous TiO_2. The X-ray absorption spectrum is derived from the two $L_3(2p_{3/2})$ (457–462 eV) and $L_2(2p_{1/2})$ (462–467 eV) parts, furthermore split into sharp t_{2g} and doublet split e_g (e_1 and e_2) states, due to slight distortion from the O_h symmetry. The e_1 peak at the lower energy side originates from the long Ti–O bonds due to a hybridization effect weaker than the short Ti–O bonds. Note that the intensity ratio of the doublet split e_g is reversed in rutile TiO_2 because of the slightly different crystal symmetry (D_{2h}) in comparison with anatase TiO_2, due to its D_{2d} crystal field [46]. Below the L_3 threshold two well-separated peaks (in 456–457 eV) are observed in the absorption spectrum. These have predominantly triplet character and are mixed through the spin–orbit interaction and the Coulomb repulsion into the main L_3 edge. [47].

For the XES spectra of both rutile and anatase TiO_2, the main peak around 451 eV to the transition from the peak of valence density of states to the Ti2$p_{3/2}$ core state; lower-intensity structures are seen at the higher energy side (at 457.5 eV) due to the Coster–Kronig process, which transfers the hole from Ti2$p_{1/2}$ to Ti2$p_{3/2}$ and to a multiple electron excitation.

O K-edge X-ray absorption and emission spectra of both rutile and anatase TiO_2 recorded in total electron yield (TEY) detection modes are depicted in Figure 5.7. The oxygen XAS spectra of both rutile and anatase can be divided into two regions. First, the well-defined pre-edge is attributed to O2p weighting of states that have predominantly transition-metal 3d character, that is, Ti3d–O2p mixing. The 3d states are split into two bands, which are related to t_{2g} and e_g symmetries, although this assignment is not strictly valid, as the crystal field is slightly distorted from octahedral symmetry. The second region, above 536 eV, is attributed to O2p character hybridized with Ti4s and 4p states. A similar trend has been observed in oxygen K-edge XAS of other 3d transition-metal oxide compounds. The relative intensity ratio of the pre-edge region to that of the second region is attributed to number of unoccupied Ti3d states and the degree of hybridization between Ti3d and O2p states. O K-edge absorption spectra clearly reveal that hybridization between Ti3d and O2p states in both rutile and anatase TiO_2 is extensive.

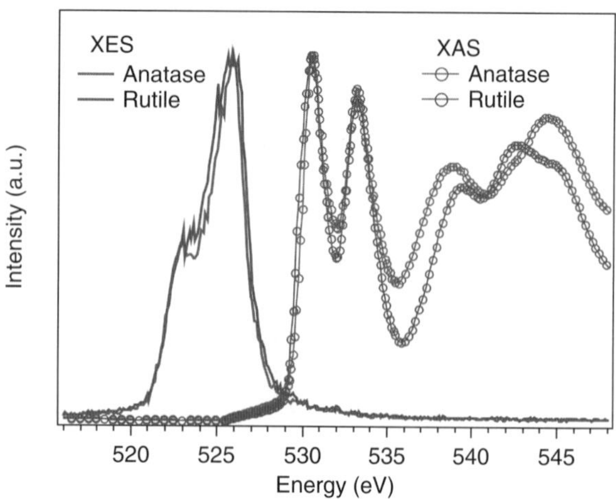

Figure 5.7 O *K*-edge XAS and XES spectra of TiO_2 in crystal structure of rutile and anatase.

A remarkable difference between rutile and anatase is the spectral structure in between 536 and 548 eV. The 538 eV peak in anatase and the 539 and 543 eV peaks in rutile are attributed to antibonding oxygen states, while the titanium 4*sp* band is related to the 544 eV peak in anatase and the 548 eV peak in rutile. The anatase crystal has an 8% less dense structure and smaller effective interactions of the titanium 4*sp* states. Hence they are antibonding and their position is at lower energy [48].

O *K*-emission spectra of TiO_2 (dotted lines) and Li_xTiO_2 (solid lines) shown in Figure 5.7. The oxygen emission spectra consist of two bands: a high-energy "main" band centered around 526 eV, and low-energy band at 522.8 eV. The assignment of each band can be easily made with reference to the other 3*d* transition-metal compounds. The main band originates from the O2*p* and Ti3*d*-derived states. Emission from the low-energy band is attributed to the O2*p* state.

The valence-core XES spectra of the rutile and anatase TiO_2 aligned together with the corresponding XAS spectra are shown in Figure 5.8. The bandgap is determined to be 1.6 for rutile and 2.1 eV for anatase TiO_2, respectively. In the case of anatase, the *absorption–emission* spectrum shows a larger bandgap. The enlarging of the bandgap in anatase TiO_2 is seen as the decrease of the valence-band maximum and increase of the conduction-band minimum.

In the lithium battery, the electrode consists of interconnected nanocrystallites forming a nanoporous structure with an extremely large inner surface allowing for electrochemical reactions to take place in almost the entire volume of the electrode. High charging capacities are reported, [41,42] when lithium is inserted into nanoporous anatase titanium dioxide. Schematically, the electrochemical insertion reaction is written as $x\mathrm{Li}^+ + \mathrm{TiO_2} = xe^-\mathrm{Li}_x\mathrm{TiO_2}$, where x is the mole fraction of lithium in the titanium dioxide.

Figure 5.9 shows the Ti *L*-emission spectra of Li-doped nanoporous TiO_2 exited at the photon energies indicated in the XAS spectrum (inset). The emission spectra mainly reflect the Ti3*d* states. Three contributions are identified in the resonantly excited XES spectra: normal emission features at constant photon emission energy, elastic scattering features at the excitation photon energy, and inelastic scattering features such as *dd* and charge-transfer (CT) excitations at energies below the excitation energy.

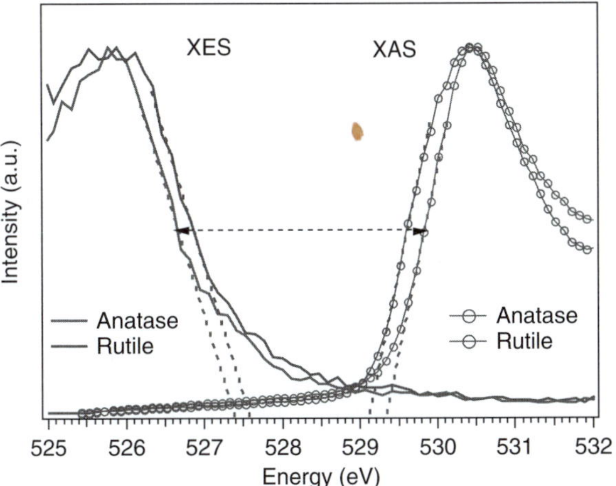

Figure 5.8 The bandgaps of rutile and anatase TiO_2 as revealed by the XAS and XES spectra.

The spectral profile of the resonant excited XES spectra Li-doped TiO_2 (Figure 5.10) shows two low energy-loss features within the bandgap below the elastic peak, which are not present in the undoped material. [49,50]. The energy-loss features are attributed to *dd* excitations that correspond to electron–hole pairs within or between the valence and conduction bands.

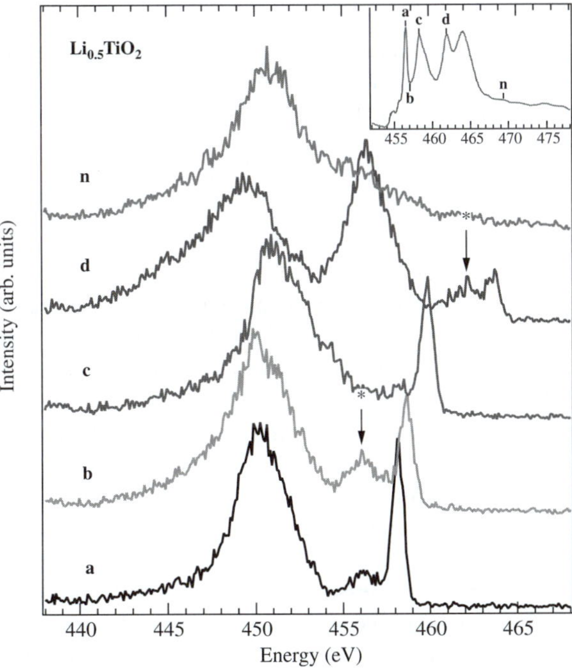

Figure 5.9 Resonant excited XES spectra of $Li_{0.5}TiO_2$ and XAS spectrum (inset).

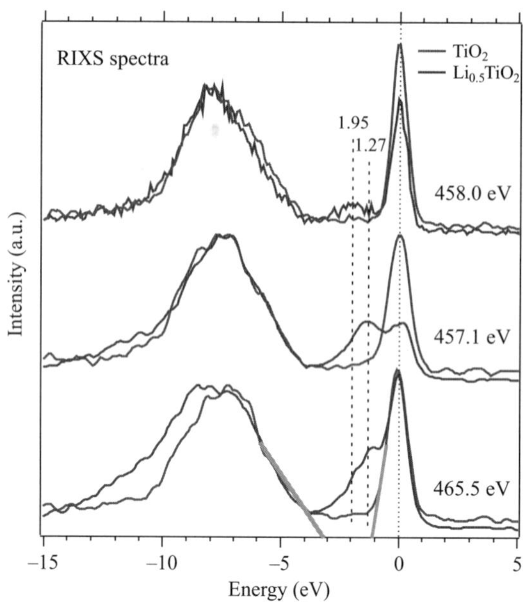

Figure 5.10 Resonant inelastic X-ray scattering of TiO_2 and Li-doped TiO_2.

The lithium insertion leads to an electron in the *d*-band (d^1), that is, t_{2g} states are occupied. In the 465.5 eV-spectrum, an asymmetry in the elastic peak is observed and at the next excitation energy (457.1 eV) this develops into a shoulder/peak. These low-energy excitations at 1.2 eV and 1.95 eV are attributed to electron–hole pairs within the t_{2g} band. The intensity of these excitations is enhanced when the excitation energy is tuned to the Ti^3-t_{2g} absorption feature, that is, the final state in the scattering process is an excited electron in the t_{2g} band. The relative intensity of the inelastic part is large, which suggests a significant electron correlation of the *d*-electrons in the system.

When the excitation energy lies between the L_3-edge t_{2g} and e_g in $Zr_{0.05}TiO_2$ (Figure 5.11) peaks in the absorption spectrum (b), there is an enhancement of the inelastic part in the resonantly excited XES spectra. This, on the other hand, is not observed at the t_{2g} (a). This could be due to the Ti^{3+} contribution to the absorption spectrum. At this particular energy (b) an enhancement of the Ti^{3+} contribution is observed where there is a dip in the Ti^{4+} spectrum (between t_{2g} and e_g). As the insertion concentration increases, this loss feature gains more intensity.

Advances in the synthesis of particles of nanometer dimensions, narrow size distribution and controlled shape have generated interest because of the potential to create novel materials with tailored physical and chemical properties [51,52]. New properties arise from quantum confinement effects and from the increasing fraction of surface atoms with unique bonding and geometrical configurations. Co nanocrystals display a wealth of size-dependent structural, magnetic, electronic and catalytic properties. The challenges in making isolated Co nanocrystals are to overcome the large attractive forces between the nanoparticles, due to surface tension and van der Waals interactions that tend to aggregate them [53,54].

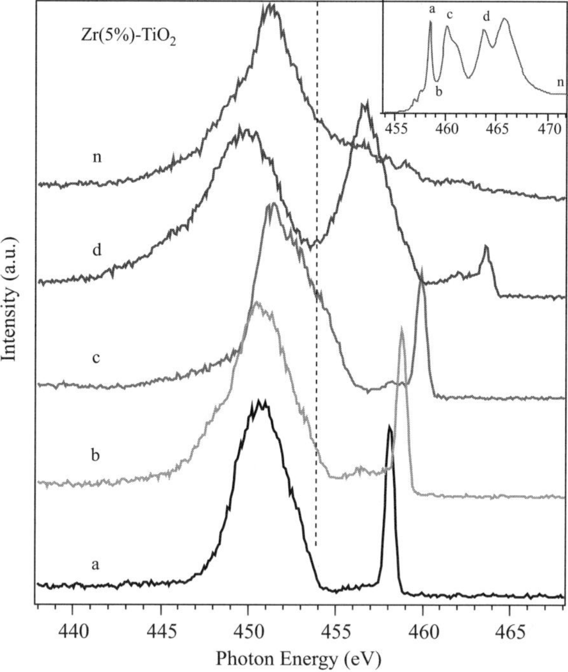

Figure 5.11 Ti 2p-XAS and L-XES spectra of Zr (5%)-doped TiO_2.

Using X-ray absorption and selectively excited X-ray emission spectroscopy to probe unoccupied and occupied electronic states, one can establish a firm interpretation for the unusual thermodynamic properties of molecular liquids. Furthermore, one can elucidate finer details of their structural properties. XAS and XES spectra reflect the local electronic structure of the various conformations; in this case, the oxygen lineshape is sensitive to the hydrogen-bonding configurations. XES spectra, emanating from the radiative decay, subsequent to core excitation, can be useful in assigning structures in XAS spectra [37,38,55,56]. Figure 5.12 shows an X-ray emission study of the local hydrogen bonding in liquid water.

We have shown that resonantly excited XES spectra of liquid water are compatible with the traditional view that three and four hydrogen bonds dominate in the structure [37]. At excitation high above threshold (545.5 eV), the experimental XES spectrum is well described by a calculation both with symmetric fourfold coordination, and conformations with one broken hydrogen bond. When tuning the excitation energy to the pre-peak at 534.7 eV, the XES spectral shape changes comply with what is expected when structures with one broken hydrogen bond at the hydrogen donor site (D-ASYM) are resonantly excited.

The electronic structure of cobalt nanocrystals suspended in liquid as a function of size has been investigated using *in situ* X-ray absorption and emission spectroscopy. A sharp absorption peak associated with the ligand molecules is found that increases in intensity upon reducing the nanocrystal size. X-ray Raman features due to *dd* and to charge-transfer excitations of ligand molecules are identified. The study reveals the local symmetry of the surface of ε-Co phase

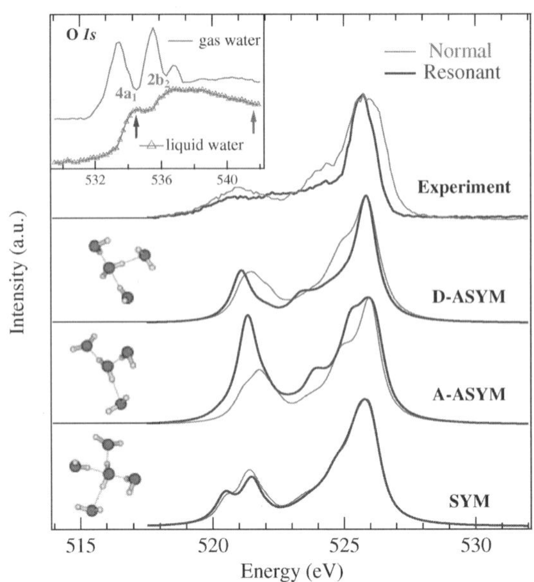

Figure 5.12 Resonant (534.7 eV) and normal (545.5 eV) excited XES spectra of liquid water in comparison with calculations for SYM, A-ASYM and D-ASYM species. Excitation energies are marked in the XAS spectrum.

nanocrystals, which originates from a dynamic interaction between Co nanocrystals and surfactant + solvent molecules [57].

In Co metal, the ground state is $4s^2 3d^7$, while for CoO, one uses the ground-state configuration $[3d^7 + 3d^8L^{-1}]$ (L^{-1}denotes a hole in the ligand level). Although this explained the CoO spectrum quite well, the $[3d^7 + 3d^8L^{-1}]$ never yielded a significant satellite contribution, as seen in Co nanocrystals. The only known octahedral systems with large satellites are cyanide complexes, where large satellites are caused by π back-bonding, that is, $[3d^7 + 3d^6L]$. The main structure is a $[2p^5 3d^8 + 2p^5 3d^7 L]$ bonding combination and the satellite is the antibonding part. It is worth noticing that the metal–ligand charge transfer (MLCT) acts mainly on the t_{2g} electrons.

Although there are extensive studies on carbon-nanotube-based gas sensors, only a few spectroscopic studies have been reported with regard to electronic structural changes upon gas interaction with SWNTs. *in situ* XAS experiments with SWNTs exposed to hydrogen gas at pressures of up to 450 Torr were performed by using a gas cell with a Si_3N_4 window of 100 nm thickness. The XAS spectra are displayed in Figure 5.13. The ambient conditions of the SWNTs were similar to that for SWNTs used as gas sensor. Spectral changes around π^* and σ^* features were observed upon introduction of H_2. It was suggested the molecular collision mechanism should be included when considering the interaction between gas and SWNTs [58].

There have been more uses of photon-in/photon-out soft X-ray spectroscopy in nanomaterial sciences. To name a few: the study of different N contents in ZnO:N thin films [59], Mg-induced increase of bandgap in $Zn_{1-x}Mg_xO$ nanorods [60], synthesis of highly water-soluble magnetite colloidal nanosrystals [61], charge transfer in nanocrystalline-Au/ZnO nanorods [62], size

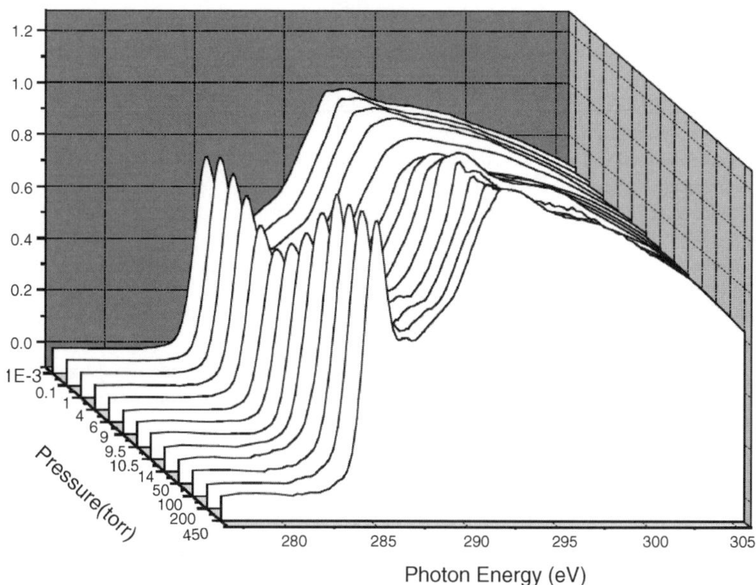

Figure 5.13 C *K*-edge XAS spectra of SWNTs under vacuum and ambient pressures.

dependence of the electronic structure of copper nanoclusters in SiC matrix [63], and electron correlation and charge transfer in $[(Ba_{0.9}Nd_{0.1})CuO_{2+\delta}]_2/[CaCuO_2]_2$ superconducting superlattices [64], one-dimensional quantum confinement effects in α-Fe_2O_3 ultrafine nanorod arrays [65] and polymerization of a confined π-system: chemical synthesis of tetrahedral amorphous carbon nanoballs from graphitic carbon nanocapsules [66].

Acknowledgments

The authors are grateful to the contribution from the collaborators, X. B. Chen, J.-W. Chiou, C. L. Dong, Y. Luo, J. Nordgren, W.-F. Pong, M. Salmeron, L. Vayssieres, Jun Zhong, Ziyu Wu, to name a few. The Advanced Light Source is supported by the Director, Office of Science, Office of Basic Energy Sciences, of the US Department of Energy under Contract No. DE-AC02-05CH11231.

References

[1] Crabtree, George W. and Lewis, Nathan S. (2007) Solar energy conversion. *Physics Today*, **60**, 37.

[2] Bolton, James R., Strickler, Stewart J., and Connolly, John S. (1985) Limiting and realizable efficiencies of solar photolysis of water. *Nature*, **316**, 495.

[3] Bard, Allen J. (1980) Photoelectrochemistry. *Science*, **207**, 139.

[4] Bolton, James R. (1978) Solar fuels. *Science*, **202**, 705.

[5] Guo, Jinghua (2004) Synchrotron radiation, soft-X-ray spectroscopy and nanomaterials. *International Journal of Nanotechnology*, **1–2**, 193.

[6] Guo, Jinghua (2006) X-ray Absorption and Emission Spectroscopy in Nanoscience and Lifesciences, in *Nanosystem Characterization Tools in the Life Sciences* (ed. Challa Kumar), Wiley-VCH Verlag GmbH & Co. KgaA, Weinheim, pp. 259–291.

[7] Zhong, Jun, Chiou, Jauwern, Dong, Chungli *et al.* (2008) Probing quantum confinement of single-walled carbon nanotubes by resonant soft-x-ray emission spectroscopy. *Applied Physics Letters*, **93**, 023107.
[8] Rubensson, J.-E., Wassdahl, N., Bray, G. *et al.* (1988) Resonant behavior in soft x-ray fluorescence excited by monochromatized synchrotron radiation. *Physical Review Letters*, **60**, 1759.
[9] Gelímukhanov, F.Kh., Mazalov, L.N., and Kondratenko, A.V. (1977) A theory of vibrational structure in the X-ray spectra of molecules. *Chemical Physics Letters*, **46**, 133.
[10] Ma, Y., Wassdahl, N., Skytt, P. *et al.* (1992) Soft-X-ray resonant inelastic scattering at the C *K* edge of diamond. *Physical Review Letters*, **69**, 2598.
[11] Gelímukhanov, F.Kh. and Ågren, H. (1994) Resonant inelastic X-ray scattering with symmetry-selective excitation. *Physical Review A*, **49**, 4378.
[12] Gelímukhanov, F.Kh. and Ågren, H. (1994) Channel interference in resonance elastic x-ray scattering. *Physical Review A*, **50**, 1129.
[13] Luo, Y., Ågren, H., and Gelímukhanov, F.Kh. (1994) Symmetry assignments of occupied and unoccupied molecular orbitals through spectra of polarized resonance inelastic x-ray scattering. *Journal of Physics B: Atomic, Molecular and Optical Physics*, **27**, 4169.
[14] Luo, Y., Ågren, H., Gelímukhanov, F.Kh. *et al.* (1995) Symmetry-selective resonant inelastic X-ray scattering of C_{60}. *Physical Review B*, **52**, 14479.
[15] Butorin, S.M., Guo, J.-H., Magnuson, M. *et al.* (1996) Low-energy *d-d* excitations in MnO studied by resonant x-ray fluorescence spectroscopy. *Physical Review B*, **54**, 4405.
[16] Kuiper, P., Guo, J.-H., Såthe, C. *et al.* (1998) Resonant X-ray raman spectra of Cu *dd* excitations in $Sr_2CuO_2Cl_2$. *Physical Review Letters*, **80**, 5204.
[17] Jia, J.J., Callcott, T.A., Shirley, Eric L. *et al.* (1996) Resonant inelastic X-Ray scattering in hexagonal boron nitride observed by soft-X-ray fluorescence spectroscopy. *Physical Review Letters*, **76**, 4054.
[18] Glans, P., Gunnelin, K., Skytt, P. *et al.* (1996) Resonant X-Ray Emission Spectroscopy of Molecular Oxygen. *Physical Review Letters*, **76**, 2448.
[19] Skytt, P., Glans, P., Guo, J.-H. *et al.* (1996) Quenching of symmetry breaking in resonant inelastic X-ray scattering by detuned excitation. *Physical Review Letters*, **77**, 5035.
[20] Skytt, P., Glans, P., Gunnelin, K. *et al.* (1997) Role of screening and angular distributions in resonant X-ray emission of CO. *Physical Review A*, **55**, 134.
[21] Warwick, T., Heimann, P., Mossessian, D. *et al.* (1995) Performance of a High-Resolution, High Flux density SGM undulator beamline at the ALS. *Review of Scientific Instruments*, **66**, 2037.
[22] Nordgren, J. and Nyholm, R. (1986) Design of a portable large spectral range grazing incidence instrument. *Nuclear Instruments and Methods in Physics Research Section A*, **246**, 242.
[23] Nordgren, J., Bray, G., Cramm, S. *et al.* (1989) Soft X-ray emission spectroscopy using monochromatized synchrotron radiation. *Review of Scientific Instruments*, **60**, 1690.
[24] Callcott, T.A., Tsang, K.-L., Zhang, C.H. *et al.* (1986) High-efficiency soft X-ray emission spectrometer for use with synchrotron radiation excitation. *Review of Scientific Instruments*, **57**, 2680.
[25] Jia, J.J., Callcott, T.A., Yurkas, J. *et al.* (1995) First experimental results from IBM/TENN/TULANE/LLNL/LBL undulator beamline at the advanced light source. *Review of Scientific Instruments*, **66**, 1394.
[26] Shin, S., Agui, A., Fujisawa, M. *et al.* (1995) Soft X-ray emission spectrometer for undulator radiation. *Review of Scientific Instruments*, **66**, 1584.
[27] Guo, J.-H., Glans, P., Skytt, P. *et al.* (1995) Resonant excitation x-ray fluorescence from C_{60}. *Physical Review B*, **52**, 10681.
[28] Ågren, H., Luo, Y., Gel'mukhanov, F. *et al.* (1995) Symmetry selective resonant inelastic X-ray scattering. *Physical Review B*, **105**, 2041.
[29] Guo, J.-H., Skytt, P., Wassdahl, N. *et al.* (1995) Resonant and non-resonant X-ray scattering from C_{70}. *Chemical Physics Letters*, **235**, 152.
[30] Nilsson, P.O., Kanski, J., Guo, J.-H. *et al.* (1995) Electronic structure of buried Si layers in GaAs(0 0 1) as studied by soft-x-ray emission. *Physical Review B*, **52**, R8643.
[31] Duda, L.-C., Isberg, P., Mirbt, S. *et al.* (1996) Soft-x-ray emission study of Fe/V (0 0 1) superlattices. *Physical Review B*, **54**, 10393.
[32] Glans, P., Skytt, P., Gunnelin, K. *et al.* (1996) Selectively excited X-ray emission spectra of N_2. *Journal of Electron Spectroscopy and Related Phenomena*, **82**, 193.

[33] Skytt, P., Glans, P., Gunnelin, K. *et al.* (1997) Lifetime-vibrational interference effects in the resonantly excited x-ray-emission spectra of CO. *Physical Review A*, **55**, 146–154.
[34] Gunnelin, Kerstin, Glans, Peter, Rubensson, Jan-Erik *et al.* (1999) Bond-length-dependent core hole localization observed in simple hydrocarbons. *Physical Review Letters*, **83**, 1315.
[35] Duda, L.-C., Isberg, P., Andersson, P.H. *et al.* (1997) Hydrogen-induced changes of the electronic states in ultrathin single-crystal vanadium layers. *Physical Review B*, **55**, 12914.
[36] Hjörvarsson, B., Guo, J.-H., Andersson, G. *et al.* (1999) Probing the local electronic structure in the H induced metal – insulator transition of Y. *Journal of Physics: Condensed Matter*, **11**, L119.
[37] Guo, J.-H., Luo, Y., Augustsson, A. *et al.* (2002) X-ray emission spectroscopy of hydrogen bonding and electronic structure of liquid water. *Physical Review Letters*, **89**, 137402.
[38] Guo, J.-H., Luo, Y., Augustsson, A. *et al.* (2003) The molecular structure of alcohol-water mixtures. *Physical Review Letters*, **91**, 157401.
[39] Guo, Jinghua, Tong, Tyler, Svec, Lukas *et al.* (2007) Soft-x-ray spectroscopy experiment of liquids. *Journal of Vacuum Science and Technology*, **25**, 1231.
[40] Forsberg, J., Duda, L.-C., Olsson, A. *et al.* (2007) System for *in situ* studies of atmospheric corrosion of metal films using soft X-ray spectroscopy and quartz crystal microbalance. *Review of Scientific Instruments*, **78**, 083110.
[41] Huang, S.Y., Kavan, L., Exnar, I., and Gratzel, M. (1995) Rocking Chair Lithium Battery Based on Nanocrystalline TiO_2 (Anatase). *Journal of the Electrochemical Society*, **142**, L142.
[42] Hagfeldt, A., Vlachopoulos, N., and Gratzel, M. (1994) Fast electrochromic switching with nanocrystalline oxide semiconductor films. *Journal of the Electrochemical Society*, **141**, L82.
[43] O'Regan, B. and Gratzel, M. (1991) A low cost, high efficiency solar cell based on dye-sensitized colloidal TiO_2 films. *Nature*, **353**, 737.
[44] Guo, J.-H., Vayssieres, L., Persson, C. *et al.* (2002) Polarization-dependent soft-x-ray absorption of highly oriented ZnO microrod arrays. *Journal of Physics: Condensed Matter*, **14**, 6969.
[45] Dong, C.L., Persson, C., Vayssieres, L. *et al.* (2004) Electronic structure of nanostructued ZnO from x-ray absorption and emission spectroscopy and local density approximation. *Physical Review B*, **70**, 195325.
[46] de Groot, F.M.F., Fuggle, J.C., Thole, B.T., and Sawatzky, G.A. (1990) $2p$ x-ray absorption of $3d$ transition-metal compounds: An atomic multiplet description including the crystal field. *Physical Review B*, **42**, 5459.
[47] de Groot, F.M.F., Fuggle, J.C., Thole, B.T., and Sawatzky, G.A. (1990) $L_{2,3}$ x-ray-absorption edges of d^0 compounds: K^+, Ca^{2+}, Sc^{3+}, and Ti^{4+} in O_h (octahedral) symmetry. *Physical Review B*, **41**, 928.
[48] de Groot, F.M.F., Faber, J., Michiels, J.J.M. *et al.* (1993) Oxygen $1s$ X-ray absorption of tetravalent titanium oxides: A comparison with single-particle calculations. *Physical Review B*, **48**, 2074.
[49] Matsubara, M., Uozumi, T., Kotani, A. *et al.* (2000) Polarization dependence of resonant X-ray emission spectra in early transition metal compounds. *Journal of the Physical Society Japan*, **69**, 1558.
[50] Harada, Y., Kinugasa, T., Matsubara, M. *et al.* (2000) Polarization dependence of soft-x-ray Raman scattering at the L edge of TiO_2. *Physical Review B*, **61**, 12854.
[51] Somorjai, G.A. and Borodko, Y.G. (2001) Research in nanosciences – Great opportunity for catalysis science. *Catalysis Letters*, **76**, 1.
[52] Konya, Zoltan, Puntes, Vitor F., Kiricsi, Imre *et al.* (2002) Novel two-step synthesis of controlled size and shape of platinum nanoparticles encapsulated in mesoporous silica. *Catalysis Letters*, **81**, 137.
[53] Puntes, Victor F., Krishnan, Kannan, M., and Alivisatos, A. Paul (2001) Colloidal Nanocrystal Shape and Size Control: The Case of Cobalt. *Science*, **291**, 2115.
[54] Puntes, Victor F., Gorostiza, Pau, Aruguete, Deborah M. *et al.* (2004) Collective behaviour in two-dimensional cobalt nanoparticle assemblies observed by magnetic force microscopy. *Nature Materials*, **3**, 263.
[55] Gunnelin, K., Glans, P., Skytt, P. *et al.* (1998) Assigning X-ray absorption spectra by means of soft-x-ray emission spectroscopy. *Physical Review A*, **57**, 864.
[56] Hellgren, N., Guo, J.-H., Såthe, C. *et al.* (2001) Nitrogen Bonding Structure in Carbon Nitride Thin Films Studied by Soft X-ray Spectroscopy. *Applied Physics Letters*, **79**, 4348–4350.
[57] Liu, Hongjian, Guo, Jinghua, Yin, Yadong *et al.* (2007) Electronic structure of cobalt nanocrystals suspended in liquid. *Nano Letters*, **7**, 1919.
[58] Zhong, Jun, Chiou, Jauwern, Dong, Chungli *et al.* (2008) Probing quantum confinement of single-walled carbon nanotubes by resonant soft-x-ray emission spectroscopy. *Applied Physics Letters*, **93**, 023107.

[59] Bär, M., Ahn, K.-S., Shet, S. *et al.* (2009) Impact of air exposure on the chemical and electronic structure of ZnO: Zn_3N_2 thin films. *Applied Physics Letters*, **94**, 012110.
[60] Chiou, J.W., Tsai, H.M., Pao, C.W. *et al.* (2008) Mg-induced increase of bandgap in $Zn_{1-x}Mg_xO$ nanorods revealed by X-ray absorption and emission spectroscopy. *Journal of Applied Physics*, **104**, 013709.
[61] Ge, Jianping, Hu, Yongxing, Biasini, Maurizio *et al.* (2007) One-Step Synthesis of Highly Water-Soluble Magnetite Colloidal Nanosrystals. *Chemistry-A European Journal*, **13**, 7153.
[62] Chiou, J.W., Ray, S.C., Tsai, H.M. *et al.* (2007) Charge transfer in nanocrystalline-Au/ZnO nanorods investigated by X-ray spectroscopy and scanning photoelectron microscopy. *Applied Physics Letters*, **90**, 192112.
[63] Shin, D.-W., Dong, C.L., Mattesini, M. *et al.* (2006) Size dependence of the electronic structure of copper nanoclusters in SiC matrix. *Chemical Physics Letters*, **422**, 543.
[64] Freelon, B., Augustsson, A., Guo, J.-H. *et al.* (2006) Electron correlation and charge transfer in $[(Ba_{0.9}Nd_{0.1})CuO_{2+\delta}]_2/[CaCuO_2]_2$ superconducting superlattices. *Physical Review Letters*, **96**, 017003.
[65] Vayssieres, L., Såthe, C., Butorin, S.M. *et al.* (2005) One-dimensional quantum confinement effect in α-Fe_2O_3 Ultrafine Nanorod Arrays. *Advanced Materials*, **17**, 2320.
[66] Chu, Cheng-Che, Hwang, Gan-Lin, Chiou, Jau-Wern *et al.* (2005) Polymerization of a confined π-system: chemical synthesis of tetrahedral amorphous carbon nanoballs from graphitic carbon nanocapsules. *Advanced Materials*, **17**, 2707.

6

X-Ray and Electron Spectroscopy Studies of Oxide Semiconductors for Photoelectrochemical Hydrogen Production

Clemens Heske[1], Lothar Weinhardt[2], and Marcus Bär[3]

[1]*Department of Chemistry, University of Nevada, Las Vegas, USA,*
Email: heske@unlv.nevada.edu
[2]*Experimentelle Physik II, Universität Würzburg, Germany,*
Email: lothar.weinhardt@physik.uni-wuerzburg.de
[3]*Helmholtz-Zentrum Berlin für Materialien und Energie, Berlin, Germany,*
Email: marcus.baer@helmholtz-berlin.de

6.1 Introduction

As outlined in various chapters throughout this book, a successful implementation of photoelectrochemical hydrogen production using sunlight (PEC) requires significant material science breakthroughs. A material needs to be found that simultaneously fulfils several requirements, among them an optimized bulk bandgap for efficient utilization of the incoming solar photon flux and its spectral distribution, an optimized electronic structure at the interface between the material and the surrounding electrolyte, and a sufficient chemical stability (lifetime) of the material under the conditions of a very high or a very low pH value in the electrolyte. While no single such material is in hand today, significant advances have been made with a variety of materials that fulfill at least one of these requirements, as outlined in this book.

How can the search for the optimal ("holy grail") material be facilitated? Ultimately, of course, a PEC candidate material will be judged by its ability to split water and to produce hydrogen in a cost-effective way. To reach this goal, however, individual properties of

On Solar Hydrogen & Nanotechnology Edited by Lionel Vayssieres

particular materials need to be understood and optimized, and, in particular, fundamental barriers in one (or more) of the requirements need to be identified. It is thus crucially important to be able to characterize candidate materials with respect to each material requirement individually, that is, independent of the other requirements. It is the purpose of this chapter to demonstrate how soft X-ray- and electron-based spectroscopic methods are a powerful and evolving "tool chest" to do just that: to focus on a specific materials requirement and to collect information about the fundamental properties of a candidate material that are of direct relevance to the ultimate performance in a PEC cell.

We will focus on two particular requirements: the understanding of the electronic structure of the material surface (and, ultimately, its interface with the electrolyte), as well as the chemical structure relevant for chemical stability, by discussing two examples.

The first example is based on the need of a PEC material to exhibit suitable positions of the conduction band minimum (CBM) at the surface of the hydrogen electrode and of the valence band maximum (VBM) at the surface of the oxygen electrode, respectively. A detailed knowledge of these levels is of large importance for the choice and optimization of an electrode material. However, a direct determination of these levels – especially at the surface of the material – is not straightforward. In most studies related to this topic, one of the energy levels (the VBM for p-type systems or the CBM for n-type systems) is determined by electrochemical methods. These techniques require specific sets of assumptions about the possibility of achieving flat-band conditions, and the position of the other band edge (i.e., the CBM for p-type systems or the VBM for n-type systems) is generally inferred from optically determined bulk bandgaps. However, in general, electronic bandgaps and band edge positions at the surface of compound semiconductors (and thus also at the interface with the electrolyte) are different from optical bulk gaps and bulk band edge positions (see, for example, [1–3] for respective studies on chalcopyrite compound semiconductors). For a correct description it is thus necessary to measure band edge positions and gaps directly with surface-sensitive techniques. This will be demonstrated in Section 6.3.

In terms of chemical stability, in particular of a multicomponent material system, a detailed understanding of the chemical composition at and near the surface is crucial. With such understanding, chemical changes during operation (or, as in our case, air exposure) can be monitored precisely, giving detailed insight into the fundamental chemical behavior of a PEC candidate material. Of particular interest is the ability to derive bond-specific composition information, that is, not just a quantification of the presence of specific elements, but rather a quantification of the presence of elements in a *particular chemical environment*. In Section 6.4, we will discuss the example of $ZnO:Zn_3N_2$ thin films and show how the local chemical environment of the oxygen atoms can be used to monitor degradation processes in such systems.

Some of the experimental methods that are capable of deriving such information potentially lend themselves to *in situ* studies, that is, soft X-ray investigations of a PEC cell *under operation*. This is an emerging and exciting new development in the field of soft X-ray spectroscopy, and a brief outlook will be given in Section 6.5.

First, however, a description of the experimental methods is given in Section 6.2. A wide variety of soft X-ray- and electron-based spectroscopies exists, of which a subset is discussed here, namely photoelectron spectroscopy, inverse photoemission spectroscopy, X-ray emission spectroscopy and X-ray absorption spectroscopy. Other techniques (e.g., Auger electron spectroscopy) can give valuable additional insights; in this text we will limit ourselves to the above-mentioned four techniques for simplicity only.

6.2 Soft X-Ray and Electron Spectroscopies

As sketched in Figure 6.1, soft X-rays and electrons interact with the electronic structure of a PEC candidate material in numerous ways. In photoelectron spectroscopy (PES), one of the fundamental techniques of surface science [4,5], an incoming soft X-ray photon excites an electron, either from a core level or the valence band. If excited with X-rays, then PES is often called "X-ray photoelectron spectroscopy" (XPS), while PES with UV excitation is often also called "UV photoelectron spectroscopy" (UPS). In PES, the emitted electron and its kinetic energy are detected, giving rise to a spectrum that shows the number of electrons emitted as a function of kinetic energy. The energy axis can be easily converted from "kinetic energy" to "binding energy," taking the excitation (photon) energy and the work function of the electron analyzer into account. While core-level spectra are element-specific and chemically sensitive, valence-band spectra give detailed insight into the band structure, in our context, in particular, the energy position of the VBM with respect to the Fermi energy (E_F). The information derived from PES is representative of the electronic structure *at the surface* of a sample, since the detected electrons have an inelastic mean free path in the nanometer range [6,7]. A photoemission spectrum can also be used to derive the work function of a sample surface (i.e., the minimal energy required to remove an electron from the sample to the vacuum level in front of the sample surface) by evaluating the secondary electron cut-off at low kinetic energies.

The inverted photoemission process, in which an incoming electron is placed into an unoccupied state above the vacuum level and subsequently decays into a lower unoccupied state, is used for inverse photoemission spectroscopy (IPES [8]). By detecting emitted UV photons as a function of electron energy, it is possible to extract a spectrum of the conduction band, in our context, in particular, to derive the energy position of the CBM with respect to E_F.

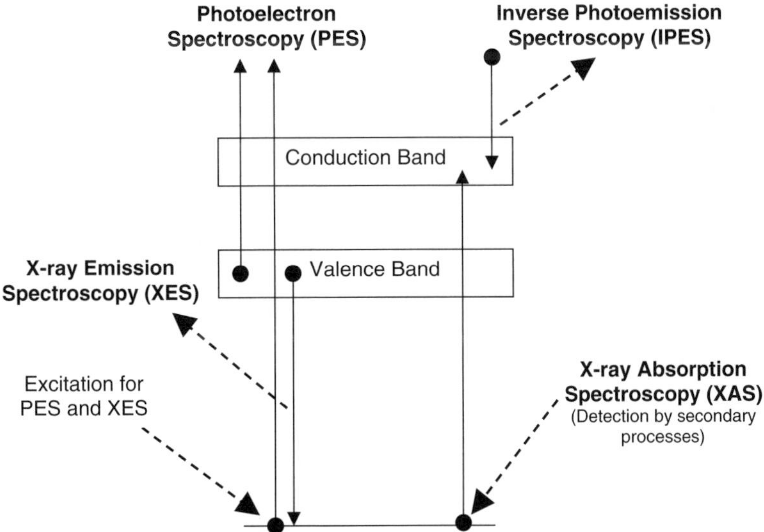

Figure 6.1 Schematic diagram of the soft X-ray spectroscopy techniques discussed in this chapter. Black dots represent electrons, solid arrows represent electronic transitions and dashed arrows represent incoming or outgoing photons.

As in the case of PES, IPES is a surface-sensitive technique due to the small inelastic mean free path of the incoming electrons. Thus, by combining PES and IPES, it is possible to derive a very detailed picture of the electronic structure at the surface of a sample, including the energies of VBM, CBM and the vacuum level with respect to the Fermi energy, as will be discussed in Section 6.3 using the example of WO_3. Note that the determination of the VBM and CBM energies also implies a determination of the electronic surface bandgap. As mentioned in the introduction, this electronic surface bandgap can deviate significantly from the optical bulk bandgap that is responsible for light absorption in a PEC candidate material.

In X-ray emission spectroscopy (XES [9,10]), a core electron is removed from the system, and one of the two possible channels for the subsequent decay of this core hole (namely the emission of a soft X-ray photon) is monitored. An XES spectrum thus depicts the number of emitted photons as a function of photon energy, which represents a partial and local density of states – "partial" because of the need to obey dipole selection rules, and "local" because of the local character of the core hole that initiates the X-ray emission process. An XES spectrum of a decay process that involves valence electrons thus gives additional insights into the valence-band structure, as will be demonstrated in Section 6.4. Furthermore, the localized character of the core hole can be used to probe the local chemical environment of the core-excited atom and thus allows XES to also reveal information about the chemical bonding.

In X-ray absorption spectroscopy (XAS or NEXAFS – near-edge X-ray absorption fine structure [11]), a core electron is excited into an unoccupied state and the ensuing core-hole decay process (Auger electron emission or X-ray fluorescence) is monitored, either in the electron channel (electron yield) and/or in the fluorescence channel (fluorescence yield – FY). XAS is thus sensitive to both chemical shifts in the ionized core level and the conduction-band structure into which the core electron is excited.

It should be noted that all of the here-described spectroscopies are not only governed by the initial state configuration of the specific technique (i.e., the ground state for PES or a core-hole state for XES), but are more precisely described by transition elements that take both initial and final states (and an operator describing the photon field) into account. In the case of XAS, for example, a determination of the energy position of the conduction-band minimum from the leading edge of the XAS spectrum can be obscured by the potential presence of a core exciton, that is, an electron–hole pair that is bound by Coulomb interaction and that is formed between the hole in the ionized core level and the electron excited into the conduction band. If a core exciton is present, the onset of the XAS spectrum is shifted to lower excitation energies, approximately by the binding energy of the core exciton.

In contrast to PES and IPES, which involve electrons in the detection or excitation process, respectively, XES and FY XAS use photon-in–photon-out processes. Consequently, the information depth is defined by the attenuation length of the employed soft X-ray photons, which is typically on the order of a few tens to a few hundred nanometers [12]. Thus, the information derived from XES and FY XAS spectra describes the near-surface bulk of the probed sample. Furthermore, the increased information depth forms the basis of exciting new developments in *in situ* XES and XAS spectroscopy using suitably designed *in situ* cells (see, e.g., [13–18]). While soft X-rays typically require an ultra-high-vacuum environment, these cells allow the use of XES and FY XAS of samples in non-vacuum environments, such as a PEC electrolyte. A brief outlook of such an approach is given in Section 6.5.

To perform such spectroscopies, sophisticated equipment is required. For IPES, a high-flux, low-energy electron source with a narrow thermal broadening is needed, combined with an efficient UV band-pass detection system. Such set-ups are rare and typically located in stationary ultra-high vacuum systems in a single-investigator lab environment.

PES set-ups, in contrast, have seen a remarkable commercial development and are frequently found in lab environments, as well as at synchrotron radiation sources. While the lab environment allows constant access, the synchrotron radiation environment has experimental advantages, in particular the tunability of the photon energy and the (potentially) higher-energy resolution compared to lab-based X-ray sources.

XES requires a high-flux excitation source and thus is best performed with high-brilliance synchrotron radiation from a third-generation synchrotron light source. For XES, the tunability of the excitation source is very important to optimize photoionization cross-sections. This is particularly true in the soft X-ray regime, since the competing relaxation process (i.e., Auger electron emission) is substantially faster and thus overwhelmingly dominant. Consequently, XES is a "photon-hungry" experiment, requiring high-flux excitation, as well as efficient detection. In order to extract detailed information about the electronic structure, however, this detection also needs to be performed with high-energy resolution (i.e., on the scale of a few tenths of an eV), which in turn calls for highly sophisticated soft X-ray spectrometers [19–22].

6.3 Electronic Surface-Level Positions of WO_3 Thin Films

6.3.1 *Introduction*

In the following, we will demonstrate how a combination of photoelectron spectroscopy (PES) and inverse photoemission (IPES) can be used to directly derive the band-edge position in vacuum and how these values can be correlated to the electrochemical energy scale relevant for a complete PEC device [23]. The discussion will exemplarily be done for WO_3, which has been discussed as a photoanode material for photoelectrochemical (PEC) hydrogen production in recent years (e.g., [24–29]).

WO_3 is a particular interesting example, since its optical and electronic properties strongly depend on the actual stoichiometry of the films. For example, it was found that O-poor films change color and have different electrical properties compared to stoichiometric films. A detailed discussion of these effects and their origin can be found in, for example, [30–32]. One advantage of WO_3 is that the films can be deposited with a large variety of different techniques, including RF- [33,34] and DC-sputtering [33,35], screen-printing [36] and thermal evaporation [37]. The WO_3 films studied in this chapter were deposited by the group of E. Miller, University of Hawaii, using reactive sputtering from a W target under an argon and oxygen ambient.

6.3.2 *Sample Handling and the Influence of X-Rays, UV-Light and Low-Energy Electrons on the Properties of the WO_3 Surface*

When employing very surface-sensitive techniques like PES and IPES, it is necessary to ensure that the measurement is not compromised by any surface contamination. This would change the properties of the surface itself and can also lead to contamination-induced spectral features that overlap with those of the actual sample. In principle, it would therefore be ideal to have sample preparation and analysis in one vacuum system. However, it is even more important that

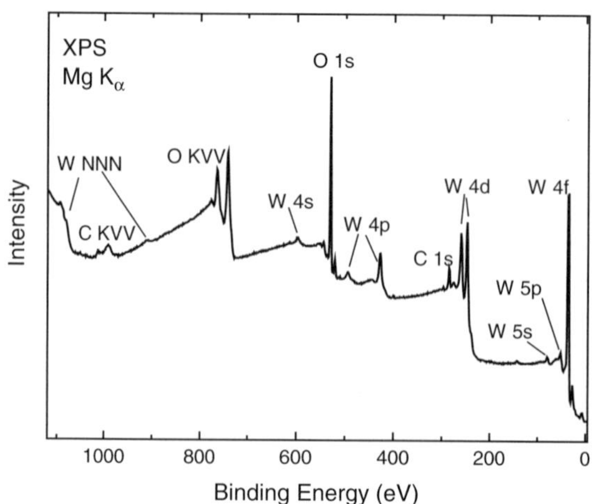

Figure 6.2 XPS survey spectrum of a WO_3 surface after an exposure time of 5 min. (Reprinted with permission from L. Weinhardt, M. Blum, M. Bär *et al.*, Electronic surface level positions of WO_3 thin films for photoelectrochemical hydrogen production, *Journal of Physical Chemistry C*, **112**(8), 3078, 2008. © 2008 American Chemical Society.)

the studied samples are of relevance, meaning that they originate from a system that is dedicated to produce optimized anode materials. In the present case, the growth system is located at the University of Hawaii, while the measurement system is situated at the University of Nevada, Las Vegas (UNLV). To be able to study a pristine WO_3 surface, sample handling thus has to be optimized. For this purpose, the samples were sealed under dry nitrogen conditions after film preparation at the University of Hawaii and immediately shipped to Las Vegas in an inert sample box equipped with desiccant. At UNLV, the samples were unpacked in a glovebox under nitrogen atmosphere and directly introduced into the ultra-high vacuum system used for the spectroscopic experiments.

The success of the careful packing and shipping method is indicated by the very low amount of C adsorbates at the WO_3 surface, as exemplarily shown by the survey spectrum in Figure 6.2. Such an initially low surface contamination level is desirable for every surface study and especially critical for the study of WO_3 surfaces. The latter is due to the fact that conventional surface-cleaning approaches such as Ar^+ ion sputtering leads to strong changes in the spectra of WO_3, as reported by Dixon *et al.* [31]. This is not only true for sputtering with high-energy ions – significant ion damage can already be observed at energies as low as 50 eV (and currents of approx. 50 nA cm^{-2}).

In addition to the careful sample handling discussed above, WO_3 requires special caution during the measurement: it shows a significant sensitivity to X-ray and electron irradiation. It is necessary to carefully monitor any changes induced by the excitation sources and to minimize the experiment duration, such that any irradiation-induced effects on the WO_3 surface can be neglected. Figure 6.3 shows the W 4f spectrum after increasing exposure time to nonmonochromatized Mg K_α radiation (1253.6 eV). The first of the recorded spectra (exposure time about 1 min) consists of two symmetric peaks (W $4f_{5/2}$ at 37.92 (±0.02) eV and W $4f_{7/2}$ at 35.80 (±0.02) eV), which are representative for W atoms with an oxidation state of +6 (as expected

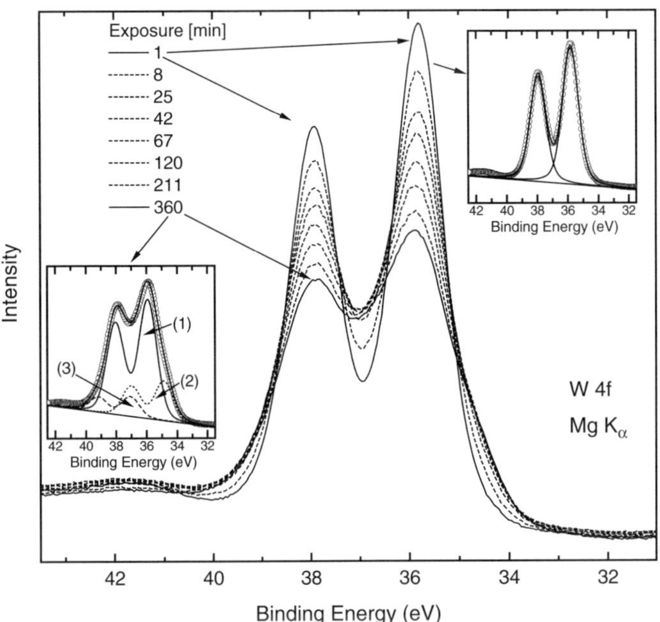

Figure 6.3 W 4f XPS spectra after increasing exposure time to Mg K_a radiation. The two insets show the fit of the W 4f line after very short exposure (top right), and the fit of the W 4f line after prolonged exposure (left). (Modified from L. Weinhardt, M. Blum, M. Bär *et al.*, Electronic surface level positions of WO_3 thin films for photoelectrochemical hydrogen production, *Journal of Physical Chemistry C*, **112**(8), 3078, 2008. © 2008 American Chemical Society.)

for the WO_3 environment) [30–32,36–39]. The presence of (only) two peaks is illustrated by the fit in Figure 6.3 (top right). Upon further exposure to Mg K_α radiation, the spectra change drastically. The fit on the bottom left in Figure 6.3 shows how the spectrum now consists of three different components. It is generally agreed that the low binding-energy component ((2) in Figure 6.3) is due to a loss of O at the surface and corresponds to W atoms in a $+5$ oxidation state [30–32,37–39], while the high binding-energy component (3) has been attributed either to surface defects [36,39] or a plasmon loss peak [38]. Further experiments show that the UV radiation used to measure the UV-PES spectra (He I and He II radiation) does not affect the WO_3 surface, while the low-energy electrons used for the IPES measurements do (on a timescale of a few tens of minutes), limiting the usable measurement time for this experiment.

6.3.3 Surface Band Edge Positions in Vacuum – Determination with UPS/IPES

The UPS spectrum of the valence band (left) is shown together with the IPES spectrum of the conduction band (right) on a common energy scale in Figure 6.4 (together with the secondary electron cut-off, bottom right, to be discussed in Section 6.3.4). For application in a PEC device, the positions of the VBM and the CBM are of interest, as discussed above. They can be determined from the spectra in Figure 6.4 by a linear extrapolation of the leading edge in the

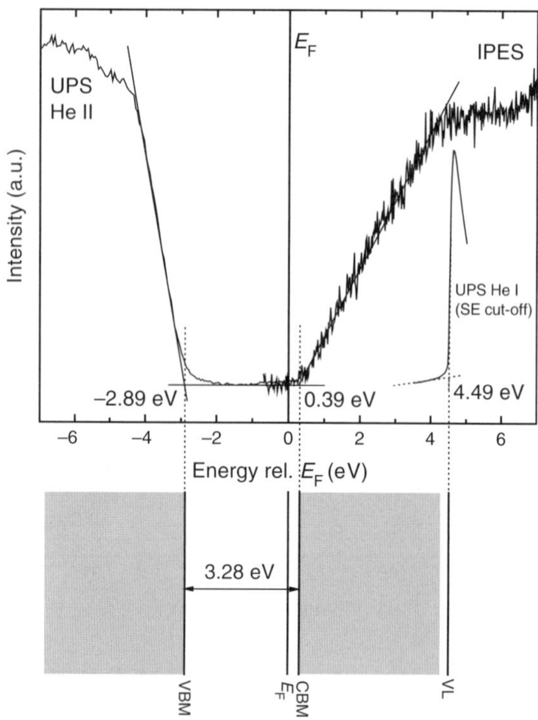

Figure 6.4 UPS and IPES spectra of an undamaged WO_3 surface after minimized exposure, together with the secondary electron (SE) cut-off measured by UPS. The linear extrapolations of the edges give the positions of the VBM, the CBM and the vacuum level (VL), respectively, as schematically shown below the panel. Error bars are ± 0.10 eV for the VBM and CBM positions, ± 0.14 eV for the derived electronic surface bandgap, and ± 0.05 eV for the VL (work function).

respective spectra. This procedure was found to yield correct results for a number of reference materials and is based on basic considerations regarding the shape of the UPS valence- and IPES conduction-band spectra [40]. Using the linear extrapolation as shown in Figure 6.4, the VBM can be determined to be -2.89 (± 0.10) eV and the CBM to be 0.39 (± 0.10) eV relative to E_F. This shows that the surface of the investigated WO_3 films is n-type. The bandgap at the surface of the WO_3 film can now simply be determined by adding up these values, yielding 3.28 (± 0.14) eV. This number is significantly larger than the bulk values of 2.6–2.9 eV commonly found by optical measurements [27,28,41,42], but coincides with the value of 3.2 eV shown in [43]. The observation of an increased bandgap at the surface of a polycrystalline semiconductor film is not unusual, as mentioned above. In the case of $CuInSe_2$, for example, the observed electronic surface bandgap of (solar cell) device-grade films (1.4 eV [1]) is significantly larger than the corresponding optical bulk bandgap (1.0 eV). In the present case, both structural, as well as compositional differences between bulk and surface could be responsible for the observed enhancement of the surface bandgap.

While we have directly determined the values of the VBM and CBM relative to E_F in vacuum, the respective positions relative to the normal hydrogen electrode (NHE) and thus relative to the redox potentials of water are of primary interest for electrochemical applications.

As we will show in the following subsection, these values can be determined from the values relative to E_F.

6.3.4 Estimated Surface Band-Edge Positions in Electrolyte

The first step to correlate the above-derived band-edge values to energy levels relevant for water splitting in an electrolyte is to express them with respect to the vacuum level. This is possible by directly measuring the work function of WO_3 using the secondary electron cutoff in the UPS spectrum. This cutoff, which is also shown in Figure 6.4 (bottom right), consists of the slowest electrons able to leave the sample (i.e., those that exactly reach the vacuum level directly in front of the sample surface) and therefore defines the position of the vacuum level E_{Vacuum} with respect to E_F (i.e., the work function Φ – note that the vacuum level is strictly defined only for a surface in vacuum). Again using a linear extrapolation, a value for Φ of 4.49 ($\pm$0.05) eV can be derived. Using this value and the absolute potential of the NHE of −4.44 eV (as recommended by the International Union of Pure and Applied Chemistry [44]), the band-edge position relative to the NHE can be calculated with:

$$-\mathrm{E(NHE)}-4.44\,\mathrm{eV} = E_{\mathrm{Vacuum}} = E_F + \Phi. \tag{6.1}$$

This equation couples our vacuum-derived values with an energy scale relative to the NHE.

Note, however, that two potentially very significant uncertainties still exist. First, a semiconductor surface in vacuum typically exihibts a surface charge (e.g., in/at surface states, defects and/or adsorbates) and thus a band bending towards the surface may exist. It is therefore, in general, incorrect to assume that flat-band conditions prevail at a semiconductor/vacuum interface. In the present case of n-type WO_3, a potential band bending, if present, would be "upward" towards the surface and smaller than 0.39 ($\pm$0.10) eV (i.e., the measured distance between CBM and E_F at the surface, see Figure 6.4). An exact determination of the magnitude of the band bending could be achieved by determining the bulk Fermi level position within the bandgap using Hall measurements. Alternatively, one could measure the surface photovoltage under illumination with sufficiently high intensity to achieve flat-band conditions.

Second, the surface band bending in the semiconductor is generally different for the solid/vacuum interface compared to the solid/electrolyte interface due to the formation of a Helmholtz layer in the latter case. To discuss the impact of this uncertainty, we will first follow the approximative approach in [45] to correlate the vacuum measurements with the electrochemical energy scale by assuming a zero surface charge (i.e., flat-band conditions) in vacuum. Then, we will discuss the numerical results in view of the surface- and Helmholtz-layer-induced band bending discussed above.

As discussed in various publications [41,43,46], the band alignment at a solid/electrolyte interface is strongly affected by the creation of the Helmholtz layer. It can induce an additional band bending of significant magnitude up to the difference in (vacuum!) work function between the semiconductor surface and the metal counter-electrode surface. In the present case, however, we speculate that this effect is small for the following reason. First, we note that the H_2O/O_2 oxidation and H^+/H_2 reduction potentials are usually given for the standard state (i.e., pH = 0). In comparison, PEC experiments of WO_3 have been performed at pH = 1.3 [27–29]. Furthermore, it was found that the isoelectric point of WO_3 (i.e., flat-band conditions at the WO_3/electrolyte interface) corresponds to a pH value of 0.5 [47]. Thus, the measured band-edge positions have to be modified to accommodate for the difference in pH between standard state and isoelectric point. For metal oxides, this shift is approximately 60 mV pH^{-1} [48], and, thus, the correction for a direct comparison of the measured band-edge positions of WO_3 with the redox potentials is very small (i.e., 0.5×60 meV = 30 meV). Finally, we note that the (vacuum) work function of WO_3 is lower than the work function of most counter-electrode metal surfaces. This suggests that not only the surface- but also the Helmholtz-layer-induced band bending should be upwards. Hence, even if there is a band bending present at the initial surface (in vacuum), the measured band positions would have to be shifted down by (at most) 0.39 (±0.10) eV if the sample is brought into contact with electrolyte at pH 0.5 (i.e., at the isoelectric point). This shift in band position would be counteracted by the formation of the Helmholtz layer, potentially leading to an upward band bending. Altogether, since the pH values of relevance for our comparisons (pH at the standard state and under operating conditions) are very close to the isoelectric point of WO_3, we speculate that the impact of Helmholtz layer formation on the relevance of the vacuum-derived energetic levels is rather small in our case.

A deliberate change of the pH value of the electrolyte will shift the redox potentials by the same amount as the electronic levels (according to the Nernst Equation). This effectively "couples" the relative electronic levels throughout the full pH range, and thus no further corrections due to pH variation are necessary.

By using Equation 6.1, we can now derive the positions of VBM and CBM *relative to the NHE*, as summarized in Table 6.1 and Figure 6.5, including the 30 meV correction discussed above. For comparison, the relevant levels for the splitting of water are also included.

The band-edge positions of a device for photoelectrochemical hydrogen production are crucial for its function, as mentioned above. Once an electron–hole pair is created by a photon and the charges are separated, the CBM at the photocathode (i.e., where the electrons are collected) has to be above that of the H^+/H_2 reduction potential. Likewise, the VBM at the photoanode (i.e., where the holes are collected) has to be below the H_2O/O_2 oxidation potential. If both conditions are met, water splitting can take place. For our case of WO_3 this implies that, in principle, water splitting should be possible: the CBM of WO_3 lies above the H^+/H_2 reduction potential. However, the separation between these two levels is (at most) 0.31 (±0.11) eV and has potentially to be corrected for the neglected surface-induced band

Table 6.1 Experimental energetic positions of the valence band maximum (VBM) and conduction band minimum (CBM) of WO_3 relative to the Fermi energy (E_F), the vacuum level (E_{Vacuum}), and the normal hydrogen electrode (NHE). The positions of the H_2O/O_2 and H^+/H_2 redox potentials are given for comparison. The values in the "E rel. NHE" column do not include a potential surface band bending in the vacuum-based experiments. (Reprinted with permission from L. Weinhardt, M. Blum, M. Bär *et al.*, Electronic surface level positions of WO_3 thin films for photoelectrochemical hydrogen production, *Journal of Physical Chemistry C*, **112**(8), 3078, 2008. © 2008 American Chemical Society.)

	E rel. E_F (eV)	E rel. E_{Vacuum} (eV)	E rel. NHE (eV)
VBM	−2.89 (±0.10)	−7.38 (±0.11)	2.97 (±0.11)
CBM	0.39 (±0.10)	−4.10 (±0.11)	−0.31 (±0.11)
H_2O/O_2 oxidation potential		−5.67	1.23
H^+/H_2 reduction potential		−4.44	0

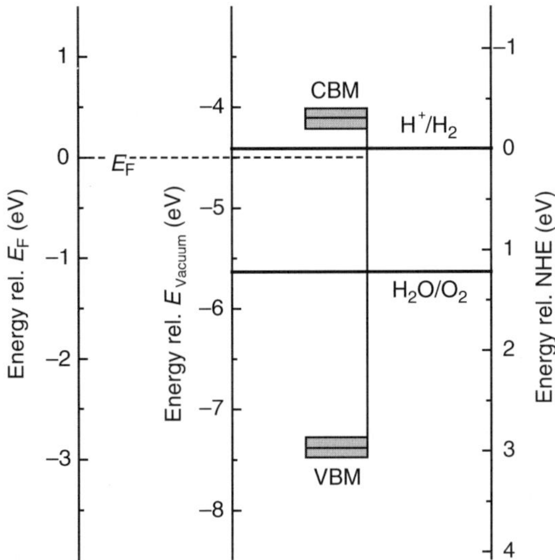

Figure 6.5 Positions of VBM and CBM of WO_3 relative to E_F, E_{Vacuum}, and the NHE. For comparison, the H_2O/O_2 oxidation and the H^+/H_2 reduction potentials are given. Gray bars for VBM and CBM indicate the error bars along the energy axis. The alignment of the "E rel. NHE (eV)" axis does not include a potential surface band bending in the vacuum-based experiments. (Reprinted with permission from L. Weinhardt, M. Blum, M. Bär *et al.*, Electronic surface level positions of WO_3 thin films for photoelectrochemical hydrogen production, *Journal of Physical Chemistry C*, **112**(8), 3078, 2008. © 2008 American Chemical Society.)

bending of the film in vacuum, which would further reduce this separation (in the extreme case even moving the CBM below the H^+/H_2 reduction potential). Furthermore, in this situation the current is zero. For a current to flow, the Fermi level of the metal has to be below the CBM of the semiconductor surface. The maximum possible current is then defined by how far the Fermi level can be shifted downwards before the metal Fermi level is pushed below the H^+/H_2 reduction potential. For pure WO_3 it is therefore necessary to add a bias for the cell to split water with a significant efficiency, in particular if additional overpotential losses need to be accommodated.

6.3.5 Conclusions

We have shown how PES and IPES can be used to derive the positions of the VBM and CBM at the surface of a compound semiconductor film. These levels play a crucial role for a PEC device, but have to be correlated with their respective positions in the electrolyte. This can be done by measuring the work function of the semiconductor, yielding energy levels that have to be compared with the positions of the relevant redox potentials for water splitting. Note that it is necessary, but not sufficient, that VBM and CBM straddle these potentials; for a current to flow, the energy separation needs to be large enough to accommodate a change in the band bending at the surface of the semiconductor (and to accommodate additional overpotentials if needed). For WO_3, this separation is too small, making an additional bias necessary for water splitting.

6.4 Soft X-Ray Spectroscopy of $ZnO:Zn_3N_2$ Thin Films[1]

6.4.1 Introduction

In the past, ZnO has been used in a large variety of applications [49]. For example, its wide bandgap ($E_g^{ZnO} = 3.3$ eV [50]) and large exciton binding energy make it a promising material for UV optoelectronic devices [49,51]. ZnO has also been suggested as a candidate material for photoelectrochemical hydrogen production (PEC) [52]. Since ZnO possesses similar or even better optical and electronic properties than TiO_2 (which has received much PEC attention in the past [[53] and chapters in this book]), it is expected to result in good PEC performance if chemical stability can be achieved. As in the case of TiO_2, the bandgap of ZnO is too large to effectively use the full spectrum of solar irradiation. It is therefore critical for PEC applications to reduce the bandgap of ZnO. One possible approach is to mix ZnO with Zn_3N_2. Since the chemical activity of oxygen is much higher than that of nitrogen [54], it is difficult to incorporate nitrogen into ZnO, even in the case of low N concentration (as used for p-type doping [49,51,55]). However, theoretical studies [56] have shown that the incorporation of nitrogen into ZnO can be enhanced by increasing the "chemical potential" of the nitrogen source. Recently, a corresponding reactive radio frequency (RF) sputter-based deposition method [57] was developed. It uses the RF sputter power as a parameter to control the incorporation of nitrogen. Such layers show an increasing N content with increasing sputter power and the (expected) corresponding E_g reduction down to 1.76 eV for an N content of approx. 35 at% [57]. Corresponding $ZnO:Zn_3N_2$ ("ZnO:N") thin films indeed show a higher photoresponse than pure ZnO [52], which is promising for their application in PEC devices. However, after prolonged storage in ambient air, a discoloration of the ZnO:N layers with high N content is observed and first tests indicate that the films are not stable in electrolyte [52]. Both observations show that ZnO:N layers have to be optimized in terms of stability, which is critical with respect to their use in PEC applications. To study the changes of the chemical and electronic (surface) structure of ZnO:N layers upon exposure to ambient air, soft X-ray emission spectroscopy (XES) was used to identify instabilities and their underlying mechanisms [58].

6.4.2 The O K XES Spectrum of ZnO:N Thin Films – Determination of the Valence Band Maximum

A set of ZnO:N layers deposited with different RF sputter powers (100, 120 and 150 W) was characterized by XES (see [58] for more details). The O K XES spectrum of the ZnO:N sample (150 W) with the nominally highest N content is compared to that of a ZnO powder reference in Figure 6.6 (left). The spectra are dominated by four distinct features A–D. A, B and C are ascribed to hybridized O 2p – Zn 4p, Zn 4s, and Zn 3d states [59,60], respectively, decaying into the O 1s core hole. Feature D, in contrast, is attributed to the Zn L_2 emission (detected in second order of the spectrometer and excited by higher harmonics of the beamline). In the following, feature C is thus used as an indicator for oxygen atoms in O–Zn bonds, while feature D is a measure for the entire Zn content, irrespective of chemical environment.

The comparison of the O K XES spectra of the ZnO reference and the 150 W ZnO:N layer reveals that the position of the maximum slightly shifts from approx. 526.6 to 526.4 eV. Furthermore, the shoulder at approx. 523.9 eV (feature B) is less pronounced in the spectrum of

[1]Reprinted with permission from M. Bär, K-S. Ahn, S. Shet, *et al.* Band gap narrowing of ZnO:N films by varying rf sputtering power in O_2/N_2 mixtures, *Applied Physics Letters*, **94**, 012110, 2009. © 2009 American Institute of Physics.

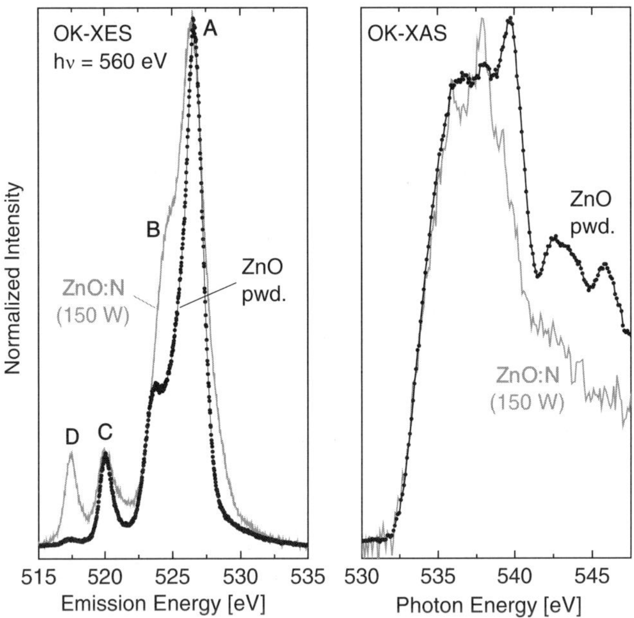

Figure 6.6 O K XES (left) and XAS (right) spectra of the 150 W ZnO:N sample, compared to those of a ZnO powder reference.

the ZnO:N sample, but increases in intensity (relative to feature A). Close inspection of features C and D, as well as of the onset at high emission energies (i.e., the region near the VBM) shows that the C/D intensity ratio of the ZnO:N sample is reduced and that the high-energy onset is shifted towards higher emission energies (i.e., opposite to their maximum). The latter is in good agreement with theoretical considerations [57], showing that the incorporation of N into ZnO (and hence the presence of N 2p states) shifts the VBM towards lower binding energies (here: higher emission energies). Theory also suggests that this VBM shift is primarily responsible for the E_g reduction, since no shift of the position of the CBM was observed in corresponding calculations [57]. An analysis of the onsets of respective XAS spectra (Figure 6.6, right) indeed shows no significant CBM shifts, corroborating the theoretical predictions.

6.4.3 The Impact of Air Exposure on the Chemical Structure of ZnO:N Thin Films

In order to quantify the impact of air exposure on the chemical structure of the ZnO:N layers, the corresponding $Zn_3N_2/(Zn_3N_2 + ZnO)$ ratio was derived based on the presented XES data. To avoid quantification uncertainties induced by variations of the excitation intensity (from the synchrotron source) and sample positions for each measurement, intensity *ratios* rather than absolute intensities were used. As mentioned above, feature C (hybridized O 2p–Zn 3d states) is directly indicative for oxygen atoms in O–Zn bonds, while feature D (Zn L_2 emission) is a measure for all Zn atoms in the probed volume. Thus, assuming that the investigated samples are exclusively composed of Zn_3N_2 and ZnO and using the C/D intensity of the pure ZnO sample [C/D(ZnO)] as reference, one can use the C/D intensity ratio of the different samples to

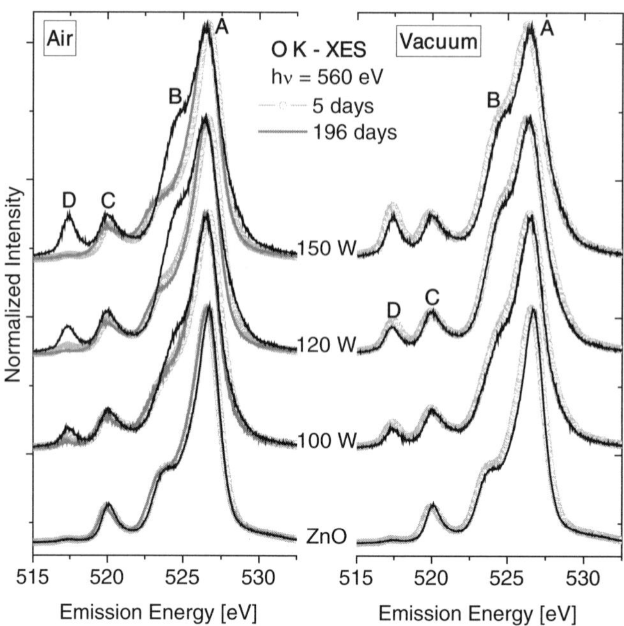

Figure 6.7 Initial O K XES spectra (black lines in both panels) of the ZnO:N sample set (together with a ZnO reference), compared to corresponding spectra of samples stored in air (left) and vacuum (right). The spectra were taken after 5 days (open spheres) and after 196 days (grey lines) of storage. (Reprinted with permission from M. Bär, K-S. Ahn, S. Shet *et al.*, Band gap narrowing of ZnO:N films by varying rf sputtering power in O_2/N_2 mixtures, *Applied Physics Letters*, **94**, 012110, 2009. © 2009 American Institute of Physics.)

determine their $Zn_3N_2/(Zn_3N_2 + ZnO)$ value as {1 – [C/D: C/D(ZnO)]}. The corresponding $Zn_3N_2/(Zn_3N_2 + ZnO)$ ratio for the 150 W sample is 0.9.

After initial characterization, the ZnO:N samples were cut into two pieces. One set was stored in ambient air and the other one was stored in vacuum (see [58] for more details). After five days, the samples were re-characterized (open spheres in Figure 6.7). While no significant spectral changes can be observed for the pure ZnO samples or the ZnO:N samples stored in vacuum (close inspection of the O K spectra before and after 5 days in vacuum reveals a slight decrease of the C/D intensity ratio, see [58] for a detailed discussion), the O K spectra of the ZnO:N samples exposed to ambient air show substantial changes. Compared to the initial O K spectra, the C/D intensity ratio is significantly increased. Furthermore, the intensity of feature B is decreased and the high-energy onset is shifted to lower emission energies.

These findings indicate a decrease of the $Zn_3N_2/(Zn_3N_2 + ZnO)$ ratio at the sample surfaces. An additional re-characterization after 196 days of air exposure (grey lines in Figure 6.7, left) shows a continuation of this trend. Furthermore, we observe that these changes are more pronounced for high RF powers (i.e., higher initial N contents). The quantification of the $Zn_3N_2/(Zn_3N_2 + ZnO)$ ratio (using the above-described C/D intensity ratio approach) reveals that the "loss" of N indeed depends on the RF power and thus on the initial $Zn_3N_2/(Zn_3N_2 + ZnO)$ ratio of the ZnO:N layer, as shown in Figure 6.8. Note that optical measurements show that the bandgap of ZnO:N samples increases if they are stored in ambient air. This indicates that the observed compositional changes are not limited to the sample surface, but eventually also affect the sample bulk.

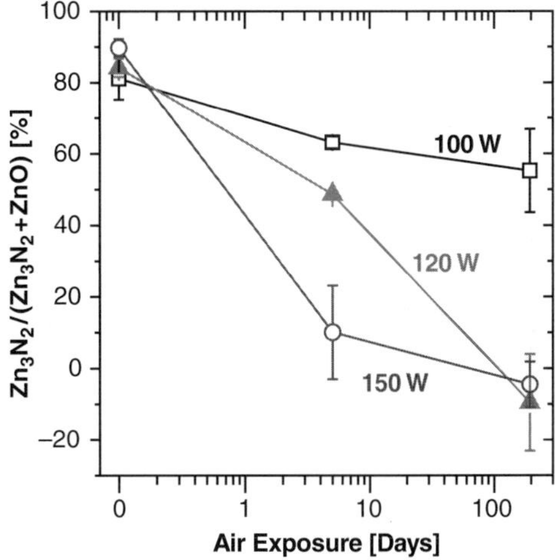

Figure 6.8 $Zn_3N_2/(Zn_3N_2 + ZnO)$ ratios of the investigated ZnO:N samples (derived from XES) as a function of air exposure time. (Reprinted with permission from M. Bär, K-S. Ahn, S. Shet *et al.*, Band gap narrowing of ZnO:N films by varying rf sputtering power in O_2/N_2 mixtures, *Applied Physics Letters*, **94**, 012110, 2009. © 2009 American Institute of Physics.)

According to the enthalpy of formation of ZnO ($-350.5\,kJ\,mol^{-1}$ [61]) and Zn_3N_2 ($-22.6\,kJ\,mol^{-1}$ [62]), ZnO is significantly more stable than Zn_3N_2. Thus, an explanation for the "loss" of N and hence the decrease of the $Zn_3N_2/(Zn_3N_2 + ZnO)$ ratio observed for the air-exposed ZnO:N samples could be based on the (thermodynamically likely) conversion of Zn_3N_2 to ZnO in ambient air, but not in vacuum. Already in 1875 [63], it was reported that Zn_3N_2 is stable under "exclusion of air" (i.e., vacuum), and that it will heavily decompose when brought into contact with H_2O (i.e., $Zn_3N_2 + 3H_2O \rightarrow 3ZnO + 2NH_3$ [64]). Thus, we speculate that the humidity in the ambient air is already sufficient to initiate this process for our air-exposed ZnO:N samples, explaining the observed decrease of the $Zn_3N_2/(Zn_3N_2 + ZnO)$ ratio and the related increase of the optical bandgap (i.e., discoloration).

6.4.4 Conclusions

Using soft X-ray emission spectroscopy for the (quantitative) investigation of the chemical and electronic surface structure of ZnO:N layers, it was found that ZnO:N layers are stable in vacuum, but degrade in ambient air (presumably due to humidity). The observed change in surface composition depends on exposure time and initial $Zn_3N_2/(Zn_3N_2 + ZnO)$ ratio of the ZnO:N layer. Thus – based on our data – the large range of reported bandgap values for Zn_3N_2 could in fact very well be explained by a different degree of degradation. Even though the N content is much higher in our case, the findings might also provide insight into the variety of findings and instabilities reported for N-doped p-type ZnO [55,65]. In the context of PEC for hydrogen production, the findings demonstrate the susceptibility of ZnO:N to ambient conditions and show that XES, especially in combination with XAS, is a uniquely suited tool to monitor such influences, in the future possibly even under *in situ* conditions.

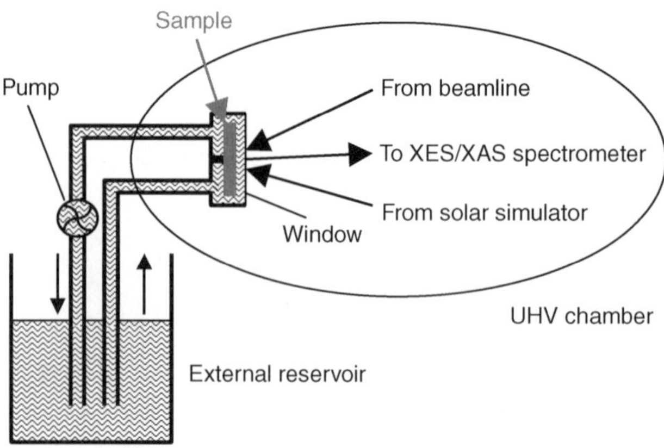

Figure 6.9 Schematic diagram of an experimental set-up for *in situ* XES and XAS studies of PEC candidate materials.

6.5 *In Situ* Soft X-Ray Spectroscopy: A Brief Outlook

As mentioned in Section 6.2, X-ray emission spectroscopy (XES) and fluorescence-yield X-ray absorption spectroscopy (FY XAS) are photon-in–photon-out spectroscopies and, thus, their information depth is governed by the attenuation length of soft X-rays, which, for many host materials, lies in the range of tens to hundreds of nanometers. This allows the excitation and detection of such spectroscopies through thin membrane windows, for example, 1 μm of polyimide or 100 nm of Si_3N_4 [13,14]. Consequently, *in situ* cells can be designed that allow the sample under study to be immersed in an environment different from vacuum. While current operational *in situ* liquid/solid interface cells are static [14], that is, do not flow or replenish the liquid, future set-ups will include a flow-through design, as schematically depicted in Figure 6.9. In the case of PEC, it is envisioned to build a cell that flows the electrolyte (and formed gases) through a compartment housing the sample. The compartment is separated from the ultra-high vacuum of the synchrotron beamline by a thin membrane window, and the sample is electrically insulated and connected to external leads for the application of an external bias. The sample is illuminated by a solar simulator and, simultaneously, by the soft X-ray excitation from the beamline, allowing monitoring of the surface of PEC candidate materials under operation. This will give insights into the electronic structure and, as described in the example in Section 6.4, into the chemical structure and stability of PEC candidate material surfaces. Furthermore, it will complement the electronic structure data that can only be obtained by the powerful electron- (and vacuum-) based techniques described in Section 6.3, thus leading to a comprehensive characterization of the electronic structure of PEC candidate materials.

6.6 Summary

In summary, this chapter has discussed four experimental soft X-ray- and/or electron-based techniques and has shed light on some of their characteristic properties that promise to be helpful in the search for the PEC material of the future. Detailed insight into the electronic and chemical structure can be gained, which: (i) aids in the verification of the "status quo," (ii) can point out shortcomings in the current systems that allow an insight-based approach to overcome the currently dominating barriers and (iii) even provides a tool to monitor whether deliberately induced modifications in the composition or preparation of PEC candidate materials indeed lead to the desired changes.

Numerous materials challenges have to be overcome to make large-scale PEC for hydrogen production a viable component of the future energy portfolio, and consequently numerous requirements for the detailed characteristics of PEC materials exist. Soft X-ray spectroscopies promise to play an important supporting role in fulfilling these requirements in an insight-based, custom-designed approach, working hand-in-hand with the materials screening and preparation activities in the PEC community.

Acknowledgments

We gratefully acknowledge very fruitful collaborations with B. Cole, B. Marsen, N. Gaillard and E. Miller, Hawaii Natural Energy Institute, School for Ocean, Earth Science, Technology, University of Hawaii, Manoa, Hawaii; K.-S. Ahn, S. Shet, Y. Yan and M. Al-Jassim, National Renewable Energy Laboratory, Golden, Colorado; and O. Fuchs and M. Blum, Experimentelle Physik II, Universität Würzburg, Germany. The contributions of K. George, S. Pookpanratana, Y. Zhang and S. Krause, Department of Chemistry, UNLV, are gratefully acknowledged, and without the help of W. Yang and J.D. Denlinger, Advanced Light Source, Lawrence Berkeley National Laboratory, none of our synchrotron-based experiments would have been possible. We gratefully acknowledge very fruitful discussion with T.F. Jaramillo, Stanford University. Finally, we acknowledge funding by the Department of Energy under subcontracts UNLV #RF-05-HFS-006 (grant no. DE-FG36-03GO13063), UNLV #RF-05-SHGR-004 (grant no. DE-FG36-03GO13062), and NFH-8-88502-01 (prime contract DE-AC36-99GO10337), as well as by the Nevada System of Higher Education under SFFA No. NSHE 07-100. M.B. also thanks the Deutsche Forschungsgemeinschaft (Emmy-Noether-Programm).

References

[1] Morkel, M., Weinhardt, L., Lohmüller, B. *et al.* (2001) Flat conduction band alignment at the CdS/$CuInSe_2$ thin film solar cell heterojunction. *Applied Physics Letters*, **79**, 4482.

[2] Weinhardt, L., Bär, M., Muffler, H.-J. *et al.* (2003) Impact of Cd^{2+}-treatment on the band alignment at the ILGAR-ZnO/CuIn(S,Se)$_2$ heterojunction. *Thin Solid Films*, **431–432**, 272.

[3] Bär, M., Weinhardt, L., Pookpanratana, S. *et al.* (2008) Depth-resolved band gap energies in Cu(In,Ga)(S,Se)$_2$ thin films. *Applied Physics Letters*, **93**, 244103.

[4] Hüfner, Stefan (2003) *Photoelectron Spectroscopy: Principles and Application*, 3rd edn, Springer.

[5] Briggs, D. and Seah, M.P. (1990) *Practical Surface Analysis, Volume 1, Auger and X-ray Photoelectron Spectroscopy*, 2nd edn, John Wiley & Sons.

[6] Seah, M.P. and Dench, W.A. (1979) Quantitative electron spectroscopy of surfaces: A standard data base for electron inelastic mean free paths in solids. *Surface and Interface Analysis*, **1**, 2.

[7] NIST Electron Inelastic-Mean-Free-Path Database: Version 1.1, C. Powell (scientific contact), http://www.nist.gov/srd/nist71.htm (accessed 3 September 2009).

[8] Smith, N.V. (1988) Inverse photoemission. *Reports on Progress in Physics*, **51**, 1227.

[9] Meisel, A., Leonhardt, G., and Szargan, R. (1989) *X-Ray Spectra and Chemical Binding, Springer Series in Chemical Physics*, **37**, Springer Verlag, Berlin, Heidelberg.

[10] (2000) See various articles in *Journal of Electron Spectroscopy and Related Phenomena*, **110/111**.

[11] Stöhr, J. (1998) *NEXAFS Spectroscopy*, Springer-Verlag, Berlin, Heidelberg, New York.

[12] To calculate an attenuation length, use the excellent web page by the Center for X-ray Optics, Lawrence Berkeley National Lab (www.cxro.lbl.gov): http://henke.lbl.gov/optical_constants/atten2.html (accessed 3 September 2009).

[13] Guo, J.-H., Luo, Y., Augustsson, A. *et al.* (2002) X-ray emission spectroscopy of hydrogen bonding and electronic structure of liquid water. *Physical Review Letters*, **89**, 137402–1.

[14] Heske, C., Groh, U., Fuchs, O. *et al.* (2003) Monitoring chemical reactions on surfaces in liquids: water on CuIn $(S,Se)_2$ thin film solar cell absorbers. *Journal of Chemical Physics*, **119**, 10467.

[15] Duda, L.C., Schmitt, T., Augustsson, A., and Nordgren, J. (2004) Resonant soft X-ray emission of solids and liquids. *Journal of Alloys and Compounds*, **362**, 116.

[16] Heske, C. (2004) Spectroscopic investigation of buried interfaces and liquids with soft x-rays. *Applied Physics A*, **78**, 829.

[17] Forsberg, J., Duda, L.-C., Olsson, A. *et al.* (2007) System for *in situ* studies of atmospheric corrosion of metal films using soft x-ray spectroscopy and quartz crystal microbalance. *Review of Scientific Instruments*, **78**, 083110.

[18] Fuchs, O., Maier, F., Weinhardt, L. *et al.* (2008) A liquid flow cell to study the electronic structure of liquids with soft x-rays. *Nuclear Instruments and Methods A*, **585**, 172–177.

[19] Nordgren, J. and Nyholm, R. (1986) Design of a portable large spectral range grazing incidence instrument. *Nuclear Instruments and Methods A*, **246**, 242.

[20] Jia, J.J., Callcott, T.A., Yurkas, J. *et al.* (1995) First experimental results from IBM/TENN/TULANE/LLNL/LBL undulator beamline at the advanced light source. *Review of Scientific Instruments*, **66**, 1394.

[21] Nordgren, J. and Guo, J. (2000) Instrumentation for soft x-ray spectroscopy. *Journal of Electron Spectroscopy and Related Phenomena*, **110–111**, 1.

[22] Fuchs, O., Weinhardt, L., Blum, M. *et al.* (2009) High-resolution, high-transmission soft x-ray spectrometer for the study of biological samples. *Review of Scientific Instruments*, **80** (6), 063103–063103-7.

[23] Weinhardt, L., Blum, M., Bär, M. *et al.* (2008) Electronic surface level positions of WO_3 thin films for photoelectrochemical hydrogen production. *Journal of Physical Chemistry C*, **112**, 3078.

[24] Santato, C., Ulmann, M., and Augustynski, J. (2001) Photoelectrochemical properties of nanostructured tungsten trioxide films. *Journal of Physical Chemistry B*, **105**, 936.

[25] Miller, E.L., Rocheleau, R.E., and Deng, X.M. (2003) Design considerations for a hybrid amorphous silicon/ photoelectrochemical multijunction cell for hydrogen production. *International Journal of Hydrogen Energy*, **28**, 615.

[26] Park, J.H., Park, O.O., and Kim, S. (2006) Photoelectrochemical water splitting at titanium dioxide nanotubes coated with tungsten trioxide. *Applied Physics Letters*, **89**, 163106.

[27] Miller, E.L., Marsen, B., Cole, B., and Lum, M. (2006) Low-temperature reactively sputtered tungsten oxide films for solar-powered water splitting applications. *Electrochemical and Solid-State Letters*, **9**, G248.

[28] Marsen, B., Cole, B., and Miller, E.L. (2007) Influence of sputter oxygen partial pressure on photoelectrochemical performance of tungsten oxide films. *Solar Energy Materials and Solar Cells*, **91**, 1954.

[29] Marsen, B., Miller, E.L., Paluselli, D., and Rocheleau, R.E. (2007) Progress in sputtered tungsten trioxide for photoelectrode applications. *International Journal of Hydrogen Energy*, **32**, 3110.

[30] Fleisch, T.H. and Mains, G.J. (1982) An XPS study of the UV reduction and photochromism of MoO_3 and WO_3. *Journal of Chemical Physics*, **76**, 780.

[31] Dixon, R.A., Williams, J.J., Morris, D. *et al.* (1998) Electronic states at oxygen deficient $WO_3(001)$ surfaces: a study by resonant photoemission. *Surface Science*, **399**, 199.

[32] Romanyuk, A. and Oelhafen, P. (2006) Evidence of different oxygen states during thermal coloration of tungsten oxide. *Solar Energy Materials and Solar Cells*, **90**, 1945.

[33] Franke, E.B., Trimble, C.L., Schubert, M. *et al.* (2000) All-solid-state electrochromic reflectance device for emittance modulation in the far-infrared spectral region. *Applied Physics Letters*, **77**, 930.

[34] Yamada, Y., Tabata, K., and Yashima, T. (2007) The character of WO_3 film prepared with RF sputtering. *Solar Energy Materials and Solar Cells*, **91**, 29.

[35] DeVries, M.J., Trimble, C., Tiwald, T.E. *et al.* (1999) Optical constants of crystalline WO_3 deposited by magnetron sputtering. *Journal of Vacuum Science and Technolology A*, **17**, 2906.

[36] Bittencourt, C., Felten, A., Mirabella, F. *et al.* (2005) High-resolution photoelectron spectroscopy studies on WO_3 films modified by Ag addition. *Journal of Physics: Condensed Matter*, **17**, 6813.

[37] Santucci, S., Cantalini, C., Crivellari, M. *et al.* (2000) X-ray photoemission spectroscopy and scanning tunneling spectroscopy study on the thermal stability of WO_3 thin films. *Journal of Vacuum Science and Technolology A*, **18**, 1077.

[38] Hollinger, G., Duc, T.M., and Deneuville, A. (1976) Charge transfer in amorphous colored WO_3 films observed by X-Ray photoelectron spectroscopy. *Physical Review Letters*, **37**, 1564.

[39] Ottaviano, L., Bussolotti, F., Lozzi, L. *et al.* (2003) Core level and valence band investigation of WO_3 thin films with synchrotron radiation. *Thin Solid Films*, **436**, 9.

[40] Gleim, Th., Heske, C., Umbach, E. *et al.* (2003) Formation of the ZnSe/(Te/)GaAs(100) heterojunction. *Surface Science*, **531**, 77.

[41] Bak, T., Nowotny, J., Rekas, M., and Sorrell, C.C. (2002) Photo-electrochemical hydrogen generation from water using solar energy. Materials-related aspects. *International Journal of Hydrogen Energy*, **27**, 991.

[42] Chandra, S. (1985) *Photoelectrochemical Solar Cells*, Gordon and Breach, New York, p. 98.

[43] Nozik, A.J. and Memming, R. (1996) Physical chemistry of semiconductor-liquid interfaces. *Journal of Physical Chemistry*, **100**, 13061.

[44] Trasatti, S. (1986) The absolute electrode potential: an explanatory note (Recommendations 1986). *Pure and Applied Chemistry*, **58**, 955.

[45] Chun, W.-J., Ishikawa, A., Fujisawa, H. *et al.* (2003) Conduction and valence band positions of Ta_2O_5, TaON, and Ta_3N_5 by UPS and electrochemical methods. *Journal of Physical Chemistry B*, **107**, 1798.

[46] Grimes, C.A., Varghese, O.K., and Ranjan, S. (2008) *Light, Water, Hydrogen*, Springer, New York.

[47] Parks, G.A. (1965) The isoelectric points of solid oxides, solid hydroxides, and aqueous hydroxo complex systems. *Chemical Reviews*, **65**, 177.

[48] Matsumoto, Y., Yoshikawa, T., and Sato, E. (1989) Dependence of the band bending of the oxide semiconductors on *p*H. *Journal of the Electrochemical Society*, **136**, 1389.

[49] Klingshirn, C. (2007) ZnO: from basics towards applications. *Physica Status Solidi B*, **244**, 3027.

[50] Pankove, J.I. (1971) *Optical Processes in Semiconductors*, Dover, New York.

[51] Özgür, Ü., Alivov, Y.I., Liu, C. *et al.* (2005) A comprehensive review of ZnO materials and devices. *Journal of Applied Physics*, **98**, 041301.

[52] Ahn, K.-S., Yan, Y., Lee, S.-H. *et al.* (2007) Photoelectrochemical properties of N-incorporated ZnO films deposited by reactive RF magnetron sputtering. *Journal of the Electrochemical Society*, **154**, B956.

[53] O'Regan, B. and Grätzel, M. (1991) A low-cost, high-efficiency solar cell based on dye-sensitized colloidal TiO_2 films. *Nature*, **353**, 737.

[54] Li, B.S., Liu, Y.C., Zhi, Z.Z. *et al.* (2003) Optical properties and electrical characterization of p-type ZnO thin films prepared by thermally oxiding Zn_3N_2 thin films. *Journal of Materials Research*, **18**, 8.

[55] Look, D.C., Reynolds, D.C., Litton, C.W. *et al.* (2002) Characterization of homoepitaxial *p*-type ZnO grown by molecular beam epitaxy. *Applied Physics Letters*, **81**, 1830.
Look, D.C. and Claflin, B. (2004) P-type doping and devices based on ZnO. *Physica Status Solidi B*, **241**, 624.

[56] Yan, Y., Zhang, S.B., and Pantelides, S.T. (2001) Control of doping by impurity chemical potentials: predictions for *p*-Type ZnO. *Physical Review Letters*, **86**, 5723.

[57] Ahn, K.-S., Yan, Y., and Al-Jassim, M. (2007) Band gap narrowing of ZnO:N films by varying rf sputtering power in O_2/N_2 mixtures. *Journal of Vacuum Science and Technolology B*, **25**, L23.

[58] Bär, M., Ahn, K.-S., Shet, S. *et al.* (2009) Impact of air-exposure on the chemical and electronic structure of ZnO: Zn_3N_2 thin films. *Applied Physics Letters*, **94**, 012110.

[59] Strunskus, T., Fuchs, O., Weinhardt, L. *et al.* (2004) The valence electronic structure of zinc oxide powders as determined by X-ray emission spectroscopy: variation of electronic structure with particle size. *Journal of Electron Spectroscopy and Related Phenomena*, **134**, 183.

[60] Dong, C.L., Persson, C., Vayssieres, L. *et al.* (2004) Electronic structure of nanostructured ZnO from x-ray absorption and emission spectroscopy and the local density approximation. *Physical Review B*, **70**, 195325.

[61] Lide, David R. (ed.) 88th Edition of the CRC Handbook of Chemistry and Physics 2007–2008.

[62] Wriedt, H.A. (1988) The N-Zn (Nitrogen-Zinc) system. *Bulletin of Alloy Phase Diagrams*, **9**, 247.

[63] Kraut, K. (1875) *Gmelin-Kraut's Handbuch der Chemie III*, Carl Winter's Universitätsbuchhandlung, Heidelberg, 33.

[64] Roscoe, H.E. and Schorlemmer, C. (1907) *A Treatise on Chemistry: Volume II, The Metals*, 4th edn, Macmillan, London, p. 650.

[65] Xiao, Z.Y., Liu, Y.C., Mu, R. *et al.* (2008) Stability of *p*-type conductivity in nitrogen-doped ZnO thin film. *Applied Physics Letters*, **92**, 052106.

7

Applications of X-Ray Transient Absorption Spectroscopy in Photocatalysis for Hydrogen Generation

Lin X. Chen[1,2]

[1]*Chemical Sciences and Engineering Division, Argonne National Laboratory, Argonne, Illinois, 60439, USA, Email: lchen@anl.gov*

[2]*Department of Chemistry, Northwestern University, Evanston, Illinois, 60208, USA*

7.1 Introduction

Solar hydrogen generation ($2H_2O$ + light = $2H_2$ + O_2) is a renewable, clean and abundant way of producing alternative fuels [1,2]. Because this reaction is uphill in energy, photocatalysis is needed, where reactants or catalysts can absorb energy from photons of sunlight. Three schemes have emerged in solar hydrogen generation from water splitting: (i) doped TiO_2 or other inorganic nanoparticles coupling water reduction and water oxidation via a Z-scheme to simultaneously generate O_2 and H_2 [1–4]; (ii) transition-metal complex-mediated artificial photosynthetic systems consisting of light-harvesting antenna, and electron donors and acceptors, as well as metal complex redox centers [5–9]; and (iii) hybrid biomimetic systems coupling photosystem II with biomimetic hydrogenase [10,11]. Although these schemes vary significantly, they share common fundamental photocatalytic processes in three basic steps: light absorption, charge separation and reduction/oxidation of reactants, such as water or other reactants, which eventually lead to hydrogen production.

The metal center can be a primary light absorber, or an electron donating or receiving entity that accumulates multiple electrons or holes for a redox reaction necessary for fuel production, such as water splitting. As the primary light absorber, solar photons will generate the excited

On Solar Hydrogen & Nanotechnology Edited by Lionel Vayssieres

state of the metal center, and elevate its energy over the reaction potential energy barrier. As an electron donor or acceptor, the metal center will change its oxidation state one or more times to facilitate multiple electron redox reactions in water splitting, such as in the well-known Joliot–Kok cycle using a manganese cluster in photosynthetic water oxidation [12,13]. In a typical photocatalytic reaction, all three basic steps would happen at the metal centers, either in metal complexes or metal oxide nanoparticles. Consequently, the oxidation state of the metal center that catalyzes the reaction would change at different stages of the reaction, accompanied by coordination geometry rearrangements. In addition, light absorption and redox reactions are strongly coupled in time and space. Therefore, we can only discover the details of the catalytic mechanisms by following the electronic and nuclear configurations of the metal center structures during photocatalytic reactions, either at the interfaces of the reactants and catalysts, in solution, or in other disordered environments.

Optical and vibrational spectroscopies have provided very useful information on reaction kinetics and intermediate reactant structural dynamics, which have advanced our understanding of reaction mechanisms [14–16]. However, they provide no or limited direct transient structural information on the metal centers with atomic resolution during photocatalysis. Moreover, optical signals involving oxidation state or nuclear geometry changes of the metal centers are often weak, or overshadowed by other strong optical signals, such as those of the π–π^* transition in conjugated aromatic systems. Frequencies of vibrational modes involving metal centers are often very low (i.e., $<100\,cm^{-1}$) because of the large reduced mass, and may be difficult to detect. Hence, new methods need to be employed to specifically probe metal-center structures at the electronic and atomic levels, in terms of their oxidation states, molecular orbital (MO) configurations and coordination geometry, all of which are important for projecting reaction coordinates along photocatalytic pathways.

One of the main stumbling blocks in the rational design of photocatalysts for effective solar hydrogen production is the lack of knowledge of the reaction intermediate structural dynamics at the atomic and molecular levels relevant to hydrogen production. We have learned that structures of metal complexes or catalytic centers in nanoparticles determine their energetics and substrate binding using static structural studies, such as X-ray diffraction (XRD), X-ray absorption spectroscopy (XAS) and scattering (SAXS, small angle X-ray scattering and WAXS, wide angle X-ray scattering). However, it is often the dynamic nature of a photocatalyst structure that is crucial for the key transformation in the reaction. Therefore, developing tools for time-resolved concomitant structure/function analyses of fundamental molecular mechanisms is very important for the rational design of catalysts with optimal efficiencies for solar hydrogen generation. In order to resolve the time evolution of photocatalytic reaction coordinates in a molecular ensemble, the same reaction needs to be synchronized among all metal centers in the ensemble on the timescale of the chemical transformation. This can be achieved using an ultrafast laser excitation pulse to trigger the photocatalysis, followed by an X-ray or electron pulse with variable time delays to the excitation pulse to probe transient structures of the catalyst. This "pump–probe" scheme has been commonly used in laser spectroscopy, where a laser pump pulse acts as a "clock" to synchronously trigger an elemental chemical event in a molecular ensemble, and a subsequent laser probe pulse interrogates the optical responses (e.g., absorption and emission) of the ensemble as a function of the delay time between the pump and the probe [17–21]. On the basis of the time evolution of characteristic optical features for different intermediate states, optical transient absorption (OTA) (or emission, OTE) spectroscopy tracks populations and energetics, as well as coherence and correlation among different transient species in photochemical reactions. With the exception of time-resolved vibrational spectroscopy, which indirectly measures transient structures [22–24], optical transient spectroscopy provides no direct structural information for reaction intermediates. These transient structures are often inferred by theoretical calculations without experimental verification.

Hence, extending the pump–probe approach to the X-ray regime offers an opportunity to shed new light on photocatalytical water splitting, as intense pulsed X-ray sources have become available during the past decade and more new light sources are being built around the world. Compared to sophisticated optical spectroscopic measurements, fast or ultrafast X-ray transient structural determination is still in its early stages because the key technical requirements for such studies have only been met in recent years, including sufficiently short and intense X-ray pulses, sophisticated optics for manipulating X-rays, as well as high sensitivity/dynamic-range and fast detectors. The third-generation synchrotron sources, including the Advanced Photon Source (APS) at Argonne National Laboratory, have provided X-ray pulses of 30–100 ps duration (full width at half maximum, fwhm) with a high photon flux within a pulse, and have enabled several pioneering time-resolved structural studies [25–31]. Meanwhile, ultrafast laser-driven hard X-ray pulses with <50 fs duration have become available [26,32–37], which offer subpicosecond time resolution with a much lower photon flux than the synchrotron sources. From now until the next decade, several ultrafast X-ray free electron laser (XFEL) facilities are either planned, under construction or have been constructed [38]. Therefore, significant progress in ultrafast X-ray applications is expected.

XAS, including X-ray absorption near edge structure (XANES) and X-ray absorption fine structure (XAFS), appears to be one of the methods of choice for studying metal-center catalysts in photocatalytic solar hydrogen generation processes, because it can be used to study disordered systems, is metal specific and provides both electronic and nuclear structural information for the metal centers in catalysts. In this chapter, we will review the current status of X-ray transient absorption (XTA) spectroscopy or laser-initiated time-resolved X-ray absorption spectroscopy (LITR-XAS, a previously used name), in the determination of structures of metal centers in the excited or transient state, which can be directly applied to the characterization of photocatalyst structures during water splitting in the above-0mentioned three schemes. The fundamental aspects of the method will be described first, followed by examples of XTA from previous studies. Finally, prospects and challenges for XTA applications in photocatalytic solar hydrogen generation will be discussed.

7.2 X-Ray Transient Absorption Spectroscopy (XTA)

Conventional XAS, including XANES and XAFS, is a local structural method based on resonant electronic transitions from core levels to vacant orbitals and to the continuum (Figure 7.1) [39–41]. XANES measures dipole-mediated transitions from a core (i.e., 1*s*, 2*s*, 2*p*, 3*d*, etc.) orbital to unoccupied orbitals, which transforms an initial state $|i\rangle$ to a final state $|f\rangle$ of the X-ray-absorbing atom. The X-ray absorption intensity $\mu(E)$ is proportional to $\Sigma|\langle f|\boldsymbol{e}_x{\cdot}\boldsymbol{r}_x|i\rangle|^2\ \delta(E)$, where $\boldsymbol{e}_x$ and $\boldsymbol{r}_x$ are the electric field and the transition dipole vectors of the X-ray, respectively [42]. The pre-edge and edge features originate from electronic transitions from a core level to empty bound states [43], and their values and intensities are

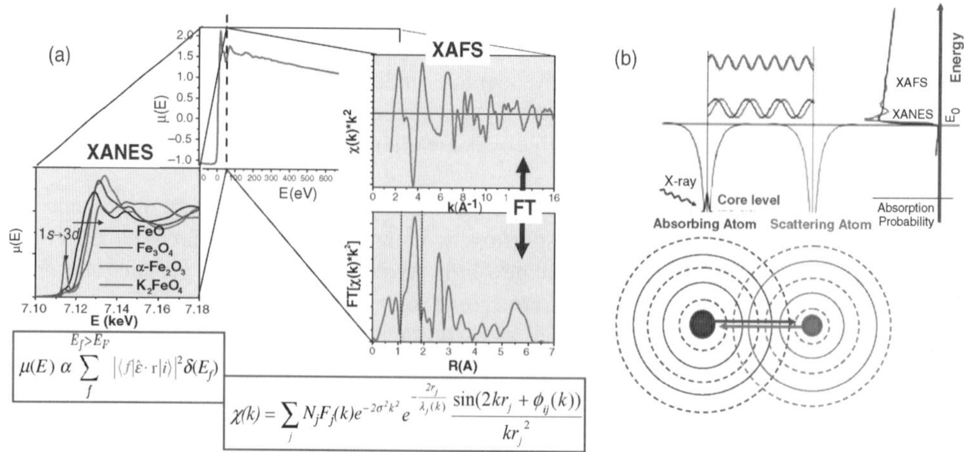

Figure 7.1 (a) Illustration of XAS with XANES and XAFS, and (b) illustration of conventional XAS (after Matthew G. Newville's lecture in XAFS summer school at the Advanced Photon Source, Argonne National Laboratory).

sensitive to energy levels and the local symmetry of the vacant MOs, following the selection rules for dipole-mediated transitions [44]. The X-ray photon energy required for ejecting a core electron to the continuum depends on the charges of the X-ray-absorbing atom. Thus, the transition-edge features and its energy in an XANES spectrum are often unambiguous characters to reveal the oxidation states of the X-ray-absorbing atom when other methods fail. The pre-edge features observed in transition-metal complexes with centrosymmetry, such as $1s \rightarrow 3d$ transitions, are quadrupole-allowed, with the intensities two orders of magnitude lower than the dipole-allowed transitions. However, these weak pre-edge features can gain extra intensity from mixing with MOs with np character when a coordination environment has no centrosymmetry [44]. Hence, these features are highly sensitive to the coordinating symmetry of the X-ray-absorbing atom and can be used to reveal both energy levels of the vacant MOs and the coordination geometry.

XAFS encompasses the oscillatory features above the transition edge from the interference between an outgoing photoelectron wave from the X-ray absorbing atom with back-scattered photoelectron waves from neighboring atoms (Figure 7.1) [39–41]. XAFS can be described by Equation 7.1 below,

$$\chi(k) \propto \sum_j N_j F_j(k) \cdot e^{-2\sigma^2 k^2} \cdot e^{-\frac{2r_j}{\lambda_j(k)}} \cdot \frac{\sin[2kr_j + \delta_{ij}(k)]}{kr_j^2}, \tag{7.1}$$

which can be successfully Fourier transformed into an atomic radial distribution centered at the X-ray-absorbing atom [39–41]. j is the index for the neighboring atom shells around the X-ray absorbing atom, $F(k)$ is the back-scattering amplitude, N, the coordination number, r, the average distance, σ, the Debye–Waller factor, λ, the electron mean free path and δ, the phase shift of the photoelectron wave. k is the photoelectron wavevector, $k = [2m(E - E_0)/\hbar^2]^{1/2}$, where m is the electron mass and E_0, the threshold energy for the transition edge. XAS provides precise local structures without limitation of sample forms. Hence, it is very useful for studying many chemical reactions that mostly take place in solution and other disordered media.

XAS has been used extensively to characterize *in situ* and *operando* catalyst structures and the oxidation states and coordination geometry of the metal center, as a function of temperature, pressure and other experimental conditions [45–47]. The development of quick XAFS or QXAFS using the third-generation synchrotron sources has made a significant contribution in understanding catalytic reaction mechanisms [48–52]. Although the time resolution of these experiments can be as fast as 100 μs, it is still much slower than the timescales of elemental chemical events or the excited state lifetimes of the photocatalysts [53–56]. In contrast, XTA presented in this chapter is analogous to the "pump–probe" OTA, where a laser pulse triggers synchronously a photochemical reaction (i.e., photocatalysis) among molecules in an ensemble, and an X-ray pulse probes changes in electronic and geometric structures of the ensemble resulting from the pump pulse (Figure 7.2). The time resolution of the experiment is independent of the response time of the detector (as long as the response time is shorter than the time intervals between two adjacent X-ray pulses), but is dependent on the duration of the pump or probe pulse, whichever is longer, convoluted with the instrumental response function (IRF) of the experiment. When fs X-ray pulses become available in the future, it will be conceivable to probe, on the fs time resolution, atomic movements as reactions take place, such as bond formation and breakage. Therefore, the XTA described here differs from other time-resolved XAS studies using synchrotron X-rays as a continuous wave source incorporating lasers or steady-state light sources [57–70], where the time resolution is independent of the X-ray pulse duration, but dependent on other components, such as the shutter or the detector.

XTA development started when intense pulsed X-rays from third-generation synchrotron sources became available. The main characteristics of these synchrotron sources that enable the XTA experiments are the sub-100 ps pulse duration and the high photon flux in each X-ray pulse (e.g., $>10^6$ photons per pulse at the sample with monochromatic X-rays). Demonstrated

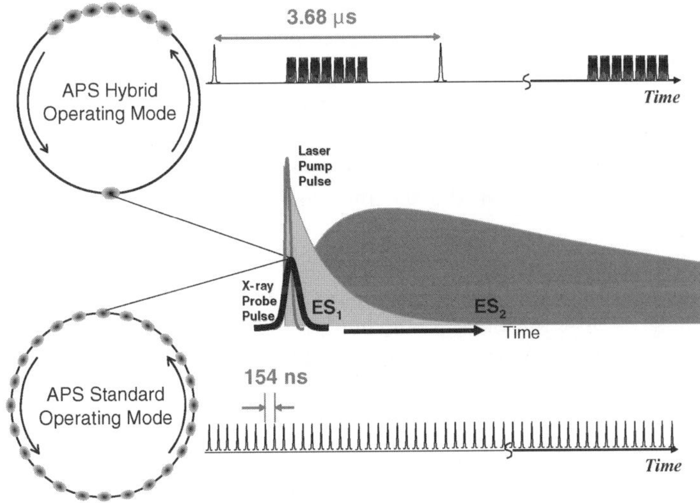

Figure 7.2 The XTA studies at the APS use the hybrid and normal timing modes. The notation of ES1 and ES2 stand for the first and the second excited states or other transient states.

by the development of the XTA methodology in the past decade, it is important to simultaneously optimize light and matter interactions for two different transitions that differ in energies and absorption extinction coefficients by three orders of magnitudes. Generally speaking, molecules have approximately 1000 times higher extinction coefficients interacting with laser photons with a few eVs in energy, than when they interact with X-ray photons with a few to tens of keV. Hence, a sample with a high concentration of X-ray-absorbing atoms is preferred to obtain intense XAS signals, while a sample with a low concentration is preferred for laser excitation to create a high fraction of the excited-state population. A multidimensional parameter-optimization approach is used to optimize experimental conditions (i.e., laser-pulse energy, beam size, sample concentration and thickness, etc.), to simultaneously achieve a relatively high fraction of the excited species (which favors dilute samples) and as high as possible signal-to-noise ratios in XAS spectra (which favors high concentration samples). From the laser-excitation aspect, the fraction of the excited state f_{ex} can be calculated according to the Beer–Lambert law and approximated by Equation 7.2 below,

$$f_{ex} = \frac{N_p \cdot e^{-kt}}{N_m} \cdot \left\{ 1 - 10^{-lC[\varepsilon_0(\lambda) - \varepsilon_{ex}(\lambda) f_{ex} e^{-kt}]} \right\}, \tag{7.2}$$

where N_p and N_m are numbers of laser photons and molecules respectively; k, t and λ are the excited-state decay rate constant, the delay time from the pump laser pulse and the laser wavelength, respectively; and l and C are the thickness and the concentration of the sample, respectively. In our previous XTA experiments, X-ray fluorescence detection was used for dilute samples due to their scarcity and the high f_{ex} requirement. From the transient XAS signal aspect, the total number of X-ray fluorescence photons from the sample can be estimated by Equation 7.3 below [71],

$$I_f = I_a \frac{\Omega}{4\pi} \eta \frac{\mu_k}{\mu_T} \eta_{Det}, \tag{7.3}$$

where I_f is the fluorescence signal, I_a is number of photons absorbed by the sample, Ω is the solid angle covered by the detector, η is the quantum yield of the fluorescence, μ_k and μ_T are the absorption cross-sections of the atoms of interest and of the whole sample, respectively and η_{Det} is the detector efficiency. Based on the beamline conditions and the intensity of the spectral features of interest, I_f generally should exceed 100 000 counts in order to have sufficient quality of XAS to be analyzed. However, the specific requirements could vary in each individual experiment [71].

Another challenge of XTA is the repetition rate differences between the laser and X-ray pulses. In order to achieve a relatively high fraction of the excited state photocatalyst, the photons in each laser pulse must exceed the number of X-ray-absorbing atoms in the sample area being probed by the X-ray pulse, which limits the repetition rate of the laser to a few kHz compared to the repetition rate of a few MHz for the X-ray pulses. Therefore, only one out of a few thousand X-ray pulses can be used, due to the kHz repetition for the pump–probe cycles. Therefore, consideration of the I_f required for the XTA needs to take the repetition rate of the pump–probe cycles into account, which implies a prolonged data acquisition time, that is, tens of hours for each spectrum. However, the X-ray photon flux has been improved by new configurations of insertion devices at the synchrotrons, which has produced strongly focused X-ray probe beams enabling the use of tightly focused laser pump beams. Such an improvement will allow the use of higher-repetition-rate laser systems as the excitation source. Meanwhile, significant improvements have been realized in data quality and data-acquisition time. It is achievable to acquire a relatively high-quality XTA spectrum (XANES and XAFS) within a few hours for a reasonably dilute sample (i.e., ~1 mM) at the APS, with presumably similar capabilities at other synchrotron sources.

Because XAFS oscillatory variations are fine features overlaid with a step function of the atomic absorption many times the oscillatory amplitudes, it is important to take at least a pair of spectra under the same experimental conditions with the exception of with and without the laser. This can be accomplished by gating the XTA detector signals with different X-ray pulses, only one of which is synchronized to the laser pulse. We have done so in two different detector systems, the solid-state germanium detector arrays (Canberra) [71] and the plastic scintillator coupled with PMT arrays with Sollar slits/filter combination. The later system incorporates a fast digitizing board and sophisticated software for extracting XTA signals from X-ray pulses before and after the laser pulses, as the reference and the time-evolution spectra, respectively.

The XTA spectra can be analyzed as conventional XAS spectra, with the exception of the consideration of multiple coexisting species, namely the excited state and the remaining ground state, as well as other possible reaction intermediates. XANES spectra are analyzed with arctan function background subtraction and peak fitting with combinations of Gaussian and Lorentzian functions. XAFS spectra have been analyzed by commercially available software program packages (i.e., Athena, WinXAS, etc.) using starting structural parameters calculated from simulated structures or crystal structures of the molecule of interest by FEFF8.0 calculation, to obtain the scattering paths with the phases and amplitudes of the ground state [72,73]. The photoexcited transient-state spectrum was extracted according the Equation 7.4,

$$\chi(k)_{ex} = \frac{\chi(k)_{\text{laser-on}} - \chi(k)_{\text{laser-off}} \cdot f_0}{1 - f_0}, \tag{7.4}$$

where k is the photon electron wave vector, $\chi(k)_{laser\text{-}off}$ and $\chi(k)_{laser\text{-}on}$ are taken from the XAS spectra without and with the synchronization of the laser pulse, the former being the ground state spectrum and the latter, the spectrum of the mixture of the photoexcited and the remaining ground state. $\chi(k)_{ex}$ is the resulting signal for the excited state and f_o, the fraction of the remaining ground state under laser pulse excitation. There are two unknowns in Equation 7.4, $\chi(k)_{ex}$ and f_o (or $1 - f_{ex}$, assuming $f_o + f_{ex} = 1$). f_{ex} has been obtained from the change in optical density due to the laser excitation by OTA measurements using the same laser as the pump source. However, it can also be simulated by theoretical approaches, such as MXAN, resolving the Equation 7.4 [74–76]. A number of theoretical modeling methods have been developed to extract the structural information beyond those structural parameters in Equation 7.1. These calculations are very useful in analyzing XTA data, as demonstrated by the recent work on the metal-to-ligand-charge-transfer (MLCT) state structure of a copper(I) complex, which extracted not only coordination numbers and bond distances, but also bond angles [77].

To accommodate XTA applications for samples with limited solubility and scarce quantity, we have been optimizing the experiments with X-ray fluorescence detection that preferentially deals with samples at low concentrations (i.e., ~1 mM) (Figure 7.3). If the laser can turn over a sufficient fraction of the excited state in a relatively concentrated sample, transmission XTA

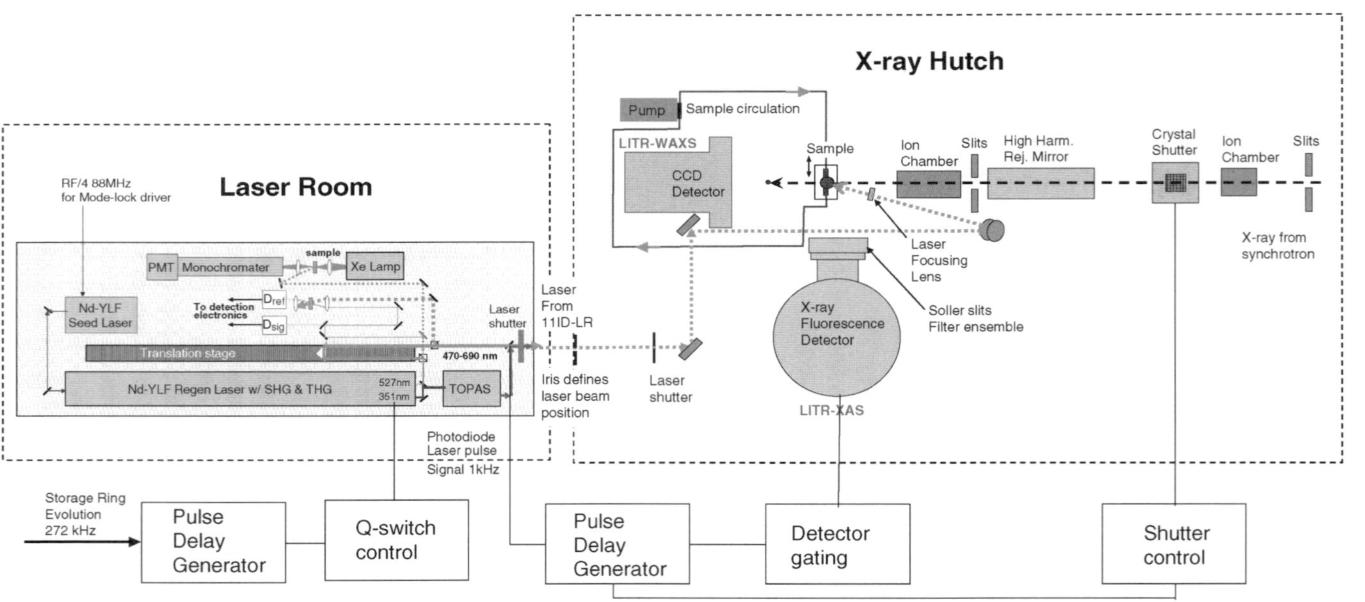

Figure 7.3 Current XTA setups at Beamline 11-ID-D of the APS. The laser room (left) has three operating modes, (i) XAS/WAXS, (ii) nanosecond photolysis and (iii) ps pump–probe transient absorption, marked by a solid line, a thick dotted line and thin dotted line, respectively. The X-ray hutch (right) can host both XAS and WAXS experiments. (LITR-XAS = XTA).

can be used, as documented in previous studies by others [78–83]. As far as the XAFS data quality is concerned, a sufficiently high fraction of the excited state population must be generated (>10–20%) in order to extract structural parameters from a mixture of the excited state and the remaining ground state. This requires a high laser-pulse energy and limits the pump-laser pulse repetition rate to a few kHz compared to a few MHz for X-ray pulses from a synchrotron. Therefore, the pump–probe cycle limits usable X-ray probe pulses to one in every 2000–6000 X-ray pulses, resulting in prolonged data acquisition time (i.e., tens of hours per XAFS spectrum). However, the X-ray photon flux can be increased by a factor of >100 using an in-line dual-undulator beamline, which significantly shortens the data acquisition time and improves the signal-to-noise ratio in the spectra. Even a regular undulator beamline at the third generation synchrotron source can significantly improve the data acquisition time and data quality. A laser-pulse pump, X-ray pulse probe, X-ray absorption/scattering facility adjacent to the X-ray experimental station at Beamline 11-ID-D at the APS is displayed in Figure 7.3, where XTA studies have been carried out on a series of metal complexes and metal/metal oxide nanoparticle systems, which will be described below, to demonstrate the potential of XTA in characterizing photocatalytic reactions for water splitting.

Like other methods, XTA alone only provides structural information in photocatalytic reactions. Our general approach is combining optical transient absorption and emission spectroscopy with the XTA studies, as well as quantum mechanical calculations to acquire a comprehensive understanding of the reaction and to provide feedback for synthetic chemistry in catalyst designs. The optical spectroscopy characterizes dynamics of excited states in terms of reaction rates, energetics, and coherence for energy/electron transfer, intersystem crossing, internal and overall rotation, as well as the fraction and stability of the excited state molecules created by the same laser system used in the XTA experiments [84–90].

In order to demonstrate the potential applications of XTA in studying structural dynamics in photocatalytic reactions for water splitting, we describe a few examples related to the currently studied photocatalytic systems mentioned in the Introduction. The possible structural changes on the metal center catalysts during the water-splitting reaction include electron occupation in MOs, oxidation states, coordination geometry and ligation. In the following, a few examples of relevant metal centers, namely metal complexes and metal nanoparticles, will be described.

7.3 Tracking Electronic and Nuclear Configurations in Photoexcited Metalloporphyrins

Metalloporphyrins have versatile functions due to their structural diversity. The metal center serves as a ligation site to bind other molecules or as a redox site to donate or accept electrons. The porphyrin macrocycles have been important chromophores in various light-driven reactions because of their high extinction coefficients in the visible region for sunlight harvesting, and their electron-donating/withdrawing capabilities. Meanwhile, metalloporphyrins are key building blocks for photoinduced charge/energy transfer, photocatalysis and molecular devices in covalently linked or self-assembled arrays [91–94]. Recently, self-assembled metalloporphyrin nanotubes were discovered to perform solar hydrogen generation [95]. The metal centers in various enzymes with analogous heme groups have also been recognized for their catalytic redox functions in nature [96–99]. Moreover, metalloporphyrins are closely related to chlorophylls in photosynthetic systems, where they carry out photoinduced charge separation and light harvesting in photosynthetic-reaction-center proteins, such as photosystems I and II [100]. Despite the promise in photocatalytic applications, details of metalloporphyrins and their composites concerning the electronic configuration and molecular geometry of excited states, and their influence on light-activated functions have mainly eluded direct experimentation. In particular, the electronic configurations and coordination geometry of the central metal ion in the excited states, such as electron occupation and energy levels of MOs, often remain vague because metal-centered electronic states are often optically "dark," or masked by other strong optical features, making it difficult to track the transition-metal *d* electrons in the excited states.

XTA measurements on a metalloporphyrin have been recently reported which simultaneously track electronic and nuclear configurations for an excited-state nickel-tetramesitylporphyrin (NiTMP) with a 200 ps lifetime in a dilute solution [101,102]. As far as we know, it is the shortest-lived excited state structure to be captured by the present XTA method using X-ray pulses from a synchrotron source so far. The electronic configuration of Ni(II) ($3d^8$) in a nearly square-planar ground state S_0 has an empty $3d_{x2-y2}$ MO and a doubly occupied $3d_{z2}$ MO [103]. The $S_0 \rightarrow S_{1,2}$ ($S_{1,2}$, the lowest and second lowest energy excited states, respectively) transitions can be induced by exciting the Q and Soret bands, respectively. Ultrafast OTA studies revealed that the S_1 or S_2 (π, π^*) state converts to an intermediate $T_1{}'$ state within 350 fs, which then undergoes vibrational relaxation in less than 20 ps to a relaxed T_1 state, with a presumed $^3(3d_{x2-y2}, 3d_{z2})$ configuration [104–107] and ~200 ps lifetime (Figure 7.4). Nevertheless, the electronic configuration and the geometric structure of the T_1 excited states are still subject to some uncertainties.

The pre-edge region of an XANES spectrum contains features that are related to the coordination geometry of the metal. For the first row transition metal, the pre-edge region represents the transitions from 1*s* to 3*d* MOs, which are dipole forbidden and quadrupole allowed. Although these features are weak, they are sensitive to *d-p* mixing and are enhanced

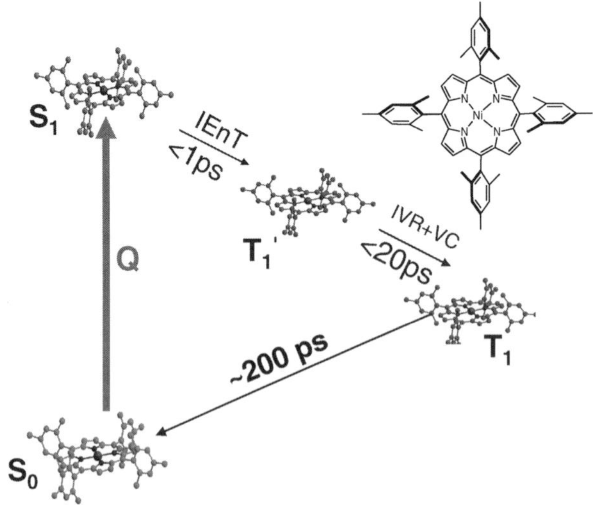

Figure 7.4 Reaction cascade and kinetics time constants of the NiTMP excited state.

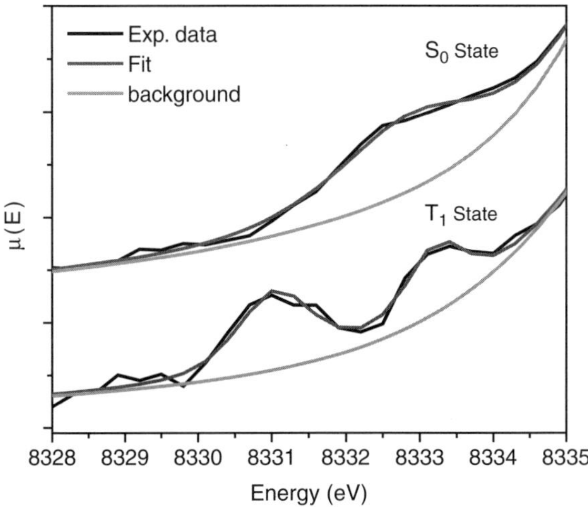

Figure 7.5 The pre-edge spectra of the ground S_0 state and the excited T_1 state. The latter shows that the energy difference for the transitions from 1s to singly occupied $3d_{z2}$ and $3d_{x2-y2}$ MO respectively in the XAS final state is 2.2 ± 0.2 eV.

when the dipole-allowed p MO component is present in a noncentrosymmetric coordination of the metal center. Using the XTA method at the Ni K-edge and laser excitation at 527 nm (which induces the S_0 to S_1 transition), the ground state NiTMP shows a single peak at 8332.8 eV, indicating one $1s \rightarrow 3d_{x2-y2}$ transition from Ni(II) $3d^8$ in a nearly square-planar coordination geometry with one empty $3d_{x2-y2}$ MO, whereas the excited T_1 state shows two pre-edge features at 8331.0 and 8333.2 eV, corresponding to $1s \rightarrow 3d_{x2\text{-}y2}$ and $1s \rightarrow 3d_{z2}$ transitions, respectively, with an energy difference of 2.2 ± 0.2 eV (Figure 7.5). This result confirms the quantum-mechanical calculations of the excited state NiTPP (nickel-tetraphenylporphrin), a molecule closely related to NiTMP, where the two highest energy $3d_{x2-y2}$ and $3d_{z2}$ MOs are singly occupied. The transition energy of $1s \rightarrow 4p_z$ in the T_1 state relative to those in the S_0 state is also mapped out (Figure 7.6A) which is 1.5 eV higher than that of the S_0 state. This example suggests the feasibility of XTA applications in solar hydrogen-generation processes, where the transient electronic configuration of the photocatalyst metal centers can be revealed during photocatalysis.

In addition, the X-ray fluorescence difference signals $\Delta\mu(E)$ (laser-on–laser-off) as functions of the delay time between the laser and the X-ray pulses at three X-ray photon energies have been used to characterize specific $1s \rightarrow 4p_z$ and $1s \rightarrow 4p_{x,y}$ electronic transitions induced by X-rays. As a result of the laser excitation at 527 nm, the $1s \rightarrow 4p_z$ transition intensity at 8338 eV for the S_0 state decreases, while a new peak emerges at 8340 eV, characterizing the same transition for the T_1 state, indicating the correlation between the two states. These signals are analogous to the ground-state bleaching and excited-state absorption signals in OTA. Although the current XTA time resolution is too low to resolve the depletion and the rise of the S_0 and T_1 states, all three time scans (Figure 7.6B) agree with the calculated convolution of a Gaussian function for the X-ray pulse with 160 ps fwhm (Note: the X-ray pulse here has a current of 16 mA, approximately four times the normal operating mode, where the current in an

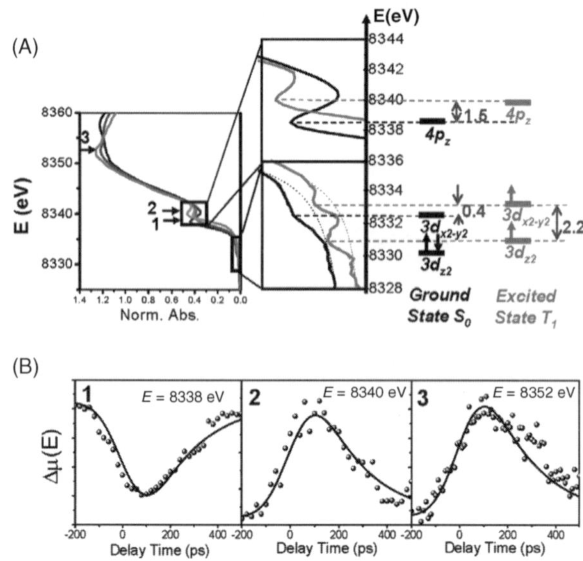

Figure 7.6 (A) XTA spectra in the Ni K-edge XANES region and their correlation with the MO energy levels (black, laseroff; green, laseron at $t = 0$; blue, laseron at $t = 100$ ps; and red, the T1 state spectrum; $t =$ delay time). (B) XTA difference signals (laseron–laseroff) as a function of the delay time taken at three X-ray photon energies labeled by arrows in XANES spectra.

X-ray pulse is 4 mA, which gives a pulse duration of about 80 ps) and an exponential decay function for the T_1 state population with a 200 ps time constant. This capability suggests that XTA can resolve the dynamics for the electronic and geometric structural changes simultaneously. When the X-ray pulse duration becomes shorter in future XFEL, it is conceivable that one could directly verify the Bohr–Oppenheimer approximation to resolve the timing of the electronic transition and the nuclear geometry change. These dynamics signals are complementary, with optical transient absorption measurements that are generally vague for metal-specific transitions.

As the excited NiTMP proceeds through the energy cascade along the path $S_1 \rightarrow T_1$, the excess energy generates vibrationally hot states accompanying the electron redistribution [105,107,108], resulting in molecular geometry rearrangements. These structural changes can be extracted from the data analysis of the XAFS spectra in Figure 7.7A, where the laser-off and laser-on at $t = 100$ ps spectra are clearly different in oscillation frequency and phase. Fourier-transformed XAFS spectra in Figure 7.7B and data fitting indicate lengthenings of average Ni–N and Ni–C_α distances (without phase corrections) in the T_1 state, by 0.08 and 0.07 Å, respectively, which is consistent with an expansion of the porphyrin ring. Apparently, the addition of one electron to the higher energy antibonding $3d_{x2-y2}$ MO in the T_1 state causes electrostatic repulsion and reduces the Ni–N bond order, resulting in longer Ni–N bond distances. As the porphyrin ring expands, the electron density on Ni(II) through σ-bonding and π-conjugation decreases, while its effective charge increases compared to that of the S_0 state, which consequently causes the energy up-shift of the $1s \rightarrow 3d_{x2\text{-}y2}$ and $1s \rightarrow 4p_z$ transitions in the T_1 state (Figure 7.5).

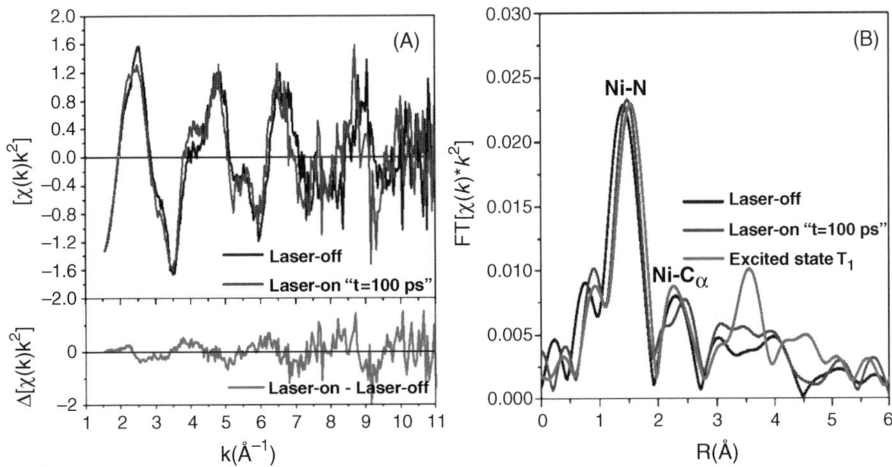

Figure 7.7 (A) XAFS spectra for NiTMP. The difference spectrum is displayed at the bottom; (B) the phase-uncorrected Fourier-transformed XAFS spectra of NiTMP.

The photoinduced axial ligation of NiTMP in its excited state in the presence of coordinating ligand molecules has been also investigated by XTA, by monitoring the intensity of the $1s \rightarrow 4p_z$ transition feature (Figure 7.6A), which is distinctive when Ni(II) in NiTMP has no axial ligand and an empty $4p_z$ MO, and vanishes or is significantly attenuated when NiTMP has one or two axial ligands that donate electron density to $4p_z$. To our surprise, XTA spectra taken from NiTMP in a solvent mixture of toluene (~75%) and pyridine (~25%, the axial ligating molecule) with 527 nm light excitation revealed that the pyridine ligation does not take place immediately after the S_1 is generated. Instead, it takes approximately 200–600 ps for the ligand to bind Ni(II) and the ground-state recovery time is prolonged to tens of ns. This observation can be explained by a required triplet state $^3(3d_{z2}, 3d_{x2-y2})$ generated from the intersystem crossing from the S_1 state and the presence of pyridine ligands. Once the NiTMP binds two pyridine axial ligands, it will take tens of ns to release them and return to the ground state. The ligation state of the metal center is crucial in photocatalytic steps, and it can be obtained by XTA, as demonstrated by the above example.

The above example clearly demonstrates the prospects of XTA for simultaneously tracking electronic configurations and atomic coordinates in metal complexes along their excited state pathways. While valence electrons of molecules are promoted to higher-energy MOs by laser photons, XTA spectra can probe electronic transitions from core to valence levels induced by X-rays. This provides an alternative means for obtaining metal excited-state electronic configurations that may otherwise be inaccessible by optical transient absorption spectroscopy. These results can directly verify and guide theoretical calculations and spectroscopic assignments for excited-state metal complexes. In particular, the XTA method can selectively probe the time evolution of the electron occupation and the energy level for a particular MO (e.g., $4p_z$) after the photoexcitation, as seen in Figure 7.6B. This capability validates the great potential of XTA using next-generation X-ray sources, namely X-ray free electron lasers with selected X-ray photon energies, in resolving excited-state kinetics and coherence with fs time resolution. Hence, accurate structural and kinetic information can be obtained, not only for relaxed excited states, but likely also for evanescent transition states (i.e., the T'_1 state) or coherent atomic motions. These capabilities will make simultaneously visualizing electronic and molecular structures during photochemical reactions a reality, enabling new insight into reaction mechanisms and validating the rational design of molecules for solar-energy conversion, catalysis and molecular devices.

7.4 Tracking Metal-Center Oxidation States in the MLCT State of Metal Complexes

The MLCT excited states of many metal complexes are precursors for photochemical reactions in water splitting, during which the metal centers will change oxidation states to donate electrons or holes for redox reactions. Therefore, knowledge of the metal-center oxidation states during photocatalysis is important in characterization of the reaction mechanisms and can provide input for the design of better photocatalysts. The MLCT transitions of metal complexes are due to $d \rightarrow \pi^*$ transitions, shifting electron density from the metal to the ligands, which is the basis for employing these complexes as building blocks in artificial photosynthetic and photocatalytic systems for water splitting [109–112]. Due to the MLCT transitions, the metal-center oxidation states will change, accompanied by changes in coordination geometry and bond distance, all of which can be probed by XTA.

Cuprous diimine compounds can act as energy and electron donors [113–118], and as functional groups of molecular machines [119,120]. The ground state Cu^{I} center has a $3d^{10}$ configuration with a pseudo-tetrahedral coordination geometry [121], whereas the Cu^{II^*} center of the thermally equilibrated MLCT state has a $3d^9$ configuration, susceptible to Jahn–Teller distortion to a flattened tetrahedral geometry (Figure 7.8) [122]. We studied $[Cu(I)(dmp)_2]^+$ (dmp, 2,9-dimethyl-1,10-phenanthroline) using both ultrafast transient optical spectroscopy

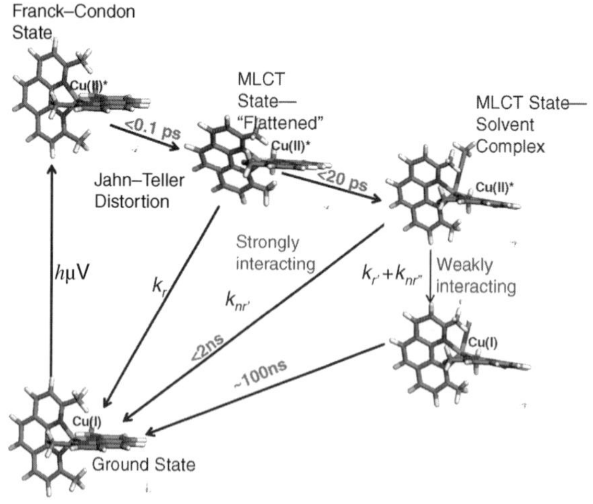

Figure 7.8 The MLCT excited-state pathways and dynamics.

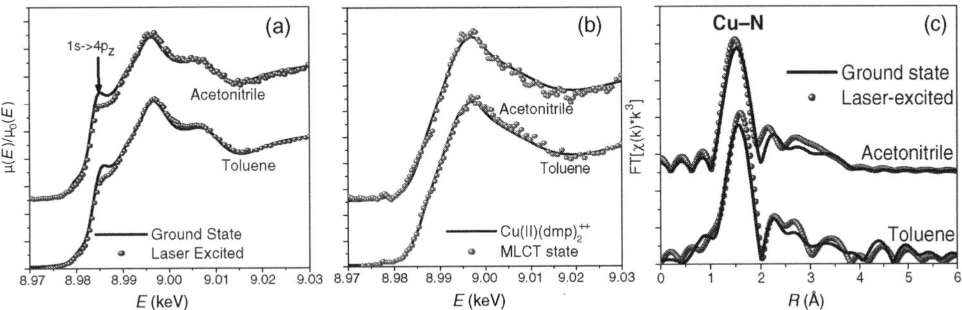

Figure 7.9 (a) XTA spectra within 200 ps of the laser excitation in XANES region $[Cu(I)(dmp)_2]^+$; (b) XANES spectra after removal of the contribution from the remaining ground state, and (c) FT-XAFS spectra with no phase correction.

and XTA [86]. The MLCT state dynamics of $[Cu(I)(dmp)_2]^+$ fit a triexponential function, with two solvent-independent fast time constants of ~0.5 ps and 15 ps, and one solvent-dependent slow time constant of 2 ns in acetonitrile (coordinating solvent), and 100 ns in toluene (non-coordinating solvent) [84,85,123]. The ultrafast transient optical spectroscopy and the DFT (density functional theory) results revealed two parallel excited-state pathways (Figure 7.8) [123,124]: (i) from the Franck–Condon (F–C) state to a flattened geometry then to the thermally equilibrated triplet MLCT state; and (ii) from the F–C state to a triplet MLCT state then flattening to reach thermal equilibrium [125].

Thermally equilibrated MLCT-state structures emerging ~20 ps after the excitation were studied in coordinating acetonitrile and in non-coordinating toluene by XTA with laser excitation at 527 nm. Differences in XANES spectra for $[Cu(I)(dmp)_2]^+$ within 200 ps of the laser excitation in both solvents [84] are clearly seen (Figure 7.9A). After subtracting the contribution from the remaining ground state, the resulting MLCT spectra in both solvents resemble the ground state of $[Cu(II)(dmp)_2]^{2+}$ (Figure 7.9B). FT-XAFS spectra (Figure 7.9C) show increased intensities of the Cu–N peak for laser-excited $[Cu(I)(dmp)_2]^+$ in both solvents, suggestive of the additional ligand binding to Cu(II)* in the transient MLCT state. Interestingly, the average Cu–N distance compared to the ground state shifts to a shorter distance in acetonitrile, but to a longer distance in toluene. The results of the experiment demonstrated: (i) light excitation of $[Cu(I)(dmp)_2]^+$ generated a formal Cu(II)* center in the excited MLCT state; (ii) the inner-sphere reorganization changed the coordination number of the MLCT state from four to five in presumed noncoordinating toluene and (iii) the average Cu–ligand bond distances lengthened in the MLCT state in toluene, but shortened in acetonitrile. According to the energy-gap law [126–128], the nonradiative rate constant increases exponentially with decreasing energy separation between the ground state and the excited state. Although the LITR-XAS results in both solvents support the formation of a pentacoordinated Cu(II)* center in the MLCT state, the respective elongation and shortening of the average Cu–N bond distance for the MLCT state of the $[Cu(I)(dmp)_2]^+$ indicate the difference in interactions of Cu(II)* with the fifth ligand. Toluene has less electron-donating capability compared to acetonitrile, hence it could only act as the fifth ligand with a longer and less-stable bond with Cu(II)*. Consequently, the MLCT state of the $[Cu(I)(dmp)_2]^+$–toluene complex will not be as stable and will not lower the energy of the MLCT state as much as the acetonitrile complex. Hence, the excited state quenching is only significant in acetonitrile, and the "exciplex" previously inferred is more appropriate for cases where strong excited-state adducts completely quench the excited state. In this work we have resolved internal atomic reorganization linked to ET for a MLCT complex, and established a foundation for the proposed futures studies of internal reorganization for energy conversion in tuned PET complexes.

This example clearly indicates that XTA measurements are sensitive to the transient oxidation state of the metal center when it undergoes a redox reaction. Because the MLCT states of metal complexes are normally the reactants for water-splitting reactions, the identities of their excited state, their intermediate state after donating or accepting electrons and their final state at the end of the reaction cycle are all important to know, in order to find a rational way of improving the reaction efficiency. For some later catalytic reaction steps that are not directly induced by light, XAS measurements with μs–ms time resolution have been carried out, such as metal-complex catalyzed water splitting, as detected by time-resolved XAS studies on the Mn cluster of the oxygen-evolving complex in photosystem II [129–131].

7.5 Tracking Transient Metal Oxidation States During Hydrogen Generation

Solar hydrogen generation catalyzed by inorganic complexes has been studied extensively since its discovery. The first photocatalytic solar hydrogen-generation system consisted of a TiO_2 electrode coupled with a Pt electrode, which facilitates the water splitting [1]. More recently, Pt/TiO_2 nanoparticles have been used for the same heterogeneous photocatalytic reaction, with the Pt directly deposited onto the TiO_2 nanoparticle surfaces (Figure 7.10) [16]. Under exposure to the radiation source such as the UV laser ($\lambda = 351$ nm) employed in this experiment, a charge transfer occurs within TiO_2 nanoparticles, leading to electron–hole separation, with the electrode potential substantially higher than that of needed for electro-decomposition of water to H_2 and O_2 ($E^0 = 1.23$ V). The platinum metal crystallites decorated over the surface of the TiO_2 act as the cathode to promote the combination of the electrons and protons to form hydrogen. Since the reduction of protons is more facile over the surface of Pt than that of TiO_2, the rate of hydrogen production is more competitive than electron–hole recombination. Thus the presence of Pt significantly improves the efficiency of photocatalytic splitting of water.

Unlike homogeneous systems, such as NiTMP in solution, there is scarcely any optical method for monitoring the dynamic structure of Pt/TiO_2 catalysts, because these systems are

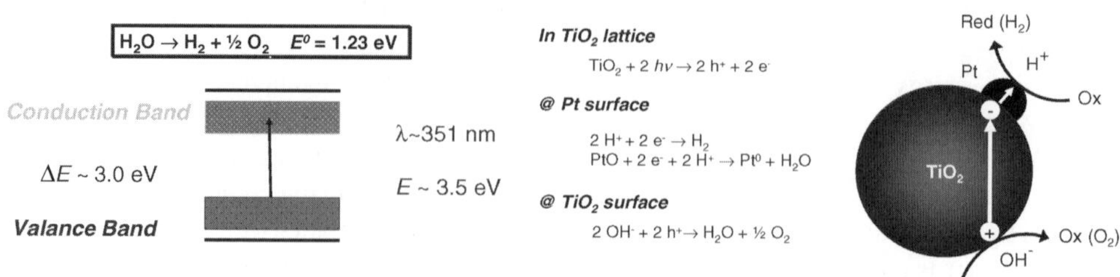

Figure 7.10 Photocatalysis of TiO_2 for hydrogen generation.

usually not transparent or have poor optical qualities. Although the charge-carrier dynamics of the TiO_2 catalyst has been investigated by fs diffuse reflectance spectroscopy [132], little information is known regarding to the change of Pt electronic and coordination structures during the process of photocatalytic water splitting. Alternative ways to capture the intermediate structures of Pt are hence desirable for the overall understanding of the molecular kinetics of hydrogen generation in these systems.

In our recent study using the same XTA facility, we conducted preliminary investigations into the dynamic variation of Pt oxidation states in Pt/TiO_2 catalysts in hydrogen production, after photoinitiation by a UV pulse, using XTA at the Pt L_{III}-edge. Highly dispersed Pt supported by TiO_2 nanoparticles (1 wt% Pt) was prepared using the incipient wetness method with $Pt(NH_3)_4(NO_3)_2$ solution over Degussa P-25, followed by drying and calcination in air at 500 °C for two hours. The catalyst powder thus produced was subsequently mixed with water and ball-milled for up to four hours to ensure maximum suspension and minimum catalyst particle size in liquid before the pump–probe experiment. A recycling water/methanol slurry jet with suspended Pt/TiO_2 was used in the XTA study. Laser pulses with 5 ps fwhm at 351 nm were used as the excitation source, promoting charge separation across the bandgap of TiO_2 nanoparticles. Only X-ray absorption near-edge structure spectra were collected at the L_{III}-edge of Pt with time delays between the pump laser and the probe X-ray pulses, of 0, 1 and 50 round trips of the X-ray pulses in the storage ring, where one round trip was equal to 3.7 μs. Shown in Figure 7.11 are the XANES spectra taken at zero delay (overlap) and one cycle delay (3.7 μs). For comparison, the XANES spectra of the reference compounds, Pt foil and

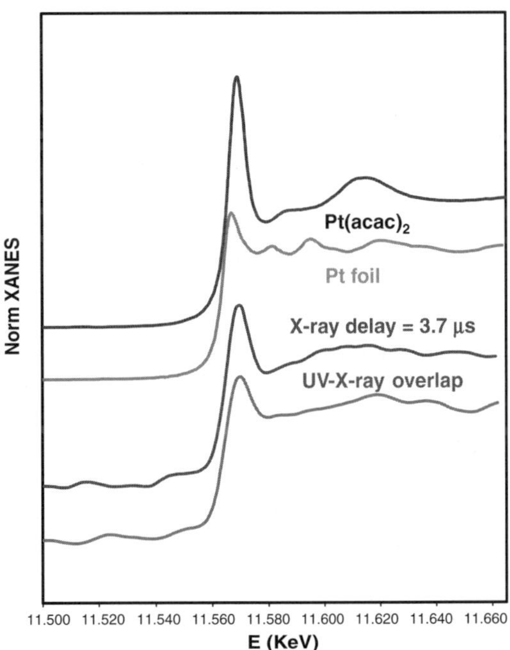

Figure 7.11 XANES spectra of Pt at L_{III}-edge for Pt coated TiO2 under 351 nm light excitation at different time delays between the laser pump pulse and the x-ray probe pulse.

Pt(II) acetoacetonate, are also plotted. The variation of the Pt white-line peak under different delays clearly demonstrated an oxidation state change for Pt when the spectra were compared with that of the reference. The prelimary results indicated the following observations: (i) Pt in Pt/TiO_2 catalyst was in the mixed state of Pt(II) and Pt(0) even though it was initially prepared in the form of Pt(II)O, (ii) the photoexcitation of TiO_2 apparently led to a small amount of reduction of Pt(II) within 100 ps of the laser excitation when compared with the baseline sample (no UV exposure), suggesting reduction resulted from electron transfer and/or by hydrogen formed during proton–electron combination and (iii) a portion of Pt was immediately oxidized to Pt(II) in the water/methonal slurry after several μs of the 351 nm excitation, possibly by the migration of oxygen formed on the surface of TiO_2 from the oxidation of the hydroxyl group. Detailed mechanism and interpretation are still in progess and require further investigation with a better signal-to-noise ratio of the data. Nevertheless, this preliminary study provides structural information that has not been accessible before, and has demonstrated great potential for the XTA in solar hydrogen-generation studies.

These examples clearly demonstrate the capability of XTA in simultaneously tracking electronic configurations and nuclear coordinates along their excited-state pathways. While OTA involves the promotion of valence electrons of molecules to higher-energy MOs by laser photons, XTA spectra reveal electronic transitions from the core to valence levels induced by X-rays. This provides an alternative and specific means for obtaining electronic configurations for the metal complexes that may otherwise be inaccessible by transient optical spectroscopy. This capability validates the great potential of XTA using next-generation X-ray sources, namely XFEL with selected X-ray photon energies, in resolving excited-state kinetics and coherence with fs time resolution.

7.6 Prospects and Challenges in Future Studies

We have shown through the above examples, the current capability of XTA in identifying transient structures of photocatalysts with metal centers during solar hydrogen generation. This method is uniquely useful when optical signals involving the key intermediates during photocatalytic reactions are weak, missing or impossible to obtain due to the poor optical quality of the samples. Moreover, transient X-ray methodology, when combined with the pulsed nature of the X-ray source, can provide direct structural information with a spatial resolution of 0.01 Å and a temporal resolution limited mainly by the X-ray pulse duration. The time resolution for the current synchrotron-based XTA is still slower than most of the elemental chemical reactions, such as bond breaking and formation, but it is an evolving parameter, depending on the available light sources. XTA has shown its capabilities in resolving structures of photoexcited states of metal centers in photocatalysts, which are first created by solar photons. However, challenges exist in applying XTA to solar hydrogen generation: (i) how to synchronously trigger catalytic steps in a time-resolution of the elemental chemical events when they are not directly triggered by light; (ii) how to probe structural changes at the interface between catalysts and reactants in heterogeneous photocatalytic reactions, where only a very small fraction of the surface sites are involved in photocatalysis; (iii) how to recycle samples fast enough to prevent sample damage by long exposure of intense laser and X-ray pulses; (iv) how to incorporate OTA or OTE during the XTA experiments to ensure sample integrity, (v) how to probe samples undergoing irreversible reactions using XTA and (vi) how to develop theoretical analysis to find the fractions of multiple species evolving with time. These are the issues to be considered in future XTA spectroscopy.

Moreover, XTA probes only local structures around the metal centers in solar hydrogen generation, which is often insufficient to understand the whole catalytic system. If larger-amplitude structural changes or structural changes involving only light atoms are involved, X-ray scattering or diffuse scattering in solution and noncrystalline media are better methods to provide the details [133–137]. Future studies include: (i) extending XTA methodology in X-ray scattering and diffraction in solar hydrogen generation photocatalytic processes, (ii) resolving evanescent excited state using the coming fourth-generation X-ray sources, such as XFEL, where either wide-angle scattering or resonance inelastic X-ray scattering will be used, due to the limited spectral range, (iii) probing semiconductor nanoparticle structural variation during photocatalytic reactions, to obtain simultaneous spatial and temporal resolution in dynamic molecular structures during photochemical reactions for solar hydrogen generation and many other photocatalytic reactions. These studies will shed light on the mechanisms for coupling single photon excitation with multiphoton reduction/oxidation reactions in solar hydrogen fuel production.

Acknowledgments

This work is supported by the Division of Chemical Sciences, Office of Basic Energy Sciences, US Department of Energy under contracts DE-AC02-06CH11357. Use of the Advanced Photon Source was supported by the US Department of Energy, Office of Science, Office of Basic Energy Sciences, under Contract No. DE-AC02-06CH11357. LXC wishes to acknowledge support from the set-up fund of Northwestern University. The author would like to acknowledge her collaborators for the past 10 years at Argonne National Laboratory, Drs. W. J. H. Jäger, G. Jennings, K. Attenkofer, T. Liu, G. B. Shaw, X. Zhang, E. C. Wasinger, J. V. Lockard, J. Guo and D. Liu. The collaboration reviewed here was with Profs. J. S. Lindsey (North Carolina State University) G. J. Meyer (Johns Hopkins University) and P. Coppens (SUNY, Baffalo) and many inspiring discussions with them are greatly appreciated.

References

[1] Fujishima, A. and Honda, K. (1972) Electrochemical photolysis of water at a semiconductor electrode. *Nature*, **238**, 37.

[2] Asahi, R., Morikawa, T., Ohwaki, T. *et al.* (2001) Visible-light photocatalysis in nitrogen-doped titanium oxides. *Science*, **293**, 269–271.

[3] Kudo, A., Kato, H., and Tsuji, I. (2004) Strategies for the development of visible-light-driven photocatalysts for water splitting. *Chemistry Letters*, **33**, 1534–1539.

[4] Zelifiska, B., Borowiak-Palen, E., and Kalenczuk, R.J. (2008) Photocatalytic hydrogen generation over alkaline-earth titanates in the presence of electron donors. *Hydrogen Energy*, **33**, 1797–1802.

[5] Falkenstrom, M., Johansson, O., and Hammarstroem, L. (2007) Light-induced charge separation in ruthenium based triads New variations on an old theme. *Coordination Chemistry Reviews*, **360**, 741–750.

[6] Sakai, K. and Ozawa, H. (2007) Homogeneous catalysis of platinum(II) complexes in photochemical hydrogen production from water. *Coordination Chemistry Reviews*, **251**, 2753–2765.

[7] Esswein, A.J. and Nocera, D.G. (2007) Hydrogen production by molecular photocatalysis. *Chemical Reviews*, **107**, 4022–4047.

[8] Kanan, M.W. and Nocera, D.G. (2008) *In situ* formation of an oxygen-evolving catalyst in neutral water containing phosphate and Co^{2+}. *Science*, **321**, 1072–1075.

[9] Hodgkiss, J.M., Krivokapic, A., and Nocera, D.G. (2007) Ligand-field dependence of the excited state dynamics of hangman bisporphyrin dyad complexes. *Journal of Physical Chemistry B*, **111**, 8258–8268.

[10] Ghrardi, M.L., Posewitz, M.C., Maness, P.-C. *et al.* (2007) Hydrogenase and Hydrogen photoproduction in oxygenic photosynthetic organisms. *Annual Review of Plant Biology*, **58**, 71–91.

[11] Sproviero, E.M., Gascon, J.A., McEnvoy, J.P. *et al.* (2008) Quantum mechanical/molecular mechanics study of the catalytic cycle of water splitting in photosysten II. *Journal of the American Chemical Society*, **130**, 3428–3442.

[12] Joliot, P., Barbieri, G., and Chabaud, R. (1969) Un nouveau modele des centres photochimiques du systeme II*. *Photochemistry and Photobiology*, **10**, 309–329.

[13] Kok, B., Forbush, B., and McGloin, M. (1970) Cooperation of charges in photosynthetic O2 evolution: I. A linear four step mechanism. *Photochemistry and Photobiology*, **11**, 457–475.

[14] Stair, P.C. (2008) Advanced synthesis for advancing heterogeneous catalysis. *Journal of Chemical Physics*, **128**, 182507.

[15] Wu, Z.L., Kim, H.S., Stair, P.C. *et al.* (2005) On the structure of vanadium oxide supported on aluminas: UV and visible Raman spectroscopy, UV-visible diffuse reflectance spectroscopy, and temperature-programmed reduction studies. *Journal of Physical Chemistry B*, **109**, 2793–2800.

[16] Moon, S., Matsumura, Y., Kitano, M. *et al.* (2003) Hydrogen production using semiconducting oxide photocatalysis. *Reearch in Chemical Intermediates*, **29**, 233–256.

[17] Lenderink, E., Duppen, K., and Wiersma, D.A. (1995) Femtosecond Twisting and Coherent vibrational motion in the excited state of tetraphenylethylene. *Journal of Physical Chemistry*, **99**, 8972–8977.

[18] Chachisvilis, M., Fidder, H., and Sundstroem, V. (1995) Electronic coherence in pseudo two-colour pump-probe spectroscopy. *Chemical Physics Letters*, **234**, 141–150.

[19] Walhout, P.K., Alfano, J.C., Kimura, Y. *et al.* (1995) Ultrafast experiments on the photodissociation, recombination, and vibrational relaxation of I_2^-: Role of solvent-induced solute charge flow, *Chemical Physics Letters*, **232**, 135–140.

[20] Whitnell, R.M., Wilson, K.R., Yan, Y., and Zewail, A.H. (1994) Classical theory of ultrafast pump-probe spectroscopy: applications to I2 photodissociation in Ar solution. *Journal of Molecular Liquids*, **61**, 153–165.

[21] Yan, Y.J., Fried, L.E., and Mukamel, S. (1989) Ultrafast pump-probe spectroscopy: femtosecond dynamics in Liouville space. *Journal of Physical Chemistry*, **93**, 8149–8162.

[22] Franzen, S., Jasaitis, A., Belyea, J. *et al.* (2006) Hydrophobic distal pocket affects NO-heme geminate recombination dynamics in dehaloperoxidase and H64V myoglobin;. *Journal of Physical Chemistry B*, **110**, 14483–14493.

[23] Dattelbaum, D.M., Kober, E.M., Papanikolas, J.M., and Meyer, T.J. (2006) Application of time-resolved near-infrared spectroscopy (TRNIR) to the metal-to-ligand charge transfer (MLCT) excited state(s) of Os(phen)(3) (2+). *Chemical Physics*, **326**, 71–78.

[24] Negrerie, M., Cianetti, S., Vos, M.H. *et al.* (2006) Ultrafast heme dynamics in ferrous versus ferric cytochrome c studied by time-resolved resonance Raman and transient absorption spectroscopy. *Journal of Physical Chemistry B*, **110**, 12766–12781.

[25] Barty, C.P.J., Ben-Nun, M., Guo, T. *et al.* (1997) Ultrafast x-ray diffraction and absorption. *Oxford Series on Synchrotron Radiation*, **2**, 44–70.

[26] Schoenlein, R.W., Chattopadhyay, S., Chong, H.H.W. *et al.* (2000) Generation of femtosecond pulses of synchrotron radiation. *Science (Washington D.C.)*, **287**, 2237–2240.

[27] Srajer, V., Crosson, S., Schmidt, M. *et al.* (2000) Extraction of accurate structure-factor amplitudes from Laue data: wavelength normalization with wiggler and undulator X-ray sources. *Journal of Synchrotron Radiation*, **7**, 236–244.

[28] DeCamp, M.F., Reis, D.A., Bucksbaum, P.H. *et al.* (2001) Coherent control of pulsed X-ray beams. *Nature*, **413**, 825–828.

[29] Neutze, R., Wouts, R., Techert, S. *et al.* (2001) Visualizing photochemical dynamics in solution through picosecond x-ray scattering. *Physical Review Letters*, **87**, 195508/1–195508/4.

[30] Techert, S., Schotte, F., and Wulff, M. (2001) Picosecond x-ray diffraction probed transient structural changes in organic solids. *Physical Review Letters*, **86**, 2030–2033.

[31] Chen, L.X., Jager, W.J.H., Jennings, G. *et al.* (2001) Capturing a photoexcited molecular structure through time-domain x-ray absorption fine structure. *Science*, **292**, 262–264.

[32] Gibson, E.A., Paul, A., Wagner, N. *et al.* (2003) Coherent soft x-ray generation in the water window with quasi-phase matching. *Science*, **302**, 95–98.

[33] Hatanaka, K., Miura, T., and Fukumura, H. (2002) Ultrafast x-ray pulse generation by focusing femtosecond infrared laser pulses onto aqueous solutions of alkali metal chloride. *Applied Physics Letters*, **80**, 3925–3927.

[34] Lee, T., Benesch, F., Jiang, Y., and Rose-Petruck, C.G. (2004) Ultrafast tabletop laser pump – x-ray probe measurement of solvated Fe(CN)6-4. *Journal of Chemical Physics*, **122**, 084506–084506-8.

[35] Chang, Z.H., Rundquist, A., Wang, H.W. *et al.* (1997) Generation of coherent soft X rays at 2.7 nm using high harmonics. *Physical Review Letters*, **79**, 2967–2970.

[36] Blome, C., Sokolowski-Tinten, K., Dietrich, C. *et al.* (2001) Set-up for ultrafast time-resolved x-ray diffraction using a femtosecond laser-plasma keV x-ray source. *Journal de Physique IV: Proceedings*, **11**, 491–494.

[37] Chen, J., Tomov, I.V., Elsayed-Ali, H.E., and Rentzepis, P.M. (2006) Hot electrons blast wave generated by femtosecond laser pulses on thin Au(111) crystal, monitored by subpicosecond X-ray diffraction. *Chemical Physics Letters*, **419**, 374–378.

[38] In Websites on XFEL, http://www.xfel.eu/en/index.php; http://www.riken.jp/XFEL/eng/index.html; http://www-ssrl.slac.stanford.edu/lcls/.

[39] Stern, E.A., Sayers, D.E., and Lytle, F.W. (1975) Extended x-ray-absorption fine-structure technique. III. Determination of physical parameters. *Physical Review B*, **11**, 4836–4846.

[40] Lytle, F.W., Sayers, D.E., and Stern, E.A. (1975) Extended x-ray-absorption fine-structure technique. II. Experimental practice and selected results. *Physical Review B*, **11**, 4825–4835.

[41] Sayers, D.E., Stern, E.A., and Lytle, F. (1971) New technique for investigating noncrystalline structures. Fourier analysis of the extended x-ray-absorption fine structure. *Physical Review Letters*, **27**, 1204–1207.

[42] Teo, B.K., and Joy, D.C. (eds) (1981) *EXAFS [Extended X-Ray Absorption Fine Structure] Spectroscopy: Techniques and Applications*, WileyBlackwell, New York.

[43] Koningsberg, D.C. and Prins, R. (1988) *X-ray Absorption: Principles, Applications, Techniques of EXAFS, SEXAFS and XANES*, John Wiley & Sons, New York.

[44] Westre, T.E., Kennepohl, P., DeWitt, J.G. *et al.* (1997) A multiplet analysis of Fe K-Edge 1s → 3d pre-edge features of iron complexes. *Journal of the American Chemical Society*, **119**, 6297–6314.

[45] Wachs, I.E. (2005) Recent conceptual advances in the catalysis science of mixed metal oxide catalytic materials. *Catalysis Today*, **100**, 79–94.

[46] Thomas, J.M. and Sankar, G. (2001) The role of synchrotron-based studies in the elucidation and design of active sites in titanium-silica epoxidation catalysts. *Accounts of Chemical Research*, **34**, 571–581.

[47] Alexeev, O. and Gates, B.C. (2000) EXAFS characterization of supported metal-complex and metal-cluster catalysts made from organometallic precursors. *Topics in Catalysis*, **10**, 273–293.

[48] Nikitenko, S., Beale, A.M., van der Eerden, A.M.J. *et al.* (2008) Implementation of a combined SAXS/WAXS/QEXAFS set-up for time-resolved in situ experiments. *Journal of Synchrotron Radiation*, **15**, 632–640.

[49] Koningsberger, D.C., Mojet, B.L., van Dorssen, G.E., and Ramaker, D.E. (2000) XAFS spectroscopy; fundamental principles and data analysis. *Topics in Catalysis*, **10**, 143–155.

[50] Park, E.D. and Lee, J.S. (1999) Effects of pretreatment conditions on CO oxidation over supported Au catalysts. *Journal of Catalysis*, **186**, 1–11.

[51] Lamberti, C., Bordiga, S., Salvalaggio, M. *et al.* (1997) XAFS, IR, and UV-vis study of the Cu-I environment in Cu-I-ZSM-5. *Journal of Physical Chemistry B*, **101**, 344–360.

[52] Yamashita, H., Ichihashi, Y., Anpo, M. *et al.* (1996) Photocatalytic decomposition of NO at 275 K on titanium oxides included within Y-zeolite cavities: The structure and role of the active sites. *Journal of Physical Chemistry*, **100**, 16041–16044.

[53] Okumura, K., Honma, T., Hirayama, S. *et al.* (2008) Stepwise growth of Pd clusters in USY zeolite at room temperature analyzed by QXAFS. *Journal of Physical Chemistry C*, **112**, 16740–16747.

[54] Shimizu, K., Sugino, K., Kato, K. *et al.* (2007) Formation and redispersion of silver clusters in Ag-MFI zeolite as investigated by time-resolved QXAFS and UV-vis. *Journal of Physical Chemistry C*, **111**, 1683–1688.

[55] Okumura, K., Kato, K., Sanada, T., and Niwa, M. (2007) *In-situ* QXAFS studies on the dynamic coalescence and dispersion processes of Pd in the USY zeolite. *Journal of Physical Chemistry C*, **111**, 14426–14432.

[56] Okumura, K., Yoshino, K., Kato, K., and Niwa, M. (2005) Quick XAFS studies on the Y-type zeolite supported au catalysts for CO-O-2 reaction. *Journal of Physical Chemistry B*, **109**, 12380–12386.

[57] Chance, B., Fischetti, R., and Powers, L. (1983) Structure and kinetics of the photoproduct of carbonylmyoglobin at low temperatures; an x-ray absorption study. *Biochemistry*, **22**, 3820–3829.

[58] Mills, D.M., Lewis, A., Harootunian, A. *et al.* (1984) Time-resolved x-ray absorption spectroscopy of carbon monoxide-myoglobin recombination after laser photolysis. *Science*, **223**, 811–813.
[59] Chance, B., Kumar, C., Korszun, Z.R. *et al.* (1984) A time-resolved modulated-carrier rapid flow system for following rapid structural changes in biological reactions. *Nuclear Instruments and Methods in Physics Research Section A*, **222**, 180–184.
[60] Tischler, J.Z., Larson, B.C., and Mills, D.M. (1985) Time-resolved x-ray studies during pulsed-laser irradiation of germanium. *Materials Research Society Symposium Proceedings*, **35**, 119–124.
[61] Matsushita, T., Oyanagi, H., Saigo, S. *et al.* (1986) Twenty-five millisecond resolution time-resolved x-ray absorption spectroscopy in dispersive mode. *Japanese Journal of Applied Physics, Part 2: Letters*, **25**, L523–L525.
[62] Chance, M.R., Wirt, M.D., Scheuring, E.M. *et al.* (1993) Time-resolved x-ray absorption spectroscopy on microsecond timescales: Implications for the examination of structural motions. *Review of Scientific Instruments*, **64**, 2035–2036.
[63] Chen, L.X., Bowman, M.K., Montano, P.A., and Norris, J.R. (1993) EXAFS studies on the structure of photoexcited cyclopentadienylnickelnitrosyl(C5H5NiNO). *Materials Research Society Symposium Proceedings*, **307**, 45–50.
[64] Chen, L.X., Bowman, M.K., Montano, P.A., and Norris, J.R. (1993) X-ray absorption structural study of a reversible, photoexcited charge-transfer state. *Journal of the American Chemical Society.*, **115**, 4373–4374.
[65] Chen, L.X., Bowman, M.K., Wang, Z. *et al.* (1994) Structural Studies of Photoinduced Intramolecular Electron Transfer in Cyclopentadienylnitrosylnickel. *Journal of Physical Chemistry*, **98**, 9457–9464.
[66] Chen, L.X., Wang, Z., Burdett, J.K. *et al.* (1995) X-ray Absorption Studies on Electronic Spin State Transitions of Fe(II) Complexes in Different Media. *Journal of Physical Chemistry*, **99**, 7958–7964.
[67] Chen, L.X., Wasielewski, M.R., Rajh, T. *et al.* (1997) Molecular structure determination for photogenerated intermediates in photoinduced electron transfer reactions using steady-state and transient XAFS. *Journal de Physique IV*, **7**, 569–572.
[68] Chen, L.X., Lee, P.L., Gosztola, D. *et al.* (1999) Time-Resolved X-ray Absorption Determination of Structural Changes following Photoinduced Electron Transfer within Bis-porphyrin Heme Protein Models. *Journal of Physical Chemistry B*, **103**, 3270–3274.
[69] Chen, L.X., Lee, P.L., Gosztola, D. *et al.* (1999) Time-resolved energy-dispersive XAS studies of photoinduced electron transfer intermediates in electron donor-acceptor complexes. *Journal of Synchrotron Radiation*, **6**, 403–405.
[70] Kleifeld, O., Frenkel, A., Martin, J.M.L., and Sagi, I. (2003) Active site electronic structure and dynamics during metalloenzyme catalysis. *Nature Structural Biology*, **10**, 98–103.
[71] Jennings, G., Jaeger, W.J.H., and Chen, L.X. (2002) Application of Multielement Ge Detector in Laser Pump/X-ray Probe XAFS. *Review of Scientific Instruments*, **72**, 362–368.
[72] Rehr, J.J. and Ankudinov, A.L. (2001) New developments in the theory of x-ray absorption and core photoemission. *Journal of Electron Spectroscopy and Related Phenomena*, **114–116**, 1115–1121.
[73] Lavrentyev, A.A., Nikiforov, I.Y., Dubeiko, V.A. *et al.* (2001) The use of the FEFF8 code to calculate the XANES and electron density of states of some sulfides. *Journal of Synchrotron Radiation*, **8**, 288–290.
[74] Benfatto, M., Della Longa, S., Hatada, K. *et al.* (2006) A full multiple scattering model for the analysis of time-resolved X-ray difference absorption spectra. *Journal of Physical Chemistry B*, **110**, 14035–14039.
[75] Benfatto, M. and Della Longa, S. (2001) Geometrical fitting of experimental XANES spectra by a full multiple-scattering procedure. *Journal of Synchrotron Radiation*, **8**, 1087–1094.
[76] Della-Longa, S., Chen, L.X., Frank, P. *et al.* (2009) Direct deconvolution of two-state pump-probe x-ray absorption spectra and the structural changes in a 100 ps transient of Ni(II)-tetramesitylporphyrin. *Inorganic Chemistry*, **48**, 3934–3942.
[77] Smolentsev, G., Soldatov, A.V., and Chen, L.X. (2008) Three-dimensional local structure of photoexcited Cu (I) diimine complex refined by quantitative XANES analysis. *Journal of Physical Chemistry A*, **112**, 5363–5367.
[78] Saes, M., Bressler, C., Abela, R. *et al.* (2003) Observing photochemical transients by ultrafast x-ray absorption spectroscopy. *Physical Review Letters*, **90**, 047403.
[79] Gawelda, W., Johnson, M., de Groot, F.M.F. *et al.* (2006) Electronic and molecular structure of photoexcited Ru-II(bpy) (3) (2 +) probed by picosecond X-ray absorption spectroscopy. *Journal of the American Chemical Society*, **128**, 5001–5009.

[80] Gawelda, W., Pham, V.T., Benfatto, M. *et al.* (2007) Structural determination of a short-lived excited iron(II) complex by picosecond x-ray absorption spectroscopy. *Physical Review Letters*, **98**, 057401.

[81] Pham, V.T., Gawelda, W., Zaushitsyn, Y. *et al.* (2007) Observation of the solvent shell reorganization around photoexcited atomic solutes by picosecond X-ray absorption spectroscopy. *Journal of the American Chemical Society*, **129**, 1530–1531.

[82] Khalil, M., Marcus, M.A., Smeigh, A.L. *et al.* (2006) Picosecond X-ray absorption spectroscopy of a photoinduced iron(II) spin crossover reaction in solution. *Journal of Physical Chemistry A*, **110**, 38–44.

[83] Cavalleri, A., Rini, M., Chong, H.H.W. *et al.* (2005) Band-selective measurements of electron dynamics in VO2 using femtosecond near-edge x-ray absorption. *Physical Review Letters*, **95**, 067405.

[84] Chen, L.X. (2002) Excited state molecular structure determination in disordered media using laser pump/X-ray probe time-domain X-ray absorption spectroscopy. *Faraday Discussions*, **122**, 315–329.

[85] Chen, L.X., Jennings, G., Liu, T. *et al.* (2002) Rapid excited-state structural reorganization captured by pulsed x-rays. *Journal of the American Chemical Society*, **124**, 10861–10867.

[86] Chen, L.X., Shaw, G.B., Novozhilova, I. *et al.* (2003) The MLCT State Structure and Dynamics of a Cu(I) Diimine Complex Characterized by Pump-probe X-ray and Laser Spectroscopies and DFT Calculations. *Journal of the American Chemical Society.*, **125**, 7022–7034.

[87] Chen, L.X. (2004) Taking snapshots of photoexcited molecules in disordered media by using pulsed synchrotron x-rays. *Angewandte Chemie International Edition*, **43**, 2886–2905.

[88] Chen, L.X., Shaw, G.B., Liu, T. *et al.* (2004) Exciplex formation of copper(II) octaethylporphyrin revealed by pulsed x-rays. *Chemical Physics*, **299**, 215–223.

[89] Chen, L.X. (2005) Probing Transient Molecular Structures in Photochemical Processes Using Laser Initiated Time-resolved X-ray Absorption Spectroscopy. *Annual Review of Physical Chemistry*, **56**, 221–254.

[90] Chen, L.X., Zhang, X., Wasinger, E.C. *et al.* (2007) Tracking electrons and atoms in a photoexcited metalloporphyrin by X-ray transient absorption spectroscopy. *Journal of the American Chemical Society*, **129**, 9616–9618.

[91] Gust, D., Moore, T.A., and Moore, A.L. (2001) Mimicking photosynthetic solar energy transduction. *Accounts of Chemical Research*, **34**, 40–48.

[92] Holten, D., Bocian, D.F., and Lindsey, J.S. (2002) Probing electronic communication in covalently linked multiporphyrin arrays. A guide to the rational design of molecular photonic devices. *Accounts of Chemical Research*, **35**, 57–69.

[93] Rosenthal, J., Bachman, J., Dempsey, J.L. *et al.* (2005) Oxygen and hydrogen photocatalysis by two-electron mixed-valence coordination compounds. *Coordination Chemistry Reviews*, **249**, 1316–1326.

[94] Holten, D., Bocian, D.F., and Lindsey, J.S. (2002) Probing electronic communication in covalently linked multiporphyrin arrays. A guide to the rational design of molecular photonic devices J. S. *Accounts of Chemical Research*, **35**, 57–69.

[95] Wang, Z.C., Ho, K.C.J., Medforth, C.J., and Shelnutt, J.A. (2006) Porphyrin nanoriber bundles from phase-transfer ionic self-assembly and their photocatalytic self-metallization. *Advanced Materials*, **18**, 2557.

[96] Collman, J.P., Sunderland, C.J., Berg, K.E. *et al.* (2003) Spectroscopic evidence for a heme-superoxide/Cu(I) intermediate in a functional model of cytochrome c oxidase. *Journal of the American Chemical Society*, **125**, 6648–6649.

[97] Dawson, J.H., Andersson, L.A., Hodgson, K.O., and Hahn, J.E. (1982) The active site structure of cytochrome P-450 as determined by extended X-ray absorption fine structure spectroscopy, *Developments in Biochemistry*, **23**, 589–596.

[98] Graige, M.S., Feher, G., and Okamura, M.Y. (1998) Conformational gating of the electron transfer reaction QA-Q $\rightarrow$ QAQB in bacterial reaction centers of Rhodobacter sphaeroides determined by a driving force assay. *Proceedings of the National Academy of Sciences USA*, **95**, 11679–11684.

[99] Halgrimson, J., Horner, J., Newcomb, M. *et al.* (2008) X-ray absorption spectroscopic characterization of the Compound II intermediate produced by reaction of a cytochrome P450 enzyme with peroxynitrite. *Proceedings of the National Academy of Sciences USA*, **105**, 8179–8184.

[100] Blankenship, R.E. (2002) *Molecular Mechanisms of Photosynthesis*, Balckwell Science, Oxford.

[101] Zhang, X.Y., Wasinger, E.C., Muresan, A.Z. *et al.* (2007) Ultrafast stimulated emission and structural dynamics in nickel porphyrins. *Journal of Physical Chemistry A*, **111**, 11736–11742.

[102] Chen, L.X., Zhang, X.Y., Wasinger, E.C. *et al.* (2007) Tracking electrons and atoms in a photoexcited metalloporphyrin by X-ray transient absorption spectroscopy. *Journal of the American Chemical Society*, **129**, 9616–9620.

[103] Ballhausen, C. J. (1962) *Introduction to Ligand Field Theory*, McGraw-Hill, New York.

[104] Kim, D., Kirmaier, C., and Holten, D. (1983) Nickel porphyrin photophysics and photochemistry. A picosecond investigation of ligand binding and release in the excited state. *Chemical Physics*, **75**, 305–322.

[105] Rodriguez, J. and Holten, D. (1989) Ultrafast vibrational dynamics of a photoexcited metalloporphyrin. *Journal of Chemical Physics*, **91**, 3525–3531.

[106] Gentemann, S., Nelson, N.Y., Jaquinod, L. *et al.* (1997) Variations and temperature dependence of the excited state properties of conformationally and electronically perturbed zinc and free base porphyrins. *Journal of Physical Chemistry B*, **101**, 1247–1254.

[107] Eom, H.S., Jeoung, S.C., Kim, D. *et al.* (1997) Ultrafast vibrational relaxation and ligand photodissociation/photoassociation processes of Nickel(II) porphyrins in the concensed phase. *Journal of Physical Chemistry A*, **101**, 3661–3669.

[108] Mizutani, Y. and Kitagawa, T. (2001) A role of solvent in vibrational energy relaxation of metalloporphyrins. *Journal of Molecular Liquids*, **90**, 233–242.

[109] Meyer, T.J. (1989) Chemical approaches to artificial photosynthesis. *Accounts of Chemical Research*, **22**, 163–170.

[110] Vlcek, A. Jr., (1998) Mechanistic roles of metal-to-ligand charge-transfer excited states in organometallic photochemistry. *Coordination Chemistry Reviews*, **177**, 219–256.

[111] Balzani, V., Credi, A., and Venturi, M. (1998) Photochemistry and photophysics of coordination compounds. An extended view. *Coordination Chemistry Reviews*, **171**, 3–16.

[112] Durr, H. and Bossmann, S. (2001) Ruthenium polypyridine complexes. On the route to biomimetic assemblies as models for the photosynthetic reaction center. *Accounts of Chemical Research*, **34**, 905–917.

[113] Ahn, B.-T. and McMillin, D.R. (1981) Photostudies of copper(I) systems. 7. Studies of bimolecular reactions involving cobalt(III) complexes, chromium(III) complexes, or oxygen following the excitation of bis(2,9-dimethyl-1,10-phenanthroline)copper(I). *Inorganic Chemistry*, **20**, 1427–1432.

[114] Blasse, G. and McMillin, D.R. (1980) On the luminescence of bis(triphenylphosphine)phenanthrolinecopper(I). *Chemical Physics Letters*, **70**, 1–3.

[115] Dietrich-Buchecker, C.O., Marnot, P.A., Sauvage, J.P. *et al.* (1983) Bis(2,9-diphenyl-1,10-phenanthroline) copper(I): a copper complex with a long-lived charge-transfer excited state. *Journal of the Chemical Society, Chemical Communications*, 513–515.

[116] Castellano, F.N., Ruthkosky, M., and Meyer, G.J. (1995) Photodriven energy transfer from cuprous phenanthroline derivatives. *Inorganic Chemistry*, **34**, 3–4.

[117] Miller, M.T., Ganzel, P.K., and Karpishin, T.B. (1998) A Photoluminescent Copper(I) Complex with an Exceptionally High CuII/CuI Redox Potential: [Cu(bfp)2] + (bfp=2,9-bis(trifluoromethyl)-1,10-phenanthroline). *Angewandte Chemie International Edition*, **37**, 1556.

[118] Ruthkosky, M., Kelly, C.A., Zaros, M.C., and Meyer, G.J. (1997) Long-lived charge-separated states following light excitation of Cu(I) donor-acceptor compounds. *Journal of the American Chemical Society.*, **119**, 12004–12005.

[119] Collin, J.-P., Dietrich-Buchecker, C., Gavina, P. *et al.* (2001) Shuttles and muscles: linear molecular machines based on transition metals. *Accounts of Chemical Research*, **34**, 477–487.

[120] Kern, J.-M., Raehm, L., Sauvage, J.-P. *et al.* (2000) Controlled molecular motions in copper-complexed rotaxanes: An XAS study. *Inorganic Chemistry*, **39**, 1555–1560.

[121] Hamalainen, R., Algren, M., Turpeinen, U., and Raikas, T. (1979) Bis(2,9-dimethyl-1,10-phenanthroline)copper (I) Nitrate. *Crystal Structure Communication*, **8**, 75–80.

[122] McMillin, D.R. and McNett, K.M. (1998) Photoprocesses of copper complexes that bind to DNA. *Chemical Reviews*, **98**, 1201–1219.

[123] Chen, L.X., Shaw, G.B., Novozhilova, I. *et al.* (2003) MLCT state structure and dynamics of a Copper(I) diimine complex characterized by pump-probe x-ray and laser spectroscopies and DFT calculations. *Journal of the American Chemical Society*, **125**, 7022–7034.

[124] Siddique, Z.A., Yamamoto, Y., Ohno, T., and Nozaki, K. (2003) Structure-dependent photophysical properties of singlet and triplet metal-to-ligand charge transfer states in copper(I) Bis(diimine) compounds. *Inorganic Chemistry*, **42**, 6366–6378.

[125] Shaw, G.B., Grant, C.D., Shirota, H., Castner, E.W. Jr., Meyer, G.J. and Chen, L.X. (2007) Ultrafast Structural Rearrangements in the MLCT Excited State for Copper(I) bis-Phenanthrolines in Solution, *Journal of the American Chemical Society*, **129**, 2147–2160.

[126] Freed, K.F. and Jortner, J. (1970) Multiphonon processes in the nonradiative decay of large molecules. *Journal of Chemical Physics*, **52**, 6272–6280.

[127] Bixon, M., Jortner, J., Cortes, J. *et al.* (1994) Energy gap law for nonradiative and radiative charge transfer in isolated and in solvated supermolecules. *Journal of Physical Chemistry*, **98**, 7289–7299.

[128] Scaltrito, D.V., Thompson, D.W., O'Callaghan, J.A., and Meyer, G.J. (2000) MLCT excited states of cuprous bis-phenanthroline coordination compounds. *Coordination Chemistry Reviews*, **208**, 243–266.

[129] Dau, H. and Haumann, M. (2008) The manganese complex of photosystem II in its reaction cycle – Basic framework and possible realization at the atomic level. *Coordination Chemistry Reviews*, **252**, 273–295.

[130] Grundmeier, A., Loja, P., Haumann, M., and Dau, H. (2007) On the structure of the manganese complex of photosystem II: extended-range EXAFS data and specific atomic-resolution models for four S-states. *Photosynthesis Research*, **91**, PS446.

[131] Dau, H. and Haumann, M. (2007) Time-resolved X-ray spectroscopy leads to an extension of the classical S-state cycle model of photosynthetic oxygen evolution. *Photosynthesis Research*, **92**, 327–343.

[132] Furube, A., Asahi, T., Masuhara, H. *et al.* (1999) Charge carrier dynamics of standard TiO_2 catalysts revealed by femtosecond diffuse reflectance spectroscopy. *Journal of Physical Chemistry B*, **103**, 3120–3127.

[133] Kong, Q.Y., Lee, J.H., Plech, A. *et al.* (2008) Ultrafast X-ray solution scattering reveals an unknown reaction intermediate in the photolysis of [Ru-3(CO)(12)]. *Angewandte Chemie-International Edition*, **47**, 5550–5553.

[134] Plech, A., Kotaidis, V., Istomin, K., and Wulff, M. (2007) Small-angle pump-probe studies of photoexcited nanoparticles. *Journal of Synchrotron Radiation*, **14**, 288–294.

[135] Wulff, M., Bratos, S., Plech, A. *et al.* (2006) Recombination of photodissociated iodine: A time-resolved x-ray-diffraction study. *Journal of Chemical Physics*, **124**, 034501.

[136] Ihee, H., Lorenc, M., Kim, T.K. *et al.* (2005) Ultrafast x-ray diffraction of transient molecular structures in solution. *Science*, **309**, 1223–1227.

[137] Plech, A., Wulff, M., Bratos, S. *et al.* (2004) Visualizing chemical reactions in solution by picosecond x-ray diffraction. *Physical Review Letters*, **92**, 125505/1–125505/4.

8

Fourier-Transform Infrared and Raman Spectroscopy of Pure and Doped TiO_2 Photocatalysts

Lars Österlund[1,2]
[1]*Totalförsvarets forskningsinstitut (FOI), Umeå, Sweden, Email: lars.osterlund@foi.se*
[2]*Dep. Engineering Sciences, The Ångström Laboratory, Uppsala University, Uppsala, Sweden*

8.1 Introduction

Photocatalysis is a broad research field which lies at the heart of modern sustainable technologies such as air and water cleaning, solar hydrogen production, wet solar cells, self-cleaning and antibacterial surface coatings. A detailed understanding of photocatalytic processes is not possible without explicit model studies of photocatalyst structure, adsorbate structures and surface reactions. Vibrational spectroscopy is an important tool to investigate these properties [1–4]. This is the subject of this chapter. Raman spectroscopy is an important technique to characterize metal-oxide nanoparticles, which complements traditional analytical techniques. In some cases, Raman gives additional advantages, such as enhanced sensitivity and information of the location of active phases. Raman spectroscopy also provides new insights into quantum-size phenomena in nanoparticles, and can be used to quantify phonon-confinement phenomena, and variations of surface stress and unit cell volume. Fourier-transform infrared spectroscopy (FTIR) is an invaluable method to probe surface acidity, adsorbate structure and surface reactions on metal-oxide nanoparticles. In particular, FTIR lends itself favorably to the study of molecular transformations *in situ* or *in vacuo*. This is utilized to study surface reactions on oxide surfaces in, for example, heterogeneous catalysis [5–7], photochemistry [8,9] and at solid–liquid interfaces [10–12]. *In situ* or *in vacuo* vibrational spectroscopy studies of photocatalytic reactions have been conducted by Yates and co-workers [13,14], Anderson and co-workers [12,15], Liao *et al.* [16,17] and Österlund and co-workers [9,18–21]. Although the emphasis here will be

On Solar Hydrogen & Nanotechnology Edited by Lionel Vayssieres

on photoinduced oxidation reactions, in particular gas–solid reactions, the results are generally applicable in a broader context, namely the reactivity–structure relationships of transition-metal oxides. Gas–solid photocatalysis connects to advancements in the surface science of metal oxides in general. Important tools such as scanning tunneling microscopy (STM), X-ray absorption and diffraction techniques, vibrational surface spectroscopy and *ab initio* quantum-mechanical calculations are today widely used to explore, predict and verify photocatalytic properties, as discussed elsewhere in this book.

The surface science of TiO_2 photocatalysis has matured considerably during the past 10–15 years [13,14,22]. In particularly, our understanding of elementary surface processes occurring on single-crystal TiO_2 has advanced considerably, albeit almost exclusively on the rutile polymorph [22]. Such studies provide suitable experimental input to *ab initio* theoretical modeling (without, for example, complications of hydration shells, solvated ions, etc.). The basic understanding has thus advanced considerably for simple adsorption systems like O_2, H_2, H_2O, CO, CO_2, NO_x, SO_2 and C_1 and to some extent C_2 organic molecules. The implication of these studies are expected to give fundamental insight into photocatalytic oxidation processes, and promise to provide a toolbox for predicting reactivity, which has been long sought for in this research field.

Advancements in materials science, notably nanoscience and nanotechnology, have facilitated fabrication of well-defined model systems that can bridge the structure gap that exists between single-crystal oxides and nanostructured materials employed in practical applications [92]. Examples of monodispersed TiO_2 photocatalysts with different structures and morphologies are shown in Figure 8.1. Each of them exhibit unique and well-defined distributions of crystal facets. Despite these advancements, there exists a structure gap in photocatalysis, which arises from the fact that most experimental studies have been conducted in colloidal suspensions employing anatase TiO_2, which is commonly considered to be the most active TiO_2 polymorph. On the other hand, much of the fundamental surface-science studies conducted during the past decade have been done on rutile single crystals, and in particular the low-surface-energy rutile (110) 1×1 surface [22], which is commercially available. In contrast, large anatase single crystals are not readily available and are difficult to grow in the laboratory. Specific anatase crystal orientations have been prepared by molecular beam epitaxy and metal organic chemical vapor deposition on various substrates [23–25]. Diebold *et al.* [26–28] have instead employed natural mineral samples for fundamental STM studies of anatase which however are difficult to implement in other experimental techniques due to their inherent small sizes. The different anatase facets that so far have been examined in detailed exhibit either bulk-terminated or reconstructed structures [23–27,29]. Analogies with the huge body of existing experimental data on TiO_2 nanoparticles are difficult, or even misleading, unless a detailed structural analysis of the nanoparticles are done in parallel and translated into the observed reactivity. Distortions of the TiO_2 lattice known to be present at the surface of the nanoparticles have different reactivity compared to perfect bulk terminated surfaces [30,31]. The literature is not consistent in this respect, and there is a large spread in reported data.

This chapter is organized as follows: in Section 8.2, a brief overview of experimental considerations in vibrational spectroscopy of photocatalysts is presented. In Section 8.3 we demonstrate the applicability of Raman spectroscopy as an important experimental technique to characterize metal oxides and, in particular, semiconductor photocatalysts. In Section 8.4 we introduce current models of surface reactions of simple organics on TiO_2 nanoparticles relevant for our discussion and show a case study of the photocatalytic oxidation of propane on different TiO_2 nanoparticles. In Section 8.5 a comprehensive overview of infrared spectroscopy

Figure 8.1 Transmission and scanning electron micrographs of TiO_2 nanoparticles prepared by different techniques; From top left to bottom right: 3×5 nm rutile (reprinted with permission from M. Andersson, A. Kiselev, L. Österlund and A.E.C. Palmqvist, Microemulsion-mediated room temperature synthesis of high surface area rutile and its photocatalytic performance, *Journal of Physical Chemistry C*, **111**, 6789, 2007. © 2007 American Chemical Society), 5×60 nm rutile (reprinted with permission from F.P. Rotzinger, J.M. Kesselman-Truttman, S.J. Hug *et al.*, Structure and vibrational spectrum of formate and acetate adsorbed from aqueous solution on the TiO_2 rutile (110) surface, *Journal of Physical Chemistry B*, **108**, 5004–5017, 2004. © 2004 American Chemical Society), top view of 25 nm 10% Nb-TiO_2 films (reprinted with permission from A. Mattsson, M. Leideborg, K. Larsson *et al.*, Adsorption and solar light decomposition of acetone on anatase TiO_2 and niobium doped TiO_2 thin films, *Journal of Physical Chemistry B*, **110**, 1210–1220, 2006. © 2006 American Chemical Society), 66 nm anatase (BDH), 25 nm anatase, 40 nm anatase and cross-sections through N-TiO_2 and TiO_2 thin films, respectively (reprinted with permission from J.M. Mwabora, T. Lindgren, E. Avendaño, *et al.*, Structure, composition, and morphology of photoelectrochemically active $TiO_{2-x}Nx$ thin films deposited by reactive DC magnetron sputtering, *Journal of Physical Chemistry B*, **108**, 20193–20198, 2004. © 2004 American Chemical Society).

studies of surface reactions of formic acid and acetone on pure and doped TiO_2 nanoparticle systems is provided. We give examples of both anion- and cation-doped TiO_2, as well as particle size and shape dependence of reactivity. From a mechanistic viewpoint, we focus on one particular issue, namely adsorbate structure of formic acid and acetone – the simplest organic acid and ketone, respectively. They represent prototype TiO_2 adsorption systems that are well studied *in vacuo* on single-crystal TiO_2 (primarily the TiO_2(110) rutile surfaces) and are known reaction intermediates and by-products formed in the course of many photocatalytic reactions. It turns out that that it is often crucial to understand these systems to understand (and predict) the overall kinetics. We include tables of compilations of measured frequencies and vibrational mode assignments to facilitate comparisons between different nanoparticles systems and also single-crystal data, when available.

8.2 Vibrational Spectroscopy on TiO_2 Photocatalysts: Experimental Considerations

Infrared (IR) and Raman vibrational spectroscopy are unique characterization techniques that provide information on: (i) metal-oxide nanoparticles (structure, lattice dynamics, anisotropy, particle size, type and properties of surface centers), (ii) adsorbate structures and (iii) mechanism of molecular surface processes occurring on metal oxides. In addition, Raman and IR spectroscopy allows for studies under a wide range of experimental conditions: from ultrahigh vacuum to high pressures, from cryogenic temperatures to ~800 K, in gas or liquids, on single crystals, powders, thin films or colloidal solutions. This allows for true *in situ* or *operando* studies, which can provide valuable information that can help bridge pressure and structure-gap problems known to hamper our understanding of, for example, catalytic reactions on solid surfaces, including photocatalytic reactions.

Raman and IR spectroscopy are complementary methods for studies of molecular vibrations. While IR spectroscopy has become a standard method in most laboratories since the introduction of interferometers and Fourier transform (FT) data processing in the 1970s [32], Raman has long been considered as a dedicated instrument typically found in physics laboratories. Raman spectroscopy has, however, been revolutionized by advancements in laser technology, with sources readily available for the whole visible and near-IR spectral range (and soon also UV). This has paved the way for bench-top instruments with user-friendly hardware and software interfaces manufactured by several companies. In general Raman has higher spatial resolution than IR and can readily access the low wavenumber spectral range where most phases of catalyst materials (metal oxides and sulfides) have their characteristic absorption bands (phonon bands). Raman has therefore been applied for characterization of bulk and supported catalysts. This can be rationalized as follows: a vibrational transition is IR active if the electric dipole moment of the molecule changes during the vibration. It is Raman active if the polarizability (i.e., the field-induced change of the dipole moment) changes during the vibration. Hence bonds having ionic character give strong IR absorption and bonds with covalent character give strong Raman absorption. The ionic character of the Ti–O bonds in TiO_2 is manifested in the strong IR absorption below about 950 cm^{-1} (Figure 8.2), while the Raman signal is weak down to 700 cm^{-1}. IR spectroscopy of adsorbed species on metal-oxide surfaces is therefore, in general, prohibited below about 1000 cm^{-1}. Metal oxides containing sufficiently large metal atoms have a weaker covalent contribution from the M–O bonds,

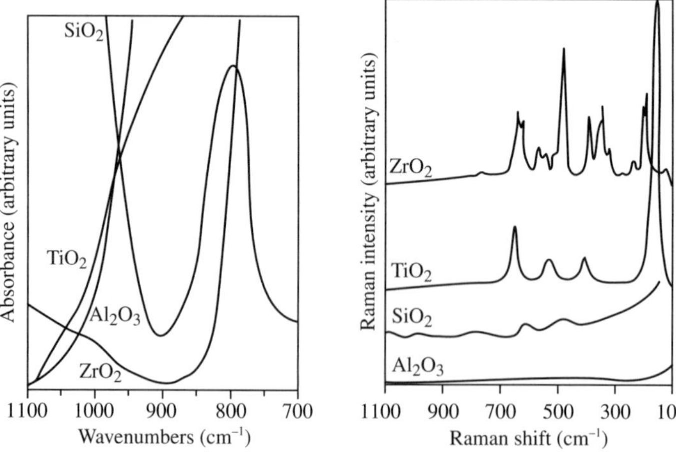

Figure 8.2 Infrared (left) and Raman spectra (right) of dehydrated metal oxides [3].

which, in the case of TiO_2, gives rise to strong, discrete absorption bands in the 700–100 cm^{-1} region due to specific symmetry of the Ti–O vibrations (Figure 8.2). As a consequence, Raman is the preferred method to detect and classify metal-oxide vibrations (M–O and M–O–M bonds) in the low-wavenumber region. At higher frequencies, IR and Raman spectroscopy can give complementary information, that is, about terminal metal-oxide bonds (M=O). Metal oxides are generally hydroxyl terminated yielding M–OH bonds that give rise to strong IR absorption, while the corresponding Raman signal is absent or very weak. Information about the hydroxyl chemistry on metal-oxide surfaces are therefore preferably gained from IR spectroscopy. The acid and basic character of hydroxyls adsorbed on the amphoteric TiO_2 surface has been the subject of several IR studies [1].

IR spectroscopy has been by far the most widely employed technique to characterize molecules adsorbed on surfaces, to probe surface acidity, and to study surface reactions on catalysts at low and high pressures, in gas and liquids (and even aqueous solutions using special cells to circumvent IR absorption by water) or at low and high temperatures (typically from cryogenic up to ~600 K). Several authors have reviewed IR spectroscopy in catalysis [1,33–35]. Recent progress in heterogeneous catalysis includes experimental set-ups that facilitate simultaneous measurements of both adsorbed and gaseous species by combined IR and mass spectrometry or gas chromatography, some of which are truly *in situ* measurements [9,15,20,36,37]. Often diffuse reflectance FTIR spectroscopy (DRIFTS) is employed for nanostructured catalysts. Today dedicated DRIFTS cells are commercially available that efficiently separate diffuse and specular components of the scattered IR light (Figure 8.3) [38]. Figure 8.4 shows a schematic set-up of an experiment employing a Praying Mantis DRIFTS cell (Harrick Scientific Corp.) capable of simultaneous *in situ* FTIR, mass spectrometry/gas chromatography and light illumination [9,19,21]. Note that in DRIFT spectroscopy, the signal is related to the surface species present in the upper layer of the catalyst bed (typically of the order of 100 μm depth for TiO_2 nanoparticles) compared to the total depth of the catalyst bed

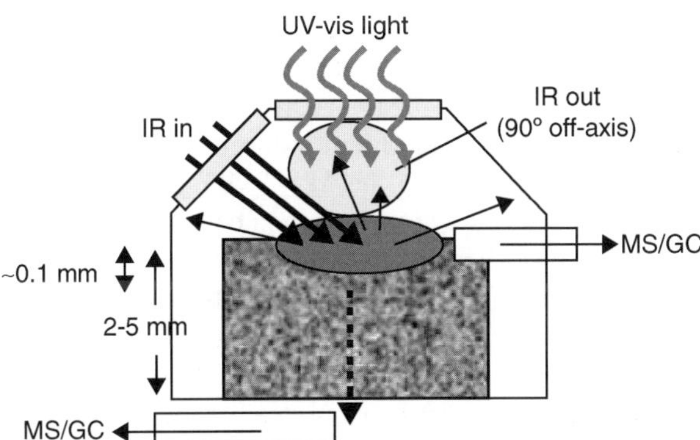

Figure 8.3 Schematic drawing of a DRIFTS reaction cell adapted to a 90° off-axis optical scattering geometry (cf. Figure8.4). The cell allows for simultaneous IR spectroscopy of adsorbed species on the catalyst surface and mass spectrometry (MS) analysis of gas-phase species. The catalyst bed and the zone probed by the infrared light are indicated. The MS data are collected at the exit of the DRIFT cell, that is, beneath the bed.

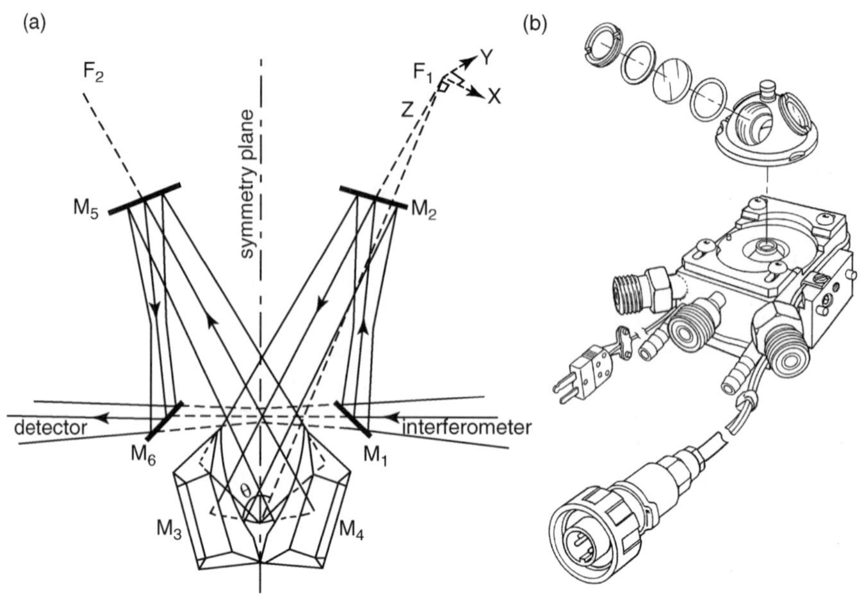

Figure 8.4 (a) Optical diagram of the Harrick 2D attachment for DRIFTS. (b) The Harrick high temperature/high pressure reaction (HTHP) cell used for DRIFTS studies of powder materials [38]. The extra window can be used to access the powder bed with an external light source. (Reprinted with permission from K. Moradi, C. Depecker and J. Corset, Diffuse reflectance infrared spectroscopy: experimental study of nonabsorbing materials and comparisions with theories, *Applied Spectroscopy*, **48**, 1491–1497, 1994. © 1994 Society for Applied Spectroscopy.)

(typically >1 mm depth). For DRIFT spectroscopy of adsorbates on TiO_2 nanoparticles, dilution of the solid by, for example, KBr is not necessary, since IR absorption by TiO_2 is negligible above $1000\,cm^{-1}$, where most necessary spectral information is located. It should, however, be noted that even though the particle size is, in general, much smaller than the irradiation wavelength for nanoparticles, the IR spectrum of TiO_2 (and metal oxides in general) is still not independent of particle size. Field-induced polarization effects and state of aggregation may have strong influence on the M–O vibrational stretching modes [39]. IR spectroscopy has been used to study the optical phonon band of nanocrystalline TiO_2 [40] and anatase TiO_2 nanoparticles [41,42]. The DRIFTS signal of adsorbed species overlaps with gas phase species, both originating from the probe volume in the catalyst bed and the free volume above the bed. This necessitates clever experimental designs and background subtraction schemes. A background spectrum is collected under steady-state conditions (constant flow, reactant pressure and temperature). Deviation from steady state due to either photon irradiation or a temperature increase, or by addition of a reactive gas molecule is readily detected in the IR spectrum. Nevertheless, both photon irradiation and temperature changes introduce spectral changes due to changes in the population of low-frequency (IR active) phonons or shallow bandgap states, which must be corrected for in the analysis. In fact, IR spectroscopy has been used to characterize formation of shallow O vacancy states in TiO_2 at elevated temperatures, as a function of water coverage and UV illumination [13,41,43,44].

Recent developments in high-pressure methods for *in situ* vibrational spectroscopy studies of catalysts have been reviewed by Rupprechter [35]. Moreover, the use of attenuated total reflection ATR-FTIR spectroscopy in heterogeneous catalysis has been demonstrated by several authors [10,45]. Burgi *et al.* [45] have showed the applicability of modulation excitation spectroscopy and phase-sensitive detection in combination with ATR for *in situ* FTIR spectroscopy of solid–liquid catalytic reactions with simultaneous irradiation capability, which enables distinction between species that have different responses to, for example, concentration changes. IR spectroscopy is today almost exclusively performed by FTIR spectroscopy, which offers several advantages, such as broad spectral range, high throughput, high spectral resolution and short acquisition time compared to other complementary vibrational techniques such as Raman, and electron energy loss spectroscopy (EELS). Nevertheless, in practice, the number of absorbents is typically notoriously small in heterogeneous catalysis and care must be taken to distinguish the desired IR signal from the background in FTIR experiments. This may also be the case for powder catalysts with surface areas exceeding 1 $m^2\ g^{-1}$, where the IR probe depth is limited (due to the optical properties of the catalysts), there is limited diffusion of reactant molecules in the porous catalyst structure, or when the IR cross-section is small for available molecular absorption bands. The extreme situation occurs on single-crystal surfaces, where the available number of absorbents is limited to the projected beam area ($\sim$1 cm^2) on the single-crystal surface. In this case the reflectance IR spectrum is dominated by the absorption of the molecules in the surrounding media (most often the gas phase). Here it is essential to eliminate this background signal, which complicates *in situ* measurements. Traditionally ultrahigh vacuum (UHV)-based infrared reflection absorption spectroscopy (IRRAS) has been employed for such (rather elaborate) studies [33,46]. Recent advancements in dual-channel spectroscopy, such as polarization-modulation infrared reflection absorption spectroscopy (PM-IRRAS), however, allow simultaneous acquisition of the static reference (or background) and the differential absorption of IR light of different polarization, where the polarization modulation is produced by a photoelastic modulator (PEM) operating at much higher frequency than the Fourier frequencies. The principle of PM-IRRAS is based on the selective absorption of p- and s-polarized light by adsorbed molecules when the light is near the grazing angle. Only the field due to p-polarized light couples efficiently with the dipolar field of the adsorbates, while the s-polarized light cancels due to interference between the incident and the reflected light. After demodulation of the data from the two channels, a differential reflectivity spectrum $\Delta R/R$ is calculated from the s- and p-polarized spectra acquired simultaneously, and constitutes the vibrational spectrum of the adsorbates on the surface. This procedure results in complete removal of spectral features of nonadsorbed species (including the ubiquitous CO_2 and H_2O) and allows *in situ* measurements of chemical interactions on surfaces on timescales much shorter than traditional IRRAS. The usefulness of PM-IRRAS has been demonstrated on various metal surfaces [35], and should also be applicable for fundamental studies of semiconductor surfaces (such as TiO_2) with appropriate consideration of the specific optical material properties [47].

8.3 Raman Spectroscopy of Pure and Doped TiO_2 Nanoparticles

Size effects, structure, location of active phases and morphology play an important role in photocatalysis [48–51]. It has been reported that many photocatalytic properties in semiconducting photocatalysts are size dependent [52,53]. Although the exact size at which quantum size effects become important is debated, and may be different for different materials and structures (e.g., rutile particles are "rod-like" due to the anisotropy in the low-index surface energies), it is reasonable to assume that most of the experimental studies of TiO_2 do not a priori exhibit pronounced electronic quantum size effects at particle sizes >5nm. In addition, these properties are modified when dopants are introduced [18]. Instead it is argued that the special nature of nanoparticle semiconductors, compared with their bulk counterparts, generally arises due to defect-related properties rather than from quantum size effects [50,51]. These defect-related sites (notably O-vacancies, interstitial Ti^{3+} and under-coordinated Ti atoms) are very reactive, and considered to be the photoactive sites in photocatalysis [54–57]. In contrast, Raman spectroscopy is a powerful tool to investigate size dependence in semiconducting photocatalysts due to phonon confinement. In fact, TiO_2 has become a model system to study size effects, phase transitions and morphology of monodisperse, as well as polydisperse, nanocrystals [58–64].

Here we outline the phonon confinement (QC) model, or the $\vec{q}$-vector relaxation model, which has been widely used to account for observed Raman frequency shifts and asymmetric line broadening on nanoparticles. This model was originally proposed by Richter *et al.* [65] to explain Raman spectra in nanocrystalline silicon, and is today frequently used to interpret Raman shifts of the low-frequency $E_g(\nu_6)$ mode in TiO_2 nanoparticles [59,61–64,66–76]. On the nanoscale, the phonon momentum selection rule $\Delta\vec{q} = 0$ for the photon–phonon interaction that is valid in bulk crystals breaks down, and phonons located outside the Brillouin zone center (up to $q \sim 1/L$, where L is the crystal diameter) contribute to the line profile [65]. A convenient way to obtain an approximate analytical solution for this problem is to assume that all phonons over the first Brillouin zone (BZ) contribute to the Raman scattering and construct a wavefunction that is a superposition of eigenfunctions weighted by Fourier coefficients, $c(0, \vec{q})$, whose weight for off-center phonons ($\vec{q} > 0$) increases as L decreases. This can be done by modifying the phonon wavefunction for an infinite crystal by introducing a confinement function, $W(\vec{r}, L)$ [65]:

$$\Psi(\vec{q}_0\vec{r}) = W(\vec{r}, L)u(\vec{q}_0\vec{r})\exp(-i\vec{q}_0\vec{r}), \tag{8.1}$$

where $\vec{q}_0$ is the wave vector in the ideal case of an infinite crystal and $u(\vec{q}_0\vec{r})$ is the periodicity of the lattice. The confinement function, $W(\vec{r}, L)$, which is superimposed on the Bloch-type wavefunction is typically taken to be a Gaussian-type function, [65,77] *viz.*

$$W(\vec{r}, L) = \exp\left(\frac{\alpha r^2}{L^2}\right), \tag{8.2}$$

where r is the radial coordinate. The constant α is chosen to be $\alpha = 8\pi^2$, based on the work by Campbell and Fauchet [77]. This form of $W(\vec{r}, L)$ is consistent with spherical particle morphology, which is pertinent to anatase particles (cf. Figure 8.1). Fourier expansion around $\vec{q} = 0$, which is appropriate for optical phonons, then yields the following functional form of the Fourier coefficients:

$$|C(0, L)|^2 \propto \exp\left(-\frac{q^2L^2}{16\pi^2}\right), \tag{8.3}$$

The Raman intensity is governed by the phonon transition matrix elements with contributions also at $q \neq 0$, which are weighted by these Fourier coefficients. Using Equations 8.1–8.3, the first-order Raman spectrum can be written as a superposition of weighted Lorentzian contributions over the first BZ, *viz.*

$$I(\omega) \propto \int \rho(L)dL \int_{BZ} \frac{|C(0,\vec{q})|^2 d\vec{q}}{(\omega - \omega(\vec{q}))^2 + \left(\frac{\Gamma_0}{2}\right)^2}, \tag{8.4}$$

where $\omega(\vec{q})$ is the phonon dispersion relation, Γ_0 represents the Raman peak line width, whose temperature dependence has been explored in detail [63], and $\rho(L)$ is the particle size distribution. Since experimental data for the phonon dispersion for anatase is lacking, either theoretical calculations [78] or data for rutile [69] is used. As shown in Ref. [78] the phonon dispersion is positive and similar in the $\Gamma \rightarrow$ X and $\Gamma \rightarrow$ N direction ([100] and [110] directions), but rather flat in the $\Gamma \rightarrow$ Z direction ([001] direction). To a first approximation, the contribution from the latter integration thus yields a multiplicative contribution to Equation 8.4. Typically the phonon dispersion is therefore described by the simplest possible tight-binding relation using a spherical BZ with isotropic dispersion, *viz.*

$$\omega(q) = \omega_0 + \Delta \times [1 - \cos(qa)], \tag{8.5}$$

where, typically, $\Delta = 20\,\text{cm}^{-1}$ is used [69]. Other values of Δ have been reported, but this has no physical significance [76], a is a unit cell parameter somewhere in the range 3.78 Å to 9.51 Å (the unit cell dimension of anatase); typically $a = 3.768$ Å is used; ω_0 is the bulk $E_g(\nu_6)$ mode frequency ($\omega_0 = 144\,\text{cm}^{-1}$ for bulk TiO_2) [79], which is also temperature dependent [63]. The particle size distribution $\rho(L)$ has, in contrast to many reports, no significant effect on the Raman shift and only a small effect on the line width [64]. Figure 8.5 shows experimental and calculated Raman spectra of the $E_g(\nu_6)$ mode using Equation 8.4. The agreement between the model and experimental data strongly support the quantum confinement model.

One fundamental physical reason that alters the lattice vibrations that has been widely neglected in the analysis of Raman spectra of nanoparticles is the hydrostatic pressure across the grains arising from surface tension. Wang *et al.* [63] and Lejon and Österlund [64] recently highlighted the importance of including hydrostatic pressure (or equivalently surface stress) in nanosized anatase grains to explain the shift in Raman peak positions, as well as peak broadening. The hydrostatic pressure gives rise to an important contribution to the Raman shift of the $E_g(\nu_6)$ mode in anatase [64]. Surface atoms have a lower coordination number than bulk atoms. Therefore the bond strengths between surface atoms are different from those in the bulk. This surface tension induces a pressure across the particle of a hydrostatic type. The pressure across the grain, P, relates to the surface stress, f, by the Kelvin equation

$$P = \frac{4f}{L}, \tag{8.6}$$

where L is the particle diameter as before. The frequency change as a function of pressure is reported to be $d\omega/dP = 2.5$ (Ref. [62]) and $f \approx 1.5\,\text{J m}^{-2}$ (Ref. [80]) for TiO_2. The effect of phonon confinement and hydrostatic pressure on peak position for different particle sizes is shown in Figure 8.6 [64]. It can be seen that the hydrostatic pressure can contribute substantially to the Raman blue-shift for small L. In fact, in these simulations it can be shown that quantum confinement and hydrostatic pressure give rise to comparable Raman *blue-shifts*, while inclusion of particle size distribution leads to a minor *red-shift*.

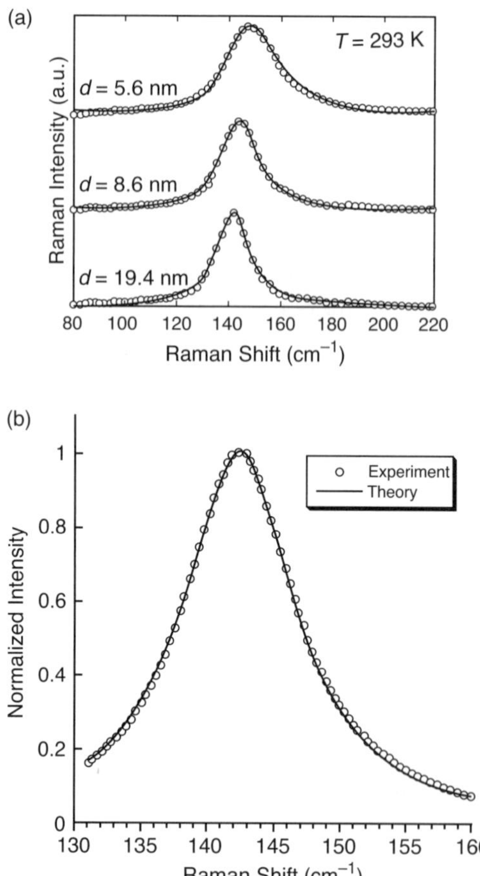

Figure 8.5 Experimental and calculated Raman spectra of the $E_g(\nu_6)$ mode (a) using Equation 8.4 [63], and (b) with additional inclusion of the hydrostatic pressure (Equation 8.6) [64]. (Reprinted with permission from D. Wang, J. Zhao, B. Chen, and C. Zhu, Lattice vibration fundamentals in nanocrystalline anatase investigated with Raman scattering, *Journal of Physics: Condensed Matter*, **20**, 085212, 2008. © 2008 IOP Publishing.)

Choi *et al.* [81] observed a contraction within the anatase TiO_2 nanoparticle due to size-induced radial pressure. The contraction in turn increases the force constants between ions in the lattice and consequently the Raman bands shift to higher wavenumber [81]. Inclusion of Zr substituent atoms relaxes the lattice contraction and leads to a Raman red-shift. It is also known that nonstoichiometry can break long-range order and also change the mode frequency [67]. This effect is counteracting the tensile strain induced by the changed surface tension. The hydrostatic pressure normally dominates, and this appears in general true for TiO_2, albeit it depends on preparation procedures and experimental conditions. Practically, the net effect can be measured by analysis of the unit cell variation as a function of particle size and used as a net change of surface tension [63].

The finding that surface tension is important and should be included in the analysis is, however, not in conflict with the QC model. The hydrostatic pressure can be incorporated in an

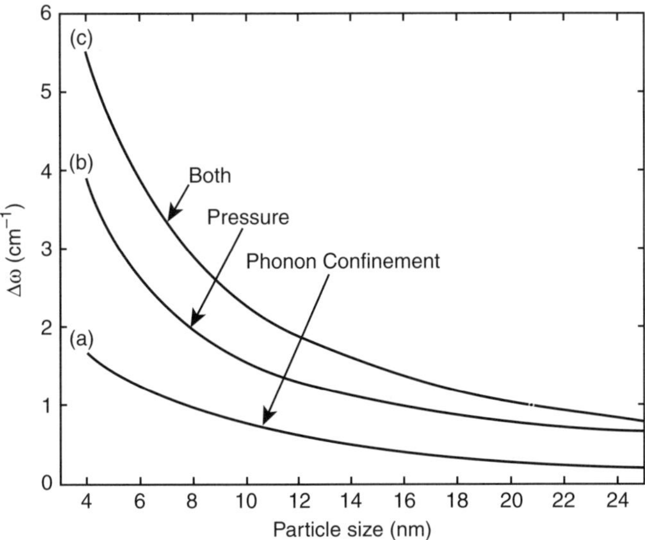

Figure 8.6 The effect of phonon confinement and hydrostatic pressure on the mode frequency change of the anatase $E_g(\nu_6)$ mode for undoped TiO_2 [64].

extension of the QC model without changing the functional form of the calculated Raman spectra, as shown in Figure 8.5b [64]. The physical interpretation of the QC model will, however, change and explicitly show that the Raman blue-shift can originate from both the positive phonon dispersion (Equation 8.5) and the tensile strain (Equation 8.6). Analysis of the symmetry of the Raman modes may give further insight into structural modifications caused by substitutional doping, for example, symmetry changes, which has been shown in the Zr–TiO_2 system [59,64].

8.4 Gas–Solid Photocatalytic Reactions Probed by FTIR Spectroscopy

Photocatalytic oxidation reactions of a wide range of organic molecules on semiconductor nanoparticles has been extensively studied in the aqueous phase and has been shown to lead to complete mineralization by means of photon-generated hydroxyl radical oxidants [14,82–88]. The corresponding reactions in the gas–solid phase are less well studied (see Refs. [22,89] and references therein). A historic account of photocatalysis has recently been given by Fujishima *et al.* [88]. The interest in gas–solid heterogeneous photocatalysis has, however, increased during the past two decades. In part this is due to an increasing interest in air-cleaning and self-cleaning devices [87,90,91]. Simultaneously the surface science of TiO_2 has advanced considerably, and photocatalysis studies on single crystals have started to appear [22,88]. It has become apparent that the gas–solid reactions can be (at least) one order of magnitude larger than corresponding liquid–solid phase reactions with quantum efficiencies (i.e., [number of reacted molecules]/[number of incident photons]) of elementary reaction steps of the order of 1% (in favorable cases) [86]. We refer the reader to the reviews by Peral and Ollis [89], Ollis [83], Diebold [22], and Kaneko and Okura [87] for a compilation of various gas–solid photocatalysis and kinetic studies that have been done on TiO_2 nanoparticles. Most experimental data so far have come from studies on semiconductor nanoparticles or colloids. It is expected that further studies on single crystals, along with detailed experiments on well-defined nanoparticle systems made by recent advancements in nanotechnology will contribute to a deepened understanding of photocatalytic gas–solid surface reactions – and also open up new possibilities [7,13,22,92].

In the most simple representation, the photocatalytic oxidation of organic molecules over TiO_2 is simply described by the overall reaction:

$$\text{Organic molecule} + O_2 \xrightarrow{h\nu > E_g;\mathrm{TiO_2}} CO_2 + H_2O + \text{mineral acid/inorganic compounds} \quad (8.7)$$

The apparent simplicity of this reaction disguises the very complex nature of the full reaction, which consists of several elementary steps occurring at timescales ranging from femtoseconds to (at least) minutes. For pure semiconductor crystals, the reaction is initiated by an electronic interband transition, or bandgap excitation, with photon energies larger than the optical band gap, $h\nu > E_g$. The chain of events leading to (8.7) is summarized in Figure 8.7. The reactions occur within the semiconductor, at the surface or in the homogenous phase above the semiconductor surface, and each step depends on a variety of parameters. It should be borne in mind that in most applications, metal oxides like TiO_2 are hydroxylated or even covered with adsorbed layers of H_2O. It poses a real challenge to the surface-science community to mimic these "real" surfaces under idealized conditions, such that deeper insight into the physico-chemical processes occurring at the surface can be obtained [7,92,93].

It is evident from Figure 8.7 that the photocatalytic reactions involve a great number of elementary steps. It is a matter of continuous discussion how these elementary reactions are initiated, which intermediate species are formed and to what extent branching of reactions occurs. Lacking detailed information of the intermediate reaction pathways, most authors report initial reaction-rate equations or final production CO_2 rates. However, it is not straightforward to translate photocatalytic degradation rates obtained in different laboratories and extract degradation efficiencies or figures of merit for different reactions and materials which are of general validity. This is due to differences in experimental factors, such as reactor designs, light sources, physical form and shape of photocatalyst material and so on. Compared to thermally activated heterogeneous gas–solid reactions, the kinetic analysis of photocatalytic reactions is complicated by the additional photoflux parameter with associated photon absorption, which is often difficult to assess. This not only complicates the quantification of quantum yield (i.e., [rate of photoreaction]/[rate of photon absorption]), but also obstructs comparisons of results obtained in different laboratories due to differences in reactor designs, modes of operation (flow or batch mode) and sample presentation (powder bed, fluidized bed, fixed bed, thin films, etc.). To remedy this problem, various "standardized" methods to extract quantum efficiencies have been proposed [54,94], including the widely used quantum efficiency, where the incident instead of absorbed photon flux is employed.

Most commonly a Langmuir–Hinshelwood (LH) reaction scheme is employed to quantify the photodegradation kinetics. As is usually found in heterogeneous catalysis [95], the LH scheme describes much of the reported data satisfactorily. This may be the case even if the LH adsorption isotherm description fails, and may in some cases be explained because only a few sites are active (which may be the case in photocatalysis [54]). Still, it must be considered rather surprising that the LH scheme is so successful, since not all reported photocatalytic reactions

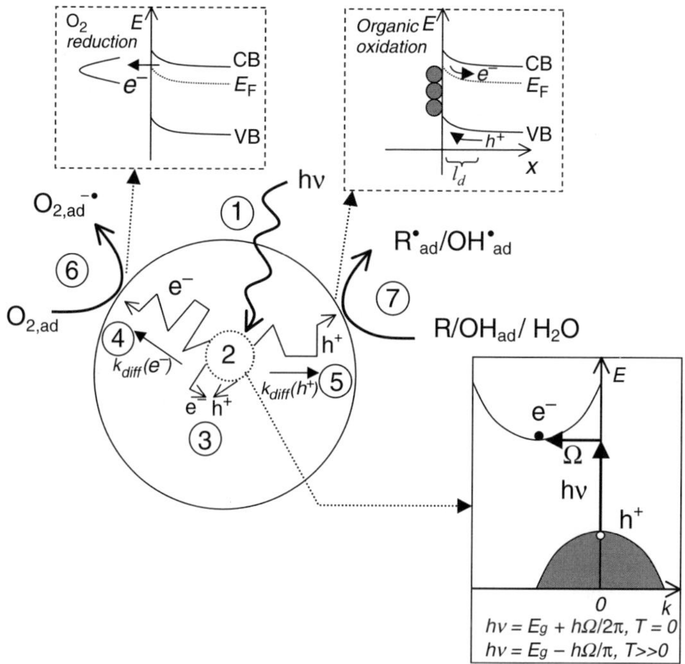

Figure 8.7 Schematic drawing of elementary processes occurring in semiconductor photocatalysis relevant for gas–solid reactions. Schematic drawing of some key elementary photoinduced reaction steps in TiO_2 nanoparticles: photon absorption in the solid (1), excitation of e–h pairs in the solid (2), bulk e–h recombination (3), e–h pair separation and subsequent scattering of electrons (4) and holes (5) towards the surface, interfacial charge transfer with electron attachment on O_2 (6) and hole attachment directly on an adsorbed organic molecules or OH groups (7). The upper insets show the energetics of the electron and hole attachment at the semiconductor surface, including adsorbate-induced band bending. The lower-right inset show an E-K diagram of the indirect bandgap excitation in a semiconductor such as TiO_2. (Reprinted with permission from L. Österlund and A. Mattsson, Surface characteristics and electronic structure of photocatalytic reactions on TiO_2 and doped TiO_2 nanoparticles, in *Solar Hydrogen and Nanotechnology* vol. 6340 (ed. L. Vayssieres) © 2006 SPIE.)

are truly catalytic (turn-over number (TON) > 1) and that the state of the catalyst actually changes over a timescale comparable to the experiments, for example, due to site inhibition. Instead it can be expected that Mars–Van Krevelen-type mechanisms better describe photocatalytic reactions on semiconductor metal oxides. The latter mechanism involves repeated surface reduction–oxidation of the metal oxide, where the catalyst actively participates in the reaction. Surface O atoms can react with adsorbates and continuously create reactive O vacancies. Experiments with isotope labeled ^{18}O introduced in the gas phase show, however, that the involvement of lattice O depends on the particular reaction being studied [16,96]. In cases when the photoreaction is carried out without further (or sufficient) supply of O from the reaction media (gas or liquid), photodegradation of organics proceeds by reactions with lattice O through grain-boundary diffusion [18]. It has been widely assumed in the literature that O vacancies on rutile $TiO_2(110)$ play a significant role in the reactivity. In particular they have been reported to greatly affect the O_2 and H_2O chemistry [22,56,97–105]. Recently, the dependence on O vacancies for formate dehydration and dehydrogenation reactions has been studied [56,106]. The dehydration reaction is substantially favored at sites adjacent to O vacancies on $TiO_2(110)$, while this is not the case for the dehydrogenation reaction. If not present initially, O vacancies can form by condensation of bridge OH groups. Similarly, lattice O plays an active role in the thermal and photocatalytic oxidation of 2-propanol and the corresponding thermal activity of rutile $TiO_2(100)$ [55]. This is interpreted in terms of the proximity of the bridged O-atoms to the fivefold coordinated Ti surface cation. This work should be contrasted to the recent reports of the importance of interstitial Ti^{3+} as an active species that facilitates O_2 dissociation on rutile (110) surface [57].

In the following, we give an example emphasizing the importance of having a detailed understanding of surface reactions in gas–solid photocatalysis, namely propane photo-oxidation over anatase and rutile TiO_2 [20], where the surface intermediates ultimately determine the overall reactivity of the catalyst. The strength of vibrational spectroscopy to unravel detailed information of surface reactions and reaction intermediates is highlighted. Teichner and co-workers reported the first mechanistic studies of photocatalytic degradation of alkanes and alkenes in the early 1970s [107,108]. Although these studies were conducted with high concentrations of the alkanes, with the interest directed towards partial oxidation, their main results are valid for photodegradation of trace amounts of hydrocarbons. In particular, small alkanes (C_nH_{2n+2}) and alkenes (C_nH_{2n}) with $n < 3$, are completely oxidized to CO_2 and H_2O. Partial oxidation products for alkanes with $n \leq 3$ are mainly ketones ($C_nH_{2n}O$) and aldehydes ($C_nH_{2m}O$), $2 < m \leq n$. The reactivity increases for longer carbon chains ($n > 3$). The reactivity decreases in the sequence $C_{tert} > C_{quat} > C_{sec} > C_{prim}$. The photo-oxidation scheme proceeds via alcohol formation and the subsequent reaction scheme should therefore be valid for alkanes, alkenes and alcohols. Acetone is a common partial oxidation product in the photocatalytic oxidation of C_3 hydrocarbon molecules or functional groups. Acetone is stable and is reacted further only under strongly oxidizing conditions (see Section 8.5.2.1). The selectivity for aldehydes is generally lower. The latter is not surprising considering that aldehydes appear reducing and are readily oxidized to carboxylic acid. Considering the protolysis of carboxylic acids to their corresponding ions (e.g., acetic acid to acetate, or formic acid to formate), which are known to form strong complexes with transition-metal cations [109], it is easy to understand why, for example, acetate and formate species are commonly observed on metal oxides, such as TiO_2 (with concomitant hydroxylation of the surface) [1,4,101,110]. The latter species are known to be very stable and are only oxidized very slowly on TiO_2.

Figures 8.8 and 8.9 show DRIFT and gas–phase mass spectra, respectively, measured simultaneously during UV illumination of anatase and rutile powders in propane/synthetic air gas mixtures in a dedicated *in situ* photocatalysis reaction cell [20]. By deconvoluting the contribution from the spectrally resolved surface species measured during the photoreaction, the overall reactivity is shown to be directly related to the oxidation of acetone on anatase, while it is oxidation of formate on the rutile phase, which is also confirmed by microkinetic modeling of the reaction kinetics (Figure 8.10) [92]. First, it should noted from Figure 8.9 that the initial reaction is not stoichiometric. Only on the large anatase particles (Degussa P25) does the reaction reach stoichiometry (1 mole propane $\rightarrow$ 3 mole CO_2) after the third illumination period. On the small anatase particles, mass balance is typically obtained only after > 60 min. This is not due to a surface-area effect, which is already included in the analysis leading to Figure 8.9; rather it is related to formation of surface species that are accumulated on the

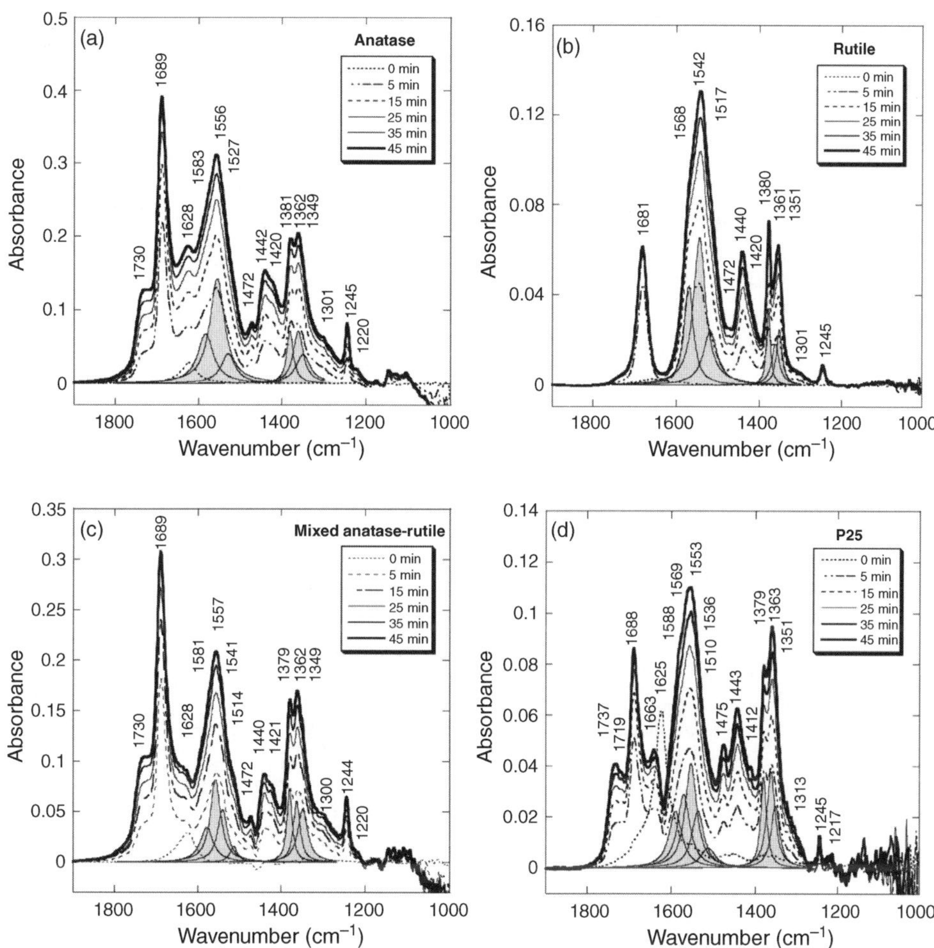

Figure 8.8 Diffuse reflectance infrared spectra obtained at different times (0–45 min) during photo-oxidation of propane on (a) 14 nm anatase, (b) 6 × 80 nm rutile, (c) mixed anatase–rutile, and (d) 25 nm anatase (P25), respectively. The gray areas show the deconvoluted Lorentzian peaks in the ν_a(OCO) and ν_s(OCO) regions due to μ-formate, μ-acetate and aqueous formate. Spectra of TiO_2 samples in synthetic air prior to illumination were used as background in each measurement. (Reprinted with permission from T. van der Meulen, A. Mattson and L. Österlund, A comparative study of the photocatalytic oxidation of propane on anatase, rutile, and mixed-phase anatase-rutile TiO_2 nanoparticles: Role of surface intermediates, *Journal of Catalysis*, **251**, 131–144, 2007. © 2007 Elsevier.)

particles during the initial illumination period, which have different total oxidation rates (different values for their rate determining steps) on each nanoparticle system (Figure 8.10). Accumulation of intermediate species can thus deactivate (or more appropriately inhibit in this case) the reactivity of the oxide and determine the overall reactivity of the catalyst. In view of this there are comparably few studies aimed at elucidating the elementary photoreaction steps of gas-phase carboxylic acids and aldehydes on different semiconductor nanoparticles [8,17,20,111–115]. In

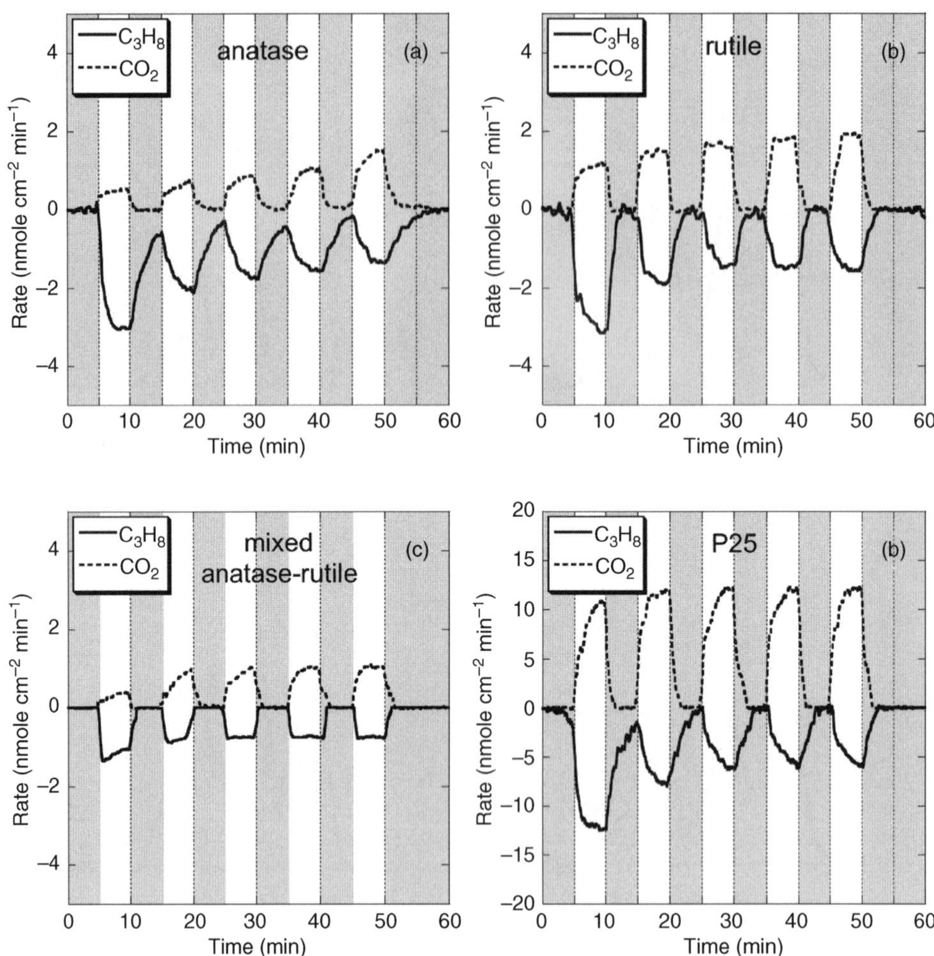

Figure 8.9 The surface area normalized reaction rate determined for (a) 14 nm anatase, (b) 6 × 80 nm rutile, (c) mixed anatase–rutile, and (d) 25 nm (P25). A positive (negative) value indicates formation (removal) of the species. (Reprinted with permission from T. van der Meulen, A. Mattson and L. Österlund, A comparative study of the photocatalytic oxidation of propane on anatase, rutile, and mixed-phase anatase-rutile TiO_2 nanoparticles: Role of surface intermediates, *Journal of Catalysis*, **251**, 131–144, 2007. © 2007 Elsevier.)

contrast, adsorption and thermal reactions of carboxylic acids, and formic acid in particular, have been extensively studied on both single crystals [22,25,101,102,116–128] (primarily rutile TiO_2(110)) and TiO_2 nanoparticles (see Refs. [1,4,8,9,89,110,129–133] and references therein). Guided by the schematic reaction scheme outlines in Figure 8.11, which highlight the importance of intermediate adsorbate structures, such as simple carboxylates and acetone, we will explore in greater detail in Section 8.5.1 the formic acid/formate and acetone adsorption systems on the various polymorphs of TiO_2 and doped TiO_2 nanoparticles.

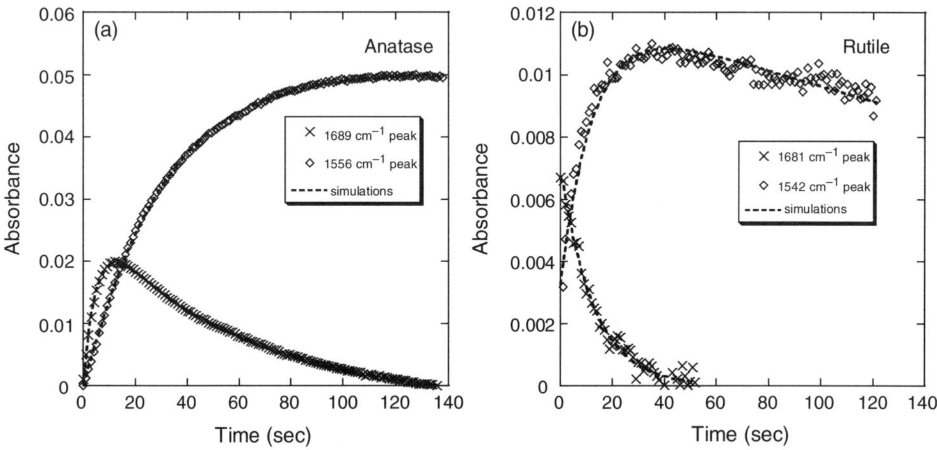

Figure 8.10 The intensity of the ν(C=O) and ν_a(OCO) vibrational bands in acetone and formate, respectively, as a function of irradiation time after propane adsorption on 14 nm anatase and 6 × 80 nm rutile TiO_2 nanoparticles, respectively. The rate-determining step is shown to be acetone oxidation for anatase ($k_d = 0.003\,min^{-1}$), and μ-formate oxidation for rutile ($k_d = 0.002\,min^{-1}$). (Reprinted with permission from L. Österlund and A. Mattsson, Surface characteristics and electronic structure of photocatalytic reactions on TiO_2 and doped TiO_2 nanoparticles, in *Solar Hydrogen and Nanotechnology* vol. 6340 (ed. L. Vayssieres) © 2006 SPIE.)

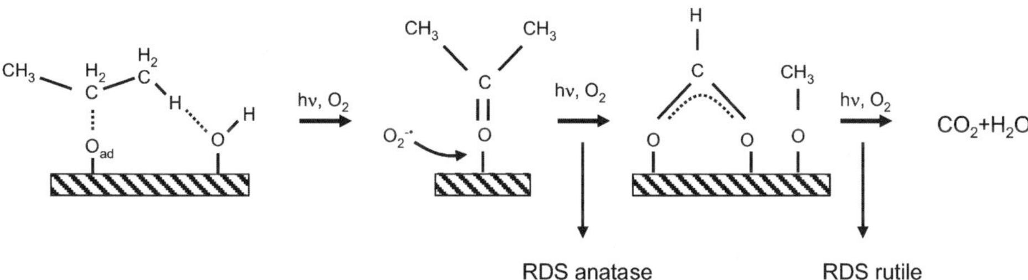

Figure 8.11 Schematic drawing of the proposed propane photo-oxidation mechanism deduced from the FTIR data, which explicitly shows the different rate-determining steps (RDS) observed on anatase and rutile TiO_2, respectively. (Reprinted with permission from T. van der Meulen, A. Mattson and L. Österlund, A comparative study of the photocatalytic oxidation of propane on anatase, rutile, and mixed-phase anatase-rutile TiO_2 nanoparticles: Role of surface intermediates, *Journal of Catalysis*, **251**, 131–144, 2007. © 2007 Elsevier.)

8.5 Model Gas–Solid Reactions on Pure and Doped TiO_2 Nanoparticles Studied by FTIR Spectroscopy

8.5.1 Reactions with Formic Acid

Vibrational spectroscopy has been extensively used to measure and classify carboxylate–metal [109,134] and carboxylate–oxide coordination on metal-oxide nanoparticles [1,4,9,110].

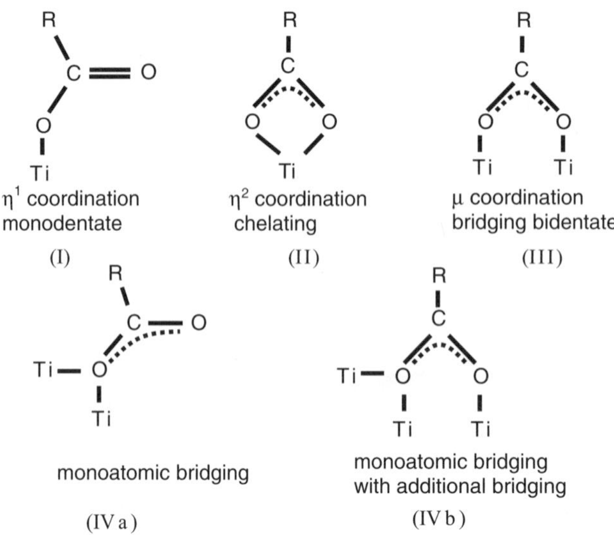

Figure 8.12 Schematic drawing of possible carboxylate ion ($R\text{-}CO_2^-$) coordination to Ti metal atoms.

Historically, most infrared spectra have been based on comparisons with spectra obtained on organotransition metal complexes with known structures. Deacon *et al.* [134] have devised a semiempirical classification scheme that enables distinction of structures I, II and III in Figure 8.12. They analyzed the symmetric and asymmetric O–C–O vibrations, v_a(OCO) and v_s(OCO) and their separation, $\Delta v_{a-s} = v_a(\text{OCO})\text{-}v_s(\text{OCO})$, for a large number of coordinated acetate compounds with known X-ray crystal structure. By comparisons of Δv_{a-s} for the free aqueous acetate (ionic specie) with transition-metal acetato-complexes they arrived at the following conclusions:

- The monodentate acetate complexes (I) have $\Delta v_{a-s}(\text{monodentate}) > \Delta v_{a-s}(\text{ionic})$.
- Chelating bidentate acetate complexes (II) have $\Delta v_{a-s}(\text{chelating}) \ll \Delta v_{a-s}(\text{ionic})$.
- Bridging bidentate acetate complexes (III) have $\Delta v_{a-s} >$ chelating bidentate complexes and close to Δv_{a-s}(ionic).

Confirmation of the last correlation requires, however, a detailed analysis which must be made on a case-by-case basis. The bonding of carboxylate to a transition metal is governed by interactions of the HOMO levels with the metal *d* molecular orbitals, which give rise to either strong or weak bonding, depending on whether bonding or antibonding levels are populated. For example population of d_{σ^*} levels in Ni^{II} and Cu^{II} carboxylato-complexes produces bond weakening, which modulates the bonding. The corresponding bonding to TiO_2 is illustrated in Figure 8.13, which explicitly demonstrates the coupling of the oxide (mainly O 2*p*-derived states) to the σ and π levels of formate [118]. It has been shown that, in particular, the v_a(OCO) vibrational mode is sensitive to the acidity of the metal center [131]. This type of correlation is important, since it also can give additional support to the detailed adsorbate structure and reactivity. For example, it has been shown that the charge of the CH hydrogen atom in HCOOH correlates strongly to adsorption to Ti^{4+} Lewis acid sites on rutile $TiO_2(110)$ [106]. A word of

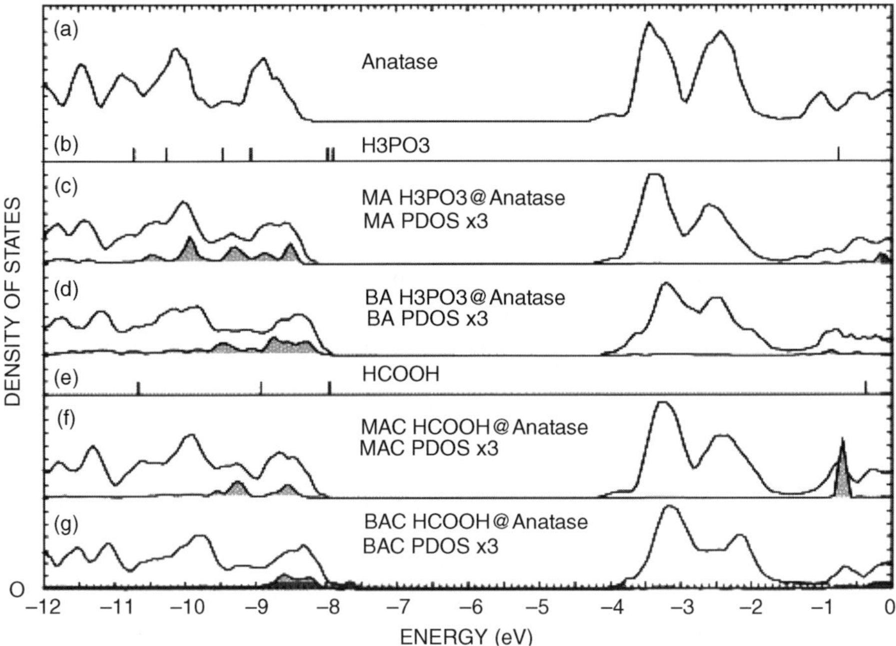

Figure 8.13 Calculated total and projected density of states for (a) anatase TiO_2, (b) phosponic acid (H_3PO_3) (c) monodenate coordinated H_3PO_3 on TiO_2, (d) bridged bidenate coordinated H_3PO_3 on TiO_2, (e) formic acid (HCOOH), (f) monodenate coordinated HCOOH on TiO_2, (g) bridged bidenate coordinated HCOOH on TiO_2. (Adapted from M. Nilsing, S. Lunell, P. Persson and L. Ojamae, Phosphonic acid adsorption at the TiO_2 anatase (101) surface investigated by periodic hybrid HF-DFT computations, *Surface Science*, **582**, 49–60, 2005 Elsevier Science.)

caution, however. Even though the rules by Deacon *et al.* [134] are useful for coarse classifications of adsorbate coordination, they should be treated with care without complementary structural information [131].

8.5.1.1 Adsorption of HCOOH on TiO_2

Adsorption of formic acid (HCOOH) on oxides from the gas phase has been extensively studied by a wide range of methods [1,4,17,20,22,25,101,102,110,114,117–133,135]. Formic acid, the simplest organic acid, is a suitable probe molecule to study the important carboxylate–TiO_2 surface interactions discussed in the previous section. Besides its well-known importance in biochemistry, where it is one of the building blocks of amino acids, it represents a basic building block in oxide-ligand chemistry to achieve binding of, for example, dyes to TiO_2 in photoelectrochemical cells. The reason for the latter is obvious considering the strong coupling of the HOMO and LUMO with the valence band (VB) and conduction band (CB) of anatase TiO_2, which is borne out by the DFT calculations for formic acid on anatase (Figure 8.13) [118]. Moreover, a thorough understanding of HCOOH/oxide systems is a prerequisite to understand a large class of reactions of organic molecules, since formate is a commonly observed intermediate due to its high stability. It is of central importance in pollution abatement technology; many organic pollutants proceed via formic-acid formation before its complete oxidation [83,136]. It is the prototype system to probe the acid–base character of oxide catalysts, since its reaction products directly reflect the acid–base properties of the oxide. For example, the dehydration reaction of formate ($HCOOH \rightarrow CO + H_2O$) occurs on acid oxides (e.g., Al_2O_3), while dehydrogenation ($HCOOH \rightarrow CO_2 + H_2$) occurs on basic oxides (e.g., MgO). Both reactions are observed to occur on TiO_2, depending on reaction conditions. This may seem to falter the typical definition of acid–base catalysts, and highlights the amphoteric nature of TiO_2. The analysis of this apparently simple adsorption system is, however, complicated by the generally low symmetry of carboxylate ions ($R{-}CO_2^-$), which prevents unambiguous spectroscopic distinction of different types of ions (in, for example, high-resolution electron energy-loss spectroscopy (HREELS) the perpendicular modes are not seen due to the surface selection rules). The species I–III in Figure 8.12 are commonly reported on TiO_2 [4,9,102,110,114,120,126,127,131], while the other types (IVa and IVb) can be found in various carboxylato-complexes [109,134] and possibly on TiO_2 nanoparticles with high defect concentrations.

Formic acid dissociates on rutile TiO_2 at temperatures above 110 K under most studied conditions [4,102,110,114,120,131]. On the perfect rutile $TiO_2(110)$ 1×1 surface, the H atom bonded to the O atom in HCOOH is abstracted and transferred to a neighboring bridging O atom to form a hydroxyl group (HO_B), while adsorbed formate ($HCOO_{ad}$) binds to the fivefold coordinated Ti^{4+} atoms (Ti(5)) along the (001) direction with each O atom bonded to adjacent Ti(5) atoms in a bridging configuration (Figure 8.14) [56,102,120]. Condensation of neighboring bridging OH groups provides a pathway to produce H_2O and an oxygen vacancy. The $HCOO_{ad}$ molecule can readily diffuse along the trough in the [001] direction and occupy the O

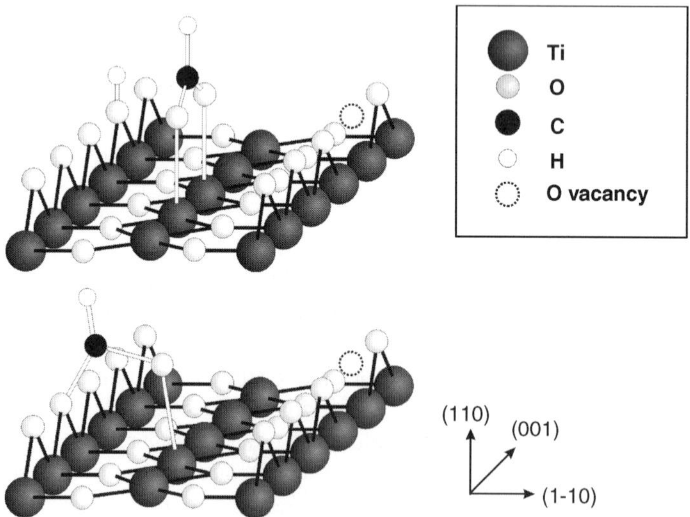

Figure 8.14 Ball and stick model of formate μ-coordination on the rutile (110) 1×1 surface. Species A are adsorbed along the [001] direction (upper panel), and species B with the molecular axis along the [1–10] direction with one O atom filling up an O vacancy site (lower panel). A hydroxyl bonded to a bridging O site (HO_B) is indicated.

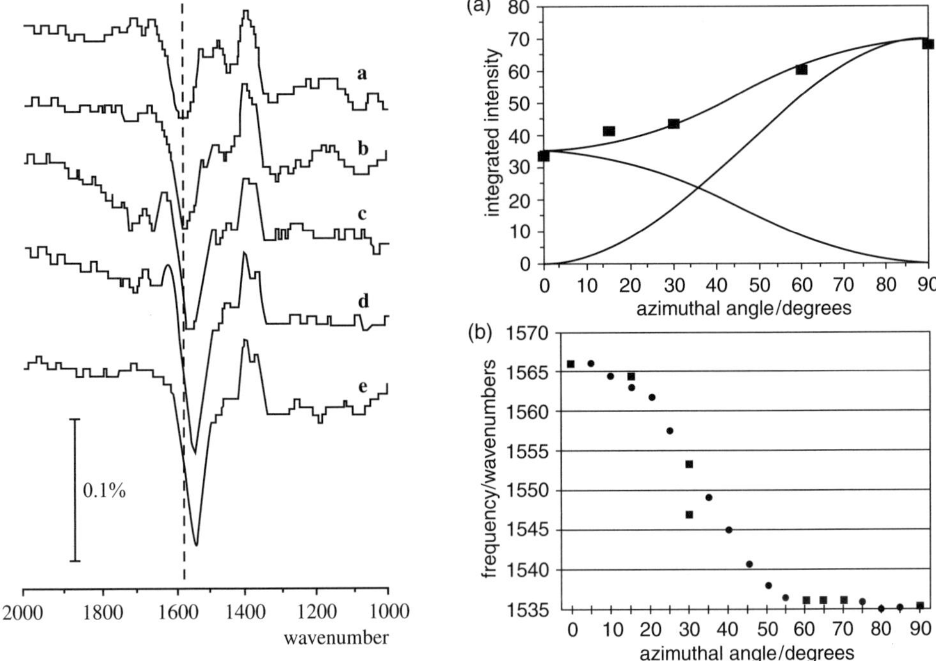

Figure 8.15 (Left) FT-RAIRS spectra obtained as a function of azimuthal angle Φ for formate on $TiO_2(110)$ adsorbed at saturation coverage by exposure of the surface to formic acid at 300 K: (a) 0°; (b) 15°; (c) 30°; (d) 60°; (e) 90°. (Right) Measured (■) integrated intensity (a) and frequency (b) of the absorption band resulting from the coupling of I_t with ν_{as}(OCO) as a function of azimuthal angle Φ together with the prediction, assuming 0.4 ML of species A and 0.2 ML of species B. The measured peak is assumed to be composed of two Gaussian components at frequencies and with FWHM obtained from measurement at $\Phi = 90°$ and $\Phi = 0°$. (Reprinted with permission from B.E. Hayden, A. King and M.A. Newton, Fourier transform reflection-absorption IR spectroscopy study of formate adsorption on $TiO_2(110)$, *Journal of Physical Chemistry B*, **103**, 203–208, 1999. © 1999 American Chemical Society.)

vacancy site, whereby the O−C−O molecular plane is rotated into the [1–10] direction. The different types of bridging bidentate (μ-coordinated) structures are denoted μ-formate A and μ-formate B, respectively. In an elegant FTIR study, Hayden *et al.* [102] proved the existence of both A- and B-type formates on single-crystal rutile $TiO_2(110)$ (Figure 8.15). The catalytic dehydration reaction mechanism providing the active O vacancy sites has been theoretically examined by Morikawa *et al.* [56] and is depicted in Figure 8.16; it involves a unimolecular reaction with an experimentally determined activation barrier of $120\,kJ\,mol^{-1}$ (1.25 eV) [117,137]. The dehydration reaction is observed at higher temperatures ($T > 500$ K) and low HCOOH pressure. The dehydrogenation is observed to be a bimolecular reaction that occurs predominately at low temperatures ($T < 450$ K) at high HCOOH pressures with apparent activation energy of $15\,kJ\,mol^{-1}$ [117,137]. Uemura *et al.* [106] have recently investigated the dehydrogenation mechanism by DFT calculations. They found that the most probable dehydration mechanism involves reaction between μ-bridging B molecules adsorbed on neighboring Ti^{4+} sites along the [001] direction. In contrast to the dehydration reaction, the

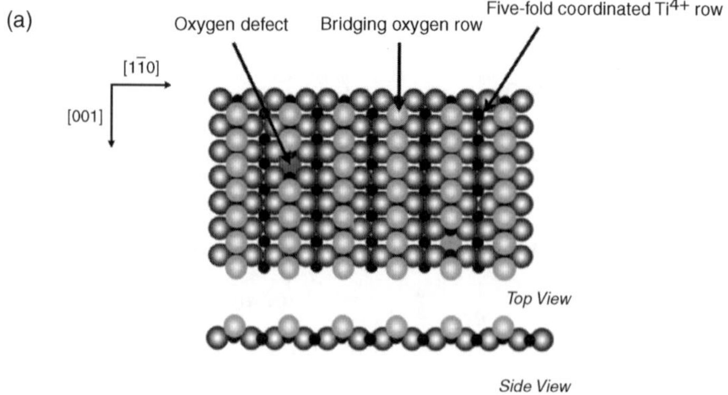

Figure 8.16 (a) Top view of the rutile $TiO_2(110)$ 1 × 1 surface structure showing O vacancy sites, Ti^{4+} cation sites and bridging O. (b) Suggested mechanism for dehydration of formic acid. (Reprinted with permission from Y. Morikawa, I. Takahashi, M. Aizawa, *et al.*, First-principles theoretical study and scanning tunneling microscopic observation of dehydration process of formic acid on a $TiO_2(110)$ surface, *Journal of Physical Chemistry B*, **108**, 14446–14451, 2004. © 2004 American Chemical Society; Y. Uemura, T. Taniike, M. Tada *et al.*, Switchover of reaction mechanism for the catalytic decomposition of HCOOH on a $TiO_2(110)$ surface, *Journal of Physical Chemistry C*, **111**, 16379–16386, 2007. © 2007 American Chemical Society.)

dehydrogenation reaction does not involve O vacancy sites, but occurs via monodentate transition states on the stoichiometric surfaces. Again the $TiO_2(110)$ catalyst confounds the traditional uniqueness rule of acid–base catalysis and contains different active sites that catalyze different reactions depending on reaction conditions [117,137].

On the rutile $TiO_2(111)$ surface, it has been suggested that formate binds as monodentate (η^1-formate) and chelating (η^2-formate) species,, based on HREELS and STM observations [121]. In addition it has been reported that upon gas-phase adsorption of HCOOH at room temperature on TiO_2 powders of anatase/rutile structure (Degussa P25) and pure rutile (Sachtleben Chemie), adsorbed molecular HCOOH and aqueous formate molecules co-exist with the dissociated molecules [17], with the nondissociated species being less stable.

Much less is known about the bonding of formate on anatase, due to lack of detailed single-crystal experiments. Recent theoretical [127,128] and FTIR spectroscopy experiments [9,114] studies suggest that the detailed surface structure greatly affects the adsorption structure. On the anatase TiO_2(101), surface a monodentate HCOOH configuration is preferred, which is hydrogen bonded to a neighboring bridging O. The stability of this configuration is, however, strongly influenced by H_2O interactions, which leads to dissociation and formation of a monodentate species, again stabilized by a hydrogen bond to a surface hydroxyl. In contrast, on the anatase TiO_2(001), dissociation is favored and either bidentate bridging, bidentate chelating or monodentate formate form, depending on reaction conditions and interactions with surface hydroxyls [128]. An important factor deciding the HCOOH adsorption structure appears to be the basicity of the undercoordinated surface O atoms, which may either stabilize the molecule by hydrogen bonding or abstract the H atom, forming a HO_B surface hydroxyl.

The database of single-crystal and theoretical studies discussed above provides an excellent background to study formic acid and formate adsorption on TiO_2 nanoparticles. Indeed vibrational spectroscopy has been widely employed to characterize the HCOOH/TiO_2 and HCOO/TiO_2 adsorption systems in gaseous [4,9,17,20,102,114,119,130,133,138] and aqueous phases [131,132,139,140]. The general features of these studies are outlined below.

Figure 8.17 shows DRIFT spectra obtained on rutile and anatase TiO_2 nanoparticles after HCOOH gas-phase adsorption [9,141]. Table 8.1 lists the observed vibrational bands of adsorbed formate species and formic acid molecules, along with the proposed mode assignments. The spectra in Figure 8.17 are representative of most IR spectra on HCOOH/TiO_2 found in the literature [17,102,130,131,138], and suggest by comparison with Table 8.1 that the surfaces contain similar formate species. It is, however, evident from Table 8.1 that the reported data and assignments differ, in particular between the different nanoparticle systems. Indeed, a systematic analysis of the spectra as a function of HCOOH dosing time, temperature and photon illumination reveals significant differences, both between anatase and rutile, and between different particle sizes [114,141]. The 1800–1200 cm^{-1} region, which is characteristic for the ν(C=O) and ν(OCO) vibrational modes of coordinated carboxylate species, is composed of many absorption bands due to the heterogeneity of adsorption sites on nanoparticles. Hydrogen-bonded species are also likely to be present, since in almost all studies of TiO_2 nanoparticles, the surface is to some extent hydrated, which further increases the complexity. This is corroborated by changes in the typical water IR δ(HOH) and ν(OH) bands. A general feature is that HCOOH adsorption displaces water upon adsorption, which is also predicted by DFT calculations [128]. The dominant IR bands around 1550 cm^{-1} and the characteristic double-peak structure in the 1380-1360 cm^{-1} region, which is the typical spectral formate signature, is composed of several bands. A consistent analysis can only be made by comparisons with particles with *known* structures, and comparisons with single-crystal data and/or calculations. By systematic deconvolution of the DRIFT spectra obtained from different known nanoparticle structures and sizes as a function of coverage, temperature and irradiation time, the following correlations can be made between the observed DRIFT spectra and specific adsorbate structures [141]. Starting with the rutile nanoparticles, which mainly exposes (110) facets [114,131] and thus should mimic the single-crystal data well, it is seen from Figure 8.17 and Table 8.1 that the large rutile particle spectra are dominated by the

Table 8.1 Compilation of vibrational frequencies and mode assignments of adsorbed formate and formic acid on TiO_2.

TiO_2 substrate	Mode assignment					Reference
	ν_a(OCO)	ν_s(OCO)	ν(C=O)	ν(CO)	δ(C-H)	
		Single crystal (UHV)				
Rutile TiO_2(110)						
μ-formate	1566 (Species B)[a]	1393 (Species B)				[102]
μ-formate	1535 (Species A)[a]	1363 (Species A)				[102]
μ-formate		1365				[120]
Rutile TiO_2(111)						
chelating or bridge	1560	1350				[121]
monodentate			1625	1235		
Nanoparticles (gas phase)						
Anatase $d = 14$ nm						[9,20,141]
μ-formate	1550	1364			~1380	
$HCOO^-$ ions	~1582	~1351			~1391	
HCOOH(a)			1683, 1655	1257	1321, 1291[f]	
Anatase $d = 23$ nm[b]						[138]
μ-formate	1550	1370				
HCOOH(a)			1665			
H-bonded HCOOH			1728		1380	
Anatase/rutile $d = 10$-nm[d]						[15]
μ-formate	1554	1360			1381	
P25 (Degussa)						[17]
μ-formate	1553	1379			1385	
HCOOH(a)			1682	1277	1325	
P25 (Degussa) $d = 30$ nm[b]						[141]
μ-formate	1552	1363			~1380	
monodentate HCOOH			1687, 1665	1250	1320, 1288[f]	
Rutile $w \times l \approx 6 \times 80$ nm[c]						[114]
μ-formate	1563 (Species B)				1377	
	1536 (Species A)	-1362 (Species A)				
Rutile $w \times l \approx 3 \times 5$ nm[c]						[114]
μ-formate	1562 (Species B)	1387 (Species B)			1373	
	1534 (Species A)	1360 (Species A)				
monodentate			1653, 1602	1261	1350, 1300[f]	

Table 8.1 (*Continued*)

TiO_2 substrate	Mode assignment					Reference
	ν_a(OCO)	ν_s(OCO)	ν(C=O)	ν(CO)	δ(C-H)	
Rutile $w \times l =$ 9×26 nm[c]						[141]
μ-formate	1560 (Species B) 1540 (Species A)	1387 (Species B) 1358 (Species A)			1375	
monodentate			1651, 1601	1258	1334, 1308[f]	
H-bonded HCOOH			1713	~1208		
		Nanoparticles (aqueous phase)				
Rutile $w \times l \approx$ 5×60 nm[c]						
Aqueous formate	1540	1349			1388	[131]
Aqueous formic acid	1580, 1585	1349, 1351	1719	1211	1383, 1383	[109,131]
Anatase/rutile $d = 10$-nm[d]						
Aqueous formic acid			~1605	~1296–1300		[132]

[a] Note the mixing of mode assignments in this latter reference: $1566 \leftrightarrow 1535\,cm^{-1}$.

[b] Diameter deduced from Scherrer analysis of XRD data.

[c] The rutile particles exhibit a rectangular morphology in TEM and are classified according to the length (*width* × *length*), which scales with exposed <110> surface area [114,131]. The width of the particles grows with increasing particle size from approximately 2 to 10 nm.

[d] In ethanol and methanol at pH = 3.1.

[e] Can also be assigned to or overlap with δ(OH..H) in bicarbonate species [110].

ν_a(OCO) peak at $1536\,cm^{-1}$ and an associated ν_s(OCO) peak at $1357\,cm^{-1}$. The peak at $1377\,cm^{-1}$ is due to the in-plane δ(CH) deformation mode, giving rise to the typical twin-peak appearance in this region on rutile TiO_2. The absorption band at $1562\,cm^{-1}$, which is evident as a shoulder on the high-energy side of the $1534\,cm^{-1}$ peak, becomes more pronounced as the particle size is decreased. Simultaneously, a peak at $1387\,cm^{-1}$ becomes clearly resolved. The 1562 and $1387\,cm^{-1}$ vibrational losses are due to ν_a(OCO) and ν_s(OCO) modes characteristic of μ-formate A species bonded to O vacancy sites (Figure 8.15). The observed trends in the rutile particles as a function of particle size can be rationalized as follows: the dominant facet on the rutile particles is the (110) facet due to its low surface energy compared to the other low-index facets. With increasing particle size, the fraction of the (110) facet thus grows at the expense of other higher-energy facets. The resulting particles can be several hundred nanometer in the [100] direction, while the width is only of the order of 10 nm and the exposed surfaces are almost exclusively (110) terminated, up to 98% [114,131]. On (110) facets, only μ-formate A and μ-formate B are observed [102]. The growth of μ-formate B with decreasing particle size reflects that the relative concentration of (110)-terminated surfaces decreases and adsorption sites associated with O vacancies (or defect sites with similar properties) increases. Formate

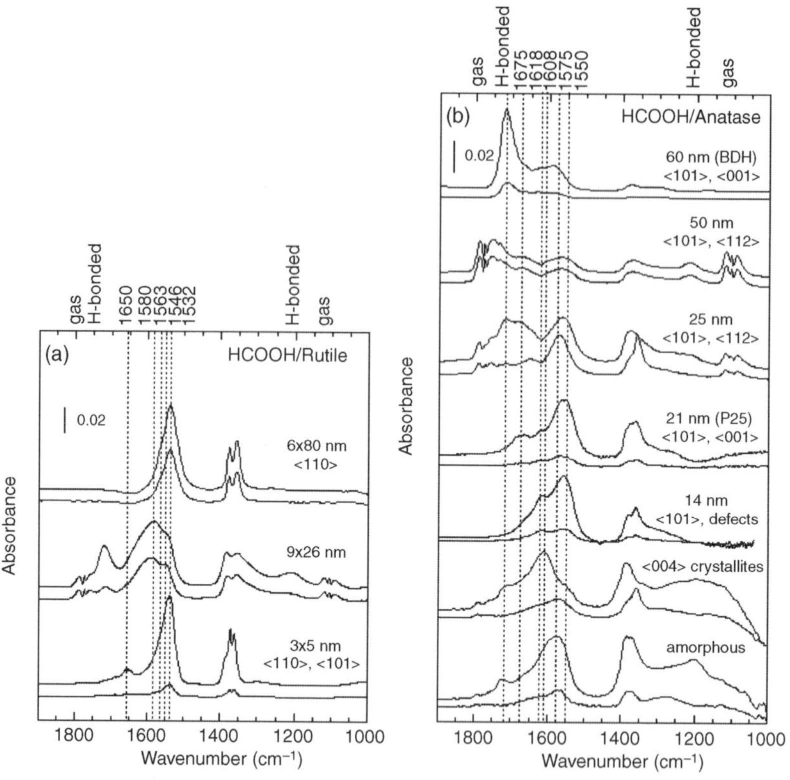

Figure 8.17 *In situ* infrared spectra of formate adsorbed on (a) rutile and (b) anatase nanoparticles with different crystallinity and prepared by different methods, showing the initial stage of formate adsorption. The bottom spectra for each sample show results at submonolayer coverage. The anatase spectra have been multiplied by a factor of (from top bottom) 0.05, 1.7, 1, 1, 0.1, 0.5 and 0.5, respectively. (Reprinted with permission from L. Österlund and A. Mattsson, Surface characteristics and electronic structure of photocatalytic reactions on TiO_2 and doped TiO_2 nanoparticles, in *Solar Hydrogen and Nanotechnology* vol. 6340 (ed. L.Vayssieres) © 2006 SPIE; L. Österlund, A. Mattsson and P.O. Andersson, What makes a good TiO_2 photocatalyts?, in *Nanostructured Materials and Nanotechnology II: Ceramic Engineering and Science Proceedings*, vol. 29 (eds S. Mathur and M. Singh) © 2008 John Wiley & Sons Inc.)

adsorption produces two combination bands in the C−H stretching region: $\nu_a(\text{COO}) + \delta(\text{CH})$ at $\sim$2950 cm^{-1} and $\nu_s(\text{COO}) + \delta(\text{CH})$ at $\sim$2740 cm^{-1}. The latter peak is often confused with the ν(CH) band due to formaldehyde which occurs in the same frequency region.

On small rutile nanoparticles, the relative surface termination by (101) facets increases and the associated formate species becomes apparent [141]. The IR spectrum for the small "rectangular" (9 × 26 nm) rutile particles has strong ν(OCO) vibrational losses at $\sim$1582 cm^{-1} and $\sim$1602 cm^{-1}, in addition to the bands due to μ-formate A and B on the (110) facet at $\sim$1540 and $\sim$1560 cm^{-1}. Even smaller 3 × 5 nm rutile nanoparticles exhibit new vibrational bands, which have been associated with μ-formate bonded to (101) facets and monodentate species [114]. In addition, absorption bands present at $\sim$1653, $\sim$1300 and $\sim$1260 cm^{-1} have been attributed to the ν(C=O), δ(CH) and ν(CO) modes, respectively, in monodentate HCOOH coordinated to Lewis acid sites [1,17]. Note that this assignment is consistent with Deacon's rule [134]. Monodentate HCOOH is also associated with a weak absorption band at $\sim$2920 cm^{-1} due to ν(CH). The weak peaks at $\sim$1713 and $\sim$1208 cm^{-1} seen in Figure 8.17 are due to small amounts of hydrogen-bonded HCOOH (similar to HCOOH dimers in the gas phase) at saturation coverages [17].

At elevated temperatures, it is reported that the relative intensity of the ν_a(OCO) and ν_s(OCO) bands associated with μ-formate A and B, respectively, changes in favor of B species [141]. Thus a gradual transformation of A species into B species occurs as O vacancies are formed and B species depleted. This can be explained by the results obtained by Morikawa *et al.* [56] and Hayden *et al.* [102], that is, that O vacancies are formed through a thermally activated process through recombination of bridge-bonded OH groups to produce H_2O and O vacancies, *viz.*

$$2O_BH \xrightarrow{T>500\,K} H_2O + O_B, \tag{8.8}$$

where O_B denotes twofold coordinated lattice O in the bridging oxygen row (cf. Figures 8.14 and 8.16). Assuming that A species can populate such new O vacancy sites via surface diffusion, this explains the redistribution of the relative A and B surface species at elevated temperatures.

The energetically favored decomposition path for formate at low pressures is reported to occur by decomposition of μ-formate B via a monodentate transition state [56]:

$$HCOO_B \rightarrow CO + O_BH \tag{8.9}$$

Indeed, CO is observed to readily desorb from the surface in vacuum studies [119]. The IR spectrum of HCOOH adsorbed on small anatase nanoparticles exhibits two pairs of asymmetric and symmetric ν(OCO) bands, but the ν_a(OCO) modes are shifted to higher energies compared to rutile (Figure 8.17). The ν_a(OCO) and ν_s(OCO) vibrational modes in μ-formate are observed at 1552 and 1363 cm^{-1}, respectively [114]. The interpretation of the 1582 and 1350 cm^{-1} absorption bands is, however, different in this case. These bands are attributed to $HCOO^-$ ions, which are H-bonded in a thin water layer on the small particles, since the temperature dependence shows that they decrease in concert with the infrared bands associated with water. Adsorbed water is always abundant in powders with small TiO_2 particles in studies conducted at atmospheric pressures. This interpretation is consistent with the detailed study by Rotzinger *et al.* [131] of formate adsorption in aqueous solutions. The appearance of peaks at 1582 cm^{-1} on small rutile particles, and the associated 1350 cm^{-1} peak suggests a similar behavior on rutile. Similar to rutile, the weak absorption bands at $\sim$1655, $\sim$1320, $\sim$1290 and $\sim$1260 cm^{-1} can be attributed to ν(C=O), δ(CH) and ν(CO) vibrational modes in (at least two pairs of) adsorbed HCOOH molecules with monodentate η^1-coordination. The thermal stability of HCOOH is also similar to rutile, and the absorption bands associated with HCOOH disappears after slight annealing, while the μ-formate remains stable. The DRIFT spectra of the Degussa P25 nanoparticles are similar to those obtained for 14 nm anatase prepared by hydrothermal treatments of microemulsions. Detailed analysis reveals, however, that P25 nanopartticles contain contributions from μ-formate bonded to the rutile phase [114].

8.5.1.2 Photoreactions of HCOOH on TiO_2

In contrast to formic acid and formate adsorption on TiO_2, much fewer photoreaction studies have been reported. An IR spectroscopy analysis of the photodegradation of HCOOH

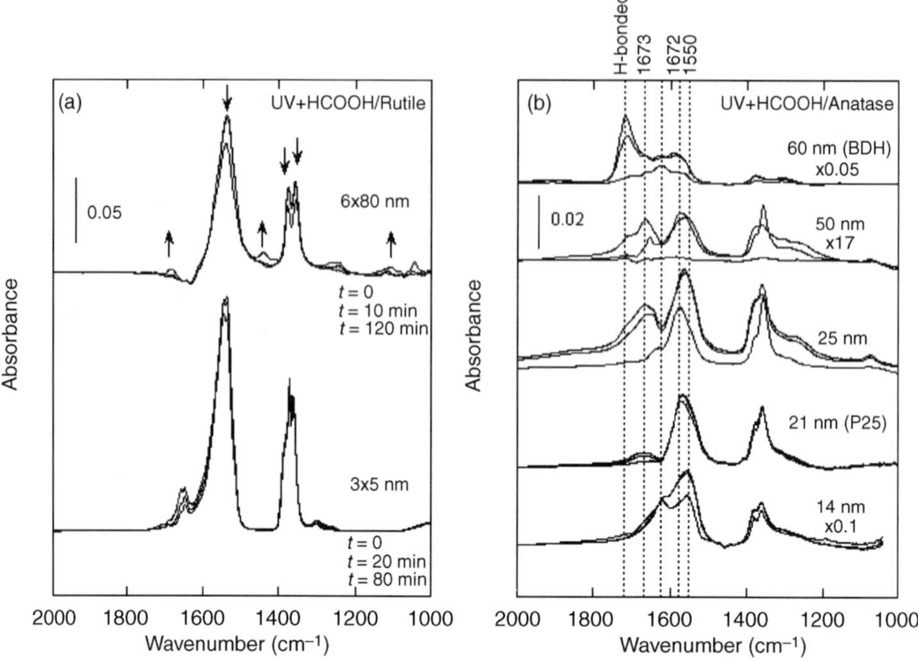

Figure 8.18 *In situ* infrared spectra of photodegradation of formate preadsorbed on (a) rutile and (b) anatase nanoparticles after different durations of solar light illumination employing a 200 W Xe lamp with AM1.5 filters. The illumination time on the large 6×80 nm particles was extended by 40 min compared to the small 3×5 nm particles to enhance the degradation effect and explicitly show the appearance of the carbonate bands at 1680 and 1100 cm^{-1} (asymmetric) and 1441 cm^{-1} (symmetric). The anatase spectra have been multiplied by the indicated factor. (Reprinted with permission from L. Österlund and A. Mattsson, Surface characteristics and electronic structure of photocatalytic reactions on TiO_2 and doped TiO_2 nanoparticles, in *Solar Hydrogen and Nanotechnology* vol. 6340 (ed. L.Vayssieres) © 2006 SPIE; L. Österlund, A. Mattsson and P.O. Andersson, What makes a good TiO_2 photocatalyts?, in *Nanostructured Materials and Nanotechnology II: Ceramic Engineering and Science Proceedings*, vol. 29 (eds S. Mathur and M. Singh) © 2008 John Wiley & Sons Inc.)

preadsorbed on TiO_2 nanoparticles reveals interesting dependencies on phase, size and structure (Figure 8.18) [9,20,114,141]. In contrast to the thermal oxidation reactions indicated in Figure 8.16, the detailed reaction pathways in photocatalytic decomposition of HCOOH are not known [117,137]. Concerning rutile TiO_2, the photodegradation rate is reported to be higher on small nanoparticles than on larger ones [114,141]. The reason is suggested to be due to stable μ-formate species, which have a relatively higher concentration on the large particles. In contrast, the monodentate and H-bonded HCOOH on the small particles are readily removed. [114,141] This is the same trend as for the thermal stability discussed above. Moreover, the rate of photodegradation of μ-formate species on the rutile particles is lower than on the anatase particles [20].

On anatase TiO_2 nanoparticles, there appears to be a correlation between the reactivity and the detailed morphology and surface structure [142–146]. T. Taguchi *et al.* [147] recently reported that photocatalytic reduction of $PtCl_6^{2-}$ in 2-propanol resulted in Pt deposition on (101) facets, while PbO_2 deposited on (112) facets upon photocatalytic reduction of Pb^{2+} in 2-propanol, suggesting that (101) facets provide reductive sites and (112) oxidative sites. Similarly, Byun and co-workers [142,147] reported different photo-oxidation rates of TiO_2 films prepared by chemical vapor deposition depending on crystal orientation, as determined by X-ray diffraction.

Given the rather detailed knowledge of HCOOH adsorbate structure on TiO_2 outlined above, IR surface spectroscopy provides means to unravel structure–reactivity relationships by comparing known adsorbate structures with observed reaction rates. The IR spectra depicted in Figure 8.18 reveal an intimate correlation between the nature of the ν_a(OCO) mode and the rate of photodegradation. We have already seen that the ν_a(OCO) mode is sensitive to the acidity of the metal ion that it binds with, which in turn is correlated with the neighboring atomic arrangement, that is, existence of O vacancy sites [106], and nearest neighbor Ti–Ti distance [126]. The latter is predicted to be decisive for the reactivity of brookite TiO_2 [126]. Since rutile exhibits the shortest Ti–Ti distance among the three TiO_2 polymorphs, anatase, rutile and brookite, and also the largest reactivity (μ-formate B formation and ν_a(OCO) frequency shift) there appears to exist a fundamental correlation between the detailed surface structure for the TiO_2 surfaces, HCOOH adsorbate structure and ν_a(OCO) frequency (or $\Delta\nu_{a-s}$). This implies that the surface energy, and thus the morphology, are important to modulate and control the reactivity of TiO_2 nanoparticles.

The reaction mechanism for photo-oxidation of formate and formic acid can be different to corresponding thermal reactions due to reactions with oxygen radicals. Possible reaction paths include degradation of μ-formate A and B species by direct reaction with an $O_2^{-\bullet}$ radical, as proposed for 2-propanol photodegradation on the rutile $TiO_2(110)$ surface [55], which simultaneously replenishes lattice O, *viz.*

$$HCOO + O_2^{-\bullet} \rightarrow OCOO + OH^{-\bullet} + O \tag{8.10}$$

Evidence for coordinated carbonate formation has been reported by several authors [18,20,132,148]. Reported band frequencies and assignments of (bi)carbonate on various TiO_2 nanoparticles are compiled in Table 8.2.

Photoinduced formation of O vacancies, which is known to promote formate dissociation [56], may also be formed by photoreactions with surface OH groups (acting as a hole acceptors), and recombination to yield water and O vacancies (Figure 8.16b). Indeed, it is observed that depletion of the ν(O–H) absorption band associated with bridging OH groups occurs simultaneously as the oxidation of small organic molecules commences, including formate species [9,15,20,149]. This would be the photochemical analog to reaction (8.8):

$$2O_BH \xrightarrow{h\nu > Eg} H_2O + O_B \tag{8.11}$$

A principal difference is that reaction (8.9) occurs at elevated temperatures, where facile diffusion of formate species may occur. Instead, in reaction (8.10) energetic $O_2^{-\bullet}$ provides the diffusion pathway.

Direct electron transfer may result in depopulation of bonding formate $\pi-$ and σ-states by charge transfer to the TiO_2 valence band, due to their strong coupling with the HOMO level extending above the VB edge (Figure 8.13). Indeed, carboxylic acids are well-known hole acceptors [150], *viz.*

Table 8.2 Observed infrared absorption bands and vibrational mode assignment of carbonate and bicarbonate complexes observed on TiO_2.

Substrate	Bicarbonate[a]	Carbonate compounds[b]				References
		Bridged bidentate carbonate	Chelating bidentate carbonate	Monodentate carbonate	Carbonate ions	
Bulk solids		1620–1670 ($\nu(CO_{II})$) 1220–1270 ($\nu_a(CO_I)$) 980–1020 ($\nu_s(CO_I)$)	1530–1620 ($\nu(CO_{II})$) 1250–1270 ($\nu_a(CO_I)$) 1020–1030 ($\nu_s(CO_I)$)	1470–1530 ($\nu_a(CO_{II})$) 1300–1370 ($\nu_s(CO_{II})$) 1080–1049 ($\nu(CO_I)$)		[1]
Dimer HCO_3^- (solid state)	1655–1615 ($\nu_a(C{=}O)$), 1400–1370 ($\nu_s(C{=}O)$), 1300 ($\delta(OH \cdots H)$)					[110]
Anatase (O_2 preads)	1431	1671, 1219, –				[188]
Anatase (H_2O preads)	1420	1676, 1245, –				[188]
Anatase[c]	1580–1555, 1443–1420	1670, 1243, –		1590–1575, 1370–1320		[110]
Rutile[d]	1586–1555, 1440–1420, 1220		1485, 1325, –			[110]
Rutile $w \times l \approx 6 \times 80$ nm[e]		~1665, ~1200, ~1040				[20]
Anatase $d = 14$ nm		~1668, ~1215, ~1160				[19,20]
Anatase/rutile $d = 10$ nm	1562, 1500, 1239	1582, 1321, 1010				[189]
P25 (Degussa)		1579, 1319, –				[16]
P25 (Degussa)					1441	[149]
		Nanoparticles (aqueous phase)				
Anatase/rutile $d = 10$ nm[f]	1582, 1500, 1239		1582, 1322, 1010			[189]

[a] Aqueous bicarbonate (HCO_3^- in H_2O): 1617, 1359 cm^{-1} [190].

[b] For noncoordinated CO_3^{2-} (D_{3h} symmetry) the ν_3 mode ($\nu_a(CO)$) at 1450–1420 cm^{-1} is the only IR active vibration >900 cm^{-1}. The symmetry of CO_3^{2-} is reduced to C_{2v} (monodentate) or C_s (bridged or chelating), upon metal coordination and the ν_3 modes splits into two bands. The following "rule" has been suggested: $\Delta\nu_3$ (bridged bidentate) > $\Delta\nu_3$ (chelating bidenate) > $\Delta\nu_3$ (monodenate) [109,110]. Where distinction between the former has not been made, the reported IR bands have been classified according this rule.

[c] Hydrated surface.

[d] Deuterated surface.

[e] The rutile particles exhibit a rectangular morphology in TEM and are classified according to the length (*width* × *length*), which scales with exposed <110> surface area [114,131]. The width of the particles grows with increasing particle size from approximately 2 to 10 nm.

[f] In ethanol and methanol at pH = 1.4–3.1.

$$HCOO^- + h^+ \xrightarrow{h\nu} CO_2 + H^{\bullet} \tag{8.12}$$

The photoelectrochemical analog to these two scenarios, that is, an indirect ((8.10) or (8.11)) or a direct pathway (8.12), has been discussed in the literature [151]. Conclusive evidence discriminating the two is, however, scarce. In these studies the indirect path is typically discussed in terms of reaction of organic compounds with surface radical intermediates such as $OH^{\bullet}$ or $Ti{-}O^{\bullet}$, which are formed, either by reduction of dissolved O_2 with conduction-band electrons [152] or by photo-oxidation of water [151,153,154]. A photoinduced dehydrogenation reaction producing CO_2 in gas–solid reactions may explain features in reported IR spectra in photocatalytic studies that are typical for asymmetric CO_2 in the $\sim$1720 cm^{-1} and $\sim$1200 cm^{-1} region (i.e., end-on Ti–O–C=O species [109,155]) and carbonate species $\sim$1650 cm^{-1} and 1000–1100 cm^{-1} [4,9,15,18,19,110,114,156]. However, at present there is no consensus on the origin of these absorption bands. For example, in vacuum-based IR photoreaction experiments of CO_2 with mixed phase TiO_2 nanoparticles (Degussa P25) [16] or upon reactive CO_2 reaction in humid atmospheres on large rutile particles [20], similar spectral bands were not seen. It cannot be excluded that particle size and structure plays a significant role.

The photodegradation proceeds faster on a hydrated TiO_2 surface [9,17]. Selloni and coworkers have reported that hydrogen bonding strongly affects the stability of HCOOH. Moreover, water ensures that the surface is fully hydroxylated and provides surface OH groups that are efficient hole acceptors to facilitate O vacancy formation (8.11).

8.5.1.3 Adsorption and Photoreactions of HCOOH on Doped TiO_2

Although studies of doped TiO_2 have dramatically increased during the past few years, spurred by encouraging reports of visible-light activity, very few studies have been published on the adsorption and reactions of carboxylic acids with doped TiO_2 nanoparticles. In the following we consider reactions of formic acid with cation- and anion-doped TiO_2.

Wetchakun *et al.* [157] synthesized transition-metal ion (V, Fe, or Cu) doped TiO_2 nanoparticles from titanium tetraisopropoxide by sol-gel methods. The photodegradation rate of several organic acids, including HCOOH was studied and was found to be highest for pure TiO_2 under UVA illumination. Under visible-light irradiation it was found that Fe-doped TiO_2 nanoparticles only could mineralize oxalic acid with appreciable yield. The authors suggested that the increased activity of Fe–TiO_2 was due to the formation of easily photolyzed ferrioxalate complexes rather than red-shift of the bandgap energy. Arana *et al.* [148,156,158] conducted a series of experiments on the interaction of HCOOH with wet impregnated Fe [156,158], Pd and Cu (Ref. [148]) anatase or mixed-phase anatase/rutile (Degussa P25) TiO_2 nanoparticles in aqueous solutions. They employed FTIR spectroscopy to characterize the adsorbates and identify reaction intermediates. Similar to the conclusion put forward by Wetchakun *et al.* [157], they also suggested that the observed photodegradation was due to efficient complexation of formic acid with Fe–TiO_2. They even inferred a complex catalytic cycle, including dissolution of Fe(III)-carboxylato-complexes and photomediated reoxidation of Fe^{2+}. High Fe loading was observed to be detrimental for the photoreactivity. Upon adsorption of HCOOH on TiO_2 impregnated with Fe, Pd and Cu nanoparticles, a strong correlation with the metal type and the reactivity was observed, which affected the HCOOH bonding to the catalyst and the vibrational frequency of the formate species [148,156].

It was suggested to be due to simultaneous formate bonding to both Ti and Fe, Pd or Cu surface cations, which facilitated hole reaction (reaction 8.12) with, simultaneously, Pd and Cu oxides acting as electron acceptors. The predicted weakening of the metal–metal bonding interaction and strengthening of the metal–oxygen bonding upon 3*d* transition-metal doping [159] is expected to decrease the acidity of the dopant atom and hence strongly affect the ν_a(OCO) vibrational mode and the $\Delta\nu_{a-s}$ value, which are sensitive indicators of the cation charge in metal-bonded formate species. Such correlations were inferred by Mattsson *et al.* [19], which correlated their vibrational data with DFT calculations (Table 8.3). In several of these studies (bi)carbonate and aldehyde complex formation was reported [19,148,158]. As noted in Section 8.5.1.1 unambiguous assignment of aldehyde species only based on IR spectra is complicated in this context since many peaks overlap with formate and carbonyl bands. In Table 8.3 it can be seen that carbonate formation is energetically favorable and will occur, provided efficient oxygenation of the formate species can occur. At present, it is, however, difficult to make any conclusive statements about the formic acid (or formate) transition-metal-doped TiO_2 systems, since most studies so far concern metal deposition on TiO_2 rather than substitutional metal doping. It is therefore expected that observed vibrational signatures of carboxylate complexes are due to superposition of oxide and metal-bonded species. It is known that the carboxyl stretching frequencies of late transition metals (e.g., Pd) [148] exhibit monodentate bonding (type I complex in Figure 8.12) with $\Delta\nu_{a-s}$(monodentate) > $\Delta\nu_{a-s}$(ionic).

A few studies of reactions of formic acid with N- [160–164] and C- [151,165] doped TiO_2 have been reported. Similar to the ideas put forward by Umebayashi *et al.* [159] to rationalize photocurrent measurements on cation-doped TiO_2, Liu *et al.* [151] showed that photocurrent measurements provide the means to distinguish direct and indirect photo-oxidation of formic acid and methanol, respectively, on C–TiO_2. Localized C-induced mid-gap states proved to be

Table 8.3 Calculated adsorption energies and bond electron population for bridging bidentate formate and carbonate on anatase TiO_2(001) surface and 9 at.% Nb:TiO_2 [19].

	$HCOO^{\cdot}$	$CO_3^{\cdot}$
	Anatase TiO_2(001)	
Adsorption energy	$-579\,kJ\,mol^{-1}$	$-724\,kJ\,mol^{-1}$
Electron population		
C-$O_{i,ii}$	0.87, 0.88	0.78, 0.76
C-O_{iii}		1.09
$T_{ii}\cdots O_i$	0.30	0.42
$T_{ii}\cdots O_{ii}$	0.32	0.42
TiO_2 (bulk)	0.70	0.70
	9% Nb-anatase TiO_2(001)	
Adsorption energy	$-579\,kJ\,mol^{-1}$	$-868\,kJ\,mol^{-1}$
Electron population		
C–$O_{i,ii}$	0.86, 0.80	0.82, 0.76
C–O_{iii}		1.10
$T_{ii}\cdots O_i$	0.35	0.39
$Nb_i\cdots O_{ii}$	0.32	0.28
NbO_2 (bulk)	0.73	0.73

responsible for the photo-oxidation of formic acid thus favoring a direct mechanism. Conflicting results concerning formic acid and formate photo-oxidation have been reported; either no visible-light photo-oxidation [160,164], or enhanced photo-oxidation [161,163] compared to pure TiO_2 has been reported. In one case the detailed adsorbate structure was reported [161]. In this study, DRIFT spectra obtained for $N-TiO_2$ in the anatase modification show evidence of formate comparable to those observed on pure anatase particles of similar size. Irokawa *et al.* [136] also showed by DRIFT spectroscopy that stable formate species formed on the surface of visible active $N-TiO_2$ catalysts upon photo-oxidation of gaseous toluene.

The influence of point defects introduced into the TiO_2 crystal structure upon N doping has been studied by FTIR spectroscopy [162]. Formation of Ti–N triple bonds in the materials was correlated to an inferior visible-light-driven photo-oxidation rate of formate, and related to a loss of crystallinity. In the studies quoted above it is likely that the $N-TiO_2$ catalyst contained only low concentrations of substitutionally bonded N atoms. Thus it appears that the reported photoresponse should be interpreted in terms of localized N 2*p*-derived mid-gap states, crystallite defects [162] or even core-shell morphologies [163]. Possibly, photoinduced defects due to Ti^{3+} states, as discussed by Emeline *et al.* [166], also contribute. From these studies it is clear that additional vibrational spectroscopy data are crucial. It is important to characterize the adsorbate structure, which can give evidence of the metal–adsorbate bonding and the acid–base character of adsorption sites. This will help to confirm the predicted oxidized states of the S, N and C ions and the increased basicity of neighboring O atoms [167], and the reduction of Ti^{4+} ions upon illumination. The work of Balcerski *et al.* [162] further shows the potential to characterize lattice bonds that form upon anion doping.

8.5.2 Reactions with Acetone

Acetone is an important VOC compound and considered as a model compound in photocatalytic air cleaning [89]. Moreover, acetone is a common intermediate in photodegradation (and thermal) reactions of organic pollutants. It was identified as the rate-determining step for photo-oxidation of propane on anatase TiO_2 nanoparticles in Section 6.4. The acetone photochemistry is much more complex compared to the formic-acid adsorption system and may result in diverse reaction pathways that range from complete oxidation to condensation reactions. Thus, a variety of surface species have been suggested, based on IR spectroscopy on powder TiO_2 samples, ranging from radicals [14,168] to condensation products [169] such as mesityl oxide [149,170–172] and polyacetone [100], or decomposition products such as carboxylate species (acetate and formate) [15,18,19,171–174] and methoxy groups [18,19,174]. Further reactions between decomposition products are likely to occur, and a wide range of possibilities exist, depending on reaction conditions. For example, it is likely that aldehydes, and carbonato- and bicarbonato-complexes form in the course of the photoreaction [15,18,19,171–174]. Again, the assignments of these species are not unambiguous, or a priori comparable between different experiments; they are often made under different conditions, with TiO_2 materials with different structures and particle sizes, employing different sample preparation procedures and so on. As for HCOOH, vibrational spectroscopy in combination with single-crystal experiments and *ab initio* calculations are invaluable to validate and confirm reaction pathways. Such data for acetone/TiO_2 is, however, scarce. Only Henderson has reported results for acetone adsorption [100,175,176] and photo-oxidation [97,174] on rutile $TiO_2(110)$.

8.5.2.1 Photoreactions with TiO_2

Photo-oxidation of acetone cannot be treated without considering the complete O_2–acetone TiO_2 system. The diverse O_2 chemistry at oxygen vacancies [104,105,177] and Ti^{3+} sites [57,174] is reported to result in both O_2 adsorption and O_2 dissociation, as well as O hopping along the [001] troughs [178], and have been shown to affect the reactivity of organics, including acetone oxidation [100]. Using temperature-programmed and isotope desorption Henderson suggested that acetone converts to an acetone–oxygen complex by thermal reactions with O_2 adsorbed on Ti^{3+} sites [174]. In addition, O_2 plays an important role as both electron acceptor and source of O for regeneration of O vacancies. The importance of oxygen radicals produced by adsorbed O_2 has been demonstrated by many authors in both the thermal [55,100], and photochemical oxidation [19,96,174] of acetone. In a recent molecular-beam study of 2-propanol, oxidation on single-crystal rutile $TiO_2(001)$ and $TiO_2(110)$, the higher thermal than photocatalytic reaction probability on the (001) facet was attributed to the proximity of bridging O atoms to the Ti(5) alcohol binding sites (which are closer on the (001) surface) [55]. Electron paramagnetic resonance (EPR) and IR spectroscopy have been extensively used in the past to study surface radical species formed over UV-irradiated TiO_2 [179]. From these studies, it has been shown that the oxidation state of the catalyst greatly influences the type of radical formed. On reduced TiO_2 at 298 K, paramagnetic oxygen radicals (O^-, O_2^- or O_3^-) form upon O_2 exposures, while at high temperatures (823 K), a reduced TiO_2 produces only diamagnetic O^{2-} lattice O due to re-oxidation of TiO_2. Thus, it is expected that the TiO_2 preparation procedure influences the detailed reaction pathway. Attwood *et al.* [168] showed that, in the absence of O_2^-, on UV illumination of dehydrated TiO_2 nanoparticles (Degussa P25) acetone photo-oxidation produced alkylperoxy radicals ($CH_3COCH_2{-}OO^{\bullet}$), which was suggested to be due to rapid hole transfer (e.g., via Ti^{3+} or O^- centers) to adsorbed acetone, yielding propanone radicals ($CH_3COCH_2^{\bullet}$) and subsequent reaction with O_2. The same propanone radical has been predicted to form by reactions with surface OH radicals [180].

Henderson has studied the acetone chemistry on single-crystal rutile $TiO_2(110)$ [97,100,174–176]. We are not aware of any single-crystal studies on anatase surfaces. Using primarily photon-stimulated and temperature-programmed desorption mass spectrometry (PSD and TPD), Henderson has, in a series of experiments, explored the thermal and photon-stimulated reactions of acetone [100], O_2 + acetone [97,174] and H_2O + acetone [175,176] with reduced and oxidized rutile $TiO_2(110)$. The general mechanism proposed by Henderson involves an initial thermal reaction pathway, whereby acetone (and other carbonyl-containing compounds [181]) reacts with adsorbed oxygen to form acetone–oxygen complexes at low temperatures ($T < 375$ K), rather than forming surface acetate or condensation products (aldol condensation). Further photo-oxidation results in acetate formation and methyl abstraction (Figure 8.19) [96,174]. Mass-spectrometric analysis of TPD data revealed that most of the adsorbed acetone desorbs molecularly below 500 K (93%), with only small amounts of other high-temperature desorption products (7%). Among these, ketene was detected ($T \sim 650$ K) and interpreted in terms of acetate decomposition (0.02 ML) associated with O vacancies [100]. Ketene desorption is also seen on the reconstructed (001) rutile surfaces at elevated temperatures [182]. In fact, it is well known that ketene is produced by pyrolysis of acetone. Back-reaction of decomposition products to yield acetone may, in principle, also occur on TiO_2. On the {114} facets of rutile $TiO_2(001)$ it is reported that acetone forms via a bimolecular reaction

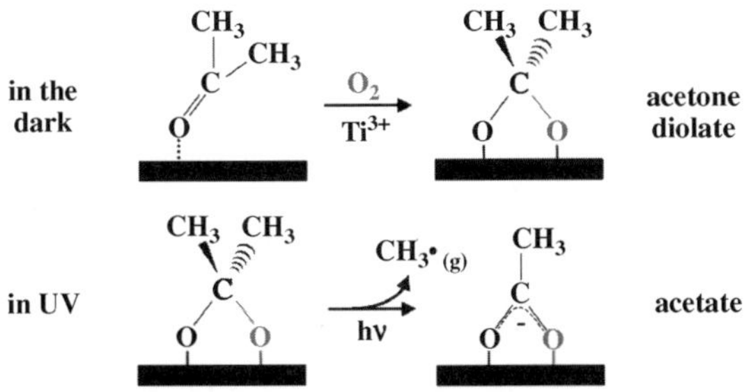

in the dark
O_2
Ti^{3+}
acetone diolate
in UV
$CH_3^{\bullet}$ (g)
hν
acetate

Figure 8.19 Schematic picture depicting the suggested reaction pathways for acetone adsorption and photodecomposition into acetate on rutile $TiO_2(110)$ deduced from UHV studies. (Reprinted with permission from M.A. Henderson, Relationship of O_2 photodesorption in photooxidation of acetone on TiO_2, *Journal of Physical Chemistry C*, **112**, 11433–11440, 2008. © 2008 American Chemical Society.)

at unsaturated Ti(4) centers ($CH_3{-}COO{-}Ti(4){-}OOC{-}CH_3 \rightarrow CH_3{-}CO{-}CH_3 + CO_2 +$ Ti(5)−O) [101]. However, the latter reaction occurs only at elevated temperatures ($T > 650$ K) in the presence of Ti(4) centers, and should be of minor importance in most photocatalysis studies. Acetate is a commonly suggested intermediate in the photodegradation of acetone, as well as alcohols [1,9,15,18,19,149,171,174]. Several studies report vibrational spectra characteristic of acetate species, albeit an unambiguous assignment is difficult [15,20,100]. The mechanism by which acetate formation occurs is, however, not known. Several authors have suggested nucleophilic attack on the carbonyl group followed by formation of bidentate acetate and formate complexes (Figure 8.20) [9,15,19,20,100]. This is supported by the simultaneous production of methyl radicals, either into the gas phase [96,174] in vacuum studies or onto the TiO_2 surface in the form of adsorbed methoxy species under high pressures [19]. In the former reaction, Ti^{3+} sites provide the active sites for O_2 adsorption and further dissociation.

In contrast, the EPR [168] and theoretical results [180] discussed above provide evidence that oxygen attack occurs on the methyl group, which may suggest a different mechanism whereby CH_3COCH_2-$OO^{\bullet}$ dissociates into either acetate or formate complexes depending on

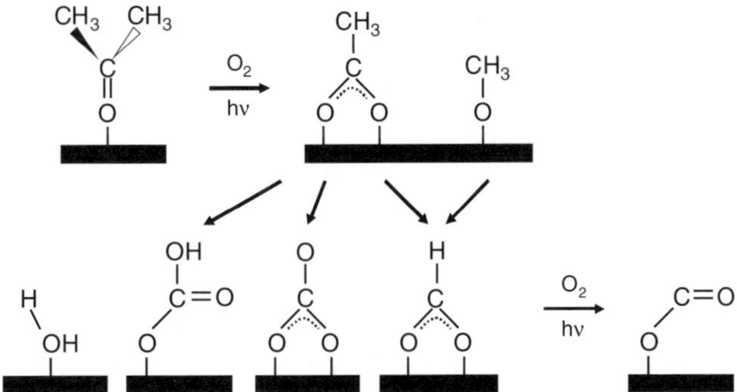

Figure 8.20 Schematic picture showing reported surface species formed in the course of photo-oxidation of acetone on TiO_2 nanoparticles, deduced from FTIR measurements.

reaction conditions. The different vibrational fingerprint of reported intermediate acetone decomposition species between vacuum studies [100] and high-pressure studies on TiO_2 nanoparticles [15,19], including isopropanol [183] (which proceeds through acetone formation) may suggest that the state of the catalyst (oxidized or reduced; Ti^{3+} surface concentration) is important. Furthermore, as shown in the preceding section, it is clear that the detailed surface structure also must be considered for a complete description of the experimental observations.

There are only a few IR spectroscopy studies of acetone photo-oxidation on TiO_2 nanoparticles [15,19,149,170,172]. Overall they all exhibit similar vibrational features, although the experimental conditions, samples and reported reaction mechanisms differ. In general, it is not possible, based only on IR spectra, to indentify adsorbate structure and reaction intermediates. Additional data is required. In order to illustrate the reactions following the initial photoinduced oxidation of acetone on TiO_2 nanoparticles, we show, in Figure 8.21, FTIR

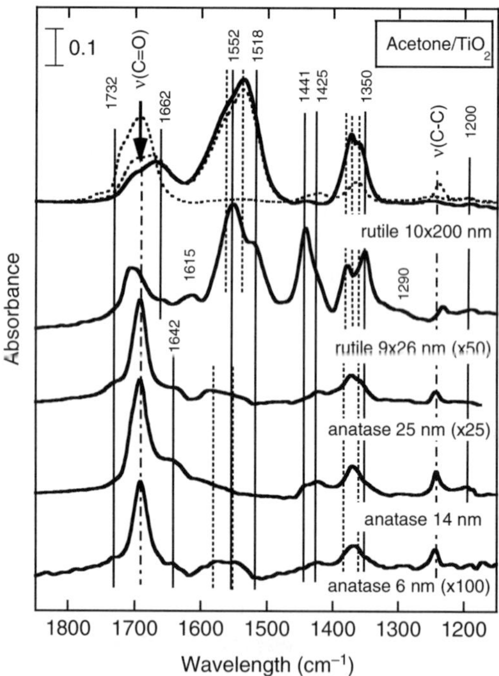

Figure 8.21 DRIFT spectra of acetone adsorbed on anatase and rutile TiO_2 nanoparticles. Note that acetone dissociates on the rutile particles (dashed spectra acquired at 10 min interval in synthetic air). (Reprinted with permission from L. Österlund and A. Mattsson, Surface characteristics and electronic structure of photocatalytic reactions on TiO_2 and doped TiO_2 nanoparticles, in *Solar Hydrogen and Nanotechnology* vol. 6340 (ed. L.Vayssieres) © 2006 SPIE; T. van der Meulen, A. Mattson and L. Österlund, A comparative study of the photocatalytic oxidation of propane on anatase, rutile, and mixed-phase anatase-rutile TiO_2 nanoparticles: Role of surface intermediates, *Journal of Catalysis*, **251**, 131–144, 2007. © 2007 Elsevier.)

spectra obtained after different durations of acetone dosing onto anatase and rutile nanoparticles [9]. In all cases, acetone binds via the carbonyl to the surface Ti atoms. It is evident that adsorption onto anatase and onto rutile nanoparticles proceeds very differently. On a pre-oxidized single-crystal rutile $TiO_2(110)$ surface it has been shown that acetone desorbs at $T > 300$ K [100]. It is thus expected that under steady-state conditions in most photocatalytic experiments conducted at room temperature, some acetone desorbs spontaneously. Indeed, on anatase nanoparticles, a desorption rate of ~0.03 ML min^{-1} is deduced from *in situ* FTIR spectroscopy [19]. In contrast, on rutile nanoparticles, acetone dissociation readily occurs (compare the dashed and solid lines in Figure 8.21). This contrasts with reported single-crystal data [100], and indicates that either the state of a vacuum-sputtered–annealed TiO_2 crystal is very different from the rutile nanoparticles prepared by calcination in synthetic air, or that surface defects (or minority facets) and OH/H_2O [128] present on the nanoparticle surfaces completely govern the adsorption kinetics. Otherwise, it is hard to explain why most acetone dissociates on the large rutile particles, where >98% of the exposed are pre-oxidized (110) facets [114,131], that is, most acetone molecules adsorb on a surface very similar to a $TiO_2(110)$ single crystal. When corrected by the evaporation rate, which is assumed to be the same as on the anatase particles, the dissociation rates as deduced by the decay of the $\nu(C{=}O)$ and $\nu(C{-}C)$ peaks in Figure 8.21, are estimated to be about 2 and 3 times as fast as the evaporation rate on the large and small rutile particles, respectively. Acetone dissociation is concerted by evolution of surface species, which forms at rates exceeding the dissociation rate of acetone by a factor of ~2–3, implying that these are C_2 and/or C_1 dissociation products. The accumulation of surface species on rutile is manifested by pronounced absorption bands that develop in the 1500–1600 cm^{-1} region and at 1441 cm^{-1} (Figure 8.21) [9]. The relative intensity of the latter peak is intense on small rutile particles. The former absorption region is characteristic of ν_a(OCO) bands in formate, acetate and bicarbonate species, as discussed in Section 8.5.1. Absorption bands at ~1565 cm^{-1} and 1382 cm^{-1} are attributed to μ-formate B species (Table 8.1). The absorption bands in the 1300–1600 cm^{-1} region contain contributions from adsorbed acetate $CH_3{-}COO$ and bicarbonate $OH{-}COO$ and carbonates, albeit assignments of these modes are not conclusive (Table 8.2). For example, the ν_s(OCO) mode in the acetate–TiO_2 complexes has been reported anywhere in the range 1420–1460 cm^{-1} [15,20,100,117]. Of these candidates, bridge-bonded acetate (1518 and 1425 cm^{-1}), bicarbonate (1615, 1441, ~1290 cm^{-1}) and bridge-bonded carbonate (1662, ~1200, and 1042 cm^{-1}) have tentatively been assigned to absorption bands observed on the rutile particles [9,20]. The former is supported by the enhancement of the 1350 cm^{-1} peak due to $\delta(CH_3)$ in acetate. In addition, the corresponding adsorbed (H-bonded) ions should be considered, for example, for acetate the ν_a(OCO) and ν_s(OCO) bands occur at 1552 and 1415 cm^{-1}, respectively [131]. Obviously, unambiguous assignments of all these species cannot be made unless additional control experiments are done. Analysis of the formation kinetics, isotopic H/D shifts, stoichiometric considerations and reactive $CO_2 + H_2O/D_2O$ adsorption experiments, have been used to scrutinize the detailed reaction pathways [20,96,174]. Several unresolved issues remain, however. For example, the observed vibrational band at ~1440 cm^{-1} has been attributed to acetate [15], or CO_3^{2-} ions [149], while recent *in situ* IR spectroscopy studies suggest that this peak is composed of several vibrational modes originating from ν_s(OCO) in bicarbonate (main peak), and acetate (weak low-frequency shoulder), respectively [9,20]. Moreover, the issue of the initial acetone adsorbate structure is not settled. Henderson reported that acetone -oxygen complex formation precedes the photo-oxidation reaxtion [96,174].

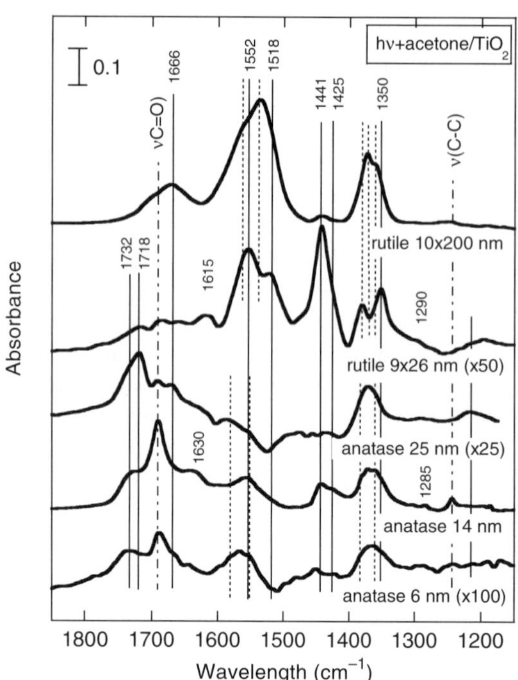

Figure 8.22 DRIFT spectra of acetone adsorbed on anatase and rutile TiO_2 nanoparticles after 45 min solar-light illumination. Note the different scales. On the 25 nm anatase particles, almost all acetone and intermediate decomposition surface species are gone after UV irradiation. (Reprinted with permission from L. Österlund and A. Mattsson, Surface characteristics and electronic structure of photocatalytic reactions on TiO_2 and doped TiO_2 nanoparticles, in *Solar Hydrogen and Nanotechnology* vol. 6340 (ed. L. Vayssieres) © 2006 SPIE.)

Figure 8.22 shows FTIR spectra after 45 min solar-light illumination of different forms of TiO_2 nanoparticles [9,18,19,141]. Again there are significant differences in reported data, which appear to depend on reaction conditions. At high acetone coverages, aldol condensation reactions have also been reported on anatase TiO_2 nanoparticles [149,170], while at low coverages [9,19] and/or under vacuum conditions on $TiO_2(110)$, decomposition reactions have been observed [96,174]. From Figure 8.22, it is evident that the photocatalytic degradation of acetone proceeds very differently on rutile and anatase, and that again the photochemical processes have distinct particle-size dependence. On the rutile particles, the total photodegradation rate is slower compared to anatase, as judged by the disappearance of intermediate surface species. Furthermore, the anatase nanoparticles in the size range 25–50 nm show the highest total oxidation rate (converting all acetone into gas-phase products after 60 min illumination). In general, the major difference between anatase and rutile appears to be the relative larger concentration of bridged bidentate species on rutile (formate, acetate and bicarbonate). On anatase, the relative concentration of carbonate species is higher, which is consistent with DFT calculations (Table 8.3). It is evident from Figure 8.22 that the intensity of the 1441 cm^{-1} (bicarbonate) peak increases and the FWHM decreases (i.e., the 1425 cm^{-1} acetate peak decreases) during the whole illumination period on both the large and small rutile particles (but with vastly different intensities), which shows that a gradual conversion to bicarbonate occurs. The bicarbonate formation is coupled to the CO_2 and H_2O production. This may explain why bicarbonate formation occurs in the case of acetone decomposition, but not for formic acid. In the latter, dehydration appears to be favored (Section 8.5.1).

A quantitative comparison between anatase and rutile is complicated by the fact that the photodegradation proceeds via different reaction pathways, and the two polymorphs contain different surface species at different reaction times. Small anatase particles in general have a higher concentration of species associated with the bands in the 1400–1600 cm^{-1} region compared to larger particles (Figure 8.22). In particular, the relative absorption at ~1560 cm^{-1} increases with decreasing anatase particle size, analogous to the findings on the rutile, where ν_a(OCO) in μ-formate B increases with decreasing particle size. The latter points to chemical similarities between the surface species coordinated to point defects on anatase and rutile, respectively. In general, the results indicate that it is desirable to suppress formation of bidentate-bonded species in pollutant-abatement applications, since the photodegradation of these species are slow. Accumulation of these species blocks the TiO_2 surface and can be viewed as a deactivation mechanism (site inhibition). The latter is quantified in Figure 8.23, which shows the (normalized) integrated 1500–1600 cm^{-1} region (characteristic for the ν_a(OCO) bands discussed above) after 45 min UVA irradiation on various anatase and rutile

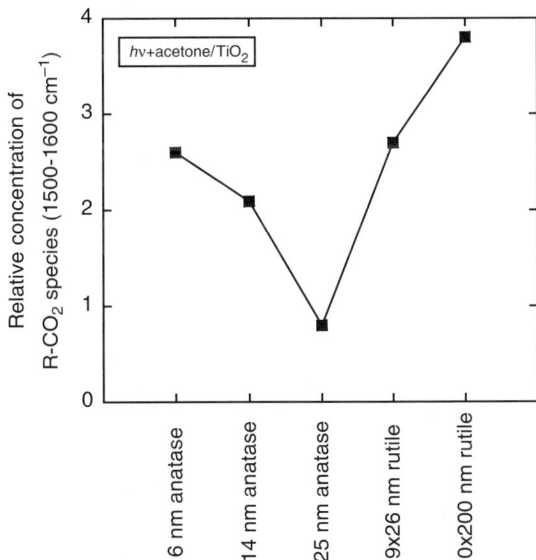

Figure 8.23 Concentration of residual formate, bicarboxylate and carboxylate surface species on different TiO_2 nanoparticles after 45 min solar-light illumination in dry gas. (Reprinted with permission from L. Österlund and A. Mattsson, Surface characteristics and electronic structure of photocatalytic reactions on TiO_2 and doped TiO_2 nanoparticles, in *Solar Hydrogen and Nanotechnology* vol. 6340 (ed. L. Vayssieres) © 2006 SPIE.)

TiO_2 nanoparticles. Similar arguments have been used to explain why surface acidity influences the quantum yield for acetone photo-oxidation on anatase TiO_2 nanoparticles [184]. The quantum yield for acetone photo-oxidation was found to be higher on H_2SO_4-treated TiO_2 nanoparticles, compared to nontreated and NaOH-treated. It was proposed that strong bonding of acidic intermediates inhibits acetone reaction by bonding to basic sites (surface OH and/or lattice O^{2-}).

The observations that large anatase particles are beneficial for photo-oxidation of gaseous acetone is at variance with the common view that small anatase particles are desirable. It does however provide a rationale why Degussa P25 is a good photocatalyst (in fact, better than most of the photocatalysts reported in the literature). The P25 catalyst consists of a large fraction of >21 nm anatase particles (about 70–80 mol%). However, the presence of rutile deteriorates the performance of P25, due to formation of μ-formate and carbonates. It thus appears that the photo-oxidation rate of carboxylates is highest on anatase particles in the size range 25–50 nm and is due to the smaller concentration of μ-formate and bicarbonate species formed on these particles. The fundamental reason for their high activity is not known, but the discussion here and in the previous section strongly suggests that structure sensitive bonding and reactivity play important roles, qualitatively similar to what has been inferred for preferentially oriented TiO_2 thin films and facetted macrocrystals [142–147].

8.5.2.2 Photoreactions with Doped TiO_2

Very few studies have reported on the photocatalytic reactions of acetone with doped TiO_2. In a ^{15}N solid-state nuclear magnetic resonance (NMR) and EPR study of N-doped TiO_2 monolayers on porous borosilicate glass ($TiO_2{-}N_x{-}TiO_2$/PVG) prepared by Reyes-Garcia *et al.* [185] showed that the acetone oxidation occurred upon illumination with visible light ($\lambda > 420$ nm) employing fairly high light flux (~1 W onto sample). The photo-oxidation of acetone was found to be less efficient than for ethanol, and in contrast to ethanol, no or very little intermediate species were detected by ^{13}C NMR during the illumination time. In an *in situ* FTIR study of photocatalytic decomposition of acetone on pure and Nb-doped anatase TiO_2 nanoparticles, it was shown that acetone adsorbs with η^1-coordination to the surface cations [19]. The carbonyl–metal bonding is stabilized on Nb sites as evidenced by a lowering of the ν(C=O) frequency by 20 cm^{-1} (Figure 8.24). Upon solar-light illumination, acetone was shown to readily decompose with concomitant formation of stable intermediate surface species. Despite a 40% more effective UV–visible absorption for Nb-doped TiO_2, the decomposition rate was an order of magnitude lower on the Nb-doped films despite an enhanced visible-light absorption in these materials. The enhanced visible-light absorption and decreased photocatalytic activity was instead proposed to be due to the presence of Nb=O clusters and cation vacancies formed during the particle synthesis in accordance with the different valence of the Ti and Nb cations. Supporting DFT calculations (Table 8.3) indicated that key surface intermediates are bidentate bridged formate and carbonate, and H-bonded bicarbonate, respectively, whose concentration on the surface can be correlated with their heats of formation and bond strength to coordinatively unsaturated surface Ti and Nb atoms. In another study, Mattsson *et al.* [18] employed *in situ* FTIR spectroscopy to monitor diffusion-limited photoreactions in inert gas atmosphere (anaerobic conditions) on Zr- and Nb-doped anatase TiO_2 nanoparticles. A coupled diffusion reaction model was developed to quantitatively determine the role of O diffusion. The results provided evidence of an oxygen surface-

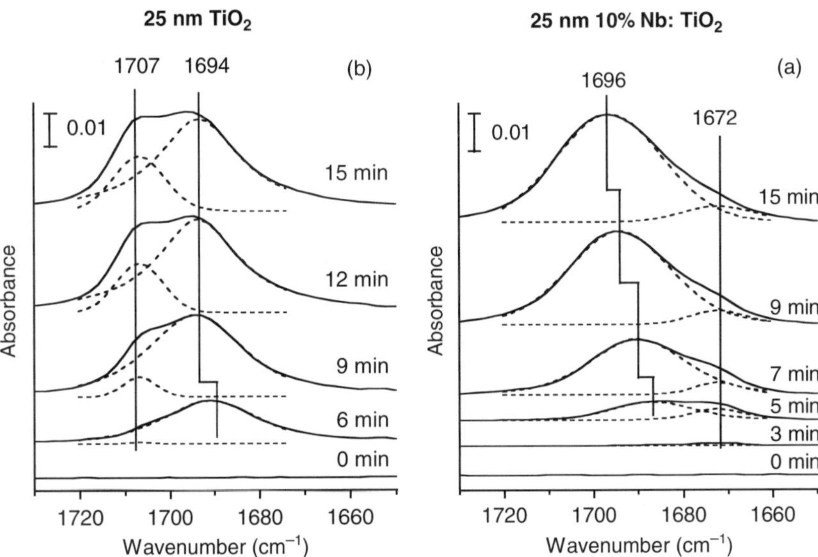

Figure 8.24 FTIR spectra of acetone adsorbed on Nb-doped TiO_2 at different dopant concentrations. (Reprinted with permission from A. Mattsson, M. Leideborg, K. Larsson *et al.*, Adsorption and solar light decomposition of acetone on anatase TiO_2 and niobium doped TiO_2 thin films, *Journal of Physical Chemistry B*, **110**, 1210–1220, 2006. © 2006 American Chemical Society.)

diffusion mechanism, which depletes the surface from oxygen and gradually deactivates the particles in absence of external oxygen supply. The diffusion-mediated photoreaction pathway was shown to be significant on the doped TiO_2 particles. The contribution of this reaction pathway was also estimated to be up to 65% of the total PID rate on Nb- and Zr-doped TiO_2 in synthetic air. Interestingly, a new parallel reaction pathway was found in inert atmosphere by FTIR surface spectroscopy that favored surface carbonate formation. This may indirectly support the different results obtained after reactive CO_2 adsorption in vacuum and atmospheric studies, as discussed above [16,18,110].

Even though many studies employing cation doping strategies to achieve enhanced UV–visible photocatalytic activity turn out to be unsuccessful, and may be related to fundamental problems associated with defect formation [166], there may be unexpected benefits associated with structural stabilization of the photocatalyst, as shown by Österlund *et al.* [186]. In this study it was shown that Zr-doped TiO_2 films exhibit an exceptionally stable activity over time and usage; much better than the pure TiO_2 films. These studies indicate means to circumvent photocatalyst deactivation and implies that there may be a balance between high activity and sustained activity for some material combinations.

8.6 Summary and Concluding Remarks

Vibrational spectroscopy provides important information relevant for photocatalysis. Raman spectroscopy is today a widely accessible technique that gives information on fundamental metal-oxide vibrations, in particular those occurring at low wavenumbers in the far-infrared region which is not readily accessible by infrared methods. Further Raman can be a useful characterization tool for metal-oxide nanoparticles that complements other techniques, such as XRD, with phase and structural information with sometimes superior sensitivity. Raman gives insight into quantum size effects and data on particle grain size. It has been demonstrated that Raman spectroscopy studies on TiO_2 and doped TiO_2 nanoparticles provide information of material properties, such as hydrostatic pressure, surface tension and thermal expansion, which can be quantified by appropriate modeling. Infrared spectroscopy is the preferred method to study surface acid properties and surface reactions. Moreover, FTIR is an ideal technique to study photoinduced surface processes on photocatalysts *in situ* at high and low pressures over a wide temperature range, which gives insight into both kinetics and dynamics of photocatalytic reactions. Even though the focus here has been on gas–solid reactions, *in situ* FTIR is suitable also for aqueous-phase studies. Infrared surface spectroscopy gives detailed information on molecular transformations on TiO_2 nanoparticles. By interplay between dedicated nanoscale synthesis, single-crystal data and *ab initio* calculations, it is possible to unravel detailed information on adsorbate structures, reaction kinetics and to identify rate-determining steps. The result that carboxylate–TiO_2 bonding can be tuned with the appropriate crystal structure, has implications for a wide range of photocatalysis applications, ranging from pollutant abatement, metallo-organic dyes or fluorophore–ligand bonding in wet solar cells, and bioimaging. The approach to correlating vibrational spectroscopy, adsorbate structures and well-defined nanostructures, where critical aspects of the structure and morphology are systematically varied, is a rather recent advancement in photocatalysis research that has been facilitated by advancements in material characterization and nanotechnology. This approach can be seen as yet another tool to be added to the experimental toolbox that must be employed to unravel details of photocatalytic reactions.

Acknowledgments

The author thanks A. Mattsson and C. Lejon for help during various parts of this work.

References

[1] Davydov, A.A. (2003) *Molecular Spectroscopy of Oxide Catalyst Surfaces*, John Wiley & Sons, Chichester.

[2] Basu, P., Ballinger, T.H., and Yates, J.T. (1988) Wide temperature-range ir spectroscopy cell for studies of adsorption and desorption on high area solids. *Review of Scientific Instruments*, **59**, 1321–1327.

[3] Wachs, I.E. (1996) Raman and IR studies of surface metal oxide species on oxide supports: Supported metal oxide catalysts. *Catalysis Today*, **27**, 437–455.

[4] Busca, G. (1996) Infrared studies of the reactive adsorption of organic molecules over metal oxides and the mechanisms of their heterogeneous-catalyzed oxidation. *Catalysis Today*, **27**, 457–496.

[5] Beebe, T.P. and Yates, J.T. (1986) An *in situ* infrared spectroscopic investigation of the role of ethylidyne in the ethylene hydrogenation reaction on Pd/A_2O_3. *Journal of the American Chemical Society*, **108**, 663–671.

[6] Somorjai, G.A. and Rupprechter, G. (1999) Molecular studies of catalytic reactions on crystal surfaces at high pressures and high temperatures by infrared-visible sum frequency generation (SFG) surface vibrational spectroscopy. *Journal of Physical Chemistry B*, **103**, 1623–1638.

[7] Freund, H.J., Baumer, M., Libuda, J. *et al.* (2003) Preparation and characterization of model catalysts: from ultrahigh vacuum to *in situ* conditions at the atomic dimension. *Journal of Catalysis*, **216**, 223–235.

[8] Chuang, C.-C., Wu, W.-C., Huang, M.-C. *et al.* (1999) FTIR study of adsorption and reactions of methyl formate on powdered TiO_2. *Journal of Catalysis*, **185**, 423–434.

[9] Österlund, L. and Mattsson, A. (2006) Surface characteristics and electronic structure of photocatalytic reactions on TiO_2 and doped TiO_2 nanoparticles, in *Solar Hydrogen and Nanotechnology*, vol. **6340** (ed. L. Vayssieres), SPIE.

[10] McQuillan, A.J. (2001) Probing solid-solution interfacial chemistry with ATR-IR spectroscopy of particle films. *Advanced Materials*, **13**, 1034–1038.

[11] Tickanen, L.D., Tejedor-Tejedor, M.I., and Anderson, M.A. (1997) Quantitative characterization of aqueous suspensions using variable-angle ATR-FTIR spectroscopy: Determination of optical constants and absorption coefficient spectra. *Langmuir*, **13**, 4829–4836.

[12] Tunesi, S. and Anderson, M.A. (1992) Surface effects in photochemistry – an *in situ* cylindrical internal-reflection fourier-transform infrared investigation of the effect of ring substituents on chemisorption onto TiO_2 ceramic membranes. *Langmuir*, **8**, 487–495.

[13] Thompson, T.L. and Yates, J.T. (2006) Surface science studies of the photoactivation of TiO_2 new photochemical processes. *Chemical Reviews*, **106**, 4428–4453.

[14] Linsebigler, A.L., Lu, G., and Yates, J.T. Jr. (1995) Photocatalysis on TiO_2 surfaces: principles, mechanisms, and selected results. *Chemical Reviews*, **95**, 735–758.

[15] Coronado, J.M., Kataoka, S., Tejedor-Tejedor, I., and Anderson, M.A. (2003) Dynamic phenomena during the photocatalytic oxidation of ethanol and acetone over nanocrystalline TiO_2: simultaneous FTIR analysis of gas and surface species. *Journal of Catalysis*, **219**, 219–230.

[16] Liao, L.F., Lien, C.F., Shieh, D.L. *et al.* (2002) FTIR study of adsorption and photoassisted oxygen isotopic exchange of carbon monoxide, carbon dioxide, carbonate, and formate on TiO_2. *Journal of Physical Chemistry B*, **106**, 11240–11245.

[17] Liao, L.-F., Wu, W.-C., Chen, C.-Y., and Lin, J.-L. (2001) Photooxidation of formic acid vs formate and ethanol vs ethoxy on TiO_2 and effect of adsorbed water on the rates of formate and formic acid photooxidation. *Journal of Physical Chemistry*, **105**, 7678–7685.

[18] Mattson, A., Leideborg, M., Persson, L. *et al.* (2009) Oxygen diffusion and photon-induced decomposition of acetone on Zr and Nb doped TiO_2 nanoparticles. *Journal of Physical Chemistry C*, **113**, 3810–3818.

[19] Mattsson, A., Leideborg, M., Larsson, K. *et al.* (2006) Adsorption and solar light decomposition of acetone on anatase TiO_2 and niobium doped TiO_2 thin films. *Journal of Physical Chemistry B*, **110**, 1210–1220.

[20] van der Meulen, T., Mattson, A., and Österlund, L. (2007) A comparative study of the photocatalytic oxidation of propane on anatase, rutile, and mixed-phase anatase-rutile TiO_2 nanoparticles: Role of surface intermediates. *Journal of Catalysis*, **251**, 131–144.

[21] Österlund, L., Stengl, V., Mattsson, A. *et al.* (2008) Effect of sample preparation and humidity on the photodegradation rate of CEES on pure and Zn doped anatase TiO_2 nanoparticles prepared by homogeneous hydrolysis. *Applied Catalysis B*, **88**, 194–203.

[22] Diebold, U. (2003) The surface science of titanium dioxide. *Surface Science Reports*, **48**, 53–229.

[23] Herman, G.S., Sievers, M.R., and Gao, Y. (2000) Structure determination of the two-domain (1 × 4) Anatase $TiO_2(001)$ surface. *Physical Review Letters*, **84**, 3354–3357.

[24] Liang, Y., Gan, S., Chambers, S.A., and Altman, E.I. (2001) Surface structure of anatase $TiO_2(001)$: Reconstruction, atomic steps, and domains. *Physical Review B*, **63**, 235402.

[25] Tanner, R.E., Liang, Y., and Altman, E.I. (2002) Structure and chemical reactivity of adsorbed carboxylic acids on anatase $TiO_2(001)$. *Surface Science*, **506**, 251–271.

[26] Diebold, U., Ruzycki, N., Herman, G.S., and Selloni, A. (2003) One step towards bridging the materials gap: surface studies of TiO_2 anatase. *Catalysis Today*, **85**, 93–100.

[27] Li, S.-C., Dulub, O., and Diebold, U. (2008) Scanning tunneling microscopy study of a vicinal anatase TiO_2 surface. *Journal of Physical Chemistry C*, **112**, 16166–16170.

[28] Hebenstreit, W., Ruzycki, N., Herman, G.S. *et al.* (2000) Scanning tunneling microscopy investigation of the TiO_2 anatase (101) surface. *Physical Review B*, **62**, R16334–R16336.

[29] Lazzeri, M. and Selloni, A. (2001) Stress-driven reconstruction of an oxide surface: The anatase $TiO_2(001)$-(1 × 4) surface. *Physical Review Letters*, **87**, 266105.

[30] Chen, L.X., Rajh, T., Jager, W. *et al.* (1999) X-ray absorption reveals surface structure of titanium dioxide nanoparticles. *Journal of Synchrotron Radiation*, **6**, 445–447.

[31] Gong, X.Q., Selloni, A., Batzill, M., and Diebold, U. (2006) Steps on anatase $TiO_2(101)$. *Nature Materials*, **5**, 665–670.

[32] Griffiths, P.R. and de Haseth, J. (1986) *Fourier Transform Infrared Spectroscopy*, John Wiley & Sons, New York.

[33] Yates, J.T. and Madey, T.E. (1992) *Vibrational Spectrsocopy of Molecules Adsorbed on Surfaces*, Plenum Press, New York.

[34] Sheppard, N. and De la Cruz, C. (1996) Vibrational spectra of hydrocarbons adsorbed on metals.1. Introductory principles, ethylene, and the higher acyclic alkenes. *Advanced Catalysis*, **41**, 1–112.

[35] Rupprechter, G. (2004) Surface vibrational spectroscopy on noble metal catalysts from ultrahigh vacuum to atmospheric pressure. *Annual Reports on the Progress of Chemistry C*, **100**, 237–311.

[36] Kataoka, S., Tejedor-Tejedor, M.I., Coronado, J.M., and Anderson, M.A. (2004) Thin-film transmission IR spectroscopy as an in situ probe of the gas-solid interface in photocatalytic processes. *Journal of Photochemistry and Photobiology A*, **163**, 323–329.

[37] Naito, K., Tachikawa, T., Fujitsuka, M., and Majima, T. (2008) Real-time single-molecule imaging of the spatial and temporal distribution of reactive oxygen species with fluorescent probes: Applications to TiO_2 photocatalysts. *Journal of Physical Chemistry C*, **112**, 1048–1059.

[38] Moradi, K., Depecker, C., and Corset, J. (1994) Diffuse reflectance infrared spectroscopy: experimental study of nonabsorbing materials and comparisions with theories. *Applied Spectroscopy*, **48**, 1491–1497.

[39] Ocana, M., Fornes, V., Ramos, J.V.G., and Serna, C.J. (1988) Factors affecting the infrared and raman-spectra of rutile powders. *Journal of Solid State Chemistry*, **75**, 364–372.

[40] Gonzalez, R.J., Zallen, R., and Berger, H. (1997) Infrared reflectivity and lattice fundamentals in anatase TiO_2. *Physical Review B*, **55**, 7014–7017.

[41] Warren, D.S. and McQuillan, A.J. (2004) Influence of adsorbed water on phonon and UV-induced IR absorptions of TiO_2 photocatalytic particle films. *Journal of Physical Chemistry B*, **108**, 19373–19379.

[42] Gesenhues, U. (2007) The effects of plastic deformation on band gap, electronic defect states and lattice vibrations of rutile. *Journal of Physical Chemistry Solids*, **68**, 224–235.

[43] Yamakata, A., Ishibashi, T., and Onishi, H. (2001) Water- and oxygen-induced decay kinetics of photogenerated electrons in TiO_2 and Pt/TiO_2: A time-resolved infrared absorption study. *Journal of Physical Chemistry B*, **105**, 7258–7262.

[44] Szczepankiewicz, S.H., Moss, J.A., and Hoffmann, M.R. (2002) Slow surface charge trapping kinetics on irradiated TiO_2. *Journal of Physical Chemistry B*, **106**, 2922–2927.

[45] Dolamic, I. and Burgi, T. (2007) Photocatalysis of dicarboxylic acids over TiO_2: An *in situ* ATR-IR study. *Journal of Catalysis*, **248**, 268–276.

[46] Tolstoy, V.P., Chernyshova, I.V., and Skryshevsky, V.A. (2003) *Handbook of Infrared Spectroscopy of Ultrathin Films*, John Wiley & Sons, Hoboken.

[47] Ozensoy, E., Meier, D.C., and Goodman, D.W. (2002) Polarization modulation infrared reflection absorption spectroscopy at elevated pressures: CO adsorption on Pd(111) at atmospheric pressures. *Journal of Physical Chemistry B*, **106**, 9367–9371.

[48] Gerischer, H. and Heller, A. (1991) The role of oxygen in photooxidation of organic-molecules on semiconductor particles. *Journal of Physical Chemistry*, **95**, 5261–5267.

[49] Kormann, C., Bahnemann, D.W., and Hoffmann, M.R. (1988) Preparation and characterization of quantum-size titanium-dioxide. *Journal of Physical Chemistry*, **92**, 5196–5201.

[50] Monticone, S., Tufeu, R., Kanaev, A.V. *et al.* (2000) Quantum size effect in TiO_2 nanoparticles: does it exist? *Applied Surface Science*, **162**, 565–570.

[51] Serpone, N., Lawless, D., and Khairutdinov, R. (1995) Size effects on the photophysical properties of colloidal anatase TiO_2 particles – size quantization or direct transitions in this indirect semiconductor. *Journal of Physical Chemistry*, **99**, 16646–16654.

[52] Brus, L.E. (1984) Electron electron and electron-hole interactions in small semiconductor crystallites – the size dependence of the lowest excited electronic state. *Journal of Chemical Physics*, **80**, 4403–4409.

[53] Höfler, H.J., Hahn, H., and Averback, R.S. (1991) Diffusion in nanocrystalline materials. *Defect and Diffusion Forum*, **75**, 195–210.

[54] Serpone, N. and Emeline, A.V. (2002) Suggested terms and definitions in photocatalysis and radiocatalysis. *International Journal of Photoenergy*, **4**, 91–131.

[55] Brinkley, D. and Engel, T. (2000) Evidence for structure sensitivity in the thermally activated and photocatalytic dehydrogenation of 2-Propanol on TiO_2. *Journal of Physical Chemistry B*, **104**, 9836–9841.

[56] Morikawa, Y., Takahashi, I., Aizawa, M. *et al.* (2004) First-principles theoretical study and scanning tunneling microscopic observation of dehydration process of formic acid on a TiO_2(110) surface. *Journal of Physical Chemistry B*, **108**, 14446–14451.

[57] Wendt, S., Sprunger, P.T., Lira, E. *et al.* (2008) The role of interstitial sites in the Ti3d defect state in the band gap of titania. *Science*, **320**, 1755–1759.

[58] Gouadec, G. and Colomban, P. (2007) Raman Spectroscopy of nanomaterials: How spectra relate to disorder, particle size and mechanical properties. *Progress in Crystal Growth and Characterization of Materials*, **53**, 1–56.

[59] Hirata, T. (1998) Pressure, temperature, and concentration dependences of phonon frequency with variable Grüneisen parameter: fits to the raman-active E_g mode in TiO_2 and $Ti_{1(x}Zr_x$ ($x \leq 0.1$). *Physica Status Solidi (b)*, **209**, 17.

[60] Pottier, A.S., Cassaignon, S., Chaneac, C. *et al.* (2003) Size tailoring of TiO_2 anatase nanoparticles in aqueous medium and synthesis of nanocomposites. Characterization by Raman spectroscopy. *Journal of Materials Chemistry*, **13**, 877–882.

[61] Swamy, V. (2008) Size-dependent modifications of the first-order Raman spectra of nanostructured rutile TiO_2. *Physical Review B*, **77**, 195414.

[62] Swamy, V., Kuznetsov, A., Dubrovinsky, L.S. *et al.* (2005) Finite-size and pressure effects on the Raman spectrum of nanocrystalline anatase TiO_2. *Physical Review B*, **71**, 184302.

[63] Wang, D., Zhao, J., Chen, B., and Zhu, C. (2008) Lattice vibration fundamentals in nanocrystalline anatase investigated with Raman scattering. *Journal of Physics: Condensed Matter*, **20**, 085212, (7pp).

[64] Lejon, C. and Österlund, L. (2009) Influence of quantum confinement, hydrostatic pressure and Zr dopant concentration on the Raman vibrational properties of Zr doped anatase TiO_2 nanoparticles. *Physical Review B*, submitted.

[65] Richter, H., Wang, Z.P., and Ley, L. (1981) The one phonon Raman spectrum in microcrystalline silicon. *Solid State Communications*, **39**, 625–629.

[66] Balaji, S., Djaoued, Y., and Robichaud, J. (2006) Phonon confinement studies in nanocrystalline anatase-TiO_2 thin films by micro Raman spectroscopy. *Journal of Raman Spectroscopy*, **37**, 1416–1422.

[67] Bassi, A.L., Cattaneo, D., Russo, V. *et al.* (2005) Raman spectroscopy characterization of titania nanoparticles produced by flame pyrolysis: The influence of size and stoichiometry. *Journal of Applied Physics*, **98**, 074305.

[68] Bersani, D., Antonioli, G., Lottici, P.P., and Lopez, T. (1998) Raman study of nanosized titania prepared by sol-gel route. *Journal of Non-Crystalline Solids*, **232–234**, 175–181.

[69] Bersani, D., Lottici, P.P., and Ding, X.-Z. (1998) Phonon confinement effects in the Raman scattering by TiO_2 nanocrystals. *Applied Physics Letters*, **72**, 73–75.

[70] Hirata, T., Asahiri, E., and Kitjima, M. (1994) Infrared and Raman Spectroscopic Studies of ZrO_2 Polymorphs doped with Y_2O_3 or CeO_2. *Journal of Solid State Chemistry*, **110**, 201–207.

[71] Ivanda, M., Musić, S., Gotić, M. *et al.* (1999) The effects of crystal size on the Raman spectra of nanophase TiO_2. *Journal of Molecular Structure*, **480–481**, 641–644.

[72] Kelly, S., Pollak, F.H., and Tomkiewicz, M. (1997) Raman spectroscopy as a morphological probe for TiO_2 aerogels. *Journal of Physical Chemistry B*, **101**, 2730–2734.

[73] Ricci, P.C., Salis, M., and Anedda, A. (2008) Phonon characterization of nano-crystals by Raman spectroscopy. *Chemical Physics Letters*, **457**, 191–193.

[74] Zhang, W.F., He, Y.L., Zhang, M.S. *et al.* (2000) Raman scattering study on anatase TiO_2 nanocrystals. *Journal of Physics D: Applied Physics*, **33**, 912–916.

[75] Iida, Y., Furukawa, M., Aoki, T., and Sakai, T. (1998) Raman spectrum of ultrafine anatase powders derived from hydrolysis of alkoxide. *Applied Spectroscopy*, **52**, 673–678.

[76] Zhu, K.-R., Zhang, M.-S., Chen, Q., and Yin, Z. (2005) Size and phonon-confinement effects on low-frequency Raman mode of anatase TiO_2 nanocrystal. *Physics Letters A*, **340**, 220–227.

[77] Campbell, I.H. and Fauchet, P.M. (1986) The effects of microcrystal size and shape on the one phonon Raman spectra of crystalline semiconductors. *Solid State Communications*, **58**, 739–741.

[78] Mikami, M., Nakamura, S., Kitao, O., and Arakawa, H. (2002) Lattice dynamics and dielectric properties of TiO_2 anatase: A first-principles study. *Physical Review B*, **66**, 155213/1–6.

[79] Ohsaka, T., Izumi, F., and Fujiki, Y. (1978) Raman spectrum of anatase, TiO_2. *Journal of Raman Spectroscopy*, **7**, 321–324.

[80] Hearne, G.R., Zhao, J., Dawe, A.M. *et al.* (2004) Effect of grain size on structural transitions in anatase TiO_2: A Raman spectroscopy study at high pressures. *Physical Review B*, **70**, 134102.

[81] Choi, H.C., Jung, Y.M., and Kim, S.B. (2005) Size effects in the Raman spectra of TiO_2 nanoparticles. *Vibrational Spectroscopy*, **37**, 33–38.

[82] Turchi, C.S. and Ollis, D.F. (1990) Photocatalytic degradation of organic water contaminants: Mechanisms involving hydroxyl radical attack. *Journal of Catalysis*, **122**, 178–192.

[83] Ollis, D.F. and Al-Ekabi, H. (eds) (1993) *Photocatalytic Purification and Treatment of Water and Air*, Elsevier, Amsterdam.

[84] Fox, M.A. and Dulay, M.T. (1993) Heterogeneous photocatalysis. *Chemical Reviews*, **93**, 341–357.

[85] Hoffmann, M.R., Martin, S.T., Choi, W., and Bahnemann, D.W. (1995) Environmental applications of semiconductor photocatalysis. *Chemical Reviews*, **95**, 69–96.

[86] Mills, A. and Le Hunte, S. (1997) An overview of semiconductor photocatalysis. *Journal of Photochemistry and Photobiology A*, **108**, 1–35.

[87] Kaneko, M. and Okura, I. (eds) (2002) *Photocatalysis: Science and Technology*, Kodansha and Springer-Verlag, Tokyo.

[88] Fujishima, A., Zhang, X., and Tryk, D.A. (2008) TiO_2 photocatalysis and related surface phenomena. *Surface Science Reports*, **63**, 515–582.

[89] Peral, J., Domenech, X., and Ollis, D.F. (1997) Heterogeneous photocatalysis for purification, decontamination and deodorization of air. *Journal of Chemical Technology and Biotechnology*, **70**, 117–140.

[90] Nakajima, A., Hashimoto, K., Watanabe, T. *et al.* (2000) Transparent superhydrophobic thin films with self-cleaning properties. *Langmuir*, **16**, 7044–7047.

[91] Mills, A., Lepre, A., Elliott, N. *et al.* (2003) Characterization of photocatalyst Pilkington ActivTM: a reference film photocatalyst? *Journal of Photochemistry and Photobiology A*, **160**, 213–224.

[92] Österlund, L., Grant, A.W., and Kasemo, B. (2006) *Lithographic Techniques in Nanocatalysis*, Springer-Verlag.

[93] Hansen, P.L., Wagner, J.B., Helveg, S. *et al.* (2002) Atom-resolved imaging of dynamic shape changes in supported copper nanocrystals. *Science*, **295**, 2053–2055.

[94] Mills, A., Wang, J.S., and McGrady, M. (2006) Method of rapid assessment of photocatalytic activities of self-cleaning films. *Journal of Physical Chemistry B*, **110**, 18324–18331.

[95] Masel, R.I. (1996) *Principles of Adsorption and Reaction on Solid Surfaces*, John Wiley & Sons, New York.

[96] Henderson, M.A. (2008) Relationship of O_2 photodesorption in photooxidation of acetone on TiO_2. *Journal of Physical Chemistry C*, **112**, 11433–11440.

[97] Di Valentin, C., Tilocca, A., Selloni, A. *et al.* (2005) Adsorption of water on reconstructed rutile $TiO_2(011)$-(2×1): Ti = O double bonds and surface reactivity. *Journal of the American Chemical Society*, **127**, 9895–9903.

[98] Tilocca, A. and Selloni, A. (2004) Structure and reactivity of water layers on defect-free and defective anatase $TiO_2(101)$ surfaces. *Journal of Physical Chemistry B*, **108**, 4743–4751.

[99] Henderson, M.A. (1996) Structural sensitivity in the dissociation of water on TiO_2 single-crystal surfaces. *Langmuir*, **12**, 5093–5098.

[100] Henderson, M.A. (2004) Acetone chemistry on oxidized and reduced $TiO_2(110)$. *Journal of Physical Chemistry B*, **108**, 18932–18941.

[101] Kim, K.S. and Barteau, M.A. (1990) Structure and composition requirements for deoxygenation, dehydration, and ketonization reactions of carboxylic acids on $TiO_2(001)$ single crystals. *Journal of Catalysis*, **125**, 353–375.

[102] Hayden, B.E., King, A., and Newton, M.A. (1999) Fourier transform reflection-absorption IR spectroscopy study of formate adsorption on $TiO_2(110)$. *Journal of Physical Chemistry B*, **103**, 203–208.

[103] Rasmussen, M.D., Molina, L.M., and Hammer, B. (2004) Adsorption, diffusion, and dissociation of molecular oxygen at defected $TiO_2(110)$: A density functional theory study. *Journal of Chemical Physics*, **120**, 988–997.

[104] Wendt, S., Schaub, R., Matthiesen, J. *et al.* (2005) Oxygen vacancies on $TiO_2(110)$ and their interaction with H_2O and O_2: A combined high-resolution STM and DFT study. *Surface Science*, **598**, 226–245.

[105] Wu, X., Selloni, A., Lazzeri, M., and Nayak, S.K. (2003) Oxygen vacancy mediated adsorption and reactions of molecular oxygen on the $TiO_2(110)$ surface. *Physical Review B*, **68**, 241402.

[106] Uemura, Y., Taniike, T., Tada, M. *et al.* (2007) Switchover of reaction mechanism for the catalytic decomposition of HCOOH on a $TiO_2(110)$ surface. *Journal of Physical Chemistry C*, **111**, 16379–16386.

[107] Formenti, M., Juillet, F., Meriaudeau, P., and Teichner, S.J. (1971) Heterogeneous photocatalysis for partial oxidation of paraffins. *Chemical Technology*, **1**, 680.

[108] Djeghri, N., Formenti, M., Juillet, F., and Teichner, S.J. (1974) Photointeraction on the surface of titanium dioxide between oxygen and alkanes. *Faraday Special Discussions of the Chemical Society*, **58**, 185.

[109] Nakamoto, K. (1997) *Infrared and Raman Spectra of Inorganic and Coordination Compounds*, 5th edn, John Wiley & Sons, New York.

[110] Busca, G. and Lorenzelli, V. (1982) Infrared spectroscopic identification of species arising from reactive adsorption of carbon oxides on metal oxide surfaces. *Materials Chemistry*, **7**, 89–126.

[111] Sato, S. (1983) Photo-Kolbe reaction at gas-solid interfaces. *Journal of Physical Chemistry*, **87**, 3531–3537.

[112] Muggli, D.S., Keyser, S.A., and Falconer, J.L. (1998) Photocatalytic decomposition of acetic acid on TiO_2. *Catalysis Letters*, **55**, 129–132.

[113] Muggli, D.S. and Falconer, J.L. (1999) Parallel pathways for photocatalytic decomposition of acetic acid on TiO_2. *Journal of Catalysis*, **187**, 230–237.

[114] Andersson, M., Kiselev, A., Österlund, L., and Palmqvist, A.E.C. (2007) Microemulsion-mediated room temperature synthesis of high surface area rutile and its photocatalytic performance. *Journal of Physical Chemistry C*, **111**, 6789.

[115] Yu, Z.Q. and Chuang, S.S.C. (2007) In situ IR study of adsorbed species and photogenerated electrons during photocatalytic oxidation of ethanol on TiO_2. *Journal of Catalysis*, **246**, 118–126.

[116] Wang, L.-Q., Ferris, K.F., Shultz, A.N. *et al.* (1997) Interactions of HCOOH with stoichiometric and defective TiO_2(110) surfaces. *Surface Science*, **380**, 352–364.

[117] Onishi, H., Aruga, T., and Iwasawa, Y. (1994) Switchover of reaction paths in the catalytic decomposition of formic acid on TiO_2(110) surface. *Journal of Catalysis*, **146**, 557–567.

[118] Nilsing, M., Lunell, S., Persson, P., and Ojamae, L. (2005) Phosphonic acid adsorption at the TiO_2 anatase (101) surface investigated by periodic hybrid HF-DFT computations. *Surface Science*, **582**, 49–60.

[119] Henderson, M.A. (1997) Complexity in the decomposition of formic acid on the TiO_2(110) surface. *Journal of Physical Chemistry B*, **101**, 221–229.

[120] Chambers, S.A., Henderson, M.A., and Kim, Y.J. (1998) Chemisorption geometry, vibrational spectra, and thermal desorption of formaic acid on TiO_2(110). *Surface Review and Letters*, **5**, 381–385.

[121] Uetsuka, H., Henderson, M.A., Sasahara, A., and Onishi, H. (2004) Formate adsorption on the (111) surface of rutile TiO_2. *Journal of Physical Chemistry B*, **108**, 13706–13710.

[122] Idriss, H., Lusvardi, V.S., and Barteau, M.A. (1996) Two routes to formaldehyde from formic acid on TiO_2(001) surfaces. *Surface Science*, **348**, 39–48.

[123] Gutiérrez-Sosa, A., Martínez-Escolano, P., Raza, H. *et al.* (2001) Orientation of carboxylates on TiO_2(110). *Surface Science*, **471**, 163–169.

[124] Bates, S.P., Kresse, G., and Gillan, M.J. (1998) The adsorption and dissociation of ROH molecules on TiO_2(110). *Surface Science*, **409**, 336–349.

[125] Bowker, M., Stone, P., Bennett, R., and Perkins, N. (2002) Formic acid adsorption and decomposition on TiO_2(110) and on Pd/TiO_2(110) model catalysts. *Surface Science*, **511**, 435–448.

[126] Li, W.K., Gong, X.Q., Lu, G., and Selloni, A. (2008) Different reactivities of TiO_2 polymorphs: Comparative DFT calculations of water and formic acid adsorption at anatase and brookite TiO_2 surfaces. *Journal of Physical Chemistry C*, **112**, 6594–6596.

[127] Vittadini, A., Selloni, A., Rotzinger, F.P., and Gratzel, M. (2000) Formic acid adsorption on dry and hydrated TiO_2 anatase (101) surfaces by DFT calculations. *Journal of Physical Chemistry B*, **104**, 1300–1306.

[128] Gong, X.Q., Selloni, A., and Vittadini, A. (2006) Density functional theory study of formic acid adsorption on anatase TiO_2(001): Geometries, energetics, and effects of coverage, hydration, and reconstruction. *Journal of Physical Chemistry B*, **110**, 2804–2811.

[129] Idriss, H. and Barteau, M.A. (2000) *Advanced Catalysis*, vol. **45** (eds B.C. Gates and H. Knozinger), Academic Press.

[130] Groff, R.P. and Manogue, W.H. (1983) An infrared study of formate formation and reactivity on TiO_2 surfaces. *Journal of Catalysis*, **79**, 462–465.

[131] Rotzinger, F.P., Kesselman-Truttman, J.M., Hug, S.J. *et al.* (2004) Structure and vibrational spectrum of formate and acetate adsorbed from aqueous solution onte the TiO_2 rutile (110) surface. *Journal of Physical Chemistry B*, **108**, 5004–5017.

[132] Brownson, J.R.S., Tejedor-Tejedor, M.I., and Anderson, M.A. (2006) FTIR spectroscopy of alcohol and formate interactions with mesoporous TiO_2 surfaces. *Journal of Physical Chemistry B*, **110**, 12494–12499.

[133] Chen, T., Wu, G.P., Feng, Z.C. *et al.* (2008) *In situ* FT-IR study of photocatalytic decomposition of formic acid to hydrogen on Pt/TiO_2 catalyst. *Chinese Journal of Catalysis*, **29**, 105–107.

[134] Deacon, G.B. and Phillips, R.J. (1980) Relationships between the carbon-oxygen stretching frequencies of carboxylates and the type of carboxylate coordination. *Coordination Chemistry Reviews*, **33**, 227–250.

[135] Capecchi, G., Faga, M.G., Martra, G. *et al.* (2007) Adsorption of CH_3COOH on TiO_2: IR and theoretical investigations. *Research on Chemical Intermediates*, **33**, 269–284.

[136] Irokawa, Y., Morikawa, T., Aoki, K. *et al.* (2006) Photodegradation of toluene over $TiO_{2-x}N_x$ under visible light irradiation. *Physical Chemistry Chemical Physics*, **8**, 1116–1121.

[137] Iwasawa, Y., Onishi, H., Fukui, K. *et al.* (1999) The selective adsorption and kinetic behaviour of molecules on TiO_2(110) observed by STM and NC-AFM. *Faraday Discussions*, **114**, 259–266.

[138] Popova, G.Y., Andrushkevich, T.V., Chesalov, Y.A., and Stoyanov, E.S. (2000) *In situ* FTIR study of the adsorption of formaldehyde, formic acid, and methyl formiate at the surface of TiO_2 (Anatase). *Kinetics and Catalysis*, **41**, 805–811.

[139] Ferri, D., Burgi, T., and Baiker, A. (2002) Probing catalytic solid-liquid interfaces by attenuated total reflection infrared spectroscopy: Adsorption of carboxylic acids on alumina and titania. *Helvetica Chimica Acta*, **85**, 3639–3656.

[140] Li, Y.X., Lu, G.X., and Li, S.B. (2003) Photocatalytic production of hydrogen in single component and mixture systems of electron donors and monitoring adsorption of donors by in situ infrared spectroscopy. *Chemosphere*, **52**, 843–850.

[141] Österlund, L., Mattsson, A., and Andersson, P.O. (2009) What makes a good TiO_2 photocatalyts?, in *Nanostructured Materials and Nanotechnology II: Ceramic Engineering and Science Proceedings*, vol. **29** (eds S. Mathur and M. Singh), Wiley & Sons.

[142] Byun, D., Jin, Y., Kim, B. *et al.* (2000) Photocatalytic TiO_2 deposition by chemical vapor deposition. *Journal of Hazardous Materials*, **73**, 199–206.

[143] Ohno, T., Sarukawa, K., and Matsumura, M. (2002) Crystal faces of rutile and anatase TiO_2 particles and their roles in photocatalytic reactions. *New Journal of Chemistry*, **26**, 1167–1170.

[144] Chu, S.Z., Inoue, S., Wada, K. *et al.* (2003) Highly porous TiO_2/Al_2O_3 composite nanostructures on glass by anodization and the sol-gel process: fabrication and photocatalytic characteristics. *Journal of Materials Chemistry*, **13**, 866–870.

[145] Taguchi, T., Saito, Y., Sarukawa, K. *et al.* (2003) Formation of new crystal faces on TiO_2 particles by treatment with aqueous HF solution or hot sulfuric acid. *New Journal of Chemistry*, **27**, 1304–1306.

[146] Tokita, S., Tanaka, N., Ohshio, S., and Saitoh, H. (2003) Photo-induced surface reaction of highly oriented anatase polycrystalline films synthesized using a CVD apparatus operated in atmospheric regime. *Journal of the Ceramic Society Japan*, **111**, 433–435.

[147] Kim, B., Byun, D., Lee, J.K., and Park, D. (2002) Structural analysis on photocatalytic efficiency of TiO_2 by chemical vapor deposition. *Japanese Journal of Applied Physics*, **41**, 222–226.

[148] Arana, J., Cabo, C.G.I., Dona-Rodriguez, J.M. *et al.* (2004) FTIR study of formic acid interaction with TiO_2 and TiO_2 doped with Pd and Cu in photocatalytic processes. *Applied Surface Science*, **239**, 60–71.

[149] El-Maazawi, M., Finken, A.N., Nair, A.B., and Grassian, V.H. (2000) Adsorption and photocatalytic oxidation of acetone on TiO_2: An *in situ* transmission FT-IR study. *Journal of Catalysis*, **191**, 138–146.

[150] Kraeutler, B. and Bard, A.J. (1978) Heterogeneous photocatalytic decomposition of saturated carboxylic-acids on TiO_2 powder – decarboxylative route to alkanes. *Journal of the American Chemical Society*, **100**, 5985–5992.

[151] Liu, H., Imanishi, A., and Nakato, Y. (2007) Mechanisms for photooxidation reactions of water and organic compounds on carbon-doped titanium dioxide, as studied by photocurrent measurements. *Journal of Physical Chemistry C*, **111**, 8603–8610.

[152] Ishibashi, K.-i., Nosaka, Y., Hashimoto, K., and Fujishima, A. (1998) Time-dependent behavior of active oxygen species formed on photoirradiated TiO_2 films in air. *Journal of Physical Chemistry B*, **102**, 2117–2120.

[153] Bickley, R.I. and Stone, F.S. (1973) Photoadsorption and photocatalysis at rutile surfaces. I Photoadsorption of oxygen. *Journal of Catalysis*, **31**, 389.

[154] Wang, C., Rabani, J., Bahnemann, D.W., and Dohrmann, J. (2002) Photonic efficiency and quantum yield of formaldehyde formation from methanol in the presence of various TiO_2 photocatalysts. *Journal of Photochemistry and Photobiology A*, **148**, 169–176.

[155] Jegat, C., Fouassier, M., Tranquille, M., and Mascetti, J. (1991) Carbon dioxide coordination chemistry. 2. Synthesis and FTIR study of $CpTi(CO_2)(PMe_3)$. *Inorganic Chemistry*, **30**, 1529–1536.

[156] Arana, J., Diaz, O.G., Saracho, M.M. *et al.* (2001) Photocatalytic degradation of formic acid using Fe/TiO_2 catalysts: the role of Fe^{3+}/Fe^{2+} ions in the degradation mechanism. *Applied Catalysis B-Environmental*, **32**, 49–61.

[157] Wetchakun, N., Chiang, K., Amal, R., and Phanichphant, S. (2008) Synthesis and characterization of transition metal ion doping on the photocatalytic activity of TiO_2 nanoparticles. Proceedings of the 2nd IEEE International Nanoelectronics Conference, January 3–8, 2008, Hong Kong, 1–3, 43–47.

[158] Arana, J., Diaz, O.G., Rodriguez, J.M.D. *et al.* (2003) Role of Fe^{3+}/Fe^{2+} as TiO_2 dopant ions in photocatalytic degradation of carboxylic acids. *Journal of Molecular Catalysis A-Chemistry*, **197**, 157–171.

[159] Umebayashi, T., Yamaki, T., Itoh, H., and Asai, K. (2002) Analysis of electronic structures of 3d transition metal-doped TiO_2 based on band calculations. *Journal of Physical Chemistry Solids*, **63**, 1909–1920.

[160] Randeniya, L.K., Bendavid, A., Martin, P.J., and Preston, E.W. (2007) Photoelectrochemical and structural properties of TiO_2 and N-doped TiO_2 thin films synthesized using pulsed direct current plasma-activated chemical vapor deposition. *Journal of Physical Chemistry C*, **111**, 18334–18340.

[161] Morikawa, T., Ohwaki, T., Suzuki, K.-i. *et al.* (2008) Visible-light-induced photocatalytic oxidation of carboxylic acids and aldehydes over N-doped TiO_2 loaded with Fe, Cu or Pt. *Applied Catalysis B*, **83**, 56–62.

[162] Balcerski, W., Ryu, S.Y., and Hoffmann, M.R. (2007) Visible-light photoactivity of nitrogen-doped TiO_2: Photo-oxidation of HCO_2H to CO_2 and H_2O. *Journal of Physical Chemistry C*, **111**, 15357–15362.

[163] Kisch, H., Sakthivel, S., Janczarek, M., and Mitoraj, D. (2007) A low-band gap, nitrogen-modified titania visible-light photocatalyst. *Journal of Physical Chemistry C*, **111**, 11445–11449.

[164] Mrowetz, M., Balcerski, W., Colussi, A.J., and Hoffmann, M.R. (2004) Oxidative power of nitrogen-doped TiO_2 photocatalysts under visible illumination. *Journal of Physical Chemistry B*, **108**, 17269–17273.

[165] Neumann, B., Bogdanoff, P., Tributsch, H. *et al.* (2005) Electrochemical mass spectroscopic and surface photovoltage studies of catalytic water photooxidation by undoped and carbon-doped titania. *Journal of Physical Chemistry B*, **109**, 16579–16586.

[166] Emeline, A.V., Sheremetyeva, N.V., Khomchenko, N.V. *et al.* (2007) Photoinduced formation of defects and nitrogen stabilization of color centers in n-doped titanium dioxide. *Journal of Physical Chemistry C*, **111**, 11456–11462.

[167] Tian, F. and Liu, C. (2006) DFT description on electronic structure and optical absorption properties of anionic s-doped anatase TiO_2. *Journal of Physical Chemistry B*, **110**, 17866–17871.

[168] Attwood, A.L., Edwards, J.L., Rowlands, C.C., and Murphy, D.M. (2003) Identification of a surface alkylperoxy radical in the photocatalytic oxidation of acetone/O_2 over TiO_2. *Journal of Physical Chemistry A*, **107**, 1779–1782.

[169] Rekoske, J.E. and Barteau, M.A. (1995) *In-situ* studies of carbonyl coupling – comparisons of liquid-solid and gas-solid reactions with reduced titanium reagents. *Industrial and Engineering Chemistry Research*, **34**, 2931–2939.

[170] Griffith, D.M. and Rochester, C. (1978) Infrared study of the adsorption of acetone on rutile. *Journal of the Chemical Society, Faraday Transactions I*, **74**, 403–417.

[171] Zaki, M.I., Hasan, M.A., and Pasupulety, L. (2001) Surface reactions of acetone on Al_2O_3, TiO_2, ZrO_2, and CeO_2: IR spectroscopic assignment of impacts of the surface acid-base properties. *Langmuir*, **17**, 768–774.

[172] Xu, W., Raftery, D., and Francisco, J.S. (2003) Effect of irradiation sources and oxygen concentration on the photocatalytic oxidation of 2-propanol and acetone studied by *in situ* FTIR. *Journal of Physical Chemistry B*, **107**, 4537–4544.

[173] Vorontsov, A.V., Kurkin, E.N., and Savinov, E.N. (1999) Study of TiO_2 deactivation during gaseous acetone photocatalytic oxidation. *Journal of Catalysis*, **186**, 318–324.

[174] Henderson, M.A. (2005) Photooxidation of acetone on $TiO_2(110)$: conversion to acetate via methyl radical ejection. *Journal of Physical Chemistry B*, **109**, 12062–12070.

[175] Henderson, M.A. (2005) Acetone and water on $TiO_2(110)$: competion for sites. *Langmuir*, **21**, 3443–3450.

[176] Henderson, M.A. (2005) Acetone and water on $TiO_2(110)$: H/D exchange. *Langmuir*, **21**, 3451–3458.

[177] Henderson, M.A., White, J.M., Uetsuka, H., and Onishi, H. (2003) Photochemical charge transfer and trapping at the interface between an organic adlayer and an oxide semiconductor. *Journal of the American Chemical Society*, **125**, 14974–14975.

[178] Du, Y., Dohnalek, Z., and Lyubinetsky, I. (2008) Transient mobility of oxygen adatoms upon O_2 dissociation on reduced $TiO_2(110)$. *Journal of Physical Chemistry C*, **112**, 2649–2653.

[179] Thompson, T. and Yates, J. (2005) TiO_2-based photocatalysis: surface defects, oxygen and charge transfer. *Topics in Catalysis*, **35**, 197–210.

[180] Lv, C., Wang, X., Agalya, G. *et al.* (2005) Photocatalytic oxidation dynamics of acetone on TiO_2: tight-binding quantum chemical molecular dynamics study. *Applied Surface Science*, **244**, 541–545.

[181] Zehr, R.T. and Henderson, M.A. (2008) Acetaldehyde photochemistry on $TiO_2(110)$. *Surface Science*, **602**, 2238–2249.

[182] Wilson, J.N. and Idriss, H. (2003) Effect of surface reconstruction of TiO_2(001) single crystal on the photoreaction of acetic acid. *Journal of Catalysis*, **214**, 46–52.

[183] Arsac, F., Bianchi, D., Chovelon, J.M. *et al.* (2006) Experimental microkinetic approach of the photocatalytic oxidation of isopropyl alcohol on TiO_2. Part 1. Surface elementary steps involving gaseous and adsorbed C_3H_xO species. *Journal of Physical Chemistry A*, **110**, 4202–4212.

[184] Kozlov, D., Bavykin, D., and Savinov, E. (2003) Effect of the acidity of TiO_2 surface on its photocatalytic activity in acetone gas-phase oxidation. *Catalysis Letters*, **86**, 169–172.

[185] Reyes-Garcia, E.A., Sun, Y., Reyes-Gil, K., and Raftery, D. (2007) ^{15}N solid state NMR and EPR characterization of N-doped TiO_2 photocatalysts. *Journal of Physical Chemistry C*, **111**, 2738–2748.

[186] Österlund, L., Mattsson, A., Leideborg, M., and Westin, G. (2007) Photodecomposition of acetone on ZrO_2-TiO_2 thin films in O_2 excess and deficit conditions, in *Nanostructured Materials and Nanotechnolog: Ceramic Engineering and Science Proceedings*, vol. **28** (eds S. Mathur and M. Singh), Wiley & Sons.

[187] Mwabora, J.M., Lindgren, T., Avendaño, E. *et al.* (2004) Structure, composition, and morphology of photoelectrochemically active $TiO_{2-x}N_x$ thin films deposited by reactive DC magnetron sputtering. *Journal of Physical Chemistry B*, **108**, 20193–20198.

[188] Tanaka, K. and White, J.M. (1982) Characterisation of species adsorbed on oxidized and reduced anatase. *Journal of Physical Chemistry*, **86**, 4708–4714.

[189] Brownson, J.R.S., Tejedor-Tejedor, M.I., and Anderson, M.A. (2005) Photoreactive anatase consolidation characterized by FTIR Spectroscopy. *Chemistry of Materials*, **17**, 6304–6310.

[190] Dobson, K.D. and McQuillan, A.J. (1997) An infrared spectroscopic study of carbonate adsorption to zirconium dioxide sol-gel films from aqueous solutions. *Langmuir*, **13**, 3392–3396.

9

Interfacial Electron Transfer Reactions in CdS Quantum Dot Sensitized TiO_2 Nanocrystalline Electrodes

Yasuhiro Tachibana

Department of Applied Chemistry, Graduate School of Engineering, Osaka University 2-1 Yamada-oka, Suita, Osaka 565-0871, Japan,
Email: y.tachibana@chem.eng.osaka-u.ac.jp

Center for Advanced Science and Innovation (CASI), Osaka University 2-1 Yamada-oka, Suita, Osaka 565-0871, Japan

School of Aerospace, Mechanical and Manufacturing Engineering, RMIT University Bundoora Campus, Building 251 level 3, Plenty Road, Bundoora VIC 3083, Melbourne, Australia

PRESTO, Japan Science and Technology Agency (JST), 4-1-8 Honcho Kawaguchi, Saitama 332-0012, Japan

9.1 Introduction

Light-driven electron transfer is a primary fundamental chemical process, converting light energy into chemical energy or electricity [1,2]. With increasing global pressure on renewable energy, chemistry-based solar energy conversion systems have become one of the most attractive energy sources. Although these are produced at low cost, as yet, there is one major drawback, which is fundamental to their progression and development, that the overall efficiency is extremely low.

The strategy of employing material design, at the nanometer level, has allowed this efficiency issue to be overcome, drastically improving solar energy conversion [2]. Thus, the key to efficiency improvements is nanomaterial design; this includes the necessary parameters

On Solar Hydrogen & Nanotechnology Edited by Lionel Vayssieres

to control the forward electron-transfer reaction, that is, charge separation. There are two main types of material developments promoting photoinduced charge-separation reactions; the first relates to the development of visible-light-sensitive semiconductors, and the second employs sensitization of wide band-gap semiconductors. The former can absorb light at visible-light wavelengths; here the generated electron and hole are separated by the built-in electric field inside the material. Intensive studies have been performed, particularly for water-splitting reactions, to improve the efficiency [3–6], although there is still further research required, with respect to the light sensitivity at longer wavelengths.

Sensitization of wide bandgap semiconductors has recently generated considerable interest for photovoltaic and water-splitting applications [7,8]. Dye sensitized metal oxide nanocrystalline solar cells are the most attractive devices to generate electricity from the sun. Recent development of these solar cells has reached solar-energy conversion efficiencies of >11% [9,10]. In the sensitization systems, the sensitizer and the metal oxide play different and distinguishable roles. The sensitizer absorbs visible light, followed by electron injection from the excited sensitizer to the conduction band of the metal-oxide semiconductor. The injected electron can diffuse away from the interface, and finally be converted to either electricity or hydrogen. In contrast to the visible-light-sensitive semiconductors, the sensitizer does not transport charges, but absorbs light, participating in electron-transfer reactions. The overall efficiency is controlled by the light harvesting efficiency (LHE) of the sensitizer and electron-transfer efficiencies at the interfaces. Therefore, material choice and interfacial design are vital to maximize the efficiency.

Recently, semiconductor quantum dots (QDs) or semiconductor nanocrystals have been considered as sensitizer materials in solar cells [11–24]. QD sensitization can also be employed for water-splitting reactions. In this chapter, the concept of sensitization of wide bandgap metal-oxide semiconductors by QDs is described, with further in-depth discussions on interfacial nanostructure design to improve the efficiency.

9.2 Nanomaterials

9.2.1 Semiconductor Quantum Dots

Metal chalcogenide QDs (e.g., CdS, CdSe, CdTe, PbS, PbSe, Ag_2S) have received broad attention due to their diverse applications in various fields, such as photoluminescence [25], biomedical sensors [26] and solar cells [22]. QDs also possess unique electronic and optical characteristics, which are rarely recognized in organic compounds. Of particular importance is modulation of the bandgap energy with size reduction to less than 10 nm in dimension, regardless of the chemical composition, that is, the smaller the size, the larger the bandgap. Such characteristics are termed "quantum size effects" and are shown in Figure 9.1. The other characteristics relating to quantum size effects are summarized as follows:

1. Conduction and valence-band potentials can be modulated by controlling the size
2. QDs can absorb light with energy higher than the bandgap energy
3. QDs exhibit large absorption coefficients ($>10^5$ dm^3 mol particle^{-1} cm^{-1})
4. Multiple exciton (electron–hole pair) generation can be achieved by one photon absorption, generating a concept of enhanced quantum yield of >100%
5. Exciton relaxation dynamics are slower, possibly allowing a hot charge-transfer reaction.

Note: Process 4 has not yet been confirmed experimentally.

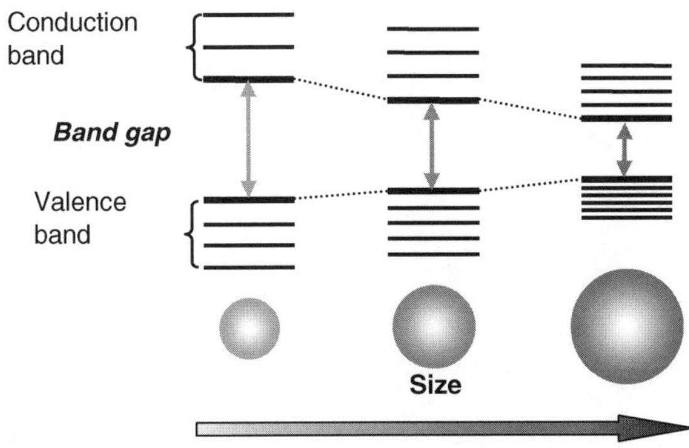

Figure 9.1 Relationship between QD size and bandgap energy.

These characteristics are extremely attractive, allowing QDs to be used as light absorbers, with control of the QD size control tuning an absorption wavelength range [27,28]. Band-edge tuning is also attractive for solar-energy-conversion devices; the donor and acceptor band levels are adjustable to favor the thermodynamics involved in interfacial electron-transfer reactions. Moreover, as shown in points 4 and 5, described above, novel physical concepts, that can be adopted in the devices, have emerged. Exciton multiplication [29] and hot electron injection [2] processes, by controlling excited state relaxation and charge transfer dynamics, have been reported.

9.2.2 Metal Oxide Nanocrystalline Semiconductor Films

Wide bandgap metal-oxide semiconductors, such as titanium dioxide (TiO_2), zinc oxide (ZnO) and tin oxide (SnO_2), are generally employed as electron acceptor materials since most of them possess an n-type character. TiO_2 is one of the most attractive metal oxides, as devices employing TiO_2, such as dye-sensitized solar cells [7], exhibit a high quantum yield, that is, efficient interfacial electron transfer occurs. As described in the previous section, both LHE and electron-transfer efficiency contribute to the efficiency improvement. Thus, in order to increase the efficiency further, an increase in surface area for light absorption was required. For photocatalytic reactions, metal-oxide nanoparticles have been developed and characterized. In contrast, for photoelectrochemical applications, recent developments have introduced nanoporous structures for the metal-oxide film. Figure 9.2 shows an example of an FE-SEM (field emission scanning electron microscopy) image of the TiO_2 nanoporous film. The film was prepared by screen printing a commercially available TiO_2 nanoparticle paste onto a slide glass, followed by a calcination process at 500 °C. The prepared film consists of TiO_2 nanoparticles with an average diameter of 15 nm, enhancing the surface area by three orders of magnitude compared to the projected area. In addition, the film becomes optically transparent owing to less light-scattering with the decrease in nanoparticle size, facilitating optical spectroscopic studies. Moreover, the calcination process introduces electronic conduction between the nanoparticles, and thus the film can be employed as a semiconducting electrode.

Figure 9.2 FE-SEM image of a nanocrystalline TiO_2 film.

9.2.3 QD Sensitized Metal Oxide Semiconductor Films

Sensitization of wide band-gap semiconductors, for example TiO_2, ZnO and SnO_2, by the QDs is one of the most advantageous structures for solar-energy conversion devices [11,17,21,23,24,30,31]. Figure 9.3 shows a photograph of a CdS QD-sensitized TiO_2 nanocrystalline film (size: $5 \times 5\,mm^2$). The small dot reflects adsorption of CdS QDs on the TiO_2 surface. FE-SEM and X-ray diffraction (XRD) measurements revealed a QD diameter of $2 \sim 4\,nm$.

The advantages of employing QD sensitization is that the potential energy levels for QD conduction and valence bands can be controlled by adjusting the QD size, that is, using quantum size effects [27,28], and thus, an appropriate size can be tuned to optimize efficient electron injection and slow charge recombination between a QD and a metal oxide. For this tuning, intensive studies have been performed to investigate the influence of QD type on the photovoltaic performance, for example, CdS [14,15,32], CdSe [13], PbS [23] and InAs [16].

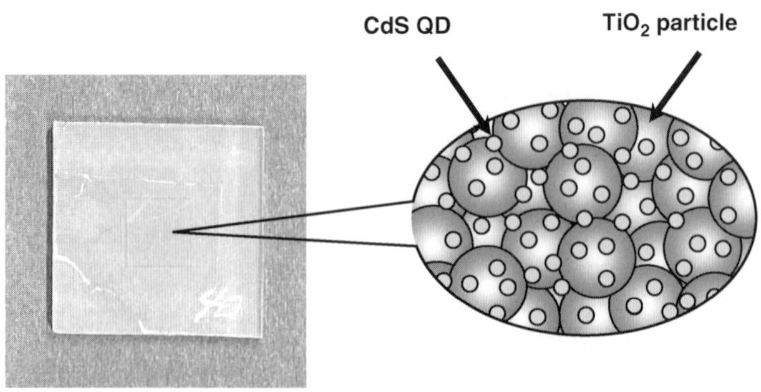

Figure 9.3 Photograph and illustration of a CdS QD-sensitized TiO_2 film.

Indeed, photocurrent response corresponds to an absorption spectrum of the dot, thereby indicating the possibility of increasing photocurrent when a low bandgap QD is employed.

The sensitization has mainly been achieved by two methods: by dipping a metal oxide nanoparticle film in a presynthesized QD solution, and by *in situ* growth of QDs or a thin film on the surface of a metal-oxide nanoparticle film. The advantage of the former method is that precise QD size control can be realized with recently developed synthetic routes in solution phase [33,34]. In contrast, facile simultaneous synthesis and sensitization can be achieved by the latter method, although the size control is relatively limited [14]. In this section, the characterization of CdS QD-sensitized TiO_2 films prepared by the latter method are discussed in detail.

CdS deposition inside the nanocrystalline film was performed using the SILAR, Successive Ionic Layer Adsorption and Reaction, technique [15,35,36]. The film was dipped in 0.1 M $Cd(ClO_4)_2$, and subsequently in 0.1 M Na_2S aqueous solutions at room temperature. After each dipping procedure, the film was rinsed with distilled water. This sequential coating was repeated 5 ~ 25 cycles for the TiO_2 film (CdS-5~25/TiO_2).

9.2.3.1 Absorption Spectra

Absorption spectra of the transparent TiO_2 films coated repeatedly by Cd^{2+} and S^{2-} are shown in Figure 9.4. The absorption edge was shifted to a longer wavelength and the absorption amplitude gradually increased with the coating cycles, consistent with the results reported by Weller *et al.* [32]. The red-shift of the absorption edge is owing to the quantum size effect, as described previously [15,32]. This absorbance increase was analyzed by plotting the absorbance at 400 nm, after subtraction of a baseline obtained for a TiO_2 film alone, as a function of the number of CdS coating cycles, shown in the inset of Figure 9.4. The linear relationship

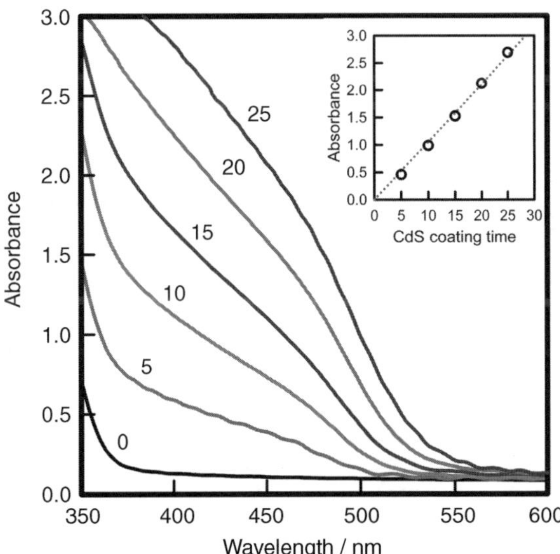

Figure 9.4 Absorption spectra of CdS/TiO_2 films. The number corresponds to CdS coating cycles. The inset shows a linear relationship between the CdS coating cycles and the absorbance at 400 nm.

suggests that the amount of CdS on the TiO_2 quantitatively increases with the number of coating cycles.

9.2.3.2 XRD Spectra

The *in situ* chemical-bath deposited CdS on the TiO_2 film was also characterized by XRD measurements. Figure 9.5 presents X-ray diffraction patterns of the CdS-5 ~ 25/TiO_2. The intense sharp peaks observed for the TiO_2 film alone were assigned typically to an anatase structure, in agreement with the previous work [37]. With the CdS coating, three new broad peaks appeared at 26, 44 and 52°, suggesting formation of CdS nanocrystals. Extracting the CdS peaks from the CdS-5 ~ 25/TiO_2, using the pattern of the TiO_2 film alone, revealed that the CdS nanocrystal actually indicates a cubic structure [38], assigned by JCPDS (Joint Committee on Powder Diffraction Standard) (10–454). Interestingly, the peak amplitude gradually grew as the coating was repeated. The nanocrystal size was estimated by fitting the line widths ((111), (220) and (311) faces) of the extracted CdS XRD spectra with a Gaussian function, using the Debye–Scherrer formula. The relationship between the mean CdS nanocrystal diameter and the number of coating cycles is shown in Figure 9.6. For <15 coating cycles, the crystallinity diameter increases almost linearly. In contrast, the diameter remains unchanged with >15 coating cycles, although the CdS amount increases linearly in this range, as indicated in Figure 9.4. These observations therefore suggest that the average number of CdS nanocrystals with a diameter of approximately 4 nm, increases with >15 coating cycles.

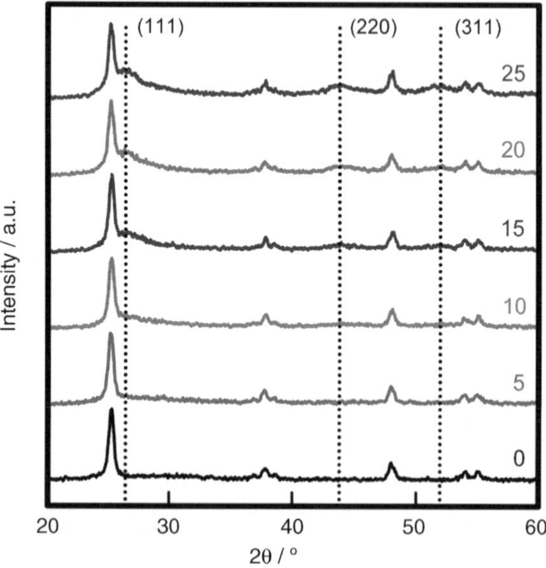

Figure 9.5 X-ray diffraction patterns of CdS/TiO_2 nanocrystalline films prepared by the SILAR technique. The number on the right-hand side corresponds to the CdS coating cycles.

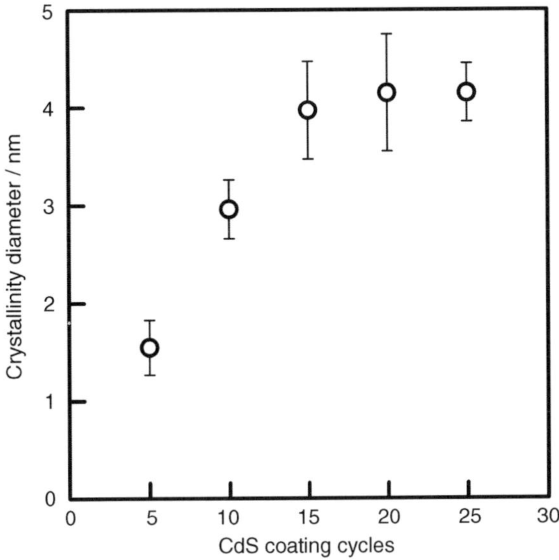

Figure 9.6 CdS crystalline diameter as a function of number of CdS coating cycles.

9.3 Transient Absorption Spectroscopy

9.3.1 Principle

The development of transient absorption spectroscopy driven by the improvement in high-power pulsed-laser systems has allowed time-resolved observation of physical and chemical reactions, since the technique of flash photolysis was introduced by Professor George Porter in the 1950s [39]. The principle of submicrosecond to millisecond transient absorption spectroscopy is that a sample is irradiated with a short, intense laser pulse known as a "pump" pulse, and then resultant changes in the optical density of the sample are monitored by a continuous probe beam. The time resolution is generally limited by the response time of the detection electronics. Reaction rates are determined by comparing transient absorption decays obtained from plotting the amplitude of transmitted light against time and by comparing transient spectra obtained from plotting the amplitude against probe wavelengths.

9.3.2 Calculation of Absorption Difference

Physical and chemical reactions are monitored by measuring the intensity of transmitted probe light passing through a sample after excitation, and presented as differences in optical density (absorption difference or ΔOD). This ΔOD is simply the difference in optical density before and after optical excitation of the sample in transient absorption spectroscopy. The optical density (OD) of the sample is defined by the Lambert–Beer Law in Equation 9.1:

$$\mathrm{OD}(\lambda) = \log_{10}\left(\frac{I_\mathrm{i}(\lambda)}{I_\mathrm{o}(\lambda)}\right) = \varepsilon(\lambda)cl, \tag{9.1}$$

where I_i is the incident light intensity, I_o is the transmitted light intensity, λ is the probe wavelength, ε is the extinction coefficient of the absorbing species at a probe wavelength, c is the concentration of the sample and l is the optical path length of the sample. Using Equation 9.2, ΔOD is given as:

$$\Delta\mathrm{OD}(\lambda, t) = \mathrm{OD}_{\mathrm{pump}}(\lambda, t) - \mathrm{OD}_{\mathrm{unpump}}(\lambda) = \log_{10}\left(\frac{I_i}{I_{o,\mathrm{pump}}(\lambda, t)}\right) - \log_{10}\left(\frac{I_i}{I_{o,\mathrm{unpump}}(\lambda)}\right) \tag{9.2}$$

Equation 9.2 is simplified to Equation 9.3:

$$\Delta\mathrm{OD}(\lambda, t) = \log_{10}\left(\frac{I_{o,\mathrm{unpump}}(\lambda)}{I_{o,\mathrm{pump}}(\lambda, t)}\right), \tag{9.3}$$

where $\mathrm{OD}_{\mathrm{pump}}(t)$ and $I_{o,\mathrm{pump}}(t)$ are the optical density and transmitted light intensity respectively at a delay time, t, after excitation. $\mathrm{OD}_{\mathrm{unpump}}$ and $I_{o,\mathrm{unpump}}$ are the optical density and transmitted light intensity without excitation of the sample.

9.3.3 System Arrangement

The arrangement of the apparatus is illustrated in Figure 9.7. Submicrosecond to millisecond transient absorption studies were conducted with a Nd/YAG laser (Spectra Physics, Quanta-Ray GCR-11) pumped dye laser (Usho Optical Systems, DL-100, ~10 ns pulse duration) as a pump source, a 100 W tungsten lamp as a probe source, a photodiode-based detection system (Costronics Electronics) and a TDS-2022 Tektronix oscilloscope [40]. The excitation wavelength can be tuned with this dye laser system. The probe light wavelength from a tungsten lamp was selected by an appropriate monochromator and by a color filter to minimize the probe light incident upon the sample. Another monochromator and a color filter after the sample were employed to reduce emission and pump laser scatter from the sample.

All transient data were collected by confirming identical excitation homogeneity and density of absorbed photons for all samples. An excitation wavelength was selected to 425, 480, 500, 515 and 520 nm for CdS-5, 10, 15, 20 and 25/TiO_2, respectively (absorbance: approximately

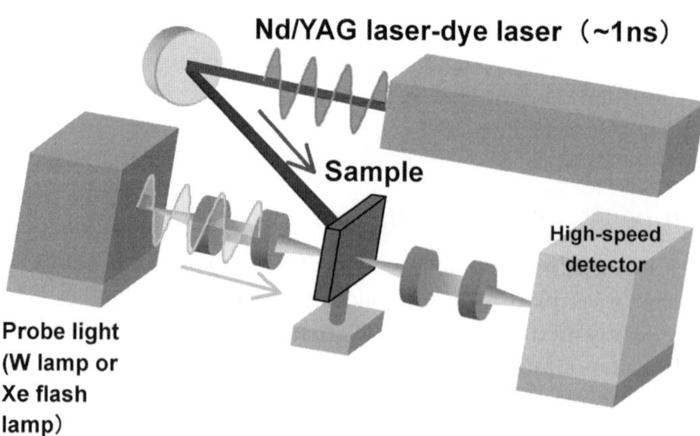

Figure 9.7 Submicrosecond to millisecond transient absorption spectrometer.

0.4 after baseline subtraction). The CdS/Al_2O_3 film was excited at 480 nm. Experiments were conducted with low pulse excitation energy densities (120 ~ 150 μJ cm^{-2}), corresponding to 1.1 absorbed photons per TiO_2 nanoparticle, with a repetition rate of 1 Hz at 25 °C. The excitation energy adjustment was finally confirmed by obtaining a transient absorption amplitude of ~2 mΔOD at 750 ~ 800 nm using TiO_2 film sensitized by (tetrabutylammonium)$_2$ *cis*-(2,2′-bipyridyl-4-COOH, 4′-COO-)$_2$(NCS)$_2$-ruthenium(II), N719 (Solaronix) for each excitation wavelength, following previous reports [41,42]. No change in the steady-state absorption spectra before and after the transient experiments was observed, suggesting that the samples are stable during the optical experiments, in agreement with the data reported by Nozik *et al.* [43].

9.4 Controlling Interfacial Electron Transfer Reactions by Nanomaterial Design

The efficiency of a solar-energy conversion system based on QD-sensitized metal oxide is dominated by electron-transfer reactions at each interface, that is, the kinetics at all interfaces have to be satisfied to optimizie the efficiency. Figure 9.8 shows the potential energy diagram of the interfaces in QD-sensitized metal oxide. For water-splitting reactions, the substrate or electrode is not necessarily employed, since hydrogen and oxygen are potentially evolved on different sites of the QD-sensitized metal-oxide nanomaterial. For applications to photovoltaics and two-electrode-based photoelectrochemical water splitting, a conducting glass is used as a supporting electrode. In this case, there are three interfaces; (i) QD/metal oxide [44], (ii) QD/electrolyte or QD/electron donor [15] and (iii) conducting glass substrate/electrolyte [35]. For an increase in efficiency, these three interfaces should be simultaneously optimized.

The mechanisms of interfacial electron-transfer reactions for QD-sensitized metal oxide are similar to those interpreted for dye-sensitized solar cells [45,46]; the differences can be described by replacing a sensitizer dye by a QD. Initially, QD excitation forms an exciton inside the QD, and subsequently, an electron is injected from the QD to the metal oxide. The injected

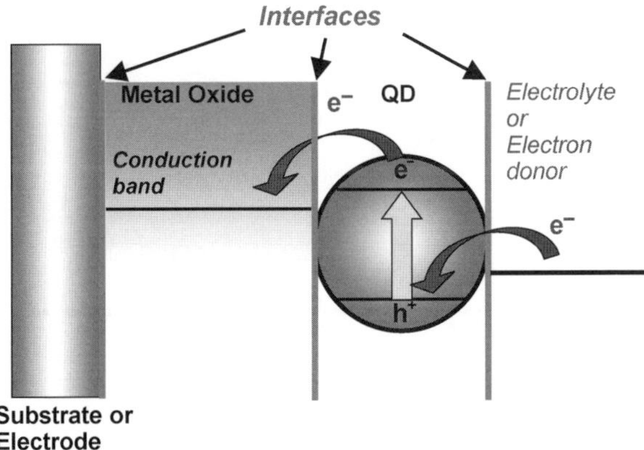

Figure 9.8 Potential energy diagram of QD-sensitized metal oxide semiconductor interfaces.

electron diffuses away from the interface through the metal-oxide nanostructure, and finally reaches a conducting glass support. The oxidized QD or the hole in the QD is reduced by a redox electrolyte. The overall efficiency is controlled by electron-transfer reactions at each step. The key to improving the efficiency is therefore to find parameters controlling these interfacial electron-transfer reactions. Below, the required points for interfaces (i ~ iii) will be discussed.

9.4.1 QD/Metal-Oxide Interface

The primary photoinduced charge-separation process occurs at the QD/metal-oxide interface. Since reactions at this interface occur on ultrafast timescales, typically picoseconds to milliseconds, the electron-transfer processes, in addition to the QD excited state dynamics (excited state relaxation and exciton dynamics), are monitored by transient absorption spectroscopy. For fundamental kinetic investigations, interfacial electron injection [11,43] and charge recombination [12] between the QD and the metal oxide have been investigated, and the initial operation mechanisms of the solar-energy conversion devices were interpreted to some extent. However, in order to improve the performance further, the parameters controlling interfacial charge transfer remain to be addressed. In this section, elucidating parameters controlling charge-recombination kinetics is discussed using a CdS/TiO_2 nanocrystalline interface as a model system [44]. *In situ*-grown chemical-bath deposited CdS on a TiO_2 nanoporous film (CdS/TiO_2) was employed owing to the superior performances of solar cells based on this film [14,35]. The sample preparation conditions are identical to those detailed in Section 9.2.3.

Figure 9.9 shows transient absorption spectra for CdS-5 ~ $25/TiO_2$, observed at 2 μs after the pulse excitation. The Al_2O_3 film was used as an inert nanoporous substrate since this insulator

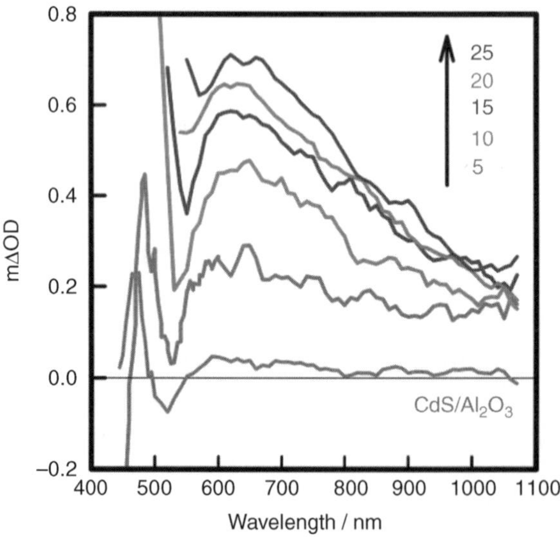

Figure 9.9 Transient absorption spectra of CdS/TiO_2 nanocrystalline films, obtained at 2 μs after CdS excitation. The number corresponds to the CdS coating cycles. A transient spectrum of CdS/Al_2O_3 is also shown as a comparison.

does not accept electrons from the excited CdS. The spectra obtained for CdS/TiO_2 clearly exhibit broad absorption, with a maximum at 600 ~ 650 nm, while weak positive absorption was observed for CdS/Al_2O_3. Thus, the broad positive absorption observed for CdS/TiO_2 indicates charge separation at the interface. This transient spectral profile can be assigned to trapped hole or S^- radical states [47–49]. Absorption by an electron in the TiO_2 is broad in the visible wavelength region, and weak, with an extinction coefficient of ~3400 $M^{-1}\,cm^{-1}$ [50,51]. The absorption spectrum is also known to exhibit a broad and featureless spectrum with a gradual increase in absorbance towards 2500 nm. Therefore, the positive transient signals presented here are mainly attributed to absorption by the S^- radical states.

Figure 9.10 shows transient decays measured at 650 nm for CdS-5 ~ $25/TiO_2$; all decays indicate multiexponential kinetics, being similar to the transient dynamics observed for dye-sensitized TiO_2 films [42,50]. For CdS/Al_2O_3, negligible transient signal was observed, since the CdS excited state decays on picosecond to nanosecond timescales [52]. These kinetics for CdS/TiO_2 are therefore associated with interfacial charge recombination between the electron in TiO_2 and the hole in CdS. Noticeably, the decays in Figure 9.10 are retarded with CdS coating cycles by more than two orders of magnitude. With respect to the potential energy levels for the CdS/TiO_2, the conduction-band edge of bulk CdS is more negative by at least 200 mV than the TiO_2 conduction-band edge (see Figure 9.8) [53–56]. This potential-energy difference between two conduction bands becomes larger with smaller number of CdS coating cycles, owing to modification of the electronic states by the quantum confinement. Following this argument, efficient electron injection occurs, irrespective of the CdS coating cycles, and thus the charge separation yield is nearly unity. This conclusion is supported by the ultrafast electron injection

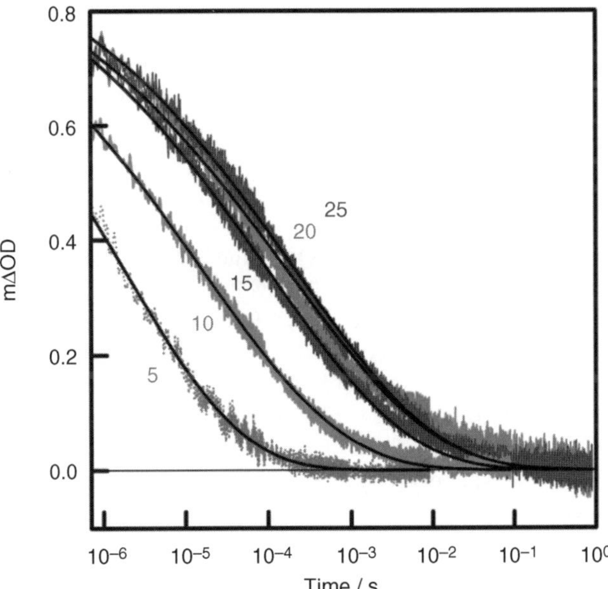

Figure 9.10 Transient absorption kinetics for the CdS-5 ~ $25/TiO_2$ monitored at 650 nm. Solid lines indicate fits to a stretched exponential function (Equation 9.4). See also color-plate section.

Table 9.1 Analytical results obtained by fitting the transient absorption decays of CdS/TiO_2. Half lifetimes are shown as $t_{50\%}$, from initial amplitude (ΔOD_0).

Coating cycles	Initial amplitude ($\Delta OD_0 \times 10^3$)	$\tau^a/\mu s$	$t_{50\%}/\mu s$	α^b
5	0.95	1.78	0.5	0.30
10	0.95	20	3.8	0.25
15	0.95	100	24	0.26
20	0.95	170	37	0.24
25	0.95	210	50	0.25

[a] Lifetime determined by fitting the decays with Equation 9.1.
[b] Stretch parameter, defined in Equation 9.4, resulted from fitting the transient decays shown in Figure 9.10.

rate ($10 \sim 50$ ps) reported for *in situ* chemical-bath-deposited CdS-sensitized TiO_2 films [43], and further justified by the data analysis described below; the initial absorption differences shown in Figure 9.10 are identical.

These decay dynamics, assigned to charge recombination, were further analyzed by fitting to a stretched exponential function, Equation 9.4 [57].

$$\Delta OD(t) = \Delta OD_0 \exp\left(-\frac{t}{\tau}\right)^{\alpha}, \tag{9.4}$$

where $\Delta OD(t)$ is the time-dependent transient absorption amplitude at 650 nm, ΔOD_0 is the initial amplitude originated mainly from the S^- radical states, and α is the stretch parameter. The fitted results are shown as the smooth black lines in Figure 9.10, and the fitted parameters are summarized in Table 9.1. All decays indicated an excellent fit, with an initial amplitude of 0.95×10^{-3}, supporting an identical electron injection efficiency for all samples. The stretch parameter resulted in a small value, that is, dispersive kinetics, being similar to those obtained for dye-sensitized TiO_2 films [57,58].

9.4.2 QD/Electrolyte Interface

Following electron injection from the excited QD to the TiO_2 film, the hole in the QD has to be re-reduced by either a redox electrolyte for photovoltaic applications or water for the water-splitting reaction. If this re-reduction process is not fast enough, the QDs may decompose. This decomposition has been typically observed under light irradiation, that is, QD photocorrosion occurred, as reported previously [24,59,60]. However, the parameters restricting this QD regeneration or the photocorrosion process have not been addressed in detail. In this section, the investigation of QD regeneration behavior is described for CdS QD-sensitized solar cells by employing several different types of electrolyte [15]. The preparation of the CdS QD-sensitized TiO_2 films is identical to that given in Section 9.2.3.

Figure 9.11a compares incident photo-to-current conversion efficiency (IPCE) spectra for CdS QD-sensitized films (five coating cycles) using various electrolytes. When the I_3^-/I^- electrolyte [61] (redox potential, E_{red}: $+0.45$ V versus NHE [45]), typically employed for Ru dye-sensitized solar cells, was used, the IPCE spectrum indicated no response in the visible wavelength range (>400 nm). A similar result was noted with the Fe^{3+}/Fe^{2+} electrolyte

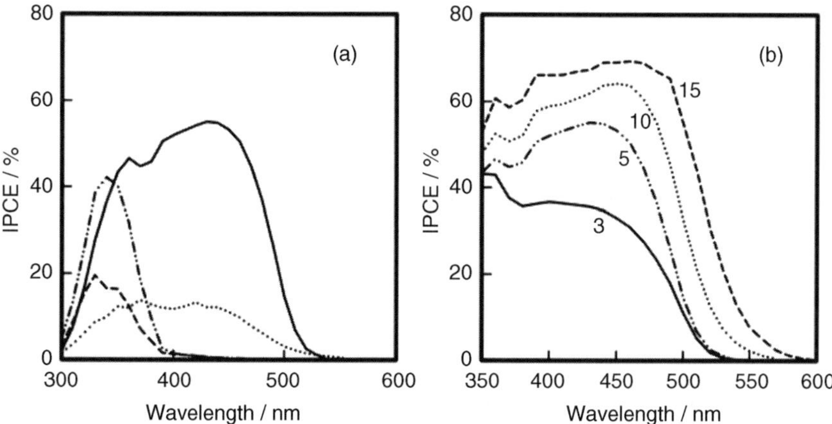

Figure 9.11 (a) IPCE spectra of CdS QD-sensitized solar cells for various electrolytes. The employed electrolytes are Na_2S_x/Na_2S (▬), I_3^-/I^- (▬ ▪ ▪), Fe^{3+}/Fe^{2+} (▬ ▬), and $Fe(CN)_6^{3-}/Fe(CN)_6^{4-}$ (▪ ▪ ▪). (b) Dependence of the QD coating cycles (indicated as the number) on IPCE spectra using the Na_2S_x/Na_2S electrolyte.

(E_{red}: +0.6 V versus NHE [62]) consisting of 0.1 M $LiClO_4$, 0.1 M $FeCl_2$ and 0.05 M $FeCl_3$ in H_2O. In contrast, a visible light spectral response was obtained with the $Fe(CN)_6^{3-}/Fe(CN)_6^{4-}$ electrolyte (E_{red}: +0.15 V versus NHE [62]) consisting of 0.1 M $LiClO_4$, 0.1 M $K_4Fe(CN)_6$, and 0.05 M $K_3Fe(CN)_6$ in H_2O. Remarkable improvement was observed with the Na_2S_x/Na_2S, polysulfide electrolyte (E_{red}: −0.45 V versus NHE [63]), consisting of 2 M Na_2S and 3 M S in water. The data clearly suggest the photocurrent response/amplitude is not dependent on the electrolyte redox potential, since the QD valence-band edge potential is located more positively than the bulk band edge (+1.6 V versus NHE).

Figure 9.11b shows dependence of the number of CdS coating cycles on IPCE spectra using the polysulfide electrolyte. The IPCE increases with the number of coating cycles due to the increase in the light-harvesting efficiency (LHE). The highest IPCE of about 70% was achieved with 15 coatings. The IPCE onset shifted to a longer wavelength with increase in the number of coating cycles. This shift can be attributed to a decrease in the bandgap resulting from the QD size increase.

CdS photocorrosion is well known, and thus the QD may be unstable in a particular electrolyte under light irradiation. In order to investigate this instability, transient photocurrents were measured for the polysulfide and I_3^-/I^- electrolytes under the continuous light irradiation at 450 nm (6.47 mW cm^{-2}), as shown in Figure 9.12. The photocurrent amplitude for the polysulfide is about 10 times greater than for I_3^-/I^-, and is maintained over several hours. Considering the electrolyte volume inside the spacer, the photocurrent observed for 10 000 s corresponds to an electrolyte turnover number (the number of collected electrons over the electrolyte moles) of about five, verifying that the polysulfide is functioning as an electron mediator. The film showed almost identical absorption spectra prior to and after the measurements, suggesting the QDs are stable under light irradiation. Note that the cell was not sealed through the stability measurement, suggesting that the photocurrent decrease may originate from electrolyte leakage. For the I_3^-/I^- electrolyte, reproducibility of the transient data proved

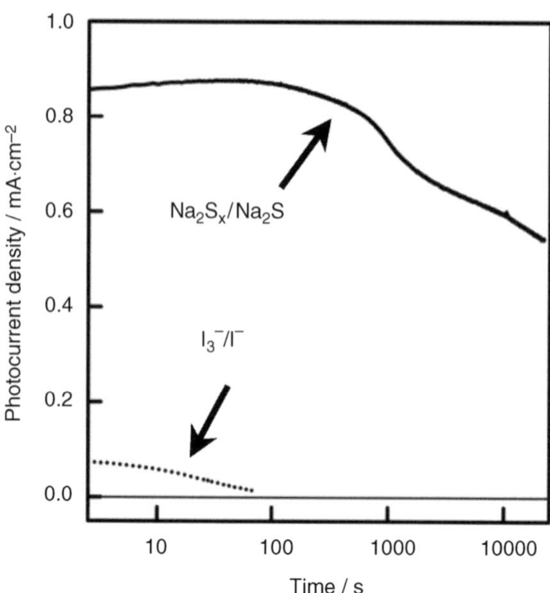

Figure 9.12 Photocurrent transients observed for CdS sensitized (five coatings) cells with Na_2S_x/Na_2S and I_3^-/I^- electrolytes at 450 nm.

difficult. The yellow CdS color occasionally disappeared immediately after the electrolyte was injected to the cell, indicative of photocorrosion [24,59,60]. The electron injection from the QD into the TiO_2 occurs on ultrafast timescales [20,43] and the charge recombination is relatively slow, similar to dye-sensitized TiO_2, as discussed above. These suggest that the photocorrosion reaction of the long-lived hole competes with QD re-reduction by I^-. Thus, QD re-reduction occurs more slowly than photocorrosion.

Figure 9.13 shows the *J–V* characteristics under 450 nm light using the same series of electrolytes. As anticipated, both the short-circuit photocurrent (J_{sc}) and open-circuit voltage (V_{oc}) are small for the I_3^-/I^- and Fe^{3+}/Fe^{2+} electrolytes. For $Fe(CN)_6^{3-}/Fe(CN)_6^{4-}$, a fast reversible reaction at the F–SnO_2/electrolyte interface was observed (see ohmic *J–V* line), being identical to the data reported by Gregg *et al.* for the ferrocene/ferrocenium couple [64]. J_{sc} and V_{oc} were remarkably higher for the Na_2S_x/Na_2S electrolyte. These results, therefore, demonstrate the polysulfide electrolyte can be used as the active redox electrolyte and for stabilizing the QDs.

9.4.3 Conducting Glass/Electrolyte Interface

The charge-separation efficiency at the QD/TiO_2 interface and the QD regeneration efficiency at the QD/electrolyte interface can simultaneously be optimized. However, even if these efficient stepwise charge-separation processes are achieved, charge recombination may still occur. For example, owing to the TiO_2 nanoporous structure, there always exists a part of the conducting glass surface exposed to the redox electrolyte. If the redox electrolyte possesses a significantly high standard rate constant at a conducting glass electrode, the electrons collected

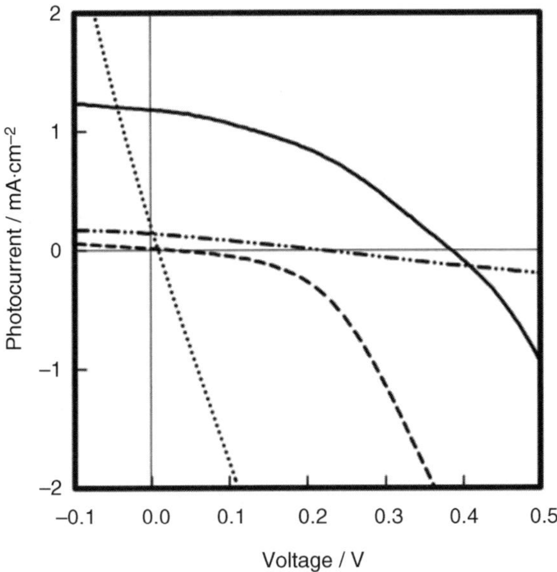

Figure 9.13 *J*—*V* characteristics obtained for QD sensitized solar cells under 450-nm light using various electrolytes: Na_2S_x/Na_2S (——), I_3^-/I^- (— ▪ ▪), $FeCl_3/FeCl_2$ (— —) or $Fe(CN)_6^{3-}/Fe(CN)_6^{4-}$ (▪ ▪ ▪ ▪).

through the TiO_2 nanoporous structure may react with the oxidized electrolyte (see Figure 9.14). This electron leakage then becomes an important cause of efficiency loss.

An electron-blocking layer on the conducting glass in ruthenium dye-sensitized wet solar cells was introduced to minimize electron leakage to the I_3^-/I^- electrolyte [65,66]. However,

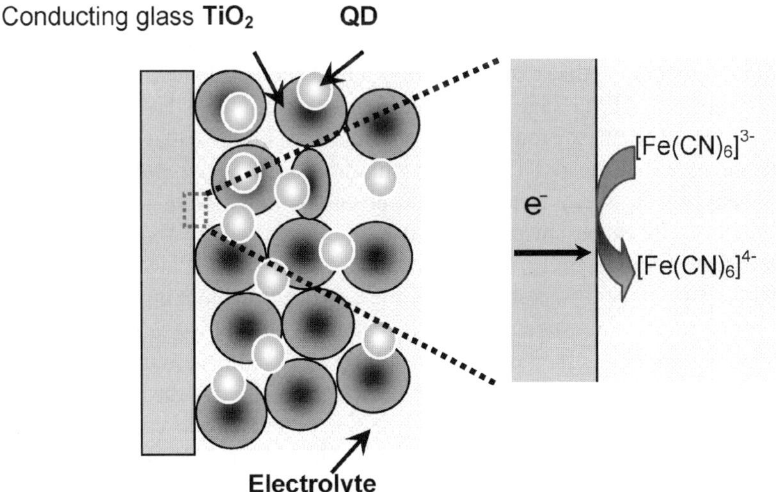

Figure 9.14 Electron-leakage reaction at the conducting glass electrode/electrolyte interface in QD-sensitized TiO_2 film.

the effect is relatively low for the optimized cell [65], since the standard rate constant of the I_3^-/I^- couple for an electrode reaction is low. Recently, cobalt-complex redox electrolytes have been employed as an alternative redox couple for dye-sensitized solar cells [67,68]. As the standard rate constant is slightly faster than the I_3^-/I^- couple [69], the electron-blocking layer effectively functions to prevent the electron leakage. In contrast, in searching for an appropriate electrolyte for QD-sensitized solar-energy conversion systems, inserting the blocking layer can be critical, since an optimum redox couple may possess a significantly high standard rate constant at a conducting glass electrode, resulting in the electron leakage reaction, as shown in Figure 9.14. Thus, clarifying the thickness and compactness of the blocking layer on cell performance is deemed essential. However, quantitative characterization of electron leakage as a function of blocking-layer thickness, using a redox electrolyte with a high standard rate constant, has not explicitly been addressed to date. In this section, the influence of an electron-blocking layer on electrolyte reactivity at the electrode/electrolyte interface is discussed [35]. A ferricyanide/ferrocyanide redox couple was employed as a model electrolyte (Figure 9.14), since it possesses a high standard rate constant for an electrode reaction.

A dense TiO_2 layer, d-TiO_2, as an electron blocking layer, was prepared on a slide glass or a fluorine-doped tin-oxide (FTO) glass, (Asahi glass, type-U, 10 Ω/square), by spray pyrolysis, following the reported method [70]. A clean glass was placed on a hot plate at 450 °C. A solution containing 0.38 M titanium di-isopropoxide bis(acetylacetonate) (Aldrich) in 2-propanol was sprayed onto the glass surface for 1 s with 0.12 M Pa nitrogen gas. This coating was repeated with a 1 min. pause after each spray. The coated substrate was heated at 450 °C for 15 min in air. The d-TiO_2 thickness was measured using a Dektak 6 M profiler, and the data were averaged over 3–7 different samples. The layer surface roughness was observed by AFM measurements (Seiko Instruments, SPI3800).

A typical example of a scanning profile for a d-TiO_2 film (eight sprays) is shown in the inset of Figure 9.15. The layer thickness was evaluated by averaging over 3–7 different samples for each spray film. Figure 9.15 shows the relationship between the spray times and d-TiO_2 layer thickness. This linear relation is consistent with the thickness-determining parameters reported by Kavan *et al.* [70]. This result indicates that a thin-film structure can be formed immediately after each spray process, so that one can readily adjust the layer thickness.

Figure 9.16 shows an AFM image for surface morphology of the d-TiO_2 layer prepared with eight spray coatings. The surface roughness is within 2 nm, similar to that of the glass substrate. This result indicates that the d-TiO_2 layer forms homogeneously, and supports the linear relationship between the spray times and the layer thickness. Note that triangle domains (~200 nm length, ~100 nm base width) were also observed, probably formed from a relatively large droplet of the solution.

Influence of d-TiO_2 thickness, that is, spray times, on electron blocking behavior at the FTO/electrolyte interface was investigated by two methods: (i) the first method monitored the current amplitude of cyclic voltammograms resulting from redox activity under bias application to the d-TiO_2/FTO electrode in a three-electrode cell; (ii) the second method was performed by measuring the diode behavior of the two-electrode cell based on a nano-TiO_2/d-TiO_2/FTO electrode. Figure 9.17 shows cyclic voltammograms of a d-TiO_2/FTO electrode as a function of d-TiO_2 thickness. A Pt plate and an Ag/AgCl electrode were used as the counter and reference electrodes, respectively, in an aqueous electrolyte containing 1 mM $K_3[Fe(CN)_6]$, 1 mM $K_4[Fe(CN)_6]$ and 0.1 M $LiClO_4$. In the absence of the d-TiO_2 layer, the anodic and cathodic current peaks of a $Fe(CN)_6^{3-}/Fe(CN)_6^{4-}$ redox couple were clearly

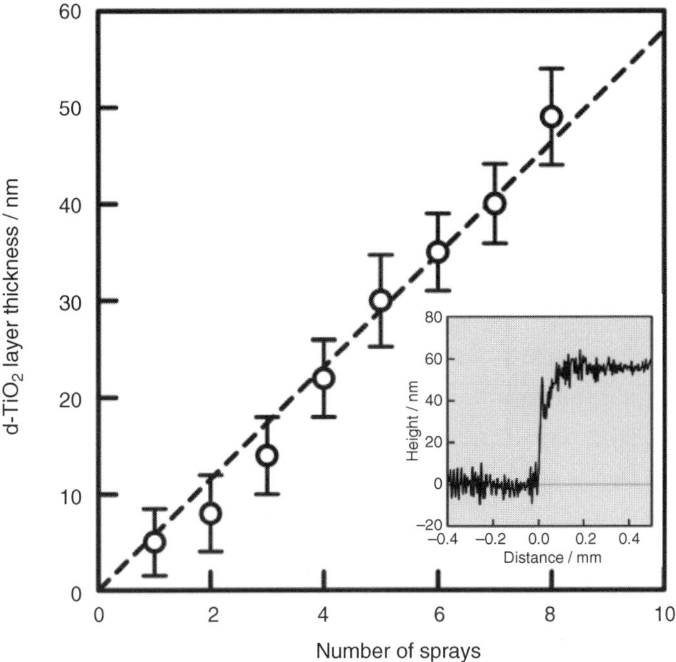

Figure 9.15 Film thickness of d-TiO_2 layer as a function of number of sprays. The inset shows a typical example of a scanning profile at the edge of the d-TiO_2 layer (eight spray on a slide glass).

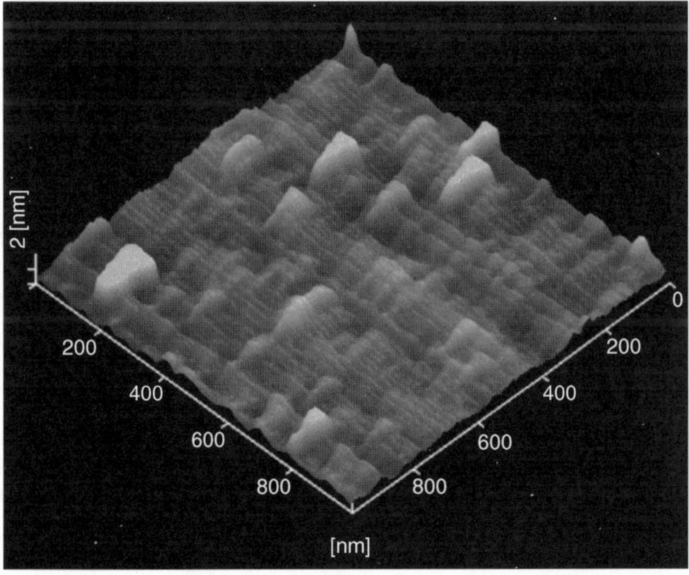

Figure 9.16 An AFM image of a d-TiO_2 layer (eight sprays) deposited on a slide glass.

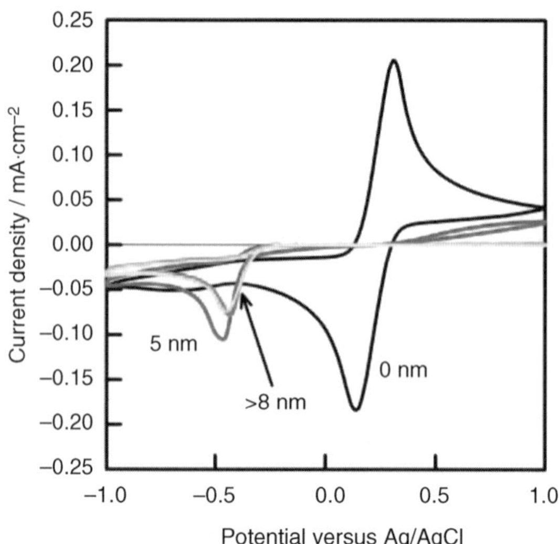

Figure 9.17 Cyclic voltammograms of a d-TiO_2/FTO electrode in an aqueous electrolyte containing 1 mM $Fe(CN)_6^{3-}$/$Fe(CN)_6^{4-}$ and 0.1 M $LiClO_4$ at a sweep rate of 20 mV s^{-1}. The number indicates the average thickness of a d-TiO_2 layer.

observed, with a peak separation of ~170 mV on the FTO, being consistent with its high standard rate constant (~0.2 cm s^{-1}) [71]. Following the formation of the d-TiO_2 layer, this redox behavior immediately disappeared, while cathodic current at <-0.3 V appeared. As the cathodic current amplitude exhibits similar size irrespective of d-TiO_2 thickness, this probably originates from $Fe(CN)_6^{3-}$ reduction at the TiO_2 surface with electron transport through the TiO_2 conduction band.

Figure 9.18 presents *J–V* characteristics of a two-electrode cell based on nano-TiO_2/d-TiO_2/FTO and a Pt counter-electrode under dark conditions. The electrolyte contains 0.1 M $Fe(CN)_6^{3-}$ and 0.1 M $Fe(CN)_6^{4-}$ in water. In the absence of the d-TiO_2 layer, the cell shows an ohmic line, indicating that the electron readily leaks from the FTO to the electrolyte. However, as the d-TiO_2 layer thickness increases, a diode behavior gradually appears. The electron leakage from the FTO to the electrolyte is gradually suppressed, while the electron transfer from the TiO_2 to $Fe(CN)_6^{3-}$ occurs with an applied voltage of <-0.7 V. The resultant *J–V* curves were fitted with a diode Equation 9.5, based on a one-diode model shown in the inset of Figure 9.18 [72,73].

$$J = -J_0\left[\exp\left(\frac{q(V+JR_s)}{nkT}\right)-1\right]-\frac{V+JR_s}{R_{sh}}, \tag{9.5}$$

where J_0 is the exchange current density, q is the elementary charge, V is the voltage, J is the current density, R_s is the series resistance, n is the ideality factor, k is the Boltzmann constant, T is the temperature, and R_{sh} is the parallel (shunt) resistance. The fitted parameters are summarized in Table 9.2. J_0 and n exhibit a slight decrease with increasing d-TiO_2 thickness. The decreased n (~3) is similar to those obtained from ruthenium dye-sensitized solar cells with I_3^-/I^- electrolyte [72]. R_s is unchanged in the range of d-TiO_2 layer thickness <40 nm,

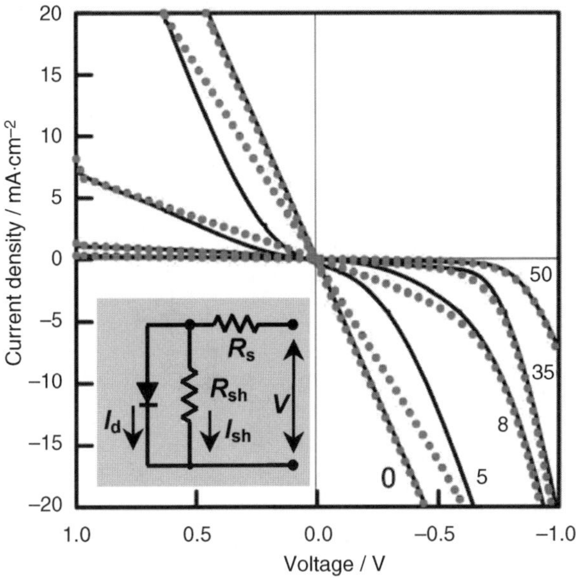

Figure 9.18 *J–V* characteristics of the cell with a nano-TiO_2/d-TiO_2/FTO electrode and a Pt electrode, observed at a sweep rate of 20 mV s^{-1}. The electrolyte is 0.1 M $Fe(CN)_6^{3-}/Fe(CN)_6^{4-}$ aqueous solution. Solid lines: measured in the dark. Dotted lines: fits of the corresponding data with Equation 9.5, based on a one-diode model shown in the inset. The number indicates average thickness (unit: nm) of a d-TiO_2 layer.

implying that the R_s originates from the resistivities of the FTO and the electrolyte. Clear increase with the d-TiO_2 thickness increase can be recognized for R_{sh}; this relation is shown in Figure 9.19. The gradual increase in R_{sh} at <30 nm may indicate the presence of pin holes (incomplete coverage) or electron leakage to the electrolyte through a thin d-TiO_2 layer. In contrast, the sharp R_{sh} increase over 30 nm indicates that the electron leakage can effectively be suppressed, and an R_{sh} increase by a factor of 200 was achieved at ~50 nm. These results suggest that a d-TiO_2 layer thickness of 30 ~ 40 nm is optimum for increasing R_{sh} to a maximum, while preserving R_s at a minimum in these experimental conditions.

Table 9.2 Thickness of d-TiO_2 layer and fit parameters resulted from *J–V* curves shown in Figure 9.18 using the one-diode model.

d-TiO_2 thickness/nm	J_0/A cm^{-2}	n	R_{sh}/Ω	R_s/Ω
0	1×10^{-7}	—	17 ± 12	—
5	5×10^{-6}	4.3	27 ± 40	5
8	3×10^{-6}	3.8	140 ± 110	6
14	1×10^{-6}	5.3	190 ± 50	5
22	1×10^{-6}	4.0	320 ± 90	4
30	2×10^{-7}	2.9	410 ± 300	2
35	6×10^{-7}	3.3	850 ± 150	5
40	2×10^{-7}	2.8	1400 ± 400	3
49	1×10^{-7}	3.3	3500 ± 400	8

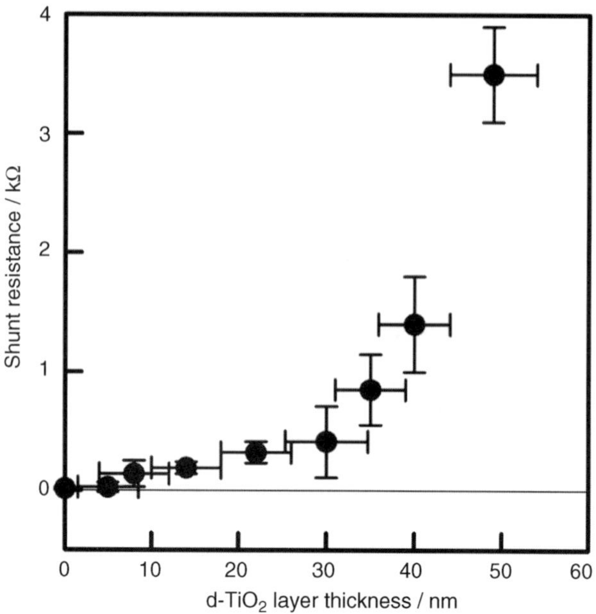

Figure 9.19 Relationship of the d-TiO_2 layer thickness with the shunt resistance.

The d-TiO_2-coated FTO electrodes were used for CdS QD-sensitized solar cells, prepared by binding a redox electrolyte with a QD/nano-TiO_2/d-TiO_2/FTO electrode and a Pt counter-electrode [61]. Figure 9.20 shows the *J–V* curves and IPCE spectrum of the solar cell using an electrode with a 30 nm d-TiO_2 layer. A CdS QD -sensitized TiO_2 film (15 coatings) was prepared, as described in Section 9.2.3, with an additional ZnS coating (two coatings). The *J–V* curves in the dark and under light exhibit negligible signs of electron leakage at the FTO surface. The J_{sc}, is 2.45 mA cm^{-2}, the V_{oc} is 0.68 V and the fill factor, *FF*, is 0.60, resulting in an energy-conversion efficiency of 1.0%. We discovered that the ZnS coating significantly improves J_{sc}, being consistent with the previous reports [74,75]. The V_{oc} and *FF* are drastically improved compared to the previous report [15]. The IPCE indicates more than 55% at 400 nm; this efficiency is similar to those reported previously [14,15,19]. Even if this high quantum efficiency was observed, integration of the IPCE spectrum with the AM 1.5 solar spectrum [61] results in 2.64 mA cm^{-2}, in close agreement with J_{sc} observed in this study. This relatively low photocurrent density under one-sun conditions limits the maximum energy-conversion efficiency, in agreement with that reported by Larramona *et al.* (1.3%) [14].

9.5 Application of QD-Sensitized Metal-Oxide Semiconductors to Solar Hydrogen Production

As well as photovoltaic devices, QD-sensitized metal-oxide semiconductors can be applied to solar hydrogen production/water-splitting reactions. Recently, a large number of studies have been reported on this issue, in addition to nanomaterials synthesis [76–78] and general photocatalytic reactions [79,80]. In this section, recent studies regarding the application of

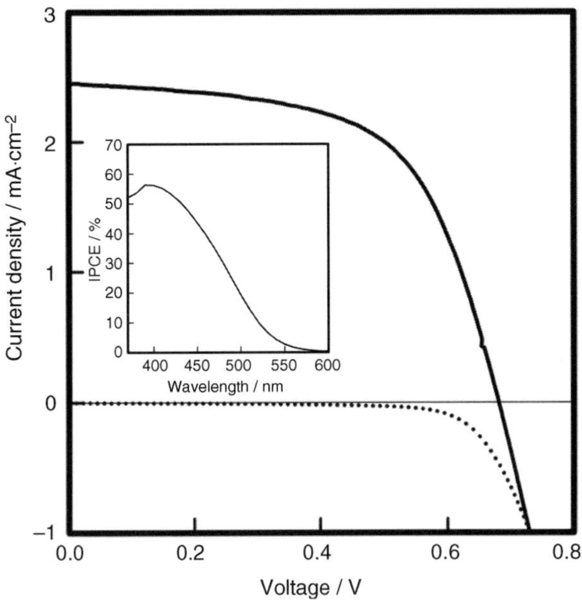

Figure 9.20 *J–V* characteristic of a CdS QD-sensitized solar cell using a 30 nm d-TiO_2 layer under AM 1.5 simulated solar light (100 mW cm^{-2}, Yamashita Denso, YSS-50A). The inset shows the IPCE spectrum, obtained using an auto-wavelength-scanned IPCE apparatus (Bunko Keiki, SM-100).

QD-sensitized metal-oxide semiconductors to solar hydrogen production systems are reviewed.

The concept of employing QD-sensitized metal-oxide semiconductors originates from hydrogen production under visible light ($\lambda > 420$ nm) irradiation, since a wide-bandgap semiconductor absorbs only the UV region. To improve hydrogen production rates, a transition metal, such as Pt, Pd or Rh, is deposited as a cocatalyst on the QD-sensitized metal-oxide semiconductor structure. The location of Pt nanoparticles is important to enhance the production rate of solar hydrogen from water, and thus synthetic methods to form QD/metal oxde/cocatalyst hybrid structures have been intensively developed. For example, Lee *et al.* have investigated three different methods, photodeposition, wet impregnation and chemical reduction, to deposit Pt nanoparticles on the CdS-sensitized TiO_2 structure [81], identifying that Pt deposition only on the TiO_2 surface provides the most efficient performance. Hoffmann also found a difference in the hydrogen production rate between Pt deposited on CdS, hybridized with TiO_2 and Pt deposited on TiO_2, hybridized with CdS, by implementing the Pt deposition process before or after the CdS deposition process [82]. They concluded that the structure with Pt deposited on TiO_2, hybridized with CdS provided 3–30 times faster production rates compared to the other hybridized structure.

Structures of QD/metal-oxide hybrid materials were modified to investigate the influence of the structure on visible-light hydrogen production ability. Zhang *et al.* showed an enhanced hydrogen production rate with the employment of TiO_2 nanotubes hybridized with CdS, compared with the $Na_2Ti_3O_7$ nanotubes [83]. Subrahmanyam *et al.* reported a CdS/ZnS hybridized structure deposited on TiO_2 nanoparticles, and modeled an efficient charge separation reaction between the CdS/ZnS and the TiO_2 [84]. Tada recently managed to fabricate CdS-Au-TiO_2 nanohybrid structures with enhanced photocatalytic processes by mimicking photosynthetic reactions [85].

9.6 Conclusion

The photoenergy conversion efficiency of the QD-sensitized metal oxide semiconductor is primarily determined by LHE and electron-transfer efficiency. The LHE is solely dependent on the amount of chromophore. If the chromophore size increases, the LHE becomes larger. However, the electron-transfer efficiency is generally forfeited by this size increase, owing to the limited diffusion length of the excited state in the chromophore to reach the interface. These requirements are suited to QD sensitization, since the absorption coefficient of the QD is large compared to organic dyes and the exciton delocalizes over the QD. Therefore, kinetic tuning of the interfacial electron-transfer reactions is vital for an overall efficiency improvement.

In QD-sensitized metal-oxide systems, three interfaces are recognized to be kinetically controlled: (i) QD/metal oxide, (ii) QD/electrolyte or QD/electron donor and (iii) conducting glass substrate/electrolyte. All these three interfaces have to be optimized simultaneously to maximize the efficiency. For example, even if the charge separation at interfaces (i) and (ii) are efficient, a low charge separation efficiency at the interface (iii) reduces the overall performance.

Nanomaterial design relates directly to the charge-separation efficiency. For example, a QD size increase retards the charge recombination rates between the electron in the TiO_2 and the hole in the CdS, being favorable for efficiency improvement. The introduction of a dense electron barrier layer at interface (iii) also retards the charge recombination between the electron in the conducting glass and the oxidized electrolyte. Thus, the interfacial nanostructural design plays a key role in maximizing the overall efficiency.

Acknowledgments

I acknowledge all of my students, colleagues and collaborators for valuable experimental assistance, helpful discussions and sample supply. This work was financially supported by TEPCO Research Foundation, by JST PRESTO program, and by Grant-in-Aid for Scientific Research, 18201022, 21550133 and 18685002, from the Ministry of Education, Culture, Sports, Science and Technology, Japan. The Association for the Progress of New Chemistry, Japan, the Murata Science Foundation, CASIO Science Promotion Foundation and the Venture Business Laboratory, Osaka University are also acknowledged for financial support.

References

[1] Bard, A.J. and Fox, M.A. (1995) Artificial photosynthesis: solar splitting of water to hydrogen and oxygen. *Accounts of Chemical Research*, **28**(3), 141–145.

[2] Nozik, A.J. and Memming, R. (1996) Physical chemistry of semiconductor-liquid interfaces. *Journal of Physical Chemistry*, **100**(31), 13061–13078.

[3] Maeda, K., Teramura, K., Lu, D. *et al.* (2006) Photocatalyst releasing hydrogen from water. *Nature (London, United Kingdom)*, **440**(7082), 295.

[4] Maeda, K., Takata, T., Hara, M. *et al.* (2005) GaN:ZnO solid solution as a photocatalyst for visible-light-driven overall water splitting. *Journal of the American Chemical Society*, **127**(23), 8286–8287.

[5] Tsuji, I., Kato, H., Kobayashi, H. and Kudo, A. (2004) Photocatalytic H_2 evolution reaction from aqueous solutions over band structure-controlled $(AgIn)_xZn_{2-1(x)}S_2$ solid solution photocatalysts with visible-light response and their surface nanostructures. *Journal of the American Chemical Society*, **126**(41), 13406–13413.

[6] Cesar, I., Kay, A., Gonzalez Martinez, J.A. and Grätzel, M. (2006) Translucent thin film Fe_2O_3 photoanodes for efficient water splitting by sunlight: nanostructure-directing effect of Si-doping. *Journal of the American Chemical Society*, **128**(14), 4582–4583.

[7] O'Regan, B. and Grätzel, M. (1991) A low-cost, high-efficiency solar cell based on dye-sensitized colloidal titanium dioxide films. *Nature (London, United Kingdom)*, **353**(6346), 737–740.

[8] Abe, R., Sayama, K. and Arakawa, H. (2002) Efficient hydrogen evolution from aqueous mixture of I^- and acetonitrile using a merocyanine dye-sensitized Pt/TiO_2 photocatalyst under visible light irradiation. *Chemical Physics Letters*, **362**(56), 441–444.

[9] Grätzel, M. (2004) Conversion of sunlight to electric power by nanocrystalline dye-sensitized solar cells. *Journal of Photochemistry and Photobiology, A: Chemistry*, **164**(1–3), 3–14.

[10] Grätzel, M. (2005) Mesoscopic solar cells for electricity and hydrogen production from sunlight. *Chemistry Letters*, **34**(1), 8–13.

[11] Robel, I., Subramanian, V., Kuno, M. and Kamat, P.V. (2006) Quantum dot solar cells. harvesting light energy with cdse nanocrystals molecularly linked to mesoscopic TiO_2 films. *Journal of the American Chemical Society*, **128** (7), 2385–2393.

[12] Lee, H.J., Yum, J.-H., Leventis, H.C. *et al.* (2008) CdSe quantum dot-sensitized solar cells exceeding efficiency 1% at full-sun intensity. *Journal of Physical Chemistry C*, **112**(30), 11600–11608.

[13] Shen, Q., Kobayashi, J., Diguna, L.J. and Toyoda, T. (2008) Effect of ZnS coating on the photovoltaic properties of CdSe quantum dot-sensitized solar cells. *Journal of Applied Physics*, **103**(8), 084304/084301–084304/084305.

[14] Larramona, G., Chone, C., Jacob, A. *et al.* (2006) Nanostructured photovoltaic cell of the type titanium dioxide, cadmium sulfide thin coating and copper thiocyanate showing high quantum efficiency. *Chemistry of Materials*, **18**(6), 1688–1696.

[15] Tachibana, Y., Akiyama, H.Y., Ohtsuka, Y. *et al.* (2007) CdS quantum dots sensitized TiO_2 sandwich type photoelectrochemical solar cells. *Chemistry Letters*, **36**(1), 88–89.

[16] Yu, P., Zhu, K., Norman, A.G. *et al.* (2006) Nanocrystalline TiO_2 solar cells sensitized with InAs quantum dots. *Journal of Physical Chemistry B*, **110**(50), 25451–25454.

[17] Peter, L.M., Riley, D.J., Tull, E.J. and Wijayantha, K.G.U. (2002) Photosensitization of nanocrystalline TiO_2 by self-assembled layers of CdS quantum dots. *Chemical Communications (Cambridge, United Kingdom)*, (10), 1030–1031.

[18] Leschkies, K.S., Divakar, R., Basu, J. *et al.* (2007) Photosensitization of ZnO nanowires with CdSe quantum dots for photovoltaic devices. *Nano Letters*, **7**(6), 1793–1798.

[19] Chang, C.-H. and Lee, Y.-L. (2007) Chemical bath deposition of CdS quantum dots onto mesoscopic TiO_2 films for application in quantum-dot-sensitized solar cells. *Applied Physics Letters*, **91**(5), 053503/053501–053503/053503.

[20] Plass, R., Pelet, S., Krueger, J. *et al.* (2002) Quantum dot sensitization of organic-inorganic hybrid solar cells. *Journal of Physical Chemistry B*, **106**(31), 7578–7580.

[21] Zaban, A., Micic, O.I., Gregg, B.A. and Nozik, A.J. (1998) Photosensitization of nanoporous TiO_2 electrodes with InP quantum dots. *Langmuir*, **14**(12), 3153–3156.

[22] McDonald, S.A., Konstantatos, G., Zhang, S. *et al.* (2005) Solution-processed PbS quantum dot infrared photodetectors and photovoltaics. *Nature Materials*, **4**(2), 138–142.

[23] Vogel, R., Hoyer, P. and Weller, H. (1994) Quantum-sized PbS, CdS, Ag_2S, Sb_2S_3 and Bi_2S_3 particles as sensitizers for various nanoporous wide-bandgap semiconductors. *Journal of Physical Chemistry*, **98**(12), 3183–3188.

[24] Kohtani, S., Kudo, A. and Sakata, T. (1993) Spectral sensitization of a titania semiconductor electrode by cadmium sulfide microcrystals and its photoelectrochemical properties. *Chemical Physics Letters*, **206**(1–4), 166–170.

[25] Lodahl, P., Floris van Driel, A., Nikolaev, I.S. *et al.* (2004) Controlling the dynamics of spontaneous emission from quantum dots by photonic crystals. *Nature (London, United Kingdom)*, **430**(7000), 654–657.

[26] Bruchez, M. Jr., Moronne, M., Gin, P. *et al.* (1998) Semiconductor nanocrystals as fluorescent biological labels. *Science (Washington, D.C.)*, **281**(5385), 2013–2016.

[27] Trindade, T., O'Brien, P. and Pickett, N.L. (2001) Nanocrystalline semiconductors: synthesis, properties and perspectives. *Chemistry of Materials*, **13**(11), 3843–3858.

[28] Alivisatos, A.P. (1996) Semiconductor clusters, nanocrystals and quantum dots. *Science (Washington, D.C.)*, **271** (5251), 933–937.

[29] Murphy, J.E., Beard, M.C., Norman, A.G. *et al.* (2006) PbTe colloidal nanocrystals: synthesis, characterization and multiple exciton generation. *Journal of the American Chemical Society*, **128**(10), 3241–3247.

[30] Fang, J., Wu, J., Lu, X. *et al.* (1997) Sensitization of nanocrystalline TiO_2 electrode with quantum sized CdSe and ZnTCPc molecules. *Chemical Physics Letters*, **270**(12), 145–151.

[31] Toyoda, T., Arae, D. and Shen, Q. (2005) Photoacoustic and photoelectrochemical characterization of CdSe quantum dots grafted onto fluorine-doped tin oxide (FTO) substrate. *Japanese Journal of Applied Physics, Part 1: Regular Papers, Brief Communications & Review Papers*, **44**(6B), 4465–4468.

[32] Vogel, R., Pohl, K. and Weller, H. (1990) Sensitization of highly porous, polycrystalline titanium dioxide electrodes by quantum sized cadmium sulfide. *Chemical Physics Letters*, **174**(3–4), 241–246.

[33] Peng, Z.A. and Peng, X. (2001) Formation of high-quality CdTe, CdSe and CdS nanocrystals using CdO as precursor. *Journal of the American Chemical Society*, **123**(1), 183–184.

[34] Murray, C.B., Norris, D.J. and Bawendi, M.G. (1993) Synthesis and characterization of nearly monodisperse CdE (E = sulfur, selenium, tellurium) semiconductor nanocrystallites. *Journal of the American Chemical Society*, **115**(19), 8706–8715.

[35] Tachibana, Y., Umekita, K., Otsuka, Y. and Kuwabata, S. (2008) Performance improvement of CdS quantum dots sensitized TiO_2 solar cells by introducing a dense TiO_2 blocking layer. *Journal of Physics D: Applied Physics*, **41**(10), 102002/102001–102002/102005.

[36] Akiyama, H.Y., Torimoto, T., Tachibana, Y. and Kuwabata, S. (2006) Quantum dot sensitized semiconductors for solar energy conversion. *Proceedings of SPIE-The International Society for Optical Engineering*, **6340**, 63400H/63401–63400H/63413, (Solar Hydrogen and Nanotechnology).

[37] Barbe, C.J., Arendse, F., Comte, P. *et al.* (1997) Nanocrystalline titanium oxide electrodes for photovoltaic applications. *Journal of the American Ceramic Society*, **80**(12), 3157–3171.

[38] Bandaranayake, R.J., Wen, G.W., Lin, J.Y. *et al.* (1995) Structural phase behavior in II-VI semiconductor nanoparticles. *Applied Physics Letters*, **67**(6), 831–833.

[39] Porter, G. (1950) Flash photolysis and spectroscopy. A new method for the study of free radical reactions. *Proceedings of the Royal Society A*, **200**, 284–300.

[40] Tachibana, Y., Otsuka, Y., Umekita, K. and Kuwabata, S. (2008) Interfacial electron transfer mechanisms in bithiophene sensitized TiO_2 based solar cells. *Transactions of the Materials Research Society of Japan*, **33**(1), 161–164.

[41] Tachibana, Y., Haque, S.A., Mercer, I.P. *et al.* (2001) Modulation of the rate of electron injection in dye-sensitized nanocrystalline TiO_2 films by externally applied bias. *Journal of Physical Chemistry B*, **105**(31), 7424–7431.

[42] Haque, S.A., Tachibana, Y., Willis, R.L. *et al.* (2000) Parameters influencing charge recombination kinetics in dye-sensitized nanocrystalline titanium dioxide films. *Journal of Physical Chemistry B*, **104**(3), 538–547.

[43] Blackburn, J.L., Selmarten, D.C. and Nozik, A.J. (2003) Electron transfer dynamics in quantum dot/titanium dioxide composites formed by in situ chemical bath deposition. *Journal of Physical Chemistry B*, **107**(51), 14154–14157.

[44] Tachibana, Y., Umekita, K., Otsuka, Y. and Kuwabata, S. (2009) Charge recombination kinetics at an *in situ* chemical bath-deposited CdS/Nanocrystalline TiO_2 interface. *Journal of Physical Chemistry C*, **113**(16), 6852–6858.

[45] Hagfeldt, A. and Grätzel, M. (1995) Light-induced redox reactions in nanocrystalline systems. *Chemical Reviews*, **95**(1), 49–68.

[46] Tachibana, Y., Akiyama, H.Y. and Kuwabata, S. (2007) Optical simulation of transmittance into a nanocrystalline anatase TiO_2 film for solar cell applications. *Solar Energy Materials & Solar Cells*, **91**(2–3), 201–206.

[47] Haase, M., Weller, H. and Henglein, A. (1988) Photochemistry of colloidal semiconductors. 26. Photoelectron emission from cadmium sulfide particles and related chemical effects. *Journal of Physical Chemistry*, **92**(16), 4706–4712.

[48] Baral, S., Fojtik, A., Weller, H. and Henglein, A. (1986) Photochemistry and radiation chemistry of colloidal semiconductors. 12. Intermediates of the oxidation of extremely small particles of cadmium sulfide, zinc sulfide and tricadmium diphosphide and size quantization effects (a pulse radiolysis study). *Journal of the American Chemical Society*, **108**(3), 375–378.

[49] Gopidas, K.R., Bohorquez, M. and Kamat, P.V. (1990) Photophysical and photochemical aspects of coupled semiconductors: charge-transfer processes in colloidal cadmium sulfide-titania and cadmium sulfide-silver(I) iodide systems. *Journal of Physical Chemistry*, **94**(16), 6435–6440.

[50] Tachibana, Y., Moser, J.E., Grätzel, M. *et al.* (1996) Subpicosecond interfacial charge separation in dye-sensitized nanocrystalline titanium dioxide films. *Journal of Physical Chemistry*, **100**(51), 20056–20062.

[51] Rothenberger, G., Fitzmaurice, D. and Grätzel, M. (1992) Spectroscopy of conduction band electrons in transparent metal oxide semiconductor films: optical determination of the flatband potential of colloidal titanium dioxide films. *Journal of Physical Chemistry*, **96**(14), 5983–5986.

[52] Sant, P.A. and Kamat, P.V. (2002) Interparticle electron transfer between size-quantized CdS and TiO_2 semiconductor nanoclusters. *Physical Chemistry Chemical Physics*, **4**(2), 198–203.

[53] Miyake, M., Torimoto, T., Sakata, T. *et al.* (1999) Photoelectrochemical characterization of nearly monodisperse cds nanoparticles-immobilized gold electrodes. *Langmuir*, **15**(4), 1503–1507.

[54] Matsumoto, H., Matsunaga, T., Sakata, T. *et al.* (1995) Size dependent fluorescence quenching of CdS nanocrystals caused by TiO_2 colloids as a potential-variable quencher. *Langmuir*, **11**(11), 4283–4287.

[55] Dewitt, R. and Kirsch-De Mesmaeker, A. (1983) Capacitance characteristics of the polycrystalline cadmium sulfide/sodium hydroxide and cadmium sulfide/cysteine interfaces. *Journal of the Electrochemical Society*, **130** (10), 1995–1998.

[56] Tachibana, Y., Haque, S.A., Mercer, I.P. *et al.* (2000) Electron injection and recombination in dye sensitized nanocrystalline titanium dioxide films: a comparison of ruthenium bipyridyl and porphyrin sensitizer dyes. *Journal of Physical Chemistry B*, **104**(6), 1198–1205.

[57] Clifford, J.N., Palomares, E., Nazeeruddin, M.K. *et al.* (2004) Molecular control of recombination dynamics in dye-sensitized nanocrystalline TiO_2 films: free energy vs distance dependence. *Journal of the American Chemical Society*, **126**(16), 5225–5233.

[58] Clifford, J.N., Yahioglu, G., Milgrom, L.R. and Durrant, J.R. (2002) Molecular control of recombination dynamics in dye sensitised nanocrystalline TiO2 films. *Chemical Communications (Cambridge, United Kingdom)*, (12), 1260–1261.

[59] Gerischer, H. (1977) On the stability of semiconductor electrodes against photodecomposition. *Journal of Electroanalytical Chemistry and Interfacial Electrochemistry*, **82**(1), 133–143.

[60] Ellis, A.B., Kaiser, S.W. and Wrighton, M.S. (1976) Optical to electrical energy conversion. Characterization of cadmium sulfide and cadmium selenide based photoelectrochemical cells. *Journal of the American Chemical Society*, **98**(22), 6855–6866.

[61] Tachibana, Y., Hara, K., Sayama, K. and Arakawa, H. (2002) Quantitative analysis of light-harvesting efficiency and electron-transfer yield in ruthenium-dye-sensitized nanocrystalline TiO_2 solar cells. *Chemistry of Materials*, **14**(6), 2527–2535.

[62] Tian, Y. and Tatsuma, T. (2005) Mechanisms and applications of plasmon-induced charge separation at TiO2 films loaded with gold nanoparticles. *Journal of the American Chemical Society*, **127**(20), 7632–7637.

[63] Allongue, P., Cachet, H., Froment, M. and Tenne, R. (1989) On the kinetics of charge transfer between an illuminated cadmium selenide electrode and polysulfide electrolyte. *Journal of Electroanalytical Chemistry and Interfacial Electrochemistry*, **269**(2), 295–304.

[64] Gregg, B.A., Pichot, F., Ferrere, S. and Fields, C.L. (2001) Interfacial recombination processes in dye-sensitized solar cells and methods to passivate the interfaces. *Journal of Physical Chemistry B*, **105**(7), 1422–1429.

[65] Ito, S., Liska, P., Comte, P. *et al.* (2005) Control of dark current in photoelectrochemical (TiO2/I–I3-) and dye-sensitized solar cells. *Chemical Communications (Cambridge, United Kingdom)*, (34), 4351–4353.

[66] Peng, B., Jungmann, G., Jager, C. *et al.* (2004) Systematic investigation of the role of compact TiO_2 layer in solid state dye-sensitized TiO_2 solar cells. *Coordination Chemistry Reviews*, **248**(13–14), 1479–1489.

[67] Nakade, S., Makimoto, Y., Kubo, W. *et al.* (2005) Roles of electrolytes on charge recombination in dye-sensitized TiO_2 solar cells (2): The case of solar cells using cobalt complex redox couples. *Journal of Physical Chemistry B*, **109**(8), 3488–3493.

[68] Nusbaumer, H., Zakeeruddin, S.M., Moser, J.-E. and Graetzel, M. (2003) An alternative efficient redox couple for the dye-sensitized solar cell system. *Chemistry–A European Journal*, **9**(16), 3756–3763.

[69] Cameron, P.J., Peter, L.M., Zakeeruddin, S.M. and Grätzel, M. (2004) Electrochemical studies of the Co(III)/Co (II)(dbbip)$_2$ redox couple as a mediator for dye-sensitized nanocrystalline solar cells. *Coordination Chemistry Reviews*, **248**(13–14), 1447–1453.

[70] Kavan, L. and Grätzel, M. (1995) Highly efficient semiconducting TiO_2 photoelectrodes prepared by aerosol pyrolysis. *Electrochimica Acta*, **40**(5), 643–652.

[71] Saji, T., Yamada, T. and Aoyagui, S. (1975) Electron-transfer rate constants for redox systems of iron(III)/iron(II) complexes with 2,2′-bipyridine and/or cyanide ion as measured by the galvanostatic double pulse method. *Journal of Electroanalytical Chemistry and Interfacial Electrochemistry*, **61**(2), 147–153.

[72] Murayama, M. and Mori, T. (2006) Equivalent circuit analysis of dye-sensitized solar cell by using one-diode model: effect of carboxylic acid treatment of TiO_2 electrode. *Japanese Journal of Applied Physics, Part 1: Regular Papers, Brief Communications & Review Papers*, **45**(1B), 542–545.

[73] Boschloo, G., Lindstrom, H., Magnusson, E. *et al.* (2002) Optimization of dye-sensitized solar cells prepared by compression method. *Journal of Photochemistry and Photobiology, A: Chemistry*, **148**(1–3), 11–15.

[74] Diguna, L.J., Shen, Q., Kobayashi, J. and Toyoda, T. (2007) High efficiency of CdSe quantum-dot-sensitized TiO_2 inverse opal solar cells. *Applied Physics Letters*, **91**(2), 023116/023111–023116/023113.

[75] Yang, S.-m., Huang, C.-h., Zhai, J. *et al.* (2002) High photostability and quantum yield of nanoporous TiO_2 thin film electrodes co-sensitized with capped sulfides. *Journal of Materials Chemistry*, **12**(5), 1459–1464.

[76] Buonsanti, R., Grillo, V., Carlino, E. *et al.* (2006) Seeded growth of asymmetric binary nanocrystals made of a semiconductor TiO_2 rodlike section and a magnetic gamma -Fe_2O_3 spherical domain. *Journal of the American Chemical Society*, **128**(51), 16953–16970.

[77] Sobczynski, A., Bard, A.J., Campion, A. *et al.* (1987) Photoassisted hydrogen generation: platinum and cadmium sulfide supported on separate particles. *Journal of Physical Chemistry*, **91**(12), 3316–3320.

[78] Spanhel, L., Weller, H. and Henglein, A. (1987) Photochemistry of semiconductor colloids. 22. Electron ejection from illuminated cadmium sulfide into attached titanium and zinc oxide particles. *Journal of the American Chemical Society*, **109**(22), 6632–6635.

[79] Xiao, M., Wang, L., Wu, Y. *et al.* (2008) Preparation and characterization of CdS nanoparticles decorated into titanate nanotubes and their photocatalytic properties. *Nanotechnology*, **19**(1), 015706/015701–015706/015707.

[80] Wu, L., Yu, J.C. and Fu, X. (2006) Characterization and photocatalytic mechanism of nanosized CdS coupled TiO_2 nanocrystals under visible light irradiation. *Journal of Molecular Catalysis A: Chemical*, **244**(1–2), 25–32.

[81] Jang, J.S., Choi, S.H., Kim, H.G. and Lee, J.S. (2008) Location and state of Pt in platinized CdS/TiO_2 photocatalysts for hydrogen production from water under visible light. *Journal of Physical Chemistry C*, **112** (44), 17200–17205.

[82] Park, H., Choi, W. and Hoffmann, M.R. (2008) Effects of the preparation method of the ternary CdS/TiO_2/Pt hybrid photocatalysts on visible light-induced hydrogen production. *Journal of Materials Chemistry*, **18**(20), 2379–2385.

[83] Zhang, Y.J., Yan, W., Wu, Y.P. and Wang, Z.H. (2008) Synthesis of TiO_2 nanotubes coupled with CdS nanoparticles and production of hydrogen by photocatalytic water decomposition. *Materials Letters*, **62**(23), 3846–3848.

[84] Tambwekar, S.V., Venugopal, D. and Subrahmanyam, M. (1999) H_2 production of (CdS-ZnS)-TiO_2 supported photocatalytic system. *International Journal of Hydrogen Energy*, **24**(10), 957–963.

[85] Tada, H., Mitsui, T., Kiyonaga, T. *et al.* (2006) All-solid-state Z-scheme in CdS-Au-TiO_2 three-component nanojunction system. *Nature Materials*, **5**(10), 782–786.

Part Three

Development of Advanced Nanostructures for Efficient Solar Hydrogen Production from Classical Large Bandgap Semiconductors

10

Ordered Titanium Dioxide Nanotubular Arrays as Photoanodes for Hydrogen Generation

M. Misra and K.S. Raja
Chemical and Materials Engineering, University of Nevada, Reno, Nevada, USA
Email: misra@unr.edu

10.1 Introduction

Titanium dioxide is the most actively investigated material for photoelectrochemical hydrogen generation by splitting water. For efficient photoelectrolysis of water the photoconversion material should satisfy three basic requirements: [1,2] (i) The electronic bandgap should be low (1.7–2.1 eV) so that most of the solar light spectrum can be used for photoexcitation; (ii) the energy level of the conduction-band minimum of the semiconductor should lie above the H_2/H_2O energy level, and the valance-band maximum should lie below the H_2O/O_2 energy level; (iii) the charge carriers generated in the photoconversion process should be transported with minimal recombination losses and be available for hydrogen and oxygen evolution reactions at the cathode and anode respectively and (iv) the semiconductor is stable against photocorrosion in the electrolyte. Among the available photosensitive semiconductor materials, TiO_2 is considered more stable against photocorrosion, even though the bandgap is about 3–3.2 eV [2,3]. This larger bandgap requires higher light energy, predominantly the UV part of solar light, for photoexcitation of electron–hole pairs. Therefore, only 3–5% of solar light can be used for conversion into photocurrent. In order to make TiO_2 more responsive to the natural solar spectrum, various approaches have been investigated. Among these approaches, bandgap modification and increasing the surface area by nanostructures are noteworthy. TiO_2 nanocrystalline anodes have been fabricated by various routes, such as coating titania slurry on conducting glass [3], spray pyrolysis and layer-by-layer colloidal coating on glass substrate

On Solar Hydrogen & Nanotechnology Edited by Lionel Vayssieres

followed by calcination at an appropriate temperature. These processes result in the formation of a 3D network of interconnected nanoparticles. It has been suggested that instead of the 3D configuration of nanoparticles, fabrication of vertical-standing nanowires of TiO_2 could improve the photoconversion efficiency [4]. Anodization of titanium metal substrate in acidified fluoride solution results in the formation of ordered arrays of vertical-standing TiO_2 nanotubes [5–7]. In this chapter, formation of ordered TiO_2 nanotubular structures using a simple electrochemical anodization process, a method for improving the light absorbance of the TiO_2 nanotubes in the visible wavelength region and a method for improving the charge transport for enhanced photoconversion efficiencies are discussed.

10.2 Crystal Structure of TiO_2

TiO_2 exists as three polymorphs, brookite, anatase and rutile. Among these, rutile is the most stable phase. Both the anatase and rutile structures are commonly used in photocatalysis applications. Figure 10.1 illustrates the crystal structures of TiO_2 in the form of TiO_6 octahedrons. Each octahedron can be considered as one Ti^{4+} cation surrounded by six O^{2-} anions. In rutile, each distorted octahedron is in contact with ten neighbor octahedrons (two sharing edge oxygen pairs and eight sharing corner oxygen atoms). The octahedron in anatase has eight neighbors, in which four share edges and four share corners. Ti–Ti distances are shorter in rutile than in anatase (0.357 and 0.296 nm versus 0.379 and 0.304 nm in anatase), whereas the Ti–O distances are shorter in anatase than in rutile (0.1934 and 0.198 nm versus 0.1949 and 0.198 nm in rutile) [8,9]. Both these phases have tetragonal lattice structures and the octahedron chains create square voids and fourfold symmetry. The rutile phase forms a linear chain and the anatase forms a zig-zag chain. As seen from Figure 10.1, the anatase has a more open structure and lower specific gravity than rutile (3.894 versus 4.25). Therefore, anatase is considered to have a larger specific surface area ($m^2\,g^{-1}$) and higher photocatalytic activity. However, the bandgap of anatase is larger than rutile (3.2 versus 3.0 eV). In the tetragonal rutile structure, metal cations are located at (0,0,0) and ($^1/_2$,$^1/_2$,$^1/_2$) and oxygen anions are located at ($1 \pm u$,* $1 \pm u$, 0) and ($^1/_2 \pm u$, $^1/_2 \pm u$, $^1/_2$), where u = oxygen internal coordinate with a value of 0.305. In anatase the internal coordinate value is 0.208.

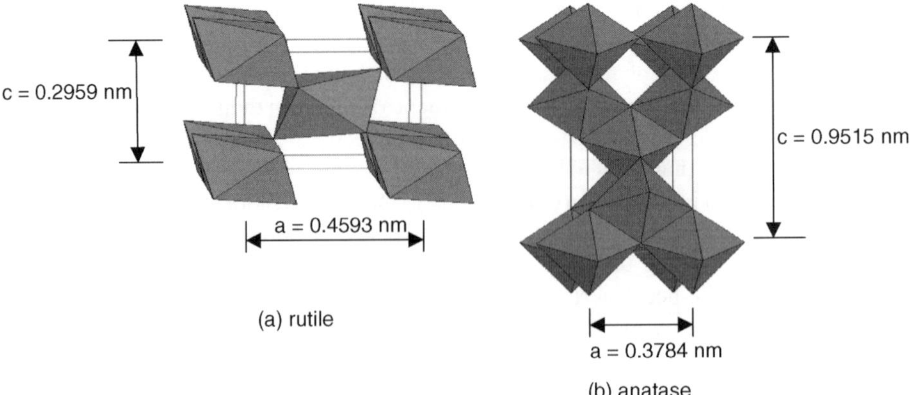

Figure 10.1 Crystal structure of TiO_2: (a) rutile and (b) anatase.

When the material is present at the nanoscale, reduction of crystallite size alters the lattice parameters of both metals and nonmetals. Lattice contraction is reported in metal nanocrystals [10], whereas either nonlinear expansion [11] or contraction [12] is observed in metal-oxide nanocrystals, depending on their composition, valence state, surface stress effects and surface hydration layer. Variations in the lattice dimensions will significantly alter the Brillouin zone, which subsequently modifies the energy band structure of the material. Li *et al.* [13] reported lattice expansion in rutile TiO_2 nanocrystals below a critical size of 54 nm. A blue-shift in bandgap is reported with decrease in the crystal size below 50 nm. In contrast to the results observed with rutile, lattice contraction is reported in pure anatase nanocrystals with sizes less than 15 nm. Depending on the purity and preparation method, nonlinear expansion is also observed in anatase with crystal sizes less than 10 nm [14]. The internal coordinate value of oxygen also increases from 0.208 to about 0.212 when the anatase crystallite size is 2 nm.

10.2.1 Electronic and Defect Structure of TiO_2

TiO_2 is an n-type semiconductor. The valence band is composed of O 2*p* orbitals and the conduction band is composed of Ti 3*d*, 4*s* and 4*p* orbitals. Following the crystal field theory of the octahedral ligand field, the metal *d* states are split into e_g (>5 eV, consisting of $d_{x^2-y^2}$ and d_{z^2} states) and t_{2g} (<5 eV, consisting of d_{yz}, d_{zx} and d_{xy} states) regions. O 2*p* has two components: (i) O p_π sates (in the plane of the Ti_3O clusters) and (ii) O p_σ states (out of the plane of the Ti_3O clusters). Ti_3O clusters are associated with the rutile structure, where one π-bond (from ligand p_y and metal t_{2g}) and two σ-bonds (from ligand p_x and p_z and metal $e_g^2sp^3$) are preferred. The ligands of the TiO_6 cluster are involved in two π-bonds and one σ-bond. Since the Ti–Ti distance in anatase is larger, the Ti d_{xy} orbitals at the bottom of the conduction band are nonbonding and isolated. In rutile, the t_{2g} orbitals provide the Ti–Ti interaction at a smaller distance. The O p_π states at the top of the valence band are nonbonding and O p_σ contributes to the bonding in the lower-energy region of the valence band [15–18]. If the TiO_2 is fully ionic, then the separation between filled oxygen 2*p* and empty Ti 3*d* levels should be about 15 eV. Since the actual bandgaps of anatase and rutile are much lower than the theoretical ionic case, significant covalent bonding is considered to be present in these crystals [19].

The electron states of TiO_2 can be described using plots, such as density of states versus energy, energy versus crystal momentum and energy versus Brillouin zone of the crystal structure [20]. The density versus energy plot gives information on the number of electron states per unit energy per unit volume in a particular energy level. In general, zero energy is given to the top of the valence band. Positive energy is given to the conduction band states and negative energy is given to the valence band states. The energy, E_n of the electron, based on a free electron model confined to a potential well of size *L*, is given by the relation:

$$E_n = \frac{(\hbar k_n)^2}{2m_e}, \tag{10.1}$$

where, k_n = the wave vector and is a quantum number given as $k_n = n\pi/L$, and $n = 1, 2, 3, \ldots$.

Therefore, the energy increases parabolically with the wave vector. The term $\hbar k_n$ gives the electron momentum. However, an electron in a crystal of size *L* having *N* atoms arranged with an interatomic distance of *a* will have a periodic potential-energy variation. Therefore, the electron wave functions will also be periodic, and are called Bloch wave functions. Bloch waves are traveling waves and given by the expression:

$$\psi_k(x) = U_k(x)\exp(jk_n x), \tag{10.2}$$

where $U_k(x)$ = a periodic function that depends on the periodic potential energy function $V(x) = V(x + m_i a), m_i = 1, 2, 3, \ldots$ and a = periodicity of the crystal. The overall wave function of the electrons inside a crystal that describes the energy states is given by the expression:

$$\Psi(x,t) = U_k(x)\exp(jk_n x)\exp\left(\frac{-jE_n t}{\hbar}\right) \tag{10.3}$$

Each wave function $\psi_k(x)$ for a particular wave vector k_n represents a state with an energy E_k. The E_k versus K_n plots give an idea about whether the electron transfer from the valence band to the conduction band occurs directly without any loss of momentum or indirectly with a change in k value. Figure 10.2 shows a schematic E_k diagram for direct and indirect bandgap semiconductors. In the direct bandgap material (Figure 10.2a), the valence-band maximum and conduction-band minimum are located at the same k vector. Therefore, an electron can be transferred from the valence band to the conduction band, conserving momentum. On the other hand, the conduction-band minimum is not directly above the valence-band maximum in the indirect bandgap materials as seen in Figure 10.2b. Therefore, transfer of an electron is associated with a momentum change from k_v to k_c. This process occurs by allowed phonon (quantum of lattice vibration)-assisted transitions at the lowest energy.

TiO_2 is an indirect bandgap material in the bulk form. The photoconversion ability of indirect bandgap materials is limited by energy loss due to phonon excitation and stronger energy dependence of the optical absorption coefficient. In other words, TiO_2 has a low optical absorption coefficient for lower-energy photons. The nature of the bandgap can be determined from the steady-state optical absorption spectra using the well-known relation:

$$\alpha h\nu = B(h\nu - E_g)^{\gamma}, \tag{10.4}$$

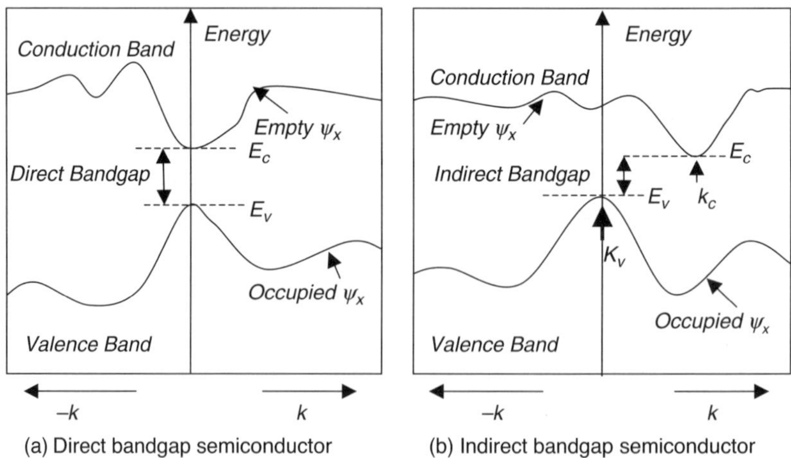

Figure 10.2 Schematic illustration of energy (E) versus crystal momentum (hk) diagram for direct and indirect bandgap semiconductors.

Where α = optical absorption coefficient, cm^{-1}, hv = photon energy, E_g = bandgap, B = absorption constant and the exponent γ takes the value of 0.5 for direct bandgap material and 2.0 for indirect bandgap material. The optical absorption coefficient, α can be determined from the absorption spectra (absorbance versus wavelength plot) using the relation:

$$\alpha = \frac{2.303 \times 10^3 A\rho}{lc} \tag{10.5}$$

Where A = experimentally determined sample absorbance from the absorption spectra, ρ = density of the semiconductor, c = loading concentration of the colloid under investigation in $g\,l^{-1}$, and l is the length of the photon path.

In nanostructured colloidal anatase particles, direct transitions are reported when the particle diameter is in the vicinity of 2 nm [19]. Size quantization in TiO_2 nanoparticles is questionable, because of the lack of accurate data on effective mass of charge carriers. A large density of surface states is envisaged in nanosized TiO_2 particles. Lattice and surface defects play a major role in electron transitions as trap levels or centers, as well as recombination centers. The difference between a trap and a recombination center in an n-type semiconductor for photoconversion applications is that the trap temporarily removes either an electron or a hole from the valence band, depending on its energy level and releases it later due to the incidence of an energetic photon, whereas a recombination center permanently removes the charge carrier without any useful photoconversion.

Defects can be classified into two main categories, depending upon the energy level at which the defect is present: (i) shallow-level defects, which are present in the vicinity of the conduction band minimum (CBM) or the valence band maximum (VBM) within the bandgap; and (ii) deep-level defects that are present at the mid bandgap. The shallow level defects are predominantly traps for charge carriers, whereas the deep-level defects are predominantly recombination centers. Most of the shallow trap levels in TiO_2 are identified with oxygen vacancies at energies from zero to one eV below the CBM. Surface state defects of Ti^{4+} ions adjacent to oxygen vacancies are identified at 0.2–0.9 eV below CBM. In addition, interstitial Ti^{3+} ions are identified as defects, presumably responsible for trapping holes. Modification of the bandgap is required in order to improve the photoconversion ability of TiO_2. Doping of various metal cations has been reported. Doping of TiO_2 with La^{3+} and Nd^{3+} has been reported [21,22]. Doping with 1.5 at% Nd was found to decrease the bandgap by 0.55 eV, which introduced electron states in the bandgap of TiO_2 and altered the conduction band (formed a new lowest unoccupied molecular orbital). Doping with W, V, Ce, Zr, Fe, Cu, Cr, Mn and Co ions has been investigated by many research groups [23–25]. In general, cation doping introduces mid-bandgap states in TiO_2 and electrons are localized around the dopants. The states due to the 3*d* dopants shift to a lower energy as the atomic number of the dopant increases. When doping with nonmetals, especially C, N, F, S, Cl and so on, the valence band of the TiO_2 is modified. Khan *et al.* [26] reported chemical modification of carbon-doped n-$TiO_{2-x}C_x$-type film synthesized by combustion of Ti metal sheet. Asahi *et al.* [27] calculated density of states (DOS) for doping with carbon and phosphorous in the anatase TiO_2 crystal and considered that the states introduced by these substitutions were too deep for overlapping with TiO_2 bandgaps. However, doping with nitrogen was found to be more effective by introducing bandgap states slightly above the O 2*p* valence band. Doping with sulfur increases the valence band width due to mixing of sulfur 3*p* states with the O 2*p* valence band and hence decreases the bandgap of TiO_2 [28]. The above discussion indicates that preparation of bandgap-modified TiO_2 nanotubes will have enhanced photoactivity in the visible-wavelength region.

10.2.2 Preparation of TiO_2 Nanotubes

TiO_2 nanotubes are generally prepared by the templated sol-gel method [29], hydrothermal processing [30,31], electrospinning, electrodeposition onto templates and electrochemical anodization methods. In sol-gel-related processes, the source material is either titanium tetraisopropoxide or titanium(IV) alkoxide. Hydrolysis of titanium tetrachloride at controlled temperatures (usually $0\,°C$) is carried out for synthesis of TiO_2 nanoparticles. Nanotubes prepared by templated sol-gel method, electrospinning or hydrothermal processing are annealed at $>350\,°C$ for crystallization. It should be noted that the predominant phases present in the nanotubes formed by hydrothermal processing are either titanate [32] or monoclinic TiO_2-B [33]. Further, these processes yield nanotubes in a loosely held powder form. Therefore, preparation of a photoelectrode involves an additional process step of coating the nanotubes onto a conducting substrate, followed by drying. The coating process gives only a random orientation to the nanotubes, with reference to the substrate, unless an electric-field-assisted coating method is employed. Vertical orientation and self-ordering of the nanotubes will enhance photoactivity because of less tortuous charge-transfer paths and a large surface area that is easily accessible to the electrolyte.

10.2.2.1 Electrochemical Anodization

Formation of vertically oriented and self-ordered TiO_2 nanotubes on a Ti substrate using a simple anodization process has been reported by several research groups [34]. In this method, Ti substrate in the form of foil, sheet or the Ti sputter-coated surface of a conducting glass is anodized in a fluoride salt containing electrolyte. Typically used electrolytes are:

- Inorganic based: 0.5 M H_3PO_4 + 0.14 M NaF, 0.5 – 1 M H_2SO_4 + 0.1 M HF, 0.5 – 1.0 M Na_2SO_4 + 0.1 M, 0.5 M $Na(NO_3)_2$ + 0.15 M NaF, 0.5 $Na_3(PO4)_2$ + 0.1 M NaF and so on
- Organic based: 0.2–0.5 wt% ammonium fluoride dissolved in ethylene glycol, glycerol, dimethyl formamide, propylene carbonate and so on. The minimum required water content in the organic electrolyte is about 0.2 wt%.

The important parameters determining the dimensions of the nanotubes are anodization potential, anodization time, pH, temperature and fluoride content of the solution. The diameter of the nanotube is essentially determined by the anodization potential and increases with an increase in the potential. The wall thickness of the nanotubes is a function of solution temperature and fluoride content. Thicker walls are observed at lower temperatures and an increase in the fluoride content decreases the wall thickness. The length of the nanotubes is a strong function of pH. Lower pH always results in shorter nanotubes, irrespective of the length of the anodization time. In organic-based electrolytes, longer anodization times result in much longer nanotubes.

A two-electrode configuration is used for anodization, as shown in Figure 10.3. A flag-shaped Pt electrode serves as a cathode. The anodization is carried out at a potential in the range of 10–60 V. Initially the potential is ramped at a rate of $0.1\,V\,s^{-1}$ from free

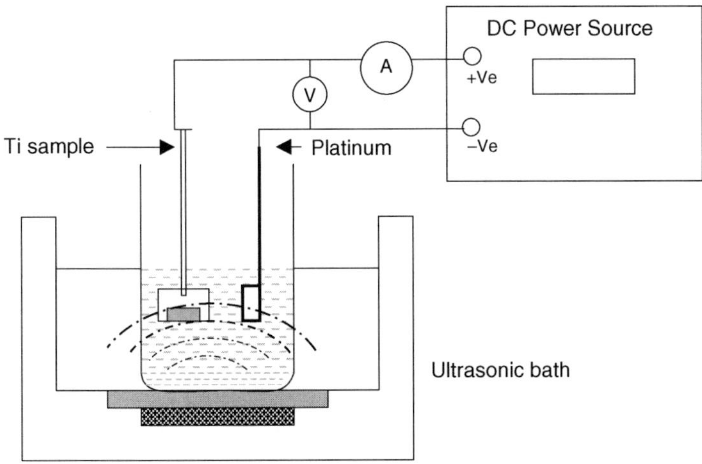

Figure 10.3 Experimental setup for anodization of Ti under ultrasonication.

corrosion potential to the final anodization potential. The electrolyte is either mechanically stirred using a magnetic bar or an ultrasonic bath/ultrasonic probe. The method of stirring significantly affects the nanotube-formation kinetics, as observed from the current transient (Figure 10.4). The anodization current is monitored continuously. After an initial increase–decrease transient, the current reaches a steady-state value at about 1.1–1.4 mA cm^{-2}, in the case of an acidified fluoride solution. The current transient shows three stages of increase–decrease behavior, as seen in Figure 10.4. The anodization is continued for 20 minutes after reaching the above current plateau values in lower-pH electrolytes.

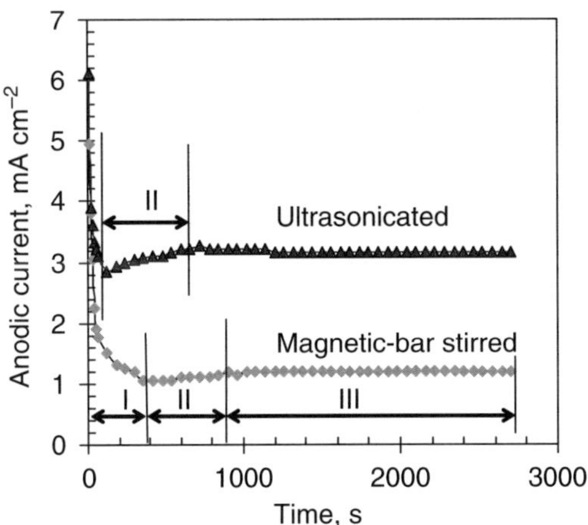

Figure 10.4 Current transients during anodization of Ti in 0.5 M H_3PO_4 + 0.14 M NaF at 20 V, with and without ultrasonication.

10.2.2.2 Stages of Nanotube Formation

Large anodic current (stage I in Figure 10.4) is measured at the instant of application of the anodization potential, indicating oxidation of Ti to Ti^{4+} ions according to the reaction:

$$Ti \rightarrow Ti^{4+} + 4e^- \tag{10.6}$$

The anodic current decays rapidly due to formation of an oxide layer. Formation of oxide could be related to the following hydrolysis reaction:

$$Ti^{4+} + 2H_2O \rightarrow TiO_2 + 4\,H^+ \tag{10.7}$$

According to the reaction (10.7), H^+ ions accumulate during hydrolysis. For electroneutrality F^- ions would have to migrate to the sites of H^+. Further, F^- ions could compete for the O^{2-} sites in the oxide. When concentration of these ions reached a critical level at local regions, dissolution of TiO_2 would occur by fluorotitanic acid following the reaction:

$$Ti^{4+} + 2H^+ + 6F^- \rightarrow H_2TiF_6(aq) \tag{10.8}$$

The reaction (10.8), dissolution of Ti cations, would create negatively charged cation vacancies in the oxide, which would migrate to the metal/oxide interface because of the potential gradient across the oxide [35]. Presence of metal-cation vacancies near the metal/oxide interface would facilitate reaction (10.6) as Ti^{4+} can easily jump to the available vacancy sites. This event is marked by a rise in anodic current (stage II). Figure 10.5 schematically illustrates the reaction steps. During the second stage of anodization, nanopores are nucleated on the oxide surface. A steady-state growth of the nanotubular oxide layer is observed when the anodic current reaches a plateau value (Region III). The mechanism of nanopore formation has been discussed, based on the surface perturbation phenomenon [6]. It is observed that the pH of the electrolyte significantly affects the rate of pore nucleation and growth. The dissolution rate of oxide increases with decrease in pH [36]. The steady-state length of the nanopores/nanotubes increases with an increase in the pH. In addition to pH, anodization potential, fluoride content and stirring method also affect the dimensions of the nanotubes and the kinetics of their formation. As the dissolution rate increases with ultrasonication, more cation vacancies are created (Figure 10.5b) and migrate to the metal/oxide interface (Figure 10.5c). Availability of a large number of cation vacancies at the interface facilitates faster oxidation of Ti (Reaction (10.4)) so that the cation can occupy the vacant sites (Figure 10.5d). This indicates that the anodic current has two components: (i) current due to the dissolution process at the oxide/electrolyte interface and (ii) current due to the oxidation of Ti at the metal/oxide interface. Therefore the increased current density under ultrasonicated conditions increases the nanotube formation kinetics.

10.2.2.3 Morphology of Nanotubes

Figure 10.6a and b shows the morphology of the nanotubes obtained in pH 2 conditions. The nanotubes are of 60–100 nm in diameter and 400–500 nm in length. Anodization in higher pH solution or ethylene glycol solution results in longer nanotubes. Anodization in ethylene glycol solution for 6 h results in nanotubes a couple of micrometers in length under steady-state conditions (Figure 10.6c and d). The diameters of the nanotubes are smaller in this case. In both the cases, the nanotubes are vertically oriented and self-ordered. It should be noted that

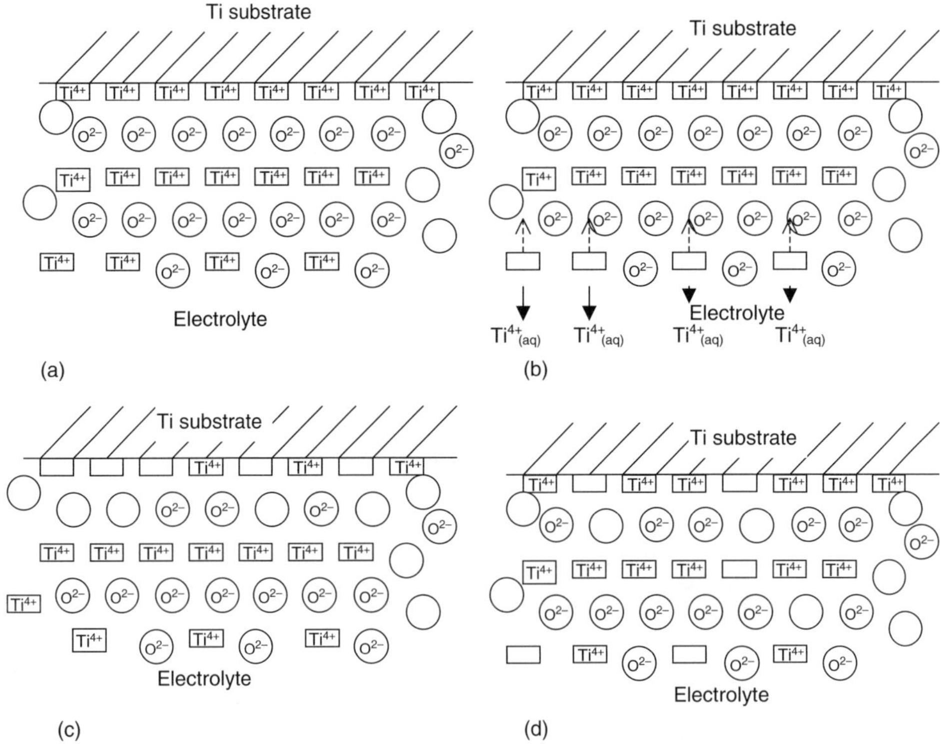

Figure 10.5 Schematic illustration of oxide-forming and -dissolution stages during anodization of Ti in fluoride solutions. (a) Schematic of stable oxide indicating Ti^{4+} and oxygen lattice positions. Oxygen vacancies are also present. (b) Dissolution stage, where cation vacancies are generated at the oxide/electrolyte interface. (c) Cation vacancies are transported to the metal/oxide interface. (d) Ti oxidation occurs, resulting in increased anodic current, and Ti^{4+} jumps to the vacancy site and migrates towards the oxide/electrolyte interface. Ti^{4+} dissolution occurs simultaneously, thus creating cation vacancies, which subsequently migrate to the metal/oxide interface. Rectangles are cation vacancies and circles are oxygen vacancies.

nanotubes formed by anodization inherently contain a large density of point defects, such as oxygen vacancies and Ti^{3+}. Under anodized conditions, the nanotubes are amorphous. The presence of defects and an amorphous phase will slow down the charge-transportation kinetics and result in diminished photoactivity. Therefore, an annealing treatment is required for crystallization.

10.2.2.4 Annealing of Nanotubes

The anodized specimens are annealed in a nitrogen or hydrogen atmosphere at 350–500 °C for 1–6 h. Annealing in reducing or inert environments does not significantly alter the bandgap states already present in the TiO_2 because of the existence of oxygen vacancies and Ti^{3+} cations. Nanotubes prepared in organic base electrolytes are modified by carbon incorporation

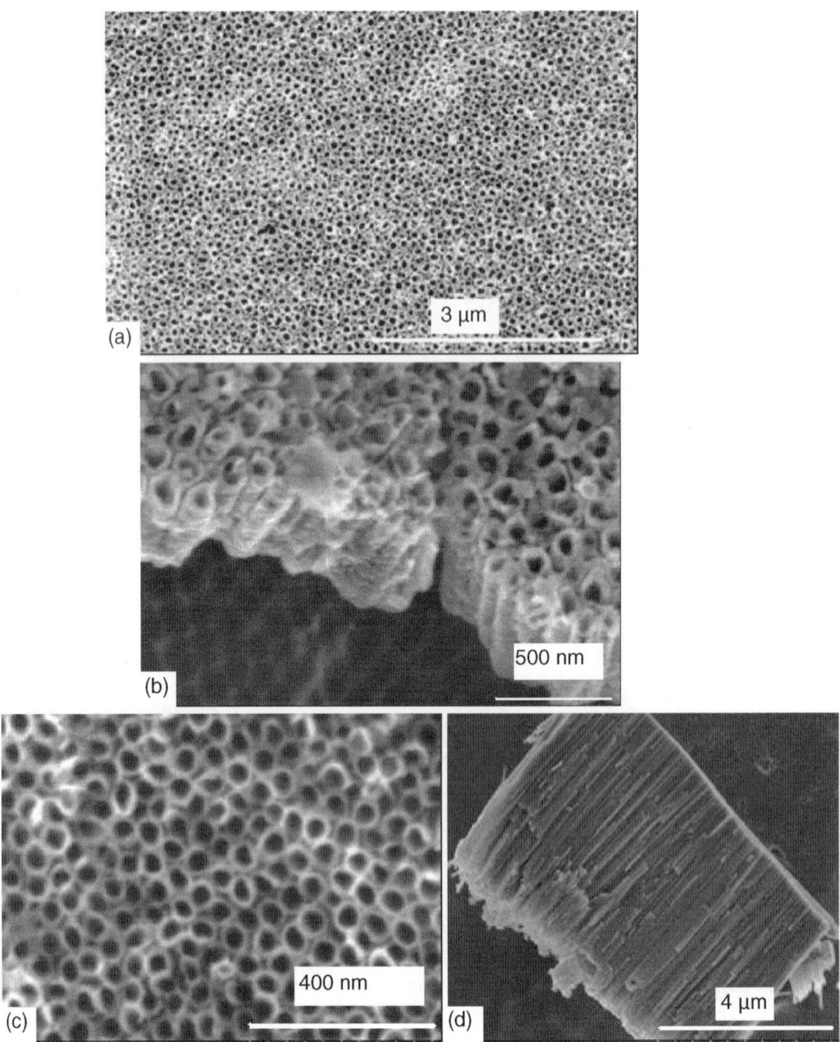

Figure 10.6 FESEM images of the TiO_2 nanotubes. (a) Front view of the nanotubes prepared by anodization in 0.5 M H_3PO_4 + 0.14 M NaF solution at 20 V for 45 minutes; (b) Side view of the nanotubes in (a); (c) Front view of the nanotubes prepared by anodization in ethylene glycol + 0.3 wt% NH_4F solution at 20 V for 6 h; and (d) Side view of the nanotubes in (c).

upon annealing in reducing or inert atmospheres. In order to incorporate carbon into TiO_2 nanotubes anodized in inorganic electrolyte, annealing is carried out at 550–650 °C for 5–20 minutes in a gas mixture of acetylene + argon + hydrogen with a typical flow rate of $20\,cm^3\,min^{-1}$, $40\,cm^3\,min^{-1}$ and $200\,cm^3\,min^{-1}$, respectively. Heat treatment of anodized specimens in a carbon-containing gas mixture results in incorporation of carbon into the nanotubes of TiO_2 arrays.

10.2.2.5 Characterization of Nanotubes

The electronic bandgap values of the TiO_2 samples are estimated from the optical absorption spectra using a UV-VIS diffuse reflectance spectrometer (Model: UV-2401 PC, Shimadzu Corporation, Kyoto, Japan). Further characterization of TiO_2 nanotubes is carried out by glancing angle X-ray diffraction (GXRD), high-resolution X-ray photoelectron spectroscopy (XPS) and Mott–Schottky analysis. The Mott–Schottky analysis is carried out by conducting standard electrochemical impedance spectroscopy at 3000 Hz in 1 M NaOH solution by scanning the potential from positive to negative in steps of 50 mV s^{-1}.

Figure 10.7a shows optical absorption spectra of nanotubular TiO_2 arrays anodized in 0.5 M H_3PO_4 + 0.14 M NaF solution at 20 V for 45 minutes under magnet-bar stirred and ultrasonicated conditions. The absorption spectra correspond to a predominant anatase phase with a possible shoulder from rutile. It can be observed that the ultrasonicated sample shows a clear red-shift in absorption take-off wavelength. The broad fluctuations in the visible region could be attributed to a plasmon resonance associated with trapped electrons and holes in the TiO_2. Lower-valance Ti and oxygen vacancies act as electron- and hole-trapping centers, respectively. Figure 10.7b is the absorption spectra of samples anodized in ethylene glycol 0.3 wt% NH4F + 3 vol% water at 20 V for 45 minutes under magnet-bar stirred and ultrasonicated conditions. The annealing of the specimens is carried out in hydrogen at 500 °C for 6 h. Because of anodization in ethylene glycol solution, the TiO_2 nanotubes incorporate carbon upon annealing in hydrogen. This carbon modification extends the absorption into the visible region. The enhanced absorption of the ultrasonically prepared sample is attributed to the Y-morphology of the nanotubes, as seen in the inset of Figure 10.7b. Nanotubes with Y-junctions are formed because of an increased perturbance at the oxide/electrolyte interface by ultrasonication. It is observed that annealing atmospheres significantly influence the optical absorption behavior of the nanotubes. This result can be linked to the volume fraction of the anatase and rutile phases of TiO_2, in addition to the defect structures.

A GXRD study of phase composition of TiO_2 nanotubes (prepared in acidified fluoride solution) is shown in Figure 10.8. For clarity, only the prominent first two peaks, assigned to (101) and (110) planes of anatase and rutile phases respectively, are given. These peaks are observed at 2θ positions of 25.3° and 27.4°. Two high-intensity peaks from (002) and (001) planes of titanium substrates are also seen at 38.4° and 40.3° (not shown in the figure). As-anodized TiO_2 nanotubes do not reveal any crystalline planes because of their amorphous state [37]. Similarly, annealing at 200 °C also does not show any crystalline peaks. Transformation of the amorphous structure into a crystalline phase is observed on the 350 °C annealed samples. Figure 10.8B shows the predominantly anatase structure of the sample annealed at 350 °C in oxygen. The samples annealed in an oxygen atmosphere at 500 °C show almost equal amounts of anatase and rutile, by considering the integrated intensity of the peaks associated with the (101) and (110) planes (Figure 10.8C), whereas the samples annealed in non-oxidizing environments contain no (Figure 10.8D) or very little rutile (Figure 10.8E). It can be seen that the nanotubes do not show any preferential growth orientation.

In order to show that the absorption peaks in the visible region are due to the presence of defect states in the bandgaps, XPS results are used, as shown in Figure 10.9a and b. XPS results show binding-energy peaks of Ti $2p_{3/2}$ and Ti $2p_{1/2}$ at around 458.8 eV and 464.74 eV, respectively, for the sample anodized in acidified fluoride solution. As-anodized TiO_2 nanotubes have an oxygen-deficient structure [38]. Annealing in an oxygen atmosphere

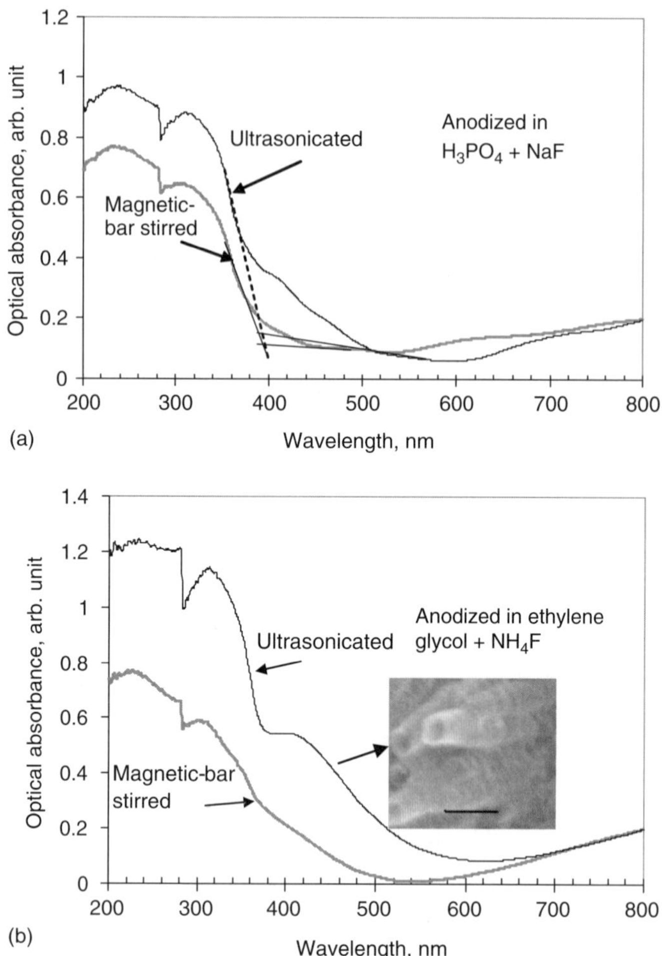

Figure 10.7 Diffuse reflectance photospectrometry results of TiO_2 nanotubes prepared by anodization in: (a) 0.5 M H_3PO_4 + 0.14 M NaF solution at 20 V for 45 minutes; and (b) in ethylene glycol + 0.3 wt% NH4F + 3 vol% water at 20 V for 45 minutes under magnet-bar stirred and ultrasonicated conditions. Inset shows Y-morphology of the nanotubes under ultrasonication. Scale bar is 150 nm.

annihilates the oxygen vacancies. A decrease in the oxygen vacancies lowers electron-cloud density, resulting in higher binding energies. Therefore, the oxygen-annealed samples show slightly higher binding energy (0.5 eV) than the as-anodized sample. The hydrogen- and nitrogen-annealed samples show lower binding energies than the oxygen-annealed sample. Defects inherited during the anodization, such as oxygen vacancies, are present even after annealing in nitrogen and hydrogen. When oxygen vacancies are present, for charge neutrality, Ti cations have a lower valence. A shoulder peak observed at a binding energy lower than 458 eV could be attributed to the presence of Ti^{3+} in the hydrogen-annealed sample [39,40]. A significantly different trend is noticeable in binding-energy values of the O 1s spectra, as

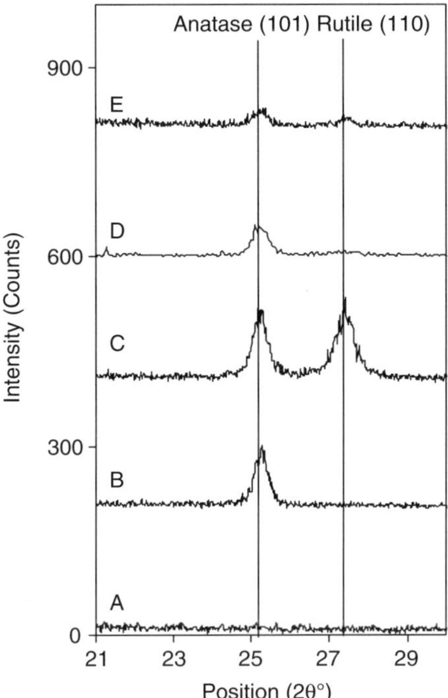

Figure 10.8 Glancing angle X-ray diffraction results of nanotubular TiO_2 templates anodized in 0.5 M H_3PO_4 + 0.14 M NaF at 20 V for 45 minutes: (A) As anodized, (B) 350 °C oxygen annealed, (C) 500 °C oxygen annealed, (D) 500 °C nitrogen annealed and (E) 500 °C hydrogen annealed.

shown in Figure 10.9b. O 1*s* spectra of all the samples are resolved into two different peaks associated with the O_2^- and hydroxyl species at 530.2 eV and 532.4 eV respectively [41–43].

10.2.3 Energetics of Photodecomposition of Water on TiO_2

Memming [44] analyzed the basic processes involved in photoelectrochemical cells and proposed models to describe the energy levels at the semiconductor/electrolyte interface and to understand the mechanism of photoelectrolysis. Based on his model, the differences in photovoltages developed on the n-type nanotubular TiO_2 photoanode in various electrolytes can be explained. When a semiconductor material comes into contact with an electrolyte, the Fermi levels of both phases are equal under equilibrium (Figure 10.10a). In order to achieve the equilibrium, band bending occurs (Figure 10.10b) in the semiconductor and there is a potential drop across the space-charge layer of the semiconductor in addition to the potential drop across the Helmholtz double layer. The amount of band bending depends on the redox potential of the electrolyte, which determines the Fermi level of the electrolyte. When the electrolyte has a specific redox system such as Ce^{4+}/Ce^{3+}, Fe^{3+}/Fe^{2+} and so on, the Fermi level of the electrolyte is fixed at the redox potential of ionic species. In the absence of a specific redox system, such as pure KOH or NaCl solutions, the Fermi level is not well defined. The Fermi level will be closer to the O_2/H_2O energy level and the valence band in the case of the KOH

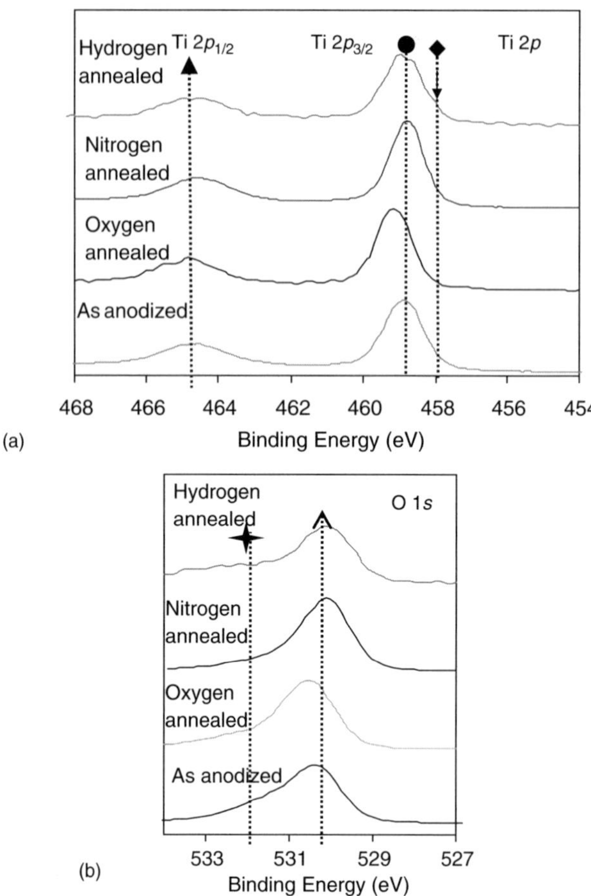

Figure 10.9 XPS spectra of TiO_2 samples anodized in 0.5 M H_3PO_4 + 0.14 M NaF at 20 V for 45 minutes and annealed in different atmospheres at 500 °C. (a) Ti $2p$ and (b) O $1s$.

electrolyte and to H_2/H_2O and conduction band in the case of 0.5 M H_2SO_4. A maximum band bending could be observed if the Fermi level of the electrolyte is closer to the valence band of the material, in the case of an n-type semiconductor. Accordingly, the band bending will be a maximum with KOH solution, as the Fermi level is closer to the O_2/H_2O energy level and the valence band. Following the same principle, the band bending will be a minimum in sulfuric acid solution and moderate in NaCl solution. The amount of band bending determines the magnitude of photopotential developed. The condition is similar to that observed in illumination of a solid semiconductor np-junction. The maximum photopotential can observed if the Fermi levels are closer to the conduction- and valence-band edges in n-type and p-type semiconductors, respectively. When the semiconductor is illuminated, the energy bands become flat under open-circuit conditions (because of the electrons reaching the conduction band and the holes reaching the surface interface), which shifts the energy level corresponding to a photopotential V_{ph} as shown in Figure 10.10c. The shift in photopotential corresponds to a negative shift in open-circuit potential in case of n-type semiconductors. It should be noted that

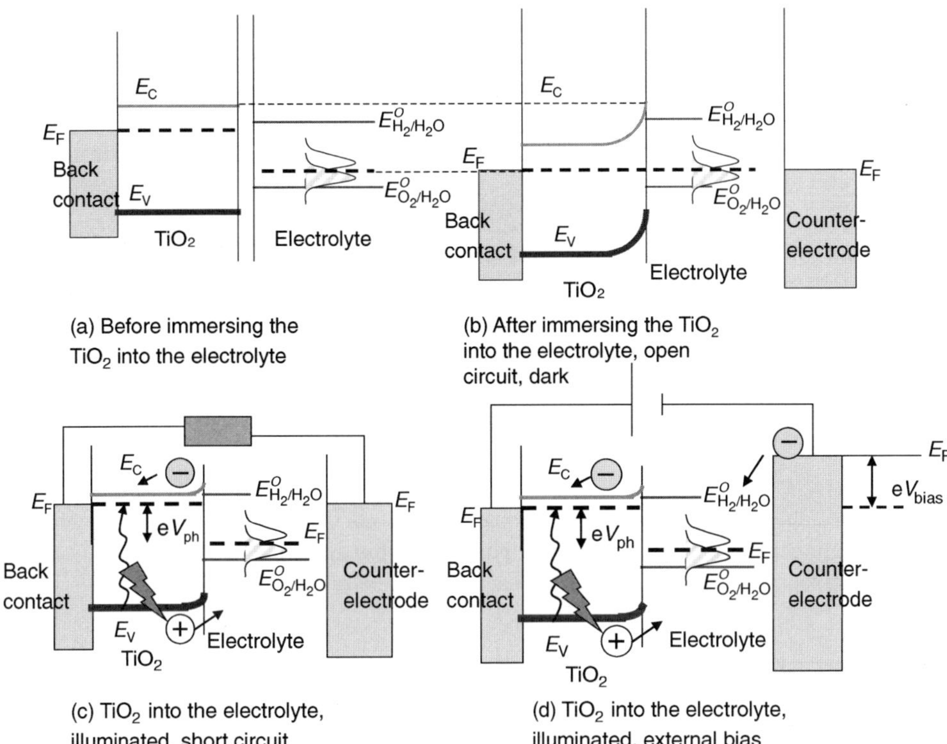

Figure 10.10 Energy scheme of photoelectrochemical cell with a TiO_2 photoanode and Pt counter-electrode in a oxygen-saturated electrolyte: (a) the energy levels of TiO_2 and electrolyte before contact; (b) after immersion, the Fermi level of TiO_2 equilibrates with that of the electrolyte, causing band bending; (c) the TiO_2 electrode is illuminated, a photopotential is developed under open-circuit conditions and the Fermi level of the counter-electrode is lower than the energy level of H_2/H_2O; (d) application of an external potential increases the Fermi level above the H_2/H_2O energy level and hydrogen evolves at the counter-electrode.

the conduction-band edge of the TiO_2 is just above the H_2/H_2O energy level. The Fermi level of the TiO_2 will be below the conduction-band edge at a distance determined by the charge-carrier density. The Fermi levels of the TiO_2 and the Pt counter-electrode will be equilibrated at the same level in an electrical circuit without any applied bias, as illustrated in Figure 10.10c. For spontaneous hydrogen evolution at the Pt surface, without any applied bias, the Fermi level of the Pt electrode should be above the H_2/H_2O energy level. This does not happen with one-sun intensity ($100\,mW\,cm^{-2}$) illumination of TiO_2. Therefore, an external bias is required to raise the Fermi level significantly above the H_2/H_2O energy level. Figure 10.10d shows the energy levels of the electrodes under an external bias.

Under an external bias, the photoanode is anodically polarized. Anodic polarization of the samples increases the electric field within the nanotube walls and the barrier oxide layer thickness by steeper band bending. This electric field is in addition to the field generated by illumination of the TiO_2. A steeper potential gradient helps separation of the electrons and holes generated by incident photons, and also minimizes recombination losses. The electrons are drifted along the length of the nanotubes to the Ti substrate because of the electric field, as well as the concentration gradient (diffusion), flow through the circuit and reach the counter-electrode to participate in the hydrogen-reduction reaction. The holes reach the surface as minority carriers and participate in the oxygen-evolution reaction. These reactions are given as:

$$TiO_2 + h\nu \rightarrow TiO_2(e^-/h^+) \rightarrow e^-_{CB} + h^+_{VB} \tag{10.9}$$

Reaction (10.9) represents generation of electron and hole pairs by energetic photons.

$$2H_2O + 4h^+ \rightarrow 4\,H^+ + O^2\uparrow \tag{10.10}$$

Reaction (10.10) occurs at the TiO_2 anode surface. Consumption of holes results in water oxidation to hydrogen ions and oxygen evolution. The protons are diffused to the cathode surface and are reduced according to the reaction:

$$4H^+ + 4e^- \rightarrow 2H_2\uparrow \tag{10.11}$$

In order to achieve higher photoconversion efficiency, not only should the available photons in the sunlight generate electron–hole pair per photon, but also the generated charge carriers should be transported to their respective reaction sites before they recombine.

10.2.3.1 Photoelectrochemical Studies

Experiments on hydrogen generation by photoelectrolysis of water are carried out in 1 M KOH using a glass cell with separated photoanode (nanotubular TiO_2 specimen) and cathode (Pt foil) compartments, as shown in Figure 10.11. The compartments are connected by a fine porous glass frit. The reference electrode (Ag/AgCl) is placed close to the anode using a salt bridge (saturated KCl)-Luggin probe capillary. The cell is provided with a 60 mm diameter quartz window for light incidence. A computer-controlled potentiostat (Model: SI 1286, Schlumberger, Farnborough, England) is employed to control the potential and record the photocurrent generated. A solar simulator (Model: 69911, Newport-Oriel Instruments, Stratford, CT, USA) is used as a light source. The light at 160 W power is passed through an AM1.5 filter. The intensity of the light is measured by a thermopile sensor (Model 70268, Newport) using a radiant power and energy meter (Model 70260, Newport Corporation, Stratford, CT, USA). The samples are anodically polarized at a scan rate of $5\,mV\,s^{-1}$ with illumination and the photocurrent is recorded. The intensity of the light falling on the electrode is in the range of 87–100 $mW\,cm^{-2}$.

Figure 10.12a shows typical potential transients recorded on the TiO_2 nanotubular samples with and without light illumination. The open-circuit potential (OCP) without illumination is generally less negative for n-type electrodes and, upon illumination, the potential moves to more negative values. In most cases, the open-circuit potential under illumination coincides with the flat band potential. In some cases, flat band potential is considered when the current during a potential scan becomes zero. This is equivalent to the OCP by definition. The magnitude of negative shift in OCP is an indication of the extent of photoactivity of an n-type semiconductor. The more the negative shift, the better will be the photoactivity. Further, the potential transient during light on–off conditions gives valuable information on the dynamics of recombination and the presence of recombination centers, traps and so on. In

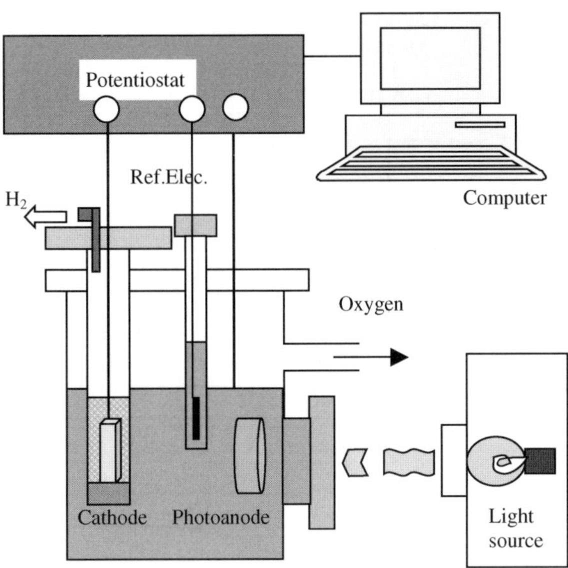

Figure 10.11 Schematic experimental set-up for conducting photoelectrochemical studies.

an extrinsic semiconductor, where the Fermi level is closer to the conduction-band minimum, the number of electrons is much higher than the number of holes. Illumination creates equal number of electrons and holes. The concentration of holes created by photoexcitation of TiO_2 is many orders of magnitude larger than the original concentration. However, the concentration of electrons created by photoexcitation is only a fraction of the original concentration. When the illumination is stopped, the concentrations of electrons and holes will go back to their original value by a recombination process. Excess holes generated during photoexcitation disappear by combining with electrons. During this process, the OCP becomes more positive, as observed in Figure 10.12a. The recombination process occurs at recombination centers. The time required to revert back to the original conditions is a function of the concentration of the recombination centers and their effectiveness in capturing holes. The recombination centers can be present as surface states,as well as mid-gap states. Further, surface absorbates also act as carrier traps, aiding recombination. Figure 10.12a indicates that TiO_2 nanotubes prepared by ultrasonic treatment show a more negative potential than nanotubes prepared under magnetic-stirred conditions. However, both materials show photoconversion and recombination behaviors.

Figure 10.12b compares the photocurrent-generation behavior of TiO_2 nanotubes prepared under magnetic-stirred and ultrasonicated conditions in acidified fluoride solution as a function of applied potential. Nanotubes prepared by ultrasonication show superior photoactivity. In spite of similar OCPs, the ultrasonically prepared sample shows a larger increase in the photocurrent density with increase in the potential. The photocurrent density of the ultrasonic sample almost reaches a plateau value for potentials more than -0.4 V versus Ag/AgCl, whereas the magnetic-stirred sample shows a continuously increasing photocurrent density with increase in the potential. These observations can be explained, based on the charge-transfer resistance and space-charge-layer thickness and wall thickness of the nanotubes. Figure 10.13a shows Nyquist plots of the TiO_2 samples. It is evident that the ultrasonically

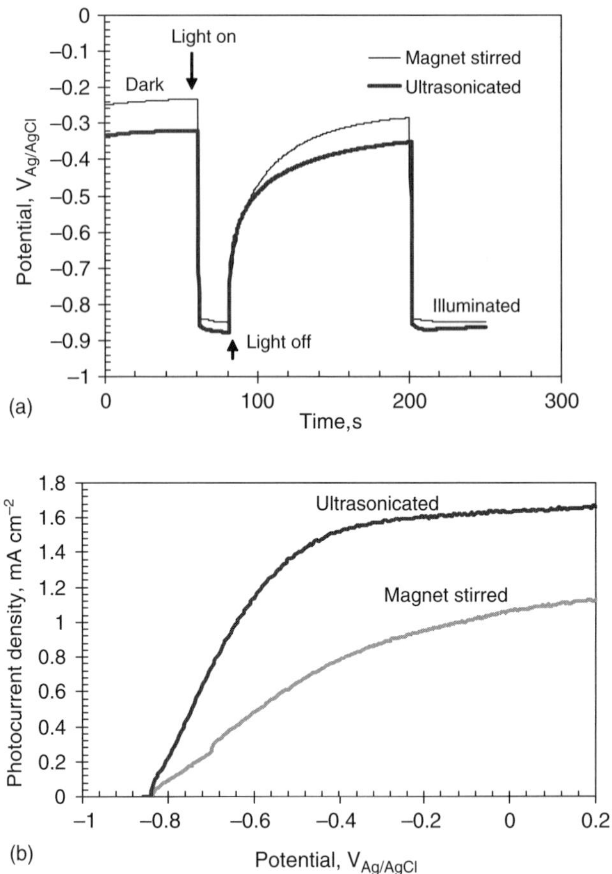

Figure 10.12 Photoelectrochemical test results of TiO_2 nanotubular samples prepared by anodization in 0.5 M H_3PO_4 + 0.14 M NaF at 20 V for 45 minutes and annealed in hydrogen atmosphere at 500 °C under magnetic-bar stirred and ultrasonicated conditions. (a) Transients of open-circuit potential with and without AM 1.5 light illumination. (b) Photocurrent generated as a function of applied potential under AM 1.5 light illumination. The dark current density was less than 12 μA cm^{-2} in both cases.

prepared sample shows about five times lower charge-transfer resistance than the magnetically prepared sample. Better charge-transport properties will result in enhanced photocurrent. Charge separation occurs more effectively in the space-charge-layer region of the nanotubes. The space-charge thickness increases with an increase in the anodic potential of the TiO_2 sample. In the case of nanotubes, band bending occurs on both sides of the wall. Therefore, the maximum allowable space-charge-layer thickness is only half of the wall thickness. If the space-layer thickness exceeds this value, then further increase in the anodic potential has no effect on band bending and charge separation. The applied potential will be realized only at the oxide/electrolyte interface. The wall thickness of the TiO_2 nanotubes varies from10–25 nm, depending on the anodization conditions. If the charge-carrier density and wall thickness are less, then the space-charge layer will extend through the entire wall of the nanotubes. A plateau

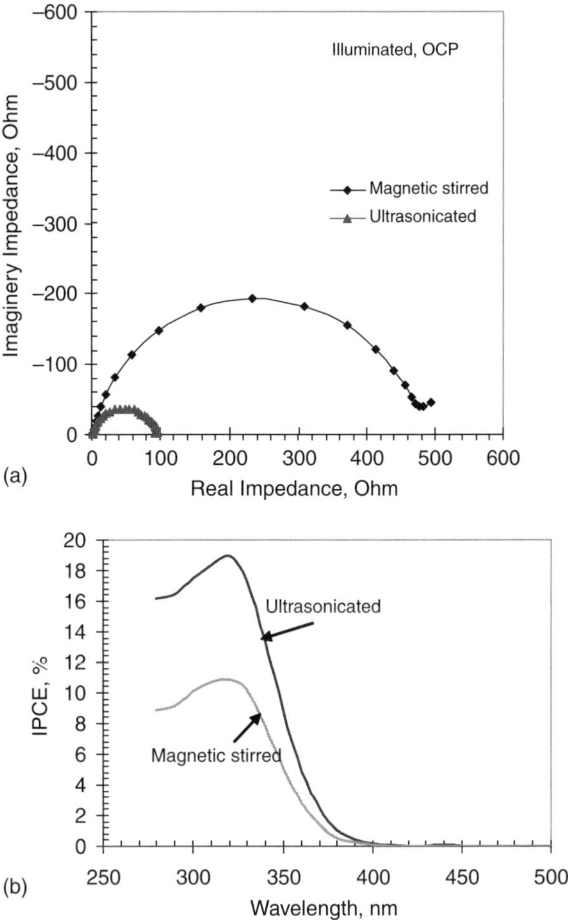

Figure 10.13 (a) Electrochemical impedance spectroscopy results (Nyquist plot) of samples prepared by anodization in 0.5 M H_3PO_4 + 0.14 M NaF at 20 V for 45 minutes and annealed in a hydrogen atmosphere at 500 °C under magnetic-bar stirred and ultrasonicated conditions. (b) Incident photon to current efficiency (IPCE, or quantum efficiency) of the samples in (a). In both the cases the samples were illuminated under open-circuit conditions.

current behavior will be observed in this case at higher applied potentials. Figure 10.13b shows the incident photon to current efficiency (IPCE) of the nanotubes. Because of the better charge-transport properties, the ultrasonically prepared sample shows higher IPCE than the samples prepared by magnetic stirring. Figure 10.14 shows the potentiostatic (at 0.2 V versus Ag/AgCl) photocurrent behavior of TiO_2 nanotubular samples prepared in inorganic and organic base electrolytes. The sample prepared in ethylene glycol base solution shows much higher photocurrent density than the sample prepared in phosphoric acid base solution. This could be attributed to the incorporation of carbon into the TiO_2 nanotubes. During anodization in ethylene glycol solution, carbon species adsorb on the surface of the nanotube walls. Upon annealing at high temperature in a hydrogen atmosphere, carbon diffuses across the nanotube

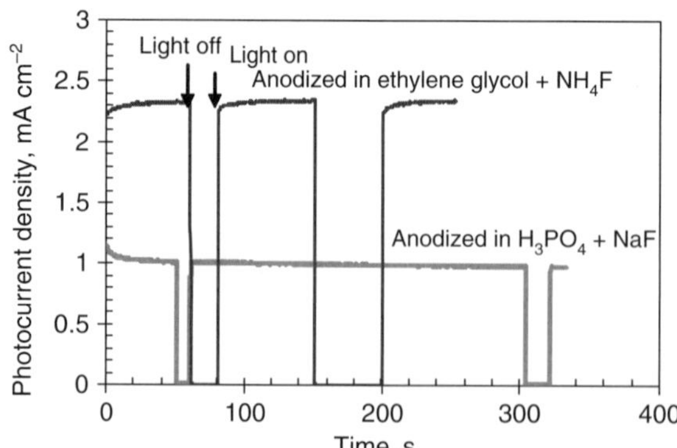

Figure 10.14 Photocurrent density of samples prepared by anodization in two different electrolytes: (i) 0.5 M H_3PO_4 + 0.14 M NaF at 20 V for 45 minutes and (ii) ethylene glycol + 0.3 wt% NH4F at 20 V for 45 minutes under magnetic-bar stirred conditions and annealed in hydrogen at 500 °C. The applied potential was 0.2 V versus Ag/AgCl.

walls as doped carbon and graphitized carbon as well. Figure 10.15 shows the high-resolution X-ray photospectrometry result of the C 1*s* spectrum. The peak at 288.4 eV could be attributed to a carbonate-type species incorporated into the nanotubes during thermal treatment.

Figure 10.16a shows the photocurrent results of carbon-modified TiO_2 samples as a function of applied potential. When the UV component is filtered out from solar light, the carbon-modified electrode shows a photocurrent density of 0.45 mA cm^{-2} under the applied anodic potential. This result clearly indicates that the carbon-modified TiO_2 nanotubes show photo-

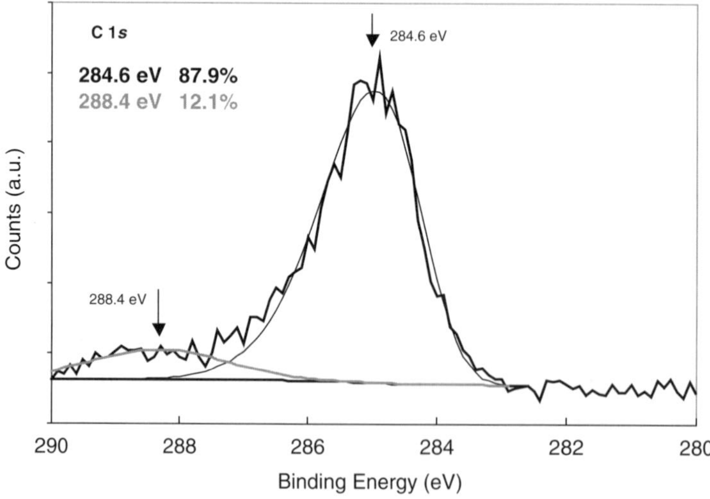

Figure 10.15 C 1*s* XPS spectrum of carbon-modified TiO_2 nanotubes. The peak at 284.6 eV is associated with adventitious elemental carbon and the peak at 288.4 eV is attributed to carbonate-type species.

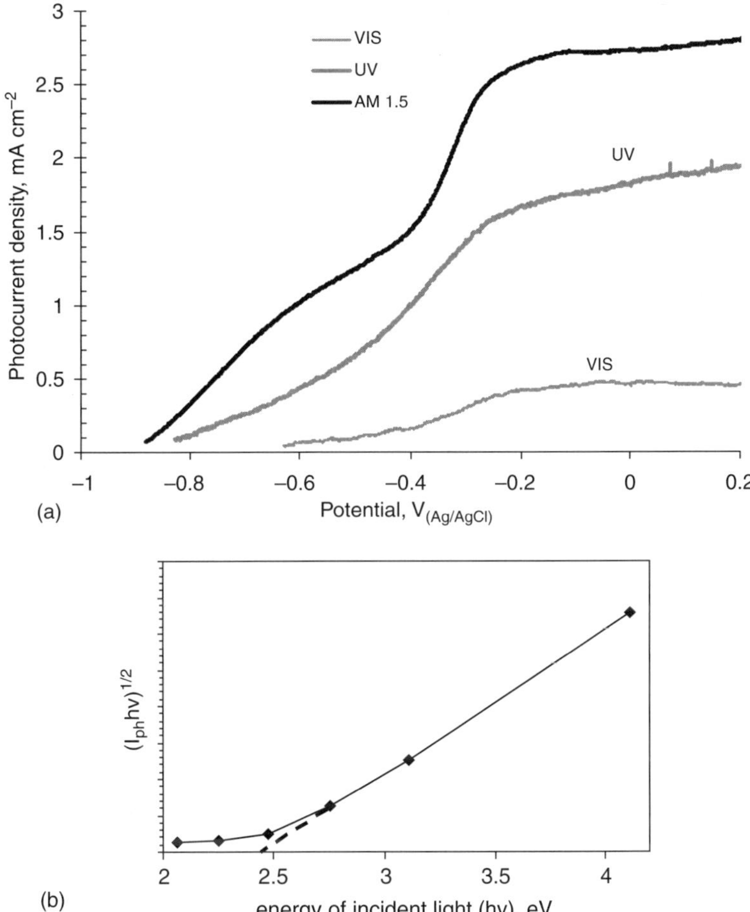

Figure 10.16 (a) Photocurrent–potential characteristics of the carbon-modified (treated at 650 °C in acytelene for 5 minutes) TiO_2 nanotubular array in AM 1.5 light, UV light (CWL: 330, FWHM: 140 nm) and visible light without UV spectrum (VIS- with center wavelength at 520 nm and FWHM of 92 nm). (b) Tauc plot of the carbon-modified TiO_2 nanotubes, indicating an indirect bandgap of ~2.4 eV.

activity in visible light without any UV component. Figure 10.16b shows the results of bandgap determination based on the photocurrent (I_{ph}) values as a function of the light energy. A linear relation could be observed between $(I_{ph}h\upsilon)^{1/2}$ and $h\upsilon$, indicating the transition is indirect. From the figure, the bandgap of the carbon-modified TiO_2 nanotubular arrays could be considered to be < 2.4 eV. The energy of the light is varied by employing bandpass filters in steps of 50 nm in the visible region. Therefore, the accuracy of the determination of the band-transition energy level is limited. The photoelectrochemical behavior of the samples is in line with the optical absorbance results, even though it has been established that bandgap modification alone does not result in increased photoactivity [45]. Yu *et al.* [46] reported enhanced photocatalytic activity of mesoporous TiO_2-carbon nanotube (CNT) composite by investigating the degradation of acetone. In addition to the increased surface area by CNT, suppression of the recombination of electron–hole pairs by CNT has been attributed to the increased photoactivity. The carbon-modified samples show better photoelectrochemical behavior than the unmodified nanotubular samples [47–57]. This improved behavior could be attributed to two possible reasons: (i) bandgap states introduced by carbon and (ii) the presence of trivalent Ti interstitials and oxygen-vacancy states introduced by the reducing environments. In this study, enhanced absorption at visible wavelengths suggests that carbon modification results in local bandgap states. High-resolution XPS studies carried out on the nitrogen/hydrogen-annealed samples and carbon-modified TiO_2 nanotubular samples suggest the presence of Ti^{3+} species. The presence of Ti^{3+} cations in the TiO_2 should be associated with oxygen vacancies in order to maintain electroneutrality.

References

[1] Khan, S.U.M., Al-Shahry, M. and Ingel, W.B. Jr. (2002) Efficient photochemical water splitting by a chemically modified n-TiO_2. *Science*, **297**, 2243–2245.

[2] Khaselev, O. and Turner, J.A. (1998) A monolithic photovoltaic-photoelectrochemical device for hydrogen production via water splitting. *Science*, **280**, 425–427.

[3] van de Lagemaat, J., Park, N.-G. and Frank, A.J. (2000) Influence of electrical potential distribution, charge transport and recombination on the photopotential and photocurrent conversion efficiency of dye-sensitized nanocrystalline TiO_2 solar cells: a study by electrical impedance and optical modulation techniques. *Journal of Physical Chemistry B*, **104**, 2044–2052.

[4] Khan, S.U.M. and Sultana, T. (2003) Photoresponse of n-TiO_2 thin film and nanowire electrodes. *Solar Energy Materials and Solar Cells*, **76**, 211–221.

[5] Raja, K.S., Misra, M. and Paramguru, K. (2005) Deposition of calcium phosphate coating on nanotubular anodized titanium. *Materials Letters*, **59**, 2137–2141.

[6] Raja, K.S., Misra, M. and Paramguru, K. (2005) Formation of self-ordered nano-tubular structure of anodic oxide layer on titanium. *Electrochimica Acta*, **51**, 154–165.

[7] Mor, G.K., Vargheese, O.K., Paulose, M. *et al.* (2003) Fabrication of tapered, conical-shaped titania nanotubes. *Journal of Materials Research*, **18**, 2588–2591.

[8] Burdett, J.K. (1985) Electronic control of the geometry of rutile and related structures. *Inorganic Chemistry*, **24**, 2244.

[9] Fahmi, A., Minot, C., Silvi, B. and Causa, M. (1993) Theoretical analysis of the structures of titanium dioxide crystals. *Physical Reviews B*, **47**, 11717.

[10] Taneda, A. and Kawazoe, Y. (1999) Structure and magnetism of small fe clusters. *Journal of the Magnetics Society of Japan*, **23**, 679.

[11] Tsunekawa, S., Ishikawa, K., Li, Z., Kawazoe, Y. and Kasuya, A. (1999) Origin of anomalous lattice expansion in oxide nanoparticles. *Nanostructured Materials*, **11**, 141; (2000) *Physical Review Letters.*, **85**, 3440; (2004) *Applied Physics Letters*, **85**, 3845; (2000) *Physical Reviews B*, **62**, 3065.

[12] Li, G., Li, L., Boerio-Goates, J. and Woodfield, B.F. (2005) High purity anatase TiO_2 nanocrystals: near room-temperature synthesis, grain growth kinetics and surface hydration chemistry. *Journal of the American Chemical Society*, **127**, 8659.

[13] Li, G., Boerio-Goates, J., Woodfield, B.F. and Li, L. (2004) Evidence of linear lattice expansion and covalency enhancement in rutile TiO_2 nanocrystals. *Applied Physics Letters*, **85**, 2059.

[14] Swamy, V., Menzies, D., Muddle, B.C. *et al.* (2006) Nonlinear size dependence of anatase TiO_2 lattice parameters. *Applied Physics Letters*, **88**, 243103.

[15] Sorantin, P.I. and Schwartz, K. (1992) Chemical bonding in rutile-type compounds. *Inorganic Chemistry*, **31**, 567.

[16] Asahi, R., Taga, Y., Mannstadt, W. and Freeman, A.J. (2000) Electronic and optical properties of anatase TiO_2. *Physical Reviews B*, **61**, 7459.

[17] Poumellect, B., Durham, P.J. and Guo, G.Y. (1991) Electronic structure and X-ray absorption spectrum of rutile TiO_2. *Journal of Physics Condensed Matter*, **3**, 8195.

[18] Umebayashi, T., Yamaki, T., Itoh, H. and Asai, K. (2002) Analysis of electronic structures of 3d transition metal-doped TiO_2 based on band calculations. *Journal of Physics and Chemistry of Solids*, **63**, 1909.

[19] Serpone, N., Lawless, D. and Khairutdinov, R. (1995) Size effects on the photophysical properties of colloidal anatase TiO_2 particles: size quantization versus direct transitions in this indirect semiconductor? *Journal of Physical Chemistry*, **99**, 16646.

[20] Kasap, S.O. (2002) *Principles of Electronic Materials and Devices*, Tata McGraw-Hill Publishing Company Limited, New Delhi, p. 363.

[21] Li, F.B., Li, X.Z. and Hou, M.F. (2004) Photocatalytic degradation of 2-mercaptobenzothiazole in aqueous La^{3+}–TiO_2 suspension for odor control. *Applied Catalysis B.*, **48**, 185.

[22] Li, W., Wang, Y., Lin, H. *et al.* (2003) Bandgap tailoring of Nd^{3+}-doped TiO_2 nanoparticles. *Applied Physics Letters*, **83**, 4143.

[23] Nagaveni, K., Hegde, M.S. and Madras, G. (2004) Structure and photocatalytic activity of $Ti_{1(x}M_xO_{2)\delta}$ (M = W, V, Ce, Zr, Fe and Cu) synthesized by solution combustion method. *Journal of Physical Chemistry B*, **108**, 20204.

[24] Anpo, M. (2000) Use of visible light. Second-generation titanium oxide photocatalysts prepared by the application of an advanced metal ion-implantation method. *Pure and Applied Chemistry*, **72**, 1787.

[25] Wang, Y., Cheng, H., Hao, Y. *et al.* (1999) Photoelectrochemical properties of metal-ion-doped TiO_2 nanocrystalline electrodes. *Thin Solid Films*, **349**, 120.

[26] Khan, S.U.M., Al-Shahry, M. and Ingel, W.B. Jr. (2002) Efficient photochemical water splitting by a chemically modified n-TiO_2. *Science*, **297**, 2243–2245.

[27] Asahi, R., Morikawa, T., Ohwaki, T. *et al.* (2001) Visible-light photocatalysis in nitrogen-doped titanium oxides. *Science*, **293**, 269.

[28] Umebayashi, T., Yamaki, T., Itoh, H. and Asai, K. (2002) Bandgap narrowing of titanium dioxide by sulfur doping. *Applied Physics Letters*, **81**, 454.

[29] Lee, S., Jeon, C. and Park, Y. (2004) Fabrication of TiO_2 tubules by template synthesis and hydrolysis with water vapor. *Chemistry of Materials*, **16**, 4292.

[30] Yao, B.D., Chan, Y.F., Zhang, X.Y. *et al.* (2003) Formation mechanism of TiO_2 nanotubes. *Applied Physics Letters*, **82**, 281.

[31] Kasuga, T., Hiramatsu, M., Hoson, A. *et al.* (1998) Formation of titanium oxide nanotube. *Langmuir*, **14**, 3160.

[32] Chen, Q., Zhou, W., Du, G. and Peng, L.M. (2002) Trititanate nanotubes made via a single alkali treatment. *Advanced Materials*, **14**, 1208.

[33] wang, Q., Wen, Z. and Li, J. (2006) Solvent-controlled synthesis and electrochemical lithium storage of one-dimensional TiO_2 nanostructures. *Inorganic Chemistry*, **45**, 6944.

[34] Grimes, C.A. (2007) Synthesis and application of highly ordered arrays of TiO_2 nanotubes. *Journal of Materials Chemistry*, **17**, 1451.

[35] Macak, J.M., Hildebrand, H., Marten-Jahns, U. and Schmuki, U.P. (2008) Mechanistic aspects and growth of large diameter self-organized TiO_2 nanotubes. *Journal of Electroanalytical Chemistry*, **621**, 254–266.

[36] Crawford, G.A. and Chawla, N. (2009) Tailoring TiO_2 nanotube growth during anodic oxidation by crystallographic orientation of Ti. *Scripta Materialia*, **60**, 874–877.

[37] Bhargava, Y.V., Nguyen, Q.A.S. and Devine, T.M. (2009) Initiation of organized nanopore/nanotube arrays in titanium oxide II. Nanopore size and spacing. *Journal of The Electrochemical Society*, **156**, E62–E68.

[38] Nguyen, Q.A.S., Bhargava, Y.V., Radmilovic, V.R. and Devine, T.M. (2009) Structural study of electrochemically synthesized TiO_2 nanotubes via cross-sectional and high-resolution TEM. *Electrochimica Acta*, **54**, 4340–4344.

[39] Macdonald, D.D. (1992) The point defect model for the passive state. *Journal of The Electrochemical Society*, **139**, 3434–3449.

[40] Ghicov, A. Tsuchiya, H., Macak, J.M. and Schmuki, P. (2005) Titanium oxide nanotubes prepared in phosphate electrolytes. *Electrochemical Communications*, **7**, 505–509.

[41] Misra, M., Paramguru, K. and Mohapatra, S.K. (2007) Growth of carbon nanotubes on nanoporous titania templates. *Journal of Nanoscience and Nanotechnology*, **7**, 2640–2646.

[42] Funk, S., Hokkanen, B., Burghaus, U. *et al.* (2007) Unexpected adsorption of oxygen on TiO_2 nanotube arrays: influence of crystal structure. *Nano Letters*, **7**(4), 1091–1094.

[43] Wu, Nae-Lih, Lee, Min-Shuei, Pon, Zern-Jin and Hsu, Jin-Zern (2004) Effect of calcination atmosphere on TiO_2 photocatalysis in hydrogen production from methanol/water solution. *Journal of Photochemistry and Photobiology A: Chemistry*, **163**, 277–280.

[44] Guimarae, J.L., Abbate, M., Betim, S.B. and Alves, M.C.M. (2003) Preparation and characterization of TiO_2 and V_2O_5 nanoparticles produced by ball-milling. *Journal of Alloys and Compounds*, **352**, 16–20.

[45] Balaur, Eugeniu, Macak, Jan M, Tsuchiya, Hiroki and Schmuki, Patrick (2005) Wetting behaviour of layers of TiO_2 nanotubes with different diameters. *Journal of Materials Chemistry*, **15**, 4488–4491.

[46] Wang, R., Hashimoto, K., Fujishima, A. *et al.* (1997) Light-induced amphiphilic surfaces. *Nature*, **388**, 431.

[47] McCafferty, E., Wightman, J.P. and Frank Cromer, T. (1999) Surface properties of hydroxyl groups in the air-formed oxide film on titanium. *Journal of The Electrochemical Society*, **146**(8), 2849–2852.

[48] Memming, R. (1980) Solar energy conversion by photoelectrochemical processes. *Electrochimica Acta*, **25**, 77–88.

[49] Nowotny, J., Sorrell, C.C., Bak, T. and Sheppard, L.R. (2005) Solar-hydrogen: Unresolved problems in solid-state science. *Solar Energy*, **78**, 593–602.

[50] Yu, Y., Yu, J.C., Yu, J.G. *et al.* (2005) Enhancement of photocatalytic activity of mesoporous TiO_2 by using carbon nanotubes. *Applied Catalysis A: General*, **289**, 186–196.

[51] Raja, K.S., Misra, M., Mahajan, V.K. *et al.* (2006) Photo-electrochemical hydrogen generation using band-gap modified nanotubular titanium oxide in solar light. *Journal of Power Sources*, **161**, 1450.

[52] Raja, K.S., Mahajan, V.K. and Misra, M. (2006) Determination of photo conversion efficiency of nanotubular titanium oxide photo-electrochemical cell for solar hydrogen generation. *Journal of Power Sources*, **159**, 1258.

[53] Mohapatra, S.K., Misra, M., Mahajan, V.K. and Raja, K.S. (2007) A novel method for the synthesis of titania nanotubes using sono-electrochemical method and its application for photo-electrochemical splitting of water. *Journal of Catalysis*, **246**, 362.

[54] Mohapatra, S.K. and Misra, M. (2007) Enhanced photoelectrochemical generation of hydrogen from water by 2,6-Dihydroxyantraquinone-functionalized titanium dioxide nanotubes. *Journal of Physical Chemistry C*, **111**, 8677.

[55] Mohapatra, S.K., Raja, K.S., Mahajan, V.K. and Misra, M. (2008) Efficient photoelectrolysis of water using TiO_2 nanotube arrays by minimizing recombination losses with organic additives. *Journal of Physical Chemistry C*, **112**, 11007.

[56] Mahajan, V.K., Misra, M., Raja, K.S. and Mohapatra, S.K. (2008) Self-organized TiO_2 nanotubular arrays for photoelectrochemical hydrogen generation: effect of crystallization and defect structures. *Journal of Physics D: Applied Physics*, **41**, 125307.

[57] Liu, Z.Y., Pesic, B., Raja, K.S. *et al.* (2009) Hydrogen generation under sunlight by self ordered TiO_2 nanotube arrays. *International Journal of Hydrogen Energy*, **34**, 3250–3257.

11

Electrodeposition of Nanostructured ZnO Films and Their Photoelectrochemical Properties

Torsten Oekermann

Institute of Physical Chemistry and Electrochemistry, Leibniz Universität Hannover, Hannover, Germany, Email: oekermann@pci.uni-hann'over.de

11.1 Introduction

While materials based on titanium dioxide (TiO_2) are generally known as the most active photocatalysts to date [1,2], zinc oxide (ZnO) has recently gained increased interest due to very similar bandgaps (3.2 eV for ZnO versus 3.0 eV for TiO_2) and band positions (conduction band of ZnO at −4.3 eV on the vacuum level versus −4.5 eV for TiO_2) [3]. In addition, it was found to have superior electrical properties compared to TiO_2 due to its higher electron mobility [4,5]. Consequently, ZnO has recently been reported to be more efficient than TiO_2 for some photocatalytic reactions, for example, the photo-oxidation of phenol and nitrophenols [6,7]. Combination of ZnO with other semiconductor oxides, such as Fe_2O_3 and WO_3, was found to further enhance photocatalytic activity in the degradation of organic compounds [8,9]. Doped ZnO is also being studied as a transparent conducting oxide (TCO) for the next generation of CIGS (copper indium gallium selenide) solar cells [10,11]. Another interesting feature of ZnO is that it can, unlike TiO_2 and most other metal oxides, be easily prepared by electrodeposition.

Fabrication of compound thin films presently relies mostly on gas-phase methods, such as vacuum evaporation, sputtering and chemical vapor deposition. While high levels of purity and structural homogeneity of the products can be achieved with such methods, they also lead to

On Solar Hydrogen & Nanotechnology Edited by Lionel Vayssieres

high production costs because of the need for expensive facilities and high energies to operate them. Low process yields from raw materials to products, and treatment of gaseous wastes are additional problems with these methods. Chemical and electrochemical solution methods are cost effective, applicable to large areas and environmentally benign, because all the wastes are confined in solutions to facilitate their treatment. Compared to chemical deposition from solution, electrochemical deposition can have additional advantages, such as good electron transport, since the deposition itself depends on it to proceed.

Until recently, industrial use of electrochemical methods has been largely limited to surface protection, lubrication or decoration of products by plating of metallic layers or anodization. It is therefore traditionally regarded as a method for mass production of cheap materials and not for advanced materials with high values. In the past decade, however, there has been a kind of renaissance in electrodeposition as a state-of-the-art technology. The most prominent example is the Damascene process, in which Cu is electrodeposited to fill up small trenches and other features on Si wafers created at submicrometer geometries by photolithography, largely replacing vacuum-deposited Al interconnects in microprocessors [12]. This has not only led to lower production costs, but also to improvements in device performance, since the electrodeposited Cu interconnects can accommodate much larger currents than the Al interconnects they replace.

In addition to advances in the electrodeposition of metals, methods for the electrodeposition of compounds such as metal oxides have also been developed since the 1980s. One of the methods best investigated is the electrodeposition of ZnO. After introducing the fundamentals of electrodeposition of compounds and especially metal oxides, this chapter will focus on the electrodeposition of nanostructured ZnO films and their properties with a view to photoelectrochemical and photocatalytic applications.

11.2 Fundamentals of Electrochemical Deposition

In general, electrodeposition is based on the local reduction or oxidation of precursors soluble in the deposition bath at an electrode. A species insoluble in the deposition bath is formed, which then precipitates at the electrode. This is opposed to electrophoretic deposition methods, where solid particles already present in the deposition bath are attracted to an electrode by an electric field. Among the most prominent examples are electrodeposition methods for metals and alloys, where the metal cations present in the solution are simply reduced to elementary metal. Methods for the electrodeposition of compounds will be discussed in more detail in Section 11.3.

The most fundamental principle of electrodeposition is Faraday's first law of electrolysis. According to this law, the amount of electrodeposited substance n is proportional to the flown charge Q, which is the product of current I and time t.

$$n = \frac{Q}{zF} \tag{11.1}$$

Here z is the number of electrons transferred per ion and F is the Faraday constant ($=96\ 485\,\mathrm{C\,mol^{-1}}$), which is the product of the elementary charge e and the Avogadro constant N_A and describes the amount of charge necessary to deposit 1 mol of substance if $z = 1$.

The thermodynamic basis of electrodeposition is the Nernst equation. It describes the redox potential E_{redox} of a redox couple, where E^0 is the standard electrode potential of the redox couple with respect to the normal hydrogen electrode (NHE) and a_{ox} and a_{red} are the activities of the oxidized and reduced form of the species, respectively:

$$E_{\mathrm{redox}} = E^0 + \frac{RT}{zF} \ln \frac{a_{\mathrm{ox}}}{a_{\mathrm{red}}} \tag{11.2}$$

The standard electrode potential of a given redox couple A/A^{n+} is related to the standard Gibbs free energy ΔG^0 of the reaction $A^{n+} + n\,e^- \rightarrow A$ by the equation

$$\Delta G^0 = -zFE^0 \tag{11.3}$$

Thermodynamically, the electrodeposition can take place if the potential applied to the working electrode is more negative than E_{redox} in case of a reduction and more positive in case of an oxidation.

In addition to thermodynamic aspects, kinetic aspects have to be considered. The latter are described by the Butler–Volmer equation [13]:

$$j = j_a - j_c = j_0 \left[e^{\frac{\alpha zF}{RT}\eta} - e^{-\frac{(1-\alpha)zF}{RT}\eta} \right] \tag{11.4}$$

Here, j is the overall current density at an electrode, j_a and j_c the anodic and cathodic current densities, j_0 the exchange current density, α the symmetry factor and η the overpotential, that is, the difference between the applied potential and the equilibrium potential. Under equilibrium conditions, that is, if $\eta = 0$, j_a and j_c are equal to j_0. In case of an overpotential >0 or <0, j_a becomes higher than j_c or vice versa, so that an overall current flow is observed. However, if j_0 is small, considerable overpotentials may be needed to achieve a noticeable electrodeposition. The overpotential according to the Butler–Volmer equation is also called activation overpotential and is required for the passage through the Helmholtz layer, which represents a potential barrier for the electroactive species to reach the electrode surface, and the charge transfer at the electrode surface itself.

Another important contribution to the overpotential is the concentration overpotential. It is due to a potential difference caused by differences in the concentrations of the electroactive species in the bulk solution and near the electrode surface. It occurs when the electrochemical reaction is sufficiently rapid to lower the surface concentration of the charge carriers to below that of the bulk solution, that is, when a concentration gradient $\mathrm{d}c/\mathrm{d}x$ exists. The overall transport of the electroactive species towards the electrode by diffusion (first term), migration (second term) and convection (third term) is described by the Nernst–Planck equation:

$$j = -D\frac{\mathrm{d}c}{\mathrm{d}x} - cD\frac{zF}{RT}\frac{\mathrm{d}\Phi}{\mathrm{d}x} + cv \tag{11.5}$$

Here D is the diffusion coefficient, $\mathrm{d}\Phi/\mathrm{d}x$ the potential gradient and v the flow speed. The migration term can usually be neglected, since the potential gradient is usually low in an electrochemical cell, especially if the conductivity of the solution is high due to the addition of a supporting salt. Therefore, convection is important to assure sufficient transport without a large concentration gradient, that is, in order to keep the diffusion overpotential low. This can be achieved by stirring the solution or by using a rotating disc electrode as a working electrode [14–17].

To assure that an appropriate potential is applied to the working electrode to fulfill the thermodynamic and kinetic conditions, the electrodeposition can be carried out potentiostatically in a three-electrode setup. In this setup a potentiostat keeps the potential of the working electrode constant with respect to a reference electrode, which is usually a solid-state electrode with a stable potential such as a calomel electrode or an Ag/AgCl electrode. To achieve a certain potential at the working electrode, a certain current, defined by the Butler–Volmer equation, has to be sent through it. This is facilitated by applying a voltage between the working electrode and the counter-electrode. The value of this voltage also depends on the current–voltage characteristics of the counter-electrode. In order to keep it low and thereby achieve good control of the working-electrode potential, electrodes with high electrocatalytic activity and therefore small overpotential, such as Pt electrodes, are usually used as counter-electrodes.

The control of the potential is also important in order to avoid undesired side reactions, for example, oxidation or reduction of the solvent or of components of the supporting salt. Once the current–voltage characteristics of the working electrode are known, however, it is also possible to carry out the electrodeposition under galvanostatic conditions, that is, by setting a constant current, in a two-electrode setup without a reference electrode.

11.3 Electrodeposition of Metal Oxides and Other Compounds

Among the first studies on the electrodeposition of compounds in the 1980s was the development of methods to prepare thin films of II–IV compound semiconductors [18,19]. For example, electrodeposition of cadmium sulfide from solutions of cadmium salts was carried out by electroreduction of elemental sulfur either dissolved in dimethylsulfoxide or released by decomposition of thiosulfate in water. Light absorbers in thin-film photovoltaic cells such as CdTe and Cu(In, Ga)Se_2 were also prepared by electrodeposition and achieved promising efficiencies [20,21]. Another example for a (p-type) compound semiconductor that can be electrodeposited is CuSCN. In this case the deposition is carried out from aqueous or ethanolic solutions containing Cu^{2+} and SCN^- ions. When Cu^{2+} is cathodically reduced to Cu^+, CuSCN, which is insoluble in ethanol and water, precipitates at the electrode [22,23].

The most traditional approach for the electrodeposition of metal oxides is based on the cathodic reduction of nitrate ions in aqueous solutions of the respective metal ion. The OH^- generated during the reduction of nitrate leads to the precipitation of metal hydroxide at the electrode. Depending on the kind of metal, the conversion of the metal hydroxide to the metal oxide can be achieved *in situ* by carrying out the deposition at elevated temperature or subsequently by calcination. The first application of this concept was the electrodeposition of redox active $Ni(OH)_2$ in 1983 [24]. The same approach was later used for the deposition of numerous different metal oxides, such as CdO [25,26], ZrO_2 [27], TiO_2 [28,29] and several mixed oxides [30,31]. For these materials, however, the electrodeposition process only yields metastable amorphous hydroxides, which can only be converted to crystalline oxides by post-annealing at high temperatures (typically $>400\,°C$). Direct formation of a nanocrystalline oxide phase was only reported for the electrodeposition of SnO_2 [32], In_2O_3 [25] and Fe_2O_3 [33]. Also in these cases, however, full crystallinity of the materials could only be achieved by annealing. ZnO is virtually the only metal oxide for which perfectly crystallized thin films can directly be electrodeposited, which is caused by its high chemical reactivity. Dissolution and recrystallization reactions occur at the surface of the growing ZnO film during electrochemical growth, so that the deposited material can reorganize into perfect ZnO crystals.

Other cathodic electrodeposition methods for metal oxides involve the substitution of NO_3^- by other oxidants, for example, O_2 or H_2O_2, in case of ZnO electrodeposition. Details will be given in the next section. Besides, it should be mentioned that anodic electrodeposition methods for oxides have also been developed. For example, TiO_2 can be electrodeposited from $TiCl_3$ solutions by oxidation of Ti(III) to Ti(IV) [34–37]. Another anodic method for the electrodeposition of TiO_2 involves stable basic solutions of Ti(IV)-alkoxides, from which Ti-oxo-hydroxide precipitates upon lowering the pH by electrochemical reduction of hydroquinone derivatives or water [38,39].

It should also be mentioned that the methods for the deposition of compounds are often not described as actual "electrodeposition" but rather as "electrochemically triggered chemical precipitation." This is because, unlike the deposition of metals, the deposition of a compound is usually not a simple Faradic reaction, but the precipitation involves reaction with other species in the solution that do not undergo an electrochemical reaction. For example, in the case of metal oxide deposition from nitrate solution, the metal ions do not undergo any electrochemical reaction, but only react with the OH^- ions locally generated at the electrode surface. Therefore, where the supply of the metal oxide ions at the electrode surface is not sufficient, some of the generated OH^- ions might diffuse away from the electrode surface. This usually leads to yields of below 100% for metal hydroxide or metal oxide generation.

11.4 Electrodeposition of Zinc Oxide

11.4.1 Electrodeposition of Pure ZnO

As already mentioned in the foregoing section, ZnO can be synthesized electrochemically from aqueous solutions of zinc salts containing NO_3^-, O_2 or H_2O_2 as oxidants. In 1996, Izaki *et al.* [40,41] and Peulon *et al.* [42,43] independently discovered methods involving cathodic reduction of nitrate ion and dissolved oxygen, respectively. While the deposition from nitrate solution was carried out with rather high (typically 0.1 M) zinc nitrate concentrations in the original work, deposition from oxygen-saturated solution was carried out using rather low (typically 5 mM) concentrations of $ZnCl_2$ with the addition of KCl as a supporting electrolyte. Later, $Zn(ClO_4)_2/LiClO_4$ solutions were also used for deposition from O_2-saturated solutions, which avoids contamination of the deposited ZnO with chloride ions caused by the predominance of $ZnCl^+$ complexes in aqueous solution containing Zn^{2+} and high concentrations of Cl^- at temperatures above 50 °C [44]. Finally, in 2001, Pauporté *et al.* proposed ZnO electrodeposition by reduction of hydrogen peroxide [45–47]. All these cathodic reductions lead to the formation of OH^- ions as follows:

$$NO_3^- + H_2O + 2e^- \rightarrow NO_2^- + 2OH^- \qquad E^0 = -0.24\ \text{V versus SCE} \tag{11.6a}$$

$$O_2 + 2H_2O + 4e^- \rightarrow 4OH^- \qquad E^0 = +0.12\ \text{V versus SCE} \tag{11.6b}$$

$$H_2O_2 + 2e^- \rightarrow 2OH^- \qquad E^0 = +0.66\ \text{V versus SCE} \tag{11.6c}$$

The hydroxyl ions then react with Zn^{2+} ions to precipitate zinc hydroxide, which is converted to fully crystalline ZnO by dehydration at elevated temperature:

$$Zn^{2+} + 2OH^- \rightarrow Zn(OH)_2 \rightarrow ZnO + H_2O \tag{11.7}$$

The overall reactions in the nitrate, oxygen and hydrogen peroxide systems can therefore be written as:

$$Zn^{2+} + NO_3^- + 2e^- \rightarrow ZnO + NO_2^- \tag{11.8a}$$

$$Zn^{2+} + {}^1\!/_2O_2 + 2e^- \rightarrow ZnO \tag{11.8b}$$

$$Zn^{2+} + H_2O_2 + 2e^- \rightarrow ZnO + H_2O \tag{11.8c}$$

Details about the temperature dependence of ZnO electrodeposition from O_2-saturated $ZnCl_2$ solution were reported by Goux *et al.* in 2005 [48]. Below 34 °C, current transients show low current values and a continuous decrease in the deposition current, suggesting that the electrode is more or less passivated by the deposition of a poorly conducting layer, probably zinc hydroxide. Consequently, very thin films of only 10–20 nm were obtained after 1 h. Oxide nucleation and film growth start above 34 °C, however optimum film transparency and full crystallinity are only obtained above 40 °C. Above 70 °C the formation of the ZnO seeds takes place only 1 or 2 s after the potential application, while it still takes 1–2 min at 40 °C and about 20 s at 50 °C. Therefore, ZnO electrodeposition is usually carried out at 70 °C.

Looking at the reduction reactions, the reduction of the nitrate ion is a kinetically slow reaction compared to the reduction of O_2 and H_2O_2. Nitrate ions are usually inert and can be reduced only in the presence of certain metal cations [49–51]. In this case, the Zn^{2+} ions serve as catalyst for the reduction of nitrate, so that the rate of ZnO growth from zinc nitrate solution was found to be influenced by the surface concentration of Zn^{2+} ions that follows Langmuir-type adsorption on ZnO [52]. Also, due to other reasons, electrodeposition from aqueous nitrate solution proved to be less advantageous, especially compared to deposition from O_2-saturated solution. The films proved to be less homogeneous and poorly reproducible. This is partly due to a hitherto unidentified change in zinc nitrate solutions with ageing. On the other hand, deposition of ZnO from nitrate solution was successfully used by Sumida *et al.* [53] as well as by Liu *et al.* [54] for the preparation of inverse opal structures of ZnO using opal layers of polystyrol spheres as templates. This method was modified by Yang *et al.* [55], who used a 1 : 1 mixture of water and ethanol as solvent for the same purpose. As for deposition from hydrogen peroxide solution, it is believed that it is only a variation of the method involving O_2, since H_2O_2 is known to decompose with formation of H_2O and O_2, especially at elevated temperature [56], so that the H_2O_2 can only be seen as an O_2-supplying agent and not as an electrochemically active species.

Due to the high overpotentials for the cathodic reductions involved, electrodeposition of ZnO has to be carried out at potentials of −0,7 V versus SCE or more negative. However, a limit is given by the reduction of Zn^{2+} to elementary Zn, which starts to occur at around −1.15 V versus SCE. If the electrodeposition is carried out on TCO substrates such as indium tin oxide (ITO) or F-doped tin oxide (FTO), activation of the substrate prior to the ZnO deposition is often done in order to obtain more seed crystals and thereby faster film growth and more homogenous films. This activation can be achieved either by dipping the substrate into 2 M nitric acid for two minutes [44] or by pre-electrolysis in O_2-satured solution without Zn^{2+} for about 15 minutes [57].

Recently, a method for anodic deposition of ZnO was also developed by Switzer and co-workers [58]. The method exploits the amphoteric nature of ZnO and starts with an alkaline solution of Zn^{2+} at pH 13, where Zn^{2+} is soluble due to formation of $Zn(OH)_3^-$ and $Zn(OH)_4^{2-}$ ions. The local pH at the working electrode is then electrochemically decreased by oxidizing ascorbate dianions, leading to the precipitation of ZnO according to the equations

$$Zn(OH)_3^- + H^+ \rightarrow ZnO + 2\,H_2O \tag{11.9a}$$

$$Zn(OH)_4^{2-} + 2H^+ \rightarrow ZnO + 3H_2O \tag{11.9b}$$

Finally, ZnO electrodeposition from nonaqueous solution has also been tested. Jayakrishnan and Hodes carried out deposition from oxygen-saturated solutions of $ZnCl_2$/LiCl or $Zn(ClO_4)_2$/$LiClO_4$ in dimethyl sulfoxide (DMSO) at temperatures between 30 and 150 °C [59]. However, films with a rather high resistivity, probably mainly consisting of $Zn(OH)_2$ or ZnO(OH) species, were obtained. The reason for the tendency towards the formation of these species, which would actually more likely be expected for deposition from aqueous solution, remains unknown.

11.4.2 Electrodeposition of Doped ZnO

Electrodeposited ZnO, like that prepared by nonelectrochemical solution methods, is usually n-type, due to the oxygen deficiencies, that is, the presence of Zn atoms, which represent electron donors. Anthony *et al.* found that the kind of electrodeposition solution can have an influence on the resultant doping level [60]. For example, ZnO deposited from $Zn(NO_3)_2$ solution showed a slightly higher bandgap than that deposited from $Zn(ClO_4)_2$ solution. This was explained by the slower OH^- formation in case of the zinc nitrate bath, leading to more oxygen defects and a higher doping level. On the other hand, ZnO deposited from zinc nitrate solution with the addition of H_2O_2 led to a smaller bandgap, caused by enhanced OH^- generation and a resultant lower density of oxygen defects and a lower doping level.

Doping of electrodeposited ZnO with other electron donors has long been inaccessible using aqueous methods. However, electrodeposition has been successfully employed for this purpose. The main purpose has been the preparation of transparent front contact layers in CIGS and CdTe solar cells. Higher conductivity of these layers has been achieved by incorporation of B [61,62], Al [63,64] or In [63,65] during electrodeposition of ZnO from aqueous zinc nitrate solution, by addition of dimethylamine borane, aluminum trichloride or indium nitrate to the electrodeposition bath. For moderate doping levels of 1–2%, the typical hexagonal crystal structure of ZnO is preserved in the obtained materials.

Doping of electrodeposited ZnO for other applications has included the incorporation of nickel and cobalt [66,67]. The crystals exhibited dilute ferromagnetic behavior, which is interesting with respect to spintronics. Doping with cobalt by deposition from mixed solutions of zinc nitrate and cobalt nitrate was also found to improve the photocatalytic activity of ZnO in photocatalytic water splitting (see Section 11.8) [68]. Goux *et al.* reported the doping of ZnO with Er^{3+} [69]. Films deposited in the presence of 5 mM Zn^{2+} and 0.15 to 0.35 mM Er^{3+} exhibited well-crystallized hexagonal columns of Er-doped ZnO with a bottom layer rich in Er, which proved to be interesting materials for electroluminescence. Similar results were also obtained for the electrochemical co-deposition of ZnO and Eu^{3+} [70].

The existence of ZnO with stable p-type doping is still the subject of debate. Some techniques have been proposed for the preparation of p-type ZnO, for example metal–organic chemical vapor deposition of Sb-doped ZnO [71], preparation of Ga-doped [72], Mg-doped [73] and P-doped [74] ZnO by pulsed laser deposition, and deposition of N-doped [75] and P-doped [76] ZnO by radio frequency magnetron sputtering. An electrochemical method was proposed by Lan *et al.*, who added bismuth nitrate in different concentrations to the zinc nitrate solutions used for ZnO electrodeposition [77]. The Bi content in the deposits could be varied between 0.09 and 1.58 atom %. The films exhibited p-type conduction, and a film with a Bi content of 1.58 atom % possessed a four orders of magnitude improvement in conductivity compared to that of undoped ZnO.

11.4.3 P-n-Junctions Based on Electrodeposited ZnO

Due to its tendency to form nanorods in solution methods (see Section 11.5), thin compact ZnO films are hard to obtain by electrodeposition. In fact, front contact layers in CIGS and CdTe solar cells are usually still made by gas-phase methods, such as electron-beam evaporation [78] or sputtering [64,79] for this reason. For the same reason, only few studies exist in which electrodeposited ZnO films have been used as a component in p-n bulk heterojunctions. Ae *et al.* reported the electrochemical formation of p-n junctions consisting of n-type ZnO and p-type CuSCN, which showed some potential for ZnO-based UV-light-emitting diodes [80]. Subsequent electrodeposition of ZnO and Cu_2O on ITO led to the formation of a p-i-n heterojunction [81]. Mukherjee *et al.* prepared p-n junctions by cathodic electrodeposition of ZnO thin films on p-type single-crystal silicon substrates from $ZnSO_4$ solutions [82].

Since reliable methods for the electrodeposition of p-type ZnO are not yet known, ZnO-based p-n-homojunctions have also not been prepared to date. Regarding possible future developments, it has to be kept in mind that p-n-homojunctions, for example, made from silicon, are usually prepared starting from either n- or p-type wavers and changing the doping in a particular part by redoping from the gas phase. If reliable gas-phase methods for the redoping of n-ZnO to p-ZnO can be found, ZnO-based p-n-homojunctions could be prepared on the basis of n-ZnO films electrodeposited by currently known methods.

Studies involving nanostructured p-n heterojunctions based on electrodeposited n-ZnO nanorod arrays or electrodeposited nanoporous n-ZnO films are discussed in the following sections.

11.5 Electrodeposition of One- and Two-Dimensional ZnO Nanostructures

11.5.1 ZnO Nanorods

Due to its high chemical reactivity and the dissolution and recrystallization reactions occurring during film growth, it is possible to obtain highly ordered arrays of ZnO crystals by electrodeposition. This is best exhibited in its heteroepitaxial growth on single crystal GaN substrates [58,83]. ZnO and GaN have the same Wurtzite structure and similar lattice constants ($a = 3.25$ Å and $c = 5.21$ Å for ZnO, $a = 3.16$–3.19 Å and $c = 5.13$–5.19 Å for GaN). Because of the small lattice mismatch, ZnO is electrodeposited epitaxially on GaN (0001) surfaces, leading to arrays of hexagonal ZnO crystals with their c-axis perpendicular to the substrate.

The nanorods present an amplified UV emission centered at 381 nm with an excitation threshold at 4.4 MW cm^{-2}, which is a promising result towards the use of this material for the preparation of UV nanolasers [84]. Heteroepitaxial electrodeposition of ZnO nanopillars was also observed on single crystalline Au substrate [85].

However, even on non-single-crystalline substrates, a preferential growth of ZnO crystals with their [0001] axis perpendicular to the substrate is observed. This is due to the faster crystal growth on the polar (0001) face compared to the faces perpendicular to the (0001) face, which are less polar. This intrinsic anisotropy in the growth rate of ZnO has already been observed for crystal growth by nonelectrochemical solution methods [86–88]. The reason for the anisotropy has been found in differences in the dissolution rates of different ZnO surfaces. Since ZnO deposition and dissolution compete with each other, higher dissolution rates lead to lower net growth rates, even if the electrodeposition process itself is assumed to be equally fast on all surfaces. About 20 years ago, Gerischer *et al.* performed a series of experiments on bulk ZnO, using a rather elegant electrochemical technique that enabled them to measure dissolution rates directly [89–91]. Their results showed a lower dissolution rate of the zinc-terminated (0001) plane compared to the oxygen-terminated (0001) and nonpolar (1010) planes. However, in the presence of complexing agents, such as ammonia and EDTA (ethylenediaminetetraacetic acid), they were able to change the dissolution rates dramatically. Anions were also found to influence the dissolution rates significantly, with an order of decreasing rates of $SO_4^{2-} > Cl^- > NO_3^- > ClO_4^-$.

Levy-Clement and co-workers have published a number of papers on the electrodeposition of ZnO nanowires on transparent conducting oxide substrates for use in extremely thin absorber (ETA) solar cells. In 2000 they reported electrodeposition of ZnO columns with diameters between 100 and 200 nm and aspect ratios >10 from oxygen-saturated aqueous solution containing 10 mM $ZnCl_2$ and 0.1 M KCl at 80 °C and −1.0 V versus SCE on SnO_2/glass substrates [92]. Later, the alignment of the nanowires was improved significantly by electrodeposition on a compact ZnO bottom layer with a thickness of 150 nm deposited on the conducting glass substrate, which also served to prevent short-circuiting inside the solar cell [93,94]. The bottom layers were deposited from a 0.1 M zinc acetate dihydrate and 0.2 M acetic acid solution in a 25 : 75 (by volume) water/ethanol mixture by spray pyrolysis on substrates preheated to 450 °C. ZnO nanowires with a length of up to 2 μm could be obtained by electrodeposition on these substrates.

The method was later modified by other authors. Anthony *et al.* electrodeposited ZnO nanowires using nitrate and H_2O_2 as oxidants on ZnO bottom layers prepared by spin coating using a saturated methanol solution of $Zn(CH_3COO)_2{\cdot}2H_2O$ on ITO glass, followed by heating at 350 °C for 30 min [60]. The nanowire growth was carried out at −1.0 V versus SCE from 5 mM aqueous zinc perchlorate solution with 2.5 mM H_2O_2 and lithium perchlorate as a supporting electrolyte, or from 5 mM aqueous zinc nitrate solution with potassium chloride as supporting electrolyte. Comparison with ZnO electrodeposited on bare ITO substrates showed that the presence of the ZnO bottom layer leads to a much better alignment, as well as a higher density of ZnO nanowires. Compared to the original method of Levy-Clement and co-workers, using O_2 as oxidant, however, shorter nanowires with only up to 500 nm length were obtained. Longer nanowires of up to 1 μm could be prepared by using lower zinc nitrate concentrations with KCl as supporting electrolyte [95].

Both groups reported the possibility of varying the thickness and length of the nanowires by variation of the concentration of the reactants and/or supporting salt in the electrodeposition bath. For instance, reducing the zinc nitrate concentration to 1 mM led to the formation of ZnO nanowires with smaller diameters of 25–50 nm, compared to 40–70 nm for 5 mM zinc nitrate. Increasing the zinc nitrate concentration led to an increase in the diameter of the nanowires. For electrodeposition from O_2-saturated $ZnCl_2$/KCl solution, variation of the KCl concentration was found to have a significant influence [96]; an increase in the KCl concentration considerably decreased the rate of O_2 reduction. The consequent decrease in OH^- production rate resulted in an increase in the ZnO deposition efficiency, from a value around 3% for 5×10^{-2} M KCl to more than 40% for 3.4 M KCl. The increase in the deposition efficiency mainly resulted in an enhancement of the longitudinal growth rate. However, very high KCl concentrations (>1 M) also favored the lateral growth of the ZnO nanowires, resulting in diameters as large as 300 nm (in comparison to a diameter of 80 nm typically obtained for concentrations <1 M). The observed effects were discussed in terms of Cl^- ion adsorption on the cathode surface, with preferential adsorption taking place on the polar (0001) ZnO surface, as also suggested by Xu *et al.* [97]. The Cl^- adsorption results in the stabilization of the (0001) ZnO surface, thereby decreasing the growth rate along the corresponding direction (*c*-axis). The ratio between the OH^- generation rate and Zn^{2+} diffusion to the cathode has been proposed as a major parameter in the electrodeposition of ZnO nanowires. For moderate increases in KCl concentration this ratio becomes lower, causing the observed increase in the deposition efficiency, but the Zn^{2+} ions are still exclusively consumed at the (0001) face. At very high KCl concentrations, not all the Zn^{2+} ions are consumed at the (0001) face any more, allowing the diffusion of some of them towards the (1010) faces, where they lead to the observed lateral growth of the nanowires.

Variation of the nature of the anions in the deposition solution was also found to influence the electrodeposition of ZnO nanowire arrays [98]. Replacement of Cl^- by ${SO_4}^{2-}$ or CH_3COO^- led to higher aspect ratios of the deposited nanowires. Nanowires with lengths of up to 3.4 μm were obtained by using CH_3COO^- instead of Cl^-. Obviously, ${SO_4}^{2-}$ and especially CH_3COO^- ions adsorb less strongly on the (0001) face of the ZnO and thereby promote longitudinal growth of the nanowires. This was directly seen in the cathodic current densities of 0.54, 0.76 and 0.85 mA cm^{-2} for the chloride, sulfate and acetate solutions, respectively, at a potential of −1.0 V versus SCE. Yet another means to influence the nanowire dimensions was found in variation of the ZnO bottom layer on which the nanowire arrays are grown. The nanowire diameter could be varied between 45 and 160 nm by using buffer layers with different crystal sizes [99]. Nanocrystalline layers provide a higher density of nucleation sites for nanowire growth, thereby leading to the deposition of more, but thinner, nanowires.

On the other hand, ZnO nanowires could also be electrodeposited without any pretreatment or seed layers, even on polished substrates, which is a significant advantage compared to nonelectrochemical wet chemical methods to grow ZnO, and makes the electrodeposition of ZnO even more interesting for industrial applications. This was demonstrated by Cui *et al.*, who electrodeposited ZnO nanowires on polished silicon surfaces [67]. The deposition was carried out using aqueous solutions of 10 mM zinc nitrate hydrate and 10 mM hexamethylenetetramine (HMTA). HMTA is a complexing agent also used in nonelectrochemical methods for the deposition of ZnO nanowires, whose detailed role is still under investigation [100,101]. ZnO nanowires with lengths of several micrometers and diameters around 200 nm were obtained.

ZnO nanorods were also electrodeposited on n-type and p-type Si wavers by Konenkamp *et al.* [102,103]. High-quality nanowires can be grown on strongly doped n-type

Si, under similar conditions as used for other highly conductive substrates. For low electron concentrations occurring in weakly n-type or in p-type wafers, nanowire growth is inhibited. This difference allows selective growth in strongly n-type areas. The inhibited growth on weakly n-type and p-type wafers can be improved by applying higher cathodic electrode potentials or by illuminating the growth area [103]. Interestingly, nanorods with a pentagonal cross-section were obtained on p-type Si substrates at smaller cathodic potentials. The pentagonal wires have a typical length of 1–2 μm and grow out of an inhomogenous nanocrystalline thin film, which is initially formed. The occurence of the pentagonal morphology was explained in terms of kinetic limitations for electron transport in the p-Si substrate and altered chemical conditions at the growth surface [102].

ZnO nanowire arrays with an even higher order, including the exact determination of the location of each nanowire, can be obtained by using electrodeposition in conjunction with electron-beam lithography [104,105]. In these studies, ZnO rods were grown on gold-coated substrates patterned with polymethylmethacrylate (PMMA). Scaling down the structure of the PMMA enabled both groups to nucleate single rods in the exposed parts of the gold film. The PMMA layers could be spin-coated from 4% solutions in anisole, followed by patterning of the layers using a scanning electron microscope (SEM). After exposure and development, a substrate with template holes was placed in a zinc nitrate solution for ZnO electrodeposition. ZnO nanorods have also been electrodeposited using porous polycarbonate or anodic alumina membranes as templates. By controlling the electrodeposition process, for example the applied potential, it was found that nanowires could be grown as single crystals of ZnO, as well as polycrystalline Zn or ZnO, or composites of both [106]. The method was later modified by Ren *et al.*, who first deposited Zn nanorods on the templates, followed by their oxidation to polycrystalline ZnO nanorods [107].

The approach of turning Zn into ZnO had already been applied before by Lopez and Choi [108]. They electrodeposited Zn nanofibers based on a type of dendritic growth mechanism, which could be initiated when the deposition rate exceeded the diffusion rate of the Zn^{2+} ions, creating diffusion-limited conditions [109]. Subsequent transformation into ZnO nanofibers was performed by heating at 450 °C for 90 min. Fibrous ZnO electrodes obtained by this method cannot be directly obtained by electrodeposition, because the current density required to fulfill this condition cannot be obtained by potentials that allow the deposition of pure ZnO. The zinc films were electrodeposited from DMSO solutions containing 0.03 M $Zn(ClO_4)_2{\cdot}6\ H_2O$ as the zinc source and 0.1 M $LiClO_4{\cdot}3\ H_2O$ as the supporting electrolyte. DMSO was chosen as solvent because it offers a wider range of deposition temperatures due to its high boiling point at 189 °C. A deposition potential of −1.5 V versus SCE was found to be optimum, since more positive potentials lead to the co-deposition of nonfibrous ZnO and more negative potentials lead to poor adhesion of the fibrous Zn film on the substrate.

11.5.2 ZnO Nanotubes

The electrodeposition of ZnO nanotubes (ZNT) instead of nanorods may be desirable in view of many applications, including photovoltaics and photocatalysis due to the potentially higher surface area. In early 2007, Tang *et al.* reported a potentiostatic deposition of ZNT on F-doped SnO_2 by first forming nanorods, which are then self-assembled into a hexagonal circular nanotube shape at a deposition bath temperature of 80 °C after deposition times of 60 min and longer [110]. Important points in this method are the very low $ZnCl_2$ concentration of 0.0001 M and the rather positive deposition potential of −0.7 V versus SCE, both leading to a slow ZnO deposition and thereby enabling the rearrangement.

Around the same time, Xu *et al.* published a two-step method, where electrodeposited nanorods are turned into nanotubes by subsequent coordination-assisted selective dissolution along the *c*-axis, based on the preferential adsorption of ethylenediamine and OH^- on different crystal faces [111]. Levy-Clement and co-workers further developed this approach into a three-step method [112]. After electrodeposition of ZnO nanorods, the core of ZnO nanowires is selectively etched in a KCl solution, where ZnO dissolution of the nanowire core occurs for KCl concentrations ≥ 1 M and the etching rate can be enhanced by increasing the temperature. Arrays of ZnO with tailored dimensions (200–500 nm external diameter and 1–5 μm length) are obtained by varying the conditions of nanowire-array deposition and dissolution. The optional third step enables even more precise control of the nanotube wall thickness and is achieved by performing a further ZnO electrodeposition step.

Another approach to the electrochemical formation of nanotubes was chosen by She *et al.*, who also first published their method in 2007 [113]. The procedure is based on the cathodic electrodeposition of ZnO nanorods, like all other methods discussed here, however, the etching of the nanorods to form nanotubes is also done by an electrochemical method, namely electrochemical etching by H^+ formed by the anodic oxidation of water. The advantage of this purely electrochemical process is that it can be completed in less than two hours, while the etching methods in the processes proposed in [111] and [112] usually take 10 hours or more.

11.5.3 Two-Dimensional ZnO Nanostructures

Vertically grown planar ZnO nanowalls, with typical dimensions of 40–80 nm thickness and several micrometers wide, were electrodeposited on glass and polymer substrates covered with ITO by Pradhan *et al.* [114,115]. The structures were generated at 70 °C from solutions containing 0.1 M zinc nitrate as well as 0.1 M KCl, the KCl obviously making the difference compared to electrodeposition from 0.1 M zinc nitrate solution without KCl, where rather compact films are obtained. The lateral growth of the nanowalls appeared to cease when one nanowall meets another with a preferred angle, between 60 °C and 90 °C, leading to rather ordered arrays of nanowalls. These ZnO nanowall arrays exhibit excellent field emission performance, with not only a considerably lower turn-on field of 3.6 V μm^{-1} (at 0.1 μA cm^{-2}), but also a higher current density of 0.34 mA cm^{-2} at 6.6 V μm^{-1} than previously observed with ZnO nanowires and other one-dimensional ZnO nanostructures.

At lower zinc nitrate concentrations of 0.05–0.06 M, two-dimensional ZnO nanoplates [97,116] and nanosheets [117] have been obtained under otherwise similar conditions. The growth of these nanostructures appeared to be more randomly oriented and to terminate without any physical obstructions. This may be due to slower growth kinetics at the lower zinc salt concentrations. The use of even lower zinc nitrate concentratrions (0.001–0.05 M) led to the formation of ZnO nanorods [115]. On the other hand, increased zinc nitrate concentrations of up to 0.5 M led to the formation of ordered arrays of ZnO walls with increased wall thickness of up to about 1 μm, enabling the tuning of the wall thickness by changing the zinc nitrate concentration [114].

11.6 Use of Additives in ZnO Electrodeposition

The disadvantage of one- and two-dimensional ZnO structures for use in solar cells and photocatalysis is their generally smaller surface area compared to three-dimensional structures at the same film thickness. The electrodeposition of such structures has been achieved by the use of organic structure-directing additives in the electrodeposition bath. The use of organic additives in electrodeposition has actually been known for a quite long time, but the additives have typically been used to make electrodeposited films smoother and more homogenous. A good example of this is again the Damascene process for the electrodeposition of Cu, where void-free filling of the narrow trenches in the patterned Si substrates is realized by certain organic catalysts, which preferably adsorb on concave surfaces and thereby accelerate the growth rate at the bottom of a trench [118]. While the adsorption of these additives takes place reversibly, additives that adsorb more strongly at the growing surface are needed to create porous ZnO. Such additives need to be incorporated into the films to form ZnO/organic hybrid films, which can then be converted into pure porous ZnO films by removal of the organic additive.

11.6.1 Dye Molecules as Structure-Directing Additives

Organic dye molecules are, interestingly, the kind of additive that give ZnO films of the highest porosity by electrodeposition to date. The approach goes back to a concept with the original goal to electrodeposit ZnO/dye hybrid films for use in dye-sensitized solar cells (DSSCs) in a single step. To achieve this, water-soluble dyes were simply added to the aqueous zinc nitrate solution used for the electrodeposition of ZnO. Among the first dyes used in this concept were water-soluble phthalocyanine dyes such as 2,9,16,23-tetrasulfophthalocyaninatometal(II) (TSPcMe, Me = Zn, Al(OH), $Si(OH)_2$), yielding blue-colored transparent ZnO/phthalocyanine films [119–121]. The presence of both the phthalocyanine dyes and crystalline ZnO in the films was confirmed by absorption spectra and X-ray diffraction patterns, respectively. However, only small photocurrents were obtained with such "as-deposited" ZnO/dye hybrid films, which was attributed to the formation of dye aggregates in the ZnO pores and the trapping of some of the dye molecules inside ZnO grains, thus making them inaccessible to the redox electrolyte of the solar cell (cf. Section 11.7) [120,121].

Despite the poor photoelectrochemical performance of the "as-deposited" films, the method caught attention, since it was soon discovered that the dye molecules could be easily extracted by aqueous KOH (pH 10.5) from some of the ZnO/dye hybrid films, yielding highly porous films purely consisting of ZnO. The highest dye contents in the hybrid films and therefore the highest porosities in the ZnO films after dye removal were achieved in films deposited with the disodium salt of eosin Y (Figure 11.1). The deposition bath typically contains 5 mM $ZnCl_2$, 0.1 M KCl supporting electrolyte and eosin Y at several tens of μM. The pore volume after extraction of eosin Y is about 50% of the total film volume. The change of the film from a dense to a porous structure by alkaline treatment was also supported by Kr sorption measurements [123,124]. While the as-deposited ZnO/eosin-Y film showed no porosity in these measurements, the roughness factor (measured surface area/projected film area) increased to values as high as 400 after dye desorption. These films proved to be highly suitable for the preparation of efficient DSSCs after readsorption of a monolayer of eosin Y, or another dye with a broader absorption band, by soaking the films in a dye solution. The dye desorption and readsorption process is illustrated in Figure 11.2. Details of the photoelectrochemical

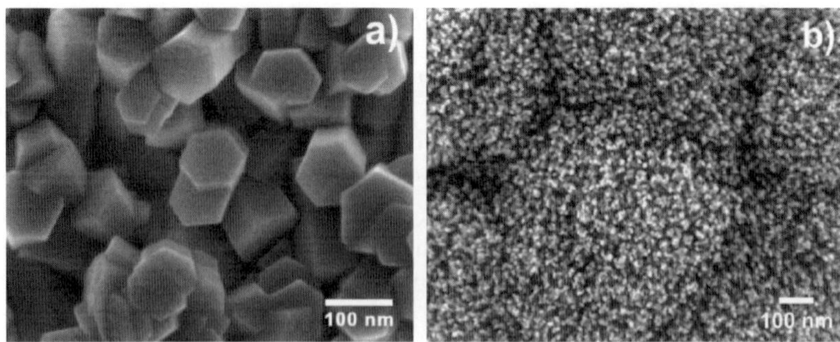

Figure 11.1 SEM of ZnO films electrodeposited from oxygen-saturated $ZnCl_2$ (5 mM)/KCl (0.1 M) solution at 70 °C and −1.0 V versus SCE for 20 min: (a) without additive, (b) with 50 μM eosin Y in the deposition solution after removal of the dye by extraction in aqueous KOH solution (pH 10.5). (Reprinted with permission from T. Oekermann, T. Yoshida, C. Boeckler *et al.*, Capacitance and field-driven electron transport in electrochemically self-assembled nanoporous ZnO/dye hybrid films, *Journal of Physical Chemistry B*, **109**(25), 12560, 2005. © 2005 American Chemical Society.)

properties of such films and their performance in DSSCs after readsorption of efficient sensitizer dyes with broad absorption bands will be given in Section 11.7.

High dye content and porosity of ZnO films deposited with eosin Y is achieved especially at deposition potentials more negative than −0.9 V versus SCE [125–127]. In this potential region the dye is reduced and forms a strong complex with Zn^{2+}, which also means that it strongly interacts with the ZnO surface. In addition, complex formation with the Zn^{2+} also seems to promote the formation of dye aggregates by formation of a polymeric structure, which can then act as template for the pores [128]. Details on eosin Y reduction and complex formation with Zn^{2+} were reported by Goux *et al.* [129]. Without reduction, eosin Y is present in aqueous solution as the EY^{2-} ion. The COO^- group in EY^{2-} forms a cycled carboxylate, explaining the

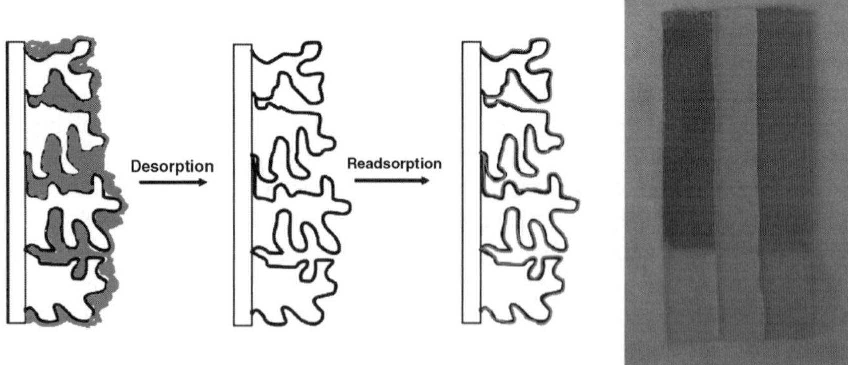

Figure 11.2 Illustration of the dye desorption/readsorption process. The photo shows three parts of the same ZnO film deposited with eosin Y from O_2-saturated solution in all three stages. The virtual absence of any dye in the film after the desorption process (middle part) and the lower dye content after the readsorption process (right part) compared to the as-deposited film (left part) are clearly seen in this photo. See also color-plate section.

lack of interaction with Zn^{2+}. At negative potentials, the EY^{3-} radical is formed, in the absence of Zn^{2+}, as proven by electron spin resonance (ESR) spectroscopy. However, in the presence of Zn^{2+}, EY^{3-} is not stable and is further reduced to EY^{4-}, which is the species that forms a stable complex with Zn^{2+}. EY^{4-} is actually colorless, so that ZnO/eosin Y films deposited at potentials < -0.9 V versus SCE also appear colorless shortly after the deposition. They become colored only after reoxidation of EY^{4-} to EY^{2-} by oxygen from the air.

Recent investigations show, however, that, at potentials more positive than -0.9 V versus SCE, porous ZnO films can also be obtained by co-deposition with eosin Y. In this case, a higher dye concentration in the electrodeposition bath is necessary in order to achieve enough dye aggregation and dye adsorption on the ZnO surface during film growth. Lower eosin Y concentrations at more positive potentials lead to the formation of nonporous ZnO/eosin Y hybrid films from which the dye cannot be desorbed, since only a few eosin Y molecules adsorb onto the ZnO surface during film growth in this case, and are eventually completely surrounded by the growing ZnO [130].

Other investigations show that the ZnO–dye and dye–dye interactions are also strongly influenced by the concentration of the supporting electrolyte in the electrodeposition bath. A decrease in the KCl concentration leads to the incorporation of more eosin Y into the hybrid films [131]. This shows competition between the interaction of eosin Y with Zn^{2+} on the one hand and K^+ on the other hand. A lower concentration of K^+ leads to more interaction with Zn^{2+} and therefore to stronger ZnO–dye and dye–dye interaction. Consequently, lower eosin Y concentrations are needed to achieve the formation of a porous ZnO structure if lower KCl concentrations are employed. Another interesting feature of ZnO electrodeposition with eosin Y is that the dye was found to catalyze film formation in O_2-saturated solution, that is, that it seems to catalyze the cathodic reduction of O_2 [132]. This is not the case for ZnO deposition from nitrate solution, where eosin Y was even found to hinder the film formation at high concentrations [127]. Especially for formation of ZnO/eosin Y hybrid films, electrodeposition from O_2-saturated zinc salt solution is therefore the preferred method.

A further feature of electrodeposition with eosin Y is that the porosity and the pore size can be tuned to a certain range by changing the eosin Y concentration in the deposition bath [128]. Higher eosin Y concentrations lead to more incorporation of eosin Y and therefore to a more porous ZnO after desorption. Furthermore, the porous structure tends to become finer, with smaller pores and thinner ZnO walls, at higher eosin Y concentrations.

The addition of dye molecules was not only found to lead to the formation of pores in the electrodeposited ZnO, it was also found to influence the crystallographic orientation of the deposited ZnO on the substrate. The most striking example is the hybrid thin film electrodeposited in the presence of 2,9,11,23-tetrasulfophthalocyaninato-dihydroxosilicon(IV) (TSPcSi), where a strong crystallographic orientation with the *c*-axis parallel to the substrate was found, making a clear contrast to the preferred orientation without additive, with the *c*-axis perpendicular to the substrate [132–134]. This led to a unique film morphology, with stacks of ZnO platelets standing on the substrate (Figure 11.3a), and XRD patterns revealing that the planes of the platelets correspond to the (001) face of ZnO. The change in the film structure and preferential crystallographic orientation on the substrate can be understood by preferential adsorption of the dye molecules on certain ZnO surfaces. In case of TSPcSi, the dye molecules adsorb preferentially on the (001) planes of ZnO through their sulfonic acid groups, so that crystal growth along the *c*-axis is hindered and takes place only in the direction of the (100) and (110) planes (Figure 11.3b). A similar effect and film morphology was found for

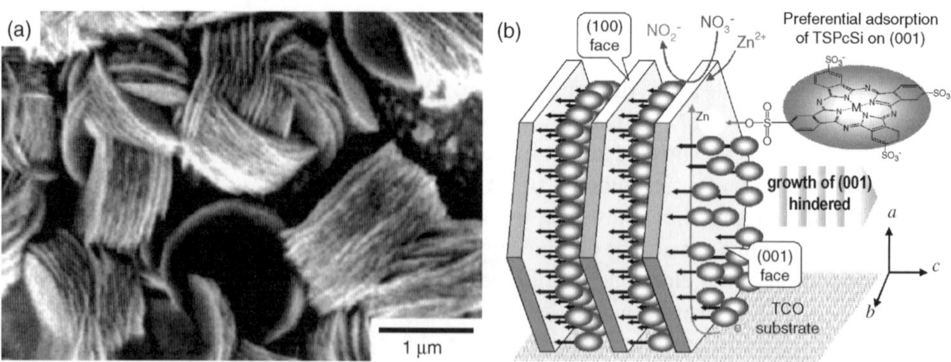

Figure 11.3 (a) SEM micrograph of a ZnO/TSPcSi film electrodeposited from 0.1 M $Zn(NO_3)_2$/50 μM TSPcSi mixed aqueous solution at −0.9 V versus SCE. (Reprinted from T. Oekermann, T. Yoshida, D. Schlettwein *et al.*, Photoelectrochemical properties of ZnO/tetrasulfophthalocyanine hybrid thin films prepared by electrochemical self-assembly, *Physical Chemistry and Chemical Physics*, **3**, 3387, 2001, PCCP Owner Societies.) (b) Growth model explaining the crystallographic orientation and formation of ZnO platelets in electrodeposited ZnO/TSPcSi films. See also color-plate section.

nonelectrochemical precipitation of ZnO from zinc salt solution containing citrate ions, which are also assumed to adsorb most strongly on the polar (001) face [135]. In addition, the presence of the two OH^- ions coordinated to the Si central metal ion on both sides of the TSPcSi structure may play an additional role in the generation of the ZnO platelets, as they may form bridges between ZnO layers. This is also proven by the fact that TSPcAl (only one OH^-) and TSPcZn (no OH-) do not lead to the formation of platelets [121]. Furthermore, it was recently shown that even with TSPcSi, no platelets were formed if OH^- was replaced by Cl^- [136].

While the ZnO/phthalocyanine films were almost nonporous with the dyes difficult to desorb, coumarin 343 (C343) was found as another example of an additive giving the same crystallographic orientation as TSPcSi, but able to be completely desorbed, yielding ZnO films almost as porous as those deposited with eosin Y. Both films, those deposited with eosin Y, as well as those deposited with C343, give a single crystalline porous structure after desorption of the dye, although with opposite crystallographic orientations [129]. This is seen in the diffractograms shown in Figure 11.4. As already discussed in Section 11.5, electrodeposition without additives (diffractogram (a)) gives preferential 001 orientation as seen in the more intense (002), but less intense (100), (101) and (110) peaks compared to ZnO powder (diffractogram (d)). Electrodeposition in the presence of eosin Y gives an even more distinct 001 orientation, as seen by the fact that the (100), (101) and (110) peaks almost disappear (diffractogram (c)). This shows that eosin Y preferentially adsorbs on the (100), (101) and (110) planes, thereby suppressing further ZnO growth on these planes and promoting growth in the direction of the (001) plane. In contrast, in the diffractogram of the film electrodeposited with C343, the (001) peak totally disappears, while the (100) peak becomes most intense, indicating preferential adsorption of the C343 molecules on the (001) plane.

It should be mentioned at this point that similar porous structures and crystallographic orientations as those obtained with eosin Y as additive were also obtained by O'Regan *et al.* in 2000 by ZnO electrodeposition from propylene carbonate solution containing 0.15 M $LiNO_3$, 0.05 M $ZnCl_2$ and either 5 mM $Zn(SO_3CF_3)_2$ or 5 mM $Zn(NO_3)_2$ hydrate [137]. However, the

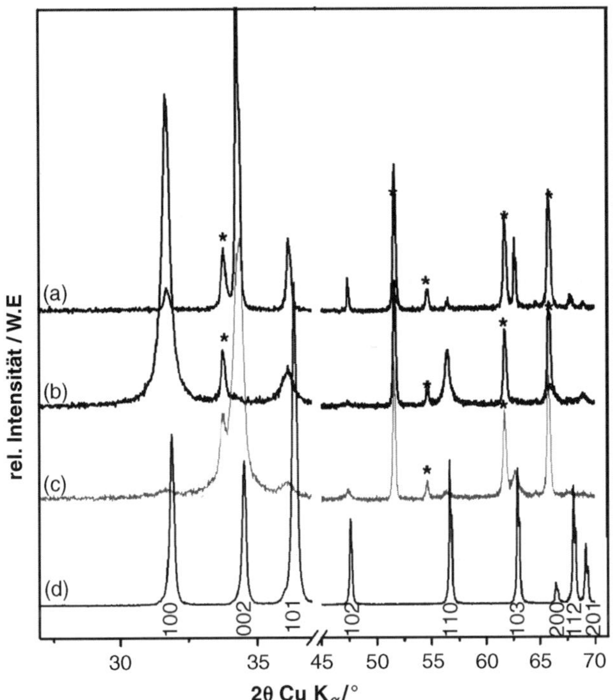

Figure 11.4 X-ray diffractograms of ZnO films electrodeposited: (a) without additive, (b) with coumarin 343 and (c) with eosin Y. A powder diffractogram of ZnO is shown for comparison (d). Miller indices refer to wurtzite ZnO, stars indicate reflexes originating from the conducting FTO substrate.

maximum thickness of these films was rather low (1 to 2 μm), the porosity was lower than in the films deposited with eosin Y and the porous single-crystalline structures were not continuous through the whole film. Nevertheless, the films achieved quite good results in solid-state DSSCs (see Section 11.7.3). The reason for the formation of the porous structure was not investigated in detail; one possible influencing factor would be strong interaction of the surface of the growing ZnO film with the solvent. The porosity, pore size and crystal orientation of the ZnO films are dependent on the deposition voltage and solution composition. For example, complete replacement of chloride with SO_3CF_3 resulted in rotation of the *c*-axis of the ZnO from perpendicular to approximately parallel to the substrate (as seen when eosin Y is replaced by coumarin 343). Varying the voltage from −0.8 to −1.0 V versus Ag/AgCl changed the average column size from ∼17 nm to ∼42 nm.

11.6.2 ZnO Electrodeposition with Surfactants

Sol-gel-based supramolecular templating methods using surfactants have been essential in producing materials with ordered nanoporous structures of uniform pore sizes [138,139]. The surfactants form micelles above a certain concentration, the critical micelle concentration (cmc), which act as templates for the formation of pores in the synthesized materials [139]. This method can be easily coupled with dip-coating methods based on evaporation-induced self-assembly (EISA) to prepare nanoporous materials as thin-film-type electrodes [140,141]. The orientation of the mesostructure in the resulting films is determined by the thermodynamically stable packing of surfactant/inorganic aggregates when the substrate is exposed to air and the solvents are evaporated [140].

When surfactants are used in electrodeposition, however, the solid–liquid interface at the working electrode provides a different environment for surfactant assembly. Due to surface forces, micelles, also called surface micelles, can form at the interface, even when the surfactant concentration is lower than the cmc in solution [142]. In this case, the formation of the surface aggregates is mainly determined by the substrate properties (i.e., hydrophobicity, surface charge density), which are frequently different from those in free solution [143]. For example, Burgess *et al.* demonstrated potential-controlled transformation of hemimicellar aggregates of sodium dodecyl sulfate (SDS) aggregates into a condensed monolayer at a gold electrode surface by inducing rearrangement of surfactant molecules that can match the surface charges on the electrode [144].

The electrodeposition of ZnO in the presence of relatively small concentrations of amphiphilic molecules was first reported by Choi and co-workers [145–147]. Using the deposition from zinc nitrate solution, the influence of amphiphilic molecules with anionic groups, such as alkyl sulfates, sulfonates and phosphates, could be found. However, on the nanometer scale, the deposition of an open porous system, as achieved in sol-gel methods could not be observed. Instead, a lamellar structure with alternating layers of ZnO and amphiphilic molecules was reported, as seen in transmission electron microscopy (TEM) images (Figure 11.5a) and also in small-angle X-ray diffraction (SAXRD) measurements, where the distance between the ZnO layers was found to decrease with decreasing length of the hydrophobic chains, as expected. The formation of the lamellar phase was observed, even for surfactant concentrations far above the cmc for micelle formation in solution (up to 20 wt%), suggesting that the interfacial structure is not significantly affected by the bulk

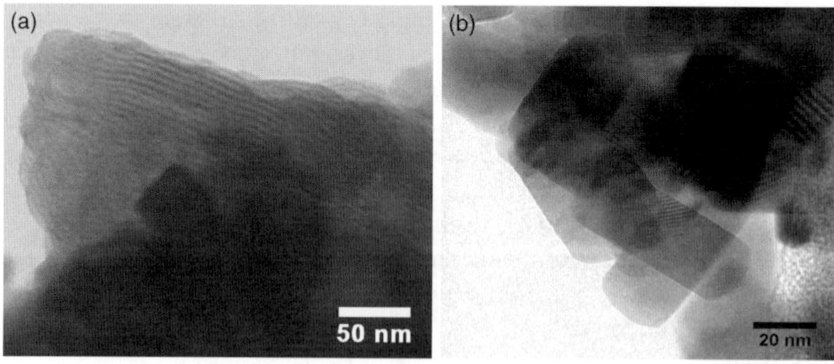

Figure 11.5 TEM images of: (a) the lamellar-structured phase and (b) the nanoparticulate phase in a ZnO film electrodeposited with decyl sulfate at a concentration of 6 mM, that is, slightly below the cmc, which was found at 8 mM. (Reprinted with permission from C. Boeckler, T. Oekermann, A. Feldhoff and M. Wark, The role of the critical micelle concentration in the electrochemical deposition of nanostructured ZnO films under utilization of amphiphilic molecules, *Langmuir*, **22** (22), 9427, 2006. © 2006 American Chemical Society.)

surfactant concentration. The quality of the lamellar structures incorporated in ZnO films was also dependent on the deposition potentials that determine the deposition rates [146]. A higher degree of ordering was achieved when a slower deposition rate ($< 0.15\,\mathrm{mA\,cm^{-2}}$) was used.

Interestingly, no influence of amphiphilic molecules with cationic groups such as quaternary alkyl ammonium as well as nonionic block co-polymers (e.g., Pluronic P123 and F127) on the electrodeposition of ZnO was observed. This observation can therefore be taken as a proof that, as already seen in the case of dye molecules, strong interactions between the structure-directing agents and the ZnO surface is a prerequisite to obtaining any influence on the film growth, and that this strong interaction can only be achieved with an anionic group. On the other hand, cationic surfactants can have an influence on the electrodepositon if an anionic surfactant is present as well. Results of ZnO electrodeposition in the presence SDS and cetyltrimethylammonium bromide (CTAB) showed an influence of the CTAB on the packing of the SDS–bilayer assemblies and their degree of ordering [147]. The best-quality lamellar ordering was achieved with 1.0 wt% SDS and 0.4 wt% CTAB. The addition of CTAB also led to a lower SDS concentration being required to generate a lamellar structure. The same effect, however, was later also achieved by using tetramethylammonium bromide (TAB), that is, CTAB without the hydrophobic chain [148]. This indicates that the amphiphilic nature of the cationic co-additive CTAB is not required in order to obtain the described effects. Investigations of the effect of various supporting cations and anions (e.g., NaCl, NaBr, NaI, Na_2SO_4) demonstrated that the effect seen with the addition of CTAB and TAB is primarily due to the cationic groups reducing the repulsion of SDS head groups and enhancing interactions between anionic Zn(II) species (i.e., $[Zn(OH)_4]^{2-}$) and anionic SDS.

To further study the influence of surfactants with cationic head groups, they were also tested as additives in the anodic electrodeposition of ZnO from basic solution, where almost all the Zn(II) is present in $Zn(OH)_4{}^{2-}$ complexes. Compared to the cathodic deposition of ZnO with SDS, this condition offers reversed polarity of the working electrode, as well as the Zn precursor and the additive, which should actually be a good match and also lead to the formation of a lamellar structure. However, addition of CTAB even at high concentrations (10 wt%) did not result in the formation of any nanostructures in the deposited ZnO films. One reason could be the much weaker interaction between the CTAB and the Zn^{2+} at the surface of the growing ZnO film.

For a more detailed understanding of the micelle formation during electrodeposition, the cmc has to be measured directly in the electrodeposition bath, since, typically, cmc values reported in the literature were obtained from measurements at room temperature in pure solvent [149]. Elevated temperatures and high concentrations of foreign ions, as used in the electrodeposition of ZnO are expected to change the cmc. The cmc values were therefore determined for several alkyl sulfates and alkyl sulfonates with different chain lengths, experimentally, directly in the electrodeposition bath by surface tension measurements. With increasing surfactant concentration the surface tension decreases until the formation of micelles starts and the surface tension remains constant. The results of the measurements showed that cmc values in the electrodeposition bath are significantly lower than those obtained in water. The rather high concentration of foreign ions more than compensates for the effect of the increased temperature on the cmc, which can be expected based on results of other authors, since an increase in the temperature of 50 K was usually found to increase the cmc by a factor $\ll 2$ [150,151], while high salt concentrations typically decreased the cmc by almost one order of magnitude [151].

Subsequently, electrodeposition of ZnO films was performed in the presence of surfactant concentrations both above and below the cmc in order to elucidate the influence of micelle formation on the structure and porosity of the deposited films [152]. Two different ZnO phases were found, depending on the surfactant concentration in the electrodeposition bath: at concentrations above the cmc a single phase with a lamellar structure was observed; at concentrations below the cmc, however, a second phase appeared. Energy-dispersive X-ray spectroscopy (EDXS) results revealed high sulfur content for the lamellar-structured phase, proving the incorporation of surfactant layers between the ZnO sheets, as found in earlier studies. The second, nanoparticulate phase (Figure 11.5b) was found to contain no sulfur in EDXS measurements, which means that no surfactant molecules were incorporated. Kr adsorption measurements at films templated with dodecyl sulfate after desorption of the surfactant with ethanol showed a porosity of $76\,cm^2\,cm^{-2}$ for films with lamellar structure, while no significant Kr uptake was detected for films consisting of nanoparticles, indicating that the particles have no inner porosity. On the other hand, the almost spherical ZnO particles of the nanoparticulate phase are surprisingly small in comparison with ZnO deposited in the absence of surfactant molecules, in which ZnO columns with lengths in the μm range and diameters of 200–300 nm are usually formed under the given conditions. In order to explain this remarkable result it is assumed that an adsorption of the surfactant molecules on the surface of the ZnO particles blocks the growth. As long as the molecules only reversibly adsorb to the surface, they cannot be detected in the deposited films by EDXS. It should also be noted that in a concentration range of about $^1/_2$ cmc $< c <$ cmc, both ZnO phases, lamellar and nanoparticular, were found, as also seen in Figure 11.5, where both TEM micrographs originate from the same film. Since surface micelles are known to start forming at electrode surfaces at around $^1/_2$ cmc, these are assumed to be the origin for the formation of the lamellar phase.

Surfactants were also tested as additives in ZnO electrodeposition from zinc nitrate solution by Jing *et al.* [153]. These authors reported similar results to Boeckler *et al.* for electrodeposition from $ZnCl_2/O_2$ solution, that is, the formation of a lamellar structure above a certain critical value, in this case 0.002 wt%, and the formation of crystalline ZnO particles below this value. From their explanation, below the critical concentration, the number of the adsorbed surfactant molecules is not large enough to fully cover the surface of the growing ZnO. The particles are islands formed at uncovered positions. Above 0.002 wt% ($0.07\,mmol\,l^{-1}$), the adsorbed surfactant molecules are dense enough to cover the whole surface so that the periodic lamellar structures can be observed.

Unfortunately, the lamellar structures were found to be not thermally stable and collapse upon heating at 500 °C for 2 h, which would be necessary for a complete removal of the surfactants. Due to the dominant solid–liquid interaction at the electrode, very high concentrations of surfactants (> 40 wt%) and inorganic salts (> 40 wt%) have to be used in order to obtain more stable three-dimensional porous structures comparable to those obtained from sol-gel methods by electrodeposition [154]. A lyotropic liquid crystalline phase is formed in the deposition solution, which is used as a physical cast to electrodeposit mesoporous films. Therefore, the electrodeposited mesoporous phase is determined by the mesophase type of the liquid crystalline phase in the bulk plating solution, and any specific interactions (e.g., coulombic interaction) between the surfactant and metal ions are not critical for the success of this method [154–160]. However, electrodeposition of ZnO using this method has only reached the stage of preliminary studies, as yet [161].

11.6.3 Other Additives

Since dye molecules had been shown to have a significant impact on the electrochemical growth of ZnO, it may also be possible to use other types of large molecules as structure-directing agents. In one of the first studies using molecules other than dyes or amphiphilic molecules for this purpose, sugar molecules were used as additives for the electrodeposition of ZnO [136]. Sugar molecules are readily available or can easily be synthesized in many variations, for example, as monomers, dimers or oligomers. Furthermore, they can easily be modified by functionalization, eventually allowing the tunability of their interaction with other additives, substrates and the growing ZnO film. The first study involved deposition of ZnO in the presence of glucose and glucuronic acid. In the presence of glucose at concentrations up to 2 mM, the typical morphology of a ZnO film deposited without additive, exhibiting hexagonal wurtzite ZnO crystals, was seen in SEM investigations, which means that there is no influence of glucose on the ZnO morphology even at high concentrations. In contrast, a strong influence of glucuronic acid was seen, even at relatively low concentrations. Since glucose and glucuronic acid molecules only differ in the carboxylic acid group present in the glucuronic acid molecule instead of an OH group in glucose, this acid group is obviously an important factor for the structure-directing effect. This is consistent with results, using dye molecules and surfactants as structure-directing agents, that organic molecules need to have acidic groups (or the respective basic anionic groups) for interaction with the growing ZnO film in order to be incorporated into the film and have any influence on its morphology.

Investigations by SEM also showed that the influence of glucuronic acid on the ZnO morphology depends on its concentration in the electrodeposition bath. At a concentration of 100 μM, disc-like particles start to form, instead of the hexagonal crystals seen without the addition of glucuronic acid. At 500 μM, the disc-like particles appear to form stacks and become more oriented perpendicular to the surface of the substrate. X-ray diffractograms show an increasing crystallographic orientation, with the *c*-axis parallel to the substrate, as observed in the electrodeposition of ZnO with the dye $TSPcSi(OH)_2$ (Figure 11.3). The similarities between glucuronic acid and $TSPcSi(OH)_2$ concerning the film morphology and crystallographic orientation, as well as the presence of several OH groups in the molecules, is further proof of the assumption that the OH groups facilitate the bridging of the ZnO discs by organic molecules and thereby contribute to the film structure. This was further supported by experiments in which cyclohexane carboxylic acid, which is very similar to a glucuronic acid molecule without any OH groups, was used as an additive. Up to a concentration of 2 mM, no significant influence on the morphology and crystallographic orientation of the ZnO films was seen in this case.

Pauporte and Yoshida successfully used 2,2′-bipyridine 4,4′-dicarboxylic acid (dcbpy), which is the ligand facilitating adsorption onto TiO_2 in the Ru complexes used as sensitizers in TiO_2-based DSSCs and a strong complexing agent for many metal ions, in the electrodeposition of ZnO from O_2-saturated $ZnCl_2$ solution [162,163]. The films obtained with dcbpy concentrations between 50 and 200 μM in the electrodeposition bath were found to be macroporous, even without removal of the additive, featuring entangled and interconnected fibers, between which pores, in the submicrometer range, were present. Due to their porosity and the presence of dcbpy, the films could subsequently be loaded with Eu^{3+} or Tb^{3+} by refluxing in ethanolic $EuCl_3$ or $TbCl_3$ solution, respectively. Intense luminescence was observed for both films under UV light illumination, which led to red emission for films loaded with Eu^{3+} and green emission for films loaded with Tb^{3+}. At higher concentrations, dcbpy also revealed an influence on the crystallographic orientation of the deposited ZnO films similar to that of coumarin 343 and glucuronic acid: concentrations of 100 and 200 μM dcbpy led to a diminished intensity of the (002) peak in the XRD, while those arising from the (100) and (101) planes became more pronounced, showing that dcbpy impinges on the ZnO growth in the *c*-axis direction.

Polymers were also found to have influence on the electrodeposition of ZnO. Addition of polyvinylpyrrolidine (PVP) led to the formation of nanoporous ZnO with grain sizes of 20–40 nm, although only with a very large PVP concentration of $4\,g\,l^{-1}$ [164]. Also in this case, film growth was hindered by higher additive concentrations and stopped altogether at $6\,g\,l^{-1}$. Another water-soluble polymer additive used in the electrodeposition of ZnO was polyvinyl alcohol (PVA) [165]. Films deposited from oxygen-saturated aqueous $ZnCl_2$ solution at PVA concentrations ranging between 2 and $10\,g\,l^{-1}$ were electrically conducting and contained well-crystallized ZnO. They were also luminescent and highly transparent in the visible wavelength range. The PVA content in the films increased continuously with the PVA concentration in the deposition bath, leading to changes in the film properties. For instance, at high PVA concentration, ZnO luminescence was dominated by the near-edge emission in the UV, revealing better stoichiometry of the ZnO. Influence on the crystallographic orientation of the ZnO was seen in the disappearance of the (002) peak in the XRD at higher PVA concentrations. The PVA could be eliminated by a subsequent annealing treatment of the films, leading to the formation of porous ZnO without changes in the crystallographic orientation.

11.7 Photoelectrochemical and Photovoltaic Properties

11.7.1 Dye-Sensitized Solar Cells (DSSCs)

The photoelectrochemical properties of electrodeposited porous ZnO films have been especially investigated in view of the use of these films in DSSCs and similar solar-cell concepts. This is due to the very similar bandgap and band positions of ZnO compared to TiO_2, which was used in the original concept for DSSCs. Electrodeposition of porous ZnO for ZnO films has the advantage of film preparation at relatively low temperatures, which allows film deposition on flexible polymer substrates for the fabrication of flexible solar cells.

The working principle of DSSCs is shown in detail in Figure 11.6. A porous film of a wide bandgap n-type semiconductor is prepared on a TCO electrode, for example, ITO or FTO on glass. The film is sensitized by dye molecules, which are adsorbed in a monolayer on the semiconductor surface. If the lowest unoccupied molecular orbital (LUMO) of the dye is energetically higher than the conduction-band edge of the semiconductor, photoexcited electrons can be injected from the dye to the semiconductor and transported to the conducting back contact. The dye is regenerated by hole injection into a redox electrolyte (usually I^-/I_3^-), whose redox potential has to be energetically higher than the highest occupied molecular orbital (HOMO) of the dye. The redox electrolyte is regenerated by electron transfer from the counter-electrode, in most cases a platinized TCO/glass electrode in order to catalyze the charge transfer. Since the Fermi level of the counter-electrode is determined by the redox potential of the electrolyte and the maximum quasi-Fermi level of electrons in the semiconductor is close to the

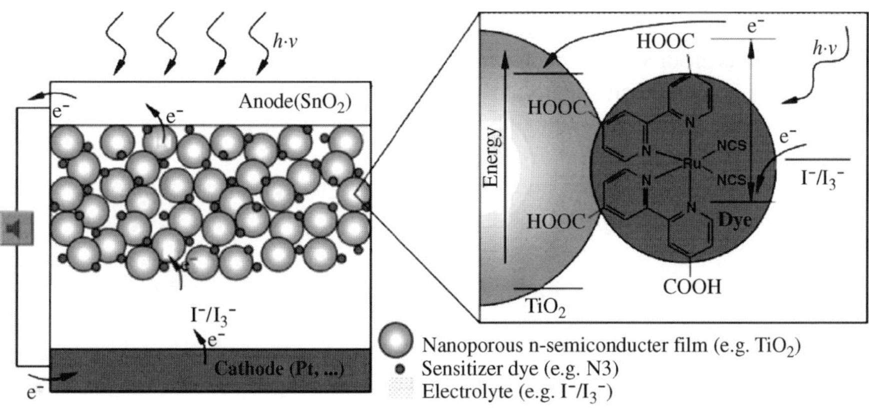

Figure 11.6 Working principle of a dye-sensitized solar cell.

conduction band, the maximum reachable photovoltage of a DSSC is the difference between the conduction-band edge and the redox potential of the electrolyte.

The overall light-to-electricity conversion efficiency η of solar cells is given by the equation

$$\eta = \frac{I_{\mathrm{MPP}} U_{\mathrm{MPP}}}{\Phi} = \frac{I_{\mathrm{sc}} U_{\mathrm{oc}} FF}{\Phi} \tag{11.10}$$

where Φ is the intensity of the incident light, I_{sc} the short-circuit current, U_{oc} the open-circuit voltage, FF the fill factor, and I_{MPP} and U_{MPP} are the current and voltage at the maximum power point (Figure 11.7). Efficiencies of solar cells are usually given for the AM1.5 sun spectrum, which has an intensity of about 1000 W m^{-2} (often referred to as one-sun light intensity).The efficiency record for DSSCs, which was held by Grätzel and co-workers for more than two decades, was recently taken over by the research group of Sharp Corp., where 11.1% was

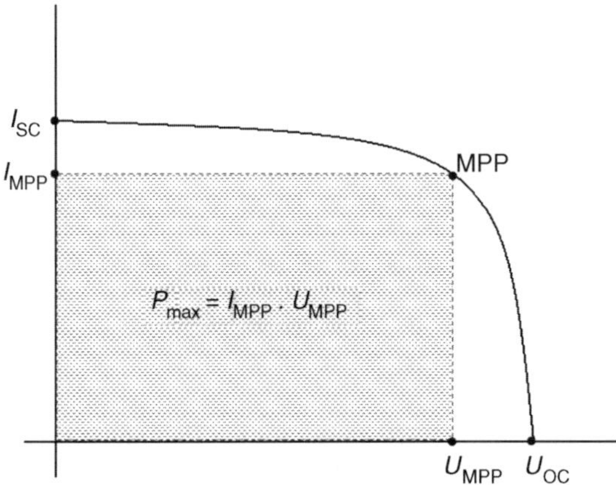

Figure 11.7 Typical *I–V* curve of a solar cell.

reached with a 0.22 cm^2 cell. The highest efficiency for the standard size of 1 cm^2 of 10.4% was also reached by this group [166,167]. This is considerably lower than efficiencies achieved with (monocrystalline) Si solar cells (up to 25%), however, DSSCs promise much lower production costs, and their partial transparency and color also make them suitable for new applications (e.g., windows, cars etc.).

Beside the overall efficiency η, one way to measure the performance of solar cells is the incident-photon-to-current conversion efficiency (IPCE), which can also be described by the equation

$$\mathrm{IPCE}(\lambda) = \mathrm{LHE}(\lambda)\Phi_{\mathrm{inj}}\eta_c \tag{11.11}$$

where LHE is the light-harvesting efficiency, Φ_{inj} the efficiency of electron injection and η_c the collection efficiency of injected electrons. Since the LHE depends on the absorption coefficient of the dye, the LHE, as well as the IPCE, changes with the wavelength λ of the incident light.

As already discussed in Section 11.6, electrodeposited ZnO films, especially those electrodeposited with eosin Y as a structure-directing additive, can have a very high surface area, so that a high LHE can be reached. Φ_{inj} has not been investigated in detail for ZnO-based DSSCs, however it has been shown to be usually high for TiO_2-based DSSCs [168]. In order to achieve a high η_c, the electron transport, which can be characterized by the electron transit time τ_D, has to be faster than the back reaction of electrons to the redox electrolyte or oxidized dye molecules, which is characterized by the electron lifetime τ_n. The electron transport and back-reaction properties of electrodeposited porous ZnO films templated by eosin Y have been investigated in detail, and also in comparison with porous ZnO films prepared from ZnO nanoparticles. Details of this study will be given in the following.

In general, electron transport has been found to be slow in nanoporous oxide semiconductor films. Since no electric field is present in the films due to screening by the redox electrolyte in the pores, electron transport takes place solely by diffusion. The electron transport can therefore be described by an electron diffusion coefficient D_n, and the electron generation (first term), electron transport (second term) and electron recombination (third term) are described by the continuity equation [169,170]

$$\frac{\mathrm{d}n}{\mathrm{d}t} = \Phi_{\mathrm{inj}}\alpha\Phi e^{-\alpha x} + D_n\frac{\mathrm{d}^2 n}{\mathrm{d}x^2} - \frac{n-n_0}{\tau_n} \tag{11.12}$$

where α is the absorption coefficient, n the electron density under illumination, n_0 the equilibrium electron concentration in the dark and τ_n the effective first-order electron lifetime (also intensity dependent). D_n is often called "efficient" or "apparent" diffusion coefficient, since it is determined by trapping and detrapping events of the electrons on their way through the film. It can be defined by

$$D_n = D_{\mathrm{cb}}\frac{k_d}{k_t} \tag{11.13}$$

where k_{t} and k_{d} are the first-order rate constants for the trapping and detrapping, respectively, and D_{cb} is the diffusion coefficient of electrons in the conduction band [171]. It has been estimated that the time spent in traps by photogenerated electrons on their way through a nanoparticulate TiO_2 film is about 100 times higher than the time spent as free electrons in the conduction band. Knowing D_n and τ_n, the diffusion lengths L_n of the electrons can be calculated according to the equation [172]

$$L_n = \sqrt{D_n \tau_n} \tag{11.14}$$

The diffusion length is a useful value to discuss the collection efficiency of DSSCs, since it has to be higher than the film thickness in order to collect all photoinjected electrons. In addition, the collection efficiency of photogenerated electrons can also be estimated by the equation [173]

$$\eta_c = 1 - \frac{\tau_D}{\tau_n} \tag{11.15}$$

The slow electron transport in nanoparticulate metal-oxide films is illustrated by the photocurrent transient of the nanoparticulate ZnO film in Figure 11.8 [130]. Upon beginning the illumination, the photocurrent increases only slowly to a constant value, since electrons farther away from the back contact need some time to be collected (curve (c)). After the end of the illumination, decrease of the photocurrent as slow as the increase is seen, which is due to photogenerated electrons remaining in the film even after the light is switched off. In contrast, electrodeposited ZnO/eosin Y film deposited at −1.0 V versus SCE exhibits an almost rectangular response to the light pulse, indicating a much faster transport of photogenerated electrons (curve (b)). Figure 11.8 also shows the transient of a nonporous ZnO/eosin Y film with eosin Y molecules trapped within ZnO crystals. These molecules can photoinject

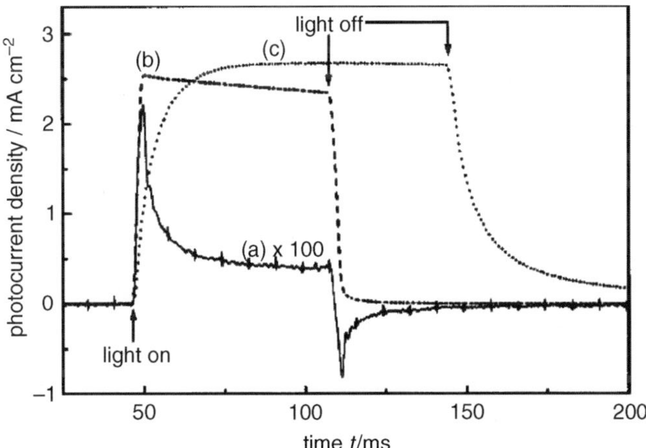

Figure 11.8 Photocurrent transients measured at ZnO/eosin Y hybrid thin-film electrodes, which were prepared by electrodeposition at: (a) −0.7 V versus SCE and (b) −1.0 V versus SCE for 60 min from oxygen-saturated aqueous $ZnCl_2$ containing 24 mM eosin Y. In comparison (c) a photocurrent transient measured at a nanoparticulate porous ZnO electrode sensitized with eosin Y is shown. (Reprinted from T. Yoshida, T. Pauporté, D. Lincot *et al.*, Cathodic electrodeposition of ZnO/eosin Y hybrid thin films from oxygen-saturated aqueous mixed solution of ZnC12 and eosinY, *Journal of the Electrochemical Society*, **150**, C608, 2003, The Electrochemical Society.)

electrons into the ZnO, but they cannot be regenerated by the redox electrolyte, leading to a fast decrease in the photocurrent after an initial peak (curve (a)).

11.7.2 Photoelectrochemical Investigation of the Electron Transport in Porous ZnO Films

A disadvantage of photocurrent transient measurements with large light steps, as shown in Figure 11.8, is that the electron concentration and therefore also the electron-transport properties change to a large extent during the experiment. This problem has been overcome by measurements under working conditions, that is, under constant illumination, using small changes (up to $\pm 10\%$) in the illumination level. Two methods based on this principle are intensity modulated photocurrent spectroscopy (IMPS) and intensity modulated photovoltage spectroscopy (IMVS), which can be used to measure the values of τ_n and τ_D in a nanoporous dye-sensitized metal-oxide film. During the IMVS and IMPS measurements, the cell is illuminated with sinusoidally modulated light with a small ac component, which can be described by the periodic illumination function

$$I(t) = I_0[1 + (\delta e^{i\omega t})] \tag{11.16}$$

where $\omega = 2\pi f$ is the variable modulation frequency and δ is $\ll 1$. The photocurrent or photovoltage response is measured in terms of its amplitude and phase shift with respect to the illumination function, and it can be represented in the IMPS and IMVS complex plane plots. Examples for electrodeposited ZnO films are shown in Figure 11.9.

The IMVS response is characterized by a semicircle in the positive/negative quadrant of the complex plane, and τ_n can be calculated directly from the IMVS response since $\tau_n = 1/\omega_{\min} = 1/2\pi f_{\min}$ where $f_{\min}$ is the frequency of the minimum of the semicircle, that is, the frequency of the lowest imaginary component in the IMVS plot [175]. Correspondingly, an electron transit time $\tau_D = 1/\omega_{\min} = 1/2\pi f_{\min}$ can be defined for IMPS, which gives a convenient estimate of the average time that photoinjected electrons need to reach the back contact. Figure 11.10 shows a comparison of values obtained from IMPS and IMVS measurements for electrodeposited as well as nanoparticulate ZnO films [174].

The decrease in τ_n and τ_D towards higher light intensities is explained by a higher trap occupancy, which leads to a faster electron diffusion because only shallow traps with faster detrapping are able to retard electrons. At the same time, less trapping of electrons also means that the back reaction can occur faster, leading to a smaller τ_n value.

Two main aspects can be extracted from the results in Figure 11.10. Firstly, τ_D for the electrodeposited films is found to be smaller than τ_D of the colloidal film by more than two orders of magnitude. This result is consistent with the photocurrent transient measurements shown in Figure 11.8. One reason for this faster charge collection can be seen in the lower thickness of the electrodeposited film. Random-walk diffusion predicts that the transit time from any point in the film at a distance x from the substrate is given by

$$\tau_D = \frac{x^2}{D_n} \tag{11.17}$$

However, the difference in the film thickness would only explain a difference in the transit times by a factor of 50. Therefore, the results suggest a higher D_n for electrons in the

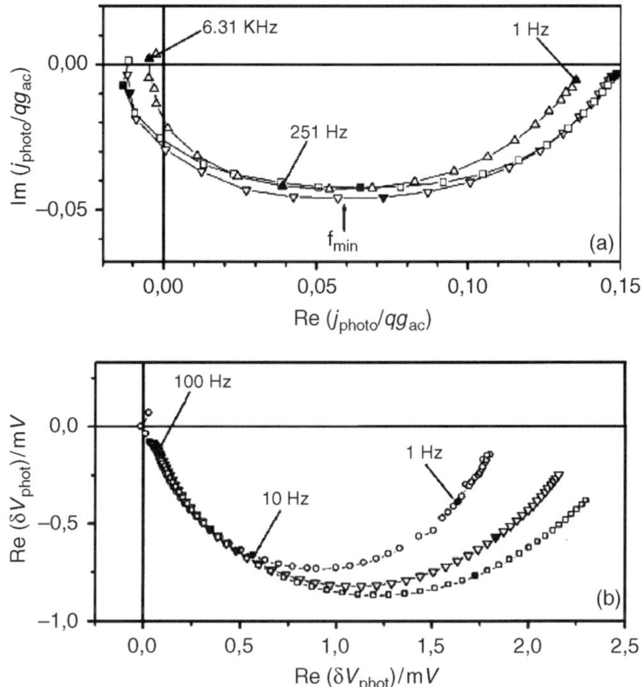

Figure 11.9 (a) Typical IMPS plots of an electrodeposited ZnO/eosin Y electrode. The film was illuminated with dc light intensities of 0.89 mW cm^{-2} (▽), 0.50 mW cm^{-2} (□) and 0.29 mW cm^{-2} (△). The solid symbols (▼,■,▲) indicate measurements at 6.31 kHz, 251 Hz and 1 Hz, respectively. The axes show normalized photocurrents (electron flux/photon flux), so that the amplitudes correspond to the ac component of the IPCE. (b) IMVS plots of the same film measured at light intensities of 2.6 mW cm^{-2} (▷), 0.8 mW cm^{-2} (○) and 0.4 mW cm^{-2} (◁) between 0.2 Hz and 200 Hz. The solid symbols (▶,●,◀) indicate measurements at 100, 10 and 1 Hz. (Reprinted with permission from T. Oekermann, T. Yoshida, H. Minoura *et al.*, Electron transport and back reaction in electrochemically self-assembled nanoporous ZnO/dye hybrid films, *Journal of Physical Chemistry B*, **108**(24), 8364, 2004. © 2004 American Chemical Society.)

electrodeposited film. Secondly, compared to the differences in τ_D, the differences in the electron lifetimes τ_n of the two kinds of film are much less significant. This leads to a much larger difference between τ_n and τ_D of the electrodeposited film and an electron collection efficiency of about 99% according to Equation 11.15. On the other hand, τ_n and τ_D of the nanoparticulate film only differ by a factor of about two, which translates into an electron collection efficiency of only 50%.

In order to compare the diffusion of photogenerated electrons in electrodeposited and nanoparticulate ZnO films in more detail, the D_n values have been calculated by fitting the IMPS results to Equation 11.12. Analytical solutions of this equation based on the illumination function in Equation 11.16 for short-circuit (IMPS) and open-circuit (IMVS) conditions have been published in the literature [172,175,176] and can be used for this purpose. The electron diffusion coefficients D_n, which were obtained from the fits of the IMPS plots, are shown in

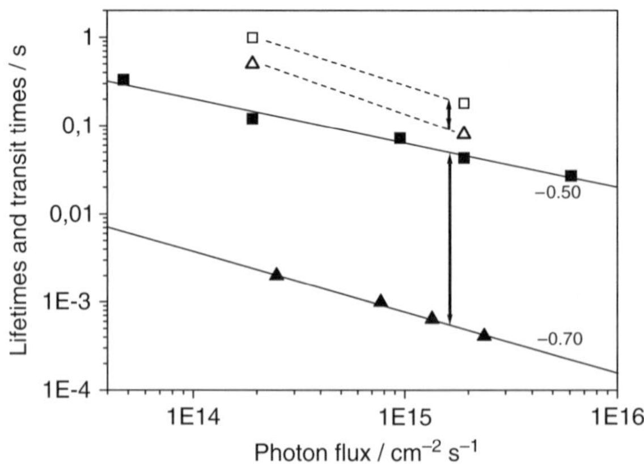

Figure 11.10 Electron transit times τ_D and lifetimes τ_n for an electrodeposited ZnO/eosin Y film at different dc light intensities ($\tau_D = ▲$; $\tau_n = ■$). Transit times and lifetimes for a colloidal ZnO/eosin Y film ($\tau_D = △$; $\tau_n = □$) are shown for comparison. Solid lines were calculated by linear regression with the slopes given in the plot. The two arrows indicate the differences between τ_D and τ_n for both kinds of films. (Reprinted with permission from T. Oekermann, T. Yoshida, H. Minoura *et al.*, Electron transport and back reaction in electrochemically self-assembled nanoporous ZnO/dye hybrid films, *Journal of Physical Chemistry B*, **108**(24), 8364, 2004. © 2004 American Chemical Society.)

Figure 11.11. The values are shown against the dc photocurrent instead of the light intensity. Taking the dc photocurrent as a measure for the concentration of excess electrons in a first approximation, this enables a better comparability of the electron-transport properties, since the occupancy of the traps in the ZnO, which has a large influence on D_n, depends on the excess

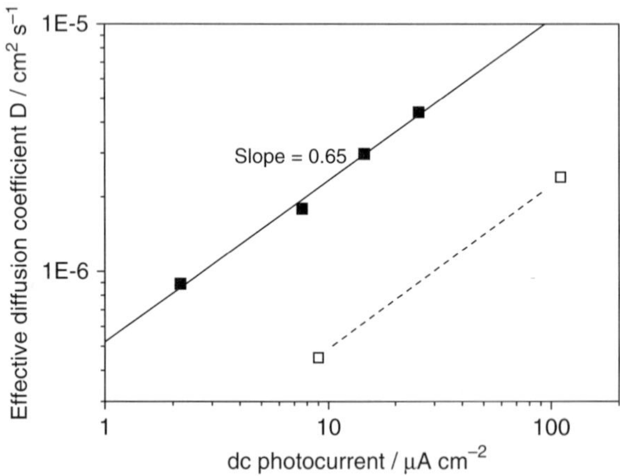

Figure 11.11 Effective electron diffusion coefficients D_n for a ZnO/eosin Y film calculated by fitting the IMPS plots (■). D_n values for a colloidal ZnO/eosin Y film (□) are shown for comparison. The solid line was calculated by linear regression. (Reprinted with permission from T. Oekermann, T. Yoshida, H. Minoura *et al.*, Electron transport and back reaction in electrochemically self-assembled nanoporous ZnO/dye hybrid films, *Journal of Physical Chemistry B*, **108**(24), 8364, 2004. © 2004 American Chemical Society.)

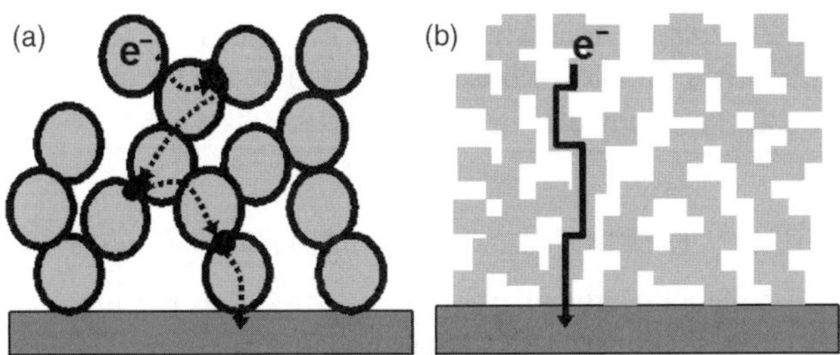

Figure 11.12 Simple qualitative model for the comparison of the electron transport in: (a) colloidal ZnO and (b) electrodeposited ZnO. The black points at the grain boundaries in (a) indicate electron traps. (Reprinted with permission from T. Oekermann, T. Yoshida, H. Minoura *et al.*, Electron transport and back reaction in electrochemically self-assembled nanoporous ZnO/dye hybrid films, *Journal of Physical Chemistry B*, **108**(24), 8364, 2004. © 2004 American Chemical Society.)

electron concentration in the film. The results indicate that electron transport in the electrodeposited ZnO film is faster by a factor of about four compared to the nanoparticulate film.

The electron diffusion coefficients found in this study are also comparable with or even higher than electron diffusion coefficients of sintered nanoporous TiO_2 films in a large number of publications. The faster electron transport in the electrodeposited film is quite remarkable, since it is known for nanoporous TiO_2 that films prepared without high-temperature sintering show significantly slower electron transport compared to films sintered at high temperatures [177,178]. However, no high temperatures are used in the electrodeposition of ZnO films, while the colloidal ZnO films can be expected to show even slower electron transport, as shown here, if sintered at a lower temperature. Figure 11.12 illustrates the likely reason for the faster electron transport in electrodeposited ZnO films. Grain boundaries and electron traps, which are often located at grain boundaries, are present in the colloidal film (Figure 11.12a). As electrons spend some time in traps, the effective diffusion coefficient D_n becomes smaller if the trap density becomes higher. The conditions for electron transport in the electrodeposited films seem to be significantly different, which can be attributed to the ordered and crystallographically highly oriented film structure consisting of porous single crystals (Section 11.6.1). It has been shown before that an ordered film structure can lead to an improvement in electron-transport properties [179], as is the case for this material. Furthermore, it can be assumed that photoinjected electrons have to pass no or much fewer grain boundaries in these films on their way to the back contact (Figure 11.12b), and a lower trap density can be assumed for the electrodeposited film, which leads to a higher effective diffusion coefficient D_n. Figure 11.12 also reflects the smaller pore size and higher surface area of the electrodeposited film.

It can be concluded at this point that, compared to nanoparticulate films, electrodeposited porous ZnO films templated with eosin Y show high electron collection efficiencies due to both very high dye concentrations in the films, which enables the use of very thin films, while still assuring efficient light harvesting, and better electron transport properties. An even slightly faster electron transport has been found in IMPS measurements for electrodeposited ZnO films templated with coumarin 343 instead of eosin Y [180]. This is due to the anisotropic structure of ZnO with higher electron mobility along the *a* and *b* axes compared to the *c* axis [181]. As discussed in Section 11.6.1, ZnO films templated with coumarin 343 are preferentially oriented with the *c*-axis parallel to the substrate, leading to faster electron transport towards the back contact, while films templated with eosin Y are preferentially oriented with the *c*-axis perpendicular to the substrate.

11.7.3 Performance of Nanoporous Electrodeposited ZnO Films in DSSCs

The higher electron collection efficiency of the electrodeposited films is reflected in their higher efficiency in DSSCs compared to nanoparticulate ZnO films. The highest efficiency for ZnO-based DSSCs to date of about 5.6% has been achieved with an electrodeposited ZnO film templated with eosin Y after readsorption of the broadly absorbing indoline dye D149 (Figure 11.13) as sensitizer [57]. Films templated with coumarin 343 have not been tested in DSSCs with broadly absorbing dyes yet; however, an increase in efficiency is not expected, since films templated with eosin Y already collect 99% of photoinjected electrons, as discussed in the foregoing section. This high collection efficiency is also seen in IPCE values of up to 90%, which can be achieved at wavelengths near the absorption maximum of the dye. Considering that about 10% of the incident light is lost by reflection and absorption by the TCO substrate, this result proves that virtually all injected electrons are collected. Furthermore, it proves the high light-harvesting efficiency of the films and a high electron-injection efficiency, both also at or near to 100%.

Due to the high internal conversion efficiency of DSSCs based on electrodeposited ZnO, it should actually be possible to achieve overall efficiencies as high as those with TiO_2 (>10%) or even higher, since typical IPCE values shown for the most efficient TiO_2-based DSSCs are still below 90%. The most important challenge for the use of ZnO in DSSCs, however, is to find a suitable sensitizer dye. Ruthenium complexes, which achieve the highest efficiency in TiO_2-based DSSCs, were actually tested with ZnO electrodes, but yielded disappointingly low efficiencies [182]. The problem in this case is that these dyes are too aggressive chemicals for ZnO, which is chemically more reactive than TiO_2. The dye molecules act as acids to dissolve ZnO and form aggregates by extended soaking of ZnO films in dye solutions [183]. A moderately high efficiency up to 5% could still be achieved by careful optimization of the electrode and dye adsorption process [184]. Concerning organic dyes, it is often seen that dyes exhibit a broader absorption spectrum after adsorption on TiO_2 as compared to ZnO. This is due to the extension of the spectral sensitivity to longer wavelengths, which is presumably caused by rather strong TiO_2–dye interactions. In fact, the adsorption of most dyes on TiO_2 surfaces seems to be more stable than on ZnO surfaces, probably due to the more covalent nature of TiO_2, which allows the formation of stronger covalent bonds to adsorbed species. Therefore, even the best sensitizer dye for ZnO to date, the D149 dye, achieves a much higher overall

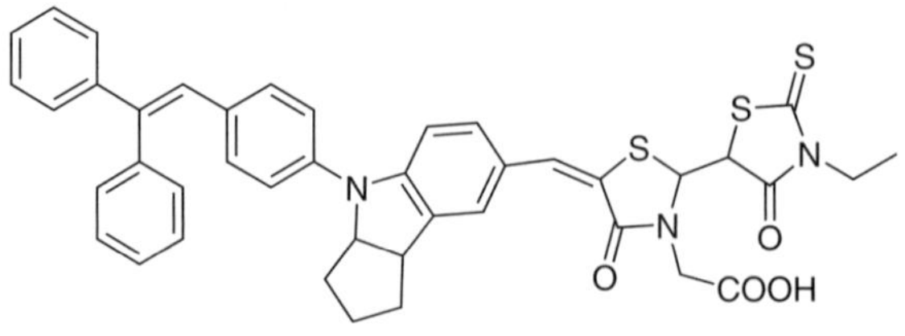

Figure 11.13 Structural formula of the indoline D149 dye.

efficiency of about 9% in combination with structurally optimized nanoparticulate TiO_2 electrodes [185]. When D149 dye is adsorbed on TiO_2, its light absorption extends to around 750 nm, while it reaches only slightly above 650 nm on ZnO. The search for a good sensitizer dye for ZnO therefore has to go on. If it can be found, highly efficient DSSCs, maybe even with higher efficiencies than TiO_2-based cells, could be fabricated without any high-temperature step, using structure-directed electrodeposition of ZnO.

Some solar cell performance data were also published for porous or nanostructured ZnO films electrodeposited by other methods. The highest efficiency was reported for films deposited from zinc nitrate solution with addition of polyvinylpyrrolidine, which achieved 5.1% with the N3 dye as sensitizer, although under illumination with only 53 mW cm^{-2} instead of 100 mW cm^{-2} [164]. Results from measurements with higher light intensities were not shown. Lower efficiencies due to ion-transport limitations can be assumed, since the SEM micrographs of the films showed rather dense films with very narrow pores.

O'Regan *et al.* used their porous electrodeposited films as n-type material in solid-state solar cells [137]. After adsorption of a monolayer of Ru dye, p-type CuSCN was electrodeposited into the pores of the ZnO film. Starting the CuSCN deposition at quite positive potentials, around +0.2 V versus NHE, nucleation of CuSCN only occurred at or near the compact ZnO underlayer by minimizing the number of electrons that flow into the porous ZnO layer, so that good pore filling could be achieved. A conversion efficiency of 1.5% was achieved, which was a record value at that time. However, no higher efficiencies have been achieved with the combination of electrodeposited ZnO and CuSCN films since, reflecting the limitations of the surface area and film thickness on the one hand and the difficulties in filling ZnO, with higher surface areas and smaller pores, with CuSCN, on the other hand.

11.7.4 Use of ZnO Nanorods in Photovoltaics

ZnO nanorod arrays have mainly been tested for applications in solid-state solar cells, since they offer less surface area than three-dimensional porous networks for DSSCs, as already mentioned, but on the other hand they offer the advantage of a very open morphology, which makes the filling of the "pores" with another solid rather easy. The first successful concept used CdTe deposited by metal organic chemical vapor deposition as the absorber material on ZnO nanowires, followed by chemical solution deposition of CuSCN as a p-type material from a saturated solution of CuSCN in propylsulfide (15 mg ml^{-1}) [186,187] on glass/FTO/CdTe samples pre-heated to 80 °C [93,94]. An overall efficiency of 2.3% with $I_{sc} = 4$ mA cm^{-2} and $U_{oc} = 500$ mV was measured at a light intensity of 360 W m^{-2}. Very similar results were also achieved with thin layers of CdSe electrodeposited on ZnO nanowires from an aqueous alkaline selenosulfate solution (0.05 M cadmium acetate, 0.1 M nitrilotriacetic acid trisodium salt and 0.05 M selenosulfate with excess sulfite) under galvanostatic conditions, with a current density in the range of −1.5 to −3 mA cm^{-2} [188–190]. Annealing of the ZnO/CdSe nanowires at 350 or 400 °C proved to be an important factor to achieve a high efficiency [191]. This effect was attributed to an increase of the CdSe particle size from 3 to 9 nm, which may lead to easier charge-carrier transport in the CdSe shell due to the loss of quantum confinement. More detailed investigations revealed a high light-trapping effect of the ZnO/CdSe nanowire layer, with an effective absorbance of ~89% and effective reflectance of ~8% in the 400–800 nm region of the AM1.5 solar spectrum [192]. Internal quantum efficiencies of up to 28% were found, the highest reported for inorganic ETA solar cells.

Konenkamp *et al.* investigated nanostructured p-n junctions based on ZnO nanowires without sensitizer dye, in view of their electroluminescent properties [193]. The nanorods were sandwiched between a transparent SnO_2 film and p-type conducting polymers. The structures showed UV electroluminescence at about 390 nm.

Electrodeposited ZnO nanorods were also investigated by electrochemical and photoelectrochemical measurements in a three-electrode setup. Photoelectrochemical measurements on ZnO/CdSe core-shell nanowire arrays in ferro-/ferricyanide solution revealed high external quantum efficiencies (>70%), demonstrating their high potential for nanostructured solar cells [191]. Electron transport in ZnO nanowires and their carrier density have been investigated by means of electrochemical impedance spectroscopy in 0.1 M $LiClO_4$ propylene carbonate electrolyte [194]. Mott–Schottky analysis is a standard technique to determine both dopant density and flatband potential at semiconductor/liquid contacts based on a plot of C^{-2} versus potential, which is described by the equation [195–197]

$$\frac{1}{C_{\mathrm{scl}}^2} = \frac{2}{q\varepsilon N_D}(U_{\mathrm{scl}} + U_0) \tag{11.18}$$

where C_{scl} is the capacitance of the space-charge layer in the semiconductor, U_{scl} the potential difference across the space-charge layer, U_0 the flatband potential corrected by the contribution of the Helmholtz layer to the capacitance, q the elementary charge and N_D the donor density. In order to extract the values of C_{scl} from the impedance measurements, an equivalent circuit model taking into account the geometry of the ZnO nanowires has been developed for this purpose. A high donor density of $6.2 \times 10^{19}\,\mathrm{cm}^{-3}$ was found for as-grown electrodeposited samples, which was dramatically reduced by two orders of magnitude after an annealing in air at 450 °C for 1 h.

Nevertheless, due to their lower surface area, ZnO nanorods reached lower efficiencies than electrodeposited ZnO films with a three-dimensional porous network in DSSCs. Baxter and Aydil reported energy conversion efficiencies of 0.5% and internal quantum efficiencies of 70% for ZnO nanorods deposited by nonelectrochemical solution methods [198]. DSSCs based on electrodeposited ZnO nanorods or nanotubes have not yet been reported in the literature.

11.8 Photocatalytic Properties

The highly porous ZnO films electrodeposited with eosin Y as pore templates showed promising results in photocatalytic decomposition of dye molecules [199]. The heteropolyaromatic dye methylene blue and the diazo dye congo red served as model substances in this study, using ZnO films electrodeposited with different eosin Y concentrations. The photodegradation rate reached a maximum for ZnO films electrodeposited with 40 μM eosin Y, while films deposited with higher eosin Y concentrations, leading to a higher surface area, but a lower pore size, exhibited lower photodegradation rates, indicating that photodegradation preferentially occurs in larger-diameter pores. This may, however, be due to the size of the dye molecules, which can be expected to have a rather low diffusion coefficient and may have some difficulties penetrating into small pores. Such difficulties would be less expected for water splitting, however, it has not been tested using the highly porous ZnO films templated with eosin Y or other dyes to date.

ZnO was also co-electrodeposited with Pt or Au for photocatalytic applications [200]. Nanoparticles of metals in contact with a semiconductor are known to enhance electron–hole separation and thereby enhance the photocatalytic activity for appropriate energy levels, that is, the position of the Fermi level of the metal is below the conduction-band edge of the semiconductor. In this case, photogenerated electrons from the semiconductor are injected into the metal particles, while photogenerated holes remain in the semiconductor, largely suppressing electron–hole recombination. The co-deposition was performed by addition of $HAuCl_4$ or H_2PtCl_6 to the aqueous zinc-nitrate-containing electrodeposition bath. Due to its higher Fermi level, which leads to faster electron transfer from the metal further onto the electrolyte, ZnO/Au showed a better performance than ZnO/Pt, as determined by the degradation of methylene blue. The highest activity was found for an Au content of 2%, since higher contents lead to increasing coverage of the ZnO surface and thereby suppress photo-oxidation at the ZnO surface.

A similar effect of increased photocatalytic activity by enhanced electron–hole separation was achieved by combination of electrodeposited ZnO and TiO_2. For this purpose, ZnO was electrodeposited into highly ordered TiO_2 nanotube arrays prepared on a Ti foil by anodization in HF solution [201]. The linear-sweep photovoltammetry response of the composite electrode showed dramatically enhanced photocurrents compared to TiO_2 nanotubes without ZnO. Enhanced photocatalytic activity in the degradation of methyl orange was also observed.

A comprehensive study on photoelectrochemical hydrogen production from water has been performed with electrodeposited $Zn_{1-x}Co_xO$ films, involving 27 different compositions with $0 < x < 0.068$ (see Section 11.4.2) [68]. All films exhibited the wurtzite structure typical of pure ZnO, where Co^{2+} appears to substitute Zn^{2+}, forming a single-phase solid solution. High-throughput photoelectrochemical screening revealed improved solar hydrogen production for the cobalt-doped films, with $Zn_{0.956}Co_{0.044}O$ exhibiting a fourfold improvement over pure ZnO.

11.9 Outlook

It has been shown that electrodeposition is a useful method for the preparation of nanostructured and nanoporous ZnO films for various applications, especially as a material for solar cells. Future research will have to further explore the possibilities of the materials described in this chapter in photocatalytic applications, including the generation of solar hydrogen. Additional research efforts are also required to further develop the preparation method by increased understanding of the interactions between reactants and additives during the electrodeposition process.

References

[1] Ohko, Y., Tatsuma, T. and Fujishima, A. (2001) Characterization of TiO_2 photocatalysis in the gas phase as a photoelectrochemical system: Behavior of salt-modified systems. *Journal of Physical Chemistry B*, **105**, 10016.

[2] Matsuoka, M., Kitano, M., Takeuchi, M. *et al.* (2007) Photocatalysis for new energy production - Recent advances in photocatalytic water splitting reactions for hydrogen production. *Catalysis Today*, **122**, 51.

[3] Memming, R. (1994) Photoinduced charge-transfer processes at semiconductor electrodes and particles, in *Topics in Current Chemistry*, vol. 169, Springer.

[4] Look, D.C. (2001) Recent advances in ZnO materials and devices. *Materials Science and Engineering*, **B80**, 383.

[5] Bellingeri, E., Marré, D., Pellegrino, L. *et al.* (2005) High mobility ZnO thin film deposition on $SrTiO_3$ and transparent field effect transistor fabrication. *Superlattices and Microstructures*, **38**, 446.

[6] Villasenor, J., Reyes, P. and Pecchi, G. (1998) Photodegradation of pentachlorophenol on ZnO. *Journal of Chemical Technology and Biotechnology*, **72**, 105.

[7] Pal, B. and Sharon, M. (2002) Enhanced photocatalytic activity of highly porous ZnO thin films prepared by sol-gel process. *Materials Chemistry and Physics*, **76**, 82.

[8] Hernandez, A., Maya, L. and Sanchez-Mora, E. (2007) Sol-gel synthesis, characterization and photocatalytic activity of mixed oxide ZnO-Fe_2O_3. *Journal of Sol-Gel Science and Technology*, **42**, 71.

[9] Zhang, Y.Y. and Mu, J. (2007) One-pot synthesis, photoluminescence and photocatalysis of Ag/ZnO composites. *Journal of Colloid and Interface Science*, **309**, 478.

[10] Minami, T. (2005) Transparent conducting oxide semiconductors for transparent electrodes. *Semiconductor Science and Technology*, **20**, S35.

[11] Dhere, N.G. (2006) Present status and future prospects of CIGS thin film solar cells. *Solar Energy Materials and Solar Cells*, **90**, 2181.

[12] Datta, M. (2003) Electrochemical processing technologies in chip fabrication: Challenges and opportunities. *Electrochimica Acta*, **48**, 2975.

[13] Bard, A.J. and Faulckner, L.R. (1980) *Electrochemical Methods*, Wiley, New York.

[14] Yamamoto, J., Tan, A., Shiratsuchi, R. *et al.* (2003) A 4% efficient dye-sensitized solar cell fabricated from cathodically electrosynthesized composite titania films. *Advanced Materials*, **15**, 1823.

[15] Atkins, P.W. (2001) *Physikalische Chemie*, Wiley-VCH, Weinheim.

[16] Hamann, C.H. and Vielstich, W. (1998) *Elektrochemie*, Wiley-VCH, Weinheim.

[17] Schmickler, W. (1996) *Grundlagen der Elektrochemie*, Vieweg, Braunschweig.

[18] Baranski, A.S., Fawcett, W.R., McDonald, A.C. *et al.* (1981) The structural characterization of cadmium sulphide films grown by cathodic electrodeposition. *Journal of the Electrochemical Society*, **128**, 963.

[19] Power, G.P., Peggs, D.R. and Parker, A. (1981) The cathodic formation of photoactive cadmium sulphide films from thiosulfate solutions. *Electrochimica Acta*, **26**, 681.

[20] Duffy, N.W., Lane, D., Özsan, M.E. *et al.* (2000) Structural and spectroscopic studies of CdS/CdTe heterojunction cells fabricated by electrodeposition. *Thin Solid Films*, **361/362**, 314.

[21] Kampmann, A., Sittinger, V., Rechid, J. and Reineke-Koch, R. (2000) Large area electrodeposition of Cu(In,Ga) Se-2. *Thin Solid Films*, **361/362**, 309.

[22] O'Regan, B. and Schwartz, D.T. (1998) Large enhancement in photocurrent efficiency caused by UV illumination of the dye-sensitized heterojunction TiO_2/RuLL'NCS/CuSCN: Initiation and potential mechanisms. *Chemistry of Materials*, **10**, 1501.

[23] O'Regan, B., Schwartz, D.T., Zakeeruddin, S.M. and Grätzel, M. (2000) Electrodeposited nanocomposite n-p heterojunctions for solid-state dye-sensitized photovoltaics. *Advanced Materials*, **12**, 1263.

[24] Tench, D. and Warren, L.F. (1983) Electrodeposition of conducting transition metal oxide hydroxide films from aqueous solution. *Journal of the Electrochemical Society*, **130**, 869.

[25] Sawatani, S., Ogawa, S., Yamada, T. *et al.* (2003) Cathodic electrodeposition of metal oxide thin films from non-aqueous solutions. *Transactions of the Materials Research Society of Japan*, **28**, 381.

[26] Seshadri, A., de Tacconi, N.R., Chenthamarakshan, C.R. and Rajeshwar, K. (2006) Cathodic electrodeposition of CdO thin films from oxygenated aqueous solutions. *Electrochemical and Solid-State Letters*, **9**, C1.

[27] Gal-Or, L., Silberman, I. and Chaim, R. (1991) Electrolytic ZrO_2 coatings. 1. Electrochemical aspects. *Journal of the Electrochemical Society*, **138**, 1939.

[28] Natarajan, C. and Nogami, G. (1996) Cathodic electrodeposition of nanocrystalline titanium dioxide thin films. *Journal of the Electrochemical Society*, **143**, 1547.

[29] Karuppuchamy, S., Amalnerkar, D.P., Yamaguchi, K. *et al.* (2001) Cathodic electrodeposition of TiO_2 thin films for dye-sensitized photoelectrochemical applications. *Chemistry Letters*, 78.

[30] Minoura, H., Naruto, K., Takano, H. *et al.* (1991) Preparation of $YBa_2Cu_3O_{7-x}$ and $Bi_2Sr_2CaCu_2O_y$ films by electrodeposition technique. *Chemistry Letters*, 379.

[31] Matsumoto, Y., Adachi, H. and Hombo, J. (1993) New preparation method for PZT films using electrochemical reduction. *Journal of the American Ceramic Society*, **76**, 769.

[32] Chang, S.T., Leu, I.C. and Hon, M.H. (2002) Preparation and characterization of nanostructured tin oxide films by electrochemical deposition. *Electrochemical and Solid-State Letters*, **5**, C71.

[33] Schrebler, R., Bello, K., Vera, F. *et al.* (2006) An electrochemical deposition route for obtaining alpha-Fe_2O_3 thin films. *Electrochemical and Solid-State Letters*, **9**, C110.
[34] Kavan, L., O'Regan, B., Kay, A. and Grätzel, M. (1993) Preparation of TiO_2 (anatase) films on electrodes by anodic oxidative hydrolysis of $TiCl_3$. *Journal of Electroanalytical Chemistry*, **346**, 291.
[35] Kavan, L., Zukalova, M., Kalbac, M. and Grätzel, M. (2004) Lithium insertion into anatase inverse opal. *Journal of the Electrochemical Society*, **151**, A1301.
[36] Wessels, K., Feldhoff, A., Wark, M. *et al.* (2006) Low-temperature preparation of crystalline nanoporous TiO_2 Films by surfactant-assisted anodic electrodeposition. *Electrochemical and Solid-State Letters*, **9**, C93.
[37] Wessels, K., Maekawa, M., Rathousky, J. and Oekermann, T. (2007) One-step electrodeposition of TiO_2/dye hybrid films. *Thin Solid Films*, **515**, 6497.
[38] Sawatani, S., Yoshida, T., Ohya, T. *et al.* (2005) Electrodeposition of TiO_2 thin film by anodic formation of titanate/benzoquinone hybrid. *Electrochemical and Solid-State Letters*, **8**, C69.
[39] Wessels, K., Maekawa, M., Rathousky, J. *et al.* (2008) Highly porous TiO_2 films from anodically deposited titanate hybrids – photoelectrochemical and photocatalytic activity. *Microporous and Mesoporous Materials*, **111**, 55.
[40] Izaki, M. and Omi, T. (1996) Transparent zinc oxide films prepared by electrochemical reaction. *Applied Physics Letters*, **68**, 2439.
[41] Izaki, M. and Omi, T. (1996) Electrolyte optimization for cathodic growth of zinc oxide films. *Journal of the Electrochemical Society*, **143**, L53.
[42] Peulon, S. and Lincot, D. (1996) Cathodic electrodeposition from aqueous solution of dense or open-structured zinc oxide films. *Advanced Materials*, **8**, 166.
[43] Peulon, S. and Lincot, D. (1998) Mechanistic study of cathodic electrodeposition of zinc oxide and zinc hydroxychloride films from oxygenated aqueous zinc chloride solutions. *Journal of the Electrochemical Society*, **145**, 864.
[44] Pauporte, T. and Lincot, D. (2000) Electrodeposition of semiconductors for optoelectronic devices: results on zinc oxide. *Electrochimica Acta*, **45**, 3345.
[45] Pauporte, T. and Lincot, D. (2001) Hydrogen peroxide oxygen precursor for zinc oxide electrodeposition. I. Deposition in perchlorate medium. *Journal of the Electrochemical Society*, **148**, C310.
[46] Pauporte, T. and Lincot, D. (2001) Hydrogen peroxide oxygen precursor for zinc oxide electrodeposition. II. Mechanistic aspects. *Journal of Electroanalytical Chemistry*, **517**, 54.
[47] Ramirez, D., Silva, D., Gomez, H. *et al.* (2007) Electrodeposition of ZnO thin films by using molecular oxygen and hydrogen peroxide as oxygen precursors: Structural and optical properties. *Solar Energy Materials and Solar Cells*, **91**, 1458.
[48] Goux, A., Pauporte, T., Chivot, J. and Lincot, D. (2005) Temperature effects on ZnO electrodeposition. *Electrochimica Acta*, **50**, 2239.
[49] Cox, J.A. and Brajter, A. (1979) Electrodeposition of ZnO thin films by using molecular oxygen and hydrogen peroxide as oxygen precursors: Structural and optical properties. *Electrochimica Acta*, **24**, 517.
[50] Ogawa, N., Kodaiku, H. and Ikeda, S. (1986) On the polarographic reduction of nitrate ion in the presence of zirconium(IV). *Journal of Electroanalytical Chemistry*, **208**, 117.
[51] Ogawa, N. and Ikeda, S. (1991) On the electrochemical reduction of nitrate ion in the presence of various metal ions. *Analytical Science*, **7**, 1681.
[52] Yoshida, T., Komatsu, D., Shimokawa, N. and Minoura, H. (2004) Mechanism of cathodic electrodeposition of zinc oxide thin films from aqueous zinc nitrate baths. *Thin Solid Films*, **451/452**, 166.
[53] Sumida, T., Wada, Y., Kitamura, T. and Yanagida, S. (2001) Macroporous ZnO films electrochemically prepared by templating of opal films. *Chemistry Letters*, 38.
[54] Liu, Z., Jin, Z., Qiu, J. *et al.* (2006) Preparation and characteristics of ordered porous ZnO films by a electrodeposition method using PS array templates. *Semiconductor Science and Technology*, **21**, 60.
[55] Yang, Y.L., Yan, H., Fu, Z. *et al.* (2006) Enhanced photoluminescence from three-dimensional ZnO photonic crystals. *Solid State Communications*, **139**, 218.
[56] Bailar, J.C. (1973) *Comprehensive Inorganic Chemistry*, vol. 2, Pergamon Press, Oxford.
[57] Minoura, H. and Yoshida, T. (2008) Electrodeposition of ZnO/dye hybrid thin films for dye-sensitized solar cells. *Electrochemistry*, **76**, 109.
[58] Limmer, S.J., Kulp, E.A. and Switzer, J.A. (2006) Epitaxial electrodeposition of ZnO on Au(111) from alkaline solution: Exploiting amphoterism in Zn(II). *Langmuir*, **22**, 10535.

[59] Jayakrishnan, R. and Hodes, G. (2003) Non-aqueous electrodeposition of ZnO and CdO films. *Thin Solid Films*, **440**, 19.

[60] Anthony, S.P., Lee, J.I. and Kim, J.K. (2007) Tuning optical band gap of vertically aligned ZnO nanowire arrays grown by homoepitaxial electrodeposition. *Applied Physics Letters*, **90**, 103107.

[61] Ishizaki, H., Izaki, M. and Ito, T. (2001) Influence of $(CH_3)_2NHBH_3$ concentration on electrical properties of electrochemically grown ZnO films. *Journal of the Electrochemical Society*, **148**, C540.

[62] Ishizaki, H., Imaizumi, M., Matsuda, S. *et al.* (2002) Incorporation of boron in ZnO film from an aqueous solution containing zinc nitrate and dimethylamine-borane by electrochemical reaction. *Thin Solid Films*, **411**, 65.

[63] Kemell, M., Dartigues, F., Ritala, M. and Leskela, M. (2003) Electrochemical preparation of In and Al doped ZnO thin films for $CuInSe_2$ solar cells. *Thin Solid Films*, **434**, 20.

[64] Wellings, J.S., Samantilleke, A.P., Warren, P. *et al.* (2008) Comparison of electrodeposited and sputtered intrinsic and aluminium-doped zinc oxide thin films. *Semiconductor Science and Technology*, **23**, 125003.

[65] Machado, G., Guerra, D.N., Leinen, D. *et al.* (2005) Indium doped zinc oxide thin films obtained by electrodeposition. *Thin Solid Films*, **490**, 124.

[66] Cui, J.B. and Gibson, U.J. (2005) Electrodeposition and room temperature ferromagnetic anisotropy of Co and Ni-doped ZnO nanowire arrays. *Applied Physics Letters*, **87**, 133108.

[67] Cui, J.B. and Gibson, U.J. (2005) Enhanced nucleation, growth rate and dopant incorporation in ZnO nanowires. *Journal of Physical Chemistry B*, **109**, 22074.

[68] Jaramillo, T.F., Baeck, S.H., Kleiman-Shwarsctein, A. *et al.* (2005) Automated electrochemical synthesis and photoelectrochemical characterization of $Zn_{1-x}Co_xO$ thin films for solar hydrogen production. *Journal of Combinatorial Chemistry*, **7**, 264.

[69] Goux, A., Pauporte, T. and Lincot, D. (2006) Oxygen reduction reaction on electrodeposited zinc oxide electrodes in KCl solution at 70 degrees C. *Journal of Electroanalytical Chemistry*, **587**, 193.

[70] Pauporte, T., Pelle, F., Viana, B. and Aschehoug, P. (2007) Luminescence of nanostructured Eu^{3+}/ZnO mixed films prepared by electrodeposition. *Journal of Physical Chemistry C*, **111**, 15427.

[71] Zhao, J.Z., Liang, H.W., Sun, J.C. *et al.* (2008) p-Type Sb-doped ZnO thin films prepared by metallorganic chemical vapor deposition using metallorganic dopant. *Electrochemical and Solid-State Letters*, **11**, H323.

[72] Oh, M.S., Hwang, D.K., Seong, D.J. *et al.* (2008) Improvement of characteristics of Ga-doped ZnO grown by pulsed laser deposition using plasma-enhanced oxygen radicals. *Journal of the Electrochemical Society*, **155**, D599.

[73] Kim, S., Kang, B.S., Ren, F. *et al.* (2004) Characteristics of thin-film p-ZnMgO/n-ITO heterojunctions on glass substrates. *Electrochemical and Solid-State Letters*, **7**, G145.

[74] Polyakov, A.Y., Smirnov, N.B., Govorkov, A.V. *et al.* (2007) Electrical properties of ZnO(P) and ZnMgO(P) films grown by pulsed laser deposition. *Journal of the Electrochemical Society*, **154**, H825.

[75] Ahn, K.S., Yan, Y., Lee, S.H. *et al.* (2007) Photoelectrochemical properties of n-incorporated ZnO films deposited by reactive RF magnetron sputtering. *Journal of the Electrochemical Society*, **154**, B956.

[76] Yang, J.H., Kim, H.S., Lim, J.H. *et al.* (2006) The effect of Ar/O-2 sputtering gas on the phosphorus-doped p-type ZnO thin films. *Journal of the Electrochemical Society*, **153**, G242.

[77] Lan, C.J., Cheng, H.Y., Chung, R.J. *et al.* (2007) Bi-doped ZnO layer prepared by electrochemical deposition. *Journal of the Electrochemical Society*, **154**, D117.

[78] Kuo, C.H., Yeh, C.L., Chen, P.H. *et al.* (2008) Low operation voltage of nitride-based LEDs with Al-doped ZnO transparent contact layer. *Electrochemical and Solid-State Letters*, **11**, H269.

[79] Chou, S.M., Hon, M.H., Leu, I.C. and Lee, Y.H. (2008) Al-doped ZnO/Cu_2O heterojunction fabricated on (200) and (111)-orientated Cu_2O substrates. *Journal of the Electrochemical Society*, **155**, H923.

[80] Ae, L., Chen, J. and Lux-Steiner, M. (2008) Hybrid flexible vertical nanoscale diodes prepared at low temperature in large area. *Nanotechnology*, **19**, 475201.

[81] Zhang, D.K., Liu, Y.C., Liu, Y.L. and Yang, H. (2004) The electrical properties and the interfaces of Cu_2O/ZnO/ITO p-i-n heterojunction. *Journal of Physics B*, **351**, 178.

[82] Mukherjee, N., Bhattacharyya, P., Banerjee, M. *et al.* (2006) Galvanic deposition of nanocrystalline ZnO thin films from a ZnO-$Zn(OH)_2$ mixed phase precursor on p-Si substrate. *Nanotechnology*, **17**, 2665.

[83] Pauporté, T., Cortes, R., Froment, M. *et al.* (2002) Electrocrystallization of epitaxial zinc oxide onto gallium nitride. *Chemistry of Materials*, **14**, 4702.

[84] Pauporte, T., Lincot, D., Viana, B. and Pelle, F. (2006) Toward laser emission of epitaxial nanorod arrays of ZnO grown by electrodeposition. *Applied Physics Letters*, **89**, 233112.

[85] Liu, R., Vertegel, A.A., Bohannan, E.W. *et al.* (2001) Epitaxial electrodeposition of zinc oxide nanopillars on single-crystal gold. *Chemistry of Materials*, **13**, 508.
[86] Li, W.J., Shi, E.W., Zhong, W.Z. and Yin, Z.W. (1999) Growth mechanism and growth habit of oxide crystals. *Journal of Crystal Growth*, **203**, 186.
[87] Choy, J.H., Jang, E.S., Won, J.H. *et al.* (2003) Soft solution route to directionally grown ZnO nanorod arrays on Si wafer; room-temperature ultraviolet laser. *Advanced Materials*, **15**, 1911.
[88] Choy, J.H., Jang, E.S., Won, J.H. *et al.* (2004) Hydrothermal route to ZnO nanocoral reefs and nanofibers. *Applied Physics Letters*, **84**, 287.
[89] Gerischer, H., Lübke, M. and Sorg, N. (1986) A study of the chemical dissolution of semiconductors in aqueous electrolytes with zinc oxide as example. *Zeitschrift für Physikalische Chemie*, **148**, 11.
[90] Gerischer, H. and Sorg, N. (1991) Chemical dissolution of oxides – Experiments with intered ZnO pellets and ZnO single crystals. *Werkstoffe und Korrosion*, **42**, 149.
[91] Gerischer, H. and Sorg, N. (1992) Chemical dissolution of zinc oxide crystals in aqueous electrolytes – An analysis of the kinetics. *Electrochimica Acta*, **37**, 827.
[92] Könenkamp, R., Boedecker, K., Lux-Steiner, M.C. *et al.* (2000) Thin film semiconductor deposition on free-standing ZnO columns. *Applied Physics Letters*, **77**, 2575.
[93] Tena-Zaera, R., Kattya, A., Bastidea, S. *et al.* (2005) ZnO/CdTe/CuSCN, a promising heterostructure to act as inorganic eta-solar cell. *Thin Solid Films*, **483**, 372.
[94] Levy-Clement, C., Tena-Zaera, R., Ryan, M.A. *et al.* (2005) CdSe-Sensitized p-CuSCN/nanowire n-ZnO heterojunctions. *Advanced Materials*, **17**, 1512.
[95] Chen, Q.P., Xue, M.Z., Sheng, Q.R. *et al.* (2006) Electrochemical growth of nanopillar zinc oxide films by applying a low concentration of zinc nitrate precursor. *Electrochemical and Solid-State Letters*, **9**, C58.
[96] Tena-Zaera, R., Elias, J., Wang, G. and Levy-Clement, C. (2007) Role of chloride ions on electrochemical deposition of ZnO nanowire Arrays from O-2 reduction. *Journal of Physical Chemistry C*, **111**, 16706.
[97] Xu, L., Guo, Y., Liao, Q. *et al.* (2005) Morphological control of ZnO nanostructures by electrodeposition. *Journal of Physical Chemistry B*, **109**, 13519.
[98] Elias, J., Tena-Zaera, R. and Levy-Clement, C. (2008) Effect of the chemical nature of the anions on the electrodeposition of ZnO nanowire arrays. *Journal of Physical Chemistry C*, **112**, 5736.
[99] Elias, J., Tena-Zaera, R. and Levy-Clement, C. (2007) Electrodeposition of ZnO nanowires with controlled dimensions for photovoltaic applications: Role of buffer layer. *Thin Solid Films*, **515**, 8553.
[100] Govender, K., Boyle, D.S., Kenway, P.B. and O'Brien, P. (2004) Understanding the factors that govern the deposition and morphology of thin films of ZnO from aqueous solution. *Journal of Materials Chemistry*, **14**, 2575.
[101] Wang, Z., Qian, X.F., Yin, J. and Zhu, Z.K. (2004) Large-scale fabrication of tower-like, flower-like and tube-like ZnO arrays by a simple chemical solution route. *Langmuir*, **20**, 3441.
[102] Koenenkamp, R., Word, R.C., Dosmailov, M. and Naclarajah, A. (2007) Pentagonal ZnO nanorods. *Physica Status Solidi RRL*, **1**, 101.
[103] Koenenkamp, R., Word, R.C., Dosmailov, M. *et al.* (2007) Selective growth of single-crystalline ZnO nanowires on doped silicon. *Journal of Applied Physics*, **102**, 056103.
[104] Cui, J. and Gibson, U. (2007) Low-temperature fabrication of single-crystal ZnO nanopillar photonic bandgap structures. *Nanotechnology*, **18**, 155302.
[105] Weintraub, B., Deng, Y. and Wang, Z.L. (2007) Position-controlled seedless growth of ZnO nanorod arrays on a polymer substrate via wet chemical synthesis. *Journal of Physical Chemistry C*, **111**, 10162.
[106] Wang, J.G., Tian, M.L., Kumar, N. and Mallouk, T.E. (2005) Controllable template synthesis of superconducting Zn nanowires with different microstructures by electrochemical deposition. *Nano Letters*, **5**, 1247.
[107] Ren, X., Jiang, C.H., Li, D.D. and He, L. (2008) Fabrication of ZnO nanotubes with ultrathin wall by electro deposition method. *Materials Letters*, **62**, 3114.
[108] Lopez, C.M. and Choi, K.S. (2005) Enhancement of electrochemical and photoelectrochemical properties of fibrous Zn and ZnO electrodes. *Chemical Communications*, 3328.
[109] Das, I., Mishra, S.S., Agrawal, N.R. and Gupta, K.S. (2003) Some aspects of pattern formation during electrochemical deposition of metals. *Journal of the Indian Chemical Society*, **80**, 351.
[110] Tang, Y.W., Luo, L.J., Chen, Z.G. *et al.* (2007) Electrodeposition of ZnO nanotube arrays on TCO glass substrates. *Electrochemical Communications*, **9**, 289.

[111] Xu, L.F., Liao, Q., Zhang, J.P. *et al.* (2007) Single-crystalline ZnO nanotube arrays on conductive glass substrates by selective dissolution of electrodeposited ZnO nanorods. *Journal of Physical Chemistry C*, **111**, 4549.

[112] Elias, J., Tena-Zaera, R., Wang, G.Y. and Levy-Clement, C. (2008) Conversion of ZnO nanowires into nanotubes with tailored dimensions. *Chemistry of Materials*, **20**, 6633.

[113] She, G.W., Zhang, X.H., Shi, W.S. *et al.* (2007) Electrochemical/chemical synthesis of highly-oriented single-crystal ZnO nanotube arrays on transparent conductive substrates. *Electrochemical Communications*, **9**, 2784.

[114] Pradhan, D., Kumar, M. Ando, Y. and Leung, K.T. (2008) Efficient field emission from vertically grown planar ZnO nanowalls on an ITO-glass substrate. *Nanotechnology*, **3**, 035603.

[115] Pradhan, D., Kumar, M. Ando, Y. and Leung, K.T. (2008) One-dimensional and two-dimensional ZnO nanostructured materials on a plastic substrate and their field emission properties. *Journal of Physical Chemistry C*, **112**, 7093.

[116] Gao, Y.F. and Nagai, M. (2006) Morphology evolution of ZnO thin films from aqueous solutions and their application to solar cells. *Langmuir*, **22**, 3936.

[117] Cao, B., Teng, X., Heo, S.H. *et al.* (2007) Different ZnO nanostructures fabricated by a seed-layer assisted electrochemical route and their photoluminescence and field emission properties. *Journal of Physical Chemistry C*, **111**, 2470.

[118] Moffat, T.P., Wheeler, D., Kim, S.-K. and Josell, D. (2006) Curvature enhanced adsorbate coverage model for electrodeposition. *Journal of the Electrochemical Society*, **153**, C127.

[119] Yoshida, T., Miyamoto, K., Hibi, N. *et al.* (1998) Self assembled growth of nano particulate porous ZnO thin film modified by 2,9,16,23-tetrasulfophthalocyaninatozinc(II) by one-step electrodeposition. *Chemistry Letters*, 599.

[120] Schlettwein, D., Oekermann, T., Yoshida, T. *et al.* (2000) Photoelectrochemical sensitization of ZnO-tetrasulfophthalocyaninatozinc composites prepared by electrochemical self-assembly. *Journal of Electroanalytical Chemistry*, **481**, 42.

[121] Oekermann, T., Yoshida, T., Schlettwein, D. *et al.* (2001) Photoelectrochemical properties of ZnO/tetrasulfophthalocyanine hybrid thin films prepared by electrochemical self-assembly. *Physical Chemistry and Chemical Physics*, **3**, 3387.

[122] Oekermann, T., Yoshida, T., Boeckler, C. *et al.* (2005) Capacitance and field-driven electron transport in electrochemically self-assembled nanoporous ZnO/dye hybrid films. *Journal of Physical Chemistry B*, **109**, 12560.

[123] Rathousky, J., Löwenstein, T., Nonomura, K. *et al.* (2005) Electrochemically self-assembled mesoporous dye-modified zinc oxide thin films. *Studies in Surface Science and Catalysis*, **156**, 315.

[124] Löwenstein, T., Nonomura, K., Yoshida, T. *et al.* (2006) Efficient sensitization of mesoporous electrodeposited zinc oxide by cis-bis(isothiocyanato)bis(2,2′-bipyridyl-4,4′-dicarboxylato)-ruthenium(II). *Journal of the Electrochemical Society*, **153**, A699.

[125] Yoshida, T., Terada, K., Schlettwein, D. *et al.* (2000) Electrochemical self-assembly of nanoporous ZnO/eosin Y thin films and its sensitized photoelectrochemical performance. *Advanced Materials*, **12**, 1214.

[126] Okabe, K., Yoshida, T., Sugiura, T. and Minoura, H. (2001) Electrodeposition of Photoactive ZnO/Xanthene Dye Hybrid Thin Films. *Transactions of the Materials Research Society of Japan*, **26**, 523.

[127] Yoshida, T., Oekermann, T., Okabe, K. *et al.* (2002) Time- and frequency-resolved photoelectrochemical investigations on nano-honeycomb TiO_2 Electrodes. *Electrochemistry*, **70**, 470.

[128] Goux, A., Pauporte, T., Yoshida, T. and Lincot, D. (2006) Mechanistic study of the electrodeposition of nanoporous self-assembled ZnO/eosin Y hybrid thin films: Effect of eosin concentration. *Langmuir*, **22**, 10545.

[129] Goux, A., Pauporte, T., Lincot, D. and Dunsch, L. (2007) In situ ESR and UV/Vis spectroelectrochemical study of eosin Y upon reduction with and without Zn(II) ions. *ChemPhysChem.*, **8**, 926.

[130] Yoshida, T., Pauporté, T., Lincot, D. *et al.* (2003) Cathodic electrodeposition of ZnO/eosinY hybrid thin films from oxygen-saturated aqueous mixed solution of $ZnCl_2$ and eosinY. *Journal of the Electrochemical Society*, **150**, C608.

[131] Boeckler, C., Oekermann, T., Saruban, M. *et al.* (2008) Influence of the supporting salt concentration on the electrodeposition of ZnO/eosin Y hybrid films. *Physica Status Solidi A*, **205**, 2388.

[132] Yoshida, T., Tochimoto, M., Schlettwein, D. *et al.* (1999) Self-assembly of zinc oxide thin films modified with tetrasulfonated metallophthalocyanines by one-step electrodeposition. *Chemistry of Materials*, **11**, 2657.

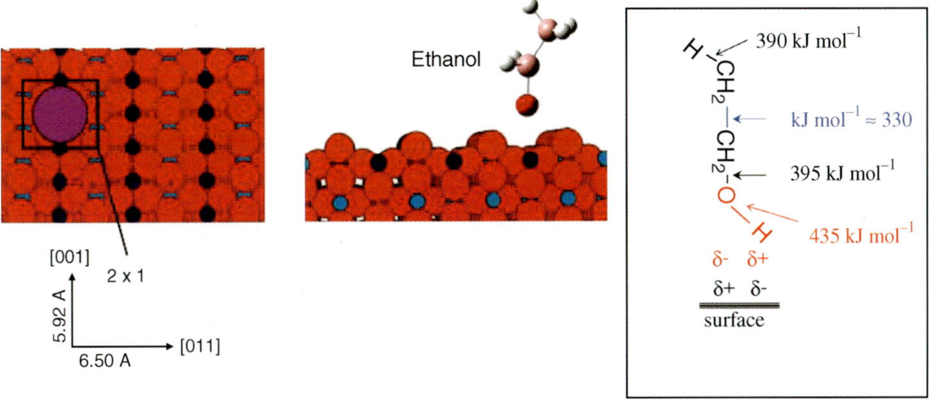

Figure 3.3 The left-hand side of the figure: top and side views of the (110) surface of the rutile TiO_2. Small (blue) circles represent Ti atoms and large (red) circles represent O atoms. The right-hand side of the figure gives the bond energies of an ethanol molecule. A representation of the adsorbed ethoxide species on top of a fivefold coordinated Ti is also seen. The H resulting from the dissociative adsorption is presumably on top of the twofold coordinated surface O atoms (bridging O atoms), not shown. The atomic scale between ethanol and TiO_2 is not unified. The very large circle (pink) represents the van der Waals dimension of an ethanol molecule. Data for this figure are provided from spectroscopic measurement. (Adapted with permission from P.M. Jayaweera *et al.*, Photoreaction of ethanol on TiO_2(110) single-crystal surface, *Journal of Physical Chemistry C*, **111**(4), 2007. © 2007 American Chemical Society.)

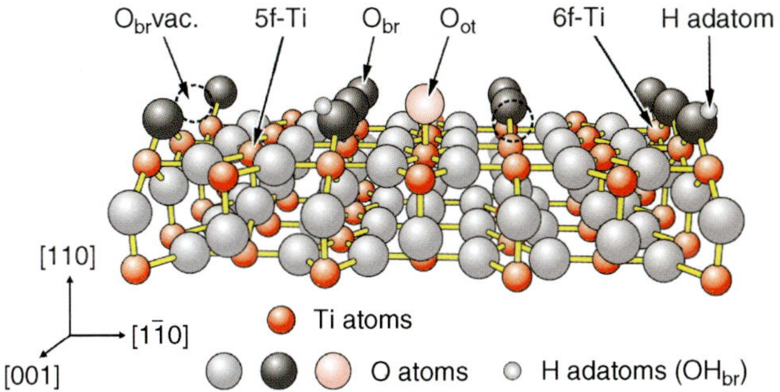

Figure 4.1 Ball-and-stick model of the TiO_2(110)–(1 × 1) surface with some of its known point defects. Large gray balls represent O atoms and medium-sized red balls sixfold coordinated Ti (6f-Ti) and fivefold coordinated surface Ti atoms (5f-Ti). Small light gray balls represent H adatoms. The bridge-bonded O species (O_{br}), single oxygen vacancies (O_{br} vac.), and the on-top bonded O species (O_{ot}) are also indicated.

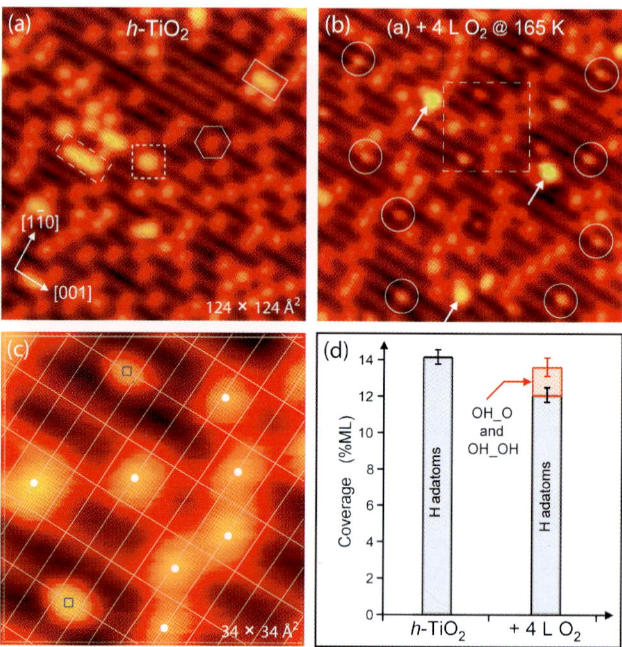

Figure 4.10 STM images of the h-$TiO_2(110)$ surface before (a) and after (b) 4 L O_2 exposure at ~165 K. Symbols in (a) indicate single H adatoms (hexagon), paired H adatoms (square, dotted line), next-nearest H adatoms (rectangle, solid line) and a chain of next-nearest H adatoms (rectangle, dashed line). The circles in (b) indicate the newly formed species in the Ti troughs and arrows indicate larger protrusions in the Ti troughs of STM heights, as typically observed for water species. The indicated square area in (b) (white, dashed line) is shown in (c), enlarged. A lattice grid was centered on top of 5f-Ti sites in (c) using single H adatoms (white small dots) for aligning in the [001] direction. Note that the z contrast in the STM image depicted in (c) is higher than in the STM images shown in (a) and (b). The STM images were acquired at ~110 K. (d) Bar graph of measured coverages in %ML of the obtained surface species before and after O_2 exposure. To extract the densities of surface species areas of ~2000 nm^2 were scanned and analyzed. (Reprinted with permission from J. Matthiesen *et al.*, Observation of all the intermediate steps of a chemical reaction on an oxide surface by scanning tunneling microscopy, *ACS Nano*, **3**, 517–526, 2009. © 2009 American Chemical Society.)

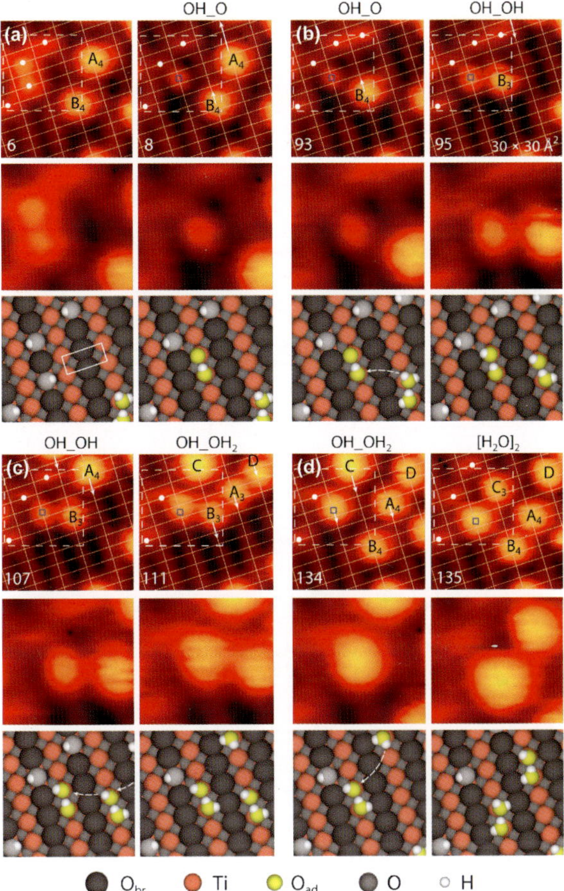

Figure 4.11 Snapshots extracted from an STM movie "O_2_h-TiO_2_reaction" recorded at ~190 K on an *h*-TiO_2(110) surface with a water coverage of ~6 %ML and in an O_2 background of 1×10^{-8} Torr. STM images were acquired with a rate of 2.76 s per frame. OH_O formation is shown in (a), and the three following H uptake reactions are identified in (b), OH_OH formation, (c), OH_OH_2 formation, and (d), $[H_2O]_2$ formation, respectively. Numbers at the lower left edges indicate the appearance of the selected STM images within the movie. White small dots indicate H adatoms, and water species in the form of dimers are denoted using the labels A_n, B_n, C_n and D, respectively, where n indicates the estimated number of containing H atoms, that is, A_n with $n = 4$ denotes a stoichiometric water dimer. In (c) and (d) the index n was omitted for some water dimers. The lattice grid is cantered on 5f-Ti sites. White arrows along the Ti troughs indicate the diffusion directions of the adsorbates. Indicated areas of 18×16.5 Å^2 size in the STM images in the first row (white dashed rectangles) are shown enlarged in the second row and explained in the third row using ball models (top views). The bent arrows indicate from which side in the STM images the H adatoms were provided. (Reprinted with permission from J. Matthiesen *et al.*, Observation of all the intermediate steps of a chemical reaction on an oxide surface by scanning tunneling microscopy, *ACS Nano*, **3**, 517–526, 2009. © 2009 American Chemical Society.)

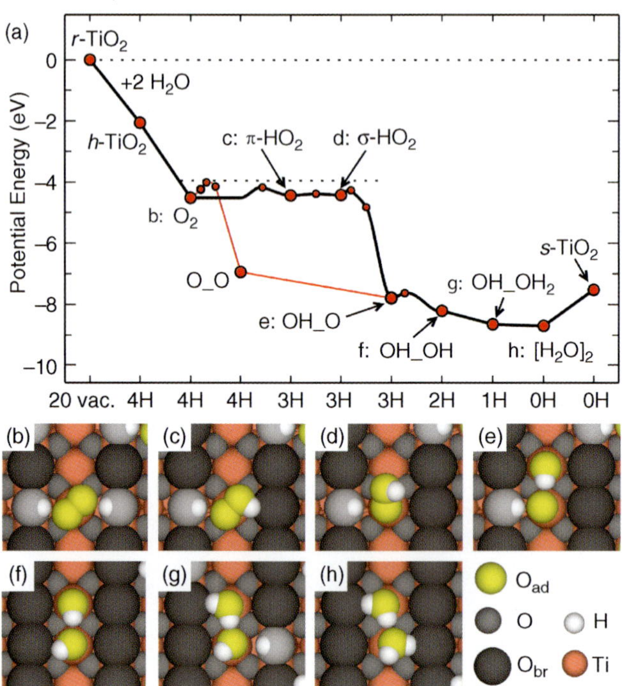

Figure 4.12 (a) DFT-based potential energy diagram. Big red dots correspond to calculated local potential energy minima, whereas small red dots indicate values deduced using the climbing NEB procedure. The initial configuration is a reduced $TiO_2(110)$ surface [r-$TiO_2(110)$] with two O_{br} vacancies and two H_2O molecules in the gas phase. The first step is dissociation of the H_2O molecules in O_{br} vacancies, thereby forming four H adatoms [h-$TiO_2(110)$] (the number of O_{br} vacancies or H adatoms on the $TiO_2(110)$ surface are denoted on the bottom axis). The second step is the adsorption of an O_2 molecule onto h-$TiO_2(110)$. The pathway in which O_2 becomes HO_2 before dissociation into OH_O is shown as a thick black line, whereas the thin red line accounts for a pathway in which O_2 dissociates ($E_a = 0.51$ eV) before the reduction. The former pathway is kinetically preferred. Desorption of the $[H_2O]_2$ leads to a clean $TiO_2(110)$ surface, s-$TiO_2(110)$, with no point defects in the uppermost surface layer. (b)–(h) Ball models (top views) of selected reaction intermediates. O_{ad} (yellow balls) indicate the oxygen atoms of the adsorbates. (Reprinted with permission from J. Matthiesen *et al.*, Observation of all the intermediate steps of a chemical reaction on an oxide surface by scanning tunneling microscopy, *ACS Nano*, **3**, 517–526, 2009. © 2009 American Chemical Society.)

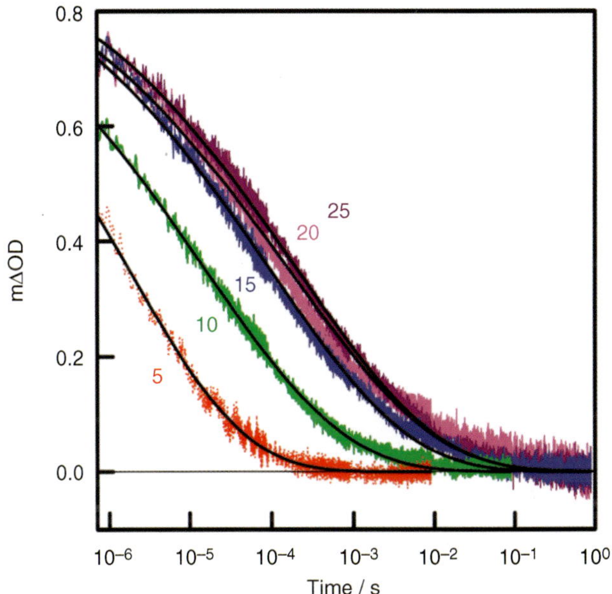

Figure 9.10 Transient absorption kinetics for the CdS-5 ~ 25/TiO_2 monitored at 650 nm. Solid lines indicate fits to a stretched exponential function (Equation 9.4).

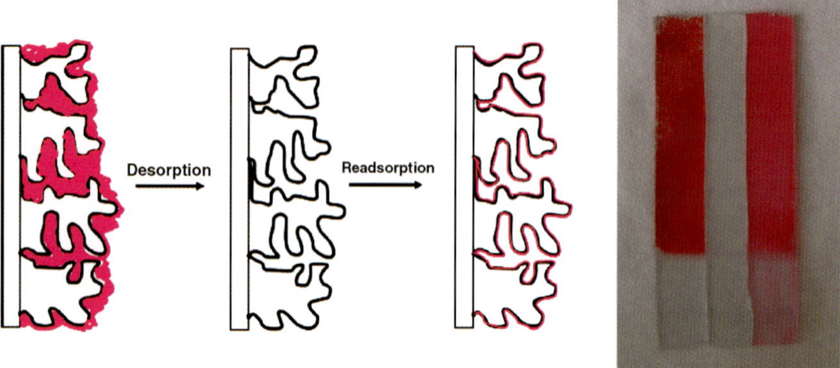

Figure 11.2 Illustration of the dye desorption/readsorption process. The photo shows three parts of the same ZnO film deposited with eosin Y from O_2-saturated solution in all three stages. The virtual absence of any dye in the film after the desorption process (middle part) and the lower dye content after the readsorption process (right part) compared to the as-deposited film (left part) are clearly seen in this photo.

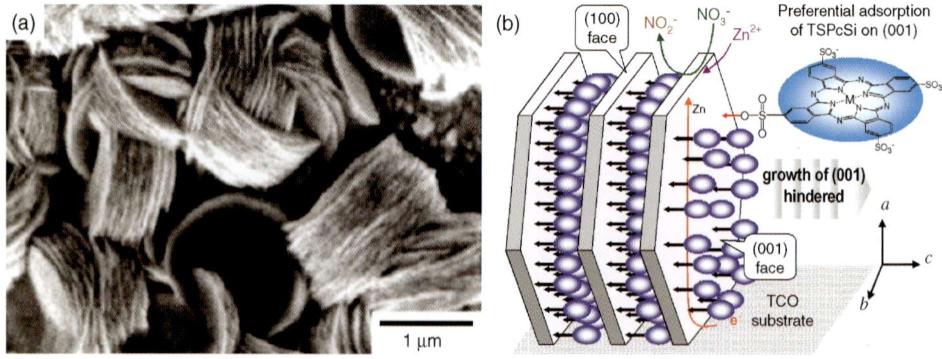

Figure 11.3 (a) SEM micrograph of a ZnO/TSPcSi film electrodeposited from 0.1 M $Zn(NO_3)_2$/50 μM TSPcSi mixed aqueous solution at −0.9 V versus SCE. (Reprinted from T. Oekermann, T. Yoshida, D. Schlettwein *et al.*, Photoelectrochemical properties of ZnO/tetrasulfophthalocyanine hybrid thin films prepared by electrochemical self-assembly, *Physical Chemistry and Chemical Physics*, **3**, 3387, 2001, PCCP Owner Societies.) (b) Growth model explaining the crystallographic orientation and formation of ZnO platelets in electrodeposited ZnO/TSPcSi films.

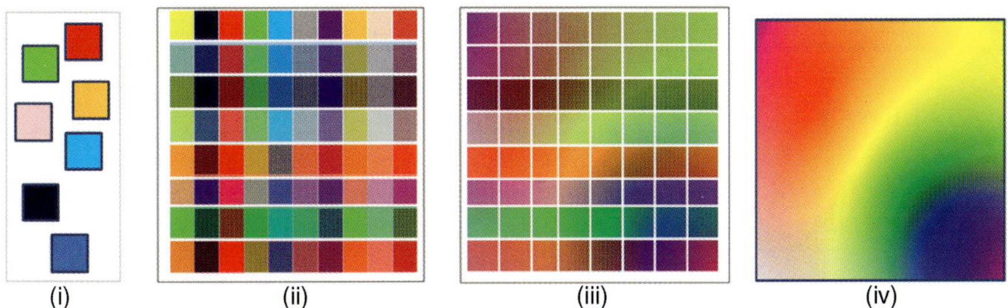

Figure 14.6 High-throughput synthesis of PEC materials in several different formats with color representing different synthesis and/or processing variables (e.g., composition, conditions, etc.). Left-to-right: Materials may be rapidly synthesized and screened as: (i) individual samples with varying properties, or arranged into two-dimensional arrays with (ii) random variations, (iii) discrete systematic variations, or (iv) continuous variations.

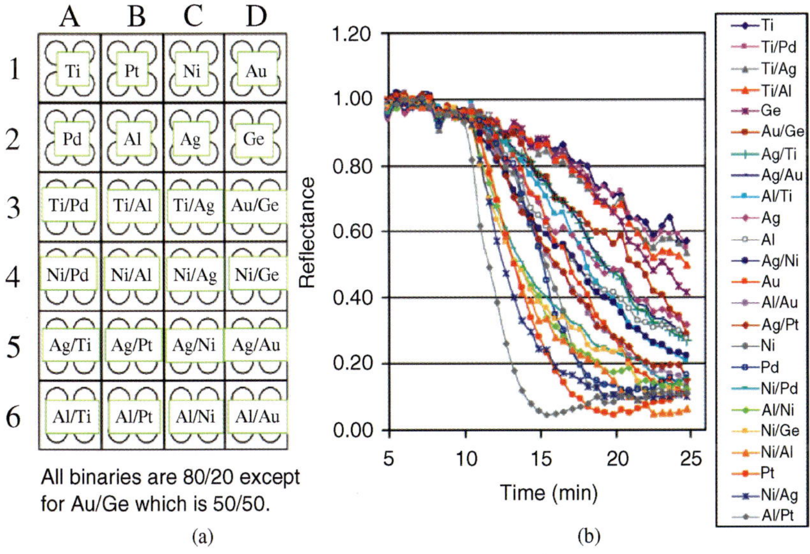

Figure 14.18 Electrocatalyst library and reflectance measurements during hydrogen production under an applied bias. (Reprinted with permission from T.F. Jaramillo, A. Ivanovskaya and E.W. McFarland, High-throughput screening system for catalytic hydrogen-producing materials, *Journal of Combinatorial Chemistry*, **4**(1), 17–22, 2002. © 2002 American Chemical Society.)

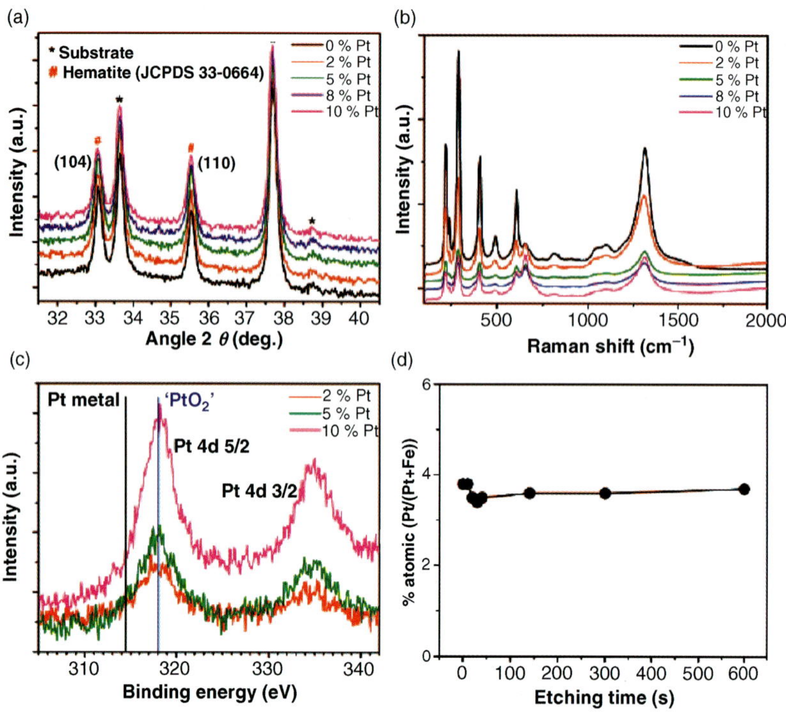

Figure 14.21 Characterizations of undoped and Pt-doped iron-oxide thin films on FTO: (a) XRD patterns, (b) Raman spectra, (c) XPS spectra and (d) XPS etching profile. (Reprinted with permission from Y.S. Hu, A. Kleiman-Shwarsctein, A.J. Forman *et al.*, Pt-doped alpha-Fe_2O_3 thin films active for photoelectrochemical water splitting, *Chemistry of Materials*, **20**(12), 3803–3805, 2008. © 2008 American Chemical Society.)

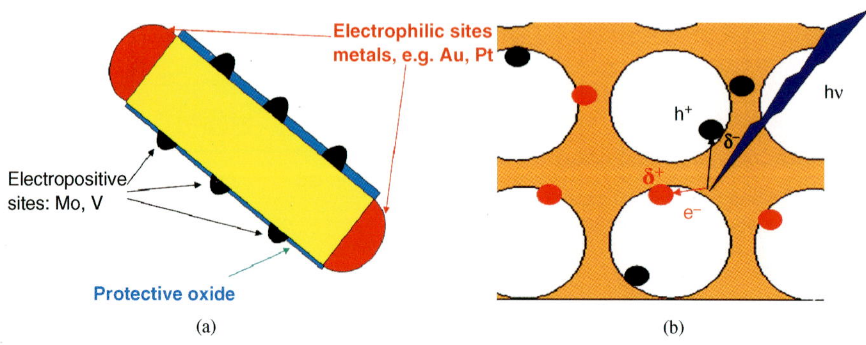

Figure 14.41 Idealized heterogeneous nanostructures for PEC systems. Red-colored particles represent electrophilic materials used for hydrogen evolution electrocatalysts, whereas black particles represent electropositive oxygen-evolution electrocatalysts.

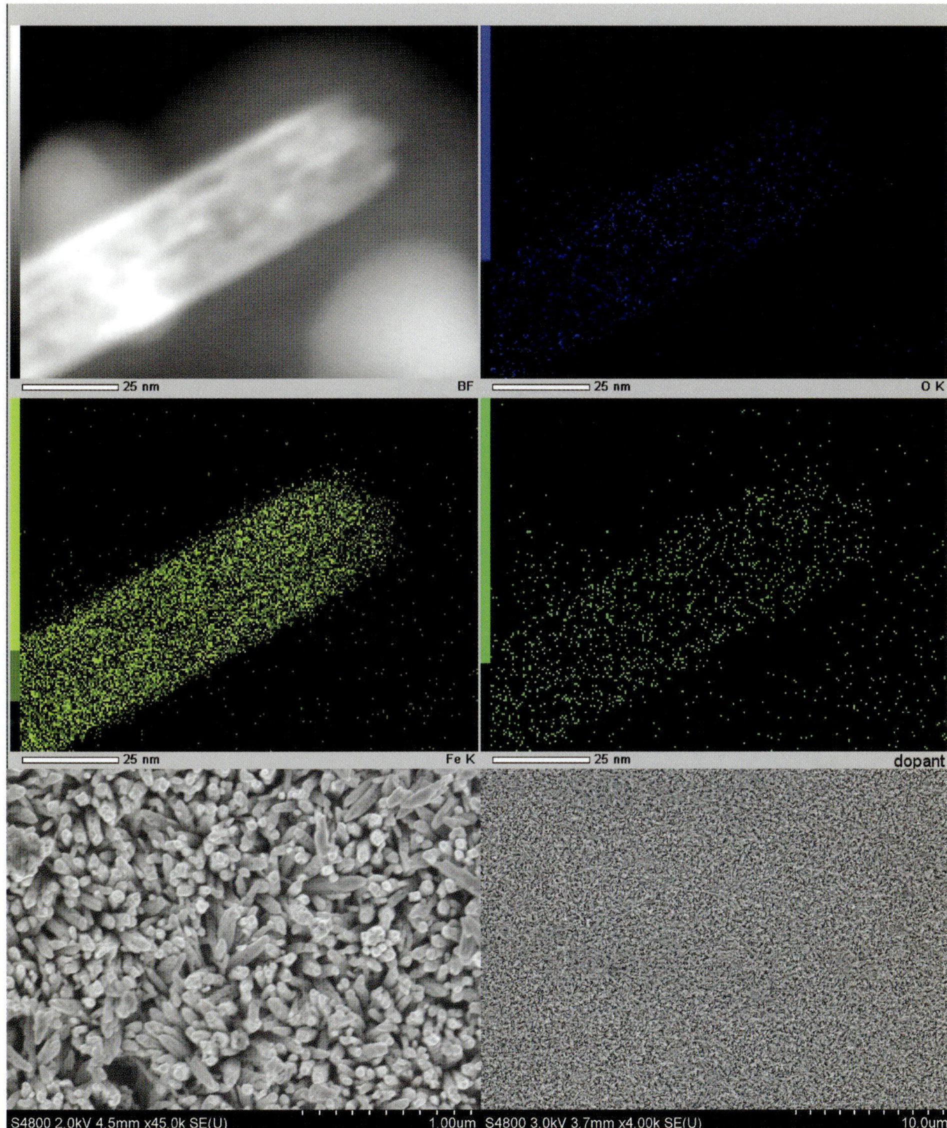

Figure 17.13 (Top) High-resolution transmission electron microscopy images of a doped iron-oxide quantum-rod bundle and the corresponding energy-dispersive elemental mapping analysis of oxygen, iron and a single dopant showing its homogeneous distribution all over the rods. (Bottom) SEM images of vertically oriented doped iron-oxide quantum-rod arrays grown onto transparent conducting oxides.

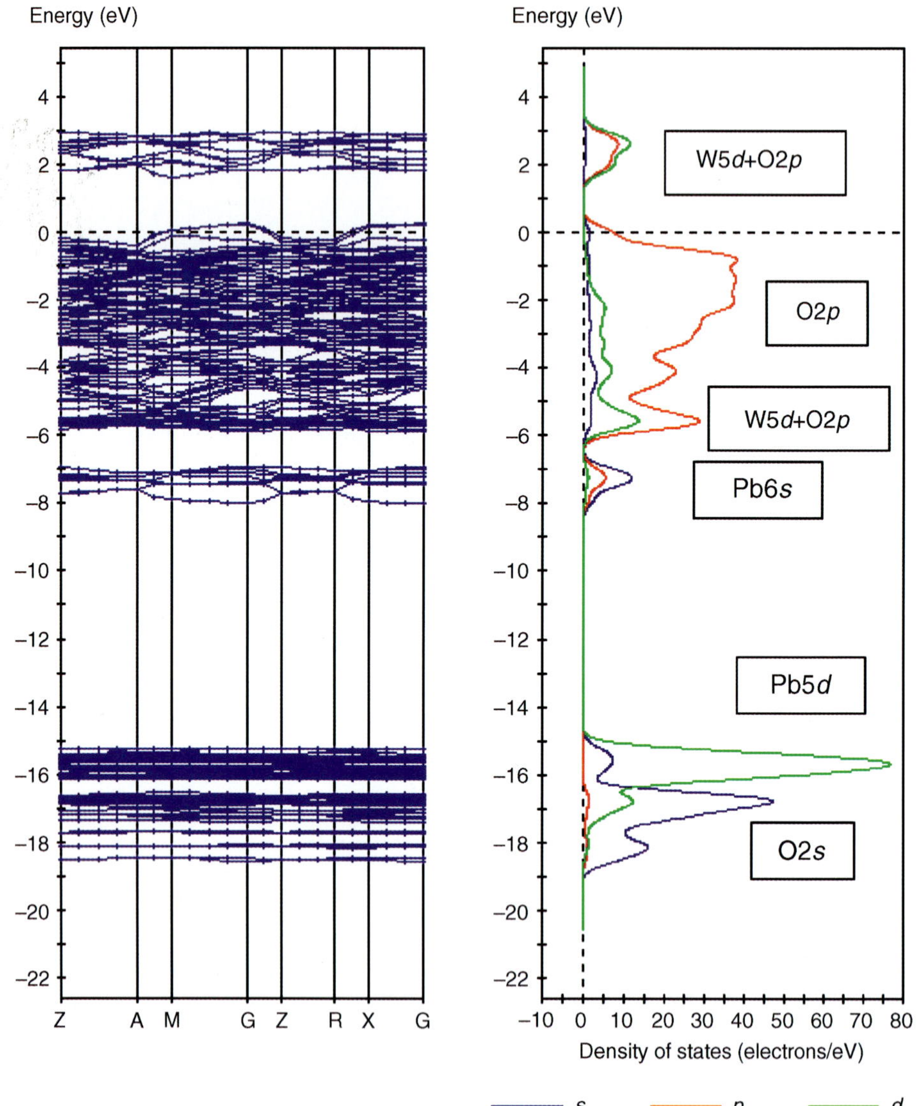

Figure 18.14 Band structure and density of state for $Pb_7W_8O_{28.8}$. Blue line; s-orbital, red; p-orbital, green; d-orbital.

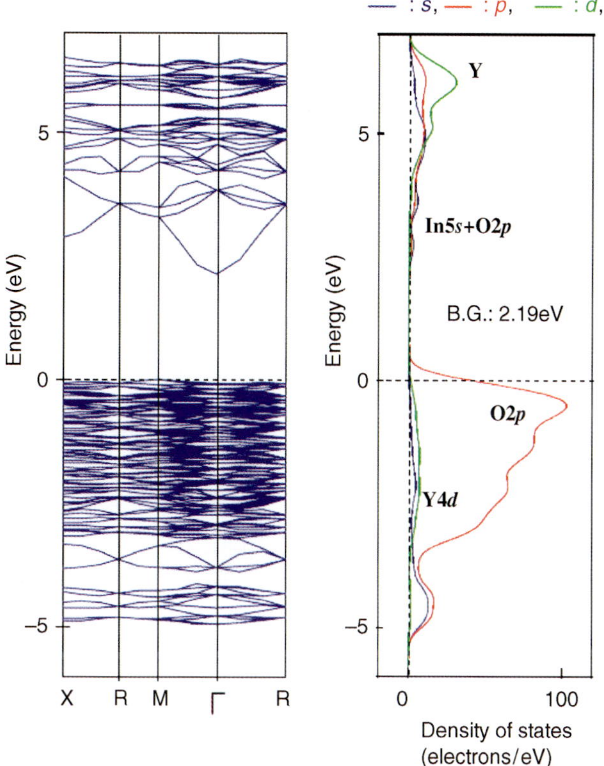

Figure 18.20 Band structure and density of states for $Y_xIn_{2-x}O_3(x = 1.0)$. Blue line: s-orbital; red: p-orbital; green: d-orbital. (Reprinted with permission from N. Arai, N. Saito, H. Nishiyama *et al.*, Photocatalytic activity for overall water splitting of RuO_2-loaded $Y_xIn_{2-x}O_3$ (x = 0.9–1.5), *Journal of Physical Chemistry C*, **112**, 5000–5005, 2008. © 2008 American Chemical Society.)

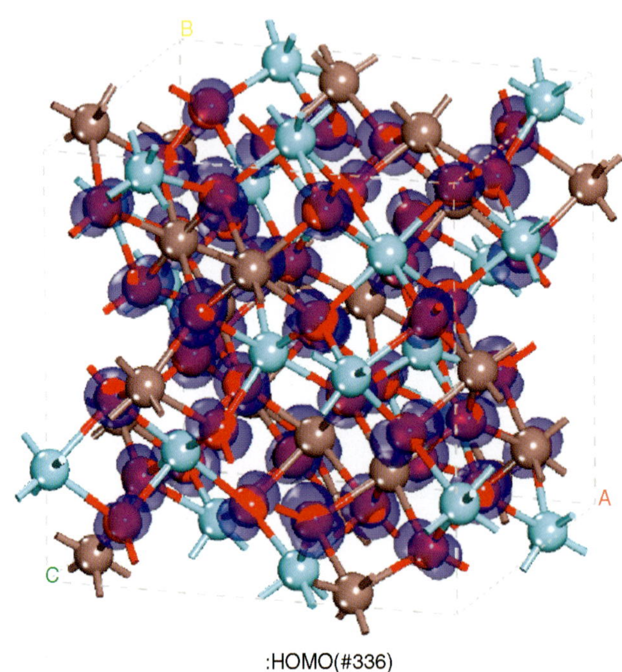

Figure 18.21 Electron density contour map for the top of the valence band (HOMO) of $Y_xIn_{2-x}O_3$-(x = 1.0). Light blue: Y atom; brown: In atom; red: O atom. (Reprinted with permission from N. Arai, N. Saito, H. Nishiyama *et al.*, Photocatalytic activity for overall water splitting of RuO_2-loaded $Y_xIn_{2-x}O_3$ (x = 0.9–1.5), *Journal of Physical Chemistry C*, **112**, 5000–5005, 2008. © 2008 American Chemical Society.)

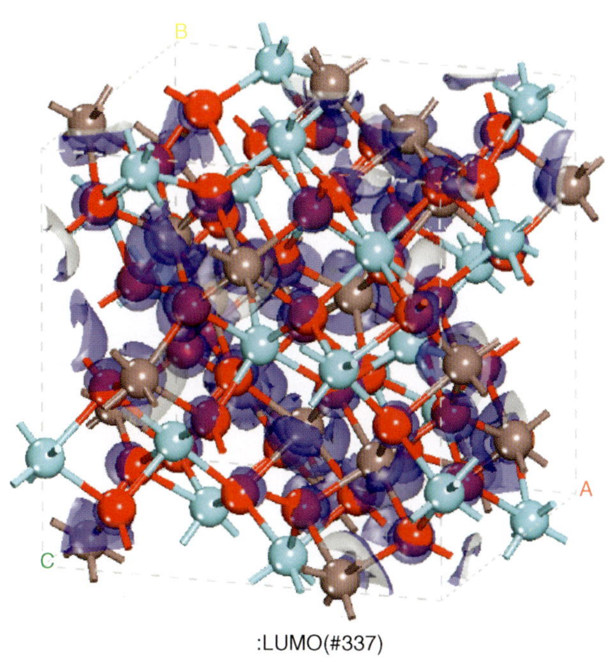

Figure 18.22 Electron density contour map for the bottom of the conduction band (LUMO) of $Y_xIn_{2-x}O_3$ ($x = 1.0$). See Figure 18.21 for the color coding. (Reprinted with permission from N. Arai, N. Saito, H. Nishiyama *et al.*, Photocatalytic activity for overall water splitting of RuO_2-loaded $Y_xIn_{2-x}O_3$ ($x = 0.9$–1.5), *Journal of Physical Chemistry C*, **112**, 5000–5005, 2008. © 2008 American Chemical Society.)

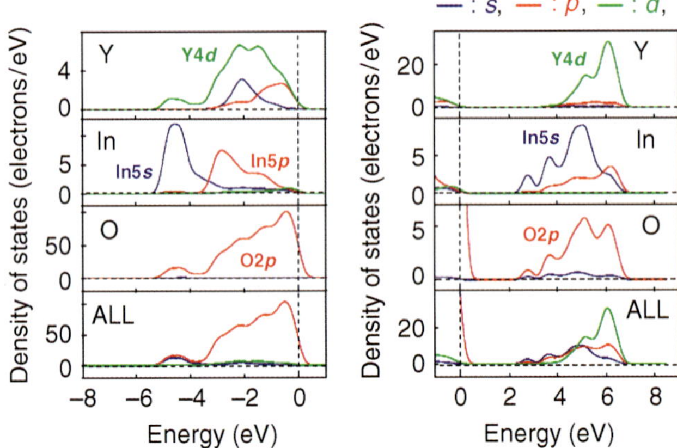

Figure 18.23 Total and atomic orbital PDOS for Y, In and O atoms for $Y_xIn_{2-x}O_3(x = 1.25)$. (Reprinted with permission from N. Arai, N. Saito, H. Nishiyama *et al.*, Photocatalytic activity for overall water splitting of RuO_2-loaded $Y_xIn_{2-x}O_3$ (x = 0.9–1.5), *Journal of Physical Chemistry C*, **112**, 5000–5005, 2008. © 2008 American Chemical Society.)

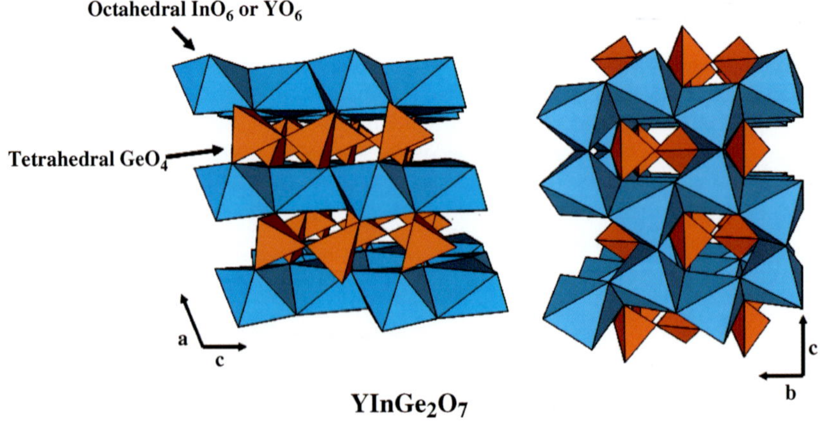

Figure 18.25 A schematic representation of the crystal structure of $YInGe_2O_7$.

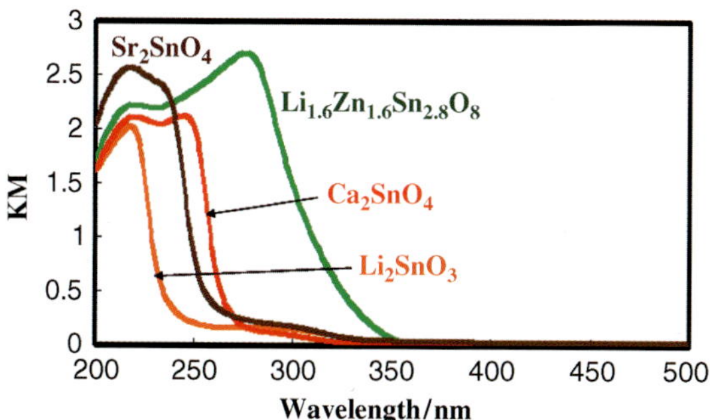

Figure 18.26 UV diffuse reflectance spectra of various Sn metal oxides.

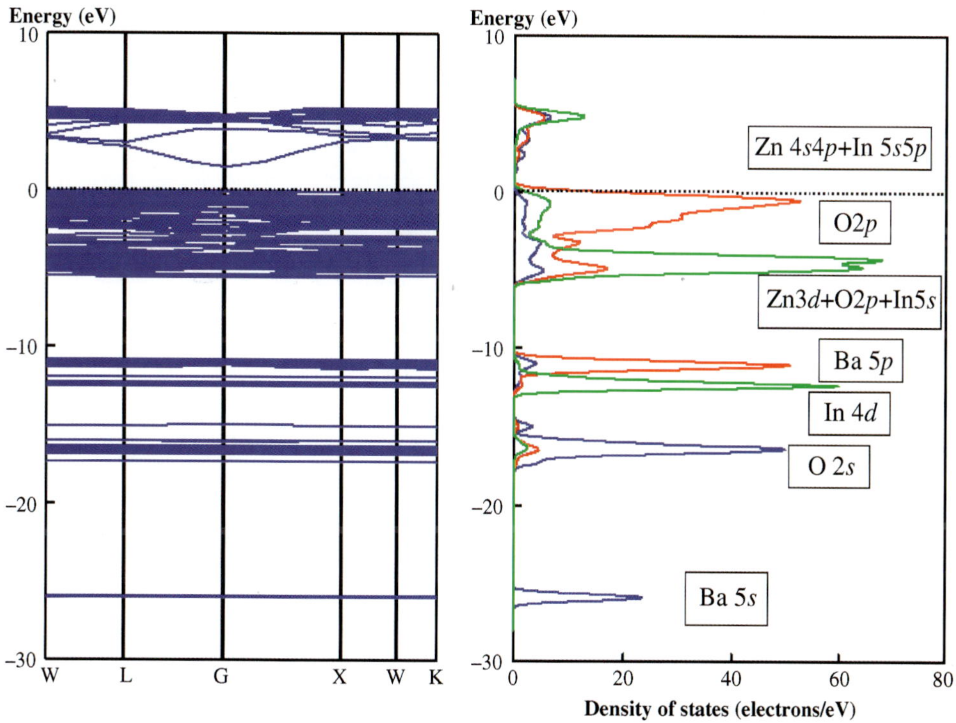

Figure 18.27 Band structure and atomic orbital PDOS for $Ba_3Zn_5In_2O_{11}$. Blue line: s orbital; red: p orbital; green: d orbital.

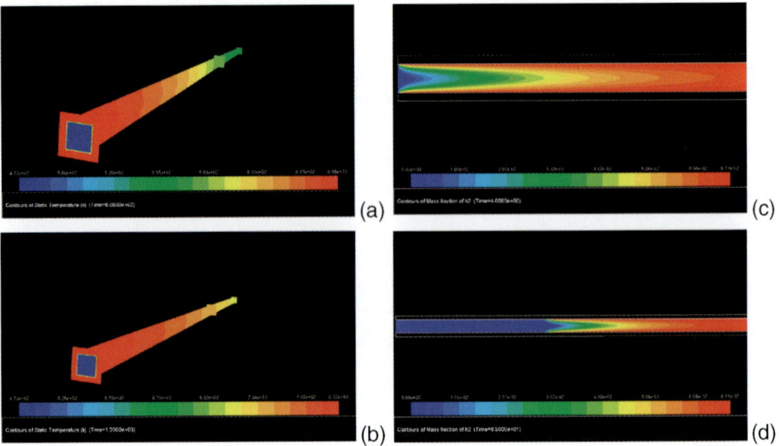

Figure 20.13 Evolution at two different time instances along a reactor's channel during the reactor's operation of: (a), (b) temperature ($t = 600$ and 1200 seconds, respectively); (c), (d) contours of H_2 mass fraction ($t = 4$ and 85 seconds, respectively). (Figure adapted from Ch. Lekkos, M. Kostoglou and A.G. Konstandopoulos, Simulation of solar water-splitting monolithic reactors, (to be submitted).)

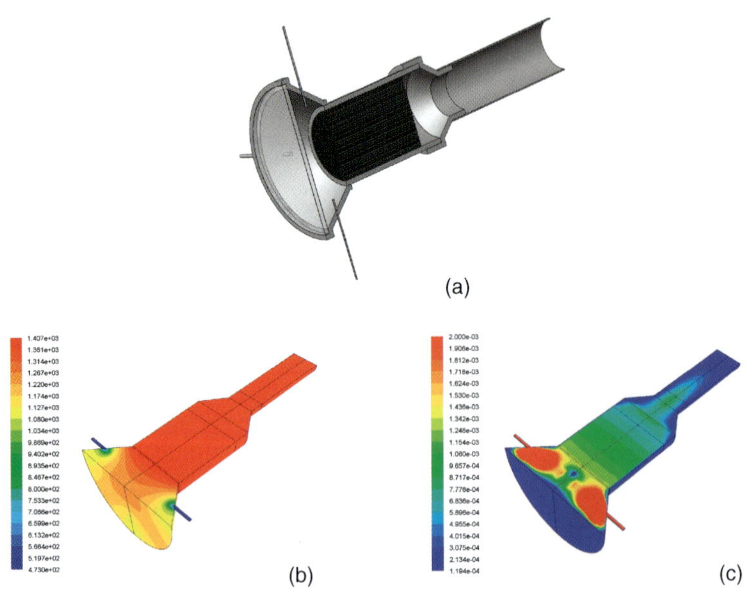

Figure 20.14 3D computational simulation of the solar water-splitting reactor shown in (a). (b) Temperature contours; (c) velocity contours. (Figure adapted from Ch. Lekkos, M. Kostoglou and A.G. Konstandopoulos, Simulation of solar water-splitting monolithic reactors, (to be submitted).)

[133] Yoshida, T. and Minoura, H. (2000) Electrochemical self-assembly of dye-modified zinc oxide thin films. *Advanced Materials*, **12**, 1219.

[134] Yoshida, T., Ide, S., Sugiura, T. and Minoura, H. (2000) Orientation Controlled Electrodeposition of Zinc Oxide Thin Films. *Transactions of the Materials Research Society of Japan*, **25**, 1111.

[135] Jang, E.S., Won, J.H., Hwang, S.J. and Choy, J.H. (2006) Fine tuning of the face orientation of ZnO crystals to optimize their photocatalytic activity. *Advanced Materials*, **18**, 3309.

[136] Boeckler, C., Feldhoff, A. and Oekermann, T. (2007) Nanostructured ZnO films electrodeposited using monosaccharide molecules as templates, in *Solar Hydrogen and Nanotechnology II*, vol. **6650** (ed. J. Guo), SPIE, Bellingham WA, (Proceedings).

[137] O'Regan, B., Schwartz, D.T., Zakeeruddin, S.M. and Grätzel, M. (2000) Electrodeposited nanocomposite n-p heterojunctions for solid-state dye-sensitized photovoltaics. *Advanced Materials*, **12**, 1263.

[138] Kresge, C.T., Leonowicz, M.E., Roth, W.J. *et al.* (1992) Ordered mesoporous molecular sieves synthesized by a liquid crystal template mechanism. *Nature*, **359**, 710.

[139] Ying, J.Y., Mehnert, C.P. and Wong, M.S. (1999) Synthesis and applications of supramolecular-templated mesoporous materials. *Angewandte Chemie*, **38**, 58.

[140] Lu, Y., Ganguli, R., Drewien, C.A. *et al.* (1997) Continuous formation of supported cubic and hexagonal mesoporous films by sol gel dip-coating. *Nature*, **389**, 364.

[141] Grosso, D., Cagnol, F., de, G.J. *et al.* (2004) Fundamentals of mesostructuring through evaporation-induced self-assembly. *Advanced Functional Materials*, **14**, 309.

[142] Manne, S., Cleveland, J.P., Gaub, H.E. *et al.* (1994) Direct visualization of surfactant hemimicelles by force microscopy of the electrical double layer. *Langmuir*, **10**, 4409.

[143] Liu, J.F. and Ducker, W.A. (1999) Surface-induced phase behavior of alkyltrimethylammonium bromide surfactants adsorbed to mica, silica and graphite. *Journal of Physical Chemistry B*, **103**, 8558.

[144] Burgess, I., Jeffrey, C.A., Cai, X. *et al.* (1999) Direct visualization of the potential-controlled transformation of hemimicellar aggregates of dodecyl sulfate into a condensed monolayer at the Au(111) electrode surface. *Langmuir*, **15**, 2607.

[145] Choi, K.S., Lichtenegger, H.C. and Stucky, G.D. (2002) Electrochemical synthesis of nanostructured ZnO films utilizing self-assembly of surfactant molecules at solid-liquid interfaces. *Journal of the American Chemical Society*, **124**, 12402.

[146] Tan, Y., Steinmiller, E.M.P. and Choi, K.S. (2005) Electrochemical tailoring of lamellar-structured ZnO films by interfacial surfactant templating. *Langmuir*, **21**, 9618.

[147] Michaelis, E., Wöhrle, D., Rathousky, J. and Wark, M. (2006) Electrodeposition of porous zinc oxide electrodes in the presence of sodium laurylsulfate. *Thin Solid Films*, **497**, 163.

[148] Steinmiller, E.M.P. and Choi, K.S. (2007) Anodic construction of lamellar structured ZnO films using basic media via interfacial surfactant templating. *Langmuir*, **23**, 12710.

[149] Dörfler, H.D. (1994) *Grenzflächen- und Kolloidchemie*, VCH, Weinheim.

[150] Kim, H.U. and Lim, K.H. (2003) Description of temperature dependence of critical micelle concentration. *Bulletin of the Korean Chemical Society*, **24**, 1449.

[151] Mukerjee, P. and Mysels, K.J. (1971) *Critical Micelle Concentrations of Aqueous Surfactant Systems*, National Bureau of Standards, Washington.

[152] Boeckler, C., Oekermann, T., Feldhoff, A. and Wark, M. (2006) The role of the critical micelle concentration in the electrochemical deposition of nanostructured ZnO films under utilization of amphiphilic molecules. *Langmuir*, **22**, 9427.

[153] Jing, H.Y., Li, X.L., Lu, Y. *et al.* (2005) Electrochemical self-assembly of highly oriented ZnO-surfactant hybrid multilayers. *Journal of Physical Chemistry B*, **109**, 2881.

[154] Attard, G.S., Bartlett, P.N., Coleman, N.R.B. *et al.* (1997) Mesoporous platinum films from lyotropic liquid crystalline phases. *Science*, **278**, 838.

[155] Attard, G.S., Leclerc, S.A.A., Maniguet, S. *et al.* (2001) Mesoporous Pt/Ru alloy from the hexagonal lyotropic liquid crystalline phase of a nonionic surfactant. *Chemistry of Materials*, **13**, 1444.

[156] Bartlett, P.N. and Marwan, J. (2003) Electrochemical deposition of nanostructured (H-1-e) layers of two metals in which pores within the two layers interconnect. *Chemistry of Materials*, **15**, 2962.

[157] Bartlett, P.N., Gollas, B., Guerin, S. and Marwan, J. (2002) The preparation and characterisation of H-1-e palladium films with a regular hexagonal nanostructure formed by electrochemical deposition from lyotropic liquid crystalline phases. *Physical Chemistry and Chemical Physics*, **4**, 3835.

[158] Bartlett, P.N., Birkin, P.R., Ghanem, M.A. *et al.* (2001) The electrochemical deposition of nanostructured cobalt films from lyotropic liquid crystalline media. *Journal of the Electrochemical Society*, **148**, C119.

[159] Elliott, J.M., Attard, G.S., Bartlett, P.N. *et al.* (1999) Nanostructured platinum (H-1-ePt) films: Effects of electrodeposition conditions on film properties. *Chemistry of Materials*, **11**, 3602.

[160] Attard, G.S., Bartlett, P.N., Coleman, N.R.B. *et al.* (1998) Lyotropic liquid crystalline properties of nonionic surfactant/H_2O/hexachloroplatinic acid ternary mixtures used for the production of nanostructured platinum. *Langmuir*, **14**, 7340.

[161] Markham, M.L., Smith, D.C., Baumberg, J.J. *et al.* (2005) Nanoporous semiconductor-based metamaterials. *Conference on Lasers and Electro-Optics (CLEO)*, **1–3**, 1435–1437.

[162] Vollhardt, K.P.C. and Schore, N.E. (2003) *Organic Chemistry. Structure and Function*, Palgrave, Berlin, New York.

[163] Pauporte, T. and Yoshida, T. (2006) Hybrid layers of ZnO/lanthanide complexes with high visible luminescences. *Journal of Materials Chemistry*, **16**, 4529.

[164] Chen, Z., Tang, Y., Zhang, L. and Luo, L. (2006) Electrodeposited nanoporous ZnO films exhibiting enhanced performance in dye-sensitized solar cells. *Electrochimica Acta*, **51**, 5870.

[165] Pauporte, T. (2007) Highly transparent ZnO/polyvinyl alcohol hybrid films with controlled crystallographic orientation growth. *Crystal Growth and Design*, **7**, 2310.

[166] Koide, N., Islam, A., Chiba, Y. and Han, L. (2006) Improvement of efficiency of dye-sensitized solar cells based on analysis of equivalent circuit. *Journal of Photochemistry and Photobiology A: Chemistry*, **182**, 296.

[167] Green, M.A., Emery, K., Hishikawa, Y. and Warta, W. (2008) Solar cell efficiency tables (version 32). *Progress in Photovoltaics: Research and Applications*, **16**, 435.

[168] Wenger, B., Grätzel, M. and Moser, J.E. (2005) Origin of the kinetic heterogeneity of ultrafast light-induced electron transfer from Ru(II)-complex dyes to nanocrystalline semiconducting particles. *Chimia*, **59**, 123.

[169] Södergren, S., Hagfeldt, A., Olsson, J. and Lindquist, S.E. (1994) Theoretical models for the action spectrum and the current-voltage characteristics of microporous semiconductor films in photoelectrochemical cells. *Journal of Physical Chemistry*, **95**, 5522.

[170] Oekermann, T., Zhang, D., Yoshida, T. and Minoura, H. (2004) Electron transport and back reaction in nanocrystalline TiO_2 films prepared by hydrothermal crystallization. *Journal of Physical Chemistry B*, **108**, 2227.

[171] Fisher, A.C., Peter, L.M., Ponomarev, E.A. *et al.* (2000) Intensity dependence of the back reaction and transport of electrons in dye-sensitized nanacrystalline TiO_2 solar cells. *Journal of Physical ChemistryB*, **104**, 949.

[172] Peter, L.M. and Vanmaekelbergh, D. (1999) Time- and frequency-resolved studies of photoelectrochemical kinetics, in *Advances in Electrochemical Science and Engineering*, vol. 6 (eds R.C. Alkire and D.M. Kolb), Wiley-VCH, Weinheim.

[173] Schlichthörl, G., Park, N.G. and Frank, A.J. (1999) Evaluation of the charge-collection efficiency of dye-sensitized nanocrystalline TiO_2 solar cells. *Journal of Physical Chemistry B*, **103**, 782.

[174] Oekermann, T., Yoshida, T., Minoura, H. *et al.* (2004) Electron transport and back reaction in electrochemically self-assembled nanoporous ZnO/dye hybrid films. *Journal of Physical Chemistry B*, **108**, 8364.

[175] Schlichthörl, G., Huang, S.Y., Sprague, J. and Frank, A.J. (1997) Band edge movement and recombination kinetics in dye-sensitized nanocrystalline TiO_2 solar cells: A study by intensity modulated photovoltage spectroscopy. *Journal of Physical Chemistry B*, **101**, 8141.

[176] Dloczik, L., Ileperuma, O., Lauermann, I. *et al.* (1997) Dynamic response of dye-sensitized nanocrystalline solar cells: Characterization by intensity-modulated photocurrent spectroscopy. *Journal of Physical Chemistry B*, **101**, 10281.

[177] Park, N.G., Schlichthörl, G., van de Lagemaat, J. *et al.* (1999) Dye-sensitized TiO_2 solar cells: Structural and photoelectrochemical characterization of nanocrystalline electrodes formed from the hydrolysis of $TiCl_4$. *Journal of Physical Chemistry B*, **103**, 3308.

[178] Nakade, S., Matsuda, M., Kambe, S. *et al.* (2002) Dependence of TiO_2 nanoparticle preparation methods and annealing temperature on the efficiency of dye-sensitized solar cells. *Journal of Physical Chemistry B*, **106**, 10004.

[179] Vayssieres, L., Hagfeldt, A. and Lindquist, S.E. (2000) Purpose-built metal oxide nanomaterials. The emergence of a new generation of smart materials. *Pure and Applied Chemistry*, **72**, 47.

[180] Nonomura, K., Komatsu, D., Minoura, H. *et al.* (2007) Dependence of the photoelectrochemical performance of sensitised ZnO on the crystalline orientation in electrodeposited ZnO thin films. *Physical Chemistry and Chemical Physics*, **9**, 1843.

[181] Wagner, P. and Helbig, R. (1974) Hall effect and anisotropy of electron mobility in ZnO. *Journal of Physical Chemistry Solids*, **35**, 327.

[182] Redmond, G., Fitzmaurice, D. and Grätzel, M. (1994) Visible light sensitization by cis-bis(thiocyanato)bis (2,2′-bipyridyl-4,4′-dicarboxylato)ruthenium(II) of a transparent nanocrystalline ZnO film prepared by sol-gel technique. *Chemistry of Materials*, **6**, 686.

[183] Keis, K., Lindgren, J., Lindquist, S.E. and Hagfeldt, A. (2000) Studies of the adsorption process of Ru complexes in nanoporous ZnO electrodes. *Langmuir*, **16**, 4688.

[184] Keis, K., Magnusson, E., Lindström, H. *et al.* (2002) A 5% efficient photo electrochemical solar cell based on nanostructured ZnO electrodes. *Solar Energy Materials and Solar Cells*, **73**, 51.

[185] Ito, S., Zakeeruddin, S.M., Humphry-Baker, R. *et al.* (2006) High-efficiency organic-dye-sensitized solar cells controlled by nanocrystalline-TiO_2 electrode thickness. *Advanced Materials*, **18**, 1202.

[186] Tennakone, K., Kumara, G.R.R.A., Kottegoda, I.R. *et al.* (1998) Nanoporous n-TiO_2/selenium/p-CuCNS photovoltaic cell. *Journal of Physics D: Applied Physics*, **31**, 2326.

[187] O'Regan, B., Lenzmann, F., Muis, R. and Wienke, J. (2003) A solid-state dye-sensitized solar cell fabricated with pressure-treated P25-TiO_2 and CuSCN: Analysis of pore filling and IV characteristics. *Journal of Materials Chemistry*, **14**, 5023.

[188] Kutzmutz, S., Lang, G. and Heusler, K. (2001) The electrodeposition of CdSe from alkaline electrolytes. *Electrochimica Acta*, **47**, 955.

[189] Skyllas-Kazacos, M. and Miller, B. (1980) Studies in selenoius acid reduction and CdSe film deposition. *Journal of the Electrochemical Society*, **127**, 869.

[190] Cocivera, M., Darkowski, A. and Love, B. (1984) Thin film CdSe electrodeposited from selenosulfite solution. *Journal of the Electrochemical Society*, **131**, 2414.

[191] Tena-Zaera, R., Katty, A., Bastide, S. and Levy-Clement, C. (2007) Annealing effects on the physical properties of electrodeposited ZnO/CdSe core-shell nanowire arrays. *Chemistry of Materials*, **19**, 1626.

[192] Tena-Zaera, R., Ryan, M.A., Katty, A. *et al.* (2006) Fabrication and characterization of ZnO nanowires/CdSe/ CuSCN eta-solar cell. *Comptes Rendus Chimie*, **9**, 717.

[193] Koenenkamp, R., Word, R.C. and Godinez, M. (2005) Ultraviolet electroluminescence from ZnO/polymer heterojunction light-emitting diodes. *Nano Letters*, **5**, 2005.

[194] Mora-Ser,ó, I., Fabregat-Santiago, F., Denier, B. *et al.* (2006) Determination of carrier density of ZnO nanowires by electrochemical techniques. *Applied Physics Letters*, **89**, 203117.

[195] Morrison, S.R. (1980) *Electrochemistry at Semiconductor and Oxidized Metal Electrodes*, Plenum, New York.

[196] Sze, S.M. (1981) *Physics of Semiconductor Devices*, Wiley, New York.

[197] Fabregat-Santiago, F., Garcia-Belmonte, G., Bisquert, J. *et al.* (2003) Mott-Schottky analysis of nanoporous semiconductor electrodes in dielectric state deposited on SnO_2(F) conducting substrates. *Journal of the Electrochemical Society*, **150**, E293.

[198] Baxter, J.B. and Aydil, E.S. (2005) Nanowire-based dye-sensitized solar cells. *Applied Physics Letters*, **86**, 053114.

[199] Pauporte, T. and Rathousky, J. (2007) Electrodeposited mesoporous ZnO thin films as efficient photocatalysts for the degradation of dye pollutants. *Journal of Physical Chemistry C*, **111**, 7639.

[200] Lee, M.K. and Tu, H.F. (2008) Au-ZnO and Pt-ZnO films prepared by electrodeposition as photocatalysts. *Journal of the Electrochemical Society*, **155**, D758.

[201] Zhang, Z.H., Yuan, Y.A., Liang, L.H. *et al.* (2008) Preparation and photoelectrocatalytic activity of ZnO nanorods embedded in highly ordered TiO_2 nanotube arrays electrode for azo dye degradation. *Journal of Hazardous Materials*, **158**, 517.

12

Nanostructured Thin-Film WO_3 Photoanodes for Solar Water and Sea-Water Splitting

Bruce D. Alexander[1] and Jan Augustynski[2]

[1]*School of Science, University of Greenwich, Kent, UK*

[2]*Department of Chemistry, Warsaw University, Warsaw, Poland,*

Email: Jan.Augustynski@unige.ch

12.1 Historical Context

Tungsten trioxide, WO_3, is a versatile material with numerous diverse applications that range from catalysis [1], to electrocatalysis [2], through gas sensors [3] and (photo)electrochromic devices [4], to photoelectrochemical water splitting [5]. In the majority of these cases, WO_3 is employed in thin films and the choice of whether it is in the amorphous or crystalline form is essential. The first extensive study of the optical and electrochromic properties of WO_3 films, by Deb [6], was motivated by the observation of the formation of color centers in WO_3, either under irradiation by UV light (photochromism) or by the application of an electric field (electrochromism). Measurement of the absorption spectra revealed a large difference between optical absorption edges of amorphous and crystalline WO_3 (about 0.38 eV); the absorption edge moves to lower energy upon crystallization, corresponding to a bandgap of 3.25 eV. In fact, the latter value is still significantly larger than the bandgap energy of 2.5 eV subsequently derived from photocurrent measurements for crystalline, and indeed nanocrystalline, WO_3 films. This difference is most likely due to the presence of some substoichiometric $W_{18}O_{49}$ in the crystalline material prepared by Deb, and the apparently smaller bandgap in nanocrystalline films which will be discussed later in this chapter. Irradiating WO_3 films with photons that have energies within the bandgap was observed to form blue color centers. This coloration occurs more efficiently in disordered amorphous films and in the presence of moisture. In what

On Solar Hydrogen & Nanotechnology Edited by Lionel Vayssieres

can be considered as the first observation of photochromism in WO_3, Deb associated film coloration with the formation of electron–hole pairs. It has been suggested that the coloration due to the photogenerated electrons (photochromism) occurs essentially in a similar way to that induced by an electric field (electrochromism) where the dissociation of adsorbed water molecules on the WO_3 surface is followed by electron/proton injection into the cathode.

The report by Deb in 1973 coincided with the notable discovery of photoelectrochemical water splitting at semiconductor TiO_2 photoanodes in the pioneering publication by Fujishima and Honda [7]. A great many authors have screened various semiconductor materials as possible alternatives to TiO_2 photoanodes, which are sensitive only to light in the near-UV range. The majority of studies have preferred to focus on n-type semiconducting oxides because of their superior chemical stability in aqueous solutions and, in most cases, the absence of photocorrosion. As it fits both of these criteria, and is sensitive to visible light, as well as near-UV radiation, WO_3 has been tested, both in the form of single crystals and as thin films. From an early stage, polycrystalline thin films have been prepared by a variety of methods, such as the thermal decomposition of $(NH_4)_2WO_4$ [8] or WCl_6 [9], by sputtering WO_3 powder [10] or by thermal oxidation of W foils [8–10].

12.2 Macrocrystalline WO_3 Films

In those early experiments, WO_3 photoanodes were illuminated either by the large spectral range output of xenon lamps or, in order to determine the spectral photoresponse, with monochromatic light of varying wavelengths whilst being polarized against a metal cathode in aqueous solutions of different pH. This resulted in photoelectrolysis of water assisted by an external bias, with hydrogen generated at the cathode and oxygen formed at the photoanode. The bandgap energies, derived from photocurrent measurements, were strongly affected by the preparation procedure and ranged from 2.7 eV for single-crystalline WO_3 [11] to values corresponding to even the near-UV range for films that were apparently only slightly substoichiometric. The close relationship between the film stoichiometry and the extent of the visible-light photoresponse was clearly demonstrated by Gissler and Memming [10] for the case of tungsten oxide films thermally grown on the W metal. Following an increase in the oxidation time, films grown in an oxygen atmosphere changed from a blue color, indicative of a partially reduced oxide, to yellow. Films grown for 2 h at 760 °C were several micrometers thick and yellow in appearance. Here the photoresponse extended to 500 nm, which seems to be the upper limit for an almost stoichiometric WO_3 film. The occurrence in these films of a secondary photocurrent maximum at 430 nm was tentatively assigned to an electron transition from the valence band to bulk intraband states located below the conduction band of WO_3 [10]. This was later clarified by a study which showed that prolonged anodic polarization in aqueous H_2SO_4 of thermally grown tungsten oxide films irradiated with intense UV/visible light resulted in a dramatic transformation of the photoresponse [12]. This is illustrated by the incident-photon-to-current efficiency (IPCE) versus wavelength curves given in Figure 12.1. The strong overall increase in the IPCE is accompanied by a suppression of the secondary maximum, which is replaced by a weak shoulder at about 430 nm. WO_3 electrodes that have been "aged" by extended photoelectrolysis in 1 M H_2SO_4 give a linear region in a Tauc plot, that is, $(\eta_{ph}h\nu)^{1/2}$ versus $h\nu$ (where η_{ph} is the IPCE close to the band edge and $h\nu$ is the energy of the incident light), which is indicative of an indirect interband transition. The intercept of the $h\nu$

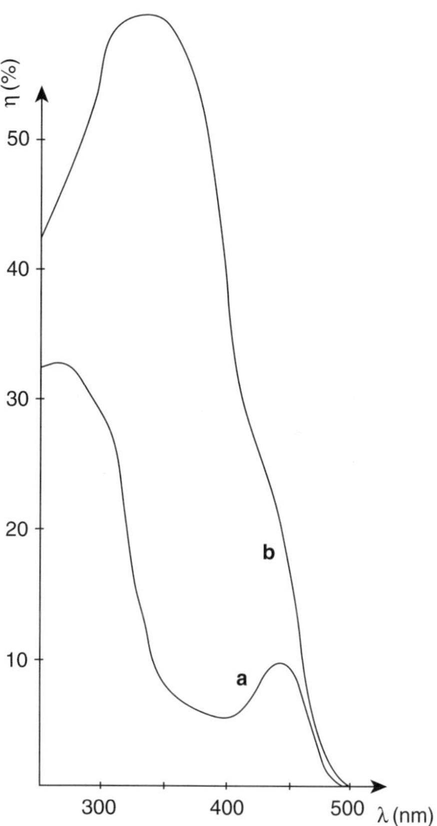

Figure 12.1 Improvement of photocurrent efficiency of a bulk WO_3 photoanode following an "aging" process: (a) as-prepared WO_3 photoanodes; (b) WO_3 photoanodes that have been subjected to 3 hours photoelectrolysis in 1 M $H_2SO_{4(aq)}$. (Reprinted with permission from M. Spichiger-Ulmann and J. Augustynski, Aging effects in n-type semiconducting WO_3 films, *Journal of Applied Physics*, **54**, 6061, 1983. © 1983 American Institute of Physics.)

axis affords the bandgap energy, about 2.5 eV in this case [12], slightly lower than that found for single-crystalline WO_3 [11]. The large enhancement of the photocurrent efficiency for the thermally grown WO_3 electrode was assigned to a partial extraction of ionic defects (interstitial W^{5+} and W^{6+}) from the film, which migrate towards the oxide/electrolyte interface due to the effect of the applied anodic bias [12]. As suggested, the outcome of the photoelectrochemical activation treatment of the WO_3 electrode was a marked decrease in the donor concentration, N_D, in the outermost part of the film, resulting in an increase in the width of the space-charge layer.[1]

[1] The width of the space charge layer, W, is given by $W = (2\epsilon\epsilon_0(V - V_{fb})/eN_D)^{1/2}$, where ϵ is the dielectric constant of the semiconductor, ϵ_0 is the permittivity of vacuum, V_{fb} is the flat-band potential and $V - V_{fb}$ is the band bending, e is the charge on the electron and N_D is the number of donors per cm^3).

12.3 Limitations of Macroscopic WO_3

The above example illustrates the intrinsic limitations in efficiency associated with the use of indirect optical transition semiconductors, such as WO_3, as sunlight-driven photoelectrodes. Given the low absorption coefficients, α, in WO_3 for photons with energies close to the bandgap, the optical penetration depth, $1/\alpha$, for light of visible wavelengths may be of the order of several micrometers [6,13]. In addition to this, the hole diffusion length, L_p, in materials with an indirect optical transition, is quite low, about 0.15 μm in WO_3 [11]: an efficient separation of the electron–hole pairs would require the establishment of a potentially wide space-charge layer in order to match $1/\alpha$. A simple estimate shows that even for the case of a semiconductor with moderate doping levels, that is, $N_D = 10^{16}\,\text{cm}^{-3}$, and a band bending as large as 1 V, the width of the space-charge layer in WO_3 does not exceed 0.8 μm [13]. This is still less than optical penetration depths for visible wavelengths. Given that the flat-band potentials for the n-type semiconducting oxide photoanodes are generally more positive than the reversible potential of the hydrogen electrode (RHE) in the same solution (with the exception of strontium titanate, $SrTiO_3$, although this responds only to UV wavelengths), the requirement of large band bending implies that even larger anodic bias has to be applied to the photoelectrode in order to reach sizeable photocurrents. In general, the smaller the bandgap energy of the n-type semiconductor oxide, the lower its conduction-band edge is in energy and the more positive its flat-band potential is. This is related to the fact that the conduction band of these oxides has mainly d-band character, whereas the valence band is mainly O $2p$ in character. In fact, the lowest photocurrent onset potential for a WO_3 photoanode (close to its flat-band potential) is about 0.4 V versus RHE [5] which leaves a rather narrow margin for band bending before reaching the thermodynamic oxygen evolution potential of 1.23 V. Clearly, the requirement of a large anodic bias to attain significant photocurrents severely limits the prospects for application of bulk semiconducting oxide photoanodes to solar-energy conversion. It was only later, following the synthesis of nanostructured WO_3 film photoanodes [14] that it was possible to show that the problems described above associated with bulk semiconductors can be largely avoided by employing semiconductors in the nanometer domain.

12.4 Nanostructured Films

It had been the search for semitransparent photoanodes, which could be employed in a photoelectrolysis cell (PEC), in conjunction with a photovoltaic (PV) cell to form a tandem device [5,15], that led to the development of nanostructured WO_3 films. Such films consist of a network of WO_3 particles with sizes in the range of tens of nanometers, as shown in a typical scanning electron micrograph (SEM) in Figure 12.2. Due to their relatively large and evenly distributed porosity, the whole films, which have thicknesses of several micrometers, are able to be permeated by electrolyte. The most significant aspect of the behavior of the nanostructured WO_3 photoanodes is that they reach saturation photocurrents under a relatively modest external bias, about 0.5 V above the onset potential. Furthermore, this is almost independent of the wavelength of incident light. Figure 12.3 shows the change in the photocurrent efficiency (IPCE) under applied anodic bias for a bulk WO_3 photoelectrode, prepared as described in Ref. [12], and a nanostructured WO_3 photoelectrode recorded under 380 nm (Figure 12.3a) and 440 nm illumination (Figure 12.3b). The unusually large IPCEs observed here are consistent

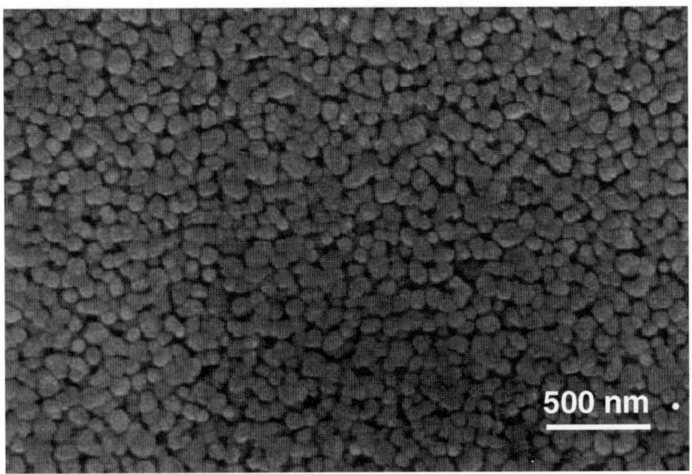

Figure 12.2 Scanning electron micrograph of a nanostructured WO_3 film annealed at 550 °C.

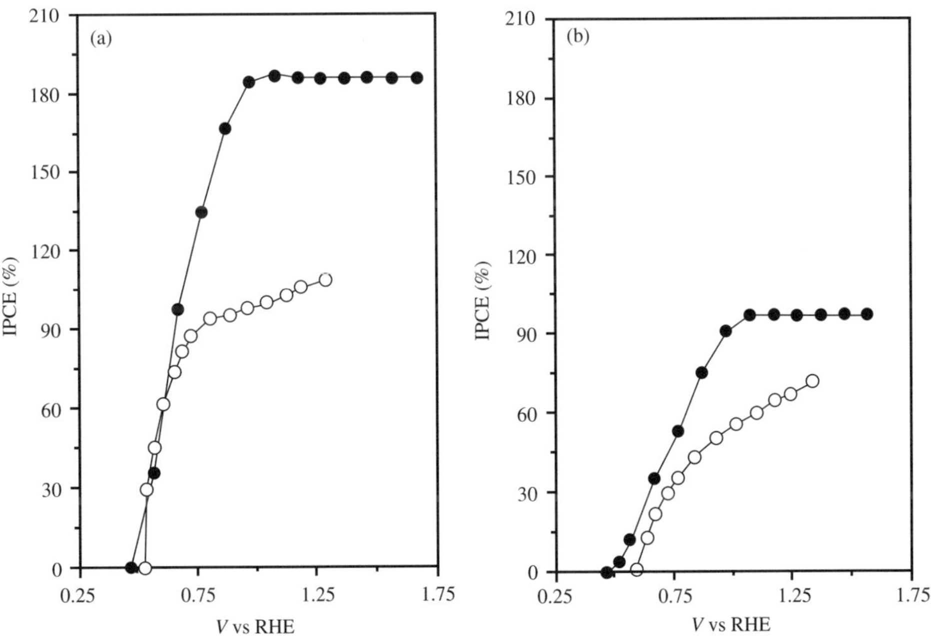

Figure 12.3 IPCE-potential curves for nanocrystalline (•) and bulk (○) WO_3 photoelectrodes recorded in a 0.1 M CH_3OH/1 M H_2SO_4 solution under: (a) 380 nm; and (b) 440 nm illumination. (Reproduced with permission from C. Santato, M. Ulmann, and J. Augustynski, Enhanced visible light conversion efficiency using nanocrystalline WO_3 films, *Advanced Materials*, **13**, 511, 2001. © 2001 Wiley VCH.)

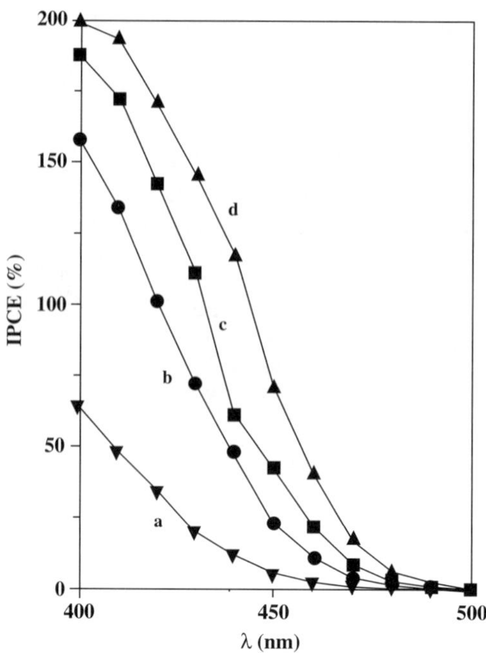

Figure 12.4 Spectral photoresponses for a series of nanocrystalline WO_3 electrodes of different thicknesses: (a) about 0.4 μm; (b) 1.2 μm; (c) 2.5 μm and (d) 4 μm in 0.1 M CH_3OH/1 M H_2SO_4 solution, at 1 V versus RHE. (Reproduced with permission from C. Santato, M. Ulmann, and J. Augustynski, Enhanced visible light conversion efficiency using nanocrystalline WO_3 films, *Advanced Materials*, **13**, 511, 2001. © 2001 Wiley VCH.)

with the preferential photo-oxidation of methanol; a reaction known to involve photocurrent doubling. The appearance of the IPCE versus potential plots for the bulk WO_3 electrodes is exactly as expected when the penetration depth of the incident light in a semiconductor exceeds the collection distance of the minority carriers within the semiconductor, given by the sum of W and L_p. The observed continuous rise of the photocurrent is consistent with the role of band bending in matching (via the increase in the width of the space-charge layer) the optical penetration depth. In contrast, given that the anodic bias required to reach the saturation photocurrent in nanostructured WO_3 photoelectrodes is relatively constant, regardless of the penetration depth of the incident light (cf. Figure 12.3), the charge separation in these materials can hardly be assigned to the presence of a conventional space-charge layer.[2] Here, it is the size of the individual nanoparticles, smaller than the hole diffusion length in WO_3, ($L_p \approx 0.15$ μm), which apparently plays an essential role in controlling the photocurrent efficiency of the nanostructured photoelectrodes. An important consequence of this, from a practical viewpoint, is that the photoresponse to longer wavelengths is directly improved by increasing the thickness of the nanostructured WO_3 film. This is demonstrated in Figure 12.4, which illustrates how the

[2] As discussed in Ref. [13], the anodic bias required to attain the saturation photocurrent might correspond to the sum of the potential drops at the substrate/WO_3 interface and across the Helmholtz layer that is formed at the interface between the network of WO_3 nanoparticles and the solution.

spectral response to visible light up to 500 nm (i.e., the portion of visible light that is absorbed by the WO_3) is affected when the film thickness increases from about 0.4 to 4 μm. It should be noted that these IPCEs still correspond to the saturation photocurrents obtained under a bias of 0.5 V above the onset potential.

12.5 Tailoring WO_3 Films Through a Modified *Chimie Douce* Synthetic Route

The nanostructured WO_3 films that have been discussed thus far were all prepared by modification of a sol-gel method [14,16]. Although sol-gel methods are frequently employed for the synthesis of metal oxides [17], a simple approach involving a freshly synthesized tungstic acid precursor [18], failed to produce satisfactory films. Evaporation of a colloidal solution of tungstic acid that had been deposited onto conducting glass substrates produced smooth xerogel films. However, attempts at densification of these films through a high-temperature heat treatment resulted in loss of both adherence and transparency. It was the addition of an organic modifier that was either polyhydroxylated or a surface active compound, such as poly(ethylene glycol) (PEG) 300 or 400, mannitol or glycerol to the tungstic acid precursor, which brought about a decisive improvement in the properties of the WO_3 films, especially with regard to their transparency and adherence to the substrate [16]. The principal effect of these compounds, which are known to form complexes with tungsten oxoanions, was to retard markedly the condensation of freshly prepared tungstic acid [18]. The use of modified precursors opens wide possibilities for tailoring both the size of the resulting WO_3 nanoparticles and the film porosity through a judicious choice of the tungstic acid/organic additive ratio and the annealing conditions. Figure 12.5 shows a series of scanning electron micrographs for the WO_3 films annealed at temperatures ranging from 400 °C to 550 °C. These films were formed by depositing a colloidal solution composed of tungstic acid and poly(ethylene glycol) 300 in a 0.5 w/w WO_3/PEG ratio onto conducting glass substrates. The formation of significant porosity in the films, as shown in SEM images in Figure 12.5c and d is the result of the combustion of the organic additive, which is present in large amounts in the precursor solution, during the high-temperature treatment. This was accompanied by the conversion of a fibrous structure visible in SEM images in Figure 12.5a and b into plate-like particles (Figure 12.5c and d). Importantly, the removal of PEG, which acts as structure-directing agent to form the mesoporous WO_3 film, does not affect the mechanical stability of the film, which is entirely preserved even after annealing at 550 °C. The complex role played by poly(ethylene glycol) in transformation of a colloidal solution of tungstic acid into a mesoporous WO_3 film was followed by Raman spectroscopy.

Films prepared with PEG and annealed below 400 °C display bands around 780 cm^{-1} and 965 cm^{-1}, both of which are comparatively broad, as shown in Figure 12.6. The former band can be attributed to stretching motions of O–W–O groups involving bridging O atoms, whereas the latter is most likely to arise from stretching of tungsten–oxygen bonds that have a pronounced double-bond character. These are typically found in tungsten-oxide hydrates where terminal $W=O$ groups give rise to bands in the region above 940 cm^{-1} [19]. Both the position of these bands and their width suggest a large amount of amorphous material is present in the films, although a small amount of hexagonal WO_3 can be detected, evidenced by bands at 244 and 322 cm^{-1} and a shoulder around 635 cm^{-1}. This is in marked contrast to films that are

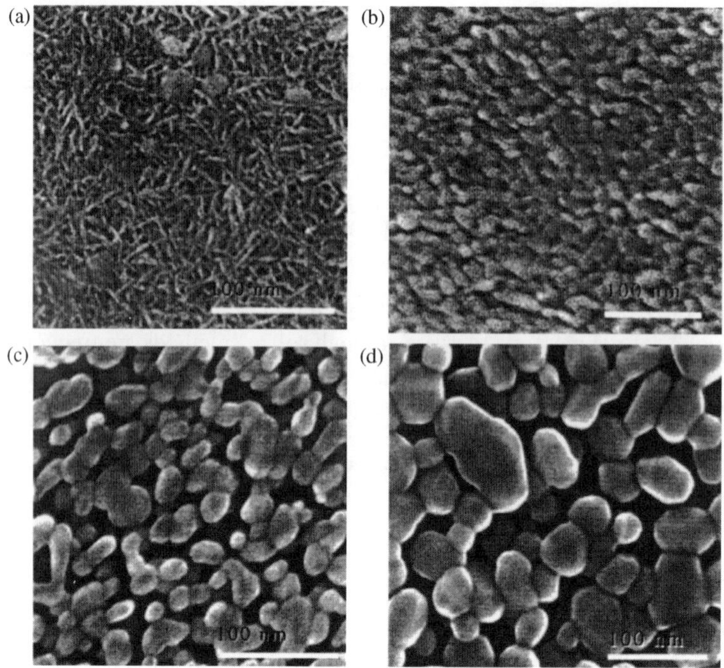

Figure 12.5 Scanning electron micrographs of WO_3 films prepared by the deposition of a tungstic acid/PEG 300 colloidal solution after annealing for 30 min at: (a) 400 °C; (b) 450 °C; (c) 500 °C; and (d) 550 °C. (Reproduced with permission from C. Santato, M. Odziemkowski, M. Ulmann and J. Augustynski, Crystallographically oriented mesoporous WO_3 films: Synthesis, characterization and applications, *Journal of the American Chemical Society*, **123**, 10639, 2001. © 2001 American Chemical Society.)

annealed below 400 °C, which do not have an organic additive present. The Raman spectra of these films show the start of the formation of monoclinic WO_3 even at 300 °C [16]. From Figure 12.6, the bands at 715 cm^{-1} and 805 cm^{-1}, along with the absence of a band due to a W = O stretch above 940 cm^{-1} show that if a film has been prepared with PEG as an additive, well-crystallized monoclinic WO_3 is only present following annealing at 500 °C. Furthermore, at annealing temperatures above 400 °C, there is an absence of bands around 1335 and 1610 cm^{-1}, which are present in films annealed at lower temperatures. These bands are due to the formation of carbon and are present due to the incomplete combustion of carbon from PEG at temperatures up to 400 °C. Thus the presence of PEG retards the crystallization of WO_3, pushing the formation of monoclinic WO_3 to higher temperatures, presumably due to the incorporation of carbon into the films. PEG also acts as a structure-directing agent as the presence of hexagonal WO_3 is not detected in films prepared without the organic additive.

Following the analysis of the Raman spectra, it was concluded that the full conversion of the tungstic acid/poly(ethylene glycol) 300 precursor into the crystalline monoclinic form of tungsten oxide required heat treatment at least at 500 °C. This was entirely consistent with a

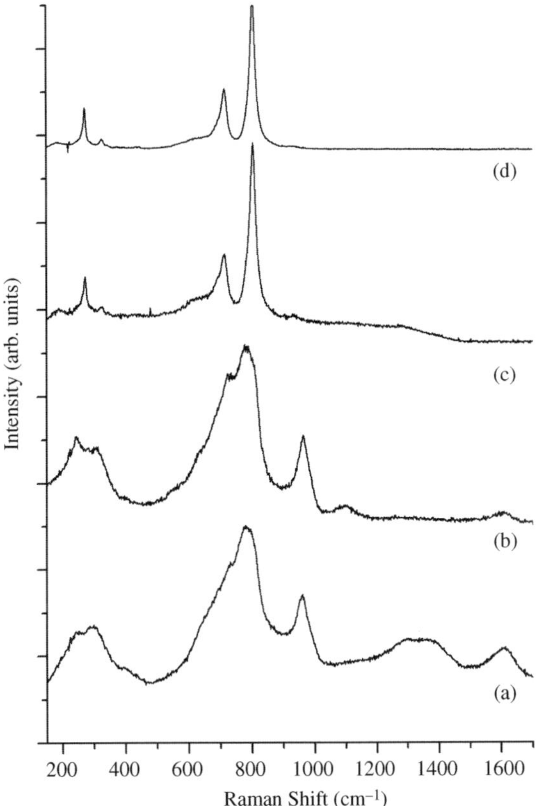

Figure 12.6 Raman spectra of WO_3 films deposited on conducting glass substrates from a tungstic acid/PEG 300 solution after annealing at: (a) 350 °C; (b) 400 °C; (c) 500 °C and (d) 550 °C.

strong increase in the intensity of X-ray diffraction (XRD) peaks for the films annealed in the range 450–550 °C. Further structural studies of such films using synchrotron radiation revealed that a heat treatment at 550 °C, in particular, induced a marked preferential orientation of the 200, 020 and 002 faces of WO_3 crystallites parallel to the substrate, shown in Figure 12.7a. Figure 12.7b illustrates the change in diffraction pattern upon decreasing the glancing angle of incident radiation. Here, the preferential orientation of WO_3 nanoparticles is maintained across a relatively thick film, about 1.2 μm. This may provide an explanation for the good transparency of the mesoporous films to wavelengths that exceed the absorption edge of monoclinic tungsten oxide [16].

Among a number of features which might affect the photoelectrochemical activity of the WO_3 films, such as the porosity, the size of nanoparticles (both of which determine the surface area that is in contact with the electrolyte), it is the crystallinity of the oxide which apparently plays an essential role. In this regard, it is particularly important to minimize the hole–electron recombination losses and to facilitate the transport of electrons across the film to the back contact and this is assisted by the crystalline structure of the boundaries between the WO_3 nanoparticles.

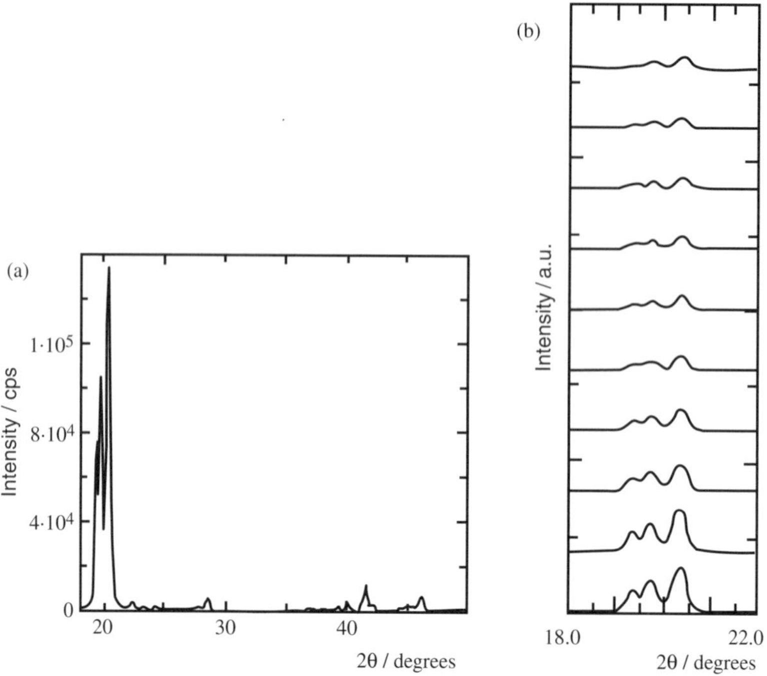

Figure 12.7 (a) Diffraction pattern of a nanostructured WO_3 film deposited on glass from a tungstic acid/PEG 300 solution after annealing at 550 °C. (b) Diffraction patterns of the same WO_3 film recorded at decreasing glancing angles, from $\alpha = 5°$ (bottom) to 0.5° (top). (Reproduced with permission from C. Santato, M. Odziemkowski, M. Ulmann, and J. Augustynski, Crystallographically oriented mesoporous WO_3 films: Synthesis, characterization and applications, *Journal of the American Chemical Society*, **123**, 10639, 2001. © 2001 American Chemical Society.)

12.6 Surface Reactions at Nanocrystalline WO_3 Electrodes

Given that the photoresponse of nanocrystalline WO_3 films is restricted to only the blue region of the visible spectrum, the significant photocurrents recorded under simulated solar irradiation are achieved due to high incident-photon-to-current efficiencies. As shown in Figure 12.8, the IPCE versus wavelength plots exhibit a maximum around 400 nm, and still display sizeable efficiencies up to 480 nm. The photocurrent efficiency of about 90%, attained for the photo-oxidation of water in the sulphuric acid solution employed here as a supporting electrolyte (the exact nature of this reaction will be discussed below) is clear evidence of a particularly low rate of hole–electron recombination. This is further confirmed by the upper plot in Figure 12.8, which shows IPCE values measured in a sulphuric acid solution that contains methanol. These values, which greatly exceed 100%, are indicative of an effective photocurrent doubling.

However, analysis of the products formed during extensive photoelectrolysis runs that last several hours revealed that oxygen is not the only species formed in acidic media at nanostructured WO_3 photoanodes. In fact, in agreement with earlier reports in the literature [20], thermodynamically unstable persulphates ($S_2O_8^{2-}$) were identified by means of

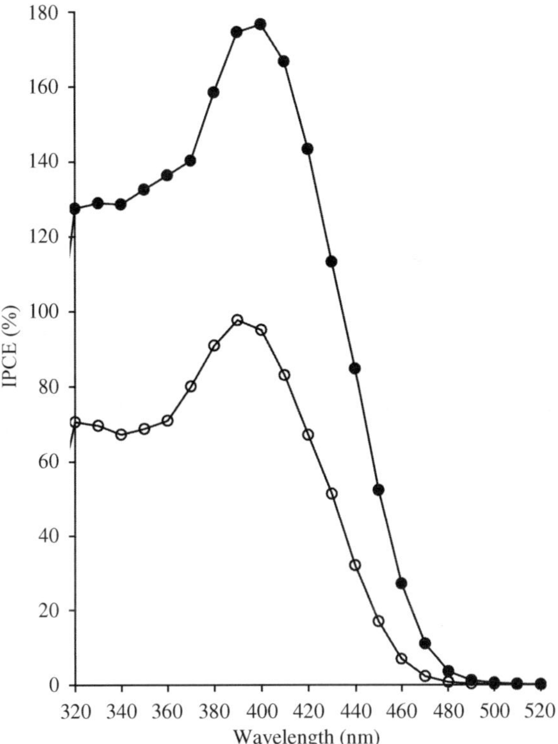

Figure 12.8 Typical spectral photoresponses for a about 2.5 μm thick WO_3 film recorded in 3 M $H_2SO_{4(aq)}$ (○) and after the addition of 0.1 M of methanol (•).

Raman spectroscopy in H_2SO_4 solutions, after prolonged photoelectrolysis under simulated solar AM 1.5 irradiation:

$$2\,SO_4^{2-} + 2\,h^+ \rightarrow S_2O_8^{2-}$$

Note that with an increase in the temperature of the electrolyte, which, for example, occurs under real solar illumination, the persulphates undergo rapid decomposition to form oxygen:

$$2\,S_2O_8^{2-} + 2\,H_2O \rightarrow 4\,SO_4^{2-} + O_2 + 4\,H^+$$

The formation of persulphates during photoelectrolysis of H_2SO_4, along with the possible formation of hydrogen peroxide in photoelectrolysis in $HClO_4$ solutions, affected the long-term stability of the photocurrents recorded at the WO_3 photoanodes. Given that tungsten trioxide exhibits excellent chemical stability and notably the absence of photocorrosion in acidic solutions, the latter behavior was assigned to the accumulation of a surface peroxide layer [21]. Indeed, the initial performance of the photoanode could be restored either by a short reductive current pulse, or by an exposure of the partially deactivated WO_3 surface to intense UV illumination, which is known to induce photodecomposition of tungsten peroxides. Interestingly, the deactivation of WO_3 photoanodes could be avoided by the addition of even small amounts of chloride or bromide ions (e.g., 0.01 M NaCl or NaBr) to the acidic electrolyte. Under such conditions oxygen remained by far the main photoelectrolysis product with chlorine or bromine generation accounting for only a small percentage of the current efficiency. Apparently, halide ions are effective promoters of oxygen evolution, most likely by causing decomposition of the surface peroxide species [21]. In light of these observations, the peroxide species appear to be key intermediates in the formation of oxygen at WO_3 photoanodes operating in acidic solutions.

Figure 12.9a shows the photocurrent–potential plot for a nanocrystalline WO_3 film photoanode recorded in a 0.5 M sodium chloride solution – a composition close to that of sea water. It can be observed that the nanostructured WO_3 electrodes deliver both significant and remarkably stable photocurrents. For comparative purposes, a similar plot obtained in the initial stage of the photoelectrolysis of a 3 M H_2SO_4 solution is shown in Figure 12.9b. The large concentration of chloride ions in the electrolyte means that diffusion control will not be the rate-limiting step in establishing the photocurrent densities of the order of 3 mA cm^{-2} that are observed under simulated solar illumination. Here, oxygen remains the main product of the photoelectrolysis, with chlorine formation

$$2\,Cl^- + 2\,h^+ \rightarrow Cl_2$$

accounting for about 20% of the current efficiency. Note that although the NaCl electrolyte is initially neutral, it does not require any preliminary acidification, since chlorine formation regulates the solution pH to about 2 close to the electrode surface. From a practical viewpoint,

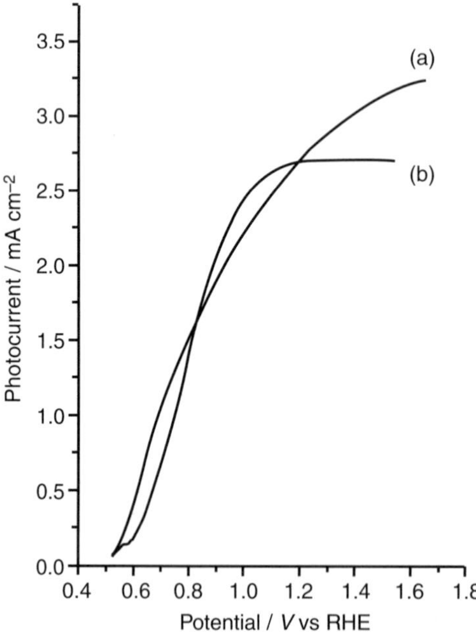

Figure 12.9 Photocurrent-potential plot (*versus* RHE in the same solution) recorded for nanocrystalline WO_3 film photoanodes under AM1.5 simulated solar irradiation in: (a) 0.5 M NaCl; and (b) 0.5 M H_2SO_4.

the production of chlorine at the WO_3 photoanode, the amount of which can be controlled by an appropriate choice of electrolyte composition, combined with hydrogen formation at the cathode, constitutes an attractive option. The evolution of both chlorine and oxygen at the photoanode enables easy conversion to hypochloric acid, an extensively used disinfectant, by washing the gases with water:

$$Cl_2 + H_2O \rightarrow HClO + H^+ + Cl^-$$

In contrast to the oxygen evolution reaction, which clearly involves the formation of peroxide species as the primary intermediates, the spectroscopic evidence [22] suggests that photo-oxidation of small organic molecules (e.g., methanol) at WO_3 is initiated by direct hole transfer to adsorbed alcoholic species rather than by the reaction of the organic species with photogenerated hydroxyl radicals

$$CH_3OH_{(ad)} + h^+ \rightarrow \bullet CH_2OH_{(ad)} + H^+$$

followed by electron injection into the conduction band of the semiconductor by the $\bullet CH_2OH_{(ad)}$ radical

$$\bullet CH_2OH_{(ad)} \rightarrow CH_2O + H^+ + e^-_{cb}$$

The combination of fast hole transfer with photocurrent doubling leads to large photoanodic currents that reach 6 mA cm^{-2} in relatively concentrated organic solutions [23]. However, the indirect oxidation pathway is likely to prevail in instances where the organic species is present in low concentrations. Thus, in the case of a supporting electrolyte that contains sulfate ions, the photo-oxidation can most probably be mediated by $\bullet SO_4^-$ radical ions.

12.7 Conclusions and Outlook

The use of an aqueous sol-gel method, involving a freshly formed tungstic acid precursor combined with a structure-directing agent belonging to the group of polyhydroxylated organic compounds, allows the fabrication of mesoporous WO_3 films of considerable thicknesses, which are able to absorb a large portion of the incident light with energies close to the bandgap of the semiconductor. Such nanostructures, consisting of several tens of stacked WO_3 nanoparticles permeated by an electrolyte, form a large number of Schottky (semiconductor/electrolyte) nano-junctions interpenetrating the whole film. A configuration of this kind allows the relaxation of the usual constraint concerning the lifetime of the minority charge carriers, provided the hole diffusion length will actually be larger than the size of the n-type semiconductor nanoparticles in contact with the electrolyte. The implication for a material such as WO_3 is that the distance between the site where e^-/h^+ pairs are generated and the electrolyte present in the pores must be smaller than 0.15 μm. The latter, as well as the essentially crystalline structure of the interparticulate boundaries, allows the preservation of high conversion efficiencies for mesoporous photoanodes under irradiation of visible wavelengths despite the films having a layer of semiconductor material of significant thickness, that is, of the range of several micrometers.

It should be noted that, when compared to, say, TiO_2, one apparent drawback to the application of nanocrystalline WO_3 films for solar hydrogen production is the unfavorable energetics associated with the conduction band. Nevertheless, the fact that nanocrystalline films of WO_3 are responsive to light well into the visible region of the electromagnetic spectrum (up to 480 nm) provides a distinct advantage over materials such as TiO_2, which is only photoresponsive to UV light, prior to the addition of dopants. Therefore, it is significant that nanocrystalline WO_3 films have shown anodic photocurrents of up to $6\,mA\,cm^{-2}$ in organic solutions. This means that devices can be constructed which couple WO_3 film electrodes with a photovoltaic cell, such as amorphous silicon or dye-sensitized TiO_2. These tandem cells [5,15] can be considered to operate by combining two photosystems. A nanocrstalline WO_3 electrode represents the first photosystem, which absorbs the UV and blue-green portions of the electromagnetic spectrum and drives the photolysis of water. The photovoltaic cell acts as the second photosystem, which absorbs the remaining electromagnetic radiation that is not absorbed by the WO_3 and thereby supplies the anodic potential required for the WO_3 electrode to photosplit water. Such a system has the benefit of being able to be employed as a "stand-alone" device and physically separates the oxygen and hydrogen that is generated at different electrodes. Despite their promise, some development of tandem cells is required to make them commercially interesting. For this to happen, the overall efficiency, that is, the portion of incident photons from solar irradiation converted into hydrogen, must be increased to 10%; currently, the maximum efficiency obtained for a tandem cell is 4%. It should be further noted that this efficiency is *not* the same as the IPCEs discussed above, or the photovoltaic efficiency associated with the photovoltaic cell, although the overall efficiency will clearly be dependent on both of these. As such, improvements in photovoltaic efficiency or the addition of organic species to the aqueous solution at the nanocrystalline WO_3 electrode, which has been shown to increase the IPCE associated with the nanocrystalline WO_3 (as shown in Figure 12.8), would potentially increase the overall efficiency. For example, initial experiments suggest that addition of organic species to a tandem cell can increase the overall efficiency to 6%. Future development would be welcomed in this area.

References

[1] Zhu, Z.Q., Bian, W., Liu, L.S. and Lu, Z. (2007) Catalytic oxidation of cyclopentene to glutaraldehyde over WO_3/Ti-HMS catalyst. *Catalysis Letters*, **117**(1–2), 79–84.

[2] Park, K.W., Choi, J.H., Ahn, K.S. and Sung, Y.E. (2004) PtRu alloy and PtRu-WO_3 nanocomposite electrodes for methanol electrooxidation fabricated by a sputtering deposition method. *Journal of Physical Chemistry B*, **108**(19), 5989–5994.

[3] Solis, J.L., Saukko, S., Kish, L. *et al.* (2001) Semiconductor gas sensors based on nanostructured tungsten oxide. *Thin Solid Films*, **391**(2), 255–260.

[4] Granqvist, C.G. (2000) Electrochromic tungsten oxide films: Review of progress 1993–1998. *Solar Energy Materials and Solar Cells*, **60**(3), 201–262.

[5] Alexander, B.D., Kulesza, P.J., Rutkowska, I. *et al.* (2008) Metal oxide photoanodes for solar hydrogen production. *Journal of Materials Chemistry*, **18**(20), 2298–2303.

[6] Deb, S.K. (1973) Optical and photoelectric properties and color centers in thin-films of tungsten oxide. *Philosophical Magazine*, **27**(4), 801–822.

[7] Fujishima, A., Honda, K. and Kikuchi, S. (1969) Photosensitized electrolytic oxidation on semiconducting n-type TiO_2 electrode. *Kogyo Kagaku Zasshi*, **72**, 108–113; Fujishima, A. and Honda, K. (1972) Electrochemical photolysis of water at a semiconductor electrode. *Nature*, **238**(5358), 37–38.

[8] Hodes, G., Cahen, D. and Manassen, J. (1976) Tungsten trioxide as a photoanode for a photoelectrochemical cell (PEC). *Nature*, **260**(5549), 312–313.

[9] Hardee, K.L. and Bard, A.J. (1977) Semiconductor electrodes. 10. Photoelectrochemical behavior of several polycrystalline metal-oxide electrodes in aqueous solutions. *Journal of the Electrochemical Society*, **124**(2), 215–224.

[10] Gissler, W. and Memming, R. (1977) Photoelectrochemical processes at semiconducting WO_3 layers. *Journal of the Electrochemical Society*, **124**(11), 1710–1714.
[11] Butler, M.A., Nasby, R.D. and Quinn, R.D. (1976) Tungsten trioxide as an electrode for photoelectrolysis of water. *Solid State Communications*, **19**(10), 1011–1014; Butler, M.A. (1977) Photoelectrolysis and physical properties of semiconducting electrode WO_3. *Journal of Applied Physics*, **48**(5), 1914–1920.
[12] Spichiger-Ulmann, M. and Augustynski, J. (1983) Aging effects in n-type semiconducting WO_3 films. *Journal of Applied Physics*, **54**(10), 6061–6064.
[13] Santato, C., Ulmann, M. and Augustynski, J. (2001) Enhanced visible light conversion efficiency using nanocrystalline WO_3 films. *Advanced Materials*, **13**(7), 511–514.
[14] Ulmann, M., Santato, C., Augustynski, J. and Shklover, V. (1997) *Electrode Materials and Processes for Energy Conversion and Storage* (eds S. Srinivasan, S. Mukerjee and J. McBreen), The Electrochemistry Society, Pennington, N.J, p. 328.
[15] Augustynski, J., Calzaferri, G., Courvoisier, J. and Grätzel, M. (1996) *Proceedings of the 11th World Hydrogen Energy Conference* (eds T.N. Veziroglu, C.J. Winter, J.P. Baselt and G. Kreysa), DECHEMA, Frankfurt, Germany, p. 2378.
[16] Santato, C., Odziemkowski, M., Ulmann, M. and Augustynski, J. (2001) Crystallographically oriented mesoporous WO_3 films: Synthesis, characterization and applications. *Journal of the American Chemical Society*, **123**(43), 10639–10649.
[17] Livage, J. and Ganguli, D. (2001) Sol-gel electrochromic coatings and devices: A review. *Solar Energy Materials and Solar Cells*, **68**(3–4), 365–381.
[18] Chemseddine, A., Morineau, R. and Livage, J. (1983) Electrochromism of colloidal tungsten oxide. *Solid State Ionics*, **9–10**(Dec), 357–361.
[19] Solarska, R., Alexander, B.D. and Augustynski, J. (2004) Electrochromic and structural characteristics of mesoporous WO_3 films prepared by a sol-gel method. *Journal of Solid State Electrochemistry*, **8**(10), 748–756.
[20] Desilvestro, J. and Grätzel, M. (1987) Photoelectrochemistry of polycrystalline n-WO_3 – Electrochemical characterization and photoassisted oxidation processes. *Journal of Electroanalytical Chemistry*, **238**(1–2), 129–150.
[21] Augustynski, J., Solarska, R., Hagemann, H. and Santato, C. (2006) Nanostructured thin-film tungsten trioxide photoanodes for solar water and sea-water splitting. *Proceedings of the Society of Photo-optical Instrumentation Engineers (The International Society for Optical Engineering)*, **6340**, 63400J–1.
[22] Léaustic, A., Babonneau, F. and Livage, J. (1986) Photoreactivity of WO_3 dispersions - spin trapping and electron spin resonance detection of radical intermediates. *Journal of Physical Chemistry*, **90**(17), 4193–4198.
[23] Santato, C., Ulmann, M. and Augustynski, J. (2001) Photoelectrochemical properties of nanostructured tungsten trioxide films. *Journal of Physical Chemistry B*, **105**(5), 936–940.

13

Nanostructured α-Fe_2O_3 in PEC Generation of Hydrogen

Vibha R. Satsangi, Sahab Dass, and Rohit Shrivastav
Dayalbagh Educational Institute (Deemed University), Dayalbagh, Agra, India, Email: vibhasatsangi@gmail.com

13.1 Introduction

This chapter presents a systematic and exhaustive literature review on nanostructured α-Fe_2O_3, from synthesis, to characterization, to its application as electrodes in photoelectrochemical cells for producing hydrogen using sunlight and water. In this context, some of the recent strategies, proven to be useful in improving the performance of PEC cells have been described and analyzed. A discussion on the efficiency of solar hydrogen production is also included with some recommendations for realizing efficient PEC systems using nanostructured α-Fe_2O_3. Section 13.2.1 describes the structural, electrical and electronic properties of α-Fe_2O_3. Section 13.2.2 covers the salient features of the research carried out on the PEC properties of α-Fe_2O_3 in the early days. In Section 13.3, various methods for the preparation of nanostructured α-Fe_2O_3 photoelectrodes and their PEC responses have been described. Section 13.4 deals with some techniques employed for enhancing the photoelectrochemical response of the material. A short discussion on the hydrogen evolution and efficiency of PEC systems using α-Fe_2O_3 has been undertaken in Section 13.5. Finally, Section 13.6 concludes the chapter by offering suggestions/recommendations to achieve efficient PEC systems for solar hydrogen production using nanostructured α-Fe_2O_3 electrodes.

On Solar Hydrogen & Nanotechnology Edited by Lionel Vayssieres

13.2 α-Fe_2O_3

13.2.1 Structural and Electrical/Electronic Properties

α-Fe_2O_3, known as hematite, has been a material of interest in PEC splitting of water for a long time. It is isostructural with $FeTiO_3$, α-Al_2O_3, Cr_2O_3 and V_2O_3. The crystal structure of α-Fe_2O_3 comprises distorted hexagonal close packing of oxygen atoms with 2/3 octahedral interstices occupied by Fe. Thus, each Fe atom has six oxygen neighbors and each oxygen atom has four Fe neighbors. In 1954, Morin [1] presented the electrical properties of α-Fe_2O_3, and suggested two alternative models. The first model assumed conduction to occur in the *d* level of Fe ions, whereas in the second, conduction was supposed to occur in the *sp* bonds of oxygen. Pure α-Fe_2O_3 is generally oxygen deficient, mainly due to oxygen vacancies created during its preparation, which is responsible for its n-type behavior.

The conduction mechanism in α-Fe_2O_3 is not fully resolved and the literature contains varied opinions. According to the scheme suggested by Gardner *et al.*, conduction in α-Fe_2O_3 sintered at temperatures in the range 450–800 °C is due to oxygen vacancies [2]. Conduction increases with the increase in density associated with a rise in sintering temperature. At temperatures below 450 °C, grain boundaries play a major role in determining electrical conduction. Likewise, at temperatures above 800 °C, intrinsic conduction due to thermal excitation becomes dominant. Considering the electroneutrality of the crystal, each oxygen vacancy is compensated by reduction of Fe^{3+} to Fe^{2+}. The n-type conduction in α-Fe_2O_3 thus occurs due to thermally activated hopping movements of electrons from 2+ to 3+ ions. However, Morin believed that this hopping of electrons occurs in localized levels, only requiring significant activation energy [3].

α-Fe_2O_3 is an extrinsic semiconductor with an indirect bandgap of 2.2 eV [4], but a direct bandgap has also been reported [5]. The bandgap of α-Fe_2O_3 measured by various methods is found to be in the range 1.9–2.3 eV [5–7]. This allows absorption of all the UV light and the blue part of the visible region, which gives α-Fe_2O_3 its brownish-red color. The most probable band structure of α-Fe_2O_3 is shown in Figure 13.1. The conduction band of ferric oxide is believed to be composed of empty *d* levels of Fe^{3+}. The valence-band edge is derived from t_{2g} Fe 3*d*

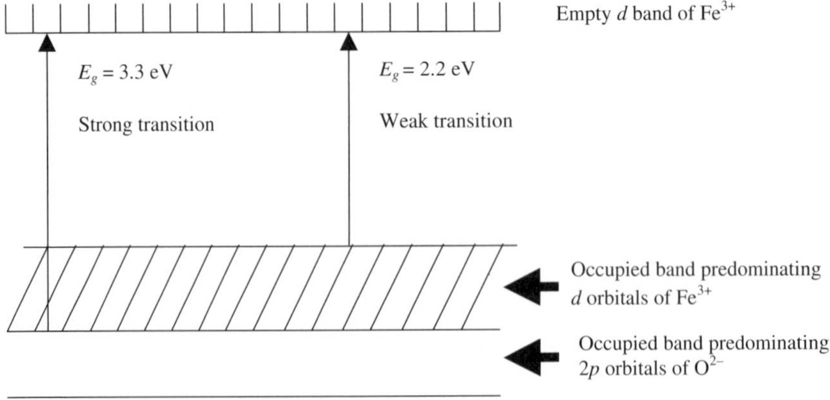

Figure 13.1 Schematic representation of the band structure of α-Fe_2O_3.

orbitals with some admixture of antibonding O 2*p* orbitals with $E_g = 2.2$ eV. It was found by ultraviolet photoemission spectroscopy (UPS) that the valence band of well-ordered, nearly stoichiometric α-Fe_2O_3 consists of overlapping O 2*p* and Fe 3*d* orbitals, which give rise to a complicated structure about 10 eV wide [8].

Gardner *et al.* observed optical absorption peaks at 2.4 eV, 3.2 eV and a strong broad absorption band centered at 5.8 eV for polycrystalline α-Fe_2O_3 [2]. Molecular-orbital studies explained these energy transitions as Fe 3*d* → 3*d*, O 2*p* → Fe 3*d* and O 2*p* → Fe 3*d*, respectively [9]. Similar energy levels were obtained with ellipsometry for both single crystals and polycrystalline films. In contrast to this, in a study based on computational analysis of the density of electronic states, using the *ab initio* periodic unrestricted Hartree–Fock approach, Catti *et al.*, in 1995, showed that the bandgap is of a *p*–*d* rather than a *d*–*d* type [10]. Optical absorption in α-Fe_2O_3 has a much wider spectrum than the photocurrent response [4]. In 1978, Kennedy and Frese explained this by suggesting that more than one optical transition occurs in α-Fe_2O_3, but only one results in the creation of an electron–hole pair, where the hole is capable of evolving oxygen from water [4].

13.2.2 α-Fe_2O_3 in PEC Splitting of Water

13.2.2.1 Advantages and Disadvantages

α-Fe_2O_3, with its bandgap of 2.2 eV, is one of the most attractive materials for harvesting solar energy for hydrogen generation through the splitting of water in PEC cells. Moreover, α-Fe_2O_3 is a stable material, which usually does not corrode in contact with most electrolytes over a wide range of pH. Being a low-cost and readily available material, it has always been the material of choice in studies related to photosplitting of water. PEC properties of α-Fe_2O_3 were extensively studied in the 1970s and 1980s and, despite several advantages, it was labeled as a poor photoelectrode, mainly due to: (i) its band edges not being properly aligned to the redox levels for hydrogen/oxygen evolution via splitting of water, (ii) large recombination rates of photogenerated carriers and (iii) short diffusion lengths of holes. The high resistance associated with α-Fe_2O_3 is another drawback leading to poor electrical conduction through the bulk material.

13.2.2.2 Summary of Earlier Work

Intensive research has been conducted to reduce the resistance of α-Fe_2O_3 and to improve its PEC properties by varying methods of preparation, dopants, sintering temperature and by using other techniques like metal–metal oxide loading, use of dyes, sensitizers etc. Research in the area up to 1984 has been very well compiled by Anderman and Kennedy in a book entitled *Semiconductor Electrodes* [11]. Most of the research was concentrated on polycrystalline α-Fe_2O_3, in the form of pellets, prepared from powder using ceramic techniques or by thermal oxidation.

Yeh and Hackerman prepared α-Fe_2O_3 by thermal oxidation [12]. McGregor *et al.* reported the preparation of polycrystalline α-Fe_2O_3 pellets and observed the onset of photocurrent at photon energies greater than 2.1 eV [7]. They further reported photocurrent saturation at higher voltages, and the predominance of the space-charge layer over surface states. Thermally grown α-Fe_2O_3 prepared by Wilhelm *et al.* exhibited bandgap variation from 2.1 to 1.9 eV [13].

Current measurement under chopped illumination allowed them to evaluate electrode parameters and reaction pathways. Single-crystal thin films with (110) preferred orientation were studied by Marusak *et al.* [14]. Redon *et al.* presented a comparison of the PEC behavior of single crystals and amorphous α-Fe_2O_3, and found that the PEC characteristics of single crystals were more favorable towards hydrogen production [15]. However, the photocurrent generated was still not very large and working with single crystals was quite expensive. Candea prepared polycrystalline α-Fe_2O_3 by thermal oxidation [16]; however, the long-term stability of films, which were too thin, was doubtful.

Doping α-Fe_2O_3 to alter its properties has interested many researchers in order to improve its PEC response. Gori *et al.* prepared doped α-Fe_2O_3 film by vacuum deposition and investigated the photoresponse under the iodine–iodide ion redox couple [17]. They found least photocorrosion in neutral solution. Polycrystalline α-Fe_2O_3 doped with Si and Group IVA elements were investigated by Kennedy *et al.*, in 1980–1981 and Si dopant was found to be more effective in improving the photoresponse, compared to Ti, exhibiting a photocurrent of 120 μA at 0.0 V versus SCE (saturated calomel electrode) with 400 nm light [18,19]. They concluded that dopants, besides introducing electronic defects, may also act as trap/recombination centers. Shinar and Kennedy, in 1982, prepared a whole range of α-Fe_2O_3 samples doped with Zr, Hf, Ce, V, Nb, W, Al and C, and they found that dopants play multiple roles [20]. Sanchez *et al.* investigated solid solubility of Ge, Si and Mg in α-Fe_2O_3 and found that Ge-doped samples exhibited a better PEC response [21].

Grätzel *et al.* reported that α-Fe_2O_3 powder may undergo visible-light-induced photodissociation/corrosion in acidic medium, especially in the presence of chloride ions [22]. Substitutional doping of α-Fe_2O_3 with Si, Ca, Nb, Cu, Ru, Mg and Zr was performed by Houlihan *et al.* in 1985 [23]. These doped samples were sintered in air at 1350 °C for 20 h, reduced in a hydrogen atmosphere at 300 °C and finally reoxidized at 700–900 °C for short periods. This resulted in samples with resistivities in the range of 0.01–5000 Ω cm. Such samples were found to have good PEC properties.

Parent *et al.* [24], in 1987, prepared Mg- and Si-doped α-Fe_2O_3 under a plasma spark and found the presence of Fe_3O_4 in the samples. They concluded that the presence of Fe_3O_4 in α-Fe_2O_3 improved the photoresponse to a great extent. Sanchez *et al.* prepared Nb-doped single crystals of α-Fe_2O_3 by chemical vapor deposition [25]. They observed that the doping results in sub-bandgap energy levels, but the low photoresponse was attributed to lower carrier mobility, and slow charge transfer across the semiconductor junction. Some attempts were also made to obtain p-type α-Fe_2O_3 using various divalent dopants, like Mg and Zn [26–28]. Studies on p/n type photoassemblies were also attempted by Leygraf *et al.* in 1982 and Turner *et al.* in 1984 [6,29,30], which could dissociate water without any external bias and showed excellent long-term stability in aqueous solution. A few studies on the use of sensitizers were also reported, without much improvement in the PEC response.

13.3 Nanostructured α-Fe_2O_3 Photoelectrodes

In recent years, the advent of nanotechnology has opened new avenues for discovering and synthesizing nanostructured metal-oxide semiconductors for efficient hydrogen production. Interest in nanosized materials in the PEC splitting of water has been stimulated on account of the small size of the building blocks (particle, grain or phase) and the high surface-to-volume ratio. Nanosized materials are expected to demonstrate unique mechanical, optical, electrical and magnetic properties. The properties of these materials depend upon: (i) fine grain size and size distribution, (ii) the chemical composition of constituent phases, (iii) the presence of grain boundaries, heterophases, interphases or free surfaces and (iv) interactions between the constituent domains. Nanocrystalline morphology is useful in increasing the photon-to-current yield by minimizing the distance that the minority carriers have to diffuse before reaching the interface. Another advantage is the large internal surface area of nanostructured electrodes, which allows larger absorption of the incoming solar radiation. Size-dependent properties, such as size quantization effects in semiconductor nanoparticles and quantization charging effects in metal nanoparticles, provide the basis for developing new and effective systems. Recent efforts in synthesizing nanostructures with well-defined geometrical shapes (solids, hollow spheres, prisms, rods and wires) and their assembly in two and three dimensions has expanded the possibility of developing new strategies for solar hydrogen production.

The use of nanostructured α-Fe_2O_3 in PEC cells is expected to deliver a more efficient system for the solar splitting of water, by overcoming the problem of recombination of photogenerated carriers, since electron–hole pairs generated near the interfaces can be separated rapidly before recombination, due to smaller dimensions. However, nanostructures of α-Fe_2O_3 have not been extensively studied with respect to PEC generation of hydrogen. Various techniques have been used by many workers to prepare α-Fe_2O_3 in nanodimensions and the PEC response is reported with respect to the method of preparation, the conditions of preparation and various other parameters, but any correlation between particle size and its impact on PEC response has not been adequately described in the literature. It is therefore not easy to describe how and why the particle size of α-Fe_2O_3 effects the PEC response of the material and so the hydrogen generation. The following section summarizes some of the nanostructured photoelectrodes described in the literature in the last two decades.

13.3.1 Preparation Techniques and Photoelectrochemical Response

13.3.1.1 Spray Pyrolysis (SP)

In this technique, the pyrolytic decomposition of iron salts, in the form of nitrate or chlorate, sprayed on the heated substrate surface, leads to the formation of oxide thin films. The electrical, optical and structural properties of thin films have strong dependence on the stoichiometry and microstructure, as well as the level of residual stress caused by deposition technique and interaction with the substrate [31]. Fine control over the spray-deposition parameters produces films with smooth and uniform thicknesses, excellent adherence to the substrate and high visible transmittance. Substrate temperature is an important factor influencing the film properties. In accordance with the mechanism of the film formation, film growth usually takes place first by nucleation, followed by growth of the nuclei. The nuclei growing on the substrate may have various crystallographic orientations and the resultant crystallite assumes the orientation of the largest one. Further increase in substrate temperature can stimulate oriented growth, facilitating crystallization.

The PEC behavior of α-Fe_2O_3 electrodes synthesized by spray pyrolysis (SP) on an SnO_2-coated glass substrate was investigated by Murthy and Reddy in 1984 [32]. The electrodes were highly stable in strongly basic medium and to applied positive bias. α-Fe_2O_3 coated on ordinary Corning glass did not give any photoresponse because of its high resistance. A bandgap of

2.0–2.2 eV was calculated from the absorption spectrum. No experimentation was reported regarding determination of particle size.

Khan *et al.*, in a series of papers [33–37], have reported the PEC response of doped/undoped nanostructured α-Fe_2O_3 thin films prepared by SP. Although particle size was not mentioned, the title of one of the paper uses the word "nanocrystalline" [35]. They prepared undoped and iodine-doped iron-oxide thin films of varying thicknesses on a conducting glass substrate at 350 °C, using a precursor of 0.01 M iodine mixed with 0.1 M $FeCl_3$ (80% ethanolic solution containing 0.1 M HCl [33]. The PEC response was studied with 0.2 M NaOH containing 0.5 M Na_2SO_4 (pH 13) as electrolyte. The photocurrent density was observed to increase with iodine doping, with a maximum of 1.0 mA cm^{-2} at 0.82 V versus SHE (standard hydrogen electrode) for a 60 nm thick film. Model calculations showed that five undoped iron-oxide thin-film electrodes of optimum thickness in stacks generated a photocurrent of 5 mA cm^{-2} at 0.82 V versus SHE, which improved to 15 mA cm^{-2} in the case of iodine-doped films. Thin films were found to be quite stable and had a mirror-like structure. No degradation of the films was observed during PEC measurements. Undoped α-Fe_2O_3 films of optimum thickness 33 nm were reported by Majumder and Khan [34], exhibiting a photocurrent density of 0.7 mA cm^{-2} at 0.82 V versus NHE, with a 200 W Xenon lamp giving out 150 mW cm^{-2} light-beam intensity. Spray time, substrate temperature, the solvent composition of the spray solution and the concentration of spray solution were optimized with respect to the best PEC response, in another paper [35]. A maximum photocurrent density of 3.7 mA cm^{-2} at 0.7 V versus SCE (Figure 13.2) was obtained with n-Fe_2O_3 nanocrystalline thin film, synthesized in optimum conditions; the spray solution concentration 0.11 M $FeCl_3$ in 100% EtOH, pyrolysis temperature 370 °C and spray time of 60 s. Previous results for iodine-doped *n*-Fe_2O_3 films prepared using HCl and EtOH in the spray solution are also included in Figure 13.2. A total conversion efficiency of 4.92% and a practical photoconversion efficiency of 1.84% at an applied potential

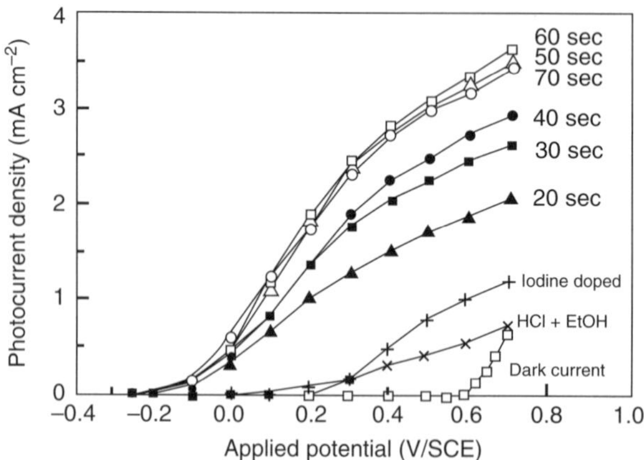

Figure 13.2 Photocurrent density: applied potential dependence at *n*-Fe_2O_3 film electrodes synthesized by the spray pyrolysis method at different spray times in optimized conditions. (Reproduced with permission from S.U.M. Khan and J. Akikusa, Photoelectrochemical splitting of water at nanocrystalline n-Fe_2O_3 thin-film electrodes, *Journal of Physical Chemistry B*, **103**, 7184–7189, 1999. © 1999 American Chemical Society.)

of 0.2 V versus SCE at pH 14 were reported for these films. Such a high photocurrent density of 3.7 mA cm^{-2} reported by the authors in undoped α-Fe_2O_3 has been questioned by many workers and could not be reproduced by others [38,39].

Zn-, Cu- and Mg-doped α-Fe_2O_3, when studied in 0.5 M H_2SO_4 as the electrolyte, exhibited p-type character [36,37,40]. The highest photocurrent density of 1.1 mA cm^{-2} at 0.8 V versus Pt was observed for optimized zinc doping of 0.0088 M, showing a maximum quantum efficiency of 21.1% at 325 nm wavelength. Mg-doped p-Fe_2O_3 thin films exhibited the highest photocurrent density of 0.21 mA cm^{-2} at 0.2 V versus SCE at a light intensity of 40 mW cm^{-2} [40]. Cu-doped Fe_2O_3 films (p-type), as reported by Ingler and Khan in 2005 [37], exhibited the highest photocurrent density of 0.94 mA cm^{-2} at zero bias versus SCE under an illumination intensity of 40 mW cm^{-2}. A maximum total conversion efficiency of 2.9% was observed with this sample, which was doped with 0.01155 M Cu^{2+}, deposited at a substrate temperature of 395 °C with a total spray time of 100 s. These films had mixed structures of α-Fe_2O_3 and $CuFe_2O_4$.

Zn-doped samples prepared by the authors' group [41] exhibited n-type conductivity in pH 13 NaOH electrolyte over a large range of doping concentrations. Films were deposited on conducting glass substrates by spraying 0.15 M precursor solution, prepared by dissolving $Fe(NO_3)_3.9H_2O$ with varying dopant concentrations (0.5, 1.0, 1.5, 5.0 and 10.0 at.%) of $Zn(NO_3)_2.6H_2O$ in double-distilled water. The solution was sprayed with air at a pressure of 10 kg cm^{-2} onto heated (350 ± 5 °C) substrate for a total of 100 s. Samples were finally sintered at 500 °C for 2 h. The average particle size in the film was found to be 60 nm and the bandgap was ~2.0 eV. The maximum photocurrent density of ~0.64 mA cm^{-2} at an applied potential of 0.7 V versus SCE (Figure 13.3) was observed with 5.0 at% Zn-doped samples, using a 150 W Xe arc lamp.

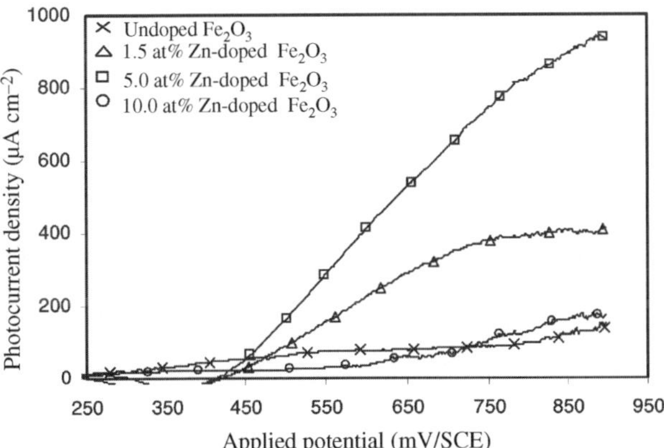

Figure 13.3 Photocurrent-density–voltage (mV versus SCE) characteristics for Zn-doped α-Fe_2O_3 using a 150 W Xe lamp as light source. Dependence of photocurrent-density on different doping concentrations (at%) in 13pH NaOH electrolyte solution. (Reproduced with permission from S. Kumari, C. Tripathi, A.P. Singh *et al.*, Characterization of Zn-doped hematite thin films for photoelectrochemical splitting of water, *Current Science*, **91**, 1062–1064, 2006. © 2006 Current Science Association.)

Detailed investigations on modifications in α-Fe_2O_3 that can affect the performance of the resulting photoanodes, such as the addition of dopants or structure-directing agents and changes in precursor species of carrier solvent, were presented by Sartoretti *et al.* in 2003 and 2005 [42,43]. A solution of 0.01 to 0.05 M iron acetylacetonate and 0.1 M $FeCl_3.6H_2O$ in absolute ethanol was sprayed with nitrogen carrier gas onto a conducting glass substrate kept at a distance of 12 cm from the nozzle, at a temperature between 400 and 440 °C. The nitrogen flow rate was approximately 7.5 l min^{-1}. A spray of 10 s was followed by a gap of 5 min to achieve the thermal hydrolysis of the freshly sprayed layer and to limit the temperature losses. Films prepared were rapidly cooled under a flow of nitrogen for 5–10 min, after which the photoelectrodes were annealed in air at 550 °C for 60 min. Finally, samples were again cooled under a flow of nitrogen for 5–10 min. SEM micrographs showed that the diameter distribution of the surface features was between 180 and 800 nm. Better adhering films to the substrate and more efficient electrodes resulted from those prepared from an ethanolic solution of ferric chloride or ferric acetylacetonate, as compared to ferric nitrate precursors. However, large photocurrents were observed by the addition of either Al^{3+} (1%) or Zn^{2+} (4%) in conjunction with Ti^{4+} (5%). Electrodes having six layers, with a precursor of 0.1 M $FeCl_3.6H_2O$ solution in absolute ethanol doped simultaneously with 5% Ti and 1% Al, were found to offer the highest photocurrent in 0.1 M NaOH, recorded as 4.38 mA cm^{-2} at 0.45 V versus NHE (normal hydrogen electrode) with a 150 W Xe lamp exhibiting an incident-photon-to-current efficiency (IPCE) of 3.7% at 550 nm and 1.07% at 600 nm. A very promising IPCE of 25% was attained at 400 nm wavelength for the same type of electrode. In Zn^{2+}/Ti^{4+}-doped samples, a negative shift in the photocurrent onset potential by ~0.22 V was observed.

Si-doped α-Fe_2O_3 films for solar water splitting prepared by spraying a solution of Fe $(AcAc)_3$ and tetraethoxysilane (TEOS) have been recently reported by Liang *et al.* 2008 [44] to exhibit a maximum photocurrent density of 0.37 mA cm^{-2} at 1.23 V versus RHE (reversible hydrogen electrode) in 0.1 M KOH solution under 80 mW cm^{-2} AM1.5 illumination. This photocurrent was achieved after depositing an interfacial layer of 5 nm SnO_2 film between the α-Fe_2O_3 film and the transparent conductive oxide (TCO) substrate for a 0.2 at% doping level.

13.3.1.2 Ultrasonic Spray Pyrolysis (USP)

Duret and Grätzel, in 2005 [45], compared the PEC response of thin films of α-Fe_2O_3 deposited on a TCO substrate by SP and ultrasonic spray pyrolysis (USP), a modified technique of spray pyrolysis, and found USP films exhibiting much better (Figure 13.4) PEC responses than SP films. In the USP method, a solution of 0.02 M ferric acetylacetonate in ethanol was used as the precursor, which was first pumped at a flow rate of 1 ml min^{-1} to the ultrasonic nozzle, connected to a 100 kHz frequency generator. The precursor solution was converted into droplets and then injected into a glass chamber to prevent the formation of big droplets. The fog of small droplets obtained at the outlet of this chamber was finally carried to a tubular oven heated to 420 °C (temperature of the substrate) using compressed air at a flow rate of 20 l min^{-1}. The optimal sprayed volume of precursor solution was 200 ml and the distance between the substrate and the outlet of the chamber was 25 cm. In the SP method, 20 ml of a 0.05 M solution of ferric acetylacetonate in ethanol was sprayed onto the TCO substrate maintained at 400 °C.

Both the samples had one indirect (2.0 eV) and one direct (3.0 eV) electronic transition. Although both types of films mainly consisted of α-Fe_2O_3 phase and at least one impurity phase

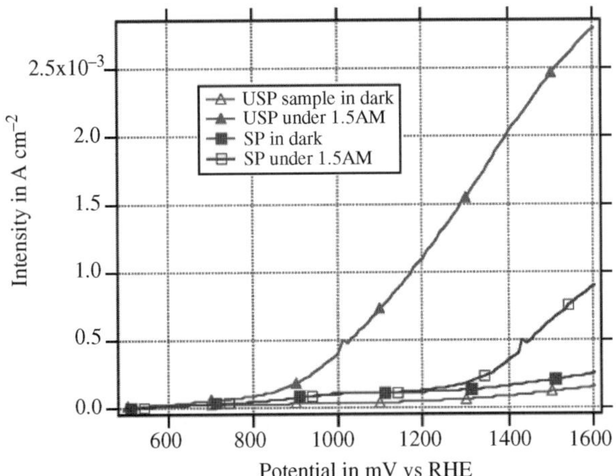

Figure 13.4 Current–potential curves for SP and USP samples measured in NaOH 1 M in the dark and under simulated AM1.5 sunlight (equivalent intensity 130 mWcm^{-2}) at a scan rate of 100 mV s^{-1}. (Reproduced with permission from A. Duret and M. Grätzel, Visible light-induced water oxidation on mesoscopic α-Fe_2O_3 films made by ultrasonic spray pyrolysis, *Journal of Physical Chemistry B*, **109**, 17184–17191, 2005. © 2005 American Chemical Society.)

(Fe_3O_4), they differed greatly in their morphology (Figure 13.5A and B). The mesoscopic α-Fe_2O_3 layers produced by USP consisted mainly of 100 nm sized platelets with a thickness of 5–10 nm (Figure 13.5A). These nanosheets were oriented mainly perpendicular to the substrate, their flat surface exposing (001) facets. The PEC response was measured with a 450 W Xe lamp adjusted with simulated global AM1.5 solar radiation in 1 M NaOH (13.6 pH)

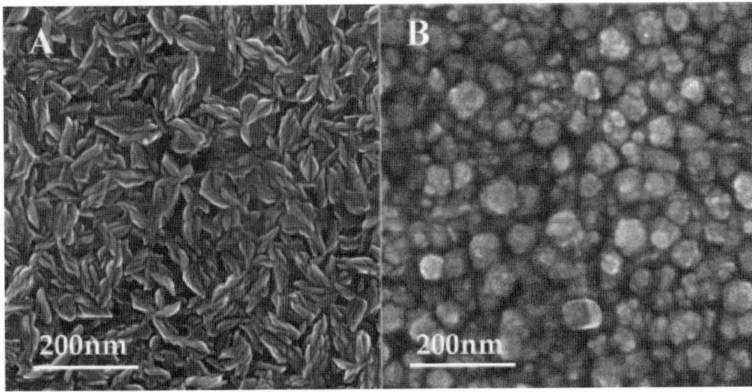

Figure 13.5 SEM pictures of (A) USP and (A) SP samples. (Reproduced with permission from A. Duret and M. Grätzel, Visible light-induced water oxidation on mesoscopic α-Fe_2O_3 films made by ultrasonic spray pyrolysis, *Journal of Physical Chemistry B*, **109**, 17184–17191, 2005. © 2005 American Chemical Society.)

as the electrolyte (Figure 13.4). The dark current of the USP sample rose gradually at potentials greater than 1.2 V versus RHE to reach 0.15 mA cm^{-2} at 1.6 V versus RHE, whereas the dark current of the SP sample increased more steeply to reach 0.25 mA cm^{-2} at 1.6 V. The photocurrent onset potential was 0.55 and 1.125 V versus RHE for the USP and the SP samples, respectively, showing that higher applied potential is needed to afford water oxidation for the SP sample than for the USP sample. The photocurrent reached 1.07 mA cm^{-2} and 14 μA cm^{-2} at 1.2 V versus RHE for the USP and the SP samples, respectively. This difference in the photocurrent was mainly attributed to the mesoscopic leaflet structure (Figure 13.5A) of α-Fe_2O_3 formed in the samples obtained by USP, which allows efficient harvesting of visible light, while offering, at the same time, the very short distance required for the photogenerated holes to reach the electrolyte interface before recombining with conduction-band electrons. Thus water oxidation by the valence-band holes may occur, even though their diffusion length is only a few nanometers. Distances were apparently longer in the particles produced by SP, favoring recombination of photoinduced charge carriers at the semiconductor/electrolyte interface. Open-circuit photovoltage measurements indicated a lower surface-state density for nanoplatelets, as compared to the spherical particles. These factors explained the much higher photoactivity of USP- compared to SP-deposited α-Fe_2O_3 layers. Addition of hydrogen peroxide to the alkaline electrolyte further improved the photocurrent–voltage characteristics of films generated by USP, indicating that hole transfer from the valence band of the semiconductor oxide to the adsorbed water is the rate-limiting kinetic step in the oxygen-generation reaction. The IPCE of the SP sample was much lower than the IPCE of the USP sample at 375 nm wavelength. For the USP electrode at 1.6 V, the IPCE quadrupled from 6% to 24% on decreasing the wavelength from 500 to 400 nm.

13.3.1.3 Atmospheric Pressure Chemical Vapor Deposition (APCVD)

Single nanocrystalline undoped and Si-doped thin films of iron oxide were prepared by atmospheric pressure chemical vapor deposition (APCVD) from $Fe(CO)_5$ and tetra-ethoxysilane (TEOS) on a fluorine-doped SnO_2 substrate at 415 °C by Kay, Cesar and co-workers [39,46]. The precursors were supplied by bubbling argon gas through 0.5 ml of each liquid at the rate of 11.3 ml min^{-1} for $Fe(CO)_5$ and at 19.4 ml min^{-1} for TEOS at 25 °C, metered by thermal mass-flow controllers. It was observed that lower TEOS concentrations yielded less efficient photoelectrodes, while higher concentrations gave inactive brown instead of red-brown films. These two vapor streams were mixed with air and directed vertically onto the substrate through a glass tube of 13 mm inner diameter from a distance of 20 mm. Within 5 min, a circular spot of iron oxide was deposited on the SnO_2 substrate, showing concentric interference fringes in reflected light, which indicated a radial thickness profile. For improved electrode performance, an interfacial layer of SiO_2 was first deposited for 5 min on SnO_2 under the same conditions, but with TEOS only, followed by α-Fe_2O_3 deposition. High-resolution scanning electron microscopy (HRSEM) studies revealed a highly developed dendritic nanostructure of 500 nm thickness, having a feature size of only 10–20 nm at the surface. Deep gaps obtained between the dendrites permitted better exposure of the internal surface of α-Fe_2O_3 into the electrolyte. The best photocurrent density offered was 2.2 mA cm^{-2} at 1.23 V versus RHE. An IPCE of 42% at 370 nm in AM1.5 sunlight of 1000 W m^{-2} was obtained in 1 M NaOH. A catalytic cobalt monolayer on the α-Fe_2O_3 surface resulted in an 80 mV cathodic shift, and the photocurrent increased to 2.7 mA cm^{-2}.

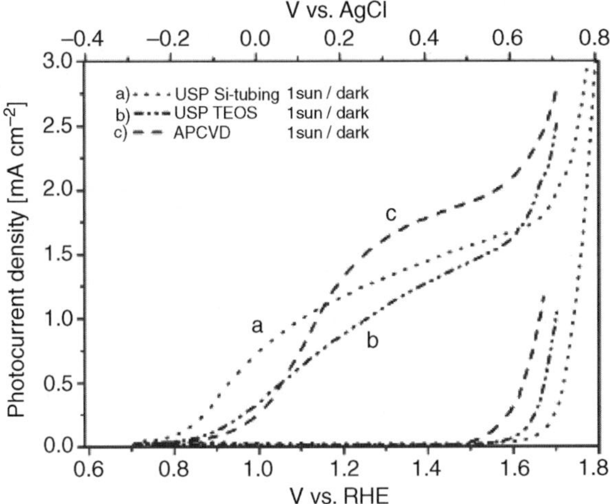

Figure 13.6 Current–potential curves of Si-doped polycrystalline α-Fe_2O_3 electrodes on TCO measured versus AgCl: (a) 370 nm thick USP film using silicone tubing; (b) 145 nm thick USP film using Tygon tubing with TEOS as Si dopant; (c) 170 nm thick APCVD film using TEOS as Si dopant. (Reproduced with permission from I. Cesar, A. Kay, J.A.G. Martinez and M. Grätzel, Translucent thin film Fe_2O_3 photoanodes for efficient water splitting by sunlight: nanostructure-directing effect of Si-doping, *Journal of the American Chemical Society*, **128**, 4582–4583, 2006. © 2006 American Chemical Society.)

Cesar *et al.* in 2006 [47] compared the PEC response of Si-doped α-Fe_2O_3 thin films prepared by USP and APCVD methods, based on thermal decomposition of iron(III)-acetylacetonate and iron pentacarbonyl, respectively. TEOS was used for silicon doping. Current–potential curves for various types of Si-doped polycrystalline α-Fe_2O_3 electrodes on TCO in 1 M NaOH (pH 13.6) in darkness and under illumination at one sun, AM 1.5 (100 mW cm^{-2}) are shown in Figure 13.6. The immersed and illuminated anode surface areas were 2.5 and 0.5 cm^2, respectively. USP-deposited film at a lower flow rate of 0.4 ml min^{-1} exhibited the best photocurrent density of 1.17 mA cm^{-2} (Figure 13.6) at 1.23 V versus RHE for a 375 nm thick film. These films consisted of stacked sheets oriented perpendicular to the substrate (Figure 13.7a and b) and had thicknesses between 15 and 25 nm, and lengths between 50 and 750 nm, depending upon the deposition time used, while APCVD Si-doped samples offered a photocurrent up to 1.45 mA cm^{-2} at 1.23 V versus RHE and showed a dendritic microstructure (Figure 13.7c and d), with 170 nm height and 20–30 nm thick branches. Photocurrents were measured at one sun (AM1.5) solar radiation, having an incident power of 100 mW cm^{-2} with 1 M NaOH (13.6 pH) as electrolyte. Si doping (n-type) in nanocrystalline films measured at one sun was found to improve the photoresponse by 50% and 90% for the USP and APCVD samples, respectively. The morphology of nanostructured α-Fe_2O_3 films for both types of samples was seen to be strongly influenced by the silicon doping (Figure 13.7b and d).

Recently, Yarahmadi *et al.* 2009 utilized ferrocene and iron pentacarbonyl as precursors to prepare undoped α-Fe_2O_3 through APCVD [48]. α-Fe_2O_3 films prepared by iron pentacarbonyl were made up of nanostructured agglomerate/clusters of around 200 nm in size, which themselves consisted of smaller particles of around 20 nm, whereas the films prepared by

Figure 13.7 Typical HRSEM images of Si-doped α-Fe_2O_3 films on TCO obtained from USP (a,b) and APCVD (c, d): (a) and (c) are side views, (b) and (d) top views. (b, Inset) α-Fe_2O_3 grains for undoped USP electrodes. (d, Inset) α-Fe_2O_3 grains for undoped APCVD electrodes. (Reproduced with permission from I. Cesar, A. Kay, J.A.G. Martinez and M. Grätzel, Translucent thin film Fe_2O_3 photoanodes for efficient water splitting by sunlight: nanostructure-directing effect of Si-doping, *Journal of the American Chemical Society*, **128**, 4582–4583, 2006. © 2006 American Chemical Society.)

ferrocene were almost free of agglomerate/cluster features and comprised bigger particles. Higher optical absorption was observed for α-Fe_2O_3 films prepared using ferrocene, which was attributed to the higher packing density. PEC studies of the films prepared using ferrocene showed superior performance to those with iron pentacarbonyl. A photocurrent density of 540 $\mu A\,cm^{-2}$ and 1.5 $\mu A\,cm^{-2}$ at 1.23 V versus RHE was achieved for α-Fe_2O_3 films prepared using ferrocene and iron pentacarbonyl, respectively, in 1 M NaOH electrolyte under illumination by solar light (300 W solar simulator).

13.3.1.4 Sol-Gel (SG)

Most commonly used method to prepare nanomaterials of metal oxide is the sol-gel (SG) method, because conversions to sol-gel occur readily with a wide variety of precursors and the process can be conducted at or near room temperatures. Additionally, gel products can be tuned to have properties ideal for desired applications. Synthesis involves co-precipitation from a solution containing inorganic salts or alkoxides of metals.

A PEC study of nanocrystalline α-Fe_2O_3 prepared by the sol-gel method was reported for the first time by Björkstén *et al.* in 1994 [49]. They prepared a thin film of α-Fe_2O_3 from a colloidal solution obtained by hydrolysis of $FeCl_3$. After depositing the α-Fe_2O_3 colloidal suspension on a conducting glass substrate using a Pasteur pipette, the electrode was dried for at least 1 h at room temperature before sintering in an oven at 560 °C for 4 h. The film prepared had a thickness of 1–1.5 μm and constituted spherical particles with diameters between 25 and 75 nm, sintered together to form a nanoporous structure. They found charge-carrier recombination to be the controlling factor for obtaining a low photocurrent. In the presence of hole scavengers, the photocurrent yields remained small, the highest value obtained being 1.7%. The reason for this is that bulk and/or grain-boundary recombination of charge carriers remains dominant, even for α-Fe_2O_3 films comprising particles in the nanometer range.

Beermann *et al.* and Lindgren *et al.* designed a nanostructured rod-like morphology of α-Fe_2O_3, with no grain boundaries with the thickness in the range ~0.1–1.5 μm, using a general thermodynamic method based on the lowering of the water–oxide interfacial tension, for direct splitting of water [5,50]. This procedure reportedly allowed control of morphology, porosity, crystallographic phase and the overall texture of the nanostructured metal-oxide thin films. Films obtained were chemically, mechanically and thermally stable with a porosity of $50 \pm 5\%$. Direct and indirect bandgap energies were calculated as 2.17 and 2.09 eV, respectively. PEC measurements were carried out in a conventional three-electrode setup, with water as the electrolyte, containing 0.1 M KI and 0.01 M K_2HPO_4/KH_2PO_4 buffers (pH 6.8) and a 450 W Xe lamp as the light source. The electrolytic solution in the vessel was either purged with nitrogen or saturated with oxygen. In two-electrode measurements, the electrolyte used was ethylene carbonate and propylene carbonate (50 : 50 by weight) containing 0.5 M LiI and 0.5 mM I_2. The IPCE obtained was fairly high, reaching 11% for the substrate/electrolyte interface (SE) illuminated at 360 nm, which was seven times higher than for sintered spherical particles reported by Björkstèn *et al.* [49]. Relatively high efficiency for the film was explained in terms of better electron transport along the nanorods, due to fewer grain boundaries (possible recombination centers) and a directed movement of electrons towards the back contact. However, the high IPCE values obtained indicated that a purpose-built nanorod of α-Fe_2O_3 is a significant way to strikingly lower the recombination rate in the case of α-Fe_2O_3 material. Using a sandwich cell, an IPCE value up to 56% was reported at 340 nm. The photocurrent onset gives the approximate position of the flatband potential and was located at around −0.5 V (−0.7 V versus NHE), in 0.1 M NaOH. *I–V* characteristics for α-Fe_2O_3 nanorods in $HClO_4$ (pH 3) and in NaOH (pH 13) were very similar in shape, indicating that the rate-limiting step could be the same in both electrolytes.

A simple wet chemical method based on the sol-gel technique, involving the use of glycerol as the key reagent, was used by Agrawal *et al.* 2003 [51] to synthesize nanoparticles of iron oxide doped with 2–9 at% titanium. These photoelectrodes, prepared in pellet form, exhibited very small photocurrent densities of 200–600 $\mu A\,cm^{-2}$ at an electrode potential of 0.8 V versus SCE. To synthesize Ti-doped α-Fe_2O_3, a mixed solution of the salts of Fe and Ti was placed on a hot plate at 60–70 °C and varying amounts of glycerol were added slowly with continuous stirring, until a gray-white solid mass was formed. The dried mass was further heated at 250 °C for about 15–20 min and then calcined at 800 °C for nearly 30 min, which resulted in Ti-doped α-Fe_2O_3 powder. The average particle size of doped α-Fe_2O_3 powder ranged from 200 to 300 nm. Doped powder was converted into pellets and then sintered at various temperatures from 1200–1350 °C. A few discs of each type were heated and cooled in a partial vacuum of 5×10^5 mbar with a soak time of 1 h, at a maximum temperature of 1350 °C, and PEC measurements were performed in NaOH and H_2SO_4 electrolytic solution at pH 4, 11 and 13, illuminating the pellets with a 100 W white-light source. All undoped/doped α-Fe_2O_3 pellets exhibited n-type conductivity. Partial-vacuum-sintered pellets exhibited much lower resistances compared to pellets sintered in air. A nearly sevenfold decrease in the resistance of α-Fe_2O_3 pellets (sintered in air at 1350 °C) was found when the Ti concentration was increased from 2 to 4 at%. Under illumination, the observed onset voltages (V_{on}), which represent the minimum bias voltages at which a significant photocurrent was observed to flow, were 100–500 mV less compared to values in the dark. There was a significant drop in V_{on} at alkaline pH and the values decreased further when the pH was raised from 11 to 13. Despite a large difference in resistance values, no significant variation was found in the PEC behavior of air-sintered and vacuum-sintered samples. The continuous use of photocatalysts in PEC cells did not show any measurable corrosion/dissolution of the material, even after 30–60 days. In another method [52] Agrawal *et al.* used hexamethyltetramine (HMTA) as the key reagent. Ferric nitrate solution mixed with a calculated quantity of dopant (tetraethoxysilane) was gradually gelled at room temperature by the slow addition of HMTA solution with continuous stirring. The ferric ions formed a hydrous-oxide gel, with the dopant ions trapped within the gel matrix. Following this, the gel was soaked in ammonia solution to cause the homogeneous precipitation of dopant ions at pH > 7. The gel was then washed with dilute ammonia solution to remove the water-soluble compounds without causing any loss of metal hydroxide by peptization. The washed gel was dried in air at ~400 °C to obtain precursor powder, which was subsequently calcined at a higher temperature ~800 °C to get doped α-Fe_2O_3 powder, free of carbonaceous matter. The sample showed n-type conductivity. Insignificant photocurrent density was reported for all the samples, even after Si doping.

Miyake and Kozuka, in 2005, prepared Fe_2O_3–Nb_2O_5 films [67] of various Nb/(Fe + Nb) mole ratios on Nesa silica glass substrates by the sol-gel method. $Fe(NO_3)_3.9H_2O$, $NbCl_5$, $CH_3(CH_2)_2CH_2OH$ and CH_3COOH, were used as the starting materials. CH_3COOH was added to $CH_3(CH_2)_2CH_2OH$ in the ambient atmosphere, and $NbCl_5$ and $Fe(NO_3)_3.9H_2O$ were dissolved in the resultant solution in flowing nitrogen gas. The solutions thus obtained were concentrated using a rotary pump until the total volume of the solutions was reduced from 30 to 20 ml. Gel films were deposited on silica glass substrates by dip-coating at a substrate withdrawal speed of 3 cm min^{-1} and fired at 600–800 °C for 10 min. Gel film deposition and heat treatment were repeated three times. The grain size increased from 50 to 150 nm with increasing Nb/(Fe + Nb) ratio. The photoanodic properties were studied in a three-electrode cell with an aqueous buffer solution at pH 7, and an aqueous solution of 0.2 M $Na_2B_4O_7$, 0.14 M H_2SO_4 and 0.3 M Na_2SO_4 as the supporting electrolyte. When the Nb/(Fe + Nb) mole ratio increased from 0 to 0.25, the crystalline phases changed from α-Fe_2O_3 to α-Fe_2O_3 + $FeNbO_4$. The photoanodic current under white-light illumination and under monochromatic-light illumination increased in both the visible and UV regions after doping. A sample consisting of α-Fe_2O_3 and $FeNbO_4$ (Nb/(Fe + Nb) = 0.25) exhibited a photoresponse extending to 600 nm and an IPCE of 18% at a wavelength of 325 nm.

Souza *et al.* prepared Si-doped and undoped α-Fe_2O_3 films by spin-coating deposition based on metallic citrate polymerization using ethylene glycol [53]. The morphology of the Si-doped and undoped α-Fe_2O_3 films showed a mesoporous structure with worm-like grains of nanometer dimensions. The 0.5% Si-doped α-Fe_2O_3 exhibited higher photocurrent response when compared with undoped films. Even though all the films strongly absorbed visible light, the poor electron mobility resulted in a maximum photocurrent of 35 μA cm^{-2} at 1.23 V versus RHE in 1 M NaOH electrolyte solution with a 300 W Xe lamp.

13.3.1.5 RF Sputtering

Schumacher *et al.* investigated reactively sputtered and plasma-oxidized iron-oxide films and the effect of thermal treatment, vacuum annealing and indium incorporation on their surface composition, and photoelectrochemical, impedance and spectroscopic properties [54]. Auger depth-profile data showed no detectable amount of indium incorporation in the plasma-oxidized films, while the reactively sputtered films contained 20 at% indium at the front surface, decreasing to 8% in the bulk, and then increasing towards the indium tin oxide (ITO) interface. A positive shift in the dc photo-onset to 1.1 V was observed for nonannealed film at the same potential compared to vacuum-annealed indium-doped film. An increase in the photoresponse was observed with increase in the thickness up to 25 nm for reactively sputtered film onto ITO at 350 °C and vacuum annealed for 10 h, beyond which it decreased, whereas indium-free films formed on ITO by plasma oxidation showed a monotonic increase in the photocurrent with increasing film thickness. A larger value of effective quantum efficiency for sputter-deposited (indium-containing) films was obtained, compared to plasma-oxidized (indium-free) films. An indirect bandgap of 2.12 eV was observed, which was comparable to the 2.10 and 2.05 eV bandgap of indium-free films formed by plasma oxidation and by sputtering onto iron metal, respectively.

A significant reduction in onset potentials was observed by Miller *et al.* in sputtered iron-oxide films [55]. Surface morphology, grain size, electrochemical and optical properties of reactively sputtered films were found to be affected by deposition parameters, such as ambient gas composition, plasma power and substrate temperature. The highest photocurrent density in 1 N KOH under outdoor sun conditions was limited to 0.1 mA cm^{-2} [56].

Glasscock *et al.* [57] used a reactive DC magnetron sputtering system for deposition of Ti- and Si-doped α-Fe_2O_3 thin films using targets: Fe for undoped, Fe with 5 at% Ti for Ti-doped and transformer steel (Fe with around 5 wt% Si and $<$1 wt% Al) for Si-doped samples. The presence of around 1% Al in this alloy was expected to be beneficial in improving the PEC response, as observed by Sartoretti *et al.* [43]. The deposition of the α-Fe_2O_3 films on a fluorine-doped tin oxide (FTO) substrate was undertaken in a vacuum chamber evacuated to a base pressure of 10^{-4} Pa. Targets were sputtered in an atmosphere of 0.2 Pa O_2 and 0.8 Pa Ar and deposited on substrates heated to 300 °C. Finally, films were moved to a furnace and annealed in air at 550 °C for 1 h. Ti-doped α-Fe_2O_3 films exhibited all the major peaks of α-Fe_2O_3, but in the case of the undoped and Si-doped α-Fe_2O_3 films, only the (110) and (300) peaks were visible. The undoped and Ti-doped α-Fe_2O_3 films had a columnar structure of irregular grains. Grain sizes of the undoped, Ti-doped and Si-doped α-Fe_2O_3 films, as estimated from SEM images of the surfaces, were in the ranges 75–150, 40–100 and 15–30 nm, respectively, whereas average crystal sizes calculated using Scherrer's equation for the undoped, Ti-doped and Si-doped α-Fe_2O_3 films were around 65, 40 and 20 nm, respectively. The PEC of α-Fe_2O_3 thin films was measured in 1 M NaOH with 1000 W Xe arc lamp fitted with a water filter. Ti-doped samples exhibited better photocurrent density compared to other dopants, with a maximum of 0.9 mA cm^{-2} at 0.6 V versus SCE.

13.3.1.6 Potentiostatic Anodization

Prakasam *et al.* first synthesized a self-organized nanoporous iron(III) oxide thin film by potentiostatic anodization of iron foil at different anodization potentials [58]. For this purpose, 0.25 mm thick pure iron foil was used and anodized in a two-electrode system with iron foil as the anode and platinum foil as the counter-electrode. The electrolytic medium used in the study consisted of 1%HF + 0.5%NH_4F + 0.2% 0.1 M HNO_3 in glycerol (pH3) at 10 °C. The current–time behavior during the potentiostatic anodization of the pure iron foil showed that a gradual and slow dissolution process led to pore formation, with the rate of current drop being an indication of the rate of oxidation. The morphology of the samples was controlled by the anodization-bath temperature and the NH_4F concentration. Depending upon the anodization potential and the electrolyte composition, the pore diameter ranged from 50 to 250 nm, with a pore depth of approximately 500 nm. X-ray diffraction (XRD) of an as-anodized sample

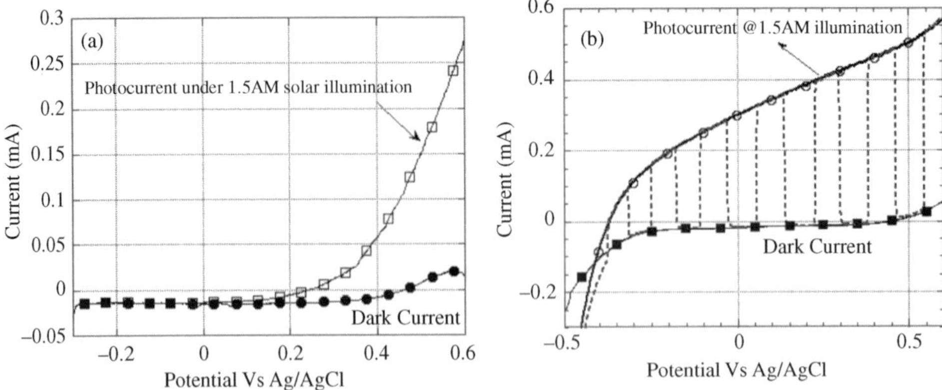

Figure 13.8 Photocurrent –potential curve for a iron(III) oxide photoanode prepared by potentiostatic anodization recorded in: (a) 1 M NaOH solution and (b) 0.5 M H_2O_2 + 1 M NaOH solution. (Reproduced with permission from H.E. Prakasam, O.K. Varghese, M. Paulose *et al.*, Synthesis and photoelectrochemical properties of nanoporous iron (III) oxide by potentiostatic anodization, *Nanotechnology*, **17**, 4285–4291, 2006. © 2006 Institute of Physics.)

showed an amorphous effect, while a sample sintered at 400 °C in a nitrogen atmosphere exhibited peaks corresponding to α-Fe_2O_3.

The crystallized nanoporous film, having a bandgap 2.2 eV, exhibited an onset potential of 0.24 mV and a net photocurrent density of 0.26 mA cm^{-2} in 1 M NaOH (Figure 13.8a) at 0.6 V versus Ag/AgCl, under simulated AM1.5 sunlight. Addition of 0.5 M H_2O_2 to the electrolyte improved the onset potential and photocurrent density to values of −0.47 V versus Ag/AgCl and 0.51 mA cm^{-2} at 0.6 V versus Ag/AgCl, respectively (Figure 13.8b). This increase in the PEC response was attributed to the addition of H_2O_2, which enhanced the reaction kinetics, as the photo-oxidation rate of H_2O_2 is much higher than water.

13.3.1.7 Electrodeposition

Synthesizing semiconductor electrodes by electrodeposition provides many advantages, which include ease of achieving electrical contact, precise control of the microstructure and the potential for co-deposition of dopants [59]. Furthermore, large-scale superstructures and nanoscale microstructures can be created economically with electrodeposition.

In 2007, Teng and Li carried out electrodeposition of iron(III) oxide films in a three-electrode system, in which a FTO glass substrate served as the working electrode and a platinum sheet and Ag/AgCl as the counter and reference electrodes, respectively [60]. Hydrated ammonium iron(II) sulfate solution [$(NH_4)_2Fe(SO_4)_2.6H_2O$; 0.1 M] was used for deposition and NaOH was used for pH adjustment to a value of 9. The deposition was conducted at 20 °C using a potentiostat at −1.6 V versus Ag/AgCl to give two accumulated charges of 7.2 and 18 C cm^{-2}. The XRD analysis of the films exhibited an amorphous nature of the film. This indicated that the crystalline size of the iron oxide was too small to be detected by X-ray diffraction. A direct bandgap of 2.2 eV was determined for both the films. The deposited films exhibited n-type semiconducting characteristics by showing photoresponse under an anodic bias in an electrolyte of 0.5 M Na_2SO_4 (pH 7). The photocurrent induced from irradiation with a 300 W Xe lamp was an increasing function of the charge applied for film deposition. Schrebler *et al.* prepared α-Fe_2O_3 films through a cyclic potential procedure in an aqueous solution of 5 mM $FeCl_3$ + 5 mM KF + 0.1 M KCl + 1 M H_2O_2 at room temperature on an SnO_2:F substrate [61]. These electrodeposited films, after annealing at 500 °C, showed good quality and crystallinity.

A similar method was utilized by Shwarsctein *et al.* to deposit α-Fe_2O_3 thin films doped with Mo and Cr on quartz Ti/Pt substrates, using $CrCl_3$ and $MoCl_5$ as precursors, for the splitting of water [62]. Electrodeposition was carried out using a standard three-electrode configuration with a graphite rod counter-electrode, an Ag/AgCl reference electrode and a Pt/Ti working electrode. The deposited samples were sintered at 700 °C. The morphologies of the undoped films were significantly different from those of the doped samples. The changes in morphology of the samples were a combined effect of changes induced by the electrodeposition conditions and those due to different rates of sintering during the high-temperature calcination. The crystallite sizes for the undoped, 5% Cr- and 15% Mo-doped samples were found to be 38, 42 and 43 nm, respectively. The optical bandgap of the samples was found to be the direct bandgap transition, with values of 2.1 eV for the undoped and 2.0 eV for both the Cr- and the Mo-doped samples. The maximum photocurrent densities were obtained for 5% Cr- and 15% Mo-doped samples, which were ~1.4 and 1.8 mA cm^{-2}, respectively, at 0.4 V versus Ag/AgCl in 1 M NaOH electrolyte. These samples had maximum IPCEs at 400 nm of 6% and 12%, which is 2.2 and four times higher than the undoped sample for the 5% Cr and 15% Mo samples, respectively.

In another paper, Hu *et al.* synthesized Pt-doped α-Fe_2O_3 thin films by electrodeposition on quartz Ti/Pt substrates, sintered at 800 °C [63]. The change in the morphology before and after Pt doping was striking. The Pt-doped films appeared more uniform and dense than the undoped. Among all the compositions, the 5% Pt-doped sample exhibited the highest relative PEC performance. The maximum photocurrent density was 1.43 mA cm^{-2} under 410 mW cm^{-2} illumination at 0.4 V versus Ag/AgCl in 1 M NaOH electrolyte. They also predicted that the best-performing Pt-doped sample, at 400 nm and 0.46 V in a two-electrode system, would yield an overall energy efficiency of ~3%, which is nearly four times higher than that of the undoped sample.

13.3.2 Flatband Potential and Donor Density

Flatband potential is an important parameter in deciding the performance of a semiconductor/electrolyte junction towards hydrogen generation. The higher the flatband potential, the greater is the ability of the photoelectrode to split water. The most frequently used method for determination of the flatband potential at a semiconductor/electrolyte interface is by Mott–Schottky plots.

If C is the junction capacitance at electrode potential V_{app}, the Mott–Schottky equation is given by

$$\frac{1}{C^2} = \left[\frac{2}{\varepsilon_o \varepsilon q N_d}\right]\left[V_{app} - V_{fb} - \left(\frac{k_B T}{q}\right)\right]$$

where ε_o is the permittivity of free space, ε is the relative permittivity of the semiconductor electrode, q is the charge on the carriers, N_d is the donor concentration, V_{fb} is the flatband potential, T is the temperature of operation and k_B is Boltzmann's constant. The intercept of the $1/C^2$ versus V_{app} curve (Mott–Schottky) on the x-axis gives the value of the flatband potential and the slope of the curve is utilized to calculate the donor density (N_d) using the formula

$$\text{slope} = \left[\frac{2}{\varepsilon_0 \varepsilon q N_d}\right].$$

For any semiconductor/electrolyte junction, the value of the flatband potential depends on the dopant, the doping concentration, the nature of the electrolyte and the pH. Using Mott–Schottky plots, Turner *et al.* reported flatband potentials of 0.2 and 2.3 V versus RHE at 1500 Hz frequency in 0.01 N NaOH electrolyte solution, for 10.0 at% Si-doped (n-type) and 5.0 at% Mg-doped (p-type) α-Fe_2O_3, respectively, prepared by conventional ceramic techniques [6]. The acceptor or donor concentrations were calculated as $5 \times 10^{16}\,cm^{-3}$ for Mg-doped samples and $2 \times 10^{18}\,cm^{-3}$ for Si-doped electrodes. For Nb-doped α-Fe_2O_3 ($Nb_xFe_{2\text{-}x}O_3$, $x = -0.03$) single crystals in 1 M NaOH solution, the flatband potential reported by Sanchez *et al.* was around -0.75 V versus SCE at different frequencies between 10 and 50 kHz [25]. The Mott–Schottky plot for Nb-doped α-Fe_2O_3 at a frequency of 30 kHz in 1 M sodium acetate solution resulted in almost the same flatband potential, both in the dark and under illumination.

Aroutiounian *et al.* studied the capacitance–voltage relationship at 100 Hz frequency for Nb- and Sn-doped α-Fe_2O_3 at different doping concentrations in 1 M NaOH solution prepared by the ceramic method [64,65]. In Nb-doped samples, the flatband potential reached a stable level of -0.94 V at Nb concentration of around 10–15 at%, which is 0.10 eV below the value required for the hydrogen evolution reaction at pH 14. However, in Sn-doped α-Fe_2O_3 samples, a maximum flatband potential of -0.90 V was obtained at 1.0–1.5 at% doping. The concentrations of "shallow"- and "deep"-donor centers in the semiconductor were calculated for Nb- and Sn-doped α-Fe_2O_3. A slower increase of the total donor density (shallow- and deep-donor density) from 0.5×10^{21} to $1.1 \times 10^{21}\,cm^{-3}$ was seen below 10 at% Nb, whereas, at 10–15 at% Nb, a sharp increase in the total donor density from 1.1×10^{21} to $3.9 \times 10^{21}\,cm^{-3}$ was obtained. On the other hand, the relationship between total donor density and Sn concentration was not simple. Deep-donor density showed a sharp increase at 0.5–1 at% Sn-doping, followed by a stable level from 1.5–2 at%. At 1.0 at% Sn doping, a maximum shallow-donor density of $4.2 \times 10^{20}\,cm^{-3}$ and a maximum total donor density of $1.0 \times 10^{21}\,cm^{-3}$ was obtained. Sn-doped α-Fe_2O_3 electrodes with 1.0 at% exhibited the best photoresponse, since they had a high donor density and a fairly reasonable flatband energy level. While comparing these two dopants, Nb offered a better photoresponse compared to Sn, due to the larger flatband potential and donor density. Aroutiounian *et al.* reported a concentration of charge carrier in a 0.5 at% Ti doped α-Fe_2O_3 of $\sim 6.82 \times 10^{18}\,cm^{-3}$. Such comparatively low donor density is connected with the formation of the located pairs Fe^{2+}-Ti^{4+}, representing donor centers in $Fe_{1.99}Ti_{0.01}O_3$ [38]. Flatband potentials obtained for 1.0 at% Ti-, Sn- and Zr- doped α-Fe_2O_3 electrodes at 1 kHz frequency in 14 pH solution were -0.79 V, -0.75 V and -0.73 V respectively [66]. Charge-carrier concentrations calculated for 1.0 at% Ti-, Sn- and Zr-doped α-Fe_2O_3 were 1.8×10^{19}, 2.9×10^{18} and $5.6 \times 10^{18}\,cm^{-3}$, respectively, at 1kHz.

Similar methods are being used in nanocrystalline α-Fe_2O_3 thin films for the determination of flatband potential and carrier density. Mott–Schottky plots for single crystals of α-Fe_2O_3 obtained by Björkstén *et al.* showed frequency dispersion and bad linearity [49]. However, flatband potential measured with this method was estimated to be -650 mV at pH 12 (10 mM

NaOH), and no significant shift was detected after adding 0.1 M LiI to the electrolyte. Spectroscopic measurement of a α-Fe_2O_3 single-crystal electrode gave an estimated flatband potential of −550 mV versus SCE at pH 10.7 and +50 mV at pH 2.7.

Lindgren *et al.* utilized the values of photocurrent onset for the measurement of flatband potential for nanorods of α-Fe_2O_3 and observed it to be located at around −0.5 V (−0.7 V versus NHE) in 0.1 M NaOH [50]. The onset for photocurrents shifted to more negative values by ~65 mV per pH over the pH range 3 to 13. This behavior was attributed to the acid–base couple action of the α-Fe_2O_3 surface, changing the flatband potential and consequently the position of the energy bands.

The flatband potentials obtained by Miyake and Kozuka, in 2005, for Fe_2O_3–Nb_2O_5 films did not show any significant change [67] on varying the doping concentrations (Nb 0, Nb 0.25, Nb 0.5, Nb 0.75 and Nb 1.0). When the Nb/(Fe + Nb) ratio increased from 0 to 0.25, the donor density increased by 2–3 orders of magnitude, due to the formation of $FeNbO_4$, which has a high donor density. The increase in donor density reduces the resistivity of the α-Fe_2O_3 film, resulting in an increase in the IPCE up to 18% at 325 nm. Glasscock *et al.*, in 2007, reported the flatband potential for undoped, Si- and Ti-doped α-Fe_2O_3 as −0.84, −0.69 and −0.99 V versus SCE, and the donor density as 1.3×10^{18}, 7.6×10^{18} and 3.3×10^{20} cm^{-3}, respectively [57]. The concentration of charge carriers in the Ti-doped α-Fe_2O_3 was over one order of magnitude larger than that of the Si-doped material.

Silicon doping from 0.01 to 10% in α-Fe_2O_3 was found to tune donor density from 10^{17} to 10^{20} cm^{-3} [44], as observed by Liang *et al.* in 2008. For undoped α-Fe_2O_3, donor density was found to be 1.2×10^{17} cm^{-3} at 30 kHz in 1 M KOH solution, whereas after introducing 0.9% Si, a donor density of 1×10^{20} cm^{-3} at 1 kHz was calculated. A maximum donor density of 4×10^{20} cm^{-3} was expected, if all the Si atoms present in the precursor solution were incorporated into the film and acted as ionized donors. A flatband potential of −0.84 V versus Ag/AgCl was measured for undoped iron oxide. In the frequency range 1–30 kHz, the flatband potential was observed to vary between −0.67 and −0.96 V versus Ag/AgCl.

Khan and Akikusa calculated a flatband potential of −0.74 V versus SCE for frequencies between 100 Hz and 1 kHz in a 1.0 M NaOH (pH 14) electrolyte solution for spray-pyrolytically deposited thin films of α-Fe_2O_3 [35]. A parallel shift of the plot towards a positive potential was observed without any change in the slope, when the pH was changed from 14.0 (1.0 M NaOH) to 0.0 (0.5 M H_2SO_4). The intercept of the lines in 0.5 M H_2SO_4 and 1 M NaOH was at 0.18 and −0.70 V versus SCE, respectively. The ac frequency dependence of the apparent donor density of n-Fe_2O_3 was relatively small, giving donor densities of 2.2×10^{20}, 2.7×10^{20} and 3.6×10^{20} cm^{-3} at frequencies 1000, 500 and 100 Hz, respectively. Ingler *et al.* reported a flatband potential of 0.0 V versus SCE and an acceptor density of 4.4×10^{18} cm^{-3} at an ac frequency of 2500 Hz in 0.5 M H_2SO_4 electrolyte solution for p-type Zn-doped iron-oxide thin films [36]. In another study on Cu-doped (0.01265 M) p-Fe_2O_3, a flatband potential of +0.08 V versus SCE at 250 and 791 Hz in 0.01 M H_2SO_4 solution was calculated [37]. They found the acceptor density to be 4.41×10^{17} and 5.57×10^{17} cm^{-3} at frequencies of 791 and 250 Hz, respectively. The acceptor densities obtained for Cu- [37], Zn- [36] and Mg-doped [6] α-Fe_2O_3 are 2–3 orders of magnitude lower than for undoped n-Fe_2O_3 [35]. These lower acceptor densities may be responsible for the lower photocurrent density obtained for p-type α-Fe_2O_3 samples. According to Khan *et al.*, of these dopants, Zn was identified as the best, as it showed improved system efficiency compared to Mg and Cu. This is due the higher acceptor density obtained for Zn-doped α-Fe_2O_3, which helps to reduce the resistivity of the film [36].

The effect of Zn-doping concentration on the flatband potential and the donor density at the α-Fe_2O_3/electrolyte interface was also studied by the authors' group, at 1 kHz in 13 pH NaOH solution [41]. It was observed that a level of doping up to 5.0 at% increases the flatband potential towards the negative from −0.58 to −0.78 V versus SCE. On the other hand, the donor density was found to decrease from 21.4×10^{19} to $11.0 \times 10^{19}\,cm^{-3}$ with increasing doping concentration.

Teng and Li measured a flatband potential of −0.58 V (versus Ag/AgCl) and a donor density of $1.02 \times 10^{22}\,cm^{-3}$ from the intercept on the x-axis and the slope of the Mott–Schottky plot, respectively, for amorphous iron(III) oxide films on a conducting glass substrate [60]. The positive value of the slope reflects the n-type semiconductor behavior of the deposited iron(III) oxide film. Cyclic voltammetry and ac impedance methods were employed to determine the band-edge levels of the deposited films. The valence-band edge of the film was located at a level of + 1.6 V versus Ag/AgCl at pH 7, which was positive enough to oxidize water into O_2 under illumination.

13.4 Strategies to Enhance Photoresponse

It is well known that the PEC response of a semiconductor is altered by the doping, the choice of electrolyte, the direction of illuminating the photoelectrode (front or back) and by many other methods. Some important factors that are being observed to affect the photoresponse of α-Fe_2O_3 are discussed in this section.

13.4.1 Doping

Impurity doping of α-Fe_2O_3 has been extensively investigated towards developing photoelectrodes with desired characteristics. Pure α-Fe_2O_3 is a highly resistive semiconductor, with resistivities of $\sim 10^{12}\,\Omega\,cm$ at room temperature. Impurity doping is largely intended to improve electrical conduction, by raising the number of charge carriers and/or by increasing the carrier mobility. Doping by higher- or lower-valent ions is known to produce mixed valence states or structure-directing agents, making it more suitable for PEC studies.

In recent times, most of the research work has been focused on the study of doped/undoped α-Fe_2O_3 in the form of nanostructured thin films, but to understand the role of dopants in improving the PEC response, α-Fe_2O_3 is still utilized in powder/pellet form prepared by conventional ceramic methods, as this is simple enough to incorporate the desired dopant.

13.4.1.1 Doping in Bulk

The effects of various dopants, such as Ti, Si, Sn, Ta, Ge, Pb, Zr, Hf, Ce, V, Nb, W and Al [11,17,19–21,27,68–71] in modifying the PEC response of α-Fe_2O_3 in PEC cells have been investigated since the 1970s. Tetravalent ions, like Ti^{4+}, Pt^{4+} and Si^{4+} act as donors in α-Fe_2O_3, while pentavalent ions like Nb^{5+} and Ta^{5+} act as double donors. On the other hand, Zn^{2+}, Cu^{2+}, Mg^{2+}, Li^{+} and Ni^{+} introduce positive charge carriers and Cr^{3+}, Al^{3+}, In^{3+} are found to act as electron traps.

Aroutiounian *et al.* extensively studied the performance of hematite in PEC cells, with variety of dopants at varying doping concentrations prepared by conventional ceramic techniques [64,65,72]. The introduction of pentavalent (Ta^{5+}, Nb^{5+}) or quadravalent (Sn^{4+}) ions resulted in the transformation of α-Fe_2O_3 into an n-type semiconductor, which is connected to the conversion of some of the trivalent Fe^{3+} ions to bivalent Fe^{2+}, due to the energy levels of the Ta^{5+}, Nb^{5+} and Sn^{4+} ions being located above the level of bivalent Fe^{2+} ions, which are not occupied in pure α-Fe_2O_3. The substitution takes place according to the formulae $Fe^{3+}_{2-3x}Ta^{5+}_{x}Fe^{2+}_{2x}O^{2-}_{3}$, $Fe^{3+}_{2-3x}Nb^{5+}_{x}Fe^{2+}_{2x}O^{2-}_{3}$ and $Fe^{3+}_{2-2x}Sn^{4+}_{x}Fe^{2+}_{x}O^{2-}_{3}$ in case of Ta-, Nb- and Sn-doped α-Fe_2O_3, respectively. The photoelectrolysis current for Ta^{5+}-doped α-Fe_2O_3 [72] at varying doping concentrations from 0.1 to 2.0 at% showed that photoelectrodes containing 0.5 at% Ta, were the most active and exhibited photocurrents of ~0.64 mA cm^{-2} at 0.7 V versus SCE. The specific resistance remained high at $\sim 8.4 \times 10^7\ \Omega$ cm for a 0.1 at% Ta-doping level. However, a drastic decrease in specific resistance up to 52 Ω cm was observed at a Ta-doping level of 1.5 at%. Specific resistance was observed to increase again at a Ta-doping level of 2.0 at%. Sn^{4+} doping in iron oxide [65] exhibited insignificant improvements in the PEC response, which was related to the recombination of photogenerated charge carriers. IPCE data indicated that α-Fe_2O_3 at lower doping levels (0.75 at%) has a better efficiency than higher Sn-doping concentrations. This was attributed to the reduction in the absorption of visible light, due to an increase in the bandgap with increasing Sn-doping level. The photocurrent onset with 0.5 at% Sn was −0.25 V, which shifted anodically with increases in doping concentration and became almost constant at −0.14 V. The specific resistance for 0.1 at% Sn-doped samples remained high ($\sim 10^{10}\ \Omega$ cm), with the energy of activation for conductivity at −0.7 eV, and reduced down to $\sim 10^2\ \Omega$ cm at 1.0 at%. Above this concentration, the specific resistance was found to increase slightly. A similar trend in the variation of specific resistance was also shown by Nb^{5+} doping in α-Fe_2O_3 [64]. It was suggested that the decrease in specific resistance with increasing amount of dopant was due to an increase in the number of ions contributing to the electrical conductivity, whereas at higher concentration of impurities, the increase in resistance was due to an increase in the carriers scattering from the impurity centers, and shedding of the second high-resistance phase. Semiconductor photoelectrodes made from solid solutions in the system Fe_2O_3–Nb_2O_5 [64] resulted in the compositions $Fe_{1.9}Nb_{0.1}O_3$ (Fe_2O_3 + 5 at% Nb), $Fe_{1.8}Nb_{0.2}O_3$ (Fe_2O_3 + 10 at% Nb), $Fe_{1.7}Nb_{0.3}O_3$ (Fe_2O_3 + 15 at% Nb) and $FeNbO_4$ (Fe_2O_3 50 at% Nb). The IPCE dependence on the Nb concentration had much greater influence at lower wavelengths than at higher wavelengths and increased with increasing Nb doping up to 10 at%, with a maximum value of ~24% at a wavelength of 450 nm at 0.5 V. A solid solution of Nb_2O_5 and Fe_2O_3 was expected to combine the best characteristics of each material set, providing spectral photosensitivity covering a large part of the visible range of the solar spectrum. In the Fe_2O_3–Nb_2O_5 solid-solution system, the top filled bands corresponding to the valence band are the filled *d* levels of FeO_6 cells, which are located more at negative energies than the 2*p* levels of O^{2-}; the empty bands are both empty *d* levels of FeO_6 cells and empty *d* levels of NbO_6 cells [64]. In these samples, larger absorption of light, compared to single material is possible through the optical transitions between the *d* levels of iron ions (~2.2 eV), as well as between *d* levels of ions of iron and niobium (~1.3 eV).

13.4.1.2 Doping in Nanostructures

Even working with nanostructures, doping has been found to be useful in improving the PEC properties of α-Fe_2O_3. With better methods available for preparation and characterization, the role of dopants is now better understood.

The effect of a wide variety of dopants on the PEC response of α-Fe_2O_3 prepared by spray pyrolysis was studied by Sartoretti *et al.* [42,43]. Multiple doping in α-Fe_2O_3 has also been studied and found to be effective in improving its photoresponse. Doping with a 5% of Ti(IV) improved photocurrents markedly from 0.78 mA cm^{-2} to 4.05 mA cm^{-2} at 0.45 V versus NHE. The increase in photocurrent was attributed to the increased conductivity of the films and the stabilization of oxygen vacancies by Ti^{4+} cations. Simultaneous doping of 1% Al and 5% Ti was found to improve the performance of the photoanodes, increasing photocurrents up to 4.38 mA cm^{-2}. These samples gave a maximum IPCE value of ~25% at a wavelength of 400 nm and a electrode potential of 0.7 V versus NHE in 0.1 M NaOH solution. These films had low-porosity surfaces containing predominately hematite, with traces of substoichiometric magnetite [43]. Adding Li impurities from 0.1 to 1.0% with 5% titanium showed comparatively low photocurrents. No significant improvement in photoresponse was obtained on doping with Cr^{3+} or In^{3+} along with 5% Ti. Photoelectrodes doped with 4% Zn^{2+} along with 5% Ti^{4+} showed a cathodic shift in the photocurrent onset potential of ~0.22 V, and a steeper rise in the photocurrent with applied voltage. Doping with Zn^{2+} at the highest doping level in iron oxide exhibited only anodic photocurrents consistent with n-type behavior. The dopants Ni, Li, Pt, Al and Cr alone had a detrimental effect on the PEC response. A significant decrease in the resistance of α-Fe_2O_3 prepared by the sol-gel route has been reported by Agrawal *et al.* on doping with Ti and Zr [51,73]. An increase in photocurrent density of 0.9 mA cm^{-2} at 0.6 V versus SCE was obtained by Glasscock *et al.* [57] on Ti doping, which was larger than the ~0.19 mA cm^{-2} obtained with Si doping. The enhancement in photocurrent with doping was attributed to improvement in the transfer-rate coefficient at the surface and passivation of the grain boundaries by the dopants. Low photocurrent density in Si-doped α-Fe_2O_3 was attributed to smaller grain size, which had a high level of surface states, enhancing recombination at the surface and at grain boundaries.

Si doping in nanocrystalline α-Fe_2O_3 prepared by APCVD [39,47] was found to modify the morphology of the films. These samples showed a dendritic microstructure (Figure 13.7c and d) about 170 nm in height with 20–30 nm thick branches [47], which could minimize the distance photogenerated holes had to diffuse to reach the α-Fe_2O_3/electrolyte interface, while still allowing light absorption, resulting in a significantly improved photoresponse. The dendritic structure shown by 1.5 at% Si-doped samples exhibited a photocurrent of ~2.2 mA cm^{-2} in 1 M NaOH solution [39], which is much larger than the photocurrent density exhibited by undoped APCVD samples of ~1.45 mA cm^{-2} at 1.23 V versus RHE. USP-deposited films consisted of stacked sheets (Figure 13.7a and b) oriented perpendicularly to the substrate, and exhibited a photocurrent density of 1.17 mA cm^{-2} under the same conditions [[47]]. Suoza *et al.* also found that Si introduction affects the grain growth, leading to a mesoporous structure in spin-coated sol-gel-deposited α-Fe_2O_3 films [53]. A preferential growth in the axial (110) plane with increase in Si concentration was predicted. The crystallographic orientation was found to be the dominant factor affecting the photocurrent. In another study by Liang *et al.*, a photocurrent of 0.37 mA cm^{-2} at 1.23 V versus RHE in a 1.0 M KOH solution was reported for spray-pyrolytically deposited α-Fe_2O_3 films doped with 0.2% Si under AM 1.5 illumination [44].

Electrodeposited iron oxide prepared by Shwarsctein *et al.* exhibited improved photoactivity after co-deposition of Mo or Cr [62]. 5% Cr- and 15% Mo-doped samples were the best performing, showing IPCEs at 400 nm of 6% and 12%, respectively, with an applied potential of 0.4 V versus Ag/AgCl. These IPCE values were 2.2 and four times higher for the 5% Cr and 15% Mo samples, respectively, than the undoped sample. In contrast to the detrimental effect of

Pt doping in α-Fe_2O_3 on the PEC response reported by Sartoretti *et al.* in 2005 [43], recently, Hu *et al.* obtained an enhanced photoreponse for Pt-doped α-Fe_2O_3 prepared by electrodeposition [63]. A maximum photocurrent density of 1.43 mA cm^{-2} was obtained at 5% Pt doping levels [63]. They reported that Pt in α-Fe_2O_3 may act in several ways. Firstly, Pt acts as an electron donor, due to the substitution of Fe^{3+} by Pt^{4+} in the α-Fe_2O_3 lattice. The increased donor concentration in n-type doping would translate into an improvement in the conductivity. Secondly, the increased donor concentration would increase the electric field across the space-charge layer, resulting in higher charge-separation efficiency. However, increasing the donor concentration would reduce the width of the space-charge layer and the Pt^{4+} would have defect-scattering/recombination properties that, at high concentrations, would negate the increased separation efficiency. The apparent optimum at 5% Pt-doping levels may balance these competing effects most effectively and yield the best PEC performance. Third, the more compact doped films may have greater electrical interconnectivity, which would facilitate charge transfer.

The addition of trivalent dopants like Al^{3+}, Cr^{3+} or In^{3+} forms oxides that are isostructural with α-Fe_2O_3 and are thereby expected to influence the crystallinity of the α-Fe_2O_3 films in forming the α-phase. Al(III)- and Cr(III)-doping in α-Fe_2O_3, as observed by Sartoretti, exhibited poor photoresponse [43]. In a study by Schumacher *et al.*, incorporation of indium into 25 nm thick iron-oxide films was shown to produce a small increase in quantum efficiency, while no change in flatband potential and bandgap was obtained [54]. Effects were explained in terms of a localized states model for α-Fe_2O_3 and it was concluded that most of the impurity is taken up in octahedral lattice sites, substituting for iron.

Interesting results have been reported by Miyake and Kozuka [67], when Fe_2O_3–Nb_2O_5 films of various Nb/(Fe + Nb) mole ratios were prepared by the sol-gel method. These films exhibited much better IPCE values compared to the pellets studied by Aroutiounian *et al.* in 2006 [64]. When the Nb/(Fe + Nb) ratio increased from 0 to 0.25, the photocurrent increased to ~1.75 mA cm^{-2} at 1.0 V versus SCE and the IPCE rose to a value of 18% at 325 nm wavelength. These samples exhibited better absorption in the visible range and an increase in donor density, due to formation of FeNbO4 phase. For ratio Nb/(Fe + Nb) = 0, the onset of absorption around 580 nm (2.1 eV) and the broad peak around 400 nm (3.1 eV) were attributed to the indirect transition and the direct transition, corresponding to charge transfer from the O 2*p* valence band to the Fe 3*d* conduction band, respectively. For Nb(Fe + Nb) = 0.25, the onset of the absorption was seen around 580 nm (2.1 eV) and 470 nm (2.6 eV). The absorption onset around 470 nm was attributed to the indirect transition in $FeNbO_4$.

13.4.1.3 Bivalent Metal-Ion Doping: p-Type α-Fe_2O_3

The introduction of bivalent metal ions, for example, Mg^{2+}, Cu^{2+}, Ca^{2+} and Zn^{2+} to α-Fe_2O_3 lattice as a substitute for Fe^{3+} ions is expected to increase the concentration of electron vacancies, resulting in p-type behavior. Hydrogen is generated at p-type electrodes in the PEC cell by the reduction of H^+ to hydrogen. In a PEC cell with a p-type electrode,oxygen is released at the Pt counter-electrode and, therefore, corrosion of photoelectrode is very small, as released oxygen is the main cause of corrosion.

Research on the synthesis of p-type α-Fe_2O_3 started as early as in 1966 [74]. Doping of Mg^{2+} and Zn^{2+} in α-Fe_2O_3 has been found to exhibit some peculiar characteristics related to p-type behavior [6,28]. These electrodes exhibited the phenomenon of transformation from negative to positive photocurrent, when the external potential was varied from negative to positive. Sieber *et al*, prepared Mg-doped α-Fe_2O_3 by the sol-gel, as well as by ceramic methods, and mentioned the formation of $Fe_{3-x}Mg_xO_4$ [26]. Transformation of pure α-Fe_2O_3 corundum phase to spinel Fe_3O_4 occurs at high temperatures, when oxygen is lost from α-Fe_2O_3 and exchange of Fe^{2+} and Mg^{2+} occurs by diffusion. Seebeck measurements indicated that samples were of the n-type, but the PEC measurement in 1 M NaOH electrolyte showed that the sintered disc electrodes exhibited photocathodic currents, characteristic of a p-type electrode. The voltammogram exhibited n-type behavior, whereas hydrogen was evolved at negative bias. Similarly, an Mg-doped, thermally formed iron-oxide electrode in boric acid/borax buffer solution, exhibited photovoltage waveforms that were partly typical of n-type and partly typical of p-type semiconductors at particular external potentials. Recently, Ingler and Khan obtained Mg^{2+}-doped p-Fe_2O_3 thin-film photoelectrodes on doping from 10 to 12.5% with respect to 0.11 M Fe concentration [40]. The photocurrent was found to increase on increasing Mg ions from 10 to 12% and a decrease was observed when concentration was increased to 12.5%. The highest photocurrent density of 0.22 mA cm^{-2} at 0.2 V versus SCE and 0.33% photoconversion efficiency, with a total conversion efficiency of 0.99%, was observed for Mg-doped p-Fe_2O_3 thin films under optimized conditions.

In a study by Cai *et al.*, Zn-doped polycrystalline iron-oxide pellets showed n-type semiconducting behavior at positive potentials more than 0.20 V, p-type behavior at negative potentials more than −0.25 V and both n-type and p-type behavior at intermediate potentials [75,76]. It was also observed that with shorter-wavelength (400 nm) incident light, a p-type photoresponse existed, while with longer wavelengths (620 nm), an n-type photoresponse was observed. Cai *et al.* explained the primary mechanism for these types of phenomena occurring in Zn-doped iron oxide [77]. SEM microregion analysis showed the presence of two kinds of crystal particles: one had a smooth surface and other was crude. The crude surface had Zn-poor crystals and the smooth surface had Zn-rich crystals. A large amount of Zn in α-Fe_2O_3 resulted in the formation of $ZnFe_2O_4$, thereby exhibiting n-type behavior. The Zn-poor crystal is some sort of solid solution of ZnO and α-Fe_2O_3 and, according to the doping principle, such material should be p-type semiconductor. When light is incident, the oxidation reaction takes place in the n-type region, producing an n-type photoresponse, while reduction takes place in the p-type region, producing a p-type photoresponse. Hence, both photochemistry and SEM microregion analysis indicate the co-existence of n-type and p-type semiconductive regions in the same electrode and the overall effect of the electrode was mainly due to competition between the n-type and p-type photoresponses.

Thin films of spray-pyrolytically deposited Zn-doped iron oxide, prepared by Ingler *et al.*, showed an increase in the photocurrent density with increasing cathodic bias, indicating p-type conductivity [36]. It was observed that the photocurrent density increased up to 0.0088 M Zn^{2+} concentration and then there was a slow, but steady, decline in photocurrent density from 0.0088 to 0.011 M Zn^{2+} doping. The highest current density of 1.1 mA cm^{-2} at −0.8 V versus Pt was obtained for optimized Zn (0.0088 M) in α-Fe_2O_3 and showed a maximum quantum efficiency of 21.1% at 325 nm, with a threshold observed at 590 nm. Formation of α-Fe_2O_3 with a $ZnFe_2O_4$ phase on Zn doping was found to show improved conductive properties when introduced in the optimum ratio [78,79]. Zn also helped in converting the indirect bandgap of α-Fe_2O_3 to the direct bandgap due to formation of $ZnFe_2O_4$.

Spray-pyrolytically deposited Zn-doped α-Fe_2O_3 was also prepared by the authors' group, which exhibited an n-type nature in pH 13 electrolyte over a large range of doping concentrations [41]. This result contradicts the p-type behavior reported by Ingler *et al.*, although the method of preparation was similar [36,37]. This discrepancy may be explained by the nature of the electrolyte used. The value of resistivity [41] was found to increase from 1.7×10^6, for undoped, to $6.8 \times 10^7\ \Omega$ cm at a doping concentration of 1.5 at%. This increase in resistivity by one order of magnitude was attributed to the substitution of Zn^{2+} ions in place of Fe^{3+} ions, thereby resulting in the loss of an electron. By increasing the doping concentration beyond 1.5 at%, the resistivity decreased. The photocurrent density increased with doping concentration upto 5.0 at% and was maximum ~0.64 mA cm^{-2} at 0.7 V versus SCE for 5.0 at% doping; further increase in doping concentration to 10.0 at% decreased the photocurrent density. Sartoretti *et al.*, in 2005, doped α-Fe_2O_3 by 5% Ti^{4+} and 8% Zn^{2+} (the same as reported by Ingler *et al.* [36] for p-type iron oxide) and noted that even at the highest Zn^{2+}-doping level, iron-oxide films exhibited only anodic photocurrents in 0.1 M NaOH solution, consistent with n-type behavior [43].

Copper-doped α-Fe_2O_3, as observed by Ingler and Khan, also exhibited p-type behavior [37]. Varying the amount of Cu from 0.011 to 0.0132 M affected the photocurrent, which increased with increasing doping concentration up to 0.01155 M, after which a decrease in the photocurrent was observed. The maximum photocurrent density of ~0.94 mA cm^{-2} at 0.0 V versus SCE was obtained at a 0.01155 M Cu^{2+}-doping level. These Cu-doped p-Fe_2O_3 thin films gave an improved photoconversion efficiency of 1.3%, with a total conversion efficiency of 2.9%, which is more than ten times that reported for Mg doping [27,28,40]. The result was on the same lines as that reported by Mohanty and Ghose in 1992 [80].

It seems that the improvement in the photoresponse of iron oxide by doping with bivalent ions offers very interesting results, establishing many new phenomena, although the observed photoresponse is still much below the expectation and the final word is yet to be said on this issue.

13.4.2 Choice of Electrolytes

The electrolyte composition and its pH play an important role in affecting the PEC response, onset potential and flatband potential of the semiconductor electrode in PEC cells. The rate of transfer of carriers at the junction is enhanced by proper band-edge matching of the semiconductor with the redox level of water. However, the positions of the conduction and valence bands of semiconductors are dependent on pH. An increase in pH results in a shift in the conduction- and valence-band positions to more cathodic potentials by 59 mV per unit pH at 25 °C (Nernstian behavior). In PEC studies on the solar splitting of water, an aqueous solution of either acid or alkali has been employed as the electrolyte. Most of the studies on iron oxide have been conducted in pH 13 NaOH solution. However, in some studies, other combinations of electrolyte have also been reported. 0.1 M KI and 0.01 M K_2HPO_4/KH_2PO_4 buffer (pH 6.8) solutions in water were employed as electrolyte by Beerman *et al.* to study the PEC response of oriented nanorod thin films of α-Fe_2O_3 [5]. When 0.1 M KI(aq) was used as the electrolyte, an increase in photoconversion efficiency (IPCE) with pH at all wavelengths was noted (Figure 13.9). Measurements showed that the IPCE (SE illumination) at 360 nm reaches 9 and 18% at pH 6.8 and 12.0, respectively. An increase in IPCE with increase in pH was also shown, when ethylene carbonate/propylene carbonate (50 : 50 by weight) with 0.5 M LiI and 0.5 mM I_2 was used as electrolyte [5]. These observations were explained by considering the surface charge of the semiconductor/electrolyte interface, where OH^- groups preferentially

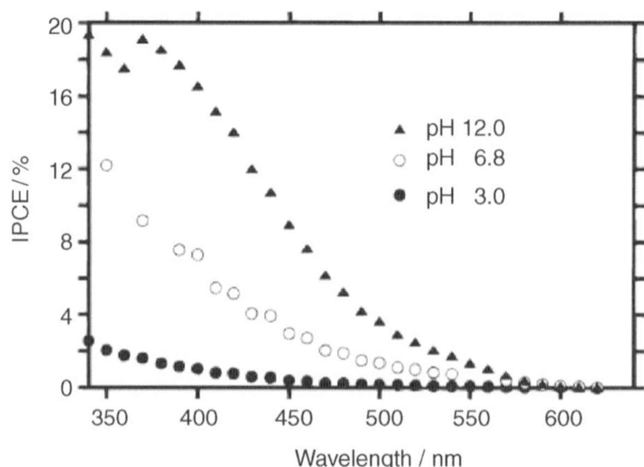

Figure 13.9 Action spectrum of a 0.6 μm thick α-Fe_2O_3 electrode, consisting of nanorods oriented in a perpendicular direction with the conducting substrate in electrolytes of different pH for SE illumination. (Reproduced with permission from N. Beermann, L. Vayssieres, S.-E. Lindquist and A. Haqfeldt, Photoelectrochemical studies of oriented nanorod thin films of hematite, *Journal of the Electrochemical Society*, **147**, 2456–2461, 2000. © 2000 Electrochemical Society.)

trap holes to form OH radicals, allowing electrons to diffuse away from the particle to the back contact. If there is efficient charge transfer from these oxidized states to the mediating redox couple in the electrolyte compared to capturing of electrons from the α-Fe_2O_3 conduction band, the surface state works in favor of a high quantum yield. The probability for a hole to be transferred into the electrolyte is controlled by its pH. Better overlap between the valence band and energy levels of the reduced form in the electrolyte, at high pH, increases the photoconversion efficiency [5]. Khan *et al.* used 0.2 M NaOH containing 0.5 M Na_2SO_4 (pH 13) as the electrolyte to study the PEC response of undoped and iodine-doped thin films of α-Fe_2O_3 because, in this electrolyte, iodine-doped films were found to be highly stable [33,34]. They also used 0.5 M H_2SO_4 as the electrolyte to study the PEC response of p-type Fe_2O_3 electrodes prepared by doping of Mg, Zn and Cu in α-Fe_2O_3 [36,37,40].

Cai *et al.* used 0.5 M Na_2SO_4 solution as the electrolyte in a boric acid and sodium borate buffer solution of pH 8.4 [75,76]. Watanabe and Kozuka [81], and Miyake and Kozuka [67] utilized an aqueous solution of 0.2 M Na_2B_2O4, 0.2 M $Na_2B_4O_7$, 0.14 M H_2SO_4 and 0.3 M Na_2SO_4 as the supporting electrolyte. Prakasam *et al.* studied the PEC properties of potentiostatically anodized nanoporous iron(III) oxide in an electrolyte containing 0.5 M H_2O_2 (50%) and 1 M NaOH (50%) and noted that the addition of H_2O_2 to the electrolyte greatly enhance the PEC response [58]. Luo *et al.*, in a study on WO_3/Fe_2O_3 nanoelectrodes, used a 0.2 mol l^{-1} Na_2SO_4 aqueous solution (pH ~7.5) as the electrolyte [82].

13.4.3 Dye Sensitizers

Apparently, not much work has been reported on the use of sensitizers/catalysts with thin films of nanostructured α-Fe_2O_3. Beerman *et al.* reported the use of dye sensitization in iron-oxide thin films consisting of nanorods [5]. Dye sensitization was carried out by soaking the film for three days in 0.5 mM solution of cis-di(thiocyanato)-bis(2,2′-bipyridyl-4,4′-dicarboxylate) ruthenium(II) (N3 dye) in ethanol. PEC studies were carried out with 0.1 M KI and 0.01 M K_2HPO_4/KH_2PO_4 buffers in a three-electrode setup, while ethylene carbonate/propylene carbonate with 0.5 M LiI and 0.5 mM I_2 was used as the electrolyte in the two-electrode setup. Use of ruthenium-based bipyridyl dye led to improvement in the photoconversion efficiency by red-shifting the photocurrent onset to 750 nm, with a maximum at 600 nm. The bare electrode exhibited higher efficiencies at shorter wavelengths compared to sensitized electrodes.

Gurunathan and Maruthamuthu, in 1995 [83], while studying the effect of loading of RuO_2, Rh and Cu on α-Fe_2O_3 in PEC splitting of water, found enhanced efficiency on addition of methyl viologen (MV^{2+}). The role of MV^{2+} was to expedite water reduction by undergoing one-electron reduction with e^-_{cb}, to form MV^+, which in turn reacts with H^+ to evolve hydrogen, returning to the dipositive state.

13.4.4 *Porosity*

The photocurrents of a semiconductor electrode in PEC cell have also been found to be affected by the porosity of the film [82]. The semiconductor electrode/electrolyte junction is the site of electrochemical reactions responsible for PEC splitting of water. The overall surface area of contact between electrodes and electrolytes depends upon the porosity of the semiconductor sample. Greater porosity is needed to raise the contact area, which in turn is expected to improve the PEC response of the material, but electrodes with very high porosity are also known to be highly resistive. Beermann *et al.* studied the PEC behavior of oriented nanorod films of α-Fe_2O_3 [5]. The procedure allowed them to control the porosity, which was reported to be $50 \pm 5\%$, independent of film thickness. It was reported that parallel orientation had a better effect on the PEC properties compared to perpendicular. Prakasam *et al.* reported the PEC properties of nanoporous α-Fe_2O_3 by potentiometric anodization [58]. They observed that anodization conditions and electrolytic composition determine the pore diameter, which ranged between 50–250 nm and 300–600 nm in length, with a pore depth of approx. 500 nm. Crystallized nanoporous films having a bandgap of 2.2 eV resulted in a net photocurrent density of 0.51 mA cm^{-2} in 0.5 M H_2O_2 + 1 M NaOH at 0.6 V versus Ag/AgCl.

13.4.5 *Forward/Backward Illumination*

Illuminating semiconductors from the back side (substrate/electrolyte interface, SE) or from the front side (electrode/electrolyte interface, EE) has also been reported to affect the PEC response of α-Fe_2O_3 thin films. Equal amounts of photocurrent density with front and back illumination have been reported by Khan and Zhou for undoped/iodine-doped iron-oxide thin-film electrodes prepared by spray pyrolysis [33]. Similar results were reported by Majumder and Khan for RuO_2-coated iron-oxide thin films [34]. This type of coating of oxides on α-Fe_2O_3 can be used to achieve stable and efficient photoelectrodes in PEC cell cells, as the front surface of the photoelectrode can be protected from photocorrosion by depositing a thick layer of stable electrocatalyst and illuminating the back surface, kept unexposed to the solution. Reactively sputtered indium-doped iron-oxide films of 25 nm thickness prepared by Schumacher *et al.* displayed double the quantum yield for the back face than for the front face [54]. A decrease in

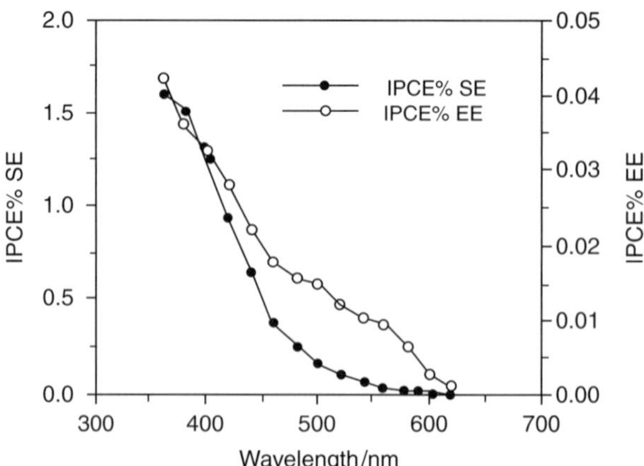

Figure 13.10 Action spectra obtained in 0.1 M NaOH and 0.1 M LiI (EtOH) electrolyte for different illuminations. (Reproduced with permission from U. Bjorksten, J. Moser and M. Gratzel, Photoelectrochemical studies on nanocrystalline hematite films, *Chemistry of Materials*, **6**, 858–863, 1994. © 1994 American Chemical Society.)

the ratio of effective quantum efficiencies between the back face and front face was observed over the wavelength range 420–600 nm, which was attributed to absorption by the ITO.

Bjorksten *et al.* investigated thin films comprising spherical nanoparticles of α-Fe_2O_3 for PEC response in ethanol with 0.1 M LiI at a potential of 0.4 V versus SCE and reported SE efficiency values of 1.7%, 40 times larger than EE illumination at 360 nm, as shown in Figure 13.10 [49]. Higher values of IPCE efficiency for SE illumination (11%) than for EE illumination (5%) at 360 nm were reported by Beermann *et al.* and Lindgren *et al.* in the case of α-Fe_2O_3 thin films comprising nanorods with controlled orientation onto a TCO substrate [5,50]. They found the IPCE to be ~8% by illumination through the substrate, with a wavelength of 350 nm and a light intensity of 0.1 mW cm^{-2}, without any applied voltage, for a rod-like α-Fe_2O_3 morphology, designed with no grain boundaries. They suggested that charge carriers produced close to the back contact are usually collected more efficiently in nanostructured systems, and collection of the photogenerated electrons in case of EE illumination across α-Fe_2O_3 films is poor. The general lowering of the IPCE for EE illumination compared to SE illumination was explained by some losses during transport of electrons from the outer part of the nanorod to the back contact. Therefore, it seems that in nanostructured thin films, back illumination is more effective, compared to front illumination.

On the other hand, Kay *et al.* have reported that nanostructured thin films prepared by APCVD exhibited better IPCE for front illumination, compared to back illumination [39]. They have explained this difference due to the dendritic morphology of the film, which had more compact stem next to the substrate and highly branched nanostructures towards the electrolyte. Since back illumination creates holes mostly in the more compact part of the α-Fe_2O_3 film near the SnO_2 substrate, and due to the short diffusion length, they have less chance of reaching the electrolyte interface and oxidizing water. Only in the long wavelength tail is light equally absorbed throughout the film thickness, where efficiency is independent of the illumination side.

13.4.6 Loading of Metal/Metal Oxide

Surface modification of α-Fe_2O_3 has been reported to be beneficial in improving the PEC response. Metal deposits on the surface of the semiconductor act as a sink for photogenerated carriers and are known to catalyze the production of hydrogen, especially in large-bandgap semiconductors [84]. Some studies on photocatalytic water splitting have been reported in the literature, using α-Fe_2O_3 loaded with metal or metal oxides. Rh and Cu loading on an α-Fe_2O_3 photocatalyst by simple heat treatment in water and further loading of RuO_2 on Rh/Fe_2O_3 and Cu (II)/Fe_2O_3 by simple mechanical mixing were tested by Gurunathan and Maruthamuthu in 1995 [83]. Rhodium doping was found to be more efficient than Cu. The rate of hydrogen production with Rh/Fe_2O_3 was studied and found to increase with doping concentration of Rh, being highest at 0.5 at%. The increase in hydrogen production rate was explained on the basis of deposition of Rh metal on the surface of α-Fe_2O_3 due to a reduction in electron–hole recombination. These photogenerated carriers form an accumulation layer at the metal–semiconductor interface and accelerate reduction of H^+ to H_2. However, at higher concentrations of Rh, the opposite effect was observed. Loading of RuO_2 on Rh/Cu/Fe_2O_3 ions was found to further enhance the efficiency. Gondal *et al.* also investigated the photocatalytic oxidation of water over pure α-Fe_2O_3 in the presence of Fe^{3+}, Ag^+ and Li^+ metal ions as electron acceptors under illumination by a strong laser beam at 355 nm [85]. An increase and decrease in the O_2 and H_2 production was observed in the presence of Fe^{3+} and Ag^+, respectively, revealing the suitability of the reduction potentials of these ions for conduction-band electron-capture processes.

Watanabe and Kozuka studied α-Fe_2O_3 films of thickness 0.04 μm, with Au and Ag particles of 5–20 nm size embedded in α-Fe_2O_3 by the sol-gel method [81]. The anodic photocurrent–potential characteristics and the action spectra of the film were investigated in a three-electrode cell in 0.2 M $Na_2B_2O_4$, 0.14 M H_2SO_4 and 0.3 M Na_2SO_4, (pH 7). Metal particles did not affect the photoanodic properties of the electrode significantly in the presence of a white-light 500 W Xe lamp, but on illuminating with weak white light, the films embedded with metal particles increased the photocurrent. When gold particles were embedded, the photoresponse in the UV region (400 nm) increased by eight times. Deposition of gold particles just on the surface of α-Fe_2O_3 film also increased the photoanodic current. This suggests that Au particles promote catalytic transfer of holes in the conduction band to the electrolyte. The flatband potential was found to be −0.3 V versus SCE for all the samples, and the donor densities were 1.3×10^{19}, 2.1×10^{19} and $1.3 \times 10^{19}\,cm^{-3}$ for films without metal particles, Au embedded particles and Ag embedded particles, respectively.

13.4.7 Layered Structures

In studies on nanostructured thin films, layered structures have also attracted the attention of researchers and are known to offer considerably higher phototoelectrochemical and photocatalytic activity. A combination of small- and large-bandgap material deposited one over the other, as shown in Figure 13.11, may absorb the full solar spectrum more efficiently. If energy-band edges match at the junction, better and efficient separation of photogenerated electrons and holes is possible.

Liou *et al.* prepared an α-Fe_2O_3/TiO_2 heterojunction electrode for photoelectrolysis of water [86] using iron foils, as early as in 1982. The heterojunction electrodes were made by the

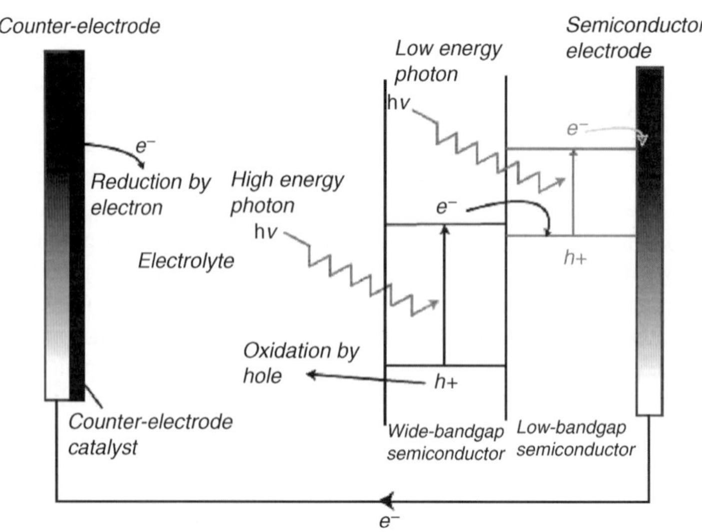

Figure 13.11 Schematic representation of layered structure of low- and high-bandgap material for PEC application.

chemical vapor deposition (CVD) of TiO_2 films on α-Fe_2O_3 substrates. The photocurrent of the α-Fe_2O_3 electrode in the positive-bias region was about half of that of a CVD TiO_2 electrode. They observed that the heterojunction electrode had the same flatband potential as that of CVD TiO_2 on Pt, though the shapes of the *I–V* curves were different. With an increase in bias potential, the photoresponse starts at a longer wavelength, and the peak of the photocurrent also shifts toward a longer wavelength, which strongly indicates a hole contribution from the α-Fe_2O_3 substrate. It appears that the holes generated in the α-Fe_2O_3 substrate do contribute to the photocurrent of the α-Fe_2O_3/TiO_2 electrode. The heterojunction thus exhibited the relatively large flatband potential of TiO_2 and the spectral threshold of α-Fe_2O_3.

An iron-oxide/n-Si heterojunction photoanode (Fe_2O_3/n-Si,) was studied by Osaka *et al.*, in 1985, for a regenerative PEC cell [87]. The electrode was prepared by vacuum evaporation of a 400–500 Å iron film onto a (100) single crystal of 5–7 Ω cm n-Si substrate, followed by vacuum-furnace heating at 400 °C for 1 hr. The addition of palladium (75–100 Å) on the hetrojunction electrode was done by vacuum evaporation, while RuO_2 loading was done by covering the electrode with $RuCl_3$ butanol solution. PEC measurements in this study were carried out photodynamically at 0.1 V s^{-1} in 0.2 M KOH solution containing 0.2 M K_4(Fe $(CN)_6$) and 0.01 M $K_3(Fe(CN)_6)$ in a purified Ar gas atmosphere, with an Hg/HgO electrode as the reference electrode. The open-circuit voltage V_{oc} for the photoanodes Fe_2O_3/n-Si, Pd/ Fe_2O_3/n-Si and RuO_2/Fe_2O_3/n-Si exhibited a constant value of 0.330 V, while the short-circuit photocurrent density I_{sc}, increased with Pd to 7.95 mA cm^{-2} and with RuO_2 to 10.5 mA cm^{-2}. The energy-conversion efficiency of the PEC cell with RuO_2/Fe_2O_3/n-Si anode was 1.6% for illumination by a 55 mW cm^{-2} Xe lamp. The heterojunction electrode showed high stability against photocorrosion. RuO_2 layering on Rh/Co/Fe_2O_3 was also found to increase the photoresponse of spray-pyrolytically deposited α-Fe_2O_3 thin films [34], whereas layering of

PbO_2 and MnO_2 on similar films was observed to decrease the photoresponse, and no such effect was observed in the case of Co_2O_3.

Miller *et al.* designed and developed a multijunction photoelectrode for hydrogen production using an α-Fe_2O_3/electrolyte as the top junction with two underlying amorphous silicon/germanium (a-Si:Ge) solid-state junctions, fabricated onto stainless-steel foil coated with a thin film of nickel–molybdenum hydrogen catalyst on the back surface [55]. The construction of the hybrid cell involved sputter deposition of an α-Fe_2O_3 layer over single-junction a-Si:Ge n-i-p device. The focus of the experiment was on the demonstration of stable operation and bias saving. The hybrid photoelectrode, on testing, showed a 0.6–0.65 V bias saving under AM1.5 illumination in potassium hydroxide.

Recently, Luo *et al.* examined the composite structure of WO_3/Fe_2O_3 by admixing both WO_3 and α-Fe_2O_3 [82]. WO_3 films were coated on an as-grown film of α-Fe_2O_3. XRD analysis confirmed the presence of α-Fe_2O_3 and WO_3 on the films, with no traces of formation of the ternary compound $Fe(WO_4)$. The bandgaps of α-Fe_2O_3 and WO_3, as calculated by transmission spectra analysis, were found to be 1.97 and 2.53 eV, respectively. The photocurrent obtained for the WO_3/Fe_2O_3 heterojunction was found to be 6 $\mu A\,cm^{-2}$ under illumination at 440 nm, which was higher than for both WO_3 and α-Fe_2O_3 films.

The energy-band diagram of α-Fe_2O_3/TiO_2, with respect to the redox level of water at pH 13 is shown in Figure 13.12, which indicates that the structure allows absorption of almost the whole range of the solar spectrum lying in the visible and UV much more efficiently, compared to a single-material electrode. A bilayered photoelectrode α-Fe_2O_3/TiO_2 system incorporating TiO_2 thin films prepared by the sol-gel technique over a spray-pyrolytically deposited thin film of α-Fe_2O_3 has been studied by the authors' group [88]. The photocurrent density exhibited by the various samples prepared was observed to vary with the thickness of the TiO_2 layers, as shown in Figure 13.13. The maximum photocurrent density, $\sim$945 $\mu A\,cm^{-2}$ at 0.6 V versus SCE, was obtained for a bilayered structure of TiO_2 (thickness 0.65 μm, particle size 24 nm) and α-Fe_2O_3 (thickness 0.96 μm, particle size 60 nm). This photoresponse was much larger than those of the single-material electrodes, which were $\sim$12 $\mu A\,cm^{-2}$ and 148 $\mu A\,cm^{-2}$ for α-Fe_2O_3 and TiO_2 thin films, respectively. An increase in the photovoltage was also observed from 0.08 for α-Fe_2O_3 to 0.26 V versus SCE in a sample of the bilayered structure of total

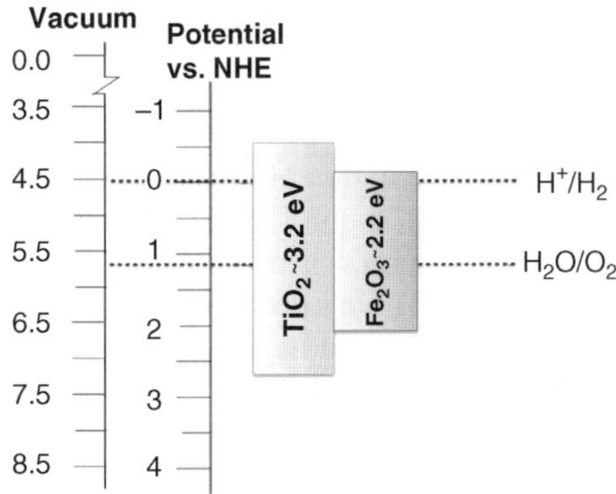

Figure 13.12 Position of energy-band edges of α-Fe_2O_3/TiO_2 at pH 13.

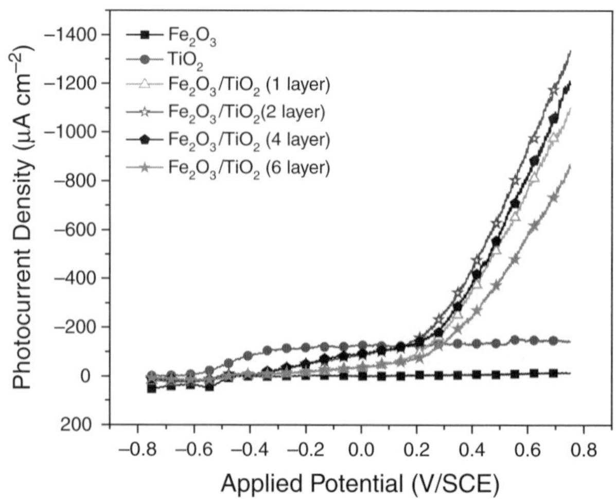

Figure 13.13 Photocurrent-density– potential curve of α-Fe_2O_3/TiO_2 bilayered thin films. (Reproduced with permission from V.R. Satsangi, A.P. Singh, P. Johri *et al.*, Bilayered photoelectrode for solar hydrogen production system, *Proceedings of XXXI National Systems Conference (CD)*, Manipal, Karnatika, India, Dec 14–15, 2007. © 2007 Systems Society of India.)

thickness ~1.68 μm. This sample exhibited a photocurrent of 92 μA cm^{-2}, even at zero bias voltage versus SCE. XRD analysis of the bilayered samples exhibited the formation of $Fe_{0.23}(Fe_{1.95}Ti_{0.42})O_4$ in addition to a hematite (α-Fe_2O_3) phase of iron oxide and an anatase phase of titanium dioxide. The bandgap energy was found to increase from 2.3 eV, for an α-Fe_2O_3 sample, to 2.7 eV, as the thickness of the TiO_2 layer increased from 0.96 μm to 2.3 μm.

13.4.8 Deposition of Zn Islands

A newer technique to modify the surface of α-Fe_2O_3 with Zn dots was attempted by the authors' group in 2007 [89]. These samples exhibited much better PEC responses, compared to doping or layering of the same metal. We modified the surface of 1.5 at% Zn-doped iron oxide (prepared by spray pyrolysis) by depositing thermally evaporated Zn through a mesh of pore diameter ~0.7 mm, using a vacuum-coating unit. The heights of the Zn dots were varied in the range 100 to 260 Å by varying the processing time of thermal evaporation. The schematic diagram explaining the method of preparation of sample is shown in Figure 13.14.

All dotted samples exhibited n-type behavior. The resistivity in Zn-doped samples was found to increase by one order of magnitude [41], whereas the presence of Zn-dotted islands on the surface of Zn-doped α-Fe_2O_3 thin films gave a decrease in resistivity, which could be due to a difference in the work functions of the two materials. A decrease in the resistivity from 2.6×10^7 to 5.7×10^6 Ω cm was observed on increasing the height of the Zn island to 230 Å on the α-Fe_2O_3 surface. However, a slight increase in resistivity up to 9.2×10^6 Ω cm was observed after loading with 260 Å thick Zn-dotted islands. This increase in resistivity at a height of 260 Å was due to interdiffusion of the Zn islands in the iron oxide near the interface, which resulted in the loss of the charge carriers.

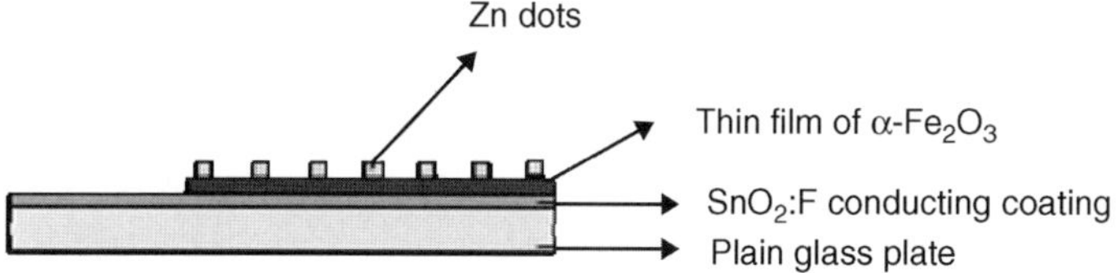

Figure 13.14 Schematic diagram of Zn-dotted islands on an α-Fe_2O_3 thin film. (Reproduced from S. Kumari, A.P. Singh, C. Tripathi *et al.*, Enhanced photoelectrochemical response of Zn-dotted hematite, *International Journal of Photoenergy*, **87467**, 1–6, 2007, Hindawi Publishing Corporation.)

Doping of 1.5 at% Zn in α-Fe_2O_3 thin films prepared by the spray-pyrolysis method enhanced the photocurrent density from 0.061 to 0.321 mA cm^{-2} at 0.7 V versus SCE. Deposition of Zn dots on 1.5 at% Zn-doped α-Fe_2O_3 thin films with increasing height of dots resulted in continuous enhancement in the photocurrent, as shown in Figure 13.15. Structures with ~230 Å thick Zn islands exhibited the maximum photocurrent density of ~1.282 mA cm^{-2}, at an applied potential of 0.7 V versus SCE. The increase in photocurrent density in this study was due to conversion of the Zn dots on the surface of α-Fe_2O_3 into wide-bandgap ZnO, at the time of annealing, which acted as an efficient catalyst for the swift migration of the photogenerated charge carriers. However, a decrease in photocurrent density with α-Fe_2O_3 deposited with thicker dots (260 Å) was correlated with an increase in resistivity and a decrease in the α-Fe_2O_3 surface area, due to interdiffusion of Zn. A minor change in donor density and flatband potential was observed on increasing the height of the dots. Observed values of photocurrent density on the surface of α-Fe_2O_3 and onset potential, along with resistivity, for various modified/unmodified thin films of α-Fe_2O_3 are given in Table 13.2.

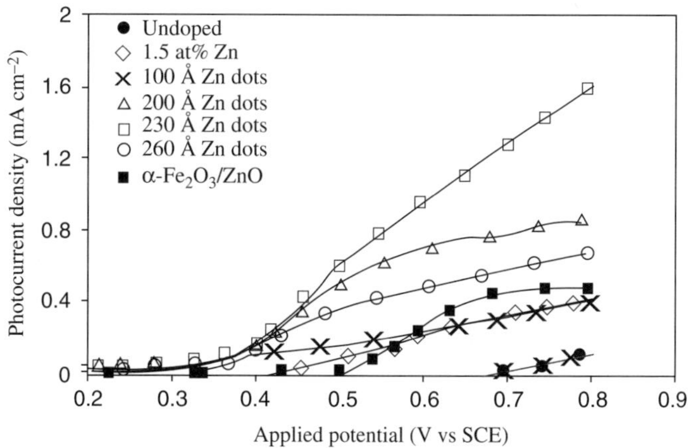

Figure 13.15 Photocurrent-density versus applied potential curves for α-Fe_2O_3 (all samples) in pH 13 NaOH electrolyte solution. (Reproduced from S. Kumari, A.P. Singh, C. Tripathi *et al.*, Enhanced photoelectrochemical response of Zn-dotted hematite, *International Journal of Photoenergy*, **87467**, 1–6, 2007, Hindawi Publishing Corporation.)

Deposition of Zn dots could also reduce the onset potential from 420.5 mV for 1.5 at% Zn-doped samples to 260.5 mV with 100 Å high Zn islands for similar reasons. Zn dots deposited on undoped α-Fe_2O_3 thin film exhibited an even better photocurrent of ~1.82 mA cm^{-2} at 0.6 V versus SCE, showing that such samples are more effective in improving the PEC response, compared to Zn dots deposited on doped α-Fe_2O_3.

13.4.9 Swift Heavy Ion (SHI) Irradiation

Swift heavy ion (SHI) irradiation plays a vital role in the field of modifications of the properties of films, foils and surfaces of bulk solids. It penetrates deep into the materials and produces a long and narrow disordered zone along its trajectory [90]. SHIs lose their energy mainly by two nearly independent processes: (i) elastic collisions known as nuclear energy loss $(dE/dx)_n$, which dominates at an energy of about 1 keV amu^{-1}, and (ii) inelastic collisions of the highly charged projectile ion with the atomic electrons of the matter, known as electronic energy loss $(dE/dx)_e$, which dominates at an energy of about 1 MeV amu^{-1} or more. SHI irradiation causes exotic effects in different classes of materials that cannot be generated by any other means. The passage of SHIs induces very rapidly developing processes, which are difficult to observe during or immediately after their occurrence. Information about these processes is stored in the resulting damage, such as size, shape and structure of defects. The degree of disorder can range from point defects to a continuous amorphized zone along the ion path, commonly called the latent track. The effect of the ion beam on the materials depends on the ion species, ion energy, ion fluence and temperature [91]. The energy loss of high-energy ions, mainly via inelastic collisions, in the target material, results in the excitation of the target electrons, which is associated with modifications of the crystallographic, optical, electrical and morphological properties [92,93]. These changes in metal-oxide semiconductors due to irradiation are expected to affect the PEC behavior of the semiconducting materials for hydrogen production.

Studies of the modification in the properties of semiconductors, especially metal oxides, using SHI irradiation were attempted for the first time by the authors' group for PEC hydrogen generation, and interesting results have been obtained.

Doped/undoped iron-oxide thin films deposited on conducting glass substrates by spray pyrolysis [41] were irradiated with 170 MeV Au^{13+}, 120 MeV Ag^{9+} and 100 Mev Si^{7+} ions at fluences ranging from 1×10^{11} to 4×10^{13} ions cm^{-2}. The irradiated and unirradiated samples were characterized for any change in their morphological, electrical, optical and PEC properties. Some of the important results are given below.

α-Fe_2O_3 irradiated with 170 MeV Au^{13+} at a fluence of 1×10^{12} to 1×10^{13} ions cm^{-2} exhibited a decrease in average particle/grain size from 58 to 35 nm [94]. Atomic force microscopy (AFM) images of the film irradiated at a fluence of 1×10^{12} ions cm^{-2} exhibited formation of a tubular structure (Figure 13.16b) with an average thickness of 450 nm, against a preirradiated image showing rough and intergranular surfaces, suggesting the possibility of surface-related effects. It is expected that irradiation using Au^{13+} ions of energy 170 MeV resulted in melting and regrowth by dissipation of energy losses in the target material leading to the formation of a tubular structure. Shifting of the absorption edge towards higher wavelengths, with a decrease in the bandgap energy was also observed in irradiated samples. In PEC studies, irradiated samples showed a significant increase in the photocurrent density at a fluence of 1×10^{12} ions cm^{-2}, which was ascribed to the formation of the tubular structure, allowing efficient photon absorption. At higher fluences of 10^{13} ions cm^{-2}, a decrease in the

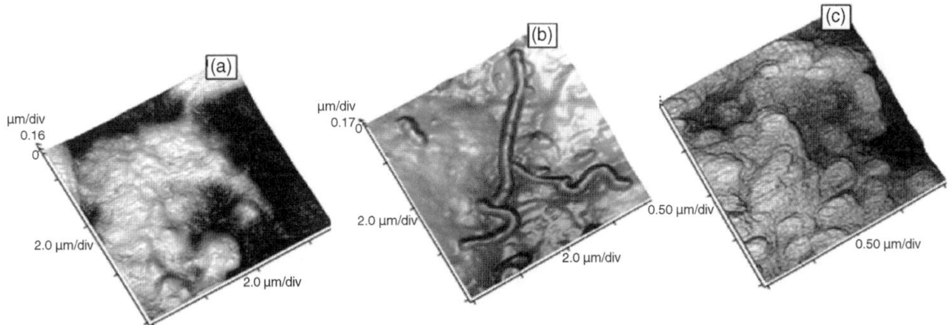

Figure 13.16 AFM images for the unirradiated and irradiated films with 170 MeV Au^{13+} ions (a) before irradiation, (b) 1×10^{12} ions cm^{-2} and (c) 1×10^{13} ions cm^{-2}. (Reproduced with permission from Y.S. Chaudhary, S.A. Khan, R. Shrivastav *et al.*, Modified structural and photoelectrochemical properties of 170 MeV Au^{13+} irradiated hematite, *Thin Solid Films*, **492**, 332–336, 2005. © 2005 Elsevier.)

photocurrent observed was due to the collapse of the tubular structure (Figure 13.16c), resulting in large number of discontinuities/dislocations that act as recombination centers for photogenerated charge carriers.

2% Cr-doped α-Fe_2O_3 films prepared by spray pyrolysis, when irradiated with 170 MeV Au^{13+} ions at 1×10^{12}, 5×10^{12} and 1×10^{13}ions cm^{-2} fluences, converted into disordered material [95], with anisotropic expansion/contraction of the lattice, as also observed and reported by other workers. Irradiation of the material also led to changes in the particle/grain size, from 53 nm, for unirradiated, to 45 nm, at 1×10^{13} ions cm^{-2} fluence, probably on account of the internal pressure of the grains upon irradiation. The value of the bandgap energy was also found to decrease from 1.99 to 1.92 eV with increase in fluence. PEC studies of the irradiated samples showed enhanced photocurrent at a fluence of 1×10^{12} ions cm^{-2}. A decrease in the photocurrent density at higher fluences was explained by the creation of a greater number of kink sites/dislocations, acting as recombination sites for photogenerated electrons.

A similar type of undoped film, when irradiated by 120 MeV Ag^{9+} at a fluence of 5×10^{11} to 1×10^{13} ions cm^{-2}, exhibited a partial phase transition from hematite (α-Fe_2O_3) to magnetite (Fe_3O_4) [96]. Irradiated samples showed a decrease in particle size up to a fluence of 5×10^{12} ions cm^{-2}; thereafter, an increase in particle size was observed. The decrease in particle size at lower fluences was attributed to grain fragmentation due to confinement of the energy in a small volume of nanoparticles during irradiation, while the increase in particle size at higher fluences was due to agglomeration of the particles. These samples exhibited improved photoresponses for all fluences and a maximum photocurrent density of ~321 µA cm^{-2} at 0.95 V versus SCE external bias, was obtained, which is five times greater than the photocurrent density obtained with the unirradiated samples (Figure 13.17).

Mott–Schottky curves (Figure 13.18) were used to evaluate the donor densities and flatband potentials for various unirradiated and irradiated samples. Flatband potentials were found to increase from −0.62 to −0.79 V versus SCE with increase in the fluence. Calculation of donor density revealed an increasing trend, on irradiation, from 1.725×10^{20} to 7.141×10^{20} cm^{-3}, apparently due to inclusion of a magnetite phase in the hematite (α-Fe_2O_3), which is highly conductive; with conduction occurring via electron transfer from Fe^{2+} to Fe^{3+} sites.

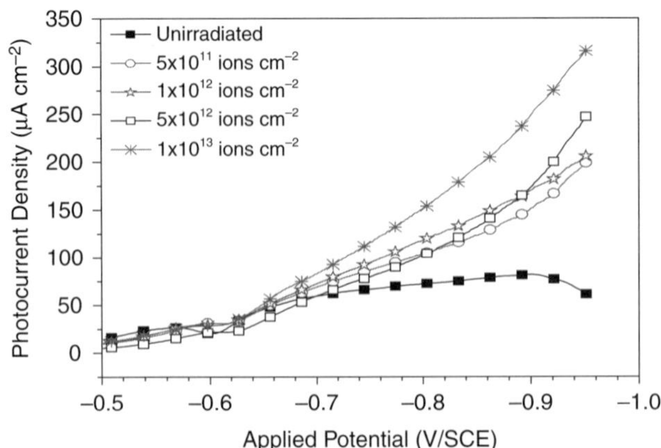

Figure 13.17 Photocurrent–potential characteristics of PEC cells comprising unirradiated and 120 MeV Ag^{9+}-irradiated α-Fe_2O_3 thin films. (Reproduced with permission from A.P. Singh, S. Kumari, R. Shrivasrav *et al.*, Improved photoelectrochemical response of haematite by high energy Ag^{9+} ions irradiation, *Journal of Physics D: Applied Physics*, **42**, 085303, 2009. © 2009 Institute of Physics.)

The photocurrent-density curves obtained for α-Fe_2O_3 irradiated with 100 MeV Si^{7+} ions at fluences ranging from 1×10^{12} to 4×10^{13} ions cm^{-2} are shown in Figure 13.19 [97]. Irradiated samples offered enhanced photocurrent densities, compared to unirradiated samples. The best photocurrent density of ~369 μA cm^{-2} at 0.95 V versus SCE external bias was obtained at a fluence of 2×10^{13} ions cm^{-2}. AFM images shown in Figure 13.20 indicate the change in surface morphology to a granular nature after irradiation. These irradiated samples also exhibited an improved crystallinity. Irradiation led to an increase in particle size from 28 to 35 nm, which was attributed to annealing of point defects and growth of small crystalline inclusions, leading to local crystallization.

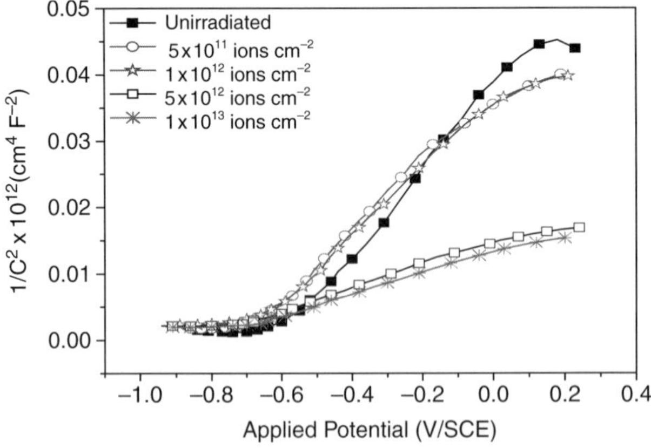

Figure 13.18 Mott–Schottky curves of PEC cells comprising unirradiated and 120 MeV Ag^{9+}-irradiated α-Fe_2O_3 thin films. (Reproduced with permission from A.P. Singh, S. Kumari, R. Shrivasrav *et al.*, Improved photoelectrochemical response of haematite by high energy Ag^{9+} ions irradiation, *Journal of Physics D: Applied Physics*, **42**, 085303, 2009. © 2009 Institute of Physics.)

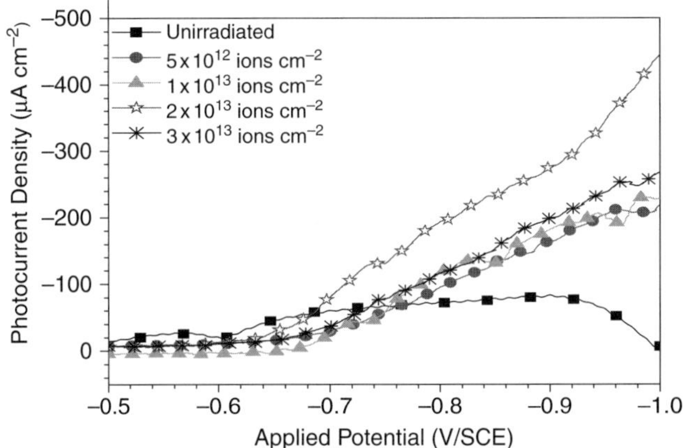

Figure 13.19 Current–potential curves for unirradiated and 100 MeV Si^{7+}-ion irradiated α-Fe_2O_3 thin films. (Reproduced with permission from A.P. Singh, A. Tripathi, R. Shrivastav *et al.*, New benchmark to improve the photoelectrochemical properties of hematite, *Proceedings of Solar Hydrogen and Nanotechnology III*, **7044** (1–8), 70440, 2008. © 2008 SPIE.)

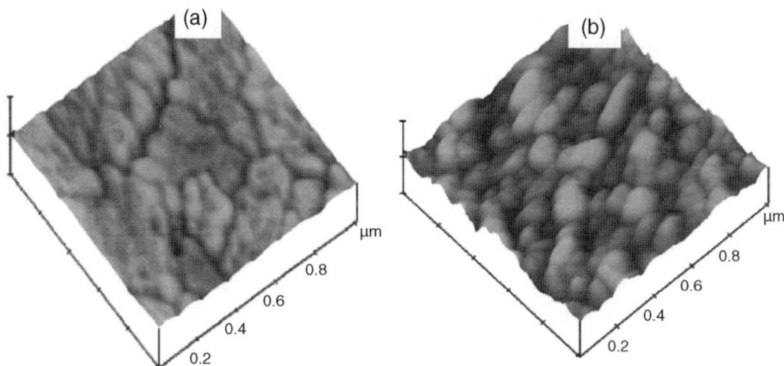

Figure 13.20 AFM images of (a) unirradiated and (b) 100 MeV Si^{7+} ion-irradiated α-Fe_2O_3 thin films at a fluence of 4×10^{13} ions cm^{-2}. (Reproduced with permission from A.P. Singh, A. Tripathi, R. Shrivastav *et al.*, New benchmark to improve the photoelectrochemical properties of hematite, *Proceedings of Solar Hydrogen and Nanotechnology III*, **7044** (1–8), 70440, 2008. © 2008 SPIE.)

Flatband potentials were observed to increase negatively from −0.62 to −0.80 V versus SCE on increasing the fluence up to 2×10^{13} ions cm^{-2}, after which a decrease in flatband potential was observed. The sample with the highest flatband potential exhibited the best photoresponse.

13.4.10 p/n Assemblies

Semiconductors with bandgaps lower than 2.2 eV lack sufficient photovoltage and need an external bias potential to split water. Iron oxide (α-Fe_2O_3) has a bandgap of 2.2 eV, which makes it an ideal semiconductor photoelectrode, due to its capability to absorb most of the photons of the solar spectrum. However, use of iron oxide alone cannot split water, mainly due to its inappropriate energy levels with respect to the redox levels of water. The use of a narrow bandgap n-type anode in combination with a p-type cathode has been proposed to eliminate the necessity of applying energy from an external source to the photoelectrolysis cell. Proper selection of both semiconductor electrode characteristics ensures that the energy necessary for water photoelectrolysis is gathered entirely from the solar light. Additionally, such a combination allows the realization of a two-quantum process that can give an additional gain to the general energetic balance. A p-n photochemical diode consisting of Zn-doped p-GaP and n-Fe_2O_3 was assembled by Mettee *et al.*, in 1981, and exposed to visible and near-UV radiation in aqueous solution [98]. Although the cell showed spontaneous water splitting under visible light illumination, the efficiency of the process remained low (~0.1%). Assembly of p-n-Fe_2O_3, having pressed pellets of Mg-doped p-type and Si-doped n-type iron(III) oxide in a PEC cell was demonstrated by Turner *et al.*, in 1984 [30]. The maximum photocurrent observed was 15 μA, which persisted undiminished throughout 320 h of continuous illumination. A pair consisting of pellets of simultaneously illuminated Ta-doped n-Fe_2O_3 and p-Cu_2O electrodes in a PEC cell was examined by Aroutiounian *et al.*, in 2002 [72,99]. A much higher photocurrent of 6.4 mA cm^{-2} was observed, indicating the possibility of water splitting using solar light as the energy source. Using n-Fe_2O_3 and Zn-doped p-Fe_2O_3 thin-film electrodes prepared by spray pyrolysis in a tandem cell to generate enough photovoltage to split water without the use of any external bias potential was reported by Ingler and Khan [100]. The low photocurrent density and consequent low solar-to-hydrogen photoconversion efficiency of 0.11% for self-driven p-n-Fe_2O_3 was attributed to the poor photoresponse of p-Fe_2O_3 films compared to n-Fe_2O_3, low conductivity and a high recombination rate of photogenerated charge carriers at the interface.

13.5 Efficiency and Hydrogen Production

Photoconversion efficiency is an important parameter to measure the performance of material in solar water splitting, defined as the ratio of the chemical potential energy stored in the form of hydrogen to the incident solar energy. Mainly, two types of conversion efficiencies are reported in the literature, *viz*, solar-to-hydrogen conversion efficiency and IPCE.

The solar-to-hydrogen conversion efficiency $h_c(\%)$ of the water-splitting reaction [101] can be determined using the equation

$$h_c = j_p \frac{(V_{WS} - V_B)}{eE_S}$$

where j_p is the photocurrent produced per unit irradiated area, $V_{WS} = 1.23$ eV is the water-splitting potential per electron, E_S is the photon flux, V_B is the bias voltage applied between the working and the counter-electrodes and 'e' is the electronic charge. According to this equation, every electron contributing to the current produces half an H_2 molecule.

The IPCE provides a measure of the efficiency of conversion of photons to current in a PEC cell flowing between the working and counter-electrodes [101]. The IPCE can be calculated using the equation

$$\text{IPCE}(\%) = \frac{J_p(\lambda)}{eE_S(\lambda)}$$

where $E_S(\lambda)$ is the incident photon flux at wavelength λ. A 100% IPCE indicates that one incident photon generates an electron–hole pair, which is effectively separated and results in current flow in the external circuit. The plot of IPCE as a function of wavelength is known as action spectra.

Maximum theoretical water-splitting efficiency for α-Fe_2O_3 is 12.9% within the thermodynamic limit, which cannot be exceeded [101]. These values are calculated using ideal conditions, but the achievable practical efficiency is expected to be significantly lower, on account of imperfect absorption, reflection losses, and recombination; putting all together these losses can reduce the efficiency by a factor of up to three.

The main objective of this section is to present collectively the reported data on IPCE and the rate of hydrogen production/evolution using α-Fe_2O_3 photoelectrodes. The IPCE values of α-Fe_2O_3 photoelectrodes under various conditions (as summarized in Table 13.4) vary from a value as low as 0.33% [40] to a value as high as 56% [5]. It can be seen that, among the various methods, USP-prepared nanostructured α-Fe_2O_3 films showed a much better IPCE value of 26%. Si-doped samples prepared by APCVD [47] and hematite prepared by spray pyrolysis, doped with two metals also exhibited good results [43]. Solar-to-hydrogen conversion efficiencies are reported in only a few papers and are very low [35,37].

In the early 1980s, the rate of hydrogen production, using α-Fe_2O_3 as photoelectrode, was very low and rarely reported [27,29,98]. In 1987, Khader *et al.* reported evolution of 380 μmol hydrogen in 400 h from photocatalytic water dissociation in visible light using α-Fe_2O_3. Assuming that all the catalyst particles were active during the experiment, the rate of hydrogen generation was 40 μmol h^{-1} g^{-1} for the first 200 hours. In a PEC measurement on Fe_2O_3 pellets, hydrogen evolution linearly increased from 0 to 17 μmol in 10 h, with increasing light intensity from 0 to 35 mW cm^{-2} under +0.5 V external bias [102]. The rate of hydrogen production in visible light by water splitting, as observed by Gurunathan and Maruthamuthu, in 1995, was 0.66 ml h^{-1} for Rh/Fe_2O_3 [83]. 1 mole% RuO_2 loading on Rh(III)/Fe_2O_3 further increased the rate of hydrogen production to 0.83 ml h^{-1}. 5.0 at% Cu(II) doping enhanced the hydrogen evolution rate, compared with undoped α-Fe_2O_3. The volume of hydrogen evolved with Cu(II) was 0.60 and 0.79 ml h^{-1} for Cu(II)/Fe_2O_3 and Cu(II)/Fe_2O_3/RuO_2, respectively.

Yields of hydrogen and oxygen production under laser-beam irradiation were determined with pure α-Fe_2O_3 and in the presence of metal ions, by Gondal *et al.* [85]. The rates of hydrogen and oxygen evolution using 0.3 g of α-Fe_2O_3 suspended in 60 ml of distilled water were ~1.7 and 4.5 mmol h^{-1}, respectively, under irradiation by a strong laser beam at 355 nm. They also studied H_2 and O_2 production rates with time in the presence of metal ions such as Fe^{3+}, Ag^{+} and Li^{+}, and reported a decrease in the rate of hydrogen production in the presence of metal ions. A comparison of hydrogen production using α-Fe_2O_3 with NiO, TiO_2 and WO_3 showed that the hydrogen production rate was in the order NiO > TiO_2 > α-Fe_2O_3 > WO_3 [103]. This order of hydrogen production corresponds well with the order of suitability of the conduction-band edges of these catalysts for production of hydrogen [104].

Recently, Aroutiounian *et al.* carried out laboratory measurements of hydrogen evolution in heterogeneous cells using an α-Fe_2O_3 photoelectrode in the form of pellets doped with Sn [65]. The rate of hydrogen evolution was equal to ~20 μmol h^{-1} cm^{-2} for 0.75–1 at% Sn-doped α-Fe_2O_3 photoelectrodes at ~780 mV chemical bias under a mercury lamp with a light intensity of 250 W. In yet another study by Aroutiounian *et al.* on the α-Fe_2O_3–Nb_2O_5 system, the rate of hydrogen evolution was found to be equal to ~19.0 and 22.4 μmol h^{-1} cm^{-2} for $Fe_{1.8}Nb_{0.2}O_3$ and $FeNbO_4$ photoelectrodes, respectively, at 780 mV external bias voltage, under a mercury lamp with a light intensity of 200 W [64].

13.6 Concluding Remarks

In the past few decades, α-Fe_2O_3, both in bulk and in nanostructures, has been studied by several researchers to obtain efficient photoelectrodes for hydrogen generation via PEC splitting of water using solar energy. In this context, various preparation techniques, modifications in material and other different approaches to improve PEC performance have been described. Table 13.1 summarizes the photocurrent density obtained with nanostructured α-Fe_2O_3 thin films prepared by various techniques. At present it seems difficult to identify the best method of preparation of nanostructured α-Fe_2O_3 with respect to efficient PEC generation of hydrogen. The data clearly reveal that a significant photocurrent with nanostructured thin films of α-Fe_2O_3 can be obtained only after doping. The desired morphology of the film and suitable electrolyte composition also play important roles. Limited reports are available on photocurrent generation under zero bias with a solar simulator or direct solar energy – a situation where solar energy actually splits water.

The values of the maximum photocurrent densities observed with α-Fe_2O_3 modified by various techniques, have been summarized in Table 13.3. Spray-pyrolytically deposited α-Fe_2O_3 thin films modified by Zn dots have shown better prospects, exhibiting a significant photoresponse in PEC cells. Deposition of metal dots on the surface of α-Fe_2O_3 could be an interesting area of research in PEC generation of hydrogen, and is expected to provide an efficient photoelectrode. The IPCE values (Table 13.4) reported in few papers at specific wavelengths may provide a comparison of the performances of different photoelectrodes, yet the solar-to-hydrogen conversion efficiencies in simulated solar conditions may be a better option to compare efficiencies, which are rarely available. Use of artificial light sources in most of the measurements on water-splitting efficiency is another issue; for example, use of a Xe lamp as the light source has the potential to overestimate the water-splitting efficiency on account of higher spectral photon flux at short wavelengths. To obtain a reliable estimate of water-splitting efficiencies using solar energy, experimental results needs to be studied under AM1.5 conditions without application of any external voltage. Therefore, at this stage, with the available data, it is difficult to compare the efficiency of hydrogen production as obtained with the different types of nanostructured α-Fe_2O_3 photoelectrodes and draw any meaningful conclusion.

Few important observations, which resulted in enhanced photoelectrochemical responses of α-Fe_2O_3 in PEC cells are listed below:

- Nanostructured α-Fe_2O_3 thin films, with modified morphologies of the photoelectrode, exhibit significant photocurrents due to efficient harvesting of solar energy and by offering short distances for the photogenerated holes to reach the electrolyte interface [39,45–47].
- The method of preparation of α-Fe_2O_3 photoelectrode greatly affects its performance.
- The sol-gel technique provides a method for preparation of purpose-built structures, which may be used to develop efficient α-Fe_2O_3 photoelectrodes [5,50].
- Thin films exhibit much better photoresponses, compared to pellets.
- Thin films of α-Fe_2O_3 with simultaneous double doping with preferred metals are more effective than with single doping [43].
- Traces of Fe_3O_4 in α-Fe_2O_3 enhance [24,43] its photoresponse, while larger amounts of Fe_3O_4 increase the conductivity, but not the photoresponse.

Table 13.1 Methods of preparation: a comparison of photoresponse.

Reported by	Method of preparation	Sample Description	Best Photocurrent
Khan and Zhou [33]	Spray Pyrolysis (SP)	Iodine-doped, n-type in 0.2 M NaOH + 0.5 M Na_2SO_4, Thickness 100 nm	1.6 mA cm^{-2} at 0.82 V versus SHE
Majumder and Khan [34]	SP	Thickness 33 nm, 0.2 M NaOH + 0.5 M Na_2SO_4,	0.7 mA cm^{-2} at 0.82 V versus NHE
Khan and Akikusa [35]	SP	Undoped nanocrystalline, 1 M NaOH	3.7 mA cm^{-2} at 0.7 V versus SCE
Ingler and Khan [40]	SP	12% Mg-doped, p-type in 0.5 M H_2SO_4	0.21 mA cm^{-2} at 0.2 V versus SCE
Ingler *et al.* [36]	SP	0.0088 M Zn-doped, p Type in 0.5 M H_2SO_4	1.1 mA cm^{-2} at 0.8 V versus Pt
Ingler and Khan [37]	SP	0.01155 M Cu-doped, p-type in 0.5 M H_2SO_4	0.94 mA cm^{-2} at 0.0 V versus SCE
Sartoretti *et al.* [43]	SP	(5%) Ti- and (1%) Al-doped, n-type in 0.1 M NaOH	4.38 mA cm^{-2} at 0.45 V versus NHE
Kumari *et al.* [41]	SP	5.0 at.% Zn-doped, n-type in 13 pH NaOH, particle size 60 nm	0.64 mA cm^{-2} at 0.7 V versus SCE
Duret and Gratzel [45]	SP	50–150 nm spherical particles, 1 M NaOH	14 μA cm^{-2}at 1.2 V versus RHE
	Ultrasonic Spray Pyrolysis (USP)	100 nm sized platelets of thickness 5–10 nm, 1 M NaOH	1.07 mA cm^{-2} at 1.2 V versus RHE
Cesar *et al.* [47]	USP	370 nm thick USP film using silicone tubing, 1 M NaOH	1.17 mA cm^{-2} at 1.23 V versus RHE
	Atmospheric Pressure Chemical vapor Deposition (APCVD)	Si doped, n-type, 375 nm thick film, dendritic microstructure of about 170 nm high with 20–30 nm thick branch, 1 M NaOH	1.45 mA cm^{-2} at 1.23 V versus RHE

(continued)

Table 13.1 (*Continued*)

Reported by	Method of preparation	Sample Description		Best Photocurrent
Kay *et al.* [39]	APCVD	Dendritic nanostructure of 500 nm thickness having a feature size of 10–20 nm	Si-doped single nanocrystalline α-Fe_2O_3 thin films, 1 M NaOH	2.2 mA cm^{-2} at 1.23 V versus RHE
			Catalytic cobalt monolayer on the Si-doped α-Fe_2O_3 film, 1 M NaOH	2.7 mA cm^{-2} at 1.23 V versus RHE
Yarahmadi *et al.* [48]	APCVD	Nanostructured particles of 20 nm prepared using ferrocene, 1 M NaOH		540 μA cm^{-2} at 1.23 V versus RHE
Cesar *et al.* [46]	APCVD	Si-doped mesoporous of 12 nm size, 1 M NaOH		2.3 mA cm^{-2} at 1.43 V versus RHE
Glasscock *et al.* [57]	Reactive DC Magnetron Sputtering	5.0 at% Ti doped α-Fe_2O_3, 40–100 nm size, 1 M NaOH		0.9 mA cm^{-2} at 0.6 V versus SCE
Beermann *et al.* [5]	Sol-gel	Nanorod with 50 nm in radius and 500 nm in length, dye sensitized		320 μA cm^{-2} at 0 V
Souza *et al.* [53]	Sol-gel	0.5% Si-doped n-type, mesoporous structure in nanodemension, 1 M NaOH		35 μA cm^{-2}at 1.23 V versus RHE
Prakasam *et al.* [58]	Potentiostatic Anodization	Nanoporous α-Fe_2O_3 with pore diameter 50–250 nm and pore depth of 500 nm, 0.5 M H_2O_2 + 1 M NaOH		0.51 mA cm^{-2} at 0.6 V versus Ag/AgCl
Shwarsctein *et al.* [62]	Electrodeposition	15% Mo-doped n-type, 1 M NaOH		1.8 mA cm^{-2} at 0.4 V versus Ag/AgCl
Hu *et al.* [63]	Electrodeposition	5% Pt-doped n-type, 1 M NaOH		1.43 mA cm^{-2} at 0.4 V versus Ag/AgCl

Table 13.2 Measured parameters of Zn dotted α-Fe_2O_3 thin films at different heights of Zn islands.

Description of samples	Resistivity (Ω cm)	Onset Potential V_{on} (mV)	Photocurrent-density at 0.7 V versus SCE (mA cm^{-2})
Undoped α-Fe_2O_3	4.0×10^6	660.5	0.061
1.5 at% Zn-doped α- Fe_2O_3	2.6×10^7	420.5	0.321
100 Å thick Zn dots	1.5×10^7	260.5	0.32
200 Å thick Zn dots	5.9×10^6	294.5	0.701
230 Å thick Zn dots	5.7×10^6	306.5	1.282
260 Å thick Zn dots	9.2×10^6	272.5	0.576

- An interfacial layer of SiO_2 on the TCO enhances the PEC response of α-Fe_2O_3 photoelectrodes [39].
- Irradiating the thin films of α-Fe_2O_3 with SHI is effective in enhancing their PEC response by modifying the morphology of the films [94–97].
- Multilayers of preferred oxides, for example, α-Fe_2O_3/TiO_2 perform better [82,86,88] than single-material electrodes.
- Deposition of dots of preferred metals on the surface of α-Fe_2O_3 thin films are much more effective in enhancing the photoresponse, compared to doping [89].
- The proper choice of electrolyte and direction of illumination are also deciding factors for photoresponse [42].

Table 13.3 Modified electrodes: A comparison of photoresponse.

Reported by	Method of preparation	Photoresponse, (as-deposited)	Modifications	Photoresponse
Luo *et al.* [82]	Sol-gel spin coating	6 μA cm^{-2} at 0.7 V (single)	100–200 nm, bilayered WO_3/α-Fe_2O_3	22 μA cm^{-2} at 0.8 V versus Ag/AgCl
Kumari *et al.* [89]	Spray pyrolysis (SP)	0.061 mA cm^{-2} at 0.7 V versus SCE	Zn dots of 0.7 mm thickness and height 230 Å deposited on 1.5 at% Zn doped Hematite surface	1.282 mA cm^{-2} at 0.7 V versus SCE
Satsangi *et al.* [88]	SP and sol-gel	12 μA cm^{-2} at 0.6 V versus SCE	α-Fe_2O_3/TiO_2	945 μA cm^{-2} at 0.6 V versus SCE
Singh *et al.* [97]	SP	50 μA cm^{-2} at 0.95 V versus SCE	100 MeV Si^{7+} irradiation α-Fe_2O_3, fluence 2×10^{13} ions cm^{-2}	369 μA cm^{-2} at 0.95 V versus SCE
Singh *et al.* [96]	SP	50 μA cm^{-2} at 0.95 V versus SCE	120 MeV Ag^{9+} irradiation α-Fe_2O_3 fluence 1×10^{13} ions/cm^2	321 μA cm^{-2} at 0.95 V versus SCE

Table 13.4 A comparative chart of efficiency of photoelectrodes synthesized by different methods.

Reported by	Sample description	Photoconversion Efficiency/IPCE
Khan and Akikusa [35]	Spray pyrolysis, n-type nanocrystalline in 0.5 M H_2SO_4	1.84% at 0.2 V versus SCE
Ingler and Khan [40]	Spray pyrolysis, Mg doped, p-type in 0.5 M H_2SO_4	0.33% at −0.61 V/V_{aoc} (V_{aoc} = 0.88 V versus SCE)
Ingler *et al.* [36]	Spray pyrolysis, Zn doped, p type in 0.5 M H_2SO_4	21.1% at 325 nm
Ingler and Khan [37]	Spray pyrolysis, Cu doped, p-type in 0.5 M H_2SO_4	2.9% at 0.68 V/V_{aoc} (V_{aoc} = 0.7 V versus SCE)
Sartoretti *et al.* [43]	Spray pyrolysis, 5%Ti and 1%Al doped, n-type, 0.1 M NaOH	IPCE 25% at 400 nm
Duret and Gratzel [45]	Spray pyrolysis, 50–150 nm spherical particles, 1 M NaOH	IPCE 7% at 400 nm (1.6 V versus RHE)
	Ultrasonic spray pyrolysis, 100 nm sized platelets of thickness 5–10 nm, 1 M NaOH	IPCE 24% at 400 nm (1.6 V versus RHE)
Beermann *et al.* [5]	Sol-gel, nanorods with 50 nm in radius and 500 nm in length, 0.1 KI	IPCE 11% for SE illumination at 360 nm
	Sandwich cell using parallel and perpendicular oriented nanorods,	IPCE 56% at 340 nm
Lindgren, et at. [50]	Sol-gel, nanorods with 700 nm in length and 50 nm in diameter, 0.5 M I^- + 0.5 mM I_2 in ethylene carbonate/propylene carbonate	IPCE 20% for SE illumination at 350 nm
Miyake and Kozuka [67]	Sol-gel, α-Fe_2O_3-Nb_2O_5 films, aq. solution of 0.2 M $Na_2B_4O_7$, 0.14 M H_2SO_4 and 0.3 M Na_2SO_4	IPCE 18% at 325 nm
Luo *et al.* [82]	Sol-gel spin coating, 100–200 nm, bilayered WO_3/Fe_2O_3	IPCE 4.8 % at 398 nm
Souza *et al.* [53]	Sol-gel, 0.5% Si-doped n-type, mesoporous structure in nanodimension, 1 M NaOH	IPCE 37% at 1.6 V versus RHE (300 nm)
Kay *et al.* [39]	APCVD, Si-doped α-Fe_2O_3, 1 M NaOH	IPCE 42% at 370 nm (1.23 V versus RHE)
Cesar *et al.* [46]	APCVD, Si-doped mesoporous of 12 nm size, 1 M NaOH	IPCE 38% at 350 nm (1.43 V versus RHE)
Glasscock *et al.* [57]	Reactive magnetron sputtering, Ti-doped film, 1 M NaOH	IPCE ~17% at 320 nm (0.5 V versus SCE)
Shwarsctein *et al.* [62]	Electrodeposition, 15% Mo-doped n-type, 1 M NaOH	IPCE 11.8% at 400 nm (0.4 V versus Ag/AgCl)
Hu *et al.* [63]	Electrodeposition, 5% Pt-doped n-type, 1 M NaOH	IPCE 14.8% at 400 nm (0.6 V versus Ag/AgCl)

- Connecting many photoelectrodes, kept one behind the other, in parallel, also produces better photocurrent than single-electrode systems [42].

It thus becomes amply evident that nanostructured, nanoporous α-Fe_2O_3 in the form of thin films, doped suitably, offers better results than bulk material. However, the values of photocurrent obtained in the present scenario are not sufficient for efficient generation of hydrogen. It is expected that nanostructured thin films of α-Fe_2O_3, modified suitably, in the light of the above mentioned observations, may yield a cost-effective and durable PEC system for producing hydrogen. Combining the above listed favorable observations and deriving inputs from the theoretical simulations, efforts can be directed to design and develop nanostructured architectures, which may result in a technically viable, reliable and cost-effective PEC system, based on nanostructured α-Fe_2O_3 photoelectrodes for efficient generation of hydrogen.

Acknowledgments

We have received Grace, Guidance and Directions from our Most Revered Professor P. S. Satsangi Sahab, Chairman, Advisory Committee on Education, Dayalbagh, Agra, at every moment. We express our most humble gratitude and reverence in His Lotus Feet.

It gives us pleasure to acknowledge all those who have contributed some way or the other in bringing this chapter to its present shape. First of all, special and sincere thanks go to our thoughtful and enthusiastic editor Dr Lionel Vayssieres for providing us with this opportunity. During the preparation of the manuscript, many discussions were held in our research group from time to time. We express our deep sense of appreciation for the contributions made by all our research scholars, in particular, Dr Saroj Kumari and Mr Aadesh P. Singh, for their continual help and sustained effort. We also take this opportunity to thankfully acknowledge the support provided by our colleague Dr Gunjan Agrawal, Department of Mathematics, through fruitful suggestions, discussions and constructive criticisms. Finally, our heartfelt thanks are due to Prof. Dr S. K. Satsangi, S.N. Medical College, Agra for his invaluable encouragement and support.

References

[1] Morin, F.J. (1954) Electrical properties of α-Fe_2O_3. *Physical Reviews*, **93**, 1195–1199.

[2] Gardner, R.F.G., Sweett, F., and Tanner, D.W. (1963) The electrical properties of alpha ferric oxide-II.: Ferric oxide of high purity. *Journal of Physics and Chemistry of Solids*, **24**, 1183–1186.

[3] Morin, F.J. (1951) Electrical properties of α-Fe_2O_3 and α-Fe_2O_3 containing titanium. *Physical Reviews*, **83**, 1005–1010.

[4] Kennedy, J.H. and Frese, K.W. Jr. (1978) Photooxidation of water at α-Fe_2O_3 electrodes. *Journal of the Electrochemical Society*, **125**, 709–714.

[5] Beermann, N., Vayssieres, L., Lindquist, S.-E., and Haqfeldt, A. (2000) Photoelectrochemical studies of oriented nanorod thin films of hematite. *Journal of the Electrochemical Society*, **147**, 2456–2461.

[6] Turner, J.E., Hendewerk, M., Parmeter, J. *et al.* (1984) The characterization of doped iron oxide electrodes for the photodissociation of the water. *Journal of the Electrochemical Society*, **131**, 1777–1783.

[7] McGregor, K.G., Calvin, M., and Otvos, J.W. (1979) Photoeffects in Fe_2O_3 sintered semiconductor. *Journal of Applied Physics*, **50**, 369–373.

[8] Kurtz, R.L. and Henrich, V.E. (1987) Surface electronic structure and chemisorption on corundum transition-metal oxides: α-Fe_2O_3. *Physical Reviews B*, **36**, 3413–3421.

[9] Debnath, N.C. and Anderson, A.B. (1983) Water adsorption on an iron oxide surface. *Surface Science*, **128**, 61–69.

[10] Catti, M., Valerio, G., and Dovesi, R. (1995) Theoretical-study of electronic, magnetic, and structural-properties of alpha-Fe_2O_3 (hematite). *Physical Reviews B*, **51**, 7441–7450.

[11] Anderman, M. and Kennedy, J.H. (1988) *Semiconductor Electrodes*, **55** (ed. Hary C. Finklea), Elsevier, pp. 147–202.

[12] Yeh, L.S.R. and Hackerman, N. (1977) Iron oxide semiconductor electrode in photoassisted elelctrolysis of water. *Journal of the Electrochemical Society*, **124**, 833–836.

[13] Wilhelm, S.M., Yun, K.S., Ballenger, L.W., and Hackerman, N. (1979) Semiconductor properties of iron oxide electrode. *Journal of the Electrochemical Society*, **126**, 419–424.

[14] Marusak, L.A., Messier, R., and White, W.B. (1980) Optical absorption spectrum of hematite, α-Fe_2O_3, near IR to UV. *Journal of Physics and Chemistry of Solids*, **41**, 981–984.

[15] Redon, A.M., Vigneron, S., Heindl, R. *et al.* (1981) Differences in the optical and photoelectrochemical behaviours of single-crystal and amorphous ferric oxide. *Solar Cells*, **3**, 179–186.

[16] Candea, R.M. (1981) Photoelectrochemical behavior of iron oxides thermally grown on Fe-Ni alloys. *Electrochemica Acta*, **26**, 1803–1808.

[17] Gori, M., Grüniger, H.R., and Calzaferri, G. (1980) Photochemical properties of sintered iron (III) oxide. *Journal of Applied Electrochemistry*, **10**, 345–349.

[18] Kennedy, J.H., Shinar, R., and Ziegler, J.P. (1980) α-Fe_2O_3 photoelectrode doped with silicon. *Journal of the Electrochemical Society*, **127**, 2307–2309.

[19] Kennedy, J.H., Anderman, M., and Shinar, R. (1981) Photoactivity of polycrystalline α-Fe_2O_3 electrodes doped with group IVA elements. *Journal of the Electrochemical Society*, **128**, 2371–2373.

[20] Shinar, R. and Kennedy, J.H. (1982) Photoactivity of doped α-Fe_2O_3 electrodes. *Solar Energy Materials*, **6**, 323–335.

[21] Sanchez, H.L., Steinfink, H., and White, H.S. (1982) Solid solubility of Ge, Si, and Mg in Fe_2O_3, and photoelectric behavior. *Journal of Solid State Chemistry*, **41**, 90–96.

[22] Gratzel, M., Kiwi, J., and Morrison, C.L. (1985) Visible-light-induced photodissolution of α-Fe_2O_3 powder in the presence of chloride anions. *Journal of the Chemical Society Faraday Transactions I*, **81**, 1883–1890.

[23] Houlihan, J.F., Pannaparayil, T., Burdette, H.L. *et al.* (1985) Substitution and defect doping effect on the photoelectrochemical properties of Fe_2O_3. *Materials Research Bulletin*, **20**, 163–177.

[24] Parent, L., Dodelet, J.P., and Dallaire, S. (1987) Phase transformation in plasma-sprayed iron oxide coatings. *Thin Solid Films*, **154**, 57–64.

[25] Sanchez, C., Sieber, K.D., and Somorjai, G.A. (1988) The photoelectrochemistry of niobium doped α-Fe_2O_3. *Journal of Electroanalytical Chemistry*, **252**, 269–290.

[26] Sieber, K.D., Sanchez, C., Turner, J.E., and Somorjai, G.A. (1985) Preparation, electrical and photoelectrochemical properties of magnesium doped iron oxide sintered discs. *Materials Research Bulletin*, **20**, 153–162.

[27] Leygraf, C., Hendewerk, M., and Somorjai, G.A. (1982) Mg- and Si- doped iron oxide for the photocatalyzed production of hydrogen from water by visisble light (2.2 eV $\leq h\upsilon \leq$ 2.7 eV). *Journal of Catalysis*, **78**, 341–351.

[28] Leygraf, C., Hendewerk, M., and Somorjai, G.A. (1983) The preparation and selected properties of Mg doped p-type iron oxide as a photocathode for the photoelectrolysis of water using visible light. *Journal of Solid State Chemistry*, **48**, 357–367.

[29] Leygraf, C., Hendewerk, M., and Somorjai, G.A. (1982) Photocatalytic production of hydrogen from water by a p- and n-type polycrystalline iron oxide assembly. *Journal of Physical Chemistry*, **86**, 4484–4485.

[30] Turner, J.E., Hendewerk, M., and Somorjai, G.A. (1984) The photodissociation of water by doped iron oxides: The unblased p/n assembly. *Chemical Physics Letters*, **105**, 581–585.

[31] Chamberlin, R.R. and Skarman, J.S. (1966) Chemical spray deposition process for inorganic films. *Journal of the Electrochemical Society*, **113**, 86–89.

[32] Murthy, A.S.N. and Reddy, K.S. (1984) Photoelectrochemical behaviour of undoped ferric oxide (α-Fe_2O_3) electrodes prepared by spray pyrolysis. *Materials Research Bulletin*, **19**, 241–246.

[33] Khan, S.U.M. and Zhou, Z.Y. (1993) Photoresponce of undoped and iodine doped iron oxide thik film electrode. *Journal of Electroanalytical Chemistry*, **357**, 407–420.

[34] Majumder, S.A. and Khan, S.U.M. (1994) Photoelectrolysis of water at bare and electrocatalyst covered thin film iron oxide electrode. *International Journal of Hydrogen Energy*, **19**, 881–887.

[35] Khan, S.U.M. and Akikusa, J. (1999) Photoelectrochemical splitting of water at nanocrystalline n-Fe_2O_3 thin-film electrodes. *Journal of Physical Chemistry B*, **103**, 7184–7189.

[36] Ingler, W.B. Jr., Baltrus, J.P., and Khan, S.U.M. (2004) Photoresponse of p-type zinc-doped iron(III) oxide thin films. *Journal of the American Chemical Society*, **126**, 10238–10239.

[37] Ingler, W.B. Jr. and Khan, S.U.M. (2005) Photoresponse of spray pyrolytically synthesized copper-doped p-Fe_2O_3 thin film electrodes in water splitting. *International Journal of Hydrogen Energy*, **30**, 821–827.

[38] Aroutiounian, V.M., Arakelyan, V.M., Shahnazaryan, G.E. *et al.* (2000) Investigation of the $Fe_{1.99}Ti_{0.01}O_3$-electrode interface. *Electrochimica Acta*, **45**, 1999–2005.

[39] Kay, A., Cesar, I., and Grätzel, M. (2006) New benchmark for water photooxidation by nanostructured α-Fe_2O_3 films. *Journal of the American Chemical Society*, **128**, 15714–15721.

[40] Ingler, W.B. Jr. and Khan, S.U.M. (2004) Photoresponse of spray pyrolytically synthesized magnesium-doped iron (III) oxide (p-Fe_2O_3) thin films under solar simulated light illumination. *Thin Solid Films*, **461**, 301–308.

[41] Kumari, S., Tripathi, C., Singh, A.P. *et al.* (2006) Characterization of Zn-doped hematite thin films for photoelectrochemical splitting of water. *Current Science*, **91**, 1062–1064.

[42] Sartoretti, C.J., Ulmann, M., Alexander, B.D. *et al.* (2003) Photoelectrochemical oxidation of water at transparent ferric oxide film electrodes. *Chemical Physics Letters*, **376**, 194–200.

[43] Sartoretti, C.J., Alexander, B.D., Solarska, R. *et al.* (2005) Photoelectrochemical oxidation of water at transparent ferric oxide film electrodes. *Journal of Physical Chemistry B*, **109**, 13685–13692.

[44] Liang, Y., Enache, C.S., and Krol, R. (2008) Photoelectrochemical cheracterization of sprayed α-Fe_2O_3 thin films: Influence of Si doping and SnO_2 interfacial layer. *International Journal of Photoenergy*, **739864**, 1–7.

[45] Duret, A. and Grätzel, M. (2005) Visible light-induced water oxidation on mesoscopic α-Fe_2O_3 films made by ultrasonic spray pyrolysis. *Journal of Physical Chemistry B*, **109**, 17184–17191.

[46] Cesar, I., Sivula, K., Kay, A. *et al.* (2009) Influence of feature size, film thickness, and silicon doping on the performance of nanostructured hematite photoanodes for solar water splitting. *Journal of Physical Chemistry C*, **113**, 772–782.

[47] Cesar, I., Kay, A., Martinez, J.A.G., and Grätzel, M. (2006) Translucent thin film Fe_2O_3 photoanodes for efficient water splitting by sunlight: nanostructure-directing effect of Si-doping. *Journal of the American Chemical Society*, **128**, 4582–4583.

[48] Yarahmadi, S.S., Tahir, A.A., Vaidhyanathan, B., and Wijayantha, K.G.U. (2009) Fabrication of nanostructured α-Fe_2O_3 electrodes using ferrocene for solar hydrogen generation. *Materials Letters*, **63**, 523–526.

[49] Bjorksten, U., Moser, J., and Gratzel, M. (1994) Photoelectrochemical studies on nanocrystalline hematite films. *Chemistry of Materials*, **6**, 858–863.

[50] Lindgren, T., Wang, H., Beermann, N. *et al.* (2002) Aqueous photoelectrochemistry of hematite nanorod array. *Solar Energy Materials and Solar Cells*, **71**, 231–243.

[51] Agrawal, A., Chaudhary, Y.S., Satsangi, V.R. *et al.* (2003) The synthesis of titanium doped photosensitive hematite by a new route. *Current Science*, **85**, 101–104.

[52] Agrawal, A., Chaudhary, Y.S., Satsangi, V.R. *et al.* (2005) Sol-gel synthesis and photoelectrochemical behaviour of Si-doped α-Fe_2O_3. *Acta Ciencia Indica*, **XXXI C** (4), 373–380.

[53] Souza, F.L., Lopes, K.P., Nascente, P.A.P., and Leite, E.R. (2009) Nanostructured hematite thin films produced by spin-coating deposition solution: Application in water splitting. *Solar Energy Materials and Solar Cells*, **93**, 362–368.

[54] Schumacher, L.C., McIntyre, N.S., Afara, S.M., and Dignam, M.J. (1990) Photoelectrochemical properties of indium doped iron oxide. *Journal of Electroanalytical Chemistry*, **277**, 121–138.

[55] Miller, E.L., Rocheleau, R.E., and Khan, S. (2004) A hybrid multijunction photoelectrode for hydrogen production fabricated with amorphous silicon/germanium and iron oxide thin films. *International Journal of Hydrogen Energy*, **29**, 907–914.

[56] Miller, E.L., Paluselli, D., Marsen, B., and Rocheleau, R.E. (2005) Development of reactively sputtered metal oxide films for hydrogen-producing hybrid multijunction photoelectrodes. *Solar Energy Materials and Solar Cells*, **88**, 131–144.

[57] Glasscock, J.A., Banner, P.R.F., Plub, I.C., and Savvides, N. (2007) The enhancement of photoelectrochemical hydrogen production from hematite thin films by the introduction of Ti and Si. *Journal of Physical Chemistry C*, **111**, 16477–16488.

[58] Prakasam, H.E., Varghese, O.K., Paulose, M. *et al.* (2006) Synthesis and photoelectrochemical properties of nanoporous iron (III) oxide by potentiostatic anodization. *Nanotechnology*, **17**, 4285–4291.

[59] Choi, K.S., Lichtenegger, H.C., Stucky, G.D., and McFarland, E.W. (2002) Electrochemical synthesis of nanostructured ZnO films utilizing self-assembly of surfactant molecules at solid-liquid interfaces. *Journal of the American Chemical Society*, **124**, 12402–12403.

[60] Teng, H. and Li, P.-S. (2007) Electrodeposited amorphous iron (III) oxides as anodes for photoelectrolysis of water. *Journal of the Chinese Institute of Chemical Engineers*, **38**, 267–273.

[61] Schrebler, R., Bello, K., Vera, F. *et al.* (2006) An electrochemical deposition route for obtaining α-Fe_2O_3 thin films. *Electrochemical and Solid State Letters*, **9**, C110–C113.

[62] Shwarsctein, A.K., Hu, Y.-S., Forman, A.J. *et al.* (2008) Electrodeposition of α-Fe_2O_3 doped with Mo or Cr as photoanodes for photocatalytic water splitting. *Journal of Physical Chemistry C*, **112**, 15900–15907.

[63] Hu, Y.-S., Shwarsctein, A.K., Forman, A.J. *et al.* (2008) Pt-doped α-Fe_2O_3 thin films active for photoelectrochemical water splitting. *Chemistry of Materials*, **20**, 3803–3805.

[64] Aroutiounian, V.M., Arakelyan, V.M., Shahnazaryan, G.E. *et al.* (2006) Photoelectrochemistry of semiconductor electrodes made of solid solutions in the system Fe_2O_3-Nb_2O_5. *Solar Energy*, **80**, 1098–1111.

[65] Aroutiounian, V.M., Arakelyan, V.M., Shahnazaryan, G.E. *et al.* (2007) Photoelectrochemistry of tin-doped iron oxide electrodes. *Solar Energy*, **81**, 1369–1376.

[66] Aroutiounian, V.M., Arakelyan, V.M., Sarkissyan, A.G. *et al.* (1999) Photoelectrochechemical characteristics of the Fe_2O_3 electrodes doped with the group IV elements. *Russian Journal of Electrochemistry*, **35**, 854–859.

[67] Miyake, H. and Kozuka, H. (2005) Photoelectrochemical properties of Fe_2O_3-Nb_2O_5 films prepared by sol-gel. *Journal of Physical Chemistry B*, **109**, 17951–17956.

[68] Kennedy, J.H. and Frese, K.W. Jr. (1978) Flatband potentials and donor densities of polycrystalline α-Fe_2O_3 determined from Mott-Schottky plots. *Journal of the Electrochemical Society*, **125**, 723–726.

[69] Fredlein, R.A. and Bard, A.J. (1979) Semiconductor electrodes. *Journal of the Electrochemical Society*, **126**, 1892–1898.

[70] Kennedy, J.H. and Anderman, M. (1983) Photoelectrolysis of water at α-Fe_2O_3 electrodes in acidic solution. *Journal of the Electrochemical Society*, **130**, 848–852.

[71] Kennedy, J.H. and Dunnwald, D. (1983) Photo-oxidation of Organic Compounds at Doped α-Fe_2O_3 Electrodes. *Journal of the Electrochemical Society*, **130**, 2013–2016.

[72] Aroutiounian, V.M., Arakelyan, V.M., Shahnazaryan, G.E. *et al.* (2002) Investigation of ceramic Fe_2O_3<Ta> photoelectrodes for solar energy photoelectrochemical converters. *International Journal of Hydrogen Energy*, **27**, 33–38.

[73] Agrawal, A., Chauhan, D., Tripathi, C. *et al.* (2006) Preparation, characterization and photoelectrochemical behaviours of titanium and zirconium doped hematite. *Proceedings of the National Academy of Sciences India*, **76A** (III), 189–195.

[74] Gardner, R.F.G., Moss, R.L., and Tanner, D.W. (1966) Electrical properties of alpha ferric oxide cotaining magnesium. *British Journal of Applied Physics*, **17**, 55–62.

[75] Cai, S., Jiang, D., Tong, R. *et al.* (1990) The n-type/p-type photoresponse transition of Mg doped and Zn-doped polycrystalline iron oxide electrode. *Corrosion Science*, **31**, 733–738.

[76] Cai, S., Jiang, D., Tong, R. *et al.* (1991) Semiconductive properties and photoelectrochemistry of iron oxide electrodes-viii. photoresponses of sintered Zn-doped iron oxide electrode. *Electrochimica Acta*, **36**, 1585–1590.

[77] Cai, S., Jiang, D., Zhang, J. *et al.* (1992) Semiconductive properties and photoelectrochemistry of iron oxide electrodes-ix. Photoresponses of sintered Zn-doped oxide electrode. *Electrochimica Acta*, **37**, 425–428.

[78] Wu, Z., Okuya, M., and Kaneko, S. (2001) Spray pyrolysis deposition of zinc ferrite films from metal nitrates solutions. *Thin Solid Films*, **385**, 109–114.

[79] Jin, Y., Li, G., Zhang, Y. *et al.* (2002) Fine structures of photoluminescence spectra of TiO_2 thin films with the addition of $ZnFe_2O_4$. *Journal of Physics D: Applied Physics*, **35**, L37–L40.

[80] Mohanty, S. and Ghose, J. (1992) Studies on some α-Fe_2O_3 photoelectrodes. *Journal of Physics and Chemistry of Solids*, **53**, 81–91.

[81] Watanabe, A. and Kozuka, H. (2003) Photoanodic properties of sol-gel derived Fe_2O_3 thin films containing dispersed gold and silver particles. *Journal of Physical Chemistry B*, **107**, 12713–12720.

[82] Luo, W., Yu, T., Wang, Y. *et al.* (2007) Enhanced photocurrent–voltage characteristics of WO_3/Fe_2O_3 nano-electrodes. *Journal of Physics D: Applied Physics*, **40**, 1091–1096.

[83] Sobas, A.G., Kusior, E., Radecka, M., and Zakrzewska, K. (2006) Visible photocurrent response of TiO_2 anode. *Surface Science*, **600**, 3964–3970.

[84] Gurunathan, K. and Maruthamuthu, P. (1995) Photogeneration of hydrogen using visible light with undoped/doped α-Fe_2O_3, in the presence of methyl viologen. *International Journal of Hydrogen Energy*, **20**, 287–295.
[85] Gondal, M.A., Hameed, A., Yamani, Z.H., and Suwaiyan, A. (2004) Production of hydrogen and oxygen by water splitting using laser induced photo-catalysis over Fe_2O_3. *Applied Catalysis A: General*, **268**, 159–167.
[86] Liou, F.-T., Yang, C.Y., and Levine, S.N. (1982) Photoelectrolysis at Fe_2O_3/TiO_2 heterojunction electrode. *Journal of the Electrochemical Society: Electrochemical Science and Technology*, **129**, 342–345.
[87] Osaka, T., Hirota, N., Hayashi, T., and Eskildsen, S.S. (1985) Characteristics of photoelectrochemical cells with iron oxide/n-Si heterojunction photoanodes. *Electrochimica Acta*, **30**, 1209–1212.
[88] Satsangi, V.R., Singh, A.P., Johri, P. *et al.* (2007) Bilayered photoelectrode for solar hydrogen production system. *Proceedings of XXXI National Systems Conference (CD)*, Manipal, Karnatika, India, Dec 14–15, 2007.
[89] Kumari, S., Singh, A.P., Tripathi, C. *et al.* (2007) Enhanced photoelectrochemical response of Zn-dotted hematite. *International Journal of Photoenergy*, **87467**, 1–6.
[90] Kanjilal, D. (2001) Swift heavy ion-induced modification and track formation in materials. *Current Science*, **80**, 1560–1566.
[91] Kucheyev, S.O., Williams, J.S., Zou, J. *et al.* (2001) The effects of ion mass, energy, dose, flux and irradiation temperature on implantation disorder in GaN. *Nuclear Instruments and Methods in Physics Research B*, **178**, 209–213.
[92] Iwase, A., Sasaki, S., Iwata, T., and Nihira, T. (1987) Anomalous reduction of stage-i recovery in nickel irradiated with heavy-ions in the energy-range 100–120 MeV. *Physical Review Letters*, **58**, 2450–2453.
[93] Bauer, P., Dufour, C., Jaouen, C. *et al.* (1997) High electronic excitations and ion beam mixing effects in high energy ion irradiated Fe/Si multilayers. *Journal of Applied Physics*, **81**, 116–125.
[94] Chaudhary, Y.S., Khan, S.A., Shrivastav, R. *et al.* (2005) Modified structural and photoelectrochemical properties of 170 MeV Au^{13+} irradiated hematite. *Thin Solid Films*, **492**, 332–336.
[95] Chaudhary, Y.S., Khan, S.A., Tripathi, C. *et al.* (2006) A study on 170 MeV Au^{13+} irradiated nanostructured metal oxide (Fe_2O_3 and CuO) thin films for PEC applications. *Nuclear Instruments and Methods in Physics Research B*, **244**, 128–131.
[96] Singh, A.P., Kumari, S., Shrivasrav, R. *et al.* (2009) Improved photoelectrochemical response of haematite by high energy Ag^{9+} ions irradiation. *Journal of Physics D: Applied Physics*, **42**, 085303.
[97] Singh, A.P., Tripathi, A., Shrivastav, R. *et al.* (2008) New benchmark to improve the photoelectrochemical properties of hematite. *Proceedings of Solar Hydrogen and Nanotechnology III*, **7044** (1–8), 70440.
[98] Mettee, H., Otvos, J.W., and Calvin, M. (1981) Solar induced water splitting with p/n hetero type photochemical diodes: n-Fe_2O_3/p-GaP. *Solar Energy Materials*, **4**, 443–453.
[99] Arutyunyan, V.M., Arakelyan, V.M., Shakhnazaryan, G.E. *et al.* (2002) Ceramic Fe_2O_3: Ta photoelectrodes for photoelectrochemical solar cells. *Russian Journal of Electrochemistry*, **38**, 378–383.
[100] Ingler, W.B. Jr. and Khan, S.U.M. (2006) A self-driven p/n-Fe_2O_3 tandem photoelectrochemical cell for water splitting. *Electrochemistry and Solid State Letters*, **9**, G144–G146.
[101] Murphy, A.B., Barnes, P.R.F., Randeniya, L.K. *et al.* (2006) Efficiency of solar water splitting using semiconductor electrodes. *International Journal of Hydrogen Energy*, **31**, 1999–2017.
[102] Khader, M.M., Vurens, G.H., Kim, I.K. *et al.* (1987) Photoassisted catalytic dissociation of H_2O to produce hydrogen on partially reduced α-Fe_2O_3. *Journal of the American Chemical Society*, **109**, 3581–3585.
[103] Gondal, M.A., Hameed, A., Yamani, Z.H., and Suwaiyan, A. (2004) Laser induced photo-catalytic oxidation/splitting of water over α-Fe_2O_3, WO_3, TiO_2 and NiO catalysts: activity comparison. *Chemical Physics Letters*, **385**, 111–115.
[104] Bard, A.J., Parsons, R., and Jordan, J. (eds) (1985) *Standard Potentials in Aqueous Solutions*, Marcel Dekker, New York.

Part Four

New Design and Approaches to Bandgap Profiling and Visible-Light-Active Nanostructures

14

Photoelectrocatalyst Discovery Using High-Throughput Methods and Combinatorial Chemistry

Alan Kleiman-Shwarsctein, Peng Zhang, Yongsheng Hu, and Eric W. McFarland

Department of Chemical Engineering, University of California, Santa Barbara, CA, Email: mcfar@engineering.ucsb.edu

14.1 Introduction

There are an enormous number of semiconductor formulations that could be created from earth-abundant non-toxic elements; however, only a relatively tiny fraction have been synthesized and studied. Unfortunately, we do not have the basic knowledge or theory to predict in advance the existence of a suitable material system or what the composition should be. Combinatorial chemistry is the synthesis and screening of large numbers of different materials from different combinations of chemical variables in a systematic and deliberate manner to explore their composition–structural-property relationships and discover, by induction, an otherwise unpredictable material with specific desirable properties. In the absence of a thorough *a priori* understanding of the many interdependent properties and relationships between components, the components are combined in a large number of different ways and from observations of specific properties of the different combinations the fundamental interdependent relationships are inductively determined. High-throughput experimentation methods using robotics and computers allows the use of combinatorial methods at high speed to increase the rate of discovery and understanding of complex materials. This chapter describes the methodology and prior work using high-throughput experimentation and combinatorial methods for potentially increasing the rate of discovery of new photoelectrocatalysts for conversion of inexpensive and abundant feedstocks into fuels, including the splitting of water into hydrogen and oxygen.

On Solar Hydrogen & Nanotechnology Edited by Lionel Vayssieres

14.2 The Use of High-Throughput and Combinatorial Methods for the Discovery and Optimization of Photoelectrocatalyst Material Systems

14.2.1 The Use of High-Throughput and Combinatorial Methods in Materials Science

Of the 117 + elements in the periodic table, less than one third might be major constituents of a cost-effective solar-conversion system, due to the enormous quantities of material required. Although, known chemical principles and experience can help narrow the number of combinations considerably, there is no theory or formal deductive selection methodology to guide us in choosing, amongst the myriad of chemically sensible possible combinations of these elements, which ones might be a semiconductor with a bandgap between 1.4 and 2.2 eV. In conventional chemical science, to obtain a new material with a specific property, experience and/or a theory is used to suggest such a material and a few carefully selected samples are painstakingly prepared and exhaustively studied until a "complete" understanding of the material's structure and properties is obtained. Thus, the discovery of new solid-state semiconducting material appropriate for solar applications must either await indefinitely a discovery "deduced" using conventional chemical methods from a not-yet-discovered theory, or proceed with an alternative approach, whereby the problem is tackled inductively through the systematic and rapid synthesis and screening of large numbers of selected combinations using methods of combinatorial chemistry and high-throughput experimentation.

Nature has successfully made use of "combinatorial methods" for the evolutionary synthesis and selection of extraordinary combinations of "living" chemical systems with remarkable functionality. In the hands of modern scientists making use of robotics and computers, high-throughput and combinatorial methods in chemistry and materials science are implemented through the deliberate, rapid creation and screening of very large numbers of new materials from different combinations of specific building-block atoms and molecules. This inductive methodology has particular utility in systems where *a priori* theoretical understanding is lacking. Through the analysis of trends in performance related to the material structure and composition obtained from large numbers of new materials, new hypotheses can be generated inductively as previously unpredicted "lead" materials are identified, and conventional analytics and characterizations used to determine why their performance was favorable and to develop new theories so that others can be predicted. High-throughput and combinatorial approaches are not "shot-gun science" or "mindless trial and error;" rather they complement traditional deductive chemistry as epidemiology complements biological and medical science. The methods follow on a rich intellectual tradition that departed early on from the purely deductive structure of Plato to a structure based on induction.

High-throughput and combinatorial methods as a discovery approach are not new and are often rather obvious. Animal and plant husbandry are basically combinatorial methods which have been used to create and select desired genetic traits of living organisms before written history. Inductive approaches to science were systematized in the fifteenth–seventeenth centuries by thinkers such as Francis Bacon, who believed that our understanding of natural phenomena was far from complete and that reductionist methods based on over-simplified models prevented complete understanding [1]. Bacon believed that it was necessary to observe a wide and diverse range of natural phenomena before real understanding could be accomplished. Indeed, it was the empirical observations of parabolic planetary motion, not a hypothesis, that led Newton to suggest his Law of Gravitation [2], and it was Balmer's empirical formula developed by observing the many line spectra of hydrogen under a variety of different conditions, without any prior understanding of the quantum nature of the atom, that led Bohr to his brilliant suggestion that electrons in an atom could only have quantized energies. Informally, bench scientists have long tried to discover new materials with special properties using inductive methods which closely resemble today's "combinatorial" methods.

Giacomo Ciamician, an Italian photochemist working in the early 1900s, was observed to place hundreds of flasks containing different mixtures of materials on the roof of his laboratory in the sunlight to search for unexpected photochemical reactions [3]. Thomas Edison believed that increasing the number of experiments performed and painstakingly recording the observations would increase the probability of a chance success, and that the accumulated information could be used for future discoveries. It was the rapid testing of more than 3000 different materials that made possible the discovery of a material suitable for an incandescent light bulb in 1879. Creating and characterizing large numbers of materials seemingly without much more than intuition has been disparagingly called "Edisonian science." Such criticisms have come from those far less imaginative or important in mankind's history of technological development than Edison. Indeed, Edison was interested in accomplishment and realized the limitations of his own predictive abilities [4]. He recognized the value of increasing the probability of chance successes by systematically increasing the number of experiments. Perhaps as importantly, each of the failed tests generated a detailed record of the unsatisfactory material's properties which was used for subsequent discoveries.

Undoubtedly, unrecognized scientists worldwide have used and continue to use high-throughput and combinatorial methodologies. The concept of laboratory-test automation and high-throughput chemical screening dates back to the 1950s and the explosive growth in medical testing [5]. High-throughput automated chemical analyses were developed to test thousands of clinical specimens each day in single-site facilities. Automated systems for synthesis, analysis and data recording for electrolytes, and specific molecular structures became commonly available [6,7]. What has become possible most recently with advanced robotics and high-speed computer systems is that automated means of performing syntheses and screening are facilitated and the capture and analysis of enormous data sets is possible, such that the age-old inductive methodology which underpins combinatorial chemistry can be more formally implemented.

In the chemical and material sciences, it is assumed that the chemical composition and structure determine the performance of a material. Using limited theory and prior knowledge, a large number of combinations of matter can be imagined, but their performance as a specific material class is not generally predictable. Furthermore, it has been realized that small details in processing have significant impact on performance. In an inductive approach, it is first postulated that a complex manifestation of nature depends upon many, as yet poorly defined, variables, which interact in as yet unknown ways. The variables are combined systematically in as many different ways as possible and observations made of the desired behavior in each case.

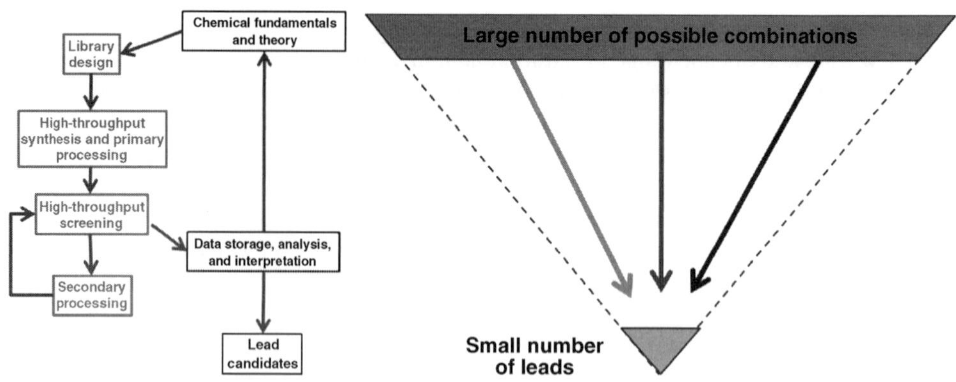

Figure 14.1 General process of combinatorial methods in material science.

Observations of systematic variations in the properties of the system behavior are correlated to the systematic variations in the variables and from these correlations physical laws and relationships are "induced." The general paradigm applied to material science is illustrated schematically in Figure 14.1.

Using the best available theory and fundamental knowledge, a potentially large set of material combinations which might provide a desired material property are selected. Combinations may be excluded due to knowledge of instability and other basic chemical synthesis rules, or due to important practical constraints (cost, toxicity, etc.), or these combinations are used to determine general trends and then excluded later. The different possible combinations are then enumerated and a search strategy defined. Ideally, the search strategy would be defined as the most efficient means of sampling the enormous composition space; however, in practice a large number of practical considerations will dictate the search (e.g., the ease of synthesis, available processing conditions). The objective is not to do many pointless or poorly conceived experiments, the objective is to do good chemistry, just faster. The frequently misunderstood feature of high-throughput and combinatorial methods is that the process always begins with known chemical fundamentals and chemical intuition. It is hoped to narrow the number of compositions whereby the precise and time-consuming traditional methods of chemical synthesis and analysis are utilized to a relatively few leads with a high probability for possessing superior properties.

Using high-throughput synthesis (HTS), an enormous number of combinations of materials are created rapidly, based on a broad hypothesis, then screened for the property of interest. Multiple screening of libraries for other potentially interesting properties may also be performed. Often, the materials are synthesized in two-dimensional $n \times m$ arrays which have become known as "libraries," where one property may be varied along one or more spatial axes. The observations of the behavior of these screening experiments are analyzed and the hypothesis re-examined and modified to guide design of the next set of high-throughput experiments (HTE's). The materials and libraries may be further processed and rescreened to examine processing variables. If the analysis of the screening results identifies materials meeting specific performance metrics, (often called "leads"), these candidates are then either further subdivided for additional HTE, or synthesized and studied using conventional chemical methods.

A formal strategy for synthesizing and testing large collections of multicomponent solid-state materials was introduced by Joseph Hanak and others in 1969 [8,9]. Hanak, while working at RCA Laboratories, published work arguing that doing experiments one sample at a time was an inefficient and expensive means of materials research. He developed physical vapor deposition (PVD) methods based on radio-frequency co-sputtering for the synthesis of diverse multicomponent compounds for use as superconductors, ceramets, photovoltaics and luminescent materials [9–15]. One of several problems Hanak tackled using high-throughput methods was to optimize doping combinations for photovoltaic (PV) semiconductors [14]. HTE led to optimized doping compositions that have been used in commercial solar cells.

Much later, Xiang and coworkers at the University of California, Berkeley demonstrated that PVD synthesis combined with masks could be used to generate small arrays of high-temperature superconducting and giant-magnetoresistive thin films [16]. This work led to the creation of the company called Symyx Technologies (SMMX: NASDAQ) which was initially devoted to the development and exploitation of combinatorial material science. Researchers at Symyx developed a variety of systems for the synthesis and characterization of functional materials now used throughout industry. In one example of work at Symyx, electron-beam evaporation was used to synthesize thin-film libraries containing thousands of potential new inorganic phosphors; unexpectedly, SrCeO3 was discovered to have unusual new luminescent properties and an unexpected structure [17,18].

Researchers at Symyx developed early the application of high-throughput methods to the discovery of electrode and electrocatalyst materials, as well as high-throughput electrochemical synthesis and screening in general [19]. Because of the many synthesis variables under direct control (e.g., voltage, current density, electrolyte composition), electrochemical methods offer a great opportunity for combinatorial exploration of inorganic materials. Later, Reddington and co-workers applied HTE methods in an electrochemical assay to screen ternary metal-alloy electrocatalyst libraries of five transition metals (Pt, Ru, Os, Rh and Pd) as potential anode materials for the electrochemical oxidation of methanol [20]. These metal-alloy fuel-cell libraries were screened in an electrochemical cell using a pH-adjusted aqueous methanol solution containing electrolyte and a pH-sensitive fluorescence indicator (acridine or quinine); electrochemical oxidation of aqueous methanol is known to proceed with the release of protons, decreasing the local pH near the anode. Optical imaging of the catalyst library during a single potential sweep under UV irradiation was used to identify library elements that initiated the oxidation of methanol at the lowest overpotentials. Preliminary results suggested that Pt–Os–Rh alloys were very effective for the anodic oxidation of methanol. In other work, an array of 64 electrodes, each with different surfaces, was evaluated using a computer-controlled multiplexer, which controlled the applied potential and performed current measurements identical to those traditionally performed as individual experiments [21].

14.2.2 HTE Applications to PEC Discovery

The first specific applications of high-throughput and combinatorial methods to solid-state photoelectrocatalysts were reported in 2001, when both high-throughput electrochemical synthesis of photoelectrocatalysts and high-throughput photoelectrochemical screening were reported on doped tungsten oxides as potential PEC candidates for hydrogen production by water splitting [22]. This work also contained the initial description of a two-dimensional WO_3/Pd based chemo-optical sensor system for directly measuring hydrogen production in libraries of photoelectrodes [23]. Shortly afterward, Nakayama and co-workers [24] screened photoanode libraries of titania doped with transition metals prepared using a two-dimensional mask system similar to those in use for high-throughput synthesis of electronic and optical materials. The photo-oxidation of water was measured using a two-dimensional potentiometric sensor system that detected the increased hydrogen-ion concentration from the water oxidation and ferric-ion reduction.

Subsequently, high-throughput investigations of doped and mixed metal oxides based on tungsten [25] and zinc [26] were performed. Like titania, oxides predominately composed of tungsten or zinc do not appear to be acceptable PEC solar spectrum absorbers due to their wide inherent bandgap. Although, the addition of significant fractions of other transition metals can provide increasing visible band sensitivity, there is no evidence that the performance will be acceptable unless other metals forming narrow-gap oxides are present as the majority species.

Recent work by Woodhouse *et al.* [27] has incorporated the versatile high-throughput liquid phase synthesis method using inkjet printers, previously used for combinatorial syntheses [28,29], into metal-oxide libraries based on the potential solar-spectrum-absorbing atoms, Fe, Co, Cu and Ni, combined with structural atoms, Al, and Nd, with charge compensation by Cs. Several libraries with wide composition ranges were created and screening performed using direct readout of the photocurrent while a laser was scanned over the library surface. Potential leads were identified; however, no quantitative performance measurements have yet been published.

Seyler *et al.* have recently published an impressive series of solid-state photocatalyst libraries synthesized and screened for hydrogen production with the greatest diversity of combinations presented to date for photocatalysts [30]. Using a liquid-phase dispensing robot, mixtures of sol-gel precursors were used to create binary and ternary oxides containing an enormous variety of metals, including Li, Na, K, Ba, Bi, Co, Mg, Mn, Nd, Ni, Sb, Sr, Y, Yb, Zn, Al, Fe, La, Zr, V, Ta and Ti. Their screening included absorption and hydrogen production. The promising, and new, lead material $Al_xBi_xPb_{x/2}O_y$ was identified; quantitative investigation of this promising material is pending.

Although still in its infancy, combinatorial synthesis and screening of solid-state materials suitable for photoelectrocatalytic applications is clearly possible, making use of exciting new automation technology for high-speed synthesis and screening of unprecedented numbers of different materials. In all cases, a material system for solar PEC requires that all of the following selection criteria are simultaneously satisfied:

1. Sufficiently abundant, safe, processible and thus cost-effective components
2. Absorbs a significant fraction of the solar photons
3. Efficient bulk carrier separation and transport
4. Efficient redox charge transfer surface reactions
5. Long-term stability in electrolyte.

The materials must come from combinations of elements with sufficient abundance to provide the required $10^3\ m^3\ yr^{-1}$ of solid material required if solar energy conversion is to have a significant impact. If we begin with the approximately 117 elements in the periodic table as potential building blocks and make the economic constraints discussed above, elements which are not at least 1/10 000 as abundant as Si may be eliminated from consideration. Further, the particularly toxic elements (e.g., Pb, Be, Sr) may also be eliminated (or used as a last resort),

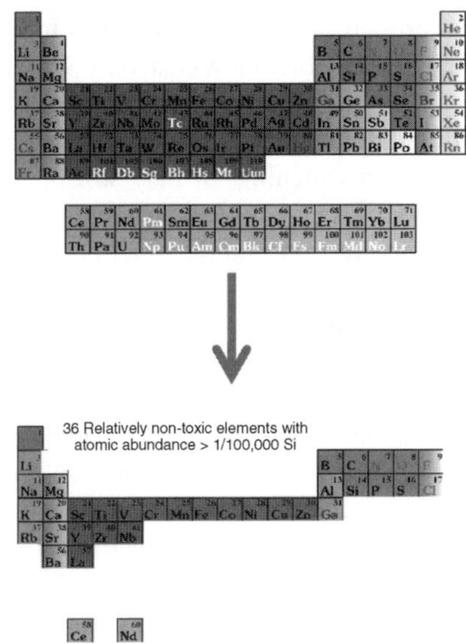

Figure 14.2 Practical components for PEC materials are a relatively small subset of the periodic table. After elimination of elements which are either too toxic or not abundant enough to supply the $10^3\ m^3\ yr^{-1}$ required to produce sufficient product quantity to have an impact on the overall energy demand, the elements of the periodic table can be reduced to a set of approximately 36.

reducing the entire periodic table to approximately 36 elements, Figure 14.2. Although several specific elements included or omitted from this list might be debated, the general approach of "elemental triage" is easily defensible. Nonetheless, the number of possible stable combinations of these elements, which might be semiconductors useful for solar-energy conversion, is enormous. No present theory can accurately predict *a priori* the bandgap or electro-optical properties of more than a tiny fraction of this enormous set of potential solids, and to synthesize and analyze them one-by-one using conventional methods would take centuries of work.

We are searching for a solid-state PEC material, m, composed of elements M_i and X_j which can be synthesized in mass quantities by a synthesis route, S, which meets the cost, \$(M,X,S), and efficiency, ε(M,X,S), requirements for cost-effective solar energy conversion, m(\$(M,X,S), ε(M,X,S)). By selecting only abundant and relatively non-toxic materials, the cost of the PEC material system will primarily be determined by the stability, synthesis and fabrication costs. The efficiency will be a complex function of the relationships of composition, synthesis and atomic-scale structure to the visible band absorbance, conductivity and macrostructure.

Most investigators work under the assumption that if a composition and structure can be identified, then, through ingenious chemistry, a practical and cost-effective synthesis route will ultimately be found; thus, the essence of the problem is to search through the combinations of acceptable elements which could, in principle, create a solid-state PEC material system and find those with a sufficiently high efficiency for solar-to-chemical conversion to meet the overall requirement of cost-effective solar-to-chemical conversion. Simply considering numerical combinations of the 36 elements alone makes little sense and, immediately, fundamental chemical principles and theory (when known) must be applied to prepare and prioritize a selected subset of combinations for potential synthesis.

Existing fundamental theories allow several material properties of elemental solids to be calculated accurately *a priori*. Of the building-block elements available at low cost, only Si is an elemental semiconductor; however, its bulk bandgap is 1.1 eV and it is highly unstable with respect to oxidation, it is thus not a single-junction candidate for water splitting. If passivated and synthesized as a nanoparticle with size-induced bandgap widening, Si may still be an excellent candidate for PEC applications. Theory provides less accuracy in the calculation of optical bandgaps and other PEC-related properties for binary, ternary and the other complex solids. Experimental data is required to reliably evaluate their properties. Thus, with relatively few exceptions, elimination of choices based on known or predicted material properties *a priori* is limited, and one would not want to eliminate a potential "hit" without very reliable evidence.

In the absence of theory or experimental evidence, prioritization based on experience, chemical intuition and extrapolations of known experimental observations is required in order to select the first combinations to be synthesized. Combinations of the elemental building blocks are synthesized using several different, high-throughput synthesis methods, guided by basic chemical reasoning (atomic size, valence, etc.), theory and chemical experience to create large numbers of material compositions, typically in n $\times$ m libraries. The new materials are screened for absorbance in the 1.6–2.2 eV range. The screening of each library may provide information on trends or unexpected performance that will result in a reprioritization of the combinations which will be synthesized next. Further, from analysis of trends, new hypotheses as to composition–structure–performance relationships may be developed and used for subsequent prioritizations (Figure 14.3).

Lead materials selected based on their absorbance will next form the basis of new libraries which may include variations in synthesis variables and impurities or dopants. Since it is known that the electronic properties of inorganic solids are exquisitely sensitive to processing conditions and impurities, the initial lead compositions are synthesized and processed differently (e.g., variable temperature of synthesis or processing, processing gases, vacancies, dopant concentrations, etc.) and the new material libraries are screened as photoelectrodes for incident photon-to-electron conversion efficiency (IPCE) from illumination in the visible spectrum, for example, at 600 nm (2 eV). IPCE will depend on the process of absorption, carrier transport and surface redox charge transfer. Electrolytes with reactant donor/acceptor pairs that possess relatively fast electron-transfer kinetics are used to minimize the contribution of the redox charge-transfer rate in this screening step. Once a material composition and set of processing conditions are identified that meet a specific criteria for IPCE performance, the material is selected as a candidate for more traditional methods of solid-state chemistry and electronic material optimization and evaluation.

14.2.3 Absorbers

In a semiconducting solid, it is understood that absorption occurs through the interaction of the electromagnetic field of the photon and the intrinsic (direct gap) or induced (indirect gap) dipole moment of the solid absorber. The details of the establishment of the dipole and the associated energy gap is not fully understood in complex solids, and thus within the group of elemental building blocks there remains an extraordinarily large number of chemically possible combinations with largely unknown absorbance properties. At present, there is no high-throughput synthesis and screening technology that would allow the actual synthesis and screening of all the possible combinations of these building blocks to fully ensure that all

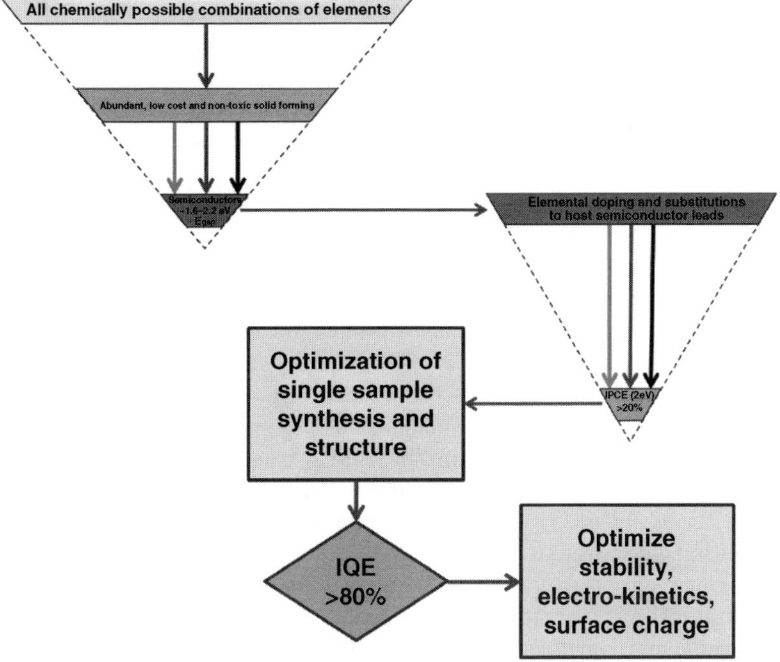

Figure 14.3 Schematic representation of the high-throughput methodology applied to solid-state PEC materials. From the Periodic table those elements that are relatively abundant, non-toxic, cost-effective, and might be combined into solid semiconductors, comprise a smaller set of approximately 35 building block elements. Libraries selected from combinations of these 35 elements are synthesized and screened for bandgaps between 1.4–2.2 eV. Lead materials satisfying this selection criteria are synthesized and optimized to achieve IQEs greater than 80%.

combinations have been checked. Instead, a combination of theory and chemical intuition is used, together with one or more high-throughput syntheses to systematically examine and exclude or include specific candidates.

We begin with the reduced set of elemental building blocks and consider what is known and understood about solid-state materials. Of the 36 possible single-element combinations, only Si forms an elemental semiconductor with visible band absorption. All two-element combinations that can form solids have been made and have known properties. Because several ionization states are possible, there are numerically greater than 36^2 two-element combinations; however, many are not stable solids. The principle classes include carbides, oxides, phosphides, nitrides, sulfides and halides.

Of the possible ternary compositions, less than 5% have been made and characterized and of the quaternaries almost nothing is known. In addition to the classes listed above, there are the more complex mixed-anion materials, oxychlorides, oxyflourides and so on, and of these, most ternary materials containing a single cationic species and two anionic species have been made and characterized. With the exception of $La_xSr_yCu_zOS$, there are no known semiconducting mixed anionic materials in these groups that have visible absorbance. This does not mean that none exist, rather only that none are known.

Although presently limited, significant progress in applying advanced theoretical methods to determining optoelectronic and electrocatalytic properties of materials possibly suitable to PEC has been made [31,32]. In particular, theory has helped explain the observations that the poor conductivity of the Mott insulators (e.g., iron oxide) may be improved by increased dopant concentrations and that mixed complex oxides such as the delafossites of the structure $Cu(X)O_2$ might be particularly efficient PEC materials.

Table 14.1 below lists most of the known combinations of the 36 target elements that have been synthesized and are known to form semiconductors with visible band absorbances and bandgaps near our target range.

There are a large number of ways of categorizing the possible material groups, (e.g., by crystal structure, etc.) and this is only one of many. In principle, there may be additional complex anion structures with three or more members, for example, oxyfluorochlorides; however, for the purposes of designing libraries of specific compositions to synthesize these will be given a lower priority since none in these classes are presently known (or predicted) to have desirable properties. Experience with more complex compositions is far less complete and based on the fact that the unpredicted, serendipitous, and best high-temperature superconductor is $Hg_{12}Tl_3Ba_{30}Ca_{30}Cu_{45}O_{125}$, there is considerable hope that a highly efficient PEC material is yet to be discovered.

A discouraging reality is that, in spite of the large financial incentives to identify low-cost solar spectrum absorbers for photovoltaic applications, only Si has shown enough promise to be produced in sufficient quantities to test the technoeconomics and, to date, Si-based solar conversion is not cost effective. Although both photovoltaics and PEC materials must absorb in the visible and have lifetimes of many years, a photovoltaic material need not be stable in an aqueous electrolyte, nor are specific redox reactions required. The bandgap for maximum extraction of energy from sunlight in a single-gap PV device is approximately 1.2 eV; thus, Si is nearly ideal. In PEC applications, Si suffers from oxidation in an aqueous environment that has been difficult to control, and thus application in electrolytes would be difficult; nonetheless, Texas Instruments developed a nearly commercial PEC process based on Si for the photoelectrochemical synthesis of hydrogen and bromine from HBr [33], and several groups remain

Table 14.1 Potential cost-effective components for PEC material systems. Material bandgap listed in eV.

>1/1000 of Si	>1/10 000 of Si	>1/100 000 of Si
Si, C, Mg, Na, H, Al, K, Ca, Ti, Mn, Fe	Li, Ba, Cu, Zn, Sr, Zr, B	Sc, V, Co, Cr, Ni, Ga, Rb, Y, Nb, Ce, Nd
Si (1.1)	**CuO(1.7)**	**Co_3O_4(2),**
SiC(2.8)	**Cu_2O(2.2),**	**$CoTiO_3$(2.2)**
FeO(2.4)	**$CuAlO_2$(1.6)**	**$LiCoO_2$(1.5)**
Fe_2O_3(2.2)	$CuGaO_2$(1.8)	NiO (3.2)
FeS_2(1.0)	Cu_5FeS_4(1.0)	$NiMoO_4$(2.0)
$FeYO_3$(2.6)	$SrCu_2O_2$(2.4)	**$NiTiO_3$(2.2)**
AlP(2.4)	Cu_2S(1.1)	Ga_2O_3 (4.7)
	ZnS_2 (2.7)	GaS(2.5)
	Zn_3P_2(1.5)	GaP(2.3)
	$ZnSiP_2$(2.1)	GaN (3.4)
	BP(2.4)	Ce_2O_3(2.4)

dedicated to passivating Si for photoelectrochemical applications [34]. Other materials that have past or present commercial photovoltaic support include Ge, GaP, CdTe, CdS, PbS, $CuIn_xGa_{(1-x)}Se_2$ (CIGS), GaAs and GaIAs. One can assume that industry has tried many other combinations and not widely publicized their negative results. Interestingly, the oxides of Fe and Cu have reasonable solar-spectrum absorbance and they are inexpensive materials, yet no commercial photovoltaic products have come from either oxide, primarily because most stable oxides of Fe and Cu have bandgaps greater than 1.5 eV and their bulk conductivity is relatively poor. Although titania has been widely studied and mixed titanium-containing materials show promise, there is no evidence that oxides with a majority of titanium will have sufficient solar-spectrum absorbance to be a candidate for cost-effective solar-energy conversion to fuels or chemicals.

Of all possible combinations from the set of approximately 36 cost-effective PEC elements, there are less than 10 binary or ternary materials that are known to have a bandgap of between 1.2 and 2.2 eV. All binary elemental materials have been made and characterized; however, thousands of possible ternary combinations remain to be investigated, as well as tens of thousands of quaternary combinations. Based on our knowledge of the stability of the general classes of materials in an electrolyte environment and the known absorption properties, an initial prioritization for library synthesis can be made. One reasonable starting point would be libraries of oxides of the classes $Fe_xM_yO_z$, $Cu_xM_yO_z$, $Co_xM_yO_z$ and $Ni_xM_yO_z$, sulfides of the classes $Fe_xM_yS_z$ and $Cu_xM_yS_z$, and phosphides based on zinc, $Zn_xM_yP_z$. Such an initial ranking is not to exclude any other possible combinations, rather to simply prioritize the order of experiments based on our knowledge today. Given enough time, all compositions not ruled out should eventually be made and the set reprioritized as new knowledge is obtained. The initial screening for a PEC host from this set will be based solely on bandgap, and if the absorbance band is not squarely in the visible, the material will not be a lead candidate for a solar PEC material system for water splitting. The approximately 10 leads which have already been identified, including the oxides of iron and copper are subject to further high-throughput experimentation and screening for other key PEC properties.

14.2.4 Bulk Carrier Transport

Upon selection of "lead" compositions based on bulk absorbance, the next step involves evaluating and optimizing the material for charge transport. The selected host absorbers are synthesized as libraries with variables controlled to influence carrier lifetime and mobility. These include the processing conditions, substitutions, vacancies and dopants, all of which will affect conductivity and recombination. Screening of PEC libraries for optimized carrier transport will typically be done with actual electrochemical measurements of the IPCE of the materials made in libraries composed of thin films. Ideally the IPCE is measured with a selected redox pair of reactants that are readily oxidized and reduced without kinetic or overpotential limitations, which may be later optimized through the use of a surface catalyst.

Libraries synthesized in thin-film formats on conducting substrates allow a bias to be applied to distinguish, to some degreee, charge-separation effects from mobility and recombination. In bulk powder samples, there may not be a sufficient asymmetric internal field to drive charge separation, even if the internal quantum efficiency (IQE) and mobility is high. Efficient charge-carrier separation and transport will be dependent on the details of the material system and micro/nanostructure. Establishing an internal electric field in which to separate the electron and hole requires a heterogeneous electronic environment. The measurement of photocurrent and IPCE during the screening process, with and without an applied bias on thin-film samples allows the materials' conductivity to be evaluated, as well as the flat-band potential; all of this information may be potentially used to make improvements in the materials. Although sample-to-sample variability in thickness, reflectance, and other variables still exists, examining trends and searching for local maximums that exceed basic criteria allows selection of lead materials. As an example of a screening criteria, a thickness-corrected IPCE of greater than 20% without an external bias at 500 nm ($\sim$ 2eV) would suggest that the material should be further optimized for PEC applications.

14.2.5 Electrocatalysts

In general, the IPCE screening for charge transport is done with donor/acceptor reactant pairs that have fast and efficient outer-sphere charge-transfer rates without the need for catalytic surface reactions (e.g., ferricyanide). The desired substrates for actual PEC systems are usually not so ideal, and often the microkinetic reaction pathway consists of multiple steps with and without electron transfer. The reaction rate on the host absorber material may be unacceptably low and only through the use of a specific electrocatalytic material on the absorber surface will the rate be acceptable. Water splitting requires both an efficient hydrogen evolution reaction (HER) pathway and an oxygen-evolution reaction (OER) pathway. Both of these processes have been well studied theoretically and experimentally, and both pathways contain both electron-transfer steps and elementary steps that do not involve net charge transfer. On many semiconducting materials (especially oxides) these pathways are not particularly efficient, and electrocatalytic materials are required. Several candidate electrocatalysts are known, including Pt and several metal alloys for the HER [35] and NiO, Pt, RuO_x, CrO_x and VO_x for the OER [36,37]. It is unlikely that Pt would be available in quantities sufficient to meet the demands of a widely deployed PEC system; however, suitable substitutions with oxides of Ni, V or Cr would be sensible materials to evaluate for the OER catalyst, and Au/Pd and Pd/Ru alloys for the HER catalyst. For other potential PEC reactions, the catalyst materials would be different. For the decomposition of biomass-derived organics, the oxidation electrocatalysts would be similar to the oxygen-evolution material; however, less work has been performed in this area and high-throughput screening might increase the number of candidates. For carbon-dioxide reduction, Cu and Pb have been used as electrocatalytic cathode materials. It is known that the semiconductor on which the electrocatalyst is deposited influences the performance of the catalyst. For a specific PEC reaction, with a selected host absorber, high-throughput screening of potential lead electrocatalysts on prepared large-area host substrates allows rapid evaluation of the substrate–catalyst interactions and integrated performance.

14.2.6 Morphology and Material System

The functional PEC material-system structure will depend in part upon the overall system design. At present there are two, fundamentally different, system configurations under consideration. The first configuration consists of photoreactors incorporating a photoelectrode structure, whereby the PEC material is supported as an electrode and oriented such that charge carriers are conducted to physically separate electrolyte compartments connected through a semi-permeable membrane. The second system configuration is based on a collection of individual PEC particles, either free floating as a slurry or supported on an inert, transparent support network in a single electrolyte tank. The configurations are shown schematically below (Figure 14.4).

Several different manifestations of the basic designs are possible, however, the concepts are the same and the design requirements and pros and cons will be similar. Photoelectrode-based designs make use of similar concepts as photovoltaic; however, their cost per unit area for the absorber is similar to PV, plus the additional cost of an ion-exchange membrane. The slurry design has the advantage of significantly lower absorber-system cost and the requirements for the absorber are less stringent, since absorption in any one particle is not critical and reactor depth is relatively inexpensive. Low-absorption indirect-gap materials can be used, provided their IQE is sufficiently high. Furthermore, the slurry designs allow replacement of the PEC material without complete disassembly of the reactor system. In both cases, appropriate surface electrocatalysts which minimize any backreactions and have high surface area with minimized distances to the sites of redox chemistry are desirable.

Ultra-high surface area mesoporous or nanoparticulate materials are an obvious means of minimizing charge-carrier pathlengths in PEC materials; however, in photoelectrode configurations, it has been shown that micron- scale features are optimal [38]. This limit does not apply to slurry-phase particulate systems and ultra-high surface area materials may be utilized. In either case, it is important to balance the decreased distance to the surface with the potential for increased numbers of surface sites for recombination. Attention to surface recombination is important, and processing and surface passivation may be important high-throughput variables for minimizing this undesirable pathway.

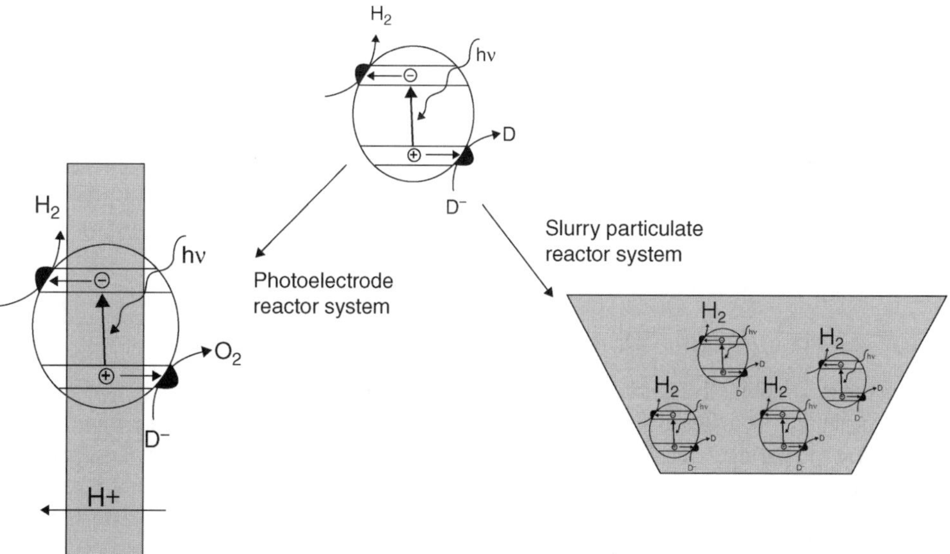

Figure 14.4 PEC reactor configurations. (Right) Photoelectrode-based reactors allow separation of products and prohibits any product back-reactions. (Left) Particulate-based photoreactors make use of low-cost tank reactors; however, secondary-product separation is necessary and rapid removal of products required to minimize back-reactions.

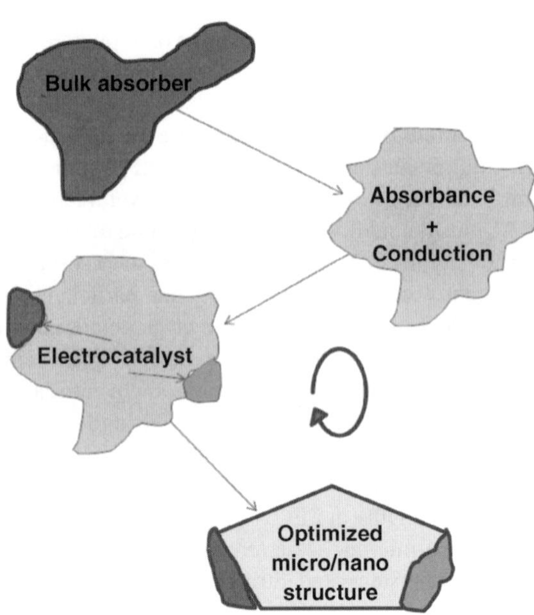

Figure 14.5 Process for optimization of PEC performance through high-throughput experimentation. First, the absorbance and conductivity are optimized; second, electrocatalysts for specific chemical conversions are identified and finally, an optimized shape suitable for a specific reactor configuration is found.

To arrive at specific structures and morphologies for a given system, high-throughput experimentation can be used in several steps after a specific host absorber with satisfactory charge transport is identified. Libraries representing different morphologies of the same host can provide insight into specific improvements and aid in optimization of the structure. The observed trends in performance properties with decreasing dimensions can point to the increasing importance of surface recombination and provide hints as to how to passivate the surface to improve the performance further (Figure 14.5).

14.2.7 Library Format, Data Management and Analysis

Among the major objectives of high-throughput experimental methods is the exploration of the largest variations in material properties in the least amount of time. There are innumerable formats for preparing, characterizing and storing the materials and information. Several common configurations are shown in Figure 14.6. Large collections of individually prepared materials may be created (left); however, for convenience, the collections of solid-state materials are typically produced in two-dimensional collections called libraries. The material(s) may be synthesized and/or arranged as an array of discrete, uniform, samples, each with different properties, or deposited over the substrate such that the variable properties change continuously as a function of position on the substrate.

Typically, the process begins with a decision as to which variable(s) will be screened within the library (e.g., atomic composition, defects, dopants, electrolytes). The library is designed

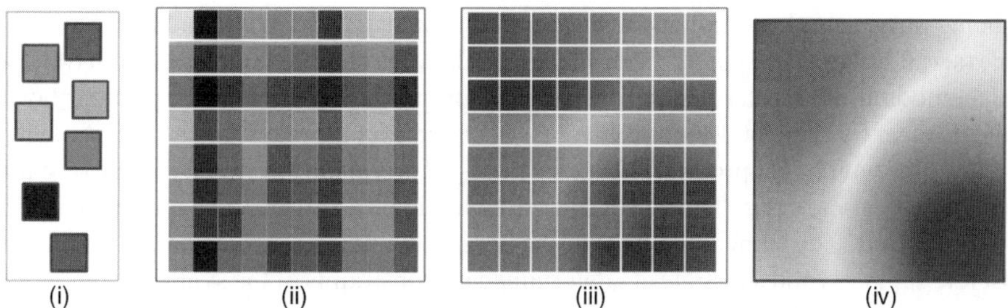

Figure 14.6 High-throughput synthesis of PEC materials in several different formats with color representing different synthesis and/or processing variables (e.g., composition, conditions, etc.). Left-to-right: Materials may be rapidly synthesized and screened as: (i) individual samples with varying properties, or arranged into two-dimensional arrays with (ii) random variations, (iii) discrete systematic variations, or (iv) continuous variations. See also color-plate section.

with both fundamental considerations regarding optimal sampling of the variables, and practical considerations, including experimental limitations. The design and layout of the libraries is coordinated with syntheses and screening, which may be performed in parallel or in series, and typically determined by the restrictions of the instrumentation and the synthesis details. The information used in the design and synthesis of the library must be recorded and attached to the data stream associated with the library. If an automation program is used for synthesis, the specific synthesis program and appropriate experimental details must be tagged to the specific library.

Screening and analysis of the PEC library may begin immediately after synthesis or await an initial processing step. The details of the processing are added to the library's data file. Initial evaluation of the library screening data is typically performed prior to subsequent syntheses so that anything learned may be incorporated into subsequent library design. Internal standards and duplicate libraries must be used to validate screening and synthesis protocols. High-throughput experimentation gives rise to enormous quantities of data, which can quickly overwhelm even the best-prepared experimental team. It is critical to design up-front a means of systematically capturing, storing and analyzing data from any high-throughput experimentation system [39].

14.3 Practical Methods of High-Throughput Synthesis of Photoelectrocatalysts

The synthesis of high-quality solid-state materials involves aspects of both art and science, and if care is not taken, HTE methods have the potential to produce enormous numbers of useless materials extremely rapidly. To develop and validate the appropriate synthesis method, such that a desired variable is indeed changing controllably within a library, can take an enormous amount of time, possibly negating any advantage of the high-throughput approach over traditional methods. The practical implementation of high-throughput methods involves tradeoffs between the number of samples, the synthetic accuracy of the samples, the quality and accuracy of the screening data, the number of different screening methods and the total time. Depending upon the specific synthesis method, this can be accomplished either by producing many samples simultaneously (parallel synthesis) or rapidly producing samples one by one (serial synthesis). Typically, parallel methods achieve more samples per unit time, while serial methods allow greater variability in sample properties at lower synthesis rates; however, there is no clear fundamental general advantages of one over the other and the specific synthesis will determine which is appropriate. A major trade-off is typically between the rate of synthesis versus the rate of screening. The rate of synthesis should be no greater than the rate of screening and data analysis. The form and format of the library, including size, demarcation (continuous or discrete samples) and substrate are determined by constraints of the synthesis and screening system, as well as details pertaining to any post-processing desired for the samples.

For creating libraries of solid-state photoelectrocatalysts, several synthetic routes are possible. The most successful work in HTS has utilized physical and chemical vapor deposition, liquid phase synthesis, electrochemical synthesis and spray pyrolysis. The selection of the synthesis method utilized for a given material class is obviously dependant on the specific material. Some materials have a limited number of practical synthetic routes (e.g., GaN) whereas others can be made utilizing a large number or synthetic pathways (e.g., metal oxides).

14.3.1 Vapor Deposition

Physical vapor deposition, PVD, is widely used for electronic and optical materials where highly controlled syntheses are needed. A solid source material(s) is evaporated under vacuum conditions onto a substrate to create a solid film. The energy to vaporize the solid may be supplied from a resistive heater (thermal evaporation), an electron beam, a laser (laser ablation), a cathodic discharge or plasma. PVD may be used to create most host materials suitable for PEC applications. Chemical vapor deposition, CVD, techniques can be performed at atmospheric pressure, low pressure or in ultra-high vacuum. The precursor used for these methods can by dosed into the system in the gas phase [40,41] by aerosols, or directly by using a liquid in which the precursors can be dissolved [42,43]. The deposition methods range from atomic-layer deposition, where the precursor reacts on the surface to methods where an external energy source is used to enhance the deposition rate, such as in the case of plasma-enhanced CVD, hotwire CVD, thermal CVD or vapor-phase epitaxy.

The major advantages of vapor deposition methods include: precise control of the film thickness, control of film stoichiometry by varying the deposition rate of each of the sources, and by using a wide variety of material types including metal oxides [44–47] and metal oxynitrides [48–51], as well as oxygen-free materials such as nitrides including GaN [52]. Vapor deposition also has several disadvantages, including the need for expensive synthesis systems, peripheral equipment and maintenance, the use of ultra-high purity source materials and the limited range of scalability due to the size and cost of the systems. Nonetheless, by adding masking systems and other automated means of high-throughput sample production, vapor deposition can be extremely rapid and solid-state materials can routinely be made at rates of more than 100 sample compositions per day; libraries of inorganic phosphors have been created with thousands of different metal oxides [17,18].

There have been few examples of PEC materials synthesized by high-throughput PVD techniques, in part due to the complexity of the systems required and the associated costs. PVD was used to produce libraries of electrocatalysts for the HER [23], which is relevant to photoelectrochemical hydrogen production. Jaramillo *et al.* [23] synthesized a library of 24 different hydrogen-evolution catalysts from eight elemental PVD sources, by electron-beam evaporation using two electron beams and masks to control which sources were used during the depositions; elemental and binary materials were readily created. Eight of the 24 materials synthesized were pure elements, while the remaining 16 were binary compounds co-evaporated in a ratio of approximately 80/20, with the exception of Au/Ge which had a ratio of 1/1. The electrocatalysts were synthesized on a single library with four replicates per composition variation, resulting in 96 different catalysts in the library. Results from this work will be discussed in Sections 14.4.1 and 14.5.3 below. While this work made use of discreet composition variations in the electrocatalyst libraries, others made use a continuous "composition spread" to produce a library in which the compositional variation of the materials varied continuously across the library [9,53,54].

Commercial systems are available for automated vapor deposition of libraries from materials that would be suitable for photocatalyst synthesis. Examples include systems made by Kurt Lesker, such as the AXXIS™ Deposition System, which can be configured for magnetron sputtering, electron beam evaporation and thermal evaporation of different sources at the same time to co-deposit different materials; or the SPECTROS® deposition system also from Kurt Lesker, which was designed for light-emitting diode synthesis and includes integrated masking capabilities. AJA International, Inc. offers custom-engineered magnetron sputters, which can include electron-beam evaporators and thermal evaporation on their ATC™ systems, which, when combined with numerous turret sources, can deposit up to 36 different binary materials. CHA Industries, Temescal, and Cambridge NanoTech Inc, among others, also offer different systems for high-throughput synthesis using electron-beam evaporation, sputtering, thermal evaporation, atomic layer deposition and other techniques.

14.3.2 Liquid Phase Synthesis

In high-throughput liquid phase synthesis of solid-state materials, liquid precursors are delivered to a library substrate where the liquid is dried and processed. Although the liquids may be delivered by hand with pipettes, two automated means of delivering the liquids have been used to produce photoelectrocatalysts: inkjet printing and liquid-dispensing robots. Both methods draw on an enormous body of knowledge from other fields for their inspiration. Including the time required to prepare the precursors, and deposit and process the library, typical synthesis rates are hundreds of samples per day.

14.3.2.1 Inkjet Printers

Commercial inkjet printers deliver liquid inks under precise control to specific locations on a paper substrate. By replacing the colored inks with different solid-state material precursors, modified inkjet printers can deliver precise amounts of specific precursors to library substrates using the software and control systems already developed for the printer. Applications to high-throughput solid-state synthesis began in 1998 [20,55] and were the first of several applications for deposition of electrocatalysts for direct methanol fuel cells [56,57]. Recently, these methods were applied to the deposition of photoelectrocatalysts [27,42]. Typically, the solid-state precursor solution is substituted into an inkjet cartridge that has been reconditioned for this purpose. Commercial inkjet printers, such as an Apple Style Writer Printer [56] or a Hewlett-Packard Deskjet [27] have been used for materials synthesis. Commercial systems for inject printing such as the Fujifilm Dimatix model DMP-2800 have also been used for "printing" photocatalysts [42]. Examples of precursors used for photoelectrocatalysts [42] include $Fe(NO_3)_3$, $Cu(NO_3)_2$, $Co(NO_3)_2$, and $Al(NO_3)_3$ and $TiCl_4$ in 2M HCl. These precursors were made suitable for the inkjet dispensers by the addition of the metal nitrate salts with 35 vol% diethylene glycol and 1 vol% diethylene glycol monobutyl ether, which serve as a viscosity agent and surfactant, respectively. A printing pattern is designed as a printable "document" and the precursors "printed" onto a substrate, such as fluorine-doped tin oxide (FTO) or Toray® carbon paper. Due to the increased interest in semiconducting polymers and flexible circuit boards manufactured by microprinting, manufacturers include ArrayIT [58] and FujiDimatix, who have developed systems that can be used directly to print diverse materials without having to adapt a commercial printer. These special-purpose systems offer many other features; however, the price is several orders of magnitude greater than the standard inkjet printers.

14.3.2.2 Liquid-Dispensing Robots

The biotechnology industry has a long experience with automated pipetting and liquid dispensing of biochemical precursors, and a large number of commercial robots for the delivery of diverse liquid compositions to multiwell plates or vials exist that can be readily adapted for delivery of precursors for solid-state PEC synthesis. Dispensing robots such as the 215 Liquid Handler/Injector from Gilson® may be used to dispense liquid precursors to a library substrate followed by drying and thermal reaction of the solution (sol-gel or high-temperature calcination). Maier *et al.* [30] have shown a combinatorial-chemistry methodology, using a liquid-dispensing robot to create metal-oxide libraries from sol-gel precursors. The precursors are dispensed into the vials where they react, gel and, after evaporation of the solvents, the gel is calcined to give metal oxides with diverse composition ranges.

Several variations of this methodology have been used, including specific applications to PEC synthesis using both homemade and commercial robotics obtained from several companies such as J-Chem Scientific and Hamilton. Specific applications have been presented by Funahashi *et al.* [59] for thermoelectrics and by Arai *et al.* [60] for photocatalysts. Their method consists of the use of a liquid-dispensing robot to prepare binary metal oxides in a 96-well microplate; after the solutions have been prepared and mixed, the dispensing robot delivers the solutions to an FTO substrate in a 14×3 array. A robotic arm is then used to transfer the substrates into an electric furnace, where the samples are calcined at 550 °C, then removed for additional depositions and recalcined.

14.3.2.3 Advantages and Disadvantages of Liquid-Phase Synthesis

The major advantage of liquid-phase synthesis is that the instrumentation required is often relatively inexpensive. In many cases, minimal modifications are needed to off-the-shelf commercial equipment commonly available to the pharmaceutical industry for high-throughput combinatorial chemistry. The synthesis components for high-throughput synthesis of powders can be particularly low cost [30]. Typically, the liquid precursors require relatively high temperatures to volatilize the unwanted components and crystallize the desired metal oxide. Unfortunately, the required high temperatures prevent the use of many desirable substrates and do not allow for metastable-phase materials to be produced, which may have superior properties.

14.3.3 Electrochemical Synthesis

Two different approaches have been developed by McFarland *et al.* [26,61] for high-throughput electrochemical synthesis of metal oxides as photoelectrocatalysts. The first method utilizes a rapid serial electrochemical deposition system where PEC samples are deposited one at a time on the library substrate, while the second method uses a parallel electrochemical deposition system, where the samples are deposited at the same time on the library. Brief descriptions of the serial and parallel electrochemical deposition systems are given below; full details of such system have been published by McFarland *et al.* [23,26,61–64].

The rapid serial electrochemical deposition system consists of five main components: (i) motion hardware, (ii) liquid delivery system, (iii) combinatorial block for sample synthesis, (iv) electrochemical/electronic equipment and (v) software. Figure 14.7 shows a photograph of the rapid serial electrochemical deposition system. The motion hardware consist of X-Y linear motion stages (Danaher Motion and Micos), which are stacked on top of each other to create an X-Y table, while a Z stage (Micos) is mounted on an custom-made arm on top of the Y stage, as shown in Figure 14.7a. The setup is controlled using a Nu-Drive® power amplifier and a motion control card PCI-4CX card (National Instruments). Diverse custom-made fastening components are needed to hold the stages and the electrode holder in position on the Z stage. The electrode holder consists of a polypropylene block, in which a working and a counter-electrode (thin metal wires) can be fastened, as shown in Figure 14.7c; alternatively a capillary tube can be used to introduce H_2O_2 into the electrodeposition well for the synthesis of iron-oxide thin films [65], prior to deposition by the use of the liquid delivery system, which consists of a computer-controlled peristaltic pump (Cole Palmer). Typically, electrolytes with varying concentrations of the metal precursors to be used for deposition are transferred to the combinatorial block by a robotic or hand-held pipetter.

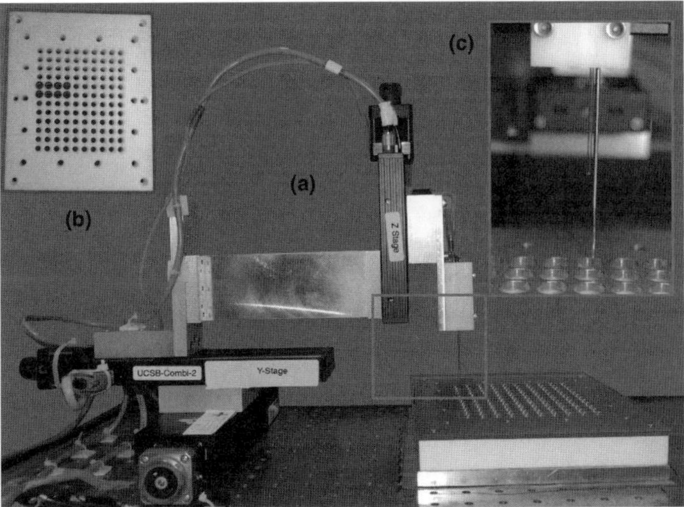

Figure 14.7 Photograph of the rapid serial electrochemical deposition system showing: (a) overall system, (b) combinatorial block, (c) electrode holder.

The combinatorial block, Figure 14.7b, consists of a perforated polypropylene-or glass-filled Teflon™ block (for nonaqueous electrolytes), which makes a seal with the substrate via O-rings in order to isolate samples from each other and to prevent leakage of the electrolyte. Hence, an array of individual working electrodes (individually isolated electrochemical cells) is created, allowing for different synthesis conditions at each library position, as each cell is filled with a compositionally unique electrolyte. The electrochemical equipment consists of an EG&G 273 Potentiostat, a PCI-6036E card, which is used to control the potentiostat for pulse-deposition experiments, and a multimeter (HP34401A), which is controlled through a GPIB interface used to read the current or voltage during the deposition. Finally, all the hardware is controlled and synchronized by a virtual instrument panel, which has been developed with LabVIEW™ (National Instruments).

The rapid serial electrochemical deposition system is versatile and permits the electrodeposition of samples by four methods: (i) galvanostatic deposition, (ii) potentiostatic deposition, (iii) pulse deposition (two or three pulses; square or sine pulse; pulse width $>25\,\mu s$) and (iv) cyclic voltammograms. The size of the library deposited is limited by the size of the combinatorial block, the travel distance of the stages or by the total deposition time needed to perform the experiment; in our case, up to 120 samples have been deposited on a single substrate.

The parallel electrochemical deposition system consists of the following main components: (i) combinatorial electrode array, (ii) combinatorial block for sample synthesis, (iii) electrochemical/electronic equipment and (iv) software. Figure 14.8 shows a sketch of the system. The main advantage of the parallel method is the rate of synthesis. The synthesis time for n samples is $\sim(n-1)^{*}t$ deposition, less than that of the serial approach. In addition, the parallel system has no moving parts. In the simplest manifestation, all deposition cells have the same potential between the working and the counter-electrodes. Control over conditions for deposition of each individual sample is possible in the parallel system with more complex (and expensive) instrumentation and multiplexing. In practical applications, library synthesis

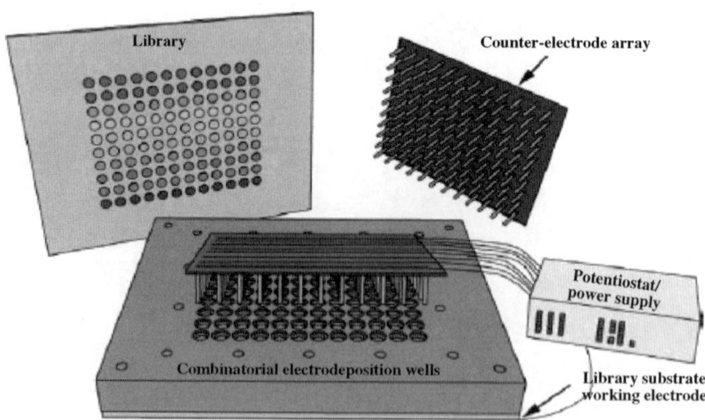

Figure 14.8 Sketch of the parallel electrochemical deposition system. (Reprinted with permission from S.H. Baeck, T.F. Jaramillo, A. Kleiman-Shwarsctein and E.W. McFarland, Automated electrochemical synthesis and characterization of TiO_2 supported Au nanoparticle electrocatalysts, *Measurement Science & Technology*, **16**(1), 54–59, 2005. © 2005 Institute of Physics.)

rates, including the time required to prepare the precursor electrolytes, deposit and process the library, are ~120 samples per day.

The combinatorial electrode array consists of series of stainless-steel counter-electrodes in which each row (12 electrodes) has the same common potential, two-electrode configuration, while each one of the eight columns can have a different applied potential, as seen in Figure 14.8. The combinatorial block used in this setup is the same as described for the serial system above. The electrochemical equipment in this case consists of a data-acquisition card, in which eight analog output channels are used to generate the applied voltage. A home-made eight-channel power amplifier is used to boost the signal from the card, such that there is enough current applied to each one of the cells to perform the electrochemical depositions. The software consists of a virtual instrument, which was programmed in LabVIEW™ (National Instruments). In this case the hardware limits the depositions to potentiostatic two-electrode depositions and to pulse depositions (in which the pulse width is in the ms range). As mentioned before, this system has the advantage of faster synthesis, since all depositions occur simultaneously; however the depositions are not as well controlled as those of the serial method.

For electrochemical synthesis of new PEC materials, the composition of the material is varied conveniently by varying the electrolyte composition, as well as the deposition parameter conditions (e.g., voltage, temperature, etc.). When establishing electrosynthetic methods for a specific class of materials, it is critical to determine the relationship between the electrolyte concentrations of precursors and the resulting solid-state material composition. Figure 14.9 shows a typical result from the preparation of doped metal oxides. It was desired to electrochemically dope iron and zinc oxides by adding to their liquid-phase precursor solutions varying amounts of Cr or Mo ions for the iron oxides, and Co ions for the zinc oxides. XPS evaluation of the films produced by electrodeposition shows the relationships between the dopant precursor concentrations in the electrolyte and the dopant resulting in the solid-phase materials formed.

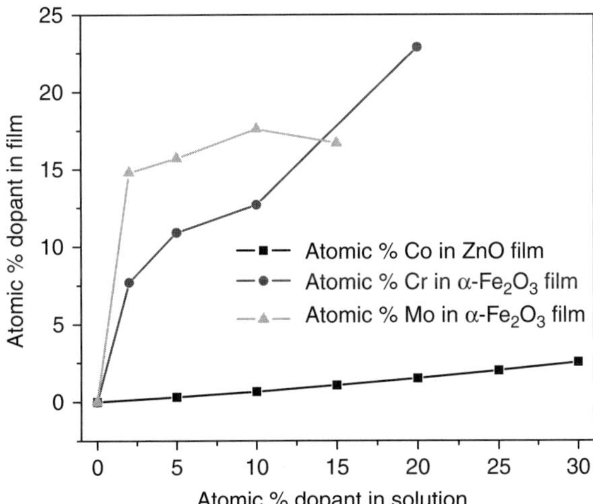

Figure 14.9 Atomic % of dopant in solution for Co in ZnO, Cr and Mo in α-Fe_2O_3 versus atomic % of dopant in electrochemically deposited film obtained from XPS data.

14.3.3.1 Advantages and Disadvantages of Liquid-Phase Synthesis

Electrochemical deposition is a versatile and cost-effective method for depositing metal and metal-oxide materials. The methods are typically scalable to industrial applications without major complications. Nonetheless, electrodeposition reactions are often quite complex and there are a limited number of materials that have been successfully deposited. Although electrodeposition has been used for many oxides, often it is the metal hydroxide that is preferentially deposited. Furthermore, often the as-deposited materials are amorphous and require high-temperature annealing or calcination to produce a photoactive material.

14.3.4 Spray Pyrolysis

Spray-pyrolysis systems have been developed to deposit a variety of thin-film materials. Semiconducting metal oxides, including ZnO [66,67] and Fe_2O_3 [68–71] have been deposited using spray pyrolysis; however, nonoxides have also been synthesized [72]. The process consists of spraying a metal-oxide precursor over a hot substrate to decompose the precursor, while the film increases in thickness. Metal salts are often used to form an oxide or hydroxide (depending on the substrate temperature). Post-deposition annealing or calcination can be performed at higher temperatures to increase the crystallinity or form the oxide from the hydroxide deposited by spray pyrolysis. Spray pyrolysis has been widely applied, due to its apparent simplicity; however, there are many challenges when using spray pyrolysis as the basis of an automated high-throughput synthesis system. To our knowledge, there has been only one application using spray pyrolysis for high-throughput synthesis of materials with variable composition [73].

The high-throughput spray-pyrolysis system developed for deposition of iron-oxide thin films was developed, consisting of: (i) spray-pyrolysis chamber, (ii) motion hardware, (iii) sprayer/liquid delivery system, (iv) nozzle/shutter, (v) heated stage-combinatorial block, (vi) gas-delivery system and (vii) control software [73]. The spray-pyrolysis chamber is an enclosure which gives mechanical support to the system and ensures that any reaction products from the pyrolysis of the solutions are vented to a fume-recovery system (See Figure 14.10). The motion hardware consists of an X-Y stage (Danaher Motion) to which a hotplate is fixed. The stage is controlled through a Nu-Drive power amplifier, using a motion control card (PCI-4CX, National Instruments). The liquid-delivery system consists of a pinch valve under software control (LabView™). The precursor can be selected from up to 19 different solutions by the use of two 10-position valves (VICI, Valco Instruments Co. Inc.) connected to syringes with solutions on the valve side, and through the pinch valve, or computer-controlled peristaltic pump, on the spray-nozzle side (Paasche, AXX-AUSDRS-000), Figure 14.12. The nozzle/shutter system consists of the nozzle, an aperture and a shutter, which blocks the spray field until a uniform spray is achieved and during cleaning of the nozzle. The heated stage consists of a custom-made hotplate, in which the substrate is sandwiched between the hotplate and a grid which restricts the sample size and eliminates any overspray which might contaminate other samples in the library. The temperature of the heated stage is controlled and monitored continuously. The gas-delivery system consists of regulated O_2 or compressed air pulsed to the nozzle through a solenoid valve, which is controlled through a data-acquisition card. The control software regulates the pulse of gas to the nozzle, and the preset composition of the precursor solution (via the multiposition valve), the solution-spray deposition interval (based

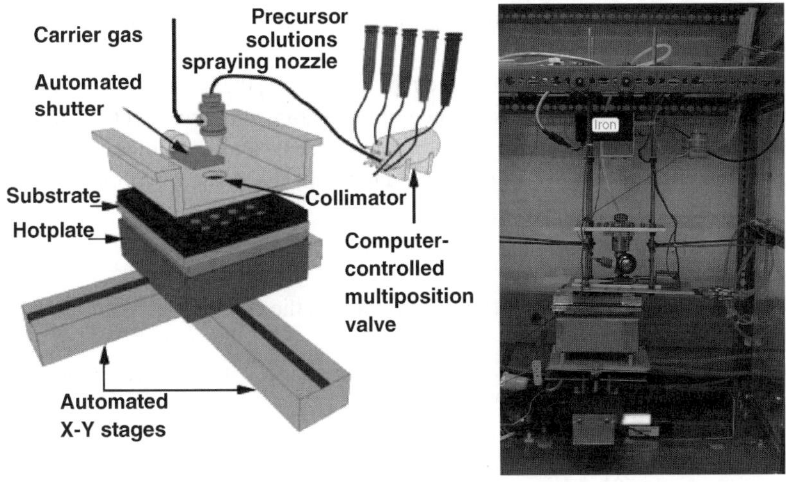

Figure 14.10 Schematic and photograph of the automated spray-pyrolysis system.

on time after the previous spray or temperature of the substrate) and the location of the sample being deposited are all controlled using software written in Labview™.

14.3.4.1 Advantages and Disadvantages of Spray Pyrolysis

The major advantage of spray pyrolysis is the inherent simplicity of the chemical synthesis by way of decomposition of the precursor. Further processing can be used to reach a thermodynamically stable phase. Mixing of precursors is relatively easy and thus synthesis of mixed metal oxides is facilitated. Nonetheless, there are many complications that can arise from the complex phenomena of solvent evaporation and gradients of temperature between the source and the heated substrate, as well as the interaction of the precursor droplets and the substrate, which will have an impact on the morphology and crystallinity of the samples. The control over the morphology of the samples is complicated; however, there have been several reports where the shape of the iron oxide can be controlled by a dopant atom such as Si [74] or by use of an ultrasonic nozzle [71].

14.4 Photocatalyst Screening and Characterization

In a high-throughput materials process there is little value in synthesizing large numbers of samples if they cannot be characterized and screened for desirable properties at a rate approximately equal to or greater than the rate of synthesis. Ideally, as much information as possible should be obtained from every sample; however, tradeoffs must usually be made between the speed of characterization and the quantitative accuracy of the specific numerical values. Often, it is valuable to simply know how a new material behaves relative to a known material, rather than having a specific numerical quantity. For example, knowing that electron donors doped into Fe_2O_3 at increasing concentrations increase the conductivity of the Fe_2O_3

relative to the undoped material is almost as valuable as knowing what the numerical value of the conductivity is.

"Primary screens" have been developed to rapidly measure the most important observable(s). For solar photoelectrocatalysts, absorbance, charge transport and electron-transfer efficiency are the critical performance variables to assess and optimize. High-throughput screening is designed to rapidly select candidates that meet one or more of the following selection criteria:

1. Bandgap between 1.6 and 2.4 eV
2. Efficiency of transfer of photoelectrons and holes to the electrolyte–solid interface greater than 90%
3. Efficiency for the production of desired redox products in the electrolyte from photogenerated electrons and holes greater than 90%.

Samples meeting the "primary" screening criteria are then selected to undergo "secondary" screening, which is typically slower, using conventional analytic methods, and resynthesized to address the fundamental questions: is the apparent improvement real and if so, what is the reason for the apparent improvement in performance?

14.4.1 High-Throughput Screening

14.4.1.1 High-Throughput UV-Visible Measurements

To screen for materials that absorb visible light, the most time and cost-efficient screen is to simply look at it. More quantitatively, digital photography is commonly used to screen materials by generating a two-dimensional map of one of the most important optical properties of the library under controlled illumination. A more quantitative high-throughput system to measure the absorbance spectrum of solid-state photocatalysts was developed, which consists of three basic subsystems: (i) motion hardware, (ii) optical setup and (iii) custom software written in LabView™ [62]. The motion hardware includes a motorized X-Y-Z stage and related hardware (See Section 14.3.2 for additional details). The optical subsystem consists of a 1 kW xenon lamp (Newport Corporation), an Ocean Optics S2000 UV-visible spectrometer, and a 7.5 cm Spectra Lab integrating sphere. Lower-intensity lamps can also be used. The integrating sphere is used in a three-port configuration (see Figure 14.11). The light is introduced into the sphere through one port by a fiber optic (Ocean Optics, 600 μm), which is oriented perpendicular to the sample being measured. The second port has an O-ring to make contact with the sample on the substrate and the third port couples to a fiber optic to the UV-visible spectrometer. The spectrometer's range is such that white light can be used for this measurement rather than scanning step-wise with a monochromator. The software component consists of OOILVD LabView™ drivers (Ocean Optics) to control the spectrometer and custom software (Labview™) to control the motion hardware. A diagram of the system is shown below in Figure 14.11.

The high-throughput UV-visible optical-screening system has been applied in the screening of $Zn_{1-x}Co_xO$ samples as photoelectrocatalysts. Co was substituted into ZnO as a means of increasing the visible light absorbance. An investigation to identify the optimal loading was performed using the high-throughput UV-visible screening system [62]. Figure 14.12a shows

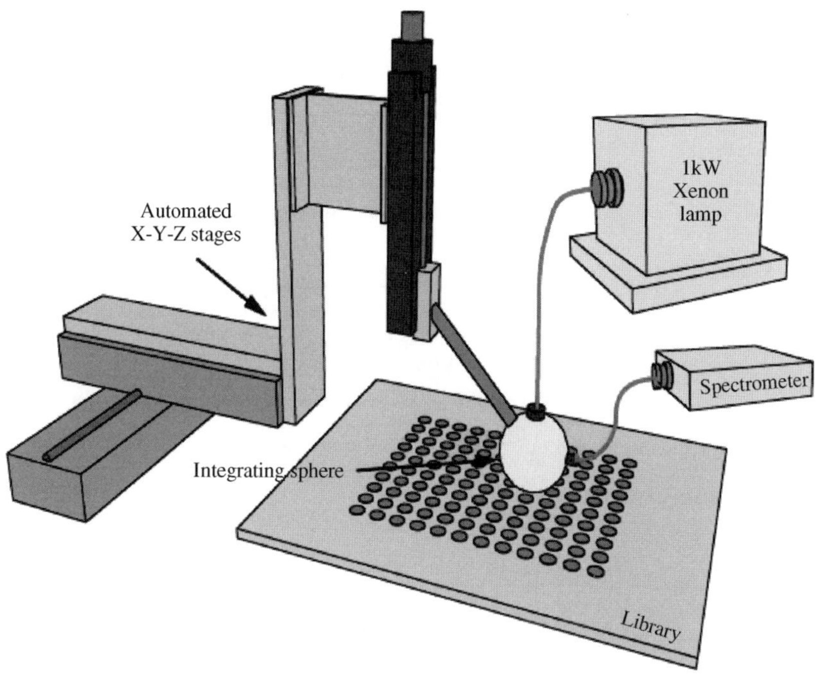

Figure 14.11 Diagram of the high-throughput UV-visible optical-screening system.

the screening results of the UV-visible absorbance of selected $Zn_{1-x}Co_xO$ thin-film samples. With an increase in the Co composition, the absorption edge shifts to longer wavelengths. The plot of $(\alpha h\nu)^2$ versus $h\nu$ in Figure 14.12b indicates the direct bandgap, E_g, as the x intercept of the linear fits. E_g values for ZnO with 2.1% and 4.4% Co dopant are 2.90 and 2.75 eV, respectively.

14.4.1.2 High-Throughput Photoelectrochemical Screening

The performance of a photoelectrocatalyst is ideally measured in an electrochemical cell with a controlled light source. Several standard electrochemical measurements, including current–voltage measurements using a rotating disc electrode and Mott–Schottky analysis may require more time than available for a primary screening. A high-throughput photoelectrochemical testing system was developed to have a screening rate of approximately 5–15 minutes per sample, which is fast enough to be equivalent to the high-throughput synthesis rate in the electrochemical synthesis system (120 samples per day) [25,26,62]. The system consists of: (i) motion hardware, (ii) an optical subsystem, (iii) an electrochemical probe head, (iv) a liquid-delivery system, (v) an electrochemical/electronic subsystem and (vi) system software. A schematic diagram of the system is shown in Figure 14.13. The motion hardware is the same as used in the rapid serial synthesis system described above (Section 14.3.2) [75,76]. The optical subsystem consists of an infrared (IR)-filtered 1 kW xenon lamp (Newport), a monochromator (CS 260 1/4M, Newport), an electronic beam chopper and a filter wheel (74010, Newport). The monochromator output beam is focused on

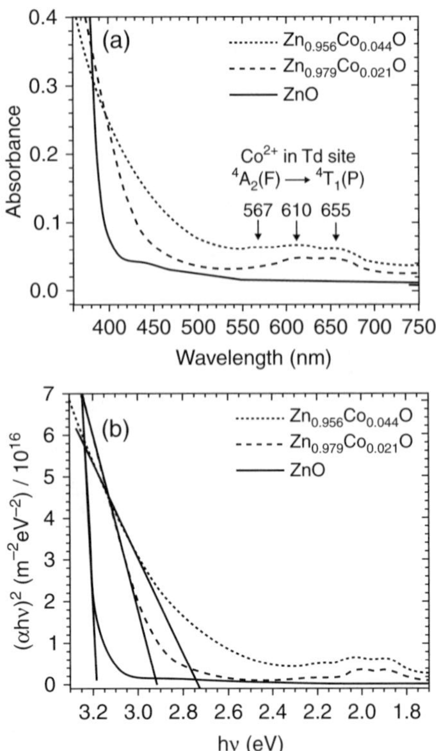

Figure 14.12 UV-visible spectroscopy of selected $Zn_{1-x}Co_xO$ samples. (Reprinted with permission from T.F. Jaramillo, S.H. Baeck, A. Kleiman-Shwarsctein *et al.*, Automated electrochemical synthesis and photoelectrochemical characterization of $Zn_{1-x}Co_xO$ thin films for solar hydrogen production, *Journal of Combinatorial Chemistry*, **7**(2), 264–271, 2005. © 2005 American Chemical Society.)

a fiber optic bundle which terminates in a borosilicate glass sleeve in the probe head. The probe head is an electrochemical cell assembled with a counter and reference electrode, and two liquid ports on the bottom of the cell where the electrolyte is pumped into and out of the cell. A seal between the probe and the library is made with an O-ring at the base of the probe head. The liquid delivery system consists of a computer-controlled pump (Masterflex), which will fill and empty the cell with electrolyte as it moves from site to site on the library (Figure 14.14) [26]. The electronic subsystem consists of a multimeter (Hewlett Packard, HP34401A), a potentiostat (PAR, EG&G 273) and a data-acquisition card (National Instruments, PCI-6036E).

The photocurrent is measured with a multimeter (Hewlett Packard, HP 4210) and any bias potential required is applied in a two- or three-electrode cell configuration using a potentiostat (EG&G 273A). For overall efficiency calculations the applied potential (E_{appl}) used is between the working and counter-electrodes (two-electrode configuration). For the current–voltage measurements, *I–V*, the experiments are performed with a voltage sweep of 100 mV s^{-1} from the potentiostat in a range from −0.5 to 0.8 V versus Ag/AgCl. The high-throughput screening system is computer controlled using custom software written in LabView™ (National

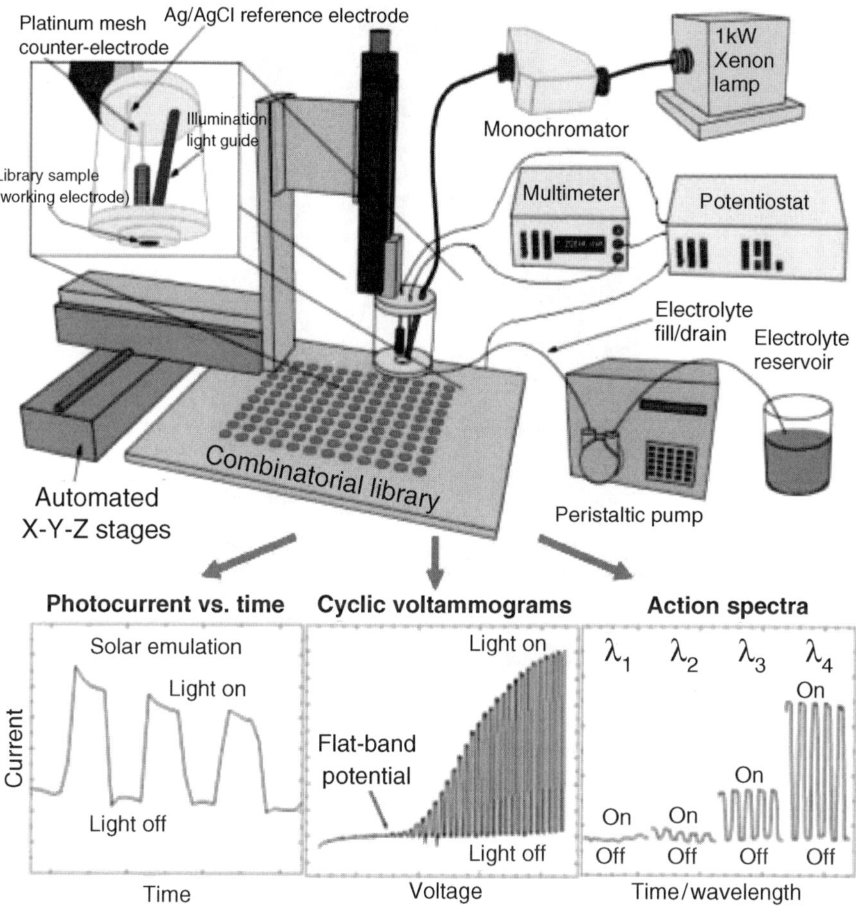

Figure 14.13 Schematic diagram of the high-throughput photoelectrochemical testing system. The system consists of motion hardware, an optical subsystem, an electrochemical probe, a liquid delivery system and an electrochemical/electronic subsystem. The schematic shows three types of performance test that can be performed with such system. (Reprinted with permission from T.F. Jaramillo, S.H. Baeck, A. Kleiman-Shwarsctein *et al.*, Automated electrochemical synthesis and photoelectrochemical characterization of $Zn_{1-x}Co_xO$ thin films for solar hydrogen production, *Journal of Combinatorial Chemistry*, **7**(2), 264–271, 2005. © 2005 American Chemical Society.)

Instruments). Figure 14.15 shows the screening results of undoped and Pt-doped iron-oxide thin films deposited on Pt/Ti-coated quartz substrates in 1M NaOH electrolyte [65]. The 5% Pt-doped sample exhibits the highest photocurrent, which increases with light intensity (Figure 14.15a and b). Figure 14.15c shows the photocurrent from a chopped light source as a function of applied bias voltage from the doped and undoped iron-oxide thin films. The behavior is similar to that observed under steady-state illumination, Figure 14.15a. As shown in Figure 14.15d, the IPCE values of the Pt-doped samples are significantly larger than those of the undoped samples.

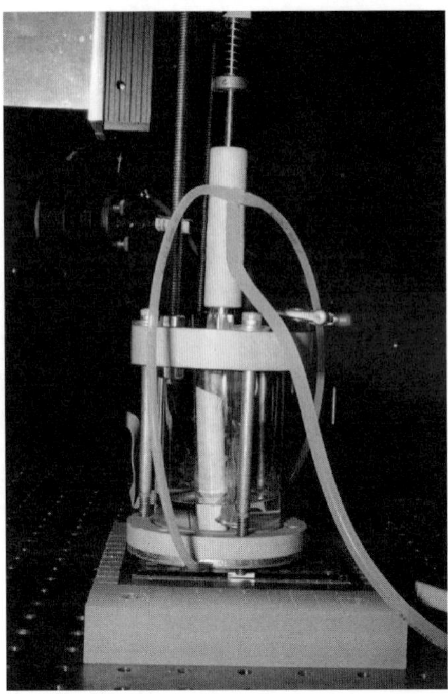

Figure 14.14 Photograph of the electrochemical probe head used for measurements, including IPCE determination. (Reprinted with permission from T.F. Jaramillo, S.H. Baeck, A. Kleiman-Schwarsctein *et al.*, Combinatorial electrochemical synthesis and screening of mesoporous ZnO for photocatalysis, *Macromolecular Rapid Communications*, **25**(1), 297–301, 2004. © 2004 Wiley-VCH Verlag GmbH & Co. KGaA).

Although it is unlikely that a cost-effective PEC system will include a material requiring an applied bias, information on the relative energy levels of the semiconductor and reactant redox or electrocatalytic barriers, which may be modified through subsequent optimizations may be obtained. For example, a material with a high solar-spectrum absorbance and an appropriate bandgap, but with a conduction band minimum (CBM) too low for hydrogen generation may be modified to shift the CBM and achieve a high overall unbiased PEC efficiency. Thus, bias is routinely applied for screening. Figure 14.16 shows the short-circuit photocurrent and photocurrents at different applied bias for Co-substituted ZnO. The photocurrent with + 1.1 V applied bias is improved by a factor of about 2.5 with comparison to the zero-bias case.

14.4.1.3 Hydrogen Production

Ultimately, the most meaningful performance characterization of PEC materials for the production of hydrogen is a measurement under solar-spectrum illumination of their rate of hydrogen production. A high-throughput screening system was developed to directly monitor the hydrogen gas produced at each location in a library of materials, which was based on the hydrogen-induced optical detection of changes in the reflectance of palladium-coated tungsten trioxide in the presence of hydrogen, Figure 14.17 [23].

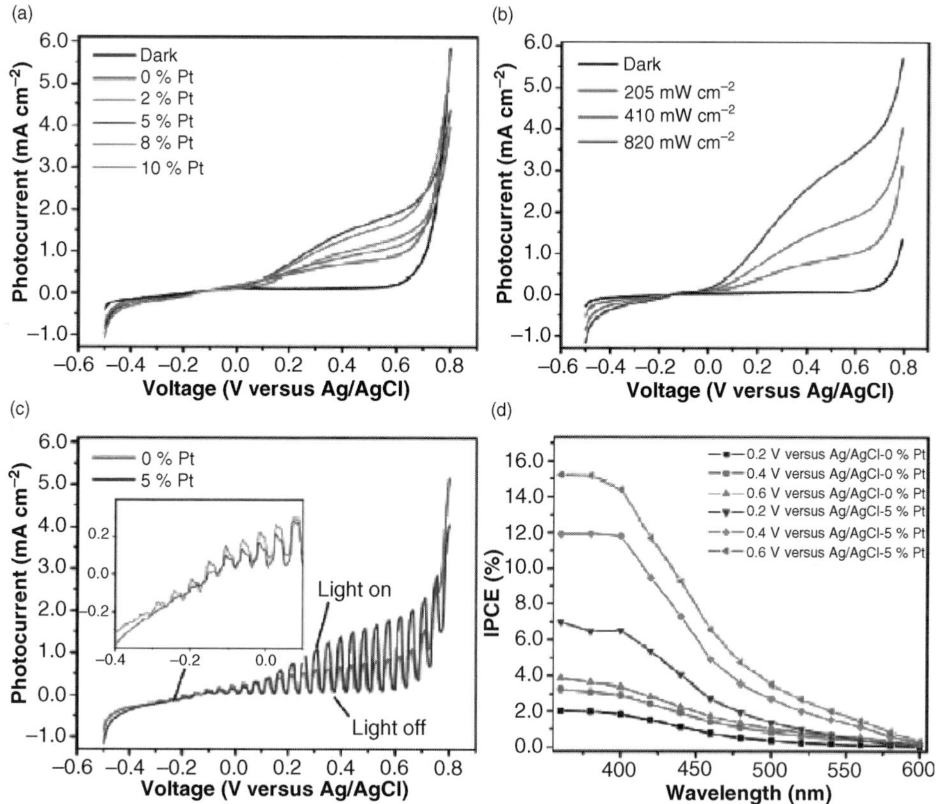

Figure 14.15 *I–V* curves and IPCE results of Pt-doped and undoped hematite. The electrolyte is 1 M NaOH solution. (Reprinted with permission from Y.S. Hu, A. Kleiman-Shwarsctein, A.J. Forman *et al.*, Pt-doped alpha-Fe_2O_3 thin films active for photoelectrochemical water splitting, *Chemistry of Materials*, **20**(12), 3803–3805, 2008. © 2008 American Chemical Society.)

The system was demonstrated by screening libraries of potential hydrogen-evolution electrocatalysts, Figure 14.18. A library of 24 different electrode materials (four samples of each for statistical comparison) was synthesized by electron-beam evaporation. Eight distinct elemental sources of 99.99% purity were used to create the library: Ti, Pt, Ni, Au, Pd, Al, Ag and Ge. Among the 24 different materials, eight were pure elements and the remaining 16 were binary compounds co-evaporated in a ratio of approximately 80/20, except Au/Ge, which was 50/50. For high-throughput screening, a polypropylene reactor block with 96 holes spaced over each of the elements of the library and sealed with individual O-rings (Figure 14.17) was used. The thin-film catalyst library served as the working cathode. A 0.2 M sodium acetate solution was used to fill each cell and a common 304 stainless-steel anode was used. Electrolysis was conducted in the sodium acetate solution at −2.8 V with respect to the 304 stainless-steel anode for 20 min. Hydrogen produced at a library cathode site would reduce the tungsten trioxide sensor layer above and cause a reflectance change. The rates of H_2 production from the libraries were measured optically by a scanning IR reflectance sensor calibrated using a known concentration of H_2.

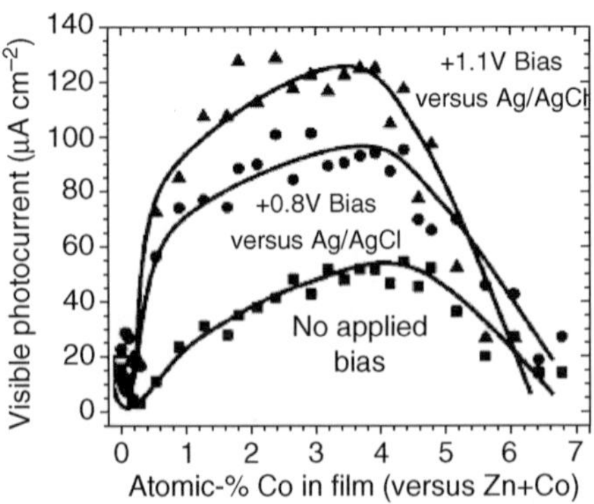

Figure 14.16 Photocurrent of $Zn_{1-x}Co_xO$ materials under visible-only illumination with and without the aid of an external bias. (Reprinted with permission from T.F. Jaramillo, S.H. Baeck, A. Kleiman-Shwarsctein *et al.*, Automated electrochemical synthesis and photoelectrochemical characterization of $Zn_{1-x}Co_xO$ thin films for solar hydrogen production, *Journal of Combinatorial Chemistry*, **7**(2), 264–271, 2005. © 2005 American Chemical Society.)

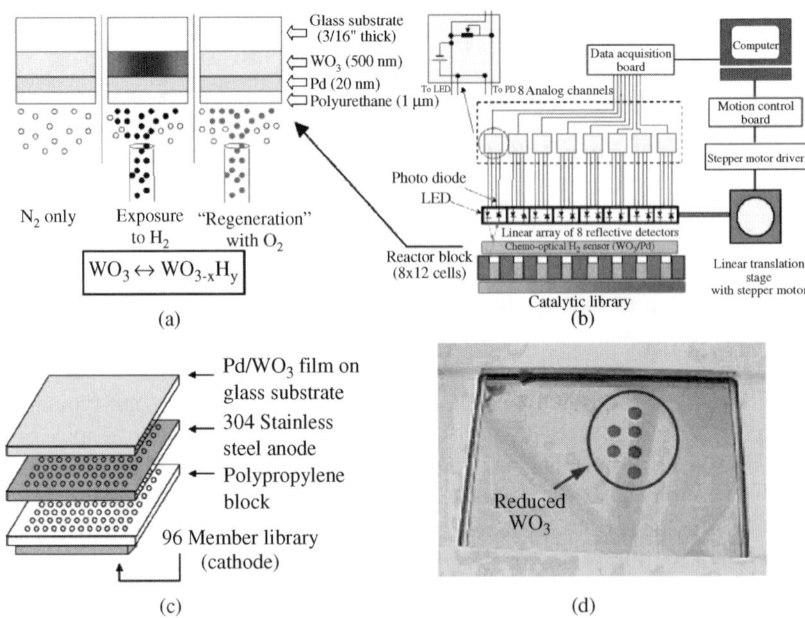

Figure 14.17 System for measurement of hydrogen production. The hydrogen-induced reflectance changes of a tungsten-oxide thin film deposited on glass are monitored by scanning an array of reflectance sensors over the library of hydrogen-producing mini-cells. (Reprinted with permission from T.F. Jaramillo, A. Ivanovskaya and E.W. McFarland, High-throughput screening system for catalytic hydrogen-producing materials, *Journal of Combinatorial Chemistry*, **4**(1), 17–22, 2002. © 2002 American Chemical Society.)

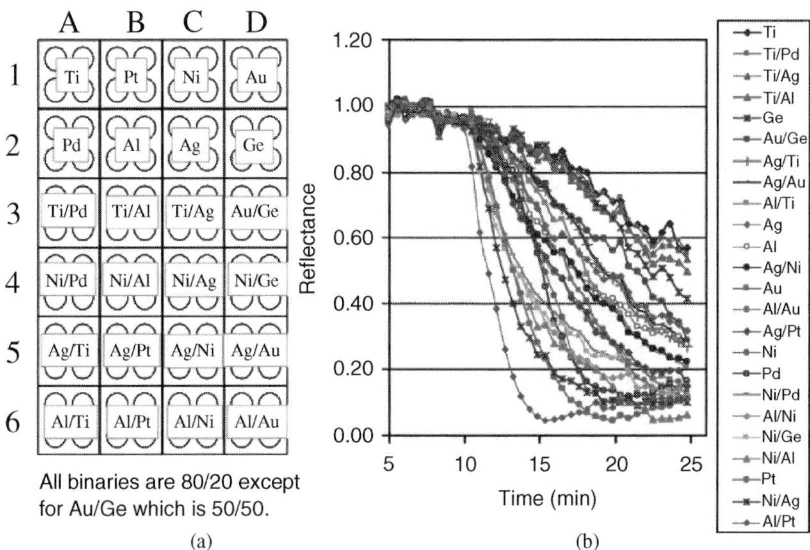

Figure 14.18 Electrocatalyst library and reflectance measurements during hydrogen production under an applied bias. (Reprinted with permission from T.F. Jaramillo, A. Ivanovskaya and E.W. McFarland, High-throughput screening system for catalytic hydrogen-producing materials, *Journal of Combinatorial Chemistry*, **4**(1), 17–22, 2002. © 2002 American Chemical Society.) See also color-plate section.

The raw reflectance data measured from the IR sensor is shown in Figure 14.18b, which illustrates the change of the reflectance versus electrolysis time. The rate of change in the reflectance indicates the hydrogen production rate. Among the 24 materials, a mixed material consisting of Al/Pt showed the highest H_2 production rate. Of the eight pure elemental materials, Pt, Pd and Ni were observed to give the highest rate of hydrogen production, which agrees with the literature and is consistent with their relatively high work functions and their ability to facilitate surface H recombination.

14.4.1.4 Other High-Throughput Screening Systems

A wide variety of high-throughput characterization methods, including fluorescence analysis, chromatography, mass spectroscopy, radioactivity assays, nuclear magnetic resonance (NMR) and IR have been developed for other areas of material science and chemistry, and many can be directly applied to screen PEC materials [77].

A laser-induced resonance-enhanced multi-photon ionization (REMPI) system was developed for screening solid catalysts [78]. In REMPI, specific products are detected from the library by selectively laser ionization and detection of the ionized products over each library element with microelectrodes. Woodhouse *et al.* [42] fabricated mixed metal oxide photoelectrocatalysts using high-throughput inkjet printing followed by pyrolysis. The oxides were then screened by scanning a 532 nm frequency doubled Nd:YAG laser (300 ms per sample) over the materials immersed in an electrolyte. The resulting photocurrent was used to generate false color photocurrent images for analysis. In other work, a modified scanning electrochemical microscope (SECM) was used for rapid screening of photocatalysts [79]. An optical fiber connected to a Xe lamp is used as the SECM tip and scanned over the photocatalyst working electrode array, which is immersed in a common electrolyte. The SECM potentiostat is used to apply a bias potential. The recorded photocurrent was processed into a color-coded two-dimensional image.

High-throughput scanning mass spectrometers have been used for the analysis of gas-phase products created from heterogeneous catalysts [80,81]. The catalyst libraries are typically mounted on a stage which translates each library element at approximately one sample per minute under a mass-spectrometer detection zone. Alternatively, a hollow fiber capillary tube "sniffer" may be scanned over a fixed library. Scanning X-ray diffractometers have also been used for screening crystalline phases of catalyst libraries. Automated XRD measurements on two-dimensional library substrates have been reported using a commercially available system (Bruker AXS D8 Discover) [82]. A map of diffraction intensity and angles was obtained by scanning a continuous thin-film library, $Bi_xY_{3-x}Fe_5O_{12}$, made by continuous-variable deposition of precursors on variably heated (410–700 °C) garnet substrates. The best substrate temperature and thin-film components for crystallization were thus screened using this method.

14.4.2 Secondary Screening and Quantitative Characterization

High-throughput experimentation does not eliminate the need for the painstaking complex quantitative analyses used conventionally in materials science and chemistry; it serves to possibly select only potentially useful materials for such detailed studies. Primary screening is performed to: (i) identify relationships between composition, structure and synthesis to relative performance trends, and (ii) identify materials with particularly promising performance as leads. Typically, trends and leads are verified from multiple libraries synthesized, when possible, with different variations of properties, but containing the same basic materials. Performance trends are used to help generate testable hypotheses as to the underlying mechanisms giving rise to the performance changes.

Lead PEC materials identified from primary screening as having suitable absorbance and electron/hole transport properties above a threshold are submitted to secondary screening [77] consisting of several levels of analysis designed to: (i) confirm observed performance trends, (ii) identify alternative explanations (e.g., rather than composition, microstructure may be responsible for activity changes), (iii) measure quantitative performance metrics of new materials to compare them to existing materials and (iv) understand their performance in the context of existing materials science. Information learned from both primary and secondary screening is fed back to help in future library design. Although all methods of solid-state quantitative analysis may be utilized, several of the more commonly utilized secondary screening techniques for photoelectrocatalysts are briefly described below.

14.4.2.1 Microscopy

The microscopic appearance of photocatalytic materials synthesized by high-throughput methods can sometimes reveal a potential explanation for observed behavior due to changes in the materials' morphology, surface area or gross variations in the interaction with the substrate (e.g., debonding). Optical microscopy can rapidly determine significant variations in film quality and rapidly provide information on film defects. Optical microscopy can be combined with Raman and fluorescence measurement equipment and used with automated stages to create high-throughput systems [83].

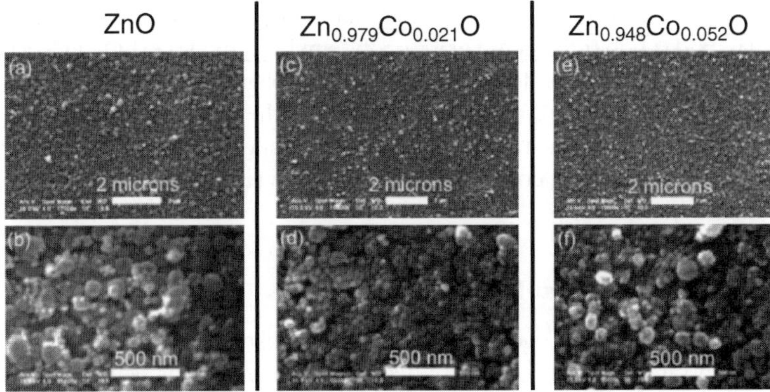

Figure 14.19 SEM images of three selected members of the $Zn_{1-x}Co_xO$ library. (Reprinted with permission from T.F. Jaramillo, S.H. Baeck, A. Kleiman-Shwarsctein *et al.*, Automated electrochemical synthesis and photoelectrochemical characterization of $Zn_{1-x}Co_xO$ thin films for solar hydrogen production, *Journal of Combinatorial Chemistry*, **7**(2), 264–271, 2005. © 2005 American Chemical Society.)

Electron microscopy is commonly used to characterize the morphologies of semiconductor photocatalysts. The surface morphology of thin-film samples is generally observed using scanning electron microscopy (SEM) at scales between microns and nanometers. Figure 14.19 shows three selected members of the $Zn_{1-x}Co_xO$ photocatalyst library described above [62]. The SEM images provide evidence that there are similar morphologies for samples with different compositions and that all the thin films consist of approximately spherical or platelet particles with sizes ranging from 20 to 200 nm. Transmission electron microscopy (TEM) requires more complex sample preparation; however, it can reveal important information on the nanoscale structure of materials at subnanometer resolution. One example from a mesoporous ZnO PEC library prepared using electrodeposition on FTO glass is shown in Figure 14.20. High-throughput

Figure 14.20 TEM morphologies of electrodeposited ZnO with different amount of $EO_{20}PO_{70}EO_{20}$. (Reprinted with permission from T.F. Jaramillo, S.H. Baeck, A. Kleiman-Schwarsctein *et al.*, Combinatorial electrochemical synthesis and screening of mesoporous ZnO for photocatalysis, *Macromolecular Rapid Communications*, **25**(1), 297–301, 2004. © 2004 Wiley-VCH Verlag GmbH & Co. KGaA).

photocurrent screening results had shown that variation of the structure-directing polymer, poly(ethylene oxide)-poly(propylene oxide)-poly(ethylene oxide), ($EO_{20}PO_{70}EO_{20}$) had a significant effect on the photocatalytic performance. Samples with 1–5 wt% showed the highest photoelectrochemical activity [26]. TEM revealed that the sample prepared with 5 wt% of $EO_{20}PO_{70}EO_{20}$ had a disordered mesoporous structure, while the 15 wt% sample had a lamellar structure. These findings led to conclusions regarding the relationship between mesostructure and activity for water splitting.

14.4.2.2 EDS, XPS, XRD, and Raman Spectroscopies

Many electron microscopes are equipped with an energy dispersive X-ray spectrometer (EDS), which provides a rapid analysis of material composition within the penetration depth of the electron beam. Morphological features can be characterized at high concentrations. Further information, such as phase separation, and the existence and distribution of impurities, and so on can then be generated. The combination of state-of-art microscopes and high-resolution EDS spectrometers provides reasonable quantitative information. Other common characterization methods include X-ray photoelectron spectroscopy (XPS), which characterizes the surface composition and ionization state, X-ray diffraction (XRD), which evaluates the bulk crystal structure and Raman spectroscopy for probing the surface chemical environment and so on. Hu *et al.* have applied several methods in their study of Pt-doped hematite thin films as candidates for photoelectrochemical water splitting [65]. In this work, XRD, XPS and Raman spectroscopes were applied to map the state and environment of Pt in the hematite thin film and at the surface. As shown in Figure 14.21a, the XRD shows pure hematite diffraction patterns beside the substrate patterns. No platinum oxide or metal diffraction was observed within the instrumental sensitivity. The Raman spectra (Figure 14.21b) show a decreasing intensity in the 657 cm^{-1} peak, which was attributed to magnetite (Fe_3O_4) or a disordered phase (such as FeO), with increasing Pt from 0% to 10%. However, the XPS in Figure 14.21c does not support the presence of magnetite or FeO. Only Pt^{4+} was observed from the XPS Pt 4*d* spectra (Figure 14.21c). It was proposed that the doped Pt had been introduced into the hematite lattice. The distribution of Pt as a function of depth in hematite thin films can also be studied by XPS performed during ion-beam etching, as shown in Figure 14.21d. The results indicate a similar concentration of Pt in the bulk and on the surface.

14.4.2.3 Internal Quantum Efficiency (IQE)

The IQE of a semiconductor photoelectrocatalyst powder is strictly defined as the number of photoelectrons produced per absorbed photon; however, for PEC applications, the IQE is frequently approximated as the number of single electron/hole pair photochemical products generated per photon adsorbed by the PEC material. An IQE measurement system for powders was developed, consisting of an integrating sphere, a photoreactor and a potentiometric measurement system, as shown in Figure 14.22. In this system, the photoreduction of $[Fe(CN)_6]^{3-}$ by hematite powders is used to determine their IQE. The ratio of $[Fe(CN)_6]^{4-}$ product to $[Fe(CN)_6]^{3-}$ reactant is quantified by measuring the potential of the ferricyanide/ferrocyanide solution at a platinum wire with respect to an Ag/AgCl electrode. The system is first calibrated with known concentrations of ferri/ferrocyanide, which are generated via bulk electrolysis of a pure ferricyanide solution in a separate cell. The concentration of the $[Fe(CN)_6]^{4-}$ can then be measured *in situ* and calculated from the calibration curve.

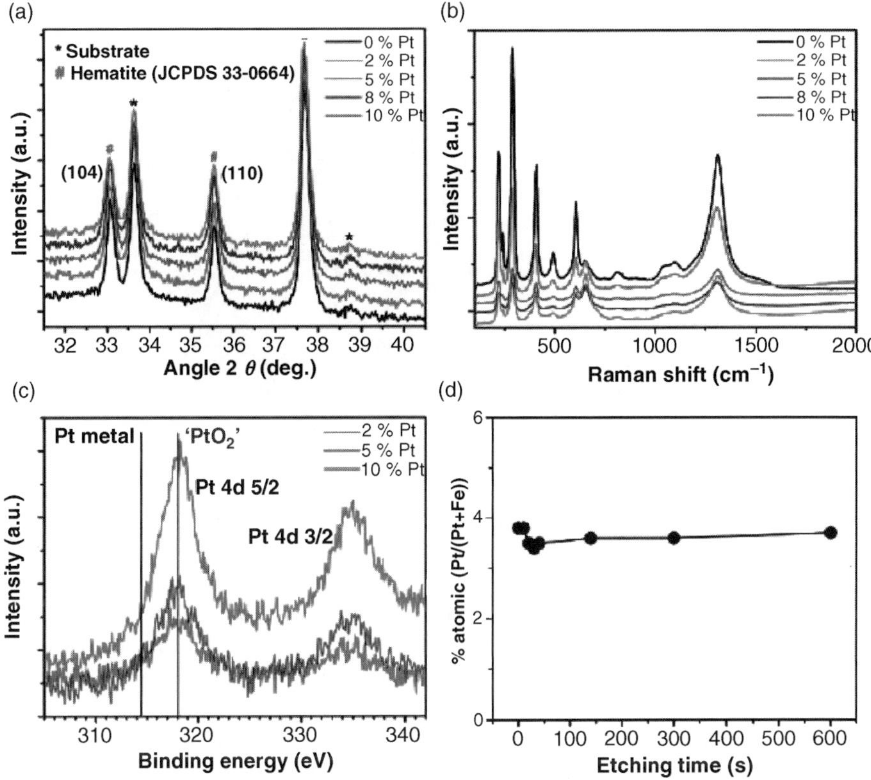

Figure 14.21 Characterizations of undoped and Pt-doped iron-oxide thin films on FTO: (a) XRD patterns, (b) Raman spectra, (c) XPS spectra and (d) XPS etching profile. (Reprinted with permission from Y.S. Hu, A. Kleiman-Shwarsctein, A.J. Forman *et al.*, Pt-doped alpha-Fe_2O_3 thin films active for photoelectrochemical water splitting, *Chemistry of Materials*, **20**(12), 3803–3805, 2008. © 2008 American Chemical Society.) See also color-plate section.

Standard integrating-sphere procedures are applied to quantify the number of photons absorbed by the sample. In general, a known input light intensity and spectrum is correlated to a power reading at the diode detector on the sphere. The ratio of the measured power with the sample in place, to the power measured when a non-absorbing sample (blank) is substituted can be used to determine the total number of photons that were absorbed by the PEC powder. The photogenerated reduction products and the number of photons absorbed are now both quantified. The IQE for any PEC powder can thus be determined in real time using this system, provided that there is no loss in transfer of charge to the sacrificial redox reactants.

In one example, hematite nanoparticles are dispersed in a ferricyanide solution and the equilibrium potential monitored as a function of exposure.

14.4.2.4 Hydrogen Production

Quantitative measurement of hydrogen production rate from powdered catalysts is the ultimate measurement of material as a photoelectrocatalyst. Powder samples may be screened in a

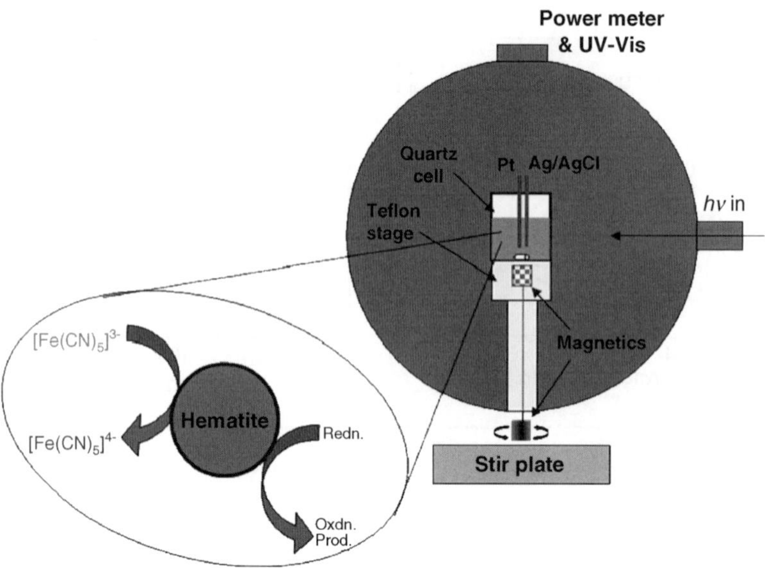

Figure 14.22 Schematic diagram of a measurement system for determining the internal quantum efficiency, IQE, of particulate photocatalysts. The system consists of an integrating sphere, a quartz cell photoreactor and a potentiometric measurement system, including Pt and Ag/AgCl electrodes.

high-throughput manner using gas-tight vials containing powdered catalysts, which are illuminated and the headspace gas periodically sampled [77], or hydrogen production can be monitored continuously in a system such as the one shown schematically in Figure 14.23. The continuous monitoring system consists of a photoreactor, a pump loop, an Ar purge system and a gas chromatograph (GC). The colloidal photoelectrocatalyst in the sample vial catalyzes water splitting and the product hydrogen will be carried by Ar purge gas into the GC loop for determination of the concentration. The rate of hydrogen production from the photoreactor can be calculated quantitatively after calibration.

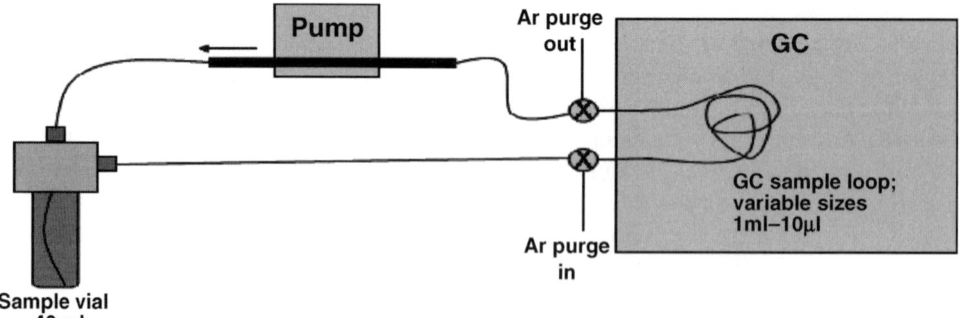

Figure 14.23 Schematic diagram of hydrogen production measurement system. The system consists of a photoreactor, a pump loop, an Ar purge system, and a GC detector.

14.5 Specific Examples of High-Throughput Methodology Applied to Photoelectrocatalysts

To date, there has been no cost-effective solar photoelectrocatalyst material system identified for the production of any product, including hydrogen; however, efforts in several groups are ongoing using high-throughput synthesis and screening in attempts to identify new candidate materials. The primary unmet challenge continues to be the identification of a cost-effective solar-photon absorber suitable for photoelectrocatalysis. Once the absorber material has been identified, optimization of charge transfer, electrocatalytic activity and structure will follow. In this section, some examples of high-throughput and combinatorial approaches are presented for optimization of three material system-design objectives, solar-spectrum absorbance, carrier-transfer efficiency and electrocatalysis. An example of the final step of assembling a complete nanostructured photocatalytic unit is also shown to illustrate the complete material system-development process.

14.5.1 Solar Absorbers

To identify new semiconductors suitable for solar-spectrum absorption, two approaches have been taken: (i) modify the absorbance of materials with known optical properties or (ii) identify entirely new materials. To date, most work has involved the former approach (with the preponderance of effort historically focused on titania). Below are described specific results for modifications to the wide-gap materials, ZnO and WO_3.

14.5.1.1 Automated Electrochemical Synthesis and Photoelectrochemical Characterization of $Zn_{1-x}Co_xO$ Thin Films for Solar Hydrogen Production

Zinc oxide is an inexpensive and non-toxic semiconducting oxide. It is not suitable in solar PEC applications because it is not stable in electrolytes and does not absorb in the visible. With the aim of improving the visible-light absorbance and stability of ZnO against photocorrosion, different atomic species were co-electrodeposited with ZnO to dope the host oxide and alter its properties. Species were selected that might form defect bands for visible-light absorption and provide stabilized surface oxides. Table 14.2 shows several of the dopant atoms tested in libraries using rapid primary screening of photocurrent under visible ($\lambda > 400$ nm) and broad UV-visible ($\lambda > 300$ nm) illumination in comparison with the undoped ZnO. A qualitative measurement of the stability of the samples with the dopant atoms is also shown in this table, as observed in the dark cell current; a rating of average would be comparable to that of the undoped ZnO. Cobalt-doped ZnO was chosen for detailed analysis by combinatorial synthesis due to its promise as a good visible photocatalyst and good stability in KNO_3 aqueous solution, compared with the undoped ZnO.

The co-electrodeposition of Co-doped ZnO was performed in a 120-well combinatorial block (see Section 14.3.2) heated on a hot plate. The electrodeposition bath consisted of a 100 mM $ZnCl_2$ and 60 mM $LiNO_3$ dissolved in dimethylsulfoxide (DMSO) while the concentrations of $Co(NO_3)_2{\cdot}6H_2O$ ranged from 0 to 150 mM across the library. The electrodeposition was carried out for two minutes for each of the samples at an applied voltage of -1.15 V versus the Ag wire; for full details consult Jaramillo *et al.* [62]. The samples were rinsed with deionized water after synthesis then calcined in air at 500 °C for eight hours.

Table 14.2 Table summarizing some of the dopants that have been tested on ZnO and the qualitative performance of the photocatalyst under visible light illumination and UV- visible light illumination, as well as a broad stability criteria, where undoped ZnO is regarded as average in terms of stability.

Dopant	Visible PC	UV-visible PC	Stability in KNO_3 (aq.)
Ag	Poor	Poor	Average
Al	Good	Average	Good
Au	Poor	Poor	Average
Ce	Average	Average	Excellent
Cd	Poor	Poor	Average
Co	Excellent	Poor	Good
Cr	Poor	Poor	Average
Cu	Poor	Poor	Average
Eu	Poor	Poor	Average
Fe	Good	Good	Average
Mn	Average	Average	Good
Mo	Poor	Poor	Poor
Ni	Excellent	Excellent	Average
Nb	Poor	Average	Good
Pd	Very Poor	Very Poor	N/A
Pt	Poor	Poor	Poor
Rh	Poor	Poor	Average
Ru	Excellent	Excellent	Average
Sb	Poor	Poor	Average
Sn	Average	Average	Good
Ti	Very Poor	Very Poor	N/A
V	Poor	Poor	Average
W	precipitated	precipitated	precipitated
Zr	Average	Average	Good

The photogenerated current under short-circuit conditions without an applied bias from the samples is shown in Figure 14.24. The photocurrent was calculated by subtraction of the current recorded under illumination and in the dark. The error bars were determined by analysis of the data from four "identical" samples. The UV + visible photocurrent was under illumination from a 1000 W xenon lamp, while the visible photocurrent was measured under illumination with a xenon lamp attenuated with a 400 nm cut-off filter. In the case of broadband (UV + visible) illumination, the undoped ZnO shows the highest photocurrent and the photocurrent rapidly decays as the cobalt concentration is increased. This decrease in photocurrent is due to increased trapping or recombination of photogenerated electrons and holes at the cobalt defect sites. Under visible light illumination there is little absorption without cobalt; however, at a cobalt doping of approximately 4.4%, a fourfold improvement in visible band photocurrent is observed. The presence of Co in the ZnO provides absorption sites, however, as the concentration increases above 4.4%, the increased absorption of visible light is overwhelmed by the recombination and trapping due to the Co defects. The apparent optimum at 4.4% Co doping may balance these competing effects and yields the best overall PEC performance.

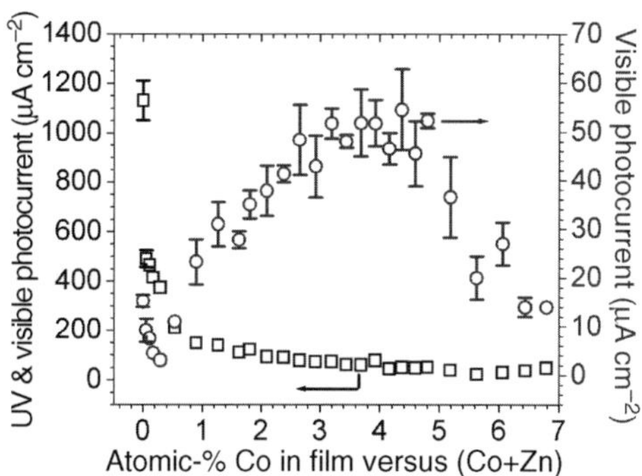

Figure 14.24 Photocurrent in 0.2 M KNO_3 as a function of $Zn_{1-x}Co_xO$ composition without the application of an external bias. Two forms of illumination were used, UV-visible (left axis, squares, 1.77 W cm^{-2}) and visible only (right axis, circles, 1.50 W cm^{-2}) which simulates solar radiation. (Reprinted with permission from T.F. Jaramillo, S.H. Baeck, A. Kleiman-Shwarsctein *et al.*, Automated electrochemical synthesis and photoelectrochemical characterization of $Zn_{1-x}Co_xO$ thin films for solar hydrogen production, *Journal of Combinatorial Chemistry*, **7**(2), 264–271, 2005. © 2005 American Chemical Society.)

Cyclic voltamograms were performed after the short-circuit photocurrent measurements of the sample were recorded using our automated screening system (See Section 14.4.1.2). Figure 14.25 shows the photocurrent of the Co-doped ZnO library obtained from the *I–V* curves at applied bias of 0.8 and 1.1 V versus Ag/AgCl in comparison with the values obtained from the short-circuit photocurrent under visible-light illumination. The expected result is that an increased photocurrent is observed at a positive applied potential for an n-type semiconductor. It can be seen that the same volcano plot shown in Figure 14.24 was also observed here, at the

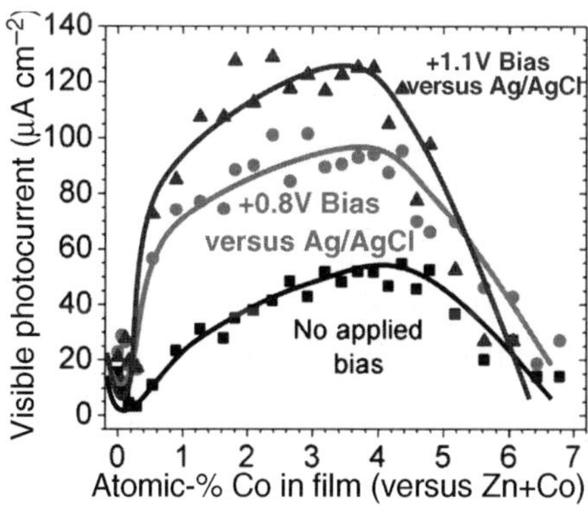

Figure 14.25 Photocurrent of $Zn_{1-x}Co_xO$ materials under visible illumination with the aid of an external bias. (Reprinted with permission from T.F. Jaramillo, S.H. Baeck, A. Kleiman-Shwarsctein *et al.*, Automated electrochemical synthesis and photoelectrochemical characterization of $Zn_{1-x}Co_xO$ thin films for solar hydrogen production, *Journal of Combinatorial Chemistry*, **7**(2), 264–271, 2005. © 2005 American Chemical Society.)

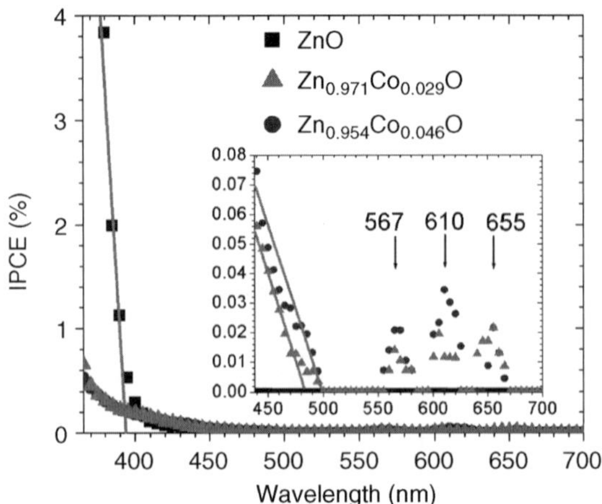

Figure 14.26 IPCE of the selected $Zn_{1-x}Co_xO$ samples. Increased cobalt loading shifts the edge of photocurrent to longer wavelengths, consistent with a decrease in the bandgap energy. The characteristic Co^{2+} *d–d* transitions observed in UV-visible spectroscopy are also responsible for increased visible-light photoelectrochemical activity. (Reprinted with permission from T.F. Jaramillo, S.H. Baeck, A. Kleiman-Shwarsctein *et al.*, Automated electrochemical synthesis and photoelectrochemical characterization of $Zn_{1-x}Co_xO$ thin films for solar hydrogen production, *Journal of Combinatorial Chemistry*, **7**(2), 264–271, 2005. © 2005 American Chemical Society.)

same concentration of 4.4% Co, suggesting that the competing effect between absorption and recombination loss even exists under applied bias conditions. The increased current under bias is likely due to electrocatalytic rate effects for the oxygen-evolution reaction on the oxide surface.

Figure 14.26 shows the action spectra (IPCE) for the selected samples. The IPCE was calculated from the following equation:

$$\text{IPCE}(\%) = \frac{(1240)(i_{\text{photocurrent}}\ \mu\text{A cm}^{-2})}{(\lambda\ \text{nm})(j_{\text{photons}}\ \mu\text{W cm}^{-2})} \times 100\%$$

The undoped ZnO sample shows the highest IPCE at 400 nm, compared to the doped samples, which is consistent with the results obtained from Figure 14.24. However, we can see an increased IPCE from the doped samples at 567, 610 and 655 nm (see inset), which may derive from characteristic *d–d* transitions of a tetrahedral coordinated Co^{2+}. The improvement observed at ~500 nm is due to a decrease in the bandgap of these $Zn_{1-x}Co_xO$ materials, where the absorption edge is shifted to a lower wavelength. Both of these features were observed by UV-visible spectroscopy and are shown in Figure 14.12.

The presence of Co also improved the stability, however, the doped material still suffered from long-term photocorrosion. The work on doped ZnO illustrates how high-throughput methods might be used to optimize the PEC performance of materials, it also suggests that even the optimized ZnO would not have performance close to that required for cost-effective solar-to-chemical applications.

14.5.1.2 Combinatorial Electrochemical Synthesis and Characterization of Tungsten-Based Mixed-Metal Oxides

Tungsten-based mixed-metal oxides, $W_nO_mM_x$ [M = Ni, Co, Cu, Zn, Pt, Ru, Rh, Pd and Ag], have been synthesized by a rapid serial automated electrochemical synthesis (R-ECEHM, see Section 14.3.2). The tungsten precursor was stabilized as a peroxo complex by dissolving 1.8 g of tungsten power in 60 ml of 30% hydrogen peroxide overnight and then decomposing the excess hydrogen peroxide using platinum black. The solution was diluted to 50 mM with a mixture of 50 : 50 2-propanol and water. Binary electrolytes were prepared by adding different volumes of a 50 mM metal chloride solution to the tungsten peroxo solution. Two different libraries of 5 × 7 elements were synthesized by applying −1.0 V versus a Pt reference for 5 minutes. The design is shown in Figure 14.27.

X-ray diffraction was performed on the selected samples and it was found that Ni formed the mixed metal oxide $WNiO_4$, whereas Co, Cu and Zn formed mixed phases of the oxides and the metal in the WO_3 films (metal–metal-oxide composites). In the case of Pt, Ru, Rh, Pd, Ag, only the WO_3 was observed (metal-doped tungsten oxide).

The photoelectrochemical activity of the libraries was screened with an electrochemical probe similar to that shown in Figure 14.13. The electrolyte was 0.1 M sodium acetate. The light source consisted of an Oriel Xe 150 W lamp coupled by an optical fiber to the scanning probe to provide an incident light intensity of 2.25 mW cm^{-2} to the sample. Figure 14.28 shows the normalized zero-bias photocurrent results. It can be seen from Figure 14.28a that the greatest improvements are observed with Ni modification, followed by Co and Cu, which showed similar trends toward improved performance. Samples prepared with 5 to 15% Ni precursor in the electrodeposited bath showed nearly twice the photocurrent, compared with the control sample. The performance of the Ni-modified samples is shown in Figure 14.29 at zero and 1 V applied bias. It can be seen that the photoactivity is significantly improved at an applied bias of 1 V versus Pt.

Tungsten–molybdenum mixed metal oxides have been synthesized by parallel automated electrochemical synthesis (see Section 14.3.2); this is the first report of atomically mixed tungsten–molybdenum oxides ($W_{1-x}Mo_xO_3$, $x = 0.0, 0.05, 0.1, 0.2, 0.3, 0.5, 0.7, 0.8, 0.9\ 0.95, 1.0$) [25].

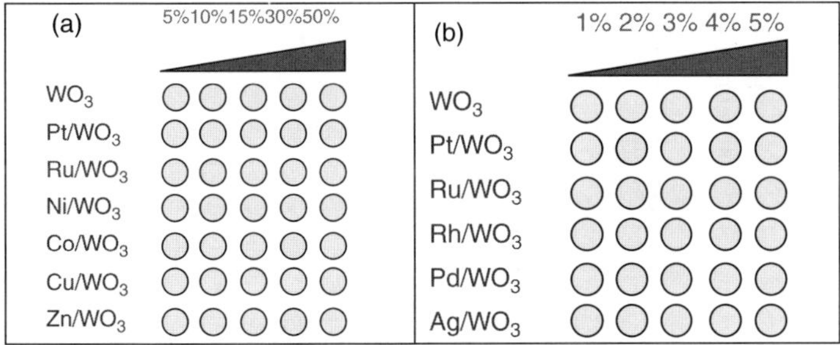

Figure 14.27 Library designs for (a) library A and (b) library B. The individual electrolytes contained peroxo-stabilized tungsten with metal salts added at the individual mole percentage for diversity. (Reprinted with permission from S.H. Baeck, T.F. Jaramillo, C. Braendli *et al.*, Combinatorial electrochemical synthesis and characterization of tungsten-based mixed-metal oxides, *Journal of Combinatorial Chemistry*, **4**(6), 563–568, 2002. © 2002 American Chemical Society.)

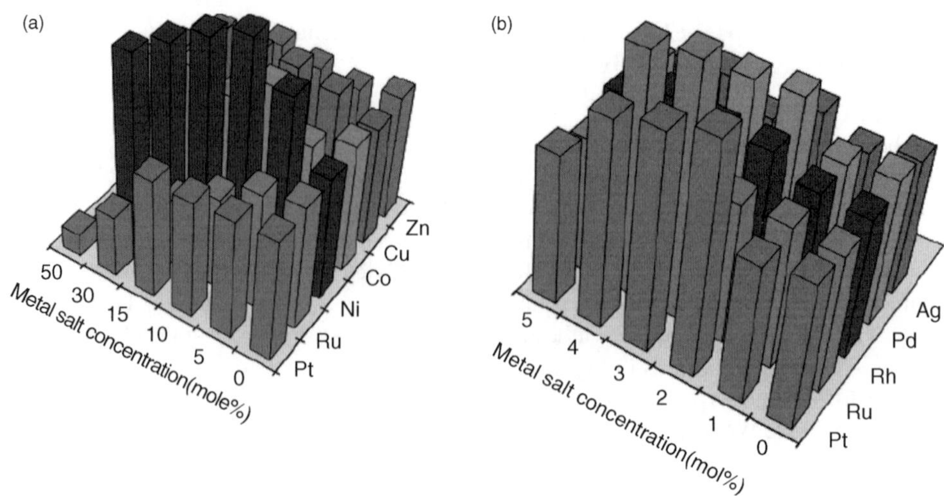

Figure 14.28 Zero-bias photocurrent of (a) metal-doped tungsten oxide in library A and (b) noble-metal-doped tungsten oxides in library B. (Reprinted with permission from S.H. Baeck, T.F. Jaramillo, C. Braendli *et al.*, Combinatorial electrochemical synthesis and characterization of tungsten-based mixed-metal oxides, *Journal of Combinatorial Chemistry*, **4**(6), 563–568, 2002. © 2002 American Chemical Society.)

The electrolyte for the electrodepositions was prepared by grinding a mixture of tungsten and molybdenum metal powders and then dissolving the metal powder mixture in 60 ml of 30% H_2O_2. The excess hydrogen peroxide was decomposed by platinum black. A 50 : 50 water and isopropanol mixture was used to dilute and stabilize the solution to 50 mM (W + Mo). The

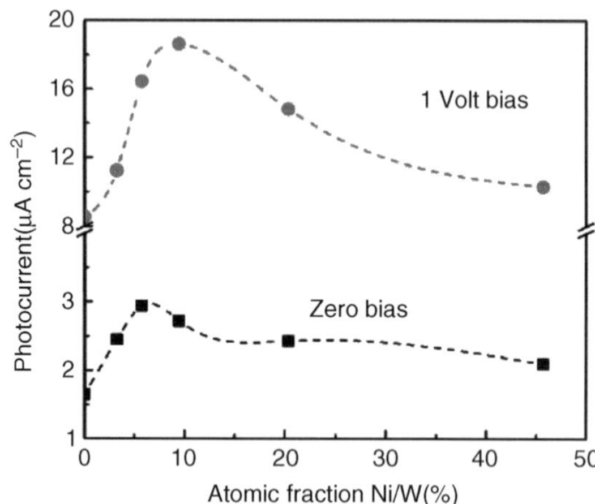

Figure 14.29 Photocurrents in 0.1 M sodium acetate solution at zero bias and 1 V bias for Ni-modified tungsten oxide as a function of Ni content. Dashed lines have been added to guide the eye. (Reprinted with permission from S.H. Baeck, T.F. Jaramillo, C. Braendli *et al.*, Combinatorial electrochemical synthesis and characterization of tungsten-based mixed-metal oxides, *Journal of Combinatorial Chemistry*, **4**(6), 563–568, 2002. © 2002 American Chemical Society.)

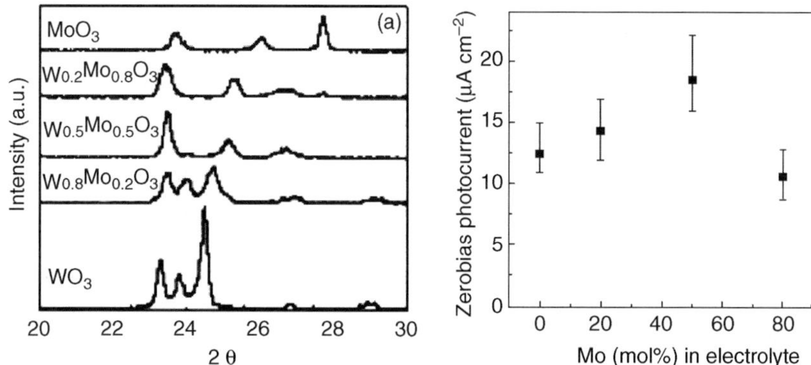

Figure 14.30 (a) X-ray diffraction patterns of tungsten–molybdenum mixed oxides and (b) zero-bias photocurrent of tungsten–molybdenum mixed metal oxide films in 0.1 M sodium acetate. Note that all samples were n-type semiconductors. Illumination was provided by a 150 W Xe lamp. (Reprinted with permission from S.H. Baeck, T.F. Jaramillo, D.H. Jeong and E.W. McFarland. Parallel synthesis and characterization of photoelectrochemically and electrochromically active tungsten–molybdenum oxides, *Chemical Communications,* (4), 390–391, 2004. © 2004 The Royal Society of Chemistry.)

concentrations for the experiment were varied from 0 to 100 mol% in a library (5 × 9 elements). The synthesis was performed using a combinatorial block with an ITO glass substrate as working electrode and a stainless steel as counter and reference electrodes. The −1.8 V versus stainless steel was applied for 10 minutes. After electrodeposition, the samples were rinsed with deionozed water and dried under air, followed by a 450 °C calcination in air for 4 hours.

Figure 14.30a shows the XRD results of selected samples in which we can clearly observe several different phases for the mixed oxide films. The XRD pattern of the $W_{0.5}Mo_{0.5}O_3$ film displays three characteristic peaks at 23.3, 25.1 and 26.6°, which are different from either pure WO_3 or pure MoO_3. The $W_{0.8}Mo_{0.2}O_3$ sample shows a mixed phase of MoO_3 and WO_3, while all the other mixed metal oxides are not simply mixed phases of the two pure oxides. Raman spectra were carried out to further understand the crystal structure of the mixed metal oxides. It was found that for the $0 \leq x \leq 0.2$ samples, the structure was similar to monoclinic WO_3, while for the $0.2 \leq x \leq 0.9$ and $0.9 \leq x \leq 1.0$ samples, the structure was similar to monoclinic β-MoO_3 and orthorhombic α-MoO_3, respectively (detailed discussion by Baeck *et al.* [25]). In Figure 14.30b, the photocurrent of tungsten oxide is 12.3 μA cm^{-2} while that of the molybdenum oxide is 5.9 μA cm^{-2}. The maximum photoresponse was found for the $W_{0.5}M_{0.5}O_3$ sample with 18.5 μA cm^{-2} which is 50% higher that that of the pure WO_3 film.

14.5.2 Improving Charge-Transfer Efficiency

14.5.2.1 Combinatorial Electrochemical Synthesis and Photoelectrochemical Characterization of Fe_2O_3 Thin Films for Photocatalytic Hydrogen Production

Hematite (α-Fe_2O_3) has many known potential advantages for photocatalytic hydrogen production [60,70,84–93]. It has a bandgap of 2–2.2 eV (allowing absorbtion of approximately

40% of solar light), it is stable in electrolytes with $pH > 3$ and is abundant, inexpensive and non-toxic. Unfortunately, it is a Mott insulator and, to date, several factors have limited the use of this material as an efficient photoelectrocatalyst, including poor charge transfer, and high recombination rates of photogenerated electrons and holes. In addition, for hydrogen production, the conduction band minimum is more positive than the redox level of H^+/H_2.

Efforts have been devoted to increasing the conductivity of thin films by increasing the amount of charge carriers or transferring electrons along the (001) planes of the hematite, which is four orders higher conductivity than transport perpendicular to this plane. Several strategies have been proposed to overcome these limitations by tailoring the hematite structure to allow more efficient transport and collection of photogenerated charge carriers, doping with heteroatoms and loading surface electrocatalysts (e.g., Pt, Au, RuO_2) [70,93]. Most work has been devoted to doping the iron oxide with heteroatoms as a means of improving PEC performance [60,84–92].

A high-throughput approach has been implemented to test the hypothesis that by deliberate doping of hematite with selected heteroatoms, the efficiency of charge transport can be improved. An automated electrochemical synthesis system was designed to allow the use of cyclic voltammogram deposition methods for the Fe_2O_3 host, investigating libraries of variable composition and structure (see Section 14.3.2 above). During the synthetic process, electrochemical parameters, including voltage range, scan rate and cycle number can be easily controlled by the software program. Figure 14.31 presents one example of a library of as-synthesized iron hydroxide thin films containing different samples.

Undoped and doped α-Fe_2O_3 thin films have been synthesized by electrochemical cyclic voltammogram deposition techniques. Results show that doping with heteroatoms can be realized by co-electrodeposition. Thirty dopants, including Al, Zn, Cu, Ni, Co, Mn, Cr, V, Ti, Mg and Pt were investigated. An automated PEC screening system was designed to measure the photocurrent, photovoltage and IPCE (see also Section 14.4.1). The photocurrent results at

Figure 14.31 Photograph of a 6×6 iron oxide thin-film library prepared by automated electrodeposition.

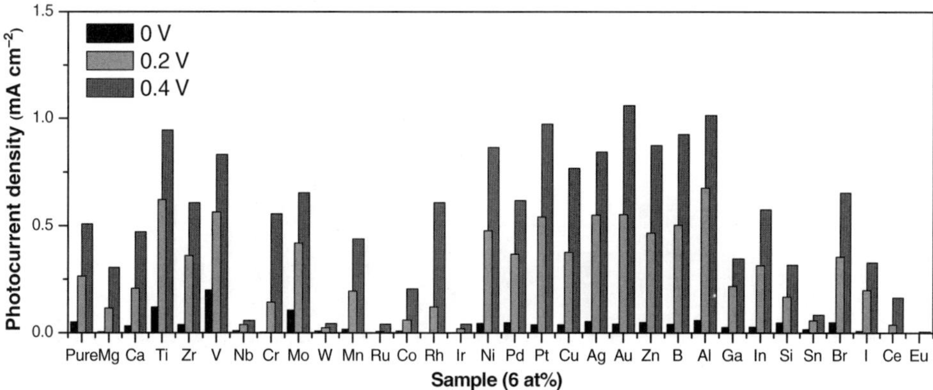

Figure 14.32 Photocurrents from samples of iron oxide electrochemically doped/substituted with various metal cations under 310 mW/cm^2 white light illumination at different bias in 1 M NaOH solution.

different bias are shown in Figure 14.32. It can be seen that doping with Al, Ti, Au and Pt exhibits the most promising results at a high bias of 0.4 V versus Ag/AgCl, whereas V doping shows the highest photocurrent at a low bias of 0 V versus Ag/AgCl.

The photovoltage measurement can indirectly reflect the relative positions of the flat-band potentials of the different photoelectrodes. For an ideal semiconductor-electrode Schottky junction, the photovoltage should follow the equation:

$$V_{OC} = V_{FB} - V_{redox}$$

where V_{OC} refers to the open-circuit photovoltage, V_{FB} refers to the flat-band potential, and V_{redox} refers to the redox potential of the redox couple present in the electrolyte in contact with the photoelectrode. The illumination intensity and other factors will significantly influence V_{OC} and the product of V_{OC} and the short-circuit photocurrent is a more useful screening quantity than either alone. Figure 14.33 shows the photovoltage of the undoped and doped samples

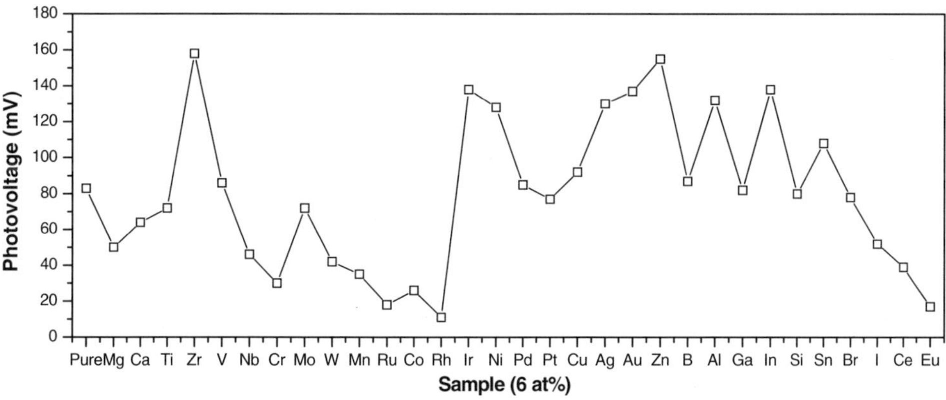

Figure 14.33 Photovoltages measured from a library of iron oxide doped/substituted with different cations under 410 mW cm^{-2} white light illumination in 1 M NaOH solution.

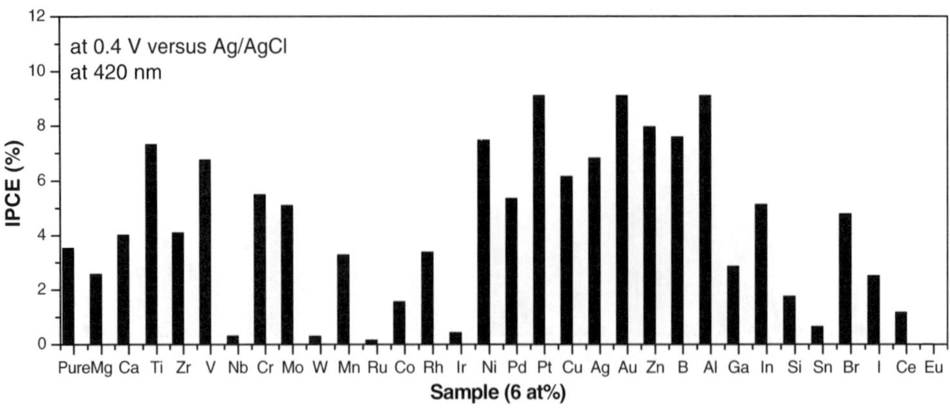

Figure 14.34 IPCE's measured from samples of iron oxide doped/substituted with different metal cations obtained under monochromatic (420 nm) light illumination at a bias of 0.4 V vs. Ag/AgCl in 1 M NaOH solution.

showing large variations in the photovoltage suggesting either changes in flat-band potentials or recombination rates.

The IPCE at 420 nm and a bias of 0.4 V versus Ag/AgCl for the undoped and doped iron-oxide thin films is displayed in Figure 14.34. The IPCEs of the Al-, Au-, Pt- and Ti-doped samples are greater than those of the other doped samples and the undoped control. Later, XPS analysis of the oxidation states of the dopant species supported a relationship between n-type doping of the hematite and an increase in IPCE likely due to conductivity increases. The trend of the IPCEs of the different samples is very similar to the photocurrents, as shown in Figure 14.31.

14.5.2.2 High-Throughput Synthesis of Hematite PEC Materials by Spray Pyrolysis

Iron-oxide thin films have been deposited by high-throughput spray pyrolysis (HTSP) described above in Section 14.3.3. Fluorine-doped tin oxide (FTO, Pilkington, Tec 15) substrates were heated to 350 °C and solutions containing 5 mM of an iron precursor ($FeCl_3$, $Fe(NO_3)_3$ and iron acetlyacetonate), were sprayed onto the hot substrate. The deposition solutions were solvated in mixtures of water and other organic solvents to control the evaporation rate and particle size of the droplets. The droplets formed in the spray head pyrolyze on impact to the FTO substrate to form iron oxide. The deposition process consists of spraying the precursor solution through the nozzle onto the substrate for a specific amount of time, followed by a dwell period to allow the substrate to reheat. The spraying time is short enough that there is less than a 10 °C drop in substrate temperature and the solution is pyrolyzed rather than forming droplets on the substrate. The spray/dwell cycle is repeated to achieve the desired thickness.

High throughput methods were used to examine different precursors and synthesis conditions for undoped Fe_2O_3 films. A library with different iron precursors solvated in different organic precursors was deposited and tested for performance. The library was prepared from

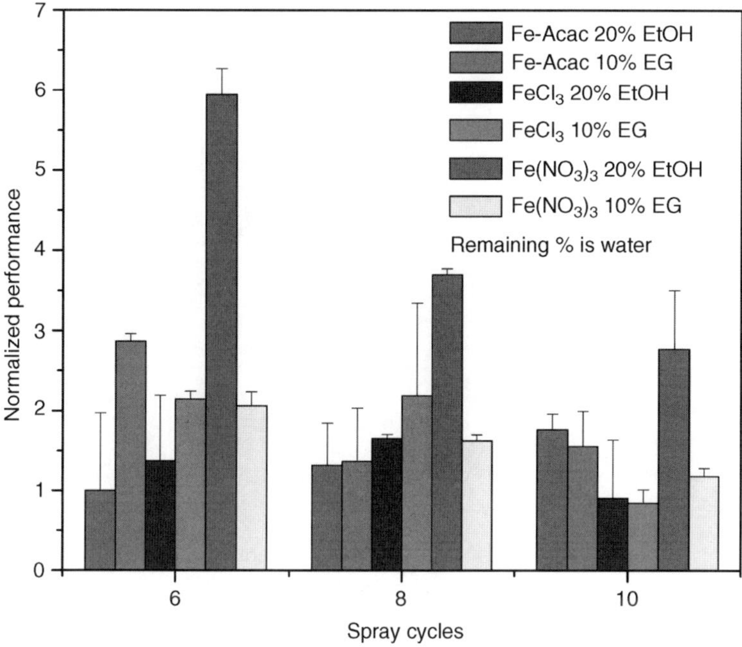

Figure 14.35 Normalized PEC performance (photocurrent) from a library of hematite films synthesized by spray pyrolysis with iron acetylacetonate, iron chloride and iron nitrate precursors, and ethanol and ethylene glycol with water as the solvating agents.

5 mM pyrolysis solutions of $FeCl_3$, $Fe(NO_3)_3$ and iron acetlyacetonate (acac) as the iron precursors, with different solvents, including 20 vol% ethanol or 10 vol% ethylene glycol in water with three different spraying cycles, 3.4 s spray time per min for 6, 8 and 10 min per sample, respectively. Two repeats were done on each sample for consistency. After synthesis at 350 °C, the samples were calcined for 6 h at 500 °C in air. The data shown in Figure 14.35 are the relative photocurrents, as measured in the high-throughput screening system for the materials on the library and suggest that the best deposition conditions consisted of $Fe(NO_3)_3$ in 20% ethanol, independent of the sample thickness. The thinnest film synthesized (six deposition cycles) had the highest relative photocurrent. Other deposition conditions have been tried (data not shown) in which 80/20 ethanol/butanol with 5 mM $Fe(NO_3)_3$ showed the highest performance.

Substrate temperature was also a synthesis variable that was optimized using the automated system. Figure 14.36 shows the performance (photocurrent) of library samples as a function of the substrate temperature (200 to 420 °C) during deposition, and as a function of precursor solutions. From the data it is clear that the substrate temperature has a significant impact on the material performance; qualitatively, the highest performance was achieved with the iron peroxo precursor at 360 °C. Titanium doping at low concentrations had little impact on the performance of the sample. However, we have observed ~9% Ti doping under optimized conditions, when iron acetlyacetonate or iron chloride is mixed with titanium ispropoxide; these observations are consistent with Augustynski *et al.* [70].

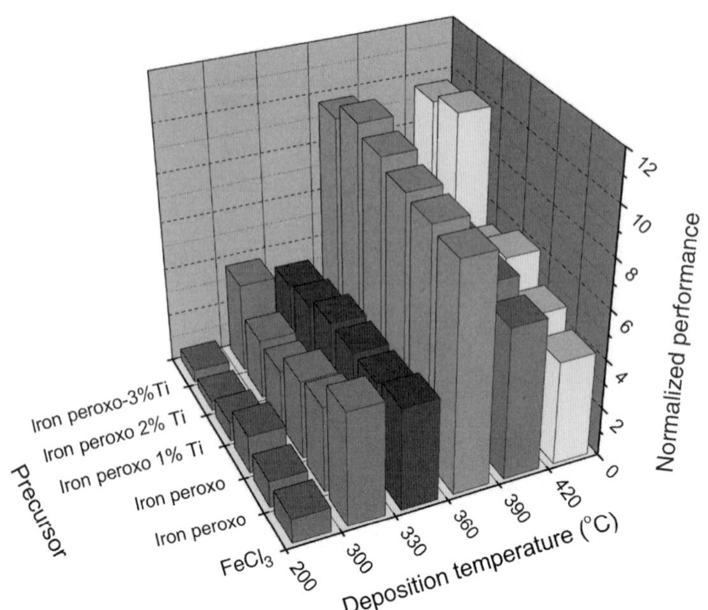

Figure 14.36 Plot of relative performance (photocurrent) of iron-oxide samples as a function of substrate deposition temperature and variations in precursor.

14.5.3 Improved PEC Electrocatalysts

In order to obtain efficient photoelectrocatalytic water splitting, electrocatalysts for decreasing the overpotential of the OER and HER reactions, as described in Section 14.2.2.4 are desired. Two examples of a strategy using high-throughput library synthesis to identify electrocatalysts are provided below.

14.5.3.1 Automated Electrochemical Synthesis and Characterization of TiO_2-Supported Au Nanoparticle Electrocatalysts

Gold is known to exhibit unusual catalytic activity when synthesized as nanoparticles and could potentially be useful for the oxygen evolution reaction. High-throughput synthesis and screening were used to deposit gold nanoparticle electrocatalysts on TiO_2 and sizes ranging from 5 to 150 nm have been demonstrated [61]. The photoelectrochemical and electrocatalytic performance was screened as a function of the coverage and particle size of the electrodeposited gold particles.

The synthesis of nanometer-size gold particles was carried out in a 32-member library by using the rapid serial synthesis system described in Section 14.3.2. TiO_2 was synthesized by thermal treatment of a Ti foil at 400 °C for one hour to produce an anatase film which was used as the substrate. Electrodeposition of the gold nanoparticles was performed using an electrolyte containing a 20 mM $HAuCl_4{\cdot}3H_2O$. Pulsed electrodeposition was used where the voltage pulse was cycled between -1 V and $+0.2$ V for 5 ms at each potential (100 pulses s^{-1}), the deposition time for the library columns were 0 s (control sample), 0.5, 1, 3, 5, 10, 30 and 60 s.

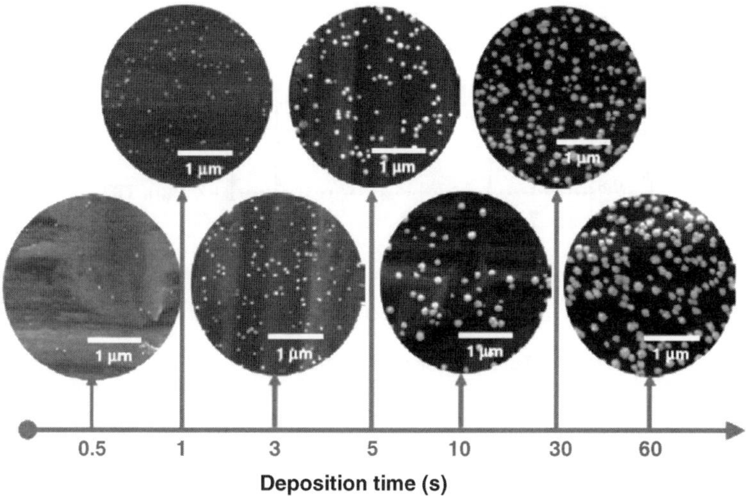

Figure 14.37 Selected SEM images of the Au/TiO_2 library as a function of total deposition time. Particle sizes varied from <5 to <150 nm. (Reprinted with permission from S.H. Baeck, T.F. Jaramillo, A. Kleiman-Shwarsctein and E.W. McFarland, Automated electrochemical synthesis and characterization of TiO_2 supported Au nanoparticle electrocatalysts, *Measurement Science & Technology*, **16**(1), 54–59, 2005. © 2005 Institute of Physics.)

Scanning electron microscopy was performed on the samples and representative images are shown in Figure 14.37. The size and loading of the samples increased with increasing deposition time. At a low deposition time, the size of the particles was 10 ± 5 nm, while at 60s the sizes increase to 200 ± 50 nm. Using TEM, Figure 14.38a, it was confirmed that nanoparticulate Au was obtained and the particles were well dispersed on the surface of the TiO_2. XPS of the Au from the electrodeposited samples, Figure 14.38b, shows the Au 4*f* peaks

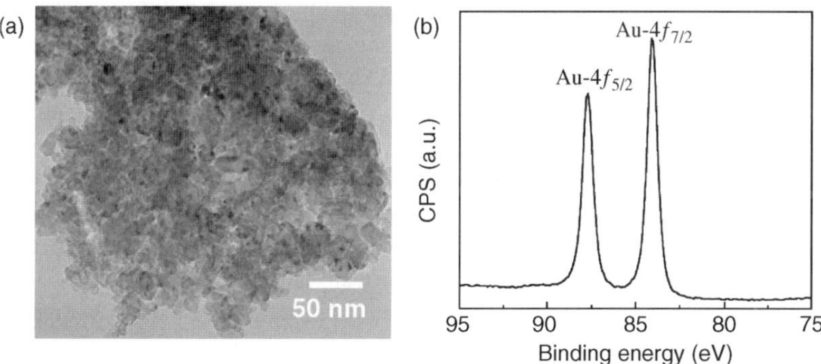

Figure 14.38 (a) TEM image of Au nanoparticles synthesized with 1 s of total deposition time and (b) X-ray photoelectron spectra (Al $K\alpha = 1486.6$ eV) corresponding to the Au 4*f* photoemission. (Reprinted with permission from S.H. Baeck, T.F. Jaramillo, A. Kleiman-Shwarsctein and E.W. McFarland, Automated electrochemical synthesis and characterization of TiO_2 supported Au nanoparticle electrocatalysts, *Measurement Science & Technology*, **16**(1), 54–59, 2005. © 2005 Institute of Physics.)

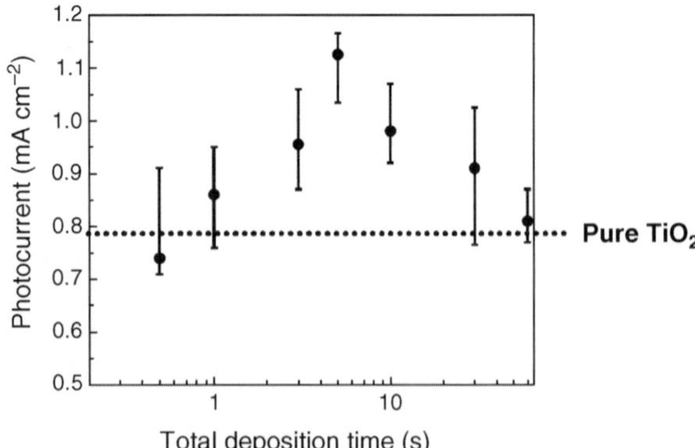

Figure 14.39 High-throughput zero-bias photocurrent screening of 32 Au/TiO_2 samples (four replicates at each deposition time). (Reprinted with permission from S.H. Baeck, T.F. Jaramillo, A. Kleiman-Shwarsctein and E.W. McFarland, Automated electrochemical synthesis and characterization of TiO_2 supported Au nanoparticle electrocatalysts, *Measurement Science & Technology*, **16**(1), 54–59, 2005. © 2005 Institute of Physics.)

at 87.7 and 84.0 eV, which are characteristic of metallic gold. The photoelectrochemical performance of the samples is shown in Figure 14.39 with the Au/TiO_2 serving as a photoanode for oxygen evolution. Very little change is observed for samples deposited for less than 3 s, while an increase of 20–40% in photocurrent is observed when the samples were deposited for between 3 and 10 s, with a peak photocurrent of 1.12 mA cm^{-12} for the sample deposited for 5 s. A decrease in photocurrent is observed with higher Au loading (and larger sizes). The maximum performance of the Au nanoparticles for the oxygen evolution reaction electrocatalyst is related to the particle size and loading.

14.5.3.2 High-Throughput Screening System for Electrocatalytic Hydrogen Production Materials

Hydrogen evolution catalysts were investigated in a library of 24 different electrocatalysts synthesized by electron-beam evaporation from eight elemental sources of Ti, Pt, Ni, Au, Pd, Al, Ag and Ge [23]. Eight of the 24 materials synthesized were single-element substances, while the remaining 16 were binary compounds co-evaporated in a ratio of approximately 80/20, with the exception of Au/Ge, which had a ratio of 1/1. Four replicas of each material were synthesized to allow determination of the statistical error on each of the material compositions and measurements. A schematic diagram of the library was shown in Figure 14.18 above. The electrocatalyst library was screened using an H_2 screening system based on chemioptical changes in tungsten oxide, as discussed above in Section 14.4.1.3.

The electrocatalytic performance of the library was based on the rate of hydrogen production, as determined from the data in Figure 14.18 (the reflectance change is a measure of the rate of hydrogen production). The results are shown in Figure 14.40. The best material in this library was an Al/Pt mixture in which the Al at the surface is certainly present as alumina, this particular combination has previously reported as an active electrocatalyst [94]. The trends of the library can be qualitatively compared to know literature trends, for example, it is well known that Pt, Pd and Ni are the most effective single-element materials for cathodic hydrogen

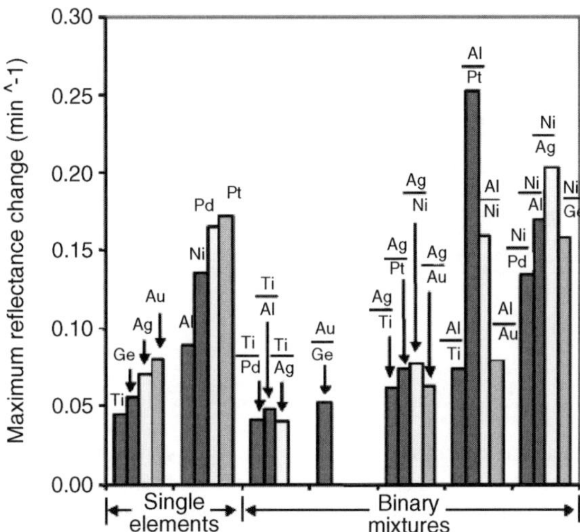

Figure 14.40 2-D bar graph representing the maximum rate of change in reflectance for each material. Expected trends are validated because materials containing Ni, Pd and Pt were the most effective H_2 producers, while materials consisting of Ti or Ge performed poorly. The Al/Pt mixture was clearly the best electrocatalyst of those studied. (Reprinted with permission from T.F. Jaramillo, A. Ivanovskaya and E.W. McFarland, High-throughput screening system for catalytic hydrogen-producing materials, *Journal of Combinatorial Chemistry*, **4**(1), 17–22, 2002, American Chemical Society.)

evolution; Ag, Al and Au would be expected to have a lower performance while Ti and Ge are known to show poor performance [95]. These trends are qualitatively the same as the experimental data shown in Figure 14.40 (see publication for full details of the other validation methods [23]). In the case of the binary mixtures, the best-performing samples were the Al/Pt followed by some of the Ni/Ag, Ni/Al, Ni/Ge and Al/Ni, which had better performance than that of the pure Ni electrocatalyst alone.

14.5.4 Design and Assembly of a Complete Nanostructured Photocatalytic Unit

The optimization of the various components of a PEC material system involves integrating an absorber, a transporter and an electrocatalyst into a stable functional unit. It is possible to make use of high-throughput methods to facilitate this process. As the lead materials are identified and optimized, the final assembly of a functional device must be taken into account. Our vision of the ideal photoelectrocatalytic material system consists of: (i) a high-efficiency absorber material made using a low-cost synthetic pathway from abundant and non-toxic elements, fabricated into (ii) a structure that maximizes charge transport from the site of creation to the site of utilization, with (iii) electrocatalytic active sites for minimal overpotential loss in performing redox chemistry and (iv) passivation against photodegradation and destruction in the working environment. Nanostructured photocatalysts have high surface area and minimized charge-transport lengths, which may be used to maximize the IQE. A balance must be struck between the improved charge transport and the potentially higher surface recombination rate.

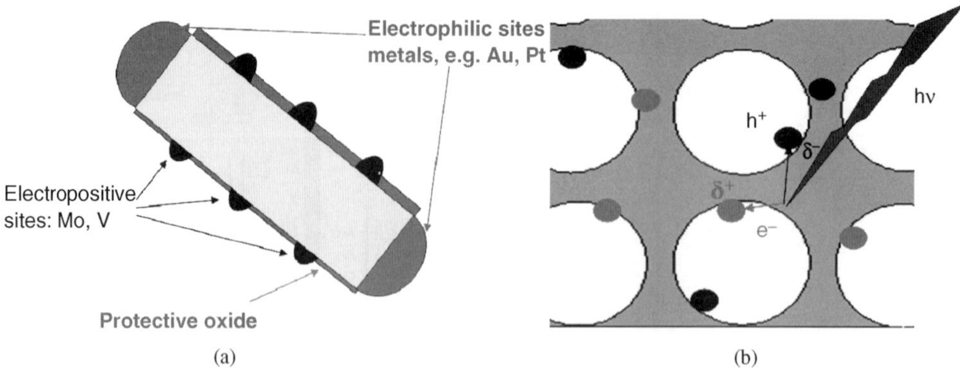

Figure 14.41 Idealized heterogeneous nanostructures for PEC systems. Red-colored particles represent electrophilic materials used for hydrogen evolution electrocatalysts, whereas black particles represent electropositive oxygen-evolution electrocatalysts. See also color-plate section.

Idealized heterogeneous nanostructures are shown above in Figure 14.41, designed to improve charge transport by minimizing the distance from absorption to utilization and improving the separation efficiency of photogenerated electrons and holes using high electron/hole affinity electrocatalysts. Along the path to create such structures, Figure 14.42A shows a micrograph of hematite nanowires/rods synthesized from a mesoporous template (SBA-15) with a diameter of approximately 6 nm and surface area of larger than $175\,m^2\,g^{-1}$. In Figure 14.42B a mesoporous iron oxide material is shown prepared by nanocasting using a KIT6 template, which was removed after synthesis. The material has a crystalline cubic

Figure 14.42 Nanoscale iron oxide: (A) hematite nanowires and rods created from casting in mesoporous SBA15 template; (B) mesoporous hematite nanostructure created from KIT6 templates. The dark particles are Pt nanoparticles deposited hydrothermally after removal of the template.

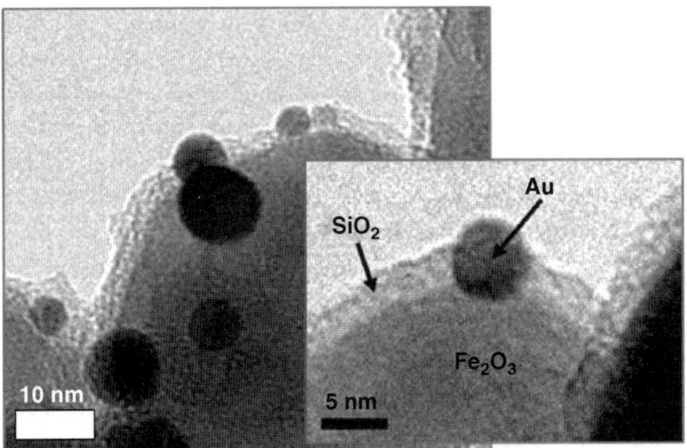

Figure 14.43 Electron micrograph of a complete PEC nanoscale heterostructure. A hematite absorber particle has been decorated with Au electrocatalyst particles and coated for stability with silica. Image courtesy of Arnold Forman.

structure (pore size ~8 nm, wall thickness ~8 nm). Pt nanoparticles deposited on the surface of the hematite are intended to assist in the separation of photogenerated charges and to function as a electrocatalyst for HER. With a higher crystallinity, the IQE of the mesoporous hematite containing Pt nanoparticles is almost two times of that of nanowires/rods, even though the surface area is less than $100\,m^2\,g^{-1}$).

The final photoelectrochemical unit may also require passivation from the electrolyte environment. Figure 14.43 shows a TEM micrograph of a hematite nanoparticle, which has had electrophilic gold nanoparticulate electrocatalysts affixed to the surface and was then coated with several nanometers of semi-porous silica. Today, wet-chemical synthesis routes exist such that these types of photosynthetic functional units can be created on enormous scales economically, once an efficient material system is identified. Unfortunately, hematite has not been shown to be efficient.

14.6 Summary and Outlook

Nature has evolved photosynthetic systems with approximately 0.5% solar-to-chemical fuel conversion efficiency. For almost 100 years, humankind has sought to identify artificial photosynthetic systems with higher efficiencies that are cost-effective. Photoelectrochemical processes are conceptually simple and the fundamental processes are well understood, yet in spite of decades of focused research there are no material systems known that could be deployed on large scales in a cost-effective manner. The primary problem is the identification of a solar-spectrum absorber that is stable in an electrochemical environment. Silicon solar cells, using one of the most abundant elements on earth, are not cost-effective today and the one or two practical alternative materials have not been shown to have the necessary stability or practical efficiencies. A PEC material has the added constraint that not only must it absorb in the visible and transport charges to an interface, it must also be stable in an electrolyte. Today, there is no known material that is satisfactory. Fortunately, there are an enormous number of semiconductor materials that have not even been synthesized and that might have favorable properties for a cost-effective PEC system. Unfortunately, we do not have the basic knowledge or theory to predict in advance what the properties of most heterogeneous semiconductors are, and thus we will either need to make and test them all one-by-one using conventional chemical methods, or we can embrace combinatorial and high-throughput experimentation methods to both increase the rate at which materials are synthesized and screened and, hopefully, discover relationships inductively that give rise to new theories that improve our predictive abilities. To date, most work in materials science has made use of high-throughput methods alone, without invoking combinatorics. Applications specifically to PEC-material discovery have been limited with little true new material-discovery experimentation underway. There is a wide range of possibilities for use of combinatorial and high-throughput methods to aid in the identification of new solar-to-chemical conversion materials, however, even with such an approach, the task is enormous and will require a long-term and sustained effort if ever a cost-effective means of splitting water into hydrogen and oxygen is to be found.

References

[1] Anderson, F.H. (1948) *The Philosophy of Francis Bacon*, University of Chicago Press.

[2] Brewster, D. (1855) *Memoirs of the Life, Writings, and Discoveries of Sir Isaac Newton*, T. Constable and Company, Edinburgh.

[3] Albini, A. and Fagnoni, F.M. (2004) Green chemistry and photochemistry were born at the same time. *Green Chemistry*, **6**, 1–6.

[4] Bryan, G.S. (1926) *Edison, the Man and his Work*, Alfred Knopf Publisher, London, New York.

[5] Mabry, C.C. Gevedon, R.E. and Gochman, N. (1965) Automated microchemistries for whole hospital. *Southern Medical Journal*, **58**(12), 1597.

[6] Lindsey, J.S., Corkan, L.A., Erb, D. and Powers, G. J. (1988) Robotic work station for microscale synthetic chemistry – online absorption-spectroscopy, quantitative automated thin-layer chromatography, and multiple reactions in parallel. *Review of Scientific Instruments*, **59**(6), 940–950.

[7] Bowie, L.J. and Gochman, N. (1983) Selection factors for automated analytic instrumentation. *Clinical Chemistry*, **29**(6), 1290–1290.

[8] Segatto, P.R. (1969) Controlled preparation of alloys by simultaneous, multitarget sputtering. *Journal of Vacuum Science & Technology*, **6**(3), 368–372.

[9] Hanak, J.J. (1970) Multiple-sample-concept in materials research – synthesis, compositional analysis and testing of entire multicomponent systems. *Journal of Materials Science*, **5**(11), 964–971.

[10] Hanak, J.J. (1971) Compositional determination of RF co-sputtered multicomponent systems. *Journal of Vacuum Science & Technology*, **8**(1), 172–175.

[11] Hanak, J.J., Wehner, R.K. and Lehmann, H.W. (1972) Calculation of deposition profiles and compositional analysis of cosputtered films. *Journal of Applied Physics*, **43**(4), 1666–1673.

[12] Faughnan, B.W. and Hanak, J.J. (1983) Photovoltaically active p layers of amorphous-silicon. *Applied Physics Letters*, **42**(8), 722–724.

[13] Hanak, J.J. (1975) Co-sputtering – its limitations and possibilities. *Vide-Science Technique et Applications*, **30**(175), 11–18.

[14] Hanak, J.J. (1985) Plasma potential gradient in dc positive-column and its effect on the deposition and photovoltaic performance of A-SI-H films. *Journal of Non-Crystalline Solids*, **77–78**, 773–776.

[15] Hanak, J.J. and Pellicane, J. (1976) DC electroluminescence in sputtered EU + 3-activated ternary metal-oxides. *Journal of the Electrochemical Society*, **123**(8), C260–C260.

[16] Xiang, X.D., Sun, X., Briceño, G. *et al.* (1995) A combinatorial approach to materials discovery. *Science*, **268** (5218), 1738–1740.

[17] Danielson, E., Golden, J.H., McFarland, E.W. *et al.* (1997) A combinatorial approach to the discovery and optimization of luminescent materials. *Nature*, **389**(6654), 944–948.

[18] Danielson, E., Devenney, M., Gaiquinta, D.M. *et al.* (1998) A rare-earth phosphor containing one-dimensional chains identified through combinatorial methods. *Science*, **279**(5352), 837–839.

[19] McFarland, E.W. (1997) Potential masking systems and methods for combinatorial library synthesis. 6468806. USPTO, Editor. Symyx: USA.

[20] Reddington, E., Sapienza, A., Gurau, B. *et al.* (1998) Combinatorial electrochemistry: A highly parallel, optical screening method for discovery of better electrocatalysts. *Science*, **280**(5370), 1735–1737.

[21] Sullivan, M.G., Utomo, H., Fagan, P.J. and Ward, M.D. (1999) Automated electrochemical analysis with combinatorial electrode arrays. *Analytical Chemistry*, **71**(19), 4369–4375.

[22] McFarland, E. (2001) Discovery of New Photocatalytic Materials for Hydrogen Synthesis. Proceedings of the 11th Canadian Hydrogen Association Conference.

[23] Jaramillo, T.F., Ivanovskaya, A. and McFarland, E.W. (2002) High-throughput screening system for catalytic hydrogen-producing materials. *Journal of Combinatorial Chemistry*, **4**(1), 17–22.

[24] Nakayama, A., Suzuki, E. and Ohmori, T. (2002) Development of high throughput evaluation for photocatalyst thin-film. *Applied Surface Science*, **189**, 260–264.

[25] Baeck, S.H., Jaramillo, T.F, Braendli, C. *et al.* (2002) Combinatorial electrochemical synthesis and characterization of tungsten-based mixed-metal oxides. *Journal of Combinatorial Chemistry*, **4**(6), 563–568.

[26] Jaramillo, T.F., Baeck, S.H., Kleiman-Schwarsctein, A. *et al.* (2004) Combinatorial electrochemical synthesis and screening of mesoporous ZnO for photocatalysis. *Macromolecular Rapid Communications*, **25**(1), 297–301.

[27] Woodhouse, M., Herman, G.S. and Parkinson, B.A. (2005) Combinatorial approach to identification of catalysts for the photoelectrolysis of water. *Chemistry of Materials*, **17**(17), 4318–4324.

[28] Lemmo, A.V., Fisher, J.T., Geysen, H.M. and Rose, D.J. (1997) Characterization of an inkjet chemical microdispenser for combinatorial library synthesis. *Analytical Chemistry*, **69**(4), 543–551.

[29] Sun, X.D., Wang, K.-A., Yoo, Y. *et al.* (1997) Solution-phase synthesis of luminescent materials libraries. *Advanced Materials*, **9**(13), 1046–1049.

[30] Seyler, M., Stoewe, K. and Maier, W.F. (2007) New hydrogen-producing photocatalysts – A combinatorial search. *Applied Catalysis B-Environmental*, **76**(1–2), 146–157.

[31] Valdes, A., Qu, Z.W., Kroes, G.-J. *et al.* (2008) Oxidation and photo-oxidation of water on TiO_2 surface. *Journal of Physical Chemistry C*, **112**(26), 9872–9879.

[32] Huda, M.N., Yan, Y., Wei, S.-H., Al-Jassim, M.M. (2008) Electronic structure of ZnO:GaN compounds: Asymmetric bandgap engineering. *Physical Review B*, **78**(19), 195204/1–195204/5.

[33] Johnson, E.L. (1981) The TI solar energy system development, in *Electron Devices Meeting International*, IEEE, Washington, D.C., pp. 2–5.

[34] Hamann, T.W. and Lewis, N.S. (2006) Control of the stability, electron-transfer kinetics, and pH-dependent energetics of Si/H_2O interfaces through methyl termination of Si(111) surfaces. *Journal of Physical Chemistry B*, **110**(45), 22291–22294.

[35] Greeley, J., Jaramillo, T.F., Bonde, J., Chorkendorff, I. and Nørskov, J.K. (2006) Computational high-throughput screening of electrocatalytic materials for hydrogen evolution. *Nature Materials*, **5**(11), 909–913.

[36] Trasatti, S. (1984) Electrocatalysis in the anodic evolution of oxygen and chlorine. *Electrochimica Acta*, **29**(11), 1503–1512.

[37] Rossmeisl, J., Qu, Z.W., Zhu, H., Kroes, G.J. and Norskov, J.K. (2007) Electrolysis of water on oxide surfaces. *Journal of Electroanalytical Chemistry*, **607**(1–2), 83–89.

[38] Kayes, B.M., Atwater, H.A. and Lewis, N.S. (2005) Comparison of the device physics principles of planar and radial p-n junction nanorod solar cells. *Journal of Applied Physics*, **97**(11), 11.

[39] Takeuchi, I., Lippmaa, M. and Matsumoto, Y. (2006) Combinatorial experimentation and materials informatics. *MRS Bulletin*, **31**(12), 999–1003.

[40] Pore, V., Ritala, M., Leskelä, M. *et al.* (2007) H_2S modified atomic layer deposition process for photocatalytic TiO_2 thin films. *Journal of Materials Chemistry*, **17**(14), 1361–1371.

[41] Pore, V., Heikklä, M., Ritala, M., Leskelä, M. and Areva, S. (2006) Atomic layer deposition of TiO_2-xNx thin films for photocatalytic applications. *Journal of Photochemistry and Photobiology A-Chemistry*, **177**(1), 68–75.

[42] Woodhouse, M. and Parkinson, B.A. (2008) Combinatorial discovery and optimization of a complex oxide with water photoelectrolysis activity. *Chemistry of Materials*, **20**(7), 2495–2502.

[43] Ahn, K.S., Shet, S., Deutsch, T. *et al.* (2008) Enhancement of photoelectrochemical response by aligned nanorods in ZnO thin films. *Journal of Power Sources*, **176**(1), 387–392.

[44] Glasscock, J.A., Barnes, P.R.F., Plumb, I.C. and Savvides, N. (2007) Enhancement of photoelectrochemical hydrogen production from hematite thin films by the introduction of Ti and Si. *Journal of Physical Chemistry C*, **111**(44), 16477–16488.

[45] Miller, E.L., Marsen, B., Cole, B. and Lum, M. (2006) Low-temperature reactively sputtered tungsten oxide films for solar-powered water splitting applications. *Electrochemical and Solid State Letters*, **9**(7), G248–G250.

[46] Miller, E.L., Paluselli, D., Marsen, B. and Rocheleau, R.E. (2005) Development of reactively sputtered metal oxide films for hydrogen-producing hybrid multijunction photoelectrodes. *Solar Energy Materials and Solar Cells*, **88**(2), 131–144.

[47] Mor, G.K., Prakasam, H.E., Varghese, O.K., Shankar, K. and Grimes, C.A. (2007) Vertically oriented Ti-Fe-O nanotube array films: Toward a useful material architecture for solar spectrum water photoelectrolysis. *Nano Letters*, **7**(8), 2356–2364.

[48] Mwabora, J.M., Lindgren, T., Avendano, E. *et al.* (2004) Structure, composition, and morphology of photoelectrochemically active $TiO_{2-x}N_x$ thin films deposited by reactive DC magnetron sputtering. *Journal of Physical Chemistry B*, **108**(52), 20193–20198.

[49] Petitjean, C., Grafoute, M., Rousselot, C. and Pierson, J.F. (2008) Reactive gas pulsing process: A method to extend the composition range in sputtered iron oxynitride films. *Surface & Coatings Technology*, **202**(19), 4825–4829.

[50] Le Dreo, H., Banakh, O., Keppner, H. *et al.* (2006) Optical, electrical and mechanical properties of the tantalum oxynitride thin films deposited by pulsing reactive gas sputtering. *Thin Solid Films*, **515**(3), 952–956.

[51] Wong, M.S., Wang, S.-H., Chen, T.-K., Weng, C.-W. and Rao, K.K. (2007) Co-sputtered carbon-incorporated titanium oxide films as visible light-induced photocatalysts. *Surface & Coatings Technology*, **202**, 890–894.

[52] Jung, H.S., Hong, Y.J., Li, Y., Cho, J., Kim, Y.J. and Yi, G.C. (2008) Photocatalysis using GaN nanowires. *ACS Nano*, **2**(4), 637–642.

[53] Rar, A., Specht, E.D., George, E.P., Santella, M.L. and Pharr, G.M. (2004) Preparation of ternary alloy libraries for high-throughput screening of material properties by means of thick film deposition and interdiffusion: Benefits and limitations. *Journal of Vacuum Science & Technology A*, **22**(4), 1788–1792.

[54] Rar, A., Frafjord, J.J., Fowlkes, J.D. *et al.* (2005) PVD synthesis and high-throughput property characterization of Ni-Fe-Cr alloy libraries. *Measurement Science & Technology*, **16**(1), 46–53.

[55] Gurau, B., Viswanathan, R., Renxuan, L. *et al.* (1998) Structural and electrochemical characterization of binary, ternary, and quaternary platinum alloy catalysts for methanol electro-oxidation. *Journal of Physical Chemistry B*, **102**(49), 9997–10003.

[56] Fenniri, H. (2000) *Combinatorial Chemistry: A Practical Approach* (ed. H. Fenniri), Oxford University Press, pp. 401–419.

[57] Chan, B.C., Renxuan, L., Jambunathan, K. *et al.* (2005) Comparison of high-throughput electrochemical methods for testing direct methanol fuel cell anode electrocatalysts. *Journal of the Electrochemical Society*, **152**(3), A594–A600.

[58] Hu, Y.-S., Kleiman-Shwarsctein, A., Forman, A.J. *et al.* (2008) Pt-Doped α-Fe_2O_3 thin films active for photoelectrochemical water splitting. *Chemistry of Materials*, **20**(12), 3803–3805.

[59] Funahashi, R., Urata, S. and Kitawaki, M. (2004) Exploration of n-type oxides by high throughput screening. *Applied Surface Science*, **223**(1–3), 44–48.

[60] Arai, T., Konishi, Y., Iwasaki, Y., Sugihara, H. and Sayama, K. (2007) High-throughput screening using porous photoelectrode for the development of visible-light-responsive semiconductors. *Journal of Combinatorial Chemistry*, **9**(4), 574–581.

[61] Baeck, S.H., Jaramillo, T.F., Kleiman-Shwarsctein, A. and McFarland, E.W. (2005) Automated electrochemical synthesis and characterization of TiO_2 supported Au nanoparticle electrocatalysts. *Measurement Science & Technology*, **16**(1), 54–59.

[62] Jaramillo, T.F., Baeck, S.H., Kleiman-Shwarsctein, A. *et al.* (2005) Automated electrochemical synthesis and photoelectrochemical characterization of $Zn_{1-x}Co_xO$ thin films for solar hydrogen production. *Journal of Combinatorial Chemistry*, **7**(2), 264–271.

[63] Baeck, S.H. and McFarland, E.W. (2002) Combinatorial electrochemical synthesis and characterization of tungsten-molybdenum mixed metal oxides. *Korean Journal of Chemical Engineering*, **19**(4), 593–596.

[64] Brandli, C., Jaramillo, T.F., Ivanovskaya, A. and McFarland, E.W. (2001) Automated synthesis and characterization of diverse libraries of macroporous alumina. *Electrochimica Acta*, **47**(4), 553–557.

[65] Hu, Y.S., Kleiman-Shwarsctein, A., Forman, A.J. *et al.* (2008) Pt-doped α-Fe_2O_3 thin films active for photoelectrochemical water splitting. *Chemistry of Materials*, **20**(12), 3803–3805.

[66] Aranovich, J., Ortiz, A. and Bube, R.H. (1979) Optical and electrical-properties of zno films prepared by spray pyrolysis for solar-cell applications. *Journal of Vacuum Science & Technology*, **16**(4), 994–1003.

[67] Eberspacher, C., Fahrenbruch, A.L. and Bube, R.H. (1986) Properties of ZnO films deposited onto INP by spray pyrolysis. *Thin Solid Films*, **136**(1), 1–10.

[68] Ingler, W.B. and Khan, S.U.M. (2004) Photoresponse of spray pyrolytically synthesized magnesium-doped iron(III) oxide (p-Fe_2O_3) thin films under solar simulated light illumination. *Thin Solid Films*, **461**(2), 301–308.

[69] Ingler, W.B. and Khan, S.U.M. (2005) Photoresponse of spray pyrolytically synthesized copper-doped p-Fe_2O_3 thin film electrodes in water splitting. *International Journal of Hydrogen Energy*, **30**(8), 821–827.

[70] Sartoretti, C.J., Ulmann, M., Alexander, B.D., Augustynski, J. and Weidenkaff, A. (2005) Photoelectrochemical oxidation of water at transparent ferric oxide film electrodes. *Journal of Physical Chemistry B*, **109**(28), 13685–13692.

[71] Duret, A. and Gratzel, M. (2005) Visible light-induced water oxidation on mesoscopic alpha-Fe_2O_3 films made by ultrasonic spray pyrolysis. *Journal of Physical Chemistry B*, **109**(36), 17184–17191.

[72] Yadav, S.P., Shinde, P.S., Rajpure, K.Y. and Bhosale, C.H. (2008) Photoelectrochemical properties of spray deposited n-ZnIn2Se4 thin films. *Solar Energy Materials and Solar Cells*, **92**(4), 453–456.

[73] Kleiman-Shwarsctein, A. (2004) Automated spray pyrolysis system for hematite synthesis as a photocatalyst for hydrogen production. *Abstracts of Papers of the American Chemical Society*, **227**, U1314–U1314.

[74] Cesar, I., Kay, A., Gonzalez Martinez, J.A. and Grätzel, M. (2006) Translucent thin film Fe_2O_3 photoanodes for efficient water splitting by sunlight: Nanostructure-directing effect of Si-doping. *Journal of the American Chemical Society*, **128**(14), 4582–4583.

[75] Chen, Z., Jaramillo, T.F., Deutsch, T.G., Kleiman-Shwarsctein, A., Forman, A., Gaillard, N., Garland, R., Takanabe, K., Heske, C., Sunkara, M., McFarland, E.W., Domen, K., Miller, E.L., Turner, J.A. and Dinh, H.N. Acceletating materials development for photoelectrochemical (PEC) hydrogen production: Standards for methods, definitions, and reporting protocols. *Journal of Materials Research*, In print.

[76] Further reference on standardized methods for photoelectrochemical research related to hydrogen production can be found in. http://www2.eere.energy.gov/hydrogenandfuelcells/photoelectrochemical_group.html.

[77] Maier, W.F., Stowe, K. and Sieg, S. (2007) Combinatorial and high-throughput materials science. *Angewandte Chemie-International Edition*, **46**(32), 6016–6067.

[78] Senkan, S.M. (1998) High-throughput screening of solid-state catalyst libraries. *Nature*, **394**(6691), 350–353.

[79] Lee, J.W., Ye, H., Pan, S. and Bard, A.J. (2008) Screening of photocatalysts by scanning electrochemical microscopy. *Analytical Chemistry*, **80**(19), 7445–7450.

[80] Bergh, S., Cong, P. Ehnebuske, B. *et al.* (2003) Combinatorial heterogeneous catalysis: oxidative dehydrogenation of ethane to ethylene, selective oxidation of ethane to acetic acid, and selective ammoxidation of propane to acrylonitrile. *Topics in Catalysis*, **23**(1–4), 65–79.

[81] Cong, P.J., Dehestani, A., Doolen, R. *et al.* (1999) Combinatorial discovery of oxidative dehydrogenation catalysts within the Mo-V-Nb-O system. *Proceedings of the National Academy of Sciences of the United States of America*, **96**(20), 11077–11080.

[82] Zhao, X.R. *et al.* (2006) High-throughput characterization of BixY3-xFe_5O_{12} combinatorial thin films by magneto-optical imaging technique. *Applied Surface Science*, **252**(7), 2628–2633.

[83] Dieing, T. and Hollricher, O. (2008) *High-Resolution, High-Speed Confocal Raman Imaging*, Elsevier Science Bv.

[84] Leland, J.K. and Bard, A.J. (1987) Photochemistry of colloidal semiconducting iron-oxide polymorphs. *Journal of Physical Chemistry*, **91**(19), 5076–5083.

[85] Shinar, R. and Kennedy, J.H. (1982) Photoactivity of doped Alpha-Fe_2O_3 electrodes. *Solar Energy Materials*, **6** (3), 323–335.

[86] Yeh, L.S.R. and Hackerman, N. (1977) Iron-oxide semiconductor electrodes in photoassisted electrolysis of water. *Journal of the Electrochemical Society*, **124**(6), 833–836.

[87] Leygraf, C., Hendewerk, M. and Somorjal, G.A. (1982) Photocatalytic production of hydrogen from water by a P-type and N-type polycrystalline iron-oxide assembly. *Journal of Physical Chemistry*, **86**(23), 4484–4485.

[88] Kay, A., Cesar, I. and Gratzel, M. (2006) New benchmark for water photooxidation by nanostructured alpha-Fe_2O_3 films. *Journal of the American Chemical Society*, **128**(49), 15714–15721.

[89] Dare-Edwards, M.P., Goodenough, J.B., Hamnett, A. and Trevellick, P.R. (1983) Electrochemistry and photoelectrochemistry of iron(III) oxide. *Journal of the Chemical Society Faraday Transactions*, **1**(79), 2027.
[90] Khan, S.U.M. and Akikusa, J. (1999) Photoelectrochemical splitting of water at nanocrystalline n-Fe_2O_3 thin-film electrodes. *Journal of Physical Chemistry B*, **103**(34), 7184–7189.
[91] Aroutiounian, V.M., Arakelyan, V.M., Shahnazaryan, G.E. *et al.* (2006) Photoelectrochemistry of semiconductor electrodes made of solid solutions in the system Fe_2O_3-Nb_2O_5. *Solar Energy*, **80**(9), 1098–1111.
[92] Lindgren, T., Wang, H., Beerman, N. *et al.* (2002) Aqueous photoelectrochemistry of hematite nanorod array. *Solar Energy Materials and Solar Cells*, **71**(2), 231–243.
[93] Watanabe, A. and Kozuka, H. (2003) Photoanodic properties of sol-gel-derived Fe_2O_3 thin films containing dispersed gold and silver particles. *Journal of Physical Chemistry B*, **107**(46), 12713–12720.
[94] Inaba, M., Kintaichi, Y. and Hamada, H. (1996) Cooperative effect of platinum and alumina for the selective reduction of nitrogen monoxide with propane. *Catalysis Letters*, **36**(3–4), 223–227.
[95] Jaksic, M.M. and Jaksic, J.M. (1994) Fermi dynamics and some structural bonding aspects of electrocatalysis for hydrogen evolution. *Electrochimica Acta*, **39**(11–12), 1695–1714.

15

Multidimensional Nanostructures for Solar Water Splitting: Synthesis, Properties, and Applications

Abraham Wolcott and Jin Z. Zhang

Department of Chemistry and Biochemistry, University of California, Santa Cruz, CA, USA, Email: zhang@chemistry.ucsc.edu

15.1 Motivation for Developing Metal-Oxide Nanostructures

15.1.1 Introduction

Metal-oxide nanostructures have attractive properties for various applications, including alternative energy conversion and generation. Photoelectrochemical (PEC) and photovoltaic (PV) cells produced from metal oxides have potential to aid in producing low-cost and environmentally benign energy. Exploiting the unique properties of metal oxides at the nanoscale has witnessed an impressive surge in recent years. Advancement in the colloidal synthesis of these nanomaterials is quickly pushing the technology in making marketable sustainable alternative energy (SAE) a reality. While scientific research commences for SAE, a growing body of evidence is confirming anthropogenic climate change and accelerating the global need to reduce CO_2 emissions [1]. Currently global energy production is dominated by fossil-fuel combustion. The subsequent release of CO_2 into the atmosphere is being theorized to cause harm to the fragile environmental balance of planet Earth. One telling piece of evidence is the drastic increase in CO_2 levels after the industrial revolution, as evidenced by ice-core samples that track the oscillatory trends of atmospheric CO_2 over the past 600 millennia [2]. Many forms of SAE exist and include wind, solar, hydro, hydrothermal, photocatalysis and photoelectrochemical (PEC) technologies. Utilization of PEC devices for water splitting and oxygen/hydrogen production is a segment of study that has received

On Solar Hydrogen & Nanotechnology Edited by Lionel Vayssieres

increasing attention. Hydrogen as an energy source in fuel cells yields electricity, heat and water. Energy sources that do not contribute to the release of CO_2 are a paramount goal for the global community. Nanotechnology and the use of nanostructured materials play an ever increasing role in this SAE push, due to their unique properties at the nanoscale.

15.1.2 PEC Water Splitting for Hydrogen Production

The utilization of metal-oxide semiconductors as water-splitting substrates was originally spawned by the work of Fujishima and Honda in 1972 [3]. Since that time, many groups have focused on using semiconductors as PEC cells to produce hydrogen and this has been reviewed in depth [4–8]. The main principles governing PEC water splitting have been thoroughly covered in Chapter 1 of this book, and will therefore not be deeply probed. In 1998, Khaselev and Turner produced a monolithic PV/PEC cell composed of planar p-GaInP_2-GaAs that established a high photon-to-hydrogen efficiency of 12.4% [9]. While very effective, the cost of producing the p-GaInP_2-GaAs cell is expensive and prevents the mass production of such a device. Photocorrosion is also a main issue in the use of III–V semiconductors and limits the long-term stability of these devices, and photochemical corrosion will be discussed below. More recently, TiO_2 nanotubes produced by potentiostatic anodization have been investigated fervently, and one report states a 16.25% photon-to-hydrogen efficiency [10]. This exciting new direction for water splitting with n-type metal-oxide one-dimensional nanostructures has been covered in depth in Chapter 10. Importantly, TiO_2, WO_3 and ZnO represent a significant segment of the work done on metal-oxide systems for PEC cells, PVcells and photocatalytic applications. The ability of metal oxides to be synthetically produced by colloidal means into zero-dimensional (0D), one-dimensional (1D) and two-dimensional (2D) morphologies is experimentally important due to lower cost, thereby making these systems more scientifically accessible to researchers around the globe.

15.1.3 Metal-Oxide PEC Cells

Three main characteristics make these metal oxides attractive: (i) ease of synthesis to produce 0D, 1D and 2D nanostructures; (ii) band structure closely matched to the reductive and oxidative potential of H_2O and (iii) resistance to photodecomposition in aqueous solutions, essential to the overall long-term stability of a marketable PEC device. The last point is worthy of discussion and explanation. Electrochemical corrosion occurs at the semiconductor–electrolyte interface (SEI) when charge is transferred across this junction. These anodic and cathodic decomposition processes with YZ semiconductors can be described by

$$\mathrm{YZ} + \mathrm{xh}^{+} \rightarrow \mathrm{Y}^{+\mathrm{x}} + \mathrm{Z} + \Delta G_{\mathrm{a}} \tag{15.1}$$

$$\mathrm{YZ} + x\mathrm{e}^{-} \rightarrow \mathrm{Z}^{-x} + \mathrm{Y} + \Delta G_{\mathrm{c}} \tag{15.2}$$

where x is the number of electrons or holes, ΔG_a is the anodic free energy change and ΔG_c is the cathodic free energy change. ΔG_a and ΔG_c relate to the enthalpy of the cathodic and anodic processes by

$$E_{\mathrm{p,d}} = \frac{\Delta G_{\mathrm{a}}}{x N_{\mathrm{A}}} \tag{15.3}$$

$$E_{n,d} = \frac{\Delta G_c}{xN_A} \tag{15.4}$$

where $E_{p,d}$ is the free enthalpy of oxidation per electron hole and $E_{n,d}$ is the free enthalpy of reduction per electron. Derived by Gerisher, the resistance to either anodic or cathodic electrochemical corrosion is defined with the free enthalpies as:

$$E(O_2/H_2O) < E_{p,d} \tag{15.5}$$

$$E(H^+/H_2) > E_{n,d} \tag{15.6}$$

where $E(O_2/H_2O)$ and $E(H^+/H_2)$ are the oxidative and reductive potentials of water, respectively [11]. Taking these thermodynamic inequalities into consideration, we see that only SnO_2, WO_3, ZnO and TiO_2 have the proper free enthalpy conditions, as seen in Figure 15.1. Simply put, the free enthalpies must bracket the oxidative and reductive potential of water in order to be stable as both a photoanode and photocathode. ZnO has an $E_{p,d}$ very close to $E(O_2/H_2O)$, and thus has stability issues with its use as a photoanode, but that has not prevented a plethora of investigations into ZnO. If one looks at the decomposition potentials of CdS, it is seen to be a stable photocathode, but would suffer from electrochemical corrosion as a photoanode. Due to this critical thermodynamic limitation, the metal oxides TiO_2, WO_3 and ZnO have received the most investigations into PEC water splitting. This is clearly seen in the coverage of these three metal oxides in Chapters 2–4 and 8–12 of this book. To further compliment this growing body of work, this chapter will review the varied and ever growing class of colloidal synthetic techniques

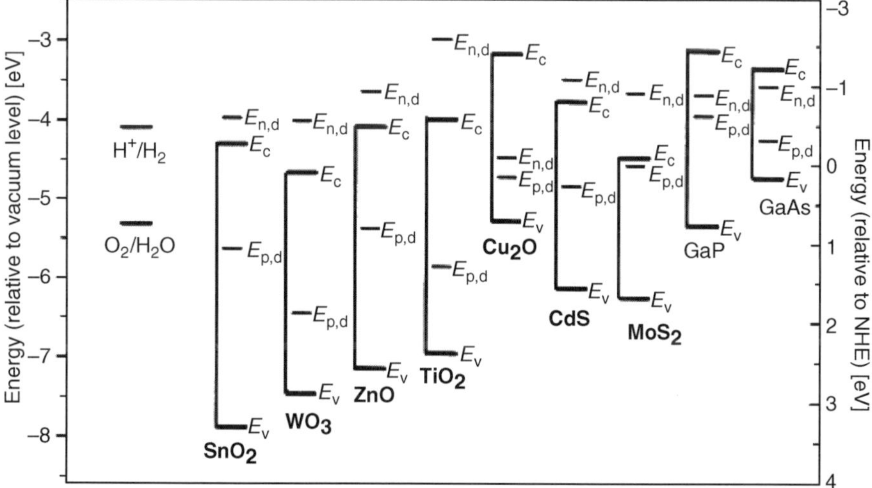

Figure 15.1 An energy level diagram of a series of semiconductors showing the positions of the decomposition potentials ($E_{n,d}/E_v$) in relation to their valence/conduction bands (E_v/E_c). (Reprinted with permission from T. Bak, J. Nowotny, M. Rekas and C. C. Sorrell, Photo-electrochemical hydrogen generation from water using solar energy. Materials-related aspects, *International Journal of Hydrogen Energy*, **27**(10), 991–1022, 2002. © 2002 Pergamon.)

to produce 0D, 1D and 2D multidimensional metal-oxide nanostructures with emphasis on synthetic methodology and structural and morphological properties.

A vital drawback to the use of metal oxides is their inherently large bandgaps and weak visible absorption. Many groups have addressed this issue through the use of doping, tandem PEC architectures and sensitization. Doping and tandem-cell technologies are thoroughly covered in Chapters 9 and 18, respectively, and will not be covered here. Sensitization through light-harvesting dyes and semiconductor nanoparticles (quantum dots or QDs) is an effective method to extend photoresponse into the visible, as exploited by O'Regan and Gratzel for their dye-sensitized solar cell (DSSC) [12].

15.1.4 Dye and QD Sensitization

Dye sensitization of metal-oxide nanostructures has been widely investigated, and it greatly enhances the light-harvesting abilities of PEC and PV systems [13–24]. Charge separation and ultrafast electron injection studies from dyes to TiO_2 reinforce one of the main kinetic features that allow dye sensitization to be successful [13,24]. Many dyes have been explored and include Ru-based dyes, safrinin, fluorescein, P3UBT and others [14,23,25–27]. A newer class of sensitization has arisen with the use of tunable QDs as inorganic light-harvesting materials. QDs have advantages over organic dyes with a tunable absorption based on quantum confinement effects, resistance to photobleaching, high absorption cross-sections and the ability to decorate the QD surface with a number of hydrophobic and hydrophilic molecules [28]. QD sensitization has been examined on a number of metal-oxide systems, as composite nanostructures for light harvesting [29–39]. Recently, a new strategy based on the simultaneous doping and QD sensitization of nanocrystalline TiO_2 has also been developed [40]. While most metal-oxide nanostructures have been produced based on low-cost colloidal techniques, as discussed in detail later, they can also be generated using deposition techniques that are often better suited for traditional thin films and multidimensional nanostructures.

15.1.5 Deposition Techniques for Metal Oxides

Chemical vapor techniques are another class of synthetic methods used to form multidimensional metal-oxide nanostructures. The term "chemical vapor" is loosely defined in this instance, and covers classic chemical vapor deposition (CVD), metal–organic chemical vapor deposition (MOCVD), molecular beam epitaxy (MBE), pulsed laser deposition (PLD), oblique angle deposition (OAD) and glancing angle deposition (GLAD). CVD, MOCVD and MBE techniques are powerful in their control over growth conditions and have been successfully applied to producing various nanostructured metal oxides [41–55]. An example of this control is a hybrid thin film → nanorod → porous thin film of ZnO that was produced by MOCVD. Morphological control was accomplished by controlling the supersaturation conditions at temperatures ranging from 520–870 K with constant diethyl zinc and oxygen concentrations. This manipulation of the supersaturation conditions allowed for a well-separated growth regime change, as seen in Figure 15.2 [44]. PLD, OAD and GLAD techniques are also versatile and can allow for multiple materials to be deposited in sequence, morphology manipulation and control over the growth angle of 1D nanostructures of various materials, including metal

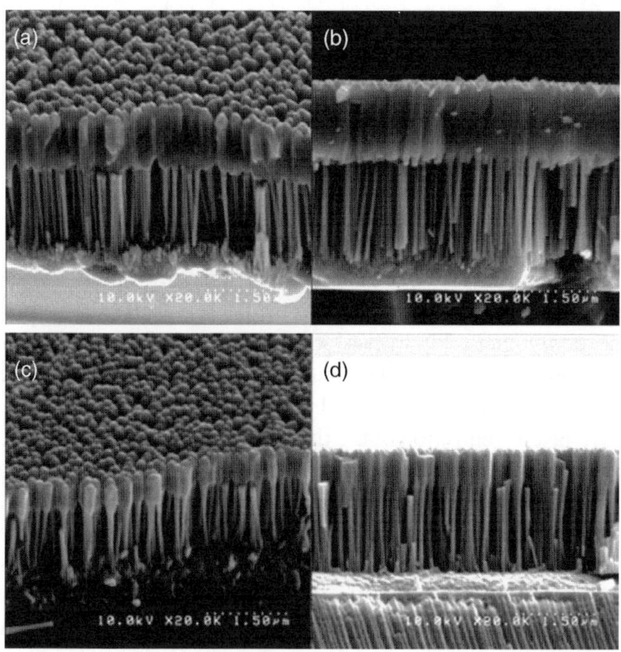

Figure 15.2 SEM images of MOCVD-prepared hybrid ZnO nanostructures on Al_2O_3 with tilted views ((a) and (c)) and cross-sectional views ((b) and (d)) with various growth conditions. (Reprinted with permission from M.C. Jeong, B.Y. Oh, O.H. Nam, T. Kim and J.M. Myoung, Three-dimensional ZnO hybrid nanostructures for oxygen sensing application, *Nanotechnology*, **17**(2), 526–530, 2006. © 2006 Institute of Physics.)

oxides [56–59]. Another benefit is the density of aligned nanostructures, as seen with TiO_2 nanorods fashioned into PEC cells and utilized as water-splitting substrates [60]. PLD, OAD and GLAD were also used to create ZnO thin films, nanoplatelets and nanoparticle PEC cells and were examined for their crystallographic, morphological and PEC properties [61]. The versatility of such techniques lends themselves quite well to the rational design of nanomaterials with specific purposes, such as active water splitting for hydrogen production.

15.2 Colloidal Methods for 0D Metal-Oxide Nanoparticle Synthesis

15.2.1 Colloidal Nanoparticles

This section will highlight the investigations of metal-oxide semiconductor (SC) nanoparticles produced by colloidal techniques. In the course of producing colloidal nanoparticles, researchers have focused on controlling porosity, nanoparticle size, size distribution and assembly properties [62–64]. Three major metal-oxide nanomaterials will be discussed, TiO_2, WO_3 and ZnO. While this is in no way a complete collection of all available nanoparticle systems, it will demonstrate the general techniques currently exploited on a well-rounded basis. The colloidal synthesis of TiO_2, WO_3 and ZnO nanoparticles spans several decades, and includes, but is not exclusive to, the utilization of sol-gel, hydrothermal, solvothermal, sonochemical and template-directed techniques. Metal-oxide nanomaterial synthesis has been expansive and its applications vary greatly, including sensors, photo- and electrochromics, photocatalysis, photocatalytic degradation, PV and PEC cells [65–67].

15.2.2 TiO_2 Sol-Gel Synthesis

Sol-gel methods for the synthesis of TiO_2 nanoparticles are based on the hydrolysis and polymerization of titanium precursors [68–74], typically metal alkoxides, inorganic metal salts and metal–organic compounds. Hydrolysis of titanium alkoxides under acidic conditions allows for the formation of Ti-O-Ti bond formation, and growth of polymerized chains leading to nanoparticle growth. The rate at which hydrolysis proceeds is dependent on water content, temperature, pH and the concentration of the titanium precursor in solution. Low, medium and high water contents lead to three regimes of Ti hydrolysis: (i) at low water content, the formation of close-packing three-dimensional polymeric skeletons of Ti–O–Ti chains; (ii) medium water content favors the formation of $Ti(OH)_4$ and (iii) high water content leads to Ti–OH formation and loosely packed first-order particles [75]. Study of the growth kinetics of TiO_2 nanoparticles with titanium(IV) isopropoxide as the precursor has found particle growth based on temperature to be in agreement with the Lifshitz–Slyozov–Wagner model for Ostwald ripening [71]; Ostwald ripening being the mechanism that describes the growth process of crystals, in which smaller particles of higher solubility become the precursors for the growth of larger particles (also known as "defocusing") [76]. Knowledge of growth mechanisms via sol-gel techniques has allowed nanocrystalline TiO_2 porosity and nanoparticle size to be used in PV applications, as discussed below.

A comparative PV study of three TiO_2 nanoparticle systems sensitized by poly(3-undecyl-2,2′-bithiophene) or P3UBT, utilized a common method for sol-gel-prepared nanoparticles [25]. In the sol-gel synthesis, 250 μl of high purity Milli-Q H_2O (18 MΩ) was added to 10 ml of 100% pure ethanol. The solution was acidified to pH 1–2 with concentrated HNO_3, and then, under inert atmosphere conditions (nitrogen glove box), 750 μl of titanium isopropoxide was added under vigorous stirring. The solution was then allowed to stir for three days in the inert atmosphere conditions. Overall nanoparticle diameter was found to be 40–60 nm in this particular synthetic route. 100 μl of the TiO_2 sol-gel solution was then spin-coated at 1000 rpm onto a conducting substrate, annealed at 450 °C, and later fashioned into functioning PV cells. The above synthesis is using control of hydrolysis and polymerization as its main nanocrystal driving force. Other sol-gel routes include the use of active ionic species such as tetramethylammonium hydroxide (TMAOH) to control growth.

In a systematic examination of TiO_2 nanocrystal growth, a sol-gel protocol utilizing the organic base TMAOH, H_2O, ethanol and titanium isopropoxide was used [68]. Highly crystalline anatase nanocrystals were produced of varying size, shape and organization, based on ratios of the aforementioned precursors. This base-catalyzed reaction had the dual purpose of promoting hydrolysis and polycondensation reactions of the titanium precursor, but also stabilized TiO_2 polyanionic cores through the organic cationic base. With high titanium concentrations, the separation of hydrolysis and condensation reactions (formation of Ti–O–Ti) is difficult to differentiate [68]. The growth control is understood by the ratio of Ti/TMAOH, and the relative control of crystal growth velocities in the [101] and [001]

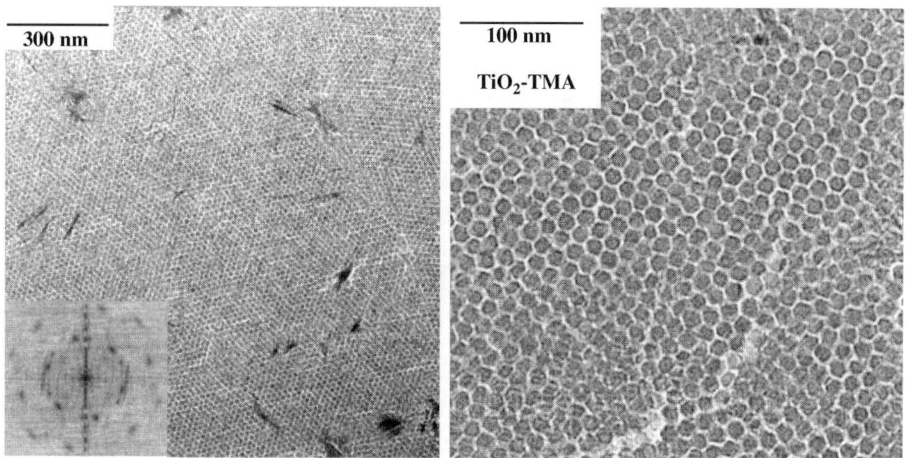

Figure 15.3 TEM images of TiO_2 nanocrystal superlattices produced with tetramethylammounium hydroxide (TMAOH) and titanium isopropoxide. (Reprinted with permission from A. Chemseddine and T. Moritz, Nanostructuring titania: Control over nanocrystal structure, size, shape, and organization *European Journal of Inorganic Chemistry*, (2), 235–245, 1999. © 1999 Wiley-VCH Verlag GmbH & Co. KGaA.)

directions. At low TMAOH concentrations, [101] was favored and at high concentrations [001] was stabilized by TME^+ cations on the nanocrystal surface. This stabilization allowed for isotropic growth that lead to high monodispersity and self-assembly, as seen in Figure 15.3.

15.2.3 TiO_2 Hydrothermal Synthesis

Hydrothermal synthesis is an approach that is typically performed in a Teflon-lined steel vessel or similar container that can be sealed tight and placed in ovens at elevated temperatures. The sealed vessels allow for aqueous solutions to be heated past their boiling point, and exceedingly high vapor pressures to be attained within the reaction vessel. Control over the temperature of the vessel and solution volume are variables which allow for the modification of the internal pressure of individual reactions. Hydrothermal techniques for TiO_2 nanoparticle synthesis have been widely applied [77–85]. In the preparation of photocatalytically active TiO_2 nanoparticles, a similar approach of using an ethanol/water mixture, titanium isopropoxide and acidification with HNO_3 in a hydrothermal vessel was utilized [79]. The modification of ethanol-to-water ratio and the concentration of titanium isopropoxide allowed the synthesis of TiO_2 nanoparticles ranging in size from 7–25 nm. Thin films were produced from particle suspensions of 7 nm, 15 nm, 25 nm and commercially available Degussa P-25, and field emission scanning electron microscopy (FESEM) images are seen in Figure 15.4. Ethanol-to-water ratio comparisons found that increasing ratios from 1:8 to 8:1 produced continually smaller particle diameters, ranging from 25 to 7 nm. Increasing the titanium isopropoxide ratio from 0.125–1.0 M also showed a similar trend of decreasing particle size from 25 to 7 nm. All solutions were hydrothermally reacted at 240 °C for four hours in a glass-lined hydrothermal bomb. Hydrothermal routes with the use

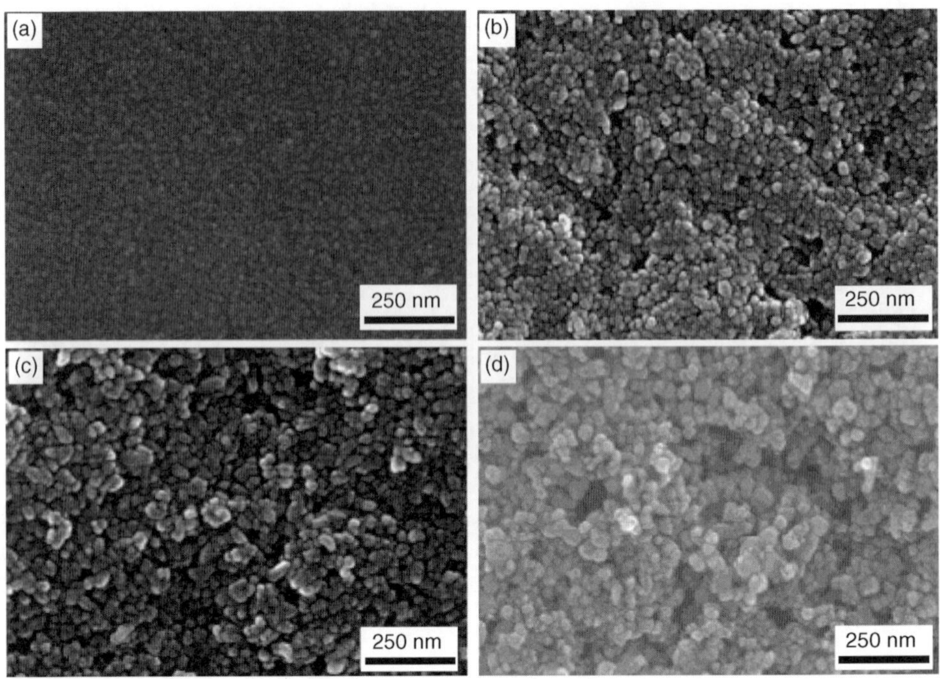

Figure 15.4 FESEM images of TiO_2 thin films produced by 7 nm, 15 nm, 25 nm and Degussa P-25 commercial powder in a hydrothermal synthesis route. (Reprinted with permission from S.Y. Chae, M.K, Park, S.K. Lee, T.Y. Kim, S.K. Kim and W.I. Lee, Preparation of size-controlled TiO_2 nanoparticles and derivation of optically transparent photocatalytic films, *Chemistry of Materials*, **15**, 3326–3331, 2003. © 2003 American Chemical Society.)

of titanium butoxide, and tetraalkylammounium hydroxide has also been explored extensively [83]. For example, TiO_2 nanoparticles synthesized using this approach were all found to be a pure anatase phase after the heat treatments. Chain length of the organic base (methyl, ethyl, butyl) had an effect on the morphology, based on concentration. A transition from spherical to rod-like was observed as the concentration increased with tetramethylammounium hydroxide. Increasing tetraethylammonium hydroxide, in contrast, caused a transition from spherical to astrerisk-like, and various concentrations of tetrabutylammonium hydroxide did not change the spherical morphology.

15.2.4 TiO_2 Solvothermal and Sonochemical Synthesis

Solvothermal reaction techniques are a close relative of hydrothermal techniques, but include the use of nonaqueous solvents. Organic solvents have the advantage of higher boiling points, and, in turn, the size and morphology in organic solvents increases control in comparison to hydrothermal routes. Consequently, the crystallinity of the materials is also increased. The solvothermal method has been effectively used to generate TiO_2 nanoparticles of narrow size

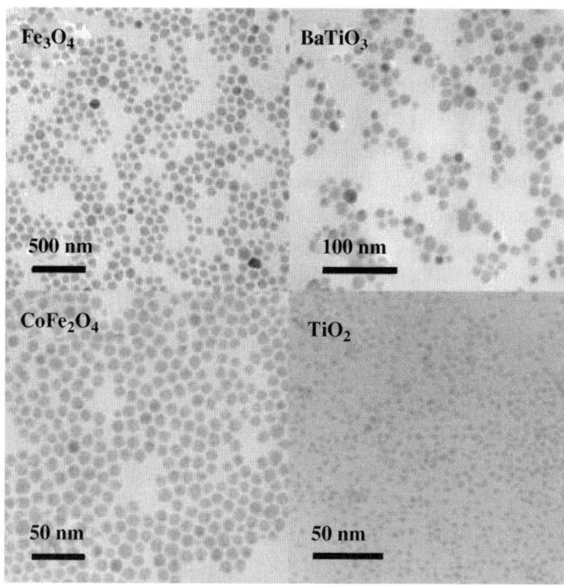

Figure 15.5 TEM images of various metal-oxide nanoparticles produced by the liquid–solid–solution (LSL) route. (Reprinted with permission from X. Wang, J. Zhuang, Q. Peng and Y.D. Li, A general strategy for nanocrystal synthesis, *Nature*, **437**(7055), 121–124, 2005. © 2005 Nature Publishing Group.)

distribution and dispersity [86–89]. A universal approach to a number of nanomaterials includes the exploitation of the liquid–solid–solution (LSS) process. The general protocol includes a metal linoleate phase (solid), an ethanol–linoleic acid liquid phase (liquid) and a water–ethanol solution (solution) at various temperatures for growth in an autoclave vessel [86]. In a typical synthesis, 0.5 g $TiCl_3$ in 20 ml of H_2O, 1.6 g sodium linoleate, 10 ml ethanol and 2 ml linoleic acid is added to an autoclave vessel and heated from 80–200 °C for 10 hours. The synthesized particles can then be collected and dispensed in a number of organic solvents, including toluene or chloroform. This particular technique was demonstrated to work for metals (Ag, Au, Rh, Ir), semiconductors (Ag_2S, PbS, ZnSe, CdSe) and metal oxides (Fe_3O_4, $BaTiO_3$, $CoFe_2O_4$, TiO_2) [86]. The metal oxides were monodisperse nanoparticles and representative TEM images can be seen in Figure 15.5.

A more traditional solvothermal route is carried out in the production of TiO_2 nanoparticles with anhydrous toluene as the solvent, with titanium isopropoxide as the titanium precursor [88]. In an inert atmosphere glove box, different ratios of titanium isopropoxide and anhydrous toluene ranging from 5 : 100–40 : 100 (Ti:Toluene) were mixed into solutions, vigorously stirred for three hours, and placed into a stainless-steel autoclave vessel with Teflon lining. The vessel was then heated to 250 °C for three hours, in which the precursors decomposed into TiO_2 and the resulting solvent. The TiO_2 nanoparticles were found to be anatase when the Ti:Toluene ratio was 10 : 100–30 : 100 and amorphous at 5 : 100 and 40 : 100. For the crystalline products, the TiO_2 nanoparticles had a median diameter of 10 nm, 14 nm and 16 nm for Ti:Toluene ratios of 10 : 100, 20 : 100 and 30 : 100, respectively. In a surfactant-aided solvothermal route, oleic acid (OA) was added to the titanium isopropoxide and toluene system, and produced monodisperse anatase TiO_2 nanocrystals ranging from nanoparticles to nanodumbells, as a function of precursor ratio [87]. The Ti:Toluene ratio of 5 : 100 produced nanoparticles of 6 nm in diameter. The oleic acid is understood to direct the nanoparticle growth, and subsequently aided in producing nanodumbells in the [001] direction with a Ti: Toluene ratio of 10 : 100, and a sliding Ti:OA ratio of 1-2-1 : 5.

Sonochemical techniques include the use of ultrasonic waves to produce microscopic bubbles through acoustic cavitation. The ultrasonic technique has been applied to nanomaterials of metals, alloys, carbides, oxides and colloids. After cavitational-bubble collapse, intense local heating is produced (~5000 K), high pressures (~1000 atm) and large heating and cooling rates ($>10^9\,K\,s^{-1}$). TiO_2 nanoparticles have been produced using the sonochemical method by various groups [90–98]. In one method, a combination of surfactants, industrially prepared nanoparticles and sonication was used to produce polyanaline-coated TiO_2 nanoparticles ranging in size from 30–200 nm [94]. Conductance measurements were made with varying degrees of polyanaline, and found that the hybrid polyanaline–TiO_2 nanoparticles increased in conductivity with increased analine wt%. The sonochemical route is rationalized to be a substitute for stirring in the case of colloidal synthesis, and can lead to better conductive behavior of these hybrid nanomaterials, in comparison to conventional stirring [94].

15.2.5 TiO_2 Template-Driven Synthesis

Template-directed growth of TiO_2 opals and photonic materials has been well studied, and includes a complimentary direction to the materials synthesis discussed previously in this section [99–104]. In general, these opal materials are produced by a templating mechanism, and will produce highly ordered crystalline structures. One synthetic protocol utilizes latex spheres laid on filter paper over a Buchner funnel, saturated in ethanol and adhered via an applied vacuum [104]. Titanium ethoxide was added dropwise to the latex beads under vacuum, allowed to dry in a vacuum desiccator for 3–24 hours, and then removed for annealing at 575 °C for 7–12 hours in open-air conditions. The TiO_2 opal produced via the templating technique shows a very well-ordered assembly with 320–360 nm voids and is seen in Figure 15.6. Templated TiO_2 also finds applications in advanced optoelectronics used as single-mode cavities for quantum electronics and quantum communications [104].

In a similar approach using a skeleton structure, a controlled CaF_2 framework of TiO_2 was produced from a template technique. The synthetic approach included the production of a polystyrene (PS) opal made by the slow evaporation of a PS solution made of 270 nm diameter particles. A 1–3 mm^3 PS piece was then saturated with titanium isopropoxide, ethanol and an argon stream. The infiltrated opal was then exposed for several days to open-air conditions to hydrolyze the titanium precursor with residual water moisture. The composite constructed of PS and titania was than annealed at 300 °C and 700 °C to remove the PS template, and the resultant material was found to be highly opalescent. Figure 15.7 shows the TiO_2 opal with its skeleton structure and polygon-like grid. The engineered material was subsequently modeled using a plane-wave expansion method and the bandgap was calculated to be 2.9 eV. This pseudogap was a product of the intrinsic atomic locations in the skeleton structure and was a compliment to the usual bandgap found for TiO_2 of 3.2 eV. This dual bandgap material was reported for the first time in this article and the designed materials can be applied to photonic crystals of dielectrics [101].

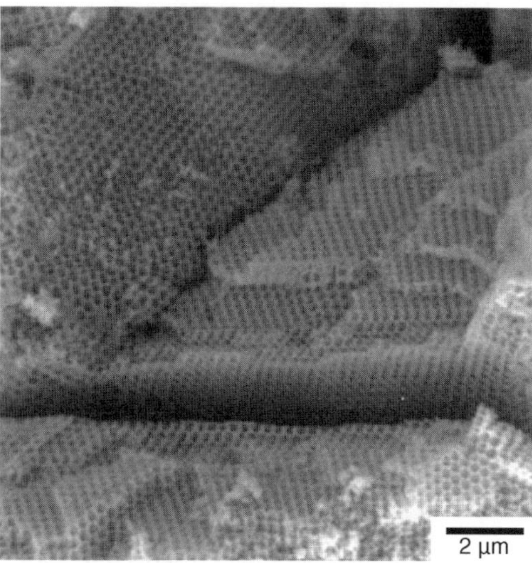

Figure 15.6 SEM image of the TiO_2 inverse opal produced by polystyrene spheres and titanium isopropoxide. (Reprinted with permission from B.T. Holland, C.F. Blanford and A. Stein, Synthesis of macroporous minerals with highly ordered three-dimensional arrays of spheroidal voids, *Science*, **281** (5376), 538–540, 1998. © 1998 American Association for the Advancement of Science.)

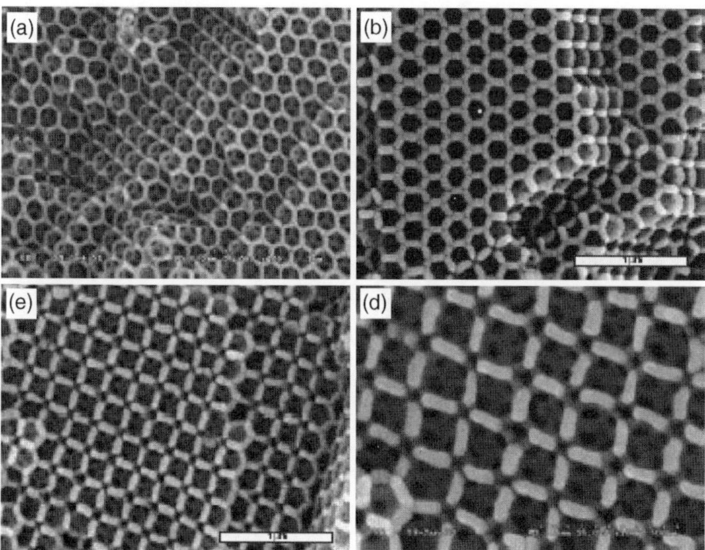

Figure 15.7 SEM image of skeleton structure of TiO_2 produced by PS template with a titanium isopropoxide precursor. (Reprinted with permission from W.T. Dong, H. Bongard, B. Tesche and F. Marlow, Inverse opals with a skeleton structure: Photonic crystals with two complete bandgaps, *Advanced Materials*, **14**(20), 1457–1460, 2002. © 2002 Wiley-VCH Verlag GmbH & Co. KGaA.)

15.2.6 Sol-Gel WO_3 Colloidal Synthesis

Synthesis of WO_3 nanoparticle systems has been of considerable interest due to the ability to use such materials in electrochromics, photocatalysts and photoelectrochemical cells [105–110]. Colloidal $WO_3{\cdot}H_2O$ can be produced by the passing of Na_2WO_4 aqueous solutions through a proton exchange column, and collected as a series of aliquots. After the acidification process, polymerization of tungsten oxoanions occurs to produce yellowish isomorphous crystalline $WO_3{\cdot}H_2O$ tungstic-acid complexes [111]. The sol-gel techniques typically employed include the addition of directing agents such as the nonionic poly(ethylene glycol) (PEG). One study demonstrated that addition of PEG led to crystallographic orientation of the WO_3 nanoparticles in the [200], [020] and [002] crystal directions, as investigated with X-ray diffraction (XRD) and Raman scattering spectroscopy [105]. Chapter 12 extensively covers WO_3 for PEC hydrogen production. In a similar tungstic-acid preparation, quantum-confined WO_3 colloids smaller than 50 Å were synthesized with the use of oxalic acid as a stabilizing agent [112]. Besides investigating the electrochromic effects via UV-visible absorption spectroscopy, energy transfer and charge trapping were explored with picosecond laser flash photolysis using the dye thionine. It was discovered that the oxalic acid acts as an efficient hole scavenger after photoexcitation. In a related synthesis, the solution of 0.25 M Na_2WO_4 was acidified by passing through a proton-exchange column. To stabilize the growth of the $WO_3{\cdot}H_2O$ colloids, 0.520 g of carbowax (PEG) was added per 10 ml of solution and one drop of Triton X-100 was added per ml of tungstic acid [113]. The solution was then stirred for six hours and stored in a refrigerator. This sol-gel nanoparticle synthesis produced 30–50 nm WO_3 nanoparticles of the monoclinic crystal phase. These particles have been extensively studied as active photoanodes in PEC cells for water splitting.

15.2.7 WO_3 Hydrothermal Synthesis

Hydrothermal synthesis of WO_3 systems has also been explored, including the use of tungsten precursors in concert with ligand additives [114–118]. In one WO_3 nanoparticle synthesis, Na_2WO_4 and L(+)-tartaric acid were used, yielding WO_3 nanoparticles of 100 nm in diameter, as shown in Figure 15.8 [118]. Another synthesis used the production of $H_2WO_4{\cdot}H_2O$ by acidic precipitation of Na_2WO_4 and the formed gels were later washed and placed inside a Parr autoclave bomb. The autoclave bomb was heated to 125 °C, and the suspensions were laid onto indium tin oxide (ITO) conducting substrates. The morphology of the nanoparticles was 100–600 nm in diameter and produced agglomerates of hexagonal WO_3 crystallites. The produced WO_3 devices were used as NH_3 sensors with concentrations ranging from 50–500 ppm [114].

15.2.8 WO_3 Solvothermal and Sonochemical Synthesis

Solvothermal routes to WO_3 nanocrystallites are limited in scope, but follow similar protocols seen in hydrothermal synthetic routes [119–121]. In general, solvothermal routes for WO_3 in the literature appear to form nanowires and nanorods preferentially over nanoparticle formation. A solvothermal approach to the production of $CaWO_4$ did produce nanoparticles in conjunction with a side product of WO_x [121].

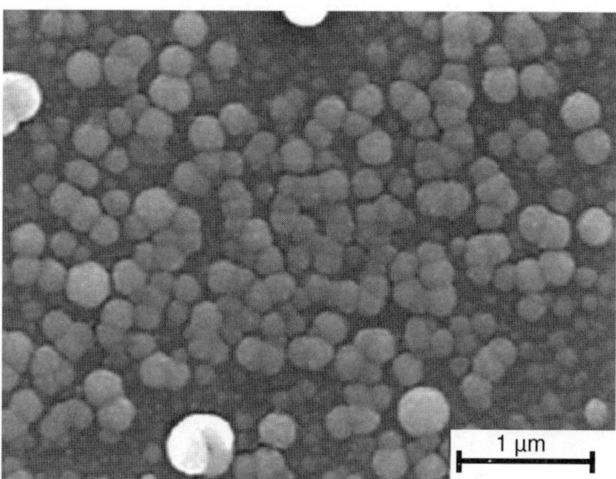

Figure 15.8 SEM image of WO_3 nanoparticles made via a hydrothermal method with Na_2WO_4 and L(+)-tartaric acid. (Reprinted with permission from Q.J. Sun, H.M. Luo, Z.F. Xie *et al.*, Synthesis of monodisperse $WO_3.2H_2O$ nanospheres by microwave hydrothermal process with L(+)-tartaric acid as a protective agent, *Materials Letters*, **62**(17–18), 2992–2994, 2008. © 2008 Elsevier.)

Sonochemical techniques using ultrasonic irradiation in the synthesis of WO_3 nanoparticles is also a rare occurrence and is seen based on a reaction of tungsten hexacarbonyl in diphenylmethane [122]. Overall, the solubility of tungsten hexacarbonyl can be aided by sonication in diphenylmethane (DPM). The residual black solid that resulted was then centrifuged and washed three times with pentane. After heating in open-air conditions at 1000 °C, the black solid became yellowish green, the color usually associated with WO_3. TEM images revealed the particles to be 50–70 nm in diameter after annealing, and of the triclinic crystal phase.

15.2.9 WO_3 Template Driven Synthesis

Template-driven synthesis of mesoporous and nanostructured WO_3 for electrochromic devices, sensing and energy production has been explored with a number of techniques [123–129]. As a gas sensor, WO_3 has been widely explored and block copolymers have been exploited to assist silica template formation for producing hexagonal and cubic mesoporous structures. The silica template and the nanoporous WO_3 produced are seen in Figure 15.9. The WO_3 nanostructures were found to be of a mixed phase of both triclinic and monoclinic, and the nanopores were in the size range of 10 nm in width, in close agreement to the silica-template pore diameter.

Utilizing natural biopolymer templates such as chitosan, which are extracted from the shells of shrimp and other sea crustaceans, WO_3 nanoparticles can also be synthesized [125]. Chitosan is a linear polysaccharide composed of randomly distributed β-(1-4)-linked D-glucosamine (deacetylated unit) and N-acetyl-D-glucosamine (acetylated unit). In the synthesis, 1 g of ammonium metatungstate was dissolved in 60 ml of acetic acid diluted with water (1 : 1), stirred until dissolved, and then 3 g of chitosan was added and left at room temperature for three hours. This chitosan/WO_3 gel was than annealed at 400–600 °C, and

Figure 15.9 HRTEM images of the silica mold (a) and its resulting WO_3 nanoporous structures (b–d). (Reprinted with permission from E. Rossinyol, J. Arbiol, F. Peiro *et al.*, Nanostructured metal oxides synthesized by hard template method for gas sensing applications, *Sensors and Actuators B, Chemical*, **109**(1), 57–63, 2005. © 2005 Elsevier.)

dispensed in water for TEM characterization. Depending on annealing temperature the WO_3 nanoparticles ranged in size from 30–50 nm and consisted of the triclinic phase at 400–500 °C; the nanoparticles formed a large agglomerated structure of the hexagonal phase on annealing at 600 °C. The WO_3 nanoparticles were fashioned into electrochemical cells and cyclic voltammetry was performed to cathodically produce hydrogen at the electrode surface.

Control over pore size and geometric positioning is a key attribute in these structures and can be accomplished by the utilization of anodized aluminum oxide (AAO) [126]. With the motivation to make a high surface-to-volume ratio gas-sensing device, an AAO was produced by the sputtering of aluminum on Si substrates, then anodizing them in malonic acid electrolytes with a an applied voltage of 95 V and a current of 6 mA cm^{-2}. In a secondary step to coat the exposed aluminum with compact alumina, a follow up anodization at 110 V followed. To increase surface-to-volume ratios with the AAO, a third preparatory step was taken, wherein a chemical bath of phosphoric and chromic acid at 50 °C was used for upwards of 9 minutes. Deposition of the WO_3 was performed using rf magnetron sputtering onto the

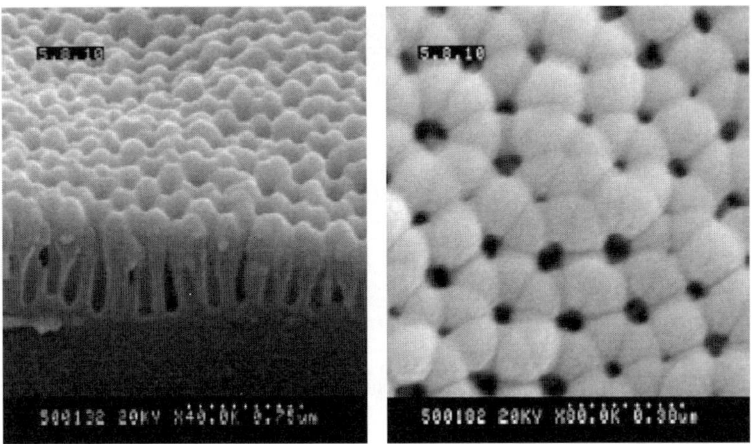

Figure 15.10 SEM images of the template WO_3 gas-sensing devices produced with anodized aluminum templates. (Reprinted with permission from V. Khatko, G. Gorokh, A. Mozalev *et al.*, Tungsten trioxide sensing layers on highly ordered nanoporous alumina template, *Sensors and Actuators B, Chemical*, **118** (1–2), 255–262, 2006. © 2006 Elsevier.)

AAO at a power of 100 W and a chamber pressure of 5×10^{-3} mbar of an Ar:O_2 mixture (1 : 1). The resulting AAO/WO_3 nanoporous structures after chemical-bath dipping in seen in Figure 15.10. Pore diameter was found to change from 50–170 nm depending on chemical etching time via SEM imaging. The nanoporous structures were originally amorphous and after annealing at 400 °C, the monoclinic WO_3 crystal phase was detected by XRD. AAO templates have been used in the synthesis of 1D nanostructures, as discussed further later.

15.2.10 ZnO Sol-Gel Nanoparticle Synthesis

Sol-gel techniques have been widely used in the course of producing ZnO nanoparticle systems for fundamental spectroscopic investigations [130–135]. In one procedure, 37.7 mg of zinc perchlorate ($Zn(ClO_4)_2{\cdot}H_2O$) in 40.5 ml of methanol, was added to a 9×10^{-3} M NaOH solution in 40.5 ml of methanol, allowed to stir overnight at 25 °C and stored in a dark environment [131]. The nanoparticles had a mean particle diameter of 3.4 nm, and were tracked for size changes by UV-visible absorption. Subsequent photoluminescence (PL) studies were conducted to probe the states of fluorescence as a function of excess Zn^{2+} ions in solution. With extremely small particles of 7–30 Å, the quantum yield of ZnO particles was found to depend on diameter, most likely due to the changes in surface trap density [135]. In a similar sol-gel synthesis, zinc acetate was used as a precursor for producing ZnO nanoparticles. SEM images of the ZnO colloids in a hexagonal packing arrangement can be seen in Figure 15.11. The mean nanoparticle diameter was found to be 60 nm and the nanoparticles were in a close-packed hexagonal structure. The growth mechanism is believed to be a two-stage process of nucleation of 3–4 nm ZnO nanoparticles, and then subsequent agglomeration that is very rapid, and this in turn allows for spherical geometries to form into larger 60 nm nanoparticles.

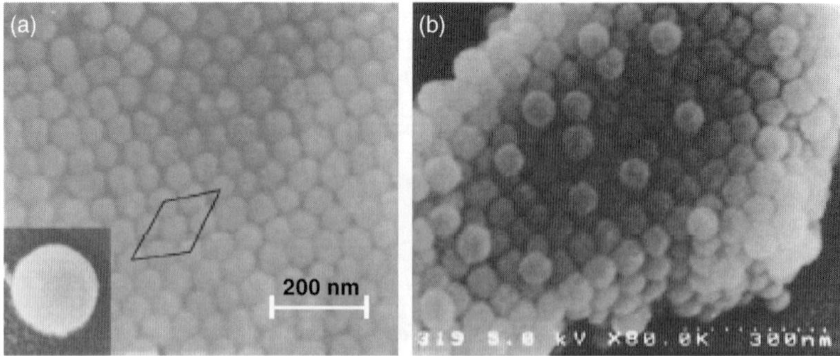

Figure 15.11 Self-assembly of ZnO spheres on silicon, illustrating hexagonal packing: (a) the well-ordered facade and (b) the flank of a cluster of ZnO spheres. (Reprinted with permission from L. Dong, Y.C. Liu, Y.H. Tong *et al.*, Preparation of ZnO colloids by aggregation of the nanocrystal subunits, *Journal of Colloid and Interface Science*, **283**(2), 380–384, 2005. © 2005 Elsevier.)

15.2.11 ZnO Hydrothermal Synthesis

Hydrothermal synthesis of ZnO nanoparticles has been investigated thoroughly, and includes the synthesis of autoclave and open-air condition synthetic routes [136–141]. Utilizing various surfactants, ions and pH conditions, control on the size of nanoparticles can be achieved, as seen in Figure 15.12. In one study, surfactant concentration, time and temperature conditions were controlled to produce a series of ZnO nanoparticles [136]. In the synthesis, equal volumes of $Zn(CH_3COO)_2{\cdot}H_2O$ and NaOH were added in the presence of varying amounts of the surfactant, hexamethylenetetramine (HMT) and left to stir at room temperature. The white colloidal suspension of $Zn(OH)_2$ was washed several times through centrifugation and decantation. $Zn(OH)_2$ suspensions were then placed into a Teflon autoclave bomb and were heated at 100, 150 and 200 °C. In general, ZnO nanoparticle diameter was modulated in the range 56–112 nm, depending on the aforementioned variables. The smallest nanoparticles synthesized were produced by a combination of no HMT, at a temperature between 125–150 °C and a growth time of five hours. Addition of HMT facilitated nanoparticle growth at lower temperatures and shorter growth times [136].

Preferential adsorption of ions onto crystal faces of growing nanocrystals is another technique exploited in concert with hydrothermal growth [139]. In the synthetic scheme, three types of zinc salts were utilized, $Zn(CH_3COO)_2$, $Zn(SO_4)$ and $Zn(NO_3)_2$, and two alkali metal hydroxides, LiOH and KOH. Solutions of the zinc salts and alkali metal hydroxides were made with distilled water at 0.05 and 0.1 mol kg^{-1}, respectively. Using a 153 cm^3 Ti alloy reaction vessel, batch reactions were completed with equal volumes of the zinc salts, and alkali metal hydroxides were added to a reaction density of 0.53 g cm^{-3}. A reaction temperature of 400 °C was utilized with a molten salt bath, and monitored by an *in situ* thermocouple. The reaction times were a total of 10 minutes, and the reaction vessel was cooled by a room temperature water bath. ZnO nanoparticles with diameters of 649–710 nm were synthesized with this technique. To control nucleation and particle growth, flow experiments were performed in the apparatus shown in Figure 15.13. Zinc and hydroxide precursors were fed

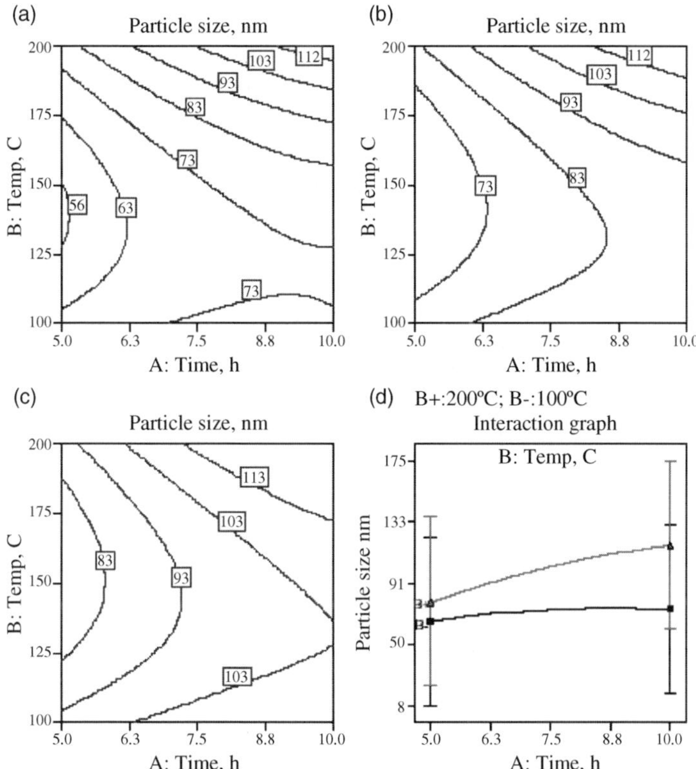

Figure 15.12 ZnO nanoparticle diameter contour plots of time versus temperature as a function of surfactant concentration: (a) 0 ppm, (b) 100 ppm and (c) 200 ppm. (d) An interaction graph of time and temperature at 0 ppm surfactant concentration. (Reprinted with permission of A.A. Ismail, A. El-Midany, E.A. Abdel-Aal and H. El-Shall, Application of statistical design to optimize the preparation of ZnO nanoparticles via hydrothermal technique, *Materials Letters*, **59**(14–15), 1924–1928, 2005. © 2005 Elsevier.)

into mixing point 1(MP1) at a rate of 3 and 1.5 g min^{-1}, and then were passed to mixing point 2 (MP2) with preheated water. At that point, the ZnO sol was injected into the reaction vessel at temperatures in the range 385–390 °C, and the resulting ZnO nanoparticles were collected with a diversion valve through a filter system. In the flow system, ZnO nanoparticle diameter was more highly controlled and allowed the synthesis of diameters ranging from 16 to 57 nm. Representative TEM images of the nanoparticles produced with the flow system are seen in Figure 15.14. Overall, ZnO nanoparticle diameter was found in the flow system to be a function of injection nozzle diameter, Li^+ concentration, temperature and pressure.

15.2.12 ZnO Solvothermal and Sonochemical Synthesis

Solvothermal routes toward ZnO nanoparticles in organic solvents have garnered a lot of attention and compliment the aqueous synthetic routes covered previously [142–148]. One synthetic procedure utilizes zinc acetylacetonate with anhydrous acetonitrile to yield ZnO

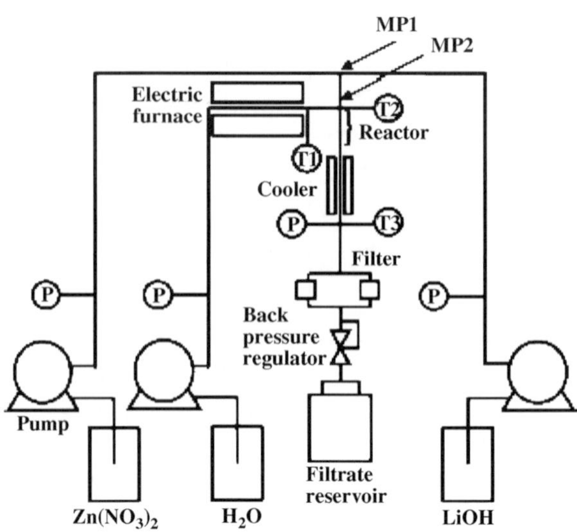

Figure 15.13 A schematic illustration of a hydrothermal flow system. (Reprinted with permission from K. Sue, K. Kimura, K. Murata and K. Arai, Effect of cations and anions on properties of zinc oxide particles synthesized in supercritical water, *Journal of Supercritical Fluids*, **30**(3), 325–331, 2004. © 2004 Elsevier.)

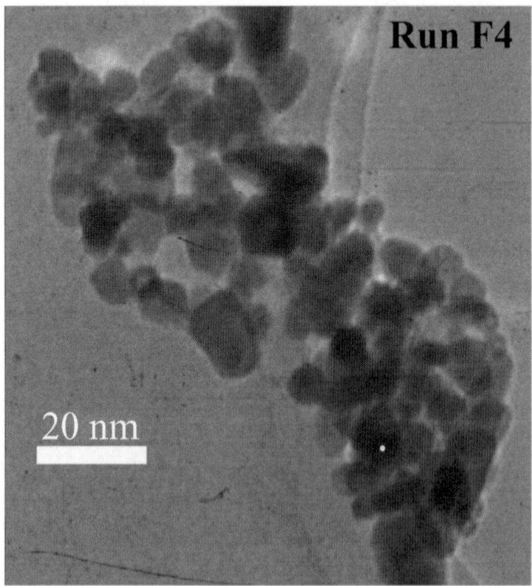

Figure 15.14 TEM image of ZnO nanoparticles produced in a hydrothermal flow through system at 386 °C with diameters of 16 ± 8 nm. (Reprinted with permission from K. Sue, K. Kimura, K. Murata and K. Arai, Effect of cations and anions on properties of zinc oxide particles synthesized in supercritical water, *Journal of Supercritical Fluids*, **30**(3), 325–331, 2004. © 2004 Elsevier.)

nanoparticles of 15–85 nm in diameter, with irregular morphology ranging from spherical to well-faceted nanocrystallites [142]. In a related synthesis, zinc acetylacetonate and benzylamine were used to produce ZnO nanoparticles. Analysis of the resulting precursors led to the proposed mechanism for the decomposition of acetylacetonate, in which: (i) aminolysis leads to the formation of N-benzyl-acetamide and acetonate/enolate binding to the Zn center; (ii) a condensation reaction follows where a benzylamine and an acetonate molecule form by a nucleophilic attack of an amine on to the electrophilic carbonyl center, and a subsequent hydroxyl release and (iii) release of an amine group [147]. ZnO nanoparticles were found to be purely of the zincite phase and were, on average 20 nm in diameter. The reaction mixture also produced ZnO nanorods of 200 nm in length and 30 nm in diameter. Another systematic synthetic approach toward ZnO nanocrystallites was explored using elemental zinc powder as a precursor, activated with dilute HCl [143]. ZnO nanoparticles were produced by the addition of 2 mmol zinc powder, 4 mmol 4-picoline N-oxide and 21 ml of an anhydrous solvent in a 30 ml Teflon-lined autoclave vessel. Experimentally, three dried solvents were explored, including toluene, ethylenediamine (EDA) and N,N,N′,N′-tetramethylenthylenediamine (TMEDA). Addition of all the reactants were carried out in an inert atmosphere glove box, and the autoclave vessel was sealed and heated to 180 °C for 91 hours. The brown precipitate that was formed was cleaned in an ultrasonic bath in acetone, and dried at 60 °C in air. While other reaction conditions produced nanorods, the above synthetic protocol specifically produced ZnO nanoparticles in the diameter range 24–60 nm. Solvent size and polarity played a factor in particle diameter, and it was found that toluene allowed the synthesis of the smallest particles, while the bulkier EDA and TMEDA produced both larger nanoparticles and nanorods [143].

Ultrasonic sonochemical synthetic techniques have been utilized for ZnO nanoparticle synthesis with the use of surfactants and ionic liquids [92,149–155]. Room temperature ionic liquids (RTILs) have been increasingly investigated due to their potential as a green alternative to volatile organic solvents in industrial processes [149]. The combination of an ionic liquid solvent and ultrasonic agitation was employed to produce ZnO nanoparticles with the RTIL, 1-hexyl-3-methylimidazolium bis-(trifluoromethylsulfonyl) imide [hmim] [NTf_2]. The resulting ZnO precipitate was separated by centrifugation and washed several times with distilled water and ethanol, and then dried in a vacuum oven at 40 °C for 10 hours. SEM images of the ZnO nanoparticles revealed that the average diameter was 60 nm, producing an interconnected porous network seen in Figure 15.15. Control experiments without sonication and RTILs revealed that only the combination of both variables produced the ZnO nanoparticles.

Ease of synthesis of nanoparticles is always desired for the scaling-up process required by industry and the production of cost-effective PEC cells. In this sonochemical synthesis of ZnO nanoparticles of 4–6 nm, 2.86 g of $Zn(CH_3COO)_2{\cdot}2H_2O$ was added to 130 ml of absolute ethanol, and allowed to stir. The resultant $ZnO(OH)_2$ solution was then concentrated to 60 ml, rediluted to 130 ml in ethanol, and then cooled to 0 °C over a 30 minute period. 0.754 g $LiOH{\cdot}H_2O$ was added to 130 mL of absolute ethanol, stirred, and then added dropwise to the $ZnO(OH)_2$ at 0 °C over 45 minutes. The ZnO suspensions were then treated with ultrasonic irradiation. Heptane was used to precipitate the ZnO nanoparticles, and centrifugation was followed by the drying of the samples in a vacuum freezer. The authors also found they could precipitate the ZnO sols with ultrasonic pulses with a water bath, centrifugation and washing of the nanoparticles with ethanol. This step simplified the cleaning process, and removed the need for expensive organic solvents such as heptane. Interestingly, they found that sonication

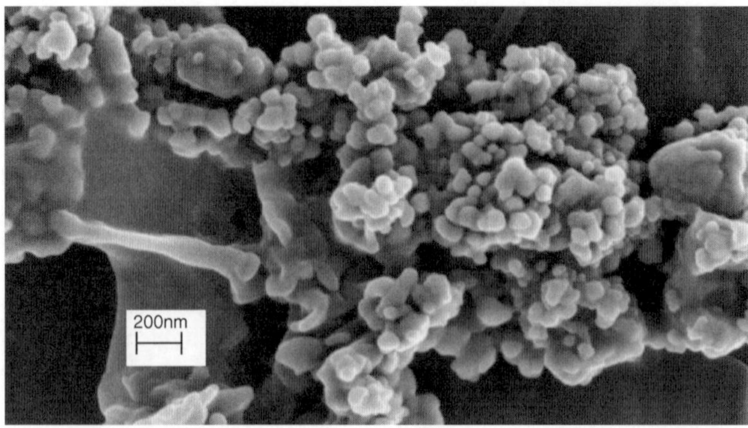

Figure 15.15 SEM image of ZnO nanoparticles produced via ultrasonic irradiation and RTILs. (Reprinted with permission from E.K. Goharshadi, Y. Ding, M.N. Jorabchi and P. Nancarrow, Ultrasound-assisted green synthesis of nanocrystalline ZnO in the ionic liquid [hmim][NTf2], *Ultrasonics Sonochemistry*, **16**(1), 120–123, 2009. © 2009 Elsevier.)

without a water bath produced precipitation the fastest (10 minutes) due to what they believed was faster heating of the solutions caused by the microbubble cavitation. In this synthesis, consistently small particles were produced without the need for a ligand system, as seen in Figure 15.16. The nanoparticles produced were highly crystalline, of the zincite crystal phase,

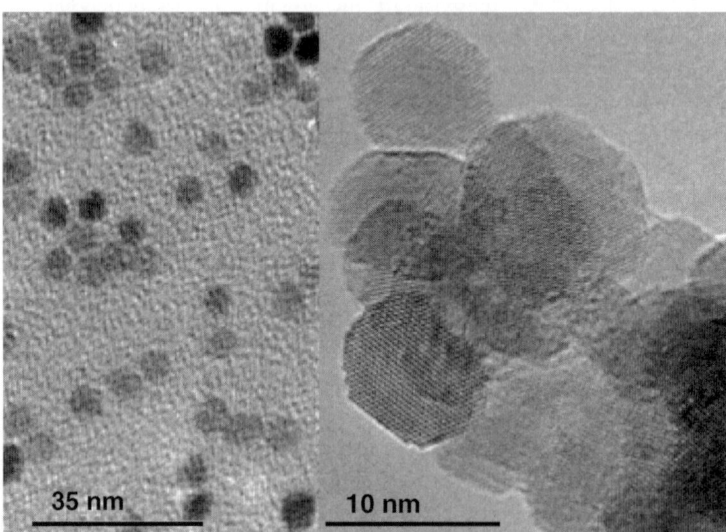

Figure 15.16 TEM and HRTEM images of ZnO nanoparticles produced via ultrasonic irradiation in absolute ethanol solutions. (Reprinted with permission from D. Qian, J.Z. Jiang and P.L. Hansen, Preparation of ZnO nanocrystals via ultrasonic irradiation, *Chemical Communications*, (9), 1078–1079, 2003. © 2003 Royal Society of Chemistry.)

and ranged in diameter between 4 and 6 nm, depending on reaction conditions, with a 10% standard deviation [153].

15.2.13 ZnO Template-Driven Synthesis

Template-driven synthesis of ZnO has also proven to be very valuable in the patterning and control of reactants, utilizing a plethora of different host materials [156–163]. Polystyrene (PS) beads, diblock copolymers and liquid crystal templating techniques have all be found to be useful for constructing ZnO materials of various architectures and pore sizes. Porous ZnO thin films have been grown directly on substrates through a seed-mediated templating technique. The overall procedure was environmentally benign and included initial dip coating of ZnO nanoparticles onto the substrate [161]. Annealing at 450 °C followed to remove organics, and then 1 μm PS beads were spincoated over the ZnO seeds and heated to 90 °C for one hour prior to hydrothermal treatment. The coated substrate was then placed in a Teflon-lined autoclave vessel with a total solution of 30 ml 0.3 M HMT, 0.03 M of $Zn(NO_3)_2{\cdot}6H_2O$ and 1.5 mg sodium citrate. The reaction vessel was heated at 60 °C for 28 hours, washed with distilled water and heated at 450 °C in open-air conditions. Sodium citrate was added to inhibit the growth of the ZnO seeds in the [001] direction, and crystal growth was then directed laterally, filling the open voids left by the PS template. Other experimentation without sodium citrate produced templated ZnO nanorod structures crowded into the gaps of the PS beads. Overall the templating was very uniform, and the pores of the ZnO porous structure were very close to those of the 1 μm PS diameter as seen in Figure 15.17. Diblock copolymers have also been found to be extremely powerful in their ability to create controlled pore diameters based on hydrophobic and hydrophilic moieties.

Diblock copolymers have attracted a lot of attention because of their ability to produce homogenous materials in a well-defined manner. The ease of preparation and the low cost of the procedure make it extremely attractive for scaling up in industrial applications. A synthetic route exploits the diblock copolymer poly(styrene)-block-poly(2-vinyl-pyridine) (PS-b-P2VP) as the templating material. The protocol calls for 10 mg of PS-b-P2VP to be added to 5 ml of toluene (2 mg ml^{-1}) and stirred for 24 hours. The polystyrene diblock copolymer is known to form micelles at this concentration, and to have a polar P2VP core and a nonpolar PS corona [162]. $Zn(NO_3)_2{\cdot}6H_2O$ and $Zn(CH_3COO)_2{\cdot}2H_2O$ were used as the zinc precursors and added to the micelle solution with a molar equivalent per pyridine unit of 0.1–0.6. The solution was stirred for 48 hours, a cleaned Si wafer was dipped into the solution at 10 mm min^{-1}, dried in air, and then cleaned with an oxygen plasma etcher to remove the polymer. Overall, particle diameter was a function of the precursor loading ratio and the Zn precursor used. Diblock copolymer micelles loaded with Zn precursors were well spaced on the Si wafer, as seen in Figure 15.18, and varied in diameter from 9 to 17 nm. After oxygen etching, the individual nanoparticles were still templated spatially, and the diameters determined by AFM varied from 0.9 to 3.5 nm. A combination of XPS and XRD data was used to determine that ZnO nanoparticles were not formed, but instead Zn_2SiO_4. This was postulated to occur during the oxygen etching on the active Si wafer surface. While the templating was successful, other substrates could be utilized to produce pure ZnO nanoparticles. The ease of production and the low cost of producing such well-defined systems are of continuing interest, and can be a great advantage in device preparation for PV, PEC, sensing and photocatalytic degradation applications.

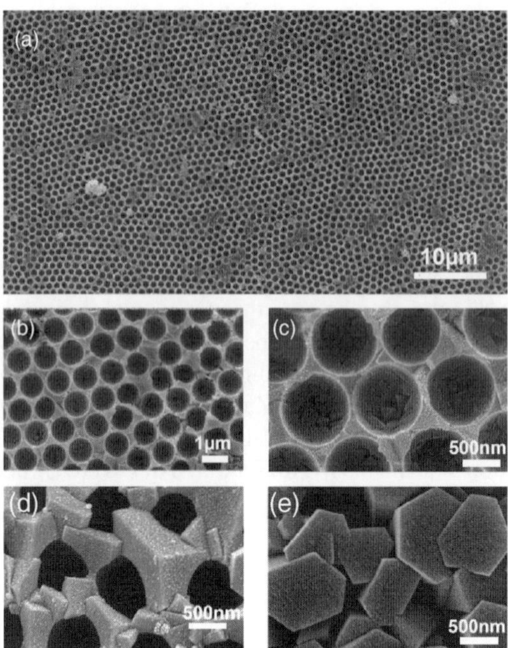

Figure 15.17 SEM images of a porous ZnO network produced directly on substrates utilizing polystyrene (PS) beads and sodium citrate as a directing agent. (Reprinted with permission from D. Lan, Y.M. Zhang and Y.R. Wang, From hexagonally arrayed nanorods to ordered porous film through controlling the morphology of ZnO crystals, *Applied Surface Science*, **254**(18), 5849–5853, 2008. © 2008 Elsevier.)

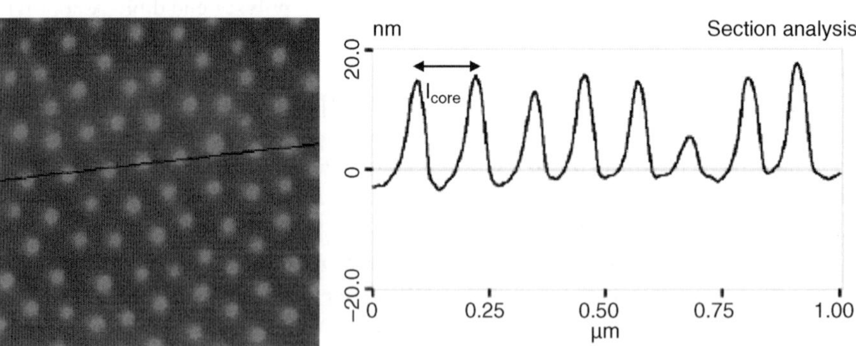

Figure 15.18 AFM image of assembled loaded Zn micelles produced with the diblock copolymer PS-b-P2VP and $Zn(NO_3)_2{\cdot}6H_2O$ at a 0.4 Zn:Pyridine molar equivalent. (Reprinted with permission from E. Pal, A. Oszko, P. Mela *et al.*, Preparation of hexagonally aligned inorganic nanoparticles from diblock copolymer micellar systems, *Colloids and Surfaces A, Physicochemical and Engineering Aspects*, **331**(3), 213–219, 2008. © 2008 Elsevier.)

15.3 1D Metal-Oxide Nanostructures

15.3.1 Colloidal Synthesis and Fabrication

Out of investigations into colloidal based nanoparticle synthesis grew exploration into 1D anisotropic nanostructures. Now commonly referred to as nanorods, nanowires, nanoneedles, nanotubes and nanowhiskers, the growth in investigation of 1D nanoarchitectures has expanded exponentially. 1D nanosystems have attracted attention because of their unique charge-transport properties [164–166], photophysics [167,168] and electronic structure [169,170]. Theoretically, 1D nanostructures should be superior as PEC and PV cells because of vectorial charge transport along the long axis of the nanorod, in contrast to trap-limited charge hopping in interconnected nanoparticle systems. A common limitation to PEC and PV cell efficiency is the diffusion of minority carriers. In 1D nanorod systems the diffusion length of minority carriers can be greatly reduced by their radial movement to the nanorod surface for their utilization to either oxidize water or produce hydrogen gas. Both of these characteristics in turn would decrease electron–hole recombination and increase the efficiency of collecting photogenerated carriers for energy production and water splitting. While this section will cover the colloidal and electrochemical techniques currently used to synthesize 1D metal-oxide nanostructures, the reader would also find of interest the use of II–VI semiconductor 1D nanostructures in energy applications. II–VI semiconductor nanorods, nanowires and nanobelts have been widely explored, and are of particular relevance for their use in light-harvesting and energy-production devices [171–179]. Similarly to the preceding section, TiO_2, WO_3 and ZnO 1D nanoarchitectures will be covered because of the general breadth of study and their applicability to energy production. Surfactant-assisted colloidal methods, seeded growth, template-driven, anodic anodization and combinatorial approaches of the aforementioned techniques are a few of the current means by which 1D nanoarchitectures have been produced and will be covered to give an overview of the state-of-the-art in synthesis and fabrication.

15.3.2 Synthesis and Fabrication of 1D TiO_2 Nanostructures

Colloidal techniques utilizing surfactants and crystallographically directing templates for nanorod synthesis, either in solution or growth onto substrates has been extensively studied due to their low cost, scalability and utilization of "green" precursors [180–193]. The synthesis of nanorods and nanotubes using a sol-gel of titania precursors with anodized aluminum templates has shown great promise for material control. AAO templates are very versatile and the pore diameter can be highly controlled and are commercially available. They are usually made by the anodization of a cleaned Al sheet with an applied potential. In a sol-gel/AAO nanorod/nanotube synthesis, titanium isopropoxide is mixed with ethanol to form a solution, and is then added to a solution of ethanol (EtOH), water and acetylacetone (ACAC) to form a TiO_2 sol [190]. Differing of the molar ratios of the above precursors optimizes the preparation of the nanorods and nanotubes. To fill the pores of a commercially available AAO template, a dipcoating procedure was followed. The loaded AAO template was air dried for one day, and then heated at 400 °C for 1 day. The AAO templates were then removed by chemical etching in a NaOH solution, and the TiO_2 nanotubes and nanorods matched the 200–250 nm pore diameter of the template. They found that a decrease in acetylacetone affects the thickness of the TiO_2 nanotubes, and high yields of nanorods can be produced when the ethanol concentration was raised. The viscosity and packing of the TiO_2 sol is of critical importance in AAO-assisted synthesis, and it is advantageous to modify a material from a solid to hollow structure through colloidal means. In general they found that molar ratios of Ti:ACAC:H_2O:EtOH affected the wall thickness of the nanotubes. Nanotubes and nanorods produced with this procedure were both found to be polycrystalline, regardless of reaction conditions. Polycrystallinity in PV and PEC cells is usually a limitation to light-harvesting energy conversion, due to grain boundaries and the energetic traps that result from such materials. In a related synthesis, TiO_2 nanotubes were utilized as dye-sensitized solar cells and achieved a power-conversion efficiency of 3.5% [184].

Another widely used template substrate is polycarbonate (PC), which, in concert with electrophoresis, has found successful application in fabricating 1D nanostructures [185,193]. Movement of sols into pores of either PC or AAO must rely on capillary forces and are thus limited by the preparation of the sol, and are highly sensitive to the viscosity and concentration of precursors. Most sols have a low solids content (~5%), and the complete filling of the pores does not produce a dense packing of material. Increasing the concentration should correct for this, but added viscosity can prevent pore filling and lead to incomplete nanorod formation [185]. Use of electrophoresis can overcome the limitations inherent in traditional sol-gel templating techniques, allowing for superior pore filling, and materials of higher quality. In electrophoresis, the use of a cathode and an anode draw charged species toward individual electrodes. For instance, positively charged species would migrate to the cathode, and deposit onto the cathodic surface, and vice versa. Species then agglomerate together and begin packing as the continued electrophoretic process continues, as seen in Figure 15.19.

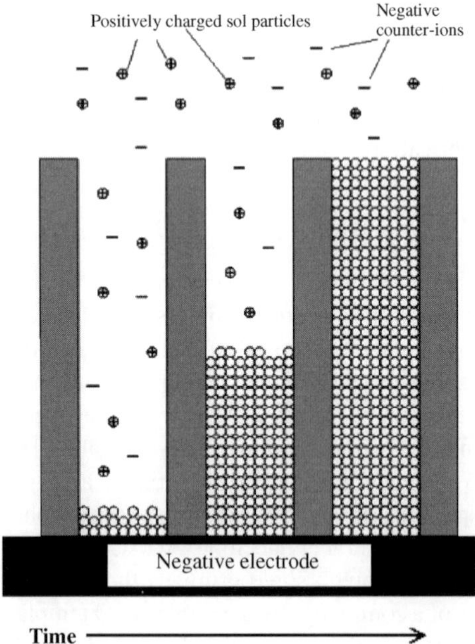

Figure 15.19 An illustration depicting the charged sol particles of TiO_2 stacking into the PC pores by electrophoresis. Sequential stacking of the charged particles is seen as time progresses. (Reprinted with permission from S.J. Limmer, S. Seraji, Y. Wu *et al.*, Template-based growth of various oxide nanorods by sol-gel electrophoresis, *Advanced Functional Materials*, **12**(1), 59–64, 2002. © 2002 Wiley-VCH Verlag GmbH & Co. KGaA.)

In one protocol, a solution of titanium isopropoxide was added to glacial acetic acid, and stirred at 400 rpm at room temperature. Deionized water was then added to the titanium isopropoxide solution and it instantly became white and turbid, but then became a clear liquid after 5 minutes of stirring. Glycerol, lactic acid and ethylene glycol were then added to stabilize the TiO_2 sol for several weeks. To form the nanorods, a commercially available PC template was connected to an aluminum electrode with conducting tape, placed in contact with only the surface of the TiO_2 sol, and the electrophoresis setup was completed with a platinum counter-electrode and power supply. The electrophoresis was performed at 5 V for three hours, excess sol was blotted from the membrane and the membrane was placed onto a Si wafer. The membranes were dried at 100 °C for several hours, and then annealed in an oven at 500 °C to remove the PC template and crystallize the deposited material. Representative SEM, TEM and XRD of the TiO_2 nanorods are seen in Figure 15.20. Diameters of the nanorods matched very well with the PC pore diameter. 180 nm diameter TiO_2 nanorods were produced with 200 nm PC pore sizes, 90 nm diameter nanorods with 100 nm pore sizes and 40 nm diameter nanorods with 50 nm pore sizes. The 10% shrinkage was attributed to the densification during annealing, and samples were found to be entirely of the anatase crystal phase. TiO_2 nanorods produced by electrophoresis were found to be polycrystalline in this study as well, which shows an intrinsic limitation of this particular procedure and of the prior template technique with AAO. For energy-related devices, single crystalline devices are usually desired for charge transport to be optimized, and have been found to perform at higher efficiency levels than their polycrystalline counterparts. While the utilization of templates is attractive, it is increasingly important to

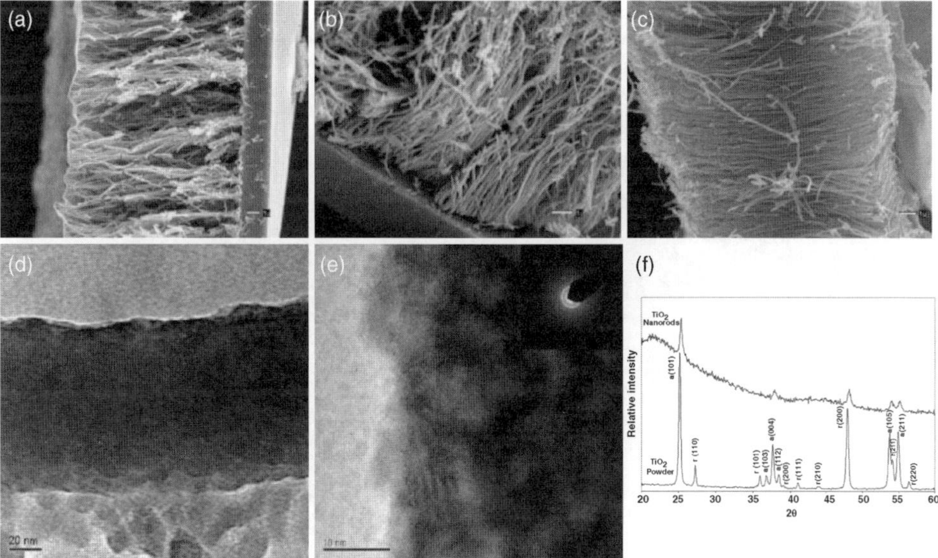

Figure 15.20 TiO_2 nanorods were produced by a polycarbonate (PC) templating technique and representative SEM (a–c), TEM (d–e) and XRD (f) data show their general growth direction and polycrystallinity. Reprinted with permission from S. J. Limmer, S. Seraji, Y. Wu *et al.*, *Advanced Functional Materials*, **12** (1), 59–64, 2002. © 2002 Wiley-VCH Verlag GmbH & Co. KGaA.

achieve single-crystal nanorods without a template, and this was recently accomplished by directly growing TiO_2 nanorods onto conducting substrates [194].

Single-crystal nanorods of aligned metal oxides have proven difficult to accomplish colloidally in the case of TiO_2. In a recent advancement, a simple procedure without a directing template or surfactants was accomplished [194], produced on the conducting substrate fluorine-doped tin oxide (FTO). Cleaned FTO was placed in a Teflon-lined autoclave vessel that contained 10 ml of toluene, 1 ml tetrabutyltitanate, 1 ml titanium tetrachloride (1 M in toluene) and 1 ml of hydrochloric acid (HCl, 37% by weight) and heated to 180 °C. The minute amounts of water present in the HCl play a critical role in nucleation at the FTO surface Phase separation of the water/toluene mixture attracts the water to the hydrophilic FTO surface wherein Ti^{4+} precursors hydrolyze at the water/FTO interface and form nucleation sites. These nucleation sites then act as the new hydrophilic TiO_2 layer where growth and crystallization persist. The single crystalline TiO_2 nanorods grew along the [001] direction and it was explained that Cl^- ions in solution preferentially adsorbed to the (110) crystal facet inhibiting growth in that direction, leading to anistropic nanocrystal formation, as shown in Figure 15.21. XRD analysis showed the TiO_2 nanorods to be of the rutile crystal phase, and SEM revealed that 2.1–4 μm long nanorods with diameters ranging from 10 to 35 nm were produced after 2–22 hours of hydrothermal treatment. Preparations to produce stable PV cells included the growth of a 20 nm dense layer of TiO_2 on cleaned FTO substrates with a solution of $TiCl_4$ to prevent electrical shorting. With a commercially available light-harvesting dye (N719), the aligned 2–3 μm TiO_2 arrays were treated to produce PV cells with a power conversion of $\eta = 5.02\%$ [194]. This is a 43% improvement over the dye-sensitized solar cells fabricated from

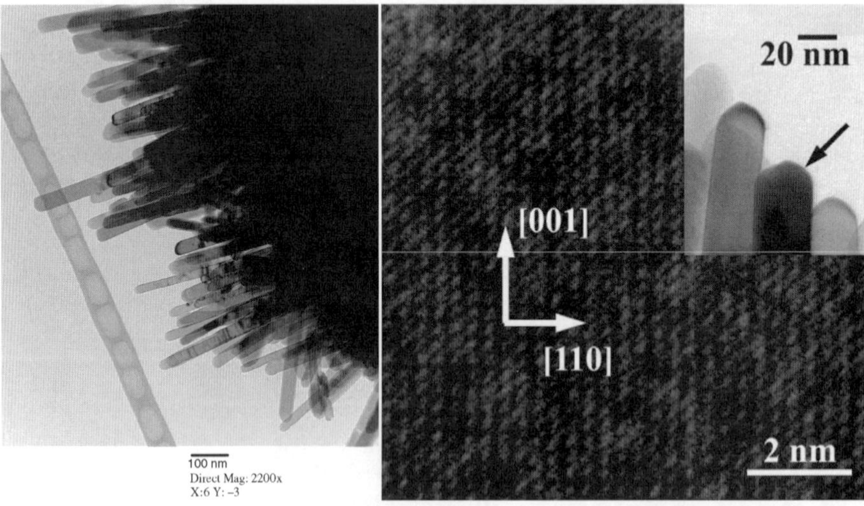

Figure 15.21 TEM (left) and HRTEM (right) of single-crystal TiO_2 nanorods growing along the [001] direction. (Reprinted with permission from X.J. Feng, K. Shankar, O.K. Varghese *et al.*, Vertically aligned single crystal TiO_2 nanowire arrays grown directly on transparent conducting oxide coated glass: synthesis details and applications, *Nano Letters*, **8**(11), 3781–3786, 2008. © 2008 American Chemical Society.)

TiO_2 nanotubes 15 μm in length using a similar commercially available dye [184]. Overall performance of the TiO_2 nanorods was found to be superior, with shorter nanorods in the 2–3 μm length regime in comparison to those in the 3–4 μm and 4–5 μm length regimes. This was attributed to nanorod bunching at the interface of the TiO_2/FTO interface when the base diameter grew, to decreased nanorod density and to increased surface area with smaller diameter nanorods. Overall, the morphological comparisons of nanotubes and nanorods are in no way complete, and continued controlled experiments of their photoelectrochemical performance are essential. Creating nanotubes through anodic anodization is another promising route to 1D nanostructures, and have allowed for longer nanotubes to be produced with high photon-to-hydrogen generation efficiencies in PEC cells.

Potentiostatic anodization is the process responsible for the production of AAO porous templates, but can also be applied directly to the formation of TiO_2 nanotubes when Ti sheets or Ti thin films are used [10,195–199]. Recent advancements have allowed for the creation of 1D nanostructures upwards of 1000 μm in length, with high levels of control over nanotube diameter and wall thickness [198]. The major changes which allowed the length of the TiO_2 nanotubes to increase from 500 nm to 1000 μm were the solvents, electrolytes and pH conditions in which the potentiostatic anodization occurred [200,201]. The anodization process is controlled by the dissolution of solid Ti metal by the applied potential and interaction of the surrounding media. A key parameter that was found to aid in TiO_2 nanotube formation is reducing the water content to below 5%, and using various organic electrolytes, in the presence of fluoride ions [10]. The procedure to produce 134 μm long TiO_2 nanotubes was performed in anhydrous ethylene glycol with NH_4F at 60 V for 17 hours in a two-component electrochemical system. The nanotubes had an inner diameter of 110 nm and an outer diameter of 160 nm and can be seen in Figure 15.22. These TiO_2 nanotube structures have been successfully applied to light-driven water splitting, and have been shown to generate gaseous hydrogen with photo-to-hydrogen efficiencies of 0.5–7% [195,202–204]. In a hybridized system, p-type Cu-Ti-O was produced by potentiostatic anodization and coupled to an n-type TiO_2 nanotube array for a

Figure 15.22 SEM (left) and HRSEM (right) of TiO_2 nanorods produced by potentiostatic anodization. (Reprinted with permission from M. Paulose, K. Shankar, S. Yoriya *et al.*, Anodic growth of highly ordered TiO_2 nanotube arrays to 134 μm in length, *Journal of Physical Chemistry B*, **110**(33), 16179–16184, 2006. © 2006 American Chemical Society.)

self-biased hydrogen production efficiency of 0.30% [197]. Potentiostatic anodization is a robust fabrication technique for 1D nanostructures and is versatile for both PEC and PV applications. The ease of fabrication and low-cost components make it an attractive alternative to the deposition techniques mentioned briefly in the introduction.

15.3.3 Colloidal Synthesis and Fabrication of 1D WO_3 Nanostructures

WO_3 has been explored as a material for electrochromic, sensing and energy applications due to its coloration at applied potentials and resistance to photocorrosion in acidic conditions. 1D nanostructures of WO_3 through colloidal and template-driven synthetic techniques have expanded considerably [205–215]. Hydrothermal routes have been shown to be robust routes for nanorod, nanobelt and sea-urchin morphologies [207]. In a typical synthesis, a 25 mM solution of Na_2WO_4 was acidified with HCl and 6.3 g of oxalic acid was added to yield a translucent and homogenous WO_3 sol. Growth of various morphologies was controlled by the addition of sulfate salts. The WO_3 sol was then added to an autoclave vessel and heated at 180 °C for 2–72 hours. Washing cycles with ethanol and water followed to remove unreacted species, and then drying at 60 °C. Addition of Na_2SO_4, K_2SO_4 and Rb_2SO_4 to the WO_3 sols produced nanorods, nanobelts and sea-urchin microspheres decorated with single-crystal WO_3 nanorods as seen in Figure 15.23.

While the exact growth mechanisms were not wholly explained, the adsorption of sulfate ions, oxalic acid and the counter-cations were surmized to preferentially bind to specific crystal facets and direct growth. An encapsulated tungsten precursor synthesis was also exploited to produce WO_3 nanorods later used to oxidize methanol for potential applications in direct methanol fuel cells [211]. In their synthesis, crystallized tetrabutylammounium decatungstate $((C_4H_9)_4N)_4W_{10}O_{32}$ was placed in a quartz container and placed in a tube furnace at 450 °C for three hours in an Ar atmosphere to generate blue crystals. Platinum decoration was carried out with a 0.01 M aqueous hexachloroplatinic acid with WO_3 crystals and then reduced in a hydrogen atmosphere. The decorated WO_3 nanorods were found to be 130–480 nm in length and 18–56 nm in width. Increased catalytic activity was found with the hybridized system over platinized bulk WO_3 for methanol oxidation, due to their enhanced electrochemical activity and reduced decomposition in acidic media [211].

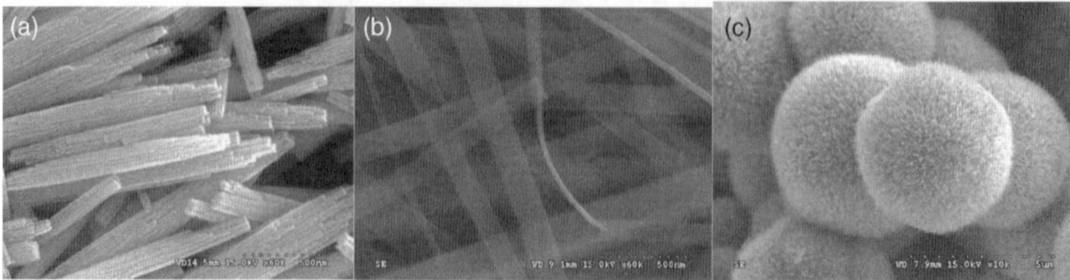

Figure 15.23 SEM images of WO_3 nanorods (a), nanobelts (b) and microspheres decorated with WO_3 nanorods (c). (Reprinted with permission from Z.J. Gu, T.Y. Zhai, B.F. Gao *et al.*, Controllable assembly of WO_3 nanorods/nanowires into hierarchical nanostructures, *Journal of Physical Chemistry B*, **110**(47), 23829–23836, 2006. © 2006 American Chemical Society.)

As a versatile means to produce 1 D nanostructures, templates produced using AAO have also been applied to WO_3 nanorod synthesis [208]. Using a commercially available AAO template with a 200 nm pore diameter, a solution of phosphotungstic ($H_3PW_{12}O_{40}$) acid in methanol was vacuumed into the membrane and dried at 95 °C for one hour. Annealing was performed at temperatures of up to 600 °C, the AAO template was removed with HF, and the WO_3 nanorods rinsed with deionized water and dried. The resulting nanostructures were of the monoclinic crystal phase, and the diameter was found to be 150 nm, which indicates that shrinkage within the pores during annealing was 25%. As a compliment to template- and surfactant-based synthesis, synthesis of a single precursor-based nanorod system is also attractive. In an inert atmosphere, the precursor tungsten oxomethoxide, $WO(OCH_3)_4$, was added to a Swagelok vessel, and placed into a tube furnace at 700 °C for three hours and then allowed to cool over a five hour period to form a black solid of $W_{18}O_{49}$ with carbon contamination. Post-annealing at 500 °C in open-air conditions led to the formation of yellow crystallites of the monoclinic phase of WO_3 with all carbon removed. The synthesized WO_3 nanorods were 80–700 nm in diameter and several micrometers in length with varied aspect ratio. One individual nanorod of 80 nm had a length of 1.5 μm with an aspect ratio of ~20 [210]. The monoclinic WO_3 nanorods grew in the [200] direction and streaking in the electron diffraction (ED) pattern was presumed to be a product of stacking faults. Overall, the diversity of synthetic techniques to produce WO_3 nanorods is not covered in its entirety here, but this is a brief description of some successful techniques that require few precursors and no expensive deposition systems.

15.3.4 Colloidal Synthesis and Fabrication of 1D ZnO Nanostructures

Due to its higher electron mobility properties in comparison to TiO_2 (1–10 $cm^2 V^{-1} s^{-1}$), ZnO (100 $cm^2 V^{-1} s^{-1}$) has received a lot of attention recently for its use in PV, PEC, sensing, emission and photocatalytic devices [216–218]. The relative positions of its valence and conduction bands are similar to TiO_2, with a bandgap of ~3.2 eV, and it has been versatile in forming nanoparticles and nanorod morphologies. Seeded growth directly on substrates of ZnO nanorods and nanowires has been found to be very successful, and been applied to fully functional PV devices [164,219–222]. Controlling the concentration of surfactants to suppress or enhance crystal growth has also shown great promise, and continues to be expanded upon [218,223–234]. ZnO nanorods and nanocolumns that were reminiscent of the structures produced by red abalone (*Haliotis rufescens*) were produced by manipulation of the citrate concentration in aqueous solutions with zinc nitrate [218,225]. ZnO nanoparticle seeds on glass substrates submerged in an aqueous solution containing 0.03 M HMT and 0.03 M $Zn(NO_3)_2$ at 60 °C for three days produced nanorods. Nanocolumn structures were then grown onto the ZnO nanorods with HMT, zinc nitrate and sodium citrate as a directing agent [225]. Individual components of the ZnO nanocolumns comprised individual ZnO nanoplatelets, as seen in Figure 15.24. These ZnO nanoplatelets grew on top of the "seed" ZnO nanorods dominated by crystal growth along the [001]. Due to the high citrate concentration that was present during growth, the [001] was inhibited and, in turn, nanoplatelets stacked preferentially. As a systematic study, it was found that a decreasing height/width ratio is seen with increasing sodium citrate concentration. In an attempt to replicate the morphology of red abalone, they were also able to switch from a nanorod growth regime to a nanoplatelet growth regime by exchanging the growth solution with varying amounts of sodium citrate. Overall, this

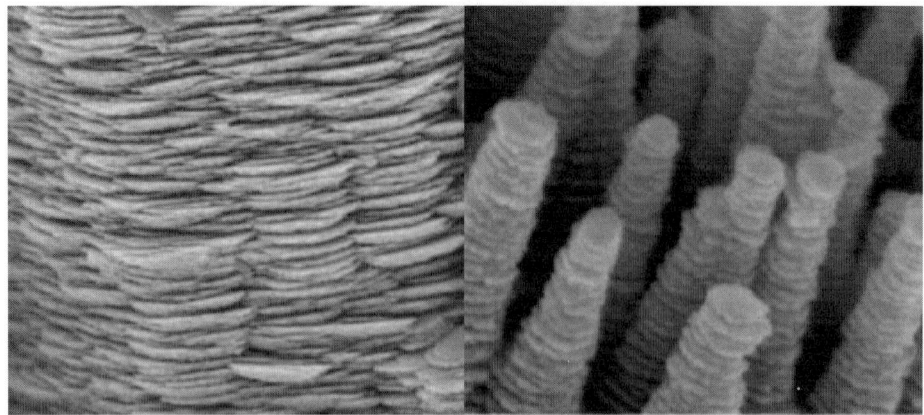

Figure 15.24 SEM images of ZnO nanocolumns grow on ZnO nanorods with a citrate-ion directing agent. (Reprinted with permission from Z.R.R. Tian, J.A. Voigt, J. Liu *et al.*, Complex and oriented ZnO nanostructures, *Nature Materials*, **2**(12), 821–826, 2003. © 2003 Nature Publishing Group.)

crystallographic control of growth in a simple manner is very powerful for creating large networks of complex nanostructures at low cost.

Directing methods are also seen with ZnO nanorod and nanowire formation with the utilization of AAO templates, plant cell adsorption, potentiostatic anodization, and microwave irradiation [235–240]. Electrodeposition of ZnO nanowires into AAO templates has been sparsely investigated because of the formation of $ZnAlO_3$ [241,242]. Zn sulfate was used as the zinc source in aqueous electrolytes, and with electrochemical conditions of 1 V, pH 3 and room temperature. After electrodeposition inside the AAO templates, annealing at a series of temperatures between 25 °C and 500 °C was performed for 10 hours. The annealing temperature was found to be important due to the formation of $ZnAlO_3$ at 400 °C, and thus annealing temperatures of 350 °C were found to produce pure ZnO crystal phases. ZnO nanowires of 50 nm were produced, consistent with the diameters of the AAO template, and over 10 μm in length. These samples were than characterized for their ferromagnetic properties, which is beyond the scope of this chapter. While these techniques have been based on colloidal or wet chemical methods, a large body of literature exists, using chemical vapor deposition in conjunction with directing templates to produce 1D ZnO nanostructures. The reader is directed to the introductory section to seek references pertaining to CVD and other related deposition techniques.

15.4 2D Metal-Oxide Nanostructures

15.4.1 Colloidal Synthesis of 2D TiO_2 Nanostructures

2D structures or those confined in their growth in only one direction have been examined due to their large surface-to-volume ratio and large aspect ratios. Taking this 2D phenomenon to the very extreme would yield a structure like single graphene sheets, which have seen broad interest over the past five years [243,244]. In our case, a brief description of colloidal techniques leading to 2D metal oxides is more appropriate for the main motivation of PEC devices as water-splitting materials. Nanostructures confined in one direction have been creatively described as nanodisks, nanosheets, nanoflakes and nanoplatelets. The examination of 2D nanomaterials is considerably less extensive than their 0D and 1D counterparts, and so there are many more avenues of exploration that remained untapped.

Colloidal synthesis of TiO_2 nanosheets and nanodisks has been widely studied using self-assembly, template, surfactant and Langmuir–Blodgett (LB) techniques [245–264]. LB deposition techniques stand out as a complimentary technique not before mentioned in this chapter. The LB technique is a deposition and assembly method based on hydrophilic/hydrophobic interactions at surfaces and the air–water interface [265,266]. Use of the LB technique can allow for precise layering of materials from hydrophilic-to-hydrophobic and systematic deposition onto substrates with a high degree of precision. Utilizing an exfoliated layered titanate precursor, $H_2Ti_{2-x/4}O_4 \cdot H_2O$, and the surfactant dioctyldimethylammonium bromide (DODAB), it is possible to alternate layers of the amphiphile and TiO_2. In order to form LB monolayers, a chloroform solution of DODAB (1.6×10^{-3} mol dm^{-3}) was spread onto the LB trough with a Hamilton syringe at 20 °C onto a TiO_2 sol already dispersed in the trough as seen in Figure 15.25 (step 2). The solvent was evaporated prior to compression, and

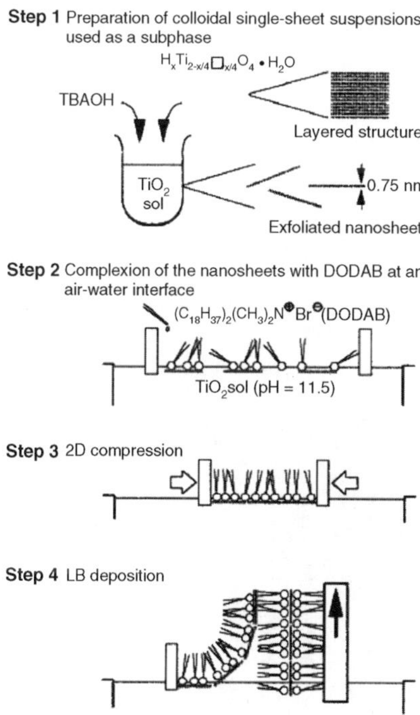

Figure 15.25 Schematic illustration depicting the deposition of TiO_2 nanosheets using the Langmuir–Blodgett (LB) deposition technique. (Reprinted with permission from T. Yamaki and K. Asai, Alternate multilayer deposition from ammonium amphiphiles and titanium dioxide crystalline nanosheets using the Langmuir–Blodgett technique, *Langmuir*, **17**(9), 2564–2567, 2001. © 2001 American Chemical Society.)

the compression of the monolayer was performed at 3000 $mm^2 min^{-1}$ and tracked via surface-area–pressure (π-A) isotherms (step 3). Compression was maintained at 30–40 mN m^{-1} and allowed to equilibrate for 60 minutes prior to a vertical dipping technique using a quartz substrate treated with trichlorosilane to make it hydrophobic. The dipping was done at 20 mm min^{-1} and stopped for 13 minutes at the top of each cycle (step 4). The packed structure was examined via XRD and transmission spectrophotometry after the deposition was completed. Absorption of the single TiO_2 nanosheets was blue-shifted considerably to 268 nm (4.68 eV), 1.5 eV from the bulk bandgap energies of anatase (3.18 eV) and rutile (3.03 eV) TiO_2. These ultra-thin structures were found to have a lattice spacing of 3.4 nm, similar to the combined thickness of one layer of TiO_2 (0.75 nm) and DODAB (2.5 nm). Control of individual TiO_2 atomic building blocks can be a powerful tool in the construction of more advanced architectures and can potentially be used in the fabrication of PEC and PV devices.

15.4.2 Colloidal Synthesis of 2D WO_3 Nanostructures

Utilization of colloidal soft-template methods for producing 2D WO_3 nanodisks and nanosheets, allow for a cost-effective route to advanced materials [119,267–271]. One investigation used the surfactant triton X-100 and the non-ionic polymer poly(ethylene glycol) (PEG) as a directing agent to produce WO_3 nanodisks [267]. The resulting WO_3 nanodisks were 350–1000 nm along the long axis, 200–750 nm along the short axis and 7–18 nm in thickness. These ultrathin WO_3 nanodisks were of the monoclinic crystal phase and the 7–18 nm of thickness was equivalent to 9–24 unit cells of the (010) facet. Representative SEM images of the nanodisks can be seen in Figure 15.26. The growth mechanism was explained by

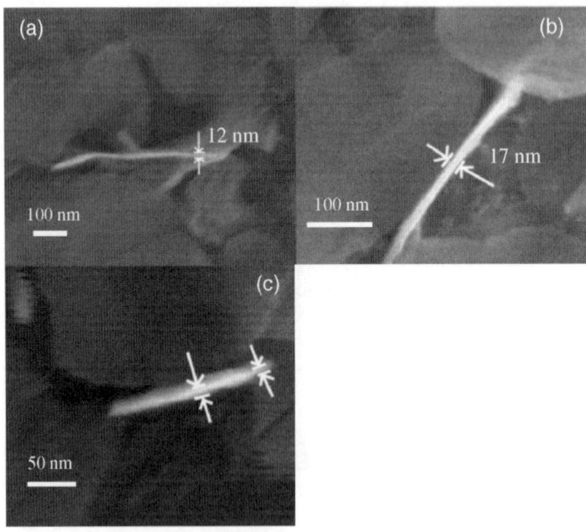

Figure 15.26 SEM images of vertically aligned WO_3 nanodisks produced by a soft-template colloidal method. (Reprinted with permission from A. Wolcott, T.R. Kuykendall, W. Chen *et al.*, Synthesis and characterization of ultrathin WO_3 nanodisks utilizing long-chain poly(ethylene glycol), *Journal of Physical Chemistry B*, **110**(50), 25288–25296, 2006. © 2006 American Chemical Society.)

the adsorption of PEG-10 000 onto the (010) crystal facets of $WO_3{\cdot}H_2O$, and thereby inhibiting crystal growth and leading to the anisotropic growth in two dimensions. The WO_3 nanodisks were fashioned into PEC cells and examined for their electrochemical properties. At negative potentials, H^+ and Na^+ intercalation lead to an electrochromic affect and a bluish tint to the films in solution. At negative potentials near −0.8 V, inefficient hydrogen generation was observed, and subsequent damage to the WO_3 nanodisk PEC cells ensued. The simple colloidal technique for high-aspect-ratio 2D nanostructures can be especially useful for catalysis, because of the large surface-to-volume ratio. The electron-transport properties of the WO_3 nanodisks were hindered by trap-limited hopping from nanodisk to nanodisk, leading to low photocurrent generation.

15.4.3 Colloidal Synthesis of 2D ZnO Nanostructures

2D ZnO nanostructures produced through colloidal means includes the use of microwave irradiation, electrodeposition, hydrothermal and surfactant-based synthesis [272–288]. Template-driven synthesis allows precise control over atomic building blocks, is a powerful methodology and can be potentially exploited in photocatalytic, PV and PEC devices. Vertically aligned ZnO nanosheets were directly applied to DSSC and were found to be successful, with a power conversion efficiency of 3.9%. The high performance level was attributed to the access of the redox electrolyte to the entirety of the ZnO array, thus allowing for increased fill thickness without diminishing the fill factor [289]. Electrodeposition of ZnO nanosheets with template-driven microsphere arrays was also accomplished with a high degree of atomic control on ITO conducting substrates. PS beads were laid onto the ITO substrate, and in a three-electrode cell with a 0.1 M $Zn(NO_3)_2$ solution, the ZnO nanosheets were electrodeposited at potentials of −0.8–1.1 V for 8 hours. The PS template was removed by annealing at 600 °C, and the resultant ordered structures were then examined, as seen in Figure 15.27. The highly crystalline porous material was found to grow along the (111) facet, and the pore diameters were 260 nm, very close to the diameter of the PS beads (270 nm). While these highly ordered ZnO nanosheets were not fashioned into a functioning PV or PEC cell, the success of the aforementioned ZnO DSSC lends a direction to the use of the templated nanosheets for alternative energy production.

Solvothermal production of 2D ZnO nanostructures is also a low-cost and scalable production technique. Using zinc foils as the elemental source and the substrate, nanosheets, along with nanorods, were produced in a autoclave bomb at elevated temperatures. ZnO nanosheets were produced by first polishing the zinc foils and cleaning them with acetone and water in an ultrasonic bath. The zinc foils were then placed into the autoclave vessel with 60 ml of ethanol, heated to 170–230 °C for 3–12 hours, removed from the vessel and then washed with water and dried in air. SEM imaging showed that nanosheets and nanorods were produced at 170 °C for three hours with four zinc foils in the reaction vessel and nanosheet thicknesses measuring 500–750 nm. Longer heating times, up to 12 hours, produced ZnO nanosheets with thicknesses of 200–250 nm. TEM and XRD revealed that the 2D structures were growing along the (0001) facet, and were of the zincite crystal phase. In general, this type of synthetic route should be regarded as "green" chemistry and seen as a fantastic opportunity to produce highly functional nanomaterials with the least environmental impact. Further studies using mild conditions and inexpensive starting materials will lower

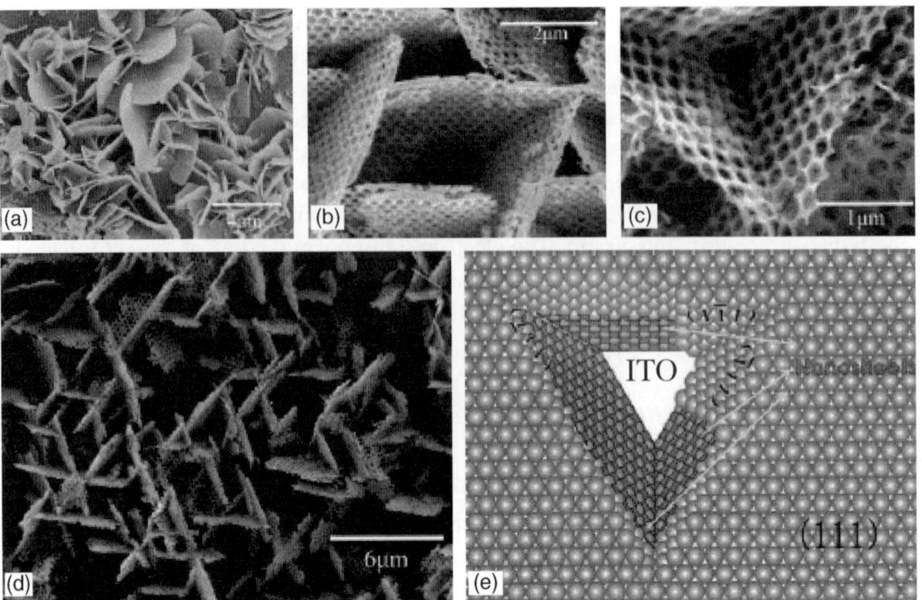

Figure 15.27 SEM images of ZnO nanosheets produced without a polystyrene bead template (a), with a closely packed polystyrene template (b–d) and a model depicting the growth direction. (Reprinted with permission of M. Fu, J. Zhou, Q.F. Xiao *et al.*, ZnO nanosheets with ordered pore periodicity via colloidal crystal template assisted electrochemical deposition, *Advanced Materials*, **18**(8), 1001–1004, 2006. © 2006 Wiley-VCH Verlag GmbH & Co. KGaA.)

the overall cost of PEC and PV devices and allow these to enter the market in a scalable fashion.

15.5 Conclusion

Since the original exploitation of nanocrystalline TiO_2 as a substrate for the now well-known Grätzel cell, many discoveries have been made and the realms of possibility greatly expanded in the synthesis, characterization and applications of metal-oxide nanostructures. This chapter has focused on surveying various common methods developed for producing various metal-oxide nanostructures with control over their size, shape, crystal structure, surface and morphology. The plethora of techniques that allow for 0D, 1D and 2D nanostructure synthesis is varied, and will no doubt continue to expand. Utilization of sol-gel, hydrothermal, solvothermal, sonochemical, surfactant driven and template-assisted synthesis show the diversity of ways in which colloidal chemistry can be used. Colloidal chemistry allows for low cost, high scalability and relatively "green" synthesis of metal oxides and is therefore a good avenue to explore for practical device applications, including solar-energy conversion and photoelectrochemical water splitting for hydrogen generation. Future research is expected to further explore the surface and interfacial properties of nanomaterials and their explications in device applications.

Acknowledgment

We wish to thank the Basic Energy Sciences Division of the US Department of Energy for financial support.

References

[1] (IPCC), I.P.o.C.C., Climate Change (2007) Synthesis Report, 2007, United Nations, New York.

[2] Brook, E.J. (2005) Tiny bubbles tell all. *Science*, **310**(5752), 1285–1287.

[3] Fujishima, A. and Honda, K. (1972) Electrochemical photolysis of water at a semiconductor electrode. *Nature*, **238**(5358), 37–111.

[4] Amouyal, E. (1995) Photochemical production of hydrogen and oxygen from water – a review and state-of-the-art. *Solar Energy Materials and Solar Cells*, **38**(1–4), 249–276.

[5] Bak, T., Nowotny, J., Rekas, M. and Sorrell, C.C. (2002) Photo-electrochemical hydrogen generation from water using solar energy. Materials-related aspects. *International Journal of Hydrogen Energy*, **27**(10), 991–1022.

[6] Bolton, J.R. (1996) Solar photoproduction of hydrogen: A review. *Solar Energy*, **57**(1), 37–50.

[7] Ni, M., Leung, M.K.H., Leung, D.Y.C. and Sumathy, K. (2007) A review and recent developments in photocatalytic water-splitting using TiO_2 for hydrogen production. *Renewable & Sustainable Energy Reviews*, **11**(3), 401–425.

[8] Rocheleau, R.E. and Miller, E.L. (1997) Photoelectrochemical production of hydrogen: Engineering loss analysis. *International Journal of Hydrogen Energy*, **22**(8), 771–782.

[9] Khaselev, O. and Turner, J.A. (1998) A monolithic photovoltaic-photoelectrochemical device for hydrogen production via water splitting. *Science*, **280**(5362), 425–427.

[10] Paulose, M., Shankar, K., Yoriya, S. *et al.* (2006) Anodic growth of highly ordered TiO_2 nanotube arrays to 134 mu m in length. *Journal of Physical Chemistry B*, **110**(33), 16179–16184.

[11] Gerisher, J. (1979) *Solar Energy Conversion (ed* B.O. Seraphin), Springer, Berlin.

[12] Oregan, B. and Gratzel, M. (1991) A low-cost, high-efficiency solar-cell based on dye-sensitized colloidal TiO_2 films. *Nature*, **353**(6346), 737–740.

[13] Cherepy, N.J., Smestad, G.P., Gratzel, M. and Zhang, J.Z. (1997) Ultrafast electron injection: Implications for a photoelectrochemical cell utilizing an anthocyanin dye-sensitized TiO_2 nanocrystalline electrode. *Journal of Physical Chemistry B*, **101**(45), 9342–9351.

[14] El Zayat, M.Y., Saed, A.O. and El-Dessouki, M.S. (1998) Photoelectrochemical properties of dye sensitized Zr-doped $SrTiO_3$ electrodes. *International Journal of Hydrogen Energy*, **23**(4), 259–266.

[15] Gratzel, M. (2005) Mesoscopic solar cells for electricity and hydrogen production from sunlight. *Chemistry Letters*, **34**(1), 8–13.

[16] Gratzel, M. and Kalyanasundaram, K. (1994) Artificial photosynthesis – efficient dye-sensitized photoelectrochemical cells for direct conversion of visible-light to electricity. *Current Science*, **66** (10), 706–714.

[17] Hagfeldt, A., Didriksson, B., Palmqvist, T. *et al.* (1994) Verification of high efficiencies for the gratzel-cell – A 7-percent efficient solar-cell based on dye-sensitized colloidal TiO_2 films. *Solar Energy Materials and Solar Cells*, **31**(4), 481–488.

[18] Huang, S.Y., Schlichthorl, G., Nozik, A.J. *et al.* (1997) Charge recombination in dye-sensitized nanocrystalline TiO_2 solar cells. *Journal of Physical Chemistry B*, **101**(14), 2576–2582.

[19] Park, J.H. and Bard, A.J. (2005) Unassisted water splitting from bipolar Pt/dye-sensitized TiO_2 photoelectrode arrays. *Electrochemical and Solid State Letters*, **8**(12), G371–G375.

[20] Park, J.H. and Bard, A.J. (2006) Photoelectrochemical tandem cell with bipolar dye-sensitized electrodes for vectorial electron transfer for water splitting. *Electrochemical and Solid State Letters*, **9**(2), E5–E8.

[21] Rensmo, H., Keis, K., Lindstrom, H. *et al.* (1997) High light-to-energy conversion efficiencies for solar cells based on nanostructured ZnO electrodes. *Journal of Physical Chemistry B*, **101**(14), 2598–2601.

[22] Schlichthorl, G., Huang, S.Y., Sprague, J. and Frank, A.J. (1997) Band edge movement and recombination kinetics in dye-sensitized nanocrystalline TiO_2 solar cells: A study by intensity modulated photovoltage spectroscopy. *Journal of Physical Chemistry B*, **101**(41), 8141–8155.

[23] Smestad, G., Bignozzi, C. and Argazzi, R. (1994) Testing of dye-sensitized TiO_2 solar-cells.1. experimental photocurrent output and conversion efficiencies. *Solar Energy Materials and Solar Cells*, **32**(3), 259–272.

[24] Tachibana, Y., Moser, J.E., Gratzel, M. *et al.* (1996) Subpicosecond interfacial charge separation in dye-sensitized nanocrystalline titanium dioxide films. *Journal of Physical Chemistry*, **100**(51), 20056–20062.

[25] Grant, C.D., Schwartzberg, A.M., Smestad, G.P. *et al.* (2002) Characterization of nanocrystalline and thin film TiO_2 solar cells with poly(3-undecyl-2,2′-bithiophene) as a sensitizer and hold conductor. *Journal of Electroanalytical Chemistry*, **522**(1), 40–48.

[26] Grant, C.D., Schwartzberg, A.M., Smestad, G.P. *et al.* (2003) Optical and electrochemical characterization of poly (3-undecyl-2,2′-bithiophene) in thin film solid state TiO_2 photovoltaic solar cells. *Synthetic Metals*, **132**(2), 197–204.

[27] Smestad, G.P., Spiekermann, S., Kowalik, J. *et al.* (2003) A technique to compare polythiophene solid-state dye sensitized TiO_2 solar cells to liquid junction devices. *Solar Energy Materials and Solar Cells*, **76**(1), 85–105.

[28] Bruchez, M., Moronne, M., Gin, P. *et al.* (1998) Semiconductor nanocrystals as fluorescent biological labels. *Science*, **281**(5385), 2013–2016.

[29] Diguna, L.J., Shen, Q., Kobayashi, J. and Toyoda, T. (2007) High efficiency of CdSe quantum-dot-sensitized TiO_2 inverse opal solar cells. *Applied Physics Letters*, **91**(2), 023116.1–023116.3.

[30] Lee, J.C., Kim, T.G., Choi, H.J. and Sung, Y.M. (2007) Enhanced photochemical response of TiO_2/CdSe heterostructured nanowires. *Crystal Growth & Design*, **7**(12), 2588–2593.

[31] Lee, J.C., Sung, Y.M., Kim, T.G. and Choi, H.J. (2007) TiO_2-CdSe nanowire arrays showing visible-range light absorption. *Applied Physics Letters*, **91**(11), 113104.1–113104.3.

[32] Leschkies, K.S., Divakar, R., Basu, J. *et al.* (2007) Photosensitization of ZnO nanowires with CdSe quantum dots for photovoltaic devices. *Nano Letters*, **7**(6), 1793–1798.

[33] Levy-Clement, C., Tena-Zaera, R., Ryan, M.A. *et al.* (2005) CdSe-Sensitized p-CuSCN/nanowire n-ZnO heterojunctions. *Advanced Materials*, **17**(12), 1512–1515.

[34] Niitsoo, O., Sarkar, S.K., Pejoux, C. *et al.* (2006) Chemical bath deposited CdS/CdSe-sensitized porous TiO_2 solar cells. *Journal of Photochemistry and Photobiology A-Chemistry*, **181**(2–3), 306–313.

[35] Shen, Q., Arae, D. and Toyoda, T. (2004) Photosensitization of nanostructured TiO_2 with CdSe quantum dots: effects of microstructure and electron transport in TiO_2 substrates. *Journal of Photochemistry and Photobiology A-Chemistry*, **164**(1–3), 75–80.

[36] Shen, Q., Katayama, K., Sawada, T. *et al.* (2006) Optical absorption, photoelectrochemical, and ultrafast carrier dynamic investigations of TiO_2 electrodes composed of nanotubes and nanowires sensitized with CdSe quantum dots. *Japanese Journal of Applied Physics Part 1*, **45**(6B), 5569–5574.

[37] Shen, Q., Sato, T., Hashimoto, M. *et al.* (2006) Photoacoustic and photo electrochemical characterization of CdSe-sensitized TiO_2 electrodes composed of nanotubes and nanowires. *Thin Solid Films*, **499**(1–2), 299–305.

[38] Shen, Q. and Toyoda, T. (2004) Characterization of nanostructured TiO_2 electrodes sensitized with CdSe quantum dots using photoacoustic and photoelectrochemical current methods. *Japanese Journal of Applied Physics Part 1*, **43**(5B), 2946–2951.

[39] Shen, Y.C., Deng, H.H., Fang, J.H. and Lu, Z.H. (2000) Co-sensitization of microporous TiO_2 electrodes with dye molecules and quantum-sized semiconductor particles. *Colloids and Surfaces a-Physicochemical and Engineering Aspects*, **175**(1–2), 135–140.

[40] Lopez-Luke, T., Wolcott, A., Xu, L.P. *et al.* (2008) Nitrogen-doped and CdSe quantum-dot-sensitized nanocrystalline TiO_2 films for solar energy conversion applications. *Journal of Physical Chemistry C*, **112** (4), 1282–1292.

[41] Han, X.H., Wang, G.Z., Jie, J.S. *et al.* (2005) Controllable synthesis and optical properties of novel ZnO cone arrays via vapor transport at low temperature. *Journal of Physical Chemistry B*, **109**(7), 2733–2738.

[42] Heo, Y.W., Varadarajan, V., Kaufman, M. *et al.* (2002) Site-specific growth of ZnO nanorods using catalysis-driven molecular-beam epitaxy. *Applied Physics Letters*, **81**(16), 3046–3048.

[43] Jeon, S. and Yong, K. (2008) Synthesis and characterization of tungsten oxide nanorods from chemical vapor deposition-grown tungsten film by low-temperature thermal annealing. *Journal of Materials Research*, **23**(5), 1320–1326.

[44] Jeong, M.C., Oh, B.Y., Nam, O.H. *et al.* (2006) Three-dimensional ZnO hybrid nanostructures for oxygen sensing application. *Nanotechnology*, **17**(2), 526–530.

[45] Kim, K.S. and Kim, H.W. (2003) Synthesis of ZnO nanorod on bare Si substrate using metal organic chemical vapor deposition. *Physica B-Condensed Matter*, **328**(3–4), 368–371.

[46] Klinke, C., Hannon, J.B., Gignac, L. *et al.* (2005) Tungsten oxide nanowire growth by chemically induced strain. *Journal of Physical Chemistry B*, **109**(38), 17787–17790.

[47] Lyu, S.C., Zhang, Y., Ruh, H. *et al.* (2002) Low temperature growth and photoluminescence of well-aligned zinc oxide nanowires. *Chemical Physics Letters*, **363**(1–2), 134–138.

[48] Mahan, A.H., Parilla, P.A., Jones, K.M. and Dillon, A.C. (2005) Hot-wire chemical vapor deposition of crystalline tungsten oxide nanoparticles at high density. *Chemical Physics Letters*, **413**(1–3), 88–94.

[49] Meda, L., Breitkopf, R.C., Haas, T.E. and Kirss, R.U. (2002) Investigation of electrochromic properties of nanocrystalline tungsten oxide thin film. *Thin Solid Films*, **402**(1–2), 126–130.

[50] Pradhan, S.K., Reucroft, P.J., Yang, F.Q. and Dozier, A. (2003) Growth of TiO_2 nanorods by metalorganic chemical vapor deposition. *Journal of Crystal Growth*, **256**(1–2), 83–88.

[51] Rao, C.N.R., Gundiah, G., Deepak, F.L. *et al.* (2004) Carbon-assisted synthesis of inorganic nanowires. *Journal of Materials Chemistry*, **14**(4), 440–450.

[52] Wu, J.J. and Yu, C.C. (2004) Aligned TiO_2 nanorods and nanowalls. *Journal of Physical Chemistry B*, **108**(11), 3377–3379.

[53] Yang, J.L., An, S.J., Park, W.I. *et al.* (2004) Photocatalysis using ZnO thin films and nanoneedles grown by metal-organic chemical vapor deposition. *Advanced Materials*, **16**(18), 1661–1664.

[54] Yoo, S. and Akbar, S.A. (2008) Gas-phase driven nano-machined TiO_2 ceramics. *Journal of Electroceramics*, **21** (1–4), 103–109.

[55] Yuan, H.T. and Zhang, Y. (2004) Preparation of well-aligned ZnO whiskers on glass substrate by atmospheric MOCVD. *Journal of Crystal Growth*, **263**(1–4), 119–124.

[56] Smith, W. and Zhao, Y.P. (2008) Enhanced photocatalytic activity by aligned WO_3/TiO_2 two-layer nanorod arrays. *Journal of Physical Chemistry C*, **112**(49), 19635–19641.

[57] He, Y.P., Zhang, Z.Y. and Zhao, Y.P. (2008) Optical and photocatalytic properties of oblique angle deposited TiO_2 nanorod array. *Journal of Vacuum Science & Technology B*, **26**(4), 1350–1358.

[58] Smith, W., Zhang, Z.Y. and Zhao, Y.P. (2007) Structural and optical characterization of WO_3 nanorods/films prepared by oblique angle deposition. *Journal of Vacuum Science & Technology B*, **25**(6), 1875–1881.

[59] Ye, D.X., Zhao, Y.P., Yang, G.R. *et al.* (2002) Manipulating the column tilt angles of nanocolumnar films by glancing-angle deposition. *Nanotechnology*, **13**(5), 615–618.

[60] Wolcott, A., Smith, W.A., Kuykendall, T.R. *et al.* (2009) Photoelectrochemical Water Splitting Using Dense and Aligned TiO_2 Nanorod Arrays. *Small*, **5**(1), 104–111.

[61] Wolcott, A., Smith, W.A., Kuykendall, T.R. *et al.* (2009) Photoelectrochemical study of nanostructured ZnO thin films for hydrogen generation from water splitting. *Advanced Functional Materials*, **19**(12), 1849–1856.

[62] Boettcher, S.W., Fan, J., Tsung, C.K. *et al.* (2007) Harnessing the sol-gel process for the assembly of non-silicate mesostructured oxide materials. *Accounts of Chemical Research*, **40**(9), 784–792.

[63] Garnweitner, G. and Niederberger, M. (2006) Nonaqueous and surfactant-free synthesis routes to metal oxide nanoparticles. *Journal of the American Ceramic Society*, **89**(6), 1801–1808.

[64] Rao, C.N.R. and Kalyanikutty, K.P. (2008) The liquid-liquid interface as a medium to generate nanocrystalline films of inorganic materials. *Accounts of Chemical Research*, **41**(4), 489–499.

[65] Mills, A. and LeHunte, S. (1997) An overview of semiconductor photocatalysis. *Journal of Photochemistry and Photobiology A-Chemistry*, **108**(1), 1–35.

[66] Linsebigler, A.L., Lu, G.Q. and Yates, J.T. (1995) Photocatalysis on TiO_2 surfaces – principles, mechanisms, and selected results. *Chemical Reviews*, **95**(3), 735–758.

[67] Hagfeldt, A. and Gratzel, M. (1995) Light-induced redox reactions in nanocrystalline systems. *Chemical Reviews*, **95**(1), 49–68.

[68] Chemseddine, A. and Moritz, T. (1999) Nanostructuring titania: Control over nanocrystal structure, size, shape, and organization. *European Journal of Inorganic Chemistry*, (2), 235–245.

[69] Moritz, T., Reiss, J., Diesner, K. *et al.* (1997) Nanostructured crystalline TiO_2 through growth control and stabilization of intermediate structural building units. *Journal of Physical Chemistry B*, **101**(41), 8052–8053.

[70] Pottier, A.S., Cassaignon, S., Chaneac, C. *et al.* (2003) Size tailoring of TiO_2 anatase nanoparticles in aqueous medium and synthesis of nanocomposites. Characterization by Raman spectroscopy. *Journal of Materials Chemistry*, **13**(4), 877–882.

[71] Oskam, G., Nellore, A., Penn, R.L. and Searson, P.C. (2003) The growth kinetics of TiO_2 nanoparticles from titanium(IV) alkoxide at high water/titanium ratio. *Journal of Physical Chemistry B*, **107**(8), 1734–1738.

[72] Reddy, K.M., Reddy, C.V.G. and Manorama, S.V. (2001) Preparation, characterization, and spectral studies on nanocrystalline anatase TiO_2. *Journal of Solid State Chemistry*, **158**(2), 180–186.

[73] Pottier, A., Chaneac, C., Tronc, E. *et al.* (2001) Synthesis of brookite TiO_2 nanoparticles by thermolysis of TiCl4 in strongly acidic aqueous media. *Journal of Materials Chemistry*, **11**(4), 1116–1121.

[74] Li, G.S., Li, L.P., Boerio-Goates, J. and Woodfield, B.F. (2005) High purity anatase TiO_2 nanocrystals: Near room-temperature synthesis, grain growth kinetics, and surface hydration chemistry. *Journal of the American Chemical Society*, **127**(24), 8659–8666.

[75] Chen, X. and Mao, S.S. (2007) Titanium dioxide nanomaterials: Synthesis, properties, modifications, and applications. *Chemical Reviews*, **107**(7), 2891–2959.

[76] Lifshitz, I.M. and Slyozov, V.V. (1961) The kinetics of precipitation from supersaturated solid solutions. *Journal of Physics and Chemistry of Solids*, **19**(35), 35–50.

[77] Buonsanti, R., Grillo, V., Carlino, E. *et al.* (2006) Seeded growth of asymmetric binary nanocrystals made of a semiconductor TiO_2 rodlike section and a magnetic gamma-Fe_2O_3 spherical domain. *Journal of the American Chemical Society*, **128**(51), 16953–16970.

[78] Yang, J., Mei, S. and Ferreira, J.M.F. (2004) Hydrothermal processing of nanocrystalline anatase films from tetraethylammonium hydroxide peptized titania sols. *Journal of the European Ceramic Society*, **24**(2), 335–339.

[79] Chae, S.Y., Park, M.K., Lee, S.K. *et al.* (2003) Preparation of size-controlled TiO_2 nanoparticles and derivation of optically transparent photocatalytic films. *Chemistry of Materials*, **15**(17), 3326–3331.

[80] Yang, J., Mei, S. and Ferreira, J.M.F. (2003) In situ preparation of weakly flocculated aqueous anatase suspensions by a hydrothermal technique. *Journal of Colloid and Interface Science*, **260**(1), 82–88.

[81] Andersson, M., Osterlund, L., Ljungstrom, S. and Palmqvist, A. (2002) Preparation of nanosize anatase and rutile TiO_2 by hydrothermal treatment of microemulsions and their activity for photocatalytic wet oxidation of phenol. *Journal of Physical Chemistry B*, **106**(41), 10674–10679.

[82] Yang, J., Mei, S. and Ferreira, J.M.F. (2002) Hydrothermal synthesis of well-dispersed TiO_2 nano-crystals. *Journal of Materials Research*, **17**(9), 2197–2200.

[83] Yang, J., Mei, S. and Ferreira, J.M.F. (2001) Hydrothermal synthesis of nanosized titania powders: Influence of tetraalkyl ammonium hydroxides on particle characteristics. *Journal of the American Ceramic Society*, **84**(8), 1696–1702.

[84] Yang, J., Mei, S. and Ferreira, J.M.F. (2000) Hydrothermal synthesis of nanosized titania powders: Influence of peptization and peptizing agents on the crystalline phases and phase transitions. *Journal of the American Ceramic Society*, **83**(6), 1361–1368.

[85] Cot, F., Larbot, A., Nabias, G. and Cot, L. (1998) Preparation and characterization of colloidal solution derived crystallized titania powder. *Journal of the European Ceramic Society*, **18**(14), 2175–2181.

[86] Wang, X., Zhuang, J., Peng, Q. and Li, Y.D. (2005) A general strategy for nanocrystal synthesis. *Nature*, **437** (7055), 121–124.

[87] Kim, C.S., Moon, B.K., Park, J.H. *et al.* (2003) Solvotherinal synthesis of nanocrystalline TiO_2 in toluene with surfactant. *Journal of Crystal Growth*, **257**(3–4), 309–315.

[88] Kim, C.S., Moon, B.K., Park, J.H. *et al.* (2003) Synthesis of nanocrystalline TiO_2 in toluene by a solvothermal route. *Journal of Crystal Growth*, **254**(3–4), 405–410.

[89] Yang, S.W. and Gao, L. (2006) Fabrication and shape-evolution of nanostructured TiO_2 via a sol-solvothermal process based on benzene-water interfaces. *Materials Chemistry and Physics*, **99**(2–3), 437–440.

[90] Meskin, P.E., Ivanov, V.K., Barantchikov, A.E. *et al.* (2006) Ultrasonically assisted hydrothermal synthesis of nanocrystalline ZrO_2, TiO_2, $NiFe_2O_4$ and Ni0.5Zn0.5Fe_2O_4 powders. *Ultrasonics Sonochemistry*, **13**(1), 47–53.

[91] Jokanovic, V., Spasic, A.M. and Uskokovic, D. (2004) Designing of nanostructured hollow TiO_2 spheres obtained by ultrasonic spray pyrolysis. *Journal of Colloid and Interface Science*, **278**(2), 342–352.

[92] Guo, W.L., Lin, Z.M., Wang, X.K. and Song, G.Z. (2003) Sonochemical synthesis of nanocrystalline TiO_2 by hydrolysis of titanium alkoxides. *Microelectronic Engineering*, **66**(1–4), 95–101.

[93] Blesic, M.D., Saponjic, Z.V., Nedeljkovic, J.M. and Uskokovic, D.P. (2002) TiO_2 films prepared by ultrasonic spray pyrolysis of nanosize precursor. *Materials Letters*, **54**(4), 298–302.
[94] Xia, H.S. and Wang, Q. (2002) Ultrasonic irradiation: A novel approach to prepare conductive polyaniline/nanocrystalline titanium oxide composites. *Chemistry of Materials*, **14**(5), 2158–2165.
[95] Yu, J.C., Zhang, L.Z. and Yu, J.G. (2002) Rapid synthesis of mesoporous TiO_2 with high photocatalytic activity by ultrasound-induced agglomeration. *New Journal of Chemistry*, **26**(4), 416–420.
[96] Yu, J.C., Zhang, L.Z. and Yu, J.G. (2002) Direct sonochemical preparation and characterization of highly active mesoporous TiO_2 with a bicrystalline framework. *Chemistry of Materials*, **14**(11), 4647–4653.
[97] Huang, W.P., Tang, X.H., Wang, Y.Q. *et al.* (2000) Selective synthesis of anatase and rutile via ultrasound irradiation. *Chemical Communications*, (15), 1415–1416.
[98] Wang, Y.Q., Tang, X.H., Yin, L.X. *et al.* (2000) Sonochemical synthesis of mesoporous titanium oxide with wormhole-like framework structures. *Advanced Materials*, **12**(16), 1183–1186.
[99] Wang, X.D., Neff, C., Graugnard, E. *et al.* (2005) Photonic crystals fabricated using patterned nanorod arrays. *Advanced Materials*, **17**(17), 2103–2106.
[100] Dong, W.T., Bongard, H.J. and Marlow, F. (2003) New type of inverse opals: Titania with skeleton structure. *Chemistry of Materials*, **15**(2), 568–574.
[101] Dong, W.T., Bongard, H., Tesche, B. and Marlow, F. (2002) Inverse opals with a skeleton structure: Photonic crystals with two complete bandgaps. *Advanced Materials*, **14**(20), 1457–1460.
[102] Holland, B.T., Blanford, C.F., Do, T. and Stein, A. (1999) Synthesis of highly ordered, three-dimensional, macroporous structures of amorphous or crystalline inorganic oxides, phosphates, and hybrid composites. *Chemistry of Materials*, **11**(3), 795–805.
[103] Melde, B.J., Holland, B.T., Blanford, C.F. and Stein, A. (1999) Mesoporous sieves with unified hybrid inorganic/organic frameworks. *Chemistry of Materials*, **11**(11), 3302–3308.
[104] Holland, B.T., Blanford, C.F. and Stein, A. (1998) Synthesis of macroporous minerals with highly ordered three-dimensional arrays of spheroidal voids. *Science*, **281**(5376), 538–540.
[105] Santato, C., Odziemkowski, M., Ulmann, M. and Augustynski, J. (2001) Crystallographically oriented Mesoporous WO_3 films: Synthesis, characterization, and applications. *Journal of the American Chemical Society*, **123**(43), 10639–10649.
[106] Santato, C., Ulmann, M. and Augustynski, J. (2001) Enhanced visible light conversion efficiency using nanocrystalline WO_3 films. *Advanced Materials*, **13**(7), 511–514.
[107] Santato, C., Ulmann, M. and Augustynski, J. (2001) Photoelectrochemical properties of nanostructured tungsten trioxide films. *Journal of Physical Chemistry B*, **105**(5), 936–940.
[108] Solarska, R., Alexander, B.D. and Augustynski, J. (2004) Electrochromic and structural characteristics of mesoporous WO_3 films prepared by a sol-gel method. *Journal of Solid State Electrochemistry*, **8**(10), 748–756.
[109] Solarska, R., Alexander, B.D. and Augustynski, J. (2006) Electrochromic and photoelectrochemical characteristics of nanostructured WO_3 films prepared by a sol-gel method. *Comptes Rendus Chimie*, **9**(2), 301–306.
[110] Spichigerulmann, M. and Augustynski, J. (1983) Aging effects in N-type semiconducting WO_3 films. *Journal of Applied Physics*, **54**(10), 6061–6064.
[111] Cotton, F.A. and Wilkinson, G. (1972) *Advanced Inorganic Chemistry*, 3rd edn, Interscience Publishers.
[112] Bedja, I., Hotchandani, S. and Kamat, P.V. (1993) Photoelectrochemistry of quantized WO_3 colloids – electron storage, electrochromic, and photoelectrochromic effects. *Journal of Physical Chemistry*, **97**(42), 11064–11070.
[113] Wang, H.L., Lindgren, T., He, J.J. *et al.* (2000) Photolelectrochemistry of nanostructured WO_3 thin film electrodes for water oxidation: Mechanism of electron transport. *Journal of Physical Chemistry B*, **104**(24), 5686–5696.
[114] Balazsi, C., Wang, L.S., Zayim, E.O. *et al.* (2008) Nanosize hexagonal tungsten oxide for gas sensing applications. *Journal of the European Ceramic Society*, **28**(5), 913–917.
[115] Huang, K., Pan, Q., Yang, F. *et al.* (2008) Controllable synthesis of hexagonal WO_3 nanostructures and their application in lithium batteries. *Journal of Physics D-Applied Physics*, **41**(15), 155417.
[116] Mo, R.F., Jin, G.Q. and Guo, X.Y. (2007) Hydrothermal synthesis of tungsten trioxides using citric acid as controlling agent. *Chinese Journal of Inorganic Chemistry*, **23**(9), 1615–1620.
[117] Song, S.W. and Kang, I.S. (2008) A facile fabrication route to tungsten oxide films on ceramic substrates. *Sensors and Actuators B-Chemical*, **129**(2), 971–976.

[118] Sun, Q.J., Luo, H.M., Xie, Z.F. *et al.* (2008) Synthesis of monodisperse WO(3 center dot)2H(2)O nanospheres by microwave hydrothermal process with L(+) tartaric acid as a protective agent. *Materials Letters*, **62**(17–18), 2992–2994.

[119] Choi, H.G., Jung, Y.H. and Kim, D.K. (2005) Solvothermal synthesis of tungsten oxide nanorod/nanowire/nanosheet. *Journal of the American Ceramic Society*, **88**(6), 1684–1686.

[120] Kominami, H., Kato, J., Murakami, S. *et al.* (2003) Solvothermal syntheses of semiconductor photocatalysts of ultra-high activities. *Catalysis Today*, **84**(3–4), 181–189.

[121] Hernandez-Sanchez, B.A., Boyle, T.J., Pratt, H.D. *et al.* (2008) Morphological and phase controlled tungsten based nanoparticles: synthesis and characterization of scheelite, wolframite and oxide nanomaterials. *Chemistry of Materials*, **20**(21), 6643–6656.

[122] Koltypin, Y., Nikitenko, S.I. and Gedanken, A. (2002) The sonochemical preparation of tungsten oxide nanoparticles. *Journal of Materials Chemistry*, **12**(4), 1107–1110.

[123] Brezesinski, T., Fattakhova-Rohlfing, D., Sallard, S. *et al.* (2006) Highly crystalline WO_3 thin films with ordered 3D mesoporosity and improved electrochromic performance. *Small*, **2**(10), 1203–1211.

[124] Cui, X.Z., Zhang, H., Dong, X.P. *et al.* (2008) Electrochemical catalytic activity for the hydrogen oxidation of mesoporous WO_3 and WO_3/C composites. *Journal of Materials Chemistry*, **18**(30), 3575–3580.

[125] Ganesan, R. and Gedanken, A. (2008) Synthesis of WO_3 nanoparticles using a biopolymer as a template for electrocatalytic hydrogen evolution. *Nanotechnology*, **19**(2), 025702.

[126] Khatko, V., Gorokh, G., Mozalev, A. *et al.* (2006) Tungsten trioxide sensing layers on highly ordered nanoporous alumina template. *Sensors and Actuators B-Chemical*, **118**(1–2), 255–262.

[127] Lai, W.H., Teoh, L.G., Su, Y.H. *et al.* (2007) Hydrolysis reaction on the characterization of wormhole-like mesoporous tungsten oxide. *Journal of Alloys and Compounds*, **438**(1–2), 247–252.

[128] Rossinyol, E., Arbiol, J., Peiro, F. *et al.* (2005) Nanostructured metal oxides synthesized by hard template method for gas sensing applications. *Sensors and Actuators B-Chemical*, **109**(1), 57–63.

[129] Sadakane, M., Sasaki, K., Kunioku, H. *et al.* (2008) Preparation of nano-structured crystalline tungsten(VI) oxide and enhanced photocatalytic activity for decomposition of organic compounds under visible light irradiation. *Chemical Communications*, (48), 6552–6554.

[130] Dong, L., Liu, Y.C., Tong, Y.H. *et al.* (2005) Preparation of ZnO colloids by aggregation of the nanocrystal subunits. *Journal of Colloid and Interface Science*, **283**(2), 380–384.

[131] Monticone, S., Tufeu, R. and Kanaev, A.V. (1998) Complex nature of the UV and visible fluorescence of Colloidal ZnO nanoparticles. *Journal of Physical Chemistry B*, **102**(16), 2854–2862.

[132] Rodriguez-Gattorno, G., Santiago-Jacinto, P., Rendon-Vazquez, L. *et al.* (2003) Novel synthesis pathway of ZnO nanoparticles from the spontaneous hydrolysis of zinc carboxylate salts. *Journal of Physical Chemistry B*, **107** (46), 12597–12604.

[133] Spanhel, L. (2006) Colloidal ZnO nanostructures and functional coatings: A survey. *Journal of Sol-Gel Science and Technology*, **39**(1), 7–24.

[134] Tokumoto, M.S., Pulcinelli, S.H., Santilli, C.V. and Briois, V. (2003) Catalysis and temperature dependence on the formation of ZnO nanoparticles and of zinc acetate derivatives prepared by the sol-gel route. *Journal of Physical Chemistry B*, **107**(2), 568–574.

[135] van Dijken, A., Makkinje, J. and Meijerink, A. (2001) The influence of particle size on the luminescence quantum efficiency of nanocrystalline ZnO particles. *Journal of Luminescence*, **92**(4), 323–328.

[136] Ismail, A.A., El-Midany, A., Abdel-Aal, E.A. and El-Shall, H. (2005) Application of statistical design to optimize the preparation of ZnO nanoparticles via hydrothermal technique. *Materials Letters*, **59**(14–15), 1924–1928.

[137] Ohara, S., Mousavand, T., Umetsu, M. *et al.* (2004) Hydrothermal synthesis of fine zinc oxide particles under supercritical conditions. *Solid State Ionics*, **172**(1–4), 261–264.

[138] Sue, K., Kimura, K. and Arai, K. (2004) Hydrothermal synthesis of ZnO nanocrystals using microreactor. *Materials Letters*, **58**(25), 3229–3231.

[139] Sue, K., Kimura, K., Murata, K. and Arai, K. (2004) Effect of cations and anions on properties of zinc oxide particles synthesized in supercritical water. *Journal of Supercritical Fluids*, **30**(3), 325–331.

[140] Viswanathan, R. and Gupta, R.B. (2003) Formation of zinc oxide nanoparticles in supercritical water. *Journal of Supercritical Fluids*, **27**(2), 187–193.

[141] Viswanathan, R., Lilly, G.D., Gale, W.F. and Gupta, R.B. (2003) Formation of zinc oxide-titanium dioxide composite nanoparticles in supercritical water. *Industrial & Engineering Chemistry Research*, **42**(22), 5535–5540.

[142] Buha, J., Djerdj, I. and Niederberger, M. (2007) Nonaqueous synthesis of nanocrystalline indium oxide and zinc oxide in the oxygen-free solvent acetonitrile. *Crystal Growth & Design*, **7**(1), 113–116.

[143] Chen, S.J., Li, L.H., Chen, X.T. *et al.* (2003) Preparation and characterization of nanocrystalline zinc oxide by a novel solvothermal oxidation route. *Journal of Crystal Growth*, **252**(1–3), 184–189.

[144] Du, H.C., Yuan, F.L., Huang, S.L. *et al.* (2004) A new reaction to ZnO nanoparticles. *Chemistry Letters*, **33**(6), 770–771.

[145] Ghosh, M., Seshadri, R. and Rao, C.N.R. (2004) A solvothermal route to ZnO and Mn-doped ZnO nanoparticles using the cupferron complex as the precursor. *Journal of Nanoscience and Nanotechnology*, **4**(1–2), 136–140.

[146] Niederberger, M., Garnweitner, G., Pinna, N. and Neri, G. (2005) Non-aqueous routes to crystalline metal oxide nanoparticles: Formation mechanisms and applications. *Progress in Solid State Chemistry*, **33**(2–4), 59–70.

[147] Pinna, N., Garnweitner, G., Antonietti, M. and Niederberger, M. (2005) A general nonaqueous route to binary metal oxide nanocrystals involving a C=C bond cleavage. *Journal of the American Chemical Society*, **127**(15), 5608–5612.

[148] Wang, C.L., Shen, E.H., Wang, E.B. *et al.* (2005) Controllable synthesis of ZnO nanocrystals via a surfactant-assisted alcohol thermal process at a low temperature. *Materials Letters*, **59**(23), 2867–2871.

[149] Goharshadi, E.K., Ding, Y., Jorabchi, M.N. and Nancarrow, P. (2009) Ultrasound-assisted green synthesis of nanocrystalline ZnO in the ionic liquid [hmim][NTf2]. *Ultrasonics Sonochemistry*, **16**(1), 120–123.

[150] Kandjani, A.E., Tabriz, M.F. and Pourabbas, B. (2008) Sonochemical synthesis of ZnO nanoparticles: The effect of temperature and sonication power. *Materials Research Bulletin*, **43**(3), 645–654.

[151] Kumar, R.V., Diamant, Y. and Gedanken, A. (2000) Sonochemical synthesis and characterization of nanometer-size transition metal oxides from metal acetates. *Chemistry of Materials*, **12**(8), 2301–2305.

[152] Kumar, R.V., Elgamiel, R., Koltypin, Y. *et al.* (2003) Synthesis and characterization of a micro scale zinc oxide-PVA composite by ultrasound irradiation and the effect of composite on the crystal growth of zinc oxide. *Journal of Crystal Growth*, **250**(3–4), 409–417.

[153] Qian, D., Jiang, J.Z. and Hansen, P.L. (2003) Preparation of ZnO nanocrystals via ultrasonic irradiation. *Chemical Communications*, (9), 1078–1079.

[154] Yadav, R.S., Mishra, P.A. and Pandey, A.C. (2008) Growth mechanism and optical property of ZnO nanoparticles synthesized by sonochemical method. *Ultrasonics Sonochemistry*, **15**(5), 863–868.

[155] Zhu, Y.C. and Qian, Y.T. (2009) Solution-phase synthesis of nanomaterials at low temperature. *Science in China Series G-Physics Mechanics & Astronomy*, **52**(1), 13–20.

[156] Mulligan, R.F., Iliadis, A.A. and Kofinas, P. (2003) Synthesis and characterization of ZnO nanostructures templated using diblock copolymers. *Journal of Applied Polymer Science*, **89**(4), 1058–1061.

[157] Polarz, S., Orlov, A.V., Schuth, F. and Lu, A.H. (2007) Preparation of high-surface-area zinc oxide with ordered porosity, different pore sizes, and nanocrystalline walls. *Chemistry-A European Journal*, **13**(2), 592–597.

[158] Sun, X.M., Deng, Z.X. and Li, Y.D. (2003) Self-organized growth of ZnO single crystal columns array. *Materials Chemistry and Physics*, **80**(1), 366–370.

[159] Yan, C.L. and Xue, D.F. (2006) Room temperature fabrication of hollow ZnS and ZnO architectures by a sacrificial template route. *Journal of Physical Chemistry B*, **110**(14), 7102–7106.

[160] Yoo, S.I., Sohn, B.H., Zin, W.C. *et al.* (2004) Self-assembled arrays of zinc oxide nanoparticles from monolayer films of diblock copolymer micelles. *Chemical Communications*, (24), 2850–2851.

[161] Lan, D., Zhang, Y.M. and Wang, Y.R. (2008) From hexagonally arrayed nanorods to ordered porous film through controlling the morphology of ZnO crystals. *Applied Surface Science*, **254**(18), 5849–5853.

[162] Pal, E., Oszko, A., Mela, P. *et al.* (2008) Preparation of hexagonally aligned inorganic nanoparticles from diblock copolymer micellar systems. *Colloids and Surfaces A-Physicochemical and Engineering Aspects*, **331**(3), 213–219.

[163] Zhou, Q., Zhao, J.J., Xu, W.W. *et al.* (2008) Formation of two-dimensional ordered cavities of zinc oxide and their confinement effect on electrochemical reactions. *Journal of Physical Chemistry C*, **112**(7), 2378–2381.

[164] Law, M., Greene, L.E., Radenovic, A. *et al.* (2006) ZnO-Al_2O_3 and ZnO-TiO_2 core-shell nanowire dye-sensitized solar cells. *Journal of Physical Chemistry B*, **110**(45), 22652–22663.

[165] Persano, A., Leo, G., Manna, L. and Cola, A. (2008) Charge carrier transport in thin films of colloidal CdSe quantum rods. *Journal of Applied Physics*, **104**(7), 074306.

[166] Tian, B.Z., Zheng, X.L., Kempa, T.J. *et al.* (2007) Coaxial silicon nanowires as solar cells and nanoelectronic power sources. *Nature*, **449**(7164), U885–U888.

[167] Knappenberger, K.L., Wong, D.B., Xu, W. *et al.* (2008) Excitation-Wavelength Dependence of Fluorescence Intermittency in CdSe Nanorods. *ACS Nano*, **2**(10), 2143–2153.

[168] Kazes, M., Lewis, D.Y. and Banin, U. (2004) Method for preparation of semiconductor quantum-rod lasers in a cylindrical microcavity. *Advanced Functional Materials*, **14**(10), 957–962.

[169] Kuykendall, T., Ulrich, P., Aloni, S. and Yang, P. (2007) Complete composition tunability of InGaN nanowires using a combinatorial approach. *Nature Materials*, **6**(12), 951–956.

[170] Steiner, D., Dorfs, D., Banin, U. *et al.* (2008) Determination of band offsets in heterostructured colloidal nanorods using scanning tunneling spectroscopy. *Nano Letters*, **8**(9), 2954–2958.

[171] Huynh, W.U., Dittmer, J.J. and Alivisatos, A.P. (2002) Hybrid nanorod-polymer solar cells. *Science*, **295**(5564), 2425–2427.

[172] Frame, F.A., Carroll, E.C., Larsen, D.S. *et al.* (2008) First demonstration of CdSe as a photocatalyst for hydrogen evolution from water under UV and visible light. *Chemical Communications*, (19), 2206–2208.

[173] Ion, L., Enculescu, I. and Antohe, S. (2008) Physical properties of CdTe nanowires electrodeposited by a template method, for photovoltaic applications. *Journal of Optoelectronics and Advanced Materials*, **10**(12), 3241–3246.

[174] Jayadevan, K.P. and Tseng, T.Y. (2005) One-dimensional semiconductor nanostructures as absorber layers in solar cells. *Journal of Nanoscience and Nanotechnology*, **5**(11), 1768–1784.

[175] Kar, S. and Chaudhuri, S. (2006) Cadmium sulfide one-dimensional nanostructures: Synthesis, characterization and application. *Synthesis and Reactivity in Inorganic Metal-Organic and Nano-Metal Chemistry*, **36**(3), 289–312.

[176] Kislyuk, V.V. and Dimitriev, O.P. (2008) Nanorods and nanotubes for solar cells. *Journal of Nanoscience and Nanotechnology*, **8**(1), 131–148.

[177] Lee, J., Kim, H.J., Chen, T. *et al.* (2009) Control of energy transfer to CdTe nanowires via conjugated polymer orientation. *Journal of Physical Chemistry C*, **113**(1), 109–116.

[178] Min, S.K., Joo, O.S., Jung, K.D. *et al.* (2006) Tubular end-capped electrodeposited CdSe nanofibers: Enhanced photochemistry. *Electrochemistry Communications*, **8**(2), 223–226.

[179] Min, S.K., Joo, O.S., Mane, R.S. *et al.* (2007) CdSe nanofiber based photoelectrochemical cells: Influence of annealing temperatures. *Journal of Photochemistry and Photobiology A-Chemistry*, **187**(1), 133–137.

[180] Attar, A.S., Ghamsari, M.S., Hajiesmaeilbaigi, F. *et al.* (2008) Synthesis and characterization of anatase and rutile TiO_2 nanorods by template-assisted method. *Journal of Materials Science*, **43**(17), 5924–5929.

[181] Attar, A.S., Ghamsari, M.S., Hajiesmaeilbaigi, F. *et al.* (2008) Study on the effects of complex ligands in the synthesis of TiO_2 nanorod arrays using the sol-gel template method. *Journal of Physics D-Applied Physics*, **41** (15), 155318.

[182] Attar, A.S., Ghamsari, M.S., Hajiesmaeilbaigi, F. *et al.* (2009) Sol-gel template synthesis and characterization of aligned anatase-TiO_2 nanorod arrays with different diameter. *Materials Chemistry and Physics*, **113**(2–3), 856–860.

[183] Cozzoli, P.D., Kornowski, A. and Weller, H. (2003) Low-temperature synthesis of soluble and processable organic-capped anatase TiO_2 nanorods. *Journal of the American Chemical Society*, **125**(47), 14539–14548.

[184] Kang, T.S., Smith, A.P., Taylor, B.E. and Durstock, M.F. (2009) Fabrication of highly-ordered TiO_2 nanotube arrays and their use in dye-sensitized solar cells. *Nano Letters*, **9**(2), 601–606.

[185] Limmer, S.J., Seraji, S., Wu, Y. *et al.* (2002) Template-based growth of various oxide nanorods by sol-gel electrophoresis. *Advanced Functional Materials*, **12**(1), 59–64.

[186] Lin, Y., Wu, G.S., Yuan, X.Y. *et al.* (2003) Fabrication and optical properties of TiO_2 nanowire arrays made by sol-gel electrophoresis deposition into anodic alumina membranes. *Journal of Physics-Condensed Matter*, **15** (17), 2917–2922.

[187] Miao, L., Tanemura, S., Toh, S. *et al.* (2004) Heating-sol-gel template process for the growth of TiO_2 nanorods with rutile and anatase structure. *Applied Surface Science*, **238**(1–4), 175–179.

[188] Miao, L., Tanemura, S., Toh, S. *et al.* (2004) Fabrication, characterization and Raman study of anatase TiO_2 nanorods by a heating-sol-gel template process. *Journal of Crystal Growth*, **264**(1–3), 246–252.

[189] Peng, T.Y., Hasegawa, A., Qiu, J.R. and Hirao, K. (2003) Fabrication of titania tubules with high surface area and well-developed mesostructural walls by surfactant-mediated templating method. *Chemistry of Materials*, **15** (10), 2011–2016.

[190] Zhang, M., Bando, Y. and Wada, K. (2001) Sol-gel template preparation of TiO_2 nanotubes and nanorods. *Journal of Materials Science Letters*, **20**(2), 167–170.

[191] Song, X.M., Wu, J.M., Tang, M.Z. *et al.* (2008) Enhanced photoelectrochemical response of a composite titania thin film with single-crystalline rutile nanorods embedded in anatase aggregates. *Journal of Physical Chemistry C*, **112**(49), 19484–19492.

[192] Suprabha, T., Roy, H.G., Thomas, J. *et al.* (2009) Microwave-assisted synthesis of titania nanocubes, nanospheres and nanorods for photocatalytic dye degradation. *Nanoscale Research Letters*, **4**(2), 144–152.

[193] Limmer, S.J. and Cao, G.Z. (2003) Sol-gel electrophoretic deposition for the growth of oxide nanorods. *Advanced Materials*, **15**(5), 427–431.

[194] Feng, X.J., Shankar, K., Varghese, O.K. *et al.* (2008) Vertically aligned single crystal TiO_2 nanowire arrays grown directly on transparent conducting oxide coated glass: synthesis details and applications. *Nano Letters*, **8**(11), 3781–3786.

[195] Allam, N.K., Shankar, K. and Grimes, C.A. (2008) Photoelectrochemical and water photoelectrolysis properties of ordered TiO_2 nanotubes fabricated by Ti anodization in fluoride-free HCl electrolytes. *Journal of Materials Chemistry*, **18**(20), 2341–2348.

[196] Mor, G.K., Varghese, O.K., Wilke, R.H.T. *et al.* (2008) p-Type Cu-Ti-O nanotube arrays and their use in self-biased heterojunction photoelectrochemical diodes for hydrogen generation (vol 8, pg 1906, 2008). *Nano Letters*, **8**(10), 3555–3555.

[197] Mor, G.K., Varghese, O.K., Wilke, R.H.T. *et al.* (2008) p-type Cu-Ti-O nanotube arrays and their use in self-biased heterojunction photoelectrochemical diodes for hydrogen generation. *Nano Letters*, **8**(7), 1906–1911.

[198] Paulose, M., Prakasam, H.E., Varghese, O.K. *et al.* (2007) TiO_2 nanotube arrays of 1000 mu m length by anodization of titanium foil: Phenol red diffusion. *Journal of Physical Chemistry C*, **111**(41), 14992–14997.

[199] Prakasam, H.E., Shankar, K., Paulose, M. *et al.* (2007) A new benchmark for TiO_2 nanotube array growth by anodization. *Journal of Physical Chemistry C*, **111**(20), 7235–7241.

[200] Cai, Q.Y., Paulose, M., Varghese, O.K. and Grimes, C.A. (2005) The effect of electrolyte composition on the fabrication of self-organized titanium oxide nanotube arrays by anodic oxidation. *Journal of Materials Research*, **20**(1), 230–236.

[201] Mor, G.K., Varghese, O.K., Paulose, M. and Grimes, C.A. (2005) Transparent highly ordered TiO_2 nanotube arrays via anodization of titanium thin films. *Advanced Functional Materials*, **15**(8), 1291–1296.

[202] Mor, G.K., Shankar, K., Paulose, M. *et al.* (2005) Enhanced photocleavage of water using titania nanotube arrays. *Nano Letters*, **5**(1), 191–195.

[203] Paulose, M., Mor, G.K., Varghese, O.K. *et al.* (2006) Visible light photoelectrochemical and water-photoelectrolysis properties of titania nanotube arrays. *Journal of Photochemistry and Photobiology A-Chemistry*, **178**(1), 8–15.

[204] Shankar, K., Mor, G.K., Prakasam, H.E. *et al.* (2007) Highly-ordered TiO_2 nanotube arrays up to 220 mu m in length: use in water photoelectrolysis and dye-sensitized solar cells. *Nanotechnology*, **18**(6), 065707.

[205] Gillet, M., Delamare, R. and Gillet, E. (2007) Growth, structure and electrical conduction of WO_3 nanorods. *Applied Surface Science*, **254**(1), 270–273.

[206] Gillet, M., Masek, K., Potin, V. *et al.* (2008) An epitaxial hexagonal tungsten bronze as precursor for WO_3 nanorods on mica. *Journal of Crystal Growth*, **310**(14), 3318–3324.

[207] Gu, Z.J., Zhai, T.Y., Gao, B.F. *et al.* (2006) Controllable assembly of WO_3 nanorods/nanowires into hierarchical nanostructures. *Journal of Physical Chemistry B*, **110**(47), 23829–23836.

[208] Maiyalagan, T. and Viswanathan, B. (2008) Catalytic activity of platinum/tungsten oxide nanorod electrodes towards electro-oxidation of methanol. *Journal of Power Sources*, **175**(2), 789–793.

[209] Mo, R.F., Jin, G.Q. and Guo, X.Y. (2007) Morphology evolution of tungsten trioxide nanorods prepared by an additive-free hydrothermal route. *Materials Letters*, **61**(18), 3787–3790.

[210] Pol, S.V., Pol, V.G., Kessler, V.G. *et al.* (2005) Synthesis of WO_3 nanorods by reacting WO(OMe)(4) under autogenic pressure at elevated temperature followed by annealing. *Inorganic Chemistry*, **44**(26), 9938–9945.

[211] Rajeswari, J., Viswanathan, B. and Varadarajan, T.K. (2007) Tungsten trioxide nanorods as supports for platinum in methanol oxidation. *Materials Chemistry and Physics*, **106**(2–3), 168–174.

[212] Song, X.C., Wang, Y., Lin, S. and Yang, E. (2007) Selected-control synthesis and photoluminescence properties of tungsten oxide nanowires and nanoparticles. *Rare Metal Materials and Engineering*, **36**, 9–12.

[213] Song, X.C., Zheng, Y.F., Yang, E. and Wang, Y. (2007) Large-scale hydrothermal synthesis Of WO_3 nanowires in the presence of K_2SO_4. *Materials Letters*, **61**(18), 3904–3908.

[214] Sun, S.B., Zhao, Y.M., Xia, Y.D. *et al.* (2008) Bundled tungsten oxide nanowires under thermal processing. *Nanotechnology*, **19**(30), 305709.

[215] Tian, Z.R.R., Voigt, J.A., Liu, J. *et al.* (2003) Large oriented arrays and continuous films of TiO_2-based nanotubes. *Journal of the American Chemical Society*, **125**(41), 12384–12385.

[216] Hendry, E., Koeberg, M., O'Regan, B. and Bonn, M. (2006) Local field effects on electron transport in nanostructured TiO_2 revealed by terahertz spectroscopy. *Nano Letters*, **6**(4), 755–759.

[217] Kaidashev, E.M., Lorenz, M., von Wenckstern, H. *et al.* (2003) High electron mobility of epitaxial ZnO thin films on c-plane sapphire grown by multistep pulsed-laser deposition. *Applied Physics Letters*, **82**(22), 3901–3903.

[218] Tian, Z.R.R., Voigt, J.A., Liu, J. *et al.* (2003) Complex and oriented ZnO nanostructures. *Nature Materials*, **2** (12), 821–826.

[219] Greene, L.E., Law, M., Yuhas, B.D. and Yang, P.D. (2007) ZnO-TiO_2 core-shell nanorod/P3HT solar cells. *Journal of Physical Chemistry C*, **111**(50), 18451–18456.

[220] Law, M., Greene, L.E., Johnson, J.C. *et al.* (2005) Nanowire dye-sensitized solar cells. *Nature Materials*, **4**(6), 455–459.

[221] Greene, L.E., Law, M., Goldberger, J. *et al.* (2003) Low-temperature wafer-scale production of ZnO nanowire arrays. *Angewandte Chemie-International Edition*, **42**(26), 3031–3034.

[222] Greene, L.E., Law, M., Tan, D.H. *et al.* (2005) General route to vertical ZnO nanowire arrays using textured ZnO seeds. *Nano Letters*, **5**(7), 1231–1236.

[223] Sounart, T.L., Liu, J., Voigt, J.A. *et al.* (2006) Sequential nucleation and growth of complex nanostructured films. *Advanced Functional Materials*, **16**(3), 335–344.

[224] Hsu, J.W.P., Tian, Z.R., Simmons, N.C. *et al.* (2005) Directed spatial organization of zinc oxide nanorods. *Nano Letters*, **5**(1), 83–86.

[225] Tian, Z.R.R., Voigt, J.A., Liu, J. *et al.* (2002) Biommetic arrays of oriented helical ZnO nanorods and columns. *Journal of the American Chemical Society*, **124**(44), 12954–12955.

[226] Vayssieres, L., Keis, K., Hagfeldt, A. and Lindquist, S.E. (2001) Three-dimensional array of highly oriented crystalline ZnO microtubes. *Chemistry of Materials*, **13**(12), 4395–4398.

[227] Vayssieres, L., Hagfeldt, A. and Lindquist, S.E. (2000) Purpose-built metal oxide nanomaterials. The emergence of a new generation of smart materials. *Pure and Applied Chemistry*, **72**(1–2), 47–52.

[228] Vayssieres, L., Keis, K., Lindquist, S.E. and Hagfeldt, A. (2001) Purpose-built anisotropic metal oxide material: 3D highly oriented microrod array of ZnO. *Journal of Physical Chemistry B*, **105**(17), 3350–3352.

[229] Guo, J.H., Vayssieres, L., Persson, C. *et al.* (2002) Polarization-dependent soft-x-ray absorption of highly oriented ZnO microrod arrays. *Journal of Physics-Condensed Matter*, **14**(28), 6969–6974.

[230] Keis, K., Vayssieres, L., Rensmo, H. *et al.* (2001) Photoelectrochemical properties of nano- to microstructured ZnO electrodes. *Journal of the Electrochemical Society*, **148**(2), A149–A155.

[231] Yang, Y., Sun, X.W., Tay, B.K. *et al.* (2008) On the fabrication of resistor-shaped ZnO nanowires. *Physica E-Low-Dimensional Systems & Nanostructures*, **40**(4), 859–865.

[232] Keis, K., Vayssieres, L., Lindquist, S.E. and Hagfeldt, A. (1999) Nanostructured ZnO electrodes for photovoltaic applications. *Nanostructured Materials*, **12**(1–4), 487–490.

[233] Wang, J.X., Sun, X.W., Yang, Y. *et al.* (2006) Hydrothermally grown oriented ZnO nanorod arrays for gas sensing applications. *Nanotechnology*, **17**(19), 4995–4998.

[234] Vayssieres, L. (2003) Growth of arrayed nanorods and nanowires of ZnO from aqueous solutions. *Advanced Materials*, **15**(5), 464–466.

[235] Yi, J.B., Pan, H., Lin, J.Y. *et al.* (2008) Ferromagnetism in ZnO nanowires derived from electro-deposition on AAO template and subsequent oxidation. *Advanced Materials*, **20**(6), 1170–1174.

[236] Cheng, C.L., Chen, C.C., Lin, H.Y. *et al.* (2008) Patterned growth of ZnO nanostructures based on the templation of plant cell walls. *Journal of Nanoscience and Nanotechnology*, **8**(12), 6344–6348.

[237] Meng, A.L., Lin, Y.S. and Wang, G.X. (2005) Preparation of ZnO nanowires by electrochemical method. *Chinese Journal of Inorganic Chemistry*, **21**(4), 583–587.

[238] Kajbafvala, A., Shayegh, M.R., Mazloumi, M. *et al.* (2009) Nanostructure sword-like ZnO wires: Rapid synthesis and characterization through a microwave-assisted route. *Journal of Alloys and Compounds*, **469**(1–2), 293–297.

[239] Wang, X.D., Summers, C.J. and Wang, Z.L. (2004) Large-scale hexagonal-patterned growth of aligned ZnO nanorods for nano-optoelectronics and nanosensor arrays. *Nano Letters*, **4**(3), 423–426.

[240] Greyson, E.C., Babayan, Y. and Odom, T.W. (2004) Directed growth of ordered arrays of small-diameter ZnO nanowires. *Advanced Materials*, **16**(15), 1348–1352.

[241] Li, Y., Meng, G.W., Zhang, L.D. and Phillipp, F. (2000) Ordered semiconductor ZnO nanowire arrays and their photoluminescence properties. *Applied Physics Letters*, **76**(15), 2011–2013.

[242] Wang, J.G., Tian, M.L., Kumar, N. and Mallouk, T.E. (2005) Controllable template synthesis of superconducting Zn nanowires with different microstructures by electrochemical deposition. *Nano Letters*, **5**(7), 1247–1253.

[243] Novoselov, K.S., Geim, A.K., Morozov, S.V. *et al.* (2005) Two-dimensional gas of massless Dirac fermions in graphene. *Nature*, **438**(7065), 197–200.

[244] Novoselov, K.S., Geim, A.K., Morozov, S.V. *et al.* (2004) Electric field effect in atomically thin carbon films. *Science*, **306**(5696), 666–669.

[245] Sasaki, T., Ebina, Y., Kitami, Y. *et al.* (2001) Two-dimensional diffraction of molecular nanosheet crystallites of titanium oxide. *Journal of Physical Chemistry B*, **105**(26), 6116–6121.

[246] Kijima, N., Takahashi, Y., Hayakawa, H. *et al.* (2008) Synthesis, characterization, and electrochemical properties of a thin flake titania fabricated from exfoliated nanosheets. *Journal of Physics and Chemistry of Solids*, **69**(5–6), 1447–1449.

[247] Yoon, M., Seo, M., Jeong, C. *et al.* (2005) Synthesis of liposome-templated titania nanodisks: Optical properties and photocatalytic activities. *Chemistry of Materials*, **17**(24), 6069–6079.

[248] Wei, M.D., Konishi, Y. and Arakawa, H. (2007) Synthesis and characterization of nanosheet-shaped titanium dioxide. *Journal of Materials Science*, **42**(2), 529–533.

[249] Wen, P.H., Itoh, H., Tang, W.P. and Feng, Q. (2007) Single nanocrystals of anatase-type TiO_2 prepared from layered titanate nanosheets: Formation mechanism and characterization of surface properties. *Langmuir*, **23** (23), 11782–11790.

[250] Sasaki, T. and Watanabe, M. (1997) Semiconductor nanosheet crystallites of quasi-TiO_2 and their optical properties. *Journal of Physical Chemistry B*, **101**(49), 10159–10161.

[251] Sakai, N., Fukuda, K., Shibata, T. *et al.* (2006) Photoinduced hydrophilic conversion properties of titania nanosheets. *Journal of Physical Chemistry B*, **110**(12), 6198–6203.

[252] Shibata, T., Sakai, N., Fukuda, K. *et al.* (2007) Photocatalytic properties of titania nanostructured films fabricated from titania nanosheets. *Physical Chemistry Chemical Physics*, **9**(19), 2413–2420.

[253] Umemura, Y., Shinohara, E., Koura, A. *et al.* (2006) Photocatalytic decomposition of an alkylammonium cation in a Langmuir-Blodgett film of a titania nanosheet. *Langmuir*, **22**(8), 3870–3877.

[254] Shibata, T., Fukuda, K., Ebina, Y. *et al.* (2008) One-nanometer-thick seed layer of unilamellar nanosheets promotes oriented growth of oxide crystal films. *Advanced Materials*, **20**(2), 231–235.

[255] Wu, B.H., Guo, C.Y., Zheng, N.F. *et al.* (2008) Nonaqueous production of nanostructured anatase with high-energy facets. *Journal of the American Chemical Society*, **130**(51), 17563–17567.

[256] Rhee, C.H., Kim, Y., Lee, J.S. *et al.* (2006) Nanocomposite membranes of surface-sulfonated titanate and Nafion (R) for direct methanol fuel cells. *Journal of Power Sources*, **159**(2), 1015–1024.

[257] Osada, M., Akatsuka, K., Ebina, Y. *et al.* (2008) Langmuir-Blodgett fabrication of nanosheet-based dielectric films without an interfacial dead layer. *Japanese Journal of Applied Physics*, **47**(9), 7556–7560.

[258] Matsuda, A., Matoda, T., Kogure, T. *et al.* (2005) Formation and characterization of titania nanosheet-precipitated coatings via sol-gel process with hot water treatment under vibration. *Chemistry of Materials*, **17**(4), 749–757.

[259] Sugimoto, W., Terabayashi, O., Murakami, Y. and Takasu, Y. (2002) Electrophoretic deposition of negatively charged tetratitanate nanosheets and transformation into preferentially oriented TiO_2(B) film. *Journal of Materials Chemistry*, **12**(12), 3814–3818.

[260] Sakai, N., Ebina, Y., Takada, K. and Sasaki, T. (2004) Electronic band structure of titania semiconductor nanosheets revealed by electrochemical and photoelectrochemical studies. *Journal of the American Chemical Society*, **126**(18), 5851–5858.

[261] Wang, F., Jiu, J., Pei, L. *et al.* (2007) Effect of nitrate ion on formation of TiO_2 nanoplate structure in hydrothermal solution. *Materials Letters*, **61**(2), 488–490.

[262] Jang, J.S., Choi, S.H., Park, H. *et al.* (2006) A composite photocatalyst of CdS nanoparticles deposited on TiO_2 nanosheets. *Journal of Nanoscience and Nanotechnology*, **6**(11), 3642–3646.

[263] Yamaki, T. and Asai, K. (2001) Alternate multilayer deposition from ammonium amphiphiles and titanium dioxide crystalline nanosheets using the Langmuir-Blodgett technique. *Langmuir*, **17**(9), 2564–2567.

[264] Peng, C.W., Ke, T.Y., Brohan, L. *et al.* (2008) (101)-exposed anatase TiO_2 nanosheets. *Chemistry of Materials*, **20**(7), 2426–2428.

[265] Ulman, A. (1991) *Introduction to Ultrathin Organic Films: From Langmuir-Blodgett to Self-Assembly*, Academic Press, Boston.

[266] Petty, M.C. (1996) *Langmuir-Blodgett Films: An Introduction*, Cambridge University Press, New York.

[267] Wolcott, A., Kuykendall, T.R., Chen, W. *et al.* (2006) Synthesis and characterization of ultrathin WO_3 nanodisks utilizing long-chain poly(ethylene glycol). *Journal of Physical Chemistry B*, **110**(50), 25288–25296.

[268] Wang, J.M., Lee, P.S. and Ma, J. (2009) Synthesis, growth mechanism and room-temperature blue luminescence emission of uniform WO_3 nanosheets with W as starting material. *Journal of Crystal Growth*, **311**(2), 316–319.

[269] Chen, D.L., Gao, L., Yasumori, A. *et al.* (2008) Size- and shape-controlled conversion of tungstate-based inorganic-organic hybrid belts to WO_3 nanoplates with high specific surface areas. *Small*, **4**(10), 1813–1822.

[270] Chen, D.L., Wang, H.L., Zhang, R. *et al.* (2008) Single-crystalline tungsten oxide nanoplates. *Journal of Ceramic Processing Research*, **9**(6), 596–600.

[271] Oaki, Y. and Imai, H. (2006) Room-temperature aqueous synthesis of highly luminescent BaWO4-polymer nanohybrids and their spontaneous conversion to hexagonal WO_3 nanosheets. *Advanced Materials*, **18**(14), 1807–1811.

[272] Cho, S., Jung, S.H. and Lee, K.H. (2008) Morphology-controlled growth of ZnO nanostructures using microwave irradiation: from basic to complex structures. *Journal of Physical Chemistry C*, **112**(33), 12769–12776.

[273] Pradhan, D. and Leung, K.T. (2008) Controlled growth of two-dimensional and one-dimensional ZnO nanostructures on indium tin oxide coated glass by direct electrodeposition. *Langmuir*, **24**(17), 9707–9716.

[274] Kim, C., Kim, Y.J., Jang, E.S. *et al.* (2006) Whispering-gallery-modelike-enhanced emission from ZnO nanodisk. *Applied Physics Letters*, **88**(9), 093104.

[275] Long, T.F., Yin, S., Takabatake, K. *et al.* (2009) Synthesis and characterization of ZnO nanorods and nanodisks from zinc chloride aqueous solution. *Nanoscale Research Letters*, **4**(3), 247–253.

[276] Reeja-Jayan, B., De la Rosa, E., Sepulveda-Guzman, S. *et al.* (2008) Structural characterization and luminescence of porous single crystalline ZnO nanodisks with sponge-like morphology. *Journal of Physical Chemistry C*, **112**(1), 240–246.

[277] Yan, Y.F., Liu, P., Wen, J.G. *et al.* (2003) In-situ formation of ZnO nanobelts and metallic Zn nanobelts and nanodisks. *Journal of Physical Chemistry B*, **107**(36), 9701–9704.

[278] Fu, M., Zhou, J., Xiao, Q.F. *et al.* (2006) ZnO nanosheets with ordered pore periodicity via colloidal crystal template assisted electrochemical deposition. *Advanced Materials*, **18**(8), 1001–1004.

[279] Pan, A.L., Yu, R.C., Xie, S.S. *et al.* (2005) ZnO flowers made up of thin nanosheets and their optical properties. *Journal of Crystal Growth*, **282**(1–2), 165–172.

[280] Cao, B.Q., Cai, W.P., Li, Y. *et al.* (2005) Ultraviolet-light-emitting ZnO nanosheets prepared by a chemical bath deposition method. *Nanotechnology*, **16**(9), 1734–1738.

[281] Kar, S., Dev, A. and Chaudhuri, S. (2006) Simple solvothermal route to synthesize ZnO nanosheets, nanonails, and well-aligned nanorod arrays. *Journal of Physical Chemistry B*, **110**(36), 17848–17853.

[282] Mo, M., Yu, J.C., Zhang, L.Z. and Li, S.K.A. (2005) Self-assembly of ZnO nanorods and nanosheets into hollow microhemispheres and microspheres. *Advanced Materials*, **17**(6), 756–760.

[283] Liu, B., Yu, S.H., Zhang, F. *et al.* (2004) Ring-like nanosheets standing on spindle-like rods: Unusual ZnO superstructures synthesized from a flakelike precursor Zn-5(OH)(8)Cl-2 center dot H_2O. *Journal of Physical Chemistry B*, **108**(14), 4338–4341.

[284] Imai, H., Iwai, S. and Yamabi, S. (2004) Phosphate-mediated ZnO nanosheets with a mosaic structure. *Chemistry Letters*, **33**(6), 768–769.

[285] Cao, J.M., Wang, J., Fang, B.Q. *et al.* (2004) Microwave-assisted synthesis of flower-like ZnO nanosheet aggregates in a room-temperature ionic liquid. *Chemistry Letters*, **33**(10), 1332–1333.

[286] Kuo, C.L., Kuo, T.J. and Huang, M.H. (2005) Hydrothermal synthesis of ZnO microspheres and hexagonal microrods with sheetlike and platelike nanostructures. *Journal of Physical Chemistry B*, **109**(43), 20115–20121.

[287] Qian, H.S., Yu, S.H., Gong, J.Y. *et al.* (2005) Growth of ZnO crystals with branched spindles and prismatic whiskers from Zn-3(OH)(2)V2O7 center dot H_2O nanosheets by a hydrothermal route. *Crystal Growth & Design*, **5**(3), 935–939.

[288] Zhan, J.H., Bando, Y., Hu, J.Q. *et al.* (2006) Fabrication of ZnO nanoplate-nanorod junctions. *Small*, **2**(1), 62–65.

[289] Hosono, E., Fujihara, S., Honna, I. and Zhou, H.S. (2005) The fabrication of an upright-standing zinc oxide nanosheet for use in dye-sensitized solar cells. *Advanced Materials*, **17**(17), 2091–2094.

16

Nanoparticle-Assembled Catalysts for Photochemical Water Splitting

Frank E. Osterloh
Department of Chemistry, University of California, Davis, California, USA, Email: osterloh@chem.ucdavis.edu

16.1 Introduction

Over 130 materials are known to catalyze the conversion of water to hydrogen and oxygen (Equation 16.1) [1], but most of them are plagued by either low energy efficiency, inadequate light absorption in the visible range, or material instability under catalytic conditions.

$$\begin{aligned} &H_2O \rightarrow {}^{1}\!/_{2}O_{2(g)} + H_{2(g)} \\ &\Delta G = +237\,\text{kJmol}^{-1}(1.3\,\text{eVe}^{-1},\ \lambda_{\text{min}} = 1100\,\text{nm}) \end{aligned} \tag{16.1}$$

These issues can potentially be solved with nanostructured catalysts that contain separate components for light absorption, water oxidation and water reduction. By building these structures in modular fashion from separate, preformed nanoparticles, it should be possible to optimize the catalytic function by adjusting the discrete properties of the various components. The working cycle of a hypothetical three-component catalyst is illustrated in Figure 16.1a. As the first step, bandgap absorption of a photon produces an electron–hole pair in the semiconductor (step 1 in Figure 16.1a). The pair becomes separated at the nano-interface, with the electron being guided (step 2) along a path of decreasing energy to the metal particle, where it enters an energy level above the Fermi level and becomes available as a reducing agent for water (step 3). If this step is fast enough ($<50\,\text{ps}$), it will prevent recombination of the electron–hole pair – a limiting factor for water-splitting photocatalysts. Subsequently, an electron from the valence band of the metal-oxide particle is injected into

On Solar Hydrogen & Nanotechnology Edited by Lionel Vayssieres

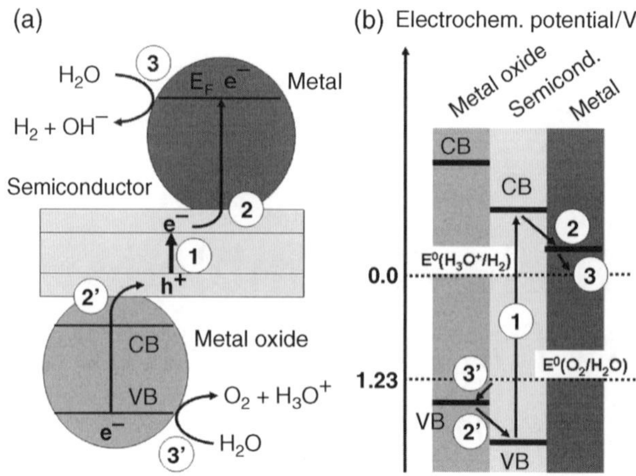

Figure 16.1 (a) Schematic structure of the catalyst showing the connectivity of the semiconductor, metal and metal-oxide nanocrystals. (b) Energy scheme for the three-component structure and desirable charge-transfer pathways (arrows). Note that electron transport from the metal-oxide particle occurs from the valence band of the particle without the need for absorption of a second photon.

the valence band of the semiconductor (step 2′). If this step is fast enough, it can potentially prevent photocorrosion of the semiconductor – a reaction that is known to decompose metal chalcogenides and nitrides. Finally, the metal-oxide nanocrystal carries out a one-electron oxidation of water (step 3′) with the concomitant formation of a water-oxidation intermediate, which is stabilized on the surface of the metal oxide. Overall, two cycles of the machine are required to split one molecule of water, and four cycles to evolve one molecule of O_2 and two molecules of H_2.

The proper function of the three-component nanostructure depends on our ability to engineer its electronic properties so that electrons and holes are guided along a path of decreasing energy (Figure 16.1b) from the semiconductor to the sites of the redox reactions. To achieve this, the energy gap of the semiconductor needs to be large enough to provide the driving force for guiding electrons and holes to the proper water reaction sites, but sufficiently small (<3.0 eV) to allow absorption of light in the visible region of the spectrum.

Potentially nanocrystals have several advantages over conventional catalysts. First, quantum size effects allow one to tailor the bandgaps and band potentials of the components, which is a condition for optimizing charge transport, as described above. Secondly, due to the small dimensions of the nanoparticles, charge-transport pathways are short, reducing energy loss due to electrical resistance. Third, defect concentrations for nanocrystals are also known to be low, which reduces the number of charge recombination sites. Finally, the surface-to-volume ratio is large, enabling maximum surface reaction rate with the minimum amount of material.

On the other hand, nanoparticulate catalysts also have significant disadvantages. Due to their higher surface energy, they are less chemically stable than bulk materials, which can reduce their lifetime as catalysts. Second, in order to be handled in solution phase, nanocrystals require surfactants that may interfere with the catalytic properties of the particles. Also, the fabrication of nanostructured composites is a challenge. For scalability, a solution-based bottom-up approach is preferable over surface-based methods, for example, lithography or vapor–liquid–solid growth, which produce smaller amounts of catalyst. However, solution assembly has lower selectivity and may lead to low yields of active catalysts.

Here we show that two-component nanostructures according to Figure 16.1a can be synthesized effectively using a bottom-up nanostructure assembly approach in solution phase, and that the resulting nanostructures are active catalysts for H_2 evolution from water or from aqueous solutions of mild reducing agents [8–12]. After describing synthesis and functional aspects of catalysts assembled from niobate nanosheets and pre-fabricated IrO_2 and Pt colloids, we demonstrate in Section 16.3, that quantum size effects can be utilized to activate CdSe as a catalyst for H_2 evolution from aqueous sulfide/sulfite solutions. Section 16.4 then summarizes problems of nanoparticle-assembled catalysts and discusses future research directions.

16.2 Two-Component Catalysts

16.2.1 Synthetic and Structural Aspects

Due to their sheet-like morphology, sheet-like nanocrystals are suitable building blocks for the construction of multicomponent nanostructured catalysts. For nanosheets derived from the layered niobate $KCa_2Nb_3O_{10}$, we have developed a protocol that allows other nanoparticles to be grown on or linked to the nanosheets (Figure 16.2A). The nanosheets are obtained by exfoliating the Dion–Jacobsen phase $KCa_2Nb_3O_{10}$ with tetrabutylammonium (TBA) hydroxide after reaction with hydrochloric acid [11]. Each nanosheet is ~1 nm thick and consists of triple stacks of edge-shared NbO_6 with intercalated Ca^{2+} ions, as shown in Figure 16.2B and C. At pH <8 protonation converts surface Nb–O^- groups into Nb–OH sites. The latter can react with alkylsubstituted alkoxysilanes, for example 3-aminopropyltrimethoxysilane, APS, or N-(2-aminoethyl)-11-aminoundecyltrimethoxysilane, AEAUS as shown in Figure 16.2A. This produces aminoalkyl-terminated nanosheets that contain 0.43 ± 0.06 linker molecules per $[HCa_2Nb_3O_{10}]$ unit [13]. The reaction is accompanied by partial restacking of the nanosheets (transmission electron microscopy (TEM) images in Figure 16.2C), probably resulting from H-bonding between the primary amines. Subsequently Pt nanoparticles can be attached to the sheets, either by photochemical growth (Figure 16.2G) or by reaction with preformed citrate-coated Pt nanoparticles (Figure 16.2H). In both cases, Pt particles are homogeneously distributed over the entire sheet surface, with ~9000 Pt nanoparticles per μm^2. Without the linker, no particle attachment to the nanosheets takes place. Photochemical deposition in the absence of the linker leads to large clustered Pt nanoparticles at the periphery of the sheets (Figure 16.2F). Nanosheets coated with iridium-dioxide particles can be synthesized similarly by direct reaction of the amino-terminated sheets with dispersion of citrate-coated IrO_2 nanoparticles [14,15]. Due to H-bonding, the IrO_2 forms clusters on the nanosheet surface (Figure 16.2F) [15].

The optical properties (Figure 16.3A) of the nanoparticle aggregates are mainly the sum of those of the building blocks. Band-edge absorption begins at 350 nm for all species, in accordance with a bandgap of 3.5 eV for $TBA[Ca_2Nb_3O_{10}]$ [16]. The compound IrO_2(cit)-APS-$[Ca_2Nb_3O_{10}]$ has a faint blue appearance due to a broad absorption band at 650 nm from the IrO_2(cit) nanoparticles (*d–d* transition) [14,17]. The compounds Pt(cit)-APS-$[Ca_2Nb_3O_{10}]$ appear brown from a broad absorption at 380–580 nm due to the Pt nanoparticles. The photochemically deposited Pt-$[Ca_2Nb_3O_{10}]$ lacks this absorption and looks metallic gray, because of the larger size of the Pt particles.

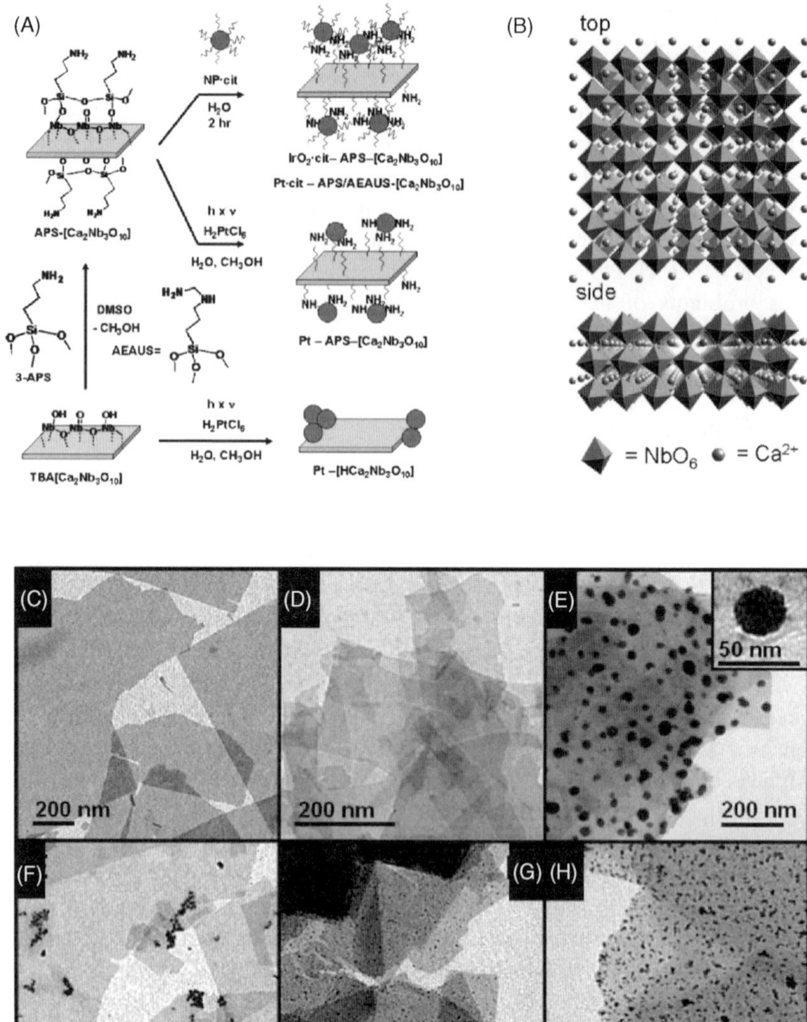

Figure 16.2 (A) Assembly of nanostructured photocatalysts. (B) Structure of $[HCa_2Nb_3O_{10}]$ nanosheet (top and side view). TEM images of (C) exfoliated $[HCa_2Nb_3O_{10}]$ nanosheets, (D) APS-$[Ca_2Nb_3O_{10}]$, (E) IrO_2(cit)-APS-$[Ca_2Nb_3O_{10}]$ with magnified IrO_2 cluster in inset, (F) Pt-$[HCa_2Nb_3O_{10}]$, (G) Pt-APS-$[Ca_2Nb_3O_{10}]$, (H) Pt(cit)-APS-$[Ca_2Nb_3O_{10}]$. (Reprinted with permission from O.C. Compton, C.H. Mullet, S. Chiang and F.E. Osterloh, A building block approach towards photochemical water splitting catalysts based on niobate nanosheets, *Journal of Physical Chemistry C*, **112** (15), 6202–6208, 2008. © 2008 American Chemical Society.)

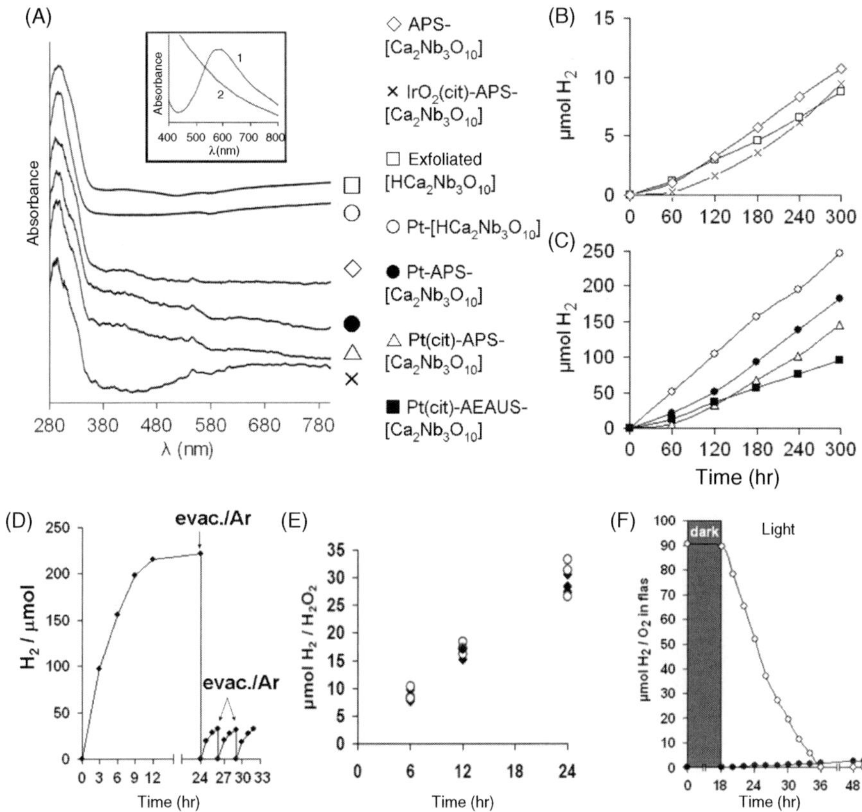

Figure 16.3 (A) Diffuse reflectance UV-visible spectra for films of the catalysts (see legend), Inset: Absorption UV-visible spectra of: (1) IrO_2(cit) colloid in solution, (2) Pt(cit) colloid in solution. (B) and (C) H_2 evolution data for various catalysts (see legend), (D) Rate of H_2 evolution from 100 mg Pt-[$HCa_2Nb_3O_{10}$] as a function of time. Arrows indicate evacuation of the sample to dispersion to 20 torr followed by purging with Ar gas. (E) Measured amounts of H_2 (◆) and H_2O_2 (○) after various irradiation times. (F) Temporal evolution of H_2 (•) and removal of added O_2 (○) under irradiation. (Reprinted with permission from O.C. Compton, C.H. Mullet, S. Chiang and F.E. Osterloh, A building block approach towards photochemical water splitting catalysts based on niobate nanosheets, *Journal of Physical Chemistry C*, **112** (15), 6202–6208, 2008; O.C. Compton and F.E. Osterloh, Niobate nanosheets as catalysts for photochemical water splitting into hydrogen and hydrogen peroxide. *Journal of Physical Chemistry C*, **113**, (1), 479–485, 2009. © 2008, 2009 American Chemical Society.)

16.2.2 Photocatalytic Hydrogen Evolution

Under irradiation with light from a 4 × 175 W Hg illumination system, all catalysts evolve hydrogen from pure water (Figure 16.3 B/C and Table 16.1). The suspension of the TBA-[$Ca_2Nb_3O_{10}$] nanosheets evolves H_2 at a rate of 1.75 $\mu mol\,h^{-1}$ (QE = 0.17%) at pH 10.8, an activity close to the parent phase $KCa_2Nb_3O_{10}$ [18].

Table 16.1 H_2 evolution data. (Reprinted with permission from O.C. Compton, C.H. Mullet, S. Chiang and F.E. Osterloh, A building block approach towards photochemical water splitting catalysts based on niobate nanosheets, *Journal of Physical Chemistry C*, **112**(15), 6202–6208, 2008. © 2008 American Chemical Society.)

	$[HCa_2Nb_3O_{10}]$	APS-$[Ca_2Nb_3O_{10}]$	Pt-$[HCa_2Nb_3O_{10}]$	Pt-APS-$[Ca_2Nb_3O_{10}]$	Pt(cit)-APS-$[Ca_2Nb_3O_{10}]$	Pt(cit)-AEAUS-$[Ca_2Nb_3O_{10}]$	IrO_2(cit)-APS-$[Ca_2Nb_3O_{10}]$
pH	10.8	10.2	10.6	9.9	9.9	10.1	9.9
H_2 [μmol] after 5 h	8.77	10.77	245.73	182.84	145.50	96.51	9.42
H_2 rate [μmol h^{-1}]	1.75	2.15	49.15	36.57	29.10	19.30	1.88
QE [%]	0.17	0.21	4.69	3.49	2.78	1.84	0.18

Conditions: 100 ml of pure water, 100 mg of catalyst, four 175 W Hg arc lamps.

Functionalization with APS slightly increases the catalytic activity to 2.15 μmol h^{-1} (QE = 0.21%) at pH 10.2. This indicates that the primary NH_2 groups of the linker molecules can act as sacrificial electron donors in this system. A large increase in activity occurs when 3 wt% Pt nanoparticles are directly grown onto the nanosheets (49.2 μmol h^{-1}, QE 4.69%) [16]. These particles act as efficient water-reduction sites. The activity of photochemically grown Pt-APS-[$Ca_2Nb_3O_{10}$] (36.6 μmol h^{-1}, QE = 3.49% at pH 9.9) is 25% lower than this, despite equal or higher Pt content. This suggests that the linker diminishes electronic contact between the metal and the semiconductor. Indeed, when APS is replaced with AEAUS (a longer chain) the activity drops further to 19.3 μmol h^{-1} (QE = 1.84%) at pH 10.1. A 20% reduction in H_2 evolution rate also occurs in Pt(cit)-APS-[$Ca_2Nb_3O_{10}$], but its activity is still higher than nanosheets without Pt. This must be attributed to the citrate surfactants, which reduce the substrate-accessible surface area on the Pt nanoparticles. IrO_2(cit)-APS-[$Ca_2Nb_3O_{10}$] produced H_2 at rates comparable to Pt-free nanosheet catalysts (Figure 16.3c). In this case, the photochemical reaction was accompanied by a color change, indicating photochemical instability of the catalyst. Indeed, X-ray photoelectron spectroscopy revealed that 10–20% of the IrO_2 was converted into metallic Ir(0) during irradiation.

16.2.3 Peroxide Formation

None of the catalysts, however, evolve any oxygen. Instead they catalyze the formation of hydrogen peroxide according to Equation 16.2. This interpretation is supported by the results of long-duration irradiation experiments, redox titrations and vibrational spectroscopy.

$$2H_2O + \text{cat} \rightarrow H_2 + \text{cat} \cdot O_2^{2-} + 2\,H^+ \qquad (16.2)$$

The timecourse of hydrogen evolution from a UV-irradiated Pt-[$HCa_2Nb_3O_{10}$] sample suspended in water is shown in Figure 16.3d. After 9 h of irradiation, the H_2 rate decreases, until it becomes zero after 24 h. This does not occur when methanol is present in solution. The activity can be restored up to 60% by flushing the catalyst suspension with Ar gas. This indicates that the catalyst is deactivated by a species that reversibly binds to the nanosheets. For TiO_2 (anatase) the deactivation has been attributed to the formation of peroxides that cover up the catalyst surface [19–24]. Such surface-bonded peroxides can be determined by titrating catalysts with the redox indicator*o*-tolidine in the presence of colloidal Pt [20,22,24]. For nanosheet catalysts, we determine that the peroxide concentration on the catalyst increases linearly with irradiation time. The measured amounts indicates 1 : 1 stoichiometry with H_2 (Figure 16.2e), in agreement with Equation 16.2.

The photogenerated peroxide was further characterized with IR and Raman spectroscopy. Figure 16.4A and B show vibrational spectra for the catalyst before and after irradiation (bottom two traces). A new band at 873 cm^{-1} was identified as the ν[O–O] stretch of a peroxo group, which is generally found in the 850–910 cm^{-1} range when bound to metal ions [25–27]. The peak at 482 cm^{-1} can be attributed to an asymmetric ν_a[Nb(O_2)] stretch of the peroxo group, given its relative weakness when compared to the strong corresponding peak at 481 cm^{-1} in the Raman spectrum. Thus, the peak at 423 cm^{-1} in the IR spectrum must be a symmetric ν_s[Nb(O_2)] stretch. Symmetric and asymmetric ν[Nb(O_2)] stretches are generally found to absorb in the range 430–520 cm^{-1} [25,27]. Similar spectral characteristics are evident in samples prepared by direct reaction of TBA[$Ca_2Nb_3O_{10}$] with 0.2 mM H_2O_2 (Figure 16.4A and B). Most of the spectral changes caused by H_2O_2 coordination can be reversed (not shown)

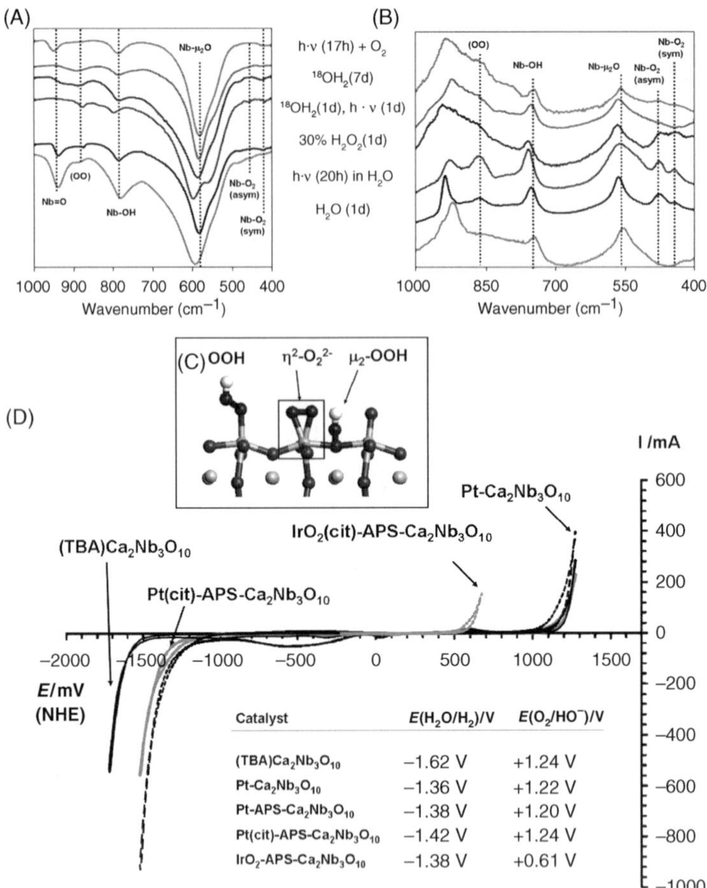

Catalyst	$E(H_2O/H_2)$/V	$E(O_2/HO^-)$/V
$(TBA)Ca_2Nb_3O_{10}$	−1.62 V	+1.24 V
$Pt\text{-}Ca_2Nb_3O_{10}$	−1.36 V	+1.22 V
$Pt\text{-}APS\text{-}Ca_2Nb_3O_{10}$	−1.38 V	+1.20 V
$Pt(cit)\text{-}APS\text{-}Ca_2Nb_3O_{10}$	−1.42 V	+1.24 V
$IrO_2\text{-}APS\text{-}Ca_2Nb_3O_{10}$	−1.38 V	+0.61 V

Figure 16.4 (A) Infrared and (B) Raman spectra of $TBA[Ca_2Nb_3O_{10}]$ before and after storage/irradiation in the presence of $H_2{}^{18}O$ or O_2, and after treatment with H_2O_2. (C) Possible and likely (square) coordination of peroxide to $HCa_2Nb_3O_{10}$. (D) Cyclic voltammograms of catalyst films mounted on Au electrode and recorded in 1.0 M NaOH, with a Pt counter and calomel reference electrode at 100 mV s^{-1}. Potentials are given versus NHE. Onset potentials for water oxidation and reduction were determined graphically from the intercept of a linear fit of the current with the x-axis. (Reprinted with permission from O.C. Compton, C.H. Mullet, S. Chiang and F.E. Osterloh, A building block approach towards photochemical water splitting catalysts based on niobate nanosheets, *Journal of Physical Chemistry C*, **112** (15), 6202–6208, 2008; O.C. Compton and F.E. Osterloh, Niobate nanosheets as catalysts for photochemical water splitting into hydrogen and hydrogen peroxide. *The Journal of Physical Chemistry C*, **113** (1), 479–485, 2009. © 2008, 2009 American Chemical Society.)

upon heating the treated nanosheets to 450 °C, indicating that the peroxide disproportionates into oxide and oxygen under these conditions. The high stability of peroxide seems to be the primary reason for the inability of the niobate nanosheets to catalyze the overall water-splitting reaction according to Equation 16.1. To elucidate the origin of the coordinated peroxide and to confirm peak assignments, we also performed a series of irradiation experiments in 98% $H_2{}^{18}O$.

Under these conditions, all peaks in the Raman either shift to lower energy or become broader, due to partial isotopic substitution (Figure 16.3A and B). Simple storage of the nanosheets in $H_2{}^{18}O$ also causes the Nb–μ_2O bands to shift to lower wavenumbers, indicating isotopic exchange of even the bridging ^{16}O ions with ^{18}O. Thermal isotope exchange of terminal and bridging O has been previously observed for the Lindqvist cluster ion $[Nb_6O_{16}]^{6-}$, which contains one μ_6O, six terminal O and 12 μ_2O sites [28].

There are several possibilities for H_2O_2 to bind to the nanosheets, including end-on, η^2, or μ_2-η^1-bridging modes (Figure 16.4c). But only the η^2-O_2 mode is consistent with the spectroscopic data, in that it produces the asymmetrical and symmetrical $\nu[Nb(O_2)]$ bands. Also, only this mode gives a nonpolar O–O bond that is weakly IR active and strongly Raman active, as observed in Figure 16.4A and B. A μ_2-η^2 mode, with one $O_2{}^{2-}$ ion connecting two adjacent metal ions has been suggested for TiO_2 [20], but is impossible for $HCa_2Nb_3O_{10}$ due to the larger Nb–Nb separation on the nanosheet surface. Side-on coordinated $O_2{}^{2-}$ is frequently observed in molecular Nb^{5+} complexes, including $[Nb(O_2)F_5]^{2-}$ [29], $[Nb(O_2)_3(phen)]$ [30], and $[Nb(O_2)_2(tart)]$ (tart = tartrate) [31] which feature up to three peroxide ligands and seven-/eight-coordinate Nb ions. Based on the relative signal intensities of the Nb=O and Nb–OH Raman bands, there seems to be a preference for the peroxide ions to coordinate to Nb ions of the Nb=O sites. Substitution at this site can occur without disturbing the charge balance of the nanosheet, and can be thought as a thermodynamically controlled acid–base equilibrium in which the more acidic H_2O_2 ($pK_a = 11.6$) displaces the less acidic water ($pK_a = 15.7$) from its salt. By contrast, incorporation of the peroxide at the Nb–OH sites would lead to a build-up of negative charge, or produce a terminal Nb–OOH, which is not observed in the vibrational spectra.

There are two possible routes for the formation of side-on-bonded peroxide in this system, either by two-electron oxidation of water or by two-electron reduction of O_2. Oxygen-uptake experiments, in which a small aliquot (2.0 ml) of O_2 was injected into the headspace of the reaction flask, were conducted to distinguish between these pathways (Figure 16.3F). After irradiation for 18 h, O_2 can no longer be detected in the headspace of the flask. During O_2 absorption, the catalyst only evolves H_2 at a 20 times reduced rate (0.084 μmol h^{-1}) when compared to O_2-free conditions. After irradiation, 6.22 μmol of peroxide was detected in the supernatant by *o*-tolidine titration, but only trace amounts comparable to nonirradiated catalyst are found associated with the niobate sheets. When the catalyst suspension was irradiated in air, weak peroxide bands could also be observed in the vibrational spectra (Figure 16.4A and B). ^{1}H nuclear magnetic resonance (NMR) spectra of the supernatant after irradiation reveal that the TBA counterion is destroyed, whereas irradiation with O_2 exclusion leaves TBA^+ intact. This leads us to conclude that the superoxide radicals are generated only when the catalyst is irradiated in air, and that these radicals react with TBA^+ under H-atom abstraction and H_2O_2 formation. Superoxide species are known to be involved in the photochemical remediation of organic waste [32–36]. Most of the H_2O_2 formed in this way would not be associated with the catalyst, while some of it could adsorb to the niobate sheets by direct reaction, as described above. Conversely, irradiation in the absence of air produces side-on-coordinated peroxide by two-electron oxidation of water (Equation 16.2), without free-radical intermediates.

16.2.4 Water Electrolysis

In order to further explore the possibility of water splitting with the catalysts, cyclic voltammetry was used to determine the overpotentials for water oxidation and reduction in aqueous 1.0 M NaOH. For this purpose, thin films of the catalysts were deposited onto an Au disk electrode by drop casting from solution, and potential scans (100 mV s^{-1}) were performed, as shown in Figure 16.4D. It can be seen that all catalysts were able to electrochemically evolve H_2 and O_2 from water, albeit at potentials exceeding the thermodynamic values ($E_{OH^-/O_2} = +0.40$ V and $E_{H_2/H_2O} = -0.83$ V at pH 14) by about 0.8 V in either direction. For example, TBA[$Ca_2Nb_3O_{10}$] had potentials at + 1.24 V and −1.62 V (NHE), respectively. All Pt-containing catalysts reduced water more easily by about 0.2 V, and oxidized water at 0.02 V lower potentials. Because Pt-containing materials were also much more active as photochemical catalysts, one can conclude that the overpotential for water reduction is a limiting factor for activity. In agreement with the photocatalytic results, Pt(cit) nanoparticles (−1.38...−1.42 V) were found to be slightly less electrochemically active than grown Pt nanoparticles (−1.36 V). This points to an inhibiting effect of citrate, which blocks surface sites on the Pt. The aminoalkyl linkers also increased the water-reduction overpotential, suggesting that charge transport across the Pt/$Ca_2Nb_3O_{10}$ interface is an activity-determining factor in these catalysts. The IrO_2 co-catalyst lowered the water-oxidation potential by 0.6 V, to a final potential close to the thermodynamic one at pH 14. However, repeated scans to negative potentials near −1.47 V diminished the water-oxidation current at +0.61 V, and suggesting that IrO_2 underwent reduction to metallic Ir(0), which is not active as a water-oxidation catalyst. This is analogous to the behavior of the catalyst under irradiation (see above).

16.3 CdSe Nanoribbons as a Quantum-Confined Water-Splitting Catalyst

One of the problems of metal-oxide semiconductors is that their bandgaps are too large to allow absorption in the visible region, which provides 45% of the solar energy. This problem can be solved with metal sulfides, whose bandgaps are much smaller [37–48]. However, because sulfides undergo photocorrosion under bandgap irradiation [49–51], sacrificial electron donors, such as Na_2S and Na_2SO_3 [40,42], are necessary to maintain catalytic activity. In an effort to circumvent this problem, we have explored a nanocystalline *metal selenide* as potential water-splitting catalyst. Metal selenides generally do not catalyze the reaction because their bandgaps are too small [52]. However, as we will show here, quantum-confinement effects can be exploited to activate the material for photochemical water splitting.

CdSe nanoribbons are accessible on a gram scale by reaction of cadmium chloride with a selenocarbamate that is generated *in situ* by reaction of CO, oleylamine and selenium [53]. The ribbons are composed of CdSe wurtzite fragments (Figure 16.5A) terminated at the top and bottom by 1120 lattice planes, and on the sides by $1\bar{1}00$ planes. TEM data (Figure 16.5B) confirms that the ribbons are approximately 1.5 nm thick and form stacks in the $11\bar{2}0$ direction. Suspensions of the ribbons are colored bright yellow, and UV-visible spectra (Figure 16.5C) reveal an absorption edge at 460 nm. Based on the optical properties, the bandgap of the material is 2.7 eV for the nanoribbons, compared to 1.7 eV for bulk CdSe. When irradiated with UV in water or methanol, the ribbons evolve only small amounts of hydrogen (Figure 16.5D, Table 16.2) and quickly turn orange, which indicates photodecomposition of the material, similar to bulk CdSe [54]. In 0.1 M aqueous Na_2S/Na_2SO_3, on the other hand, catalytic H_2 evolution takes place at 107 μmol h^{-1} and with an apparent quantum efficiency (QE) of ~10%. The amount of H_2 produced after 5h exceeds the molar amount of CdSe by a factor of 10,

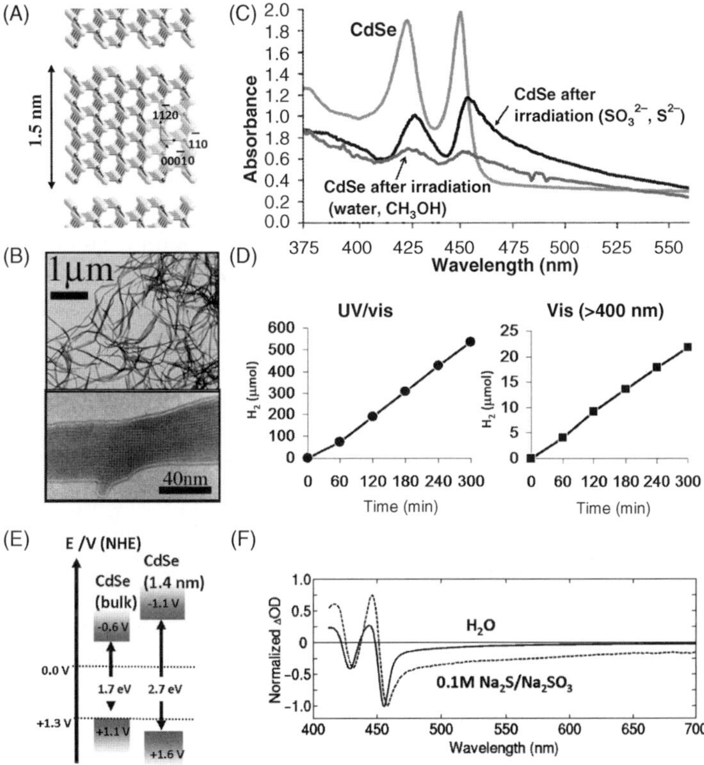

Figure 16.5 (A) Atomic structure and (B) morphology of the CdSe ribbons (TEM). Arrows indicate crystallographic directions of the lattice. (C) UV/visible spectra of CdSe before and after irradiation. (D) Hydrogen evolution from 10 mg of CdSe nanoribbons in 0.1 M Na_2S/Na_2SO_3 aqueous solution. (E) Postulated band structure of CdSe nanoribbons compared to bulk CdSe. Normalized transient spectra of CdSe ribbons in water and 0.1M Na_2S/Na_2SO_3, 1 ps following 400 nm excitation.

Table 16.2 Photochemical hydrogen generation with CdSe nanoribbons.

Experiment	CdSe used (µmol)	Initial/final pH	total H_2 produced (µmol)	H_2 produced (µmol h^{-1})	QE (%)
Pure water	45.92	10.30/9.55	4.62	0.92	0.09
20% CH_3OH (aq)	50.34	10.37/8.24	18.36	3.67	0.35
0.10 M Na_2SO_3: 0.11 M Na_2S(aq)	48.15 (53.61)[a]	13.22/13.30	533.9 (21.8)[a]	106.79 (4.36)[a]	10.19 (13.4)[a]

4 × 175 W Hg lamps, ~10 mg of catalyst, 50 ml of solvent.
[a]Separate experiment with 1.0 M $NaNO_2$ (aq) longpass filter $\lambda > 400$ nm.

proving that the reaction is catalytic. However, even under these conditions, the CdSe undergoes changes, as shown by a shift in the optical absorption maxima and band broadening (Figure 16.5C). These changes could be due to photodecomposition or surface doping of CdSe with sulfur. The activity for H_2 evolution prevails in visible light (>400 nm, 1.0M aqueous $NaNO_2$ filter). Even though the H_2 rate is lower, (4.36 μmol h^{-1}), the apparent QE of 13.4% exceeds that under UV irradiation. This shows that the CdSe ribbons are an active catalyst for hydrogen evolution from water under visible-light irradiation.

As mentioned previously, bulk CdSe is not catalytically active for the reaction, even though the reported flatband potentials for this semiconductor (−0.6 V (NHE) at pH 7 [55]; −0.2 V (NHE) at pH 0) [56] are sufficient for proton reduction in water. Apparently, proton reduction on the CdSe surface is kinetically inhibited. The overpotential for proton reduction can be overcome by the increased bandgap of the CdSe nanoribbons. Based on the larger bandgap, we estimate the flatband potential for the nanoribbons at −1.1 V (NHE) (Figure 16.5E).

The charge carriers responsible for water reduction can be directly observed with femtosecond transient absorption (TA) spectroscopy. The spectrum acquired after 1 ps (Figure 16.5F) reveals a bleach of the ground-state absorptions and two new photoinduced absorptions at 420 nm and at 440 nm. In addition, there is a broad stimulated gain component extending beyond 700 nm, which represents the population of trapped electrons responsible for the long-wavelength tail of the photoluminescence spectra (not shown). Unlike CdSe nanoparticles [57], the CdSe nanoribbons exhibit deeply trapped electrons, which we believe are responsible for the observed photocatalytic activity. As expected, their population is increased when 0.1 M Na_2S/Na_2SO_3 is added to the aqueous catalyst dispersion.

16.4 Conclusion and Outlook

In conclusion, we have demonstrated that two-component nanoparticle structures can be synthesized from metal-oxide nanosheets and metal/metal-oxide particles via controlled linkage or photochemical reactions in liquid phase. Under UV irradiation, these nanostructures are active for photocatalytic water splitting with quantum efficiencies up to 4.7%, depending on catalyst composition, method of assembly and length of the organic linkers. Interestingly, these activities do not exceed those of the bulk materials, even though the nanostructures have a much larger surface area. As has been noted in previous publications [1], surface area is not a limiting factor in photocatalysts. Unexpectedly, the catalysts produce not oxygen, but hydrogen peroxide which binds to surface Nb sites as a η^2-ligand. The coating with peroxide deactivates the catalyst over time, but partial reactivation is possible after purging the catalyst suspension with Ar gas or heating. Since Nb=O sites are responsible for the formation of the peroxide, nanosheets without such sites may enable O_2 evolution. For $HCa_2Nb_3O_{10}$ nanosheets, the O_2 evolution problem persists for the IrO_2-containing catalysts, even though IrO_2 is an excellent water-oxidation catalyst, as revealed by the electrochemical data. Apparently, there is no efficient hole transport pathway from the semiconductor to the IrO_2 particles (step 2′ in Figure 16.1A). Instead IrO_2 accepts electrons from the niobate, causing reduction to iridium metal. The platinum cocatalyst, on the other hand, enhances both the photocatalytic and the electrocatalytic activity of the nanostructures, suggesting that directional electron transport to the cocatalyst is possible here. The difference may be due to the higher mobility of electrons compared to holes, and to the specific surface reactivity of the niobate nanosheets, which favor direct reaction with water over hole transport. Improved two-component catalysts may be accessible through smart assembly, by attaching cocatalysts directly to those sites on the semiconductor that have high hole concentrations. Alternatively, the observed electron transport from the niobate to IrO_2 might be due to a mismatch of the band energies of the two materials. To settle this issue, it will be important to measure the band-edge potentials for all components. Once the potentials are known, they can be adjusted using quantum size effects or by attaching charged surfactants to the particle surfaces. As we have shown, the organic linkers that hold together the nanostructures, do inhibit internanoparticle charge-transport considerably, causing reduced catalytic activities. Thus, organic linkers should be avoided if possible. The same applies to surfactants, whose detrimental effects must be attributed to the blockage of active sites on the particle surfaces, as shown for citrate on the Pt cocatalyst. It may be possible to remove them at elevated temperature. Finally, the big fundamental problem of the catalysts discussed here concerns the inability of the niobate nanosheets to absorb visible light. This issue can potentially be overcome with metal chalcogenide nanosheets, whose bandgaps are smaller. However due to the photochemical instability, these materials will need to be protected against photocorrosion by employing cocatalysts that can remove the photogenerated holes quickly enough. The identification of suitable cocatalysts and proper attachment methods will be a major challenge. On the positive side, the possibility of activating noncatalytic materials for photocatalytic water reduction, as demonstrated here for CdSe nanoribbons, is very promising. It will be interesting to see if this effect can be applied to other materials that are both more stable under irradiation, and that allow clean water splitting, according to Equation 16.1.

Acknowledgment

F. Osterloh thanks the National Science Foundation for supporting this work with an 'Energy for Sustainability' grant (CBET 0829142).

References

[1] Osterloh, F.E. (2008) Inorganic Materials as catalysts for photochemical splitting of water. *Chemistry of Materials*, **20**(1), 35–54.

[2] Bard, A.J. and Fox, M.A. (1995) Artificial photosynthesis – solar splitting of water to hydrogen and oxygen. *Accounts of Chemical Research*, **28**(3), 141–145.

[3] Kamat, P.V. (2007) Meeting the clean energy demand: Nanostructure architectures for solar energy conversion. *Journal of Physical Chemistry C*, **111**(7), 2834–2860.

[4] Gurevich, Y.Y. and Pleskov, Y.V. (1986) *Semiconductor Photoelectrochemistry*, Consultants Bureau, New York, xxv 422.

[5] Ohashi, K., McCann, J. and Bockris, J.O.M. (1977) Stable photoelectrochemical cells for splitting of water. *Nature*, **266**(5603), 610–611.

[6] Kamat, P.V. and Meisel, D. (2003) Nanoscience opportunities in environmental remediation. *Comptes Rendus Chimie*, **6**(8–10), 999–1007.

[7] Nowotny, J., Sorrell, C.C., Bak, T. and Sheppard, L.R. (2005) Solar-hydrogen: Unresolved problems in solid-state science. *Solar Energy*, **78**(5), 593–602.

[8] Frame, F.A., Carroll, E.C., Larsen, D.S. *et al.* (2008) First demonstration of CdSe as a photocatalyst for hydrogen evolution from water under UV and visible light. *Chemical Communications*, 2206–2208.

[9] Compton, O.C., Mullet, C.H., Chiang, S. and Osterloh, F.E. (2008) A building block approach towards photochemical water splitting catalysts based on niobate nanosheets. *Journal of Physical Chemistry C*, **112** (15), 6202–6208.

[10] Carroll, E.C., Compton, O.C., Madsen, D. *et al.* (2008) Ultrafast carrier dynamics in exfoliated and functionalized calcium niobate nanosheets in water and methanol. *Journal of Physical Chemistry C*, **112** (7), 2394–2403.

[11] Compton, O.C., Carroll, E.C., Kim, J.Y. *et al.* (2007) Calcium niobate semiconductor nanosheets as catalysts for photochemical hydrogen evolution from water. *Journal of Physical Chemistry C*, **111**(40), 14589–14592.

[12] Compton, O.C. and Osterloh, F.E. (2009) Niobate Nanosheets as catalysts for photochemical water splitting into hydrogen and hydrogen peroxide. *Journal of Physical Chemistry C*, **113**(1), 479–485.

[13] Kim, J.Y., Osterloh, F.E., Hiramatsu, H. *et al.* (2005) Synthesis Real-time magnetic manipulation of a biaxial superparamagnetic colloid. *Journal of Physical Chemistry B*, **109**(22), 11151–11157.

[14] Harriman, A., Thomas, J.M. and Millward, G.R. (1987) Catalytic and structural properties of iridium-iridium dioxide colloids. *New Journal of Chemistry*, **11**, 757.

[15] Hoertz, P.G., Kim, Y., Youngblood, W.J. and Mallouk, T.E. (2007) Bidentate dicarboxylate capping groups and photosensitizers control the size of IrO_2 nanoparticle catalysts for water oxidation. *Journal of Physical Chemistry B*, **111**, 6485.

[16] Compton, O.C., Carroll, E.C., Kim, J.Y. *et al.* (2007) Calcium niobate semiconductor nanosheets as catalysts for photochemical hydrogen evolution from water. *Journal of Physical Chemistry C*, **111**, 14589.

[17] Nahor, C.S., Hapiot, P., Neta, P. and Harriman, A. (1991) Changes in the redox state of iridium oxide clusters and thelr relation to catalytic water oxidation. radiolytic and electrochemical studles. *Journal of Physical Chemistry*, **95**, 616.

[18] Ebina, Y., Sakai, N. and Sasaki, T. (2005) Photocatalyst of lamellar aggregates of RuOx-loaded perovskite nanosheets for overall water splitting. *Journal of Physical Chemistry B*, **109**(36), 17212–17216.

[19] Yesodharan, E., Yesodharan, S. and Graetzel, M. (1984) Photolysis of water with supported noble-metal clusters, the fate of oxygen in titania based water cleavage systems. *Solar Energy Materials*, **10**(3–4), 287–302.

[20] Kiwi, J. and Graetzel, M. (1987) Specific analysis of surface-bound peroxides formed during photoinduced water cleavage in titanium dioxide-based microheterogeneous systems. *Journal of Molecular Catalysis*, **39**(1), 63–70.

[21] Kalyanasundaram, K., Gratzel, M. and Pelizzetti, E. (1986) Interfacial electron-transfer in colloidal metal and semiconductor dispersions and photodecomposition of water. *Coordination Chemistry Reviews*, **69**, 57–125.

[22] Gu, B., Kiwi, J. and Gratzel, M. (1985) Photochemical water cleavage in suspensions of Pt-loaded titania particles with 0.7-percent overall light to chemical conversion efficiency. *New Journal of Chemistry*, **9**(8–9), 539–543.

[23] Kiwi, J. and Gratzel, M. (1984) Optimization of conditions for photochemical water cleavage – aqueous Pt/TiO_2 (Anatase) dispersions under ultraviolet-light. *Journal of Physical Chemistry*, **88**(7), 1302–1307.

[24] Duonghong, D. and Gratzel, M. (1984) Colloidal TiO_2 particles as oxygen carriers in photochemical water cleavage systems. *Journal of the Chemical Society, Chemical Communications*, 23, 1597–1599.

[25] Nakamoto, K. (1997) *Infrared Raman Spectra of Inorganic and Coordination Compounds*, 5 edn, Wiley-Interscience, New York.

[26] Bayot, D., Tinant, B. and Devillers, M. (2003) Water-soluble niobium peroxo complexes as precursors for the preparation of Nb-based oxide catalysts. *Catalysis Today*, **78**, 439.

[27] Bayot, D., Devillers, M. and Peeters, D. (2005) Vibrational spectra of eight-coordinate niobium and tantalum complexes with peroxo ligands: A theoretical simulation. *European Journal of Inorganic Chemistry*, **20**, 4118.

[28] Black, J.R., Nyman, M. and Casey, W.H. (2006) Rates of oxygen exchange between the [HxNb6O19]((aq))(8-x) Lindqvist ion and aqueous solutions. *Journal of the American Chemical Society*, **128**(45), 14712–14720.

[29] Stomberg, R. (1983) The disordered structure of Bis(8-Hydroxyquinolinium) Pentafluoroperoxoniobate(V) Trihydrate, $(C_9H_8NO)_2[NbF_5(O_2)].3H_2O$ – a Redetermination at 170-K and 275-K. *Acta Chemica Scandinavica A*, **37**(6), 523–530.

[30] Mathern, G. and Weiss, R. (1971) Structure of transition metal peroxide complexes .2. Crystal structure of potassium Triperoxo-(Ortho Phenanthroline)Niobate trihydrate and its hydrogen peroxide adduct $(C_{12}H_8N_2).3H_2O$ and $KNb(O_2)_3(C_{12}H_8N_2).3H_2O.H_2O_2$. *Acta Crystalographica B Structure B*, **27**(AUG 15), 1582–1588.

[31] Bayot, D., Tinant, B. and Devillers, M. (2005) Homo- and heterobimetallic niobium(v) and tantalum(v) peroxo-tartrate complexes and their use as molecular precursors for Nb-Ta mixed oxides. *Inorganic Chemistry*, **44**(5), 1554–1562.

[32] Fox, M.A. and Dulay, M.T. (1993) Heterogeneous Photocatalysis. *Chemical Reviews*, **93**(1), 341–357.

[33] Carp, O., Huisman, C.L. and Reller, A. (2004) Photoinduced reactivity of titanium dioxide. *Progress in Solid State Chemistry.*, **32**(1–2), 33–177.

[34] Halmann, M.M. (1996) *Photodegradation of Water Pollutants*, CRC Press, Boca Raton, pp. 301.

[35] Kamat, P.V. (2002) Photophysical, photochemical and photocatalytic aspects of metal nanoparticles. *Journal of Physical Chemistry B*, **106**(32), 7729–7744.

[36] Vinodgopal, K., Hotchandani, S. and Kamat, P.V. (1993) Electrochemically assisted photocatalysis – TiO_2 particulate film electrodes for photocatalytic degradation of 4-chlorophenol. *Journal of Physical Chemistry*, **97** (35), 9040–9044.

[37] Darwent, J.R. (1981) H-2 production photosensitized by aqueous semiconductor dispersions. *Journal of the Chemical Society Faraday Transactions II*, **77**, 1703–1709.

[38] Darwent, J.R. and Porter, G. (1981) Photochemical hydrogen-production using cadmium-sulfide suspensions in aerated water. *Journal of the Chemical Society, Chemical Communications*, (4), 145–146.

[39] Mills, A. and Porter, G. (1982) Photosensitized dissociation of water using dispersed suspensions of N-type semiconductors. *Journal of the Chemical Society Faraday Transactions I*, **78**, 3659–3669.

[40] Buhler, N., Meier, K. and Reber, J.F. (1984) Photochemical hydrogen-production with cadmium-sulfide suspensions. *Journal of Physical Chemistry*, **88**(15), 3261–3268.

[41] Yanagida, S., Azuma, T. and Sakurai, H. (1982) Photocatalytic hydrogen evolution from water using zinc-sulfide and sacrificial electron-donors. *Chemistry Letters*, (7), 1069–1070.

[42] Reber, J.F. and Meier, K. (1984) Photochemical Production of Hydrogen with Zinc-Sulfide Suspensions. *Journal of Physical Chemistry*, **88**(24), 5903–5913.

[43] Zheng, N., Bu, X.H., Vu, H. and Feng, P.Y. (2005) Open-framework chalcogenides as visible-light photocatalysts for hydrogen generation from water. *Angewandte Chemie, International Edition*, **44**(33), 5299–5303.

[44] Lei, Z.B., Ma, G.J., Liu, M.Y. *et al.* (2006) Sulfur-substituted and zinc-doped In(OH)(3): A new class of catalyst for photocatalytic H-2 production from water under visible light illumination. *Journal of Catalysis*, **237**(2), 322–329.

[45] Kobayakawa, K., Teranishi, A., Tsurumaki, T. *et al.* (1992) Photocatalytic activity of $CuInS_2$ and $CuIn_5S_8$. *Electrochimica Acta*, **37**(3), 465–467.

[46] Kudo, A., Nagane, A., Tsuji, I. and Kato, H. (2002) H-2 evolution from aqueous potassium sulfite solutions under visible light irradiation over a novel sulfide photocatalyst NaInS2 with a layered structure. *Chemistry Letters*, (9), 882–883.

[47] Sobczynski, A., Yildiz, A., Bard, A.J. *et al.* (1988) Tungsten disulfide – a novel hydrogen evolution catalyst for water decomposition. *Journal of Physical Chemistry*, **92**(8), 2311–2315.

[48] Bessekhouad, Y., Mohammedi, M. and Trari, M. (2002) Hydrogen photoproduction from hydrogen sulfide on Bi2S3 catalyst. *Solar Energy Materials and Solar Cells*, **73**(3), 339–350.

[49] Platz, H. and Schenk, W. (1936) Concerning the discolouration of zinc sulphides in light. *Angewandte Chemie*, **49**, 0822–0826.

[50] Gloor, K. (1937) Photolysis with zinc sulphide. *Helvetica Chimica Acta*, **20**, 853–877.

[51] Gerische, H. and Meyer, E. (1971) Mechanism for electron transfer and corrosion on cadmium sulfide electrode in redox system I3(−)/I. *Zeitschrift für Physikalische Chemie Neue Folge*, **74**(3–6), 302–318.

[52] Kambe, S., Fujii, M., Kawai, T. *et al.* (1984) Photocatalytic Hydrogen-Production with Cd(S, Se) Solid-Solution Particles - Determining Factors for the Highly Efficient Photocatalyst. *Chemical Physics Letters*, **109**(1), 105–109.

[53] Joo, J., Son, J.S., Kwon, S.G. *et al.* (2006) Low-temperature solution-phase synthesis of quantum well structured CdSe nanoribbons. *Journal of the American Chemical Society*, **128**(17), 5632–5633.

[54] Marcu, V., Tenne, R. and Rubinstein, I. (1986) Electrochemical Characterization of Photoetching Products of Cdse. *Journal of the Electrochemical Society*, **133**(6), 1143–1148.

[55] Fujii, M., Kawai, T. and Kawai, S. (1984) Photocatalytic activity and the energy-levels of electrons in a semiconductor particle under irradiation. *Chemical Physics Letters*, **106**(6), 517–522.

[56] Nozik, A.J. and Memming, R. (1996) Physical chemistry of semiconductor-liquid interfaces. *Journal of Physical Chemistry*, **100**(31), 13061–13078.

[57] Burda, C. and El-Sayed, M.A. (2000) High-density femtosecond transient absorption spectroscopy of semiconductor nanoparticles. A tool to investigate surface quality. *Pure and Applied Chemistry*, **72**(1–2), 165–177.

17

Quantum-Confined Visible-Light-Active Metal-Oxide Nanostructures for Direct Solar-to-Hydrogen Generation

Lionel Vayssieres

International Center for Materials NanoArchitectonics, National Institute for Materials Science, Tsukuba, Japan, Email: vayssieres.lionel@nims.go.jp

17.1 Introduction

With global warming seriously endangering our environment, the recent dramatic worldwide increase in gasoline price, the shortening in fossil fuels and the appeal and political will for a future implementation of the hydrogen economy, a substantial renewed interest in the field of materials for photoconversion such as solar cells, fuel cells and solar hydrogen generation has occurred within the last few years. The latter has triggered the fabrication and study of a plethora of new (as well as revisited) doped and composite-based oxide [1–72] and non-oxide [73–84] structures aiming at finding the ideal photocatalysts for water splitting under visible-light irradiation. Our strategy to fulfill the drastic requirements of materials development for direct solar water splitting is the ability to design metal-oxide semiconductors based on vertically oriented anisotropic nanostructures (i.e., nanorods) with intermediate bands and highly quantized band structures, such as quantum rods and dots, to enable high efficiency in the visible range, as well as tuning bandgap and band edges by quantum-confinement effects. Such unique characteristics, combined with in-depth investigation and modeling of their electronic structure, and large-scale and low-cost fabrication methods provide this research a substantial advance and (r)evolutionary prospect in the field of semiconductor technology and

On Solar Hydrogen & Nanotechnology Edited by Lionel Vayssieres

materials for solar-energy conversion. In addition, such advanced knowledge of the electronic structure, structural properties and surface chemistry of metal-oxide nanostructures and heterostructures has direct implications for fundamental understanding of the structure–property relationships as well as efficiency optimization of novel nanodevices based on metal-oxide nanostructures and heterostructures for renewable hydrogen generation [85–90].

Materials synthesis by chemical methods has emerged as one of the most versatile fabrication techniques for the in-depth study and rational delivery of high-purity functional nanomaterials, engineered from the molecular to the microscopic scale at low cost and large scale. The latter attributes make this approach very attractive for industrial applications and represents the immediate answer to the enormous need for new materials research to develop technologies for renewable energy. Our approach is to use aqueous materials chemistry to fabricate innovative semiconductor nanostructures and nanodevices for efficient photoelectrochemical production of hydrogen under visible-light irradiation. Nanomaterials based on metal oxides, the most stable and versatile class of compounds in terms of physical and chemical properties, are purposely designed at multilength scales (nano, meso and micro) from low-cost synthesis methods, such as aqueous chemical growth onto various substrates, including the industrially required transparent conducting substrates, and subsequently doped by chemical and/or vapor deposition techniques. Two- and three-dimensional nanostructures based on advanced and functional zero-and one-dimensional building blocks, such as quantum dots, quantum rods and nanowires, are subsequently optimized by selected processes consisting of chemical coating, specific heat treatment and interfacial electrochemistry to fabricate efficient and stable nanodevices with the optimum bandgap for visible-light absorption. Their optical, structural and electronic properties are thoroughly investigated at synchrotron radiation facilities. Simultaneously their photoelectrochemical and photocatalytic properties are probed at real scale with solar simulator bench set-ups to record their photoconversion efficiency to achieve the ideal bandgap profiling for efficient visible-light absorption. A constant and direct feedback between materials development and in-depth fundamental knowledge and experimental investigation allows selecting, tuning and optimizing of the properties of the materials.

This chapter will present the concepts, design and cost-effective fabrication, by aqueous chemical growth, of advanced nanostructures consisting of vertically oriented metal-oxide quantum-rod arrays. These nanostructures show high surface area, oriented pathways for photogenerated electron/hole carriers, new morphology with functionalized surfaces and engineered bandgap for stability against photocorrosion, as well as optimized band edges and solar absorption profile for direct solar hydrogen generation. In-depth investigation of their electronic structure and photoelectrochemical properties is also presented to demonstrate the ability to fabricate at low cost, stable visible-light-active semiconductors and understand bandgap profiling and quantum-confinement concepts for a full absorption of the solar radiation.

17.2 Design of Advanced Semiconductor Nanostructures by Cost-Effective Technique

17.2.1 Concepts and Experimental Set-Up of Aqueous Chemical Growth

A relatively new thin-film growth technique has emerged as a simple and powerful tool to fabricate, at low cost and low temperatures (below 100 °C), large areas of metaloxide nano- to microparticulate thin films [91,92]. Two- and three-dimensional arrays consisting of oriented anisotropic nanoparticles (i.e., nanorods, nanowires, nanotubes) are easily generated with enhanced control over orientations and dimensions. The synthesis involves the controlled heteronucleation of metal oxides onto substrates from the hydrolysis-condensation of metal salts in aqueous solutions [93,94]. The most pertinent parameter to control the nucleation and growth, and therefore the overall design and architecture, of a thin film is the interfacial free energy of the system [95,96].

Indeed, by applying thermodynamic stability concepts [97], that is to reduce, chemically and electrostatically, the interfacial tension of the system to the thin-film processing technology, an inexpensive and effective aqueous growth technique at mild temperatures has been developed to produce functionalized coating of metal-oxide materials onto various substrates (amorphous, polycrystalline, single crystalline, plastics, fabrics...). Such a technique allows the generation of advanced nano/microparticulate thin films without any template, membrane, surfactant, applied external fields or specific requirements in substrate activation, thermal stability or crystallinity. Given that the crystallites grow from the substrate, a large choice of thin film/substrate combinations is offered, which provides, consequently, a better flexibility and a higher degree of materials engineering and design. To understand the possibility of growing nano- and microparticulate thin films from aqueous solution, as well as the ability to grow and align anisotropic nanoparticles into large arrays on a substrate, one has to observe the differences between homogeneous and heterogeneous nucleation phenomena. The ability to design materials with different orientations stimulates the study of the influence of such parameters on the physical properties of materials and gives further opportunities for materials design.

This method consists of heating homogeneous aqueous solutions of metal salts (e.g., $FeCl_3$, $Zn(NO_3)_2$...) in the presence of a substrate (e.g., glass, Si/SiO_2 wafers, sapphire, glass, coated glass, polypropylene, teflon...) at moderate temperatures (below 100 °C) in a closed vessel. Therefore, such a technique does not require high-pressure containers and is also entirely recyclable, safe and environmentally friendly, since only water is used as solvent. Such a process avoids the safety hazards of organic solvents and their eventual evaporation, flammability and potential toxicity. In addition, since no organic solvents or surfactants are present, the purity of the materials is dramatically improved. Any residual salts (from the precursors' decomposition and/or from the added ionic strength) are easily washed away by water, due to their high solubility in aqueous solutions. In most cases, no additional heat or chemical treatments are necessary, which represent a significant improvement compared to surfactant-, template- or membrane-based synthesis methods.

A full coverage of the substrate is obtained within few hours (and up to a couple of days in certain cases) provided that the heat capacity of water and surface coverage control are achieved by monitoring the synthesis time in the early stages of the thin-film growth. Partial coverage is obtained within the first few hours, which may be necessary for certain applications to adjust and tune the overall physical properties of devices (e.g., optical properties of multibandgap thin films). Such a technique is a multideposition technique and the layer-by-layer growth of thin films is readily obtainable. The development of multilayer thin films of various morphologies and/or of various chemical compositions – that is, composite, multi-bandgap and doped thin films – is reached. The complete thin-film architecture may thus be modeled, designed and monitored to match the requirements of the application. In most cases, it does improve the physical and/or chemical properties of the devices. It also gives the capacity to create novel and/or improved thin-film integrated devices. Growing thin layers directly from the substrate does substantially improve the adherence and the mechanical stability of the thin films compared to the standard solution and colloidal deposition techniques, such as spin and/or dip coating, chemical baths, screen printing or doctor blading. Moreover, given that such materials precipitate in homogeneous solution from molecular-scale compounds (i.e., condensed metal complexes), they will grow on virtually any substrate (providing that the substrates are stable in aqueous solution). It goes without saying that the overall mechanical stability of the thin films does vary from substrate to substrate, but in most cases, strong adhesion is observed. Scaling up is easily feasible, and this concept and synthesis method is potentially applicable to all water-soluble metal ions likely to precipitate in aqueous solutions. Large-scale manufacturing at low cost is therefore achievable with such a technique. In addition to all these industrially related advantages, such a technique is also potentially very interesting due to the compatibility of water and aqueous solution with biological compounds. For instance, 3D arrays of composite bionanomaterials have been obtained using aqueous chemical growth. Such concepts and thin-film processing techniques have been applied successfully to oxides and oxyhydroxides of several transition metals.

17.2.1.1 Tailoring the Morphology

At low interfacial tension, not only the size [98], but also the shape of nanoparticles does not necessarily require being spherical; indeed very often nanoparticles develop a spherical morphology to minimize their surface energy because the sphere represents the smallest surface for a given volume. However, if the synthesis (and/or dispersion) conditions are suitable (i.e., yielding to the thermodynamic stabilization of the system), the shape of the crystallites will be driven by the symmetry of the crystal structure, as well as by the chemical environment, and various morphologies (other than spherical) may therefore be developed. Manipulating and controlling the interfacial tension enables the growth of nanoparticles with sizes and shapes tailored for their applications. Applying the appropriate solution chemistry (precursors and precipitation/dispersion conditions) to the investigated transition-metal ion, along with the natural crystal symmetry and anisotropy, or by forcing the material to grow along a certain crystallographic direction by controlling chemically the specific interfacial adsorptions of ions and/or ligand- and crystal-field stabilization, one can gain the ability to develop purpose-built crystal morphology.

17.2.1.2 Tailoring the Orientation

To develop the capability of controlling the orientation of large arrays of anisotropic nanoparticles on a substrate, one has to consider the energetic differences between homogeneous and heterogeneous nucleation phenomena. In most cases, homogenous nucleation of solid phases from solution requires a higher activation-energy barrier and, therefore, heteronucleation is promoted and energetically more favorable. Indeed, the interfacial energy between two solids is generally smaller than the interfacial energy between a solid and a solution, and, therefore, nucleation takes place at a lower saturation ratio onto a substrate than in solution. Nuclei grow by heteronucleation on the substrate, and various morphology and orientation monitoring can be obtained by experimental control of the chemical composition of the precipitation medium. To illustrate and demonstrate such

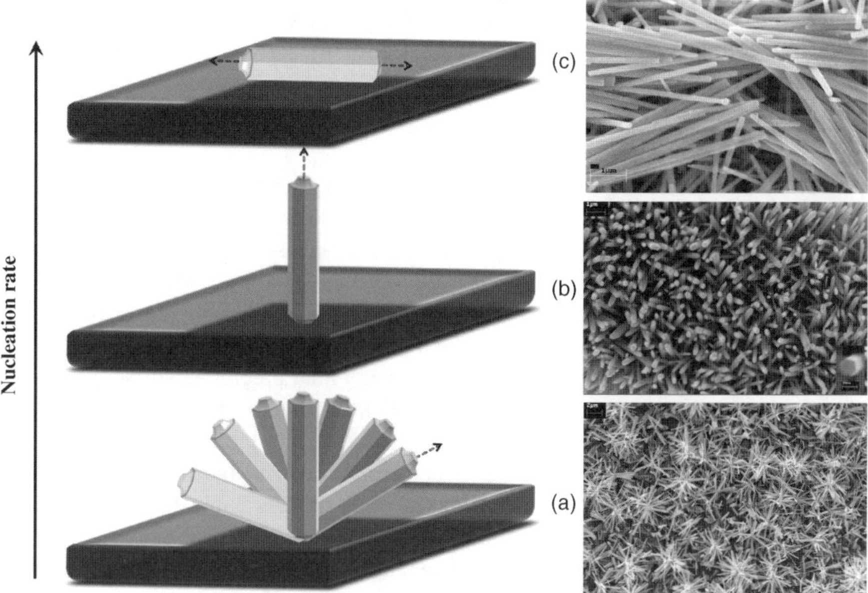

Figure 17.1 Schematic representation (left) and scanning electron microscope (SEM) images (right) of the effect of (hetero)nucleation rate on the orientation of (ZnO) nanorods grown with (a) multi-directional; (b) perpendicular; and (c) parallel orientation onto substrates by aqueous chemical growth at $T < 100\,°C$.

capabilities, the example of an anisotropic building block (e.g., nanorods) is taken and the possibilities of experimentally monitoring its orientations onto a substrate using the aqueous chemical growth thin film processing technique are exposed hereafter, and illustrated in Figure 17.1:

- Nanorods with a parallel orientation to substrates can be obtained when the nucleation rate is enhanced by the precipitation conditions (pH, ionic strength, concentration, *T*). Indeed, the fast appearance of a large number of nuclei will result in rapid two-dimensional growth. The stacking of anisotropic nanoparticles with random orientations to each other, but with an overall perpendicular orientation with respect to the substrate normal (i.e., parallel orientation), is therefore obtained (Figure 17.1c);
- Nanorods with a perpendicular orientation to substrates may be obtained when the nucleation is slightly limited by the precipitation conditions. The slow appearance of a limited number of nuclei will allow slow growth (according to the crystal symmetry and relative face velocities) along the easy direction of crystallization. As a result, a condensed phase of nanorods parallel to the substrate normal (i.e., perpendicular/vertical orientation) is generated (Figure 17.1b)
- Nanorods with multiangular orientations to substrates may be generated when the number of nuclei is exceedingly limited by the precipitation conditions (through the chemical and electrostatic monitoring of the interfacial tension). The system will promote twinning and the preferential epitaxial three-dimensional growth of the rods along their easy axis from a very

limited number of nuclei will induce a type of star shape (or flower-shape) morphology, where the nanorods grow with multidirectional morphology (Figure 17.1a)

17.2.1.3 Tailoring the Crystal Structure

In addition to the control of particle size, shape and orientation, the synthesis of nanoparticles by aqueous chemical growth at low interfacial tension allows the stabilization of oxide and oxyhydroxide metastable crystal structures [99]. Crystal-phase transitions in solution generally operate through a dissolution–recrystallisation process to comply with the surface-energy minimization requirement of the system. Indeed, when a solid offers several allotropic phases and polymorphs, it is typically the one with the highest solubility and consequently the lowest stability, (i.e., the crystallographic metastable phase) which precipitates first. This is understood by considering the nucleation kinetics of the solids. At a given supersaturation ratio, the germ size is as small (and the nucleation rate as high) as the interfacial tension of the system is low. Thus, given that the solubility is inversely proportional to the interfacial tension, the precipitation of the most soluble phase and consequently the less stable thermodynamically, is therefore kinetically promoted. Due to its solubility and metastability, this particular phase is more sensitive to secondary growth and ageing, which leads to crystallographically more stable phases, essentially by heteronucleation. Secondary growth and aging processes are delicate to control and the phase transformations appear within a few hours to a few days in solution, whence the resulting undesired mixing of allotropic phases and polymorphs. However, by careful consideration of the precipitation conditions (i.e., at thermodynamically stable conditions), such phenomena may be avoided when nanosystems are precipitated at low interfacial tension. In summary, by adjusting the experimental conditions to reach the thermodynamic stability of a system, the nanoparticle size, shape and crystal structure may be tuned and optimized. It allows the functionalized design of nanomaterials and the ability to quantitatively probe the influence of such parameters on the physical and chemical properties of metal-oxide nanoparticles and nanoparticulate materials.

17.2.2 Achievements in Aqueous Design of Highly Oriented Metal-Oxide Arrays

The ability to design materials consisting of anisotropic nanoparticles of different orientations enables experimental studies of the angular-dependence influence on the physical/chemical properties and electronic structure of materials, and gives further opportunities for materials design and engineering (e.g., increased dimensionality). The outcome allows the fabrication of innovative and functional nano- to microparticulate thin films of metal oxides directly onto substrates without membrane, template, surfactant, undercoating or applied external fields. Such purpose-built nanomaterials are prepared in such a way that particle nucleation is separated from growth, giving exquisite control of particle size. In many cases, the differences between interfacial energies among crystal facets can be exploited to produce controlled anisotropic shapes. In addition, since the purpose-built nanomaterials are fabricated at low temperature and can be deposited on almost any substrate, it is possible to deposit one phase on another. This opens up the possibility of using the anisotropy of one material to generate a high-surface-area substrate upon which a second, active, material is deposited. In effect, this produces nanocomposites – materials combining properties and architectures from multiple phases to achieve results not available from any single phase. Such a method has been successfully applied to develop a new generation of functional materials, the so-called *purpose-built materials* [91,92]. Indeed, when thermodynamic stabilization is applied to heterogeneous nucleation, rather than homogeneous nucleation, not only the size of spherical nanoparticles can be controlled, but also the shape and the orientation of anisotropic building blocks onto various substrates can be tailored to build smart and cost-effective nanostructures. Indeed, the fabrication of highly oriented crystalline arrays of large physical area consisting of nanorods of iron oxides [100], iron–chromium sesquioxide nanocomposites [101], iron oxyhydroxides and iron metal [102] have been successfully designed by following such ideas. In addition, ZnO nanorods and nanowires [103], microrods [104], microtubes [105], along with other morphologies [106], nanowires and nanoparticles of manganese oxides [107]. More recently, and for the first time, ordered arrays of c-axis-elongated nanorods of rutile SnO_2 [108] with a square cross-section have also been successfully fabricated following the same concept and aqueous thin-film processing method. It is worth mentioning that the c-axis orientation goes against crystal symmetry (centrosymmetric structure, presence of mirror planes preventing the growth along the c-axis). Yet, by providing the chemical growth conditions that allow {110} faces to grow, the c-axis elongation is naturally promoted. Only aqueous chemistry has enabled such growth where all other gas-phase techniques failed. Such purpose-built nanomaterials have been designed, for instance, to improve the photovoltaic [109], photoelectrochemical [110] and gas-sensing [111–113] properties of large bandgap semiconductors [114]. Better fundamental understanding of the bandgap, electronic structure [115] and orbital symmetry of II–VI semiconductor nanomaterials [116], as well as confinement effects on the orbital character of bandgap states of TiO_2 quantum-dot arrays [117], on the bandgap of α-Fe_2O_3 quantum rod-arrays [118] and on the interfacial chemistry (i.e., point of zero charge) of γ-Fe_2O_3 [119] (changing its surface acidity from slightly acidic to neutral to basic by changing the size of the nanoparticle from 12 to 7.5 to 3.5 nm) have also been demonstrated with materials fabricated by such concepts and experimental methods. Finally, the direct application of such models has allowed the demonstration of the thermodynamic stabilization of metastable crystal phases such as hydroxides and oxyhydroxides of transition metals in aqueous solutions [99], as well as the fabrication of novel bionanocomposites by tailoring the conformation of bioactive molecules adsorbed on metal-oxide nanoarrays [120]. Figure 17.2 illustrates the capability of aqueous chemical growth in designing vertically oriented nano-to-microrod arrays of common, large-bandgap semiconductors.

17.3 Quantum Confinement Effects for Photovoltaics and Solar Hydrogen Generation

The attempt to improve the performance of existing materials by further development along the same lines (incremental increase), is reaching its limit. Indeed, very little possibility of further improvement and innovations is foreseen, especially considering the environment, when aiming at even higher efficiency. Given this situation, it is now imperative to develop novel functional materials which can produce real innovations with the aim of realizing true renewable energy production. The necessity of materials development which is not limited to materials that can achieve their theoretical limits, but makes it possible to raise

Figure 17.2 Field-emission SEM images of oriented arrays of SnO_2 (a), ZnO (b) and α-Fe_2O_3 (c) fabricated onto various substrates by aqueous chemical growth at temperatures below 100 °C.

theoretical limits by changing the fundamental underlying physics and chemistry has become a critical issue. Materials R&D, fully exploiting nanoscience and nanotechnology, has the greatest potential to reach such highly challenging goals. Indeed, recent reports of such advanced materials using quantum-confinement effects have surfaced and triggered a renewed interest and new hopes for higher efficiency and visible-light-active materials for photovolatics and photocatalysis. The basic principles of such effects are briefly presented in this section. More detailed reviews of the subject can be found in the literature [121,122].

17.3.1 Multiple Exciton Generation

One of the major limiting steps to the development of materials for solar photoconversion with high efficiency is the 1:1 ratio between the absorption of photons and the creation of electron–hole pairs in the materials. A process called impact ionization, known in the semiconductor field since the late 1950s [123–125], allows the creation of more than one electron–hole pair per absorbed photon. Although, such a process is not efficient for bulk semiconductors (due to the high photon energy threshold), it becomes very attractive and very efficient when dealing with nanoscale semiconductors, where the threshold photon energy to generate multiple electron–hole pairs could occur in the visible or near-infrared (IR) spectral region. Such a process is now called "multiple exciton generation (MEG)" and has been shown to occur (at least from time-resolved spectroscopic measurements) in PbS, PbSe, CdSe, PbTe, InAs and Si quantum dot-arrays [126–129].

The main reason why such a process is so effective in quantum-dot-based semiconductors lies in the relatively slow rate of electron relaxation (via electron–phonon interactions), which originates from the quantized nature of the energy levels and subsequent carrier confinement

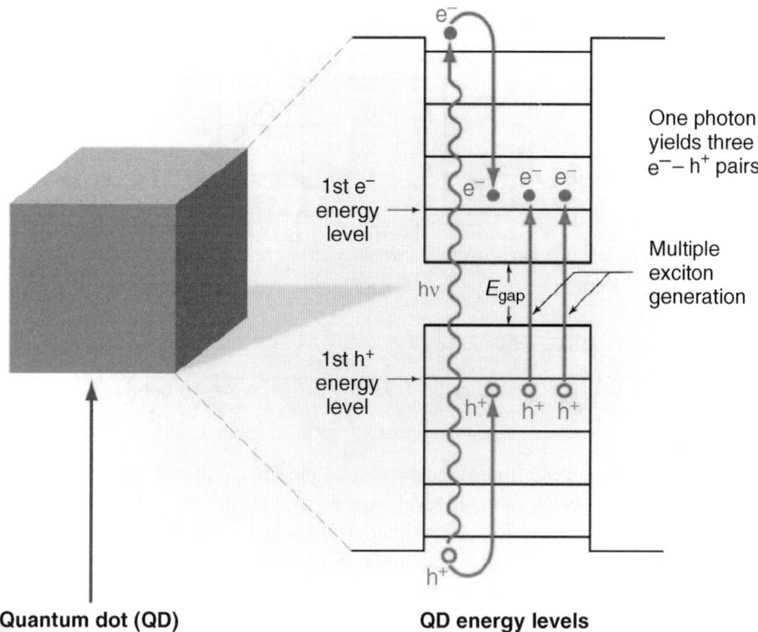

Figure 17.3 Schematic representation of the multiple exciton generation process observed in quantum dots. A single absorbed photon whose energy is at least three times the energy difference between the first energy levels for electrons and holes can create three excitons. (Reprinted from *Basic Research Needs For Solar Energy Ultilization*, Report of the Basic Energy Sciences Workshop on Solar Energy Ulitilization, April 18–21, 2005, US Department of Energy.)

and enhanced Coulomb interaction. Figure 17.3 shows the fundamental mechanism involved in multiple exciton generation in low-dimensional materials (i.e., quantum dots).

17.3.2 Quantum-Well Structures

Another possible path to increase the efficiency and/or enabling a better match between solar radiation and the absorption profile of the materials (i.e., visible-light-active materials) is found in the formation of quantum-well-based structures [131–140]. Indeed, the formation of discrete energy levels allows photons of lower energy (visible to IR range) to be efficiently absorbed, thus contributing to enhancing the efficiency of the systems (Figure 17.4).

17.3.3 Intermediate Band Materials

To efficiently absorb photons with energies lower than the bandgap and thus create electron-–hole pairs to increase the efficiency in the visible part of the solar spectrum of photovoltaics and photocatalytic devices, the fabrication of the so-called "intermediate-band" materials [141–174] becomes necessary. Figure 17.5 illustrates such a concept in quantum-dot-based materials. The intermediate band arises from the confined electronic states of the

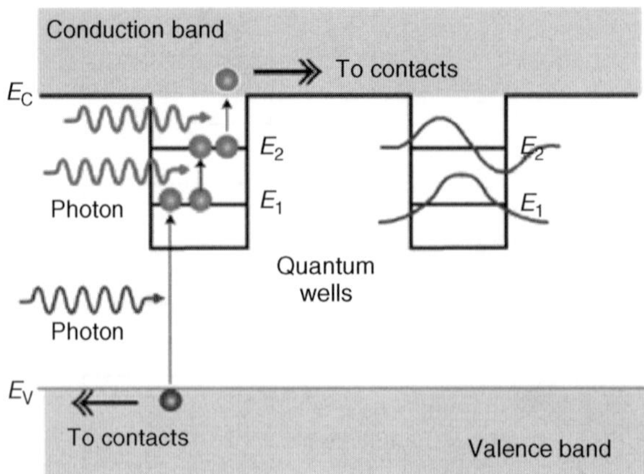

Figure 17.4 Schematic representation of the energetics and optical transitions of quantum-well-based structures. The formation of localized energy levels enables the absorption of photons of multiple energies in addition to the bandgap absorption. (Reprinted from *Basic Research Needs For Solar Energy Ultilization*, Report of the Basic Energy Sciences Workshop on Solar Energy Ulitilization, April 18–21, 2005, US Department of Energy.)

electrons in the potential wells of the conduction band. Delocalization of the electrons within the intermediate band is achieved by increasing their density until the electron wave functions have significant overlap and thus become delocalized (i.e., Mott transition). Indeed, by creating an (intermediate) band between the valence band and the conduction band, interband carrier relaxation is substantially reduced (limited probability of multiphonon interactions) and

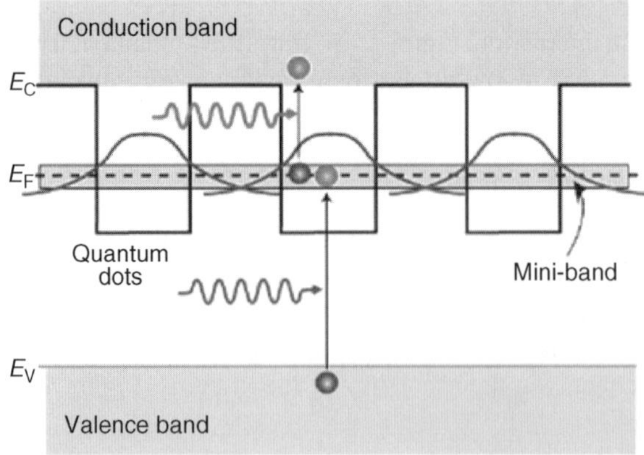

Figure 17.5 Schematic representation of the energetics and optical transitions of quantum-dot-based structures. The intermediate band arises from the confined electronic states of the electrons in the potential wells of the conduction band, which enables the absorption of photons of two different energies, in addition to the bandgap absorption. (Reprinted from *Basic Research Needs For Solar Energy Ultilization*, Report of the Basic Energy Sciences Workshop on Solar Energy Ulitilization, April 18–21, 2005, US Department of Energy.)

sub-bandgap photons can contribute to the overall photocurrent generation. It is important to understand that, with this concept, the introduction of energy levels within the bandgap of the semiconductor is not supposed to create nonradiative recombination centers (which usually reduce the performance instead of improving it). The main fundamental explanation is that the wave function associated with the electrons in the intermediate band has to be delocalized to improve efficiency. This is contrary to conventional nonradiative recombination centers, which are tightly bound.

The concept of an intermediate band seems to be the most appropriate as far as metal-oxide semiconductors are concerned to develop visible-light-active materials for efficient photoconversion.

17.4 Novel Cost-Effective Visible-Light-Active (Hetero)Nanostructures for Solar Hydrogen Generation

17.4.1 Iron-Oxide Quantum-Rod Arrays

17.4.1.1 Improving the Charge Collection by Vertical Design

The fabrication of novel and smart iron-oxide materials was demonstrated by the production of quantum rods arranged in vertical/parallel oriented bundles of nanorods onto various substrates [100,118]. Such purpose-built nanomaterials were originally designed to produce low-cost photoelectrochemical devices, such as photovoltaic cells [109], as well as photocatalytic cells for direct water splitting and solar hydrogen generation [110]. Hematite is, at least theoretically, the ideal candidate for photoelectrochemical applications. Its low cost, abundance, low toxicity, structural and chemical thermodynamic stability and relatively good stability against photocorrosion, as well as its "low" bandgap (around 2 eV) for a large-bandgap semiconductor make a very attractive candidate for such applications. Unfortunately, after many studies over several decades in the twentieth century, which included highly porous nanomaterials [102], the photoefficiencies remained at very low values, mostly due to the high level of electron–hole recombinations in the solid and at grain boundaries.

In the late 1990s, we proposed a new concept to overcome the unsuccessful reputation of iron-oxide materials for photovoltaics and photocatalytic applications. Our strategy involved the design of hematite nanomaterials consisting of anisotropic nanoparticles (presently called nanorods and nanowires) rather than spherical nanoparticles (Figure 17.6).

The purpose of such a design intended to lower the recombination processes by eliminating the grain boundaries generated by isotropic nanoparticles. In addition, designing nanorods with a perpendicular orientation onto the substrate would facilitate the transport of photogenerated electrons to the back contact through a direct pathway rather than random transport across spherical nanoparticles.

17.4.1.2 Improving the Charge Separation of Photogenerated Carriers by Structural Design

Another strategy to improve the efficiency of hematite was to tailor the diameter of the rods to match the hole (minority carrier) diffusion length, to drastically reduce the electron–hole recombinations at the interface. In hematite, the hole diffusion length is very small (ca. a few

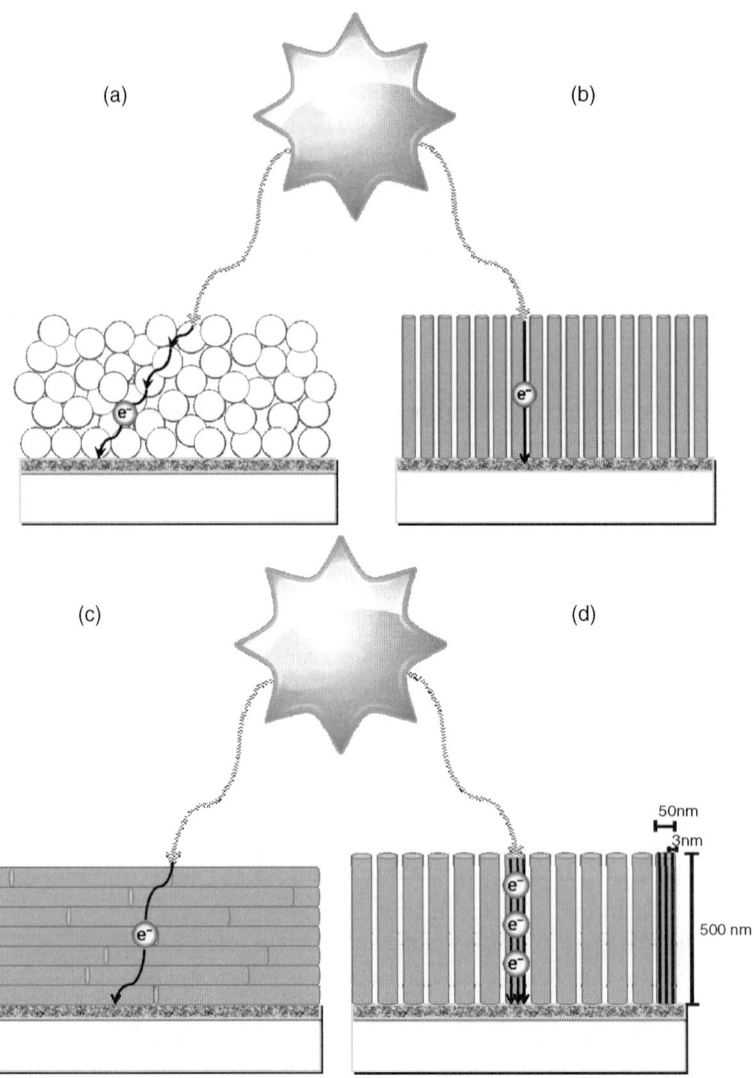

Figure 17.6 Scheme of the conventional (a) spherical nanoparticulate and the proposed (b, c) nanorod-based and (d) quantum rod-based n-type semiconductor solar cells grown onto transparent conducting oxide substrate. (Adapted from N. Beermann, L. Vayssieres, S.-E. Lindquist and A. Hagfeldt Photoelectrochemical studies of oriented nanorod thin films of hematite, *Journal of the Electrochemical Society*, **147**(7), 2456–2461, 2000, The Electrochemical Society.)

nm [177]), which has been identified as the major cause of recombination and low efficency. However, the shorter the distance the holes have to travel to reach the interface, the faster the interfacial recombination of the holes with the redox species in solution and the longer the life of the photogenerated electrons. Such a process will enable better generation and collection of photogenerated electrons, and therefore, a higher photoefficiency will be reached (Figure 17.6d).

The synthesis involved the perpendicular growth and thermodynamic stabilization of the anisotropic and metastable phase of ferric oxyhydroxide (akaganeite, β-FeOOH) according to theoretical concepts and chemical approach developed by the author. Subsequently, a phase transition was induced by heat treatment in air. As a result, hematite designed with the required anisotropic-oriented configuration (i.e., 3D nanorod array) was obtained. Stabilization of the akaganeite phase is required because the growth of β-FeOOH from solution is subject to a structural and morphological change from rod-like nanoparticles to α-Fe_2O_3 spherical or cubic ones [178]. However, lowering the interfacial tension significantly reduced the aging process and suppressed the phase and morphological transformation into hematite. According to the general concept, by lowering the pH of precipitation to a minimum value (i.e., where the precipitation is barely occurring), that is, far from the point of zero charge (PZC) and by setting the ionic strength of the synthetic medium at a high concentration of noncomplexant ions, stabilized and highly charged anisotropic nanoparticles of akaganeite were obtained with controlled orientations. According to differential scanning calorimetry (DSC)/differential thermal analysis (DTA), a subsequent heat treatment in air above 390 °C provided the thermodynamically stable crystallographic phase of ferric oxide (i.e., hematite) skillfully designed with the required nanorod morphology [100].

Matching the diameter of the nanorods to the hole diffusion length of hematite was obtained by taking advantage of the structural properties of akaganeite. β-FeOOH crystallizes in the tetragonal system (space group I4/m, $a = 10.44$, $c = 3.01$ Å). The structure can be described as a tunnel structure (similar to α-MnO_2) hosting H_2O or Cl^- and based on a defected close-packed oxygen lattice with three different kinds of oxygen layers. Every third layer is only two-thirds occupied, with rows of oxygens missing along the c-axis. The cation occupation of the octahedral sites between the other anion layers is in double rows, but separated by single rows of empty sites along the c-axis. The octahedral cation sites remaining between the third anion layer and its neighboring layers are completely filled. This structural configuration produces di-octahedral chains, which are arranged about the fourfold symmetry c-axis. The chains share vertices along their edges, forming square-cross-section tunnels, some 5 Å along the edge. The crystals are elongated along the c-axis and of rod-like morphology. These crystals have empty cores that produce a square channel of about 3 nm on a side. The anisotropic crystals form a bundle called a somatoid. Such structural particularities can be used to generate quantum nanorods of hematite with diameters matching the hole diffusion length. Dehydration of β-FeOOH at high temperature leads to the α-Fe_2O_3 phase, which crystallizes in the trigonal crystal system.

The synthesis was performed with 0.15 M iron, pH 1.5 and 1 M sodium nitrate at 95 °C [98]. Aligned single-crystalline nanorods of typically 5 nm in diameter, self-assembled as bundle of about 50 nm in diameter were obtained in fairly perpendicular fashion on the transparent conducting substrate and arranged in very large uniform arrays (Figure 17.7). The length of the nanorods, which essentially represented the thickness of a homogeneous monolayer, could be experimentally tailored to any required dimension of up to about 1 μm by varying the time and/or the temperature of aging (i.e., nanorods of 100 nm in length were produced after about an hour at 95 °C). The aspect ratio is 1 to 20 for the nanorod bundle and 1–200 for the individual nanorods.

As previously stated, such quantum rods arranged in oriented 3D bundles of several tens of cm^2 patterned arrays have been designed for photoelectrochemical applications. Indeed, since the array features oriented and direct grain-boundary-free electron pathways, and given

Figure 17.7 Electron microscopy images of iron oxide (αFe_2O_3 or $\beta FeOOH$) quantum rods and highly oriented (multidirectional, parallel and perpendicular) arrays consisting of bundles of quantum rods grown onto transparent conducting oxides by aqueous chemical growth.

that the fine diameter of the nanorods matches the minority carrier diffusion length, fast electron–hole recombination is prevented, allowing fast generation, transfer and collection of photogenerated electrons.

Compared to previously reported data [175] on nanostructured hematite, consisting of a highly porous network of spherical nanoparticles measured in similar operating conditions (light, electrolyte, bias), the vertically oriented arrays showed an incident photon-to-electron conversion efficiency (IPCE) as high as 20% at 350 nm, obtained in a three-electrode set-up [109], which represents an improvement of more than 500 times. Moreover, sandwich-type cells (two-electrode set-up) were fabricated which showed IPCEs as high as 55% at 350 nm, which means that more than half of the incident photons are used to create electron–hole pairs. An open-circuit voltage of 0.3 V and short-circuit currents of several hundreds of μA were recorded at an illumination of $100\,W\,cm^{-2}$. Although the overall efficiency remained low (0.2%), due partly to the very low film thickness (0.5 μm) and hence the absorption of the thin film, such cells constituted the first demonstration of an iron-oxide photovoltaic cell (Figure 17.8).

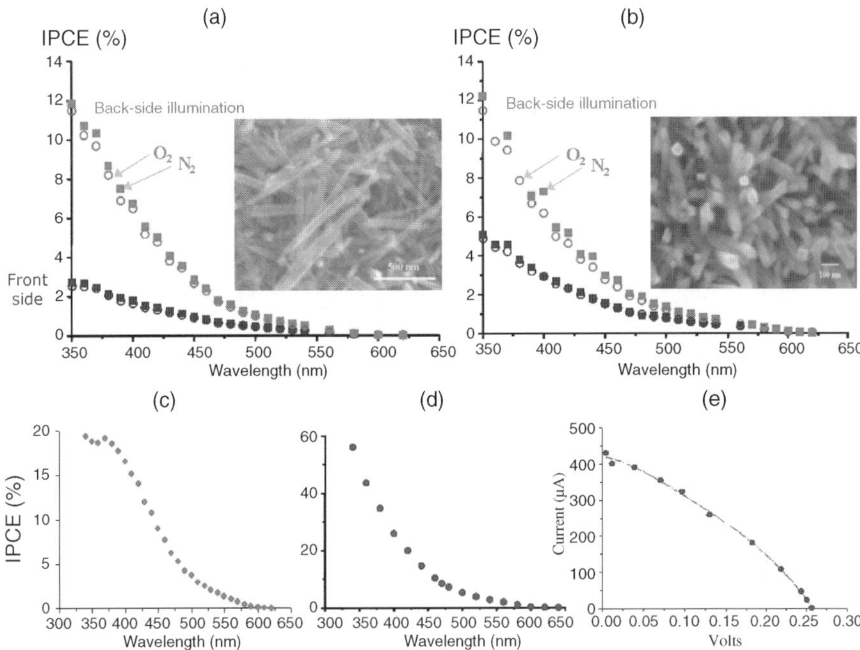

Figure 17.8 Photoelectrochemical characterization of 0.5 μm thick highly oriented nanorod-arrays of hematite (α-Fe_2O_3) grown on F-SnO_2 coated glass; Incident photon to electron conversion efficiency (IPCE) of nanorods with parallel (a) and perpendicular (b) orientation with respect to the substrate in 0.1 M KI aqueous solution and 0.01 M K_2HPO_4/KH_2PO_4 buffers (pH 6.8) purged in oxygen (○) or nitrogen (■) gas with an applied bias of +0.3 V vs Ag/AgCl under 450 W Xenon lamp irradiation (with 80 mm water filter to avoid IR) and high intensity monochromator. (c) IPCE of vertically oriented nanorod-array at pH 12 (same condition as above). (d) Two-electrode IPCE measurements (back side illumination) of nanorod-arrays in 0.5 M LiI + 50 mM I_2 in ethylene carbonate/propylene carbonate (50: 50% by weight). (e) I-V characteristics (overall efficiency) of the highly oriented nanorod-arrays under 1000 W xenon solar simulator (with 10 cm water filter).

This vertically oriented nanorod concept, along with its chemical design was first presented at international conferences in 1997 [176], 1998 [179] and 1999 [180] and the full article was published in 2000 [109].

Since then, other vertically oriented metal-oxide nanostructures have been fabricated (and shown successful), using the same concept and chemical methods for photovoltaic, as well as other applications, such as magnetics [105] and sensors, by the author [111–113], as well as by many others.

In addition to architecture design and, more importantly, bandgap and band-edge engineering, is the crucial issue scientists are facing to improve the stability and efficiency of new materials. Various methods, approaches and strategies are being explored, from calculations, simulations and imaging of model surfaces (see Chapters 2–4), to doping (see Chapter 18), to new morphological design of classical semiconductors (see Chapters 10–13), to supramolecular, nanoparticle and multidimensional assemblies (see Chapters 19, 16 and 15, respectively) to combinatorial approaches (see Chapter 14), to achieve such challenging goals.

17.4.1.3 Profiling Bandgap and Band Edges by Quantum Confinement

Indeed, the bandgap and band-edge positions, as well as the overall band structure of semiconductors are of crucial importance in photovoltaic and photocatalytic cells and even more drastically for direct water-splitting applications. The energy position of the band-edge level can be controlled by the electronegativity of the dopants, solution pH (flatband potential variation of 60 mV per pH unit), as well as by quantum-confinement effects. Accordingly, band edges and the bandgap can be tailored to achieve specific electronic structure and optical properties, as well as photocatalytic reactions. The most important application is found in the generation of H_2 from direct photo-oxidation of water with (and more importantly without) external bias. Indeed, to succeed in directly splitting water via solar irradiation at the interface of a semiconductor, the top of the valence band of the semiconductor has to be located at a lower energy level than the chemical potential of dioxygen evolution (H_2O/O_2), and the bottom of the conduction band has to be located at a higher energy level than the chemical potential of dihydrogen evolution (H_2/H^+). If the positions of the energy levels of the valence and conduction bands are not fulfilled, an external bias has to be applied to induce the photocatalytic process, which in turn substantially reduces the overall efficiency.

According to thermodynamics alone, the necessary potential to split water is found to be of 1.23 V; however, a semiconductor which possess such a bandgap would very quickly photocorrode upon illumination. It is clearly established in the literature that a suitable bandgap is around 2 eV to avoid degradation. Indeed, it has been reported that a band gap of 2.46 eV [181] is necessary for water photolysis without an external bias. Although the bandgap of hematite, reported to be around 1.9–2.2 eV (depending on its crystalline status and methods of preparation), and its valence-band edge are suitable for oxygen evolution, the conduction-band edge of hematite is too low to generate hydrogen. Therefore, a blue-shift of the bandgap of hematite of about 0.3 eV and an upward shift of the conduction-band edge would make hematite an ideal anode material for photocatalytic devices for the photo-oxidation of water.

In-depth investigations of the electronic structure of such novel nanostructures have been carried out at synchrotron radiation facilities. Such studies include polarization-dependent [115] soft X-ray absorption and emission measurements [116], as well as energy-dependent resonant inelastic X-ray scattering [118]. Both synchrotron-radiation-based soft-X-ray absorption spectroscopy (XAS) which maps the unoccupied states of the materials (conduction band) and soft-X-ray emission spectroscopy (XES) which maps the occupied state (valence band) of a variety of nanostructures (see Chapters 5–7) have brought important information on bandgap, Fermi energy and work function, as well as other band-edge characteristics, such as conduction orbital character (Figure 17.9).

With high-resolution monochromatized synchrotron-radiation excitation, resonant inelastic X-ray scattering (RIXS) has emerged as a new source of information for electronic structure and excitation dynamics of nanomaterials. The selectivity of the excitation, in terms of energy and polarization, has also facilitated studies of emission anisotropy, as well as for probing the optical transitions in transition-metal oxides, especially *d–d* transitions, which are in other spectroscopies (e.g., electron energy loss spectroscopy (EELS) or UV-visible spectrophotometry), dipole forbidden and thus of very low intensity. Moreover, angle-resolved XAS and XES spectroscopies along with density functional theory (DFT) calculations have also been carried out to probe the orbital character and symmetry of the conduction band of such important large-bandgap semiconductors (Figure 17.9). The results reveal

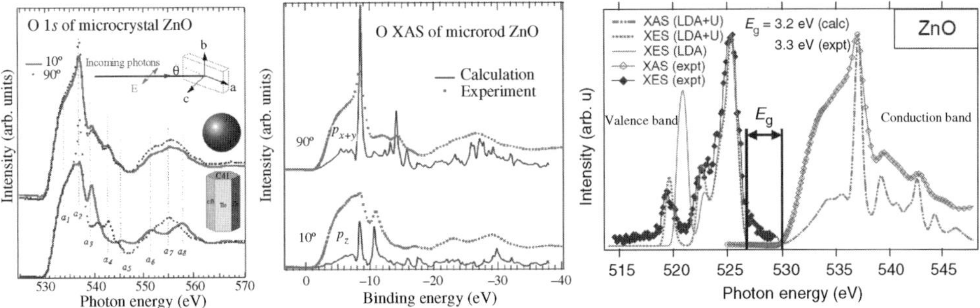

Figure 17.9 (left) O K-edge polarization dependent X-ray absorption spectra (XAS) of crystalline 3D arrays of ZnO of isotropic and oriented anisotropic morphology. The inset illustrates the experimental XAS geometry, where the *a*- and *b*-axes define the sample-surface plane, the *c*-axis is along the growth direction of the ZnO rods, E is the polarization of the incoming photons and θ represents the incident angle to the sample surface: 10° ($E // c$, lines) and 90° ($E \perp c$, dots); (center) calculated polarization dependent O 1*s* XAS of wurzite bulk ZnO, where energy is referred to valence-band maximum; experimental (dots) and calculated (lines) O K-edge XAS spectra of ZnO as a function of the polarization angle. The calculated XAS spectra feature the partial density of states (DOS) and the transition probability to O $2p_{x+y}$ and $2p_z$ states with a Lorentzian broadening equivalent to the experimental spectral resolution (0.2 eV); (right) Experimental O 1s XAS and XES spectra of ZnO nanostructures and LDA + U calculation of bulk ZnO describes *d*-state localization, and thereby also the Zn-*d*–O-*p* hybridization. E_g is the electronic bandgap.

important fundamental and applied knowledge of direct relevance for semiconductor physics and technologies.

The Fe 2*p* absorption spectrum of an α-Fe_2O_3 quantum-rod array is displayed in (Figure 17.10a). The spectral shape appears very similar to previous XAS measurements conducted on polycrystalline [182] or single-crystal [183] samples (Figure 17.10c). The typical spectrum shows the spin–orbit interaction of the 2*p* core level that splits the L_2 ($2p_{1/2}$) and L_3 ($2p_{3/2}$) edges, and the *p–d* and *d–d* Coulomb and exchange interactions that cause multiplets within the edges. The ligand-field splitting of 3*d* transition metals, being of the same order of magnitude as *p–d* and *d–d* interactions (1–2 eV), gives a 1.4 eV energy splitting between the t_{2g} (xy, yz, xz) and e_g (x^2-y^2, $3z^2-r^2$) orbitals. The charge transfer has two main effects on the spectral shape [184]. It splits *d* levels by the formation of molecular orbitals, as well as giving rise to an asymmetric shape (tail). Such effects can clearly be observed on the higher-energy side of the edges, especially in the 711–718 eV regions (L_3) of the experimental spectra (Figure 17.10).

Figure 17.10b,d shows the Fe *L*-emission spectra recorded with a high-photon-energy excitation for the quantum rods, as well as for the single crystal. The spectral shape shows two peaks originating from the transitions of 3*d* orbitals to $2p_{1/2}$ and $2p_{3/2}$ core levels. A branching ratio (L_β/L_α) of 0.8 is found for the α-Fe_2O_3 bundled nanorod arrays, which appears substantially larger than that of a α-Fe_2O_3 single crystal [183]. It has been demonstrated that the intensity ratio $I(L_\beta)/I(L_a)$ varies due to the occupancy of L_2 and L_3 levels, which depends on the chemical state of the elements [185]. Skinner *et al.* [186] showed that the ratio $I(L_\beta)/I(L_\alpha)$ of 3*d* transition metals and alloys is very small due to the non-radiative Coster–Kronig transition $L_2L_3M_{4,5}$. The probability of such a transition is distinctly lower for 3*d* oxides than for metals, due to the presence of an energy gap. Recently, Kurmaev *et al.* [187] found that the ratio $I(L_\beta)/I$

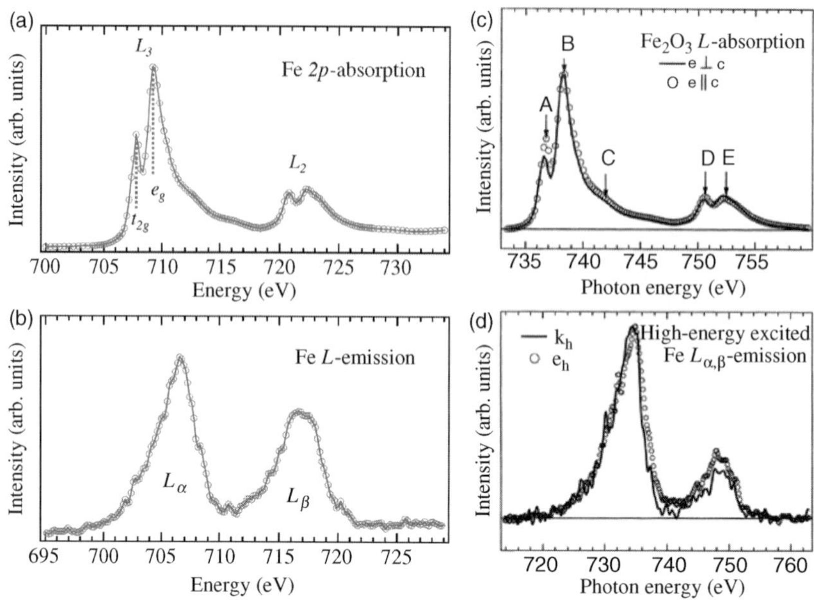

Figure 17.10 (a, c) Fe 2*p* absorption and (b, d) normal *L*-emission spectra of α-Fe_2O_3 nanorod array and single crystal. Reprinted with permission from L. Vayssieres, C. Sathe, S.M. Butorin *et al.*, One-dimensional quantum-confinement effect in α-Fe_2O_3 ultrafine nanorod arrays. *Advanced Materials*, **17** (19), 2320–2323, 2005. © 2005 Wiley-VCH and from L.-C. Duda, J. Nordgren, G. Drager *et al.*, Polarized resonant inelastic X-ray scattering from single crystal transition metal oxides. *Journal of Electron Spectroscopy and Related Phenomena*, **110–111**(1–3), 275–285, 2000. © 2000 Elsevier.

(L_α) of the molecular superconductor $(ET)_4[(H_3Ofe(C_2O_4)_3)C_6H_5CN$ was twice as big as that of iron oxides because of the highly ionic and insulating character of the oxalate layers, as well as the localization of the 3*d* electron density. The present experimental observation of a higher branching ratio $I(L_\beta)/I(L_\alpha)$ suggests the possibility of a larger bandgap in hematite nanorods. In the RIXS process, final states probed via such a channel are related to eigenvalues of the ground-state Hamiltonian. The core-hole lifetime does not limit the resolution of this spectroscopy. According to the many-body picture, the energy of a photon, scattered on a certain low-energy excitation, should change by the same amount as a change in an excitation energy of the incident beam (see the decay route of core-excitations e_g versus that of t_{2g} in the inset of Figure 17.9), so that RIXS features have constant energy losses and follow the elastic peak on the emitted-photon energy scale. In such an octahedral symmetry, a d^5-configuration is found to have well-separated *d–d* excitations. Optical absorption spectroscopy of α-Fe_2O_3 has revealed many transitions ranging from the infrared to the ultraviolet regions. In a typical spectrum, well-defined features are observed at 2.1 eV, 3.3 eV and 5.6 eV [188,189], which are assigned to the following electronic transitions ${}^6A_1 \rightarrow {}^4A_1$, ${}^6t_{1u} \rightarrow {}^2t_{2g}$ and ${}^6t_{1u} \rightarrow e_g$, respectively. These features are also found in the RIXS spectrum of an α-Fe_2O_3 single crystal [183]. In such a process, the *d–d* excitations in α-Fe_2O_3 are specifically probed by the transition sequence $2p^63d^5 \rightarrow 2p^53d^6 \rightarrow 2p^63d^5$. These *d–d* transitions become fully allowed, and their intensity can be more easily calculated compared to optical spectroscopy and EELS.

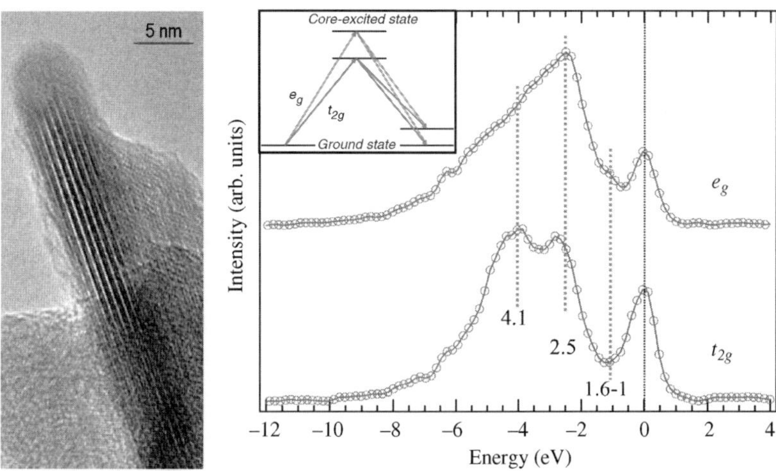

Figure 17.11 High-resolution transmission electron microscopy (HRTEM) (left) and energy-dependent resonant inelastic X-ray scattering spectra (right) of α-Fe_2O_3 quantum rod-array. The inset shows a schematic representation of the radiative de-excitation process for the two core excitations e_g and t_{2g}. Reprinted from L. Vayssieres, C. Sathe, S.M. Butorin *et al.*, One-dimensional quantum-confinement effect in α-Fe_2O_3 ultrafine nanorod arrays. *Advanced Materials*, **17**(19), 2320–2323, 2005, Wiley-VCH.)

The RIXS spectrum recorded at the Fe *L*-edge of α-Fe_2O_3 nanorods is shown in Figure 17.11. Several energy-loss features are clearly resolved. The low-energy excitations, such as the strong *d–d* and charge-transfer excitations, are identified in the region from 1 to 5 eV. The 1 eV and 1.6 eV energy-loss features originate from multiple excitation transitions. The 2.5 eV excitation, which corresponds to the bandgap transition, appears significantly blue-shifted compared to the reported 2.1∼2.2 eV bandgap of single-crystal and polycrystalline hematite samples, as suggested above by the higher L_β/L_α branching ratio observed in the emission spectrum.

Synchrotron-based spectroscopic investigations of 1D nanomaterials consisting of designed bundled quantum rods of hematite reveal a significant increase in the bandgap compared to bulk (polycrystalline and single-crystal) samples. Such a finding is successfully attributed to a 1D (lateral dimension) quantum-confinement effect. Such conclusions strongly suggest that such designed nanomaterials would meet the bandgap [181] and band-edge [190] criteria (Figure 17.12) for the photocatalytic oxidation of water without an applied bias. Indeed, an increase in bandgap of 0.3 eV found in the quantum-rod arrays would bring the top of the conduction-band edge of hematite to an energy level very similar to the level of H^+/H_2. The outcome of such a result is of great importance for the direct solar production of hydrogen, as it may thus be possible to evolve hydrogen with a limited applied bias of a few hundred mV compared to several hundred in regular samples.

17.4.2 Doped Iron-Oxide Quantum-Rod Arrays

After successfully: (i) developing cost-effective fabrication at low temperature of oriented nanorods of iron oxide and (ii) increasing the bandgap of hematite to 2.5 eV to ensure enhanced stability against photocorrosion and concomitantly change the conduction band edge to a more favorable position compared to H_2/H_2O levels (therefore reducing the value of the applied

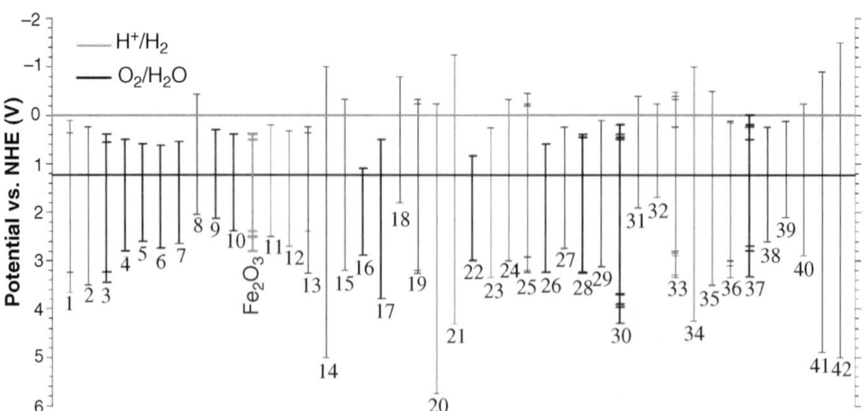

Figure 17.12 Literature survey of band edge positions of hematite and other metal oxide semiconductors: (1) $BaTiO_3$; (2) $Bi_{12}GeO_{20}$; (3) Bi_2O_3; (4) Cd_2SnO_4; (5) $Cd_2Fe_2O_4$; (6) CdO (7) $CdSnO_3$; (8) $Cr_2Ti_2O_7$; (9) Cu_3WO_6; (10) $CuWO_4$; (11) Fe_2TiO_5; (12) $FeTa_2O_6$; (13) Fe-TiO_3; (14) HfO_2; (15) $Hg_2Nb_2O_7$; (16) $Hg_2Ta_2O_7$; (17) In_2O_3; (18) $InTaO_4$; (19) $KtaO_3$; (20) La_2O_3; (21) $LaAlO_3$; (22) $LuRhO_3$; (23) $MnTiO_3$; (24) $NaNbO_3$; (25) Nb_2O_5; (26) NbO_2; (27) $PbFe_{12}O_{19}$; (28) PbO; (29) Sb2O3; (30) SnO_2; (31) Sr_2NbFeO_6; (32) $Sr_3FeNb_2O_9$; (33) $SrTiO_3$; (34) $SrZrO_3$; (35) Ta_2O_5; (36) TiO_2; (37) WO_3; (38) YFe_2O_3; (39) $ZnFe_2O_4$; (40) ZnO; (41) ZrO_2; (42) $ZrSiO_4$ with respect to the redox potential of H_2O/H_2 and O_2/H_2O (Adapted from J.A. Glasscock, Nanostructured materials for photoelectrochemical hydrogen production using sunlight, PhD Thesis, University of New South Wales, Australia, 1–220, 2008).

bias), a final step needed to be reached to improve the optical absorption of the materials with regards to the solar spectrum to full visible-light absorption.

The strategy is to use new concepts based on quantum-confinement effects to develop novel semiconductor nanostructures whose absorption profiles offer a better match with the solar spectrum, for optimal solar conversion; that is absorbing in the visible range, as well as part of the infrared yet maintaining good stability against photocorrosion. Indeed, semiconductors with low bandgaps (less than 2 eV) will absorb in the visible range, but will degrade very quickly under illumination. Our strategy to reach ideal visible-light-active and stable photocatalysts with high efficiency at low cost and large scale included the unique combination of the two major routes to large-scale fabrication of materials, that is, aqueous solution and gas-phase synthesis for the single purpose of fabricating *visible-light-active* materials. Indeed, such a merge allows the large-scale fabrication of virtually any structure, including quantum-confined, hybrid, mixed, doped and composite materials.

As we have seen earlier, various possibilities are available, using quantum effects, as well as various device architectures, such as tandem and multi-junctions, as well as plasmonic structures, to improve efficiency and absorption of the photovoltaic/photocatalytic devices. MEG is still controversial and does not really apply to metal oxides, considering the nature of the band, as well as the bandwidth (*d* bands are very narrow). We chose to use the concept of intermediate bands, given that the bandgap of iron oxide is sufficiently large to accommodate impurity levels/intermediate bands to fully absorb visible light (while maintaining stability against photocorrosion).

Figure 17.13 (Top) High-resolution transmission electron microscopy images of a doped iron-oxide quantum-rod bundle and the corresponding energy-dispersive elemental mapping analysis of oxygen, iron and a single dopant showing its homogeneous distribution all over the rods. (Bottom) SEM images of vertically oriented doped iron-oxide quantum-rod arrays grown onto transparent conducting oxides. See also color-plate section.

Based on the abovementioned fundamentals of quantum-confinement effects and synthesis strategy, excellent groundwork results have been obtained by the author, showing excellent performance of these novel nanostructures. Before doping, the typical color of the samples is orange-red. After doping, the samples are black. Figure 17.13 shows the high-resolution electron microscopy energy-dispersive elemental mapping analysis of the doped iron-oxide

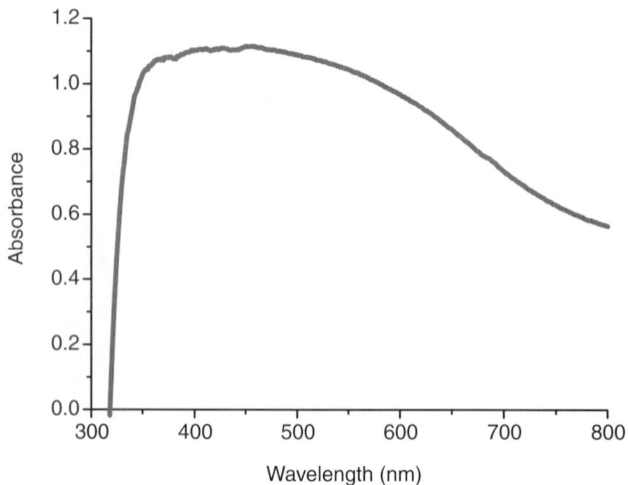

Figure 17.14 UV-visible optical absorption of a 0.5 μm thin film consisting of vertically oriented doped iron-oxide quantum-rod arrays grown on F-SnO_2 substrate. The contribution of the substrate has been subtracted.

quantum-rod arrays. The results show that the dopant is homogeneously distributed all over the rods. Figure 17.14 shows the optical absorption spectrum, which reveals very strong absorption of the nanostructures in the visible region. Indeed, an optical absorbance of 1 is recorded between 350 and 600 nm and up to 0.6 at 800 nm. The materials also absorb in the near-IR. Clearly, the goal has been achieved in terms of full visible absorption.

The photoelectrochemical properties of such samples have also been recorded. Figure 17.15 shows the dynamic action spectrum upon illumination (one sun) in aqueous phosphate buffer

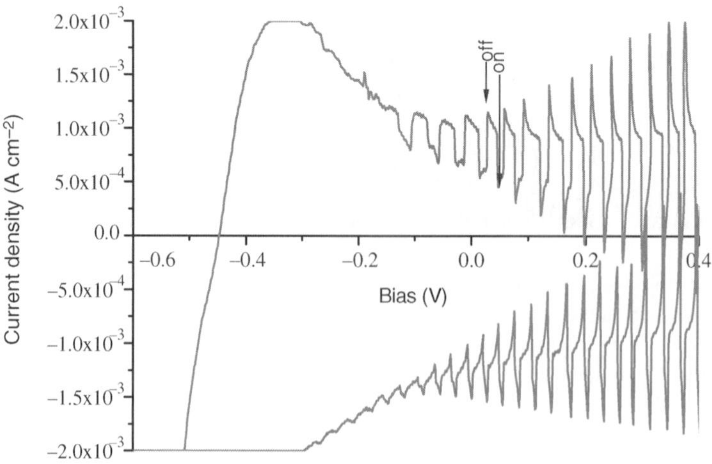

Figure 17.15 Photoelectrochemical characterization of the doped iron-oxide quantum-rod arrays at room temperature in aqueous phosphate buffer solution (pH 7.2) under chopped one-sun illumination.

electrolyte (pH 7.2) at room temperature. A very decent photocurrent (around 1–2 mA cm^{-2}) is recorded at relatively low applied bias (from 0 to 0.4 V versus SCE). The samples show good stability in water (although long-term stability tests upon illumination remain to be performed) and surprisingly fast interfacial kinetics. Indeed, when chopping the light (manually, by approximately 1 s intervals), the photocurrent recovery is very fast, reaching the same values within a second or so. Such results undoubtedly prove that these new visible-light-active nanostructures, consisting of homogeneously doped iron-oxide oriented bundles of quantum rods, are stable in air, water and upon illumination and show a full visible-light absorption profile, fast carrier injection and interfacial kinetics, in aqueous solution at neutral pH and room temperature, good photocurrent at low bias, and are one of the best candidates for large-scale production of hydrogen by direct conversion of solar energy. The dopant is not disclosed due to intellectual property issues. It is worth pointing out that a strong basic medium (i.e., concentrated NaOH for iron oxide) or strong acidic medium (i.e., concentrated H_2SO_4 for tungsten oxide) is no longer necessary to obtain relatively good efficiency. The use of a noncorrosive electrolyte, and possibly seawater, becomes available and is of crucial importance for the development of economical and safe devices capable of direct solar-to-hydrogen generation without wasting sacrificial agent or freshwater supplies.

17.4.3 Quantum-Dot–Quantum-Rod Iron-Oxide Heteronanostructure Arrays

Quantum-dot solar cells [191–196], as well as inorganic [197–201] and metal sensitization [202–224] of semiconductors is an efficient way to tailor optical absorption and photoelectrochemical efficiency of semiconductors (see Chapter 9). To develop additional visible-light-active materials, the fabrication of heteronanostructures of iron oxide was carried out. These novel structures consist of quantum dots grown onto quantum rods of iron oxide. The development of such advanced materials is still in the preliminary stage, but shows very promising results in terms of visible-light absorption. Figure 17.16 shows the UV-visible

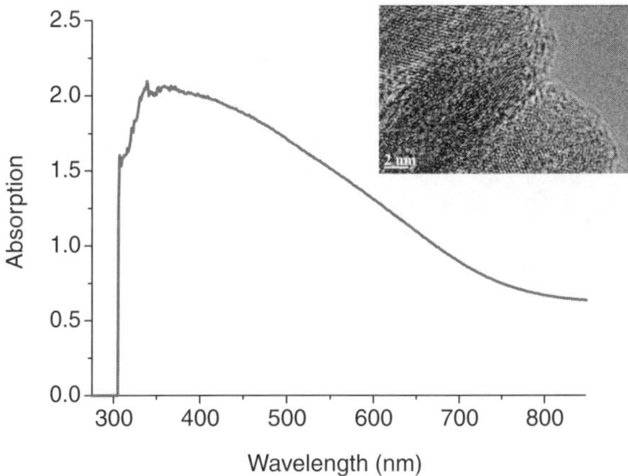

Figure 17.16 UV-visible absorption of an iron-oxide quantum-dot–quantum-rod heteronanostructure array. The inset shows an HRTEM image of such an array.

optical properties of a quantum-dot–quantum-rod iron-oxide heteronanostructure array. A very strong absorbance is found in the visible range and the absorption profile is an excellent match to the solar spectrum.

17.4.4 Iron Oxide Oriented Porous Nanostructures

Recent developments in ordered porous nanostructures (see Chapter 10) consisting of metal-oxides nanotube (doped TiO_2, TaON, etc.) have demonstrated such materials as very

Figure 17.17 High-resolution electron microscopy images of novel iron-oxide-based porous nanostructures fabricated by aqueous chemical growth at 95 °C for efficient solar hydrogen generation: (a) porous nanotubes with 10 and 50 nm inner and outer diameters, respectively. The walls of the tubes are highly porous, with pore diameters in the order of 1–3 nm (b). Bright-field image of the cross-section of highly oriented iron oxide porous nanostructures (c).

promising candidates for cost-effective solar hydrogen generation. Following the same design (but with a different fabrication technique to those structures typically created by anodization of metals), we have developed advanced oriented porous nanostructures consisting of ordered nanotubes of typically 10–15 nm in inner diameter and about 40–50 nm in outer diameter, as well as other structures where additional porosity is found in the walls of the nanotubes. Besides the always beneficial overall increase in surface area, the idea behind it is to increase the kinetics of the water-splitting reaction at the water–oxide interface by providing faster/easier access of water to the interface by fast release of H_2/O_2 gases at the pore sites, and thus better gas collection, avoiding, therefore, the limiting step of gas release. Indeed, tailoring the pore size allows us to play on the pressure of the gas bubble and thus, on its kinetics of release. The faster the gas bubble is released, the faster the next molecule of water can reach the interface. Such a design should significantly increase the kinetics of the overall process and thus the efficiency of the device. Figures 17.17 and 17.18 show the different type of porous nanostructures fabricated by aqueous chemical growth. Such porous nanostructures are currently being tested for water oxidation and excellent preliminary results confirm the relevance of such a highly porous and oriented design to increasing the efficiency of direct solar-to-hydrogen generation.

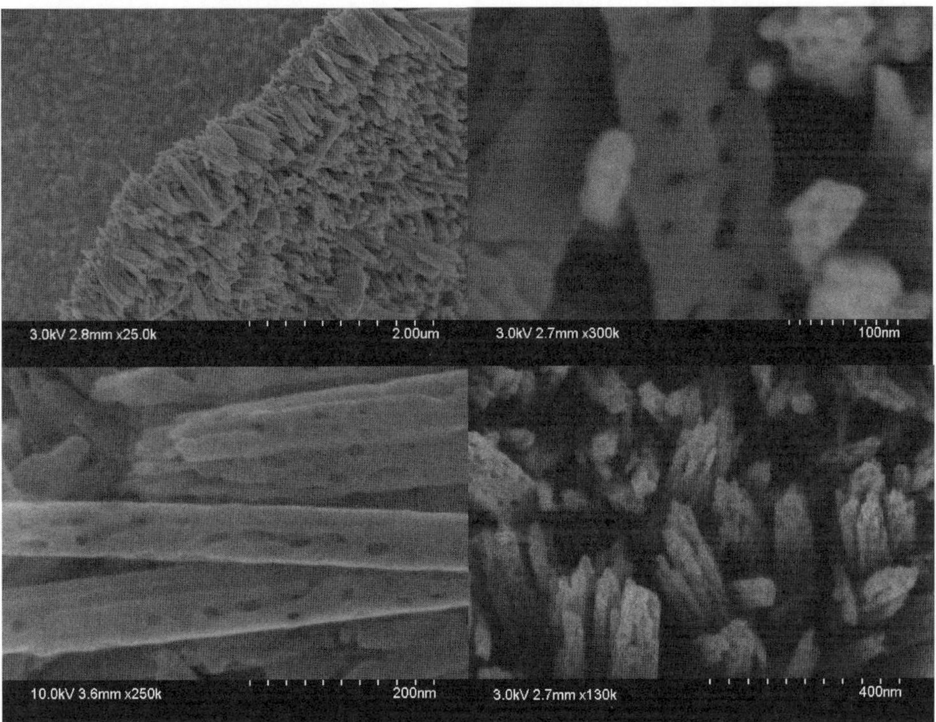

Figure 17.18 Field-emission scanning electron micrographs of highly oriented porous nanorod arrays of hematite grown on F-SnO_2 transparent conducting substrates at 90 °C by aqueous chemical growth.

17.5 Conclusion and Perspectives

A new approach using smart materials design and quantum-confinement effects along with simple and cost-effective fabrication of advanced metal-oxide materials for direct solar-to-hydrogen generation by photocatalytic oxidation of water under visible-light conditions was described in this chapter. This new generation of visible-light-active iron-oxide-based semiconducting nanostructures represents an excellent candidate for mass production of efficient photocatalysts. They comply with the drastic requirements of photocatalysts for water oxidation, in terms of stability (in water and under illumination), in terms of light absorption (visible-light active), in terms of safety (nontoxic) and in terms of cost (low fabrication and operating costs). Moreover, it demonstrates once again the advances in the application of nanoscience and nanotechnology in enabling a future hydrogen economy by fabricating novel visible-light-active materials using novel concepts and simultaneously reducing their cost of fabrication. The use of such new materials may render solar hydrogen generation commercially viable (e.g., the US department of Energy (DOE) has set a price goal of US$ 2–3 per kg of hydrogen, which includes production, delivery and dispensing for hydrogen to compete with gasoline for passenger vehicles). [88].

It goes without saying that additional academic and industrial R&D is required for the complete optimization of those nanostructures in terms of overall efficiency, doping density, rod aspect ratio and density, and long-term stability, as well as their large-scale fabrication and indoor/outdoor testing as 10×10 cm modules and/or 30×30 cm panels. Nevertheless, they will certainly pave the way for a new generation of low-cost materials and devices from which one can easily imagine the creation of off-shore power plants dedicated to renewable direct solar-to-hydrogen fuel generation (using existing facilities for gas and oil), where seawater would be used as the sole electrolyte feedstock, and hydrogen and oxygen would be generated and transported to the main land via gas pipelines. After all, the most abundant and freely available resources of this blue planet are sunlight and seawater.

References

[1] Yoshida, M., Takanabe, K., Maeda, K. *et al.* (2009) Role and function of Noble-Metal/Cr-Layer Core/Shell structure cocatalysts for photocatalytic overall water splitting studied by model electrodes. *Journal of Physical Chemistry C*, **113**(24), 10151–10157.

[2] Ang, T.P., Toh, C.S. and Han, Y.-F. (2009) Synthesis, characterization, and activity of visible-light-driven nitrogen-doped TiO_2-SiO_2 mixed oxide photocatalysts. *Journal of Physical Chemistry C*, **113**(24), 10560–10567.

[3] Yang, X., Wolcott, A., Wang, G. *et al.* (2009) Nitrogen-doped ZnO nanowire arrays for photoelectrochemical water splitting. *Nano Letters*, **9**(6), 2331–2336.

[4] Mohapatra, S.K., John, S.E., Banerjee, S. and Misra, M. (2009) Water photooxidation by smooth and ultrathin α-Fe_2O_3 nanotube arrays. *Chemistry of Materials*, **21**(14), 3048–3055.

[5] Amano, F., Nogami, K. and Ohtani, B. (2009) Visible light-responsive bismuth tungstate photocatalysts: effects of hierarchical architecture on photocatalytic activity. *Journal of Physical Chemistry C*, **113**(4), 1536–1542.

[6] Wei, W., Dai, Y. and Huang, B. (2009) First-principles characterization of bi-based photocatalysts: $Bi_{12}TiO_{20}$, $Bi_2Ti_2O_7$, and $Bi_4Ti_3O_{12}$. *Journal of Physical Chemistry C*, **113**(14), 5658–5663.

[7] Higashi, M., Abe, R., Takata, T. and Domen, K. (2009) Photocatalytic overall water splitting under visible light using $ATaO_2N$ (A = Ca, Sr, Ba) and WO_3 in a IO_3^-/I^- shuttle redox mediated system. *Chemistry of Materials*, **21** (8), 1543–1549.

[8] Saremi-Yarahmadi, S., Wijayantha, K.G.U., Tahir, A.A. and Vaidhyanathan, B. (2009) Nanostructured α-Fe_2O_3 electrodes for solar driven water splitting: effect of doping agents on preparation and performance. *Journal of Physical Chemistry C*, **113**(12), 4768–4778.

[9] Zhong, D.K., Sun, J., Inumaru, H. and Gamelin, D.R. (2009) Solar water oxidation by composite catalyst/ α-Fe_2O_3 photoanodes. *Journal of the American Chemical Society*, **131**(17), 6086–6087.

[10] Chen, H., Wen, W., Wang, Q. *et al.* (2009) Preparation of $(Ga_{1-x}Zn_x)(N_{1-x}O_x)$ photocatalysts from the reaction of NH_3 with Ga_2O_3/ZnO and $ZnGa_2O_4$: *in situ* time-resolved XRD and XAFS studies. *Journal of Physical Chemistry C*, **113**(9), 3650–3659.

[11] Lin, W.-C., Yang, W.-D., Huang, I.-L. *et al.* (2009) Hydrogen production from methanol/water photocatalytic decomposition using Pt/$TiO_{2-x}N_x$ catalyst. *Energy and Fuels*, **23**(4), 2192–2196.

[12] Cesar, I., Sivula, K., Kay, A. *et al.* (2009) Influence of feature size, film thickness, and silicon doping on the performance of nanostructured hematite photoanodes for solar water splitting. *Journal of Physical Chemistry C*, **113**(2), 772–782.

[13] Hwang, Y.J., Boukai, A. and Yang, P. (2009) High density n-Si/n-TiO_2 core/shell nanowire arrays with enhanced photoactivity. *Nano Letters*, **9**(1), 410–415.

[14] Long, R. and English, N.J. (2009) Synergistic effects of Bi/S codoping on visible light-activated anatase TiO_2 photocatalysts from first principles. *Journal of Physical Chemistry C*, **113**(19), 8373–8377.

[15] Jang, J.S., Lee, J., Ye, H. *et al.* (2009) Rapid screening of effective dopants for Fe_2O_3 photocatalysts with scanning electrochemical microscopy and investigation of their photoelectrochemical properties. *Journal of Physical Chemistry C*, **113**(16), 6719–6724.

[16] Periyat, P., McCormack, D.E., Hinder, S.J. and Pillai, S.C. (2009) One-pot synthesis of anionic (Nitrogen) and cationic (Sulfur) codoped high-temperature stable, visible light active, anatase photocatalysts. *Journal of Physical Chemistry C*, **113**(8), 3246–3253.

[17] Xu, S., Feng, D. and Shangguan, W. (2009) Preparations and photocatalytic properties of visible-light-active zinc ferrite-doped TiO_2 photocatalyst. *Journal of Physical Chemistry C*, **113**(6), 2463–2467.

[18] Livraghi, S., Chierotti, M.R., Giamello, E. *et al.* (2008) Nitrogen-doped titanium dioxide active in photocatalytic reactions with visible light: A multi-technique characterization of differently prepared materials. *Journal of Physical Chemistry C*, **112**(44), 17244–17252.

[19] Liu, H., Yuan, J., Shangguan, W. and Teraoka, Y. (2008) Visible-light-responding $BiYWO_6$ solid solution for stoichiometric photocatalytic water splitting. *Journal of Physical Chemistry C*, **112**(23), 8521–8523.

[20] Ding, Q.-P., Yuan, Y.-P., Xiong, X. *et al.* (2008) Enhanced photocatalytic water splitting properties of $KNbO_3$ nanowires synthesized through hydrothermal method. *Journal of Physical Chemistry C*, **112**(48), 18846–18848.

[21] Allendorf, M.D., Diver, R.B., Siegel, N.P. and Miller, J.E. (2008) Two-step water splitting using mixed-metal ferrites: thermodynamic analysis and characterization of synthesized materials. *Energy and Fuels*, **22**(6), 4115–4124.

[22] Arai, N., Saito, N., Nishiyama, H. *et al.* (2008) Photocatalytic activity for overall water splitting of RuO_2-Loaded $Y_xIn_{2-x}O_3$ ($x = 0.9$–1.5). *Journal of Physical Chemistry C*, **112**(13), 5000–5005.

[23] Bae, S.-T., Shin, H., Kim, J.Y. *et al.* (2008) Roles of MgO coating layer on mesoporous TiO_2/ITO electrode in a photoelectrochemical cell for water splitting. *Journal of Physical Chemistry C*, **112**(26), 9937–9942.

[24] Cole, B., Marsen, B., Miller, E. *et al.* (2008) Evaluation of nitrogen doping of tungsten oxide for photoelectrochemical water splitting. *Journal of Physical Chemistry C*, **112**(13), 5213–5220.

[25] Hu, Y.-S., Kleiman-Shwarsctein, A., Forman, A.J. *et al.* (2008) Pt-Doped α-Fe_2O_3 thin films active for photoelectrochemical water splitting. *Chemistry of Materials*, **20**(12), 3803–3805.

[26] Kleiman-Shwarsctein, S., Hu, Y.-S., Forman, A.J. *et al.* (2008) Electrodeposition of α-Fe_2O_3 doped with Mo or Cr as photoanodes for photocatalytic water splitting. *Journal of Physical Chemistry C*, **112**(40), 15900–15907.

[27] Kadowaki, H., Saito, N., Nishiyama, H. *et al.* (2007) Overall splitting of water by RuO_2-loaded $PbWO_4$ photocatalyst with d10s2-d0 configuration. *Journal of Physical Chemistry C*, **111**(1), 439–444.

[28] Mohapatra, S.K., Misra, M., Mahajan, V.K. and Raja, K.S. (2007) Design of a highly efficient photoelectrolytic cell for hydrogen generation by water splitting: application of $TiO_{2-x}C_x$ nanotubes as a photoanode and Pt/TiO_2 nanotubes as a cathode. *Journal of Physical Chemistry C*, **111**(24), 8677–8685.

[29] Maeda, K., Teramura, K., Lu, D. *et al.* (2007) Roles of Rh/Cr_2O_3 (Core/Shell) nanoparticles photodeposited on visible-light-responsive $(Ga_{1-x}Znx)(N_{1-x}Ox)$ solid solutions in photocatalytic overall water splitting. *Journal of Physical Chemistry C*, **111**(20), 7554–7560.

[30] Lee, Y., Teramura, K., Hara, M. and Domen, K. (2007) Modification of $(Zn_{1+x}Ge)(N_2O_x)$ solid solution as a visible light driven photocatalyst for overall water splitting. *Chemistry of Materials*, **19**(8), 2120–2127.

[31] Reyes-Gil, K.R., Reyes-García, E.A. and Raftery, D. (2007) Nitrogen-doped In_2O_3 thin film electrodes for photocatalytic water splitting. *Journal of Physical Chemistry C*, **111**(39), 14579–14588.

[32] Chiaramonte, T., Cardoso, L.P., Gelamo, R.V., Fabreguette, F., Sacilotti, M., Marco de Lucas, M.C., Imhoff, L., Bourgeois, S., Kihn, Y. and Casanove, M.-J. (2006) Characterization of ruthenium oxide nanocluster as a cocatalyst with $(Ga_{1-x}Zn_x)(N_{1-x}O_x)$ for photocatalytic overall water splitting. *Journal of Physical Chemistry B*, **110**(9), 4500–4501.

[33] Abe, R., Higashi, M., Sayama, K. *et al.* (2006) Photocatalytic activity of R_3MO_7 and $R_2Ti_2O_7$ (R = Y, Gd, La; M = Nb, Ta) for water splitting into H_2 and O_2. *Journal of Physical Chemistry B*, **110**(5), 2219–2226.

[34] Maeda, K., Teramura, K., Lu, D. *et al.* (2006) Characterization of Rh-Cr mixed-oxide nanoparticles dispersed on $(Ga_{1-x}Zn_x)(N_{1-x}O_x)$ as a cocatalyst for visible-light-driven overall water splitting. *Journal of Physical Chemistry B*, **110**(28), 13753–13758.

[35] Maeda, K., Teramura, K., Masuda, H. *et al.* (2006) Efficient overall water splitting under visible-light irradiation on $(Ga_{1-x}Zn_x)(N_{1-x}O_x)$ dispersed with Rh-Cr mixed-oxide nanoparticles: effect of reaction conditions on photocatalytic activity. *Journal of Physical Chemistry B*, **110**(26), 13107–13112.

[36] Cesar, A., Kay, J.A., Martinez, Gonzalez and Grätzel, M. (2006) Translucent thin film Fe_2O_3 photoanodes for efficient water splitting by sunlight: nanostructure-directing effect of Si-doping. *Journal of the American Chemical Society*, **128**(14), 4582–4583.

[37] Park, J.H., Kim, S. and Bard, A.J. (2006) Novel carbon-doped TiO_2 nanotube arrays with high aspect ratios for efficient solar water splitting. *Nano Letters*, **6**(1), 24–28.

[38] Wang, H. and Lewis, J.P. (2006) Second-generation photocatalytic materials: anion-doped TiO_2. *Journal of Physics C*, **18**, 421–434.

[39] Batzill, M., Morales, E.H. and Diebold, U. (2006) Influence of nitrogen doping on the defect formation and surface properties of TiO_2 rutile and anatase. *Physical Review Letters*, **96**, 026103.

[40] Ghicov, A., Macak, J.M., Tsuchiya, H. *et al.* (2006) Ion implantation and annealing for an efficient N-doping of TiO_2 nanotubes. *Nano Letters*, **6**, 1080–1082.

[41] Park, J.H., Kim, S. and Bard, A.J. (2006) Novel carbon-doped TiO_2 nanotube arrays with high aspect ratios for efficient solar water splitting. *Nano Letters*, **6**, 24–28.

[42] Kikuchi, H., Kitano, M., Takeuchi, M. *et al.* (2006) Extending the photoresponse of TiO_2 to the visible light region: photoelectrochemical behavior of TiO_2 thin films prepared by RF-magnetron sputtering deposition method. *Journal of Physical Chemistry B*, **110**, 5537–5541.

[43] Maeda, K., Teramura, K., Lu, D. *et al.* (2006) Photocatalyst releasing hydrogen from water. *Nature*, **440**, 295.

[44] Ryu, Abe, Sayama, K. and Sugihara, H. (2005) Development of new photocatalytic water splitting into H_2 and O_2 using two different semiconductor photocatalysts and a shuttle redox mediator IO_3^-/I^-. *Journal of Physical Chemistry B*, **109**(33), 16052–16061.

[45] Hur, S.G., Kim, T.W., Hwang, S.-J. *et al.* (2005) Synthesis of new visible light active photocatalysts of $Ba(In_{1/3}Pb_{1/3}M'_{1/3})O_3$ (M′ = Nb, Ta): A band gap engineering strategy based on electronegativity of a metal component. *Journal of Physical Chemistry B*, **109**(36), 17346–17346.

[46] Maeda, K., Takata, T., Hara, M. *et al.* (2005) GaN:ZnO solid solution as a photocatalyst for visible-lightdriven overall water splitting. *Journal of the American Chemical Society*, **127**, 8286–8287.

[47] Teramura, K., Maeda, K., Saito, T. *et al.* (2005) Characterization of ruthenium oxide nanocluster as a cocatalyst with $(Ga_{1-x}Zn_x)(N_{1-x}O_x)$ for photocatalytic overall water splitting. *Journal of Physical Chemistry B*, **109**, 21915–21921.

[48] Kitano, M., Takeuchi, M., Matsuoka, M. *et al.* (2005) Preparation of visible light-responsive TiO_2 thin film photocatalysts by an RF magnetron sputtering deposition method and their photocatalytic reactivity. *Chemistry Letters*, **34**, 616–617.

[49] Maeda, A., Takata, T., Hara, M. *et al.* (2005) GaN:ZnO solid solution as a photocatalyst for visible-light-driven overall water splitting. *Journal of the American Chemical Society*, **127**(23), 8286–8287.

[50] Teramura, K., Maeda, K., Saito, T. *et al.* (2005) Characterization of ruthenium oxide nanocluster as a cocatalyst with $(Ga_{1-x}Zn_x)(N_{1-x}O_x)$ for photocatalytic overall water splitting. *Journal of Physical Chemistry B*, **109**(46), 21915–21921.

[51] Maeda, K., Teramura, K., Takata, T. *et al.* (2005) Overall water splitting on $(Ga_{1-x}Zn_x)(N_{1-x}O_x)$ solid solution photocatalyst: relationship between physical properties and photocatalytic activity. *Journal of Physical Chemistry B*, **109**(43), 20504–20510.

[52] Shimizu, K., Itoh, S., Hatamachi, T. *et al.* (2005) Photocatalytic water splitting on Ni-intercalated ruddlesden-popper tantalate $H_2La_2/3Ta_2O_7$. *Chemistry of Materials*, **17**(20), 5161–5166.

[53] Ebina, Y., Sakai, N. and Sasaki, T. (2005) Photocatalyst of lamellar aggregates of RuOx-loaded perovskite nanosheets for overall water splitting. *Journal of Physical Chemistry B*, **109**(36), 17212–17216.

[54] Galińska, J.W. (2005) Photocatalytic water splitting over Pt-TiO_2 in the presence of sacrificial reagents. *Energy and Fuels*, **19**(3), 1143–1147.

[55] Wang, D., Zou, Z. and Ye, J. (2005) Photocatalytic water splitting with the Cr-doped $Ba_2In_2O_5/In_2O_3$ composite oxide semiconductors. *Chemistry of Materials*, **17**(12), 3255–3261.

[56] Kim, S., Hwang, S.-J. and Choi, W. (2005) Visible light active platinum-ion-doped TiO_2 photocatalyst. *Journal of Physical Chemistry B*, **109**(51), 24260–24267.

[57] Kuroda, Y., Mori, T., Yagi, K. *et al.* (2005) Preparation of visible-light-responsive $TiO_{2-x}N_x$ photocatalyst by a sol-gel method: analysis of the active center on TiO_2 that reacts with NH_3. *Langmuir*, **21**(17), 8026–8034.

[58] Yang, Q., Xie, C., Xu, Z. *et al.* (2005) Synthesis of highly active sulfate-promoted rutile titania nanoparticles with a response to visible light. *Journal of Physical Chemistry B*, **109**(12), 5554–5560.

[59] Bacsa, R., Kiwi, J., Ohno, T. *et al.* (2005) Preparation, testing and characterization of doped TiO_2 active in the peroxidation of biomolecules under visible light. *Journal of Physical Chemistry B*, **109**(12), 5994–6003.

[60] Thompson, T.L. and Yates, J.T. (2005) TiO_2-based photocatalysis: Surface defects, oxygen and charge transfer. *Topics in Catalysis*, **35**, 197–210.

[61] Sakthivel, S., Janczarek, M. and Kisch, H. (2004) Visible light activity and photoelectrochemical properties of nitrogen-doped TiO_2. *Journal of Physical Chemistry B*, **108**, 19384–19387.

[62] Abe, R., Higashi, M., Zou, Z. *et al.* (2004) Photocatalytic water splitting into H_2 and O_2 over R_3TaO_7 and R_3NbO_7 (R = Y, Yb, Gd, La): effect of crystal structure on photocatalytic activity. *Journal of Physical Chemistry B*, **108** (3), 811–814.

[63] Chiaramonte, T., Cardoso, L.P., Gelamo, R.V. *et al.* (2003) Structural characterization of TiO_2/TiN_xO_y (delta-doping) heterostructures on (110)TiO) substrates. *Applied Surface Science*, **212**, 661–666.

[64] Ihara, T., Miyoshi, M., Iriyama, Y. *et al.* (2003) Visible-light-active titanium oxide photocatalyst realized by an oxygen deficient structure and by nitrogen doping. *Applied Catalysis B, Environment*, **42**, 403–409.

[65] Burda, C., Lou, Y.B., Chen, X.B. *et al.* (2003) Enhanced nitrogen doping in TiO_2 nanoparticles. *Nano Letters*, **3**, 1049–1051.

[66] Yamakata, A., Ishibashi, T.-A., Kato, H. *et al.* (2003) Photodynamics of $NaTaO_3$ catalysts for efficient water splitting. *Journal of Physical Chemistry B*, **107**(51), 14383–14387.

[67] Kato, H., Asakura, K. and Kudo, A. (2003) Highly efficient water splitting into H_2 and O_2 over lanthanum-doped $NaTaO_3$ photocatalysts with high crystallinity and surface nanostructure. *Journal of the American Chemical Society*, **125**(10), 3082–3089.

[68] Kato, H. and Kudo, A. (2002) Visible-light-response and photocatalytic activities of TiO_2 and $SrTiO_3$ photocatalysts codoped with antimony and chromium. *Journal of Physical Chemistry B*, **106**, 5029–5034.

[69] Morikawa, T., Asahi, R., Ohwaki, T. *et al.* (2001) Bandgap narrowing of titanium dioxide by nitrogen doping. *Japanese Journal of Applied Physics Part 2, Letters*, **40**, L561–L563.

[70] Kato, H. and Kudo, A. (2001) Water splitting into H_2 and O_2 on alkali tantalate photocatalysts $ATaO_3$ (A = Li, Na and K). *Journal of Physical Chemistry B*, **105**(19), 4285–4292.

[71] Kudo, A., Kato, H. and Nakagawa, S. (2000) Water splitting into H_2 and O_2 on New $Sr_2M_2O_7$ (M = Nb and Ta) photocatalysts with layered perovskite structures: factors affecting the photocatalytic activity. *Journal of Physical Chemistry B*, **104**(3), 571–575.

[72] Zang, L., Macyk, W., Lange, C. *et al.* (2000) Visible-light detoxification and charge generation by transition metal chloride modified titania. *Chemistry-A European Journal*, **6**, 379–384.

[73] Le Paven-Thivet, C., Ishikawa, A., Ziani, A. *et al.* (2009) Photoelectrochemical properties of crystalline perovskite lanthanum titanium oxynitride films under visible light. *Journal of Physical Chemistry C*, **113**(15), 6156–6162.

[74] Tessier, F., Maillard, P., Lee, Y. *et al.* (2009) Zinc germanium oxynitride: influence of the preparation method on the photocatalytic properties for overall water splitting. *Journal of Physical Chemistry C*, **113**(19), 8526–8531.

[75] Maeda, K., Saito, N., Inoue, Y. and Domen, K. (2007) Dependence of activity and stability of germanium nitride powder for photocatalytic overall water splitting on structural properties. *Chemistry of Materials*, **19**(16), 4092–4097.

[76] Maeda, K. and Domen, K. (2007) New non-oxide photocatalysts designed for overall water splitting under visible light. *Journal of Physical Chemistry C*, **111**(22), 7851–7861.

[77] Maeda, K., Saito, N., Lu, D. *et al.* (2007) Photocatalytic properties of RuO_2-loaded β-Ge_3N_4 for overall water splitting. *Journal of Physical Chemistry C*, **111**(12), 4749–4755.

[78] Lee, Y., Terashima, H., Shimodaira, Y. *et al.* (2007) Zinc germanium oxynitride as a photocatalyst for overall water splitting under visible light. *Journal of Physical Chemistry C*, **111**(2), 1042–1048.

[79] Deutsch, T.G., Koval, C.A. and Turner, J.A. (2006) III–V nitride epilayers for photoelectrochemical water splitting: GaPN and GaAsPN. *Journal of Physical Chemistry B*, **110**(50), 25297–25307.

[80] Lee, Y., Watanabe, T., Takata, T. *et al.* (2006) Effect of high-pressure ammonia treatment on the activity of Ge_3N_4 photocatalyst for overall water splitting. *Journal of Physical Chemistry B*, **110**(35), 17563–17569.

[81] Kapoor, M.P., Inagaki, S. and Yoshida, H. (2005) Novel zirconium-titanium phosphates mesoporous materials for hydrogen production by photoinduced water splitting. *Journal of Physical Chemistry B*, **109**(19), 9231–9238.

[82] Sato, J., Saito, N. Yamada, Y. *et al.* (2005) RuO_2-loaded β-Ge_3N_4 as a non-oxide photocatalyst for overall water splitting. *Journal of the American Chemical Society*, **127**(12), 4150–4151.

[83] Licht, S., Wang, B., Mukerji, S. *et al.* (2000) Efficient solar water splitting, exemplified by RuO_2-catalyzed AlGaAs/Si photoelectrolysis. *Journal of Physical Chemistry B*, **104**(38), 8920–8924.

[84] Kudo, A. and Miseki, Y. (2009) Heterogeneous photocatalyst materials for water splitting. *Chemical Society Reviews*, **38**, 253–278.

[85] Tributsch, H. (2008) Photovoltaci hydrogen generation. *International Journal of Hydrogen Energy*, **33**(21), 5911–5930.

[86] Alexander, B.D., Kulesza, P.J., Rutkowska, I. *et al.* (2008) Metal oxide photoanodes for solar hydrogen production. *Journal of Materials Chemistry*, **18**, 2298–2303.

[87] Osterloh, F.E. (2008) Inorganic materials as catalysts for photochemical splitting of water. *Chemistry of Materials*, **20**(1), 35–54.

[88] Turner, J., Sverdrup, G., Mann, M.K. *et al.* (2008) Renewable hydrogen production. *International Journal of Energy Research*, **32**, 379–407.

[89] Sahaym, U. and Norton, M.G. (2008) Advances in the application of nanotechnology in enabling a hydrogen economy. *Journal of Materials Science*, **43**, 5395–5429.

[90] Kamat, P.V. (2007) Meeting the clean energy demand: nanostructure architectures for solar energy conversion. *Journal of Physical Chemistry C*, **111**(7), 2834–2860.

[91] Vayssieres, L., Hagfeldt, A. and Lindquist, S.-E. (2000) Purpose-built metal oxide nanomaterials. The emergence of a new generation of smart materials. *Pure and Applied Chemistry*, **72**(1–2), 47–52.

[92] Vayssieres, L. (2006) Designing ordered nano-arrays from aqueous solutions. *Pure and Applied Chemistry*, **78** (9), 1745–1751.

[93] Vayssieres, L. (2004) On the design of advanced metal oxide nanomaterials. *International Journal of Nanotechnology*, **1**(1–2), 1–41.

[94] Vayssieres, L. (2007) An aqueous approach to advanced metal oxide arrays on substrates. *Applied Physics A*, **89** (1), 1–8.

[95] Vayssieres, L. (1995) *Précipitation en Milieu Aqueux de Nanoparticules D'oxydes: Modélisation de L'interface et Contrôle de la Croissance, Thèse de Doctorat CHIMIE 1995PA066747*, Université Pierre et Marie Curie (Paris 6), Paris, France, pp. 1–145.

[96] Vayssieres, L. (2007) On aqueous interfacial thermodynamics and the design of metal oxide nanostructures, in *Synthesis, Properties and Applications of Oxide Nanomaterials* (eds J.A. Rodriguez and M. Fernandez-Garcia), Wiley, pp. 49–78, Chapter 2.

[97] Vayssieres, L. (2005) On the thermodynamic stability of metal oxide nanoparticles in aqueous solutions. *International Journal of Nanotechnology*, **2**(4), 411–439.

[98] Vayssieres, L., Chaneac, C., Tronc, E. and Jolivet, J.P. (1998) Size tailoring of magnetite particles formed by aqueous precipitation: An example of thermodynamic stability of nanometric oxide particles. *Journal of Colloid and Interface Science*, **205**(2), 205–212.

[99] Vayssieres, L. (2007) On the aqueous stabilization of metastable crystalline nanostructures. *International Journal of Nanotechnology*, **4**(6), 750–775.

[100] Vayssieres, L., Beermann, N., Lindquist, S.-E. and Hagfeldt, A. (2001) Controlled aqueous chemical growth of oriented three-dimensional nanorod Arrays: Application to iron(III) oxides. *Chemistry of Materials*, **13**(2), 233–235.

[101] Vayssieres, L., Guo, J.-H. and Nordgren, J. (2001) Aqueous chemical growth of αFe_2O_3-αCr_2O_3 nanocomposite thin films. *Journal of Nanoscience and Nanotechnology*, **1**(4), 385–388.

[102] Vayssieres, L., Rabenberg, L. and Manthiram, A. (2002) Aqueous chemical route to ferromagnetic 3D arrays of iron nanorods. *Nano Letters*, **2**(12), 1393–1395.

[103] Vayssieres, L. (2003) Growth of arrayed nanorods and nanowires of ZnO from aqueous solutions. *Advanced Materials*, **15**(5), 464–466.

[104] Vayssieres, L., Keis, K., Lindquist, S.E. and Hagfeldt, A. (2001) Purpose-built anisotropic metal oxide material: 3D highly oriented microrod-array of ZnO. *Journal of Physical Chemistry B*, **105**(17), 3350–3352.

[105] Vayssieres, L., Keis, K., Hagfeldt, A. and Lindquist, S.-E. (2001) Three-dimensional array of highly oriented crystalline ZnO microtubes. *Chemistry of Materials*, **13**(12), 4395–4398.

[106] Yang, Y., Sun, X.W., Tay, B.K. *et al.* (2008) On the fabrication of resistor-shaped ZnO nanowires. *Physica E*, **40** (4), 859–886.

[107] Rabenberg, L. and Vayssieres, L. (2003) Multiple orientation relationships among nanocrystals of Mn oxides. *Microscopy and Microanalysis*, **9**(2), 402–403.

[108] Vayssieres, L. and Graetzel, M. (2004) Highly ordered SnO_2 nanorod-arrays from controlled aqueous growth. *Angewandte Chemie International Edition*, **43**(28), 3666–3670.

[109] Beermann, N., Vayssieres, L., Lindquist, S.-E. and Hagfeldt, A. (2000) Photoelectrochemical studies of oriented nanorod thin films of Hematite. *Journal of the Electrochemical Society*, **147**(7), 2456–2461.

[110] Lindgren, T., Wang, H., Beermann, N. *et al.* (2002) Aqueous photoelectrochemistry of hematite nanorod-array. *Solar Energy Materials and Solar Cells*, **71**(2), 231–243.

[111] Vayssieres, L. and Sun, X.W. (2008) Nanorod-based sensors. *Sensor Letters*, **6**(6), 787–791.

[112] Wang, J.X., Sun, X.W., Yi, Y. *et al.* (2006) Hydrothermally grown ZnO nanorod arrays for gas sensing applications. *Nanotechnology*, **17**(19), 4995–4998.

[113] Vayssieres, L. (2004) Advanced metal oxide based structures for sensor technologies. *Chemical Sensors*, **20**(B), 324–325.

[114] Vayssieres, L. (2006) Advanced semiconductor nanostructures. *Comptes Rendus Chimie*, **9**(5–6), 691–701.

[115] Dong, C.L., Persson, C., Vayssieres, L. *et al.* (2004) The electronic structure of nanostructured ZnO from x-ray absorption and emission spectroscopy and the local density approximation. *Physical Review B*, **70**(19), 195325.

[116] Guo, J.-H., Vayssieres, L., Persson, C. *et al.* (2002) Polarization-dependent soft-x-ray absorption of highly oriented ZnO microrods. *Journal of Physics, Condensed Matter*, **14**(28), 6969–6974.

[117] Vayssieres, L., Persson, C. and Guo, J.-H., to be published.

[118] Vayssieres, L., Sathe, C., Butorin, S.M. *et al.* (2005) One-dimensional quantum-confinement effect in α-Fe_2O_3 ultrafine nanorod arrays. *Advanced Materials*, **17**(19), 2320–2323.

[119] Vayssieres, L. (2009) On the effect of nanoparticle size on water-oxide interfacial chemistry. *Journal of Physical Chemistry C*, **113**(12), 4733–4736.

[120] Vayssieres, L. (2004) 3-D bio-inorganic arrays, on *Chemical Sensors VI: Chemical and Biological Sensors and Analytical Methods* (eds C. Brukner-lea, P. Vanysek, G. Hunter *et al.*), The Electrochemical Society, Pennington, NJ, pp. 322–343.

[121] Luque, A., Marti, A. and Nozik, A.J. (2007) Solar cells based on quantum dots: multiple exciton generation and intermediate bands. *MRS Bulletin*, **32**(3), 236–241.

[122] Solanki, C.S. and Beaucarne, G. (2007) Advanced solar cell concepts. *Energy for Sustainable Development*, **11** (3), 17–23.

[123] Tauc, J. (1959) Electron impact ionization in semiconductors. *Journal of Physics and Chemistry of Solids*, **8**, 219–223.

[124] Vavilov, V.S. (1959) On photo-ionization by fast electrons in germanium and silicon. *Journal of Physics and Chemistry of Solids*, **8**, 223–226.

[125] Koc, S. (1957) The quantum efficiency of the photo-electric effect in germanium for the 0.3–2 μ wavelength region. *Czechoslovak Journal of Physics*, **7**, 91–95.

[126] Li, S., Steigerwald, M.L. and Brus, L.E. (2009) Surface States in the photoionization of high-quality CdSe core/shell nanocrystals. *ACS Nano*, **3**(5), 1267–1273.

[127] Kim, J., Wong, C.Y. and Scholes, G.D. (2009) Exciton fine structure and spin relaxation in semiconductor colloidal quantum dots. *Accounts of Chemical Research*, **42**(8) 1037–1046.

[128] Ji, M., Park, S., Connor, S.T. *et al.* (2009) Efficient multiple exciton generation observed in colloidal PbSe quantum dots with temporally and spectrally resolved intraband excitation. *Nano Letters*, **9**(3), 1217–1222.

[129] Beard, M.C., Midgett, A.G., Law, M. *et al.* (2009) Variations in the quantum efficiency of multiple exciton generation for a series of chemically treated PbSe Nanocrystal films. *Nano Letters*, **9**(2), 836–845.

[130] (2005) Basic research needs for solar energy ultilization, Report of the Basic Energy Sciences Workshop on Solar Energy Ulitilization, April 18–21.

[131] Roelver, R., Berghoff, B., Baetzner, D. *et al.* (2008) Si/SiO_2 multiple quantum wells for all silicon tandem cells: Conductivity and photocurrent measurements. *Thin Solid Films*, **516**(20), 6763–6766.

[132] Dahal, R., Pantha, B., Li, J. *et al.* (2009) InGaN/GaN multiple quantum well solar cells with long operating wavelengths. *Applied Physics Letters*, **94**(6), 063505/1–063505/3.

[133] Derkacs, D., Chen, W.V., Matheu, P.M. *et al.* (2008) Nanoparticle-induced light scattering for improved performance of quantum-well solar cells. *Applied Physics Letters*, **93**(9), 091107/1–091107/3.

[134] Magnanini, R., Tarricone, L., Parisini, A. *et al.* (2008) Investigation of GaAs/InGaP superlattices for quantum well solar cells. *Thin Solid Films*, **516**(20), 6734–6738.

[135] Aeberhard, U. and Morf, R.H. (2008) Microscopic nonequilibrium theory of quantum well solar cells. *Physical Review B*, **77**(12), 125343/1–125343/9.

[136] Wu, Pei-Hsuan, Su, Yan-Kuin, Tzeng, Yen F C. *et al.* (2007) A novel GaAsN/InGaAs strain-compensated multi-quantum wells solar cell. *Semiconductor Science and Technology*, **22**(5), 549–552.

[137] Johnson, D.C., Ballard, I.M., Barnham, K.W.J. *et al.* (2007) Observation of photon recycling in strain-balanced quantum well solar cells. *Applied Physics Letters*, **90**(21), 213505/1–213505/3.

[138] Rimada, J.C., Hernandez, L., Connolly, J.P. and Barnham, K.W.J. (2007) Conversion efficiency enhancement of AlGaAs quantum well solar cells. *Microelectronics Journal*, **38**(4–5), 513–518.

[139] Freundlich, A., Fotkatzikis, A., Bhusal, L. *et al.* (2007) III–V dilute nitride-based multi-quantum well solar cell. *Journal of Crystal Growth*, **301–302**, 993–996.

[140] Fox, M. (2006) Quantum wells, superlattices and band-gap engineering, in *Springer Handbook of Electronic and Photonic Materials* (eds Kasap Safa and Capper Peter), Springer, New York, NY, pp. 1021–1040.

[141] Antolin, E., Marti, A., Olea, J. *et al.* (2009) Lifetime recovery in ultrahighly titanium-doped silicon for the implementation of an intermediate band material. *Applied Physics Letters*, **94**(4), 042115/1–042115/3.

[142] Tablero, C. (2008) Correlation effects in Cr-Zinc chalogenides. *Computational Materials Science*, **44**(2), 303–309.

[143] Tablero, C. (2007) Correlation and nuclear distortion effects of Cr-substituted ZnSe. *Journal of Chemical Physics*, **126**(16), 164703/1–164703/7.

[144] Tablero, C. (2006). Survey of intermediate band materials based on ZnS and ZnTe semiconductors. *Solar Energy Materials and Solar Cells*, **90**(5), 588–596.

[145] Tablero, C. (2006) Optical properties for Ga32P31Cr and Ga31P32Cr intermediate band materials. *Solar Energy Materials and Solar Cells*, **90**(2), 203–212.

[146] Tablero, C. (2005) Analysis of the electronic properties of intermediate band materials as a function of impurity concentration. *Physical Review B*, **2**(3), 035213/1–035213.

[147] Tablero, C. (2005) Survey of intermediate band material candidates. *Solid State Communications*, **133**(2), 97–101.

[148] Palacios, P., Aguilera, I., Sanchez, K. *et al.* (2008) Transition-metal-substituted indium thiospinels as novel intermediate-band materials: prediction and understanding of their electronic properties. *Physical Review Letters*, **101**(4), 046403/1–046403/4.

[149] Canovas, E., Marti, A., Lopez, N. *et al.* (2008) Application of the photoreflectance technique to the characterization of quantum dot intermediate band materials for solar cells. *Thin Solid Films*, **516**(20), 6943–6947.

[150] Palacios, P., Aguilera, I., Wahnon, P. and Conesa, J.C. (2008) Thermodynamics of the formation of Ti- and Cr-doped $CuGaS_2$ intermediate-band photovoltaic materials. *Journal of Physical Chemistry C*, **112**(25), 9525–9529.

[151] Pablo, P., Wahnon, P., Pizzinato, S. and Conesa, J.C. (2006) Energetics of formation of $TiGa_3As_4$ and $TiGa_3P_4$ intermediate band materials. *Journal of Chemical Physics*, **124**(1), 014711/1–014711/5.

[152] Tablero, C., Palacios, P., Fernandez, J.J. and Wahnon, P. (2005) Properties of intermediate band materials. *Solar Energy Materials and Solar Cells*, **87**(1–4), 323–331.

[153] Lin, A.S., Wang, W. and Phillips, J.D. (2009) Model for intermediate band solar cells incorporating carrier transport and recombination. *Journal of Applied Physics*, **105**(6), 064512/1–064512.

[154] Marti, A., Tablero, C., Antolin, E. *et al.* (2009) Potential of Mn doped In1- xGaxN for implementing intermediate band solar cells. *Solar Energy Materials and Solar Cells*, **93**(5), 641–644.

[155] Antolin, E., Marti, A., Stanley, C.R. *et al.* (2008) Low temperature characterization of the photocurrent produced by two-photon transitions in a quantum dot intermediate band solar cell. *Thin Solid Films*, **516**(20), 6919–6923.

[156] Marti, A., Antolin, E., Canovas, E. *et al.* (2008) Elements of the design and analysis of quantum-dot intermediate band solar cells. *Thin Solid Films*, **516**(20), 6716–6722.

[157] Marti, A., Marron, D.F. and Luque, A. (2008) Evaluation of the efficiency potential of intermediate band solar cells based on thin-film chalcopyrite materials. *Journal of Applied Physics*, **103**(7), 073706/1–073706/6.

[158] Levy, M.Y. and Honsberg, C. (2008) Nanostructured absorbers for multiple transition solar cells. *IEEE Transactions on Electron Devices*, **55**(3), 706–711.

[159] Kechiantz, A.M., Kocharyan, L.M. and Kechiyants, H.M. (2007) Band alignment and conversion efficiency in Si/Ge type-II quantum dot intermediate band solar cells. *Nanotechnology*, **18**(40), 405401/1–405401/12.

[160] Shao, Q., Balandin, A.A., Fedoseyev, A.I. and Turowski, M. (2007) Intermediate-band solar cells based on quantum dot supracrystals. *Applied Physics Letters*, **91**(16), 163503/1–163503/3.

[161] Lopez, N., Marti, A., Luque, A. *et al.* (2007) Experimental analysis of the operation of quantum dot intermediate band solar cells. *Journal of Solar Energy Engineering*, **129**(3), 319–322.

[162] Yu, K.M., Scarpulla, M.A., Farshchi, R. *et al.* (2007) Synthesis of highly mismatched alloys using ion implantation and pulsed laser melting. *Nuclear Instruments and Methods in Physics Research B*, **261**(1–2), 1150–1154.

[163] Marti, A., Lopez, N., Antolin, E. *et al.* (2007) Emitter degradation in quantum dot intermediate band solar cells. *Applied Physics Letters*, **90**(23), 233510/1–233510/3.

[164] Wei, G. and Forrest, S.R. (2007) Intermediate-band solar cells employing quantum dots embedded in an energy fence barrier. *Nano Letters*, **7**(1), 218–222.

[165] Marti, A., Antolin, E., Stanley, C.R. *et al.* (2006) Production of photocurrent due to intermediate-to-conduction-band transitions: A demonstration of a key operating principle of the intermediate-band solar cell. *Physical Review Letters*, **97**(24), 247701/1–247701/4.

[166] Luque, A., Marti, A., Lopez, N. *et al.* (2006) Operation of the intermediate band solar cell under nonideal space charge region conditions and half filling of the intermediate band. *Journal of Applied Physics*, **99**(9), 094503/1–094503/9.

[167] Marti, A., Lopez, N., Antolin, E. *et al.* (2006) Novel semiconductor solar cell structures: The quantum dot intermediate band solar cell. *Thin Solid Films*, **511–512**, 638–644.

[168] Yu, K.M., Walukiewicz, W., Ager, J.W. III *et al.* (2006) Multiband GaNAsP quaternary alloys. *Applied Physics Letters*, **88**(9), 092110/1–092110/3.

[169] Luque, A., Marti, A., Lopez, N. *et al.* (2005) Experimental analysis of the quasi-Fermi level split in quantum dot intermediate-band solar cells. *Applied Physics Letters*, **87**(8), 083505/1–083505/3.

[170] Lucas, C., Marti, A. and Luque, A. (2004) Influence of the overlap between the absorption coefficients on the efficiency of the intermediate band solar cell. *IEEE Transactions on Electron Devices*, **51**(6), 1002–1007.

[171] Luque, A., Marti, A., Stanley, C. *et al.* (2004) General equivalent circuit for intermediate band devices: Potentials, currents and electroluminescence. *Journal of Applied Physics*, **96**(1), 903–909.

[172] Marti, A., Cuadra, L. and Luque, A. (2004) Intermediate-band solar cells, in *Next Generation Photovoltaics* (eds A. Marti and A. Luque), pp. 140–164, Taylor & Francis.

[173] Marti, A., Cuadra, L. and Luque, A. (2002) Quasi-drift diffusion model for the quantum dot intermediate band solar cell. *IEEE Transactions on Electron Devices*, **49**(9), 1632–1639.

[174] Luque, A. and Marti, A. (1997) Increasing the efficiency of ideal solar cells by photon induced transitions at intermediate levels. *Physical Review Letters*, **78**(26), 5014–5017.

[175] Bjoerksten, U., Moser, J. and Graetzel, M. (1994) Photoelectrochemical studies on nanocrystalline hematite films. *Chemistry of Materials*, **6**(6), 858–863.

[176] Vayssieres, L. (1997) Photoelectrochemical properties of anisotropic hematite nanostructured electrode. Lecture presented at the 3rd International Conference on New Trends in Photoelectrochemistry, Estes Park, CO, USA, May 1997.

[177] Kennedy, J.H. and Frese, K.W. Jr. (1978) Photooxidation of water at alpha-iron(III) oxide electrodes. *Journal of the Electrochemical Society*, **125**(5), 709–714.

[178] Bailey, J.K., Brinker, C.J. and Mecartney, M.L. (1993) Growth mechanisms of iron oxide particles of differing morphologies from the forced hydrolysis of ferric chloride solutions. *Journal of Colloid and Interface Science*, **157**(1), 1–13.
[179] Vayssieres, L. (1998) New purpose-built nanostructured metal oxide materials for photoelectrochemical applications. Lecture presented at the 194th Meeting of the Electrochemical Society, Symposium on Photoelectrochemistry and Solar Energy Conversion, Boston, MA, USA, November 1998.
[180] Vayssieres, L. (1999) New generation of designed metal oxides for photoelectrochemical devices. 4th International Symposium on new trends in photoelectrochemistry, fundamentals, photovoltaic & environmental aspects, Sophia Antipolis, France, June 1999.
[181] Matsumoto, Y. (1996) Energy positions of oxide semiconductors and photocatalysis with iron complex oxides. *Journal of Solid State Chemistry*, **126**(2), 227–234.
[182] Kuiper, P., Searle, B.G., Rudolf, P. *et al.* (1993) X-ray magnetic dichroism of antiferromagnet Fe_2O_3: The orientation of magnetic moments observed by Fe 2p x-ray absorption spectroscopy. *Physical Review Letters*, **70**, 1549.
[183] Duda, L.-C., Nordgren, J., Drager, G. *et al.* (2000) Polarized resonant inelastic X-ray scattering from single-crystal transition metal oxides. *Journal of Electron Spectroscopy and Related Phenomena*, **110–111**(1–3), 275–285.
[184] Crocombette, J.P., Pollak, M., Jollet, F. *et al.* (1995) X-ray-absorption spectroscopy at the Fe L2,3 threshold in iron oxides. *Physical Review B*, **52**, 3143.
[185] Holliday, J.E. (1973) *Band Structure Spectroscopy of Metals and Alloys* (eds D.J. Fabian and L.M. Watson), Academic Press, London, pp. 713.
[186] Skinner, H.W., Bullen, T.G. and Jonston, J. (1954) Notes on soft X-ray spectra, particularly of the Fe group elements. *Philosophical Magazine*, **45**, 1070.
[187] Kurmaev, E.Z., Galakhov, V.R., Moewes, A. *et al.* (2000) Electronic structure of molecular superconductors containing paramagnetic 3d ions. *Physical Review B*, **62**, 11380.
[188] Marusak, L.A., Messier, R. and White, W.B. (1980) Optical absorption spectrum of hematite from near-IR to UV. *Journal of Physics and Chemistry of Solids*, **41**, 981–984.
[189] Akl, A.A. (2004) Optical properties of crystalline and non-crystalline iron oxide thin films deposited by spray pyrolysis. *Applied Surface Science*, **233**, 307–319.
[190] Glasscock, J.A. (2008) Nanostructured materials for photoelectrochemical hydrogen production using sunlight, PhD Thesis, University of New South Wales, Australia, 1–220.
[191] Kamat, P.V. (2008) Quantum dot solar cells. Semiconductor nanocrystals as light harvesters. *Journal of Physical Chemistry C*, **112**(48), 18737–18753.
[192] Bang, J.H. and Kamat, P.V. (2009) Quantum dot sensitized solar cells. A tale of two semiconductor nanocrystals: CdSe and CdTe, ACS Nano, ASAP; Kirchartz, Thomas; Rau, Uwe, Modeling charge carrier collection in multiple exciton generating PbSe quantum dots. *Thin Solid Films*, **517**(7), 2438–2442.
[193] Oshima, R., Takata, A. and Okada, Y. (2008) Strain-compensated InAs/GaNAs quantum dots for use in high-efficiency solar cells. *Applied Physics Letters*, **93**(8), 083111/1–083111/3.
[194] Brown, P. and Kamat, P.V. (2008) Quantum dot solar cells. Electrophoretic deposition of CdSe-C60 composite films and capture of photogenerated electrons with nC60 cluster shell. *Journal of the American Chemical Society*, **130**(28), 8890–8891.
[195] Kongkanand, A., Tvrdy, K., Takechi, K. *et al.* (2008) Quantum dot solar cells. Tuning photoresponse through size and shape control of CdSe-TiO_2 architecture. *Journal of the American Chemical Society*, **130**(12), 4007–4015.
[196] Laghumavarapu, R.B., El-Emawy, M., Nuntawong, N. *et al.* (2007) Improved device performance of InAs/GaAs quantum dot solar cells with GaP strain compensation layers. *Applied Physics Letters*, **91**(24), 243115/1–243115/3.
[197] Gerischer, H. and Luebke, M. (1986) A particle size effect in the sensitization of TiO_2 electrodes by a CdS deposit. *Journal of Electroanalytical Chemistry*, **204**, 225–227.
[198] Spanhel, L., Weller, H. and Henglein, A. (1987) Photochemistry of semiconductor colloids. 22. Electron injection from illuminated CdS into attached TiO_2 and ZnO particles. *Journal of the American Chemical Society*, **109**, 6632–6635.
[199] Hotchandani, S. and Kamat, P.V. (1992) Charge-transfer processes in coupled semiconductor systems. Photochemistry and photoelectrochemistry of the colloidal CdS-ZnO system. *Journal of Physical Chemistry*, **96**, 6834–6839.

[200] Zaban, A., Micic, O.I., Gregg, B.A. and Nozik, A.J. (1998) Photosensitization of nanoporous TiO_2 electrodes with InP quantum dots. *Langmuir*, **14**, 3153–3156.

[201] Vogel, R., Hoyer, P. and Weller, H. (1994) Quantum-sized PbS, CdS, Ag2S. Sb2S3 and Bi2S3 particles as sensitizers for various nanoporous wide-bandgap semiconductors. *Journal of Physical Chemistry*, **98**, 3183–3188.

[202] Lahiri, D., Subramanian, V., Bunker, B.A. and Kamat, P.V. (2006) Probing photochemical transformations at TiO_2/Pt and TiO_2/Ir interfaces using x-ray absorption spectroscopy. *Journal of Chemical Physics*, **124**, 204720.

[203] Chen, M.S. and Goodman, D.W. (2004) The structure of catalytically active gold on titania. *Science*, **306**, 252–255.

[204] Subramanian, V., Wolf, E.E. and Kamat, P.V. (2003) Influence of metal/metal-ion concentration on the photocatalytic activity of TiO_2-Au composite nanoparticles. *Langmuir*, **19**, 469–474.

[205] Cozzoli, P.D., Comparelli, R., Fanizza, E. *et al.* (2004) Photocatalytic synthesis of silver nanoparticles stabilized by TiO_2 nanorods: A semiconductor/metal nanocomposite in homogeneous nonpolar solution. *Journal of the American Chemical Society*, **126**, 3868–3879.

[206] Cozzoli, P.D., Fanizza, E., Comparelli, R. *et al.* (2004) Role of metal nanoparticles in TiO_2/Ag nanocomposite-based microheterogeneous photocatalysis. *Journal of Physical Chemistry B*, **108**, 9623–9630.

[207] Kamat, P.V., Flumiani, M. and Dawson, A. (2002) Metal-metal and metal-semiconductor composite nanoclusters. *Colloids and Surfaces A*, **202**, 269–279.

[208] Subramanian, V., Wolf, E. and Kamat, P.V. (2001) Semiconductor-metal composite nanostructures. To what extent metal nanoparticles (Au, Pt, Ir) improve the photocatalytic activity of TiO_2 films? *Journal of Physical Chemistry B*, **105**, 11439–11446.

[209] Dawson, A. and Kamat, P.V. (2001) Semiconductor-metal nanocomposites. Photoinduced fusion and photocatalysis of gold-capped TiO_2 (TiO_2/Au) nanoparticles. *Journal of Physical Chemistry B*, **105**, 960–966.

[210] Chandrasekharan, N. and Kamat, P.V. (2000) Improving the photoelectrochemical performance of nanostructured TiO_2 films by adsorption of gold nanoparticles. *Journal of Physical Chemistry B*, **104**, 10851–10857.

[211] Hiesgen, R. and Meissner, D. (1998) Nanoscale photocurrent variations at metal-modified semiconductor surfaces. *Journal of Physical Chemistry B*, **102**, 6549–6557.

[212] de Tacconi, N.R., Carmona, J. and Rajeshwar, K. (1997) Chemically modified Ni/TiO_2 nanocomposite films. Charge transfer from photoexcited TiO_2 particles to hexacyanoferrate redox centers within the film and unusual photoelectrochemical behavior. *Journal of Physical Chemistry B*, **101**, 10151–10154.

[213] Amouyal, E. (1995) Photochemical production of hydrogen and oxygen from water: A review and state of the art. *Solar Energy Materials and Solar Cells*, **38**, 249–276.

[214] Henglein, A. (1993) Physicochemical properties of small metal particles in solution: Microelectrode reactions, chemisorption, composite metal particles and the atom-to-metal transition. *Journal of Physical Chemistry*, **97**, 5457–5471.

[215] Kamat, P.V. (1993) Photochemistry on nonreactive and reactive (semiconductor) surfaces. *Chemical Reviews*, **93**, 267–300.

[216] Anpo, M., Chiba, K., Tomonari, M. *et al.* (1991) Photocatalysis on native and platinum-loaded TiO_2 and ZnO catalysts. Origin of different reactivities on wet and dry metal oxides. *Bulletins of the Chemical Society of Japan*, **64**, 543–551.

[217] Nakato, Y., Ueda, K., Yano, H. and Tsubomura, H. (1988) Effect of microscopic discontinuity of metal overlayers on the photovoltages in metalcoated semiconductor-liquid junction photoelectrochemical cells for efficient solar energy conversion. *Journal of Physical Chemistry*, **92**, 2316–2324.

[218] Domen, K., Sakata, Y., Kudo, A. *et al.* (1988) The photocatalytic activity of a platinized titanium dioxide catalyst supported over silica. *Bulletins of the Chemical Society of Japan*, **61**, 359–362.

[219] Heller, A. (1986) Optically transparent metallic catalysts on semiconductors. *Pure and Applied Chemistry*, **58**, 1189–1192.

[220] Smotkin, E., Bard, A.J., Campion, A. *et al.* (1986) Bipolar titanium dioxide/platinum semiconductor photoelectrodes and multielectrode arrays for unassisted photolytic water splitting. *Journal of Physical Chemistry*, **90** (19), 4604–4607.

[221] Baba, R., Nakabayashi, S., Fujishima, A. and Honda, K. (1985) Investigation of the mechanism of hydrogen evolution during photocatalytic water decomposition on metal-loaded semiconductor powders. *Journal of Physical Chemistry*, **89**, 1902–1905.

[222] Nakato, Y. and Tsubomura, H. (1985) Structures and functions of thin metal layers on semiconductor electrodes. *Journal of Photochemistry*, **29**, 257–266.

[223] Nakato, Y. and Tsubomura, H. (1982) The photoelectrochemical behavior of an n-TiO_2 electrode coated with a thin metal film, as revealed by measurements of the potential of the metal film. *Israel Journal of Chemistry*, **22**, 180–183.

[224] Nakato, Y., Shioji, M. and Tsubomura, H. (1982) Photoeffects on the potentials of thin metal films on a n-TiO_2 crystal wafer. The mechanism of semiconductor photocatalysts. *Chemical Physics Letters*, **90**, 453–456.

18

Effects of Metal-Ion Doping, Removal and Exchange on Photocatalytic Activity of Metal Oxides and Nitrides for Overall Water Splitting

Yasunobu Inoue
Department of Chemistry, Nagaoka University of Technology, Nagaoka,
Email: inoue@analysis.nagaokaut.ac.jp

18.1 Introduction

Hydrogen production by photocatalytic water splitting using solar energy has been attracting interest from the viewpoint of an ecofriendly and renewable energy source, and the new development of photocatalysts is among the key issues. In the past three decades, efforts toward photocatalytic water splitting have been focused on metal oxides. There are two major groups of metal oxides (NiO_x and RuO_2 are mostly loaded as co-catalysts) that have the ability to split water into hydrogen and oxygen. As shown in Table 18.1, one group is composed of transition-metal oxides. Representative transition-metal oxides include various kinds of titanates (TiO_2 [1], $SrTiO_3$ [2], $A_2Ti_6O_{13}$ (A = Na, K, Rb) [3,4], $BaTi_4O_9$ [5,6], $A_2La_2Ti_3O_{10}$(A = K, Rb, Cs) [7,8], $Na_2Ti_3O_7$ [9], $K_2Ti_4O_9$ [10]), zirconium oxide (ZrO_2 [11]) niobates ($A_4Nb_6O_{17}$ (A = K, Rb) [12], $Sr_2Nb_2O_7$ [13], $Cs_2Nb_4O_{11}$ [14], $Ba_5Nb_4O_{15}$ [15]), tantalates ($ATaO_3$ (A = Na, K) [16,17], MTa_2O_6 (M = Ca, Sr, Ba) [18,19], $Sr_2Ta_2O_7$ [13], $ACa_2Ta_3O_{10}$ (A = H, Na, Ca) [20], $A_2SrTa_2O_7{\cdot}nH_2O$ (A = H, K, Rb) [21], $ALnTa_2O_7$ (A = H, Na, Rb, Cs, Ln = La, Pr, Nd, Sm) [22], $K_3Ta_3B_2O_{12}$ [23], $Ba_5Ta_4O_{15}$ [24]) and tungstates ($PbWO_4$ [25,26], Pb_xWO_4 [27]). These transition-metal oxides are composed of a core metal ion with d^0

On Solar Hydrogen & Nanotechnology Edited by Lionel Vayssieres

Table 18.1 A list of metal-oxide photocatalysts with d^0 and d^{10} electronic configurations for overall water splitting.

	d^0 electronic configuration		d^{10} electronic configuration
Ti^{4+}	TiO_2, $SiTiO_3$, $A_2Ti_6O_{13}$ (A = Na, K, Rb), $BaTi_4O_9$, $A_2La_2Ti_3O_{10}$ (A = K, Rb, Cs), $K_2Ti_4O_9$, $Na_2Ti_3O_7$	Ga^{3+} In^{3+} Ge^{4+}	MGa_2O_4 (M = Ca, Sr, Zn), Ga_2O_3 MIn_2O_4 (M = Ca, Sr), $LaInO_3$, $AInO_2$ (A = Li,Na) Zn_2GeO_4
Zr^{4+}	ZrO_2	Sn^{4+}	M_2SnO_4 (M = Ca, Sr)
Nb^{5+}	$A_4Nb_6O_{17}$ (A = K, Rb), $Sr_2Nb_2O_7$, $Cs_2Nb_4O_{11}$	Sb^{5+}	$NaSbO_3$, MSb_2O_6, $M_2Sb_2O_7$ (M = Ca, Sr)
Ta^{5+}	$ATaO_3$ (A = Na, K), MTa_2O_6 (M = Ca, Sr, Ba), $Sr_2Ta_2O_7$, $ACa_2Ta_3O_{10}$ (A = H, Na, Ca), $A_2SrTa_2O_7{\cdot}nH_2O$ (A = H, K, Rb), $ALnTa_2O_7$ (A = H, Na, Rb; Ln = La, Nd), $Ba_5Ta_4O_{15}$, $K_3Ta_3B_2O_{12}$		
W^{6+}	$PbWO_4$		

electronic configuration, such as Ti^{4+}, Zr^{4+}, Nb^{5+}, Ta^{5+} and W^{6+}. The other group contains typical metal oxides that consist of various kinds of indates (MIn_2O_4 (M = Ca, Sr) [28,29], $AInO_2$ (A = Li, Na) [30], $LaInO_3$ [31]), gallates (Ga_2O_3 [32], $ZnGa_2O_4$ [33], MGa_2O_4 (M = Ca, Sr), germanate (Zn_2GeO_4 [34]), stannates (M_2SnO_4 (M = Ca, Sr) [28]) and antimonates ($NaSbO_3$ [28,35], MSb_2O_6 (M = Ca, Sr) [35], $M_2Sb_2O_7$ (M = Ca, Sr) [35]). The core metal ions are Ga^{3+}, In^{3+}, Ge^{4+}, Sn^{4+} and Sb^{5+} with d^{10} electronic configuration. The kinds and numbers of d^{10} photocatalysts discovered in recent years are almost comparable to those of the d^0 metal oxides established in the past three decades. This is partly associated with an advantage relating to the conduction bands of typical metal oxides with d^{10} configurations, since the bands are composed of the *sp* orbitals, with large band dispersions, and permit the generation of photoexcited electrons with high mobility, compared to the flat conduction bands of transition-metal oxides with d^0electronic configurations [9–22].

The essential steps for water splitting in a solid photocatalyst are photoexcitation of electrons from the valence band to the conduction band, transfer of the electrons and holes to the surface without recombination, and surface reactions of electrons with H^+ and holes with OH^-. The efficiency of photocatalysts is evidently associated with the geometric and electronic structures that control the behavior of photoexcited charges. Although there is a wide variety of metal oxides with different crystal structures and electronic structures, the key factor is the symmetry of the metal–oxygen octahedral/tetrahedral coordination, and the valence- and conduction-band structures. For example, a composite metal oxide, $LiInGeO_4$ [36], consisting of two *p*-block metal ions of Ir^{3+} and Ge^{4+}, with a d^{10}–d^{10} electronic configuration, has the capability of splitting water under UV irradiation, when combined with RuO_2 as a co-catalyst. The activity is considerably larger than that of simple metal oxides, such as $AInO_2$ (A = Li, Na) [36] and $NaGeO_4$ [36], with d^{10} configurations. This indicates that the mixing of different core metal ions has a significant effect, not only on the geometric structures of tetrahedral and octahedral coordination, but also the density of states and band dispersion due to the hybridization of two *sp* orbitals. Metal-ion doping and the removal and exchange of constituent metal ions of metal oxides are thought to cause changes in the geometric and electronic structures of the metal oxides, and their effects on photocatalytic performance of transition, typical and lanthanide metal oxides and nitrides for overall water splitting are reviewed.

18.2 Experimental Procedures

Details of the experimental procedure for photocatalytic water splitting described here have been reported elsewhere [5,28]. Briefly, powdered metal oxides were prepared by the calcination of constituent metal oxides, carbonates or nitrates at high temperatures, and metal nitrides were prepared by nitridation of metal sulfides in a NH_3 atmosphere at moderate temperatures. Ruthenium oxide, RuO_2, was loaded on metal oxides and nitrides as a cocatalyst. For the loading, metal oxides and nitrides were impregnated up to incipient wetness with the ruthenium carbonyl complex, $Ru_3(CO)_{12}$, in THF, dried at 353 K, and oxidized in air at 673 K for 5 h in order to convert the loaded Ru complex into RuO_2 particles [28,33]. RuO_2-loaded powder was placed in distilled, ion-exchanged water in an inner- or outer-type reaction cell. The photocatalyst powder was dispersed in the water by continuous bubbling of Ar gas during photocatalytic reaction in a closed gas-circulating apparatus and irradiated with an Xe or Hg–Xe lamp. The amounts of H_2 and O_2 produced in the gas phase were analyzed using an on-line gas chromatograph.

18.3 Effects of Metal Ion Doping

It is quite interesting to see that a small amount of doping with different metal ions converts inactive chemical compounds to be phtocatalytically active for water splitting. Although such a marked dopant effect on the photocatalytic performance has been limited, we can see good examples with a lanthanide metal oxide of CeO_2 and a metal nitride of GaN.

Compared to the transition-metal and typical-metal oxides, rare-earth-metal oxides have not been fully investigated as photocatlaysts. CeO_2 has $[Xe]f^0d^0$ electronic structure, and the core metal ion of Ce^{4+} has a kind of d^0 electronic configuration. The metal oxide was reported to have the ability to produce oxygen from a suspending solution of CeO_2 containing Ce^{4+} ions as electron acceptors [37], but the photocatalytic activity for overall water splitting is not known yet. Very recently, doping of Sr onto CeO_2 converted it to be photocatalytically active [38]. This is the first example of a lanthanide metal oxide as a photocatalyst for overall water splitting. As for metal nitrides, GaN [39,40] is inactive, but doping with a small amount of divalent metal ions change it to have remarkably high activity. Their detailed photocatalytic performance is described in the following sections.

18.3.1 Sr^{2+} Ion-Doped CeO_2

For the preparation of Sr^{2+}-doped CeO_2, CeO_2 and $SrCO_3$ were used as starting materials and calcined in air at 1273 K for 10 h [38]. For comparison, undoped CeO_2 was subjected to the same temperature. The UV-visible diffuse reflectance spectra showed that Sr^{2+}-doped CeO_2 had light absorption characteristics starting at around 400 nm and leveling off at 350 nm, which was similar to that of undoped CeO_2. Undoped CeO_2 exhibited negligible photocatalytic activity for water splitting under UV irradiation in the presence of RuO_2 cocatalyst. However, RuO_2-loaded Sr^{2+}-doped CeO_2 produced both H_2 and O_2 immediately upon irradiation. Although the production

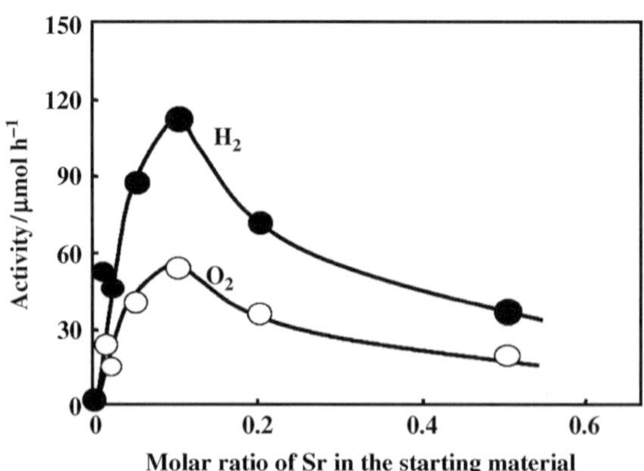

Figure 18.1 Photocatalytic activity as a function of molar ratio of doped Sr^{2+}. RuO_2-loading was 1.0 wt%. ●: H_2; ○: O_2. (Reprinted with permission from H. Kadowaki, N. Saito, H. Nishiyama, and Y. Inoue, RuO_2-loaded Sr^{2+}-doped CeO_2 with d^0 electronic configuration as a new photocatalyst for overall water splitting, *Chemistry Letters*, **36**(3), 440–441, 2007. © 2007 The Chemical Society of Japan.)

gradually decreased over an initial period, it remained at a constant level after prolonged irradiation. During the reaction, the ratio of H_2 to O_2 production was maintained at 2.0, indicating that water splitting on RuO_2-loaded Sr^{2+}-doped CeO_2 proceeded photocatalytically.

Figure 18.1 shows photocatalytic activity for water splitting as a function of the amount of doped Sr^{2+}, which is defined as the molar ratio of Sr to Sr + Ce atoms. With increasing molar ratio of Sr^{2+}, both the activities for H_2 and O_2 production increased markedly, reached a maximum at a doping ratio of 0.1, and decreased monotonically. A change in the activity might be associated with the appearance of $SrCeO_3$ and/or Sr_2CeO_4. Separately, the photocatalytic activities of stoichiometric cerium compounds such as $SrCeO_3$ and Sr_2CeO_4 were examined, but little activity was observed. Thus, it is evident that a small amount of dopant converts the photocatalytically inactive metal oxide to an efficient photocatalyst. As shown in Figure 18.2, the single-phase X-ray diffraction patterns due to fluorite structure are maintained without significant peak shifts by Sr^{2+}-doping up to a doping ratio of 0.2. The small shift in the peak is due to a small difference in ionic sizes between Ce^{4+} (114 nm) and Sr^{2+} ions (125 nm) in tetrahedral coordination. At a doping ratio of 0.5, however, new phases associated with the stoichiometric compounds of $SrCeO_3$ and Sr_2CeO_4 appear. Calculation of the electronic structure of CeO_2 by a density functional theory (DFT) method showed that the occupied valence band was composed of the O2*p* atomic orbital (AO), and the unoccupied band consisted of the Ce4*f* AO band at the lower level and Ce5*d* AO at the upper level. For the calculation for Sr-doped CeO_2, no significant changes in the band structures were observed, except for the hybridization of the Ce5*d* AO with the Sr5*s*5*p* AO in the unoccupied band. Thus, the threshold absorption at around 400 nm observed in the UV spectrum is assigned to electron transfer from the O2*p* AO to the Ce4*f* AO. However, there is a question as to whether the electron transferred to the Ce4*f* level contributes to the photocatalysis, since the *f* orbital is narrow and isolated. When Sr-doped CeO_2 photocatalysts were irradiated by light with wavelength longer than 300 nm, no water splitting occurred at all. This indicates that the electrons transferred to the Ce4*f* orbital are useless for water splitting, and that those at the Ce5*d* AO are responsible. A photocatalysis model is shown in Figure 18.3.

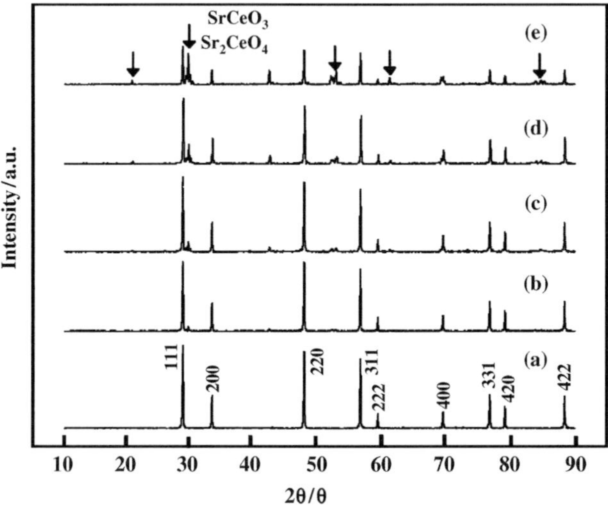

Figure 18.2 XRD patterns of undoped(a) and Sr^{2+}-doped CeO_2 with different molar ratios Sr/(Sr + Ce) = 0.05(b), 0.10(c), 0.20(d) and 0.50(e). (Reprinted with permission from H. Kadowaki, N. Saito, H. Nishiyama, and Y. Inoue, RuO_2-loaded Sr^{2+}-doped CeO_2 with d^0 electronic configuration as a new photocatalyst for overall water splitting, *Chemistry Letters*, **36**(3), 440–441, 2007. © 2007 The Chemical Society of Japan.)

If one considers the ionic size ratio $r(M^{n+})/r(O^{2-})$, where $r(M^{n+})$ and $r(O^{2-})$ are the ion sizes of a metal ion and an oxygen anion in metal oxides, respectively, the ratio for CeO_2 fluorite is calculated to be 0.703. Since the ideal ionic size ratio of $r(M^{n+})/r(O^{2-})$ for a fluorite structure of an MO_8 decahedron is 0.732, the ratio of CeO_2 deviates considerably from the ideal value, which could cause instability in the fluorite structure. However, stability is maintained by converting some of the Ce^{4+} ions to larger Ce^{3+} ions. This results in the presence of a recombination center for photoexcited charges and leads to poor photocatalytic performance. Thus, it is required to depress the formation of Ce^{3+} in CeO_2. Yabe *et al.* [41] showed that the

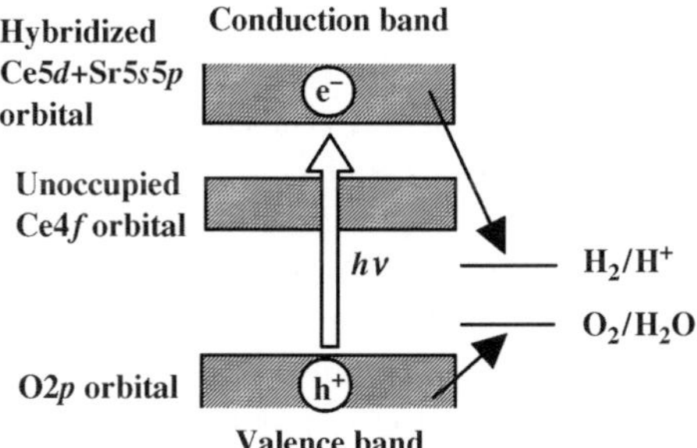

Figure 18.3 A photocatalysis model for Sr^{2+}-doped CeO_2. (Reprinted with permission from H. Kadowaki, N. Saito, H. Nishiyama, and Y. Inoue, RuO_2-loaded Sr^{2+}-doped CeO_2 with d^0 electronic configuration as a new photocatalyst for overall water splitting, *Chemistry Letters*, **36**(3), 440–441, 2007. © 2007 The Chemical Society of Japan.)

partial replacement of Ce^{4+} by larger and/or less positively charged cations permitted maintenance of the fluorite structure without reduction of Ce^{4+} to Ce^{3+}. Sr^{2+} has a slightly larger ionic size and less positive charge than Ce^{4+}, indicating that Sr^{2+}-doping yields Ce^{3+}-free CeO_2. Thus, the doping effect of Sr^{2+} on the activation of CeO_2 is explained in terms of the formation of a d^0f^0 electronic configuration in the absence of recombination centers.

18.3.2 Metal-Ion Doped GaN

Metal oxide photocatalysts useful for water splitting have wide bandgaps, since the energy levels of valence bands due to O2*p* bands are deep enough to generate holes for oxidizing water, while the conduction bands must be higher than the reduction level of H^+. Thus, recent photocatalyst research into visible-light-driven water splitting has shifted to oxynitrides and/or metal nitrides in aiming at obtaining narrow bandgap photocatalysts, since the potential of the N2*p* orbital is higher than that of the O2*p* orbital.

In metal nitrides with d^{10} electronic configurations, β-Ge_3N_4 has become a strong photocatalyst for overall water splitting to produce H_2 and O_2 when RuO_2 is loaded as a co-catalyst [43]. β-Ge_3N_4 was the first example of a photocatalytically active d^{10} metal nitride. However, GaN with a d^{10} electronic configuration exhibited negligible activity for water splitting, even in the presence of RuO_2. Divalent and tetravalent metal-ion doping onto GaN, forming p-type and n-type GaN, respectively, is thought to be effective for activating GaN [39,40].

Experimentally, nitridation was performed in a NH_3 flow at 1273 K for 15 h using a rotary kiln-type electric furnace [39,40]. For example, undoped GaN was synthesized by nitridation of Ga_2S_3, and Mg^{2+}-doped GaN was prepared under similar conditions by the nitridation of a mixture of Ga_2S_3 and MgS with different molar ratios of Mg/(Ga + Mg). SEM images showed that the powdered GaN had round shapes with an average particle size of 1.2 µm, and their morphology remained nearly unchanged upon doping.

Figure 18.4 shows photocatalytic water splitting on RuO_2-dispersed Mg^{2+}-doped GaN under light irradiation [39,40]. The production of H_2 and O_2 occurred in proportion to irradiation time from the first run. The activity was fairly stable over the runs, and the ratio of H_2 to O_2 was close to the stoichiometric value of 2.0. Little evolution of N_2 occurred in repeated runs. The total amounts of H_2 produced were larger by a factor of approximately 280, 100 and 290 for Mg^{2+}-, Be^{2+}-and Zn^{2+}-doped GaN, respectively, than the estimated amount of Ga^{3+} ion present at the GaN surface, whereas the total amounts of N_2 evolved were 240–340 times lower than that of H_2 in all cases. The activation effects of the divalent metal ions doped increased in the order $Mg^{2+} > Be^{2+} > Zn^{2+}$. In GaN thin-film photoelectrodes deposited on a sapphire substrate grown by metalorganic vapor-phase epitaxy, H_2 and a trace amount of O_2, together with a considerable amount of N_2, were reported to be produced from H_2O by UV illumination under an applied voltage of + 1.0 V [43,44]. It should be noted that divalent metal ion (Zn^{2+}, Mg^{2+} or Be^{2+})-doped GaN powder combined with RuO_2 is able to photocatalytically produce both H_2 and O_2 in a stoichiometric ratio without any external force.

Figure 18.5 shows the photocatalytic activity of undoped, divalent metal-ion (Zn^{2+},Mg^{2+}, Be^{2+})-doped and tetravalent metal-ion (Si^{4+},Ge^{4+})-doped GaN with the deposition of RuO_2 as a co-catalyst. The divalent metal-ion-doped GaN had marked photocatalytic activity. However, the tetravalent metal-ion-doped GaN exhibited little production of H_2 and O_2. These results clearly indicate that only divalent metal-ion-doping is able to activate GaN to be an efficient and stable photocatalyst.

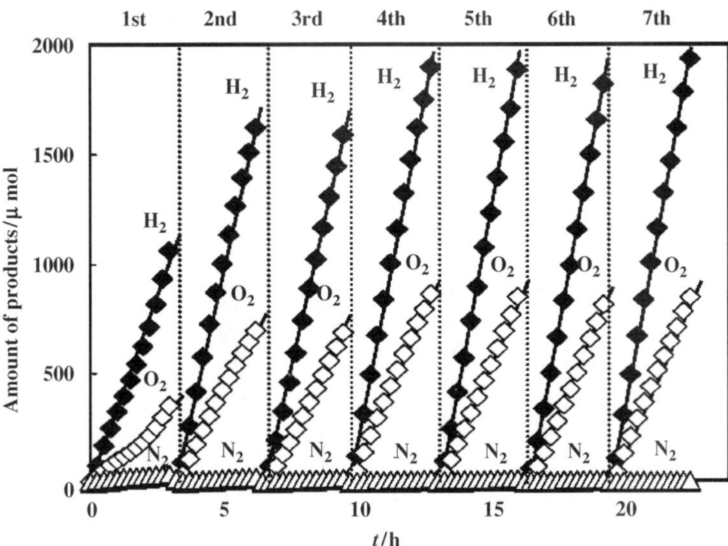

Figure 18.4 Overall water splitting on Mg^{2+}-doped GaN. The RuO_2-loading was 3.5 wt%. □: H_2; ◇: O_2; Δ: N_2. (Reprinted with permission from N. Arai, N. Saito, H. Nishiyama, and Y. Inoue, Overall water splitting by RuO_2-dispersed p-type GaN photocatalyts with d^{10} electronic configuration, *Chemistry Letters*, **35**(7), 796–797, 2006. © 2006 The Chemical Society of Japan.)

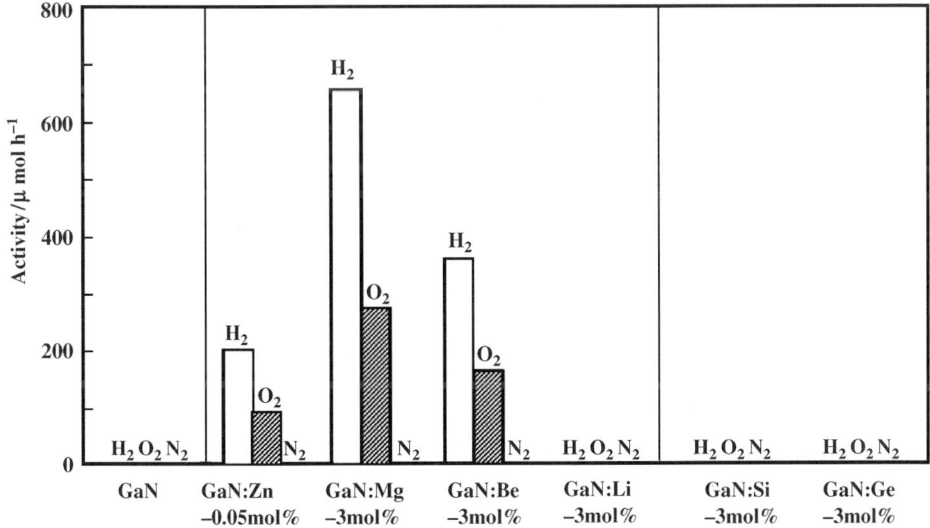

Figure 18.5 Photocatalytic activity of undoped, monovalent metal-ion (Li^+), divalent metal-ion (Zn^{2+}, Mg^{2+}, Be^{2+}) and tetravalent metal-ion (Si^{4+}, Ge^{4+})-doped GaN. The amount of dopant was 3 mol%, except for Zn, which was 0.05 mol%. The RuO_2-loading was 3.5 wt%. (Reprinted with permission from N. Arai, N. Saito, H. Nishiyama, and Y. Inoue, Overall water splitting by RuO_2-dispersed p-type GaN photocatalysts with d^{10} electronic configuration, *Chemistry Letters*, **35**(7), 796–797, 2006). © 2006 The Chemical Society of Japan.)

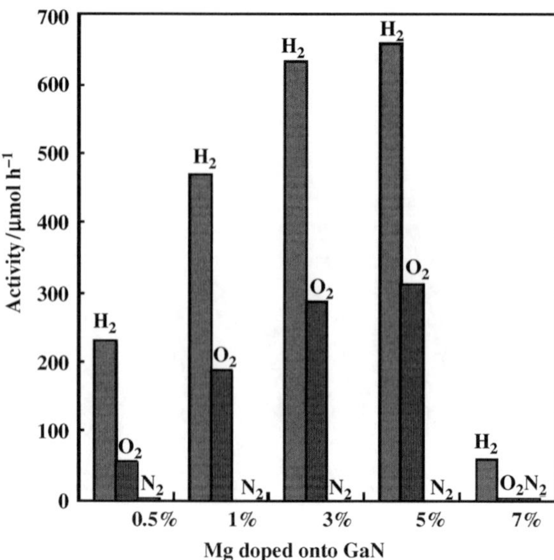

Figure 18.6 Photocatalytic activity of Mg^{2+}-doped GaN as a function of Mg^{2+} content in the mixture of starting materials. The RuO_2-loading was 3.5 wt%.

Figure 18.6 shows photocatalytic activity as a function of the mole percent ($= Mg^{2+}/(M^{2+} + Ga^{3+}) \times 100\%$) in the starting material. The activity for H_2 and O_2 production increased sharply with addition of Mg^{2+}, reaching the highest level at around 5 mol% Mg^{2+}, above which it decreased sharply [40]. The photocatalytic activity was stable independent of the amount of Mg^{2+} doped. The lowest N_2 evolution was observed at 5 mol% Mg^{2+}, at which the photocatalytic activity became the highest. The XRD patterns of undoped and doped GaN exhibited single-phase diffraction patterns characteristic of a wurtzite structure. No significant changes in the diffraction peak positions were observed. Electron probe microanalysis (EPMA) showed that the doped Zn content was around 0.05 mol%. As the size of Mg^{2+} is smaller by 15% than that of Zn^{2+} (ion radii size for Mg^{2+} and Zn^{2+} for tetrahedral coordination is 0.063 and 0.074 nm, respectively), it is thought that the amount of Mg^{2+} ion doped was similar to or slightly larger than that of the Zn^{2+} ion. The UV-visible diffuse reflectance spectra showed that the absorption of undoped GaN occurred at around 390 nm, increasing sharply with decreasing wavelength, and leveled off at 370 nm, which coincided with a reported bandgap of approximately 3.4 eV. The doping of Zn^{2+} shifted the absorption band to a slightly longer wavelength (375 nm). The Mg^{2+} doping also caused red-shifts, the extent of which was smaller than that with Zn^{2+}-doping.

Calculation by a DFT method showed that N2*s* and Ga3*d* AOs exist in the inner level. The lower and upper part of the valence band is composed of N2*p* mixed with Ga4*p* and Ga4*s* orbitals, respectively. The conduction band consists of hybridized Ga4*s*4*p* mixed with N2*p* orbitals. The conduction bands of GaN have a large dispersion characteristic of a d^{10} electronic configuration. The large dispersion is able to produce photoexcited electrons with high mobilities that are advantageous for photocatalytic reactions. The fact that tetravalent

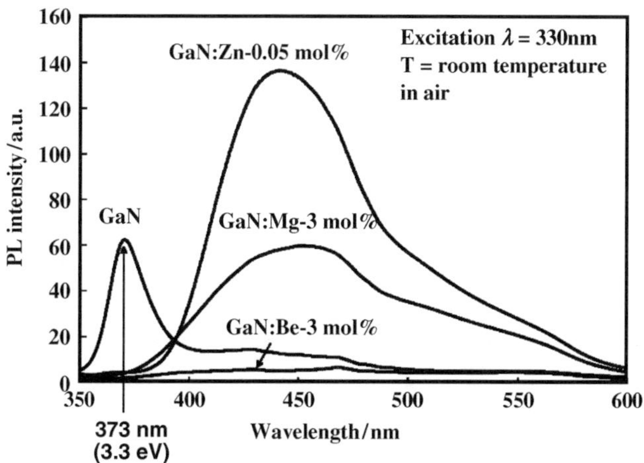

Figure 18.7 Photoluminescence spectra of undoped and divalent metal-ion-doped GaN. Measurement at room temperature, excitation wavelength = 330 nm. The amount of dopant was 3 mol% except for Zn, which was 0.05 mol%.

Si^{4+} and Ge^{4+} ions enhancing the n-type character had no activation effect means that the improvement in electron behavior has nothing to do with the activation. Namely, the key is the improvement of poor hole behavior in the valence band consisting of the N2*p* AO.

Figure 18.7 shows the photoluminescence spectra of undoped and doped GaN. The undoped GaN exhibited emission at 373 nm. This is nearly the same as the bandgap and is due to electron transfer from the conduction to the valence band. The photoemission for Si^{4+}- or Ge^{4+}-ion-doped GaN was analogous to that for undoped GaN, indicating that the donor levels were lying immediately beneath the conduction band. Doping with Zn^{2+} and Mg^{2+} changed the bandgap-type emission to broad emissions that had a maximum at around 450 nm and a tail extending to longer wavelengths up to 600 nm. In the excitation spectra, the 450 nm emission band appeared, with an excitation wavelength of 400 nm, and became strongest with a wavelength of 350 nm (not shown). The broad photoluminescence emission bands were reported to appear at 2.87 eV, 2.95 eV and 2.2 eV at 2 K for Zn^{2+}-, Mg^{2+}- and Be^{2+}-doped thin GaN films deposited on a sapphire substrate, respectively. The photoluminescence was explained in terms of a simple free-to-bound mechanism, that is transfer of free electrons in conduction bands to the acceptor levels due to doped Zn_{Ga}, Mg_{Ga} and Be_{Ga}, respectively [45,46]. The acceptor levels, Zn_{Ga}, Mg_{Ga} and Be_{Ga} were calculated to be higher by 0.34, 0.25 and 0.7 eV above the edge of the valence band. Thus, photoluminescence observed in the present study of doped GaN is related to transfer of electrons from the conduction band/donor levels to acceptor levels, such as Zn_{Ga} and Mg_{Ga}, formed in the forbidden band. Namely, the electronic structure is converted to that of p-type GaN. Figure 18.8 shows the photoluminescence spectra of Mg^{2+}-doped GaN with different Mg contents. The highest emission intensity appeared at a molar ratio of 0.5 mol%. With increasing Mg content, the intensity decreased monotonously, accompanied by a red-shift of the center of the emission bands, and almost disappeared at

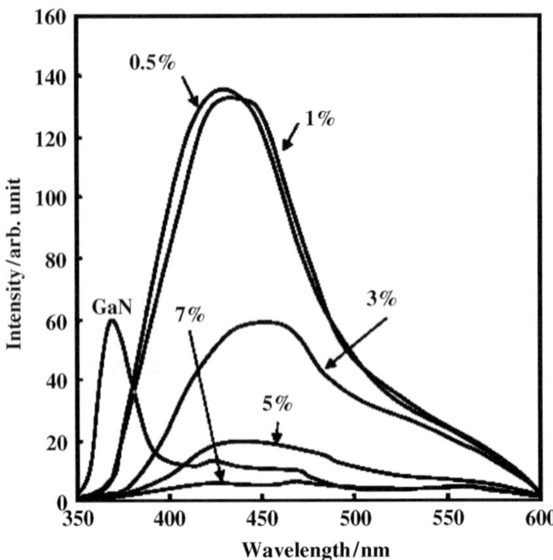

Figure 18.8 Changes in photoluminescence spectra of Mg^{2+}-doped GaN with Mg^{2+} content in the mixture of starting materials. Measurement at room temperature, excitation wavelength = 330 nm. Mg/(Mg + Ga) × 100% = 0, 0.5, 1, 3, 5 and 7. (Reprinted with permission from N. Arai, N. Saito, H. Nishiyama *et al.*, Effects of divalent metal Ion (Mg^{2+}, Zn^{2+} and Be^{2+}) doping on photocatalytic activity of ruthenium oxide-loaded gallium nitride for water splitting, *Catalysis Today*, **129**, 407–413, 2007. © 2007 Elsevier.)

7 mol%. Meanwhile, photocatalytic activity increased with increasing Mg, reaching a maximum at around 5 mol%, and sharply decreasing at 7 mol%. Thus, with increasing Mg content, the change in photoluminescence intensity was opposite to that in photocatalytic activity, although both decreased to negligible levels at large Mg contents. Photoluminescence intensity is associated with various factors, such as the density of photoexcited charges, the extent of nonradiation processes, the mobility of photoexcited charges and the density of recombination centers, and it is suggested that an increase in Mg content enhances the density of the acceptor levels and increases the concentration of holes and their mobility. The red-shift of the center of the photoluminescence peak with increasing Mg content indicates upward shifts in the acceptor level and strengthening of p-type properties. This increases the mobility and concentration of holes. Figure 18.9 shows a model of the band structure [45] and the mechanism of photocatalysis. These findings indicate that an improvement in hole behavior leads to the photocatalytic performance.

The light-absorption threshold of doped GaN is slightly shifted to longer wavelengths, compared to that of undoped GaN. This is ascribable to an absorption increase at around 400–450 nm, which is plausibly due to the formation of defects. Thus, weakening of the photoluminescence and disappearance of photocatalytic activity by a large amount of Mg doping are associated with an increase in the density of defects.

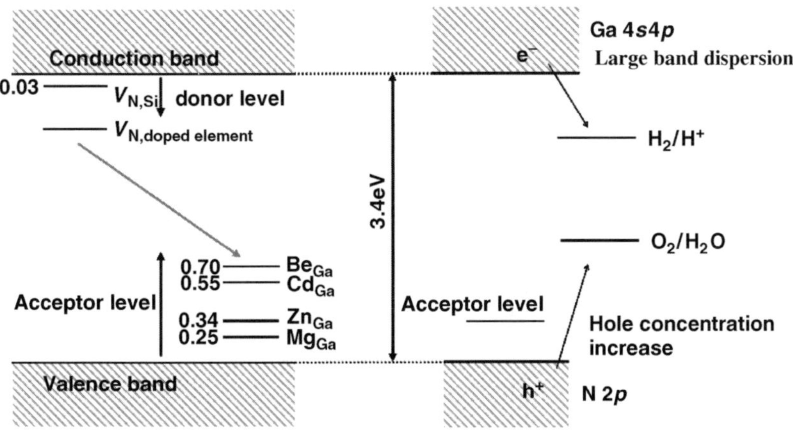

Figure 18.9 A model for the band structure of doped GaN and the mechanism of photocatalytic water splitting.

18.4 Effects of Metal-Ion Removal

The partial deficiency of one component metal ion from a stoichiometric compound would usually result in a significant, or at least considerable, deactivation of photocatalytic performance in view of the formation of defect sites as recombination centers for excited electrons and holes. However, an exception is observed for a lead tungstate $PbWO_4$; the phenomenon whereby the activity for water splitting increases when some part of the Pb is removed from stoichiometric $PbWO_4$ is interesting for understanding of the mechanism of photocatalysis by metal oxides.

Pb-deficient and Pb-rich lead tungstates, Pb_xWO_4 ($x = 0.2 \sim 1.1$), were obtained by changing the ratio of PbO to WO_3 in the starting mixture using a solid-state reaction at high temperatures in air and under vacuum-sealed conditions. Figure 18.10 shows the X-ray diffraction patterns of Pb_xWO_4 ($x = 0.6 \sim 1.1$). The main diffraction pattern for $x = 1.0$ was assigned to a single phase of tetragonal $PbWO_4$ [47]. Decreasing from $x = 1.0$ to $x = 0.60$, the diffraction patterns were maintained, although small peaks due to WO_3 appeared at $x = 0.875$ and grew with decreasing x. On the other hand, small peaks due to Pb_2WO_5 were observed at $x = 1.1$. In the UV-visible diffuse reflectance spectra, in addition to the main light absorption of $PbWO_4$ that occurred at a wavelength of 330 nm and leveled off at 300 nm, a broad bump due to WO_3 appeared in the range 330–450 nm with decreasing x. When the concentrations of Pb and W atoms in Pb_xWO_4 prepared in air and under vacuum conditions were measured using EPMA, the ratios of Pb to W atoms in $PbWO_4$ and $Pb_{0.75}WO_4$ (prepared in air) were 0.90 and 0.68, and those in $PbWO_4$ and $Pb_{0.75}WO_4$ (prepared under vacuum) were 0.93 and 0.70, respectively. The deficiency in Pb ions was close to that expected. The SEM images showed similar morphologies between $PbWO_4$ and $Pb_{0.75}WO_4$, although the particle sizes were considerably larger for the latter.

In water splitting on RuO_2-loaded $Pb_{0.75}WO_4$, both H_2 and O_2 were evolved upon UV irradiation and increased almost proportionally to irradiation time, as shown in Figure 18.11. The photocatalytic activity remained nearly unchanged and stable over runs. Figure 18.12 shows the photocatalytic activity of RuO_2-loaded Pb_xWO_4 (prepared in air) as a function of x. The photocatalytic activity for $x = 1.1$ was approximately one-half that for $x = 1.0$. An excess

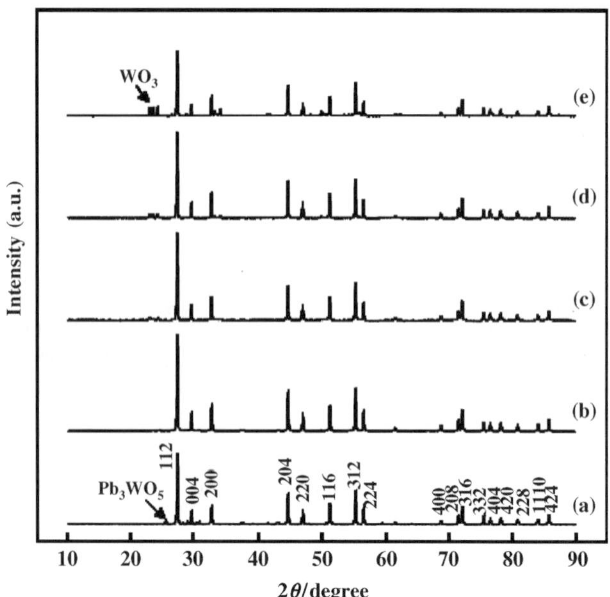

Figure 18.10 X-ray diffraction patterns of Pb_xWO_4 for x = (a) 1.1, (b) 1.0, (c) 0.875, (d) 0.75, (e) 0.6. (Reprinted with permission from H. Kadowaki, N. Saito, H. Nishiyama *et al.*, Photocatalytic activity of RuO_2-loaded Pb_xWO_4(x=0.2–1.1) for water decomposition, *Chemistry Letters*, **36**(3), 424–425, 2007. © 2007 The Chemical Society of Japan.)

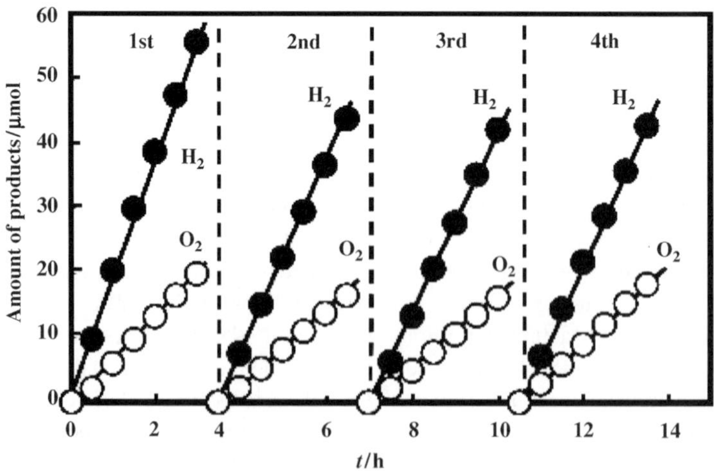

Figure 18.11 Stable production of H_2 (●) and O_2 (○) in repeated runs of overall water splitting on 1 wt % RuO_2-loaded $Pb_{0.75}WO_4$. (Reprinted with permission from H. Kadowaki, N. Saito, H. Nishiyama *et al.*, Photocatalytic activity of RuO_2-loaded Pb_xWO_4(x=0.2–1.1) for water decomposition, *Chemistry Letters*, **36**(3), 424–425, 2007. © 2007 The Chemical Society of Japan.)

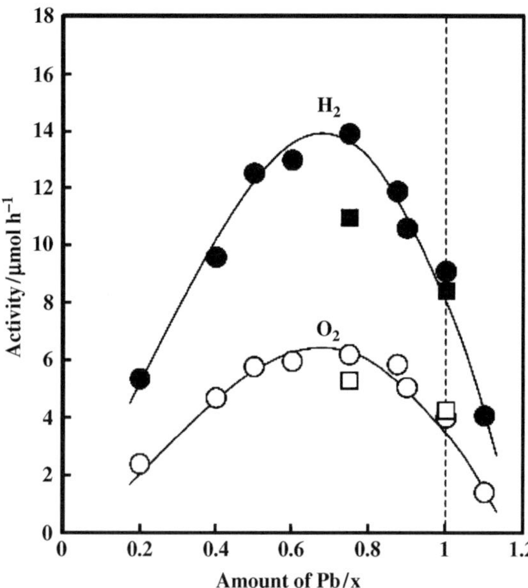

Figure 18.12 Photocatalytic activity of 1 wt% RuO_2-loaded Pb_xWO_4 as a function of x. Pb_xWO_4 prepared in air (●: H_2; ○: O_2) and under vacuum in a sealed quartz (■: H_2; □: O_2). (Reprinted with permission from H. Kadowaki, N. Saito, H. Nishiyama *et al.*, Photocatalytic activity of RuO_2-loaded Pb_xWO_4(x=0.2–1.1) for water decomposition, *Chemistry Letters*, **36**(3), 424–425, 2007. © 2007 The Chemical Society of Japan.)

of Pb lowered the activity. Furthermore, the activity of Pb_2WO_5 with excess Pb was negligible. Meanwhile, with decreasing x, the activity increased, reached a maximum at around $x = 0.75$, and the decreased markedly. The highest activity, for $x = 0.75$, was 1.5-fold larger than that for $x = 1.0$. For $PbWO_4$ and $Pb_{0.75}WO_4$ (prepared under vacuum), the photocatalytic activity was 1.3-fold larger for the latter than for the former. One might suggest that the activity enhancement is due to the formation of new metal-oxide interfaces and phases. One possibility is that the co-existence of WO_3 and $PbWO_4$ is responsible for the enhancement of photocatalytic activity, since the WO_3 phase remained unreacted for $x < 0.875$ (cf. Figure 18.10). However, in the case where a small amount of WO_3 was deliberately added to $PbWO_4$ and was subjected to calcination, the photocatalytic activity of the $PbWO_4$ decreased significantly.

As Pb-deficient lead tungstates, $Pb_{0.9375}WO_4$ and $Pb_7W_8O_{28.8}$ are known [48,49]. These provide nearly the same X-ray diffraction patterns as that of $PbWO_4$, and it is difficult to confirm the formation of $Pb_{0.9375}WO_4$ and $Pb_7W_8O_{28.8}$ based on the diffraction patterns alone. A possibility that the activity enhancement of Pb_xWO_4 with decreasing x is attributable to the formation of $Pb_{0.9375}WO_4$ and/or $Pb_7W_8O_{28.8}$ cannot completely excluded.

Figure 18.13 shows the energy-band structure and density of states (DOS) for $PbWO_4$. For comparison, the results for photocatalytically inactive $CaWO_4$ are also shown [26]. For $PbWO_4$, the O2*p* AO mainly formed the valence band, in which the lower part of the O2*p* AO was mixed with the W5*d* AO. The top of the valence band was composed of the O2*p* AO, largely hybridized with the Pb6*s* AO. On the other hand, the bottom of the conduction band consists of the W5*d* orbital hybridized with the O2*p* and Pb6*p* AOs. The narrow bandgap for

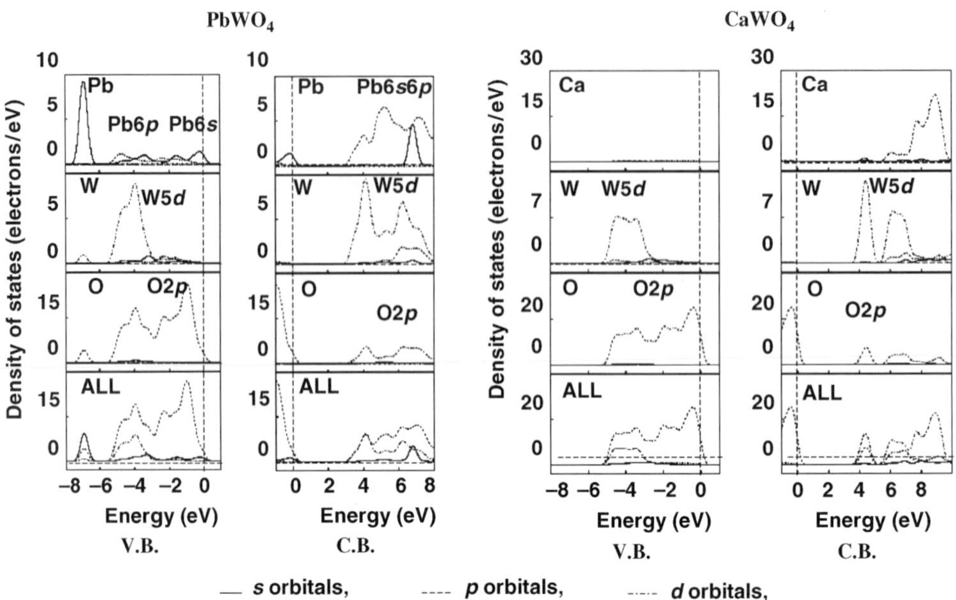

Figure 18.13 Total and atomic orbital projected density of state for Pb, W, O and Ca atom of $PbWO_4$ and $CaWO_4$. (Reprinted with permission from H. Kadowaki, N. Saito, H. Nishiyama *et al.*, Overall splitting of water by RuO_2-loaded $PbWO_4$ photocatalyst with $d^{10}s^2$-d^0 configuration, *Journal of Physical Chemistry C*, **111**, 439–444, 2007. © 2007 American Chemical Society.)

$PbWO_4$ is due to the hybridization of the Pb6*s* with the O2*p* AOs, causing a split-off state. Furthermore, it should be noted that both the valence and conduction bands of $PbWO_4$ have large dispersions. This is due to the contribution of the Pb6*s* AO to the valence band and that of the Pb6*p* to the conduction band. Thus, a large band dispersion generates electrons and holes with high mobilities, which is advantageous in photocatalysis. On the other hand, the band dispersion of inactive $CaWO_4$ is small for both the valence and conduction bands, since the top of the valence band is mainly composed of the O2*p* AO only, and the bottom of the conduction band consists of the W5*d* and the O2*p* AOs. These results clearly demonstrate that both the Pb6*s* and 6*p* AOs play an important role in photocatalytic performance.

Figure 18.14 shows the electronic structures for $Pb_7W_8O_{28.8}$ as Pb-deficient Pb_xWO_4. The contributions of the Pb6*s* to the O2*p* AO in the valence band and the Pb 6*p* to the W5*d* AO in the conduction bands, respectively, are increased, compared to those for $PbWO_4$, which indicates that Pb deficiency in Pb_xWO_4 enhances the interactions between Pb^{2+} and W^{6+} ions. This is responsible for increases in the photocatalytic activity. The intensity of photoluminescence for Pb_xWO_4 ($x = 0.6 \sim 1.0$) upon UV excitation at 250 nm considerably increased with decreasing x, becoming strongest at around $x = 0.75 \sim 0.875$, and significantly decreasing at $x = 0.6$. Furthermore, the Raman peak due to A_{1g} appearing at 904 cm^{-1} caused broadening for $Pb_{0.75}WO_4$, compared to that for $PbWO_4$. These spectroscopic changes are considered to reflect stronger interactions between the Pb^{2+} and W^{6+} metal ions, which are due to changes in both the geometric and electronic structures of $PbWO_4$ by partial loss of Pb ions.

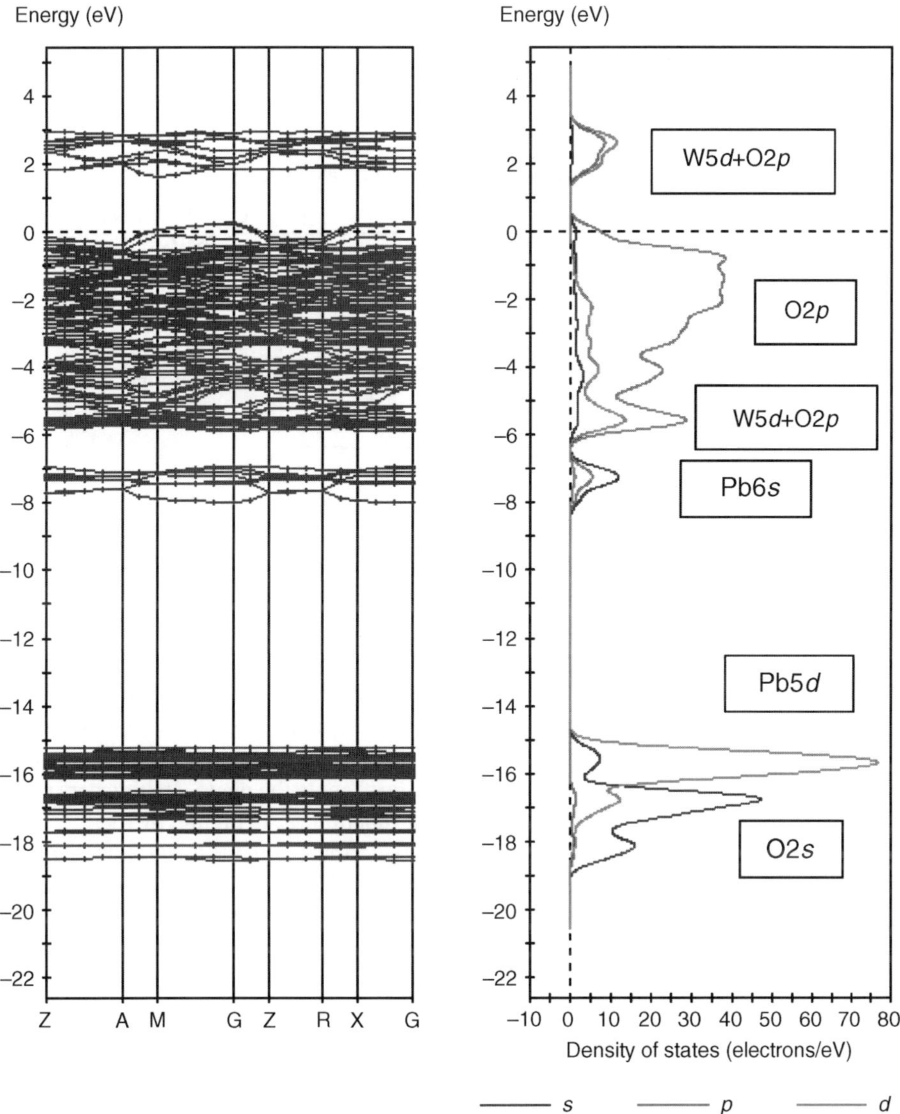

Figure 18.14 Band structure and density of state for $Pb_7W_8O_{28.8}$. Blue line; s-orbital, red; p-orbital, green; d-orbital. See also color-plate section.

18.5 Effects of Metal-Ion Exchange on Photocatalysis

18.5.1 $Y_xIn_{2-x}O_3$

In_2O_3 and Y_2O_3 have cubic structures with a lattice constant of $a = b = c = 1011$ and 1060 pm, respectively [49,50]. Because of the similar ionic radii of Y^{3+} (103 pm) and In^{3+} (97 pm), two cubic oxides make a single-phase solid solution of $Y_xIn_{2-x}O_3$, in which a melt of the octahedral

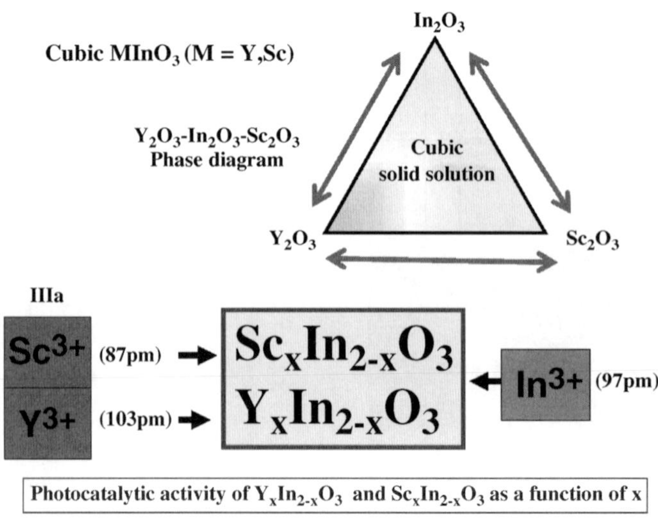

Figure 18.15 A schematic representation of a phase diagram for Y_2O_3-In_2O_3-Sc_2O_3.

InO_6 and YO_6 are randomly distributed. In_2O_3 is photocatalytically inactive for the water-splitting reaction, because of a lower conduction-band level than the H^+/H energy level. However, the partial replacement of In by Y atom is able to activate In_2O_3 to undergo water splitting when RuO_2 is loaded as a co-catalyst. This is related to the interactions of two metal ions of Y^{3+} and In^{3+} in a unit cell, and it is interesting to study the photocatalytic performance of a single-phase solid solution of $Y_xIn_{2-x}O_3$ as a function of x (Figure 18.15).

Although the X-ray diffraction pattern due to cubic $YInO_3$ remained nearly unchanged for $Y_xIn_{2-x}O_3$ in the range from x = 0.9 to x = 1.5, each peak of $Y_xIn_{2-x}O_3$ shifted to a lower angle nearly in proportion to x, as shown in Figure 18.16; for example, the main peak due to the (222) plane at $2\theta = 29.86°$ for x = 1.0 changed to 29.50° for x = 1.5. This is due to an increase in the lattice constant: a 50% increase in Y atoms brought about a 1.2% enhancement in the lattice constant. These results confirm that the prepared $Y_xIn_{2-x}O_3$ (x = 0.9–1.5) has a single-phase solid solution with randomly distributed octahedral YO_6 and InO_6.

Figure 18.17 shows the UV diffuse reflectance spectra of $Y_xIn_{2-x}O_3$ as a function of x. For x = 1.0 ($YInO_3$), light absorption started at around 350 nm and reached a maximum level at around 320 nm. For x < 0.9 (e.g., $Y_{0.9}In_{1.1}O_3$), the absorption curve shifted by 10 nm to longer wavelength, whereas for x > 1.0, the shift occurred in the opposite direction. The maximum absorption wavelength was 296, 282 and 254 nm for x = 1.1, 1.3 and 1.5, respectively.

The photocatalytic activity of 1 wt% RuO_2-loaded In_2O_3 was negligible for water splitting under Hg-Xe lamp illumination, but 1.0 wt % RuO_2–loaded $Y_xIn_{2-x}O_3$ (1.0 < x < 1.5) showed stable production of both H_2 and O_2 with the onset of the reaction at constant rates. Figure 18.18 shows the photocatalytic activity of 1 wt% RuO_2-loaded $Y_xIn_{2-x}O_3$ as a function of x. For x = 0.5, a small amount of hydrogen was evolved, but the activity was small. Significant activity was observed for x = 1.0. The activity increased gradually in the range x = 1.0–1.3 with increasing x, passed through a maximum at x = 1.3, and significantly decreased at x = 1.5. It might be possible that an increase in the concentration of Y atom promotes the crystallization of

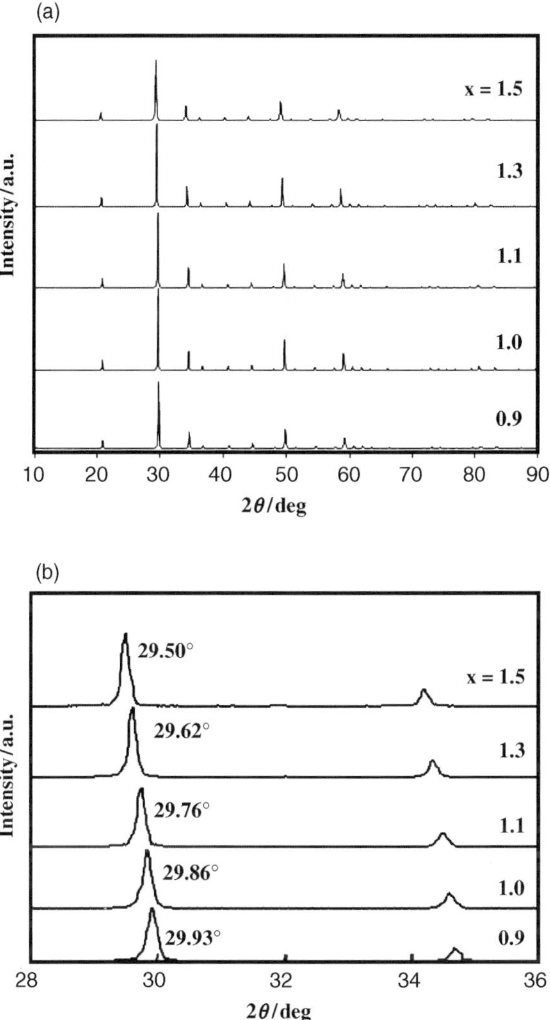

Figure 18.16 X-ray diffraction patterns for $Y_xIn_{2-x}O_3$ (x = 0.9–1.5) over a wide range (a) and a narrow range (b). (Reprinted with permission from N. Arai, N. Saito, H. Nishiyama *et al.*, Photocatalytic activity for overall water splitting of RuO_2-loaded $Y_xIn_{2-x}O_3$ (x = 0.9–1.5), *Journal of Physical Chemistry C*, **112**, 5000–5005, 2008. © 2008 American Chemical Society.)

$Y_xIn_{2-x}O_3$ and thus eliminates impurities and structural imperfections that frequently work as traps for charge recombination, leading to better photoctalytic performance. However, this is unlikely, because no significant narrowing in the X-ray diffraction peaks was observed with increasing x. Furthermore, in the SEM images for $Y_xIn_{2-x}O_3$ (x = 1.0, 1.3, 1.5), the morphological features of irregular-shaped particles with uneven surfaces remained nearly unchanged, irrespective of x. Thus, macrocrystal structures are not responsible for the activity changes.

Figure 18.19 shows the Raman spectra for $Y_xIn_{2-x}O_3$(x = 0.0–2.0). In_2O_3 (x = 0.0 in $Y_xIn_{2-x}O_3$) has peaks at 112, 118, 121, 134, 156, 172, 215, 309, 370, 393, 499, and 631 cm^{-1},

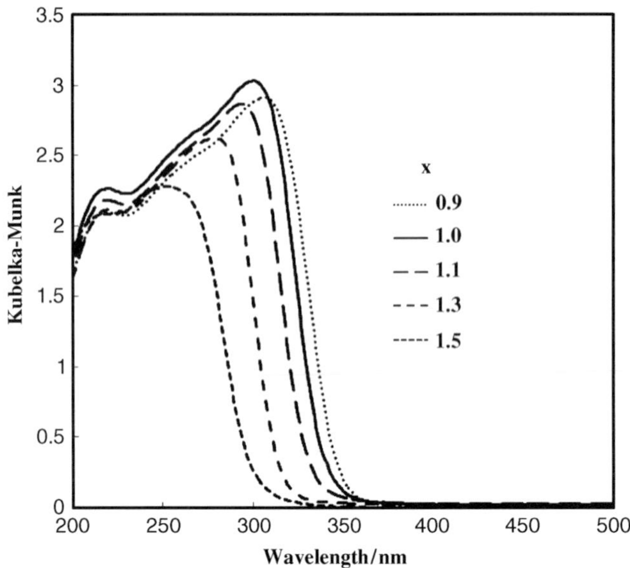

Figure 18.17 UV diffuse reflectance spectra of $Y_xIn_{2-x}O_3$(x = 0.9–1.5). (Reprinted with permission from N. Arai, N. Saito, H. Nishiyama *et al.*, Photocatalytic activity for overall water splitting of RuO_2-loaded $Y_xIn_{2-x}O_3$ (x = 0.9–1.5), *Journal of Physical Chemistry C*, **112**, 5000–5005, 2008. © 2008 American Chemical Society.)

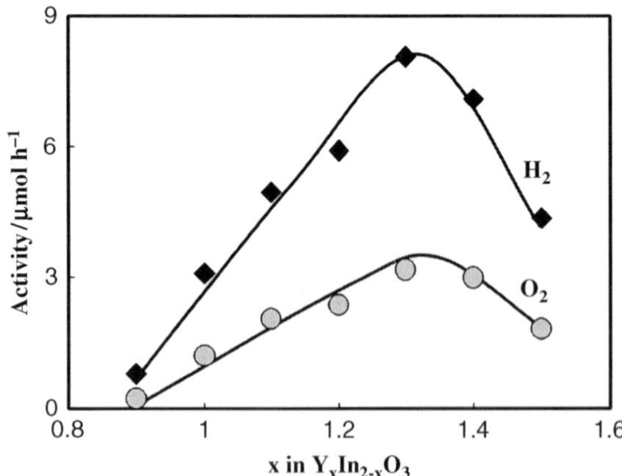

Figure 18.18 Photocatalytic activity of 1.0 wt% RuO_2-loaded $Y_xIn_{2-x}O_3$ (x = 0.9–1.5) for H_2 and O_2 production as a function of x. (Reprinted with permission from N. Arai, N. Saito, H. Nishiyama *et al.*, Photocatalytic activity for overall water splitting of RuO_2-loaded $Y_xIn_{2-x}O_3$ (x = 0.9–1.5), *Journal of Physical Chemistry C*, **112**, 5000–5005, 2008. © 2008 American Chemical Society.)

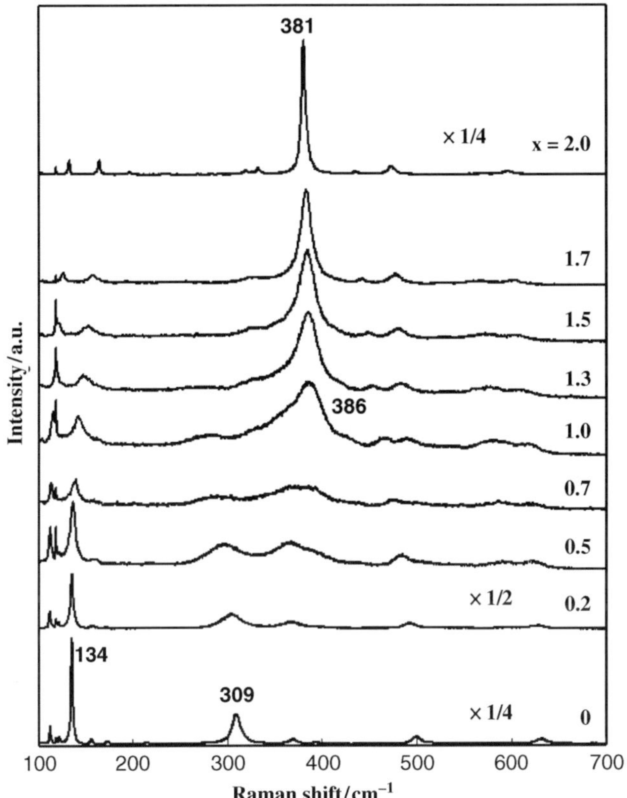

Figure 18.19 Raman spectra of $Y_xIn_{2-x}O_3(x = 0.0–2.0)$. Excitation wavelength: 514.54 nm. (Reprinted with permission from N. Arai, N. Saito, H. Nishiyama *et al.*, Photocatalytic activity for overall water splitting of RuO_2-loaded $Y_xIn_{2-x}O_3$ (x = 0.9–1.5), *Journal of Physical Chemistry C*, **112**, 5000–5005, 2008. © 2008 American Chemical Society.)

in which two strong peaks appear at 134 and 309 cm^{-1}. With increasing x, the two peaks become broader and their intensity decreases. The peak at 309 cm^{-1} almost disappears at x = 0.7. At x = 1.0 ($YInO_3$), an extremely broad peak newly appears at 386 cm^{-1} and becomes sharper with increasing x. At x = 1.5–1.7, the Raman spectra changes to characteristic peak patterns due to Y_2O_3 that has peaks at 118, 132, 165, 184, 197, 320, 333, 381, 435, 473, 569 and 596 cm^{-1}. A remarkably strong peak is observed at 381 cm^{-1}. The Raman spectra of Y_2O_3 have $4E_g + 4A_g + 14F_g$ modes; the eight modes of F_g have the same frequencies as those of E_g and A_g [52]. The peaks at 596, 473, 381 and 165 cm^{-1} are associated with $F_g + A_g$ modes, whereas the peaks at 569, 435, 333, 320 and 197 cm^{-1} belong to the $E_g + A_g$ modes. The peaks at 118, 132 and 184 cm^{-1} were assigned to the F_g mode (the other weak peaks due to F_g at 399, and 526 cm^{-1} are missing in these spectra). The strongest peak at 381 cm^{-1} for Y_2O_3 and 309 cm^{-1} for In_2O_3 deviates from a relation [53] of metal–oxygen bond length versus Raman wavenumber. This indicates that the strongest peak is not due to the stretching vibration only, as demonstrated by the presence of the $F_g + A_g$ mode for Y_2O_3. By the replacement of In by Y atom, the strongest peak for Y_2O_3 becomes broader and shifts to higher wavenumber by 5 cm^{-1}. Since the octahedral InO_6 and YO_6 in $Y_xIn_{2-x}O_3$ are randomly distributed by edge connections sharing two oxygen atoms, the changes in the Raman peaks with the partial replacement of the octahedral InO_6 by octahedral YO_6 (vice versa) are indicative of significant deformation of both the octahedral InO_6 and YO_6. A correlation between the photocatalytic activity and the distortion of octahedral XO_6 (X = metal ion) has been demonstrated for various kinds of metal-oxide photocatalysts [5,31,54–56]. For instance, among alkaline-metal and alkaline-earth-metal indates, $SrIn_2O_4$ and $Sr_{0.93}Ba_{0.07}In_2O_4$, consisting of two kinds of distorted octahedral InO_6 were photocatalytically active for water splitting when RuO_2 was loaded as a co-catalyst, whereas $LiInO_2$ with undistorted octahedral InO_6 was inactive under similar reaction conditions [31]. Thus, the activity enhancement of $Y_xIn_{2-x}O_3$ with increasing x in the range x = 1.0–1.3 is ascribable to the effects of distortion of the octahedral InO_6. A decrease in the activity above $x > 1.3$ is due to the decrease in the concentration of InO_6.

In the DFT calculation for $YInO_3$, the optimized $Y_{16}In_{16}O_{48}$ structure was used as a unit cell. For $Y_xIn_{2-x}O_3$, $Y_{1.25}In_{0.75}O_3$ (x = 1.25) was taken as a typical example, in which $Y_{20}In_{12}O_{48}$ was used, where four Y atoms in $Y_{16}In_{16}O_{48}$ were replaced with In atoms. The valence atomic configurations for Y, In, and O atoms are $4s^24p^65s^24d^1$, $5s^24d^{10}5p^1$, and $2s^22p^4$, respectively. For the inner bands of $YInO_3$, the Y4*s* and Y4*p* bands appear at around −40 and −20 eV, respectively. The O2*s* band is observed at around −15 and −17 eV, and the In4*d* band appears at −12.3 eV as a single peak without mixing with other AO. Figure 18.20 shows the valence and conduction bands. The DOS breaks down into the projected DOS (PDOS). The valence band is mainly composed of the O2*p* AO. In the lower region of the valence band, the O2*p* AO is mixed with In5*s* AO, and the upper part consists of the O2*p* AO mixed to a small extent with the Y4*d* AO. This type of mixing occurs in the in-phase (bonding). The bottom of the conduction band is formed by the In5*s* + O2*p* AOs, whereas the upper band consists of the In5*p* and Y4*d* AOs. Figures 18.21 and 18.22 show the electron density contour maps for the HOMO and LUMO levels, respectively. For the former, the electron density is solely localized on the O atoms, indicating that the HOMO levels are solely localized on the O2*p* AO, whereas for the latter, it is largely localized on In atoms with a small portion on the O atoms. The band dispersion is remarkably large for the bottom of the conduction band. This permits the generation of photoexcited electrons with large mobilities and leads to good photocatalytic performance.

A comparison of the electronic structures for $YInO_3$ with those for Y_2O_3 and In_2O_3 is interesting. The valence band of In_2O_3 is composed of the O2*p* AO with a small contribution from the In5s AO in the lower part of the valence band. The conduction band has the In5*s*5*p* AOs. The valence band of Y_2O_3 is primarily formed by the O2*p* AO, while the Y4*d* AO is involved over the entire range of the valence band. The conduction band is composed of the Y4*d* and Y5*p* AOs. Interestingly, the valence bands of $YInO_3$ are similar to those of Y_2O_3, whereas the conduction bands are analogous to those of In_2O_3, although the conduction band of $YInO_3$ involves the contribution of the Y4*d* AO.

Figure 18.23 shows the band structure and PDOS of $Y_xIn_{2-x}O_3$ (x = 1.25). The valence and conduction band structures for $Y_xIn_{2-x}O_3$(x = 1.25) are similar to those of $YInO_3$. However, the contribution of the Y4*d* AO to the valence and conduction band increases with $Y_xIn_{2-x}O_3$ (x = 1.25), compared to $YInO_3$. The bandgap calculated is larger for $Y_xIn_{2-x}O_3$ (x = 1.25) (2.42 eV) than for $YInO_3$ (1.19 eV). The contribution of the Y4*d* AO to the lower region of the conduction band raises the LUMO level, whose effects are larger for $Y_xIn_{2-x}O_3$ (x = 1.25). This result is in good agreement with the shift of the primary absorption bands to shorter

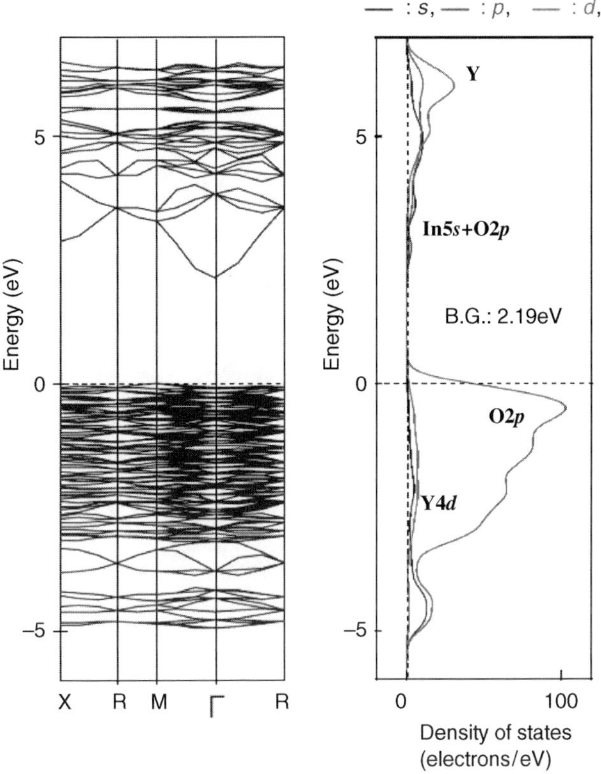

Figure 18.20 Band structure and density of states for $Y_xIn_{2-x}O_3(x=1.0)$. Blue line: s-orbital; red: p-orbital; green: d-orbital. (Reprinted with permission from N. Arai, N. Saito, H. Nishiyama *et al.*, Photocatalytic activity for overall water splitting of RuO_2-loaded $Y_xIn_{2-x}O_3$ (x = 0.9–1.5), *Journal of Physical Chemistry C*, **112**, 5000–5005, 2008. © 2008 American Chemical Society.) See also color-plate section.

wavelengths with increasing Y content in $Y_xIn_{2-x}O_3$ (cf. Figure 18.17). The shift of the LUMO to a higher energy level is advantageous for the production of high-energy electrons and is also responsible for the increased photocatalytic activity.

From these findings, the activity increases of a single-phase solid solution of $Y_xIn_{2-x}O_3$ with increasing x can be attributed to the deformation of InO_6/YO_6 octahedral units and also to upward shifts in the conduction band levels. A composite metal oxide, $LiInGeO_4$, with the d^{10}–d^{10} electronic configuration, showed higher photocatalytic performance than single core metal oxides such as $AInO_2$ (A = Li,Na) and Li_2GeO_3, with d^{10} configurations. This is explained in terms of the distortion effects of the octahedral InO_6 and tetrahedral GeO_4 and the hybridization effects of the *sp* orbitals of In^{3+} and Ge^{4+} metal ions in conduction bands [36]. Thus, solid solutions of metal oxides is pointed out to have the advantage that they readily cause the distortion of metal–oxygen octahedral/tetrahedral units and orbital hybridization between different atoms.

As mentioned above, both GaN and ZnO were photocatalytically inactive for water splitting even when RuO_2 or Cr_2O_3/Rh was loaded as a cocatalyst. However, a solid solution of GaN

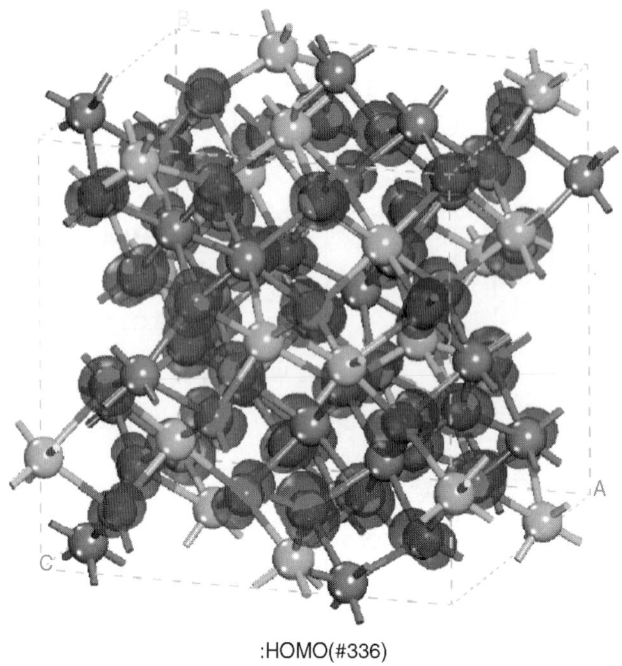

Figure 18.21 Electron density contour map for the top of the valence band (HOMO) of $Y_xIn_{2-x}O_3$-(x = 1.0). Light blue: Y atom; brown: In atom; red: O atom. (Reprinted with permission from N. Arai, N. Saito, H. Nishiyama *et al.*, Photocatalytic activity for overall water splitting of RuO_2-loaded $Y_xIn_{2-x}O_3$ (x = 0.9–1.5), *Journal of Physical Chemistry C*, **112**, 5000–5005, 2008. © 2008 American Chemical Society.) See also color-plate section.

with ZnO, $(Ga_{1-x}Zn_x)(N_{1-x}O_x)$, was remarkably active. Interestingly its light absorption wavelength shifted to longer wavelength, extending to 550 nm from 380–400 nm of GaN and ZnO, and water splitting occurred even under the visible light region at around 550 nm [57,58]. This clearly shows the solid-solution effects on photocatalysis.

18.5.2 $Sc_xIn_{2-x}O_3$

In_2O_3 and Sc_2O_3 form a solid solution of $Sc_xIn_{2-x}O_3$. A difference from $Y_xIn_{2-x}O_3$ is the smaller ionic radii of Sc^{3+} (87 pm) than that of Y^{3+} (103pm). The X-ray diffraction patterns of $Sc_xIn_{2-x}O_3$ showed a single–phase cubic pattern of $ScInO_3$ in the range of x = 1.0–1.3. However, the main peak shifted to a larger angle with increasing x; for example, the (222) plane at $2\theta = 30.98°$ for x = 1.0 shifted to 31.12° for x = 1.3. Light absorption for x = 1.0 ($ScInO_3$) started at around 370 nm and reached a maximum level at around 290 nm. For x = 1.3, the absorption maximum shifted to 260 nm. The blue-shift with increasing x occurred in a manner similar to $Y_xIn_{2-x}O_3$.

Figure 18.24 shows the activity of $ScInO_3$ (x = 1.0 in $Sc_xIn_{2-x}O_3$) and $Sc_{1.3}In_{0.7}O_3$. The activity is 4.4-fold larger for x = 1.3 than for x = 1.0. The extent of activity enhancement is nearly the same as those observed for $Y_xIn_{2-x}O_3$(x = 1.0 and 1.3). In the Raman spectra for

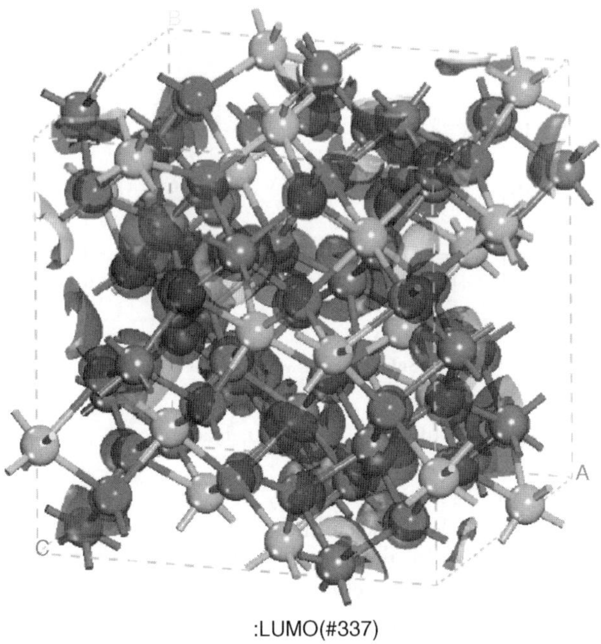

Figure 18.22 Electron density contour map for the bottom of the conduction band (LUMO) of $Y_xIn_{2-x}O_3$ (x = 1.0). See Figure 18.21 for the color coding. (Reprinted with permission from N. Arai, N. Saito, H. Nishiyama *et al.*, Photocatalytic activity for overall water splitting of RuO_2-loaded $Y_xIn_{2-x}O_3$ (x = 0.9–1.5), *Journal of Physical Chemistry C*, **112**, 5000–5005, 2008. © 2008 American Chemical Society.) See also color-plate section.

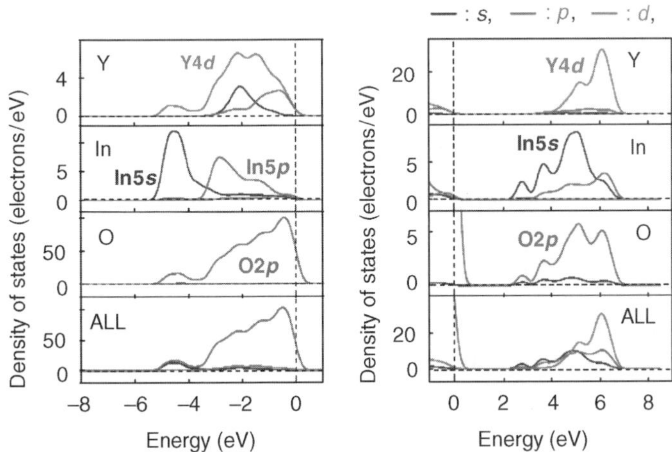

Figure 18.23 Total and atomic orbital PDOS for Y, In and O atoms for $Y_xIn_{2-x}O_3$(x = 1.25). (Reprinted with permission from N. Arai, N. Saito, H. Nishiyama *et al.*, Photocatalytic activity for overall water splitting of RuO_2-loaded $Y_xIn_{2-x}O_3$ (x = 0.9–1.5), *Journal of Physical Chemistry C*, **112**, 5000–5005, 2008. © 2008 American Chemical Society.) See also color-plate section.

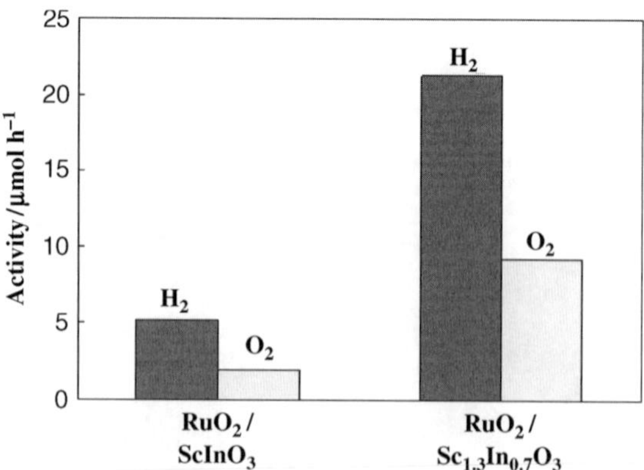

Figure 18.24 Photocatalytic activity of 1 wt% RuO_2-loaded $Sc_xIn_{2-x}O_3$ for x = 1.0 and 1.30.

$Sc_xIn_{2-x}O_3$(x = 0.0–2.0), Sc_2O_3 (x = 2.0 in $Sc_xIn_{2-x}O_3$) showed a strong peak at 421 cm^{-1}due to the ScO_6 stretching vibration. With the addition of an In atom, that is, decreasing x, the peak shifted to shorter wavenumber with broadening and appeared at 397 cm^{-1} for x = 1.0–1.3. Significant changes in the Raman peaks with the partial replacement of the octahedral InO_6 by octahedral ScO_6 (vice versa) in $Sc_xIn_{2-x}O_3$ provide evidence for the deformation of octahedrally coordinated InO_6 and ScO_6. Thus, the activity enhancement of $Sc_xIn_{2-x}O_3$with increasing x in the range x = 1.0–1.3 is quite similar to that observed in $Y_xIn_{2-x}O_3$, with a decrease in the activity above x > 1.3.

Furthermore, the positive effects of metal-ion exchange on the activity were observed for β-$Ga_{2-x}In_xO_3$, in which the activity increased by two-fold when Ga atoms were replaced by 5 mol% In atoms.

18.5.3 $Y_xIn_{2-x}Ge_2O_7$

It is interesting to see how the photocatlytic activity of a composite metal oxide varies with the replacement of the component metal ions. In a composite metal oxide of $YInGe_2O_7$ with d^{10}–d^{10} electronic configuration, Y and In can replace each other in different ratios. The X-ray diffraction patterns for $Y_xIn_{2-x}Ge_2O_7$ remained unchanged in the range x = 0.0–2.0 except for shifts of each peak to a lower diffraction angle. This indicates that the single crystal phase is maintained, and the unit cell is expanded by an increase in the content of a larger Y ion. The main light-absorption band of $Y_xIn_{2-x}Ge_2O_7$ (x = 1.0) was observed at around 280 nm, and blue-shifts occurred with increasing x. The photocatalytic activity of RuO_2-loaded Y_xIn_{2-x}-Ge_2O_7 increased gradually in the range x = 0–0.5 with increasing x, reached a maximum at x = 1.25, and markedly decreased with further increase in x. The activity was nearly zero at x = 2.0. In a manner similar to the case of $Y_xIn_{2-x}O_3$, the maximum activity was observed at x = 1.25, not at x = 1.0.

$YInGe_2O_7$ [59] has octahedral YnO_6 and tetrahedral GeO_4 linked to each other through the corner oxygen atoms, as shown in Figure 18.25. Both the InO_6 and GeO_4 units are so heavily distorted that the position of the In cation in InO_6 deviates from the center of gravity of the surrounding six oxygen anions. This produces a dipole moment of 3.0 D in the InO_6 unit. The GeO_4 tetrahedron is also distorted, and has a dipole moment of 3.4 D. Thus, the high

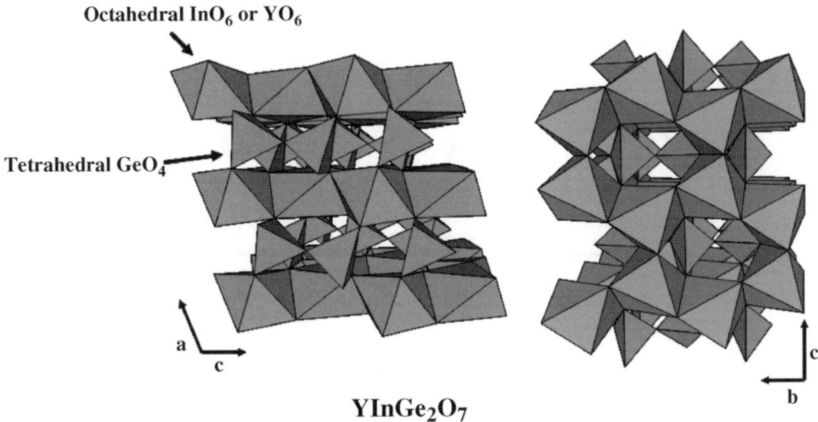

Figure 18.25 A schematic representation of the crystal structure of $YInGe_2O_7$. See also color-plate section.

photocatalytic performance of RuO_2-dispersed $YInGe_2O_7$ is in line with the correlation between activity and distortion [31]. With increasing x in $Y_xIn_{2-x}Ge_2O_7$, the distortion of InO_6 and GeO_4 varies with mixing of metal ions of different sizes, which is effective for the enhancement of activity.

In the DFT calculation, the optimized $(YInGe_2O_7)_2$ structure is used as a unit cell where half of a Y atom in $Y_2Ge_2O_7$ is replaced by an In atom. The valence atomic configurations of $YInGe_2O_7$ are $1s^22s^1$ for Y, $5s^24d^{10}5p^1$ for In, $4s^24p^2$ for Ge and $2s^22p^4$ for O atom. The lower part of the valence band is composed of the O2*p* AO hybridized with the Ge4*s* and In5*s* AOs, and the middle part is hybridized with the Ge4*p* and In5*p* AOs. The upper part of the valence band consists of the O2*p* AO, slightly mixed with Y4*d* AO. On the other hand, the bottom of the conduction band is composed of the Ge4*s*, In5*s* and O2*p* AOs. With increasing energy in the conduction band, Ge4*p*, In5*p* and Y4*d* AOs appear in that order. When compared with the band structure of $YInO_3$ (cf. Figure 18.22), the valence band, consisting of the O2*p* AO slightly mixed with the Y4*d* AO, is similar, whereas there is a difference in the conduction band: for $YInGe_2O_7$, the Ge4*s*4*p* AO is added to the In5*s* + O2*p* AOs mixed with the Y4*d* AO. The calculation also shows that the higher photocatalytic activity of $Y_xIn_{2-x}Ge_2O_7$ (x = 1.25), compared to $YInGe_2O_7$, is due to the increased contribution of the Y4*d* AO to the conduction bond, as observed in $Y_xIn_{1-x}O_3$.

18.6 Effects of Zn Addition to Indate and Stannate

A metal ion of Zn^{2+} has a $d^{10}s^0$ electronic structure and offers an interesting electronic effect on the band structures of metal oxides, because the Zn3*d* AO is involved in the O2*p* valence band region, and the Zn4*s* AO contributes to the conduction band. This induces *p*–*d* repulsion in the valence band, which raises the energy level of the valence band, causing band narrowing [33]. For example, light absorption occurred at longer wavelength for $ZnGa_2O_4$ than for $SrGa_2O_4$ (the photocatalytic activities of two RuO_2-loaded gallates for water splitting were similar). Thus, chemical compounds involving Zn^{2+} are interesting from the viewpoint of band narrowing.

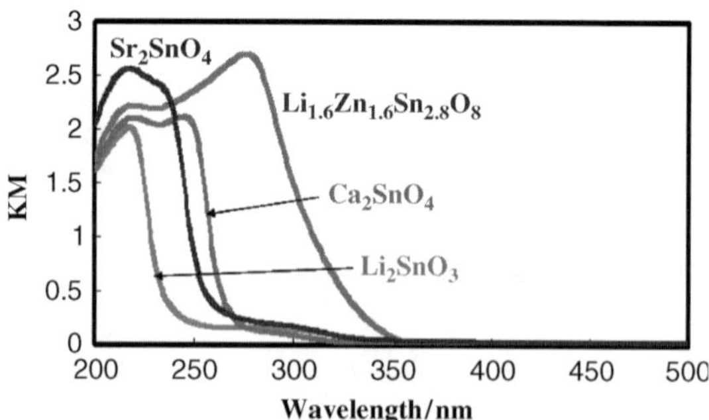

Figure 18.26 UV diffuse reflectance spectra of various Sn metal oxides. See also color-plate section.

18.6.1 *$Li_{1.6}Zn_{1.6}Sn_{2.8}O_8$*

Figure 18.26 shows the UV diffuse reflectance spectra of Li_2SnO_4, M_2SnO_4 (M = Ca, Sr) and $Li_{1.6}Zn_{1.6}Sn_{2.8}O_8$. Li_2SnO_3 has a main absorption band at around 230 nm. Sr_2SnO_4 shows a threshold band at around 260 nm and a main band at 240 nm, whereas Ca_2SnO_4 has an absorption band shifted by 10 nm to longer wavelength than that of Sr_2SnO_4. On the other hand, $Li_{1.6}Zn_{1.6}Sn_{2.8}O_8$ shows a threshold band at around 350 nm and a maximum band at 280 nm. It is evident that the inclusion of Zn causes a shift of absorption band to longer wavelength.

In water splitting on RuO_2-loaded $Li_{1.6}Zn_{1.6}Sn_{2.8}O_8$, the production of H_2 was considerably lower than the stoichiometric ratio in an earlier period of reaction, but the ratio of H_2 to O_2 approached 2.0 on prolonged irradiation. The photocatalytic behavior of $Li_{1.6}Zn_{1.6}Sn_{2.8}O_8$ was sensitive to the amount of RuO_2 loaded and the oxidation temperature to convert the surface Ru carbonyl complex to RuO_2. Suitable amounts and temperature were 1.0 wt% and around 673 K, respectively. Thus, the anomalous behavior in the initial reaction stage might be associated with the sensitive role of RuO_2 on $Li_{1.6}Zn_{1.6}Sn_{2.8}O_8$ surface. Under similar reaction conditions, the activity of $Li_{1.6}Zn_{1.6}Sn_{2.8}O_8$ was larger by a factor of about two than M_2SnO_4(M = Ca,Sr). M_2SnO_4 (M = Ca,Sr) has deformed octahedral SnO_6; for Ca_2SnO_4, SnO_6 has an elongated octahedral structure, in which two Sn–O bonds are longer than the rest of the Sn–O bonds, whereas for Sr_2SnO_4, octahedral SnO_6 has a compressed structure, in which two SnO bonds ae shorter than the other four bonds. However, the core metal ion is located at the center of gravity of six oxygen atoms, and the dipole moment is zero for both the stannates. On the other hand, $Li_{1.6}Zn_{1.6}Sn_{2.8}O_8$ has heavily distorted SnO_6 with a dipole moment of 3.5 D. Thus, the higher activity of $Li_{1.6}Zn_{1.6}Sn_{21.8}O_8$ is related to the characteristic structures of octahedral SnO_6.

18.6.2 *$Ba_3Zn_5In_2O_{11}$*

The effects of Zn were also found for Ba–In–O metal oxides. In RuO_2-loaded MIn_2O_4 (M = Ca, Sr, Ba) metal oxides, activity for water splitting was observed for M = Ca and Sr, but not for M = Ba. Furthermore, a series of Ba–In–O metal oxides consisting of $(BaO)_n(In_2O_3)_m$, where n = 2, 3,4,5, 8 or m = 1, 3 were photocatalytically inactive even in the presence of RuO_2. However, a metal oxide involving Zn^{2+} such as $Ba_3Zn_5In_2O_{11}$ that consisted of $(BaO)_3$ $(ZnO)_5(In_2O_3)_1$ was photocatalytically active. As shown in Figure 18.27, the DFT calculation for $Ba_3Zn_5In_2O_{11}$ shows that the Zn3*d* AO contributes to the valence band consisting of the O2*p*

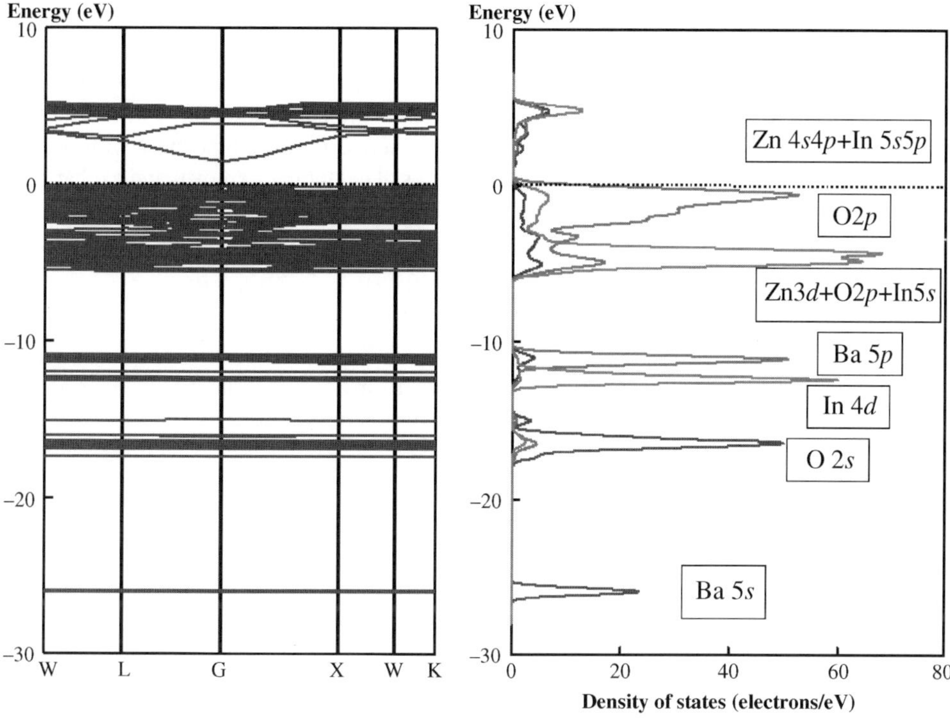

Figure 18.27 Band structure and atomic orbital PDOS for $Ba_3Zn_5In_2O_{11}$. Blue line: s orbital; red: p orbital; green: d orbital. See also color-plate section.

AO, whereas Zn4*s* AO is mixed with In 5*s*5*p* AOs in the conduction band. It is evident that the electronic contribution of Zn^{2+} induces the photocatalytic activity.

18.7 Conclusions

The effects of metal ion doping and the removal and exchange of constituent metal ions on the photocatalytic performance for water splitting of transition, typical and lanthanide metal oxides and nitrides were demonstrated. Although neither undoped CeO_2 nor stoichiometric compounds such as $SrCeO_3$ and Sr_2CeO_4 were active, Sr^{2+} doping activated CeO_2 to produce H_2 and O_2 when RuO_2 was loaded as a co-catalyst. An inactive metal nitride, GaN, with d^{10} configuration became active by doping with divalent metal ions such as Zn^{2+}, Mg^{2+} or Be^{2+}. Undoped GaN showed a photoluminescence peak at 350 nm due to the bandgap transition, whereas Zn^{2+} and Mg^{2+}-doped GaN provided broad photoluminescence peaks at around 450 nm. This is indicative of the formation of p-type GaN, by which the concentration and mobility of holes are increased, and hence the photocatalytic activity is invoked. Pb-deficient lead tungstate, $Pb_xWO_4(x < 1.0)$, showed the highest photocatalytic activity at $x = 0.75$ in the range $x = 0.2 \sim 1.1$, which was also related to changes in the local structure and band structures. The photocatalytic activity of RuO_2-loaded $Y_xIn_{2-x}O_3$ increased with increasing x, reached a maximum at $x = 1.3$ and decreased with further increase in x. Similar changes in the activity with x were observed for the solid solutions of $Sc_xIn_{2-x}O_3$ and $Y_xIn_{2-x}Ge_2O_7$. The Raman spectra for $Y_xIn_{2-x}O_3$ showed that both the octahedral InO_6 and YO_6 were deformed with increasing x. Furthermore, the DFT calculation showed that the contribution of the Y4d AO to the O2p AO in the valence band and to the In5s AO in the conduction band increased with increasing x. Activity enhancement of $Y_xIn_{2-x}O_3$ with increasing x was associated with changes in the local structure by distortion and band structures. The doping, removal and exchange of constituent metal ions are concluded to be an interesting approach to the activation of the metal oxides and nitrides, because of easier changes in the geometric and electronic structures.

Acknowledgments

This work was partially supported by a Grant-in-Aid for Basic Research (No. 5) from The Ministry of Land, Infrastructure and Transport Government of Japan.

References

[1] Fujishima, A. and Honda, K. (1972) Electrochemical photolysis of water at a semiconductor electrode. *Nature*, **238**, 37–38.

[2] Domen, K., Kudo, A. and Onishi, T. (1986) Mechanism of photocatalytic decomposition of water into H_2 and O_2 over NiO-$SrTiO_3$. *Journal of Catalysis*, **102**, 92–98.

[3] Inoue, Y., Kubokawa, T. and Sato, K. (1991) Photocatalytic activity of alkali-metal titanates combined with Ru in the decomposition of water. *Journal of Physical Chemistry*, **95**, 4059.

[4] Ogura, S., Kohno, M., Sato, K. and Inoue, Y. (1997) Photocatalytic activity for water decomposition of RuO_2 combined with $M_2Ti_6O_{13}$ (M = Na,K,Rb,Cs). *Applied Surface Science*, **121/122**, 521.

[5] Inoue, Y., Asai, Y. and Sato, K. (1994) Photocatalysts with tunnel structures for decomposition of water. $BaTi_4O_9$ having a pentagonal prism tunnel structure and various promoters. *Journal of the Chemical Society Faraday Transactions*, **90**, 797.

[6] Kohno, M., Kaneko, T., Ogura, S. *et al.* (1998) Photocatalysts with tunnel structures for decomposition of water. $BaTi_4O_9$ having a pentagonal prism tunnel structure and various promoters. *Journal of the Chemical Society Faraday Transactions*, **94**, 89.

[7] Takata, T., Furumi, Y., Shinohara, K. *et al.* (1997) Photocatalytic decomposition of water on spontaneously hydrated layered perovskites. *Chemistry of Materials*, **9**, 1063–1064.

[8] Takata, T., Shinohara, K., Tanaka, A. *et al.* (1997) A Highly active photocatalyst for overall water splitting with a hydrated layered perovskite structure. *Journal of Photochemistry and Photobiology A, Chemistry*, **106**, 45–49.

[9] Ogura, S., Kohno, M., Sato, K. and Inoue, Y. (1998) Effects of RuO_2 on activity for water decomposition of a $RuO_2/Na_2Ti_3O_7$ photocatalyst with a zigzag layer structure. *Journal of Materials Chemistry*, **8**, 2335.

[10] Ogura, S., Sato, K. and Inoue, Y. (2000) Effects of RuO_2 dispersion on photocatalytic activity for water decomposition of $BaTi_4O_9$ with a pentagonal prism tunnel and $K_2Ti_4O_9$ with a zigzag layer structure. *Physical Chemistry and Chemical Physics*, **2**, 2449.

[11] Sayama, K. and Arakawa, H. (1993) Photocatalytic decomposition of water and photocatalytic reduction of carbon dioxide over zirconia catalyst. *Journal of Physical Chemistry*, **97**, 531–533.

[12] Kudo, A., Tanaka, A., Domen, K. *et al.* (1988) Photocatalytic decomposition of water over NiO-$K_4Nb_6O_{17}$ catalyst. *Journal of Catalysis*, **111**, 67–76.

[13] Kudo, A., Kato, H. and Nakagawa, S. (2000) Water splitting into H_2 and O_2 on New $Sr_2M_2O_7$ (M = Nb and Ta) photocatalysts with layered perovskite structures – factors affecting the photocatalytic activity. *Journal of Physical Chemistry B*, **104**, 571–575.

[14] Miseki, Y., Kato, H. and Kudo, A. (2005) Water splitting into H_2 and O_2 over $Cs_2Nb_4O_{11}$ photocatalyst. *Chemistry Letters*, **34**, 54–55.

[15] Miseki, Y., Kato, H. and Kudo, A. (2006) Water splitting into H_2 and O_2 over $Ba_5Nb_4O_{15}$ photocatalysts with layered perovskite structure prepared by polymerizable complex method. *Chemistry Letters*, **35**, 1052–1053.

[16] Kato, H. and Kudo, A. (1999) Highly efficient decomposition of pure water into H_2 and O_2 over $NaTaO_3$ photocatalysts. *Catalysis Letters*, **58**, 153–155.

[17] Ishihara, T., Nishiguchi, H., Fukamachi, K. and Takita, Y. (1999) Effects of acceptor doping to $KTaO_3$ on photocatalytic decomposition of pure H_2O. *Journal of Physical Chemistry B*, **103**, 1–3.

[18] Kato, H. and Kudo, A. (1998) New tantalate photocatalysts for water decomposition into H_2 and O_2. *Chemical Physics Letters*, **295**, 487–492.

[19] Kato, H. and Kudo, A. (1999) Photocatalytic decomposition of pure water into H_2 and O_2 over $SrTa_2O_6$ prepared by a flux method. *Chemistry Letters*, 1207–1208.

[20] Machida, M., Mitsuyama, T., Ikeue, K. *et al.* (2005) Photocatalytic property and electronic structure of triple-layered perovskite tantalates, $MCa_2Ta_3O_{10}$ (M = Cs, Na, H, and $C_6H_{13}NH_3$). *Journal of Physical Chemistry B*, **109**, 7801–7802.

[21] Shimizu, K., Tsuji, Y., Kawakami, M. *et al.* (2002) Photocatalytic water splitting over spontaneously hydrated layered tantalate $A_2SrTa_2O_7{\cdot}nH_2O$ (A = H, K, Rb). *Chemistry Letters*, **31**, 1158.

[22] Machida, M., Yabunaka, J., Kijima, T. *et al.* (2001) Electronic structure of layered tantalates photocatalysts, $RbLnTa_2O_7$ (Ln = La, Pr, Nd, and Sm). *International Journal of Inorganic Materials*, **3**, 545–550.

[23] Machida, M., Yabunaka, J. and Kijima, T. (2000) Synthesis and Photocatalytic property of layered perovskite tantalates, $RbLnTa_2O_7$ (Ln = La, Pr, Nd, and Sm). *Chemistry of Materials*, **12**, 812–817.

[24] Kurihara, T., Okutomi, H., Miseki, Y. *et al.* (2006) Highly efficient water splitting over $K_3Ta_3B_2O_{12}$ photocatalyst without loading cocatalyst. *Chemistry Letters*, **35**, 274–275.

[25] Otsuka, H., Kim, K., Kouzu, A. *et al.* (2005) Photocatalytic performance of $Ba_5Ta_4O_{15}$ to decomposition of H_2O into H_2 and O_2. *Chemistry Letters*, **34**, 822.

[26] Saito, N., Kadowaki, H., Kobayashi, H. *et al.* (2004) A new photocatalyst of RuO_2-loaded $PbWO_4$ for overall splitting of water. *Chemistry Letters*, **133**, 1452.

[27] Kadowaki, H., Saito, N., Nishiyama, H. *et al.* (2007) Overall splitting of water by RuO_2-Loaded $PbWO_4$ photocatalyst with $d^{10}s^2$-d^0 configuration. *Journal of Physical Chemistry C*, **111**, 439.

[28] Sato, J., Saito, N., Nishiyama, H. and Inoue, Y. (2001) New photocatalytic group for water decomposition of RuO_2-Loaded p-Block metal (In, Sn and Sb) oxides with d^{10} configuration. *Journal of Physical Chemistry*, **105**, 6061.

[29] Sato, J., Saito, N., Nishiyama, H. and Inoue, Y. (2001) Photocatalytic activity for water decomposition of RuO_2-loaded $SrIn_2O_4$ with d^{10} configuration. *Chemistry Letters*, 868.

[30] Sato, J., Kobayashi, H., Saito, N. *et al.* (2003) Photocatalytic activities for water decomposition of RuO_2–loaded $AInO_2$ (A = Li, Na) with d^{10} configuration. *Journal of Photochemistry and Photobiology A: Chemistry*, **158**, 139.

[31] Sato, J., Saito, N., Nishiyama, H. and Inoue, Y. (2003) Photocatalytic activity for water decomposition of indates with octahedrally coordinated d^{10} configuration. I. influences of preparation conditions on activity. *Journal of Physical Chemistry B*, **107**, 7965.

[32] Yanagida, T., Sakata, Y. and Imamura, H. (2004) Photocatalytic decomposition of H_2O into H_2 and O_2 over Ga_2O_3 loaded with NiO. *Chemistry Letters*, **33**, 726.

[33] Ikarashi, K., Sato, J., Kobayashi, H. *et al.* (2002) Photocatalysis for Water decomposition by RuO_2–dispersed $ZnGa_2O_4$ with d^{10} configuration. *Journal of Physical Chemistry*, **106**, 9048.

[34] Sato, J., Kobayashi, H., Ikarashi, K. *et al.* (2004) Photocatalytic activity for water decomposition of RuO_2-dispersed Zn_2GeO_4. *Journal of Physical Chemistry B*, **108**, 4369.

[35] Sato, J., Saito, N., Nishiyama, H. and Inoue, Y. (2002) Photocatalytic water decomposition by RuO_2-Loaded antimonates, $M_2Sb_2O_7$ (M = Ca, Sr), $CaSb_2O_6$ and $NaSbO_3$, with d^{10} configuration. *Journal of Photochemistry and Photobiology A: Chemistry*, **148**, 85.

[36] Kadowaki, H., Sato, J., Kobayshi, H. *et al.* (2005) Photocatalytic activity of the RuO_2-dispersed composite p-block metal oxide $LiInGeO_4$ with d^{10}-d^{10} configuration for water decomposition. *Journal of Physical Chemistry*, **109**, 22995.

[37] Bamwenda, G.R., Sayama, K. and Arakawa, H. (1999) The photoproduction of O_2 from a suspension containing CeO_2 and Ce^{4+} cations as an electron acceptor. *Chemistry Letters*, 1047.

[38] Kadowaki, H., Saito, N., Nishiyama, H. and Inoue, Y. (2007) RuO_2-loaded Sr^{2+}-doped CeO_2 with d^0 electronic configuration as a new photocatalyst for overall water splitting. *Chemistry Letters*, **36**, 440.

[39] Arai, N., Saito, N., Nishiyama, H. *et al.* (2006) Overall water splitting by RuO_2-dispersed p-type GaN photocatalyts with d^{10} electronic configuration. *Chemistry Letters*, **35**, 796.

[40] Arai, N., Saito, N., Nishiyama, H. *et al.* (2007) Effects of divalent metal Ion (Mg^{2+}, Zn^{2+} and Be^{2+}) doping on photocatalytic activity of ruthenium oxide-loaded gallium nitride for water splitting. *Catalysis Today*, **129**, 407.

[41] Yabe, S., Yamashita, M., Momose, S. *et al.* (2001) Synthesis and UV-shielding properties of metal oxide doped ceria via soft solution chemical processes. *International Journal of Inorganic Materials*, **3**, 1003–1008.

[42] Sato, J., Saito, N., Yamada, Y. *et al.* (2005) RuO_2-loaded β-Ge_3N_4 as a Non-oxide Photocatalysts for overall water splitting. *Journal of the American Chemical Society*, **127**, 4150–4151.

[43] Fujii, K., Kusakabe, K. and Ohkawa, K. (2005) Photoelectrochemical properties of InGaN for H_2 generation from aqueous water. *Japanese Journal of Applied Physics*, **44**, 7433–7435.

[44] Fujii, K., Karasawa, T. and Ohkawa, K. (2005) Hydrogen gas generation by splitting aqueous water using n-type GaN photoelectrode with anodic oxidation. *Japanese Journal of Applied Physics*, **44**, L543–L545.

[45] Bergman, P., Ying, Gao, Monemar, B. and Holtz, P.O. (1987) Time-resolved spectroscopy of Zn- and Cd-doped GaN. *Journal of Applied Physics*, **61**, 4589.

[46] Monemar, B., Bergman, J.P. and Buyanova, I.A. (1997) *Optoelecronic Properties of Semiconductors and Superlattices*, vol. **2** (ed. M.O. Manasreh), Gordon and Breach Science Publishers, Amsterdam, p. 85, Chap. 4.

[47] Plakhov, G.F., Pobedimskaya, E.A., Simonov, M.A. and Belov, N.V. (1970) The crystal structure of $PbWO_4$. *Soviet Physics, Crystallography*, **15**, 928.

[48] Moreau, J.M., Galez, Ph., Peigneux, J.P. and Korzhik, M.V. (1996) Structural characterization of $PbWO_4$ and related new phase $Pb_7W_8O_{(32-x)}$. *Journal of Alloys and Compounds*, **238**, 46–48.

[49] Moreau, J.M., Gladyshevskii, E., Galez, Ph. *et al.* (1999) A new structural model for Pb-deficient $PbWO_4$. *Journal of Alloys and Compounds*, **284**, 104–107.

[50] ICSD No.50846.

[51] ICSD No.153500.

[52] Repelin, Y., Proust, C., Husson, E. and Beny, J.M. (1995) Vibrational spectroscopy of the C-form of yttrium sesquioxide. *Journal of Solid State Chemistry*, **118**, 163–169.

[53] White, W.B. and Keramidas, V.G. (1972) Vibrational spectra of oxides with the *C*-type rare earth oxide structure. *Spectrochimica Acta Part A*, **28**, 501–509.

[54] Kohno, M., Ogura, S., Sato, K. and Inoue, Y. (1997) Properties of photocatalysts with tunnel structures: formation of a lattice O radical by the UV irradiation of $BaTi_4O_9$ with a pentagonal-prism tunnel structure. *Chemical Physics Letters*, **267**, 72.

[55] Inoue, Y., Kubokawa, T. and Sato, K. (1991) Photocatalytic activity of alkali-metal titanates combined with Ru in the decomposition of water. *Journal of Physical Chemistry*, **95**, 4059.

[56] Ogura, S., Kohno, M., Sato, K. and Inoue, Y. (1999) Photocatalytic properties of $M_2Ti_6O_{13}$ (M=Na,K,Rb,Cs) with rectangular tunnel structures: behavior of a surface radical produced by UV irradiation and photocatalytic activity for water decomposition. *Physical Chemistry and Chemical Physics*, **1**, 179.

[57] Maeda, K., Takata, T., Hara, M. *et al.* (2005) GaN:ZnO solid solution as a photocatalyst for visible-light-driven overall water splitting. *Journal of the American Chemical Society*, **127**, 8286–8287.

[58] Maeda, K., Teramura, K., Lu, D. *et al.* (2006) Photocatalyst releasing hydrogen from water – Enhancing catalytic performance holds promise for hydrogen production by water splitting in sunlight. *Nature*, **440**, 295.

[59] Juarez-Arellano, E.A., Bucio, L., Ruvalcaba, J.L. *et al.* (2002) The crystal structure of $InGe_2O_7$ germanate. *Zeitschrift für Kristallographie*, **217**, 201.

19

Supramolecular Complexes as Photoinitiated Electron Collectors: Applications in Solar Hydrogen Production

Shamindri M. Arachchige and Karen J. Brewer

Department of Chemistry, Virginia Polytechnic Institute and State University, Blacksburg, Virginia, USA, Email: Kbrewer@vt.edu

19.1 Introduction

The urgent need to explore alternative energy sources has stimulated solar energy research globally. Solar energy that reaches the southern United States has an instantaneous maximum intensity of 1 kW m^{-2} and an average 24 h intensity of 250 W m^{-2} in a year [1]. It is a clean, abundant, and renewable energy source. Solar energy research is concentrated on its direct conversion to electricity in photovoltaic devices, conversion to heat in solar thermal devices, or conversion to chemical energy to produce fuels via artificial photosynthesis. Hydrogen has been proposed as an energy solution for the future due to its high energy content per gram (120 kJ g^{-1}) and burning hydrogen results in only energy and water, having no impact on our carbon foot print.

19.1.1 Solar Water Splitting

Solar hydrogen production through water splitting is an attractive alternative energy solution for the future [1–7]. Solar water splitting uses light energy from the sun to convert water into hydrogen and oxygen. Water splitting is a challenging, energetically uphill process. The reactions involved in water splitting require bond breaking, bond formation and multielectron

On Solar Hydrogen & Nanotechnology Edited by Lionel Vayssieres

transfer reactions. The overall reaction for water splitting is represented in Equation 19.1.

$$2H_2O_{(l)} \rightarrow 2H_{2(g)} + O_{2(g)} \quad (19.1)$$

The reduction and oxidation half reactions for water splitting are represented in Equations 19.2 and 19.3:

$$4H^+_{(aq)} + 4e^- \rightarrow 2H_2 \quad (19.2)$$

$$2H_2O_{(l)} \rightarrow O_{2(g)} + 4e^- + 4H^+_{(aq)} \quad (19.3)$$

The free energy change for the reaction (19.1) corresponds to 1.23 V versus NHE [1]. This energy is less compared to the ca. 5 V energy required for water splitting via a single electron mechanism [8,9]. Multielectron processes offer a lower-energy mechanism for water splitting. Although much of the solar spectrum (>1.23 V) possesses sufficient energy for water splitting, photochemical agents that absorb in the visible region of the solar spectrum are necessary to catalyze this complex reaction. The need to understand multielectron catalysis and multi-electron photochemistry is critical for efficient harnessing of solar energy through water splitting.

19.1.2 Supramolecular Complexes and Photochemical Molecular Devices

The use of photochemical devices at the molecular scale to collect reducing equivalents and deliver them to their substrates provides attractive means to make fuels [10–12]. In this regard, design and development of photochemical molecular devices capable of light- and/or redox-induced processes is an active area of research [13]. A comprehensive review of photochemical molecular devices constructed from supramolecular complexes was presented by Balzani *et al.* [11]. Supramolecular complexes as described therein are large molecular assemblies, constructed by the covalent attachment of smaller building block units, designed to perform complex functions in which the individual subunits perform specific tasks [11]. Common building blocks used to construct supramolecular complexes for light- and/or redox-induced processes are light-absorbing units (LA) to harvest energy, bridging ligands (BL) to act as connectors at the atomic scale, electron donors (ED) to supply electrons, and electron collectors (EC) to collect reducing equivalents. The coupling of multiple LAs to a single EC with appropriate orbital energetics should generate devices for photoinitiated electron collection, a process where light energy is used to collect reducing equivalents. The development of supramolecular assemblies that function as photoinitiated electron collectors allows for the development of systems for multielectron photocatalysis and solar hydrogen production [11,13,14]. Despite the introduction of these concepts several decades ago and a wide array of laboratories working in this arena, functional multielectron photocatalysts are rare. This chapter will focus on the recent progress in the use of supramolecular complexes in multielectron photocatalysis and solar hydrogen production. The electrochemical, photochemical and photophysical properties of these molecules will be discussed. Applications of supramolecular complexes in solar hydrogen production will be summarized.

19.1.3 Polyazine Light Absorbers

Upon optical excitation by absorption of a photon, a LA forms an electronically excited state, *LA, Equation 19.4, that possesses properties unique compared to the electronic ground state, often able to undergo electron or energy transfer.

$$LA + h\nu \xrightarrow{I_a} {}^*LA \tag{19.4}$$

Relaxation of the *LA can occur in a nonradiative fashion, Equation 19.5, or in a radiative fashion (emission of a photon of light), Equation 19.6 (k_{nr} and k_r are rate constants for nonradiative and radiative decay, respectively).

$${}^*LA \xrightarrow{k_{nr}} LA + \text{heat} \tag{19.5}$$

$${}^*LA \xrightarrow{k_r} LA + h\nu' \tag{19.6}$$

The *LA is both a better oxidizing and a better reducing agent than the ground state LA. The *LA undergoes energy and/or electron-transfer reactions to a quencher molecule (Q) providing means of application of Qs in solar-energy conversion schemes, Equations (19.7)–(19.9). Electron acceptors (EAs) or EDs can act as Qs (k_{en}, k_{et} and k'_{et} are rate constants for energy-transfer quenching, oxidative quenching and reductive quenching, respectively). Typically, the processes of sensitization are bimolecular and limited by the need for intermolecular collision during the excited state lifetime (τ) of the *LA.

$${}^*LA + Q \xrightarrow{k_{en}} LA + {}^*Q \tag{19.7}$$

$${}^*LA + EA \xrightarrow{k_{et}} LA^+ + EA^- \tag{19.8}$$

$${}^*LA + ED \xrightarrow{k'_{et}} LA^- + ED^+ \tag{19.9}$$

The driving force for excited-state electron-transfer quenching (E_{redox}) depends on the excited-state reduction potential of the LA ($E({}^*LA^{n+}/LA^{(n-1)+})$) and the ground state oxidation potential of the ED ($E(ED^{0/+})$) for a reductively quenching event, Equation 19.10, or the excited-state oxidation potential of the LA ($E({}^*LA^{n+}/LA^{(n+1)+})$) and the ground-state reduction potential of the EA ($E(EA^{0/-})$) for an oxidatively quenching event, Equation 19.11. The excited-state potentials, $E({}^*LA^{n+}/LA^{(n-1)+})$ and $E({}^*LA^{n+}/LA^{(n+1)+})$ are calculated based on the ground-state reduction or oxidation potential of the LA, respectively, and the energy gap between the ground vibronic state of the electronic ground and excited states (E^{0-0}), Equations 19.12 and 19.13.

$$E_{redox}(\text{red}) = E({}^*LA^{n+}/LA^{(n-1)+}) - E(ED^{0/+}) \tag{19.10}$$

$$E_{redox}(\text{oxd}) = E({}^*LA^{n+}/LA^{(n+1)+}) - E(EA^{0/-}) \tag{19.11}$$

$$E({}^*LA^{n+}/LA^{(n-1)+}) = E(LA^{n+}/LA^{(n-1)+}) + E^{0-0} \tag{19.12}$$

$$E({}^*LA^{n+}/LA^{(n+1)+}) = E(LA^{n+}/LA^{(n+1)+}) - E^{0-0} \tag{19.13}$$

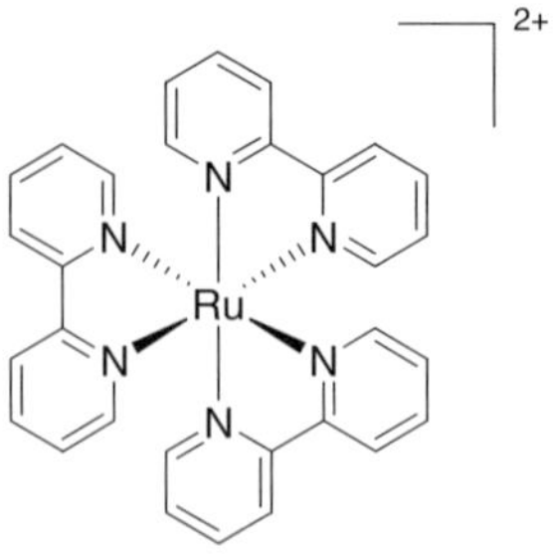

Figure 19.1 Representation of $[Ru(bpy)_3]^{2+}$ (bpy = 2,2′-bipyridine).

Considerable research has been conducted on the excited-state electron- [15–17] and energy-transfer reactions [18,19] of the prototypical LA $[Ru(bpy)_3]^{2+}$, Figure 19.1. $[Ru(bpy)_3]^{2+}$ absorbs light with high extinction coefficients in the ultraviolet and visible regions of the spectrum [11,20–23]. The UV region is dominated by intense absorptions that are intraligand (IL) $\pi \rightarrow \pi^*$ in nature, while the visible region is dominated by Ru(dπ) $\rightarrow$ bpy(π^*) metal-to-ligand charge-transfer (^{1}MLCT) transitions with $\lambda_{max}^{abs} = 450$ nm. Absorption of a photon of light with optical excitation at $\lambda_{max}^{abs} = 450$ nm, followed by electron spin inversion, populates the lowest energy ^{3}MLCT excited state. Jablonski (state) diagrams are typically used to demonstrate the conversion between electronic excited states of molecules. A Jablonski diagram depicting the low-energy excited states of $[Ru(bpy)_3]^{2+}$ populated upon optical excitation is shown in Figure 19.2. The ^{3}MLCT state of $[Ru(bpy)_3]^{2+}$ is emissive ($\lambda_{max}^{em} =$ 605 nm), relatively long-lived ($\tau = 860$ ns in room temperature (RT) acetonitrile solution), and can undergo electron- or energy-transfer quenching or radiatively and nonradiatively decay to the ground state [21]. The emissive ^{3}MLCT excited state of $[Ru(bpy)_3]^{2+}$ provides a probe to

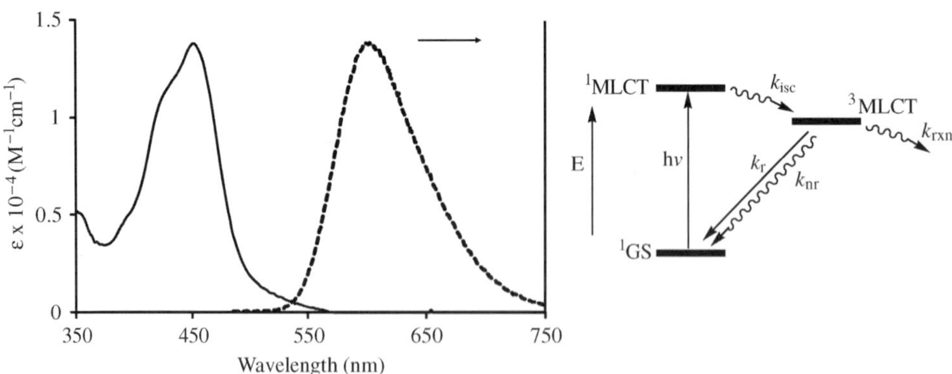

Figure 19.2 The Ru(dπ) $\rightarrow$ bpy(π^*) metal-to-ligand charge transfer (^{1}MLCT) transition (—) and RT-^{3}MLCT emission spectra (- - -) for $[Ru(bpy)_3](PF_6)_2$ in acetonitrile (left) and a Jablonski diagram for $[Ru\text{-}(bpy)_3]^{2+}$ (right) (bpy = 2,2′-bipyridine, GS = ground state, MLCT = metal-to-ligand charge-transfer, k_r = rate constant for radiative decay, k_{nr} = rate constant for nonradiative decay, k_{isc} = rate constant for intersystem crossing, k_{rxn} = rate constant for rate of reaction).

study excited-state dynamics. The τ is the inverse of the sum of all the rate constants for deactivation of an electronic excited state in the absence of a quencher. For $[Ru(bpy)_3]^{2+}$, τ is represented in Equation 19.14:

$$\tau = \frac{1}{k_r + k_{nr}} \tag{19.14}$$

The quantum yield, Φ, for an excited-state process can be described as the number of defined events which occur per photon of light absorbed. This can be calculated by considering the ratio between the rate constants for the process of interest and the sum of all the rate constants for the deactivation of a state. For an indirectly populated state, this ratio is multiplied by the fraction of light that populates this state, Equation 19.15. For $[Ru(bpy)_3]^{2+}$, Φ^{em} from the ^{3}MLCT state is represented in Equation 19.15:

$$\Phi^{em} = \Phi_{^3MLCT} \frac{k_r}{k_r + k_{nr}} \tag{19.15}$$

$\Phi_{^3MLCT}$ is the quantum efficiency for generation of the ^{3}MLCT state and is unity for $[Ru(bpy)_3]^{2+}$ and most Ru(II) polyazine LAs.

The relationship between Φ of photophysical processes and concentration of a Q is described using Stern–Volmer kinetics, Equation 19.16:

$$\Phi^{\circ}/\Phi = 1 + K_{sv}[Q] \tag{19.16}$$

Φ^{o} and Φ are quantum yields in the absence and presence of Q, respectively, and K_{sv} is the Stern–Volmer constant which can be used to determine rate of quenching by Q. For $[Ru(bpy)_3]^{2+}$, the ^{3}MLCT state is reductively quenched by the sacrificial electron donor, N,N-dimethylaniline (DMA), at a rate of $7.1 \times 10^7\ M^{-1}\ s^{-1}$ [24]. The efficiency for reductive quenching of DMA can be modulated through ligand substitution. For example, replacing the TL bpy with 4,7-diphenyl-1,10-phenanthroline (Ph_2phen) provides for more emissive molecules with longer excited-state lifetimes. Figure 19.3 represents some common TL ligands used. Replacing ruthenium with osmium provides a means to tune the energetics between the highest occupied molecular orbital (HOMO) and lowest unoccupied molecular orbital (LUMO). Table 19.1 provides the variation of the spectroscopic and redox properties through ligand and/or metal substitution.

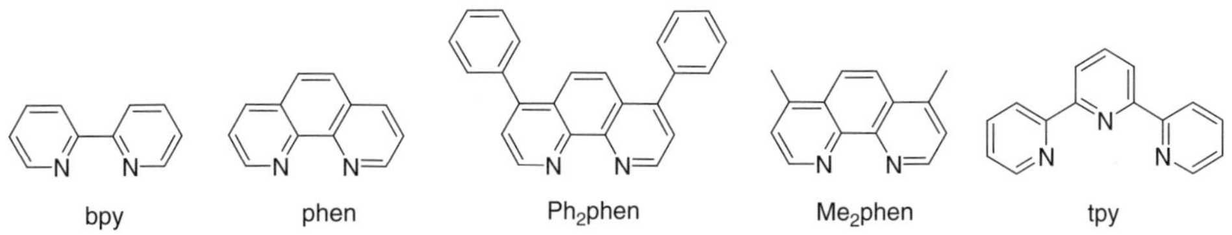

bpy phen Ph_2phen Me_2phen tpy

Figure 19.3 Representation of common polyazine terminal ligands.

Table 19.1 Spectroscopic and redox properties of Ru^{II} and Os^{II} polyazine complexes.

Complex	λ_{max}^{abs} (RT) (nm)	λ_{max}^{em} (RT) (nm)	τ (RT) (ns)	$E_{1/2}^{ox}$ (V vs SCE)^b	$E_{1/2}^{red}$ (V vs SCE)^b	Ref.
$[Ru(bpy)_3]^{2+a}$	450	605	860	1.27 ($Ru^{II/III}$)	−1.34 ($bpy^{0/-}$)	[21,25]
$[Ru(phen)_3]^{2+a}$	443	604	400	1.27 ($Ru^{II/III}$)	−1.35 ($phen^{0/-}$)	[25]
$[Ru(Ph_2phen)_3]^{2+c}$	460	613	4680	1.26 ($Ru^{II/III}$)	−1.24 ($Ph_2phen^{0/-}$)	[26,27]
$[Os(bpy)_3]^{2+a}$	640	723	20^c	0.81($Os^{II/III}$)	−1.29 ($bpy^{0/-}$)	[28]
$[(bpy)_2Ru(dpp)]^{2+a}$	475	691	240	1.31 ($Ru^{II/III}$)	−1.06 ($dpp^{0/-}$)	[29–31]
$[(phen)_2Ru(dpp)]^{2+a}$	465	652	252	1.39 ($Ru^{II/III}$)	−1.07 ($dpp^{0/-}$)	[32]
$[(bpy)_2Os(dpp)]^{2+a}$	486	798	37	0.91 ($Os^{II/III}$)	−0.99 ($dpp^{0/-}$)	[33,34]
$[Ru(tpy)_2]^{2+}$	476^d	629^a	0.25^c	1.30 ($Ru^{II/III}$)^a	−1.23 ($tpy^{0/-}$)^a	[35,36]
$[Os(tpy)_2]^{2+}$	477^d	718^d	269^a	0.97 ($Os^{II/III}$)^a	−1.23 ($tpy^{0/-}$)^a	[35,36]

^{a} In acetontrile solution at room temperature.
^{b} Potentials versus SCE.
^{c} In aqueous solutions.
^{d} In ethanol-methanol (4/1).

19.1.4 Polyazine Bridging Ligands to Construct Photochemical Molecular Devices

Replacing one or more of the bpy of $[Ru(bpy)_3]^{2+}$ with polyazine BLs allows these prototypical LAs to be incorporated into large molecular assemblies, such as photochemical molecular devices [29–33,37–39]. The most studied of such systems is $[Ru(bpy)_2(dpp)]^{2+}$, Figure 19.4 [29–31]. The remote nitrogen atoms of the BL allows for the construction of polymetallic systems using this LA. Optical excitation of $[Ru(bpy)_2(dpp)]^{2+}$ affords a Ru ($d\pi$) $\rightarrow$ dpp(π^*) MLCT state, at $\lambda_{max}^{abs} = 475$ nm, with the promoted electron formally located on the dpp used for attachment to additional metal centers. The ^{3}MLCT state is emissive ($\lambda_{max}^{em} = 691$ nm) with $\tau = 240$ ns in RT aerated acetonitrile solution [29–31]. The redox chemistry shows a $Ru^{II/III}$ oxidation at 1.31 V versus SCE (saturated calomel electrode) with the $dpp^{0/-}$ couple occurring at −1.06 V versus SCE, prior to the $bpy^{0/-}$ couple. The electrochemical data illustrate a Ru($d\pi$)-based HOMO and a dpp(π^*)-based LUMO in

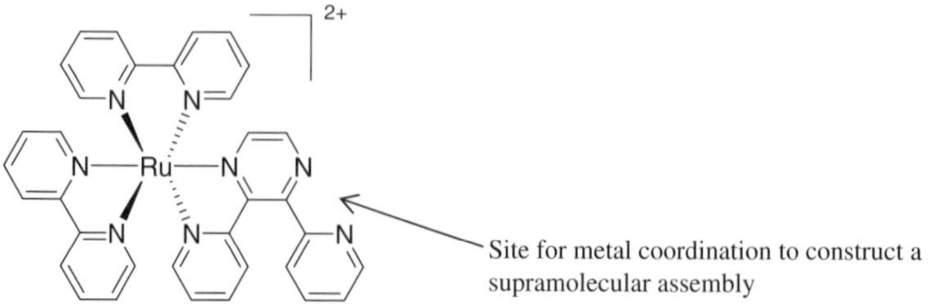

Site for metal coordination to construct a supramolecular assembly

Figure 19.4 Representations of $[(bpy)_2Ru(dpp)]^{2+}$.

$[Ru(bpy)_2(dpp)]^{2+}$. The attachment of a metal to the remote nitrogen atoms of dpp affords perturbations to the redox and excited-state properties, as evidenced by $[(bpy)_2Ru(dpp)Ru(bpy)_2]^{4+}$. The redox chemistry of $[(bpy)_2Ru(dpp)Ru(bpy)_2]^{4+}$ in dimethylformamide shows two $Ru^{II/III}$ couples at 1.47 and 1.69 V versus SCE, suggesting electronic coupling of the ruthenium centers across the dpp bridge [31]. The $dpp^{0/-}$ and $dpp^{-/-2}$ couples of $[(bpy)_2Ru(dpp)Ru(bpy)_2]^{4+}$ occur at −0.64 and −1.08 V versus SCE, respectively, prior to the $bpy^{0/-}$ couple [31]. This redox behavior is consistent with a more stabilized dpp(π^*) orbital with coordination of a second metal to the remote nitrogens of dpp. The light-absorbing properties of $[(bpy)_2Ru(dpp)Ru(bpy)_2]^{4+}$ are also consistent with the electrochemical behavior leading to a substantial red-shift of the Ru(dπ) → dpp(π^*) MLCT at $\lambda_{max}^{abs} = 525$ nm versus the 475 nm in $[Ru(bpy)_2(dpp)]^{2+}$ [29]. The lower-energy ^{3}MLCT excited state exhibits a shortened lifetime of 80 ns for $[(bpy)_2Ru(dpp)Ru(bpy)_2]^{4+}$, consistent with the energy-gap law [40].

19.1.5 Multi-Component System for Visible Light Reduction of Water

The ^{3}MLCT state of $[Ru(bpy)_3]^{2+}$ has sufficient energy to drive water splitting. Direct photocatalysis does not occur. A multicomponent system uses an $[Ru(bpy)_3]^{2+}$ LA, an $[Rh(bpy)_3]^{3+}$ EA, a triethanolamine (TEOA) ED and a metallic platinum catalyst for photochemical hydrogen production from water, Figure 19.5 [41,42]. In the excited state, $[Ru(bpy)_3]^{2+}$ undergoes electron transfer to $[Rh(bpy)_3]^{3+}$ to produce $[Ru(bpy)_3]^{3+}$ and

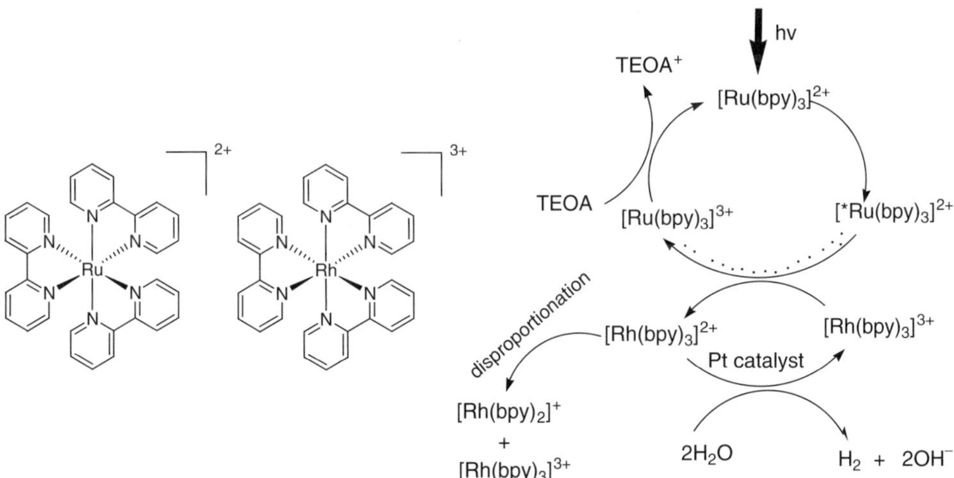

Figure 19.5 Mechanism of solar hydrogen production from water using an $[Ru(bpy)_3]^{2+}$ LA, an $[Rh(bpy)_3]^{3+}$ EA, a TEOA ED and a metallic platinum catalyst (bpy = 2,2′-bipyridine). (Figure adapted from G.M. Brown, S.-F. Chan, C. Creutz, *et al.*, Mechanism of the formation of dihydrogen from the photoinduced reactions of Tris(bipyridine)ruthenium(II) with Tris(bipyridine)-rhodium(III), *Journal of the American Chemical Society*, **101**, 7638, 1979; C. Creutz, A.D. Keller, N. Sutin, and A.P. Zipp, Poly (pyridine)ruthenium(II)-Photoinduced redox reactions of bipyridinium cations, Poly(pyridine)rhodium complexes, and osmium ammines, *Journal of the American Chemical Society*, **104**, 3618, 1982, American Chemical Society.)

$[Rh(bpy)_3]^{2+}$. The photogenerated $[Rh(bpy)_3]^{2+}$ rapidly disproportionates to $[Rh(bpy)_2]^{+}$ and $[Rh(bpy)_3]^{3+}$. In addition, the Pt catalyst can accept an electron from the reduced rhodium species to catalyze the reduction of water to hydrogen. Energy- and electron-transfer efficiency in such multicomponent systems is limited by the need for diffusional contact during the excited state of the LA and concentration of the EA. The inherent inefficiencies of such bimolecular quenching can be avoided by designing complex polymetallic supramolecular assemblies capable of multielectron photochemistry. This key fundamental work on multicomponent systems has paved the way for the systems described herein.

19.1.6 Photoinitiated Charge Separation

Photoinitiated electron transfer to afford long-lived charge-separated states is key to efficient solar-energy harnessing and photosynthesis. Synthetic multicomponent molecules that mimic this process would be pivotal for conversion of solar energy into fuels and artificial photosynthesis. Extension of the excited-state lifetime of the charge-separated state can be accomplished in these assemblies. A basic device for photoinitiated charge separation would consist of an ED–LA–EA assembly, Figure 19.6. The ED supplies electrons to the *LA to prevent back electron transfer and decay of the *LA to the ground state. The optically excited electron can be transferred to an EA to generate a charge-separated state with a positively charged ED and negatively charged EA.

Polyazine BLs are widely used to connect ED, LA and EA units. Polyazine type BLs separated by short rigid spacers promote directional electron or energy transfer within the molecular assembly and often provide molecular architectures that display observable emissions at room temperature [13]. Some common polyazine BLs are represented in Figure 19.7.

Component modification to modulate basic chemical, excited-state and catalytic properties is a very attractive feature of supramolecular complexes. In this regard, supramolecular assemblies can be designed so that at least two ED–LA assemblies are coupled to a single EA. If this EA can accept multiple electrons, optical excitation of such systems would provide a simple means of multiple charge collection at an EA site, allowing the EA to function as an EC,

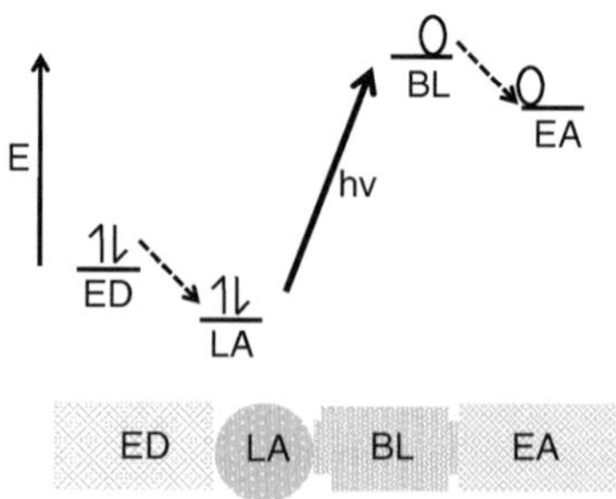

Figure 19.6 Photoinduced charge separation in an ED–LA–EA molecular device (ED = electron donor, LA = light absorber, EA = electron acceptor). (Figure adapted from M. Elvington, J.R. Brown, D.F. Zigler, and K.J. Brewer, Supramolecular complexes as photoinitiated electron collectors: applications in solar hydrogen production, *Proceedings of SPIE*, **6340**, 63400W-1, 2006, SPIE.)

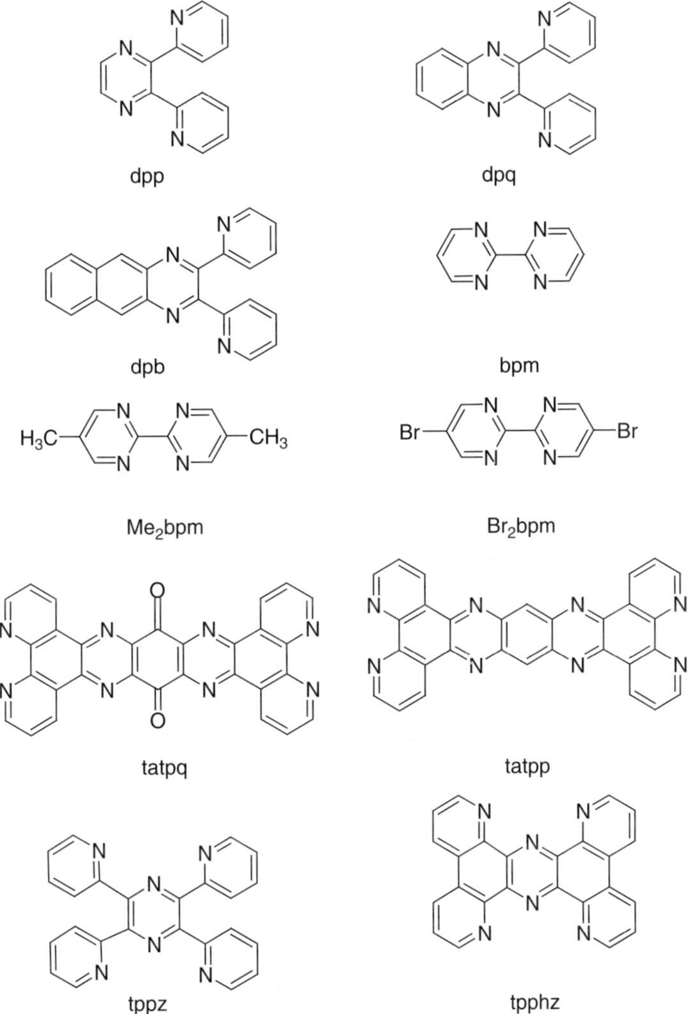

Figure 19.7 Representative polyazine bridging ligands.

Figure 19.8. Following the absorption of two photons (*hv*) the system would produce two oxidized electron donors (ED^+) and a doubly reduced electron collector (EC^{2-}). The collection of electrons provide for a means to use low-energy visible, as well as ultraviolet, light in solar-energy conversion schemes. Despite the significance of collecting electrons as a means to harness solar energy, preparation of functioning systems that undergo photoinitiated electron collection (PEC) has been challenging and only a few such systems exist. Very few systems undergo PEC and can be used for solar hydrogen production. This chapter will be limited to the recent progress in supramolecular systems incorporating polyazine LAs for photoinitiated electron collection. The application of some of these supramolecules in solar hydrogen production will be discussed.

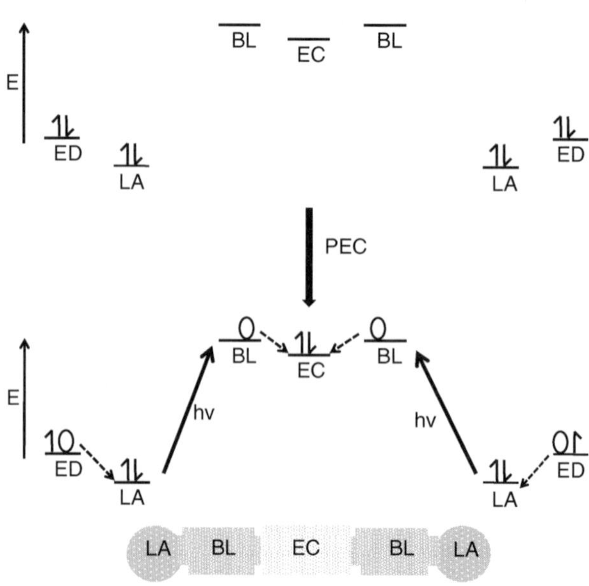

Figure 19.8 An orbital energy diagram of a photoinitiated electron collector (BL = bridging ligand, LA = light absorber, ED = electron donor, and EC = electron collector).

19.2 Supramolecular Complexes for Photoinitiated Electron Collection

Photochemical molecular devices for PEC may provide systems for efficient solar energy conversion. The lack of fundamental understanding of multielectron photochemistry greatly impedes the design and development of systems for this purpose. The degree of perturbation of subunit properties upon assembly of sumpramolecules, intercomponent coupling and relative orbital energetics of the components to promote directional charge transfer are some factors to be considered in designing systems for PEC. Coupling of reactive metals is essential to conversion of solar energy to transportable fuels [43].

19.2.1 Photoinitiated Electron Collection on a Bridging Ligand

19.2.1.1 Ruthenium Polyazine Light Absorbers Coupled to an Iridium Core

A series of supramolecular complexes of the type, $[\{(bpy)_2Ru(BL)\}_2IrCl_2]^{5+}$ (BL = 2,3-bis-(2-pyridyl)pyrazine (dpp), 2,3-bis(2-pyridyl)quinoxaline (dpq) or 2,3-bis(2-pyridyl)benzo-quinoxaline (dpb)), that couple and separate two $(bpy)_2Ru^{II}(BL)$ LA subunits by a catalytically active $(BL)_2Ir^{III}Cl_2$ (BL = dpp, dpq or dpb) core were reported by Brewer *et al.* [44]. Closely related $[(TL)_2Ru(BL)Ru(TL)_2]^{4+}$ systems have been well studied [29–31,45,46]. The direct coupling of two LA subunits by a polyazine BL prohibits PEC in these systems.

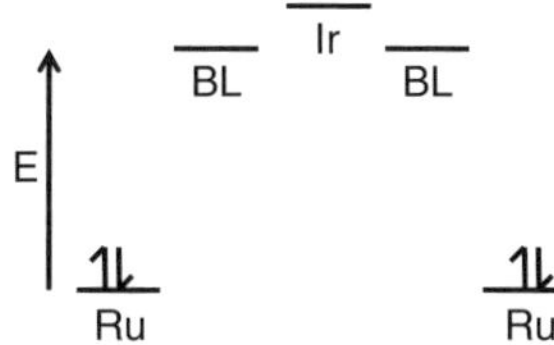

Figure 19.9 Orbital energy diagram for Ir-centered supramolecular complex, $[\{(bpy)_2Ru\text{-}(BL)\}_2IrCl_2]^{5+}$ (BL = 2,3-bis(2-pyridyl)pyrazine (dpp), 2,3-bis(2-pyridyl)quinoxaline (dpq), or 2,3-bis(2-pyridyl)benzoquinoxaline (dpb)).

19.2.1.2 Redox Properties of Ruthenium Polyazine Light Absorbers Coupled to an Iridium Core

The redox properties can provide useful information on the orbital energetics of supramolecular complexes. The electrochemical data for $[\{(bpy)_2Ru(BL)\}_2IrCl_2]^{5+}$ (BL = dpp, dpq, or dpb) are given in Table 19.2. The oxidative electrochemistry reveals overlapping $Ru^{II/III}$ oxidations at 1.53 V versus SCE, suggesting weak electronic communication of the Ru LA units through the BLs (dpp, dpq or dpb) [44]. The reductive electrochemistry reveals the BL reductions prior to the bpy reductions. The two BL units of each complex are electronically coupled through Ir and reduce separately. The reductive processes are tuned by the BL employed. The first four reductions are reversible and represent two $BL^{0/-}$ couples followed by two $BL^{-/-2}$ couples [44]. The BL reductions move to more positive potentials from (−0.43 and −0.58 V for $dpp^{0/-}$, −0.16 and −0.30 V for $dpq^{0/-}$, and −0.005 and −0.16 V for $dpb^{0/-}$ versus SCE), consistent with a more stabilized dpb(π^*) acceptor orbital. The electrochemical properties suggest a Ru(dπ)-based HOMO and a BL(π^*)-based (BL = dpp, dpq or dpb) LUMO in this structural motif, Figure 19.9.

19.2.1.3 Spectroscopic and Photophysical Properties of Ruthenium Polyazine Light Absorbers Coupled to an Iridium Core

The trimetallic complexes coupling two ruthenium LAs to a central Ir core absorb efficiently throughout the UV and visible regions of the spectrum [44]. The energies of the lowest-lying electronic transitions for the trimetallic complexes are given in Table 19.2. The electronic absorption spectra of these complexes display bpy- and BL-based $\pi \rightarrow \pi^*$ transitions in the UV region and Ru(dπ) $\rightarrow$ bpy(π^*) and Ru(dπ) $\rightarrow$ BL(π^*) MLCT transitions in the visible region, with the Ru(dπ) $\rightarrow$ BL(π^*) MLCT transitions occurring at the lowest energy. This lowest-energy absorption shifts to the red as the easier to reduce BL unit dpb (666 nm) or dpq (616 nm) is substituted for dpp (522 nm).

The ^{3}MLCT excited states of Ru polyazine complexes are often emissive. Polymetallic systems with BLs such as dpp and dpq often display greatly stabilized ^{3}MLCT states that are shorter lived than the monometallic LA subunits consistent with their redox properties. The dpp and dpq bridged systems of $[\{(bpy)_2Ru(BL)\}_2IrCl_2]^{5+}$ exhibit emissions from the Ru(dπ) $\rightarrow$ BL(π^*) ^{3}MLCT state at λ^{em}_{max} = 794 and 866 nm, respectively, in deoxygenated acetonitrile solutions at room temperature [44]. The Φ^{em} and τ are 1.2×10^{-4} and 32 ns for the dpp system and $<10^{-6}$ and <5 ns for the dpq system, respectively. The dpb system does not display a

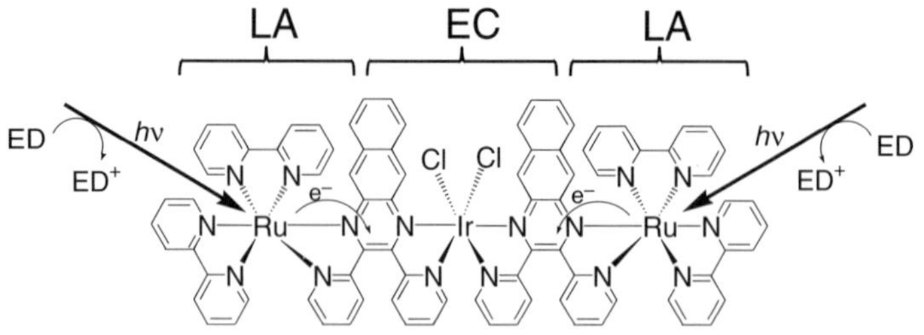

Figure 19.10 Electron collection in Ir-centered complexes.

detectable emission under the conditions used. The redox and excited-state properties of $[\{(bpy)_2Ru(BL)\}_2IrCl_2]^{5+}$ (BL = dpp, dpq, or dpb) have been modulated by variation of the BL. Changing the BL from dpp or dpq to dpb has little effect on the energy of the Ru($d\pi$)-based HOMO, while the relative energy of the BL(π^*)-based LUMO varies dramatically. This is evidenced by the similar UV spectra and varied visible spectra, as well as the constant oxidation potential and varied $BL^{0/-}$ reduction potentials of these complexes.

19.2.1.4 $[\{(bpy)_2Ru(dpb)\}_2IrCl_2]^{5+}$, The First Molecular System for Photoinitiated Electron Collection

Complexes having the general formula $[\{(bpy)_2Ru(BL)\}_2IrCl_2]^{5+}$ provide ideal structural motifs to study PEC. Brewer *et al.* have established $[\{(bpy)_2Ru(dpb)\}_2IrCl_2]^{5+}$ as the first functioning photoinitiated electron collector [47]. Photolysis of this complex in the presence of an electron donor, DMA, affords the doubly reduced $[\{(bpy)_2Ru(dpb^-)\}_2IrCl_2]^{3+}$, Figure 19.10. Bulk electrolysis to generate $[\{(bpy)_2Ru^{II}(dpb^-)\}_2IrCl_2]^{4+}$, $[\{(bpy)_2Ru^{II}(dpb^{2-})\}_2IrCl_2]^{3+}$, $[\{(bpy)_2Ru^{II}(dpb^{3-})\}_2IrCl_2]^{2+}$, $[\{(bpy)_2Ru^{II}(dpb^{4-})\}_2IrCl_2]^{+}$, and $[\{(bpy)_2Ru^{III}(dpb)\}_2IrCl_2]^{7+}$ has been achieved with greater than 95% reversibility [47]. Studies indicate the spectroscopy of the doubly reduced electrochemically generated product is consistent with that of the photoproduct. This implies the ability of $[\{(bpy)_2Ru^{II}(dpb)\}_2IrCl_2]^{5+}$ to store two electrons on the orbitals of the dpb(π^*) to form $[\{(bpy)_2Ru^{II}(dpb^{2-})\}_2IrCl_2]^{3+}$. This complex has not been demonstrated to deliver the collected electrons to a substrate.

19.2.2 Ruthenium Polyazine Light Absorbers Coupled Through an Aromatic Bridging Ligand

Systems that couple two ruthenium polyazine LAs through extended, aromatic BLs have been reported by MacDonnell and Campagna *et al.* [48–50]. These systems are of the form $[(phen)_2Ru(tatpq)Ru(phen)_2]^{4+}$ (phen = 1,10-phenanthroline, tatpq = 9,11,20,22-tetraazatetrapyrido[3,2-a:2′3′-c:3″,2″-1 : 2,3−n]pentacene-10,21-quinone) and $[(phen)_2Ru(tatpp)Ru(phen)_2]^{4+}$ (tatpp = 9,11,20,22- tetraazatetrapyrido[3,2-a:2′3′-c:3″,2″-1:2−,3−n]pentacene). Photolyses of these complexes in the presence of an ED, triethylamine (TEA), leads to two or four electrons being collected on the polyazine tatpp or tatpq BL unit, respectively.

19.2.2.1 Redox Properties of Ruthenium Polyazine Light Absorbers Coupled Through an Aromatic Bridging Ligand

Electrochemistry of the ruthenium bimetallic complexes provides insight into the HOMO and LUMO energies. The complexes $[(phen)_2Ru(tatpq)Ru(phen)_2]^{4+}$ and $[(phen)_2Ru(tatpp)Ru(phen)_2]^{4+}$ show two overlapping $Ru^{II/III}$ oxidations at about 1.37 V versus SCE [48]. The oxidative electrochemistry is suggestive of weak electronic interaction between the LA components. The complex $[(phen)_2Ru(tatpq)Ru(phen)_2]^{4+}$ displays a reversible $tatpq^{0/-}$ couple and a quasi-reversible $tatpq^{-/-2}$ couple at −0.23 and −0.60 V versus SCE, respectively. The reductive electrochemistry of $[(phen)_2Ru(tatpp)Ru(phen)_2]^{4+}$ displays two reversible couples at −0.18 and −0.56 V versus SCE, consistent with the formation of $tatpp^{0/-}$ and $tatpp^{-/-2}$. Differential pulse voltammetry was used to confirm that the reductions were one-electron processes. The electrochemistry of the two complexes suggests a Ru(dπ)-based HOMO and a BL(π^*)-based (BL = tatpq or tatpp) LUMO in this structural motif.

19.2.2.2 Spectroscopic and Photophysical Properties of Ruthenium Polyazine Light Absorbers Coupled Through an Aromatic Bridging Ligand

The ruthenium bimetallic complexes are efficient LAs throughout the UV and visible regions of the spectrum [48]. The UV region is dominated by ligand ($\pi \rightarrow \pi^*$)-based absorptions, while the visible region is dominated by Ru(dπ) $\rightarrow$ phen(π^*) and Ru(dπ) $\rightarrow$ BL(π^*) (BL = tatpq or tatpp) MLCT transitions, $\lambda_{max}^{abs} = 440$ nm in acetonitrile solution [48]. The excited state LAs can decay back to the ground state, emitting light, providing a probe to study the excited-state dynamics. The complexes $[(phen)_2Ru(tatpq)Ru(phen)_2]^{4+}$ and $[(phen)_2Ru(tatpp)Ru(phen)_2]^{4+}$ do not display detectable emissions from the Ru(dπ) $\rightarrow$ phen(π^*) ^{3}MLCT states, upon optical excitation, at room temperature in acetonitrile solution or at 77 K in butyronitrile rigid matrix [48]. The nonluminescence suggests that the emission from the Ru(dπ) $\rightarrow$ phen(π^*) ^{3}MLCT state is quenched by electron transfer to populate a charge-separated state with an oxidized ruthenium and a reduced tatpp or tatpq at room temperature and at 77 K. The tatpp and tatpq BLs are somewhat unique in that the lowest-lying π^* acceptor orbital is localized on the central part of the BLs. The optically populated acceptor orbital is higher in energy and based on the two phen-type subunits of the BLs.

19.2.2.3 Photoinitiated Electron Collection of Ruthenium Polyazine Light Absorbers Coupled Through an Aromatic Bridging Ligand

Electronic isolation between multiple polyazine LA units has been achieved using extended aromatic BL units [48–50]. Photolyses of $[(phen)_2Ru(tatpq)Ru(phen)_2]^{4+}$ and $[(phen)_2Ru(tatpp)Ru(phen)_2]^{4+}$ in the presence of the ED, TEA, generate four- and two-electron reduced complexes, respectively, as shown in Figure 19.11. The mechanisms of action for the electron-collection processes have been determined through spectral changes observed during controlled potential electrolysis experiments. Each electron-transfer process is suggested to be coupled with protonation of the reduced site, avoiding negative charge build-up in the system [51]. The key to the functioning of this system is that electrons added to the complex are localized on a π^* acceptor orbital on the center of the BLs. This allows the phen-type spectroscopic orbital on the BLs to still be active for absorption of light, even in the reduced

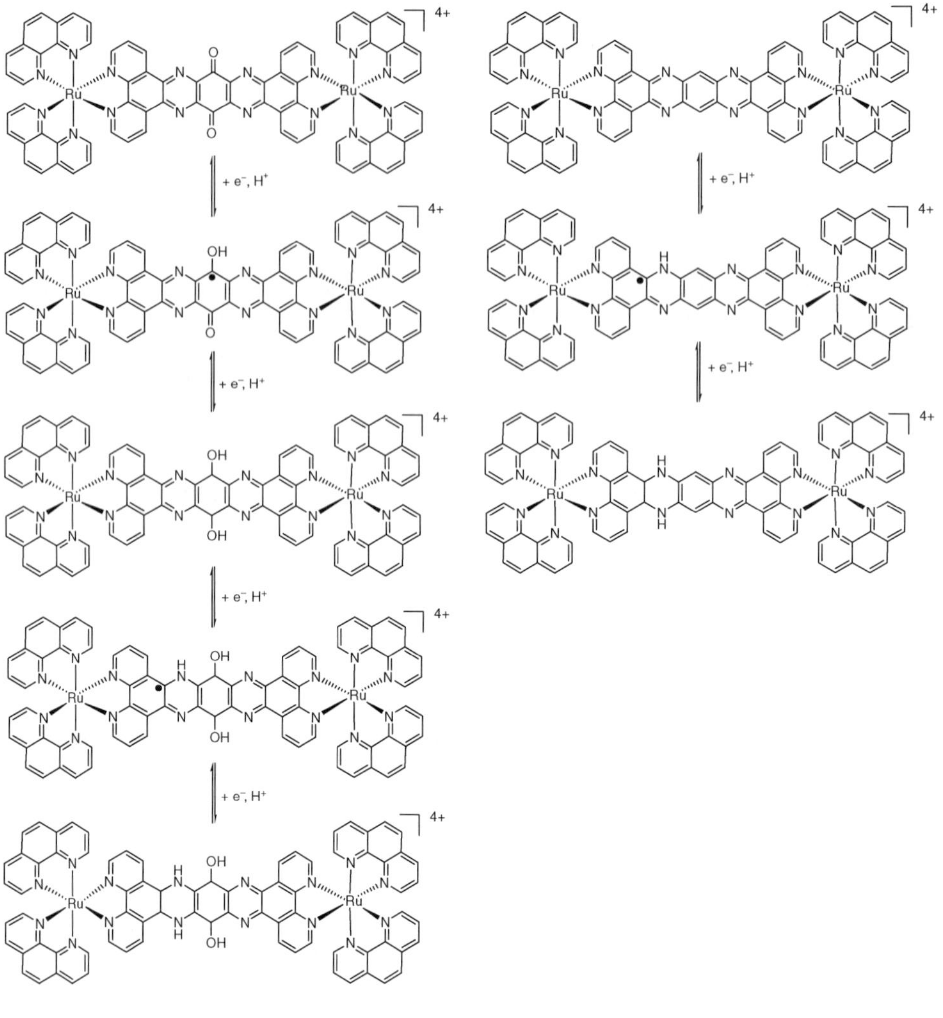

Figure 19.11 Redox reactions for $[(phen)_2Ru(tatpq)Ru(phen)_2]^{4+}$ (left) and $[(phen)_2Ru(tatpp)Ru(phen)_2]^{4+}$ (right) that are involved for photoinitiated electron collection. (Figure adapted from R. Konduri, H. Ye, F.M. MacDonnell, *et al.*, Ruthenium photocatalysts capable of reversibly storing up to four electrons in a single acceptor ligand: A step closer to artificial photosynthesis, *Angewandte Chemie International Edition*, **41**, 3185, 2002, German Chemical Society.)

forms of the complex. These systems represent the second molecular systems shown to undergo PEC. The collected electrons have not been delivered to a substrate and this system does not catalyze multielectron reduction of water.

19.2.3 Photoinitiated Electron Collection on a Platinum Metal

Photoinitiated electron collection at a metal center was first demonstrated by Bocarsly *et al.* [52–54]. A series of trimetallic complexes of the form $[(NC)_5M^{II}(CN)Pt^{IV}(NH_3)_4(NC)$

$M^{II}(CN)_5]^{4-}$ (M = Fe, Ru, or Os), where the Pt^{IV} center is attached within the molecular assemblies to act as an electron collector, were reported by Bocarsly *et al.* [52–54]. The approach is to use a single photon to transfer multiple electrons. For M = Fe, absorption of one photon of light results in two electrons being transferred to the central Pt^{IV} via intermediate formation of a Fe^{III}, Pt^{III} complex [52–54]. Electron transfer leads to dissociation of $[(NC)_5Fe^{II}(CN)Pt^{IV}(NH_3)_4(NC)Fe^{II}(CN)_5]^{4-}$ into two Fe^{III} complexes and a reduced Pt^{II} complex, Figure 19.12.

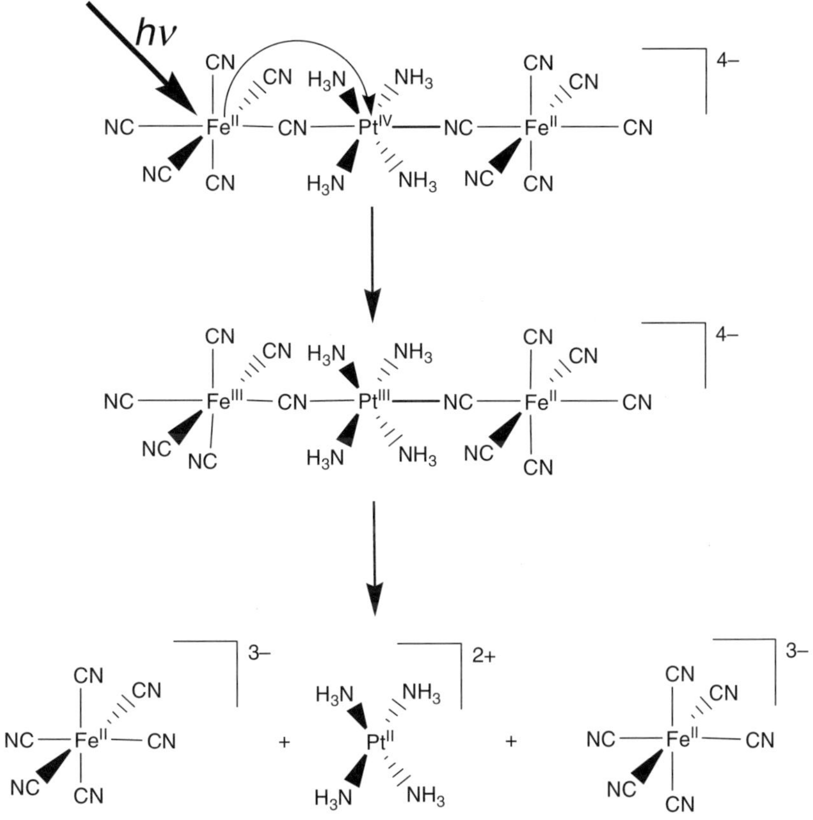

Figure 19.12 Photoinitiated electron collection at a metal center followed by dissociation of $[(NC)_5Fe^{II}(CN)Pt^{IV}(NH_3)_4(NC)Fe^{II}(CN)_5]^{4-}$. (Figure adapted from M. Zhou, B.W. Pfennig, J. Steiger, *et al.*, Multielectron transfer and single-crystal X-Ray structure of a trinuclear cyanide-bridged platinum-iron species, *Inorganic Chemistry*, **29**, 2456, 1990; C.C. Chang, B. Pfennig, and A.B. Bocarsly, Photoinduced multielectron charge transfer processes in group 8-platinum cyanobridged supramolecular complexes, *Coordination Chemistry Review*, **208**, 33, 2000; D.F. Watson, J.L. Wilson, and A.B. Bocarsly, Photochemical image generation in a cyanogel system synthesized from Tetrachloropalladate(II) and the trimetallic mixed-valence complex [(NC)5FeII-CN-PtIV(NH3)4-NC-FeII(CN) 5]4: Consideration of photochemical and dark mechanistic pathways of prussian blue formation, *Inorganic Chemistry*, **41**, 2408, 1990, 2000, 2002, American Chemical Society, Elsevier.)

19.2.3.1 Redox Properties of $[(NC)_5Fe^{II}(CN)Pt^{IV}(NH_3)_4(NC)Fe^{II}(CN)_5]^{4-}$

The electrochemistry of $[(NC)_5Fe^{II}(CN)Pt^{IV}(NH_3)_4(NC)Fe^{II}(CN)_5]^{4-}$ suggests an Fe-based HOMO. The oxidative electrochemistry $[(NC)_5Fe^{II}(CN)Pt^{IV}(NH_3)_4(NC)Fe^{II}(CN)_5]^{4-}$ shows two overlapping $Fe^{II/III}$ oxidations at 0.60 V versus SCE [52–54]. The overlapping oxidations suggest weak intercomponent coupling between the Fe units within the molecular assembly. The electrochemistry is consistent with the formation of a $[(NC)_5Fe^{III}(CN)Pt^{IV}(NH_3)_4(NC)Fe^{III}(CN)_5]^{4-}$ species upon electrochemical oxidation.

19.2.3.2 Spectroscopic and Photophysical Properties of $[(NC)_5Fe^{II}(CN)Pt^{IV}(NH_3)_4(NC)Fe^{II}(CN)_5]^{4-}$

The trimetallic complex, $[(NC)_5Fe^{II}(CN)Pt^{IV}(NH_3)_4(NC)Fe^{II}(CN)_5]^{4-}$, absorbs in the UV and visible regions of the spectrum. The absorption at $\lambda_{max}^{abs} = 424\,nm$ is assigned to a metal-to-metal charge-transfer (MMCT) transition from Fe^{II} to Pt^{IV} [52–54]. This absorption feature occurs at lower energy relative to the $Fe(d\pi) \rightarrow CN(\pi^*)$ MLCT transitions of $[Fe^{II}(CN)_5]^{3-}$, which occur at $\lambda_{max}^{abs} = 416\,nm$. Excitation at 424 nm yields a yellow solution that matches the spectrum of $[Fe^{II}(CN)_5]^{3-}$. The Φ for the formation of $[Fe^{II}(CN)_5]^{3-}$ was determined to be 0.02. Photolysis was done at low intensity to eliminate multiphoton events. Photoinduced dissociation of $[(NC)_5Fe^{II}(CN)Pt^{IV}(NH_3)_4(NC)Fe^{II}(CN)_5]^{4-}$ through multielectron transfer is suggested to occur through an unstable Pt^{III} oxidation state.

19.2.4 Two-Electron Mixed-Valence Complexes for Multielectron Photochemistry

Metal–metal bonded mixed-valence systems that undergo photochemical multielectron photochemistry, which are able to convert hydrohalic acids to hydrogen using light and a halogen trap have been reported by Nocera *et al.* [55–58]. The approach takes advantage of two-electron mixed-valence compounds, $M^{n+}\cdots M^{n+2}$, to drive multielectron chemistry. The systems are designed so that the single-electron-transfer products are unstable with respect to the two-electron-transfer products, promoting multielectron chemistry.

19.2.4.1 Dirhodium Photocatalysts

The mixed-valence dirhodium system, $[Rh_2^{0,II}(dfpma)_3X_2(L)]$ (dfpma = $MeN(PF_2)_2$, X = Cl or Br, L = CO, PR_3, or CNR) has been generated by irradiating $[Rh_2(dfpma)_3X_4]$ containing solutions at excitation wavelengths between 300–400 nm in the presence of excess L and halogen-atom traps, including tetrahydrofuran, dihydroanthracene or 2,3-dimethylbutadiene. Further irradiation of the $[Rh_2^{0,II}(dfpma)_3X_2(L)]$ complex resulted in the activation of Rh^{II}-X and generation of $[Rh_2^{0,0}(dfpma)_3L_2]$ [55–57]. The dfpma ligand consists of a π-acceptor and a π-donor to stabilize the mixed-valence oxidation states of $[Rh_2^{0,II}(dfpma)_3X_2(L)]$. Photolysis of $[Rh_2^{0,0}(dfpma)_3(PPh_3)(CO)]$ in the presence of HCl results in photolabilization of CO, allowing for attack by HCl at both metal centers to form an intermediate Rh^{II}, Rh^{II} dihydride, dihalide. Upon photolysis of the dihydride, one equivalent of hydrogen is evolved, yielding a blue intermediate, which is attributed to be the valence-symmetric $[Rh_2^{I,I}(dfpma)_3Cl_2]$ complex. This product is unstable with respect to internal disproportionation to the catalyti-

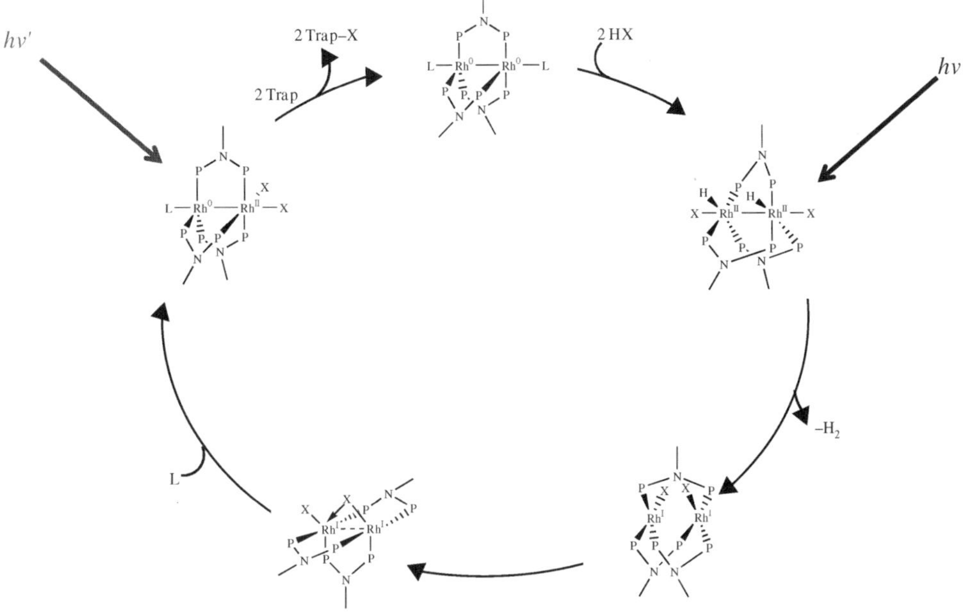

Figure 19.13 Mechanism for the photocatalytic generation of hydrogen from hydrohalic acids using a dirhodium mixed-valence photocatalyst. (Figure adapted from A.J. Esswein and D.G. Nocera, Hydrogen production by molecular photocatalysis, *Chemical Reviews*, **107**, 4022, 2007, American Chemical Society.)

cally inactive species $[Rh_2^{0,II}(dfpma)_3(PPh_3)Cl_2]$. In the presence of a halogen-atom trap, photolysis leads to regeneration of the catalytically active species $[Rh_2^{0,0}(dfpma)_3L]$ with an overall quantum yield for hydrogen of $\Phi \approx 0.01$, Figure 19.13. The Rh^{II}, Rh^{II} dihydride, dihalide complex has been isolated by using the more sterically demanding and less electron-withdrawing ligand tfepma (tfepma = $MeN(P(OCH_2CF_3)_2)_2$) [57]. The Rh^{II}–X bond activation is the rate-determining step in hydrogen production. Heterobimetallic complexes, $[Rh^IAu^I(tfepma)_2(CN^tBu)_2]^{2+}$ and $[Pt^{II}Au^I(dppm)_2PhCl]^+$ (dppm = $CH_2(PPh_2)_2$), have been synthesized to increase the rate of M–X bond activation. These complexes undergo two-electron oxidation upon photolysis to the Rh^{II}–Au^{II} and Pt^{III}–Au^{II} complexes respectively [58]. The Rh^{II}–Au^{II} complex $[Rh^{II}Au^{II}(tfepma)_2(CN^tBu)_2Cl_2]^{2+}$ is unstable toward internal disproportionation to yield Rh^{III} and $Au_2^{I,I}$ products, but the Pt^{III}–Au^{II} complex $[Pt^{III}Au^{II}(dppm)_2PhCl_3]^+$ is robust, with a 10-fold increase in the efficiency of metal–halide bond activation with respect to the dirhodium complexes.

19.2.5 Rhodium-Centered Electron Collectors

Trimetallic supramolecular assemblies that combine two Ru(II) or Os(II) LAs to a single Rh^{III} acceptor having potentially labile halide ligands have been reported by Brewer *et al.* [59–66]. Polyazine BLs covalently couple device components within the molecular architecture. The supramolecular assemblies provide a LA–BL–RhX_2–BL–LA structural motif (LA = Ru^{II} or Os^{II} polyazine chromophore, X = Cl or Br, BL = dpp). Electron collection at the rhodium center is followed by loss of the labile ligands, providing photoreactivity to the molecule. The electrochemical, photochemical properties and photocatalytic activity of the trimetallic complexes having the general formula $[\{(TL)_2M(dpp)\}_2RhX_2]^{5+}$ (TL = bpy or phen, M = Ru or Os, X = Cl or Br) and $[\{(tpy)MCl(dpp)\}_2RhCl_2]^{3+}$ (M = Ru or Os) have been investigated [59–66]. The redox, spectroscopic, and photochemical properties are dictated by component identity. Studies have established $[\{(bpy)_2Ru(dpp)\}_2RhX_2]^{5+}$ (X = Cl or Br) and $[\{(phen)_2Ru(dpp)\}_2RhCl_2]^{5+}$ as photochemical molecular devices for electron collection at a metal center [59,63,66]. These complexes have also been established as photocatalysts for solar hydrogen production from water with a hydrogen yield of $\Phi \approx 0.01$, being the first known PECs that function as photocatalysts to produce hydrogen [59,60,63,66]. Variation of the LA metal to produce $[\{(bpy)_2Os(dpp)\}_2RhCl_2]^{5+}$ or the TL to produce $[\{(tpy)MCl(dpp)\}_2RhCl_2]^{3+}$ (M = Ru or Os) tunes the energy of the LA metal ($d\pi$) orbital. The complexes $[\{(bpy)_2Os\text{-}(dpp)\}_2RhCl_2]^{5+}$ and $[\{(tpy)RuCl(dpp)\}_2RhCl_2]^{3+}$ also function as photocatalysts for hydrogen production, but with lower quantum efficiencies.

19.2.5.1 Redox Properties of Rhodium-Centered Electron Collectors

The redox properties of the supramolecular complexes of the form LA–BL–RhX_2–BL–LA are dictated by the components used. Table 19.2 summarizes the electrochemical properties of the trimetallic complexes. The electrochemistry of the trimetallic complexes of the general formula $[\{(TL)_2M(dpp)\}_2RhX_2]^{5+}$ (TL = bpy or phen, M = Ru or Os, X = Cl or Br) and $[\{(tpy)MCl(dpp)\}_2RhCl_2]^{3+}$ (M = Ru or Os) demonstrates metal-based oxidations and ligand-based reductions that are tuned over large potential ranges by subunit identity. All the trimetallics show two overlapping $Ru^{II/III}$ or $Os^{II/III}$ based oxidations, indicating the absence of significant intercomponent coupling between the ruthenium or osmium subunits. A representative cyclic voltammogram of $[\{(phen)_2Ru(dpp)\}_2RhCl_2]^{5+}$ is given in Figure 19.14. Oxidative electrochemistry of $[\{(bpy)_2Ru(dpp)\}_2RhX_2]^{5+}$ (X = Cl or Br) or $[\{(phen)_2Ru(dpp)\}_2RhCl_2]^{5+}$ shows overlapping $Ru^{II/III}$ couples at about 1.60 V versus SCE. The reductive electrochemistry shows irreversible $Rh^{III/II/I}$ reductions, followed by two reversible $dpp^{0/-}$ reductions [59,63,66]. The $Rh^{III/II/I}$ reduction is followed by the loss of halides, similar to that reported for $[Rh(bpy)_2Cl_2]^+$ [67]. The variation of the identity of the halides on the rhodium impact the reductive electrochemistry. The $Rh^{III/II/I}$ reduction occurs 40 mV more positively for $[\{(bpy)_2Ru(dpp)\}_2RhBr_2]^{5+}$ than the chloride analog, −0.37 V versus SCE. This is consistent with a rhodium center that is more electron deficient, due to the weaker σ-donor ability of Br^- versus Cl^- [59,63]. Replacing the LA metal with Os to form $[\{(bpy)_2Os(dpp)\}_2RhCl_2]^{5+}$ generates a destabilized Os($d\pi$) orbital relative to the ruthenium analogs, with the $Os^{II/III}$ couple occurring at 1.12 V versus SCE [65].

Replacing bpy or phen with tpy provides some stereochemical control in supramolecular complexes by eliminating Δ and Λ isomeric mixtures associated with the tris(bidentate) metal centers. The tpy-based systems are easier to oxidize relative to the bpy analogs, consistent with a more electron-rich ruthenium center due to Cl coordination in place of a pyridine ring. The $Ru^{II/III}$ oxidation of $[\{(tpy)RuCl(dpp)\}_2RhCl_2]^{3+}$ occurs at 1.09 V versus SCE [64]. The $Os^{II/III}$ couple occurs at an even more positive potential of 0.82 V versus SCE, consistent with

Table 19.2 Electrochemical data for supramolecular complexes that undergo multi-electron photochemistry.

Complex[a]	$E_{1/2}^{ox}$ (V vs SCE)	$E_{1/2}^{red}$ (V vs SCE)			Ref.
		$Rh^{III/II/I}$	$BL^{0/-}$	$BL^{-/2-}$	
$[\{(bpy)_2Ru(dpp)\}_2IrCl_2]^{5+\,b,c}$	1.53 ($2Ru^{II/III}$)		−0.43 −0.58	−1.10 −1.26	[44]
$[\{(bpy)_2Ru(dpq)\}_2IrCl_2]^{5+\,b,c}$	1.53 ($2Ru^{II/III}$)		−0.16 −0.30	−0.94 −1.26	[44]
$[\{(bpy)_2Ru(dpb)\}_2IrCl_2]^{5+\,b,c}$	1.53 ($2Ru^{II/III}$)		−0.005 −0.16	−0.90 −1.02	[44]
$[(phen)_2Ru(tatpq)Ru(phen)_2]^{4+\,b}$	1.37 ($2Ru^{II/III}$)		−0.23	−0.60[d]	[48]
$[(phen)_2Ru(tatpp)Ru(phen)_2]^{4+\,b}$	1.36 ($2Ru^{II/III}$)		−0.18	−0.56	[48]
$[(NC)_5Fe^{II}(CN)Pt^{IV}(NH_3)_4(NC)Fe^{II}(CN)_5]^{4-\,e}$	0.60 ($2Fe^{II/III}$)				[52–54]
$[\{(bpy)_2Ru(dpp)\}_2RhCl_2]^{5+\,b,c}$	1.60 ($2Ru^{II/III}$)	−0.41[d]	−0.80 −1.04		[59]
$[\{(tpy)OsCl(dpp)\}_2RhCl_2]^{3+\,b,c}$	0.82 ($2Os^{II/III}$)	−0.55[d]	−0.90 −1.24		[62]
$[\{(bpy)_2Ru(dpp)\}_2RhBr_2]^{5+\,b,c}$	1.57 ($2Ru^{II/III}$)	−0.37[d]	−0.76 −1.06		[63]
$[\{(tpy)RuCl(dpp)\}_2RhCl_2]^{3+\,b,c}$	1.09 ($2Ru^{II/III}$)	−0.51[d]	−0.91 −1.24		[64]
$[\{(bpy)_2Os(dpp)\}_2RhCl_2]^{5+\,b,c}$	1.12 ($2Os^{II/III}$)	−0.43[d]	−0.80 −1.04		[65]
$[\{(phen)_2Ru(dpp)\}_2RhCl_2]^{5+\,b,c}$	1.54 ($2Ru^{II/III}$)	−0.43[d]	−0.84 −1.10		[66]

[a] Potentials reported vs SCE.
[b] In acetonitrile with 0.1 M Bu_4NPF_6.
[c] Converted Ag/AgCl to SCE by subtracting 35 mV from the potential vs Ag/AgCl.
[d] Only the cathodic peak was observed.
[e] In 1 M $NaNO_3$.

the higher-energy Os($d\pi$) orbitals [62]. These changes in LA metal and TLs allow for about 1 V tuning in metal oxidation potential. The reductive electrochemistry of $[\{(tpy)MCl(dpp)\}_2RhCl_2]^{3+}$ (M = Ru or Os) shows irreversible $Rh^{III/II/I}$ reductions at −0.51 and −0.55 V versus SCE, respectively, for the ruthenium- and osmium-based systems, occurring at more negative potentials relative to the bpy analogs. The lower cationic charge and the presence of more electron-rich TLs make the $Rh^{III/II/I}$ reduction more difficult, shifting the potential to more negative values. The $Rh^{III/II/I}$ reductions are followed by $dpp^{0/-}$ couples for each dpp [62,64]. The oxidative electrochemistry of the trimetallic supramolecular complexes predicts a Ru (dπ) or Os(dπ)-HOMO with energy tuned by the TL or LA metal. The reductive electrochemistry predicts a Rh(dσ^*)-based LUMO that can accept two electrons, allowing the rhodium to function as an electron collector with higher-energy BL(π^*) orbitals, Figure 19.15. The electrochemistry predicts a lowest lying Ru(dπ) $\rightarrow$ Rh(dσ^*) metal-to-metal charge-transfer (^{3}MMCT) excited state in the LA–BL–RhX_2–BL–LA structural motif.

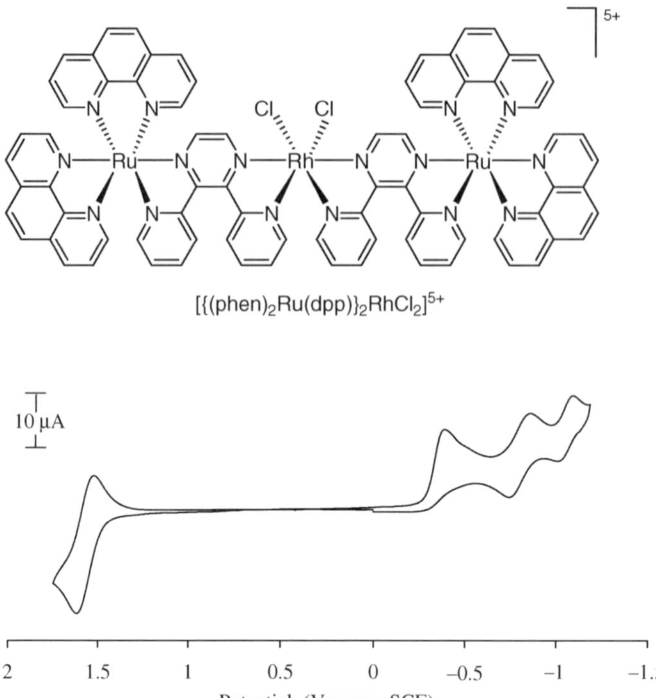

Figure 19.14 Structure and cyclic voltammogram of $[\{(phen)_2Ru(dpp)\}_2RhCl_2]^{5+}$ (phen = 1,10-phenanthroline, dpp = 2,3-bis(2-pyridyl)pyrazine). Electrochemistry conducted in 0.1 M Bu_4NPF_6 at RT in CH_3CN.

19.2.5.2 Spectroscopic Properties of Rhodium-Centered Electron Collectors

Electronic absorption spectroscopy demonstrates that the trimetallic supramolecular assemblies, LA–BL–RhX_2–BL–LA, are efficient light absorbers throughout the UV and visible regions of the spectrum, with transitions characteristic of each subunit of the LA–BL

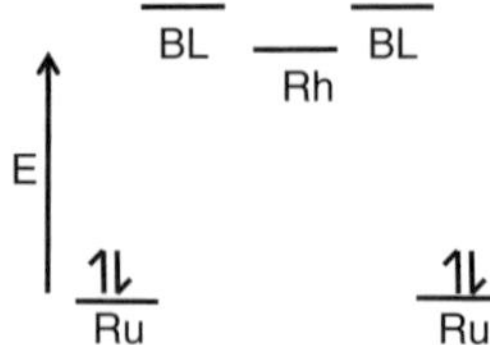

Figure 19.15 Orbital energy diagram of Rh centered photoinitiated electron collection of the form LA-BL-RhX_2-BL-LA (LA = bpy or phen, BL = dpp, X = Cl or Br). (Figure adapted from M. Elvington, J.R. Brown, D.F. Zigler, and K.J. Brewer, Supramolecular complexes as photoinitiated electron collectors: applications in solar hydrogen production, *Proceedings of SPIE*, **6340**, 63400W-1, 2006, SPIE.)

Table 19.3 Spectroscopic and photophysical data for supramolecular complexes that undergo multi-electron photochemistry.

Complex	λ_{max}^{abs} (nm)	λ_{max}^{em} (nm)	Φ^{em} (RT)	τ (RT) (ns)	Ref.
$[\{(bpy)_2Ru(dpp)\}_2IrCl_2]^{5+a}$	522	794	1.2×10^{-4}	32	[44]
$[\{(bpy)_2Ru(dpq)\}_2IrCl_2]^{5+a}$	616	866	$<10^{-6}$	<5	[44]
$[\{(bpy)_2Ru(dpb)\}_2IrCl_2]^{5+a}$	666				[44]
$[(phen)_2Ru(tatpq)Ru(phen)_2]^{4+a}$	440				[48]
$[(phen)_2Ru(tatpp)Ru(phen)_2]^{4+a}$	443				[48]
$[(NC)_5Fe^{II}(CN)Pt^{IV}(NH_3)_4(NC)Fe^{II}(CN)_5]^{4-b}$	424				[53,54]
$[\{(bpy)_2Ru(dpp)\}_2RhCl_2]^{5+a}$	520	760	7.3×10^{-5}	32	[59]
$[\{(tpy)OsCl(dpp)\}_2RhCl_2]^{3+a}$	538				[62]
$[\{(bpy)_2Ru(dpp)\}_2RhBr_2]^{5+a}$	520	760	1.5×10^{-4}	26	[63]
$[\{(tpy)RuCl(dpp)\}_2RhCl_2]^{3+a}$	540				[64]
$[\{(bpy)_2Os(dpp)\}_2RhCl_2]^{5+a}$	530				[65]
$[\{(phen)_2Ru(dpp)\}_2RhCl_2]^{5+a}$	520	746	1.8×10^{-4}	27	[66]

[a] Room temperature spectra obtained in deoxygenated acetonitrile.
[b] In distilled water.

unit [59–66]. The energies of the lowest-lying transitions for the trimetallic complexes are given in Table 19.3. The electronic absorption spectra of $[\{(bpy)_2Ru(dpp)\}_2RhX_2]^{5+}$ (X = Cl or Br) and $[\{(phen)_2Ru(dpp)\}_2RhCl_2]^{5+}$ are similar, exhibiting intense $\pi \rightarrow \pi^*$ TL and dpp transitions in the UV and Ru(dπ) $\rightarrow$ TL(π^*) (TL = bpy or phen) and Ru(dπ) $\rightarrow$ dpp(π^*) MLCT transitions in the visible [59,63,66]. The Ru(dπ) $\rightarrow$ bpy(π^*) MLCT transitions occur between 410–420 nm, while the Ru(dπ) $\rightarrow$ dpp(π^*) MLCT transitions occur at $\lambda_{max}^{abs} = 520$ nm for all complexes. The electronic absorption spectra of $[\{(bpy)_2Ru(dpp)\}_2RhX_2]^{5+}$ (X = Cl or Br) are nearly identical [59,63], which suggest that the type of halide on the rhodium center has no observable impact on the light-absorbing properties of the supramolecules. The electronic absorption spectra for the tpy-based systems are similar to the bpy analogues showing changes consistent with the decreased HOMO–LUMO gap with the Ru(dπ) $\rightarrow$ dpp(π^*) MLCT transitions at $\lambda_{max}^{abs} = 540$ [59,62–64,66]. The complexes $[\{(bpy)_2Os(dpp)\}_2RhCl_2]^{5+}$ and $[\{(tpy)OsCl(dpp)\}_2RhCl_2]^{3+}$ have similar spectroscopies to their ruthenium analogues. The Os(dπ) $\rightarrow$ dpp(π^*) MLCT transitions occur at slightly lower energies, reflective of the higher-energy Os($d\pi$) orbitals observed in electrochemistry [62,65]. The electronic absorption spectra of the osmium complexes also show higher intensities in their low-energy tails due to the higher spin–orbit coupling in Os enhancing the ^{3}MLCT absorption.

19.2.5.3 Photophysical Properties of Rhodium-Centered Electron Collectors

Supramolecular complexes incorporating polyazine LA units typically display observable emissions at room temperature, providing a means to probe the excited-state dynamics of these molecules. Optical excitation of the Ru/Os(dπ) ^{1}MLCT states in the LA–BL–RhX_2–BL–LA structural motif leads to intersystem crossing populating the ^{3}MLCT states that are often emissive. In systems which possess Ru/Os(dπ)-based HOMOs and Rh(dσ^*)-based LUMOs, the emissions from the ^{3}MLCT states are quenched by electron transfer to low lying ^{3}MMCT states. The complexes $[\{(bpy)_2Ru(dpp)\}_2RhCl_2]^{5+}$, $[\{(bpy)_2Ru(dpp)\}_2RhBr_2]^{5+}$, and $[\{(phen)_2Ru(dpp)\}_2RhCl_2]^{5+}$ display weak emissions from the $Ru(d\pi) \rightarrow dpp(\pi^*)$ ^3MLCT state at $\lambda_{max}^{em} = 760\,nm$ for $[\{(bpy)_2Ru(dpp)\}_2RhCl_2]^{5+}$ and $[\{(bpy)_2Ru(dpp)\}_2RhBr_2]^{5+}$ [59,63] and $\lambda_{max}^{em} = 746\,nm$ for $[\{(phen)_2Ru(dpp)\}_2RhCl_2]^{5+}$ with $\Phi^{em} = 7.3 \times 10^{-5}$, 1.5×10^{-4} and 1.8×10^{-4}, respectively, at RT in deoxygenated acetonitrile solutions following excitation at 520 nm [59,63,66]. The emission from the ^3MLCT states for these complexes are about 10–15 % of $[(bpy)_2Ru(dpp)Ru(bpy)_2]^{4+}$ that lacks a rhodium electron acceptor ($\lambda_{max}^{em} = 744\,nm$, $\Phi^{em} = 1.38 \times 10^{-3}$) [59]. The excited state lifetimes of the ^3MLCT states in deoxygenated acetonitrile solution at RT for $[\{(bpy)_2Ru(dpp)\}_2RhCl_2]^{5+}$, $[\{(bpy)_2Ru(dpp)\}_2RhBr_2]^{5+}$, and $[\{(phen)_2Ru(dpp)\}_2RhCl_2]^{5+}$ are similar and have values 32, 26 and 27 ns, respectively [66]. The rate constant for electron transfer to populate the ^3MMCT state, k_{et}, can be estimated by considering k_r and k_{nr} of these complexes to be identical to $[(bpy)_2Ru(dpp)Ru(bpy)_2]^{4+}$. Based on this assumption, similar rates of electron transfer with $k_{et} = 1.2 \times 10^8\,s^{-1}$, $5.2 \times 10^7\,s^{-1}$ and $4.4 \times 10^7\,s^{-1}$, respectively, for $[\{(bpy)_2Ru(dpp)\}_2RhCl_2]^{5+}$, $[\{(bpy)_2Ru(dpp)\}_2RhBr_2]^{5+}$ and $[\{(phen)_2Ru(dpp)\}_2RhCl_2]^{5+}$ are obtained [66]. The expected lower energy ^3MLCT states of the other trimetallics render their emissions even weaker, and are not observed within the detection limit of the emission spectrometer. Table 19.3 summarizes the electrochemical and photophysical properties of the mixed-metal trimetallic complexes.

19.2.5.4 Photoinitiated Electron Collection on a Rhodium Center

Supramolecular complexes can be designed to collect multiple electrons at a single site, upon optical excitation, by connecting two or more LA units to a single EC. Electronic isolation between the LA units is desirable for the design of functioning PECs. Within this perspective, supramolecular complexes of the general form $LA–BL–RhX_2–BL–LA$ have been designed [59–66]. The complex $[\{(bpy)_2Ru(dpp)\}_2RhCl_2]^{5+}$ has been established as the first photoinitiated electron collector that undergoes photoreduction and collects multiple electrons on a metal center with the supramolecular architecture remaining intact [59]. Photoreduction is followed by halide loss to produce the coordinatively unsaturated Rh^I species, $[\{(bpy)_2Ru(dpp)\}_2Rh^I]^{5+}$ through a Rh^{II} intermediate, Figure 19.16.

Electrochemical reduction of $[\{(bpy)_2Ru(dpp)\}_2RhCl_2]^{5+}$ by two electrons just negative of the $Rh^{III/II/I}$ couple leads to a spectroscopic shift in which the $Ru(d\pi) \rightarrow dpp(\pi^*)$ ^1MLCT shifts to higher energy [59]. This shift to higher energy is consistent with the destabilization of the $dpp(\pi^*)$ orbital upon formation of an electron-rich Rh^I species. The spectroscopy of the photochemically reduced product obtained by photolysis in the presence of an electron donor is identical to the electrochemically reduced product, establishing that $[\{(bpy)_2Ru(dpp)\}_2RhCl_2]^{5+}$ is reduced by two electrons photochemically with the electrons

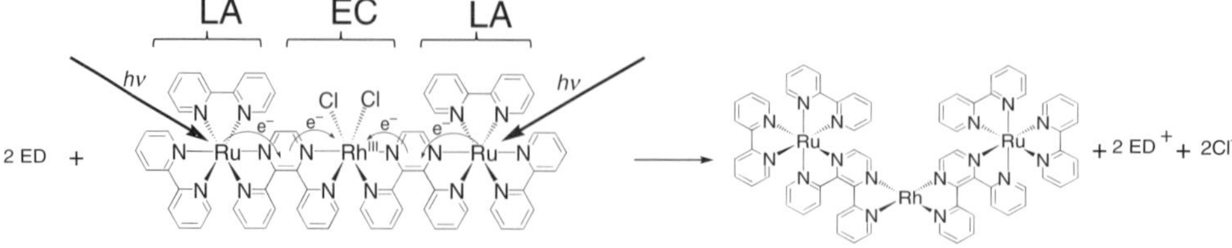

Figure 19.16 Representation of photoinduced electron transfer to generate $[\{(bpy)_2Ru(dpp)\}_2Rh^I]^{5+}$.

being collected at the rhodium center. A Stern–Volmer kinetic investigation was done to probe the photoreduction of $[\{(bpy)_2Ru(dpp)\}_2RhCl_2]^{5+}$ in the presence of DMA [59]. The rate of reductive quenching to generate the Rh^{II} species was found to be $1.9 \times 10^9\,M^{-1}\,s^{-1}$ [59]. Product formation can occur from the ^{3}MLCT state or the ^{3}MMCT state or a combination of both. The excited state reduction potentials for the ^{3}MLCT and ^{3}MMCT states are estimated as 1.23 and 0.84 V, respectively [66]. Based on the oxidation potential, $E_{1/2} = 0.81$ V versus SCE, DMA has the necessary driving force to reductively quench both the ^{3}MLCT and ^{3}MMCT states [66]. A Stern–Volmer quenching analysis of the emission from the ^{3}MLCT state demonstrates that this process is efficient, with a rate of $2 \times 10^{10}\,M^{-1}\,s^{-1}$ [59]. The complexes $[\{(bpy)_2Ru(dpp)\}_2RhBr_2]^{5+}$ and $[\{(phen)_2Ru(dpp)\}_2RhCl_2]^{5+}$ also undergo photoinitiated electron collection on the rhodium center [66]. Photoreduction of $[\{(bpy)_2Ru(dpp)\}_2RhBr_2]^{5+}$ or $[\{(phen)_2Ru(dpp)\}_2RhCl_2]^{5+}$ in the presence of DMA affords spectroscopic shifts consistent with the formation of $[\{(TL)_2Ru(dpp)\}_2Rh^I]^{5+}$ (TL = bpy or phen). The excited-state reduction potentials of the ^{3}MLCT and ^{3}MMCT states for $[\{(bpy)_2Ru(dpp)\}_2RhBr_2]^{5+}$ or $[\{(phen)_2Ru(dpp)\}_2RhCl_2]^{5+}$ are similar to $[\{(bpy)_2Ru(dpp)\}_2RhCl_2]^{5+}$. Thus, the driving forces for reductive quenching of the ^{3}MLCT and ^{3}MMCT states of $[\{(bpy)_2Ru(dpp)\}_2RhBr_2]^{5+}$ and $[\{(phen)_2Ru(dpp)\}_2RhCl_2]^{5+}$ by DMA are also similar to $[\{(bpy)_2Ru(dpp)\}_2RhCl_2]^{5+}$ [66].

19.2.5.5 Photocatalysis Using Rhodium-Centered Electron Collectors

Photochemical reduction of $[\{(bpy)_2Ru(dpp)\}_2RhCl_2]^{5+}$, $[\{(bpy)_2Ru(dpp)\}_2RhBr_2]^{5+}$ or $[\{(phen)_2Ru(dpp)\}_2RhCl_2]^{5+}$ in the presence of DMA leads to reduction of Rh^{III} and displacement of two halides to form a coordinately unsaturated $[\{(TL)_2Ru(dpp)\}_2Rh^I]^{5+}$ (TL = bpy or phen) species, which can interact with substrates [59,63,66]. The complexes $[\{(bpy)_2Ru(dpp)\}_2RhX_2]^{5+}$ (X = Cl or Br) and $[\{(phen)_2Ru(dpp)\}_2RhCl_2]^{5+}$ are the first photoinitiated electron collectors to photochemically produce hydrogen from water [59,63,66]. Photolysis of acetonitrile solutions of $[\{(bpy)_2Ru(dpp)\}_2RhX_2]^{5+}$ (X = Cl or Br) or $[\{(phen)_2Ru(dpp)\}_2RhCl_2]^{5+}$ in the presence of DMA and water at 470 nm, produces hydrogen with Φ 0.01 [59,60,63,66]. Hydrogen production appears linear within the 4 h investigated. The halide on rhodium impacts the photcatalytic activity, as shown by higher hydrogen yields for $[\{(bpy)_2Ru(dpp)\}_2RhBr_2]^{5+}$ [63]. Photoreduction to form the Rh^I product is followed by halide loss that may be a kinetically important step in photocatalysis and may occur more rapidly in $[\{(bpy)_2Ru(dpp)\}_2RhBr_2]^{5+}$ due to the weak σ-donor ability of Br^- versus Cl^- leading to higher photcatalytic activity. A mercury test [68,69] suggests that Rh decomplexation was not an operating pathway for hydrogen generation using the LA–BL–RhX_2–BL–LA structural motif.

Photocatalytic activity was observed for all three complexes when a 520 nm excitation source was used [66]. This process was less efficient. This lower photocatalytic activity may be attributed to competition between the Rh^I species and the photocatalyst for light absorption as photolysis proceeds. The Rh^I photoproduct does not absorb well at 520 nm, so if excitation of this complex is important, then a lower yield at excitation at 520 nm would be expected.

The impact on the photocatalytic efficiency when the LA metal is replaced by Os and the TL, bpy, is replaced by tpy was investigated [66]. The complexes $[\{(bpy)_2Os(dpp)\}_2RhCl_2]^{5+}$, $[\{(tpy)RuCl(dpp)\}_2RhCl_2]^{3+}$ and $[\{(tpy)OsCl(dpp)\}_2RhCl_2]^{3+}$ were analyzed for photochemical hydrogen production from water. The two complexes $[\{(bpy)_2Os(dpp)\}_2RhCl_2]^{5+}$

and $[\{(tpy)RuCl(dpp)\}_2RhCl_2]^{3+}$ yield similar amounts of hydrogen when irradiated at 470 nm in the presence of DMA and water, an amount that is lower than the $[\{(bpy)_2Ru(dpp)\}_2RhCl_2]^{5+}$ photocatalyst. The excited-state reduction potentials of the ^{3}MLCT and ^{3}MMCT were predicted as 0.91 and 0.54 vs. SCE, respectively, for $[\{(bpy)_2Os(dpp)\}_2RhCl_2]^{5+}$, and 1.01 and 0.61 vs. SCE respectively, for $[\{(tpy)RuCl(dpp)\}_2RhCl_2]^{3+}$ [66]. These potentials are quite similar, suggesting similar driving forces for reductive quenching by DMA, predicting similar photocatalytic efficiency. The small driving force for reductive quenching of the ^{3}MLCT for these systems relative to $[\{(bpy)_2Ru(dpp)\}_2RhCl_2]^{5+}$ may be a significant contributor to the lower hydrogen yields. The lower-energy excited states of $[\{(bpy)_2Os(dpp)\}_2RhCl_2]^{5+}$ and $[\{(tpy)RuCl(dpp)\}_2RhCl_2]^{3+}$ likely result in shorter excited-state lifetimes. This limits bimolecular quenching by DMA during the excited-state lifetime of the catalyst, further impeding the photocatalytic efficiency. The complex $[\{(tpy)OsCl(dpp)\}_2RhCl_2]^{3+}$ does not produce detectable hydrogen under the conditions investigated. The driving forces to reductively quench the ^{3}MLCT or ^{3}MMCT states by DMA is thermodynamically unfavorable, as evidenced by the excited-state reduction potentials of these states (0.71 and 0.37 V versus SCE, respectively, for the ^{3}MLCT and ^{3}MMCT states). The even shorter ^{3}MLCT lifetime relative to $[\{(tpy)RuCl(dpp)\}_2RhCl_2]^{3+}$, and the thermodynamically unfavorable driving force for reductive quenching both predict a lack of photocatalytic activity for $[\{(tpy)OsCl(dpp)\}_2RhCl_2]^{3+}$.

The complex $[\{(bpy)_2Ru(dpb)\}_2IrCl_2]^{5+}$ undergoes multielectron photochemistry and collects electrons on the dpb BL [47]. This complex was evaluated for photochemical hydrogen production from water in the presence of DMA by excitation at 470 and 520 nm [66]. Based on the excited-state reduction potential of the ^{3}MLCT state of $[\{(bpy)_2Ru(dpb)\}_2IrCl_2]^{5+}$ (1.13 V versus SCE), DMA has sufficient driving force to reductively quench this ^{3}MLCT state. Thus, photoreduction of $[\{(bpy)_2Ru(dpb)\}_2IrCl_2]^{5+}$ in the presence of DMA is expected. This iridium-based complex does not produce detectable amounts of hydrogen under the conditions studied. The lack of photochemical hydrogen production by the Ir complex implies that electron collection on the rhodium center is key for the photochemical reduction of water to hydrogen and signifies the importance of the rhodium center for hydrogen photocatalysis.

The impact of the electron donor was investigated for the most efficient photocatalysts, $[\{(bpy)_2Ru(dpp)\}_2RhCl_2]^{5+}$, $[\{(bpy)_2Ru(dpp)\}_2RhBr_2]^{5+}$ and $[\{(phen)_2Ru(dpp)\}_2RhCl_2]^{5+}$ [66]. Significantly, photocatalysis is seen using the electron donors DMA, TEA and TEOA, illustrating the general applicability of this process with varied supramolecules and electron donors. The variation of electron donor showed the photocatalysis efficiency to decrease in the order DMA > TEA > TEOA. Variation of the electron donor provides a method to study factors that impact photocatalytic activity, including driving force, effective pH and so on. The oxidation potentials of TEA and TEOA are 0.96 and 0.90 V versus SCE, respectively, in acetonitrile. Therefore much lower driving forces for reductive quenching of the ^{3}MLCT and ^{3}MMCT are expected for all three photocatalysts, using either TEA or TEOA as electron donor. Consistent with this, considerably lower hydrogen yields are obtained when TEA or TEOA are used as the electron donor. TEOA produces the lowest hydrogen yield, despite the slightly larger driving force with respect to TEA. This suggests that factors other than driving force impact photocatalytic efficiency. The effective pH values of the photolysis solutions using DMA, TEA or TEOA were determined as about 9.1, 14.7 and 11.8, respectively, on the assumption that the pK_a values of their conjugated acids remain unchanged in the photocatalytic solutions relative to aqueous conditions (pK_a = 5.07 ($DMAH^+$), 10.75 ($TEAH^+$) and 7.76 ($TEOAH^+$)) [70]. The lower hydrogen production with more basic EDs implies that pH effects may play a role in photocatalytic efficiency observed for DMA. This is not unexpected, as the reduction potential of water is pH dependant, being more facile at lower pH. In addition, DMA can form electron-donor–catalyst adducts through π-stacking, providing a higher hydrogen yield.

The photochemical properties of $[\{(bpy)_2Ru(dpp)\}_2RhBr_2]^{5+}$ in aqueous medium have been investigated [71]. Studies have established this complex to function as a photoinitiated electron collector in water. Studies also show this complex to photocatalyze the production of hydrogen from water in the presence of TEOA buffered with triflic acid, hydrobromic acid or phosphoric acid. The photocatalytic efficiency is lower in the aqueous medium, possibly due to the large excess of amines impeding the catalyst function.

19.2.5.6 Photolysis System for Photochemical Hydrogen Production

High throughput of photolysis experiments is key to being able to uncover the key factors impacting PEC and photocatalysis. Development of an LED array with the LEDs wired in series provides for high throughput. Reproducible light delivery is maintained by power control for each LED [72]. The LEDs have been evaluated by colorimetric measurements, as well as chemical actinometry. Studies show no statistical difference in hydrogen production, irrespective of the LED used. This allows the study of multiple experiments simultaneously under identical conditions. A schematic of the experimental design using the LED array is represented in Figure 19.17.

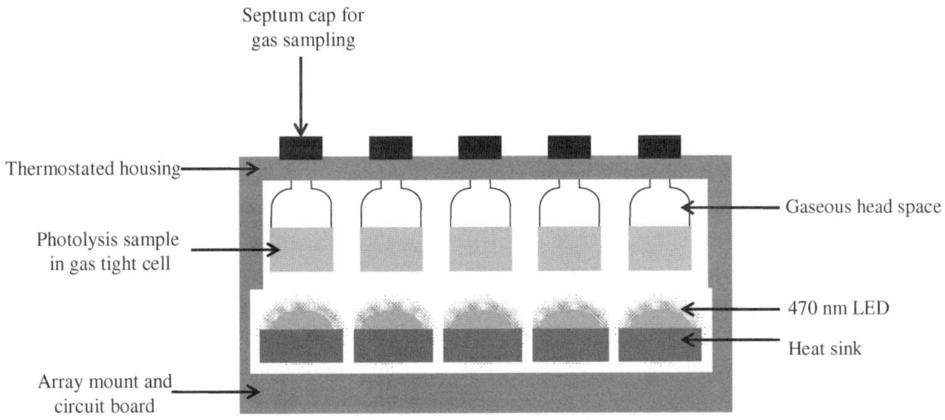

Figure 19.17 Experimental design for photocatalytic hydrogen production.

19.2.6 *Mixed-Metal Systems for Solar Hydrogen Production*

Mixed-metal polyazine complexes incorporating reactive metals have been recently explored as solar hydrogen catalysts. Sakai *et al.* recently investigated an Ru–Pt bimetallic system capable of photochemically producing hydrogen from water with $\Phi \approx 0.01$ in the presence of ethylenediaminetetraacetic acid [73,74]. Rau *et al.* reported an Ru–Pd bimetallic system that

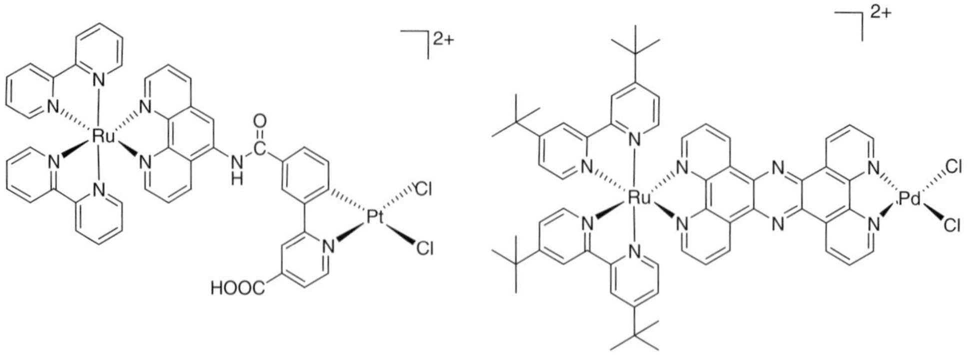

Figure 19.18 Bimetallic Ru-Pt and Ru-Pd systems as solar hydrogen catalysts.

photochemically produces hydrogen in the presence of TEA, Figure 19.18 [75]. Studies by Eisenberg and Castellano *et al.* [76] and Hammarström *et al.* [77] have shown that decomplexation of similar systems incorporating reactive metals generate metal colloids, which act as the hydrogen-generation catalysts.

19.3 Conclusions

Multielectron photochemistry is key to efficient solar-energy conversion schemes. Although much has been understood concerning single-electron photochemistry, the field of multielectron photochemistry is still in its infancy. Development of photocatalytic systems that are capable of solar hydrogen production from water requires a thorough understanding of this field. Systems that collect and deliver multiple electrons to substrates, upon optical excitation, provide an attractive means of producing many fuels, including hydrogen via water splitting. The impediments to this process include the small number of molecular systems capable of photochemically collecting reducing equivalents, the lack of fundamental understanding of multielectron photochemistry and the lack of large degrees of structural diversity of LA subunits coupled to catalytically active metal centers.

Mixed-metal supramolecular complexes have been designed to couple multiple LA units to a single reactive metal. The LA units are electronically isolated, critical to multielectron photochemistry. The energy of the acceptor orbitals for electron collection has been modulated by the type of ligand used, as well as the central metal. In the systems which incorporate Ir, $[\{(bpy)_2Ru(BL)\}_2IrCl_2]^{5+}$, or functional Ru bimetallic complexes, such as $[(phen)_2Ru(tatpq)Ru(phen)_2]^{4+}$, electron collection occurs at the BL. In systems which incorporate rhodium, electron collection occurs at the Rh site. The Rh center contains labile ligands that can be lost following photoreduction of the rhodium center, imparting reactivity at the metal site and facilitating reaction with substrates.

Rhodium centered supramolecular systems have been shown to photocatalyze water reduction to hydrogen, unprecedented in molecular photoinitiated electron collectors. The design considerations for a functioning system for photoactivated multielectron reduction of water to produce hydrogen have been investigated. Although the Ir-centered complex undergoes PEC to produce the two-electron-reduced complex $[\{(bpy)_2Ru(dpb^-)\}_2IrCl_2]^{3+}$, photocatalytic activity was not observed. This illustrates the importance of the Rh center, which is able to accept electrons and potentially bind substrates, important to chemical transformations of substrates. The coordination environment on the Rh center impacts the photocatalytic activity, as evidenced by the greater hydrogen yields when weaker σ-donors are present on Rh. The bromide complex, $[\{(bpy)_2Ru(dpp)\}_2RhBr_2]^{5+}$, provides a system with a lower-lying Rh(dσ^*) acceptor orbital, with a larger driving force for intramolecular electron transfer to produce the ^{3}MMCT state and/or promotes halide loss to generate the Rh^I system.

The complexes $[\{(bpy)_2Ru(dpp)\}_2RhCl_2]^{5+}$, $[\{(bpy)_2Ru(dpp)\}_2RhBr_2]^{5+}$ and $[\{(phen)_2Ru(dpp)\}_2RhCl_2]^{5+}$ all function as PEC devices and catalyze the reduction of water to hydrogen. The variation of electron donor was explored with the three photocatalysts that provide the highest yield of hydrogen. Lower hydrogen production was observed when the driving force for excited-state reduction is lower using the electron donors TEA and TEOA. The study using TEOA provides evidence that the driving force for excited-state reduction alone does not provide an explanation for all the experimental results, and other factors, such as ED–catalyst interactions and pH effects, have significant impact on the photocatalytic activity.

Investigation of the factors impacting hydrogen production from water using Rh-centered supramolecules has been undertaken. The TL variation or LA metal variation can be used to tune the HOMO energy and thus modulate the excited state reduction potential of the complex, as illustrated through the study of $[\{(tpy)RuCl(dpp)\}_2RhCl_2]^{3+}$ and $[\{(bpy)_2Os(dpp)\}_2RhCl_2]^{5+}$. These complexes show similar photocatalytic activity consistent with the similar excited-state reduction potentials. However, lower quantum efficiency for hydrogen production is observed for these complexes, consistent with the much lower driving force for excited-state reduction by the electron donor. These systems possess ^{3}MLCT states that should be reductively quenched by DMA, while this reaction is prohibited from the ^{3}MMCT state, implying that the ^{3}MLCT state can function for PEC and photochemical hydrogen production. The system $[\{(tpy)OsCl(dpp)\}_2RhCl_2]^{3+}$, does not produce hydrogen under the conditions investigated, consistent with the even lower ^{3}MLCT energy in this molecular architecture, inhibiting excited-state reduction by DMA.

Mixed-metal Ru–Pd and Ru–Pt systems have been designed that reduce water to hydrogen. The bimetallic complexes studied to date appear to undergo colloid Pd(s) or Pt(s) formation. The colloids formed under these conditions may have enhanced photocatalytic activity relative to typical metal colloids. Recently, our group has explored tetratmetallic complexes with Pt reactive metals that undergo PEC. These systems catalyze the reduction of water to hydrogen with high turnovers. These new systems are not greatly impacted by the addition of mercury, suggestive that a colloidal pathway is not operative. The tetrametallic systems have pathways to store multiple reducing equivalents not available in the bimetallic complexes reported to date.

The complexity of the PEC and the water-splitting processes requires significant basic science advancement to adequately address these timely issues. The development of molecular devices for PEC have concentrated on Ru and Os polyazine light absorbers. New developments point to electronic isolation of multiple light absorbers being essential for functioning PEC devices. The lack of a large array of functioning PEC devices makes application to reduction of water more challenging. The coupling of reactive metals has recently led to functioning photocatalysts for water reduction. Detailed studies of the photochemistry and photophysics of these systems will lead to a development of the knowledge base in multielectron photochemistry. It is this knowledge base that is essential to the molecular-based harvesting of the vast energy stored and delivered to our planet by the sun.

List of Abbreviations

bpy	2,2′-bipyridine
dpp	2,3-bis(2-pyridyl)pyrazine
EA	electron acceptor
ED	electron donor
en	excited state energy transfer
et	excited state electron transfer
GS	ground state light absorber
HOMO	highest occupied molecular orbital
ic	internal conversion
IL	internal ligand
isc	intersystem crossing
k_x	rate constant of process "x"
LA	ground state light absorber
*LA	excited state light absorber
LUMO	lowest unoccupied molecular orbital
Me_2bpy	4,4′-dimethy-2,2′-bipyridine
MLCT	metal-to-ligand charge-transfer
MMCT	metal-to-metal charge-transfer
nr	non-radiative decay
Ph_2phen	4,7-diphenyl-1,10-phenanthroline
phen	1,10-phenanthroline
q	bimolecular deactivation
Q	quencher
rxn	photochemical reaction
tpy	2,2′:6′,2″-terpyridine
Φ^{em}	quantum yield of emission
TL	terminal ligand
BL	bridging ligand

Acknowledgments

Acknowledgment is made of all the students and research scientists who have worked in this area in the Brewer Group. Special thanks to Ms. Kacey McCreary, Ms. Jessica Knoll and Mr. Travis White for their help with this manuscript. Acknowledgment is made to the Chemical Sciences, Geosciences and Biosciences Division, Office of Basic Energy Sciences, Office of Sciences, US Department of Energy for their generous support of our research. Acknowledgment is made to the financial collaboration of Phoenix Canada Oil Company which holds long-term license rights to commercialize our Rh-based technology. Acknowledgment is also made to H Gencorp Inc. for their generous support of our research.

References

[1] Bard, A.J. and Fox, M.A. (1995) Artificial photosynthesis: solar splitting of water to hydrogen and oxygen. *Accounts of Chemical Research*, **28**, 141.

[2] Nocera, D.G., (2009) Living healthy on a dying planet. *Chemical Society Reviews*, **38**, 13.

[3] Dempsey, J.L., Esswein, A.J., Manke, D.R. *et al.* (2005) Molecular chemistry of consequence to renewable energy. *Inorganic Chemistry*, **44**, 6879.

[4] Rosenthal, J., Bachman, J., Dempsey, J.L. *et al.* (2005) Oxygen and hydrogen photocatalysis by two-electron mixed-valence coordination compounds. *Coordination Chemistry Reviews*, **249**, 1316.

[5] Lubitz, W., Reijerse, E.J. and Messinger, J. (2008) Solar water-splitting into H_2 and O_2: design principles of photosystem II and hydrogenases. *Energy and Environmental Science*, **1**, 15.

[6] Barbaro, P.(ed.) (2009) *Catalysis for Sustainable Energy Production* (ed. C. Bianchini), Wiley-VCH.

[7] Wicks, G. (ed.) (2009) *Materials Innovations in an Emerging Hydrogen Economy: Ceramic Transactions*, vol. **202**, Wiley-VCH.

[8] Lehn, J.-M. and Sauvage, J.-P., (1977) Chemical storage of light energy. catalytic generation of hydrogen by visible light or sunlight. Irradiation of neutral aqueous solutions. *Nouveau Journal de Chimie*, **1**, 449.

[9] Kirch, M., Lehn, J.-M. and Sauvage, J.-P. (1979) Hydrogen generation by visible light irradiation of aqueous solutions of metal complexes. An approach to the photochemical conversion and storage of solar energy. *Helvetica Chimica Acta*, **62**, 1345.

[10] Kamat, P.V. (2007) Meeting the clean energy demand: nanostructure architectures for solar energy conversion. *Journal of Physical Chemistry C*, **111**, 2834.

[11] Balzani, V., Moggi, L. and Scandola, F. (1987) *Supramolecular Photochemistry* (ed. V. Balzani), D. Reidel, Dordrecht, p. 1.

[12] Lewis, N.S. and Nocera, D.G. (2006) Powering the planet: chemical challenges in solar energy utilization. *Proceedings of the National Academy of Sciences, USA*, **103**, 15729.

[13] Balzani, V., Juris, A., Venturi, M. *et al.* (1996) Luminescent and redox-active polynuclear transition metal complexes. *Chemical Reviews*, **96**, 759.

[14] Elvington, M., Brown, J.R., Zigler, D.F. and Brewer, K.J. (2006) Supramolecular complexes as photoinitiated electron collectors: applications in solar hydrogen production. *Proceedings of SPIE*, **6340**, 63400W–1.

[15] Gafney, H.D. and Adamson, A.W. (1972) Excited state $Ru(bipyr)_3{}^{2+}$ as an electron-transfer reductant. *Journal of the American Chemical Society*, **94**, 8238.

[16] Bock, C.R., Meyer, T.J. and Whitten, D.G. (1974) Electron transfer quenching of the luminescent excited state of Tris(2,2′-bipyridine)ruthenium(II). Flash photolysis relaxation technique for measuring the rates of very rapid electron transfer reactions. *Journal of the American Chemical Society*, **96**, 4710.

[17] Bock, C.R., Connor, J.A., Gutierrez, A.R. *et al.* (1979) Estimation of excited-state redox potentials by electron-transfer quenching. Application of electron-transfer theory to excited-state redox processes. *Journal of the American Chemical Society*, **101**, 4815.

[18] Demas, J.N. and Adamson, A.W. (1971) A new photosensitizer. Tris(2,2′-bipyridine)ruthenium(II) chloride. *Journal of the American Chemical Society*, **93**, 1800.

[19] Balzani, V., Moggi, L., Manfrin, M.F. *et al.* (1975) Quenching and sensitization processes of coordination compounds. *Coordination Chemistry Reviews*, **15**, 321.

[20] Kalyanasundaram, K. (1982) Photophysics, photochemistry and solar energy conversion with Tris(bipyridyl) ruthenium(II) and its analogues. *Coordination Chemistry Reviews*, **46**, 159.

[21] Durham, B., Caspar, J.V., Nagle, J.K. and Meyer, T.J. (1982) Photochemistry of $Ru(bpy)_3{}^{2+}$. *Journal of the American Chemical Society*, **104**, 4803.

[22] Juris, A., Balzani, V., Barigelletti, F. *et al.* (1988) Ru(II)-Polypyridine complexes: Photophysics, photochemistry, eletrochemistry, and chemiluminescence. *Coordination Chemistry Reviews*, **84**, 85.

[23] Anderson, P.A., Strouse, G.F., Treadway, J.A. *et al.* (1994) Black MLCT absorbers. *Inorganic Chemistry*, **33**, 3863.

[24] Anderson, C.P., Salmon, D.J., Meyer, T.J. and Young, R.C. (1977) Photochemical generation of $Ru(bpy)_3{}^{+}$ and $O_2{}^{-}$. *Journal of the American Chemical Society*, **99**, 1980.

[25] Kawanishi, Y., Kitamura, N. and Tazuke, S. (1989) Dependence of spectroscopic, electrochemical, and excited-state properties of tris chelate ruthenium(II) complexes on ligand structure. *Inorganic Chemistry*, **28**, 2968.

[26] Lin, C.-T., Böttcher, W., Chou, M. *et al.* (1976) Mechanism of the quenching of the emission of substituted Polypyridineruthenium(II) complexes by Iron(III), Chromium(III), and Europium(III) ions. *Journal of the American Chemical Society*, **98**, 6536.

[27] Sutin, N. and Creutz, C. (1978) Properties and reactivities of the luminescent excited states of polypyridine complexes of Ruthenium(II) and Osmium(II). *Advances in Chemistry Series*, **168**, 1.

[28] Kober, E.M., Sullivan, B.P., Dressick, W.J. *et al.* (1980) Highly luminescent polypyridyl complexes of Osmium (II). *Journal of the American Chemical Society*, **102**, 7383.

[29] Braunstein, C.H., Baker, A.D., Strekas, T.C. and Gafney, H.D., (1984) Spectroscopic and electrochemical properties of the dimer tetrakis (2,2′- bipyridine)(μ-2,3- bis (2-pyridyl) pyrazine)diruthenium(II) and its monomeric analogue. *Inorganic Chemistry*, **23**, 857.

[30] Fuchs, Y., Lofters, S., Dieter, T., Shi, W., Morgan, R., Strekas, T.C., Gafney, H.D. and Baker, A.D. (1987) Spectroscopic and electrochemical properties of dimeric Ruthenium(II) diimine complexes and determination of their excited state redox properties. *Journal of the American Chemical Society*, **109**, 2691.

[31] Denti, G., Campagna, S., Sabatino, L. *et al.* (1990) Luminescent and redox-reactive building blocks for the design of photochemical molecular devices: Mono-, Di-, Tri-, and tetranuclear Ruthenium(II) polypyridine complexes. *Inorganic Chemistry*, **29**, 4750.

[32] Wallace, A.W., Murphy, W.R. and Petersen, J.D. (1989) Electrochemical and photophysical properties of mono- and bimetallic Ruthenium(II) complexes. *Inorganica Chimica Acta*, **166**, 47.

[33] Richter, M.M. and Brewer, K.J. (1991) Synthesis and characterization of Osmium(II) complexes incorporating polypyridyl bridging ligands. *Inorganica Chimica Acta*, **180**, 125.

[34] Abdel-Shafi, A.A., Worrall, D.R. and Ershov, A.Y. (2004) Photosensitized generation of singlet oxygen from Ruthenium(II) and Osmium(II) bipyridyl complexes. *Journal of the Chemical Society, Dalton Transactions*, 30.

[35] Winkler, J.R., Netzel, T.L., Creutz, C. and Sutin, N., (1987) Direct observation of metal-to-ligand charge-transfer (MLCT) excited states of Pentaammineruthenium(II) complexes. *Journal of the American Chemical Society*, **109**, 2381.

[36] Beley, M., Collin, J.-P., Sauvage, J.-P., Sugihara, H., Heisel, F. and Miehé, A. (1991) Photophysical and photochemical properties of ruthenium and osmium complexes with substituted terpyridines. *Journal of the Chemical Society, Dalton Transactions*, 3157.

[37] Rillema, D.P. and Mack, K.B. (1982) The low-lying excited state in ligand π-acceptor complexes of Ruthenium (II): mononuclear and binuclear species. *Inorganic Chemistry*, **21**, 3849.

[38] Rillema, D.P., Taghdiri, D.G., Jones, D.S. *et al.* (1987) Structure and redox and photophysical properties of a series of ruthenium heterocycles based on the ligand 2,3-bis(2-pyridyl)quinoxaline. *Inorganic Chemistry*, **26**, 578.

[39] Berger, R.M. (1990) Excited-state absorption spectroscopy and spectroelectrochemistry of Tetrakis(2,2′-bipyridine)(μ-2,3-bis(2-pyridyl)pyrazine)diruthenium(II) and its mononuclear counterpart: a comparative study. *Inorganic Chemistry*, **29**, 1920.

[40] Caspar, J.V., Kober, E.M., Sullivan, B.P. and Meyer, T.J. (1982) Application of the energy gap law to the decay of charge-transfer excited states. *Journal of the American Chemical Society*, **104**, 630.

[41] Brown, G.M., Chan, S.-F., Creutz, C. *et al.* (1979) Mechanism of the formation of dihydrogen from the photoinduced reactions of Tris(bipyridine)ruthenium(II) with Tris(bipyridine)rhodium(III). *Journal of the American Chemical Society*, **101**, 7638.

[42] Creutz, C., Keller, A.D., Sutin, N. and Zipp, A.P. (1982) Poly(pyridine)ruthenium(II)-Photoinduced redox reactions of bipyridinium cations, Poly(pyridine)rhodium complexes, and osmium ammines. *Journal of the American Chemical Society*, **104**, 3618.

[43] Arachchige, S.M. and Brewer, K.J. (2009) Mixed-metal supramolecular complexes coupling polyazine light absorbers and reactive metal centers, in *Macromolecules Containing Metal and Metal Like Elements: Supramolecular Structures*, vol. **9**, Wiley and Sons, in press.

[44] Bridgewater, J.S., Vogler, L.M., Molnar, S.M. and Brewer, K.J. (1993) Tuning the spectroscopic and electrochemical properties of polypyridyl bridged mixed-metal trimetallic Ruthenium(II), Iridium(III) complexes: a spectroelectrochemical study. *Inorganica Chimica Acta*, **208**, 179.

[45] Murphy, W.R., Brewer, K.J., Gettliffe, G. and Petersen, J.D. (1989) Luminescent tetrametallic complexes of ruthenium. *Inorganic Chemistry*, **28**, 81.

[46] Kalyanasundaram, K. and Nazeeruddin, M.K. (1990) Photophysics and photoredox reactions of ligand-bridged binuclear polypyridyl complexes of Ruthenium(II) and of their monomeric analogs. *Inorganic Chemistry*, **29**, 1888.

[47] Molnar, S.M., Nallas, G., Bridgewater, J.S. and Brewer, K.J. (1994) Photoinitiated electron collection in a mixed-metal trimetallic complex of the form $[\{(bpy)_2Ru(dpb\}_2IrCl_2](PF_6)_5$ (bpy = 2,2′-Bipyridine and dpb = 2,3-Bis(2-pyridyl)benzoquinoxaline). *Journal of the American Chemical Society*, **116**, 5206.

[48] Kim, M.-J., Konduri, R., Ye, H. *et al.* (2002) Dinuclear Ruthenium(II) polypyridyl complexes containing large, redox-active, aromatic bridging ligands: synthesis, characterization, and intramolecular quenching of MLCT excited states. *Inorganic Chemistry*, **41**, 2471.

[49] Konduri, R., Ye, H., MacDonnell, F.M. *et al.* (2002) Ruthenium photocatalysts capable of reversibly storing up to four electrons in a single acceptor ligand: A step closer to artificial photosynthesis. *Angewandte Chemie International Edition*, **41**, 3185.

[50] Konduri, R., de Tacconi, N.R., Rajeshwar, K. and MacDonnell, F.M. (2004) Multielectron photoreduction of a bridged ruthenium dimer, $[(phen)_2Ru(tatpp)Ru(phen)_2][PF_6]_4$: Aqueous reactivity and chemical and spectroelectrochemical identification of the photoproducts. *Journal of the American Chemical Society*, **126**, 11621.

[51] de Tacconi, N.R., Lezna, R.O., Chitakunye, R. and MacDonnell, F.M. (2008) Electroreduction of the ruthenium complex $[(bpy)_2Ru(tatpp)]Cl_2$ in water: insights on the mechanism of multielectron reduction and protonation of the tatpp acceptor ligand as a function of pH. *Inorganic Chemistry*, **47**, 8847.

[52] Zhou, M., Pfennig, B.W., Steiger, J. *et al.* (1990) Multielectron transfer and single-crystal X-Ray structure of a trinuclear cyanide-bridged platinum-iron species. *Inorganic Chemistry*, **29**, 2456.

[53] Chang, C.C., Pfennig, B. and Bocarsly, A.B. (2000) Photoinduced multielectron charge transfer processes in group 8-platinum cyanobridged supramolecular complexes. *Coordination Chemistry Reviews*, **208**, 33.

[54] Watson, D.F., Wilson, J.L. and Bocarsly, A.B. (2002) Photochemical image generation in a cyanogel system synthesized from Tetrachloropalladate(II) and the trimetallic mixed-valence complex $[(NC)_5Fe^{II}\text{-CN-}Pt^{IV}(NH_3)_4\text{-NC-}Fe^{II}(CN)_5]^{4-}$: Consideration of photochemical and dark mechanistic pathways of prussian blue formation. *Inorganic Chemistry*, **41**, 2408.

[55] Heyduk, A.F., Macintosh, A.M. and Nocera, D.G. (1999) Four-electron photochemistry of dirhodium fluorophosphine compounds. *Journal of the American Chemical Society*, **121**, 5023.

[56] Esswein, A.J. and Nocera, D.G. (2007) Hydrogen production by molecular photocatalysis. *Chemical Reviews*, **107**, 4022.

[57] (a) Heyduk, A.F. and Nocera, D.G., (2001) Hydrogen produced from hydrohalic acid solutions by a two-electron mixed-valence photocatalyst. *Science*, **293**, 1639; (b) Esswein, A.J., Veige, A.S. and Nocera, D.G. (2005) A photocycle for hydrogen production from two-electron mixed-valence complexes. *Journal of the American Chemical Society*, **127**, 16641.

[58] (a) Esswein, A.J., Dempsey, J.L. and Nocera, D.G., (2007) A Rh^{II}-Au^{II} bimetallic core with a direct metal-metal bond. *Inorganic Chemistry*, **46**, 2362; (b) Cook, T.R., Esswein, A.J. and Nocera, D.G. (2007) Metal-halide bond photoactivation from a Pt^{III}-Au^{II} complex. *Journal of the American Chemical Society*, **129**, 10094.

[59] Elvington, M. and Brewer, K.J. (2006) Photoinitiated electron collection at a metal in a rhodium-centered mixed-metal supramolecular complex. *Inorganic Chemistry*, **45**, 5242.

[60] Elvington, M., Brown, J., Arachchige, S.M. and Brewer, K.J. (2007) Photocatalytic hydrogen production from water employing A Ru, Rh, Ru molecular device for photoinitiated electron collection. *Journal of the American Chemical Society*, **129**, 10644.

[61] Molnar, S.M., Jensen, G.E., Vogler, L.M. *et al.* (1994) Photochemical properties of mixed-metal supramolecular complexes. *Journal of Photochemistry and Photobiology A, Chemistry*, **80**, 315.

[62] Zigler, D.F., Mongelli, M.T., Jeletic, M. and Brewer, K.J. (2007) A trimetallic supramolecular complex of Osmium(II) and Rhodium(III) displaying MLCT transitions in the near-IR. *Inorganic Chemistry Communications*, **10**, 295.

[63] Arachchige, S.M., Brown, J. and Brewer, K.J. (2008) Photochemical hydrogen production from water using the new photocatalyst $[\{(bpy)_2Ru(dpp)\}_2RhBr_2](PF_6)_5$. *Journal of Photochemistry and Photobiology A, Chemistry*, **197**, 13.

[64] Swavey, S. and Brewer, K.J. (2002) Synthesis and study of Ru, Rh, Ru triads: modulation of orbital energies in a supramolecular architecture. *Inorganic Chemistry*, **41**, 4044.

[65] Holder, A.A., Swavey, S. and Brewer, K.J. (2004) Design aspects for the development of mixed-metal supramolecular complexes capable of visible light induced photocleavage of DNA. *Inorganic Chemistry*, **43**, 303.

[66] Arachchige, S.M., Brown, J.R., Chang, E. *et al.* (2009) Design considerations for a system for photocatalytic hydrogen production from water employing mixed-metal photochemical molecular devices for photoinitiated electron collection. *Inorganic Chemistry*, **48**, 1989.

[67] Kew, G., DeArmond, K. and Hanck, K. (1974) Electrochemistry of rhodium-dipyridyl complexes. *Journal of Physical Chemistry*, **78**, 727.

[68] Anton, D.R. and Crabtree, R.H., (1983) Dibenzo[a,e]cyclooctatetraene in a proposed test for heterogeneity in catalysts formed from soluble platinum-group metal complexes. *Organometallics*, **2**, 855.

[69] Baba, R., Nakabayashi, S., Fujishima, A. and Honda, K. (1985) Investigation of the mechanism of hydrogen evolution during photocatalytic water decomposition on metal-loaded semiconductor powders. *Journal of Physical Chemistry*, **89**, 1902.

[70] Lide, D.R. (ed.) (2008) *CRC Handbook of Chemistry and Physics,* 88th Edition (Internet Version 2008), CRC Press/Taylor and Francis, Boca Raton, FL.

[71] Rangan, K., Arachchige, S.M., Brown, J.R. and Brewer, K.J. (2009) Solar energy conversion using photochemical molecular devices: photocatalytic hydrogen production from water using mixed-metal supramolecular complexes. *Journal of Energy and Environmental Science*, **2**, 410.

[72] Brown, J.R., Elvington, M., Mongelli, M.T. *et al.* (2006) Analytical methods development for supramolecular design in solar hydrogen production. *Proceedings of SPIE, Solar Hydrogen and Nanotechnology*, **6340**, 634007W1.

[73] Ozawa, H., Haga, M. and Sakai, K., (2006) A photo-hydrogen-evolving molecular device driving visible-light-induced EDTA-reduction of water into molecular hydrogen. *Journal of the American Chemical Society*, **128**, 4926.

[74] Ozawa, H., Yokoyama, Y., Haga, M. and Sakai, K. (2007) Syntheses, characterization, and photo-hydrogen-evolving properties of Tris(2,2-bipyridine)ruthenium(II) derivatives tethered to a Cis-Pt(II)Cl_2 unit: insights into the structure-activity relationship. *Journal of the Chemical Society, Dalton Transactions*, 1197.

[75] Rau, S., Schäfer, B., Gleich, D. *et al.* (2006) A supramolecular photocatalyst for the production of hydrogen and the selective hydrogenation of tolane. *Angewandte Chemie International Edition*, **45**, 6215.

[76] Du, P., Schneider, J., Li, F. *et al.* (2008) Bi- and terpyridyl Platinum(II) chloro complexes: molecular catalysts for the photogeneration of hydrogen from water or simply precursors for colloidal platinum? *Journal of the American Chemical Society*, **130**, 5056.

[77] Lei, P., Hedlund, M., Lomoth, R. *et al.* (2008) The role of colloid formation in the photoinduced H_2 production with a Ru^{II}-Pd^{II} supramolecular complex: A study by GC, XPS, and TEM. *Journal of the American Chemical Society*, **130**, 26.

Part Five

New Devices for Solar Thermal Hydrogen Generation

20

Novel Monolithic Reactors for Solar Thermochemical Water Splitting

Athanasios G. Konstandopoulos[1,2] and Souzana Lorentzou[1,2]
[1]*Aerosol and Particle Technology Laboratory, CERTH/CPERI, PO Box 60361, 6th km Charilaou-Thermi Rd, 57001, Thessaloniki, Greece, Email: agk@cperi.certh.gr*
[2]*Department of Chemical Engineering, Aristotle University, PO. Box 1517, 54006, Thessaloniki, Greece*

20.1 Introduction

20.1.1 Energy Production and Nanotechnology

Energy production was acknowledged [1] "...as the single most important challenge facing humanity today...," by the late Nobel Laureate in Chemistry Richard Smalley. Professor Smalley believed that nanotechnology would provide the breakthrough solutions that are needed to surpass the limits of conventional materials. In his testimony to the Senate Committee on Energy and Natural Resources [1] on April 2004, Professor Smalley stated that there is a need for finding "...'the New Oil', the new technology that provides massive clean, low cost energy necessary for advanced civilization...." He coined the term "Terawatt Challenge" [2], to describe the problem of providing adequate, carbon-free energy, of the level of 10–30 TW by 2050, to the growing number of people on our finite planet to the same level as that consumed in the developed world, in ways that are less harmful to the environment and to human health, preferably avoiding potential geopolitical conflicts and price volatility.

Therefore, a high-priority goal worldwide in the immediate future is to increase renewable energy penetration. However, not all renewable sources have the potential of producing the necessary energy. As presented in [3], exploiting biomass would block the entire nonpopulated land of the Earth. Exploiting wind power would require covering all of the available land (where

On Solar Hydrogen & Nanotechnology Edited by Lionel Vayssieres

wind speeds of 5.1 m s^{-1} at 10 m above ground occur) with windmills. The use of hydroelectric power plants for the production of the necessary energy would require the damming of all available rivers [3]. Geothermal energy could be exploited for power generation, but there are great difficulties in accessing it at the increased depths under the Earth's surface. Producing the necessary amount of energy by nuclear fission would mean the construction of 1000 new nuclear power plants of 1 GW_e for every 1 TW of energy required [3].

One virtually inexhaustible renewable energy source with an immense potential is solar power. The amount of power that falls on Earth from the sun every day is $\sim 1.7 \times 10^5$ TW. The exploitable quantity of solar power that the earth receives amounts to 600–1000 TW, while only a minute area of land needs to be covered with almost 10% efficient solar conversion systems to produce 20 TW of power [3].

20.1.2 *Application of Solar Technologies*

Solar power can be exploited in several ways, such as in architecture and in solar-thermal applications (water heating, heating, cooling and ventilation, water treatment, process heat and electrical generation, solar-chemical processes etc.).

The enhancement of the penetration of solar technologies will need advances in material technologies and nanotechnology holds a large promise in this regard: the unique properties that nanostructured materials have, may enable breakthrough solutions in several areas that are key for the development of solar technologies, such as photovoltaics, fuel cells, batteries and supercapacitors, high-current cables and quantum conductors, materials with increased tolerance to harsh conditions (e.g., extremely high temperatures) or antiwear properties (e.g., advanced heliostats with enhanced optical properties, decreased size and resistance to atmospheric conditions, wind, dust etc.).

20.2 Solar Hydrogen Production

However, there is also another challenge. How to efficiently exploit solar energy when it is not available everywhere and at anytime? The solution is to "store" this energy and a "storage" medium that has been heralded as the "energy carrier" of the future is hydrogen. Utilization of solar energy for the production of hydrogen would be a means of "storing" this energy to a more manageable and available form.

Solar energy can be employed for the renewable production of hydrogen via several solar-to-hydrogen paths, such as photoelectrochemical, photocatalytic, photobiological and thermochemical processes and so on. In Figure 20.1, adapted from [4], some sustainable solar-to-hydrogen paths are depicted.

A particularly attractive technology for the supply of electricity for electrolytic hydrogen production or the necessary heat for thermochemically produced hydrogen is that of concentrated solar power (CSP) systems. In these systems, special mirror assemblies (such as parabolic troughs, heliostats or parabolic dishes) provide concentrated solar radiation that can heat a fluid medium, generate steam that drives a steam turbine and produce renewable electricity, for example, for the electrolysis process.

Among CSP systems, solar tower systems (which are based on concentrating solar radiation by an array of heliostats on a heat-absorbing body, termed a "volumetric receiver" located at the top of a solar tower) can provide the necessary heat to reactors for the thermochemical reactions. In both cases, renewable hydrogen is produced.

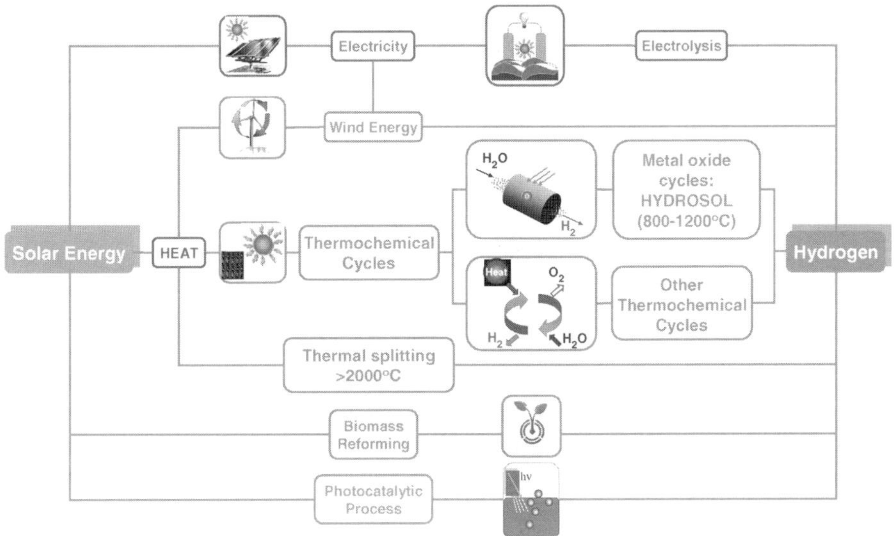

Figure 20.1 Schematic diagram of the solar-to-hydrogen production paths. (Figure adapted from J.A. Turner, M.C. Williams and K. Rajeshwar, Hydrogen economy based on renewable energy resources, *The Electrochemical Society Interface*, **13**(3), 24–31, 2004, The Electrochemical Society.)

20.2.1 Solar Hydrogen Production: Thermochemical Processes

Hydrogen from solar water splitting is a long-sought target of renewable hydrogen production. Direct water thermolysis, into its constituents, hydrogen and oxygen, requires temperatures higher than 2000 °C. These temperatures can be reached with the aid of concentrated solar radiation, however maintaining the durability of reactor materials at these temperatures is very challenging. To lower the water dissociation temperature, multistep thermochemical cycles have been suggested to enable the splitting of the water molecule at more "material-friendly" temperatures. These multistep schemes also allow the generation of hydrogen without the need for separation from oxygen, since both gases are produced at different steps.

There have been several significant works that have reviewed thermochemical cycles for the production of hydrogen [5–8]. One of the earliest investigations for multistep hydrogen production from water splitting was by Funk and Reinstorm in the 1960s [9]. These authors evaluated the energy requirements and the possibility of employing two-step processes for water dissociation and hydrogen production by oxides and hydrides.

Multistep cycles reviewed in [10] have been investigated by several researchers during the last 40 years. Carty *et al.* [11] and McQuillan *et al.* [12] examined over 200 multistep thermochemical cycles for hydrogen production. The most promising reactions involved cycles based on metal sulfates, iodine, chlorine and bromide cycles, as well as metal oxides. Most of the cycles have either increased complexity or involve management of highly toxic and corrosive reactants. An attractive way is the two-step solar thermochemical water-splitting cycle, with the aid of redox-pair metal oxides, since it involves less complex chemical steps and noncorrosive materials.

The concept behind the redox-pair metal-oxide systems is the exploitation of metals with multiple oxidation states for the removal of oxygen from water. A metal or metal oxide (with the metal existing in its reduced state) is oxidized by removing oxygen from water (water-splitting step) and therefore releasing hydrogen (Equation 20.1). This step is exothermic and can take place at temperatures as low as 600–700 °C. In the second step, the oxidized material has to be reduced (regenerated) in order to reach its initial state (regeneration step, Equation 20.2). This step is endothermic and in order to be achieved it needs addition of external thermal energy to reach the much higher temperatures than the water-splitting step that are required.

$$MO_{reduced} + H_2O(g) \rightarrow MO_{oxidised} + H_2 \tag{20.1}$$

$$MO_{oxidised} \rightarrow MO_{reduced} + {}^1\!/_2\, O_2 \tag{20.2}$$

The first investigations of two-step thermochemical cycles involving redox-pair metal oxides were conducted by Nakamura [13]. His research was based on the employment of iron-oxide redox pairs (Fe_3O_4/FeO), the so-called "ferrite process." Several groups investigated other redox pairs that involved oxides of Mn, Zn, Sn, Co, Nb, In and so on. The research on these redox pairs revealed that some might have good hydrogen yields at relatively low water-splitting temperatures, but need significantly high regeneration temperatures and vice versa. These facts lead to the concept of modifying the redox materials, for example, in the case of iron oxides by partial substitution of iron by other metals, such as Mn, Ni, Co, Mg or Zn, and formation of mixed iron–metal oxides [5,14,15].

The aforementioned research has been conducted at the laboratory scale usually with the use of reactors inside electrically heated or infrared kilns [5,16,17], or in solar-heated systems [18]. The redox materials would be in the form of powders, pellets or porous substrates [5,18,19] coated with the redox materials inside quartz fixed-bed reactors.

20.2.2 Solar Chemical Reactors

Examples of early investigations of the possibilities of the hydrogen production via thermochemical cycles with the aid of solar energy, include the work of Bilgen *et al.* [20], Sibieude *et al.* [21] and Tofighi [22]. Later studies on the potential use of solar thermochemical reactors containing redox materials are presented in [5–8,23].

One of the types of solar chemical reactors investigated was the "rotating-cavity" solar reactor that was applied for the thermal dissociation of ZnO [24], while configurations such as the "two-cavity" and the "vortex" solar reactors were employed for carbothermal processes [6]. The reactor that produced for the first time [19] solar hydrogen from the dissociation of water vapors via thermochemical cycles was the HYDROSOL reactor, as described in more detail later.

Researchers at Sandia National Laboratories [25] are currently developing a "rotating"-type reactor, the CR5, which is based on the concept of a receiver/reactor/recuperator with counter-rotating rings or disks that periodically expose the redox material to solar radiation and water vapor in order to drive a cyclic process. The redox material is specially molded into rod-like stacked structures that are carried upon the rotating rings [25]. Another "rotary"-type system for two-step thermochemical water splitting, with separate reaction chambers for the water splitting and the thermal reduction was developed at the Tokyo Institute of Technology [5,26]. It consists of a cylindrical rotor upon which foam structures coated with the redox material are fixed.

Other configurations that were considered for possible use in the production of hydrogen involve direct solar irradiation of particles, usually metal powders, either as falling particles or in the form of a suspension in a gaseous stream (e.g., particle "cloud" and fluidized bed-type solar reactors) and were reviewed in [5].

The HYDROSOL reactor was developed during the homonymous European projects "HYDROSOL" and "HYDROSOL-II" [27]. This reactor was conceived by the senior author of this chapter who serves as the coordinator of the HYDROSOL projects. The concept was based on our past experience at the Aerosol and Particle Technology Laboratory with automobile catalytic converters that employ a honeycomb porous structure coated with an active material for catalytic applications and on volumetric receivers for concentrated solar radiation [28,29]. The idea was put into practice during the HYDROSOL project with the cooperation of experts in concentrating solar technologies (Deutsches Zentrum für Luft und Raumfahrt, DLR), automotive catalysis (Johnson Matthey, JM) and ceramic manufacturing (Stobbe Tech Ceramics, STC), while in "HYDROSOL-II," the research group was enlarged with the contribution of the Centro de Investigationes Energéticas, Medioambientales y Tecnológicas (CIEMAT) with its solar-tower facilities in Almeria, Spain, where the work continues at the pilot scale.

20.3 HYDROSOL Reactor

20.3.1 The Idea

The reactor is based on two concepts: the synthesis of active iron-oxide-based redox pairs and their incorporation into multichanneled ceramic monolithic honeycomb structures capable for achieving and sustaining high temperatures [19].

The "HYDROSOL" reactor contains no moving parts and is constructed from special refractory (silicon-carbide-based) ceramic thin-wall, multichanneled (honeycomb) monoliths, optimized to absorb solar radiation and develop the required high temperatures. The monolith channels are coated with an active water-splitting material and the overall reactor looks very similar to the familiar catalytic converter of modern automobiles. When steam passes through the solar reactor, the coating material splits water vapor by "trapping" its oxygen and leaving in the effluent gas stream, pure hydrogen.

In a subsequent step the oxygen "trapping" coating is regenerated by increasing the amount of solar heat absorbed by the reactor and hence a cyclic operation is established in a single solar receiver-reactor (Figure 20.2). The proof of concept was demonstrated on the solar receiver-reactor, at the solar facilities in DLR, where quasi-continuous solar-operated water-splitting–regeneration cycles were achieved, producing the first ever solar hydrogen with monolithic honeycomb reactors [19,30–33].

20.3.2 Redox Materials

The redox water-splitting materials applied in the HYDROSOL project were based on doped/mixed iron oxides ($M_xFe_yO_z$) and were synthesized via different synthesis routes, that involved conventional solid synthesis techniques (solid-state synthesis, SSS), combustion methods in the liquid and the solid phase (liquid and solid phase self-propagating high temperature

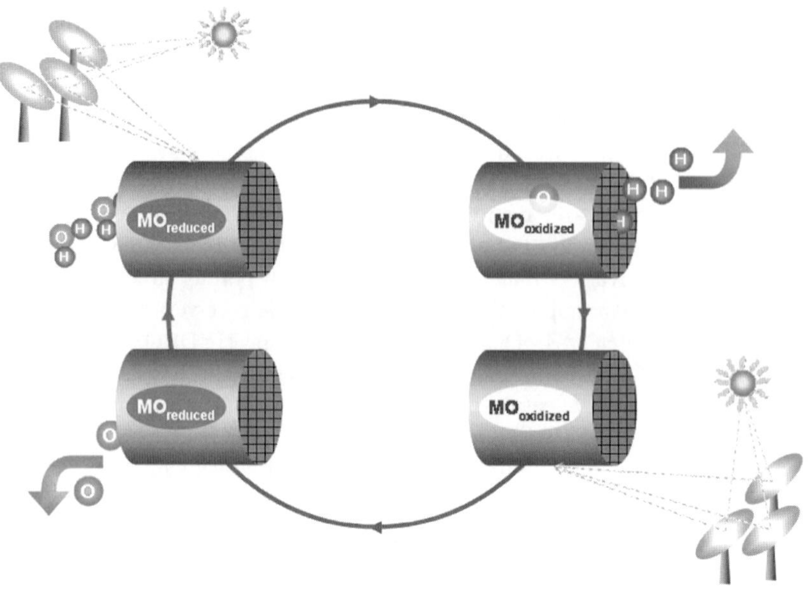

Figure 20.2 Schematic of the two-stage, water-splitting/hydrogen production (top) and regeneration/oxygen release (bottom) HYDROSOL process.

synthesis, LPSHS and SPSHS) and an aerosol synthesis technique (aerosol spray pyrolysis, ASP).

Among all synthesis methods, SSS resulted in highly crystallized phases (sharper peaks in the X-ray diffraction diagram of Figure 20.3) while ASP produced materials with the smallest

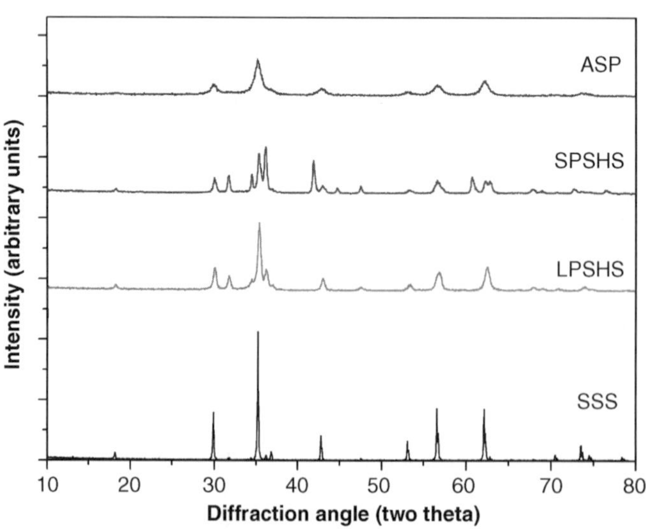

Figure 20.3 XRD patterns of the SSS-, SHS- and ASP-synthesized redox materials.

Table 20.1 Surface area and crystallite size of materials synthesized with different methods.

Synthesis route	Specific Surface Area ($m^2 g^{-1}$)	Crystallite size (nm)
SSS	1–2	1421
SPSHS	4–5	205
LPSHS	20–40	194
ASP	80–120	91

crystallite size (Table 20.1). As can be seen in Figure 20.3, all materials synthesized by SSS, LPSHS and ASP were crystallized in the spinel phase (MFe_2O_4), but the product of the SPSHS method was crystallized in the (more oxygen deficient) wustite phase (M,Fe)O. The highest-surface-area materials were produced by the ASP method, while the lowest-surface-area materials were produced by the SSS method (Table 20.1).

Water-splitting activity depends on oxygen deficiency, crystallite size and surface area, as presented in [18,32].

Typical SEM photographs of the different morphologies of powders synthesized by the four methods are shown in Figure 20.4.

20.3.3 Water Splitting: Laboratory Tests

The first screening of the redox materials with respect to their water-splitting performance was conducted in a laboratory unit consisting of a quartz reactor placed in a temperature-programmable furnace [17,18].

Figure 20.4 (a) SSS-, (b) SPSHS-, (c) LPSHS- and (d) ASP-synthesized materials.

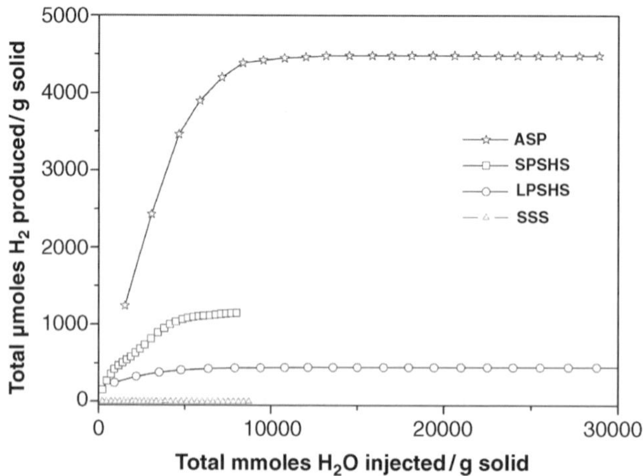

Figure 20.5 Performance of redox materials synthesized by four different methods.

Figure 20.5 illustrates an example of the hydrogen yield evaluation during water-splitting of redox materials with the same chemical composition synthesized via all four methods [33]. As it can be observed, the materials synthesized via ASP have the highest water-splitting activity followed by the materials synthesized by SPSHS. Materials synthesized with these techniques were subsequently deposited on monolithic honeycomb substrates in order to be evaluated on the HYDROSOL solar reactor. Production of SPSHS-based materials is currently possible at our laboratory at a scale of a few kg per day while production of ASP-based nanopowders is being scaled up [34] from the g per day figures to 100–300 g per day.

20.3.4 HYDROSOL Reactors

Several series of small-scale (Ø25 × 50 mm) and large-scale (Ø144 × 50 mm, Ø144 × 200 mm) SiC-based monoliths (Figure 20.6) were coated with redox materials and evaluated with respect

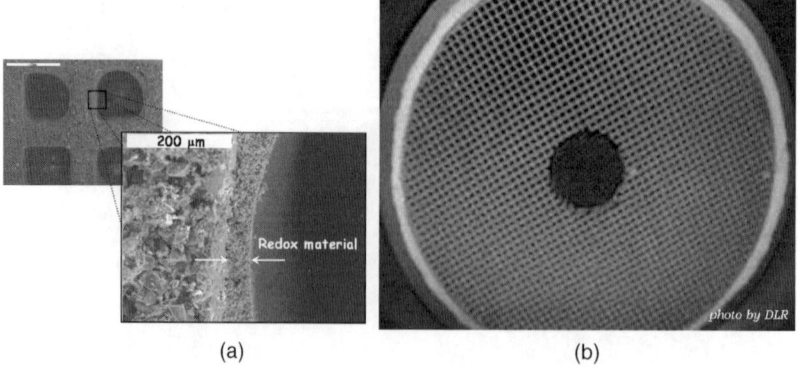

Figure 20.6 SiC-based monolith coated with redox material illustrated (a) with a scanning electron microscope and (b) inside the reactor.

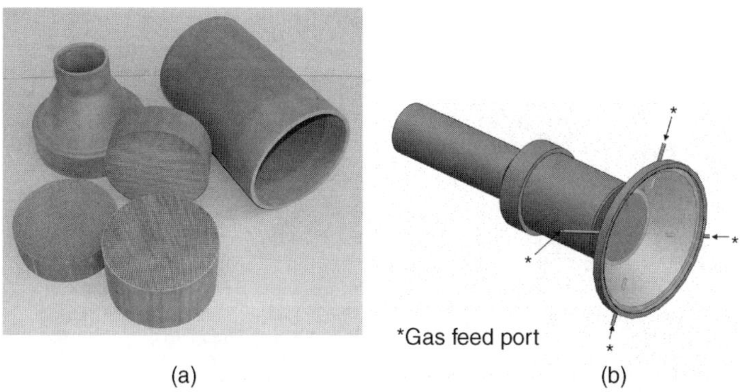

Figure 20.7 (a) Honeycomb monoliths [18] and (b) schematic of the chamber hosting the monolith.

to their water-splitting activity and regeneration capability in solar campaigns performed during the HYDROSOL project.

The proof of concept of the technology was achieved with the first solar receiver-reactor, HYDROSOL-I (Figure 20.7). The reactor consists of a ceramic SiC-based monolith, enclosed in a ceramic canister (Figure 20.6) which is in turn embedded in a metallic funnel shaped container, the front face of which has a quartz window that allows the concentrated solar radiation to pass through and heat the ceramic monolith. Reactor feed gas (steam and nitrogen) is fed through radial ports at the front face of the reactor Figure 20.7. All testing of the HYDROSOL-I reactor was performed at the solar-furnace facilities of DLR [19]. Initially small-size (Ø25 × 50 mm) coated honeycombs were embedded inside a larger similar uncoated SiC-based monolith (Ø144 × 50 mm) and several water-splitting and regeneration cycles were performed to enable the screening of the different redox materials.

The "offspring" of the HYDROSOL-I reactor was the HYDROSOL-I dual-chamber reactor [30,35]. The modular set-up of the HYDROSOL-I dual-chamber reactor allowed continuous production of solar hydrogen, with the one part of the modules splitting water while the other is being regenerated. After completion of the reactions, the regenerated module was switched to the splitting mode and vice versa, by switching the feed gas. Control of the heat demands for water splitting and regeneration on the modules was achieved by providing two focal spots (one on each module), with different flux density, via realignment of the facets of a faceted solar concentrator at the solar-furnace facilities of DLR. All the reactors constructed within the two HYDROSOL projects can be seen in Figure 20.8, in nonoperational mode, as well as during operation under solar radiation. HYDROSOL-I and the dual chamber reactor were operated at the solar facilities of DLR, while HYDROSOL-II operated at the Plataforma Solar de Almeria in Spain [36,37] Figure 20.9.

20.3.5 Solar Testing

The hydrogen production profile during water splitting in the laboratory unit and on the solar reactor is depicted in Figure 20.10.

In both cases the profile of the hydrogen evolution is similar, with the hydrogen concentration initially increasing with the introduction of water vapor up to a peak value. After that the

Figure 20.8 HYDROSOL reactors: (a) HYDROSOL-I, (b) HYDROSOL-I dual-chamber, (c) HYDROSOL-II in operation.

Figure 20.9 The thermochemical solar hydrogen plant, with the solar tower where the scaled-up HYDROSOL-II reactor is installed at the Plataforma Solar de Almeria, in Spain.

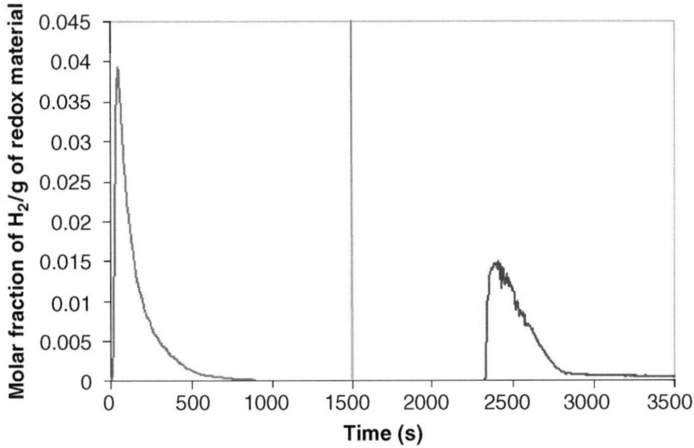

Figure 20.10 Hydrogen production in the laboratory reactor (powder-scale) (left) and the solar reactor (monolith-scale) (right).

hydrogen concentration decreases, signifying the saturation of the oxygen storage capacity of the material and the need for the regeneration step to occur.

Results from continuous hydrogen production from the HYDROSOL-I reactors (Figure 20.11a) have already appeared in the literature [30,35,38] while results from the HYDROSOL-II reactor (Figure 20.11b) will be reported in a forthcoming publication [37].

20.3.6 Simulation

The design and optimization of the operation strategy of the solar reactor necessitates the development of models for reactor simulation. In parallel to the experimental work, simulation models of different complexities are being developed by the HYDROSOL consortium partners

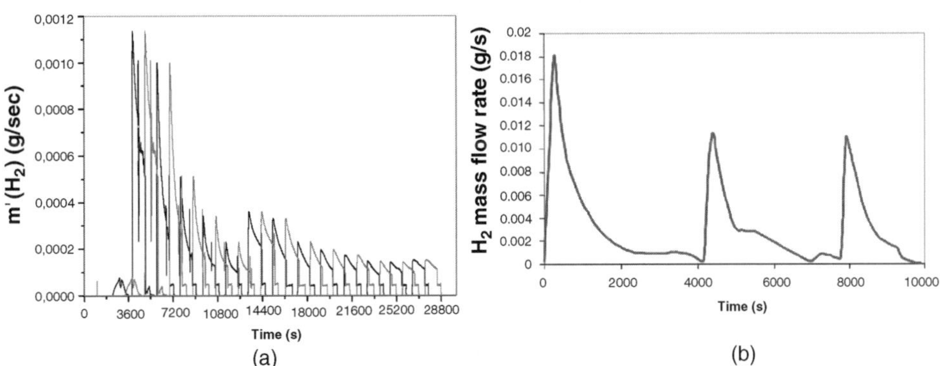

Figure 20.11 (a) cyclic solar hydrogen production with the HYDROSOL-I dual chamber reactor, (b) cyclic solar hydrogen production with the HYDROSOL-II reactor.

for description of the cyclic water-splitting/regeneration process as well as heliostat field tuning [16,19,39].

The water-splitting process towards the production of hydrogen, inside a honeycomb monolithic reactor under the influx of solar radiation on its front face has some similarities, as well as important differences, from the emission control processes that occur in automotive catalytic converters. For example, heating of the catalytic converter is effected by convection from the flowing exhaust gases, while in the solar monolithic reactor, solar radiation is absorbed on the monolith face and transmitted by conduction and radiation to the rest of the monolith body, while at the same time heating the much colder feed gas that enters the reactor channels. An additional difference stems from the fact that in the HYDROSOL reactor, the molar fraction of the reactive species can be quite large (as the reactor could be fed with 100% steam) and therefore the dilute limit approximation (typically employed in the simulation of automotive catalytic converters) is not valid.

The HYDROSOL reactor simulation model includes the basic mass, momentum and energy-transport processes, as well as the heterogeneous surface reactions of water vapor with the redox coating on the channels of the reactor. At the first level of approach, a single channel of the monolithic reactor is simulated, as described in [16]. Figure 20.12 depicts the hydrogen production curve, which has a good similarity with the hydrogen-evolution profile of Figure 20.10, leading credence to the simulation model adopted.

Single-channel models (in their perimeter averaged, 1D form as well as in 3D form) are the building blocks (Figure 20.13) of full 3D computational fluid dynamics (CFD) simulations (Figure 20.14) and provide significant insight into the process. For example the hydrogen-evolution profile of Figure 20.10 is seen to correspond to a traveling reaction front along the channel of the monolith Figure 20.13 while the 3D simulation (Figure 20.14) allows study of the placement of the feed gas ports in order to achieve a uniform flow distribution at the inlet of the monolith.

Reactor simulations can aid the identification of the design factors and parameters that affect the thermal response during the cyclic operation of the reactor. For example, the simulations identify the important effect of the thermal conductivity of the reactor material in distributing,

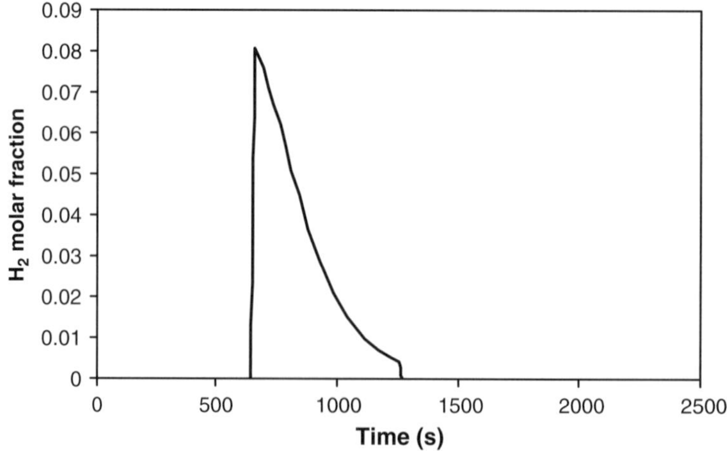

Figure 20.12 Results from a single channel simulation of the HYDROSOL reactor.

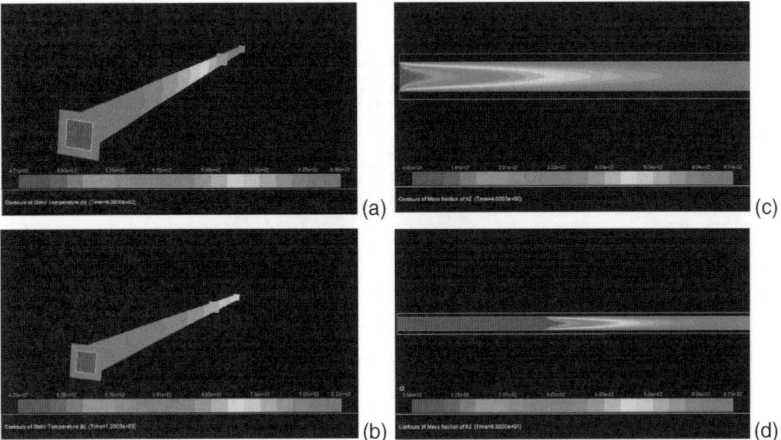

Figure 20.13 Evolution at two different time instances along a reactor's channel during the reactor's operation of: (a), (b) temperature ($t = 600$ and 1200 seconds, respectively); (c), (d) contours of H_2 mass fraction ($t = 4$ and 85 seconds, respectively). (Figure adapted from Ch. Lekkos, M. Kostoglou and A.G. Konstandopoulos, Simulation of solar water-splitting monolithic reactors, (to be submitted).) See also color-plate section.

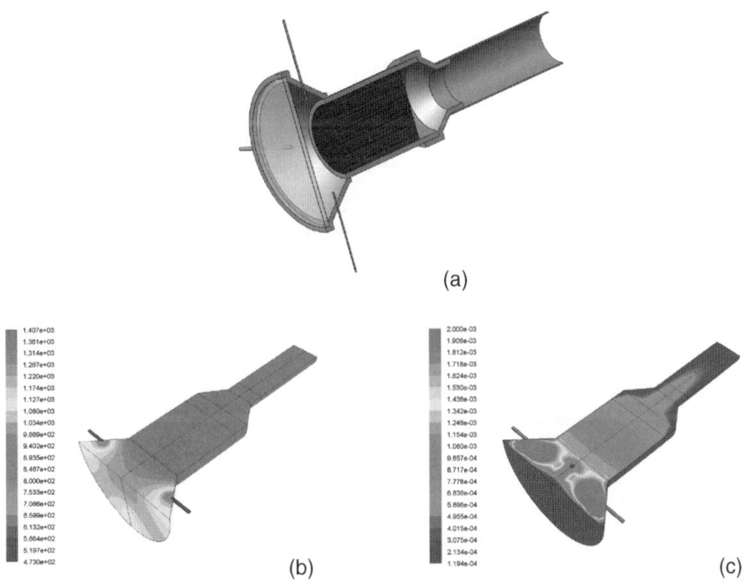

Figure 20.14 3D computational simulation of the solar water-splitting reactor shown in (a). (b) Temperature contours; (c) velocity contours. (Figure adapted from Ch. Lekkos, M. Kostoglou and A.G. Konstandopoulos, Simulation of solar water-splitting monolithic reactors, (to be submitted).) See also color-plate section.

Figure 20.15 High cell density (300 cells in^{-2}) SiC-based monolithic structures for future deployment as HYDROSOL reactor supports.

not only the incoming solar energy, but also the heat of the water-splitting reaction along its length (Figure 20.14). These factors have to be taken into account to prevent the regeneration reaction taking place at any location along the reactor channels simultaneously with the water-splitting reaction, and to optimize the overall reactor design and operation cycle for an efficient scale-up of the process.

20.3.7 Future Developments

There are several process parameters that could be optimized to enhance the efficiency of the process. Initially the monolithic materials used in the HYDROSOL reactors were constructed from low cell density (100 cells in^{-2}), coarse-grained SiC-based ceramics. New developments are leading in the manufacturing of fine-grained ceramic monoliths with higher cell density by a factor of three with respect to the current HYDROSOL technology (Figure 20.15).

Regarding the redox coating optimization, a number of newer redox materials (not limited to the ferrite family) is being developed that show increased performance with respect to the state-of-the-art while nonuniform coating distributions are also investigated. Another development towards increasing the yield of the system is to replace the coated SiC-based monoliths with monolithic elements consisting entirely of the redox material. Foam-type monolithic elements are readily constructed. For the future, the possibility for combining the produced hydrogen with sequestered CO_2 in order to generate hydrocarbon fuels, such as methane/methanol based on the Sabatier reaction [40] cannot be stressed enough. Ongoing research is aimed at identifying the parameter space under which CO_2 conversion into hydrocarbons can be achieved within a HYDROSOL-based thermochemical plant.

20.4 HYDROSOL Process

The HYDROSOL reactor technology involves a modular arrangement, including individual reaction chambers with fixed monolithic honeycomb absorbers. The modularity allows an area-wide arrangement of modules. A fixed absorber avoids movable parts of the receiver, as well as the devices driving those movable parts. The honeycomb structure is enclosed in a direct absorbing receiver, ensuring the availability of higher temperatures and lower re-radiation losses than corresponding indirect absorbing ones. The modular set-up itself allows for a continuous provision of solar hydrogen, because one part of the modules splits water, while the rest is being regenerated.

The modularity of HYDROSOL technology allows easy replication. In addition, it can advantageously be combined with future planned high-temperature solar towers, since it only represents a marginal added cost to the installation. At the same time it enables the chemical storage of solar energy into hydrogen and in this way it resolves the known supply–demand, temporal mismatch of solar technologies. Therefore it is expected that deployment of the HYDROSOL process may proceed in the future, along with the commercialization of high-temperature solar tower plants.

State-of-the-art volumetric receivers can extract with high (>70%) efficiency factors, the incident solar energy into the heat-transfer medium [41]. The need for thermal-energy storage and the use of steam turbines to produce electricity, however, reduce the overall solar-to-electricity conversion efficiency down to ~15% [42]. The HYDROSOL process, however, retains the high efficiencies of the solar reactor and its combination with solar tower systems eliminates efficiency draining stages, and can more than double the overall solar-to-electricity conversion efficiency. This is achieved using chemical storage (instead of thermal storage) of solar energy via hydrogen, which can be further fed into a fuel cell (which is a much more efficient energy converter than the gas turbine) to generate electricity on demand, either on-site after transportation to another location for use (as a fuel or feedstock).

Cost comparisons of technologies in development depend on various assumptions and scenarios/methodologies, including those based on so-called "experience curves" or detailed technoeconomic evaluations. All of the current hydrogen mass production processes require fossil fuels as feedstock and generate CO_2 and other pollutants. Natural gas steam reforming and coal gasification deliver the majority of hydrogen produced today at a cost of ~0.8 € kg^{-1} (0.024 € kWh^{-1}) and ~1.8 € kg^{-1} (0.054 € kWh^{-1}), respectively [41]. Future cost projections for hydrogen need to consider fuel price volatility due to shortage of resources, as well as the cost of CO_2 taxes/cost of carbon capture and storage, which is estimated to add 0.02–0.03 € kWh^{-1} to the cost. The cost of solar hydrogen produced by the HYDROSOL technology, based on its current level of performance, has been estimated to decrease as follows: starting with performance data from the HYDROSOL-I reactor, the hydrogen cost was estimated at 0.23 € kWh^{-1}; taking into account improvements in material performance and reactor design in 2008 the cost was estimated to drop to 0.12 € kWh^{-1}, while following the completion of the HYDROSOL-II operation and considering advances in mass production of solar component and materials technologies in 2020, a cost of 0.06 € kWh^{-1} or lower is expected.

It is stressed that contrary to fossil-fuel-based technologies, the HYDROSOL cost structure is not affected by volatility and shortage of fuel prices, while all materials needed for a HYDROSOL plant are fully recyclable.

20.5 Conclusions

A novel monolithic reactor has been developed for the solar thermochemical splitting of water, producing renewable hydrogen, within the HYDROSOL projects. A key enabling element for the process was the development of active, redox nanomaterials using combustion and aerosol synthesis methods. The HYDROSOL process was successfully put into practice at the pilot scale, achieving hydrogen production exclusively at the expense of solar energy, by the thermochemical splitting of water. Due to the fact that the HYDROSOL process employs entirely renewable and abundant energy sources and raw materials – solar energy and water, respectively – it holds a significant potential for large-scale, emissions-free hydrogen production, particularly for regions of the world that lack indigenous resources, but are endowed with ample solar energy.

Acknowledgments

The authors would like to thank the European Commission for partial funding of this work within Projects HYDROSOL (ENK6-CT-2002-00629) and HYDROSOL-II (FP6-2002-Energy-1, 020030), and their partners (DLR, JM, STC, CIEMAT) in the projects. We also thank our colleagues from APTL: C. Agrafiotis, C. Pagkoura, A. Zygogianni and M. Kostoglou, for their contributions.

References

[1] www.AmericanEnergyIndependence.com/energychallenge.aspx, accessed 22 September 2009).
[2] Smalley, R.E. (2005) Future global energy prosperity: the terawatt challenge. *Material Matters Bulletin*, **30**, 412–417.
[3] Krishnan, R., McConnell, R. and Licht, S. (2008) *Solar Hydrogen Generation: Toward a Renewable Energy Future*, Springer, New York, USA.
[4] Turner, J.A., Williams, M.C. and Rajeshwar, K. (2004) Hydrogen economy based on renewable energy resources. *The Electrochemical Society Interface*, **13**(3), 24–31.
[5] Kodama, T. and Gokon, N. (2007) Thermochemical cycles for high-temperature solar hydrogen production. *Chemical Reviews*, **107**(10), 4048–4077.
[6] Steinfeld, A. (2005) Solar thermochemical production of hydrogen – a review. *Solar Energy*, **78**, 603–615.
[7] Abanades, S., Charvin, P., Flamant, G. and Neveu, P. (2006) Screening of water-splitting thermochemical cycles potentially attractive for hydrogen production by concentrated solar energy. *Energy*, **31**, 2805–2822.
[8] Perkins, C. and Weimer, A.W. (2004) Likely near-term solar-thermal water splitting technologies. *International Journal of Hydrogen Energy*, **29**, 1587–1599.
[9] Funk, J.E. and Reinstorm, R.M. (1966) Industrial and engineering chemistry process design and dvelopment. *Industrial and Engineering Chemistry Process Design and Development*, **5**(3), 336–342.
[10] Funk, J.E. (2001) Thermochemical hydrogen production: past and present. *International Journal of Hydrogen Energy*, **26**(3), 185–190.
[11] Carty, R.H., Mazumder, M.M., Schreider, J.D. and Panborn, J.B. (1981) *Thermochemical Hydrogen Production, GRI-80/0023*, Gas Research Institute for the Institute of Gas Technology, Chicago, IL, pp. 1–4.
[12] McQuillan, B.W., Brown, L.C., Besenbruch, G.E. *et al.* (2002) High efficiency generation of hydrogen fuels using solar thermochemical splitting of water, *Report GA-A24972, General Atomics*, San Diego, CA.
[13] Nakamura, T. (1977) Hydrogen production from water utilizing solar heat at high temperatures. *Solar Energy*, **19** (5), 467–475.
[14] Charvin, P., Abanades, S., Flamant, G. and Lemort, F. (2007) Two-step water splitting thermochemical cycle based on iron oxide redox pair for solar hydrogen production. *Energy*, **32**, 1124–1133.
[15] Abanades, S., Charvin, P. and Flamant, G. (2007) Design and simulation of a solar chemical reactor for the thermal reduction of metal oxides: Case study of zinc oxide dissociation. *Chemical Engineering Science*, **62**, 6323–6333.
[16] Agrafiotis, C., Pagkoura, C., Lorentzou, S. *et al.* (2007) Hydrogen production in solar reactors. *Catalysis Today*, **127**, 265–277.
[17] Nalbandian, L., Zaspalis, V.T., Evdou, A. *et al.* (2004) Redox materials for hydrogen production from the water decomposition reaction. *Chemical Engineering Transactions*, **4**, 43–48.
[18] Agrafiotis, C., Roeb, M., Konstandopoulos, A.G. *et al.* (2005) Solar water splitting for hydrogen production with monolithic reactors. *Solar Energy*, **79**(4), 409–421.

[19] Roeb, M., Sattler, C., Klüser, R. *et al.* (2006) Solar hydrogen production by a two-step cycle based on mixed iron oxides. *Journal of Solar Energy Engineering – Transactions of the ASME*, **128**, 125–133.
[20] Bilgen, E., Ducarroir, M., Foex, M. *et al.* (1977) Use of solar energy for direct and two-step water decomposition cycles. *International Journal of Hydrogen Energy*, **2**, 251–257.
[21] Sibieude, F., Ducarroir, M., Tofighi, A. and Ambriz, J. (1982) High-temperature experiments with a solar furnace: the decomposition of Fe_3O_4, Mn_3O_4, CdO. *International Journal of Hydrogen Energy*, **7**, 79–88.
[22] Tofighi, A. (1982) Contribution à l'étude de la decomposition des oxydes de fer au foyer d'un four solaire, *Ph.D. Thesis*, Institut National Polytechnique de Toulouse, France.
[23] Moeller, S. and Palumbo, R. (2001) The development of a solar chemical reactor for the direct thermal dissociation of zinc oxide. *ASME-Journal of Solar Energy Engineering*, **123**, 83–90.
[24] Haueter, P., Moeller, S., Palumbo, R. and Steinfeld, A. (1999) The production of zinc by thermal dissociation of zinc oxide-solar chemical reactor design. *Solar Energy*, **67**, 161–167.
[25] Diver, R.B., Miller, J.E., Allendorf, M.D. *et al.* (2008) Solar thermochemical water-splitting ferrite-cycle heat engines. *Journal of Solar Energy Engineering*, **130**(4), 041001-1–041001-8.
[26] Kaneko, H., Miura, T., Fuse, A. *et al.* (2007) Rotary-type solar reactor for solar hydrogen production with two-step water splitting process. *Energy and Fuels*, **21**, 2287–2293.
[27] http://www.hydrosol-project.org/, accessed 22 September 2009).
[28] Konstandopoulos, A.G., Papaioannou, E., Zarvalis, D. *et al.* Catalytic Filter Systems with Direct and Indirect Soot Oxidation Activity, *SAE Tech. Paper* No. 2005-01-0670, **SP-1942**, 2005.
[29] Agrafiotis, C., Mavroidis, I., Konstandopoulos, A.G. *et al.* (2007) valuation of porous silicon carbide monolithic honeycombs as volumetric receivers/collectors of concentrated solar radiation. *Journal of Solar Energy Materials and Solar Cells*, **91**, 474–488.
[30] Roeb, M., Monnerie, N., Schmitz, M. *et al.* (2006) Thermo-chemical production of hydrogen from water by metal oxides fixed on ceramic substrates. Proceedings of the 16th World Hydrogen Energy Conference, Lyon, France, 13–16 June.
[31] Agrafiotis, C., Lorentzou, S., Pagkoura, C. *et al.* (2006) Advanced monolithic reactors for hydrogen generation from solar water splitting. Proceedings of SolarPACES 13th International Symposium on Concentrated Solar Power and Chemical Energy Technologies, Seville, Spain, 20–23 June.
[32] Agrafiotis, C., Pagkoura, C., Lorentzou, S. *et al.* (2006) Material technologies developments for solar hydrogen. Proceedings of the 16th World Hydrogen Energy Conference, Lyon, France, 13–16 June.
[33] Agrafiotis, C., Lorentzou, S., Pagkoura, C. *et al.* (2007) The HYDROSOL process solar-aided thermochemical production of Hydrogen from water with innovative honeycomb reactors. IWH2, 27–29 October 2007, Ghardaïa, Algeria.
[34] Lorentzou, S., Agrafiotis, C.C. and Konstandopoulos, A.G. (2008) Aerosol spray pyrolysis synthesis of water-splitting ferrites for solar hydrogen production. *Granular Matter*, **10**(2), 113–122.
[35] Roeb, M., Neises, M., Säck, J.-P. *et al.* (2008) Operational strategy of a two-step thermochemical process for solar hydrogen production. *International Journal of Hydrogen Energy*, **34**(10) 4537–4545.
[36] Van Noorden, R. (2008) Cracking water with sunlight. *Chemistry World, March 28* http://www.rsc.org/chemistryworld/News/2008/March/28030801.asp accessed 22 September 2009).
[37] Roeb, M., Säck, J.-P., Rietbrock, P. *et al.* (2009) Test operation of a 100-KW pilot plant for solar Hydrogen production from water on a solar tower. to be submitted at the 15th International SolarPACES Concentrating Solar Power Symposium, 15–18 September 2009, Berlin, Germany.
[38] Roeb, M., Säck, J.P., Rietbrock, P. *et al.* (2008) Development and verification of a two-step thermochemical process for solar Hydrogen production from water. 14th International SolarPACES Concentrating Solar Power Symposium, 4–7 March 2008, Las Vegas, Nevada, USA.
[39] Roeb, M., Gathmann, N., Neises, M. *et al.* (2009) Thermodynamic analysis of two-step solar water splitting with mixed iron oxides. *International Journal of Energy Research*, **33**(10), 893–902.
[40] Sabatier, P. (1922) *Catalysis in Organic Chemistry* (ed. E.E., Reid), Van Nostrand D, New York.
[41] Graf, D., Monnerie, N., Roeb, M. *et al.* (2008) Economic comparison of solar Hydrogen generation by means of thermochemical cycles and electrolysis. *International Journal of Hydrogen Energy*, **33**(17), 4511–4519.
[42] http://www.euro-energy.net/success_stories/52.html, accessed 22 September 2009).

21

Solar Thermal and Efficient Solar Thermal/Electrochemical Photo Hydrogen Generation

Stuart Licht
Department of Chemistry, The Renewable Energy Institute, The George Washington University, Washington, DC, USA, Email: slicht@gwu.edu

21.1 Comparison of Solar Hydrogen Processes

Actualization of a hydrogen, rather than fossil fuel, economy requires H_2 storage, utilization and generation processes; the latter is the least developed of these technologies. Solar-energy driven water splitting combines several attractive features for energy utilization. The energy source (sun) and reactive media (water) for solar water splitting are readily available and are renewable, and the resultant fuel (generated H_2) and its discharge product (water) are each environmentally benign. This chapter presents an efficient renewable energy approach to produce H_2 fuel, the hybrid thermochemical solar generation of hydrogen. This energy source is capable of sustaining the highest solar-energy conversion efficiencies and fits well into a clean hydrogen energy cycle.

To better understand the significance of an efficient solar hybrid hydrogen-formation process, it is useful to introduce it in the context of other solar hydrogen processes. Solar water splitting can provide clean, renewable sources of hydrogen fuel without greenhouse-gas evolution. A variety of approaches have been studied to achieve this important goal, for example, photosynthetic, direct or indirect photothermochemical, and photovoltaic or photoelectrochemical water splitting. As summarized in Table 21.1, each of these processes has exhibited a limited conversion of solar energy to hydrogen; photosynthetic, biological and photochemical solar water splitting [1–5] have exhibited solar energy to hydrogen conversion efficiencies, η_{solar}, on the order of only 1%; single [6–9] or multistep [10–14] photothermal

On Solar Hydrogen & Nanotechnology Edited by Lionel Vayssieres

Table 21.1 Overview of solar water splitting processes effectuating generation of hydrogen.

Water-splitting process	Limitations, potential
Photosynthetic, biological and photochemical	Demonstrated efficiencies very low, generally $<1\%$ solar-energy conversion
Photothermal, direct single-step	Gas recombination limitations, high-temperature material limitations, generally $<1\%$ solar-energy conversion
Photothermal, indirect multistep	Lower temperature and less recombination than single step, although stepwise reaction inefficiencies lead to losses, generally $<10\%$ solar conversion
PV and photoelectrochemical electrolysis	10–20% solar-energy conversion, *in situ* interfacial instability limitations
Photothermal electrochemical	Theory was unavailable, potential for $\gg 20\%$ solar conversion, requires solar concentration

processes have been reported in the $\eta_{solar} = 1\text{–}10\%$ range, and photovoltaic or photoelectrochemical solar water splitting has reached $\eta_{solar} = 18\%$ [15–23]. The highest solar efficiencies have been observed recently with a hybrid process, which unlike the other processes, incorporates full utilization of the solar spectrum, further enhancing achievable solar conversion efficiencies [24]. This hybrid process, combines solar photo and solar thermal energy conversion, with high-temperature electrochemical water electrolysis for generation of hydrogen fuel.

At high temperatures ($T > 2000\,°C$), water chemically disproportionates to H_2 and O_2 (without electrolysis). Hence, in principal, using solar energy to directly heat water to these temperatures, hydrogen can be spontaneously generated. This is the basis for all direct thermochemical solar water-splitting processes [6–9]. However, catalysis, gas recombination and containment materials limitations above 2000 °C have led to very low solar efficiencies for direct solar thermal hydrogen generation. Other thermal approaches are either indirect [10–14] or hybrid processes [24–26]. Multistep, indirect, solar thermal reaction processes to generate hydrogen at lower temperatures have been studied, and a variety of pertinent reaction processes considered. These reactions are conducted in a cycle to regenerate and reuse the original reactions (ideally, with the only net reactant water, and the only net products hydrogen and oxygen). Such cycles suffer from challenges often encountered in multistep reactions [11]. While these cycles can operate at lower temperatures than the direct thermal chemical generation of hydrogen, efficiency loses can occur at each of the steps in the multistep sequence, resulting in low overall solar-to-hydrogen energy-conversion efficiencies.

Electrochemical water splitting, generating H_2 and O_2 at separate electrodes, largely circumvents the gas recombination and high-temperature limitations occurring in thermal hydrogen processes. The UV and visible energy-rich portion of the solar spectrum is transmitted through H_2O. Therefore sensitization, such as via semiconductors, is required to drive the electrical charge for the water-splitting process. In photoelectrochemical processes, illuminated semiconductors drive redox processes in solution [27,28]. The principal advantages of photoelectrochemical (PEC) compared to solid-state photovoltaic (PV), charge transfer, are the possibility for internal electrochemical charge storage [29–31], and that solution-phase processes can be used to influence the energetics of photodriven charge transfer [32–36]. The principal disadvantage is exposure to the electrolyte, which can lead to semiconductor deterioration. PV water-splitting processes utilize a photoabsorber connected *ex situ* by an electronic conductor into the electrolyte, to electrochemically drive water splitting, for example, an illuminated solar cell wired to an electrolyzer. While PEC water-splitting processes utilize *in situ* immersion of a photoabsorber in a chemical solution, such as an illuminated semiconductor in water for electrochemically driven water splitting [28].

The photosensitizer and electrode components of PV and PEC hydrogen generation can be identical, but from a pragmatic viewpoint the PV process seems preferred, as it isolates the semiconductor from contact with, and corrosion by, the electrolyte. Illuminated semiconductors, such as TiO_2 and InP can split water, but their wide bandgap, E_g, limits the photoresponse to a small fraction of the incident solar energy. Various studies had focused on diminishing the high E_G for solar water splitting, by tuning (decreasing) the E_G of the photosensitizers to better match the water-splitting potential, E_{H_2O}. These studies change the composition or size (nanoparticles or quantum dots) of the semiconductor, but have not yet demonstrated effective solar-to-hydrogen efficiencies [37–39]. Another response to this challenge is to couple together semiconductors.

Whereas a single small-bandgap material, such as silicon, cannot generate the minimum potential of 1.23 V needed to split water at room temperature, a bipolar (series) arrangement of these multiple-bandgap semiconductors can generate a larger potential to generate hydrogen and oxygen (Figure 21.1). These semiconductors can be the same (e.g., series-connected Si cells), or semiconductors which can respond to different portions of the solar spectrum (multiple bandgap semiconductors).

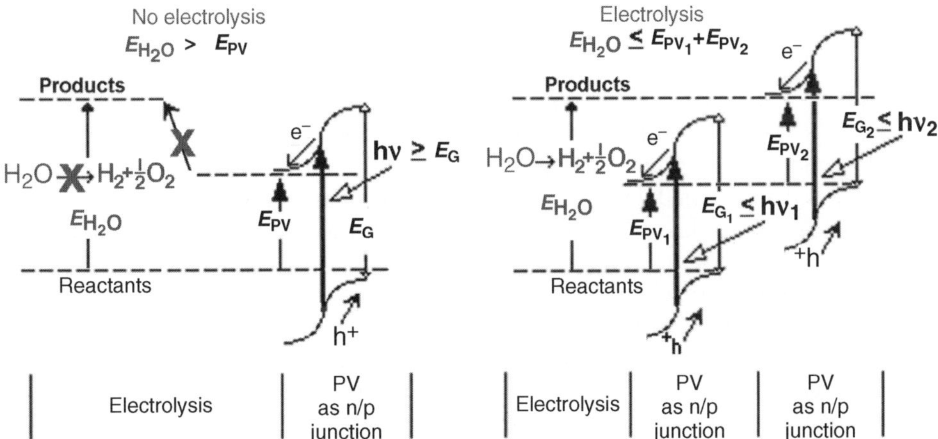

Figure 21.1 Generally, a single small-bandgap semiconductor cannot drive charge transfer at a potential sufficient to electrolyze water at room temperature. However, the combined potential of two separate semiconductors driven by two photons can drive electrolysis towards the formation of hydrogen and oxygen. Single wide-bandgap semiconductors can generate a higher photopotential, but are only photoexcited by a small fraction (towards the UV) of insolation.

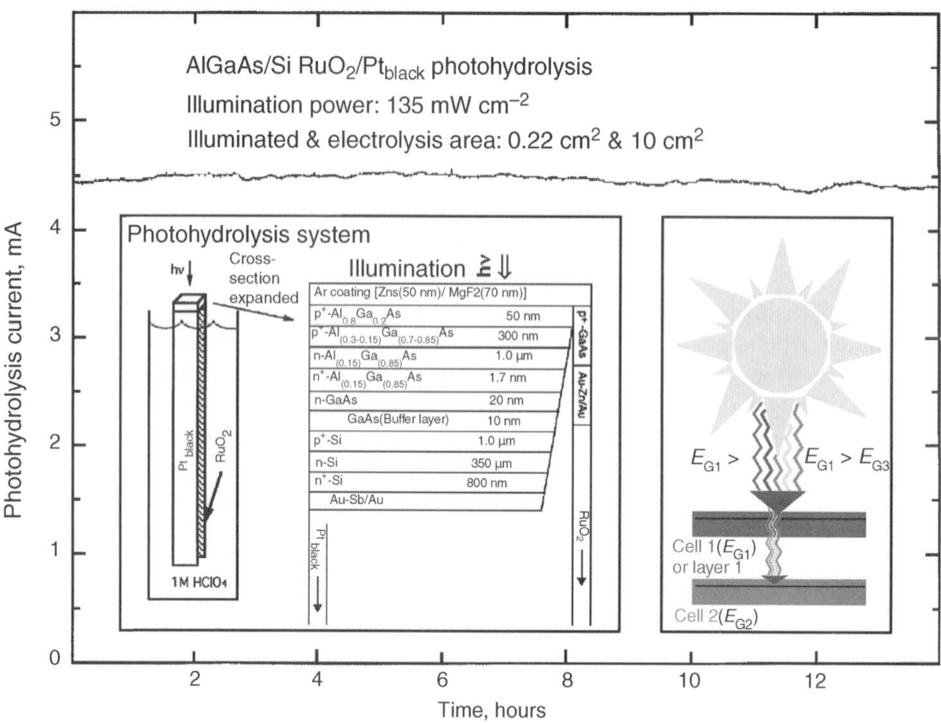

Figure 21.2 Representation (inset) and measured characteristics of the illuminated AlGaAs/Si RuO_2/Pt_{black} $\eta_{solar} = 18\%$ room-temperature photoelectrolysis cell [9]. Right inset. In stacked multi-junction solar cells, the top layer or cell converts higher-energy photons and transmits the remainder onto subsequent smaller bandgap layer(s), for more effective utilization of the solar spectrum.

Multiple-bandgap semiconductors can be combined to generate a single photovoltage well matched to the electrolysis cell, as illustrated in the right inset of Figure 21.2. Using multiple-bandgap semiconductor photovoltaics, over 18% conversion energy efficiency of solar to hydrogen was demonstrated, Figure 21.2, albeit at room temperature (without the benefit of higher-efficiency solar thermal processes) [40]. As illustrated in Figure 21.3, multiples of electrolyzers and photovoltaics can be combined to produce an efficient match between the generated and consumed power.

There has been ongoing theoretical interest in utilizing solar-generated electrical charge to drive electrochemical water splitting (electrolysis) to generate hydrogen [23,32,41–44]. Early photoelectrochemical models had underestimated low solar water-splitting conversion efficiencies, predicting a maximum of ~15%, would be attainable. This was increased to ~30% by eliminating: (i) the linkage of illumination surface area to electrolysis surface area, (ii) nonideal matching of photopotential and electrolysis potential, and incorporating the effectiveness of contemporary (iii) electrolysis catalysts and (iv) efficient multiple-bandgap photosensitizers [32]. Our experimental water-splitting cell incorporating these features achieved over 18% solar-energy-to-hydrogen conversion efficiency, Figure 21.2. However, these models and experiments did not incorporate solar-heat-effects improvements on the electrolysis energetics

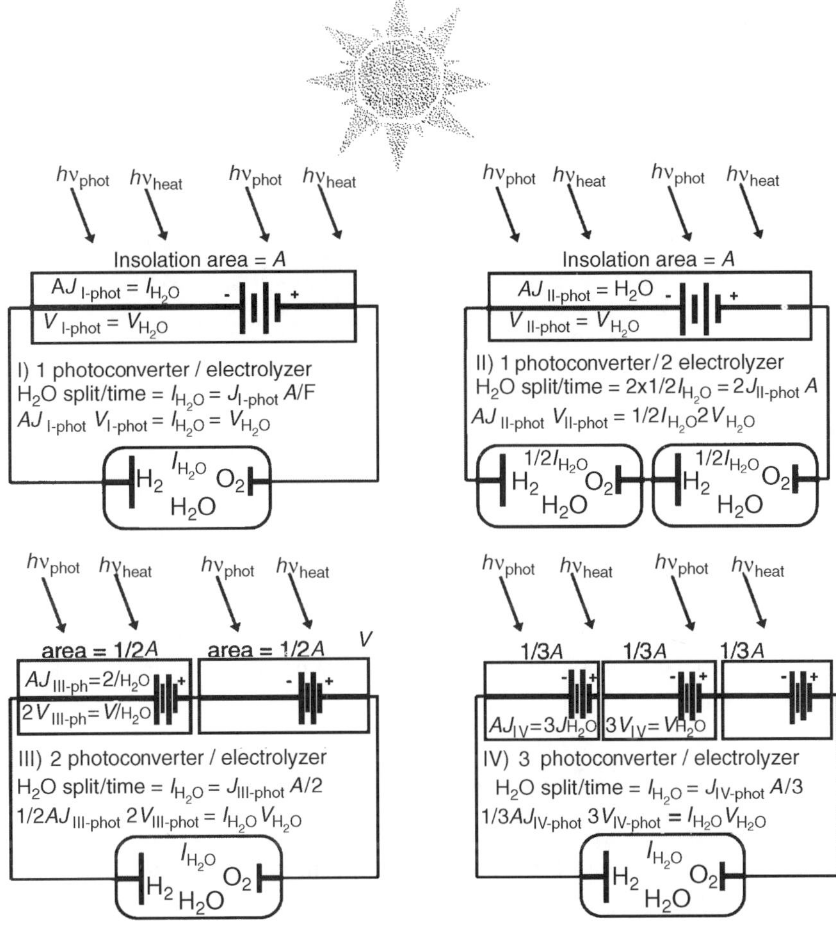

Figure 21.3 Alternate configurations varying the number of photoharvesting units and electrolysis units for solar water splitting [3]. The photoconverter in the first system generates the requisite water electrolysis potential; that in the second system generates twice that potential, while the photoconverter in the third and fourth units generate respectively only half or a third this potential.

of charge utilization, or semiconductor-imposed heat-utilization limitations. This limitation is exemplified in Figure 21.4 which presents the photoaction spectrum of a layered semiconductors which had been used in an efficient multiple band-gap semiconductor. The wide bandgap layer responds to visible radiation of energy 1.8 eV or greater, and the smaller bandgap layer responds through 1.4 eV. However, solar radiation of energy of less than 1.4 eV does not induce electronic charge transfer. This limits the use of solar energy. Energy less than 1.4 eV is thermal radiation, and comprises one third of all incident sunlight energy. As will be shown in

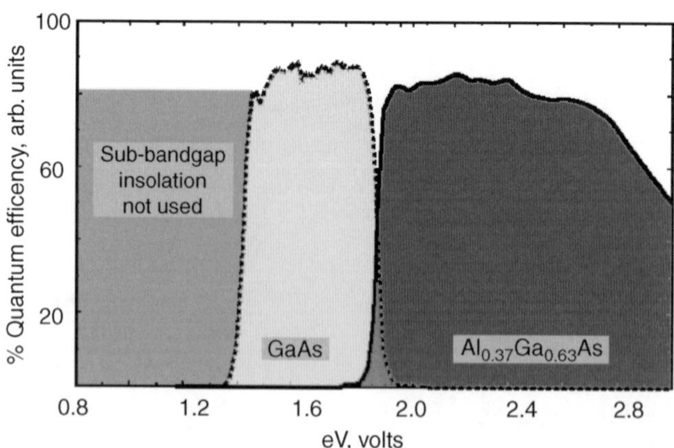

Figure 21.4 The photoaction of an efficient multiple bandgap photovoltaic illustrates that sub-bandgap sunlight does not have sufficient energy to induce photoexciation, and will not drive electronic charge transfer. In this specific example, a 28% efficient solar photovoltaic consists of wide bandgap AlGaS on smaller band gap GaAs, and sub-bandgap sunlight is radiation at lower energies than the 1.4 eV bandgap of GaAs. (Photoaction data from reference [45].)

the next sections, by applying this excess energy to the formation of hydrogen, a hybrid process increases solar efficiency and combines the advantages of photothermal and PV or PEC water-splitting processes.

21.2 STEP (Solar Thermal Electrochemical Photo) Generation of H_2

Infrared (IR) radiation, is energetically insufficient to drive conventional solar cells. Hence, PV and photoelectrochemical solar electrolysis discard (by reflectance or as re-radiated heat), solar thermal radiation. However, the hybrid process can utilize the full solar-spectrum energy, leading to substantially higher solar-energy efficiencies. As seen in Figure 21.5, and as described in a latter section of this chapter, in the hybrid process, the solar IR is not discarded, but instead separated and utilized to heat water, which substantially decreases the necessary electrochemical potential to split the water, and substantially increases the solar hydrogen energy-conversion efficiencies.

This hybrid solar hydrogen process is delineated in this section. Fundamental water thermodynamics and solar photosensitizer constraints will be shown consistent with hybrid solar-to-hydrogen fuel-conversion efficiencies in the 50% range, over a wide range of insolation, temperature, pressure and photosensitizer bandgap conditions. Nicholson and Carlisle first generated hydrogen by water electrolysis in 1800. Modifications, such as steam electrolysis, or illuminated semiconductor electrolysis [37], had been reported by the 1970s. The electrolysis of water can be substantially enhanced by heating the water with excess solar thermal energy.

With increasing temperature, the quantitative decrease in the electrochemical potential necessary to split water to hydrogen and oxygen was well known by the 1950s [46], and as early as 1980 it was noted, from this relationship, that solar thermal energy could decrease the necessary energy for the electrolytic generation of hydrogen [47]. However, the process combines elements of solid-state physics, insolation and electrochemical theory, complicating

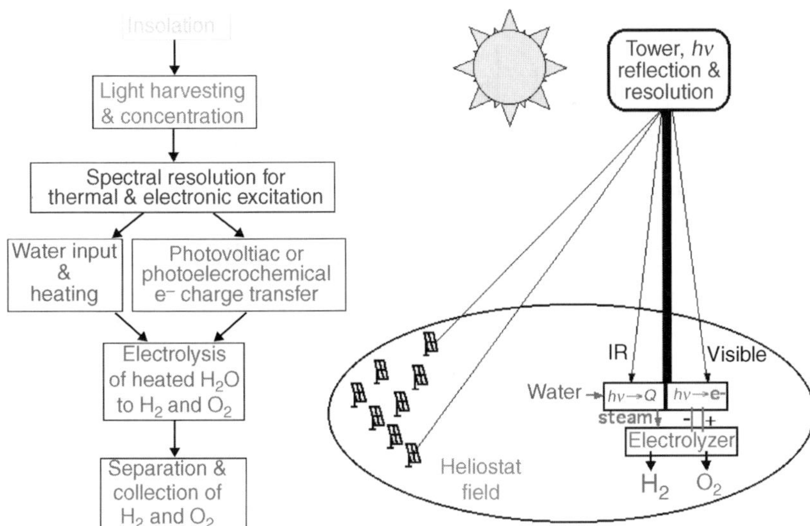

Figure 21.5 Solar water electrolysis improvement through excess solar-heat utilization via thermal electrochemical hybrid H_2 generation. (Reprinted with permission from S. Licht, Thermochemical solar hydrogen generation, *Chemical Communications*, **37**, 4623–4464, 2005. © 2005 Royal Society of Chemistry.)

rigorous theoretical support of the process. Our fundamental thermodynamic feasibility of the solar thermal electrochemical generation of hydrogen was initially derived in 2002 [48,49]. The novel theory combines photodriven charge transfer, with excess sub-bandgap insolation to lower the water potential, and derives rigorous semiconductor bandgap-restricted, thermal-enhanced solar water-splitting efficiencies in excess of 50%. In 2003, experimentation, which is described in the latter sections, was provided in support of the theory [26].

The STEP (Solar Thermal Electrochemical Photo) process is intrinsically more efficient than other solar-energy conversion processes, as it utilizes, not only the visible sunlight used to drive PVs, but also utilizes the previously detrimental (due to PV thermal degradation) thermal component of sunlight, for the electrolysis of water. The electrochemical water-splitting potential decreases with increasing temperature, and higher temperature also facilitates charge transfer (i.e., decreases kinetic overpotential losses), which arise during electrolysis. Electrochemical water splitting, generating H_2 and O_2 at separate electrodes, circumvents the gas recombination limitations or multistep repeated Carnot losses of solar thermochemical H_2 formation.

Figure 21.6 presents the energy diagram for the STEP formation of hydrogen. The STEP process distinguishes radiation that is intrinsically energy sufficient (super-bandgap, $h\nu \geq E_G$), or insufficient (sub-bandgap), to drive PV charge transfer, and applies this excess solar thermal energy to heat the water-splitting electrolysis reaction chamber. On the right-hand side, excess solar thermal energy provides heat to decrease the redox potential, while super-bandgap photons generate electronic charge with sufficient energy to drive the electrolytic formation of energetic molecules. In the top left, without this solar heat, the same solar-driven electronic charge is insufficient to drive electrolysis. The extent of the decrease in the water-electrolysis potential will vary with temperature.

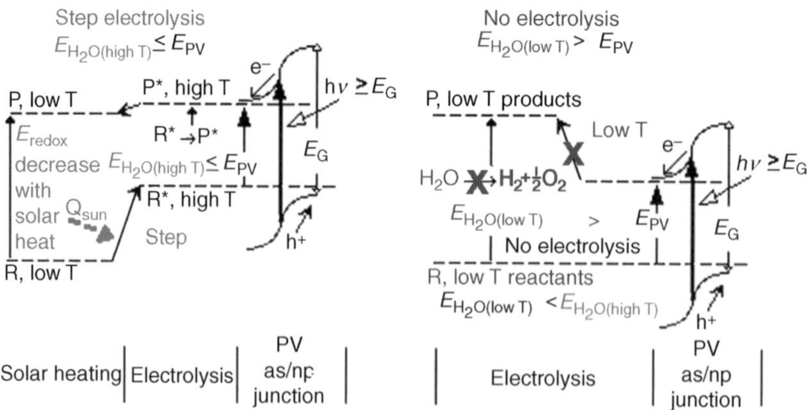

Figure 21.6 Energy diagram of solar-to-hydrogen conversion via the STEP (Solar Thermal Electrochemical Photo) process. An illuminated single small-bandgap semiconductor cannot drive charge transfer at a potential sufficient to electrolyze water at room temperature. However, the same illuminated semiconductor can drive electrolytic water splitting when superfluous solar thermal energy is used to heat and decrease the water-splitting potential.

As illustrated on the left-hand side of Figure 21.5, thermally assisted solar electrolysis consists of: (i) light harvesting, (ii) spectral resolution of thermal (sub-bandgap) and electronic (super-bandgap) radiation, the latter of which (iiia) drives photovoltaic or photoelectrochemical charge transfer $V(iH_2O)$, while the former (iiib) elevates water to temperature T, and pressure, p; finally (iv) $V(iH_2O)$-driven electrolysis of $H_2O(T,p)$. This solar thermal water electrolysis (photothermal electrochemical water splitting) is presented in Figure 21.5. Rather than a field of concentrators shown in the right-hand side of Figure 21.5, systems may use individual solar concentrators. The light harvesting can use various optical configurations; for example, in lieu of parabolic or Fresnel concentrators, the heliostat/solar tower with secondary optics can achieve higher STEP-process temperatures (>1000 °C) with concentrations of ~2000 suns. Beam splitters can redirect sub-bandgap radiation away from the PV for direct heat exchange with the eiectrolyzer.

21.3 STEP Theory

This section provides a derivation of enhanced solar water-splitting efficiencies, as determined with appropriate bandgap restrictions and incorporating excess solar thermal energy. Earlier models had described solar hydrogen energy-conversion processes with a predicted a limit of ~30% efficiency at room temperature, but the advantages (and constraints) of available excess heat had not been incorporated.

Charge transfer through a semiconductor junction cannot be driven by photons that have energy below the semiconductor bandgap. Hence a silicon photovoltaic device does not use radiation below its bandgap of ~1.1 eV, while an AlGaAs–GaAs multiple-bandgap photovoltaic does not use radiation of energy less than the 1.43 eV bandgap of GaAs. This unused, long-wavelength insolation represents a significant fraction of the solar spectrum (see later) and can be filtered and applied to heat water prior to electrolysis. The thermodynamics of heated-water dissociation are more favorable than that at room temperature. This is expressed by a shift in chemical free energy and by a decrease in the requisite water-electrolysis voltage. This decrease can considerably enhance solar water-splitting efficiencies.

The spontaneity of the water-splitting reaction, Equation 21.1, is given by the standard free energy of formation, ΔG_f^0, and is related to the voltage for water electrolysis standard potential, $E^0_{H_2O}$, as $\Delta G_f^0(H_2O)/2F$ (where F is the Faraday constant).

$$H_2O \rightarrow H_2 + \tfrac{1}{2}O_2 \tag{21.1}$$

Reaction 21.1 is endothermic, and electrolyzed water will cool unless external heat is supplied. The enthalpy balance is related to the thermoneutral potential, E_{tneut}, $\Delta H_f(H_2O_{liq})/2F$, where $E^0_{tneut}(25\,°C) = 1.481$ V. The water electrolysis rest potential varies with water, H_2 and O_2 activity as:

$$E(H_2O_{liq}) = E°(H_2O_{liq}) + (RT/2F)\ln(a_{H_2}(a_{O_2})^{1/2}/a_w); \quad E^0_{H_2O}(25°C) = 1.229\,V \tag{21.2}$$

As shown in Figure 21.7, $E^0_{H_2O}(T)$ diminishes with increasing temperature, as calculated using contemporary thermodynamic values. For solar photothermal water electrolysis, a

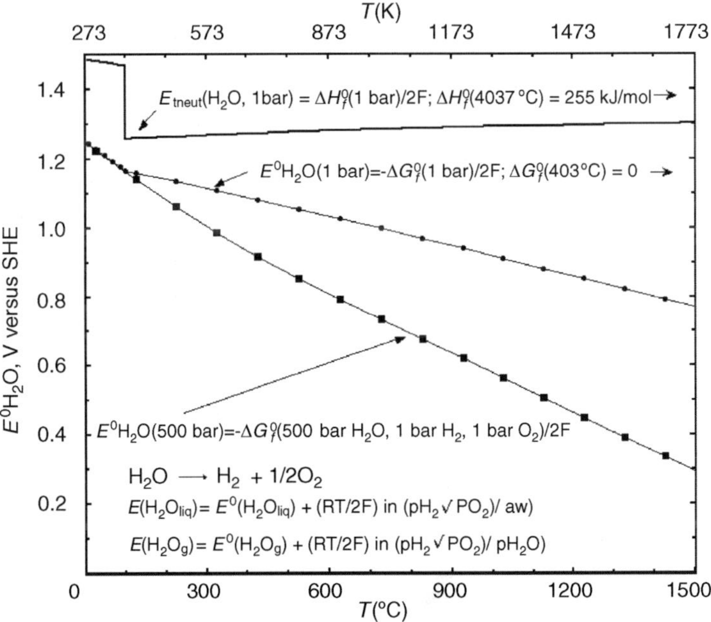

Figure 21.7 Thermodynamic and electrochemical values for water dissociation to H_2 and O_2 as a function of temperature, as calculated in ref. [3]. The $pH_2O = 1$ bar curves without squares are for liquid water through 100 °C, and for steam at higher temperatures. The high-pressure values curve ($pH_2O = 500$ bar; $pH_2 = pO_2 = 1$ bar) occur at potentials lower than that of 1 bar water. Note, the density of the high-pressure fluid is similar to that of the liquid and may be generated in a confined space by heating or electrolyzing liquid water.

portion of the solar spectrum will be used to drive charge transfer, and an unused, separate portion of the insolation will be used as a thermal source to raise ambient water to a temperature T. The overall solar-energy conversion efficiency of solar photothermal water splitting is constrained by the product of the available solar energy electronic conversion efficiency, η_{phot}, and the water-electrolysis energy-conversion efficiency:

$$\eta_{solar}(T,p) = 1.229\text{V}\ \eta_{phot}/E^0_{H_2O}(T,p) \tag{21.3}$$

Figure 21.8 presents the available power, $P_{\lambda max}$ (mW cm^{-2}) of solar radiation up to a minimum electronic excitation frequency, ν_{min} (eV) This is determined by integrating the solar spectral irradiance, S (mW cm^{-2} nm^{-1}), as a function of a maximum insolation wavelength, λ_{max}(nm). This $P_{\lambda max}$ is calculated for the conventional terrestrial insolation spectrum either above the atmosphere, AM0, or through a 1.5 atmosphere pathway, AM1.5. Relative to the total power, P_{sun} the fraction of this power available through the insolation edge is designated $P_{rel} = P_{\lambda max}/P_{sun}$. In the hybrid solar-energy electrolysis, excess heat is available primarily as

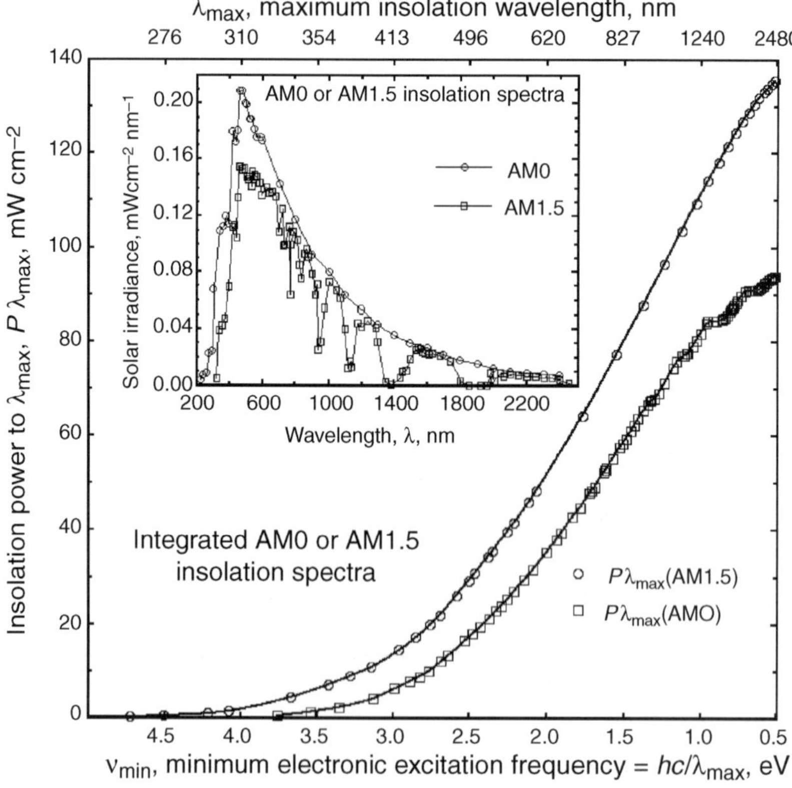

Figure 21.8 Solar irradiance (mW cm^{-2} nm^{-1}) is in the figure inset, and total insolation power (mW cm^{-2}) is in the main figure of the solar spectrum. (Reprinted with permission from S. Licht, Thermochemical solar hydrogen generation, *Chemical Communications*, **37**, 4623–4464, 2005. © 2005 Royal Society of Chemistry.)

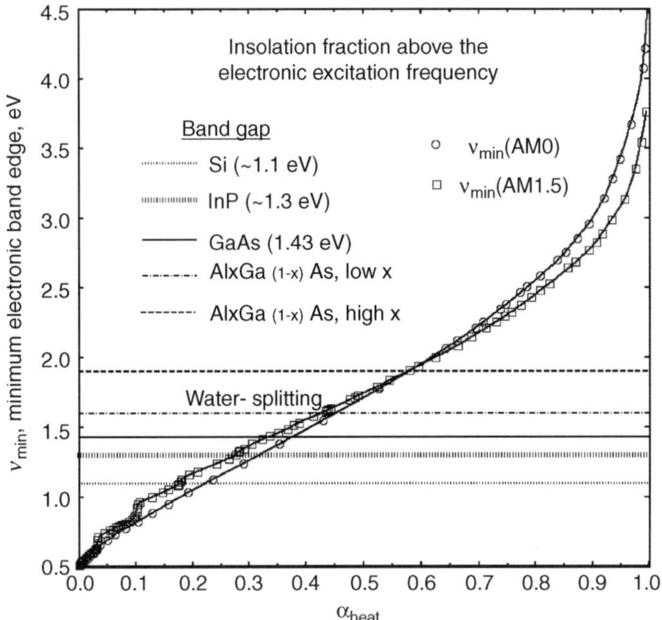

Figure 21.9 α_{heat}, $= 1 - P_{rel}$, the fraction of solar energy available below n_{min} [3], with $P_{rel} = P_{\lambda max}/P_{sun}$. Available incident power below ν_{min} is $\alpha_{heat}P_{sun}$. Various semiconductor bandgaps are superimposed as vertical lines. (Reprinted with permission from S. Licht, Thermochemical solar hydrogen generation, *Chemical Communications*, **37**, 4623–4464, 2005. © 2005 Royal Society of Chemistry.)

photons without sufficient energy for electronic excitation. The fraction these sub-bandgap photons in insolation is $\alpha_{heat} = 1 - P_{rel}$, and comprises an incident power of $\alpha_{heat}P_{sun}$.

Figure 21.9 presents the variation of the minimum electronic excitation frequency, n_{min} with α_{heat}, determined from P_{rel} using the values of $P_{\lambda max}$ summarized in Figure 21.8. A semiconductor in a photovoltaic cannot use incident energy below the band-gap. As seen in Figure 21.9 by the intersection of the solid line with ν_{min}, over one-third of insolation power occurs at $n_{min} < 1.43$ eV (>867 nm). GaAs or wider-bandgap materials cannot absorb this IR. For AM1.5 insolation spectra, $\nu_{min}(\alpha_{heat})$ is given by:

$$\nu_{min}, \text{eV} = 0.52827 + 4.0135\,\alpha_{heat} - 5.1286\,\alpha_{heat}^2 + 3.8980\,\alpha_{heat}^3 \tag{21.4}$$

When captured at a thermal efficiency of h_{heat}, the sub-bandgap insolation power is $\eta_{heat}\alpha_{heat}P_{sun}$. Other available system heating sources include absorbed super-bandgap photons, which do not effectuate charge separation, P_{recomb}, and noninsolation sources, b, such as heat recovered from process cycling Together these comprise the power for heat-balanced electrolysis, which yields α_{heat}:

$$\begin{aligned} P_{heat} &= \eta_{heat}\alpha_{heat}P_{sun} + \beta \\ \alpha_{heat} &= [\eta_{phot}/\eta_{heat}][\{E^0_{tneut}(25\,°\text{C})/(1+\zeta)E^0_{H_2O}) - 1\} - \beta/(\eta_{heat}P_{sun})]; \\ \alpha_{heat} &\cong [\eta_{phot}/\eta_{heat}][\{1.481\text{ V}/(E^0_{H_2O} - 1\} \end{aligned} \tag{21.5}$$

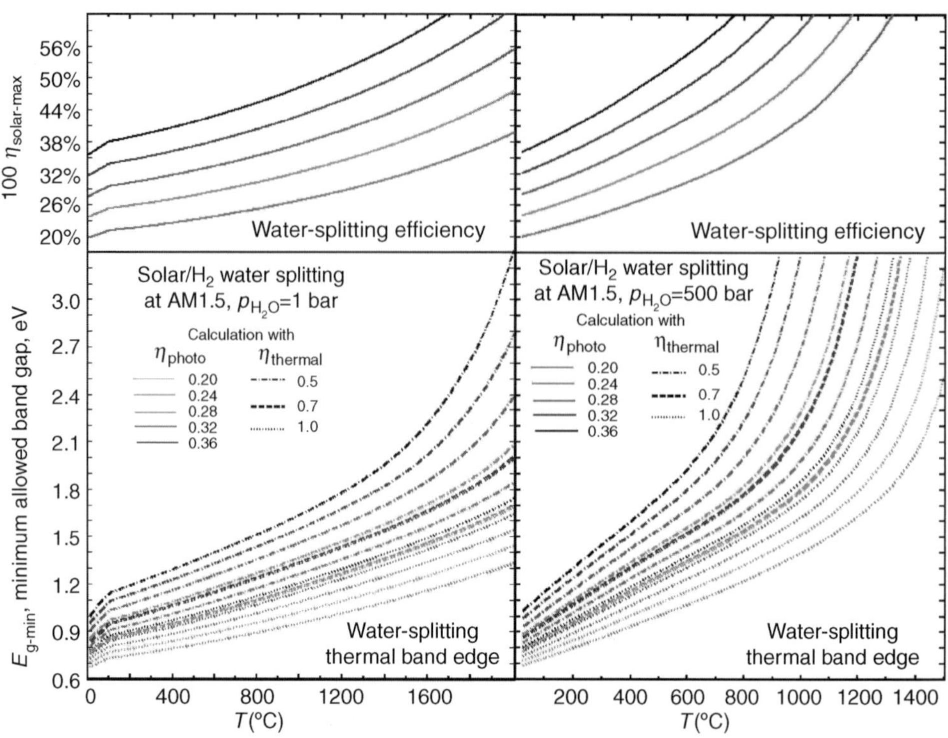

Figure 21.10 STEP solar-to-hydrogen conversion efficiency calculated at AM1.5 and at $p_{H_2O} = 1$ bar (left-hand side), or at $p_{H_2O} = 500$ bar (right-hand side). (Reprinted with permission from S. Licht, Thermochemical solar hydrogen generation, *Chemical Communications*, **37**, 4623–4464, 2005. © 2005 Royal Society of Chemistry.)

For solar electrolysis at T and p, a minimum insolation energy, ν_{min}, is required. This constrains the minimum electronic excitation energy and the bandgap, $E_{g\text{-}min}$. This is determined using $E_{g\text{-}min} = \nu_{min}(\alpha_{heat})$, where ν_{min} and α_{heat} are respectively from Equations 21.4 and 21.5.

Figure 21.10 summarizes determinations of the solar thermal–electrochemical hybrid water-splitting energy-conversion efficiency, as from Equation 21.3 using the $E^0_{H_2O}(T, p)$ data in Figure 21.9, and calculated for various solar water-splitting systems' minimum allowed bandgap, $E_{g\text{-}min}(T, p)$. The left-hand side of Figure 21.10 is calculated at various temperatures for $p_{H_2O} = 1$ bar (10^5 Pa), while the right-hand side repeats these calculations for $p_{H_2O} = 500$ bar. Comparing the top left and right sides of the figure, the rate of increase in temperature of $\eta_{solar\text{-}max}$ is significantly greater for higher pressure photoelectrolysis ($p_{H_2O} = 500$ bar). However, this higher rate of efficiency increase with temperature is offset by lower accessible temperatures (for a given bandgap).

Figure 21.10 describes the hybrid solar hydrogen conversion efficiency, η_{solar} as constrained for various values of photovoltaic conversion efficiency, η_{phot}. The high end of contemporary experimental high solar conversion efficiencies ranges from $100\%\eta_{phot} = 19.8\%$ for multi-crystalline single-junction photovoltaics, to 27.6% and 32.6% for single-junction and multiple-junction photovoltaics [56]. The efficiency of solar–thermal conversion tends to be higher than solar–electrical conversion, η_{phot}, particularly in the case of the restricted spectral range absorption used here; and values of $\eta_{heat} = 0.5$, 0.7 or 1 are used for Equation 21.3.

Representative results from Figure 21.10 for solar water-splitting to H_2 systems include a 50% solar-energy conversion for a photoelectrolysis system at 638 °Cm with $p_{H_2O} = 500$ bar, $p_{H_2} = 1$ bar and $\eta_{phot} = 0.32$. However, this high H_2O-partial-pressure system requires separation of a low partial pressure of H_2. Efficient photoelectrolysis has also been determined for high relative H_2, such as for systems of $p_{H_2} = p_{H_2O} = 1$ bar, $\eta_{heat} = 0.7$ and with an $E_{g\text{-}min} = E_g(\text{GaAs}) = 1.43$ eV, in which efficiencies improve in Figure 21.9 from 28% (at 25 °C) to 42% at 1360 °C, or from 32% (25 °C) to 46% at 1210 °C, or from 36% (25 °C) to 49% at 1060 °C. The case of the GaAs bandgap is of interest, as efficient multiple-bandgap photovoltaics have also been demonstrated using GaAs as the semiconductor minimum-bandgap component.

21.4 STEP Experiment: Efficient Solar Water Splitting

Fletcher, repeating the fascinating suggestion of Brown that saturated aqueous NaOH will never boil, hypothesized that a useful medium for water electrolysis might be very high temperature saturated aqueous NaOH solutions. These do not reach a temperature at which they boil at 1 atm due to the high salt solubility, binding solvent and changing saturation vapor pressure, as reflected in their phase diagram [50]. We measured this domain, and also electrolysis in an even higher temperature domain above which NaOH melts (318 °C), creating a molten electrolyte with dissolved water and resulting in unexpected V_{H_2O}.

Figure 21.11 summarizes measured V_{H_2O} (T) in aqueous saturated and molten NaOH electrolytes. As seen in the inset, Pt exhibits low overpotentials to H_2 evolution, and is used as

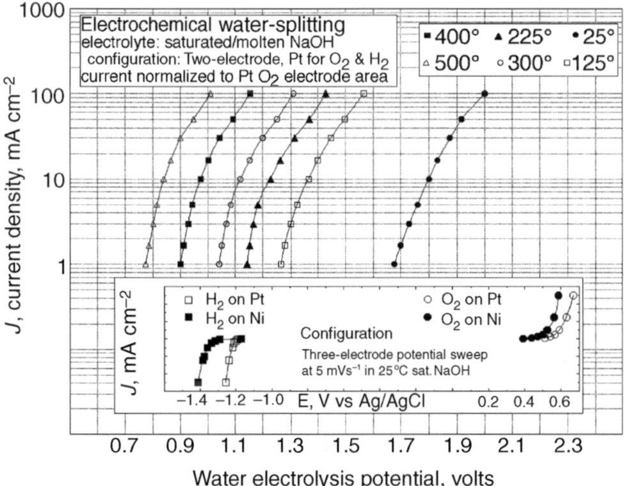

Figure 21.11 V_{H_2O}, measured in aqueous saturated or molten NaOH, at 1 atm. The molten electrolyte is prepared from heated, solid NaOH with steam injection. The O_2 anode is 0.6 cm^2 Pt foil. IR and polarization losses are minimized by sandwiching, 5 mm from each side of the anode, two interconnected Pt cathodes. Figure inset: Three-electrode values at 25°, 5 mV s^{-1} versus Ag/AgCl, with either 0.6 cm^2 Pt or Ni foil. (Reprinted with permission from S. Licht, Thermochemical solar hydrogen generation, *Chemical Communications*, **37**, 4623–4464, 2005. © 2005 Royal Society of Chemistry.)

a convenient quasi-reference electrode in the measurements which follow. As also seen in the inset, Pt exhibits a known large overpotential to O_2 evolution, as compared to a Ni electrode or to $E^0_{H_2O}(25\,°C) = 1.23\,V$. This overpotential loss diminishes at moderately elevated temperatures and, as seen in the main portion of the figure, at 125 °C there is a 0.4 V decrease in the O_2 activation potential at a Pt surface. Through 300 °C in Figure 21.11, measured V_{H_2O} remains greater than the calculated thermodynamic rest potential. Unexpectedly, V_{H_2O} at 400 °C and 500 °C in molten NaOH occurs at values substantially smaller than those predicted. These measured values include voltage increases due to IR and hydrogen overpotentials, and hence provide an upper bound to the unusually small electrochemical potential. Even at relatively large rates of water splitting (30 mA cm^{-2}) at 1 atm, a measured V_{H_2O} in Figure 21.11 is observed to be below that predicted by theory in Figure 21.7 at temperatures above the 318 °C NaOH melting point. As comparing the figures, the observed V_{H_2O} values at high temperatures approach those calculated for a thermodynamic system of 500 bar, rather than 1 bar, H_2O.

A source of the nominally less than thermodynamic water-splitting potentials is described in Figure 21.12. Shown on the left-hand side is the single-compartment cell utilized here. Cathodically generated H_2 is in close proximity to the anode, while anodic O_2 is generated near the cathode. Their presence will facilitate the water-forming back reaction, and, at the electrodes, this recombination will diminish the potential. In addition to the observed low potentials, two observations support this recombination effect. The generated H_2 and O_2 is collected, but is consistent with a coulombic efficiency of ≈50% (varying with T, J and interelectrode separation). Consistent with the right-hand side of Figure 21.12, when conducted in separate anode/cathode compartments, this observed efficiency is 98–100%. Here, however, all cell open-circuit potentials increase to beyond the thermodynamic potential, and at $J = 100$ mA cm^{-2} yields measured V_{H_2O} values of 1.45 V, 1.60 V and 1.78 V at 500 °C, 400 °C and 300 °C, which are approximately 450 mV higher than the equivalent values for the single configuration cell.

The recombination phenomenon offers advantages (low V_{H_2O}), but also disadvantages (H_2 losses), requiring study to balance these competing effects to optimize energy efficiency. In

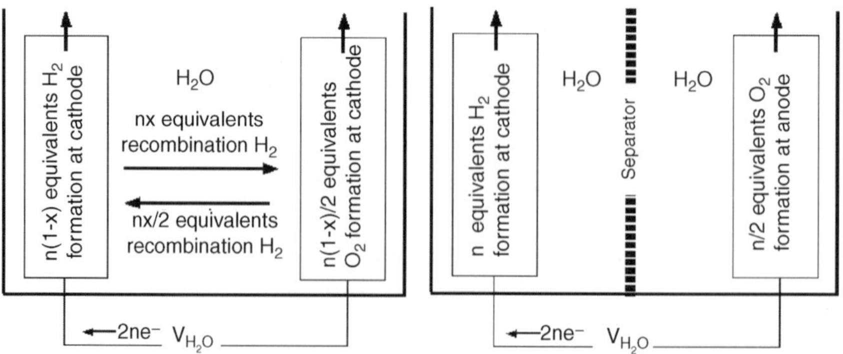

Figure 21.12 Interelectrode recombination can diminish V_{H_2O} and occurs in open (left), but not in isolated (right), configurations; such as with or without a Zr_2O mixed fiber separator between the Pt electrodes. (Reprinted with permission from S. Licht, Thermochemical solar hydrogen generation, *Chemical Communications*, **37**, 4623–4464, 2005. © 2005 Royal Society of Chemistry.)

molten NaOH, temperature-variation effects of $\Delta G_f^0(H_2O)$ and the recombination of the water-splitting products can have a pronounced effect on solar-driven electrolysis. As compared to 25 °C data in Figure 21.11, only half the potential is required to split water at 500 °C, over a wide current-density range.

The unused thermal photons which are not required in semiconductor photodriven charge generation, can contribute to heating water to facilitate electrolysis at an elevated temperature. The characteristics of one, two or three series-interconnected solar visible-efficient photosensitizers, in accordance with the manufacturer's calibrated standards, are presented in Figure 21.13. These silicon photovoltaics are designed for efficient photoconversion under concentrated insolation ($\eta_{solar} = 26.3\%$ at 50 suns). Superimposed on the photovoltaic response curves in the figure are the water electrolysis current densities for one or two series-interconnected 500 °C molten NaOH single-compartment cell configuration electrolyzers.

Constant illumination generates, for the three series cells, a constant photopotential for stability measurements at sufficient power to drive two series molten NaOH electrolyzers. At this constant power, and as presented in the lower portion of Figure 21.13, the rate of water splitting appears fully stable over an extended period. In addition, as measured and summarized in the upper portion of the figure, for the overlapping region between the solid triangle and open square curves, a single Si photovoltaic can drive 500 °C water splitting, albeit at an energy beyond the maximum power-point voltage, and therefore at diminished efficiency. This

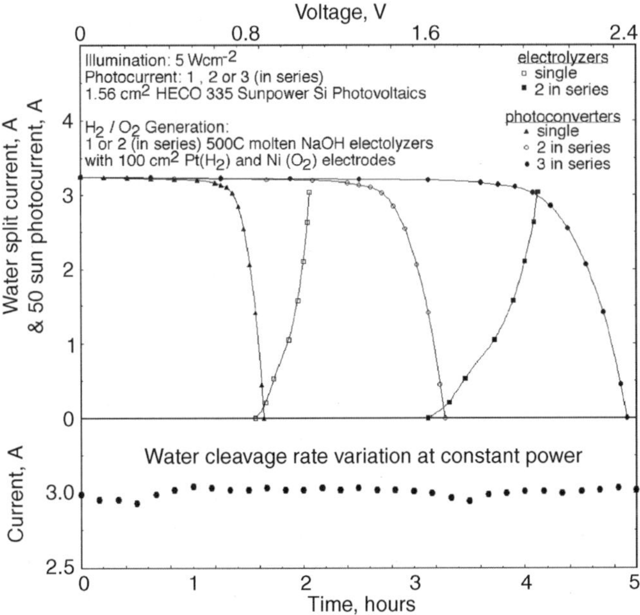

Figure 21.13 Photovoltaic and electrolysis charge transfer for thermal electrochemical solar driven water splitting. Photocurrent is shown for 1, 2 or three 1.561 cm^2 HECO 335 Si photovoltaics in series at 50 suns which drive 500 °C molten NaOH steam electrolysis using Pt gauze anode and cathodes. Inset: electrolysis current stability. (Reprinted with permission from S. Licht, Thermochemical solar hydrogen generation, *Chemical Communications*, **37**, 4623–4464, 2005. © 2005 Royal Society of Chemistry.)

appears to be the first case in which an external, single small-bandgap photosentizer has been shown to cleave water, and is accomplished by tuning the water-splitting electrochemical potential to below the Si open-circuit photovoltage. $V_{H_2O\text{-tuned}}$ is accomplished by two phenomena: (i) the thermodynamic decrease of E_{H_2O} with increasing temperature and (ii) a partial recombination of the water-splitting products. $V_{H_2O\text{-tuned}}$ can drive system efficiency advances, for example, AlGaAs/GaAs, transmits more insolation, $E_{IR} < 1.4$ eV, than Si to heat water, and with η_{photo} over 30%, prior to system engineering losses, calculates to over 50% η_{solar} to H_2.

Fundamental details of splitting of the thermal and visible insolation, with the former to heat water for electrolysis and the latter to drive electrical charge formation, have been presented. In addition, experimental components, of the representation described in Figure 21.5, of efficient solar-driven generation of H_2 fuel at 40–50% solar-energy conversion efficiencies appear to be technologically available. Without the inclusion of high-temperature effects, we had already experimentally achieved $\eta_{solar} > 0.18$, using an $\eta_{phot} = 0.20$ AlGaAs/Si system [23]. Our use of more efficient PVs, ($\eta_{phot} = 26.3\%$ at 50 suns) and inclusion of heat effects and the elevated temperature decrease of the water electrolysis potential, substantially enhances η_{solar} [26]. Existing higher $\eta_{phot} = 0.28$ to 0.33 systems should achieve proportionally higher results.

Photoelectrochemical cells tend to be unstable, which is likely to be exacerbated at elevated temperatures. Hence, the hybrid solar/thermal hydrogen process will be particularly conducive to photovoltaic-driven, rather than photoelectrochemical-driven, electrolysis. The photovoltaic component is used for photodriven charge in the electrolysis component, but does not contact the heated electrolyte. In this case the high efficiencies appear accessible; stable photovoltaics are commonly driven with concentrated insolation and, specific to the system model here, heat will be purposely filtered from the insolation prior to incidence on the photovoltaic component.

Dielectric filters used in laser optics split insolation without absorption losses. For example, in a system based on a parabolic concentrator, a casegrain configuration may be used, with a mirror made from fused silica glass with a dielectric coating acting as band-pass filter. The system will form two focal spots with different spectral configurations, one at the focus of the parabola and the other at the focus of the casegrain [51]. The thermodynamic limit of concentration is 46 000 suns, the brightness of the surface of the sun. In a medium with a refractive index greater than one, the upper limit is increased by two times the refractive index, although this value is reduced by reflective losses and surface errors of the reflective surfaces, the tracking errors of the mirrors and dilution of the mirror field. Specifically designed optical absorbers, such as parabolic concentrators or solar towers, can efficiently generate a solar flux with concentrations of ~2000 suns, generating temperatures in excess of 1000 °C [52].

Commercial alkaline electrolysis occurs at temperatures up to 150 °C and pressures up to 30 bar, and supercritical electrolysis to 350 °C and 250 bar [53]. Although less developed than their fuel-cell counterparts, which have 100 kW systems in operation and were developed from the same oxides [54], zirconia and related solid oxide-based electrolytes for high temperature steam electrolysis can operate efficiently at 1000 °C [55], and approach the operational parameters necessary for efficient solar-driven water splitting. Efficient multiple bandgap solar cells absorb light up to the bandgap of the smallest-bandgap component. Thermal radiation is assumed to be split off (removed and utilized for water heating) prior to incidence on the semiconductor and hence will not substantially effect the bandgap. Highly efficient photovoltaics have been demonstrated at a solar flux with a concentration of several hundred suns. AlGaAs/GaAs has yielded an η_{phot} efficiency of 27.6%, and a GaInP/GaAs cell 30.3% at 180 suns concentration, while GaAs/Si has reached 29.6% at 350 suns, InP/GaInAs 31.8% and GaAs/GaSb 32.6% with concentrated insolation, and approaches using semiconductor nanoparticles for photovoltaic cells have been reported [24,56].

21.5 NonHybrid Solar Thermal Processes

21.5.1 Direct Solar Thermal Hydrogen Generation

Direct thermochemical water splitting consists of heating water to a high temperature and separating the spontaneously formed hydrogen from the equilibrium mixture [6–9]. Although conceptually simple, this process has been impeded by high-temperature material limitations and the need to separate H_2 and O_2 to avoid ending up with an explosive mixture. Unfortunately, for thermal water splitting, $T > 2500$ K is necessary to achieve a significant degree of hydrogen dissociation. The free energy, ΔG, of the gas reaction $H_2O \leftrightarrows H_2 + 1/2O_2$, does not become zero until the temperature is increased to 4310 K at 1 bar pressure of H_2O, H_2 and O_2, smaller amounts of product are barely discernable at 2000 K [3]. The entropy (ΔS), driving the negative of the temperature derivative of the free energy of water, is simply too small to make direct decomposition feasible at this time [11].

The production of hydrogen by direct thermal splitting of water generated a considerable amount of research during the period 1975–1985. Fletcher and co-workers in the USA stressed the thermodynamic advantages of a one-step process with heat input at as high a temperature as possible [57]. The theoretical and practical aspects were examined by Lede and by others [6,58]. The main emphases in these investigations were the thermodynamics and the demonstration of feasibility of the process. However, no adequate solution to the crucial problem of separation of the products of water splitting has been worked out so far, although effort was devoted to demonstrate the possibility of product separation at low temperature, after quenching the hot gas mixtures by heat-exchange cooling, by immersion of the irradiated, heated target in a reactor of water liquid, by rapid turbulent gas jets, or rapid quenching by injecting a cold gas [10].

In order to attain efficient collection of solar radiation in a solar reactor operating at the requisite 2500 K, it is necessary to reach a high radiation concentration of the order 10 000. For example, a 3 MW solar-tower facility consists of a field of 64 50x concentrating heliostats. By directing all the heliostats to reflect the sun rays towards a common target, a concentration ratio of approximately 3000 may be obtained, and, to further enhance this concentration, a secondary concentration optical system is used [59].

Ordinary steels can't resist temperatures above a few hundred degrees centigrade, while the various stainless steels, including the more exotic ones, fail at less than 1300 K. In the range 3000–1800 °C, alumina, mullite or fused silica may be used. A temperature range of about 2500 K requires use of special materials for the solar reactor. Higher-melting-point materials can have additional challenges; carbide or nitride composites are likely to react with water-splitting products at the high temperatures needed for the reaction.

Water equilibria in dilute and concentrated alkaline solutions have been well described up to the critical point [60–62]. When the high-temperature gas-phase water equilibrium of water occurs, in addition to H_2O, H_2 and O_2, the atomic components H and O need to be considered. The fraction of these species is relatively insignificant at temperatures below 2500 K, as the pressure equilibrium constants for either diatomic hydrogen or oxygen formation from their atoms are each greater than 10^3 at $T \leq 2500$ K. However, the atomic components become increasingly significant at higher temperatures. The pressure equilibrium constants of the water-dissociation reaction over a range of temperatures [63] are summarized in Equation 21.6 for the relevant four equilibria, and their associated equlibrium constants K_i, considered for water splitting at temperatures in which significant, spontaneous formation of H_2 occurs:

$$\begin{array}{ll} H_2O \leftrightarrows HO + H; & K_1 \\ HO \leftrightarrows H + O; & K_2 \\ 2H \leftrightarrows H_2; & K_3 \\ 2O \leftrightarrows O_2; & K_4 \end{array} \tag{21.6}$$

	$T = 2500$ K	$T = 3000$ K	$T = 3500$ K
K_1	1.34×10^{-4}	8.56×10^{-3}	1.68×10^{-1}
K_2	4.22×10^{-4}	1.57×10^{-2}	2.10×10^{-1}
K_3	1.52×10^{3}	3.79×10^{1}	2.67×10^{0}
K_4	4.72×10^{3}	7.68×10^{1}	4.01×10^{0}

Kogan has calculated that, at a pressure of 0.05 bar, water dissociation is barely discernible at 2000 K. By increasing the temperature to 2500 K, 25% of water vapor dissociates at the same pressure. A further increase in temperature to 2800 K under constant pressure causes 55% of the vapor to dissociate [6]. These basic facts reflect the challenges that must be overcome for practical hydrogen production by a solar thermal water-splitting process: (a) attainment of very high solar-reactor temperatures, (b) solution of the materials problems connected with the construction of a reactor that can contain the water-spitting products at the reaction temperature and (c) development of an effective method of *in situ* separation of hydrogen from the water-splitting mixture.

Separation of the generated hydrogen from the mixture of water-splitting products, to prevent explosive recombination, is another challenge for thermochemically generated water-splitting processes. From the perspective of the high molecular-weight ratio of oxygen and hydrogen, separation of the thermochemically generated hydrogen from the water-splitting product mixture by gas diffusion through a porous ceramic membrane could be relatively effective. Membranes that have been considered include commercial and specially prepared porous zirconias, although sintering was observed to occur under thermal water-splitting conditions [6,64], and ZrO_2–TiO_2–Y_2O_3 oxides [65]. In such membranes, it is necessary to maintain a Knudsen flow regime across the porous wall [63]. The molecular mean free path λ in the gas must be greater than the average pore diameter ϕ [1]. A double-membrane configuration has been suggested to be superior to a single-membrane reactor [63].

In recent times, there have been relatively few studies on the direct thermochemical generation of hydrogen by water splitting [6,63–67] due to continuing high-temperature material limitations. Principal recent experimental work has been performed by Kogan and associates [1,28]. In 2004, Bayara reiterated that conversion rates in direct thermochemical processes are still quite low, and new reactor designs, operation schemes and materials are needed for a new breakthrough in this field [67].

21.5.2 Indirect (Multistep) Solar Thermal H_2 Generation

Indirect solar thermal splitting of water utilizes a reaction sequence, whose individual steps require lower temperatures than the direct solar thermal process. Historically, the reaction of reactive metals and reactive metal hydrides with water or acid was the standard way of producing pure hydrogen. These reactions involved sodium metal or calcium hydride with water, or zinc metal with hydrochloric acid, or metallic iron or ferrous oxide with steam, to produce H_2. All these methods are quite outdated and expensive.

Multireaction processes to produce hydrogen from water with a higher thermal efficiency have been extensively studied. As summarized in Figure 21.14, a variety of pertinent, spontaneous processes can be considered, which have a negative reaction free energy at temperatures considerably below that for water. These reactions are conducted in a cycle to regenerate and reuse the original reactions (ideally, with the only net reactant water, and the only net products hydrogen and oxygen. While these cycles operate at much lower temperatures than the direct thermal chemical generation of hydrogen, conversion efficiencies are insufficient and interest in these cycles has waned. Efficiency losses can occur at each of the steps in the multiple step sequence, resulting in low overall solar-to-hydrogen energy-conversion efficiencies. Interest in indirect thermal chemical generation of hydrogen started approximately 40 years ago. An upsurge in interest occurred with an average of over 70 papers per year from 1975 through 1985. Following that time, and the lack of clear success, publications have to diminished to approximately 10 per year [11]. An overview of indirect thermochemical processes for hydrogen generation has been presented by Funk, with two to six steps in the total reaction cycle, each operating at a maximum temperature of 920 to 1120 K [11].

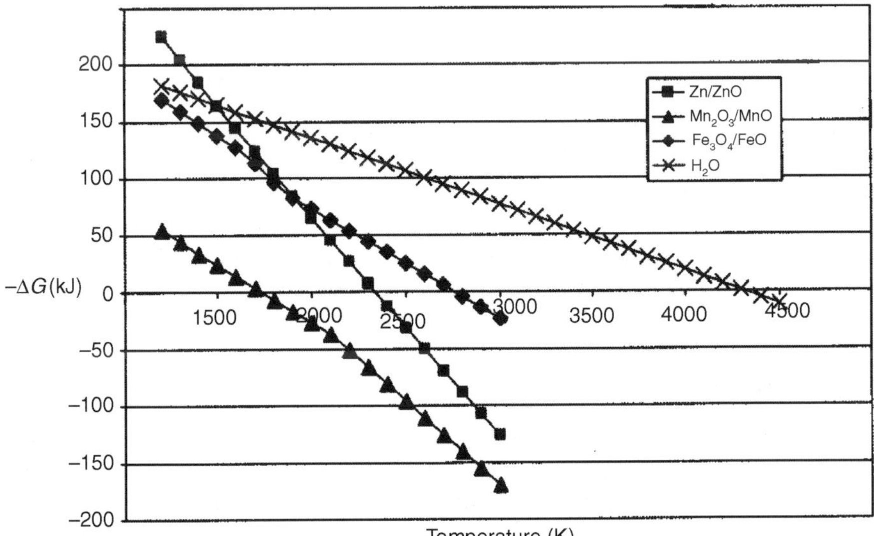

Figure 21.14 Temperature variation of the free energy for several decomposition reactions pertinent to hydrogen generation. (Reprinted with permission from S. Licht, Thermochemical solar hydrogen generation, *Chemical Communications*, **37**, 4623–4464, 2005. © 2005 Royal Society of Chemistry.)

Status reviews on multiple-step cycles have been presented [10–14,68], and leading candidates include a three-step cycle, based on the thermal decomposition of H_2SO_4 at 1130 K, and a four-step cycle, based on the hydrolysis of $CaBr_2$ and $FeBr_2$ at 1020 and 870 K; this letter process, involving two Ca and two Fe compounds at $T \leq 1050$ K has received some attention [11]. The process is operated in a cyclic manner in which the solids remain in their reaction vessels and the flow of gases is switched when the desired reaction extent is reached, and is summarized as:

$$\begin{aligned} &CaBr_2(s) + H_2O(l) \rightarrow CaO(s) + 2HBr(g) \\ &CaO(s) + Br_2(g) \rightarrow CaBr_2(s) + O_2(g) \\ &Fe_3O_4(s) + 8HBr(g) \rightarrow 3FeBr_2(s) + 4H_2O(g) + 2Br_2(g) \\ &3FeBr_2(s) + 4H_2O(g) \rightarrow Fe_3O_4(s) + 6HBr(g) + H_2(g) \end{aligned} \tag{21.7}$$

One of the most actively studied candidate metal oxide redox pairs for the two-step cycle reactions is ZnO/Zn. Several chemical aspects of the thermal dissociation of ZnO have been investigated, including reaction rates, Zn separation and heat recovery in the presence of O_2. Cycles incorporating ZnO continue to be of active research interest [10,69–76].

Higher temperatures are needed for more efficient indirect solar thermal processes, such as two-step thermal chemical cycles using metal-oxide reactions [10]. The first step is solar thermal, the metal-oxide endothermic dissociation to a lower valence or to the metal. The second step is nonsolar, and is the exothermic hydrolysis of the metal to form H_2 and the corresponding metal oxide. The net reaction remains $H_2O = H_2 + 1/2O_2$, but as H_2 and O_2 are formed in different steps, the need for high-temperature gas separation is thereby eliminated.

$$\begin{aligned} \text{1st step(solar)}: &\ M_xO_y \rightarrow xM + y/2O_2 \\ \text{2nd step(non-solar)}: &\ xM + yH_2 \rightarrow M_xO_y + yH_2 \end{aligned} \tag{21.8}$$

where M is a metal and M_xO_y is the corresponding metal oxide. Studies have been conducted on the redox pairs Fe_3O_4/FeO, TiO_2/TiO_x, Mn_3O_4/MnO Co_3O_4/CoO and MnO with NaOH, and mixed metal oxides of the type $(Fe_{1-x}M_x)_3O_4/(Fe_{1-x}M_x)_{1-y}O$ [10,69,71]. Reports include ferrite [77–83] and Ca–Br [84] cycles, sulfur–iodine [85] cycles, and tin [86,87], cerium [88,89] and magnesium oxide [90] cycles. These processes retain many of the $T > 2000$ K material challenges faced by direct thermal water splitting.

21.6 Conclusions

Solar-energy-driven water splitting combines several attractive features for energy utilization. The energy source (sun) and reactive media (water) for solar water splitting are readily available and are renewable, and the resultant fuel (generated H_2) and its discharge product (water) are each environmentally benign. An overview of solar thermal processes for the generation of hydrogen has been presented, and compared to alternate solar hydrogen processes. In particular, a hybrid solar thermal/electrochemical process combines efficient photovoltaics and concentrated excess sub-bandgap heat into highly efficient elevated-temperature solar electrolysis of water and generation of H_2 fuel. Efficiency is further enhanced by excess super-bandgap and nonsolar sources of heat, but diminished by losses in polarization and photoelectrolysis power matching. Solar concentration can provide the high temperature and diminish the requisite surface area of efficient electrical-energy conversion components, and high-temperature electrolysis components are available, suggesting that their combination for highly efficient solar generation of H_2 will be attainable.

References

[1] Kruse, O., Rupprecth, J., Mussgnug, J.R. *et al.* (2005) Photosynthesis: a blueprint for solar energy capture and biohydrogen production technologies. *Photochemical and Photobiological Sciences*, **4**, 957.

[2] Allakhverdiev, S.I., Kreslavski, V.D., Thavasi, V. *et al.* (2009) Hydrogen photoproduction by use of photosynthetic organisms and biomimetic systems. *Photochemical and Photobiological Sciences*, **8**, 148.

[3] Barber, J. (2007) Biological solar energy. *Philosphical Transactions of the Royal Society A, Mathematical, Physical and Engineering Sciences*, **365**, 1007.

[4] Esper, B., Badura, A. and Rogner, M. (2006) Photosynthesis as a power supply for (bio-) hydrogen production. *Trends in Plant Science*, **11**, 543.

[5] Smith, J.R., Van Steenkiste, T.H. and Wang, X.G. (2009) Thermal photocatalytic generation of H-2 over $CuAlO_2$ nanoparticle catalysts in H_2O. *Physical Review B*, **79**, 041403.

[6] Kogan, A. (1998) Direct solar thermal splitting of water and on-site separation of the products. II. Experimental feasibility study. *International Journal of Hydrogen Energy*, **23**, 89.

[7] Baykara, S.Z. (2004) Hydrogen production by direct solar thermal decomposition of water, possibilities for improvement of process efficiency. *International Journal of Hydrogen Energy*, **29**, 1451.

[8] Baykara, S.Z. (2004) Experimental solar water thermolysis. *International Journal of Hydrogen Energy*, **29**, 1459–1469.

[9] Ohya, H., Yatabe, M., Aihara, M. *et al.* (2002) Feasibility of hydrogen production above 2500 K by direct thermal decomposition reaction in membrane reactor using solar energy. *International Journal of Hydrogen Energy*, **27**, 369.

[10] Steinfeld, A. (2005) Solar thermochemical production of hydrogen-a review. *Solar Energy*, **78**, 603.

[11] Funk, J.E. (2001) Thermochemical hydrogen production: past and present. *International Journal of Hydrogen Energy*, **26**, 185.

[12] Ernst, F.O., Pratsinis, S.E. and Steinfeld, A. (2009) Hydrolysis rate of submicron Zn particles for solar H-2 synthesis. *International Journal of Hydrogen Energy*, **34**, 1166.

[13] Abanades, S., Chavin, P., Flamant, G. and Neveu, P. (2006) Screening of water-splitting thermochemical cycles potentially attractive for hydrogen production by concentrated solar energy. *Energy*, **31**, 2805.

[14] Perkins, C. and Weimer, A.W. (2009) Producing hydrogen using solar-thermal energy. *Chemical Engineering Progress*, **105**, 10.

[15] Dai, K., Peng, T.Y., Ke, D.N. and Wei, B.Q. (2009) Photocatalytic hydrogen generation using a nanocomposite of multi-walled carbon nanotubes and TiO_2 nanoparticles under visible light irradiation. *Nanotechnology*, **20**, 125603.

[16] Park, J.H. and Bard, A.J. (2005) Unassisted water splitting from bipolar Pt/dye-sensitized TiO_2 photoelectrode arrays. *Electrochemical and Solid State Letters*, **8**, G371.

[17] Kelly, N.A. and Gibson, T.L. (2006) Design and characterization of a robust photoelectrochemical device to generate hydrogen using solar water splitting. *International Journal of Hydrogen Energy*, **31**, 1658.

[18] Murphy, A.B., Barnes, P.R.F., Randeniya, L.K. *et al.* (2006) Efficiency of solar water splitting using semiconductor electrodes. *International Journal of Hydrogen Energy*, **31**, 1999.

[19] Kunkely, H. and Vogler, A. (2009) Water splitting by light with osmocene as photocatalyst. *Angewandte Chemie International Edition*, **48**, 1685.

[20] Tributsh, H. (2008) Photovoltaic hydrogen generation. *International Journal of Hydrogen Energy*, **33**, 5911.

[21] Park, H., Vecitis, C., Choi, W. *et al.* (2008) Solar-powered production of molecular hydrogen from water. *Journal of Physical Chemistry C*, **112**, 885.

[22] Perharz, G., Dimroth, F. and Wittstadt, U. (2007) Solar hydrogen production by water splitting with a conversion efficiency of 18%. *International Journal of Hydrogen Energy*, **32**, 3248.

[23] Licht, S., Wang, B., Mukerji, S. *et al.* (2001) Over 18% solar energy conversion to generation of hydrogen fuel; theory and experiment for sefficient solar water splitting. *International Journal of Hydrogen Energy*, **26**, 653.

[24] Licht, S. (2005) Solar water splitting to generate hydrogen fuel – a photothermal electrochemical analysis. *International Journal of Hydrogen Energy*, **30**, 459.

[25] Licht, S. (2005) Thermochemical solar hydrogen generation. *Chemical Communications*, 4635.

[26] Licht, S., Halperin, L., Kalina, M. *et al.* (2003) Electrochemical potential tuned solar water splitting. *Chemical Communications*, 3006.

[27] Licht, S. (1987) A description of energy conversion in photoelectrochemical solar cells. *Nature*, **330**, 148.

[28] Licht, S. (ed.) (2002) *Semiconductor Electrodes and Photoelectrochemistry*, Wiley-VCH, Weinheim.

[29] Licht, S., Hodes, G., Tenne, R. and Manassen, J. (1987) A light variation insensitive high efficiency solar cell. *Nature*, **326**, 863.

[30] Licht, S., Wang, B., Soga, T. and Umeno, M. (1999) Light invariant, efficient, multiple bandgap AlGaAs/Si/metal hydride solar cell. *Applied Physics Letters*, **74**, 4055.

[31] Wang, B., Licht, S., Soga, T. and Umeno, M. (2000) Stable cycling behavior of the light invariant AlGaAs/Si/metal hydride solar cell. *Solar Energy Materials and Solar Cells*, **64**, 311.

[32] Licht, S. (2001) Multiple band gap semiconductor/electrolyte solar energy conversion. *Journal of Physical Chemistry B*, **105**, 6281.

[33] Licht, S. and Peramunage, D. (1990) Efficient photoelectrochemical solar cells from electrolyte modification. *Nature*, **345**, 330.

[34] Licht, S. and Peramunage, D. (1991) Efficiency in a liquid solar cell. *Nature*, **354**, 440.

[35] Licht, S. (1995) Electrolyte modified photoelectrochemical solar cells. *Solar Energy Materials and Solar Cells*, **38**, 305.

[36] Licht, S. and Peramunage, D. (1996) Flat band variation of n-cadmium chalcogenides in aqueous cyanide. *Journal of Physical Chemistry*, **100**, 9082.

[37] Fujishima, A. and Honda, K. (1972) Electrochemical photolysis of water at a semiconductor electrode. *Nature*, **37**, 238.

[38] Heller, A., Asharon-Shalom, E., Bonner, W.A. and Miller, B. (1982) Hydrogen-evolving semiconductor photocathodes: nature of the junction and function of the platinum group metal catalyst. *Journal of the American Chemical Society*, **104**, 6942.

[39] Zou, Z., Ye, Y., Sayama, K. and Arakawa, H. (2001) Direct splitting of water under visible light irradiation with an oxide semiconductor photocatalyst. *Nature*, **414**, 625–627.

[40] Licht, S., Wang, B., Mukerji, S. *et al.* (2000) Efficient solar water splitting; exemplified by ruo_2 catalyzed AlGaAs/Si photohydrolysis. *Journal of Physical Chemistry B*, **104**, 8920.

[41] Ohmori, T., Go, H., Yamaguchi, N. *et al.* (2001) Photovoltaic water electrolysis using the sputter-deposited a-Si/c-Si solar cells. *International Journal of Hydrogen Energy*, **26**, 661.

[42] Tani, T., Sekiguchi, N., Sakai, M. and Otha, D. (2000) Optimization of solar hydrogen systems based on hydrogen production cost. *Solar Energy*, **68**, 143.

[43] Rzayeva, M.P., Salamov, O.M. and Kerimov, M.K. (2001) Modeling to get hydrogen and oxygen by solar water electroclysis. *International Journal of Hydrogen Energy*, **26**, 195.

[44] Licht, S., Tributsch, H., Ghosh, S. and Fiechter, S. (2002) High efficiency solar energy water splitting to hydrogen fuel; probing RuS_2 enhancement of multiple band electrolysis. *Solar Energy Materials and Solar Cells*, **70/4**, 471.

[45] Chung, B., Virshup, G., Hikadl, S. and Kramer, N. (1989) 27.6% efficiency (1 sun, air mass 1.5) monolithic $Al_{0.37}Ga_{0.63}As$/GaAs two-junction cascade solar cell with prismatic cover glass. *Applied Physics Letters.*, **55**, 1741.

[46] DeBethune, A.J., Licht, T.S. and Swendemna, N.S. (1959) The temperature coefficient of electrode potentials. *Journal of the Electrochemical Society*, **106**, 618.

[47] Bockris, J.O'M. (1980) *Energy Options*, Halsted Press, NY.

[48] Licht, S. (2002) Efficient solar generation of hydrogen fuel – a fundamental analysis. *Electrochemical Communications*, **4**, 790.

[49] Licht, S. (2002) Solar water splitting to generate hydrogen fuel: Photothermal electrochemical analysis. *Journal of Physical Chemistry B*, **107**, 4253.

[50] Fletcher, E. (2001) Some considerations on the electrolysis of water from sodium hydroxide solutions. *Journal of Solar Energy Engineering*, **123**, 143.

[51] Yogev, A. (1996) Quantum processes for solar energy conversion. Weizmann Sun Symp. Proc., Rehovot, Israel.

[52] Segal, E. and Epstein, M. (2001) The optics of the solar tower reflector. *Solar Energy*, **69**, 229.

[53] Misch, B., Firus, A. and Brunner, G. (2000) An alternative method of oxidizing aqueous waste in supercritical water: oxygen supply by means of electrolysis. *Journal of Supercritical Fluids*, **17**, 227.

[54] Yamamoto, O. (2000) Solid oxide fuel cells: fundamental aspects and prospects. *Electrochimica Acta*, **45**, 2423.

[55] Eguchi, K., Hatagishi, T. and Arai, H. (1996) Power generation and steam electrolysis characteristics of an electrochemical cell with a zirconia- or ceria-based electrolyte. *Solid State Ionics*, **86–8**, 1245.

[56] Green, M.A., Emery, K., Hishikawa, Y. and Warta, W. (2009) Solar cell efficiency tables (Version 33). *Progress in Photovoltaics*, **17**, 85.

[57] Fletcher, E.A. and Moen, R.L. (1977) Hydrogen and oxygen from water. *Science*, **197**, 105.

[58] Lede, J., Villermaux, J., Ouzane, R. *et al.* (1987) Production of hydrogen by simple impingement of a turbulent jet of steam upon a high temperature zirconia surface. *International Journal of Hydrogen Energy*, **12**, 3.

[59] Yogev, A., Kribus, A., Epstein, M. and Kogan, A. (1998) Solar "Thermal Reflector" systems: A new approach for high-temperature solar plants. *International Journal of Hydrogen Energy*, **26**, 239.

[60] Licht, S. (1987) pH measurement in conentrated alkaline solutions. *Analytical Chemistry*, **57**, 514.

[61] Light, T.S., Licht, S., Bevilacqua, A.C. and Morash, K.R. (2005) Conductivity and resistivity of ultrapure water. *Electrochemistry and Solid State Letters*, **8**, E16.

[62] Licht, S. (1998) Analysis in highly concentrated solutions: potentiometric, conductance, evanescent, densometric, and spectroscopic methodolgies, in *Electroanalytical Chemistry*, **20** (eds A. Bard and I. Rubinstein), Marcel Dekker, NY, p. 87.

[63] Ohya, H., Yatabe, M., Aihara, M. *et al.* (2002) Feasibility of hydrogen production above 2500 K by direct thermal decomposition reaction in membrane reactor using solar energy. *International Journal of Hydrogen Energy*, **27**, 369.

[64] Kogan, A. (2000) Direct solar thermal splitting of water and on-site separation of the products. IV. Development of porous ceramic membranes for a solar thermal water-splitting reactor. *International Journal of Hydrogen Energy*, **25**, 1043.

[65] Naito, H. and Arashi, H. (1995) Hydrogen production from direct water splitting at high temperatures using a ZrO_2-TiO_2-Y_2O_3 membrane. *Solid State Ionics*, **79**, 366.

[66] Omorjan, R.P., Paunovic, R.N., Tekic, M.N. and Antov, M.G. (2001) Maximal extent of an isothermal reversible gas-phase reaction in single- and double-membrane reaction; direct thermal splitting of water. *International Journal of Hydrogen Energy*, **26**, 203.

[67] Baykara, S.Z. (2004) Experimental solar water thermolysis. *International Journal of Hydrogen Energy*, **29**, 1459.

[68] Serpone, N., Lawless, D. and Terzian, R. (1992) Solar fuels: status and perspectives. *Solar Energy*, **49**, 221.

[69] Palumbo, R., Lede, J., Boutin, O. *et al.* (1998) The production of Zn from ZnO in a single step high temperature solar decomposition process. *Chemical Engineering Science*, **53**, 2503.

[70] Perkins, C. and Weimer, A.W. (2004) Likely near-term solar-thermal water splitting technologies. *International Journal of Hydrogen Energy*, **29**, 1587.

[71] Kaneko, H., Gokon, N., Hasewaga, N. and Tamaura, Y. (2005) Solar thermochemcial process for hydrogen production using ferrites. *Energy*, **30**, 2171.

[72] Gislon, P., Monteleone, G. and Prosini, P.P. (2009) Hydrogen production from solid sodium borohydride. *International Journal of Hydrogen Energy*, **34**, 929.

[73] Meichior, T., Platkowski, N. and Steinfeld, A. (2009) H_2 production by steam-quenching of Zn vapor in a hot-wall aerosol flow reactor. *Chemical Engineering Science*, **64**, 1095.

[74] Muller, R. and Steinfeld, A. (2008) H_2O-splitting thermochemical cycle based on ZnO/Zn-redox: Quenching the effluents from the ZnO dissociation. *Chemical Engineering Science*, **63**, 217.

[75] Weis, R.J., Ly, H.C., Wegner, K. *et al.* (2005) H_2 production by Zn hydrolysis in a hot-wall aerosol reactor. *AICHE Journal*, **51**, 1966.

[76] Perkins, C. and Weimer, A.W. (2009) Solar-Thermal Production of Renewable Hydrogen. *AICHE Journal*, **55**, 286.

[77] Ishihara, H., Kaneko, H., Hasegawa, N. and Tamaura, Y. (2008) Two-step water splitting process with solid solution of YSZ and Ni-ferrite for solar hydrogen production. *Journal of Solar Energy Engineering, Transactions of the ASME*, **130**, 044501.

[78] Gokon, N., Mizuno, T. and Nakamuro, Y. (2008) Iron-containing yttria-stabilized zirconia system for two-step thermochemical water splitting. *Journal of the Solar Energy Engineering, Transactions of the ASME*, **130**, 011018.

[79] Diver, R.B., Miller, J. and Allendorf, M.D. (2008) Solar thermochemical water-splitting ferrite-cycle heat engines. *Journal of the Solar Energy Engineering, Transactions of the ASME*, **130**, 041001.

[80] Kodama, T., Goon, N. and Yamamoto, R. (2008) Thermochemical two-step water splitting by ZrO_2-supported Ni$x$$Fe_{3(x}O_4$ for solar hydrogen production. *Solar Energy*, **82**, 73.

[81] Charvin, P., Abanades, S., Flamant, G. and Lemort, F. (2007) Two-step water splitting thermochemical cycle based on iron oxide redox pair for solar hydrogen production. *Energy*, **32**, 1124.

[82] Gokon, N. (2009) Monoclinic zirconia-supported Fe_3O_4 for the two-step water-splitting thermochemical cycle at high thermal reduction temperatures of 1400–1600 °C. *International Journal of Hydrogen Energy*, **34**, 1208.

[83] Gokon, N. (2009) Particles in an internally circulating fluidized bed. *Journal of the Solar Energy Engineering, Transactions of the ASME*, **131**, 011007.

[84] Simpson, M.F., Utgikar, V., Sachdev, P. and McGrady, C. (2007) A novel method for producing hydrogen based on the Ca–Br cycle. *International Journal of Hydrogen Energy*, **32**, 505.

[85] Prosini, P.P., Cento, C., Giaconia, A. *et al.* (2009) A modified sulphur-iodine cycle for efficient solar hydrogen production. *International Journal of Hydrogen Energy*, **34**, 1218.

[86] Charvin, P., Ababades, S., Lemont, F. and Flamant, G. (2008) Experimental study of SnO_2/SnO/Sn thermochemical systems for solar production of hydrogen. *AICHE Journal*, **54**, 2759.

[87] Abandes, S., Charvin, P., Lemont, F., Flamant, G. (2008) Novel two-step SnO_2/SnO water-splitting cycle for solar thermochemical production of hydrogen. *International Journal of Hydrogen Energy*, **33**, 6021.

[88] Abandes, S. and Flamant, G. (2006) Thermochemical hydrogen production from a two-step solar-driven water-splitting cycle based on cerium oxides. *Solar Energy*, **80**, 1611.

[89] Kaneko, H., Miura, T., Ishihara, H. *et al.* (2007) Reactive ceramics of CeO_2–MO_x (M = Mn, Fe, Ni, Cu) for H_2 generation by two-step water splitting using concentrated solar thermal energy. *Energy*, **32**, 656.

[90] Galvez, M.E., Frei, A., Albisetti, G., Lunardi, G. and Steinfeld, A. (2008) Solar hydrogen production via a two-step thermochemical process based on MgO/Mg redox reactions–Thermodynamic and kinetic analyses. *International Journal of Hydrogen Energy*, **33**, 2880.

Index

On Solar Hydrogen & Nanotechnology Edited by Lionel Vayssieres